ABOUT THE COVER

The cover image depicts a bridge in the Public Garden in Boston. Our offices look out onto this well-known Boston landmark. PWS, a Boston-based publisher, enjoys an environment in which innovative ideas are developed and applied in technical fields, in both education and industry. PWS conveys these new approaches to a worldwide audience through its core engineering series.

Statistics for Engineering Problem Solving

Statistics for Engineering Problem Solving

Stephen B. Vardeman

Iowa State University

PWS Publishing Company

Boston

PWS PUBLISHING COMPANY
20 Park Plaza, Boston, MA 02116-4324

PWS Publishing Company is a division of Wadsworth, Inc.

™

International Thomson Publishing
The trademark ITP is used under license

Library of Congress Cataloging-in-Publication Data
Vardeman, Stephen B.
    Statistics for engineering problem solving / Stephen B. Vardeman.
       p.   cm.
    Includes index.
    ISBN 0-534-92871-4
    1. Engineering — Statistical methods.   I. Title.
TA340.V37   1993
$620'.001'5195$ — dc20   93-5851
                                            CIP

Sponsoring Editor: Tom Robbins
Editorial Assistant: Cynthia Harris
Assistant Editor: Mary Thomas
Production Coordinator: Susan M.C. Caffey
Production: Hoyt Publishing Services
Interior Designer: Susan M.C. Caffey
Interior Illustrator: Hayden Graphics
Cover Designer: Books by Design, Inc./Sally Bindari
Cover Illustration: Carol Schweigert
Cover Illustration Referenced from: Stock Boston/Bill Horsman
Marketing Manager: Nathan Wilbur
Manufacturing Coordinator: Lisa M. Flanagan
Compositor: Integre Technical Publishing Co., Inc.
Cover Printer: Henry N. Sawyer Co.
Text Printer and Binder: Arcata Graphics/Hawkins

Printed and bound in the United States of America.
93 94 95 96 97 98 — 10 9 8 7 6 5 4 3 2 1

Contents

v

Preface

T he past decade has seen widespread recognition by educators and professionals that the *methods* and *thought patterns* of applied statistics are powerful tools for engineering problem solving. The purpose of this book is to present those tools in a way that both university engineering students and working engineers will find understandable and useful.

Instructors will find this book different from existing texts in several ways. No formal background in statistics is expected of the reader, but the philosophical orientation of the book is unlike that of traditional beginning texts in engineering statistics. Rather than emphasizing mathematical theory, I have tried to stress the engineering implications of statistical inferences, and to foster the development of scientific/statistical thought processes in the reader. Mathematical precision has not been compromised. But the material in this book has been treated as a means to more effective engineering practice, not as an end in itself.

I have made some nonstandard choices of content and order of presentation. This has been done to compile into a single volume what "one-shot" engineering audiences will most typically need in their professional practice. Practical issues in engineering data collection receive early and serious consideration, as do descriptive and graphical methods and the ideas of least squares surface-fitting and factorial analyses. More emphasis is given to the making of statistical intervals (including prediction and tolerance intervals) than to significance testing. Important topics in engineering statistics — such as Shewhart control charts, 2^p factorials and 2^{p-q} fractional factorials, response surface methods, mixture analyses, and variance component estimation — are given thorough, central treatment, rather than being included as supplemental topics intended to make a general statistics book into an "engineering" statistics book. Topics that seem less relevant to common engineering practice, such as axiomatic probability and counting, have been placed in appendices. There they are available for those instructors who have class time and wish to teach them, but they do not interrupt the book's main story line.

The approach and order of presentation found in this book have been used for a number of years in two different introductory engineering statistics courses at Iowa State University and in some industrial short-course teaching. In my experience, the route taken here has been more successful in demonstrating the usefulness of statistics to university engineering students and working engineers than the conventional approach of a little descriptive statistics, a lot of probability, and some one- and two-sample methods, followed by regression analysis.

An important feature of the book is its final chapter. Chapter 11 consists of an account of a highly successful industrial project in which engineering statistics played a key role. This case study integrates material from many different parts of the book and illustrates how effective the tools introduced in the book can be in engineering problem solving. I am grateful to Dow Chemical Company for providing the information necessary to write the case study and for permission to incorporate it in the text.

The discussions in this book are carried almost exclusively by examples involving real data and/or real scenarios. Some of these have been drawn from published engineering and engineering statistics literature, and some are from my own work with both academic and corporate engineers. A large proportion are from projects completed by students in my courses in engineering statistics and quality control at Iowa State University. The examples and exercises bearing only name citations (no article, book, or journal references) are based on student projects. I am grateful to those students for the use of their interesting data sets and scenarios.

I believe strongly that students in one-term engineering statistics courses need the experience of actually "doing statistics" — that is, carrying through a data collection and analysis project from the problem formulation stage all the way to the writing of a professional technical report based on the study. This book is full of examples of possible projects. The project descriptions that I give to students are included in the book's Solutions Manual, which the publisher makes available to qualified adopters.

There is more material in this book than can be readily covered in a single university course. (If I had the luxury of teaching a two-course, six-semester-hour sequence, I would have no trouble making the material of this book last through such a format.) I have attempted to make accessible to students even those topics not covered in class, either as outside reading or in subsequent professional practice. I believe it is wisest to give one-term engineering statistics students a feeling for the flavor of the subject and a wide view of what is possible, and to provide them with additional, thorough reading material that is presented at their level of sophistication. If this requires trimming some standard details from lectures and examinations, and rethinking the typical order in which topics are presented in class, I personally favor doing so. Regardless, I believe that instructors will find the book to be complete and understandable in its treatment of the subjects addressed, and flexible enough to fit a variety of syllabi.

The schedule listed in Table 1 is one that I use in a three-semester-hour, junior-level course. It maps a path culminating in the final week of class with a look at the 2^{p-q} fractional factorials. (This course meets for 75 minutes twice per week for 15 weeks.) The "**Q**" entries listed are half-period quizzes.

Parts of the book that are completely or nearly completely omitted from lecture in this fast-paced course are Sections 4-4 and 5-3, most of Section 5-4, part of Section 7-4, Sections 7-6, 8-2, 8-4, 9-3, 9-4, 10-3, 10-4, and 10-5, Appendix A, and Appendix B. If I could squeeze another topic or two into this course, the material in Sections 9-3 and 10-4 would be my first choice for inclusion.

At ISU, we also teach a three-semester-hour, freshman-level course from this text. That course covers less ground than the one described above and is somewhat more traditional in its outlook. Covered in that course are Chapters 1, 2, 3, and 4 (except for Section 4-4); Appendix B; Sections 5-1, 5-2, 5-3, and 5-5; Chapter 6; and Sections 7-5 and 7-6. Clearly, other outlines can be constructed around the ample material in this book.

Readers familiar with the engineering statistics literature will have no trouble recognizing the influence of many other authors. I have learned much from the work of Box, Hunter, and Hunter; of Guttman, Wilks, and Hunter; of Duncan; and of Daniel. My views on the teaching and practice of engineering statistics owe much to valued interaction with David Moore, H.T. David, Bill Meeker, Gerry Hahn, Bob Kasprzyk, George Kalemkarian, Bill Fulkerson, and Harvey Arnold, to name a few who come quickly to mind. Any originality on my part can only be originality of execution (if there is such a thing), not of basic outlook.

There are many others who deserve sincere thanks as I begin to see the light at the end of the tunnel on this project. Iowa State University has been a great environment in which to work and develop this material. A one-semester Faculty Improvement Leave

TABLE 1

Session	Topic(s) and/or Quiz	Reading
1	Introduction	Chapter 1
2	Data Collection	Chapter 2
3	Data Collection	Chapter 2
4	Descriptive Statistics	Chapter 3
5	Descriptive Statistics	Chapter 3
6	Curve Fitting/**Q1**	Chapter 4
7	Curve & Surface Fitting	Chapter 4
8	Surface Fitting/Factorials	Chapter 4
9	Fitted Factorial Effects	Chapter 4
10	Random Variables	Chapter 5
11	Random Variables	Chapter 5
12	Random Variables/**Q2**	Chapter 5
13	Random Variables	Chapter 5
14	Random Variables/Simple Inference	Chapter 6
15	Simple Inference	Chapter 6
16	Simple Inference	Chapter 6
17	Simple Inference	Chapter 6
18	Simple Inference/**Q3**	Chapter 6
19	One-Factor Analyses	Chapter 7
20	One-Factor Analyses	Chapter 7
21	One-Factor Analyses	Chapter 7
22	Control Charts	Chapter 7
23	Two-Factor Analyses	Chapter 8
24	2^p Analyses/**Q4**	Chapter 8
25	Inference in Curve & Surface Fitting	Chapter 9
26	Inference in Curve & Surface Fitting	Chapter 9
27	Inference in Curve & Surface Fitting	Chapter 9
28	Inference in Curve & Surface Fitting/**Q5**	Chapter 9
29	Fractional Factorials	Chapter 10
30	Fractional Factorials	Chapter 10

granted by ISU in the spring of 1989 was essential to getting me genuinely under way with the writing. Excellent teaching opportunities in the ISU Statistics and Industrial and Manufacturing Systems Engineering Departments have been both the principal motivation for putting this material down on paper and the laboratory in which many of the examples have been developed and the text debugged. I particularly appreciate the support of Professor Dean Isaacson, who has been the head of the ISU Statistics Department throughout the writing of this book. He has been a real advocate of efforts in engineering statistics and treated this project as important to the efforts of both the ISU department and the statistics profession to increase effective interaction with the engineering community. Dean David Kao of the ISU Engineering College has been another important supporter of the engineering/statistics interface; I appreciate his encouragement as well.

A number of instructors have used this book in class note form and have given me valuable suggestions for its improvement. These include especially Alan Zimmermann and Chuck Lerch, and also Peter Jones, Dean Isaacson, Jerry Hall, Ann Russey, Jean Pelkey, Peter Peterka, and Todd Sanger. Over the years, John Patterson has sent me a number of useful examples from engineering journals. I also gratefully acknowledge the help of Thomas Fischer in finding several other telling examples.

Chuck Lerch not only taught from this book in note form, providing valuable suggestions and catching many typos, but he also meticulously prepared the answer section of the book and the Solutions Manual. In the process, he identified and helped eliminate many rough spots in the exercise sets. The value of this book as a university teaching tool will owe much to his kind and careful help.

The manuscript has benefited from the comments and suggestions of "volunteer" reviews made by interested colleagues Kim Erland, Marjorie Green, and Noel Artiles. In addition, I most gratefully acknowledge the help of formal reviewers Bill Fulkerson, Dan Wardrop, Bob Kasprzyk, Harry Wadsworth, Craig Van Nostrand, John Boyer, J. Peter Jones, John Ramberg, Phillip Beckwith, and especially Harvey Arnold. (Harvey caught a number of potentially embarrassing errors in the production manuscript and page proofs, some that I and generations of students had failed to identify.) Of course, not every suggestion made by these reviewers was adopted — but many were, and all have influenced the final product.

Minitab Inc. provided software that allowed me to conveniently generate the print-outs in the book. I am grateful for their kind help in this regard.

Sharon Shepard of the ISU Statistics Department spent many hours flawlessly converting an early electronic version of the manuscript into one that was easily edited into a production version. Without her patient help, my hair would be considerably grayer.

The people of PWS and the companies they have enlisted to work on the production of the book have been first-rate professionals, and I am glad to have had this project in their hands. Special thanks are due to Michael Payne, who signed the book for PWS, championed it, and gave me substantial freedom to make it what I think it needs to be. Tom Robbins stepped in where Michael left off, and I have greatly appreciated his advocacy of treating the book as an engineering title. I am most grateful as well to Mary Thomas, who helped organize the production effort, and Susan Caffey, who handled many of the production details for PWS, including the excellent interior design. David Hoyt of Hoyt Publishing Services was wonderfully meticulous in handling day-to-day production matters and did a first-rate job of copy editing and making my sometimes arcane prose readable. Integre Technical Publishing did a superb, especially clean job of technical typesetting. Thanks also go to Hayden Graphics for excellent art and putting up with my incessant fiddling with some of the figures. As I say, this was a most professional crew and it has been a real pleasure to be associated with them.

Finally, and most importantly, there is my family. The text represents what now seems to me a huge investment of time and effort over nearly a seven-year period. During that time, my wife Jo Ellen and sons Micah and Andrew have patiently put up with too many excuses: "I'm sorry, I've got to work on the book . . . ." Jo Ellen cheerfully typed the original version of the manuscript, and I seriously doubt whether I would have ever finished even a first draft if she hadn't taken on that task in addition to all the other demands on her time. Thanks, Vardemans!

I will be grateful for feedback from users of this book in both the academic and corporate worlds. I hope that it proves genuinely helpful in teaching and practice, and I am anxious to know how the various parts of the presentation "work" in different environments. Written comments can be directed to me at the Statistics Department, Iowa State University, Snedecor Hall, Ames, Iowa 50011, or at my e-mail address: vardeman@iastate.edu

Steve Vardeman
Ames, Iowa

Statistics for Engineering Problem Solving

1 Introduction

This chapter lays a foundation for all that follows. Because it contains a road map for the study of the subject of engineering statistics, Chapter 1 should be read carefully and referred to frequently as you make your way through this text. In it, the subject of engineering statistics is defined, and its importance to an engineer is described. Some basic terminology is introduced, and the important subject of measurement is discussed. Finally, the role of mathematical models in achieving the objectives of engineering statistics is investigated.

1-1 Engineering Statistics: What and Why

In general terms, what a working engineer does is to design, build, operate, and/or improve physical systems and products. Chemical engineers design and operate systems that produce products ranging from fertilizers to fuels; civil engineers build highways, waterworks, large buildings, etc.; aeronautical engineers design and improve a myriad of different aircraft; industrial engineers design and operate manufacturing facilities. On and on and on the list could go.

An engineer is often guided by basic mathematical and physical theories (typically learned in those trying days spent negotiating a demanding undergraduate engineering curriculum). As the engineer's experience grows, these quantitative and scientific principles form the basis for, and work alongside, sound engineering judgment. But as technology advances and new systems and products are encountered, the working engineer is inevitably faced with questions and situations for which available theory and experience provide little or no help. When this happens, what is to be done?

On occasion, consultants can be called in, but most often the engineer must independently find out "what makes things tick." It is necessary to collect and interpret data that will help in understanding how the new system or product works. Without proper training, the engineer's attempts at data collection and analysis can be haphazard and poorly conceived. When this is the case, valuable time and resources are wasted, and sometimes erroneous (or at least ambiguous) conclusions are reached. To avoid such circumstances, it is vital for a working engineer to have a tool kit that includes the best possible principles and methods for gathering and interpreting data.

DEFINITION 1-1

Engineering statistics is the study of how best to

1. collect engineering data,
2. summarize or describe engineering data, and
3. draw practical conclusions on the basis of engineering data,

all recognizing the reality of variation.

The subject of engineering statistics has as its goal to provide the concepts and tools needed by an engineer who faces a problem for which his or her background or theory do not serve as adequate guides to a solution. It supplies principles of how to efficiently acquire and process empirical information for use in understanding and manipulating engineering systems.

To better understand the definition of engineering statistics, it is helpful to consider a real situation and how the elements of the subject enter a real problem.

EXAMPLE 1-1

Heat Treating Gears. The article "Statistical Analysis: Mack Truck Gear Heat Treating Experiments" by P. Brezler, (*Heat Treating*, November, 1986) describes an illustrative situation. A process engineer was faced with the question, "How should untreated gears be loaded into a continuous carburizing furnace in order to minimize distortion during heat treating?" Various people had various semi-informed opinions about how it should be done — in particular, about whether the gears should be laid flat in stacks or hung on rods passing through the gear bores. But no one really knew the consequences of laying versus hanging.

In order to settle the question, the engineer decided to get the facts — to collect some data on "thrust face runout," a measure of gear distortion, for gears laid and gears hung. Notice that exactly how this data collection should be performed required careful thought and planning. There were possible differences in gear raw material lots; machinists and machines that produced the gears; furnace conditions at different times and positions within the furnace; technicians and measurement devices that would produce the final runout measurements; etc. The engineer did not want these variations either to be mistaken for differences between the two loading techniques or to unnecessarily cloud the picture. Avoiding this required care.

In fact, the engineer conducted a well thought-out and executed study. Table 1-1 shows the runout values obtained for 38 gears laid and 39 gears hung after heat treating.

TABLE 1-1

Thrust Face Runouts (.0001 in.)

Gears Laid	Gears Hung
5, 8, 8, 9, 9	7, 8, 8, 10, 10
9, 9, 10, 10, 10	10, 10, 11, 11, 11
11, 11, 11, 11, 11	12, 13, 13, 13, 15
11, 11, 12, 12, 12	17, 17, 17, 17, 18
12, 13, 13, 13, 13	19, 19, 20, 21, 21
14, 14, 14, 15, 15	21, 22, 22, 22, 23
15, 15, 16, 17, 17	23, 23, 23, 24, 27
18, 19, 27	27, 28, 31, 36

In raw form, the runout values are hardly understandable. They lack organization; it is not possible to simply look at Table 1-1 and tell what is going on. The data needed to be summarized. One thing that was done was to compute some numerical summarizations of the data. For example, the process engineer found

$$\text{Mean laid runout} = 12.6$$

$$\text{Mean hung runout} = 17.9$$

Further, a simple graphical summarization was made of the data, as shown in Figure 1-1.

From these summaries of the runout data, several points are obvious. One is that there is variation in the runout values, even within a particular loading method. Variability is an omnipresent fact of life when one sets about to use data, and all statistical methodology and philosophy is conceived and applied in recognition of the reality of variation. In the case of the gears, it appears from Figure 1-1 that there is somewhat more variation in the hung values than in the laid values.

But in spite of the variability that complicates comparison between the loading methods, Figure 1-1 and the two group means calculated by the process engineer carry the message that the laid runouts are on the whole smaller than the hung runouts. By how much? One answer is

$$\text{Mean hung runout} - \text{Mean laid runout} = 5.3$$

FIGURE 1-1

Dot Diagrams of Runouts

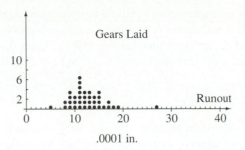

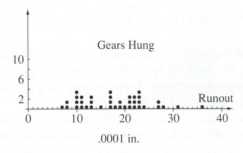

But how "precise" is this figure? It is already clear that runout values are variable. Is there any assurance that the difference seen in the present means would hold up and reappear in further testing? Or is it possibly explainable as simply "stray background noise?" Laying gears is more expensive than hanging them. Can one know whether the extra expense is justified?

These questions point to the need for methods of formal statistical inference from data and subsequent translation of those inferences into practical conclusions. Methods studied later in this text can, for example, be used to support the following statements about hanging and laying gears.

1. One can be roughly 90% sure that the difference in mean runouts that would be produced under conditions like those of the engineer's study is in the range

3.2 to 7.4

2. One can be roughly 95% sure that 95% of runouts for gears laid under conditions like those of the engineer's study would fall in the range

3.0 to 22.2

3. One can be roughly 95% sure that 95 % of runouts for gears hung under conditions like those of the engineer's study would fall in the range

.8 to 35.0

These are formal quantifications of what was learned from the study of laid and hung gears. To derive practical benefit from statements like these, the process engineer had to combine them with other information, such as the consequences of a given amount of runout and the comparative costs for hanging and laying gears, and apply sound engineering judgment to form a rational plan of action. Depending on the specifics of quality needs and costs, the engineer might or might not have been led to adopt the laying method for furnace loading. But as it turned out, the runout improvement was great enough to justify some extra expense, and the laying method was implemented.

The example shows how, in a fairly simple situation, the elements of the subject of statistics were helpful in solving an engineer's problem. Throughout this text, the intention is to convey the understanding that the topics discussed are not ends in themselves, but rather tools that engineers can use to help them do their jobs effectively.

1-2 *Basic Terminology*

A first step when entering any new subject area is to learn some basic terminology. Like most subjects, engineering statistics has both new words to be learned and new technical meanings for familiar words to be absorbed. This section introduces some common statistical jargon for types of statistical studies, types of data that can arise in those studies, and types of structures those data can have.

Types of Statistical Studies

When an engineer sets about to observe a system or process and gather data from it, it must be decided how active the observer's role will be. Will there be turning of knobs and manipulation of process variables, or will one simply let things happen and record the salient features of what transpires?

DEFINITION 1-2

An **observational study** is one in which the investigator's role is basically passive. A process or phenomenon is watched and data are recorded, but there is no intervention on the part of the person conducting the study.

DEFINITION 1-3

An **experimental study** (or more simply, an *experiment*) is one in which the investigator's role is active. Process variables are manipulated, and the study environment is regulated.

Most real statistical studies have both observational and experimental features, and the two definitions above should be thought of as representing opposite ends of a continuum. On this continuum, the experimental end is usually thought of as providing the most efficient and reliable ways to collect engineering data. This is true for several reasons. First, when one adopts a passive posture in data collection (hoping simply to observe one or more instances of favorable process behavior and then try to identify a cause from circumstances accompanying that good behavior), a long wait may be involved. It is typically much quicker to manipulate process variables and watch how a system responds to the changes.

In addition, it is far easier and safer to infer *causality* from an experiment than from an observational study. The problem with passive observation in this regard is the complexity of real systems. One may observe several instances of good process performance and note that they were all surrounded by circumstances X, without being safe in assuming that circumstances X cause good process performance. There may be important variables in the background — ones that are not recognized as part of circumstances X and perhaps not even observed — that are changing and are the true reason for instances of favorable system behavior. These so-called *lurking variables* may govern both process performance and circumstances X. Or it may simply be that many variables change haphazardly without appreciable impact on the system and that by chance, during one's limited period of observation, some of these happen to produce X at the same time that good performance occurs. In either case, an engineer's effort to create situation X, as a means of making things work well, will be wasted effort.

On the other hand, in an experiment, where the study environment is largely regulated except for a few variables the engineer changes in a purposeful way, an inference of causality is much stronger. If circumstances created by the investigator are consistently accompanied by favorable results, one can be reasonably sure that they cause the favorable results.

EXAMPLE 1-2

Pelletizing Hexamine Powder. Cyr, Ellson, and Rickard attacked the problem of reducing the number of nonconforming fuel pellets being produced in the compression of a raw hexamine powder in a pelletizing machine. This was a problem of long standing. (In fact, it has served as the basis for a number of student projects.) There were many possible factors influencing the percentage of nonconforming pellets: among others, Machine Speed, Die Fill Level, Percent Paraffin added to the hexamine, Room Temperature, Humidity at manufacture, Moisture Content, "new" versus "reground" Composition of the mixture being pelletized, and the Roughness of the chute entered by the freshly stamped pellets. Trying to passively keep track of these factors, to notice when a high percentage of the pellets produced happened to be acceptable and then to infer how best to run the pelletizing machine, was a hopeless proposition.

In fact, the students were able to make significant progress by conducting an experiment. They chose three of the factors above that seemed most likely to be important and purposely changed their values, while holding the other nonexperimental factors as close to constant as possible. The important changes they observed in the percentage of acceptable fuel pellets were appropriately attributed to the influence of the system variables they had manipulated.

Besides the distinction between observational and experimental statistical studies, it is helpful to distinguish between studies on the basis of the intended breadth of application of the results. Two relevant terms, popularized by W.E. Deming, are defined next.

DEFINITION 1-4

An **enumerative study** is one in which there is a concrete, well-defined group of objects under study. Data are collected on some or all of these objects, and conclusions are intended to apply only to these objects.

DEFINITION 1-5

An **analytical study** is one in which a process or phenomenon is investigated at one point in space and time, with the hope that the data collected will be representative of system behavior at other places and times where similar conditions hold. In this kind of study, there is rarely, if ever, a concrete group of objects to which conclusions are thought to be limited.

Most engineering studies tend to be of the second type, although some important engineering applications involve enumerative work. One such example is the reliability testing of one (or a few) of a kind of critical components — e.g., for use in a space shuttle. The interest is in the one (or the few) components actually in hand and how well they can be expected to perform, rather than on any broader problem like "the behavior of all components of this type." *Acceptance sampling* (where incoming lots of raw materials or vendor-produced components are checked before taking formal receipt), can be thought of as another important kind of enumerative study. But as indicated, most engineering studies are analytical in nature.

EXAMPLE 1-2
(continued)

The students working on the pelletizing machine were not interested in any particular batch of pellets, but rather in the question of how to make the machine work effectively. They hoped (or tacitly assumed) that what they learned in the spring of 1986 about making fuel pellets would remain valid at later times, at least under shop conditions like those they were facing. Their experimental study was analytical in nature.

Particularly when discussing enumerative studies, the next two definitions are helpful.

DEFINITION 1-6

A **population** is the entire group of objects about which one wishes to gather information in a statistical study.

DEFINITION 1-7

A **sample** is the group of objects on which one actually gathers data in a statistical study. In the case of an enumerative investigation, the sample is a subset of the population (and can in some cases include the entire population).

If a crate of 100 machine parts is delivered to a loading dock and five are examined in order to verify the acceptability of the lot, the 100 parts constitute the population of interest, and the five parts make up a (single) sample of size 5 from the population. (Notice the word usage here. There is one sample, not five samples.)

There are several ways in which the meanings of the words *population* and *sample* are often extended. For one, it is common to use the words in reference not only to objects under study but also to data values associated with those objects. For example, if one thinks of Rockwell hardness values associated with 100 crated machine parts, the 100 hardness values might be called a population (of numbers). Five hardness values corresponding to the parts examined in acceptance sampling could be termed a sample from that population.

EXAMPLE 1-2
(continued)

Cyr, Ellson, and Rickard identified eight different sets of experimental conditions under which to run the pelletizing machine. Several production runs of fuel pellets were made under each set of conditions, and each of these produced its own percentage of conforming pellets. These eight sets of percentages can be referred to as eight different samples (of numbers).

Also, although strictly speaking there is no concrete population being investigated in an analytical study, it is common to talk in terms of a *conceptual population* in such cases. Phrases like "the population consisting of all widgets that could be produced under these conditions" are sometimes used. This language is unfortunate, because it tends to encourage fuzzy thinking. But it is a common usage, and it is supported by the fact that one typically uses the same mathematics when drawing inferences from samples in enumerative and analytical contexts.

Types of Data

The word *data* has already been used in this text many times. It is time to reflect briefly on the kinds of data one can encounter. One useful distinction is provided by the degree to which data are intrinsically numerical.

DEFINITION 1-8

Qualitative or **categorical** data are the values of basically nonnumerical characteristics associated with items in a sample. There can be an order associated with qualitative data, but aggregation and counting are really required to produce any meaningful numerical values from such data.

Consider again five machine parts, constituting a sample from 100 crated parts. If each part can be classified into one of the (ordered) categories 1) conforming, 2) rework, and 3) scrap, and one knows the classifications of the five parts, one has five qualitative data points. If one then aggregates across the five and finds three conforming, one reworkable, and one scrap, then numerical summaries have been derived from the original categorical data.

In contrast to categorical data are numerical data.

DEFINITION 1-9

Quantitative or **numerical** data are the values of numerical characteristics associated with items in a sample. These are typically either counts of the number of occurrences of a phenomenon of interest or measurements of some physical property of the items.

Returning to the crated machine parts, Rockwell hardness values for five selected parts would constitute a set of quantitative measurement data. Tallies of visible blemishes on a machined surface for each of the five selected parts would make up a set of quantitative count data.

The next section considers the topic of measurement in some detail, but one point regarding measurement deserves to be made in this introduction of basic terminology. That has to do with how precisely physical quantities can be measured. It is sometimes mathematically very convenient to act as if infinitely precise measurement were possible. In that case, measured variables are continuous in the sense that their sets of possible values are whole (continuous) intervals of numbers. For example, a convenient idealization might be that the Rockwell hardness of a machine part can lie anywhere in the interval $(0, \infty)$.

But of course this is only an idealization. All real measurements are to the nearest unit (whatever that unit may be). This is especially obvious in a time when measurement instruments are increasingly equipped with digital rather than analog displays. So in reality, when looked at under a strong enough magnifying glass, all numerical data (both measured and count alike) are discrete in the sense that they have isolated possible values rather than a continuum of available outcomes. Although $(0, \infty)$ may be mathematically convenient and completely adequate for practical purposes, the real set of possible values for the measured Rockwell hardness of a machine part may be more like $\{.1, .2, .3, \ldots\}$ than like $(0, \infty)$.

It is well-known conventional wisdom (and generally sound advice) that where at all possible, numerical data are preferable to categorical data and measurements to counts. On the whole, statistical methods for quantitative data (in particular, measurement data) are simpler and more informative than methods for qualitative data. Further, there is typically far more to be learned from appropriate quantitative data than from qualitative data on the same number of physical objects. However, this must sometimes be balanced against the fact that measurement can be more time-consuming (and thus expensive) than the gathering of qualitative data.

EXAMPLE 1-3

Pellet Mass Measurements. As a preliminary to their experimental study on the pelletizing process (discussed in Example 1-2), Cyr, Ellson, and Rickard collected data on a number of aspects of machine behavior. Included was the mass of pellets produced under standard operating conditions. Since a nonconforming pellet is typically one from which some material has broken off during production, pellet mass is indicative of system performance. Informal specifications for pellet mass ranged from 6.2 to 7.0 grams.

Information on 200 pellets was collected. The students could have simply observed and recorded whether or not a given pellet had mass within the informal specifications, thereby producing qualitative data. Instead, they took time enough to actually measure pellet mass to the nearest .1 gram — thereby collecting quantitative data. A graphical summary of their findings is shown in Figure 1-2.

FIGURE 1-2

Pellet Mass Measurements

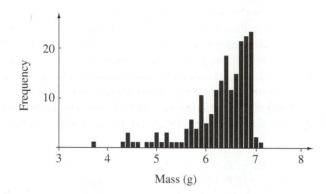

Notice that one can recover from the measurement data all that could be learned from the conformity/nonconformity information — namely, that about 28.5% (57 out of 200) of the pellets had masses that were out of specifications. But there is much more in Figure 1-2 besides this. The shape of the display can give insights into how the machine is operating and the likely consequences of simple modifications to the pelletizing process.

For example, note the *truncated* or chopped-off appearance of the figure. Masses do not trail off on the high side as they do on the low side. Upon reflection, the students realized that this feature of their data set has its origin in the fact that after powder is dispensed into a die, it passes under a paddle that wipes off excess material before a cylinder compresses the powder in the die. The amount dispensed to a given die may have a fairly symmetric mound-shaped distribution, but the paddle probably introduces the truncated feature of the display.

Also, from the numerical data displayed in Figure 1-2, one can find a percentage of pellet masses in any interval of interest, not just the interval [6.2, 7.0]. And by mentally sliding the figure to the right, it is even possible to project the likely effects of increasing die size by various amounts.

It is typical in engineering studies to have several response variables of interest. The next definitions present some jargon that is useful in specifying how many variables are involved and how they are related.

DEFINITION 1-10

Univariate data arise when only a single characteristic of each sampled item is observed.

DEFINITION 1-11

Multivariate data arise when observations are made on more than one characteristic of each sampled item. A special case of this notion involves two characteristics — **bivariate data**.

DEFINITION 1-12

When multivariate data consist of several determinations of basically the same characteristic (e.g., made with different instruments or at different times), the data are called **repeated measures data**. In the special case of bivariate responses, the term **paired data** is often used.

It is important to recognize the multivariate character of data when it is present. Having Rockwell hardness values for 5 of 100 crated machine parts and determinations of the percentage of carbon for five other parts is not at all equivalent to having both hardness and carbon content values for a single sample of five parts. One has two samples of five univariate data points in the first case and a single sample of five bivariate data points in the second. The statistical methods appropriate for use in the two situations are different. The second situation is preferable to the first, because it allows analysis and exploitation of any relationships that might exist between the variables recorded.

EXAMPLE 1-4

Paired Distortion Measurements. In the furnace-loading scenario discussed in Example 1-1, radial runout measurements were actually made on all $38 + 39 = 77$ gears both before and after heat treating. (Only after-treatment values were given in Table 1-1.) Therefore, the process engineer had two samples (of respective sizes 38 and 39) of paired data. Because of the pairing, the engineer was in the position of being able (if desired) to analyze how post-treatment distortion was correlated with pre-treatment distortion.

Types of Data Structures

Statistical engineering studies are sometimes conducted with the purpose of comparing process performance at one set of conditions to a stated standard. Such investigations involve only one sample. But it is more common for several sets of conditions to be compared with each other, in which case several samples are involved. In multisample studies, a variety of standard notions of structure or organization can be employed to advantage. Three of these are briefly discussed in the remainder of this section.

DEFINITION 1-13

A (**complete**) **factorial study** is one in which several process variables (and settings of each) are identified as being of interest, and data are collected under each possible combination of settings of the process variables. The process variables are usually called **factors**, and the settings of each variable that are studied are termed **levels** of the given factor.

For example, suppose there are four factors of interest — and call them A, B, C, and D for convenience. If A has 3 levels, B has 2, C has 2, and D has 4, a study that includes samples collected under each of the $3 \times 2 \times 2 \times 4 = 48$ different possible sets of conditions would be called a $3 \times 2 \times 2 \times 4$ factorial study.

EXAMPLE 1-2
(continued)

Experimentation with the pelletizing machine produced data with a $2 \times 2 \times 2$ (or, for shorthand, 2^3) factorial structure. The factors and respective levels studied were

Die Volume	low volume vs. high volume
Material Flow	current method vs. manual filling
Mixture Type	no binding agent vs. with binder

Combining these then produced eight sets of conditions under which data were collected (see Table 1-2).

TABLE 1-2

Combinations in a 2^3 Factorial Study

Condition Number	Volume	Flow	Mixture
1	low	current	no binder
2	high	current	no binder
3	low	manual	no binder
4	high	manual	no binder
5	low	current	binder
6	high	current	binder
7	low	manual	binder
8	high	manual	binder

When many factors and/or levels are involved, the number of samples required to constitute a full factorial study quickly reaches a size that is impractical by most standards. Engineers often find that they want to collect data for only some of the combinations that would make up a complete factorial study.

DEFINITION 1-14

A **fractional factorial study** is one in which data are collected for only some of the combinations that would make up a complete factorial study.

A little reflection ought to make clear two points concerning this definition. In the first place, it should be obvious that (at least for samples of a given size) one cannot hope to learn as much about how a response is related to a given set of factors from a fractional factorial study as from the corresponding full factorial study. Some information must be lost when only part of all possible sets of conditions are studied.

Secondly, some fractional factorial studies will be potentially more informative than others. If only a fixed number of samples can be taken, which samples to take is an issue that needs careful consideration. In fact, there is an entire subfield of statistics that concerns itself with identifying good choices of fractional factorials. A major part of Chapter 10 deals with fractional factorials in detail and concerns how to choose good ones, taking into account what part of the potential information from a full factorial study they can provide.

EXAMPLE 1-2
(continued)

The experiment actually carried out on the pelletizing process was, as indicated in Table 1-2, a full factorial study. For the sake of illustration, however, Table 1-3 contains a listing of four experimental combinations, forming what turns out to be a well-chosen fractional factorial study that includes half of the possible combinations. (These are the combinations numbered 2, 3, 5, and 8 in Table 1-2.)

TABLE 1-3

Half of the 2^3 Factorial

Volume	Flow	Mixture
high	current	no binder
low	manual	no binder
low	current	binder
high	manual	binder

One other basic type of data structure occurs in cases where groups of samples are somehow related, as are groups of groups, as are groups of groups of groups, etc. That is, there are studies involving a tree-type structure.

DEFINITION 1-15

A **hierarchical (or nested) study** is one in which samples aggregate naturally into groups of samples, which themselves aggregate naturally into groups of groups of samples, and so on. An alternative way of stating this is to say that the study involves a series of factors A, B, C, etc., where levels of B actually represent sublevels of factor A, levels of C represent sublevels of factor B, etc.

A schematic of the structure of a typical hierarchical study is given in Figure 1-3. In engineering studies, hierarchical structures arise naturally in a variety of ways. One is via subsampling. If a metallurgical engineer wishes to monitor manganese content in 1040 steel, the engineer might select five heats of steel, pick three ingots from each heat, and make two manganese determinations for each ingot. The samples of two determinations aggregate naturally according to ingots, which in turn aggregate naturally according to heats. Differently put, levels of the factor Ingot are really sublevels of the factor Heat, and levels of the factor Determination are really sublevels of the factor Ingot.

FIGURE 1-3

Schematic of a Typical
Hierarchical Data Structure

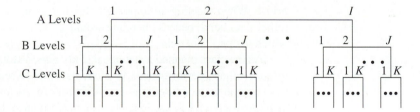

Another way in which nested structure often arises in engineering studies is through aggregation/subdivision according to the variable Time.

EXAMPLE 1-5

Hierarchical Structure in a Skiving Study. Bailey, Goodman, and Scott studied a so-called skiving process at a company that produces metal-reinforced rubber hydraulic hoses. Skiving consists of removing the outer layer of rubber on the ends of a hose before attaching couplings. The length of this skive is important, so the company routinely monitors this variable. The students collected skive lengths for five hoses, four times per shift, for two shifts per day for the first three days in December. They ended up with a data set that can be thought of as having hierarchical structure, samples of size 5 being aggregated by shift and in turn aggregated by day.

Unwanted variation is both a principal cause and principal symptom of poor process performance. Engineers are often called upon to track down and eliminate the sources of such variation. Hierarchical studies are a useful tool in that pursuit; they are sometimes called *components of variance studies* for this reason. For example, the students in Example 1-5 were in a position to investigate the issue, "Where does the most of the variation in skive length arise? Is it within a shift? Between shifts in a day? Or between days?"

1-3 *Measurement: Its Importance and Difficulty*

A prerequisite to success in statistical engineering studies is the ability to measure important features of processes and objects. For example, a chemical engineer might suspect that in order to improve a polymer production process, a pH needs adjustment. But this would hardly be helpful if there were no meter for measuring pH or if the existing meter were unreliable. Similarly, it would make no sense for a mechanical engineer to set specifications on the dimensions of a machine part if the process engineers overseeing manufacture cannot obtain gaging to monitor it, or to set tolerances on a dimension that are more stringent than the precision of the gages that will be used.

For some physical properties like length, mass, temperature, and so on, methods of measurement are commonplace and obvious. Often the behavior of an engineering system can be adequately characterized in terms of such properties. But when it cannot, engineers must carefully define what it is about the system that needs observing and then apply ingenuity and the physical and mathematical theory at their disposal to create a suitable method of measurement.

EXAMPLE 1-6

Measuring Brittleness. A senior class in metallurgical engineering design took on the project of helping a heavy-machinery manufacturer improve the performance of a spike-shaped metal part. In its intended application, this part needed to be strong but very brittle. When meeting an important obstruction in its path of motion, it must break off rather than bend, because bending would in turn cause other damage to the machine in which the part functions.

As the class planned a statistical study aimed at finding what variables of manufacture affect part performance, the students came to realize that the manufacturer didn't have a good way of assessing part performance. As a necessary step in their study, they developed a device to use in measuring performance. It looked roughly as in Figure 1-4.

FIGURE 1-4

A Device for Measuring Brittleness

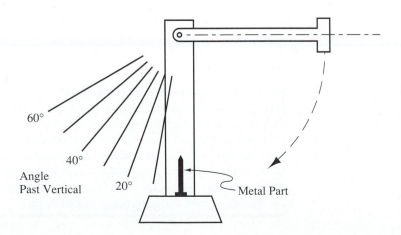

A swinging arm with a large mass at its end was brought to a horizontal position, released, and allowed to swing through a test part firmly fixed in a vertical position at the bottom of its arc of motion. The number of degrees past the vertical that the arm traversed after impact with the part provided a measure of brittleness that allowed the class to proceed effectively in its study.

EXAMPLE 1-7

Measuring Wood Joint Strength. Dimond and Dix wanted to conduct a factorial study comparing joint strengths for combinations of three different woods and three glues. But they didn't have access to strength-testing equipment. Not being easily dissuaded, they invented their own method of strength testing. To test a joint, they suspended a large container from one of the pieces of wood involved and poured water into the container until its weight was sufficient to break the joint. Knowing the volume of water poured into the container and the density of water, they could determine the force required to break the joint.

Regardless whether an engineer uses off-the-shelf measurement technology or must fabricate a new measurement device, a number of issues concerning measurement must be considered. These include notions of validity, measurement variation/error, accuracy, and precision.

DEFINITION 1-16

A measurement or measuring method is called **valid** if it usefully or appropriately represents the feature of an object or system that is of engineering importance.

It is impossible to overstate the importance of facing the question of validity before plunging ahead in a statistical engineering study. Collecting data (like all engineering activities) costs money. Expending substantial resources collecting data, only to later decide they don't really help address the problem at hand, is unfortunately all too common.

For example, consider the case of a major auto manufacturer interested in customer satisfaction. Substantial company effort was focused on the reduction of the (company) cost due to warranty repairs, which was taken as a likely indicator of customer satisfaction. It was only some time later that company engineers realized that the frequency of warranty repair requests was probably a much better index of customer approval.

The point was made in Section 1-1 that when using data, one is quickly faced with the fact that variation is omnipresent. Some of that variation comes about because the objects studied are never exactly alike. But some of it has its origin in the fact that measurement processes also have their own inherent variability. Given a fine enough scale of measurement, no amount of care will produce exactly the same value over and over in repeated measurement of even a single object. And it is naive to attribute all variation in repeat measurements to bad technique or sloppiness. (Of course, bad technique and sloppiness can increase measurement variation beyond that which is unavoidable.)

An exercise suggested by W.J. Youden in his book *Experimentation and Measurement* is helpful in making clear the reality of measurement error. Consider the problem of measuring the thickness of the paper in this book. The technique to be used is as follows. The book is to be opened to a page somewhere near the beginning and one somewhere near the end. The stack of pages in between the two is to be grasped firmly between the thumb and index finger and the stack thickness read to the nearest .1 mm. Dividing the stack thickness by the number of sheets in the stack and recording the result to the nearest .0001 mm will then produce the thickness measurement. Try this whole process (say, ten times), closing the book between measurements, so that presumably different stacks of pages will be encountered.

EXAMPLE 1-8

Book Paper Thickness Measurements. This author regularly assigns the previous measurement exercise in engineering statistics classes as homework. Presented below are ten measurements of the thickness of the paper in Box, Hunter, and Hunter's *Statistics for Experimenters*, made one semester by students Wendel and Gulliver.

Wendel: .0807, .0826, .0854, .0817, .0824, .0799, .0812, .0807, .0816, .0804
Gulliver: .0972, .0964, .0978, .0971, .0960, .0947, .1200, .0991, .0980, .1033

A graphical presentation of these data is given in Figure 1-5. The figure shows clearly that even repeat measurements by one person on one book will vary, and also that the patterns of variation for two different individuals can be quite different. (Wendel's values are both smaller and more consistent than Gulliver's.) However, note that it is impossible to know, from the particular data shown here, exactly how much of the difference in the two data patterns is due to the fact that the students were working with different copies of the text and how much is due to basic differences in the two students as measuring systems.

FIGURE 1-5

Dot Diagrams of Paper Thickness Measurements

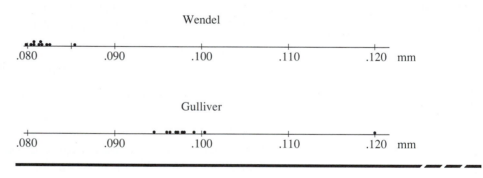

The variability or error that is inevitable in measuring systems can be thought of as having both internal and external components.

DEFINITION 1-17

A measurement system is called **precise** if it produces small variation in repeated measurement of the same object.

Precision is the internal consistency of a measurement system; typically, it can be improved only with basic changes in the configuration of the system.

EXAMPLE 1-8
(continued)

Ignoring the remote possibility that some property of Gulliver's book was responsible for his values showing more spread than those of Wendel, it appears that Wendel's measuring technique was more precise than Gulliver's.

Notice also that the precision of both students' measurements could probably have been improved by giving each a binder clip and a micrometer. The binder clip would provide a relatively constant pressure on the stacks of pages being measured, thereby eliminating the subjectivity and variation involved in grasping the stack firmly between thumb and index finger. And for obtaining stack thickness, a micrometer is clearly a more precise instrument than a ruler.

━━━━━━━━━━━━━━━━━━━━━━━━━ ▬ ▬ ▬

Precision of measurement is important, but for many purposes it alone is not adequate.

DEFINITION 1-18

A measurement system is called **accurate** (or sometimes, **unbiased**) if on average it produces the true or correct value of a quantity being measured.

Accuracy is the agreement of a measuring system with some external standard of measurement. It is a property that can typically be changed without extensive physical change in a measurement method. *Calibration* of a system against a standard (bringing it in line with the standard) can be as simple as comparing system measurements to a standard, developing an appropriate conversion scheme, and thereafter using converted values in place of raw readings from the system.

EXAMPLE 1-8
(continued)

This author doesn't know what the industry's standard measuring methodology would have produced for paper thickness in Wendel's copy of the text. But for the sake of example, suppose that a value of .0850 mm/sheet was appropriate. The fact that Wendel's measurements averaged about .0817 mm/sheet suggests that her future accuracy in measuring book paper thickness might be improved by proceeding as before but then multiplying any figure obtained by the ratio of .0850 to .0817 — i.e., multiplying by 1.04.

━━━━━━━━━━━━━━━━━━━━━━━━━ ▬ ▬ ▬

Maintaining the United States reference sets for physical measurement is, of course, the business of the National Institute of Standards and Technology (formerly called the National Bureau of Standards). It is important business. Poorly calibrated measuring devices may be sufficient for local purposes of comparing local conditions. But if one is to establish the values of quantities in any absolute (rather than relative) sense, or to expect local values to have meaning at other places and other times, it is essential to calibrate (and indeed recalibrate) measurement systems against a constant standard. A millimeter must be the same today in Iowa as it was last week in Alaska.

The possibility of bias or inaccuracy in measuring systems has at least two important implications for planning statistical engineering studies. First, the fact that measurement systems can lose accuracy over time demands that their performance be monitored over time, and that they be recalibrated when appropriate. The well-known phenomenon of *instrument drift* can ruin an otherwise flawless statistical study.

Second, it is clear that whenever possible, a single system should be used to measure responses. If several measurement devices or technicians are used, it is hard to know whether the differences observed originate with the variables under study or from differences in devices or technician biases. If the use of several measurement systems is unavoidable, they must be calibrated against a standard (or at least against each other). The following example admirably illustrates the role that human differences play in this matter.

EXAMPLE 1-9

Differences between Technicians in their Use of a Gage. Cowan, Renk, Vander Leest, and Yakes worked with a company on the monitoring of a critical dimension of a high-precision metal part that was machined on computer-controlled equipment. They encountered large, initially unexplainable variation in this dimension between different shifts at the plant. This variation was eventually traced not to any real difference in the parts produced on different shifts, but to an instability in the company's measuring system. A single gage was in use on all shifts, but different technicians turned out to use it quite differently when measuring the critical dimension. The company needed to train the technicians in a single, standardized method of using the gage.

An analogy that is sometimes helpful in understanding the difference between the concepts of precision and accuracy involves comparing measurement to target shooting. In target shooting, one can be on or off target (accurate or inaccurate), with a large or small cluster of shots (showing precision or imprecision). Figure 1-6 shows this analogy.

FIGURE 1-6

Measurement/Target
Shooting Analogy

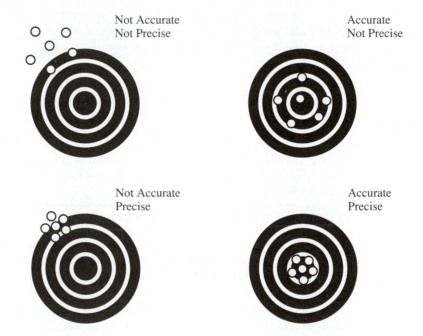

Good measurement is hard work, but without it data collection is futile. To learn how to manipulate processes to advantage, engineers must obtain valid measurements, taken by methods whose precision and accuracy are sufficient to let them see changes in system behavior that are of engineering importance. Usually, this means that measurement inaccuracy and imprecision must be an order of magnitude smaller than the variation in measured response caused by naturally occurring system fluctuations.

1-4 *Mathematical Models, Reality, and Data Analysis*

This is not a book on mathematics. Nevertheless, it contains a fair amount of mathematics, which most readers will find to be reasonably elementary — if unfamiliar and initially puzzling. Therefore, it seems wise to try to put the mathematical content of the book in perspective early. In this section, the relationships of mathematics to the physical world and to engineering statistics are discussed.

Mathematics is a construct of the human mind. Its organization of concepts and symbols into logically coherent equations, theorems, and theories exists for two different reasons. For one, people have found the development of mathematics a challenging and rewarding intellectual pursuit. However, such a perspective rarely motivates an engineer's study of mathematics. Rather, engineers generally approach the subject from the point of view that some mathematics has proved useful in describing and predicting how physical systems behave.

Although they exist only in our minds, appropriate mathematical theories are guides in every branch of modern engineering. Throughout this text, we will frequently use the phrase *mathematical model*.

DEFINITION 1-19

A **mathematical model** is a description or summarization of salient features of a real-world system or phenomenon in terms of symbols, equations, numbers, and the like.

Mathematical models are themselves not reality, but they can be extremely effective descriptions of reality. This effectiveness hinges on two somewhat opposing properties of a mathematical model: 1) its degree of simplicity and 2) its predictive ability. The most powerful mathematical models are (not surprisingly) those that simultaneously are simple *and* generate good predictions. A model's simplicity (or *tractability*) allows one to maneuver mathematically within the model's framework, deriving consequences of model assumptions that translate into predictions of process behavior. When these are largely correct in an empirical sense, one has an effective engineering tool.

The elementary "laws" of mechanics are an outstanding example of effective mathematical modeling. The simple mathematical statement that the acceleration due to gravity is constant,

$$a = g$$

yields, after one easy mathematical maneuver (an integration), the prediction that beginning with 0 velocity, after a time t in free fall an object will have

$$\text{Velocity} = gt$$

And a second integration gives the prediction that beginning with 0 velocity, a time t in free fall produces

$$\text{Displacement} = \frac{1}{2}gt^2$$

The beauty of this is that for most practical purposes, these easy predictions are quite adequate. That is, they agree well with what is observed empirically. They are relationships that can more or less be counted on as an engineer designs, builds, operates, and/or improves physical processes or products.

But then, how does the notion of mathematical modeling interact with the subject of engineering statistics? There are several ways. For one, data collection and analysis are essential in fitting or estimating parameters of mathematical models. To understand this point, consider again the example of a body in free fall. If one postulates that the acceleration due to gravity is constant, there remains the question of what numerical value that constant should have. The quantity g is a parameter of the model that must be evaluated before the model can be used for practical purposes. One does this by gathering data and using them to estimate the parameter.

This author still has (not entirely pleasant) memories of his first college physics lab, where the object was to evaluate g. The method used was to release a steel bob down a

FIGURE 1-7

A Device for Measuring *g*

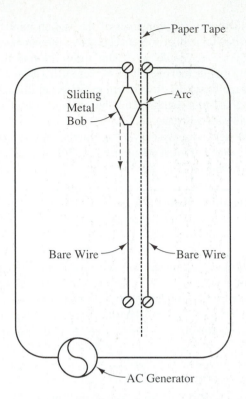

vertical wire that ran through a hole in its center and allow 60-cycle current to arc from the bob, through a paper tape, to another vertical wire, burning the tape slightly with every arc. A schematic diagram of the apparatus used is shown in Figure 1-7.

The vertical positions of the burn marks were of course bob positions, at intervals of $\frac{1}{60}$ of a second. Table 1-4 gives measurements of these positions taken from a (new) tape by this author, nearly 20 years after his first experience with this demonstration. (The author is grateful to Dr. Frank Peterson of the ISU Physics and Astronomy Department for supplying the tape used to produce the values in Table 1-4.) Plotting the bob positions in the table on graph paper at equally spaced intervals produces the approximately quadratic plot shown in Figure 1-8.

Picking a parabola to fit the plotted points amounts to identifying an appropriate value for *g*. A method of curve fitting (discussed in Chapter 4) called *least squares* produces a value for *g* of 9.79 m/sec^2, not far from the commonly quoted value of 9.8 m/sec^2.

Notice that (at least before Newton) the data in Table 1-4 might also have been used in another way. The parabolic shape of the plot in Figure 1-4 could have in fact suggested the form of an appropriate model for the motion of a body in free fall. That is, a careful observer viewing the plot of position versus time should be led to an approximately quadratic relationship between position and time (and from that via two differentiations to the conclusion that the acceleration due to gravity is roughly constant). This text is full of examples of how helpful it can be in engineering problems to use data both to identify potential forms for empirical models for the behavior of systems and to then estimate parameters of such models (preparing them for use in prediction and system optimization).

This discussion has concentrated on the fact that statistics provides raw material for developing realistic mathematical models of real systems. But there is another related and important way in which the subjects of statistics and mathematics interact. The

TABLE 1-4

Measured Displacements of a
Bob in Free Fall

Point Number	Displacement (mm)
1	.8
2	4.8
3	10.8
4	20.1
5	31.9
6	45.9
7	63.3
8	83.1
9	105.8
10	131.3
11	159.5
12	190.5
13	223.8
14	260.0
15	299.2
16	340.5
17	385.0
18	432.2
19	481.8
20	534.2
21	589.8
22	647.7
23	708.8

FIGURE 1-8

Bob Positions in Free Fall

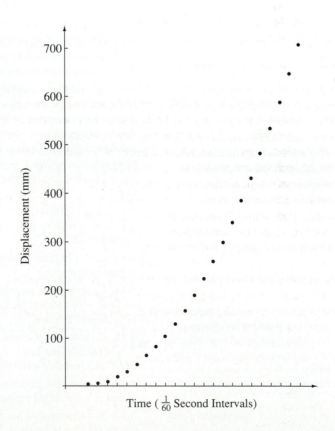

mathematical theory of probability provides a framework that guides statisticians in developing effective methods of dealing with data and in quantifying the uncertainty associated with inferences drawn from data.

DEFINITION 1-20

Probability is the mathematical theory intended to describe situations and phenomena that one would colloquially describe as involving chance.

If, for example, five students arrive at the five different laboratory values of *g*,

$$9.78, \ 9.82, \ 9.81, \ 9.78, \ 9.79$$

questions naturally arise as to how to use them to state both a best value for *g* and some measure of precision for the value. The theory of probability provides guidance in addressing these issues. Material in Chapter 6 shows that probability considerations support using the class average of 9.796 to represent these five values and attaching to it a precision on the order of plus or minus .02 m/sec^2.

Since it is not assumed that the reader has studied probability, this text will have to supply at least a minimal introduction to the subject. But do not lose sight of the fact that probability is not statistics — or vice versa. Rather, probability is a branch of mathematics and a useful subject in its own right. It is met in a statistics course as a tool, because the variation that one sees in real data is closely related conceptually to the notion of chance modeled by the theory of probability.

EXERCISES

1-1. (§1-1) Explain in your own words why engineering practice is an inherently statistical enterprise.

1-2. (§1-1) Explain in your own words why the concept of variability has a central place in the subject of engineering statistics.

1-3. (§1-1) Describe in your own words the difference between descriptive and (formal) inferential statistics.

1-4. (§1-2) Describe a situation in your field where an observational study might be used to answer a question of real importance. Describe another situation where an experiment might be used.

1-5. (§1-2) Describe two different contexts in your field where, respectively, qualitative and quantitative data might arise.

1-6. (§1-2) What kind of information can be derived from a single sample of *n* bivariate data points (x, y) that can't be derived from two separate samples of, respectively, *n* data points *x* and *n* data points *y*?

1-7. (§1-2) Describe a situation in your field where paired data might arise.

1-8. (§1-2) Consider the context of a study of making paper airplanes, where two different Designs (say, delta versus t wing), two different Papers (say, construction versus typing) and two different Loading Conditions (with a paper clip versus without a paper clip) are of interest in terms of their effects on flight distance. Describe a full factorial and then a fractional factorial data structure that might arise from such a study.

1-9. (§1-2) Describe a nested or hierarchical data structure that might arise in the quality control of concrete used in the construction of a large building.

1-10. (§1-3) Why might it be argued that in terms of producing useful measurements, one must deal first with the issue of validity, then the issue of precision, and then the issue of accuracy?

1-11. (§1-3) Often, in order to get an idea of the value of a physical quantity (for example, the mean yield of a batch chemical process run according to some standard plant operating procedures), a large number of measurements of the quantity are made and then averaged. Explain which of the three aspects — validity, precision, and accuracy — this averaging of many measurements can be expected to improve and which it cannot be expected to improve.

1-12. (§1-3) Explain the importance of the stability of the measurement system to the real-world success of a statistical engineering study.

1-13. (§1-4) Explain the importance of mathematical models to engineering practice.

1-14. Calibration of measurement equipment is most clearly associated with which of the following concepts: validity, precision, or accuracy? Explain.

1-15. If factor A has levels 1, 2, and 3, factor B has levels 1 and 2, and factor C has levels 1 and 2, list the combinations of A, B, and C that make up a full factorial arrangement.

1-16. Explain how paired data might arise in a heat treating study aimed at determining the best way to heat treat a certain alloy.

1-17. A 44-point data set consists of copper content measurements on bronze castings. Eleven castings were involved; two pieces were cut from each of the castings, and the copper contents of these pieces were each measured twice. What single term most succinctly describes the structure of the 44-point data set?

1-18. Losen, Cahoy, and Lewis purchased eight spanner bushings of a particular type from a local machine shop and measured a number of characteristics of these bushings, including their outside diameters. Each of the eight outside diameters was measured once by two student technicians, with the following results. (The units are inches.)

Bushing	1	2	3	4
Student A	.3690	.3690	.3690	.3700
Student B	.3690	.3695	.3695	.3695
Bushing	5	6	7	8
Student A	.3695	.3700	.3695	.3690
Student B	.3695	.3700	.3700	.3690

Considering both students' measurements, what type of data are given here? Explain.

1-19. Describe situations from your field where a full factorial study might be conducted (name at least three factors, and the levels of each, that would appear in the study).

1-20. Example 1-8 concerns the measurement of the thickness of book paper. Variation in measurements is a fact of life. To observe this reality, measure the thickness of the paper used in this book ten times. Use the following method. Open the book to pages near the beginning and end. Hold the stack of pages together and measure the thickness of this stack to the nearest .1 mm. Then count up the number of sheets involved and divide the measured thickness by the number of sheets. (Carry out your division to 4 decimal places.) Record the measured thickness, the number of sheets, and the quotient. Close the book and start over. If you are using this book in a formal course, be prepared to hand in your results and compare what you obtained with the values obtained by others in your class.

1-21. Exercise 1-20 illustrates the reality of variation in physical measurement. Another exercise that is similar in spirit, but leads to qualitative data, involves the spinning of U.S. pennies. Spin a U.S. penny on a hard surface 20 different times; for each trial, record whether the penny comes to rest with heads or tails showing. Did all the trials have the same outcome? Is the pattern you actually observed the one you expected to see? If not, do you have any possible explanations for what you actually observed?

1-22. Consider a situation like that of Example 1-1 (involving the heat treating of some gears). Suppose that the original gears can be purchased from a variety of vendors, they can be made out of a variety of materials, they can be heated according to a variety of regimens (involving different times and temperatures), they can be cooled in a number of different ways, and the furnace atmosphere can be adjusted to a variety of different conditions. A number of features of the final gears are of interest, including their flatness, their concentricity, their hardness (both before and after heat treating), and their surface finish.

(a) What kind of data arise if, for a single set of conditions, the Rockwell hardness of several gears is measured both before and after heat treating? (Use the terminology of Section 1-2.) In the same context, suppose that engineering specifications on flatness require that measured flatness not exceed .40 mm. If flatness is measured for several gears and each gear is simply marked Acceptable or Not Acceptable, what kind of data are generated?

(b) Describe a 3-factor full factorial study that might be carried out in this situation. Name the factors that will be used and describe the levels of each that will be involved. Write out a list of all the different combinations of levels of the factors that will be studied.

2 Data Collection

ata collection was identified in Definition 1-1 as one of the principal activities of engineering statistics. In fact, it is the most important activity of engineering statistics. Often, properly collected data will essentially speak for themselves, making formal summarization and inference rather like frosting on the cake. On the other hand, no amount of cleverness in the post-facto processing of data will salvage a badly done study. So it makes sense to consider carefully how to go about gathering data.

This chapter begins with a discussion of some general considerations in the collection of engineering data. It turns next to concepts and methods applicable specifically in enumerative contexts. Then there follows a discussion of both general principles and some more or less specific formal plans for engineering experimentation. Finally, the chapter concludes with advice for the step-by-step planning of a statistical engineering study.

2-1 General Principles in the Collection of Engineering and Industrial Data

Regardless of the purpose and constraints associated with a particular statistical engineering study, a number of general considerations must be kept in mind during the planning of the study. In this section, some of these are discussed. The discussion is organized around the topics of measurement, sampling, and recording.

Measurement

Good measurement is an indispensable ingredient of any statistical engineering study. An engineer planning a study ought to ensure that data on relevant variables (with operational definitions) will be collected by well-trained people using measurement equipment of known and adequate quality.

When choosing variables to observe in a statistical study, the concepts of measurement validity and precision discussed in Section 1-3 must be remembered. One practical point in this regard concerns how directly a measure represents a system property. When a *direct measure* exists, it is typically preferable (even at increased expense) to an indirect measure, because it will usually give much better precision. A surrogate variable is second choice for data collection purposes.

EXAMPLE 2-1

Exhaust Temperature versus Weight Loss. An engineer working on a drying process for a bulk material was having difficulty determining when a target dryness had been reached. The method being used was to monitor the temperature of hot air being exhausted from the drying device, and exhaust temperature was a valid but very imprecise indicator of moisture content.

Someone suggested the possibility of measuring the weight loss of the material, instead of exhaust temperature. The engineer developed an ingenious method of measuring this, at only slightly greater expense. This much more direct measurement greatly improved the quality of the engineer's information.

It is often easier to identify appropriate measures of the behavior of a system than to carefully and unequivocally define those measures so that they can be used. For example, a metal cylinder is to be turned on a lathe, and it is agreed that cylinder diameter is of engineering importance. What is meant by the word *diameter*? Should it be measured

22

on one end of the cylinder (and if so, which?) or in the center, or where? In practice, these locations will differ somewhat. Further, when a cylinder is gaged at some chosen location, should it be rolled in the gage to get a maximum (or minimum) reading, or should it simply be measured as first put into the gage? The cross sections of real-world cylinders are not exactly circular, and how the measurement is done will affect how the resulting data look.

It is especially necessary — and difficult — to make careful *operational definitions* to use in gathering data where qualitative and counted variables are involved. To see this, consider the case of a process engineer responsible for the operation of an injection molding machine producing plastic auto grills. If the number of abrasions appearing on these is of concern and data are to be gathered, how does one define an *abrasion*? There are certainly locations on a grill where a flaw is of no consequence. Should those areas be counted? How big should an abrasion be in order to be included in a count? How (if at all) should an inspector distinguish between abrasions and other imperfections that might appear on a grill? All of these questions must be addressed in the form of an operational definition of abrasion before consistent data collection can take place.

Once developed, operational definitions and standard measurement procedures must be communicated to those who will use them. Training (and often regular retraining) of technicians has to be taken seriously. Workers need to understand the importance of adhering to the standard definitions and methods in order to provide consistency, even in cases where they think the conventions in use are less than ideal. For example, if instructions call for zeroing an instrument before each measurement, it must be consistently done, even if some think it is an unnecessary step.

The performance of any measuring equipment used in a study must be known to be adequate — both before beginning and throughout the study. Most large industrial concerns have regular programs for both recalibrating and monitoring the precision of their numerous measuring devices. The second of these activities sometimes goes under the name of *gage R and R studies* — the two R's being *repeatability* and *reproducibility*. Repeatability is variation observed when a single operator uses the gage to measure and remeasure one item. Reproducibility is variation in measurement attributable to differences among operators. Section 8-4 includes a discussion of some statistical methods useful in gage R and R studies.

Calibration and precision studies should assure the engineer that instrumentation is adequate at the beginning of a statistical study. If the time span involved in the study is appreciable, the *stability* of the instrumentation must be maintained over the study period through checks on calibration and precision and (where needed) adjustments.

Sampling

Once it is established how measurement/observation will proceed, the engineer can consider the particulars of how much to do, who is to do it, where and under what conditions it is to be done, etc. Sections 2-2, 2-3, and 2-4 consider the question of choosing what observations to make, first in enumerative and then in experimental studies. But first, a few general comments about the issues of "How much?", "Who?", and "Where?".

Without a doubt, the most common question engineers ask applied statisticians is "How many observations do I need in order to . . .?" Unfortunately, the proper answer to the question is typically "It depends." Proceeding through this book, the reader should begin to develop some intuition and some rough guides for choosing sample sizes in statistical engineering studies. But for the time being, it will suffice to point out the only factor on which the answer to the sample size question really depends. That is the variation in response that one expects (coming both from unit-to-unit variation and measurement variation).

This makes sense. If objects to be observed were all alike and perfect measurement were possible, then a single observation would suffice for any purpose. But if there is either increase in the noise in the measurement system or increase in the variation in the system or population under study, the sample size necessary to get a clear picture of reality becomes larger.

However, one feature of the matter of sample size sometimes catches people a bit off guard. That is the fact that in enumerative studies (provided the population size is large), sample size requirements do not depend on the population size. That is, sample size requirements are not relative, but rather absolute with regard to population size. If a sample size of 5 is adequate to characterize compressive strengths of a lot of 1,000 red clay bricks, then a sample of size 5 would be adequate to characterize compressive strengths for a lot of 100,000 bricks, with similar brick-to-brick variability.

The "Who?" question of data collection cannot be effectively answered without reference to human nature and behavior. This is true even in a time when so called automatic data collection devices are proliferating. (Humans will continue to supervise these and process the information they generate.)

Those who collect engineering data must not only be well trained; they must also be convinced that the data they collect will be used and in a way that is in their best interests. Good data must be seen as a help in doing a good job, benefiting an organization, and remaining employed, rather than as pointless or even threatening. If those charged with collecting or releasing data believe that the data will be used against them, it is unrealistic to expect them to obtain useful data.

EXAMPLE 2-2

Data — an Aid or a Threat? This author once toured a facility, with a company industrial statistician as guide. That person proudly pointed out evidence that data were being collected and effectively used in the facility. Upon entering a certain department, the tone of the conversation changed dramatically. Apparently, the workers in that department had been asked to collect data on job errors. The data had pointed unmistakably to poor performance by a particular individual, who was subsequently fired frkm the company. Thereafter, convincing other workers that data collection is a helpful activity was, needless to say, a challenge.

Perhaps all the alternatives in this situation (like retraining or assignment to a different job) might already have been exhausted. But that's not the point in any case. Rather, the point is that circumstances were allowed to create an atmosphere that was not conducive to the collection and use of data.

Even where those who will gather data are convinced of its importance and are eager to cooperate, care must be exercised. Personal biases (whether conscious or subconscious) must not be allowed to enter the data collection process. Sometimes in a statistical study, hoped-for or predicted best conditions are deliberately or unwittingly given preference over others. If this is a concern, one kind of technique that can be used is to have measurements made *blind*, (i.e., without personnel knowing what set of conditions led to an item being measured). Other techniques for ensuring fair play, having less to do with human behavior, will be discussed in the next two sections.

The "Where?" question of engineering data collection can be answered in general terms: "As close as possible in time and space to the phenomenon being studied." The importance of this principle is most obvious in the observation of complex manufacturing processes. The performance of one operation in such a process is most clearly and efficiently monitored at the operation, rather than at some later point. If items being produced turn out to be unsatisfactory at the end of the line, it is rarely easy to backtrack and locate the responsible operation or operations. Even if that is accomplished, unnecessary waste has occurred during the time lag between the onset of operation sickness and its later discovery.

EXAMPLE 2-3

IC Chip Manufacturing Process Improvement. The preceding point was illustrated when this author visited a "clean room" where integrated circuit chips are manufactured. These are produced in groups of 50 or so, making up so-called wafers. Wafers are made by successively putting down a number of appropriately patterned, very thin layers of material on an inert background disk. The person conducting the tour said that at one point, a huge fraction of wafers produced in the room had been found (at the end of the line) to be nonconforming. After a number of false starts, it was discovered that by appropriate testing (data collection) at the point of application of the second layer, a majority of the eventually nonconforming wafers could be identified and eliminated, thus saving the considerable extra expense of further processing. What's more, the need for adjustments to the process was signaled in a timely manner.

Recording

The object of engineering data collection is to get data used. How they are recorded has a major impact on how successfully this objective will be met. A carefully planned data recording format can make the difference between success and failure.

EXAMPLE 2-4

A Data Collection Disaster. A group of students (whose names will be suppressed to protect the guilty) worked with a maker of molded plastic business signs in an effort to learn what factors affect the amount of shrinkage a sign undergoes as it cools. They considered factors such as Operator, Heating Time, Mold Temperature, Mold Size, Ambient Temperature, and Humidity. Then they planned a partially observational and partially experimental study of the molding process. After spending two days collecting data, they set about to analyze them. The students discovered to their dismay that although they had recorded many features of what went on during their study, they had neglected to record either the size of the plastic sheets before molding or the size of the finished signs. Their considerable effort was entirely wasted. It is likely that this gaffe could have been prevented by careful pre-collection development of a data collection form.

When data are going to be collected in a routine, ongoing process-monitoring context (as opposed to a one-shot study of limited duration), it is important that they be used to provide effective, timely feedback of information. Increasingly, computer-made graphical displays of data, in real time, are used for this purpose. But it is often possible to achieve this much more cheaply through clever design of a manual data collection form, if one keeps in sight the goal of making data recording convenient and immediately useful (without the need for transfer to another form or device).

As an example of this point, consider the essential uselessness of the little black expense books many of us keep in the glove compartments of our cars for recording gasoline purchases. Typically, odometer reading, gallons purchased, and cost are dutifully recorded and never used. How much more useful it would be to have a recording format that included a computation of miles per gallon and a plot of such values. One would then be in a position to actually track gas mileage over time and (for example) see when a tune-up is needed.

EXAMPLE 2-5

Recording Bivariate Data on PVC Bottles. Table 2-1 presents some bivariate data on bottle mass and width of bottom piece resulting from blow molding of PVC plastic bottles, (taken from *Modern Methods for Quality Control and Improvement* by Wadsworth, Stephens, and Godfrey). Six consecutive samples of size 3 are represented. (These are only part of a larger data set given in the reference.)

TABLE 2-1

Mass and Bottom Piece Widths of PVC Bottles

Sample	Item	Bottle Mass (g)	Width of Bottom Piece (mm)
1	1	33.01	25.0
1	2	33.08	24.0
1	3	33.24	23.5
2	4	32.93	26.0
2	5	33.17	23.0
2	6	33.07	25.0
3	7	33.01	25.5
3	8	32.82	27.0
3	9	32.91	26.0
4	10	32.80	26.5
4	11	32.86	28.5
4	12	32.89	25.5
5	13	32.73	27.0
5	14	32.57	28.0
5	15	32.65	26.5
6	16	32.43	30.0
6	17	32.54	28.0
6	18	32.61	26.0

Such bivariate data could be recorded in much the same way as they are listed in Table 2-1. But if it is important to have immediate feedback of information (say, to the operator of a machine), it would be much more effective to use a well-thought-out bivariate check sheet like the one in Figure 2-1.

On such a sheet, it is easy to see how the two variables recorded are related. If, in the figure, the recording symbol is varied over time, it is also easy to track changes in the characteristics over time. In the present case, width seems to be inversely related to mass, which appears to be decreasing over time.

To be useful (regardless whether data are recorded on a routine basis or in a one-shot mode, automatically or by hand), the recording must carry enough *documentation* that the important circumstances surrounding the data collection can be reconstructed. In a one-shot experimental study, someone must record responses and values of the experimental variables, and it is also wise to keep track of other nonexperimental variables that might later prove to be of interest. In the context of routine process-monitoring data collection, past data records will be useful in discovering differences in raw material

FIGURE 2-1

Check Sheet for the
PVC Bottle Data

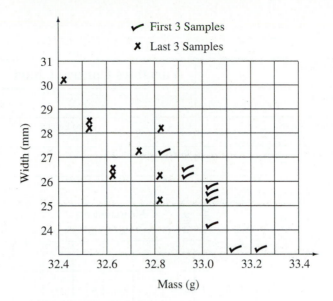

lots, machines, operators, etc., only if information on these is recorded along with the responses of interest. Figure 2-2 is a reproduction of a commonly used form for the routine collection of measurements for purposes of process monitoring. Notice how thoroughly the user is invited to document any data collection efforts.

2-2 Sampling in Enumerative Studies

Recall from the discussion in Section 1-2 that an enumerative study is one in which there is an identifiable, concrete population of items about which one wishes to gain information. This section will cover methods of selecting a sample of the items to include in a statistical investigation.

Using a sample to represent a (typically much larger) population has obvious advantages in terms of the resources required to do the necessary data collection. Measuring some characteristics of, say, 30 electrical components from an incoming lot of 10,000 such components can often be feasible in cases where it would not be feasible to perform a census (a study that attempts to include every member of the population).

Besides resource considerations, there can be other reasons for preferring a sample to a complete enumeration of a population. Sometimes testing is destructive, and studying an item renders it unsuitable for subsequent use. A fireworks manufacturer can hardly test an entire outgoing lot of firecrackers to verify its quality. Sometimes the timeliness and data quality of a sampling investigation far surpasses anything that could be achieved in a census. A moderate amount of data, collected under close supervision and put to immediate use, can be very valuable — more so than data from a study that might appear more complete but in fact takes too long. Moreover, data collection techniques can become lax or sloppy in a lengthy study.

If one is going to use a sample to stand for a population, how that sample is chosen becomes very important. The sample should somehow be representative of the population. The question to be addressed here is how to achieve this.

A variety of *systematic* and *judgment-based* methods can in some circumstances yield samples that faithfully portray the important features of a population. If a lot of items is manufactured in a known order, it may be reasonable to select, say, every 20th one for inclusion in a statistical engineering study. Or it may be effective to force the

FIGURE 2-2

Variables Control Chart Form

Part and Drawing	**Variables Control Chart**		Machine
Dimension	Gage		Operation
Production Lot(s)	Units	Zero Equals	Operator
Period Covered __	Specifications __		Raw Material Lot

	Mean	Sum	Measurements					Range		Date / Time
			5	4	3	2	1			

sample to be balanced — in the sense that every operator, machine, and raw material lot (for example) appears in the sample. Or an old hand may be able to look at a physical population and fairly accurately pick out a representative sample.

But there are problems with such methods of sample selection. Humans are subject to conscious and subconscious preconceptions and biases that can produce consistent distortions in the pictures of populations provided by judgment-based samples. Systematic methods can fail badly when unexpected cyclical patterns are present. For example, suppose one examines every 20th item in a lot, according to the order in which the items come off a production line. Suppose further that the items are at one point processed on a machine having five similar heads, each performing the same operation on every fifth item. Examining every 20th item only gives a picture of how one of the heads is behaving. The other four heads could be terribly misadjusted, and there would be no way to find this out.

Even beyond these problems with judgment-based and systematic methods of sampling, there is the additional difficulty that it is not possible to quantify their properties in any useful way. Strictly speaking, there is no way to take information from samples drawn via these methods and make reliable statements of likely margins of error.

The method introduced next avoids the deficiencies of systematic and judgment-based sampling.

DEFINITION 2-1

A **simple random sample of size *n*** from a population is a sample selected in such a manner (that most would agree) that every collection of *n* items in the population is *a priori* equally likely to compose the sample.

Probably the easiest way to think of simple random sampling is that it is conceptually equivalent to drawing *n* slips of paper out of a hat containing one for each member of the population.

EXAMPLE 2-6

Random Sampling Dorm Residents. C. Black did a partially enumerative and partially experimental study comparing student reaction times under two different lighting conditions. He decided to recruit subjects from his co-ed dorm floor, selecting a simple random sample of 20 (of the probably 80 or so) of these students to recruit. In fact, the selection method he used (which will be discussed shortly) involved the use of a table of so-called random digits. But he could have just as well written the names of all those living on his floor on standard-sized slips of paper, put them in a bowl, mixed thoroughly, closed his eyes, and selected 20 different slips from the bowl.

Methods for actually carrying out the selection of a simple random sample include mechanical methods and those involving the use of tables of *random digits*. Mechanical methods rely for their effectiveness on symmetry and/or thorough mixing in a physical randomizing device. So to speak, the slips of the paper in the hat need to be of the same size and well scrambled before sample selection begins.

The first Vietnam-era U.S. draft lottery was a famous case in which adequate care was not taken to assure appropriate operation of a mechanical randomizing device. Birthdays were supposed to be assigned priority numbers 1 through 366 in a "random" way. However, it was clear after the fact that balls representing birth dates were placed into a bin by months, and the bin was poorly mixed. When the balls were drawn out, birth dates near the end of the year received a disproportionately large share of the low draft numbers. In the present terminology, the first five dates out of the bin should *not* have been thought of as a simple random sample of size 5. Those who operate games of chance more routinely make it their business to know (via the collection of appropriate data) that their mechanical devices are operating in a more random manner.

When using a table of random digits to do sampling, one is relying for randomness on the appropriateness of the method that someone used to make up the particular table in the first place. Such tables are typically based on observation of physical random processes (like radioactive decay), or else *pseudo-random number generators* (complicated recursive numerical algorithms) are implemented on computers. In either case, the intention is to use a method guaranteeing that *a priori*

1. each digit 0 through 9 has the same chance of appearing at any particular location in the table one wants to consider, and
2. knowledge of which digit will occur at a given location provides no help in predicting which one will appear at another.

If one looks at a random digit table after the fact of its creation, condition 1 should typically be reflected in roughly equal representation of the 10 digits, and condition 2 in the lack of obvious internal patterns in the table.

Appendix D-1 contains an example of a table of random digits; the first five rows of this table are reproduced in Table 2-2 for use in this section.

TABLE 2-2

Random Digits

12159	66144	05091	13446	45653	13684	66024	91410	51351	22772
30156	90519	95785	47544	66735	35754	11088	67310	19720	08379
59069	01722	53338	41942	65118	71236	01932	70343	25812	62275
54107	58081	82470	59407	13475	95872	16268	78436	39251	64247
99681	81295	06315	28212	45029	57701	96327	85436	33614	29070

For populations that can easily be labeled with consecutive numbers, the following steps can be used to employ a random digit table to synthetically draw items out of a hat one at a time — to draw a simple random sample.

Step 1 For a population of N objects, determine the number of digits in N. (For example, $N = 1291$ is a 4-digit number.) Call this number M and assign each item in the population a different M-digit label.

Step 2 Move through the table left to right, top to bottom, M digits at a time, beginning from where one left off in last using the table, choosing objects from the population by means of their associated labels until n have been selected.

Step 3 In moving through the table according to Step 2, ignore labels that have not been assigned to items in the population and any that would indicate repeat selection of an item already included in the sample.

As an example of how this works, consider using the method to select a simple random sample of 25 members of a hypothetical population of 80 objects. One first determines that 80 is an $M = 2$ digit number and therefore labels items in the population as $01, 02, 03, 04, \ldots, 77, 78, 79, 80$. (Labels 00 and 81 through 99 are not assigned.) Then, if Table 2-2 is being used for the first time, one begins in the upper left corner and proceeds as indicated in Figure 2-3. Circled numbers represent selected labels, X's indicate that the corresponding label has not been assigned, and slash marks indicate that the corresponding item has already entered the sample.

FIGURE 2-3

Use of a Random Digit Table

12159 66144 05091 13446 45653 13684 66024 91410 51351 22772
30156 90519 95785 47544 66735 35754 11088 67310 19720 08379

As the final item enters the sample, the stopping point is marked with a penciled hash mark. Movement through the table is resumed at that point the next time the table is used.

Any predetermined systematic method of moving through the table could be substituted in place of Step 2. One could move down columns instead of across rows, for example. Pedagogically, it is useful to make the somewhat arbitrary choice of method in Step 2, for the sake of classroom consistency.

When it is not so easy to do the job of labeling the items in a population, it may still be possible to devise a method of using a random digit table in a way that a mathematician could argue would conform to Definition 2-1.

EXAMPLE 2-7

Selecting a Simple Random Sample from a Directory. Broene, Longest, Malstaff, Mathews, Moss, Roggenkamp, Rolf, Wang, and Woods worked with a local radio station on a survey of students' listening habits. The group determined that it could afford to contact a simple random sample of 100 students. It had at its disposal a student directory with student names, addresses, and phone numbers arranged in double columns on many pages. (At the time, the student body size was about 35,000.) But how was the directory to be used? Not every column had the same number of names in it, and it would have been a very unpleasant task to count the names in all the columns or actually number the entries in the directory.

Physical limitations on the number of lines on a page guaranteed that no column had more than 130 entries. The group happened across an appropriate method in a survey sampling textbook. The students used a table of random digits and

1. picked a column at random (by picking a page at random and then randomly choosing one of the two columns) and
2. chose a random number between 001 and 130, taking that numbered individual out of the chosen column.

In cases where Step 2 produced a label larger than the number of individuals in the column, they returned to Step 1. The two steps were repeated enough times to produce a list of 100 different students.

It is not obvious at this point, but probability calculations can be used to show that the method used by the students in fact conforms to Definition 2-1.

The method of executing simple random sampling described in Example 2-7 could well find engineering application in cases where a population of objects is divided into a number of possibly odd-sized lots (corresponding to columns above) and one wants a simple random sample of those items. Notice that it is not even necessary to know *a priori* how many items are in the lots, as long as one can put an upper bound on the size of the largest lot.

Regardless of the details of how Definition 2-1 is implemented, several comments about the method are in order. First, it must be admitted that simple random sampling meets the original objective of providing representative samples only in some average or long-run sense. It is possible for the method to produce particular realizations that are horribly unrepresentative of the corresponding population. A simple random sample of 20 out of 80 axle shafts could turn out to consist of those with the smallest diameters. But this doesn't happen often. On the average, a simple random sample will faithfully portray the population. Definition 2-1 is a statement about a method, not a guarantee of success on a particular application of the method.

Second, it must also be admitted that there is no guarantee that it will be an easy task to make the physical selection of a simple random sample. Imagine the pain of retrieving five out of a production run of 1,000 television sets stacked high on pallets in a warehouse for some kind of consumer testing. It would probably be a most unpleasant job to locate and gather five televisions corresponding to randomly chosen serial numbers to carry to a testing lab.

But the virtues of simple random sampling usually outweigh its drawbacks. For one thing, it is an objective method of sample selection. An engineer using it is protected from conscious and subconscious human bias. In addition, the method interjects probability into the selection process in what turns out to be a manageable fashion. As a result, the quality of information one gets from a simple random sample can be quantified. Methods of formal statistical inference, with their resulting conclusions ("I am 95% sure that . . ."), can be applied when simple random sampling is used.

There are other methods of random (or probability-based) sampling besides simple random sampling. For example, textbooks sometimes call simple random sampling *random sampling without replacement* and also discuss "random sampling with replacement." This is conceptually the same as drawing slips of paper out of a hat, but replacing each slip as it is drawn. At each draw, the hat contents are the same, and it is possible for an item to be counted more than once in a sample. But with-replacement sampling is more of a mathematical convenience than a practical method, so nothing more will be said about it here.

Another type of probability-based sampling is sequential random sampling. The tacit assumption in Definition 2-1 is that the sample size *n* is fixed in advance of sample selection. In sequential sampling, one makes each observation and, based on its value, decides (according to well-defined criteria) whether enough information has been gathered or another item should be sampled. Engineers have found this kind of methodology particularly useful in the context of acceptance sampling in quality control. However, unless warning is given to the contrary, the assumption in this text will be that sampling done in an enumerative study is simple random sampling.

It should be clear from this discussion that there is nothing mysterious or magical about simple random sampling. One sometimes gets the feeling while reading student projects (and even some textbooks) that the phrase *random sampling* is used with very little thinking (even in analytical rather than enumerative contexts) to mean "magically OK sampling" or "sampling with magically universally applicable results." Instead, simple random sampling is a concrete methodology for enumerative studies. It is generally about the best one available without *a priori* having intimate knowledge of the population.

2-3 *Principles for Effective Experimentation*

Purposely introducing changes into an engineering system and observing what happens as a result (i.e., experimentation) is a principal way of learning how such a system works and can be manipulated to advantage. Engineers meet such a great variety of experimental situations that it is impossible to give advice that will be relevant in all cases. But it is possible to raise some general issues, many of which should be considered in any given application of engineering experimentation. This is done in the present section. The discussion is organized under the headings of 1) taxonomy of variables, 2) handling extraneous variables, 3) comparative study, 4) replication, and 5) allocation of resources. Then Section 2-4 discusses a few generic experimental plans that can form the framework for planning a specific experiment.

Taxonomy of Variables

One of the realities an engineer faces when planning a first experiment is the decidedly multidimensional nature of the world. There are typically many characteristics of a system that the engineer would like to direct in appropriate directions and many variables that might well influence the characteristics of interest. To aid the process of sorting through these to determine how to deal with them in an experiment, some terminology will be introduced here

One basis of distinguishing between variables in an experiment is the question of how much management they are to be given.

DEFINITION 2-2

A **supervised (or managed) variable** in an experiment is one over which an investigator exercises power, choosing a setting or settings for use in the study. When a supervised variable is held constant (has only one setting), it will be called a **controlled variable**. And when a supervised variable is given several different settings in a study, it will be called an **experimental variable**.

Some of the variables that are not managed in an experiment will nevertheless be observed.

DEFINITION 2-3

A **response variable** in an experiment is one that is of primary interest to the investigator and is monitored in the hope that it characterizes the system's reaction to changes in the experimental variables.

DEFINITION 2-4

A **concomitant variable** in an experiment is one that is observed but is not treated as a primary response variable. Such a variable can change in reaction to either experimental or unobserved causes and may or may not itself have an impact on a response variable.

EXAMPLE 2-8

(Example 1-7 revisited) **Variables in a Wood Joint Strength Experiment.** Dimond and Dix experimented with three different woods and three different glues, investigating joint strength properties. Their primary interest was in the effects of experimental variables Wood Type and Glue Type on two observed response variables, joint strength in a tension test and joint strength in a shear test.

In addition, they recognized that strengths were probably related to the variables Drying Time and Pressure applied to the joints while drying. Their method of treating the nine wood/glue combinations fairly with respect to the Time and Pressure variables was to manage them as controlled variables, trying to hold them essentially constant for all the joints produced.

Some of the variation the students observed in strengths could also have originated in properties of the particular specimens glued, such as moisture content and surface roughness of the glued wood blocks. In fact, these were not observed in the study.

But if the students had had some way of measuring it, observation of, say, moisture content might have provided extra insight into how the wood/glue combinations behave. Moisture content would then have been termed a concomitant variable.

Handling Extraneous Variables

In planning an experiment, there are always variables that could influence the responses but which are not of practical interest to the experimenter. The investigator may recognize some of them as influential but not even think of others. Those that are recognized may well fail to be of primary interest because the experimenter has no realistic way of exercising control over them or compensating for their effects outside of the (perhaps somewhat artificial) experimental environment. So it is of little practical use to know exactly how changes in them affect the system.

But completely ignoring the existence of such *extraneous variables* in experiment planning can needlessly cloud one's perception of the effects of factors that *are* of interest. Several methods can be used in an active attempt to avoid this loss of information. These are to manage them (for experimental purposes) as controlled variables or as *blocking* variables, or to attempt to balance their effects amongst process conditions of interest through the use of the technique of *randomization*.

When one chooses to control an extraneous variable in an experiment, both the positive and negative implications of that choice should be recognized. On one hand, the control produces a homogeneous environment in which to study the effects of the primary experimental variables. In some sense, a portion of the background noise has been eliminated, allowing a clearer view of how the system reacts to changes in factors of interest.

On the other hand, system behavior at other values of the extraneous variable cannot be projected on the relatively firm basis of data. Instead, projections must be based on the somewhat shakier basis of *expert opinion* that what is seen experimentally will prove true more generally. Engineering experience is replete with examples where what worked fine in a laboratory (or even a pilot plant) setting was much less dependable in subsequent experience with a full-scale facility.

EXAMPLE 2-8
(continued)

The choice Dimond and Dix made to control drying time and the pressure under which the glue dried provided a uniform environment for comparing the nine wood/glue combinations. In fact, this gave them rather definitive results. But strictly speaking, they learned only about joint behavior under the particular experimental time and pressure conditions studied.

To make projections for other conditions, they would need to rely on their experience and knowledge of material science to decide how far the patterns they observed are likely to extend. For example, it may have been reasonable for them to expect what they observed to also hold up for any drying time at least as long as the experimental one, because of expert knowledge that the experimental time was sufficient for the joints to fully set. But the point here is that such extrapolation is based on other than statistical grounds.

An alternative to controlling extraneous variables is to handle them as experimental variables, including them in study planning at several different levels. Notice that this really amounts to applying the notion of control locally by creating not one but several (possibly quite different) homogeneous environments in which to compare levels of the primary experimental variables. The term *blocking* is often used to refer to this technique.

DEFINITION 2-5

A **block** of experimental units, experimental times of observation, experimental conditions, etc. is a homogeneous group within which different levels of primary experimental variables can be applied and compared in a relatively uniform environment.

EXAMPLE 2-8
(continued)

Returning again to the context of the gluing study of Dimond and Dix, consider embellishing on it a bit. Imagine that the students were somewhat uneasy about two issues, the first being the possibility that surface roughness differences in the pieces to be glued might mask the wood/glue combination differences of interest. Suppose also that because of unfortunate constraints on schedules, the strength testing was going to have to be done in two different sessions a day apart. Measuring techniques or background variables like humidity might vary somewhat between such periods. How might such potential problems have been handled?

Blocking is one way. If the wood specimens of each wood type were separated into relatively rough and relatively smooth groups, the factor Roughness could have then served as an experimental factor. Each of the glues could have been used the same number of times to join both rough and smooth specimens of each species. This would set up possible comparison of wood/glue combinations for rough and separately for smooth surfaces.

In a similar way, half the testing for each wood/glue/roughness combination might have been done in each testing session. Then any consistent differences between sessions could be identified and prevented from clouding the comparison of levels of the primary experimental variables. Thus, Testing Period could have also served as a blocking variable in the study.

Not all extraneous variables can be supervised. Indeed, one might argue that there are essentially an infinite number, most of which cannot even be named. Even if they could all be named, at some point one would have to leave off considering them in order to ever get any data collection done. Experimenters hope that by careful planning they account for all important extraneous variables, using control and blocking. But there is a way to take out insurance against the possibility that major extraneous variables that have been overlooked will produce effects that get mistaken for those of the primary experimental variables.

DEFINITION 2-6

Randomization is the use of a randomizing device or table of random digits at some point where experimental protocol is not already dictated by the specification of values of the supervised variables. Often this means that experimental objects (or units) are divided up between the various sets of experimental conditions at random. It can also mean that the order of experimental testing is randomly determined.

The goal of randomization is to average between sets of experimental conditions the effects of all unsupervised variables. To put it differently, one wants to treat sets of experimental conditions fairly, giving them equal opportunity to shine.

EXAMPLE 2-9

(Example 1-1 revisited) **Randomization in a Heat Treating Study.** P. Brezler, in his 1986 *Heat Treating* article, describes a very simple randomized experiment for comparing the effects on thrust face runout of laying versus hanging gears during heat treating. The variable Loading Method was, of course, the primary experimental variable. Extraneous variables Steel Heat and Machining History were controlled by experimenting on 78 gears from the same heat code, machined as a lot. The 78 gears were broken at random into two groups of 39, one to be laid and the other to be hung. (Note that Table 1-1 gives only 38 data points for the laid group. For reasons not given in the article and not important here, one laid gear was dropped from the study.)

Although there is no explicit mention of this in the original article, the principle of randomization could have been (and perhaps was) carried a step further by making the runout measurements in a random order. (This means choosing gears 01 through 78 one at a time at random, without replacement, to measure.) The effect of this randomization would have been to protect the investigator from clouding the comparison of heat treating methods with possible unexpected and unintended changes in measurement techniques. Failing to randomize and, for example, making all the laid measurements before the hung measurements, would allow changes in measurement techniques to appear in the data as differences between the two loading methods. Practice with measurement equipment might, for example, increase precision and make later runouts appear to be more uniform than early ones.

EXAMPLE 2-8
(continued)

Dimond and Dix took the notion of randomization to heart in the planning of their gluing study and, so to speak, randomized everything in sight. In the tension strength testing for a given type of wood, they glued $.5'' \times .5'' \times 3''$ blocks to a $.75'' \times 3.5'' \times 31.5''$ board of the same wood type, as illustrated in Figure 2-4.

FIGURE 2-4

Gluing Method for a Single Wood Type

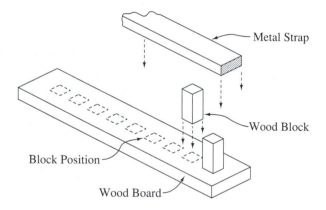

Each glue was used for three joints on each type of wood. In order to deal with any unpredicted differences in material properties (e.g., over the extent of the board) or unforeseen differences in loading by the steel strap used to provide pressure on the joints, etc., the students randomized the order in which glue was applied and the blocks placed along the extent of the base board. In addition, when it came time to do the strength testing, that was carried out in a randomly determined order.

Much as simple random sampling in enumerative studies can only be considered guaranteed effective in an average or long-run sense, randomization in experiments will not prove effective in averaging the effects of extraneous variables between settings of experimental variables every time it is used. Sometimes an experimenter will be unlucky.

But the methodology is objective, effective on the average, and about the best one can do in accounting for those variables that will not be managed.

Comparative Study

When discussing types of data structures in Section 1-2, it was mentioned that statistical engineering studies often involve more than a single sample. That is, they involve comparison of a number of settings of process variables. This is especially true in experimental contexts, not only because there may be many options open to an engineer in a given situation, but for other reasons as well.

Even in experiments where there is only a single new idea or variation on standard practice to be tried out, it is a good idea to make the study comparative and therefore to involve more than one sample. Unless this is done, there is no really firm basis on which to say that any effects observed come from the new conditions under study rather than from unexpected extraneous sources.

If standard yield for a chemical process is 63.2% and a few runs of the process with a supposedly improved catalyst produce a mean yield of 64.8%, it is not completely safe to attribute the difference to the catalyst. It could be caused by a number of things, including miscalibration of the measurement system being used. But suppose a few experimental runs (perhaps in a random order) for both the standard and the new catalysts are taken. If these produce two samples with small variation within these samples and (for example) a difference of 1.6% in mean yields, that difference is more safely attributed to differences in the performances of the catalysts.

EXAMPLE 2-9
(continued)

In the gear loading study, hanging was the standard method in use at the time of the study. From its records, the company could probably have located some values for thrust face runout that could have been used as a baseline for evaluating the laying method. But the choice to run a comparative study, including both laid and hung gears, put the engineer on firm ground for drawing conclusions about the new method.

In a potentially confusing use of language, the word *control* is sometimes used to mean the practice of including a standard or no-change sample in an experiment, for comparison purposes. (Notice that this is not the usage in Definition 2-2.) When a *control group* is included in a medical study to verify the effectiveness of a new drug, that group is either a standard treatment or no-treatment group, included to provide a solid basis of comparison for the new treatment.

Replication

In much of what has been said thus far, it has been implicit that having more than one observation for a given setting of experimental variables is a good idea.

DEFINITION 2-7

Replication of a setting of experimental variables means carrying through the whole process of adjusting values for supervised variables, making an experimental "run," and observing the results of that run — more than once. Values of the responses from replications of a setting form the (single) sample corresponding to the setting, which one hopes represents typical process behavior at that setting.

The idea of replication is fundamental in experimentation. *Reproducibility* of results is important in both science and engineering practice. Replication helps establish this, protecting the investigator from unconscious blunders and validating or confirming experimental conclusions.

But replication is not only important for establishing that experimental results are reproducible. It is also essential to quantifying the *limits* of that reproducibility — that is, getting an idea of the size of experimental error. Even under a fixed setting of supervised variables, repeated experimental runs typically will not produce exactly the same observations. The effects of unsupervised variables and measurement errors produce a kind of baseline variation or background noise, which will exist in any monitoring of the system under study. Establishing the magnitude of this variation is important. It is only against this background that one can judge whether an observed effect of an experimental variable is big enough to establish it as clearly real, rather than explainable in terms of variation in unmanaged system variables.

When planning an experiment, the engineer must think carefully about what kind of repetition will be included. Definition 2-7 was written specifically to suggest that (for example) simply remeasuring an experimental unit does not amount to real replication. Such repetition will capture measurement error, but it ignores the effects of changing unsupervised system variables. It is a common mistake in logic to seriously underestimate the size of experimental error by failing to adopt a broad enough view of what should be involved in replication, settling instead for what amounts to remeasurement.

EXAMPLE 2-10

Replication and Steel Making. A former colleague once related a consulting experience that went approximately as follows. In studying the possible usefulness of a new additive in a type of steel, a metallurgical engineer had one heat of steel made with the additive and one without. Each of these was poured into ingots. The metallurgist then selected some ingots from both heats, had them cut into pieces, and selected some pieces from the ingots, ultimately measuring a property or properties of interest on these pieces and ending up with a reasonably large amount of data. The data from the heat with additive showed it to be clearly superior to the no-additive heat. As a result, the existing production process was altered (at significant expense) and the new additive incorporated. Unfortunately, it soon became apparent that the alteration to the process had actually degraded the properties of the steel being produced.

The statistician was (only at this point) called in to figure out what had gone wrong. After all, the experimental results, based on a large amount of data, had been quite convincing, hadn't they?

The key to understanding what had gone awry was the issue of replication. In a sense, there was none. The metallurgist had essentially just remeasured the same physical objects (the heats) many times. In the process, he had learned quite a bit about the two particular heats in the study, but very little about all heats of the two types. Apparently, extraneous and uncontrolled foundry variables were producing large heat-to-heat variability. The metallurgist had mistaken an effect of this fluctuation for an improvement due to the new additive. The metallurgist had no notion of this possibility because he had not replicated the with-additive and without-additive settings of the experimental variable.

EXAMPLE 2-11

Replication and Paper Airplane Testing. Experimenting on the making and flying of paper airplanes is a favorite student project in engineering statistics classes. For example, students Beer, Dusek, and Ehlers completed a project comparing the Kline-Fogelman and Polish Frisbee designs on the basis of flight distance under a number of different conditions. In general, it was carefully done.

However, replication was a point on which their experimental plan was extremely weak. They made a number of trials for each plane under each set of experimental conditions, but only *one* Kline-Fogelman prototype and *one* Polish Frisbee prototype were used throughout the study. The students learned quite a bit about the prototypes in hand, but possibly much less about the two plane designs. If their purpose was to pick a winner between the two prototypes, to enter in a contest, then perhaps the design of their study was appropriate. But if the purpose was to make conclusions about planes like the two used in the study, they needed to make and test several (even essentially identical) prototypes.

ISU Professor Emeritus L. Wolins calls the problem of identifying what constitutes replication in an experiment the *unit of analysis* problem. There must be replication of the basic experimental unit or object. The agriculturalist who, in order to study pig blood chemistry, takes hundreds of measurements per hour on one pig, has a (highly multivariate) sample of size 1. The pig is the unit of analysis.

Without proper replication, one can only hope to be lucky. If in fact experimental error is small, then accepting conclusions suggested by samples of size 1 will lead to correct conclusions. But the problem is that without replication, one usually has little (or an overly optimistic) idea of the size of experimental error.

Allocation of Resources

Experiments are done by people and organizations that have finite resources of time and money. Allocating those resources and living within the constraints they impose is part of experiment planning. The rest of this section makes several points in this regard.

First, in spite of the impression conveyed in most engineering statistics (and indeed general engineering) courses, real-world investigations are often most effective when approached *sequentially*, the planning for each stage building upon what has been learned before. The classroom model of planning and/or executing a single experiment is more a result of constraints inherent in our methods of teaching than a realistic representation of how engineering problems are solved. The reality is most often more iterative in nature, involving a series of related experiments.

This being the case, one can not use an entire experimental budget on the first pass of a statistical engineering study. Conventional statistical wisdom on this matter is that no more than 20–25% of an experimental budget should be allocated to the first stage of a sequential study. This leaves adequate resources for follow-up work built on what is learned initially.

Second, one cannot let what is easy to do (and therefore usually cheap to do) dictate completely what is done in an experiment. In the context of the steel formula development study of Example 2-10, it seems almost certain that one reason the metallurgist chose to get his "large sample sizes" from pieces of ingots, rather than from heats, is that it was easy and cheap to get many measurements in that way. But in addition to failing to get absolutely crucial replication and thus botching the study, he probably also grossly overmeasured the two (internally relatively homogeneous) heats.

A final remark is an amplification of the discussion of sample size in Section 2-1. That is, minimum experimental resource requirements are dictated in large part by the magnitude of effects of engineering importance in comparison to the magnitude of experimental error. The larger the effects in comparison to the error (in electrical engineering parlance, the larger the signal-to-noise ratio), the smaller the sample sizes required, and thus the fewer the resources needed.

2-4 *Some Common Experimental Plans*

In previous sections, experimentation has been discussed in general terms, and the subtlety of considerations that enter the planning of an effective experiment has been illustrated. It should be obvious that any exposition of standard experimental "plans" can amount only to a discussion of standard "skeletons" around which real plans can be built. Nevertheless, it is useful to know something about such skeletons. In this section, so-called completely randomized, randomized complete block, and incomplete block experimental plans are considered.

Completely Randomized Experiments

DEFINITION 2-8

A **completely randomized experiment** is one in which all experimental variables are of primary interest (i.e., none are included only for purposes of blocking), and randomization is used at every possible point of choosing the experimental protocol.

Notice that this definition says nothing about how the settings of experimental variables included in the study are structured. In fact, they may be essentially unstructured or produce data with any of the structures discussed in Section 1-2. That is, there are completely randomized one-factor, factorial, fractional factorial, and hierarchical experiments. The essential point is that all else is randomized except what is restricted by consideration of which settings of experimental variables are to be used in the study.

Although it doesn't really fit every situation (or perhaps even most) in which the term *complete randomization* is appropriate, language like the following is commonly used to capture the intent of Definition 2-8. "Experimental units (objects) are allocated at random to the treatment combinations (settings of experimental variables). Experimental runs are made in a randomly determined order. And any *post facto* measuring of experimental results is also carried out in a random order."

EXAMPLE 2-12

Complete Randomization in a Glass Restrengthening Study. Bloyer, Millis, and Schibur carried out a class project to study the restrengthening of damaged glass through etching. They studied the effects of two experimental factors — the Concentration of hydrofluoric acid in an etching bath and the Time spent in the etching bath — on the resulting strength of damaged glass rods. (The rods had been purposely scratched in a 1-inch region near their centers by sandblasting.) Strengths were measured using a three-point bending method on a 20 kip MTS machine.

The students decided to run a 3×3 factorial experiment. The experimental levels of the factor Concentration were 50%, 75%, and 100% HF, and the levels of the factor Time employed were 30 sec, 60 sec, and 120 sec. There were thus nine treatment combinations, as illustrated in Figure 2-5.

The students decided that 18 scratched rods would be allocated — two apiece to each of the nine treatment combinations — for testing. Notice that this could be done at random by labeling the rods 01–18, placing numbered slips of paper in a hat, mixing, drawing out two for 30 sec and 50% concentration, then drawing out two for 30 sec and 75% concentration, etc.

FIGURE 2-5

Nine Combinations of Three
Levels of Concentration and
Three Levels of Time

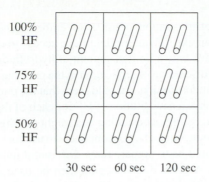

Having determined at random which rods would receive which experimental conditions, the students could again have used the slips of paper to randomly determine an etching order. And a third use of the slips of paper to determine an order of strength testing would have given the students what most people would agree was a completely randomized 3 × 3 factorial experiment.

In fact, for reasons that are not clear, the students departed from complete randomization and etched both rods for a given treatment combination simultaneously. The advisability of that maneuver might be questioned, from the point of view of the effects of possible slight timing errors on the experiment's outcome.

EXAMPLE 2-13

Complete Randomization and a Study of the Flight of Golf Balls. In spring semesters, golfing experiments are favorite projects in statistics courses. For example, G. Gronberg studied drive flight distances for 80, 90, and 100 compression golf balls, using 10 balls of each type in his experiment. Consider what complete randomization would entail in such a study involving the single factor Compression.

Notice that the amplification of Definition 2-8 (immediately preceding the last example) is not particularly appropriate to this experimental situation. The levels of the experimental factor are an intrinsic property of the experimental units (balls). There is no way to randomly divide the 30 test balls into three groups and "apply" the treatment levels 80, 90, and 100 compression to them.

In fact, about the only obvious point at which randomization could be employed in this scenario is in the choice of an order for hitting the 30 test balls. If one numbered the test balls 01 through 30 and used a table of random digits to pick a hitting order (by choosing balls one at a time without replacement), most people would be willing to call the resulting test a completely randomized one-factor experiment.

Randomization is a good idea; its virtues have been discussed at some length. So it would be wise to point out that using it can sometimes lead to practically unworkable experimental plans. Dogmatic insistence on complete randomization can in some cases be quite foolish and unrealistic. Changing experimental process variables according to a completely randomly determined schedule can be exceedingly inconvenient (and therefore expensive). If the inconvenience is great and one's fear of being misled by the effects of extraneous variables is relatively small, then backing off from complete to partial randomization may be the only reasonable course of action. But when choosing not to randomize, one must consider carefully the implications of that choice.

EXAMPLE 2-12
(continued)

For the sake of illustration, consider an embellishment on the glass strengthening scenario, where an experimenter might have access to only a single container to use for a bath and/or have only a limited amount of hydrofluoric acid.

From the discussion of replication in the previous section and present considerations of complete randomization, it would seem that the purest method of conducting the study would be to make a new dilution of HF for each of the rods as its turn comes for testing. But this would be time-consuming and might require more acid than was available.

If the investigator had three containers to use for baths but limited acid, an alternative possibility would be to prepare three different dilutions, one 100%, one 75%, and one 50% dilution. A given dilution could then be used in testing all rods assigned to that concentration. Notice that this alternative allows for a randomized order of testing, but it introduces some question as to whether there is "true" replication.

Taking the resource restriction idea one step further, notice that even if an investigator could afford only enough acid for making one bath, there is a way of proceeding. One could do all 100% concentration testing (probably with randomly ordered times), then dilute the acid and do all 75% testing (again with a random order of times), then dilute the acid again and do all 50% testing (with randomly ordered times). The resource restriction would not only affect the "purity" of replication but also prevent complete randomization of the experimental order. Thus, for example, any unintended effects of increased contamination of the acid (as more and more tests were made using it) would show up in the experimental data as indistinguishable from effects of differences in acid concentration.

To choose intelligently between complete randomization (with "true" replication) and the two plans discussed above, one would have to weigh the real severity of resource limitations against a judgment of the likelihood that extraneous factors would jeopardize the usefulness of experimental results.

Randomized Complete Block Experiments

DEFINITION 2-9

A **randomized complete block experiment** is one in which at least one experimental variable is a blocking factor (not of primary interest to the investigator); and within each block, every setting of the primary experimental variables appears at least once; and randomization is employed at all possible points where the exact experimental protocol is determined.

This definition (like Definition 2-8) says nothing about the structure of the settings of primary experimental variables included in the experiment. Nor does it say anything about the structure of the blocks. It is possible to design experiments where experimental combinations of primary variables have one-factor, factorial, fractional factorial, or hierarchical structure, and at the same time the experimental combinations of blocking variables also have one of these standard structures. The essential points of Definition 2-9 are the *completeness* of each block (in the sense that it contains each setting of the primary variables) and the *randomization* within each block.

A helpful way to think of a randomized complete block experiment is as a study consisting of a collection of similar completely randomized studies. Each of the blocks yields one of the component studies. Blocking provides the simultaneous advantages of homogeneous environments for studying primary factors and breadth of applicability of the results.

EXAMPLE 2-13
(continued)

As actually run, Gronberg's golf ball flight study amounted to a randomized complete block experiment. This is because he hit and recorded flight distances for all 30 balls on six different evenings (over a six-week period). Note that this allowed him to have (six sets of) homogeneous conditions under which to compare the flight distances of balls having 80, 90, and 100 compression. This also allowed for changes over time in his physical condition and skill level, as well as environmental conditions.

Notice the structure of the data set that resulted from the study. The settings of the single primary experimental variable Compression combined with the levels of the single blocking factor Day to produce a 3×6 factorial structure for 18 samples of size 10, as pictured in Figure 2-6.

FIGURE 2-6

18 Combinations of Compression and Day

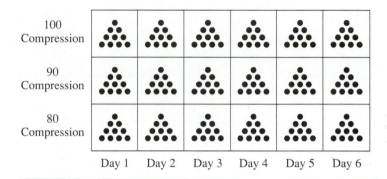

EXAMPLE 2-14

(Example 1-2 revisited) **Blocking in a Pelletizing Experiment.** Near the end of Section 1-2, the notion of a fractional factorial study was illustrated in the context of a hypothetical experiment on a pelletizing machine. The factors Volume, Flow, and Mixture were of primary interest. Table 1-3 is reproduced here as Table 2-3, listing four (out of eight possible) combinations of two levels each of the primary experimental variables, forming a fractional factorial arrangement.

TABLE 2-3

Half of a 2^3 Factorial

Volume	Flow	Mixture
high	current	no binder
low	manual	no binder
low	current	binder
high	manual	binder

Strictly for illustrative purposes (not because the final result in this example necessarily represents a wise expenditure of experimental effort), consider a situation where two different operators can make four experimental runs each on two consecutive days. Suppose further that Operator and Day are blocking factors, their combinations giving four blocks, within which the four combinations listed in Table 2-3 are run in a random order. One then ends up with a randomized complete block experiment in which the blocks have 2×2 factorial structure and the four combinations of primary experimental factors have fractional factorial structure.

There are several ways to think of this plan. For one, by temporarily ignoring the structure of the blocks and combinations of primary experimental factors, it can be considered a 4×4 factorial arrangement of samples of size 1, as is illustrated in Figure 2-7.

FIGURE 2-7

16 Combinations of Blocks
and Treatments

	High Current No Binder	Low Manual No Binder	Low Current Binder	High Manual Binder
Block 1 { Operator 1 Day 1	1 Run	1 Run	1 Run	1 Run
Block 2 { Operator 2 Day 1	1 Run	1 Run	1 Run	1 Run
Block 3 { Operator 1 Day 2	1 Run	1 Run	1 Run	1 Run
Block 4 { Operator 2 Day 2	1 Run	1 Run	1 Run	1 Run
	Combination 1	Combination 2	Combination 3	Combination 4

But from another point of view, the combinations under discussion (listed in Table 2-4) have fractional factorial structure of their own, representing a (not particularly clever) choice of 16 out of $2^5 = 32$ different possible combinations of the two-level factors Operator, Day, Volume, Flow, and Mixture. (The dotted lines in Table 2-4 separate the four blocks.)

TABLE 2-4

Half of a 2^3 Factorial Run Once
in Each of Four Blocks

Operator	Day	Volume	Flow	Mixture
1	1	high	current	no binder
1	1	low	manual	no binder
1	1	low	current	binder
1	1	high	manual	binder
2	1	high	current	no binder
2	1	low	manual	no binder
2	1	low	current	binder
2	1	high	manual	binder
1	2	high	current	no binder
1	2	low	manual	no binder
1	2	low	current	binder
1	2	high	manual	binder
2	2	high	current	no binder
2	2	low	manual	no binder
2	2	low	current	binder
2	2	high	manual	binder

Soon to be discussed will be a better use of 16 experimental runs in this situation (at least from the perspective that the combinations in Table 2-4 have their own fractional factorial structure).

⎯⎯⎯

The purpose of the preceding two examples is to illustrate that depending upon the specifics of a scenario, Definition 2-9 can describe a variety of experimental plans.

Incomplete Block Experiments

In many experimental situations, though blocking seems attractive, physical constraints make it impossible to satisfy Definition 2-9. This leads to the notion of incomplete blocks.

DEFINITION 2-10

An **incomplete** (usually randomized) **block experiment** is one in which at least one experimental variable is a blocking factor, and the assignment of combinations of primary experimental factors to blocks is such that not every combination appears in every block.

EXAMPLE 2-14
(continued)

In Section 1-2, the pelletizing machine study examined all eight possible combinations of the primary experimental variables Volume, Flow, and Mixture. These were listed in Table 1-2, which is reproduced here as Table 2-5.

TABLE 2-5

Combinations in a 2^3 Factorial Study

Combination Number	Volume	Flow	Mixture
1	low	current	no binder
2	high	current	no binder
3	low	manual	no binder
4	high	manual	no binder
5	low	current	binder
6	high	current	binder
7	low	manual	binder
8	high	manual	binder

For the sake of illustration, imagine that only half of these eight combinations could be run on a given day, and there was some fear that daily environmental conditions might strongly affect process performance. How might one have proceeded?

There are then two blocks (days), each of which will accommodate four runs. Some possibilities for assigning runs to blocks would clearly be poor. For example, running combinations 1 through 4 on the first day and 5 through 8 on the second would make it impossible to distinguish the effects of Mixture from any environmental effects.

What turns out to be a far better possibility is to run, say, the four combinations listed in Table 2-3 (combinations 2, 3, 5 and 8) on one day and the others on the next. This is illustrated in Table 2-6.

TABLE 2-6

A 2^3 Factorial Run in Two Incomplete Blocks

Day	Volume	Flow	Mixture
2	low	current	no binder
1	high	current	no binder
1	low	manual	no binder
2	high	manual	no binder
1	low	current	binder
2	high	current	binder
2	low	manual	binder
1	high	manual	binder

In a well-defined sense (explained in Chapter 10), this choice of incomplete block plan minimizes the unavoidable clouding of inferences caused by the fact all eight combinations of the primary experimental factors cannot be run on a single day.

EXAMPLE 2-14
(continued)

As one final variation on the pelletizing scenario, consider an alternative that is superior to the experimental plan outlined in Table 2-4: one that involves incomplete blocks. That is, once again suppose that two-level primary factors Volume, Flow, and Mixture are to be studied in four blocks of four observations, created by combinations of the two-level blocking factors Operator and Day.

Since a total of 16 experimental runs can be made, all eight combinations of primary experimental factors can be included in the study twice (instead of including only four combinations four times apiece). To do this, incomplete blocks are required, but Table 2-7 shows a good incomplete block plan. (Again, blocks are separated by dotted lines.)

TABLE 2-7

A Once-Replicated 2^3 Factorial Run in Four Incomplete Blocks

Operator	Day	Volume	Flow	Mixture
1	1	high	current	no binder
1	1	low	manual	no binder
1	1	low	current	binder
1	1	high	manual	binder
2	1	low	current	no binder
2	1	high	manual	no binder
2	1	high	current	binder
2	1	low	manual	binder
1	2	low	current	no binder
1	2	high	manual	no binder
1	2	high	current	binder
1	2	low	manual	binder
2	2	high	current	no binder
2	2	low	manual	no binder
2	2	low	current	binder
2	2	high	manual	binder

Notice the symmetry present in this choice of half of the $2^5 = 32$ different possible combinations of the five experimental factors. For example, a full factorial in the three primary variables (all eight possible combinations of the Volume, Flow, and Mixture variables) is run on each day, and similarly, each operator runs a full factorial in the primary experimental variables.

It turns out that the study outlined in Table 2-7 gives far more potential for learning about the behavior of the pelletizing process than the one outlined in Table 2-4. But again, a complete discussion of this must wait until Chapter 10.

At this point, there may be some uneasiness and frustration with the "rabbit out of a hat" nature of the last two examples of incomplete block experiments, since there has been no discussion of how one goes about making up a good incomplete block plan. Both the choosing of an incomplete block plan and corresponding techniques of data analysis are more advanced topics that will not be developed in any depth until Chapter 10.

The primary purpose here is to simply introduce the possibility of incomplete blocks as a useful option in experimental planning. As one more means to that end, this section will close with a brief exposition of a very specialized but well-known type of fractional factorial experimental plan, often used as an incomplete block design.

DEFINITION 2-11

A $k \times k$ **Latin square experiment** is one in which three experimental factors each have k levels, and in k^2 experimental combinations run, there is a full factorial in every pair of factors. Most often, two of the three factors are blocking factors and the other is a primary experimental variable.

The name *Latin square* is easily related to a type of diagram that is often used to illustrate the structure of this type of study. One thinks of two of the factors as defining a (square) $k \times k$ arrangement of cells. The numbers 1 through k are placed into the cells so that each number appears only once in any row or column. The resulting picture gives (via row number, column number, and cell entry) a prescription for k^2 out of k^3 possible combinations of the factor levels, forming a Latin square. This is illustrated in Figure 2-8 and Table 2-8 for the 3×3 case.

FIGURE 2-8

3×3 Latin Square

Numbers in Cells Are Levels of C

	1	2	3
Factor A 1	1	3	2
2	2	1	3
3	3	2	1

Factor B

TABLE 2-8

Factor-Level Combinations in a 3×3 Latin Square

Factor A Level	Factor B Level	Factor C Level
1	1	1
1	2	3
1	3	2
2	1	2
2	2	1
2	3	3
3	1	3
3	2	2
3	3	1

The classical textbook example of a Latin square used as an incomplete block plan is a tire mileage study. Suppose four tire Brands of primary interest are to be tested on four cars (the factor Cars not being of primary interest), and the four Mounting Positions (left front, etc.) are treated as levels of a second blocking factor. A Latin square (incomplete block) experiment for this situation is outlined in Figure 2-9. Notice that in cases like this, when a Latin square is used as the basis of an incomplete block plan, the notion of incompleteness is taken to its extreme. Blocks are of size 1.

FIGURE 2-9

4 × 4 Latin Square Tire
Wear Study

	1	2	3	4
Left Front	Brand A	Brand D	Brand C	Brand B
Left Rear	Brand B	Brand A	Brand D	Brand C
Right Front	Brand C	Brand B	Brand A	Brand D
Right Rear	Brand D	Brand C	Brand B	Brand A

Position

Car

2-5 *Preparing to Collect Engineering Data*

This chapter has raised many of the issues that engineers must address when planning a statistical study. What is still lacking, however, is a discussion of where to get started. This section first lists and then briefly discusses a series of steps that can be followed in preparing for engineering data collection. Most of these require as background only the first two chapters of this book. This section can thus form the basis for an early start on a class project on engineering data collection and analysis.

A Series of Steps to Follow

The following is a list of steps that can be used to organize the planning of a statistical engineering study.

Problem Definition

Step 1 Identify the problem to be addressed in general terms.

Step 2 Understand the context of the problem.

Step 3 State in precise terms the objective and scope of the study.

Study Definition

Step 4 Identify the response variable(s) and appropriate instrumentation.

Step 5 Identify possible factors influencing responses.

Step 6 Decide whether (and if so how) to manage factors that are likely to have effects on the response(s).

Step 7 Develop a detailed data collection protocol and timetable for the first phase of the study.

Physical Preparation

Step 8 Assign responsibility for careful supervision.

Step 9 Identify technicians and provide necessary training and instruction in the objectives and methods to be used.

Step 10 Prepare data collection forms and/or equipment.

Step 11 Do a dry run analysis on fictitious data.

Step 12 Write up a "best guess" prediction of the results of the actual study.

These 12 points are listed in a reasonably rational order, but planning any real study may involve departures from the listed order, as well as a fair amount of iterating amongst the steps before they are all accomplished. The need for other steps (like finding funds to pay for a proposed study) will also be apparent in some contexts. Nevertheless, Steps 1 through 12 form a framework for "getting going."

Problem Definition

Identifying the general problem to work on is, at least for the working engineer, largely a matter of prioritization. (The situation is admittedly somewhat different for students of engineering who are looking for a statistical engineering class project.) An individual engineer's job description and place in an organization usually dictate what problem areas need attention. And far more things could always be done than resources of time and money will permit. So some choice has to be made among the different possibilities.

It is only natural to make the choice of a general topic on the basis of the perceived importance of a problem and the likelihood of solving it (given the available resources). These criteria are, of course, somewhat subjective. So, particularly when a project team or other working group must come to consensus before proceeding, even this initial planning step is a nontrivial task. Sometimes it is possible to remove part of the subjectivity and reliance on personal impressions involved in picking an area for investigation, by either examining existing data or commissioning a statistical study of the current state of affairs. For example, suppose members of an engineering project team can name several types of flaws that occur in a mechanical part but disagree about the frequencies or dollar impacts of the types of flaws. The natural place to begin is not to try to reduce the occurrence rate of one of these, but rather to search company records or collect some new data aimed at determining the occurrence rates and/or dollar impacts.

A particularly effective and popular way of summarizing the findings of such a preliminary look at the current situation is through the use of a *Pareto diagram*. This is a bar chart whose vertical axis delineates frequency (or some other measure of impact of system misbehavior) and whose bars, representing problems of various types, have been placed left to right in decreasing order of importance.

EXAMPLE 2-15

Maintenance Hours for a Flexible Manufacturing System. Figure 2-10 is an example of a Pareto diagram. It is a representation of a breakdown (by craft classification) of the total maintenance hours required in one year on four particular machines in a company's flexible manufacturing system. (This information is excerpted from M.S. thesis work of M. Patel.)

FIGURE 2-10

Pareto Diagram of Maintenance Hours by Craft Classification

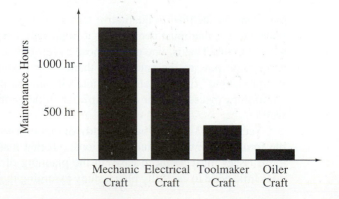

A diagram like Figure 2-10 can be an effective tool for helping to focus a project group's attention on the most important problems in an engineering system. Figure 2-10 highlights the fact that (in terms of maintenance hours required) mechanical problems required the most attention, followed by electrical problems.

In a statistical engineering study, it is essential to understand the context of the problem to be attacked. Statistics is no magic substitute for good, hard work in learning how a process is configured; what its inputs and environment are; what applicable engineering, scientific, and mathematical theory has to say about its likely behavior; etc. A statistical study is an engineering tool, not a crystal ball. When an engineer has assiduously studied and asked questions in order to gain expert knowledge about a system, he or she is then in a position to decide intelligently what is not known about the system — and thus what data will be of help.

It is often helpful at Step 2 to make *flowcharts* describing an ideal process and/or the process as it is currently operating. (Sometimes the comparison of the two is enough in itself to show an engineer how a process should be modified.) As such a display is produced, data needs and potential variables of interest can be identified in an organized manner.

EXAMPLE 2-16

Work Flow in a Printing Shop. Drake, Lach, and Shadle conducted a statistical engineering study in cooperation with a printing shop. Before collecting any data, they set about to understand the flow of work through the shop; they produced a flowchart similar to Figure 2-11. The flowchart facilitated clear thinking about what might go wrong in the printing process and at what points what kinds of data could be gathered in order to monitor process performance.

After determining the general arena and physical context of a statistical engineering study, it is necessary to agree on a statement of purpose and scope for the study. An engineering project team assigned to work on a wave soldering process for printed circuit boards must understand the various steps in that process and then begin to define what part(s) of the process will be included in the study and what the goal(s) of the study will be. Will flux formulation and application, the actual soldering, subsequent cleaning and inspection, and touchup all be studied? Or will only some part of this list be investigated? Is system throughput the primary concern, or is it instead some aspect of quality or cost? The sharper a statement of purpose and scope can be made at this point, the easier subsequent planning steps will be.

Study Definition

Once one has defined in qualitative terms what it is about an engineering system that is of interest, one must decide how to represent that property (or those properties) in precise terms. That is, one must choose a well-defined response variable (or variables) and decide how to measure it (or them). For example, in a manufacturing context, if "throughput" of a system is of interest, should it be measured in pieces/hour, or conforming pieces/hour, or net profit/hour, or net profit/hour/machine, or in some other way?

Sections 1-3 and 2-1 have already discussed issues that arise in measurement and the formation of operational definitions. All that really needs to be added here is that these issues must be faced early in the planning of a statistical engineering study. It does little good to carefully plan a study assuming the existence of an adequate piece of

FIGURE 2-11

Flowchart of a Printing Process

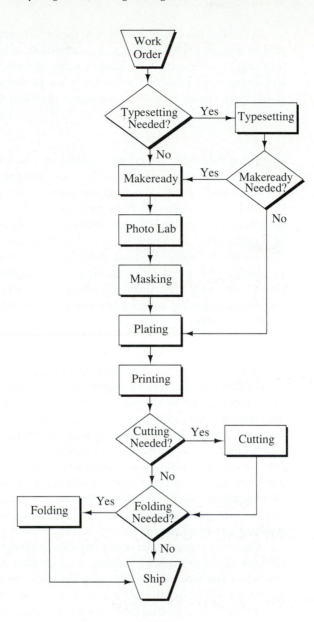

measuring equipment, only to later determine that the organization doesn't own a device with adequate precision and that the purchase of one would cost more than the entire project budget.

Identification of variables that may affect system response variables requires expert knowledge of the process under study. It can't be done completely synthetically by outsiders. Engineers who do not have hands-on experience with a system can sometimes contribute insights gained from experience with similar systems and from basic theory. But it is also wise (in most cases, essential) to include on a project team several people who have first-hand knowledge of the particular process and to talk extensively with those who work with the system on a regular basis. (To digress slightly, it is interesting to note that one competitive advantage supposedly enjoyed by Japanese manufacturers of consumer goods is that unlike American engineers, Japanese engineers typically spend the first quarter of their careers on the shop floor, learning how things work.)

Typically, the job of identifying factors of potential importance in a statistical engineering study is a group activity, carried out in brainstorming sessions. It is therefore helpful to have tools for lending order to what might otherwise be an inefficient and disorganized process. One tool that has proved extremely effective in this regard is variously known as a *cause-and-effect diagram*, or *fishbone diagram*, or *Ishikawa diagram*.

EXAMPLE 2-17

Identifying Potentially Important Variables in a Molding Process. Figure 2-12 shows a cause-and-effect diagram from a study of a molding process for polyurethane automobile steering wheels. It is taken from the paper "Fine Tuning of the Foam System and Optimization of the Process Parameters for the Manufacturing of Polyurethane Steering Wheels Using Reaction Injection Molding by Applying Dr. Taguchi's Method of Design of Experiments," by Vimal Khanna, which appeared in 1985 in the *Third Supplier Symposium on Taguchi Methods*, published by the American Supplier Institute, Inc.

Notice how the diagram in Figure 2-12 organizes the huge number of factors possibly affecting wheel quality. Without some kind of organization, it would be all but impossible to develop anything like an exhaustive list of important factors in a complex situation like this.

Armed with 1) a list of variables that might influence the response(s) of interest and some guesses at their relative importance, 2) a solid understanding of the issues raised in Section 2-3, and 3) knowledge of resource and physical constraints and time-frame requirements, one can begin to make decisions about which (if any) variables are to be managed. As has already been indicated, experiments have some real advantages over purely observational studies (see Section 1-2). Those must be weighed against possible extra costs and difficulties associated with managing both variables that are of interest and those that are not. The hope is to choose a physically and financially workable set of managed variables in such a way that the aggregate effects of variables not of interest and not managed are not so large as to mask the effects of those variables that *are* of interest (both primary experimental variables and important concomitant variables).

Choosing experimental levels and then combinations for managed variables is part (but not all) of the task of deciding on a detailed data collection protocol. Levels of controlled and blocked variables should usually be chosen to be representative of the values that will be met in routine system operation. For example, suppose the amount of contamination in a transmission's hydraulic fluid is thought to affect the transmission's time to failure when it is subjected to stress testing, where operating speed and pressure are the primary experimental variables. It only makes sense to see that the contamination level(s) during testing are representative of the level(s) that will be typical when the transmission is used in the field.

With regard to primary experimental variables, one should also choose typical levels and combinations — with a couple of provisos. Sometimes the goal in an engineering experiment is to compare an innovative, nonstandard way of doing things to current practice. In such cases, it is not good enough simply to look at system behavior with typical settings for primary experimental variables. Also, in cases where primary experimental variables are suspected to have relatively small effects on a response, in the presence of large background variation, it may be necessary to choose ranges of levels for the primary variables that are wider than normal, to see how they act on the response.

Other physical realities and constraints on data collection may also make it appropriate to use atypical values and subsequently extrapolate experimental results to

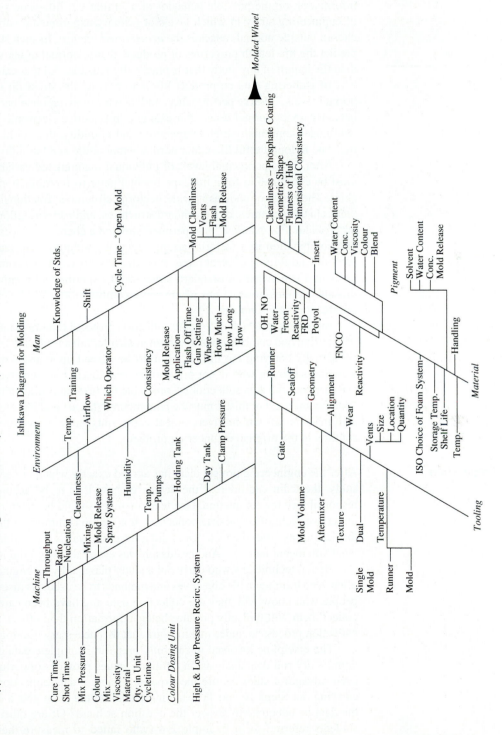

FIGURE 2-12

Cause and Effect Diagram for a Molding Process. From the Third Symposium on Taguchi Methods. © Copyright, American Supplier Institute, Dearborn, Michigan (U.S.A.). Reproduced by permission under License No. 930403.

"standard" circumstances. For example, it is costly enough to run studies on pilot plants using small quantities of chemical reagents and miniature equipment, but much cheaper than experimentation on a full-scale facility. Pilot studies are often used to gather information on the possible behavior of a similar but full-scale system. Another kind of engineering study in which levels of primary experimental variables are purposely chosen outside normal ranges is the *accelerated life test*. In such studies, one wishes to predict the life length properties of products that in normal usage would last far longer than the length of any study that is practically feasible. All that can then be done in the way of gathering data on product life is to turn up the stress on sample units beyond normal levels, observe performance, and then try to extrapolate back to a prediction for behavior under normal usage. For example, if sensitive electronic equipment performs well under abnormally high temperature and humidity, this could well be expected to translate to long useful life under normal temperature and humidity conditions.

After the experimental levels of individual manipulated variables are chosen, they must be combined to form the experimental patterns (combinations) of managed variables. The range of choices is wide: factorial structures, fractional factorial structures, hierarchical structures, other standard structures, and patterns tailor-made for a particular problem. For example, in experimentation on a chemical process, if it is known that one combination in a factorial arrangement of conditions will lead to an undesirable reaction and fouling of plant pipes, it is advisable to avoid that combination even if this leads to a somewhat nonstandard structure of experimental combinations. Obviously, this also applies in situations where some combination could be expected to create an unsafe work environment.

But developing a detailed data collection protocol requires more than these choices of experimental combinations. Experimental order must be decided. Explicit instructions for actually carrying out the testing must be agreed upon and written down in such a way that someone who is not privy to the oral discussions involved in study planning can carry out the data collection. A timetable for first data collection must be developed. With regard to all of this, it must be remembered that several iterations of data collection and analysis (all within given budget constraints) may be required in order to find a solution to the original engineering problem.

Physical Preparation

After the engineering project team has agreed exactly what is to be done in a statistical study, the team can address the details of how it is to be accomplished and assign responsibility for completion. One team member should be given responsibility for the direct oversight of actual data collection. It is all too common for the one who collected the data to say, after the fact, "Oh, I did it the other way .... I couldn't figure out exactly what you meant here .... And besides, it was easier the way I did it."

Again, technicians who carry out a study planned by an engineering project group often need training in the objectives and methods to be used. As discussed in Section 2-1, people who know why they are collecting data and have been carefully shown how to collect them will typically produce better information than others. Overseeing the data collection process includes making sure that this necessary training takes place.

The discipline involved in carefully preparing complete data collection forms and doing a dry run data analysis on fictitious values provides opportunities to refine (and even salvage) a study before the expense of actual data collection is incurred. When carrying out Steps 10 and 11, each individual on the team gets a chance to ask, "Will the data be adequate to answer the question at hand? Or are other data needed?" The students referred to in Example 2-4 (who failed to measure their primary response variable) learned the importance of these steps the hard way.

The final step in the 12-point plan, writing up a best guess at what the study will show, is one that has not been discussed thus far. This author first came across it in *Statistics for Experimenters* by Box, Hunter, and Hunter, and the motivation for it is sound. After a study is complete, it is only human to say, "Of course that's the way things are. We knew that all along." When a careful before-data statement is available to compare to an after-data summarization of findings, it is much easier to see what has been learned in the conduct of a statistical study and appreciate the value of that learning.

EXERCISES

2-1. (§2-1) Consider the context of a study on making paper airplanes where two different Designs (say delta versus t wing), two different Papers (say construction versus typing) and two different Loading Conditions (with a paper clip versus without a paper clip) are of interest with regard to their impact on flight distance. Give an operational definition of *flight distance* that you might use in running such a study.

2-2. (§2-1) Explain how training operators in the proper use of measurement equipment might affect both the repeatability and the reproducibility of measurements made by an organization.

2-3. (§2-1) What would be your response to another engineer's comment, "We have great information on our product — we take 5% samples of every outgoing order, regardless of order size"?

2-4. (§2-2) For the sake of exercise, treat the runout values for 38 laid gears (given in Table 1-1) as a population of interest, and using the random digit table (Table D-1), select a simple random sample of five of these runouts. Repeat this selection process a total of four different times. (Begin the selection of the first sample at the upper left of the table and proceed left to right and top to bottom in the table.) Are the four samples identical? Are they each what you would call "representative" of the population?

2-5. (§2-3) Consider again the paper airplane study scenario from Exercise 2-1. Describe some variables that you would want to control in such a study. What are the response and experimental variables that would be appropriate in this context? Name a potential concomitant variable here.

2-6. (§2-3) In general terms, what is the tradeoff that must be weighed as one decides whether or not to control a variable in a statistical engineering study?

2-7. (§2-3) In the paper airplane scenario of Exercise 2-1, if (e.g., because of schedule limitations) two different team members will make the flight distance measurements, discuss how the notion of blocking might be used.

2-8. (§2-3) Again using the paper airplane scenario of Exercise 2-1, suppose that two students are each going to make and fly one airplane of each of the $2^3 = 8$ possible types of airplane once. Employ the notion of randomization and Table D-1 and develop schedules for Tom and Juanita to use in their flight testing. Explain how the table was used.

2-9. (§2-3) Continuing the paper airplane scenario of Exercise 2-1, discuss the pros and cons of Tom and Juanita flying each of their own eight planes twice, as opposed to making and flying two planes of each of the eight types, one time each.

2-10. (§2-4) What standard name might be applied to the experimental plan you developed for Exercise 2-8?

2-11. (§2-4) Again in the paper airplane scenario of Exercise 2-1, suppose that Tom and Juanita each have time to make and test only four airplanes apiece, but that *in toto* they still wish to test all eight possible types of planes. Develop a sensible plan for doing this. (Which planes should each person test?) You will probably want to be careful to assure that each person tests two delta wing planes, two construction paper planes, and two paper clip planes. Why is this? Can you arrange your plan so that each person tests each Design/Paper combination, each Design/Loading combination, and each Paper/Loading combination once?

2-12. (§2-4) What standard name might be applied to the plan you developed in Exercise 2-11?

2-13. (§2-4) Once more referring to the paper airplane scenario of Exercise 2-1, suppose that only the factors Design and Paper are of interest (all planes will be made without paper clips) but that Tom and Juanita can make and test only two planes apiece. Use the 2×2 Latin square idea and assign Tom and Juanita each two planes to test.

2-14. (§2-5) Either take an engineering system and response variable that you are familiar with from your field, or consider the United Airlines passenger flight system and the response variable Customer Satisfaction, and make a cause-and-effect diagram showing a variety of variables that may potentially affect the response. How might such a diagram be practically useful?

2-15. Use Table D-1 and choose a simple random sample of $n = 8$ out of $N = 491$ widgets. Describe carefully how you label the widgets. Begin in the upper left corner of the table.

2-16. Random number tables are sometimes used in the planning of both enumerative and analytical/experimental studies. What are the two different terminologies employed in these different contexts, and what are the different purposes behind the use of the tables?

2-17. What does one hope to accomplish via blocking in an engineering experiment?

2-18. What are some purposes of replication in a statistical engineering study?

2-19. Consider a potential student project concerning the making of popcorn. Possible factors affecting the outcome of popcorn making include at least the following: Brand of corn, Temperature of corn at beginning of cooking, Popping Method (e.g., frying versus hot air popping), Type of Oil used (if frying), Amount of Oil used (if frying), Batch Size, initial Moisture Content of corn, and Person doing the evaluation of the results of a single batch. Using these factors and/or any others that you can think of, answer the following questions about such a project.

(a) What is a possible response variable in a popcorn project?

(b) Pick two possible experimental factors in this context and describe a 2×2 factorial data structure in those variables that might arise in such a study.

(c) Describe how the concept of randomization might be employed in such a popcorn study.

(d) Describe how the concept of blocking might be employed in such a study.

2-20. Explain briefly why it is safer to try to infer causality from an experiment than from an observational study.

2-21. An experiment is to be performed to compare the effects of two different methods for loading gears in a carburizing furnace on the amount of distortion produced in a heat treating process. Thrust face runout will be measured for gears laid and for gears hung while treating.

(a) 20 gears are to be used in the study. Use Table D-1 and randomly divide the gears into a group to be laid and a group to be hung. Describe carefully how you label the gears. Begin in the upper left corner of the table.

(b) What are some purposes of the randomization used in part (a)?

2-22. State briefly why it is critical to make careful operational definitions for response variables in the collection of engineering data.

2-23. Explain briefly why in an enumerative study, a simple random sample is or is not guaranteed to be representative of the population from which it is drawn.

2-24. Comment briefly on the notion that in order for a statistical engineering study to be statistically proper, one should know before beginning data collection exactly how an entire experimental budget is to be spent. (Is this, in fact, a correct idea?)

2-25. A sanitary engineer wishes to compare two methods for determining chlorine content of Cl_2-demand-free water. To do this, 8 quite different water samples are split in half, and one determination is made using the MSI method and another using the SIB method. Explain why it could be said that the principle of blocking was used in the engineer's study. Also argue that the resulting data set could be described as consisting of paired measurement data.

2-26. A research group is testing three different methods of electroplating widgets (say, methods A, B, and C). On a particular day, 18 widgets are available for testing. It is thought that the effectiveness of electroplating may be strongly affected by the surface texture of the widgets. The engineer running the experiment is able to divide the 18 available widgets into 3 groups of 6 on the basis of surface texture. (Assume that widgets 1–6 are

rough, widgets 7–12 are normal, and widgets 13–18 are smooth.)

(a) Use Table D-1 in an appropriate way, and assign each of the treatments to 6 widgets. Carefully explain exactly how you do the assignment of treatments A, B, and C to the widgets.

(b) If equipment limitations are such that only one widget can be electroplated at once, but it is possible to complete the plating of all 18 widgets on a single day, in exactly what order would you have the widgets plated? Explain where you got this order.

(c) If, in contrast to the situation in part (b), it is possible to plate only nine widgets in a single day, make up an appropriate plan for plating nine on each of two consecutive days.

(d) If measurements of plating effectiveness are made on each of the 18 plates, what kind of data structure will result from the scenario in part (b)? From the scenario in part (c)?

2-27. A company wishes to increase the light intensity of its photoflash cartridge. Two wall thicknesses ($\frac{1}{16}$ inch and $\frac{1}{8}$ inch) and two ignition point placements are under study. Two batches of the basic formulation used in the cartridge are to be made up, each batch large enough to make 12 cartridges. Discuss how you would recommend running this initial phase of experimentation if all cartridges can be made and tested in a short time period by a single technician. Be explicit about any randomization and/or blocking you would employ. Say exactly what kinds of cartridges you would make and test, in what order. Describe the structure of the data that would result from your study.

2-28. Use Table D-1 and

(a) Select a simple random sample of five widgets from a production run of 354 such widgets (begin at the upper left corner of the table and move left to right, top to bottom in the table). Tell how you labeled the widgets and name which ones make up your sample.

(b) Select a random order of experimentation for a context where an experimental factor A has two levels, a second factor, B, has three levels, and two experimental runs are going to be made for each of the $2 \times 3 = 6$ different possible combinations of levels of the factors. Carefully describe how you use the table to do this.

2-29. Return to the situation of Exercise 1-22.

(a) Name factors and levels that might be used in a 3-factor, full factorial study in this situation. Also name two response variables for the study. Suppose that in accord with good engineering data collection practice, one wishes to include some replication in the study. Make up a data collection sheet, listing all the combinations of levels of the factors to be studied, and include blanks where the corresponding observed values of the two responses could be entered for each experimental run.

(b) Suppose that it is feasible to make the runs listed in your answer to part (a) in a completely randomized order. Use a mechanical method (like slips of paper in the hat) to arrive at a random order of experimentation for your study described in part (a). Carefully describe the physical steps you follow in developing this order for data collection.

2-30. Use Table D-1 and

(a) Select a simple random sample of seven widgets from a production run of 619 widgets (begin at the upper left corner of the table and move left to right, top to bottom in the table). Tell how you labeled the widgets and name which ones make up your sample.

(b) Beginning in the table where you left off in (a), select a second simple random sample of seven widgets. Is this sample the same as the first? Is there any overlap at all?

2-31. Consider an experimental situation where three factors A, B, and C each have two levels, and it seems desirable to make three experimental runs for each of the possible combinations of levels of the factors.

(a) Select a completely random order of experimentation. Carefully describe how you use Table D-1 to do this. Make an ordered list of combinations of levels of the three factors, prescribing which combination should be run first, second, etc.

(b) Suppose that because of physical constraints, only eight runs can be made on a given day. Carefully discuss how the concept of blocking could be used in this situation, when planning which experimental runs to make on each of three consecutive days. What possible purpose would blocking serve?

(c) Use Table D-1 to randomize the order of experimentation within the blocks you described in part (b). (Make a list of what combinations of levels of the factors are to be run on each day, in what order.) How does the method you used here differ from what you did in part (a)?

2-32. Consider a study comparing the lifetimes (measured in terms of numbers of holes drilled before failure) of two different brands of 8 mm drills in drilling 1045 steel. Suppose that steel bars from three different heats (batches) of steel are available for use in the study, and it is possible that the different heats have differing physical properties. The lifetimes of a total of 15 drills of each brand will be measured, and each of the bars available is large enough to accommodate as much drilling as will be done in the entire study.

(a) Describe how the concept of control could be used to deal with the possibility that different heats might have different physical properties (such as hardnesses).

(b) Name one advantage and one drawback to controlling the heat.

(c) Describe how one might use the concept of blocking to deal with the possibility that different heats might have different physical properties.

3

Computationally Simple Descriptive Statistics

*E*ngineering data are always variable. Given precise enough measurement, even constant process conditions (if such actually exist in the real world) produce differing responses. Therefore, it is not individual data values that demand an engineer's attention as much as the pattern or *distribution* of those responses. The task of summarizing a data set is to describe its important distributional characteristics. This chapter and the next discuss first computationally simple and then computationally intensive methods that are helpful in this task.

Chapter 3 begins with some elementary graphical and tabular methods of data summarization. The notion of quantiles of a distribution is then introduced and used to make other useful graphical displays. Next, standard numerical summary measures of location and spread for quantitative data are introduced. Finally comes a brief look at some elementary methods for summarizing qualitative and count data.

3-1 Elementary Graphical and Tabular Treatment of Quantitative Data

Almost always, the place to begin in the analysis of data from a statistical engineering study is to make appropriate graphical and/or tabular displays. Indeed, in simple situations where only a few samples are involved, a good picture or table can often tell most of the story about what information the data carry. This section discusses the usefulness of dot diagrams, stem-and-leaf plots, frequency tables, histograms and scatterplots, and run charts as simple methods of descriptive statistics.

Dot Diagrams and Stem-and-Leaf Plots

When an engineering statistics study produces a small or moderate amount of univariate quantitative data, a dot diagram, easily made with pencil and paper, is often quite illuminating. One makes a *dot diagram* by representing each observation as a dot, placed at a position corresponding to its numerical value along a number line.

EXAMPLE 3-1

(Example 1-1 revisited) **Portraying Thrust Face Runouts.** Section 1-1 considered a heat-treating situation, where distortion for gears laid and gears hung was studied. Figure 1-1 has been reproduced here as Figure 3-1. It consists of two dot diagrams, one showing thrust face runout values for gears laid and the other the corresponding values for gears hung. It clearly shows that the laid values are both generally smaller and more consistent than the hung values.

EXAMPLE 3-2

Portraying Bullet Penetration Depths. Sale and Thom compared penetration depths for several types of .45 caliber bullets fired into oak wood from a distance of 15 feet. Table 3-1 gives the penetration depths (in mm from the target surface to the back of the bullets) they obtained for two bullet types. Figure 3-2 presents a corresponding pair of dot diagrams.

The dot diagrams show the penetrations of the 200 grain bullets to be both larger and more consistent than those of the 230 grain bullets. (Interestingly, the students had predicted larger penetrations for the lighter bullets, on the basis of greater muzzle velocity and smaller surface area on which friction can act. The different consistencies of penetration were neither expected nor explained.)

FIGURE 3-1

Dot Diagrams of Runouts

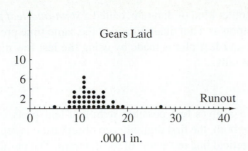

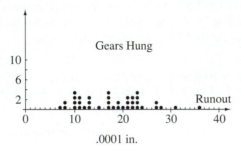

TABLE 3-1

Bullet Penetration Depths (mm)

230 Grain Jacketed Bullets	200 Grain Jacketed Bullets
40.50, 38.35, 56.00, 42.55	63.80, 64.65, 59.50, 60.70
38.35, 27.75, 49.85, 43.60	61.30, 61.50, 59.80, 59.10
38.75, 51.25, 47.90, 48.15	62.95, 63.55, 58.65, 71.70
42.90, 43.85, 37.35, 47.30	63.30, 62.65, 67.75, 62.30
41.15, 51.60, 39.75, 41.00	70.40, 64.05, 65.00, 58.00

FIGURE 3-2

Dot Diagrams of Penetration
Depths

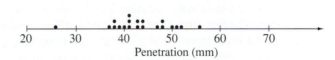

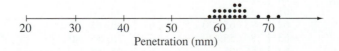

   Dot diagrams give the feel of a data set but do not always allow the user to recover exactly the values used to make them. Because of difficulty in determining exactly where a dot rests on the number line, as well as any rounding applied in order to make a workable display, it is sometimes impossible simply to read the original data values off the diagram. For example, the penetrations in Table 3-1 were rounded to the nearest millimeter to make Figure 3-2, and it is not possible to recover Table 3-1 exactly from Figure 3-2.

Another kind of diagram, called a *stem-and-leaf plot*, carries much the same visual information as a dot diagram and at the same time preserves the original values exactly. A stem-and-leaf plot is made by using the last few digits of each data point to indicate where it falls.

EXAMPLE 3-1
(continued)

Figure 3-3 gives two possible stem-and-leaf plots for the thrust face runouts of laid gears. In both, the first digit of each observation is represented by the number to the left of the vertical line or "stem" of the diagram. The numbers to the right of the vertical line make up the "leaves" of the diagram and give the second digits of the observed runouts. The difference between the two displays is that the second shows somewhat more detail by providing "0–4" and "5–9" leaf positions for each possible leading digit, instead of only a single "0–9" leaf for each leading digit.

FIGURE 3-3

Stem-and-Leaf Plots of Laid Gear Runouts

```
0 | 5 8 8 9 9 9 9
1 | 0 0 0 1 1 1 1 1 1 1 2 2 2 2 3 3 3 3 4 4 4 5 5 5 5 6 7 7 8 9
2 | 7
3 |
```

```
0 |
0 | 5 8 8 9 9 9 9
1 | 0 0 0 1 1 1 1 1 1 1 2 2 2 2 3 3 3 3 4 4 4
1 | 5 5 5 5 6 7 7 8 9
2 |
2 | 7
3 |
3 |
```

EXAMPLE 3-2
(continued)

Figure 3-4 gives two possible stem-and-leaf diagrams for the penetrations of 200 grain bullets in Table 3-1. On these, it was convenient to use two digits to the left of the decimal point to make the stem and the two following the decimal point to create the leaves. The first display was made by recording the leaf values reading directly from the table (in order from left to right and top to bottom). The second display is a somewhat improved representation, obtained by ordering the values that make up each leaf. Notice that both plots give essentially the same visual impression as the second dot diagram in Figure 3-2.

FIGURE 3-4

Stem-and-Leaf Plots of the 200 Grain Penetration Depths

```
58 | .65, .00          58 | .00, .65
59 | .50, .80, .10     59 | .10, .50, .80
60 | .70               60 | .70
61 | .30, .50          61 | .30, .50
62 | .95, .65, .30     62 | .30, .65, .95
63 | .80, .55, .30     63 | .30, .55, .80
64 | .65, .05          64 | .05, .65
65 | .00               65 | .00
66 |                   66 |
67 | .75               67 | .75
68 |                   68 |
69 |                   69 |
70 | .40               70 | .40
71 | .70               71 | .70
```

When one is comparing two data sets, a useful way to use the stem-and-leaf idea is to make two of them *back-to-back*.

EXAMPLE 3-1
(continued)

Figure 3-5 gives back-to-back stem-and-leaf plots for the data of Table 1-1. It shows clearly the differences in location and spread of the two data sets.

FIGURE 3-5

Back-to-Back Stem-and-Leaf Plots of Penetration Depths

Laid Runouts		Hung Runouts
9 9 9 9 8 8 5	0	7 8 8
4 4 4 3 3 3 3 2 2 2 2 1 1 1 1 1 1 1 0 0 0	1	0 0 0 0 1 1 1 2 3 3 3
9 8 7 7 6 5 5 5 5	1	5 7 7 7 7 8 9 9
	2	0 1 1 1 2 2 2 3 3 3 3 4
7	2	7 7 8
	3	1
	3	6

Frequency Tables and Histograms

Dot diagrams and stem-and-leaf plots are useful devices when mulling over a data set. But they are not commonly used when creating formal presentations and reports on engineering studies. In this more formal context, frequency tables and histograms are more often used.

A *frequency table* is made by first breaking an interval containing all the data into an appropriate number of smaller intervals of equal length. Then tally marks can be recorded to indicate the number of data points falling into each interval, and frequencies, relative frequencies, and cumulative relative frequencies can be added.

EXAMPLE 3-1
(continued)

Table 3-2 gives one possible presentation of the laid gear runouts in frequency table form.

TABLE 3-2

Frequency Table for Laid Gear Thrust Face Runouts

Runout (.0001 in.)	Tally	Frequency	Relative Frequency	Cumulative Relative Frequency
5–8	|||	3	.079	.079
9–12	₩ ₩ ₩ |||	18	.474	.553
13–16	₩ ₩ ||	12	.316	.868
17–20	||||	4	.105	.974
21–24		0	0	.974
25–28			1	.026	1.000
		38	1.000	

Notice that the relative frequency values are obtained by dividing the entries in the frequency column by 38, the number of data points (which is itself the sum of the values in the frequency column). The entries in the cumulative relative frequency column are the ratios of the totals in a given class and all preceding classes to the total number of data points. Except for round-off, this is the sum of the relative frequencies on the same row and above a given cumulative relative frequency. The tally column gives the same kind of information about distributional shape that is provided by a dot diagram or a stem-and-leaf plot.

The choice of intervals to use in making a frequency table is to some degree a matter of judgment. Two people given the same data set will not necessarily choose the same set of intervals. However, there are a number of simple points to keep in mind when choosing them. First, in order to avoid visual distortion when using the tally column of the table to gain an impression of distributional shape, intervals of equal length should be employed. Also, for aesthetic reasons, round numbers are preferable as interval endpoints. Since there is aggregation (and therefore some loss of information) involved in the reduction of raw data to tallies, the larger the number of intervals used, the more detailed the information portrayed by the table. On the other hand, if a frequency table is to have value as a data summarization, it can't be too "busy," cluttered with too many intervals. Besides, unless a data set is itself large, a large number of intervals will often lead to many empty or nearly empty categories.

After making a frequency table, it is common to use the organization provided by the table to create a *histogram*. A (frequency or relative frequency) histogram is a kind of bar chart used to portray the shape of a distribution of data points.

EXAMPLE 3-2
(continued)

Table 3-3 is a frequency table for the 200 grain bullet penetration depths, and Figure 3-6 is a translation of that table into the form of a histogram.

TABLE 3-3

Frequency Table for 200 Grain Penetration Depths

Penetration Depth (mm)	Tally	Frequency	Relative Frequency	Cumulative Relative Frequency
58.00–59.99	ЖtІ	5	.25	.25
60.00–61.99	|||	3	.15	.40
62.00–63.99	ЖtІ		6	.30	.70
64.00–65.99	|||	3	.15	.85
66.00–67.99			1	.05	.90
68.00–69.99		0	0	.90
70.00–71.99	||	2	.10	1.00
		20	1.00	

FIGURE 3-6

Histogram of the 200 Grain Penetration Depths

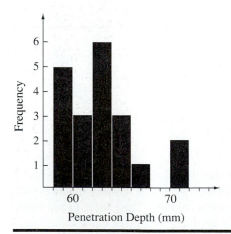

The vertical scale in Figure 3-6 is a frequency scale, and the histogram could be termed a *frequency histogram*. By changing to relative frequency on the vertical scale, one can produce a *relative frequency histogram*.

In making Figure 3-6, care was taken to 1) continue to use intervals of equal length, 2) show the entire vertical axis beginning at zero, 3) avoid breaking either axis, 4) keep a uniform scale across a given axis, and 5) center bars of appropriate heights at the midpoints of the (penetration depth) intervals. The end result of following these guidelines is a display in which equal enclosed areas correspond to equal numbers of data points. Further, data point positioning is clearly indicated by bar positioning on the horizontal axis. If these guidelines are not followed, the resulting bar chart will in one way or another fail to faithfully represent the shape of its corresponding data set.

It is useful to have terminology for some of the common distributional shapes that are encountered when making and using dot diagrams, stem-and-leaf plots, and histograms. Figure 3-7 shows a number of such distributional shapes.

FIGURE 3-7

Distributional Shapes

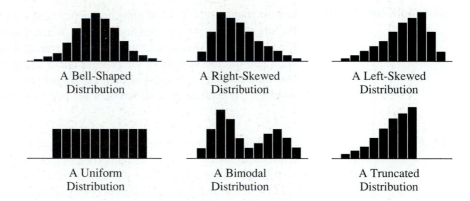

The graphical and tabular devices that have been discussed to this point are almost deceptively simple methods. When routinely and intelligently used, they are powerful engineering tools. The information on location, spread, and shape that is portrayed so clearly on a histogram can give a user strong hints as to the functioning of the physical process that is generating the data. It can also help suggest both the nature of and possible improvements for the physical mechanisms at work in the process.

For example, if one has data on the diameters of machined metal cylinders purchased from a vendor, and a histogram of the data is decidedly *bimodal* (or *multimodal*, having several clear humps), this suggests that the machining of the parts was done on more than one machine, or by more than one operator, or at more than one time. The practical consequence of such multi-channel machining is a distribution of diameters that has more variation than is intrinsic to a production run of cylinders that involves just a single machine, a single operator, and single setup.

As another possibility, if the histogram of diameters is truncated, this suggests that the lot of cylinders has been 100% inspected and sorted, thus removing all cylinders with excessive diameters.

Or, upon marking engineering specifications for cylinder diameter on the histogram, one may get a picture like that in Figure 3-8. It then becomes obvious that the lathe

FIGURE 3-8

Histogram (with Engineering Specifications Marked)

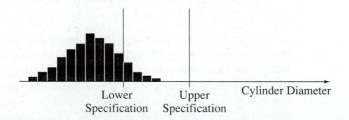

turning the cylinders needs adjustment in order to increase the typical diameter. But it also becomes clear that the basic process variation is so large that even this adjustment will fail to bring essentially all diameters into specifications. Armed with this realization and a knowledge of the economic consequences of parts failing to meet specifications, an engineer can intelligently weigh alternative courses of action: sorting of all incoming parts, demanding that the vendor use more precise equipment, seeking a new vendor, etc.

Investigating the shape of a data set is useful not only because it can lend insight into physical mechanisms behind the shape, but also because shape can be important when determining the appropriateness of applying a particular method of formal statistical inference. Various theoretical distributional shapes have their own appropriate methodologies. What is appropriate for one shape may not be appropriate for another.

Scatterplots and Run Charts

Dot diagrams, stem-and-leaf plots, frequency tables, and histograms are univariate tools, but engineering data are often multivariate. When that is the case, relationships between the variables are usually of interest. If discovered, these relationships can provide insight into the functioning of the process under study. The familiar device of making a two-dimensional *scatterplot* of data as ordered pairs is a simple and effective way of displaying potential relationships between two variables.

EXAMPLE 3-3

Bolt Torques on a Face Plate. Brenny, Christensen, and Schneider measured the torques required to loosen 6 distinguishable bolts holding the front plate on a type of heavy equipment component. Table 3-4 contains the torques (in ft lb) required for bolts number 3 and 4, respectively, on 34 different components.

TABLE 3-4

Torques Required to Loosen Two Bolts on Face Plates (ft lb)

Component	Bolt 3 Torque	Bolt 4 Torque	Component	Bolt 3 Torque	Bolt 4 Torque
1	16	16	18	15	14
2	15	16	19	17	17
3	15	17	20	14	16
4	15	16	21	17	18
5	20	20	22	19	16
6	19	16	23	19	18
7	19	20	24	19	20
8	17	19	25	15	15
9	15	15	26	12	15
10	11	15	27	18	20
11	17	19	28	13	18
12	18	17	29	14	18
13	18	14	30	18	18
14	15	15	31	18	14
15	18	17	32	15	13
16	15	17	33	16	17
17	18	20	34	16	16

Figure 3-9 is a scatterplot of the bivariate data from Table 3-4. In this figure, where several points must be plotted at a single location, the number of points needing to occupy the location has been plotted instead of a single dot.

Notice that the plot gives at least a weak indication that large torques at position 3 are accompanied by large torques at position 4. In practical terms, this is comforting; otherwise, unwanted differential forces might act on the face plate. It is also quite

FIGURE 3-9

Scatterplot of Bolt 3 and
Bolt 4 Torques

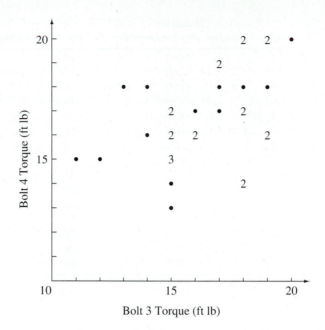

reasonable that bolt 3 and bolt 4 torques be related, when one realizes that the bolts were tightened by different heads of a single pneumatic wrench, operating on a single source of compressed air. It stands to reason that slight variations in air pressure to the wrench might affect the tightening of the bolts at the two positions similarly, producing the big-together, small-together pattern seen in Figure 3-9.

The previous example illustrates well the point that relationships seen on scatterplots suggest a common physical cause for the behavior of variables and can serve as an aid in discovering that cause.

In the most pervasive version of the scatterplot, the variable on the horizontal axis is a time variable. In this text, a scatterplot where univariate data are plotted against time order of observation will be called a *run chart* or *trend chart*. Making run charts (routinely plotting quantitative data against time) is one of the most beneficial statistical habits an engineer can develop. Seeing patterns on a run chart, one is led to think about what process variables were changing in concert with the pattern and to thus develop a keener understanding of how process behavior is affected by those variables that are changing over time.

EXAMPLE 3-4

Diameters of Consecutive Parts Turned on a Lathe. Williams and Markowski studied the behavior of a process for rough turning of the outer diameter on the outer race of a constant velocity joint. Table 3-5 gives the diameters (in inches above nominal) for 30 consecutive joints turned on a particular automatic lathe.

Figure 3-10 gives both a dot diagram and a run chart for the data in the table. In keeping with somewhat standard practice, consecutive points on the run chart have been connected with line segments.

Here the dot diagram is not particularly suggestive of the physical mechanisms that are at work in generating the data. But the time information added when moving to the run chart is revealing. Moving along in time, the outer diameters tend to get smaller until part 16, where there is a large jump in diameter, followed again by a pattern of diameter

TABLE 3-5

30 Consecutive Outer
Diameters Turned on a Lathe

Joint	Diameter (inches above nominal)	Joint	Diameter (inches above nominal)
1	−.005	16	.015
2	.000	17	.000
3	−.010	18	.000
4	−.030	19	−.015
5	−.010	20	−.015
6	−.025	21	−.005
7	−.030	22	−.015
8	−.035	23	−.015
9	−.025	24	−.010
10	−.025	25	−.015
11	−.025	26	−.035
12	−.035	27	−.025
13	−.040	28	−.020
14	−.035	29	−.025
15	−.035	30	−.015

FIGURE 3-10

Dot Diagram and Run Chart of
Consecutive Outer Diameters

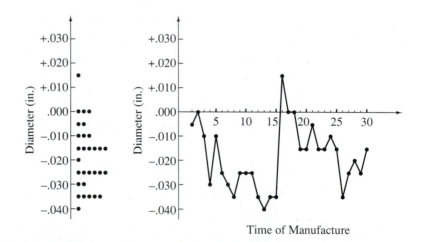

generally decreasing in time. In fact, upon checking production records, Williams and Markowski found that the lathe in question had been turned off and allowed to cool down between parts 15 and 16. It is quite likely that the pattern seen on the run chart is related to the behavior of the lathe's hydraulics. When cold, the hydraulics probably don't do as good a job pushing the cutting tool into the part being turned as when they are warm. Hence, the turned parts become smaller as the lathe warms up. In order to get parts closer to nominal, it would thus seem sensible to adjust the aimed-for diameter up by about .020 inch and run parts only after warming up the lathe.

3-2 *Quantiles and Related Graphical Tools*

Most readers will be familiar with the concept of a percentile. The notion is most famous in the context of reporting scores on educational achievement tests. For example, if a person has scored at the 80th percentile, roughly 80% of those taking the test had worse scores, and roughly 20% had better scores. This concept also turns out to be quite useful in the descriptive analysis of engineering data. However, because it is often more convenient to work in terms of fractions between 0 and 1 rather than in percentages between 0 and 100, slightly different terminology will be used here. Quantiles, rather than percentiles, will be discussed.

In this section, after the quantiles of a data set are carefully defined, they are used to create a number of useful tools of descriptive statistics. The making and use of quantile plots, boxplots, and (empirical) Q-Q plots are discussed, and probability plots (theoretical Q-Q plots) are introduced.

Quantiles and Quantile Plots

Roughly speaking, for a number p between 0 and 1, the p quantile of a distribution is a number such that a fraction p of the distribution lies to the left and a fraction $1 - p$ of the distribution lies to the right. (That is, the p quantile has the same rough interpretation as the $100p$ percentile.) However, because of the basic discreteness of finite data sets, it is necessary to adopt some convention as to exactly what will be meant by the quantiles of a data set. Definition 3-1 gives the convention that will be used in this text.

DEFINITION 3-1

For a data set consisting of n values that when ordered are $x_1 \leq x_2 \leq \cdots \leq x_n$,

1. for any number p of the form $\frac{i-.5}{n}$ where i is an integer from 1 to n, the **p quantile** of the data set will be taken to be x_i. (The ith smallest data point will be called the $\frac{i-.5}{n}$ quantile.)
2. for any number p between $\frac{.5}{n}$ and $\frac{n-.5}{n}$ that is not of the form $\frac{i-.5}{n}$, the **p quantile** of the data set will be obtained by linear interpolation between the quantiles corresponding to the two values of $\frac{i-.5}{n}$ that bracket p.

In both cases, the notation $Q(p)$ will be used to symbolize the p quantile.

This definition is easier to illustrate by means of an example than it is to discuss in the abstract.

EXAMPLE 3-5

Quantiles for Dry Breaking Strengths of Paper Towel. Consumer product testing is a popular form of student project in engineering statistics courses. Lee, Sebghati, and Straub did a study of the dry breaking strength of several brands of paper towel. Included in the study were ten strength determinations for generic towel. Table 3-6 shows the breaking strengths (in grams) reported by the students.

TABLE 3-6

Ten Paper Towel Breaking Strengths

Test	Breaking Strength (g)
1	8,577
2	9,471
3	9,011
4	7,583
5	8,572
6	10,688
7	9,614
8	9,614
9	8,527
10	9,165

By ordering the strength data and computing values of $\frac{i-.5}{10}$, one can easily find the .05, .15, .25, ..., .85, and .95 quantiles of the breaking strength distribution, as shown in Table 3-7.

TABLE 3-7

Quantiles of the Paper Towel
Breaking Strength Distribution

i	$\frac{i-.5}{10}$	ith Smallest Data Point
1	.05	$7,583 = Q(.05)$
2	.15	$8,527 = Q(.15)$
3	.25	$8,572 = Q(.25)$
4	.35	$8,577 = Q(.35)$
5	.45	$9,011 = Q(.45)$
6	.55	$9,165 = Q(.55)$
7	.65	$9,471 = Q(.65)$
8	.75	$9,614 = Q(.75)$
9	.85	$9,614 = Q(.85)$
10	.95	$10,688 = Q(.95)$

Notice that since there are $n = 10$ data points in this example, each one accounts for 10% of the data set. Applying convention 1) in Definition 3-1 to find (for example) the .35 quantile, one counts the smallest 3 data points and half of the 4th smallest as lying to the left of the desired number, and the largest 6 data points and half of the 7th largest as lying to the right. Thus, the 4th smallest data point must be the .35 quantile, as is shown in Table 3-7.

To illustrate convention 2) of Definition 3-1, consider finding the .5 and .93 quantiles of the paper towel strength distribution. Noticing first that .5 is $\frac{.5-.45}{.55-.45} = .5$ of the way from .45 to .55, linear interpolation gives

$$Q(.5) = (1 - .5)\,Q(.45) + .5\,Q(.55) = .5(9,011) + .5(9,165) = 9,088 \text{ g}$$

Then, observing that .93 is $\frac{.93-.85}{.95-.85} = .8$ of the way from .85 to .95, linear interpolation gives

$$Q(.93) = (1 - .8)\,Q(.85) + .8Q(.95) = .2(9,614) + .8(10,688) = 10,473.2 \text{ g}$$

Other authors adopt slightly different conventions for quantiles than the ones given in Definition 3-1. Any choice one makes is somewhat arbitrary, but the present one has intuitive appeal. When the size of a data set is large, the differences between the various conventions given in various texts have little numerical importance.

Particular round values of p give quantiles $Q(p)$ that are known by special names.

DEFINITION 3-2

$Q(.5)$ is called the **median** of a distribution.

DEFINITION 3-3

$Q(.25)$ and $Q(.75)$ are called the **first (or lower) quartile** and **third (or upper) quartile** of a distribution, respectively.

DEFINITION 3-4

$Q(.1), Q(.2), \ldots, Q(.8)$, and $Q(.9)$ are called the **deciles** of a distribution.

EXAMPLE 3-5
(continued)

Referring again to Table 3-7 and the value of $Q(.5)$ previously computed, one sees that for the breaking strength distribution

$$\text{Median} = Q(.5) = 9{,}088 \text{ g}$$
$$\text{1st quartile} = Q(.25) = 8{,}572 \text{ g}$$
$$\text{3rd quartile} = Q(.75) = 9{,}614 \text{ g}$$

Further, upon interpolation,

$$\text{3rd decile} = Q(.3) = .5(8{,}572) + .5(8{,}577) = 8{,}574.5 \text{ g}$$

A way of representing the quantile idea graphically is to make a *quantile plot*.

DEFINITION 3-5

A **quantile plot** is a plot of $Q(p)$ versus p. For an ordered data set of size n containing values $x_1 \leq x_2 \leq \cdots \leq x_n$, such a display is made by first plotting the points $(\frac{i-.5}{n}, x_i)$ and then connecting consecutive plotted points with straight-line segments.

It is because convention 2) in Definition 3-1 calls for linear interpolation that straight-line segments enter the picture in making a quantile plot.

EXAMPLE 3-5
(continued)

Referring again to Table 3-7 for the $\frac{i-.5}{10}$ quantiles of the breaking strength distribution, it is clear that a quantile plot for these data will involve plotting and then connecting consecutive points of the following.

$(.05, 7{,}583)$	$(.15, 8{,}527)$	$(.25, 8{,}572)$
$(.35, 8{,}577)$	$(.45, 9{,}011)$	$(.55, 9{,}165)$
$(.65, 9{,}471)$	$(.75, 9{,}614)$	$(.85, 9{,}614)$
$(.95, 10{,}688)$		

Figure 3-11 gives such a quantile plot for the breaking strength distribution.

FIGURE 3-11

Quantile Plot of Paper
Towel Strengths

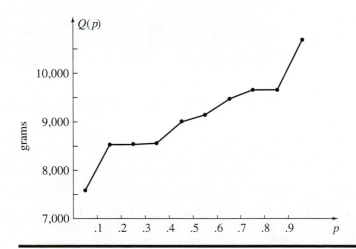

Besides allowing quick graphical evaluation of quantiles, a quantile plot allows the user to do some informal visual smoothing of the plot, to compensate for any jaggedness.

(The tacit assumption is that the underlying data-generating mechanism would itself produce smoother and smoother quantile plots for larger and larger samples.)

Boxplots

Familiarity with the quantile idea is the principal prerequisite for making *boxplots*. This type of display can sometimes be used to advantage as an alternative to dot diagrams or histograms. The boxplot carries somewhat less information than do the other types of displays but nevertheless conveys a fair amount of information about distribution location, spread, and shape. An advantage is that many boxplots can be placed side by side on a single page for comparison purposes.

For the making of boxplots, a number of possible conventions have been suggested. The one used here is borrowed from *Graphical Methods for Data Analysis* by Chambers, Cleveland, Kleiner, and Tukey; it is illustrated in generic fashion in Figure 3-12.

FIGURE 3-12

Generic Boxplot

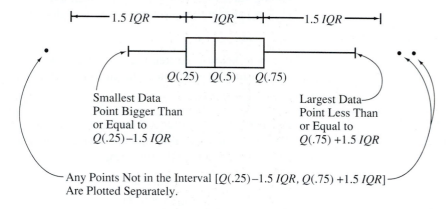

A box is made to extend from the first to the third quartiles and is divided by a line at the median. One then calculates the *interquartile range* (IQR)

$$IQR = Q(.75) - Q(.25)$$

and determines the smallest data point within $1.5IQR$ of $Q(.25)$ and the largest data point within $1.5IQR$ of $Q(.75)$. Lines (called "whiskers") are made to extend out from the box to these values. Typically, most data points will be within the interval $[Q(.25) - 1.5IQR, Q(.75) + 1.5IQR]$, but any that are not then get plotted individually. Points plotted separately are thereby identified as *outlying* or unusual.

EXAMPLE 3-5
(continued)

Consider making a box plot for the paper towel breaking strength data. To begin with,

$$Q(.25) = 8,572 \text{ g}$$
$$Q(.5) = 9,088 \text{ g}$$
$$Q(.75) = 9,614 \text{ g}$$

So

$$IQR = Q(.75) - Q(.25) = 9,614 - 8,572 = 1,042 \text{ g}$$

and

$$1.5IQR = 1,563 \text{ g}$$

Then

$$Q(.75) + 1.5IQR = 9,614 + 1,563 = 11,177 \text{ g}$$

and

$$Q(.25) - 1.5IQR = 8,572 - 1,563 = 7,009 \text{ g}$$

Since all the data points lie in the range 7,009 g to 11,177 g, the resulting boxplot is as shown in Figure 3-13.

FIGURE 3-13

Boxplot of the Paper Towel Strengths

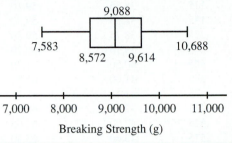

A boxplot shows distributional location through the placement of the box and whiskers along a number line. It shows distributional spread through the extent of the box and the whiskers, with the box enclosing the middle 50% of the distribution. Some elements of distributional shape are indicated by the symmetry (or lack thereof) of the box and of the whiskers. And a gap between the end of a whisker and a point plotted outside the whisker serves as a reminder that no data values fall in that interval.

Two or more boxplots drawn to the same scale and side by side provide an effective way of comparing samples.

EXAMPLE 3-6

(Example 3-2 revisited) **More on Bullet Penetration Depths.** Table 3-8 (page 72) contains the raw information needed to find the $\frac{i-.5}{20}$ quantiles for the two distributions of bullet penetration depth introduced in the previous section.

For the 230 grain bullet penetration depths, interpolation yields

$$Q(.25) = .5Q(.225) + .5Q(.275) = .5(38.75) + .5(39.75) = 39.25 \text{ mm}$$
$$Q(.5) = .5Q(.475) + .5Q(.525) = .5(42.55) + .5(42.90) = 42.725 \text{ mm}$$
$$Q(.75) = .5Q(.725) + .5Q(.775) = .5(47.90) + .5(48.15) = 48.025 \text{ mm}$$

So

$$IQR = 48.025 - 39.25 = 8.775 \text{ mm}$$
$$1.5IQR = 13.163 \text{ mm}$$
$$Q(.75) + 1.5IQR = 61.188 \text{ mm}$$
$$Q(.25) - 1.5IQR = 26.087 \text{ mm}$$

Similar calculations for the 200 grain bullet penetration depths yield

$$Q(.25) = 60.25 \text{ mm}$$
$$Q(.5) = 62.80 \text{ mm}$$
$$Q(.75) = 64.35 \text{ mm}$$
$$Q(.75) + 1.5IQR = 70.50 \text{ mm}$$
$$Q(.25) - 1.5IQR = 54.10 \text{ mm}$$

TABLE 3-8

Quantiles of the Bullet
Penetration Depth Distributions

i	$\frac{i-.5}{20}$	ith Smallest 230 Grain Data Point = $Q(\frac{i-.5}{20})$	ith Smallest 200 Grain Data Point = $Q(\frac{i-.5}{20})$
1	.025	27.75	58.00
2	.075	37.35	58.65
3	.125	38.35	59.10
4	.175	38.35	59.50
5	.225	38.75	59.80
6	.275	39.75	60.70
7	.325	40.50	61.30
8	.375	41.00	61.50
9	.425	41.15	62.30
10	.475	42.55	62.65
11	.525	42.90	62.95
12	.575	43.60	63.30
13	.625	43.85	63.55
14	.675	47.30	63.80
15	.725	47.90	64.05
16	.775	48.15	64.65
17	.825	49.85	65.00
18	.875	51.25	67.75
19	.925	51.60	70.40
20	.975	56.00	71.70

Figure 3-14 then shows boxplots for the two data sets, placed side by side on the same scale. The plots show the larger and more consistent penetration depths of the 200 grain bullets. They also show the existence of one particularly extreme data point in the 200 grain data set. Further, the relative lengths of the whiskers hint at some skewness (recall the terminology introduced with Figure 3-7) in the data. And all of this is done in a way that is quite uncluttered and compact. Many more of these boxes could be added to Figure 3-14 (to compare other bullet types) without visual overload.

FIGURE 3-14

Side-by-Side Boxplots for the
Bullet Penetration Depths

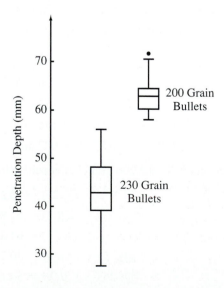

It is worthwhile to compare Figure 3-14 to Figure 3-2 and make the correspondence between the same features of the data portrayed via two different types of summary display.

Q-Q Plots and Comparing Distributional Shapes

It is often important to compare the shapes of two distributions. Comparing histograms or boxplots is one way of doing this. Another, more sensitive, way is to make a single plot based on the quantile functions for the two distributions and exploit the fact that "equal shape" is equivalent to "linearly related quantile functions." Such a plot will be called a *quantile-quantile plot* or, more briefly, a *Q-Q plot*.

In order to motivate the *Q-Q* plot idea, consider the two small artificial data sets given in Table 3-9. Dot diagrams of these two data sets are given in Figure 3-15.

TABLE 3-9

Two Small Artificial Data Sets

Data Set 1	Data Set 2
3, 5, 4, 7, 3	15, 7, 9, 7, 11

FIGURE 3-15

Dot Diagrams for Two Small Data Sets

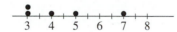

Data Set 1

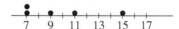

Data Set 2

On an intuitive basis, one would have to say that the two data sets have the same shape. But what is it about the two that makes this the case? One way to look at the matter is to realize that after ordering the two data sets, the gaps between consecutive data values are in the same proportions. That is, for data set 1,

$$3 - 3 = 0$$
$$4 - 3 = 1$$
$$5 - 4 = 1$$
$$7 - 5 = 2$$

and the gaps between consecutive values are in the proportions $0 : 1 : 1 : 2$. On the other hand, for data set 2,

$$7 - 7 = 0$$
$$9 - 7 = 2$$
$$11 - 9 = 2$$
$$15 - 11 = 4$$

and the gaps between the consecutive values are in the proportions $0 : 2 : 2 : 4$ — which is equivalent to the situation for data set 1.

Another way to look at the equality of the two shapes is to note that after ordering the two data sets, the data values are related by the relationship

$$\genfrac{}{}{0pt}{}{i\text{th smallest value}}{\text{in data set 2}} = 2\left(\genfrac{}{}{0pt}{}{i\text{th smallest value}}{\text{in data set 1}}\right) + 1 \tag{3-1}$$

(The 2 multiplier explains why the gaps between ordered values in the second data set are twice those in the first.) Then, recognizing ordered data values as quantiles and letting Q_1 and Q_2 stand for the quantile functions of the two respective data sets, it is clear from display (3-1) that

$$Q_2(p) = 2Q_1(p) + 1 \qquad\qquad \textbf{(3-2)}$$

That is, the two data sets have quantile functions that are linearly related. Looking at either display (3-1) or (3-2), it is obvious that a plot of the points

$$\left(Q_1\left(\frac{i - .5}{5}\right), Q_2\left(\frac{i - .5}{5}\right)\right)$$

(for $i = 1, 2, 3, 4, 5$) is exactly linear. Figure 3-16 illustrates this. The plot in Figure 3-16 is in fact a Q-Q plot for the two artificial data sets of Table 3-9.

FIGURE 3-16

Q-Q Plot for the Data of Table 3-9

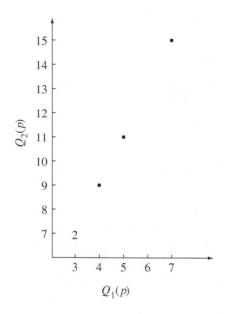

DEFINITION 3-6

An **(empirical)** Q-Q **plot** for two data sets with respective quantile functions Q_1 and Q_2 is a plot of ordered pairs $(Q_1(p), Q_2(p))$ for appropriate values of p. When two data sets of size n are involved, the values of p used to make the plot will be $\frac{i-.5}{n}$ for $i = 1, 2, \ldots, n$. When two data sets of unequal sizes are involved, the values of p used to make the plot will be $\frac{i-.5}{n}$ for $i = 1, 2, \ldots, n$, where n is the size of the smaller set.

From Definition 3-6, note that when making a Q-Q plot for two data sets of the same size, operationally one may simply 1) order each from the smallest observation to the largest, 2) pair off corresponding values in the two data sets, and 3) plot ordered pairs, with the horizontal coordinates coming from the first data set and the vertical ones from the second. When data sets of unequal size are involved, the ordered values from the smaller data set must be paired with quantiles of the larger data set obtained by interpolation.

When a Q-Q plot is reasonably linear, one may conclude that the two distributions involved have similar shapes. When there are marked departures from linearity, the character of those departures can reveal the ways in which the shapes differ.

EXAMPLE 3-6
(continued)

Returning again to the study on bullet penetration depths, notice that Table 3-8 gives the raw material for making a *Q-Q* plot. One need only pair and plot the depths on each row of that table in order to make the plot given in Figure 3-17.

FIGURE 3-17

Q-Q Plot for the Bullet Penetration Depths

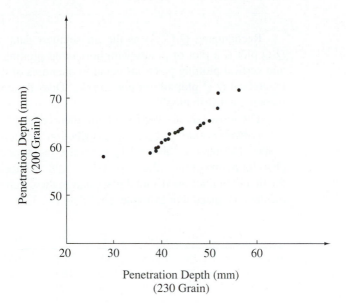

Penetration Depth (mm)
(230 Grain)

The scatterplot in Figure 3-17 is not terribly linear when looked at as a whole. However, the points corresponding to the 2nd through 13th smallest values in each data set do look fairly linear, indicating that (except for the *extreme* lower ends) the lower ends of the two distributions have similar shapes.

The horizontal jog the plot takes between the 13th and 14th plotted points indicates that the gap between 43.85 mm and 47.30 mm (for the 230 grain data) is out of proportion to the gap between 63.55 and 63.80 mm (for the 200 grain data). This perhaps hints that there was some kind of basic physical difference in the mechanisms that produced the smaller and larger 230 grain penetration depths. Once this kind of indication is discovered, it is of course a task for subject matter specialists (ballistics experts, materials people, etc.) to explain the phenomenon, if the matter is to be pursued.

Because of the marked departure from linearity produced by the 1st plotted point, (27.75, 58.00), there is a drastic difference in the shapes of the extreme lower ends of the two distributions. In order to move that point back on line with the rest of the plotted points, one would need to move it to the right or down (i.e., increase the smallest 230 grain observation or decrease the smallest 200 grain observation). That is, relative to the 200 grain distribution, the 230 grain distribution is long-tailed to the low side. (Or to put it differently, relative to the 230 grain distribution, the 200 grain distribution is short-tailed to the low side.) Note that the difference in shapes was already evident in the boxplot in Figure 3-14. Again, it would remain for a subject matter specialist to explain this difference in distributional shapes, if the matter were to be pursued.

The *Q-Q* plotting idea is useful when applied to two data sets, and it is easiest to explain the notion in such an "empirical versus empirical" context. But its greatest usefulness is really when applied to one quantile function that represents a data set and a second that represents a *theoretical distribution*.

DEFINITION 3-7

A **theoretical Q-Q plot** or **probability plot** for a data set and a theoretical distribution, with respective quantile functions Q_1 and Q_2, is a plot of ordered pairs $(Q_1(p), Q_2(p))$ for appropriate values of p. In this text, the values of p of the form $\frac{i-.5}{n}$ for $i = 1, 2, \ldots, n$ will be used.

Recognizing $Q_1(\frac{i-.5}{n})$ as the ith smallest data point, one sees that a theoretical Q-Q plot is a plot of points with horizontal plotting positions equal to observed data and vertical plotting positions equal to quantiles of the theoretical distribution. Such a theoretical Q-Q plot allows one to ask, "Does the data set have a shape similar to the theoretical distribution?"

The most famous version of the theoretical Q-Q plot occurs when quantiles for the *standard normal* or Gaussian distribution are employed. This is the archetypal bell-shaped distribution. Table 3-10 gives some quantiles of this distribution. In order to find $Q(p)$ for p, one of the values $.01, .02, \ldots, .98, .99$, locate the entry in the row labeled by the first digit after the decimal place and in the column labeled by the second digit after the decimal place. For example, $Q(.37) = -.33$.

TABLE 3-10

Standard Normal Quantiles

	.00	.01	.02	.03	.04	.05	.06	.07	.08	.09
.0		−2.33	−2.05	−1.88	−1.75	−1.65	−1.55	−1.48	−1.41	−1.34
.1	−1.28	−1.23	−1.18	−1.13	−1.08	−1.04	−.99	−.95	−.92	−.88
.2	−.84	−.81	−.77	−.74	−.71	−.67	−.64	−.61	−.58	−.55
.3	−.52	−.50	−.47	−.44	−.41	−.39	−.36	−.33	−.31	−.28
.4	−.25	−.23	−.20	−.18	−.15	−.13	−.10	−.08	−.05	−.03
.5	0.00	.03	.05	.08	.10	.13	.15	.18	.20	.23
.6	.25	.28	.31	.33	.36	.39	.41	.44	.47	.50
.7	.52	.55	.58	.61	.64	.67	.71	.74	.77	.81
.8	.84	.88	.92	.95	.99	1.04	1.08	1.13	1.18	1.23
.9	1.28	1.34	1.41	1.48	1.55	1.65	1.75	1.88	2.05	2.33

The origin of the values in Table 3-10 is not obvious at this point. It will be explained in Section 5-2, but for the time being consider the following crude argument to the effect that the quantiles in the table correspond to a bell-shaped pattern. Imagine that each entry in Table 3-10 corresponds to a data point in a set of size $n = 99$. A possible frequency table for those 99 data points is given as Table 3-11.

The tally column in Table 3-11 shows clearly the bell shape of the set of 99 theoretical quantiles listed in Table 3-10. The theoretical quantiles in Table 3-10 can be

TABLE 3-11

A Frequency Table for the Standard Normal Quantiles

Value	Tally	Frequency
−2.80 to −2.30			1
−2.29 to −1.79	||	2
−1.78 to −1.28	HHT ||	7
−1.27 to −.77	HHT HHT ||	12
−.76 to −.26	HHT HHT HHT ||	17
−.25 to .25	HHT HHT HHT HHT		21
.26 to .76	HHT HHT HHT ||	17
.77 to 1.27	HHT HHT ||	12
1.28 to 1.78	HHT ||	7
1.79 to 2.29	||	2
2.30 to 2.80			1

used (with some interpolation if desired) to make a theoretical Q-Q plot, as a way of assessing how bell-shaped a data set looks.

EXAMPLE 3-5
(continued)

Consider again the paper towel strength testing scenario and assess how bell-shaped the data set in Table 3-6 is. Note that without the theoretical Q-Q plotting idea, this would be a difficult, if not impossible, task. With only $n = 10$ data points, making a histogram (or dot diagram after some rounding of values) would probably fail to provide much help in making the assessment. Table 3-12 was made using Tables 3-7 and 3-10; it gives the information needed to produce the theoretical Q-Q plot in Figure 3-18.

TABLE 3-12

Breaking Strength and Standard Normal Quantiles

i	$\frac{i-.5}{10}$	$\frac{i-.5}{10}$ Breaking Strength Quantile	$\frac{i-.5}{10}$ Standard Normal Quantile
1	.05	7,583	-1.65
2	.15	8,527	-1.04
3	.25	8,572	$-.67$
4	.35	8,577	$-.39$
5	.45	9,011	$-.13$
6	.55	9,165	.13
7	.65	9,471	.39
8	.75	9,614	.67
9	.85	9,614	1.04
10	.95	10,688	1.65

FIGURE 3-18

Theoretical Q-Q Plot for the Paper Towel Strengths

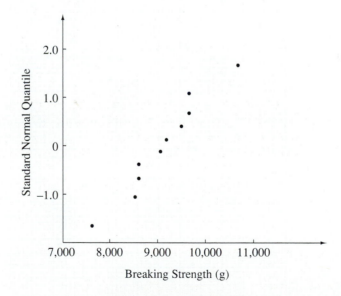

Considering the small size of the data set involved, the plot in Figure 3-18 is fairly linear, and so the data set is reasonably bell-shaped. As a practical consequence of this judgment, it is then possible to use the normal probability models discussed in Section 5-2 to describe breaking strength. Probabilities calculated using the normal models could be used to make breaking strength predictions, and methods of formal statistical inference based on normal model assumptions could be used in the analysis of breaking strength data.

Theoretical Q-Q plots made using standard normal quantiles (as in Example 3-5) are usually called *normal plots* or *normal probability plots*. Special graph paper, called *normal probability paper* (or just *probability paper*), is available as an alternative way of making normal plots. Instead of plotting points on regular graph paper using vertical plotting positions taken from Table 3-10, one plots points on probability paper using vertical plotting positions of the form $\frac{i-.5}{n}$. Figure 3-19 is a normal plot of the breaking strength data from Example 3-5, made on probability paper. Observe that this is virtually identical to the plot in Figure 3-18.

FIGURE 3-19

Normal Plot for the Paper Towel Strengths (Made on Probability Paper, used with permission of Keuffel and Esser Company)

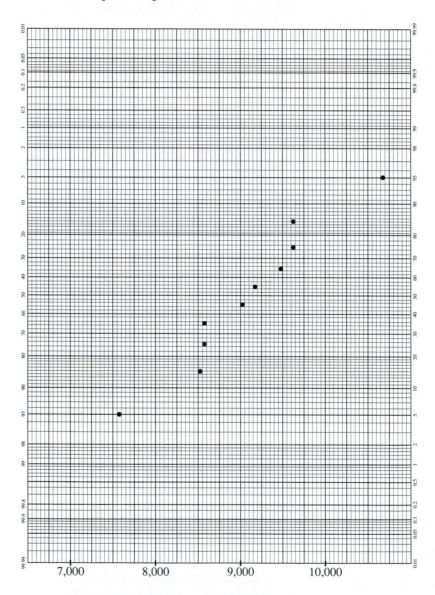

Normal plots are not the only kind of theoretical Q-Q plots useful to engineers. Many other types of theoretical distributions are of engineering importance, and each can be used to make theoretical Q-Q plots. This point is discussed in more detail in Section 5-3, after some probability has been presented. But the introduction of theoretical Q-Q plotting at this point makes it possible to emphasize the relationship between probability plotting and empirical Q-Q plotting.

3-3 *Standard Numerical Summary Measures*

The smooth functioning of most systems of modern technology (and indeed modern society at large) depends on the reduction of large amounts of data to a few numerical summary values, meant to describe the salient features of those data. For example, over the period of a month, a quality control lab doing compressive strength testing for a manufacturer's concrete blocks may make hundreds or even thousands of such measurements. But for some purposes, it may be adequate to know that those strengths average 4,618 psi with a range of 2,521 psi (from smallest to largest).

In this section several standard summary measures for quantitative data are discussed, including the mean, median, range, and standard deviation. Measures of location are considered first, then measures of spread. After introducing these and digressing briefly to discuss the difference between statistics and parameters, there will be illustrations of how numerical summaries can be effectively used to make simple plots and illuminate the results of statistical engineering studies.

Measures of Location

Most people are more or less familiar with the concept of an average as being representative of, or in the center of, a data set. Temperatures may vary between different locations in a blast furnace, but an average temperature tells something about a middle or representative temperature. Scores on a statistics exam may vary, but one is relieved to score at least above average.

The word *average*, as used in colloquial speech, has several potential technical meanings. One is the median, $Q(.5)$, which was introduced in the last section. The median divides a data set in half. Its graphical interpretation is that roughly half of the area enclosed by the bars of a well-made histogram will lie to either side of it. As a measure of center, it is completely insensitive to the effects of a few extreme or outlying observations. For example, the small set of artificial data

$$2, 3, 6, 9, 10$$

has median 6, and this remains true even if the value 10 is replaced by 10,000,000 and/or the value 2 is replaced by $-200,000$.

The previous section discussed the usefulness of the median as a summary measure and used it in the making of boxplots. But it is not the technical meaning most often attached to the notion of average in statistical analyses. Instead, it is more usual to employ the (arithmetic) *mean*.

DEFINITION 3-8

The (**arithmetic**) **mean** of a sample of quantitative data (say, $x_1, x_2, \ldots, x_n$) is

$$\bar{x} = \frac{1}{n} \sum_{i=1}^{n} x_i$$

The mean is sometimes called the *first moment* or *center of mass* of a distribution, drawing on an analogy to mechanics. If one thinks of placing a unit mass along the number line at the location of each value in a data set, the balance point of the mass distribution is at $\bar{x}$.

EXAMPLE 3-7

Waste on Bulk Paper Rolls. Hall, Luethe, Pelszynski, and Ringhofer worked with a company that cuts paper from large rolls, purchased in bulk from several suppliers. The company was interested in determining the amount of waste (by weight) on rolls

TABLE 3-13

Percent Waste by Weight on
Bulk Paper Rolls

Supplier 1	Supplier 2
.37, .52, .65,	.89, .99, 1.45, 1.47,
.92, 2.89, 3.62	1.58, 2.27, 2.63, 6.54

obtained from the various sources. Table 3-13 gives percent waste data, which the students obtained for 6 and 8 rolls, respectively, of a type of paper purchased from two different sources.

The medians and means for the two data sets are easily obtained. For the supplier 1 data,

$$Q(.5) = .5(.65) + .5(.92) = .785\% \text{ waste}$$

and

$$\bar{x} = \frac{1}{6}(.37 + .52 + .65 + .92 + 2.89 + 3.62) = 1.495\% \text{ waste}$$

For the supplier 2 data,

$$Q(.5) = .5(1.47) + .5(1.58) = 1.525\% \text{ waste}$$

and

$$\bar{x} = \frac{1}{8}(.89 + .99 + 1.45 + 1.47 + 1.58 + 2.27 + 2.63 + 6.54) = 2.228\% \text{ waste}$$

Figure 3-20 shows dot diagrams with the medians and means marked on them. Notice that a comparison of either medians or means for the two suppliers shows the supplier 2 waste to be larger than the supplier 1 waste. But there is a substantial difference between the median and mean values for a given supplier. In both cases, the mean is quite a bit larger than the corresponding median. This is a reflection of the *right-skewed* nature of both data sets. In both cases, the center of mass of the distribution is pulled strongly to the right by a few extremely large values.

FIGURE 3-20

Dot Diagrams for the Waste
Percentages

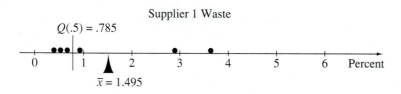

Supplier 1 Waste

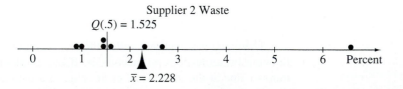

Supplier 2 Waste

Example 3-7 shows clearly that, in contrast to the median, the mean is a measure of center that can be strongly affected by a few extreme data values. Authors sometimes say that because of this, one or the other of the two measures is "better." Such statements lack sense. Neither measure of location is better; they are simply measures with different properties. And the difference in properties is one that intelligent consumers of statistical information do well to keep in mind. The average income of employees at a company paying nine line workers each $10,000/year and a president $110,000/year can be described as $10,000/year or $20,000/year, depending upon whether the median or mean is being used.

Measures of Spread

Quantifying the variation that is intrinsic in a data set can be as important as measuring its location (or even more so). In industrial contexts, for example, if a characteristic of parts coming off a particular machine is being measured and recorded, the spread of the resulting data gives information about the intrinsic precision or *capability* of the machine. The location of the resulting data is often a function of machine setup or settings of adjustment knobs. Setups can fairly easily be changed, but improvement of intrinsic machine precision usually means a capital expenditure for a new piece of equipment, or at least the overhaul of an existing one.

Although the point wasn't stressed in Section 3-2, the interquartile range, $IQR = Q(.75) - Q(.25)$, that was introduced there is one possible measure of spread for a distribution. It measures the spread of the middle half of a distribution. Therefore, it is insensitive to the possibility of a few extreme values occurring in a data set.

A measure related to the interquartile range is the *range*, which measures the spread of the entire distribution.

DEFINITION 3-9

The **range** of a data set consisting of ordered values $x_1 \leq x_2 \leq \cdots \leq x_n$ is

$$R = x_n - x_1$$

Notice the word usage here. The word *range* could be used as a verb to say, "The data range from 3 to 21." But to use the word as a noun, one says, "The range is $(21 - 3) = 18$."

Since the range depends only on the values of the smallest and largest points in a data set, it is necessarily extremely sensitive to extreme (or outlying) values. Because it is easily calculated, it has enjoyed long-standing popularity in industrial settings, particularly as a tool in statistical quality control.

However, most methods of formal statistical inference are based on a way of measuring spread that is somewhat different from the simple high-minus-low idea employed to find the range and interquartile range. Rather, a notion of "mean squared deviation" or "root mean squared deviation" is employed to produce measures that are called the *variance* and the *standard deviation*, respectively.

DEFINITION 3-10

The **sample variance** of a data set consisting of values $x_1, x_2, \ldots, x_n$ is

$$s^2 = \frac{1}{n-1} \sum_{i=1}^{n} (x_i - \bar{x})^2$$

The **sample standard deviation**, s, is the nonnegative square root of the sample variance.

Notice that apart from an exchange of $n - 1$ for n in the divisor, s^2 is an average squared distance of the data points from the central value $\bar{x}$. s^2 is thus nonnegative and is 0 only when all data points are exactly alike. s^2 has units that are the squares of the units in which the original data are expressed. Taking the square root of s^2 to obtain s then produces a measure of spread expressed in the units of the original data.

EXAMPLE 3-7
(continued)

The spreads in the two sets of percentage wastes recorded in Table 3-13 can be expressed in any of the above terms. For the supplier 1 data,

$$Q(.25) = .52$$
$$Q(.75) = 2.89$$

and so

$$IQR = 2.89 - .52 = 2.37\% \text{ waste}$$

Also,

$$R = 3.62 - .37 = 3.25\% \text{ waste}$$

Further,

$$s^2 = \frac{1}{6-1}((.37 - 1.495)^2 + (.52 - 1.495)^2 + (.65 - 1.495)^2 + (.92 - 1.495)^2$$
$$+ (2.89 - 1.495)^2 + (3.62 - 1.495)^2)$$
$$= 1.945(\% \text{ waste})^2$$

so that

$$s = \sqrt{1.945} = 1.395\% \text{ waste}$$

Similar calculations for the supplier 2 data yield the values

$$IQR = 1.23\% \text{ waste}$$

and

$$R = 6.54 - .89 = 5.65\% \text{ waste}$$

Further,

$$s^2 = \frac{1}{8-1}((.89 - 2.228)^2 + (.99 - 2.228)^2 + (1.45 - 2.228)^2 + (1.47 - 2.228)^2$$
$$+ (1.58 - 2.228)^2 + (2.27 - 2.228)^2 + (2.63 - 2.228)^2 + (6.54 - 2.228)^2)$$
$$= 3.383(\% \text{ waste})^2$$

so

$$s = 1.839\% \text{ waste}$$

Supplier 2 has the smaller IQR but the larger R and s. This is consistent with Figure 3-20. The central portion of the supplier distribution is tightly packed. But the single extreme data point makes the overall variability larger for the second supplier than for the first.

It is clear from their definitions that the measures of variation, IQR, R, and s, are not directly comparable. Although it is somewhat out of the main flow of this discussion, it is worth interjecting at this point that it is possible (and quite common in statistical quality control) to "put R and s on the same scale," so to speak. This is done by dividing R by an appropriate conversion factor, known to quality control engineers as d_2. The table in Appendix D-2 contains *control chart constants*; it can be used to find values of d_2 for various sample sizes n. For example, to get R and s on the same scale for the supplier 1 data, division of R by 2.534 is in order, since $n = 6$. (The origin of the entries in Table D-2 is not at all obvious at this point.)

Often, the calculation of s^2 or s is not particularly pleasant work if Definition 3-10 must be used directly. It is sometimes easier to use an algebraically equivalent alternative formula:

$$s^2 = \frac{n}{n-1}\left(\frac{1}{n}\sum_{i=1}^{n} x_i^2 - \bar{x}^2\right) \tag{3-3}$$

Using formula (3-3) in place of the formula in Definition 3-10 allows one to square before doing any subtraction, which can be a help if the raw data are "round" but $\bar{x}$ is not.

Of course, the most sensible way to calculate sample variances these days is using a hand-held electronic calculator with a preprogrammed variance function. One slight caution is in order, however. Calculator manufacturers are not completely consistent in their understanding of how variances should be calculated. Some brands use an n divisor in place of the $n-1$ divisor in Definition 3-10. Users should determine for themselves which choice was made in the design of any calculators they employ.

Students often find the statement "s is a root mean squared deviation" less than illuminating and have some initial difficulty developing a feel for the standard deviation. One possible help in this effort is a famous theorem of a Russian mathematician.

PROPOSITION 3-1

Chebyschev's Theorem. For any data set and any number k larger than 1, a fraction of at least $1 - (1/k^2)$ of the data are within ks of $\bar{x}$.

This little theorem says, for example, that at least $\frac{3}{4}$ of a data set is within 2 standard deviations of its mean. And at least $\frac{8}{9}$ of a data set is within 3 standard deviations of its mean. So the theorem promises that if a data set has a small standard deviation, it will be tightly packed about its mean.

EXAMPLE 3-7
(continued)

Returning to the waste data, consider using formula (3-3) and then illustrating the meaning of Chebyschev's Theorem with the supplier 1 values. First,

$$s^2 = \frac{6}{6-1}\left(\frac{1}{6}\left((.37)^2 + (.52)^2 + (.65)^2 + (.92)^2 + (2.89)^2 + (3.62)^2\right) - (1.495)^2\right)$$

$$= 1.945(\% \text{ waste})^2$$

as before. Then, for example, taking $k = 2$ and using Chebyschev's Theorem, it is clear that at least $\frac{3}{4} = 1 - (\frac{1}{2})^2$ of the 6 data points (i.e., at least 4.5 of them) must be within 2 standard deviations of $\bar{x}$. But

$$\bar{x} - 2s = 1.495 - 2(1.395) = -1.294\% \text{ waste}$$

and

$$\bar{x} + 2s = 1.495 + 2(1.395) = 4.284\% \text{ waste}$$

so simple counting shows that in fact all (a fraction of 1.0) of the data are between these two values.

Statistics and Parameters

At this point in the exposition, it is important to introduce some more basic terminology. General numerical description of distributions has been under discussion. But as it turns out, somewhat different terminology and symbolism are used in reference to distributions of samples than in reference to population distributions and theoretical distributions.

DEFINITION 3-11

Numerical summarizations of sample data are called (sample) **statistics**. Numerical summarizations of population and theoretical distributions are called (population or model) **parameters**. Typically, Roman letters are used as symbols for statistics, while Greek letters are used to stand for parameters.

As an example of the usage outlined in this definition, consider the mean. Definition 3-8 refers specifically to a calculation for a sample. If a data set represents an entire population, then it is common to use the lowercase Greek letter mu (μ) to stand for the *(population) mean* and write

$$\mu = \frac{1}{N} \sum_{i=1}^{N} x_i \qquad (3\text{-}4)$$

In comparing this expression to the one in Definition 3-8, not only is a different symbol used for the mean, but also N is used in place of n. To denote a population size as N and a sample size as n is reasonably standard notation. Chapter 5 gives a definition for what is meant by the mean of a theoretical distribution. But it is worth saying now that the symbol μ will be used in that context as well as in the context of equation (3-4).

As another pair of instances of the usage suggested by Definition 3-11, consider the variance and standard deviation. Definition 3-10 refers specifically to the sample variance and standard deviation. If a data set represents an entire population then it is common to use the lower case Greek sigma squared (σ^2) to stand for the *population variance* and to define

$$\sigma^2 = \frac{1}{N} \sum_{i=1}^{N} (x_i - \mu)^2 \qquad (3\text{-}5)$$

One then calls the nonnegative square root of σ^2 the *population standard deviation, σ*. Notice that (apparently inexplicably) the division in equation (3-5) is by N, and not the $N - 1$ that might be expected on the basis of Definition 3-10. There are reasons for this change, but they are not accessible at this point, so you will simply have to endure the apparent caprice of the situation. Chapter 5 defines a variance and standard deviation for theoretical distributions, and the symbols σ^2 and σ will be used there as well as in the context of equation (3-5).

On one point, this text will deviate from the Roman/Greek symbolism convention laid out in Definition 3-11: the notation for quantiles. $Q(p)$ will stand for the pth quantile of a distribution, whether it is from a sample, a population, or a theoretical model.

Plots of Summary Statistics

To promote the digestion of information obtained in a statistical engineering study, plotting numerical summary measures in various sensible ways is helpful. For example, plots of summary statistics against time are frequently revealing.

EXAMPLE 3-8

(Example 1-9 revisited) **Monitoring a Critical Dimension of Machined Parts.** Cowan, Renk, Vander Leest, and Yakes worked with a company that machines precision metal parts. A critical dimension of one such part was monitored by occasionally selecting and measuring 5 consecutive pieces and then plotting the sample mean and range values. Table 3-14 gives the $\bar{x}$ and R values for 25 consecutive samples of 5 parts. The values reported are in .0001 in.

Figure 3-21 is a plot of both the means and ranges against order of observation. Looking first at the plot of ranges, no strong trends are obvious, which suggests that the

TABLE 3-14

Means and Ranges for a
Critical Dimension on
Samples of $n = 5$ Parts

Sample	Time		$\bar{x}$	R
1	10/27	7:30 AM	3509.4	5
2		8:30	3509.2	2
3		9:30	3512.6	3
4		10:30	3511.6	4
5		11:30	3512.0	4
6		12:30 PM	3513.6	6
7		1:30	3511.8	3
8		2:30	3512.2	2
9		4:15	3500.0	3
10		5:45	3502.0	2
11		6:45	3501.4	2
12		8:15	3504.0	2
13		9:15	3503.6	3
14		10:15	3504.4	4
15		11:15	3504.6	3
16	10/28	7:30 AM	3513.0	2
17		8:30	3512.4	1
18		9:30	3510.8	5
19		10:30	3511.8	4
20		6:15 PM	3512.4	3
21		7:15	3511.0	4
22		8:45	3510.6	1
23		9:45	3510.2	5
24		10:45	3510.4	2
25		11:45	3510.8	3

FIGURE 3-21

Plots of $\bar{x}$ and R over Time

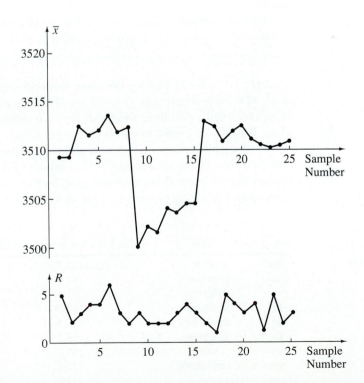

basic short-term variation measured in this critical dimension is stable. The combination of process and measurement precision is neither improving nor degrading with time.

The plot of means, however, does suggest some kind of physical change. The average dimensions from the second shift on November 27 (samples 9 through 15) are markedly smaller than the rest of the means. As discussed in Example 1-9, it turned out to be the case that the parts produced on that shift were not really systematically any different from the others. Instead, it was ultimately determined that the person making the measurements for samples 9 through 15 used the gage in a fundamentally different way than other employees. The pattern in the $\bar{x}$ values was caused by this change in measurement technique. (Essentially, the employee in question wasn't calibrated with his peers.)

Patterns revealed in the plotting of sample statistics against time ought to be enough to set the alert engineer to looking for a physical cause and (typically) a cure. *Systematic variations* or *cycles* in a plot of means can often be related to process variables that come and go on a more or less regular basis. Examples might include seasonal or diurnal variables like ambient temperature or those caused by rotation of gages or fixtures. *Instability* or variation in excess of that understandable in terms of basic equipment precision can sometimes be traced to mixed lots of raw material or overadjustment of equipment by operators. *Changes in level* of a process mean can originate in the introduction of new machinery, raw materials, or employee training and (for example) tool wear. *Mixtures* of several patterns of variation on a single plot of some summary statistic against time can sometimes (as in Example 3-8) be traced to changes in measurement calibration. They are also sometimes produced by consistent differences in machines or streams of raw material.

Plots of summary statistics against time are of course not the only useful ones. Plots of summary statistics against process variables can also be quite informative.

EXAMPLE 3-9

(Example 1-7 revisited) **Plotting the Mean Shear Strength of Wood Joints.** In their study of glued wood joint strength, Dimond and Dix obtained the values given in Table 3-15 as mean strengths over three shear tests for each combination of three woods and three glues. Figure 3-22 gives a revealing plot of these $\bar{x}$'s.

From the plot, it is obvious that the gluing properties of pine and fir are quite similar, with pine joints averaging around 40–45 lb stronger. For these two soft woods, cascamite appears slightly better than carpenter's glue, both of which make much better joints than white glue. The gluing properties of oak (a hardwood) are quite different from those of

TABLE 3-15

Mean Joint Strengths for Nine Wood/Glue Combinations

Wood	Glue	$\bar{x}$ Mean Joint Shear Strength (lb)
pine	white	131.7
pine	carpenter's	192.7
pine	cascamite	201.3
fir	white	92.0
fir	carpenter's	146.3
fir	cascamite	156.7
oak	white	257.7
oak	carpenter's	234.3
oak	cascamite	177.7

FIGURE 3-22

Plot of Mean Joint Strength
versus Glue Type for
Three Woods

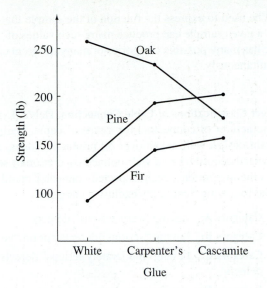

pine and fir. In fact, the glues perform in exactly the opposite ordering for the strength of oak joints. All of this is displayed quite eloquently by the simple plot in Figure 3-22.

The two previous examples have illustrated the usefulness of plotting sample statistics against time and against levels of an experimental variable. There are surely other possibilities in specific engineering situations. Plots against any variable that comes to mind in the process of engineering problem solving can potentially help the working engineer understand and manipulate the engineering system on which he or she works.

3-4 *Descriptive Statistics for Qualitative and Count Data*

The techniques presented thus far in this chapter are primarily relevant to the analysis of measurement data. As noted in Section 1-2, statistical conventional wisdom is that where they can be obtained, measurement data (or as they are sometimes called, *variables data*) are generally preferable to count and qualitative data (sometimes called *attributes data*). Nevertheless, in some situations, qualitative or count data will constitute the primary information available on the behavior of a process. It is therefore worthwhile to consider their summarization.

This section will cover the reduction of qualitative and count data to per-item or per-inspection unit figures and the display of those ratios in simple bar charts and plots.

**Numerical Summarization
of Qualitative and
Count Data**

At this point, it is advisable to review Definitions 1-8 and 1-9 and recall that aggregation and counting are required to produce numerical values from qualitative data. Beginning with count data (or after counting the number of observations falling in a given qualitative category), it is often helpful to express the counts as rates on a per-item or per-inspection unit basis.

When each item in a sample of n either does or does not have a characteristic of interest, the notation

$$\hat{p} = \frac{\text{The number of items in the sample with the characteristic}}{n}$$ (3-6)

will be used to express the fraction of the sample that possess the characteristic. Notice that a given sample can produce many such values of "p hat" if either a single characteristic has many possible categories or many different characteristics are being monitored simultaneously.

EXAMPLE 3-10

Defect Classifications of Cable Connectors. Delva, Lynch, and Stephany worked with a manufacturer of connectors for a type of cable. Daily samples of 100 connectors of a certain design were taken over 30 production days, and each sampled connector was inspected according to a well-defined (operational) set of rules. Using the information from the inspections, each inspected connector could be classified as belonging to one of the following 5 mutually exclusive categories.

Category A: having "very serious" defects

Category B: having "serious" defects but no "very serious" defects

Category C: having "moderately serious" defects but no "serious" or "very serious" defects

Category D: having only "minor" defects

Category E: having no defects

Table 3-16 gives counts of sampled connectors falling into the first 4 categories over the 30-day period.

TABLE 3-16

Counts of Connectors Classified into Four Defect Categories

Category	Number of Sampled Connectors
A	3
B	0
C	11
D	1

Then, using the fact that $30 \times 100 = 3,000$ connectors were inspected over this period,

$$\hat{p}_A = 3/3000 = .0010$$
$$\hat{p}_B = 0/3000 = .0000$$
$$\hat{p}_C = 11/3000 = .0037$$
$$\hat{p}_D = 1/3000 = .0003$$

Notice that here $\hat{p}_E = 1 - (\hat{p}_A + \hat{p}_B + \hat{p}_C + \hat{p}_D)$, because categories A through E represent a set of nonoverlapping and exhaustive classifications into which an individual connector must fall, so that the $\hat{p}$'s must total to 1.

EXAMPLE 3-11

Pneumatic Tool Manufacture. Kraber, Rucker, and Williams worked with a manufacturer of pneumatic tools. Each tool produced is thoroughly inspected before shipping. The students collected some data on several kinds of possible problems uncovered at final inspection. Table 3-17 gives counts of tools having these problems in a particular production run of 100 tools.

Notice that Table 3-17 is a summarization of highly multivariate qualitative data. In contrast to the situation in the previous example, the categories listed in Table 3-17 are not mutually exclusive; a given tool can fall into more than one of them. Instead of

TABLE 3-17

Counts and Fractions of Tools
with Various Problems

Problem	Number of Tools	$\hat{p}$
Type 1 Leak	8	.08
Type 2 Leak	4	.04
Type 3 Leak	3	.03
Missing Part 1	2	.02
Missing Part 2	1	.01
Missing Part 3	2	.02
Bad Part 4	1	.01
Bad Part 5	2	.02
Bad Part 6	1	.01
Wrong Part 7	2	.02
Wrong Part 8	2	.02

representing different possible values of a single categorical variable (as was the case with the connector categories in Example 3-10), the categories listed above each amount to 1 (present) of 2 (present and not present) possible values for a different categorical variable. For example, for type 1 leaks, $\hat{p} = .08$, so $1 - \hat{p} = .92$ for the fraction of tools *without* type 1 leaks. However, one does not necessarily expect the $\hat{p}$ values to total to the fraction of tools requiring rework at final inspection. A given faulty tool could be counted in several $\hat{p}$ values.

Another kind of per-item ratio, also based on counts, is sometimes confused with $\hat{p}$, the sample fraction possessing an attribute of interest. Such a rate arises when every item in a sample provides opportunity for a phenomenon of interest to occur, but multiple occurrences are possible and counts are kept of the total number of occurrences. In such cases, the notation

$$\hat{u} = \frac{\text{The total number of occurrences}}{\text{The total number of inspection units or sampled items}} \tag{3-7}$$

is used to express the *mean occurrences per item* (or *inspection unit*). Notice that $\hat{u}$ is really closer in meaning to $\bar{x}$ than to $\hat{p}$, even though it often turns out to be a number between 0 and 1 and is sometimes expressed as a percentage and called a *rate*.

Notice also that although the counts totaled in the numerator of expression (3-7) must all be integers, the values totaled to create the denominator need not be. For instance, suppose vinyl floor tiles are being inspected for serious blemishes. If on one occasion 1 case yields a total of 2 blemishes, on another occasion .5 case yields 0 blemishes, and on still another occasion 2.5 cases yield a total of 1 blemish, then

$$\hat{u} = \frac{2 + 0 + 1}{1 + .5 + 2.5} = .75 \text{ blemishes/case}$$

Depending on exactly how one defines terms, it may be appropriate to calculate either $\hat{p}$ values or $\hat{u}$ values or both in the descriptive analysis of a single set of count or qualitative data.

EXAMPLE 3-10
(continued)

The cable connector example can be used to illustrate the use of the notion of occurrences per item. It was possible for a single connector to have more than one defect of a given severity and, in fact, defects of different severities. For example, Delva, Lynch, and Stephany's records indicate that in the 3,000 connectors inspected, 1 connector had exactly 2 moderately serious defects (along with a single very serious defect), 11 connectors had exactly 1 moderately serious defect (and no others), and 2,988 had no moderately serious defects. So one could report the observed rate of moderately serious defects as

$$\hat{u} = \frac{2 + 11}{1 + 11 + 2988} = .0043 \text{ moderately serious defects/connector}$$

This is an occurrence rate for moderately serious defects, but not a fraction of connectors having moderately serious defects.

It may seem that a trivial point is being belabored in emphasizing the difference between the statistics $\hat{p}$ and $\hat{u}$. But it must be done, because this is a point that constantly causes students confusion. Methods of formal statistical inference based on $\hat{p}$ are not the same as those based on $\hat{u}$, although they both concern rates in some sense. The distinction between the two must be kept in mind if one is to apply those methods appropriately.

To carry this caveat a step further, note that not every quantity called a *percentage* is even of the form $\hat{p}$ or $\hat{u}$. For example, in a laboratory analysis, a specimen may be declared to be "30% carbon." The 30% cannot be thought of as having the form of $\hat{p}$ in equation (3-6) or $\hat{u}$ in equation (3-7). It is really a single, essentially continuous, measurement, not a summary statistic at all. Methods of formal inference based on $\hat{p}$ or $\hat{u}$ have nothing to say about such percentages. (Rather, methods of inference for measurement data should be applied to samples of such percentages.)

Bar Charts and Plots for Qualitative and Count Data

Often, a statistical engineering study will produce several values of $\hat{p}$ or $\hat{u}$ that need to be compared. This can come about when

1. one has a single sample, involving either one characteristic with many possible categories or several characteristics, or
2. one has several samples.

In any case, bar charts and simple bivariate plots can be a great aid in summarizing the results of such studies.

EXAMPLE 3-10
(continued)

Figure 3-23 is a bar chart representation of the fractions of connectors in the categories A through D. It shows clearly that most connectors with defects fall into category C, having moderately serious defects but no serious or very serious defects. This bar chart is a presentation of the behavior of a single categorical variable.

FIGURE 3-23

Bar Chart of Connector Defects

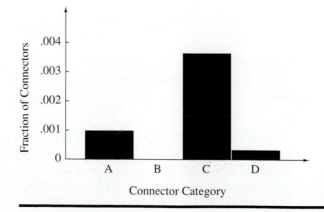

EXAMPLE 3-11
(continued)

Figure 3-24 is a bar chart representation of the information on tool problems that was given in Table 3-17. Notice that this bar chart shows leaks to be the most frequently occurring problems on this production run.

FIGURE 3-24

Bar Chart of Assembly Problems

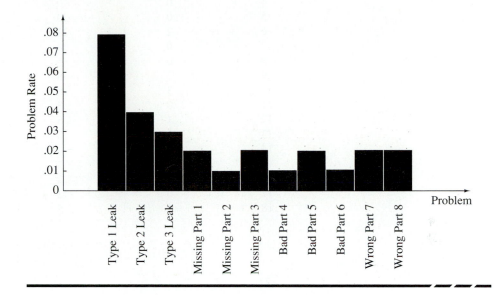

Figures 3-23 and 3-24 are both bar charts, but they differ considerably in character. The first concerns the behavior of a single (ordered) categorical variable — namely, Connector Class. The second concerns the behavior of 11 different present–not present categorical variables, like Type 1 Leak, Missing Part 3, etc. There is perhaps some reason to attach significance to the shape of Figure 3-23, since categories A through D are arranged in decreasing order of defect severity, and this order was used in the making of the figure. But the shape of Figure 3-24 is essentially arbitrary, since the particular ordering of the tool problem categories used to make the figure is arbitrary. Other equally sensible orderings would give quite different shapes.

The notion of segmenting bars on a bar chart and letting the segments stand for different categories for a single qualitative variable can be a helpful one, particularly where several different samples are to be compared.

EXAMPLE 3-12

Scrap and Rework in a Turning Operation. The article "Statistical Analysis: Roadmap for Management Action," by H. Rowe (*Quality Progress*, February 1985), describes a statistically based quality-improvement project in the turning of steel shafts. Table 3-18 gives the percentages of reworkable and scrap shafts produced in 18 production runs made during the study. (The values were read from a graph in Rowe's paper.)

Figure 3-25 is a corresponding segmented bar graph, with the jobs ordered in time, showing the behavior of both the scrap and rework rates over time. (The total height of any bar represents the sum of the two rates.)

It is interesting to note that the sharp reduction in both scrap and rework between jobs 10 and 11 was produced by overhauling one of the company's lathes. That lathe was identified as needing attention through engineering data analysis early in the plant project.

TABLE 3-18

Percents Scrap and Rework
in a Turning Operation

Job Number	Percent Scrap	Percent Rework
1	2	25
2	3	11
3	0	5
4	0	0
5	0	20
6	2	23
7	0	6
8	0	5
9	2	8
10	3	18
11	0	3
12	1	5
13	0	0
14	0	0
15	0	3
16	0	2
17	0	2
18	1	5

FIGURE 3-25

Segmented Bar Chart of Scrap
and Rework Rates

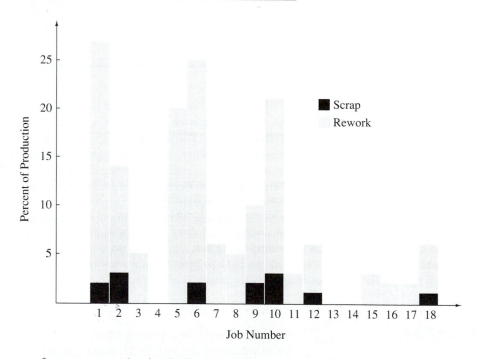

In many cases, the simple plotting of $\hat{p}$ or $\hat{u}$ values against time or process variables can make clear the essential message in a set of qualitative or count data.

EXAMPLE 3-13

Defects Per Truck Found at Final Inspection. In his text *Engineering Statistics and Quality Control*, I. W. Burr illustrates the usefulness of plotting $\hat{u}$ versus time with a set of data on defects found at final inspection at a truck assembly plant. From 95 to 130 trucks were produced daily at the plant, and Table 3-19 gives part of Burr's daily defects/truck values.

These statistics are plotted in Figure 3-26. The graph shows a marked decrease in quality (increase in $\hat{u}$) over the third and fourth weeks of December, culminating with

	Date	$\hat{u}$ = Defects/Truck
TABLE 3-19	12/2	1.54
Defects Per Truck on 26	12/3	1.42
Production Days	12/4	1.57
	12/5	1.40
	12/6	1.51
	12/9	1.08
	12/10	1.27
	12/11	1.18
	12/12	1.39
	12/13	1.42
	12/16	2.08
	12/17	1.85
	12/18	1.82
	12/19	2.07
	12/20	2.32
	12/23	1.23
	12/24	2.91
	12/26	1.77
	12/27	1.61
	12/30	1.25
	1/3	1.15
	1/6	1.37
	1/7	1.79
	1/8	1.68
	1/9	1.78
	1/10	1.84

FIGURE 3-26

Plot of Daily Defects per Truck

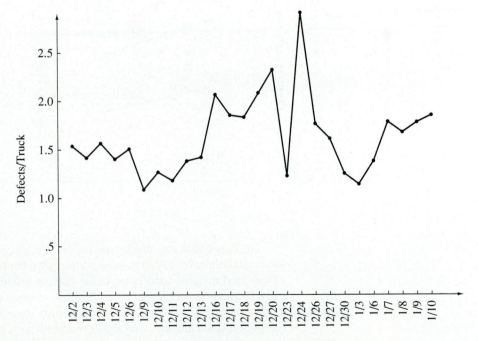

a rate of 2.91 defects/truck on Christmas Eve. Apparently, this situation was largely corrected with the passing of the holiday season.

This section will close with the observation that plots of $\hat{p}$ or $\hat{u}$ against levels of manipulated variables from an experiment are often helpful in understanding the results of that experiment.

EXAMPLE 3-14

Plotting Fractions of Conforming Pellets. Greiner, Grim, Larson, and Lukomski experimented with the same pelletizing machine studied by Cyr, Ellson, and Rickard (see Example 1-2). In one part of their study, they ran the machine at an elevated speed and varied the shot size (amount of powder injected into the dies) and the composition of that powder (in terms of the relative amounts of new versus reground material used). Table 3-20 lists the numbers of conforming pellets produced in a sample of 100 at each of $2\times2 = 4$ sets of process conditions. A simple plot of $\hat{p}$ values versus shot size is given in Figure 3-27.

TABLE 3-20

Numbers of Conforming Pellets for Four Shot Size/Mixture Combinations

Sample	Shot Size	Mixture	Number Conforming
1	small	20% reground	38
2	small	50% reground	66
3	large	20% reground	29
4	large	50% reground	53

FIGURE 3-27

Plot of Shot Sizes and Conforming Pellets

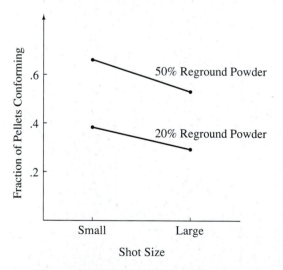

The figure indicates that increasing the shot size is somewhat detrimental but that a substantial improvement in process performance could be had by increasing the amount of reground material used in the pellet-making mixture. This could make sense physically. Reground material had been previously compressed into (nonconforming) pellets; in the process, it had been allowed to absorb some ambient humidity. Both the prior compression and the increased moisture content were potential reasons why this material improved the ability of the process to produce solid, properly shaped pellets.

EXERCISES

3-1. (§3-1) The following are percent yields from 40 runs of a chemical process, taken from J. S. Hunter's article "The Technology of Quality" (*RCA Engineer*, May/June 1985).

65.6, 65.6, 66.2, 66.8, 67.2, 67.5, 67.8, 67.8, 68.0, 68.0,
68.2, 68.3, 68.3, 68.4, 68.9, 69.0, 69.1, 69.2, 69.3, 69.5,
69.5, 69.5, 69.8, 69.9, 70.0, 70.2, 70.4, 70.6, 70.6, 70.7,
70.8, 70.9, 71.3, 71.7, 72.0, 72.6, 72.7, 72.8, 73.5, 74.2

Make a dot diagram, a stem-and-leaf plot, a frequency table, and a histogram for displaying these data.

3-2. (§3-1) Make back-to-back stem-and-leaf plots for the two samples in Table 3-1.

3-3. (§3-1) Osborne, Bishop, and Klein collected some data on the torques required to loosen bolts holding an assembly on machinery of a particular model that was being manufactured by an industrial concern. The accompanying table shows a small portion of their data with regard to two particular bolts. The torques recorded (in ft lb) were taken from 15 different pieces of equipment of the model in question, as they were assembled.

Top Bolt	Bottom Bolt
110	125
115	115
105	125
115	115
115	120
120	120
110	115
125	125
105	110
130	110
95	120
110	115
110	120
95	115
105	105

(a) Make a scatterplot of these paired data. Are there any obvious patterns in the plot?

(b) A trick often employed in the analysis of paired data such as these is to reduce the pairs to differences by subtracting the values of one of the variables from the other. Compute differences (top bolt − bottom bolt) here. Then make and interpret a dot diagram for these values.

3-4. (§3-2) The following are data (from *Introduction to Contemporary Statistical Methods* by L. H. Koopmans) on the impact strengths of sheets of insulating material cut in two different ways. (The values are in ft lb.)

(a) Make quantile plots for these two samples. Find the medians, the quartiles, and the .37 quantiles for the two data sets.

(b) Draw (to scale) carefully labeled side-by-side boxplots for comparing the two cutting methods. Discuss what these show about the two methods.

Lengthwise Cuts	Crosswise Cuts
1.15	.89
.84	.69
.88	.46
.91	.85
.86	.73
.88	.67
.92	.78
.87	.77
.93	.80
.95	.79

(c) Make and discuss the appearance of a *Q-Q* plot for comparing the shapes of these two data sets.

3-5. (§3-2) Make a *Q-Q* plot for the two small samples in Table 3-13.

3-6. (§3-2) Make and interpret a normal plot for the yield data of Exercise 3-1.

3-7. (§3-3) Calculate and compare the means, medians, ranges, interquartile ranges, and standard deviations of the two data sets introduced in Exercise 3-4. Discuss the interpretation of these values in the context of comparing the two cutting methods.

3-8. (§3-3) Are the numerical values you produced in Exercise 3-7 most naturally thought of as statistics or as parameters? Explain.

3-9. (§3-3) Add 1.3 to each of the lengthwise cut impact strengths given in Exercise 3-4, and then recompute the values of the mean, median, range, interquartile range, and standard deviation. How do these compare with the values obtained earlier? Repeat this exercise after multiplying each lengthwise cut impact strength by 2 instead of adding 1.3.

3-10. (§3-4) From your field, give an example of a variable that is a rate of the form $\hat{p}$, then an example of one that is a rate of the form $\hat{u}$, and then one of a rate that is of neither form.

3-11. (§3-4) Because gaging is easier, it is sometimes tempting to collect qualitative data related to measurements, rather than the measurements themselves. For example, in the context of Example 1-1, if gears with runouts exceeding 15 were considered to be nonconforming, it would be possible to derive fractions nonconforming, $\hat{p}$, from simple "go–no go" checking of gears. For the two sets of gears represented in Table 1-1, what would have been the sample fractions nonconforming $\hat{p}$? Give a practical reason why one might prefer having the values in Table 1-1 rather than knowing only the corresponding $\hat{p}$ values.

3-12. Explain in your own words the usefulness of theoretical *Q-Q* plotting.

3-13. Consider a laboratory situation in which one will measure the percentage copper in brass specimens. The resulting data will be a kind of rate data. Are the rates that will be obtained of the type $\hat{p}$, of the type $\hat{u}$, or of neither type? Explain.

3-14. The accompanying values are gains measured on 120 amplifiers designed to produce a 10 dB gain. These data were originally from the *Quality Improvement Tools* workbook set (published by the Juran Institute). They were then used as an example in the article "The Tools of Quality" (*Quality Progress*, September 1990).

8.1, 10.4, 8.8, 9.7, 7.8, 9.9, 11.7, 8.0, 9.3, 9.0, 8.2, 8.9, 10.1, 9.4, 9.2, 7.9, 9.5, 10.9, 7.8, 8.3, 9.1, 8.4, 9.6, 11.1, 7.9, 8.5, 8.7, 7.8, 10.5, 8.5, 11.5, 8.0, 7.9, 8.3, 8.7, 10.0, 9.4, 9.0, 9.2, 10.7, 9.3, 9.7, 8.7, 8.2, 8.9, 8.6, 9.5, 9.4, 8.8, 8.3, 8.4, 9.1, 10.1, 7.8, 8.1, 8.8, 8.0, 9.2, 8.4, 7.8, 7.9, 8.5, 9.2, 8.7, 10.2, 7.9, 9.8, 8.3, 9.0, 9.6, 9.9, 10.6, 8.6, 9.4, 8.8, 8.2, 10.5, 9.7, 9.1, 8.0, 8.7, 9.8, 8.5, 8.9, 9.1, 8.4, 8.1, 9.5, 8.7, 9.3, 8.1, 10.1, 9.6, 8.3, 8.0, 9.8, 9.0, 8.9, 8.1, 9.7, 8.5, 8.2, 9.0, 10.2, 9.5, 8.3, 8.9, 9.1, 10.3, 8.4, 8.6, 9.2, 8.5, 9.6, 9.0, 10.7, 8.6, 10.0, 8.8, 8.6

(a) Make a stem-and-leaf plot and a boxplot for these data. How would you describe the shape of this data set? Does the shape of your stem-and-leaf plot (or a corresponding histogram) give you any clue as to the method by which a high fraction within specifications was probably achieved?

(b) Make a normal plot for these data and interpret its shape. (Standard normal quantiles for $p = .0042$ and $p = .9958$ are approximately -2.64 and 2.64, respectively.)

(c) Although the nominal gain for these amplifiers was to be 10 dB, the engineering design allowed gains from 7.75 dB to 12.2 dB to be considered acceptable. About what fraction, p, of such amplifiers do you expect to meet these engineering specifications?

3-15. The article "The Lognormal Distribution for Modeling Quality Data When the Mean is Near Zero" by S. Albin (*Journal of Quality Technology*, April 1990) described the operation of a plastics recycling pilot plant run by Rutgers University. The most important material reclaimed from beverage bottles is PET plastic. A serious impurity is aluminum, which later can clog the filters in extruders when the recycled material is used. The following are the amounts (in ppm by weight of aluminum) found in bihourly samples of PET recovered at the plant over roughly a two-day period.

291, 222, 125, 79, 145, 119, 244, 118, 182, 63, 30, 140, 101, 102, 87, 183, 60, 191, 119, 511, 120, 172, 70, 30, 90, 115

(Apparently, the data here are recorded in the order in which they were collected if one reads left to right, top to bottom.)

(a) Make a run chart for these data. Are there any obvious time trends in the data? What practical engineering reason is there for looking for such trends?

(b) Ignoring the time order information, make a stem-and-leaf diagram for these data. Use the hundreds digit to make the stem and the other two digits (separated by commas to indicate the different data points) to make the leaves. After making an initial stem-and-leaf diagram by recording the data in the (time) order given above, make a second one in which the values have been ordered.

(c) How would you describe the shape of the stem-and-leaf diagram? Is the data set bell-shaped?

(d) Find the median and the first and third quartiles for the aluminum contents, and then find the .58 quantile of the data set.

(e) Make a boxplot for these data.

(f) Make a normal plot for these data, using regular graph paper. List the coordinates of the 26 plotted points. Interpret the shape of your normal plot.

(g) Try transforming the data by taking natural logarithms, and again assess the shape. Is the transformed data set more bell-shaped than the raw data set?

(h) Find the sample mean, the sample range, and the sample standard deviation for both the original data and the log-transformed values from (g). Is the mean of the transformed values equal to the natural logarithm of the mean of the original data?

3-16. The accompanying data are three hypothetical samples of size 10 that are supposed to represent measured manganese contents in specimens of 1045 steel (the units are points, or .01%). Suppose that these measurements were made on standard specimens having "true" manganese contents of 80, using three different analytical methods. (Thirty different specimens were involved.)

Method 1	87, 74, 78, 81, 78, 77, 84, 80, 85, 78
Method 2	86, 85, 82, 87, 85, 84, 84, 82, 82, 85
Method 3	84, 83, 78, 79, 85, 82, 82, 81, 82, 79

(a) Make (on the same coordinate system) side-by-side boxplots that you can use to compare the three analytical methods.

(b) Discuss the apparent effectiveness of the three different methods in terms of the appearance of your diagram from (a) and in terms of the concepts of accuracy and precision discussed in Section 1-3.

(c) An alternative method of comparing two such analytical methods is to use both methods of analysis once on each of (say) 10 different specimens (10 specimens and 20 measurements). In the terminology of Section 1-2, what kind of data would be generated by such a plan? If one simply wishes to compare the average measurements produced by two analytical methods, which data collection plan (20 specimens and 20 measurements, or 10 specimens and 20 measurements) seems to you most likely to provide the better comparison? Explain your reasoning.

3-17. Gaul, Phan, and Shimonek measured the resistances of 15 resistors of $2 \times 5 = 10$ different types. Two different wattage ratings were involved, and five different nominal resistances were used. All measurements were reported to three significant digits. Their data follow.

(a) Make back-to-back stem-and-leaf plots for comparing the $\frac{1}{4}$ watt and $\frac{1}{2}$ watt resistance distributions for each nominal resistance. In a few sentences, summarize what these show.

(b) Make pairs of boxplots for comparing the $\frac{1}{4}$ watt and $\frac{1}{2}$ watt resistance distributions for each nominal resistance.

(c) Make normal plots for the $\frac{1}{2}$ watt nominal 20 ohm and nominal 200 ohm resistors. Interpret these in a sentence or two. From the appearance of the second of these, does it seem that if one treated the nominal 200 ohm resistances as if they had a bell-shaped distribution, would one tend to overestimate or to underestimate the fraction of resistances near the nominal value?

$\frac{1}{4}$ WATT RESISTORS

20 ohm	75 ohm	100 ohm	150 ohm	200 ohm
19.2	72.9	97.4	148	198
19.2	72.4	95.8	148	196
19.3	72.0	97.7	148	199
19.3	72.5	94.1	148	196
19.1	72.7	95.1	148	196
19.0	72.3	95.4	147	195
19.6	72.9	94.9	148	193
19.2	73.2	98.5	148	196
19.3	71.8	94.8	148	196
19.4	73.4	94.6	147	199
19.4	70.9	98.3	147	194
19.3	72.3	96.0	149	195
19.5	72.5	97.3	148	196
19.2	72.1	96.0	148	195
19.1	72.6	94.8	148	199

$\frac{1}{2}$ WATT RESISTORS

20 ohm	75 ohm	100 ohm	150 ohm	200 ohm
20.1	73.9	97.2	152	207
19.7	74.2	97.9	151	205
20.2	74.6	96.8	155	214
24.4	72.1	99.2	146	195
20.2	73.8	98.5	148	202
20.1	74.8	95.5	154	211
20.0	75.0	97.2	149	197
20.4	68.6	98.7	150	197
20.3	74.0	96.6	153	199
20.6	71.7	102	149	196
19.9	76.5	103	150	207
19.7	76.2	102	149	210
20.8	72.8	102	145	192
20.4	73.2	100	147	201
20.5	76.7	100	149	257

(d) Compute the sample means and sample standard deviations for all 10 samples. Do these values agree with your qualitative statements made in answer to part (a)?

(e) Make a plot of the 10 sample means computed in part (d), similar to the plot in Figure 3-22. Comment on the appearance of this plot.

3-18. Blomquist, Kennedy, and Reiter studied the properties of three scales by each weighing a calibrated 5 g weight, a calibrated 20 g weight, and a calibrated 100 g weight twice on each scale. Their data are presented in the accompanying table.

Using whatever graphical and numerical data summary methods you find helpful, make sense of these data. Write a several-page discussion of your findings. You will probably want to consider both accuracy and precision and (to the extent possible) make comparisons between scales and between students. Part of your discussion might be phrased in terms of the concepts of repeatability and reproducibility introduced in Section 2-1. Are the pictures you get of the scale and student performances consistent across the different weights?

TABLE For Exercise 3-18

5-GRAM WEIGHINGS	Scale 1	Scale 2	Scale 3
Student 1	5.03, 5.02	5.07, 5.09	4.98, 4.98
Student 2	5.03, 5.01	5.02, 5.07	4.99, 4.98
Student 3	5.06, 5.00	5.10, 5.08	4.98, 4.98

20-GRAM WEIGHINGS	Scale 1	Scale 2	Scale 3
Student 1	20.04, 20.06	20.04, 20.04	19.94, 19.93
Student 2	20.02, 19.99	20.03, 19.93	19.95, 19.95
Student 3	20.03, 20.02	20.06, 20.03	19.91, 19.96

100-GRAM WEIGHINGS	Scale 1	Scale 2	Scale 3
Student 1	100.06, 100.35	100.25, 100.08	99.87, 99.88
Student 2	100.05, 100.01	100.10, 100.02	99.87, 99.88
Student 3	100.00, 100.00	100.01, 100.02	99.88, 99.88

3-19. The accompanying values are the lifetimes (in numbers of 24 mm deep holes drilled in 1045 steel before tool failure) for $n = 12$ D952-II (8 mm) drills. These were read from a graph in "Computer-assisted Prediction of Drill-failure Using In-process Measurements of Thrust Force" by A. Thangaraj and P. K. Wright (*Journal of Engineering for Industry*, May 1988).

$$47, 145, 172, 86, 122, 110, 172, 52, 194, 116, 149, 48$$

Write a short report to your engineering manager summarizing what these data indicate about the lifetimes of drills of this type in this kind of application. Make use of whatever graphical and numerical data summary tools make clear the main features of the data set.

3-20. Losen, Cahoy, and Lewis purchased eight spanner bushings of a particular type from a local machine shop and measured a number of characteristics of these bushings, including their outside diameters. Each of the eight outside diameters was measured once by each of two student technicians, with the following results. (The units are inches.)

Bushing	1	2	3	4
Student A	.3690	.3690	.3690	.3700
Student B	.3690	.3695	.3695	.3695

Bushing	5	6	7	8
Student A	.3695	.3700	.3695	.3690
Student B	.3695	.3700	.3700	.3690

A common device when dealing with paired data like these is to take differences and analyze the differences. Subtracting B measurements from A measurements gives the following eight values.

$$.0000, -.0005, -.0005, .0005, .0000, .0000, -.0005, .0000$$

(a) Find the first and third quartiles for these differences, and their median.

(b) Find the sample mean and standard deviation for the differences.

(c) Your mean in part (b) should be negative. Interpret this in terms of the original measurement problem.

(d) Suppose that one wishes to make a normal plot of the differences on regular graph paper. Give the coordinates of the lower left point on such a plot.

3-21. The accompanying data are the times to failure (in millions of cycles) of high-speed turbine engine bearings made out of two different compounds. These were taken from "Analysis of Single Classification Experiments Based on Censored Samples from the Two-parameter Weibull Distribution" by J. I. McCool (*The Journal of Statistical Planning and Inference*, 1979).

Compound 1   3.03, 5.53, 5.60, 9.30, 9.92, 12.51, 12.95, 15.21, 16.04, 16.84

Compound 2   3.19, 4.26, 4.47, 4.53, 4.67, 4.69, 5.78, 6.79, 9.37, 12.75

(a) Find the .84 quantile of the Compound 1 failure times.

(b) Give the coordinates of the two lower left points that would appear on a normal plot of the Compound 1 data.

(c) Make back-to-back stem-and-leaf plots for comparing the life length properties of bearings made from Compounds 1 and 2.

(d) Make (to scale) side-by-side boxplots for comparing the life lengths for the two compounds. Mark on the plots numbers indicating the locations of their main features.

(e) Compute the sample means and standard deviations of the two sets of lifetimes.

(f) Describe what your answers to parts (c), (d), and (e) above indicate about the life lengths of these turbine bearings.

3-22. Heyde, Kuebrick, and Swanson measured the heights of 405 steel punches of a certain type, purchased by a company from a

Punch Height (.001 in.)	Frequency
482	1
483	0
484	1
485	1
486	0
487	1
488	0
489	1
490	0
491	2
492	0
493	0
494	0
495	6
496	7
497	13
498	24
499	56
500	82
501	97
502	64
503	43
504	3
505	1
506	0
507	0
508	0
509	2

single supplier. The stamping machine in which these are used is designed to use .500 in. punches. Frequencies of the measurements they obtained are shown in the accompanying table.

(a) Summarize these data, using appropriate graphical and numerical tools. How would you describe the shape of the distribution of punch heights? The specifications for punch heights were in fact .500 in. to .505 in. Does this fact give you any insight as to the origin of the distributional shape observed in the data? Does it appear that the supplier has equipment capable of meeting the engineering specifications on punch height?

(b) In the manufacturing application of these punches, several had to be placed side-by-side on a drum and be used to cut the same piece of material. In this context, why is having small variability in punch height perhaps even more important than having the correct mean punch height?

3-23. The article "Watch Out for Nonnormal Distributions" by D. C. Jacobs (*Chemical Engineering Progress*, November 1990) contains 100 measured daily purities of oxygen delivered by a certain supplier. These are as follows, listed in the time order of their collection. (Read left to right, top to bottom.) The values given are in hundredths of a percent purity above 99.00%.

63, 61, 67, 58, 55, 50, 55, 56, 52, 64, 73, 57, 63, 81, 64, 54, 57, 59, 60, 68, 58, 57, 67, 56, 66, 60, 49, 79, 60, 62, 60, 49, 62, 56, 69, 75, 52, 56, 61, 58, 66, 67, 56, 55, 66, 55, 69, 60, 69, 70, 65, 56, 73, 65, 68, 59, 62, 58, 62, 66, 57, 60, 66, 54, 64, 62, 64, 64, 50, 50, 72, 85, 68, 58, 68, 80, 60, 60, 53, 49, 55, 80, 64, 59, 53, 73, 55, 54, 60, 60, 58, 50, 53, 48, 78, 72, 51, 60, 49, 67

You will probably want to make use of a statistical package on a computer to help you do the following.

(a) Make a run chart for these data. Are there any obvious time trends in the data? What would be the practical engineering usefulness of early detection of any such time trend?

(b) Now ignore the time order of data collection and represent these data with a stem-and-leaf plot and a histogram. (Use .02% class widths in making your histogram.) Mark on these the supplier's lower specification limit of 99.50% purity. Describe the shape of the purity distribution.

(c) The author of the article found it useful to reexpress the purities by subtracting 99.30 (remember that the values above are in units of .01% above 99.00%) and then taking natural logarithms. Do this with the raw data and make a second stem-and-leaf diagram and a second histogram for portraying the shape of the transformed data. Do these figures look more bell-shaped than the ones you made in part (b)?

(d) Make a normal plot for the transformed values from part (c). What does it indicate about the shape of the distribution of the transformed values? (Standard normal quantiles for $p = .005$ and $p = .995$ are approximately -2.58 and 2.58, respectively.)

3-24. The following are some data taken from the article "Confidence Limits for Weibull Regression with Censored Data" by J. I. McCool (*IEEE Transactions on Reliability*, 1980). They are the ordered failure times (the time units are not given in the paper) for hardened steel specimens subjected to rolling contact fatigue tests at four different values of contact stress.

$.87 \times 10^6$ psi	$.99 \times 10^6$ psi	1.09×10^6 psi	1.18×10^6 psi
1.67	.80	.012	.073
2.20	1.00	.18	.098
2.51	1.37	.20	.117
3.00	2.25	.24	.135
3.90	2.95	.26	.175
4.70	3.70	.32	.262
7.53	6.07	.32	.270
14.7	6.65	.42	.350
27.8	7.05	.44	.386
37.4	7.37	.88	.456

(a) Make side-by-side boxplots for these data. Does it look as if the different stress levels produce life distributions of roughly the same shape? (Engineering experience suggests that different stress levels often change the scale but not the basic shape of life distributions.)
(b) Make *Q-Q* plots for comparing all six different possible pairs of distributional shapes. Summarize in a few sentences what these indicate about the shapes of the failure time distributions under the different stress levels.

3-25. Riddle, Peterson, and Harper studied the performance of a rapid-cut industrial shear in a continuous cut mode. They cut nominally 2-inch and 1-inch strips of 14 gauge and 16 gauge steel sheet metal and measured the actual widths of the strips produced by the shear. Their data follow, in units of 10^{-3} in. above nominal.

		MATERIAL THICKNESS	
		14 Gauge	16 Gauge
MACHINE SETTING	1 Inch	2, 1, 1, 1, 0, 0, −2, −10, −5, 1	−2, −6, −1, −2, −1, −2, −1, −1, −1, −5
	2 Inch	10, 10, 8, 8, 8, 8, 7, 7, 9, 11	−4, −3, −4, −2, −3, −3, −3, −3, −4, −4

(a) Compute sample means and standard deviations for the four samples represented above. Plot the means in a manner similar to the plot in Figure 3-22. Make a separate plot of this kind for the standard deviations.
(b) Write a short report to an engineering manager to summarize what these data and your summary statistics and plots show about the performance of the industrial shear. How do you recommend that the shear be set up in the future in order to get strips cut from these materials with widths as close as possible to specified dimensions?

3-26. The accompanying data are some measured resistivity values from *in situ* doped polysilicon specimens, taken from the article "LPCVD Process Equipment Evaluation Using Statistical Methods" by R. Rossi (*Solid State Technology*, 1984). (The units were not given in the article.)

5.55, 5.52, 5.45, 5.53, 5.37, 5.22, 5.62, 5.69, 5.60, 5.58, 5.51, 5.53

(a) Make a dot diagram and a boxplot for these data and compute the statistics $\bar{x}$ and s.
(b) Make a normal plot for these data. How bell-shaped does this data set look? If you were to say that the shape departed from a perfect bell shape, in what specific way does it depart? (Refer to characteristics of the the normal plot to support your answer.)

3-27. The article "Thermal Endurance of Polyester Enameled Wires Using Twisted Wire Specimens" by H. Goldenberg (*IEEE Transactions on Electrical Insulation*, 1965) contains some data on the lifetimes (in weeks) of wire specimens tested for thermal endurance according to AIEE Standard 57. Several different laboratories were used to make the tests, and the results from two of the laboratories, using a test temperature of 200°C, follow.

Laboratory 1
14, 16, 17, 18, 20, 22, 23, 25, 27, 28
Laboratory 2
27, 28, 29, 29, 29, 30, 31, 31, 33, 34

Consider first only the Laboratory 1 data.
(a) Find the median and the first and third quartiles for the lifetimes, and then find the .64 quantile of the data set.
(b) Make and interpret a normal plot for these data. Would you describe this distribution as bell-shaped? If not, in what way(s) does it depart from being bell-shaped? Give the the coordinates of the 10 points you plot on regular graph paper.
(c) Find the sample mean, the sample range, and the sample standard deviation for these data.

Now consider comparison of the work of the two different laboratories (i.e., consider both data sets).
(d) Make back-to-back stem-and-leaf plots for these two data sets. (Use two leaves for the observations 10–19, two for the observations 20–29, etc.)
(e) Make side-by-side boxplots for these two data sets. (Draw these on the same scale.)
(f) Based on your work in parts (d) and (e), which of the two labs would you say produced the more precise results?
(g) Is it possible to tell from your plots in (d) and (e) which lab produced the more accurate results? Why or why not?

3-28. Agusalim, Ferry, and Holloway made some measurements on the thickness of sheetrock boards during their manufacture. The accompanying table shows thicknesses (in inches) of 12 different 4 ft × 8 ft boards (at a single location on the boards) both before and after drying in a kiln. (These boards were nominally .500 inch thick.)

Board	1	2	3	4	5	6
Before Drying	.514	.505	.500	.490	.503	.500
After Drying	.510	.502	.493	.486	.497	.494
Board	7	8	9	10	11	12
Before Drying	.510	.508	.500	.511	.505	.501
After Drying	.502	.505	.488	.486	.491	.498

(a) Make a scatterplot of these data. Does there appear to be a strong relationship of after-drying thickness to before-drying thickness? How might the existence of a relationship between these variables be of practical engineering importance in the manufacture of sheetrock?

(b) Calculate the 12 before–after differences in thickness. Find the sample mean and sample standard deviation of these values. How might one use the mean value in running the sheetrock manufacturing process? (Based on the mean value, what is an ideal before-drying thickness for the boards?) If one could somehow eliminate all variability in before-drying thickness, would substantial after-drying variability in thickness remain? Explain in terms of your calculations.

3-29. The accompanying values are representative of data summarized in a histogram appearing in the article "Influence of Final Recrystallization Heat Treatment on Zircaloy-4 Strip Corrosion" by Foster, Dougherty, Burke, Bates, and Worcester (*Journal of Nuclear Materials*, 1990). Given are $n = 20$ particle diameters observed in a bright-field TEM micrograph of a Zircaloy-4 specimen. The units are $10^{-2} \mu m$.

1.73, 2.47, 2.83, 3.20, 3.20, 3.57, 3.93, 4.30, 4.67, 5.03,
5.03, 5.40, 5.77, 6.13, 6.50, 7.23, 7.60, 8.33, 9.43, 11.27

(a) Compute the mean and standard deviation of these particle diameters.

(b) Make both a dot diagram and a boxplot for these data. Sketch the dot diagram on a ruled scale and make the boxplot below it.

(c) Based on your work in (b), how would you describe the shape of this data set?

(d) Make a normal plot of these data. In what specific way does the distribution depart from being bell-shaped?

(e) It is sometimes useful to find a scale of measurement on which a data set is reasonably bell-shaped. To that end, take the natural logarithms of the raw particle diameters. Normal-plot the log diameters. Does this plot appear to be more linear than your plot in (d)?

3-30. The data in the accompanying table are measurements of the latent heat of fusion of ice, taken from *Experimental Statistics* (NBS Handbook 91) by M. G. Natrella. The measurements were made (on specimens cooled to $-.072°C$) using two different methods. The first was an electrical method, and the second was a method of mixtures. The units are calories per gram of mass.

Method A (Electrical)
79.98, 80.04, 80.02, 80.04, 80.03, 80.03, 80.04, 79.97, 80.05, 80.03, 80.02, 80.00, 80.02

Method B (Mixtures)
80.02, 79.94, 79.98, 79.97, 79.97, 80.03, 79.95, 79.97

(a) Make side-by-side boxplots for comparing the two measurement methods. Does there appear to be any important difference in the precision of the two methods? Is it fair to say that at least one of the methods must be somewhat inaccurate? Explain.

(b) Compute and compare the sample means and the sample standard deviations for the two methods. How are the comparisons of these numerical quantities already evident on your plot in (a)?

4

Computationally Intensive Descriptive Statistics

*T*he methods of data summarization discussed in Chapter 3 require only relatively simple calculations that are easily carried out with only the help of an inexpensive hand-held electronic calculator. This chapter continues the study of descriptive statistics but turns to methods that can require considerably more computation. The necessary calculations are typically carried out using statistical analysis packages that are widely available for computers (personal through mainframe size).

Besides usually involving substantial computations, the methods discussed in this chapter also have somewhat more subtle and advanced conceptual bases than the methods of Chapter 3. This chapter opens discussion of the notions of fitting a simple or *parsimonious* model or equation to an engineering data set and making some preliminary evaluations of the appropriateness of that model or equation.

The chapter begins with a discussion of fitting a line to bivariate quantitative data, using the principle of least squares. Also treated is the assessment of the goodness of that fit using the sample correlation and coefficient of determination and through the analysis of residuals. Then the line-fitting ideas are generalized to the fitting of curves to bivariate data and surfaces to multivariate quantitative data. The next topic is the summarization of data from multifactor engineering studies in terms of so-called factorial effects. This technique sometimes makes it possible to conceive relatively simple equations describing multifactor data sets even where the factor levels are not quantitative, but rather qualitative. Next, the notion of data transformations is discussed, along with the idea that using some scales of measurement can allow simpler data summarizations than using others. Finally, the chapter closes with a short transitional section. That section makes the case that, however powerful methods of descriptive statistics may be, the addition of methods of formal (probability-based) inference is still necessary in order to provide a complete engineering tool.

4-1 Fitting a Line by Least Squares

One of the most commonly used techniques of descriptive statistical analysis is the fitting of a straight line to a set of bivariate data. Often, particularly in engineering contexts, such data arise because a quantitative experimental variable x has been varied between several different settings, producing a number of samples of a response variable y. In such cases, several considerations become important: 1) summarization, 2) interpolation (to values of x not included in the study, but within the range of x values used), 3) limited extrapolation (to values of x slightly outside the range of those included in the study), and sometimes 4) process optimization or adjustment based on the message carried by the data. For these purposes, it is extremely helpful to have an equation relating y to x in at least some approximate sense. Upon making a simple y-versus-x scatterplot, a linear equation

$$y \approx \beta_0 + \beta_1 x \tag{4-1}$$

relating y to x is about the simplest potentially useful equation one might be led to consider.

In this section, the principle of least squares is applied to the fitting of a line to (x, y) data. The appropriateness of that fit will be assessed using the sample correlation and the coefficient of determination. The plotting of residuals will be introduced as an important method for further investigation of possible problems with the fitted equation. Next will be a discussion of some caveats regarding the practical interpretation and use of a fitted line and the sample correlation. Finally, the use of statistical analysis computer packages in the fitting of equations to data will be briefly introduced.

Applying the Least Squares Principle

EXAMPLE 4-1

Pressing Pressures and Specimen Densities for a Ceramic Compound. Benson, Locher, and Watkins studied the effects of varying pressing pressures on the density of cylindrical specimens made by dry pressing a ceramic compound. A mixture of Al_2O_3, polyvinyl alcohol, and water was prepared, dried overnight, crushed, and sieved to obtain 100 mesh size grains. These were pressed into cylinders at pressures from 2,000 psi to 10,000 psi, and cylinder densities were calculated. Table 4-1 gives the data that were obtained; and a simple scatterplot of these data is given in Figure 4-1.

TABLE 4-1

Pressing Pressures and
Resultant Specimen Densities

x Pressure (psi)	y Density (g/cc)
2,000	2.486
2,000	2.479
2,000	2.472
4,000	2.558
4,000	2.570
4,000	2.580
6,000	2.646
6,000	2.657
6,000	2.653
8,000	2.724
8,000	2.774
8,000	2.808
10,000	2.861
10,000	2.879
10,000	2.858

It is very easy to imagine sketching a straight line through the plotted points in Figure 4-1. Such a line might then be used as a summarization of how density depends upon pressing pressure for specimens of this material. But it would seem sensible to have some analytical way of fitting a line to such linear-looking (x, y) data. The principle of least squares provides one such method. (Other methods exist but won't be discussed in this text.)

DEFINITION 4-1

To apply the **principle of least squares** in the fitting of an equation for y containing some parameters to an n-point data set, values of the parameters are chosen so as to minimize

$$\sum_{i=1}^{n}(y_i - \hat{y}_i)^2 \tag{4-2}$$

where $y_1, y_2, \ldots, y_n$ are the observed responses and $\hat{y}_1, \hat{y}_2, \ldots, \hat{y}_n$ are corresponding responses predicted or fitted by the equation.

FIGURE 4-1

Scatterplot of Density versus
Pressing Pressure

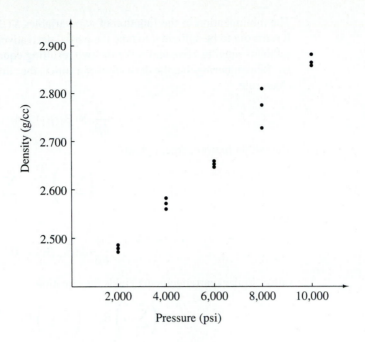

In the context of fitting a line to a set of (x, y) data, the prescription offered by Definition 4-1 amounts to choosing a slope and intercept so as to minimize the sum of squared vertical distances from (x, y) data points to the line in question. This notion is shown in generic fashion in Figure 4-2 for a fictitious five-point data set. (It is the *squares* of the five indicated differences that must be added and minimized.)

FIGURE 4-2

Five Data Points (x, y) and a
Possible Fitted Line

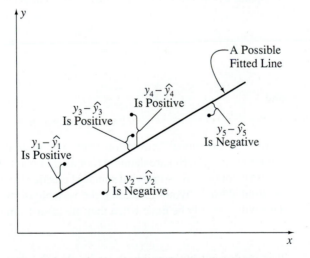

Looking at the form of the linear equation given in display (4-1), it is clear that, for the fitting of a line,

$$\hat{y} = \beta_0 + \beta_1 x$$

and therefore that the expression to be minimized by choice of slope (β_1) and intercept (β_0) is

$$S(\beta_0, \beta_1) = \sum_{i=1}^{n} (y_i - (\beta_0 + \beta_1 x_i))^2 \tag{4-3}$$

The minimization of the function of two variables $S(\beta_0, \beta_1)$ is an exercise in calculus. It turns out to be sufficient to take the partial derivatives of S with respect to β_0 and β_1, set them equal to zero, and solve the two resulting equations simultaneously for β_0 and β_1. Remembering that the derivative of a sum is the sum of the derivatives and using the chain rule,

$$\frac{\partial}{\partial \beta_0} S(\beta_0, \beta_1) = 0$$

after slight rearrangement yields

$$n\beta_0 + \left(\sum_{i=1}^{n} x_i\right)\beta_1 = \sum_{i=1}^{n} y_i \tag{4-4}$$

and

$$\frac{\partial}{\partial \beta_1} S(\beta_0, \beta_1) = 0$$

(after similar rearrangement) yields the equation

$$\left(\sum_{i=1}^{n} x_i\right)\beta_0 + \left(\sum_{i=1}^{n} x_i^2\right)\beta_1 = \sum_{i=1}^{n} x_i y_i \tag{4-5}$$

For reasons that are not obvious or important at this point, equations (4-4) and (4-5) are sometimes called the *normal* (as in perpendicular) *equations* for fitting a line. What is important for present purposes is that they are two linear equations in two unknowns and can be fairly easily solved for β_0 and β_1 (provided there are at least two different x_i's in the data set). In fact, it is fairly easy to show that a general simultaneous solution of equations (4-4) and (4-5) produces values of β_0 and β_1 given by

$$b_1 = \frac{\sum x_i y_i - \dfrac{(\sum x_i)(\sum y_i)}{n}}{\sum x_i^2 - \dfrac{(\sum x_i)^2}{n}} \tag{4-6}$$

and

$$b_0 = \bar{y} - b_1 \bar{x} \tag{4-7}$$

Notice the notational convention here. The particular numerical slope and intercept minimizing $S(\beta_0, \beta_1)$ are denoted (not as β's but) as b_1 and b_0.

In display (4-6), somewhat standard practice has been followed (and the summation notation abused) by not indicating the variable or range of summation (i, from 1 to n). This will typically be done when they are clear from the context.

EXAMPLE 4-1
(continued)

It is unpleasant but possible to verify that the data in Table 4-1 yield the following summary statistics.

$$\sum x_i = 2{,}000 + 2{,}000 + \cdots + 10{,}000 = 90{,}000$$

$$\sum x_i^2 = (2{,}000)^2 + (2{,}000)^2 + \cdots + (10{,}000)^2 = 660{,}000{,}000$$

$$\sum y_i = 2.486 + 2.479 + \cdots + 2.858 = 40.005$$

$$\sum y_i^2 = (2.486)^2 + (2.479)^2 + \cdots + (2.858)^2 = 106.982701$$

$$\sum x_i y_i = (2{,}000)(2.486) + (2{,}000)(2.479) + \cdots + (10{,}000)(2.858) = 245{,}870$$

Then the normal equations (4-4) and (4-5) are

$$15\beta_0 + 90,000\beta_1 = 40.005$$

and

$$90,000\beta_0 + 660,000,000\beta_1 = 245,870$$

Further, their solution for the least squares intercept and slope, b_0 and b_1, is given via equations (4-6) and (4-7) as

$$b_1 = \frac{245,870 - \dfrac{(90,000)(40.005)}{15}}{660,000,000 - \dfrac{(90,000)^2}{15}} = \frac{245,870 - 240,030}{120,000,000} = .0000486\overline{6} \text{ (g/cc)/psi}$$

and

$$b_0 = \frac{40.005}{15} - (.0000486\overline{6})(6,000) = 2.667 - .292 = 2.375 \text{ g/cc}$$

Figure 4-3 shows the least squares line

$$\hat{y} = 2.375 + .0000487x$$

sketched on a scatterplot of the (x, y) points from Table 4-1.

FIGURE 4-3

Scatterplot of the
Pressure/Density Data and
the Least Squares Line

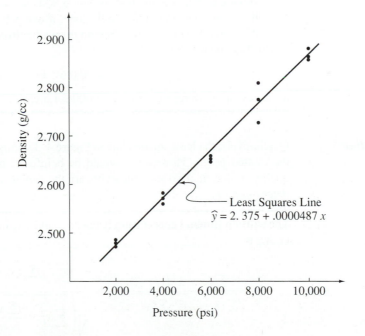

Note that the slope on this plot, $b_1 \approx .0000487$ (g/cc)/psi, has physical meaning as the (approximate) increase in y (density) that accompanies a unit (1 psi) change in x (pressure). The intercept on the plot, $b_0 = 2.375$ g/cc, so to speak positions the line vertically and is the value at which the line cuts the y axis. But it should probably not be interpreted as the density that would accompany a pressing pressure of $x = 0$ psi. The point is that the same reasonably linear-looking relation that the students found for pressures between 2,000 psi and 10,000 psi could well (in fact almost certainly does)

break down at extremely small and extremely large pressures. Before thinking of b_0 as a 0 pressure density value, one ought to have some $x = 0$ data to determine whether in fact the observed linearity holds up for x all the way down to 0.

As indicated in Definition 4-1, the value of y on the least squares line through a set of (x, y) data corresponding to a given x can be termed a *fitted* or *predicted* value. It can be used to represent likely y behavior at that x.

EXAMPLE 4-1
(continued)

Consider the problem of determining a typical density corresponding to a pressure of 4,000 psi and one corresponding to 5,000 psi.

First looking at $x = 4,000$, a simple way of representing a typical y is to note that for the three data points having $x = 4,000$,

$$\bar{y} = \frac{1}{3}(2.558 + 2.570 + 2.580) = 2.5693 \text{ g/cc}$$

and so to use this as a representative value. But assuming that y is indeed approximately linearly related to x, it is also possible to think that the fitted value

$$\hat{y} = 2.375 + .0000486(4,000) = 2.5697 \text{ g/cc}$$

might be even better than the $\bar{y}$ based on three data points for representing average density for 4,000 psi pressure.

Looking then at the situation for $x = 5,000$ psi, one finds that there are no data with such an x value. The only thing to be done, if one is to try and represent density at that pressure, is to ask whether interpolation is sensible from a physical viewpoint. If so, one might state the fitted value

$$\hat{y} = 2.375 + .0000486(5,000) = 2.6183 \text{ g/cc}$$

as representing average density for 5,000 psi pressure.

The Sample Correlation and Coefficient of Determination

Qualitatively, the least squares line in Figure 4-3 seems to do a pretty good job of fitting the plotted points. However, it would be helpful to have methods of quantifying the quality of that fit. Two such measures are the *sample correlation* and the *coefficient of determination*.

DEFINITION 4-2

The **sample (linear) correlation** between x and y exhibited in a sample of n data pairs (x_i, y_i) is

$$r = \frac{\sum (x_i - \bar{x})(y_i - \bar{y})}{\sqrt{\sum (x_i - \bar{x})^2 \cdot \sum (y_i - \bar{y})^2}} = \frac{\sum x_i y_i - \dfrac{(\sum x_i)(\sum y_i)}{n}}{\sqrt{\left(\sum x_i^2 - \dfrac{(\sum x_i)^2}{n}\right)\left(\sum y_i^2 - \dfrac{(\sum y_i)^2}{n}\right)}} \tag{4-8}$$

The sample correlation always lies in the interval from -1 to 1. Further, it is -1 or 1 only when all (x, y) data points fall on a single straight line. Comparison of formulas (4-6) and (4-8) shows the right-hand sides to have the same numerators, and both denominators are positive. Hence, b_1 and r have the same sign. So a sample correlation of -1 means that y decreases linearly in increasing x, while a sample correlation of $+1$ means that y increases linearly in increasing x.

Real data sets do not often (if ever) exhibit perfect ($+1$ or -1) correlation. Instead, $-1 < r < 1$ is typically realized. But drawing on the above facts about how it behaves, people take r as a measure of the strength of an apparent linear relationship: r near $+1$ or -1 is interpreted as indicating a relatively strong linear relationship and r near 0 is taken as indicating a lack of linear relationship. The sign of r is thought of as indicating whether y tends to increase or decrease with increased x.

EXAMPLE 4-1
(continued)

For the pressure/density data, using the summary statistics in the example following display (4-7), one obtains

$$r = \frac{245{,}870 - \dfrac{(90{,}000)(40.005)}{15}}{\sqrt{\left(660{,}000{,}000 - \dfrac{(90{,}000)^2}{15}\right)\left(106.982701 - \dfrac{(40.005)^2}{15}\right)}} = .9911$$

This value of r is near $+1$ and indicates clearly the strong positive linear relationship evident in Figures 4-1 and 4-3.

The coefficient of determination is another quantification of the quality of a fitted equation that can be applied not only in the present case of the simple fitting of a line to (x, y) data, but more widely as well.

DEFINITION 4-3

The **coefficient of determination** for an equation fitted to an n-point data set via least squares and producing fitted y values $\hat{y}_1, \hat{y}_2, \ldots, \hat{y}_n$ is the quantity

$$R^2 = \frac{\sum(y_i - \bar{y})^2 - \sum(y_i - \hat{y}_i)^2}{\sum(y_i - \bar{y})^2} \tag{4-9}$$

Once it is realized that the $\sum(y_i - \bar{y})^2$ term appearing twice in equation (4-9) is the sum of squared deviations of the observed y's from their sample mean, several points about the coefficient of determination become clear. One is that alternative formulas exist. $\sum(y_i - \bar{y})^2$ can be replaced in equation (4-9) by $(n - 1)s^2$, by $\sum y_i^2 - (\sum y_i)^2/n$, or by $\sum y_i^2 - n\bar{y}^2$.

But perhaps more important is the fact that R^2 has an interpretation as a *fraction of the raw variation in y accounted for using the fitted equation*. Provided the fitted equation includes a constant term, $\sum(y_i - \bar{y})^2 \geqslant \sum(y_i - \hat{y}_i)^2$. Further, $\sum(y_i - \bar{y})^2$ is a measure of raw variability in y, while $\sum(y_i - \hat{y}_i)^2$ is a measure of variation in y remaining after fitting the equation. So the nonnegative difference $\sum(y_i - \bar{y})^2 - \sum(y_i - \hat{y}_i)^2$ is a measure of the variability in y accounted for in the equation-fitting process. R^2 then expresses this on the basis of a fraction (of the total raw variation).

EXAMPLE 4-1
(continued)

Using the fitted line, one can find $\hat{y}$ values for all $n = 15$ data points in the original data set. These are given in Table 4-2.

Then, referring again to Table 4-1,

$$\sum(y_i - \hat{y}_i)^2 = (2.486 - 2.4723)^2 + (2.479 - 2.4723)^2 + (2.472 - 2.4723)^2$$
$$+ (2.558 - 2.5697)^2 + \cdots + (2.879 - 2.8617)^2 + (2.858 - 2.8617)^2$$
$$= .005153$$

TABLE 4-2

Fitted Density Values

x Pressure	$\hat{y}$ Fitted Density
2,000	2.4723
4,000	2.5697
6,000	2.6670
8,000	2.7643
10,000	2.8617

Further, since

$$\bar{y} = \frac{1}{15}\left(\sum y_i\right) = \frac{1}{15}(40.005) = 2.667$$

one has

$$\sum (y_i - \bar{y})^2 = (2.486 - 2.667)^2 + (2.479 - 2.667)^2 + \cdots + (2.858 - 2.667)^2$$

$$= (15 - 1)s^2$$

$$= \sum y_i^2 - \frac{\left(\sum y_i\right)^2}{15}$$

$$= 106.982701 - \frac{(40.005)^2}{15}$$

$$= .289366$$

Thus,

$$R^2 = \frac{.289366 - .005153}{.289366} = .9822$$

and the fitted line accounts for over 98% of the raw variability in density, reducing the "unexplained" variation from .289366 to .005153.

Besides the interpretation as the fraction of variation accounted for, the coefficient of determination has a second useful interpretation. For the fitting of equations that are linear in the fitted parameters (which are the only ones seriously considered in this text), R^2 turns out to be a squared correlation. It is in fact the squared (sample) correlation between the observed values y_i and the fitted values $\hat{y}_i$. Since in the present situation of fitting a line, the $\hat{y}_i$ values are perfectly correlated with the x_i values, it also turns out to be the case that R^2 is the squared correlation between the y_i and x_i values.

EXAMPLE 4-1

(continued)

For the pressure/density data, the correlation between x and y is

$$r = .9911$$

Since $\hat{y}$ is perfectly correlated with x, this is also the correlation between $\hat{y}$ and y. But notice as well that

$$r^2 = (.9911)^2 = .9822 = R^2$$

so R^2 is indeed the squared correlation between y and $\hat{y}$.

The fact that R^2 is the squared correlation between x and y for fitting a line is special to that situation. When fitting more complicated equations, this will no longer be the case. But the interpretation of R^2 as the squared correlation between y and $\hat{y}$ will remain in general.

Computing and Using Residuals

When one fits an equation to a set of data, the hope is usually that the equation so to speak extracts the main message of the data, leaving behind (unpredicted by the fitted equation) only the variation in y that is uninterpretable. That is, one hopes that the y_i's will look like the $\hat{y}_i$'s except for relatively small fluctuations that are explainable only as random variation. A way of trying to assess whether this view is sensible for a particular fitted equation is through the computation and plotting of residuals.

DEFINITION 4-4

If the fitting of an equation or model to a data set with responses $y_1, y_2, \ldots, y_n$ produces fitted values $\hat{y}_1, \hat{y}_2, \ldots, \hat{y}_n$, then the corresponding **residuals** are the values

$$e_i = y_i - \hat{y}_i$$

The notion behind using residuals is that if a fitted equation is telling the whole story contained in a data set, then its residuals ought to be patternless. So when they're plotted against time order of observation, values of experimental variables, fitted values, or any other sensible quantities, the plots should look randomly scattered. When they don't, the patterns can themselves suggest what has gone unaccounted for in the fitting and/or how one's summarization of the data might be improved.

EXAMPLE 4-2

Compressive Strength of Fly Ash Cylinders As a Function of Amount of Ammonium Phosphate Additive. As an exaggerated example of the previous point, consider the naive fitting of a line to some data of B. Roth. Roth studied the compressive strength of concrete-like fly ash cylinders. These were made using varying amounts of ammonium phosphate as an additive to the water mixed with the fly ash before molding. Part of Roth's data are given in Table 4-3. The ammonium phosphate values are the amounts used, expressed as a percentage by weight of the amount of fly ash employed.

TABLE 4-3

Additive Concentrations and Compressive Strengths for Fly Ash Cylinders

x Ammonium Phosphate (%)	y Compressive Strength (psi)
0	1221
0	1207
0	1187
1	1555
1	1562
1	1575
2	1827
2	1839
2	1802
3	1609
3	1627
3	1642
4	1451
4	1472
4	1465
5	1321
5	1289
5	1292

Using formulas (4-6) and (4-7), it is possible to show that the least squares line through the (x, y) data in Table 4-3 is

$$\hat{y} = 1498.4 - .6381x \tag{4-10}$$

Then straightforward substitution into equation (4-10) produces fitted values $\hat{y}_i$ and residuals $e_i = y_i - \hat{y}_i$, as given in Table 4-4. The residuals for this straight-line fit are plotted against x in Figure 4-4.

TABLE 4-4

Residuals from a Straight-Line Fit to the Fly Ash Data

x	y	$\hat{y}$	e
0	1221	1498.4	−277.4
0	1207	1498.4	−291.4
0	1187	1498.4	−311.4
1	1555	1497.8	57.2
1	1562	1497.8	64.2
1	1575	1497.8	77.2
2	1827	1497.2	329.8
2	1839	1497.2	341.8
2	1802	1497.2	304.8
3	1609	1496.5	112.5
3	1627	1496.5	130.5
3	1642	1496.5	145.5
4	1451	1495.8	−44.8
4	1472	1495.8	−23.8
4	1465	1495.8	−30.8
5	1321	1495.2	−174.2
5	1289	1495.2	−206.2
5	1292	1495.2	−203.2

FIGURE 4-4

Plot of Residuals versus x for a Linear Fit to the Fly Ash Data

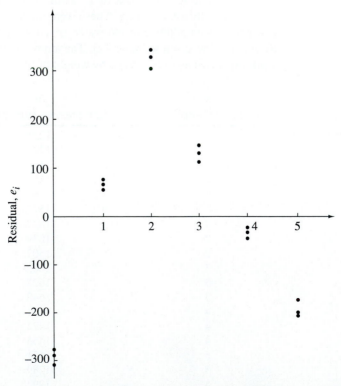

Percent Ammonium Phosphate, x_i

The distinctly "up, then back down again" curvilinear pattern of the plot in Figure 4-4 is not typical of random scatter. Something has been missed in the fitting of a line to Roth's data. Figure 4-5 is a simple scatterplot of Roth's data, which in practice should be made before fitting a line or any curve to such data. It is obvious from the scatterplot that the relationship between the amount of ammonium phosphate and compressive strength is decidedly nonlinear. In fact, a quadratic function would come much closer to fitting the data in Table 4-3 (a point that will be pursued in the next section).

FIGURE 4-5

Scatterplot of Compressive Strength versus Additive Concentration

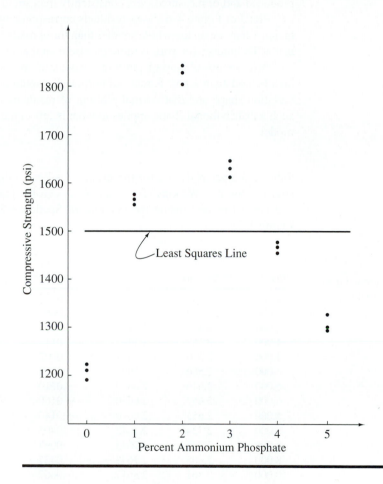

Figure 4-6 shows several patterns that might occur in plots of residuals against various variables.

FIGURE 4-6

Patterns in Residual Plots

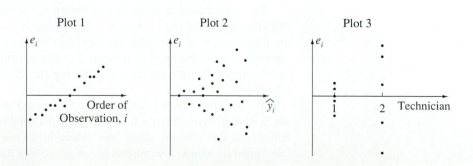

Plot 1 of Figure 4-6 shows a trend on a plot of residuals versus time order of observation. The pattern suggests that some variable changing in time is acting on y and has not been accounted for in fitting $\hat{y}$ values. For example, instrument drift (where an instrument reads higher late in a study than it did early on) could produce a pattern like that in Plot 1.

Plot 2 of Figure 4-6 shows a fan-shaped pattern on a plot of residuals versus fitted values. Such a pattern indicates that large responses are fitted (and quite possibly produced and/or measured) less consistently than small responses.

Plot 3 of Figure 4-6 shows residuals corresponding to observations made by Technician 1 that are on the whole smaller than those made by Technician 2. The suggestion is that Technician 1's work is more precise than that of Technician 2.

Another useful way of plotting residuals is to make a normal plot of them. The idea behind such a plot is that the normal distribution shape is an archetypal random variation shape and that normal-plotting of residuals is a way to investigate whether such a distributional shape applies to what is left in the data after fitting an equation or model.

EXAMPLE 4-1
(continued)

Table 4-5 gives residuals for the fitting of a line to the pressure/density data of Benson, Locher, and Watkins. The residuals e_i can be treated as a sample of 15 numbers and normal-plotted (using the methods of Section 3-2) to produce a plot like that in Figure 4-7.

TABLE 4-5

Residuals from the Linear Fit to the Pressure/Density Data

x Pressure	y Density	$\hat{y}$	$e = y - \hat{y}$
2,000	2.486	2.4723	.0137
2,000	2.479	2.4723	.0067
2,000	2.472	2.4723	−.0003
4,000	2.558	2.5697	−.0117
4,000	2.570	2.5697	.0003
4,000	2.580	2.5697	.0103
6,000	2.646	2.6670	−.0210
6,000	2.657	2.6670	−.0100
6,000	2.653	2.6670	−.0140
8,000	2.724	2.7643	−.0403
8,000	2.774	2.7643	.0097
8,000	2.808	2.7643	.0437
10,000	2.861	2.8617	−.0007
10,000	2.879	2.8617	.0173
10,000	2.858	2.8617	−.0037

The central portion of the plot in Figure 4-7 is fairly linear, indicating a generally bell-shaped distribution of residuals. But certainly the plotted point corresponding to the largest residual, and probably (this is a matter of subjective judgment) the one corresponding to the smallest residual, fail to conform to the linear pattern established by the others. Those residuals seem too big (in absolute value) in comparison to the others.

Tracking those large residuals back to Table 4-5 and to the scatterplot in Figure 4-3, one sees that they both arise from the 8,000 psi condition. And the spread for the three densities at that pressure value does indeed look considerably larger than those at the other pressure values. The normal plot suggests that the pattern of variation at 8,000 psi

FIGURE 4-7

Normal Plot of Residuals
from a Linear Fit to the
Pressure/Density Data

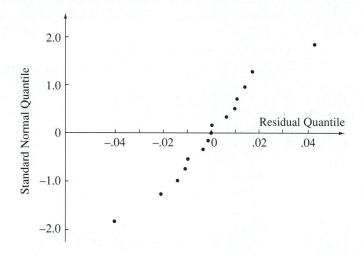

is genuinely different from those at other pressures. It may be that a different physical compaction mechanism was acting at 8,000 psi than at the other pressures. But it seems more likely that there was a problem with the students' laboratory technique or recording or with the test equipment that they were using when the 8,000 psi tests were made. At any rate, the normal plot of residuals helps draw attention to an idiosyncrasy in the data of Table 4-1 that merits further investigation, and perhaps some further data collection.

Some Caveats

The methods that have been discussed in this section are useful engineering tools when thoughtfully applied. But a few additional comments are in order, warning against some errors in logic that often accompany the use of these tools.

The first caveat regards the correlation. It must be remembered that r measures only linear relationship between x and y. It is perfectly possible to have a strong nonlinear relationship between x and y and yet have a value of r near 0. In fact, Example 4-2 provides an excellent example of this. Compressive strength is strongly related to the ammonium phosphate content. But it is possible to verify that for the data set in Table 4-3, $r = -.005$, very nearly 0.

The second warning is essentially a restatement of one implicit in the early part of Section 1-2: Correlation is not necessarily causation. One may observe a large correlation between x and y in an observational study without it necessarily being true that x drives y or vice versa. It may in fact be the case that another variable, (say, z) drives the system under study and causes simultaneous changes in both x and y.

The last of the caveats is that both $R^2(r)$ and the results of a least squares fit of an equation can be drastically affected by a very few unusual data points. This point was driven home forcefully to this author a number of years ago when teaching an elementary statistics course. As raw material for a class example, ages to the nearest year and heights to the nearest inch were collected for the 36 students taking the course. These are plotted in Figure 4-8.

By the time people reach college age, there is little useful relationship between age and height. But the correlation between ages and heights corresponding to Figure 4-8 is .73, large by most standards. The fact of the matter is that this fairly large value is produced by essentially a single data point. If one removes the data point corresponding to the 30-year-old adult student who happened to be 6 feet 8 inches from the data set, the correlation drops to .03.

FIGURE 4-8

Scatterplot of Ages and Heights
of 36 Students

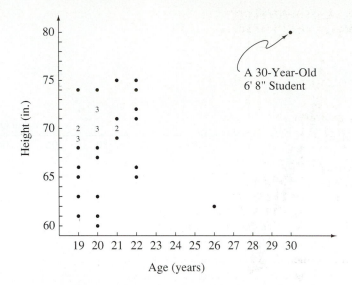

An engineer's primary insurance against being misled by the kind of phenomenon evident in Figure 4-8 is the habit of plotting the data in as many different ways as are necessary to get a feel for how it is structured. Even a simple boxplot of the age data or height data alone would have been sufficient to identify the 30-year-old student in Figure 4-8 as unusual. That would have raised the possibility of that data point strongly influencing both r and any curve that might be fit to those data via least squares.

Computing

The examples in this section have, in the interest of illustrating the meaning of various formulas, left the impression that computations were done by hand, perhaps using a simple 4-function calculator. In fact, these days it is most common to do such computations using one of many statistical analysis computer packages that are widely available. The fitting of a line by least squares is done using a *regression* program. Such a program usually also computes a value for R^2 and has an option that allows computing and later plotting of residuals.

It is not the purpose of this text to teach or recommend the use of any one of these statistical packages. But when necessary, it will be assumed that the reader has access to such computing resources. Further, annotated printouts will occasionally be given to show how Minitab, a typical statistical package, formats its output for various computations. Printout 4-1 is such a printout for an analysis of the pressure/density data in Table 4-1, paralleling the discussion in this section.

Minitab gives its user much more in the way of analysis for least squares curve fitting than has been discussed to this point, so your understanding of Printout 4-1 will be incomplete. But it should be possible to locate values of the major summary statistics discussed in this section. One clarification of what went on in creating Printout 4-1 is that the commands following the sequence "MTB >" were entered by the user. The rest of the output is the result of what Minitab did in response to those commands.

PRINTOUT 4-1

```
MTB > read c1 c2
DATA> 2000 2.486
DATA> 2000 2.479
DATA> 2000 2.472
DATA> 4000 2.558
DATA> 4000 2.570
DATA> 4000 2.580
DATA> 6000 2.646
```

(continued)

PRINTOUT 4-1
(continued)

```
DATA> 6000 2.657
DATA> 6000 2.653
DATA> 8000 2.724
DATA> 8000 2.774
DATA> 8000 2.808
DATA> 10000 2.861
DATA> 10000 2.879
DATA> 10000 2.858
DATA> end
    15 ROWS READ
MTB > name c1 'pressure' c2 'density'
MTB > plot c2 vs c1
```

```
         -
         -                                                      *
   2.85+                                                        2
         -                                              *
density  -
         -                                            *
         -                                            *
   2.70+
         -                                   *
         -                                   2
         -
         -                      2
   2.55+                        *
         -
         -     2
         -     *
         -
      --------+---------+---------+---------+---------+---------+--------
           3000      4500      6000      7500      9000  pressure
```

```
MTB > regress c2 1 c1 std res c3 yhat c4;
SUBC> residuals c5.
```

```
The regression equation is
density = 2.37 +0.000049 pressure
```

Predictor   Coef b_0 Stdev   t-ratio   p
Constant $\boxed{2.37500}$ 0.01206   197.01   0.000
pressure $\boxed{0.00004867}$ 0.00000182   26.78   0.000
b_1

s = 0.01991 $\boxed{\text{R-sq = 98.2\%}}$ R-sq(adj) = 98.1%

```
Analysis of Variance
```

SOURCE	DF	SS	MS	F	p
Regression	1	0.28421	0.28421	717.06	0.000
Error	13	0.00515	0.00040		
Total	14	0.28937			

```
Unusual Observations
```

Obs.	pressure	density	Fit	Stdev.Fit	Residual	St.Resid
10	8000	2.72400	2.76433	0.00630	-0.04033	-2.14R
12	8000	2.80800	2.76433	0.00630	0.04367	2.31R

```
R denotes an obs. with a large st. resid.
```

```
MTB > name c4 'yhat' c5 'residual'
MTB > print c1 c2 c4 c5
```

(continued)

PRINTOUT 4-1
(continued)

ROW	pressure	density	yhat	residual
1	2000	2.486	2.47233	0.0136666
2	2000	2.479	2.47233	0.0066667
3	2000	2.472	2.47233	-0.0003335
4	4000	2.558	2.56967	-0.0116665
5	4000	2.570	2.56967	0.0003333
6	4000	2.580	2.56967	0.0103333
7	6000	2.646	2.66700	-0.0210001
8	6000	2.657	2.66700	-0.0100000
9	6000	2.653	2.66700	-0.0139999
10	8000	2.724	2.76433	-0.0403333
11	8000	2.774	2.76433	0.0096667
12	8000	2.808	2.76433	0.0436668
13	10000	2.861	2.86167	-0.0006666
14	10000	2.879	2.86167	0.0173333
15	10000	2.858	2.86167	-0.0036666

**Compare to
Table 4-5**

MTB > plot c5 vs c1

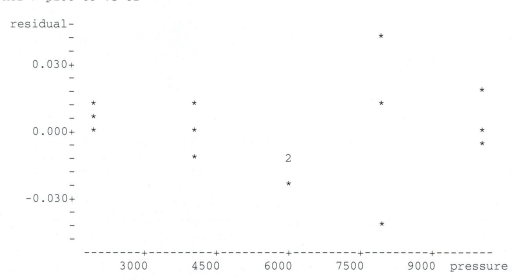

MTB > mplot c2 vs c1   c4 vs c1

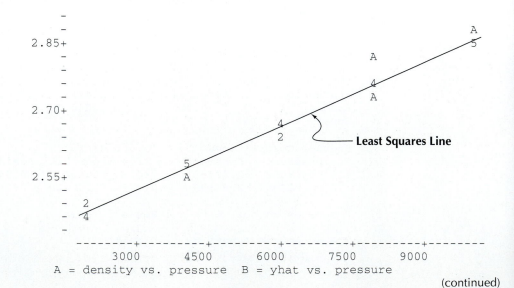

A = density vs. pressure   B = yhat vs. pressure

(continued)

PRINTOUT 4-1
(continued)

```
MTB > nsco c5 c6
MTB > name c6 'normal'
MTB > plot c6 vs c5
```
} **Create a normal plot of the residuals.**
(The MINITAB vertical plotting position is slightly
different from the one introduced in Section 3-2.)

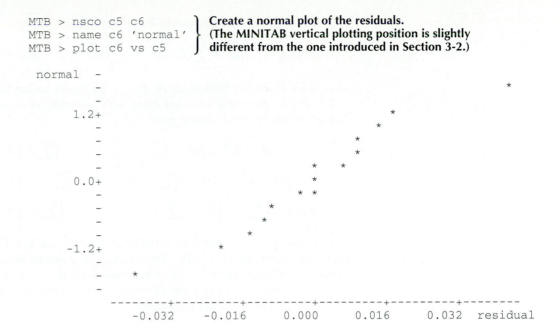

4-2 *Fitting Curves and Surfaces by Least Squares*

The basic ideas introduced in Section 4-1 generalize to produce a powerful engineering tool: *multiple linear regression*, which is introduced in this section. (However, since the term *regression* often seems obscure to students, the more descriptive terms *curve* fitting and *surface fitting* will be used here, at least initially.)

This section first covers fitting curves defined by polynomials and other functions that are linear in their parameters to a set of (x, y) data. Next comes the fitting of surfaces defined by functions that are linear in their parameters to data where a response y is thought of as depending upon the values of several process variables $x_1, x_2, \ldots, x_k$. In both cases, the discussion will stress how useful the R^2 and residual plotting ideas are and will consider the question of choosing between possible fitted equations. Lastly, some additional caveats will be given regarding the practical use of least squares curves and surfaces.

Curve Fitting by
Least Squares

In the examples of the previous section, a straight line did a reasonable job of describing the pressure/density data. But in the fly ash study, the ammonium phosphate/compressive strength data were very poorly described by a straight line. The first order of business in this section is to investigate the possibility of fitting curves more complicated than a straight line to (x, y) data. As an example, an attempt will be made to find a better equation for describing the fly ash data.

A natural generalization of the equation

$$y \approx \beta_0 + \beta_1 x \tag{4-11}$$

is the *polynomial equation*

$$y \approx \beta_0 + \beta_1 x + \beta_2 x^2 + \cdots + \beta_k x^k \tag{4-12}$$

Happily, the task of fitting equation (4-12) to a set of n pairs (x_i, y_i) via least squares is only slightly more difficult conceptually than the task of fitting equation (4-11). One wishes to minimize the function of $k + 1$ variables

$$S(\beta_0, \beta_1, \beta_2, \ldots, \beta_k) = \sum_{i=1}^{n}(y_i - \hat{y}_i)^2 = \sum_{i=1}^{n}(y_i - (\beta_0 + \beta_1 x_i + \beta_2 x_i^2 + \cdots + \beta_k x_i^k))^2$$

by choice of the coefficients $\beta_0, \beta_1, \beta_2, \ldots, \beta_k$. Upon setting the partial derivatives of $S(\beta_0, \beta_1, \ldots, \beta_k)$ equal to 0 and doing some simplification, one obtains the *normal equations* for this least squares problem:

$$\left.\begin{aligned}
n\beta_0 + \left(\sum x_i\right)\beta_1 + \left(\sum x_i^2\right)\beta_2 + \cdots + \left(\sum x_i^k\right)\beta_k &= \sum y_i \\
\left(\sum x_i\right)\beta_0 + \left(\sum x_i^2\right)\beta_1 + \left(\sum x_i^3\right)\beta_2 + \cdots + \left(\sum x_i^{k+1}\right)\beta_k &= \sum x_i y_i \\
\left(\sum x_i^k\right)\beta_0 + \left(\sum x_i^{k+1}\right)\beta_1 + \left(\sum x_i^{k+2}\right)\beta_2 + \cdots + \left(\sum x_i^{2k}\right)\beta_k &= \sum x_i^k y_i
\end{aligned}\right\} \quad \textbf{(4-13)}$$

The equations (4-13) may look formidable, but they are $k + 1$ linear equations in the $k + 1$ unknowns $\beta_0, \beta_1, \ldots, \beta_k$. And typically, they can be solved simultaneously for a single set of values $b_0, b_1, \ldots, b_k$ minimizing $S(\beta_0, \beta_1, \ldots, \beta_k)$. The mechanics of that solution, though in theory possible by hand, are usually carried out using a computer program for multiple linear regression.

EXAMPLE 4-3

(Example 4-2 continued) **More on the Fly Ash Data.** Return to the fly ash study of B. Roth. The comment was made in the last section that a quadratic equation might fit the ammonium phosphate/compressive strength data considerably better than the equation linear in x. So consider fitting the $k = 2$ version of equation (4-12)

$$y \approx \beta_0 + \beta_1 x + \beta_2 x^2 \tag{4-14}$$

to the data of Table 4-3, reproduced and extended here as Table 4-6.

TABLE 4-6

Some Calculations with the Fly Ash Data

Ammonium Phosphate x	Compressive Strength y	x^2	x^3	x^4	xy	$x^2 y$
0	1,221	0	0	0	0	0
0	1,207	0	0	0	0	0
0	1,187	0	0	0	0	0
1	1,555	1	1	1	1,555	1,555
1	1,562	1	1	1	1,562	1,562
1	1,575	1	1	1	1,575	1,575
2	1,827	4	8	16	3,654	7,308
2	1,839	4	8	16	3,678	7,356
2	1,802	4	8	16	3,604	7,208
3	1,609	9	27	81	4,827	14,481
3	1,627	9	27	81	4,881	14,643
3	1,642	9	27	81	4,926	14,778
4	1,451	16	64	256	5,804	23,216
4	1,472	16	64	256	5,888	23,552
4	1,465	16	64	256	5,860	23,440
5	1,321	25	125	625	6,605	33,025
5	1,289	25	125	625	6,445	32,225
5	1,292	25	125	625	6,460	32,300
45	26,943	165	675	2,937	67,324	238,224

From Table 4-6 and display (4-13), the normal equations for fitting the quadratic (4-14) to these data are

$$18\beta_0 + 45\beta_1 + 165\beta_2 = 26{,}943$$
$$45\beta_0 + 165\beta_1 + 675\beta_2 = 67{,}324$$
$$165\beta_0 + 675\beta_1 + 2{,}937\beta_2 = 238{,}224$$

Solving these equations is the heart of what a regression program does to fit equation (4-14). Printout 4-2 shows part of the results of a Minitab run made to accomplish this task.

The fitted quadratic equation is

$$\hat{y} = 1242.9 + 382.7x - 76.7x^2$$

Figure 4-9 shows the fitted curve sketched on a scatterplot of the (x, y) data. Notice that while the quadratic curve in Figure 4-9 is not altogether satisfactory as a summary of Roth's data, it looks as if it does a much better job of following the trend of the data than does the line sketched in Figure 4-5.

FIGURE 4-9

Scatterplot and Fitted Quadratic for the Fly Ash Data

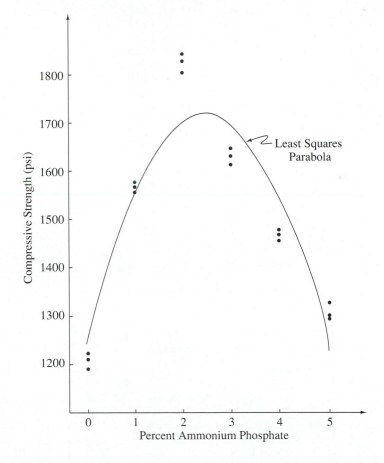

PRINTOUT 4-2

```
MTB > read c1 c2
DATA> 0 1221
DATA> 0 1207
DATA> 0 1187
DATA> 1 1555
DATA> 1 1562
DATA> 1 1575
```

(continued)

PRINTOUT 4-2
(continued)

```
DATA> 2 1827
DATA> 2 1839
DATA> 2 1802
DATA> 3 1609
DATA> 3 1627
DATA> 3 1642
DATA> 4 1451
DATA> 4 1472
DATA> 4 1465
DATA> 5 1321
DATA> 5 1289
DATA> 5 1292
DATA> end
      18 ROWS READ
MTB > mult c1 by c1 put into c3
MTB > name c1 'x'  c2 'y'  c3 'x**2'
MTB > print c1 c3 c2

  ROW     x    x**2        y

    1     0       0     1221
    2     0       0     1207
    3     0       0     1187
    4     1       1     1555
    5     1       1     1562
    6     1       1     1575
    7     2       4     1827
    8     2       4     1839
    9     2       4     1802
   10     3       9     1609
   11     3       9     1627
   12     3       9     1642
   13     4      16     1451
   14     4      16     1472
   15     4      16     1465
   16     5      25     1321
   17     5      25     1289
   18     5      25     1292


MTB > regress c2 2 c1 c3


The regression equation is
y = 1243 + 383 x - 76.7 x**2


Predictor       Coef     Stdev    t-ratio        p
Constant     1242.89     42.98      28.92    0.000
x             382.67     40.43       9.46    0.000
x**2         -76.661      7.762     -9.88    0.000

s = 82.14      R-sq = 86.7%    R-sq(adj) = 84.9%

Analysis of Variance

SOURCE        DF         SS          MS        F        p
Regression     2     658230      329115    48.78    0.000
Error         15     101206        6747
Total         17     759437

SOURCE        DF     SEQ SS
x              1         21
x**2           1     658209
```

Create the x^2 Variable

b_0 (pointing to 1242.89)
b_1 (pointing to 382.67)
b_2 (pointing to -76.661)

The previous section illustrated that when fitting a line to (x, y) data, it is helpful to quantify the goodness of that fit using the linear correlation between x and y, r, or the coefficient of determination R^2. When fitting a polynomial of form (4-12) to (x, y) data, the coefficient of determination can again be used to measure how well the equation fits the data. Recall once more from Definition 4-3 that

$$R^2 = \frac{\sum(y_i - \bar{y})^2 - \sum(y_i - \hat{y}_i)^2}{\sum(y_i - \bar{y})^2} \qquad (4\text{-}15)$$

is the fraction of the raw variability in y accounted for by the fitted equation. Its calculation by hand from formula (4-15) is possible (using the fitted equation to produce $\hat{y}$'s), but of course the easiest way to obtain R^2 is to use a computer package.

EXAMPLE 4-3
(continued)

Consulting Printout 4-2, it can be seen that for the ammonium phosphate/compressive strength data, the equation $\hat{y} = 1242.9 + 382.7x - 76.7x^2$ produces $R^2 = .867$. So 86.7% of the raw variability in compressive strength is accounted for using the fitted quadratic, and the sample correlation between the observed strengths y_i and fitted strengths $\hat{y}_i$ is $\sqrt{.867} = .93$.

Comparing what has been done with these data in the present section to what was done in Section 4-1, it is interesting to note that for the fitting of a line to the ammonium phosphate/compressive strength data, R^2 obtained there was only .000 (to 3 decimals). The present quadratic is a remarkable improvement over a linear equation for summarizing these data.

A natural question to raise, once it is realized that a quadratic equation fits the ammonium phosphate/compressive strength data much better than a linear one, is "What about a cubic version of equation (4-12)?" Printout 4-3 shows part of the results of a Minitab run made to investigate this possibility, and Figure 4-10 (page 122) shows a scatterplot of the data and a plot of the fitted cubic equation.

R^2 for the cubic equation is .952, somewhat larger than for the quadratic, but it is fairly clear from Figure 4-10 that even a cubic polynomial is not totally satisfactory as a summary of these data. In particular, both the fitted quadratic in Figure 4-9 and the fitted cubic in Figure 4-10 fail to fit the data adequately near an ammonium phosphate level of 2%. Unfortunately, this is where compressive strength is greatest — precisely the area of greatest practical interest.

PRINTOUT 4-3

```
MTB > print c1-c4

ROW     x     x**2    x**3       y

  1     0       0       0      1221
  2     0       0       0      1207
  3     0       0       0      1187
  4     1       1       1      1555
  5     1       1       1      1562
  6     1       1       1      1575
  7     2       4       8      1827
  8     2       4       8      1839
  9     2       4       8      1802
 10     3       9      27      1609
 11     3       9      27      1627
 12     3       9      27      1642
 13     4      16      64      1451
 14     4      16      64      1472
 15     4      16      64      1465
 16     5      25     125      1321
 17     5      25     125      1289
 18     5      25     125      1292
```

(continued)

PRINTOUT 4-3
(continued)

```
MTB > regress c4 3 c1-c3
* NOTE *  x**2 is highly correlated with other predictor variables
* NOTE *  x**3 is highly correlated with other predictor variables

The regression equation is
y = 1188 + 633 x - 214 x**2 + 18.3 x**3

Predictor        Coef       Stdev     t-ratio        p
Constant      1188.05       28.79       41.27    0.000
x              633.11       55.91       11.32    0.000
x**2          -213.77       27.79       -7.69    0.000
x**3           18.281       3.649        5.01    0.000

s = 50.88        R-sq = 95.2%      R-sq(adj) = 94.2%

Analysis of Variance

SOURCE         DF          SS          MS        F        p
Regression      3      723197      241066    93.13    0.000
Error          14       36240        2589
Total          17      759437

SOURCE         DF      SEQ SS
x               1          21
x**2            1      658209
x**3            1       64967
```

Coef labels: b_0 = 1188.05, b_1 = 633.11, b_2 = -213.77, b_3 = 18.281

FIGURE 4-10

Scatterplot and Fitted Cubic for the Fly Ash Data

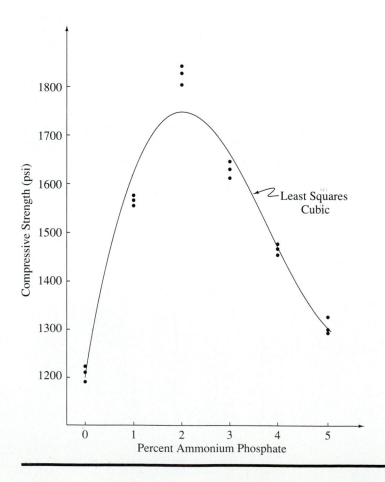

Least Squares Cubic

Compressive Strength (psi)

Percent Ammonium Phosphate

The previous example illustrates that R^2 (as useful as it is) is not the only consideration when it comes to judging the appropriateness of a fitted polynomial. The examination of plots is also important. One should not only look at scatterplots of y versus x with superimposed fitted curves; plots of residuals can also be helpful. The helpfulness of such plots can be illustrated on a data set where y can be expected to be nearly perfectly quadratic in x.

EXAMPLE 4-4

Analysis of the Bob Drop Data of Section 1-4. Consider again the experimental determination of the acceleration due to gravity (through the dropping of the steel bob) data given in Table 1-4 and reproduced here as Table 4-7.

TABLE 4-7

Measured Displacements of a Bob in Free Fall

x Point Number	y Displacement (mm)
1	.8
2	4.8
3	10.8
4	20.1
5	31.9
6	45.9
7	63.3
8	83.1
9	105.8
10	131.3
11	159.5
12	190.5
13	223.8
14	260.0
15	299.2
16	340.5
17	385.0
18	432.2
19	481.8
20	534.2
21	589.8
22	647.7
23	708.8

Recall that the positions y were recorded at $\frac{1}{60}$ sec intervals beginning at some unknown time t_0 (less than $\frac{1}{60}$ sec) after the bob was released. Since Newtonian mechanics predicts the bob displacement to be

$$\text{displacement} = \frac{gt^2}{2}$$

one expects

$$y \approx \frac{1}{2} g \left(t_0 + \frac{1}{60} (x - 1) \right)^2$$

$$= \frac{g}{2} \left(\frac{x}{60} \right)^2 + g \left(t_0 - \frac{1}{60} \right) \left(\frac{x}{60} \right) + \frac{g}{2} \left(t_0 - \frac{1}{60} \right)^2$$

$$= \frac{g}{7200} x^2 + \frac{g}{60} \left(t_0 - \frac{1}{60} \right) x + \frac{g}{2} \left(t_0 - \frac{1}{60} \right)^2 \qquad \text{(4-16)}$$

That is, y is expected to be approximately quadratic in x; indeed, the plot of (x, y) points in Figure 1-8 appears to have that character.

As a slight digression, note that expression (4-16) shows that if one fits a quadratic

$$\hat{y} = b_0 + b_1 x + b_2 x^2 \tag{4-17}$$

to the data in Table 4-7 via least squares, an experimentally determined value of g (in mm/sec^2) will be $7200 b_2$. This is in fact how the value 9.79 m/sec^2, quoted in Section 1-4, was obtained.

The use of a multiple linear regression program to fit equation (4-17) to the data in Table 4-7 gives a fitted equation

$$\hat{y} = .0645 - .4716x + 1.3597x^2$$

(from which $g \approx 9790$ mm/sec^2) with an R^2 that is 1 to 6 decimal places. Residuals for this fit can be calculated using Definition 4-4; these are given in Table 4-8.

TABLE 4-8

Calculation of Residuals for a Quadratic Fit to the Bob Displacement

x Point Number	y Displacement	$\hat{y}$ Fitted Diplacement	e Residual
1	.8	.95	−.15
2	4.8	4.56	.24
3	10.8	10.89	−.09
4	20.1	19.93	.17
5	31.9	31.70	.20
6	45.9	46.19	−.29
7	63.3	63.39	−.09
8	83.1	83.31	−.21
9	105.8	105.96	−.16
10	131.3	131.32	−.02
11	159.5	159.40	.10
12	190.5	190.21	.29
13	223.8	223.73	.07
14	260.0	259.97	.03
15	299.2	298.93	.27
16	340.5	340.61	−.11
17	385.0	385.01	−.01
18	432.2	432.13	.07
19	481.8	481.97	−.17
20	534.2	534.53	−.33
21	589.8	589.80	.00
22	647.7	647.80	−.10
23	708.8	708.52	.28

Figure 4-11 is a normal plot of the residuals. It is reasonably linear, and thus not remarkable (except for some small suggestion that the largest residual or two may not be as extreme as might be expected, a circumstance that suggests no obvious physical explanation to this author).

However, a plot of residuals versus x (the time variable) is intriguing. Figure 4-12 is such a plot, where successive plotted points have been connected with line segments.

FIGURE 4-11

Normal Plot of the Residuals
from a Quadratic Fit to the Bob
Drop Data

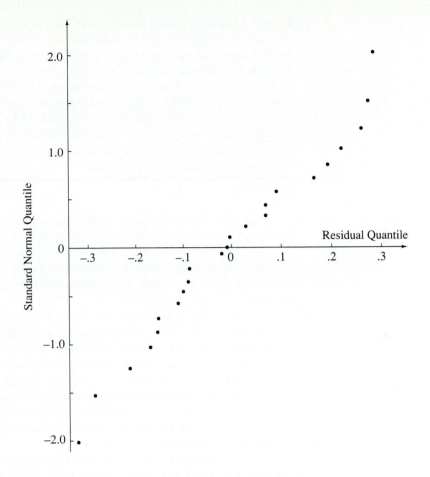

FIGURE 4-12

Plot of the Residuals from the
Quadratic Fit to the Bob Drop
Data versus x

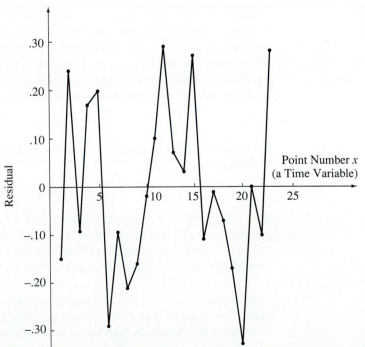

It may be a case of seeing a pattern where none really exists, but there is at least a hint in Figure 4-12 of a cyclical pattern in the residuals. Observed displacements are alternately too big, too small, too big, etc. It would be a good idea to look at several more tapes from the same apparatus, to see if a cyclical pattern appears consistently, before seriously pondering its origin. But should the pattern suggested by the plot in Figure 4-12 reappear consistently, it would be an indication that something in the mechanism generating the nominal 60 cycle current may cause the cycles to be alternately slightly shorter then slightly longer than $\frac{1}{60}$ sec. The practical implication of this would be that if a better determination of g were desired, the regularity of the waveform of the AC current used would be one problem to be addressed.

Examples 4-3 and 4-4 are (respectively) illustrations of only partial success and then great success in describing an (x, y) data set by means of a polynomial equation like (4-12). Situations like that of the first example obviously do sometimes occur in statistical engineering studies. It is reasonable to ask "Where does one go from here?" when they arise. Two simple observations can be made in response to this legitimate question.

For one, while a polynomial of form (4-12) may be unsatisfactory as a global description of a relationship between x and y, it may be quite adequate *locally* — i.e., for a relatively restricted range of x values. For example, in the fly ash study, the quadratic representation of compressive strength as a function of percent ammonium phosphate is not appropriate over the range 0 to 5%. But having identified the region around 2% as being of practical interest, it would make good sense to conduct a follow-up study concentrating on (say) 1.5 to 2.5% ammonium phosphate. It is quite possible, then, that a quadratic fit only to data with $1.5 \leq x \leq 2.5$ would be both adequate and helpful as a summarization of the follow-up data.

The second observation that can be made about using a polynomial relationship like equation (4-12) is that the factors $x, x^2, x^3, \ldots, x^k$ can be replaced by any (known) functions of x and the message(s) of this section will remain essentially unchanged. The normal equations (4-13) will look slightly different, but they will still be $k + 1$ linear equations in $\beta_0, \beta_1, \ldots, \beta_k$, and a multiple linear regression program will still produce least squares values $b_0, b_1, \ldots, b_k$. This second point can be quite useful when one has theoretical reasons to expect a particular (nonlinear but) simple functional relationship between x and y. For example, Taylor's equation for tool life is of the form

$$y \approx \alpha x^\beta$$

for y tool life (e.g., in minutes) and x the cutting speed used (e.g., in sfpm). Taking logarithms,

$$\ln(y) \approx \ln(\alpha) + \beta \ln(x)$$

This is an equation for $\ln(y)$ that is linear in the parameters $\ln(\alpha)$ and β involving the variable $\ln(x)$. So presented with a set of (x, y) data, one could determine empirical values for α and β by taking logs of both x's and y's, fitting the linear version of (4-12), and identifying $\ln(\alpha)$ with β_0 (and thus α with $\exp(\beta_0)$) and β with β_1.

Surface Fitting by Least Squares

Having generalized the ideas of fitting a line using least squares to fitting a polynomial curve, it is a small step to realize that essentially the same methods can be used to summarize the results of multifactor statistical engineering studies. In such studies, one typically wishes to describe the effects of several different quantitative variables $x_1, x_2, \ldots, x_k$ on some response y. One can think geometrically of fitting a surface described by an equation

$$y \approx \beta_0 + \beta_1 x_1 + \beta_2 x_2 + \cdots + \beta_k x_k \qquad (4\text{-}18)$$

to the data using the least squares principle. This is pictured for a $k = 2$ case in Figure 4-13, where six fictitious (x_1, x_2, y) data points are pictured in three dimensions, along with a possible fitted surface of the form (4-18).

FIGURE 4-13

Six Data Points (x_1, x_2, y) and a Possible Fitted Plane

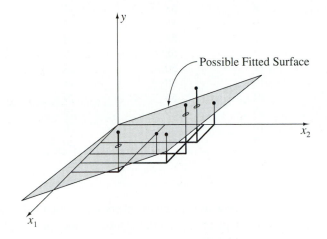

To fit a surface defined by equation (4-18) to a set of n data points $(x_{1i}, x_{2i}, \ldots, x_{ki}, y_i)$ via least squares, one must minimize the function of $k + 1$ variables

$$S(\beta_0, \beta_1, \beta_2, \ldots, \beta_k) = \sum_{i=1}^{n} (y_i - \hat{y}_i)^2 = \sum_{i=1}^{n} (y_i - (\beta_0 + \beta_1 x_{1i} + \cdots + \beta_k x_{ki}))^2$$

by choice of the coefficients $\beta_0, \beta_1, \ldots, \beta_k$. Setting partial derivatives with respect to the β's equal to 0 gives the normal equations

$$n\beta_0 + \left(\sum x_{1i}\right)\beta_1 + \left(\sum x_{2i}\right)\beta_2 + \cdots + \left(\sum x_{ki}\right)\beta_k = \sum y_i$$

$$\left(\sum x_{1i}\right)\beta_0 + \left(\sum x_{1i}^2\right)\beta_1 + \left(\sum x_{1i}x_{2i}\right)\beta_2 + \cdots + \left(\sum x_{1i}x_{ki}\right)\beta_k = \sum x_{1i}y_i$$

$$\left(\sum x_{2i}\right)\beta_0 + \left(\sum x_{1i}x_{2i}\right)\beta_1 + \left(\sum x_{2i}^2\right)\beta_2 + \cdots + \left(\sum x_{2i}x_{ki}\right)\beta_k = \sum x_{2i}y_i$$

$$\left(\sum x_{ki}\right)\beta_0 + \left(\sum x_{1i}x_{ki}\right)\beta_1 + \left(\sum x_{2i}x_{ki}\right)\beta_2 + \cdots + \left(\sum x_{ki}^2\right)\beta_k = \sum x_{ki}y_i$$

The solution of these $k + 1$ linear equations in the $k + 1$ unknowns $\beta_0, \beta_1, \ldots, \beta_k$ is the first task of a multiple linear regression program. The fitted parameters $b_0, b_1, \ldots, b_k$ that it produces are a set that minimizes $S(\beta_0, \beta_1, \beta_2, \ldots, \beta_k)$.

EXAMPLE 4-5

Surface Fitting and Brownlee's Stack Loss Data. Table 4-9 contains part of a set of real data on the operation of a plant for the oxidation of ammonia to nitric acid that appeared first in Brownlee's *Statistical Theory and Methodology in Science and Engineering*. These

TABLE 4-9

Brownlee's Stack Loss Data

i Observation Number	x_{1i} Air Flow	x_{2i} Cooling Water Inlet Temperature	x_{3i} Acid Concentration	y_i Stack Loss
1	80	27	88	37
2	62	22	87	18
3	62	23	87	18
4	62	24	93	19
5	62	24	93	20
6	58	23	87	15
7	58	18	80	14
8	58	18	89	14
9	58	17	88	13
10	58	18	82	11
11	58	19	93	12
12	50	18	89	8
13	50	18	86	7
14	50	19	72	8
15	50	19	79	8
16	50	20	80	9
17	56	20	82	15

data have been used by many others since, notably Daniel and Wood in *Fitting Equations to Data*. In plant operation, the nitric oxides produced are absorbed in a countercurrent absorption tower.

The air flow variable, x_1, represents the rate of operation of the plant. The acid concentration variable, x_3, is the percent circulating minus 50 times 10. The response variable, y, is 10 times the percentage of ingoing ammonia that escapes from the absorption column unabsorbed, (i.e, an inverse measure of overall plant efficiency). For the engineering purposes of understanding, predicting, and possibly ultimately manipulating/optimizing plant performance, it would be useful to have an equation describing how y depends on x_1, x_2, and x_3. Surface fitting via least squares is a method of coming up with an empirical equation describing this relationship.

Printout 4-4 shows part of the results of a Minitab run made to obtain a fitted equation of the form

$$\hat{y} = b_0 + b_1 x_1 + b_2 x_2 + b_3 x_3$$

The equation produced by the program is

$$\hat{y} = -37.65 + .80 x_1 + .58 x_2 - .07 x_3 \tag{4-19}$$

with $R^2 = .975$. The coefficients in this equation can be thought of as rates of change of stack loss with respect to the individual variables x_1, x_2, and x_3, holding the others fixed. For example, $b_1 = .80$ can be interpreted as the increase in stack loss y that accompanies a one unit increase in air flow x_1 if inlet temperature x_2 and acid concentration x_3 are held fixed. The signs on the coefficients indicate whether y tends to increase or decrease with increases in the corresponding x. For example, the fact that b_1 is positive indicates that the higher the rate at which the plant is run, the larger y tends to be (i.e., the less efficiently the plant operates). The large value of R^2 is a preliminary indicator that the equation (4-19) is an effective summarization of the data.

PRINTOUT 4-4

```
MTB > print c1-c4

ROW   air  water   acid  stack

  1    80    27     88    37
  2    62    22     87    18
  3    62    23     87    18
  4    62    24     93    19
  5    62    24     93    20
  6    58    23     87    15
  7    58    18     80    14
  8    58    18     89    14
  9    58    17     88    13
 10    58    18     82    11
 11    58    19     93    12
 12    50    18     89     8
 13    50    18     86     7
 14    50    19     72     8
 15    50    19     79     8
 16    50    20     80     9
 17    56    20     82    15

MTB > regress c4 3 c1-c3

The regression equation is
stack = - 37.7 + 0.798 air + 0.577 water - 0.0671 acid
```

Predictor	Coef	Stdev	t-ratio	p
Constant	b_0 -37.652	4.732	-7.96	0.000
air	b_1 0.79769	0.06744	11.83	0.000
water	b_2 0.5773	0.1660	3.48	0.004
acid	b_3 -0.06706	0.06160	-1.09	0.296

s = 1.253   R-sq = 97.5%   R-sq(adj) = 96.9%

```
Analysis of Variance
```

SOURCE	DF	SS	MS	F	p
Regression	3	795.83	265.28	169.04	0.000
Error	13	20.40	1.57		
Total	16	816.24			

SOURCE	DF	SEQ SS
air	1	775.48
water	1	18.49
acid	1	1.86

```
Unusual Observations
```

Obs.	air	stack	Fit	Stdev.Fit	Residual	St.Resid
10	58.0	11.000	13.506	0.552	-2.506	-2.23R

```
R denotes an obs. with a large st. resid.
```

While the mechanics of fitting equations of the form (4-18) to multivariate data are relatively straightforward, the choice and interpretation of appropriate equations can be a subtle matter. The problem is that where many x variables are involved, the number of potential equations of form (4-18) is huge. To make matters worse, in contrast to the case where a curve is being fitting to bivariate data, there is no completely satisfactory way to plot multivariate $(x_1, x_2, \ldots, x_k, y)$ data to "see" how an equation is fitting. About all that can be done at this point is to offer the broad advice that one wants the simplest equation that adequately fits the data. Given below are examples of how R^2 and residual plotting can be helpful tools in negotiating one's way through the subtleties that arise.

EXAMPLE 4-5
(continued)

In the context of the nitrogen plant, it is sensible to ask whether in fact all three variables x_1, x_2, and x_3 are required to adequately account for the observed variation in y. One might, for example, be able to adequately explain the behavior of stack loss using only one or two of the three x variables. There would be several consequences of practical engineering importance if this were so.

For one, in such a case, a simple or parsimonious version of equation (4-18) could be used in describing the data in Table 4-9 (and hence the behavior of the oxidation process). And if a variable is not needed to predict y, then it is possible that the expense of measuring it might be saved. Or, if a variable doesn't seem to have much impact on y (because it doesn't seem to be essential to include it when writing an equation for y), it may be possible to choose its level on purely economic grounds, without fear of degrading process performance.

As a means of investigating whether indeed some subset of x_1, x_2, and x_3 is adequate to explain stack loss behavior, a computer program to calculate R^2 values for equations based on all possible subsets of x_1, x_2, and x_3 was used. Table 4-10 gives the values that were obtained.

TABLE 4-10

R^2's for Equations Predicting Stack Loss

Equation Fit	R^2
$y \approx \beta_0 + \beta_1 x_1$	.950
$y \approx \beta_0 + \beta_2 x_2$	.695
$y \approx \beta_0 + \beta_3 x_3$	.165
$y \approx \beta_0 + \beta_1 x_1 + \beta_2 x_2$	.973
$y \approx \beta_0 + \beta_1 x_1 + \beta_3 x_3$	.952
$y \approx \beta_0 + \beta_2 x_2 + \beta_3 x_3$	.706
$y \approx \beta_0 + \beta_1 x_1 + \beta_2 x_2 + \beta_3 x_3$	.975

The figures in Table 4-10 show, for example, that 95% of the raw variability in y can be accounted for using a linear equation in only the air flow variable x_1. Use of both x_1 and the water temperature variable x_2 can account for 97.3% of the raw variability in stack loss. Inclusion of x_3, the acid concentration variable, in an equation already involving x_1 and x_2, increases R^2 only from .973 to .975.

If one has in mind identifying a simple equation for stack loss that seems to fit the data well, the message in Table 4-10 would seem to be "Consider an x_1 term first, and then possibly an x_2 term." On the basis of R^2, inclusion of an x_3 term in an equation for y seems unnecessary. And in retrospect, this is entirely consistent with the character of the fitted equation (4-19). x_3 varies from 72 to 93 in the original data set, and this means that $\hat{y}$ changes only a total amount

$$.07(93 - 72) \approx 1.5$$

based on changes in x_3. (Remember that $.07 = b_3 =$ the fitted rate of change in y with respect to x_3.) 1.5 is relatively small in comparison to the range in the observed y values.

Once R^2 values have been used to identify potential simplifications of the equation

$$\hat{y} = b_0 + b_1x_1 + b_2x_2 + b_3x_3$$

these can and should be subjected to thorough residual analyses before they are adopted as data summaries. As an example, consider calculating and plotting residuals for a fitted equation involving x_1 and x_2. A multiple linear regression program can be used to produce the fitted equation

$$\hat{y} = -42.00 - .78x_1 + .57x_2 \tag{4-20}$$

(Notice in passing that b_0, b_1, and b_2 in equation (4-20) differ somewhat from the corresponding values in equation (4-19). That is, equation (4-20) was *not* obtained from equation (4-19) by simply dropping the last term in the equation. In general, the values of the coefficients b will change depending on which x variables are and are not included in the fitting of a surface.)

Residuals for equation (4-20) can be computed and plotted in any number of potentially useful ways. Figure 4-14 shows a normal plot of the residuals and three other plots of the residuals, against respectively x_1, x_2, and $\hat{y}$. There are no really strong messages carried by the plots in Figure 4-14 except that the data set contains one unusually large x_1 value and one unusually large $\hat{y}$ (which in fact corresponds to the large x_1). But there is enough of a curvilinear "up, then down, then back up again" pattern in the plot of residuals against x_1 to suggest the possibility of adding an x_1^2 term to the fitted equation (4-20).

You might want to verify that use of a multiple regression program to fit the equation

$$y \approx \beta_0 + \beta_1x_1 + \beta_2x_2 + \beta_3x_1^2$$

to the data of Table 4-9 yields approximately

$$\hat{y} = -15.409 - .069x_1 + .528x_2 + .007x_1^2 \tag{4-21}$$

with corresponding $R^2 = .980$ and residuals that show even less of a pattern than those for the fitted equation (4-20). In particular, the hint of curvature on the plot of residuals versus x_1 for equation (4-20) is not present in the corresponding plot for equation (4-21). Interestingly, looking back over this example, one sees that fitted equation (4-21) has a better R^2 value than even fitted equation (4-19), in spite of the fact that equation (4-19) involves the process variable x_3 and equation (4-21) does not.

Equation (4-21) is somewhat more complicated than equation (4-20). But because it still really only involves two different input x's and also eliminates the slight pattern seen on the plot of residuals for equation (4-20) versus x_1, it seems an attractive choice for summarizing the stack loss data. A two-dimensional representation of the fitted surface defined by equation (4-21) is given in Figure 4-15.

The slight curvature on the plotted curves is a result of the x_1^2 term appearing in equation (4-21). Since most of the data have x_1 from 50 to 62 and x_2 from 17 to 24, the curves carry the message that over these ranges, changes in x_1 seem to produce larger changes in stack loss than do changes in x_2. This conclusion is consistent with the discussion centered around Table 4-10.

━━━━━━━━━━━━━━━━━━━━━━━━━━━━━━━━ ▬ ▬ ▬

The plots of residuals used in Example 4-5 are typical of the kinds of plots made to help assess the appropriateness of surfaces fitted to multivariate data. Normal plots, plots against all x variables, plots against $\hat{y}$, plots against time order of observation, and

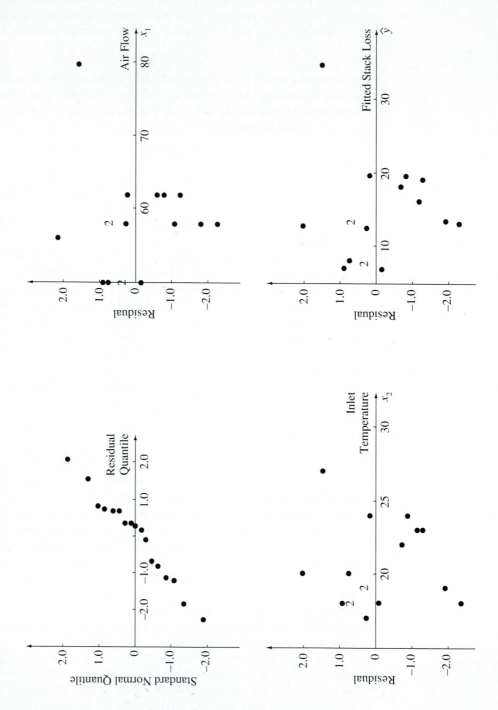

FIGURE 4-14

Plots of Residuals from a Two-Variable Equation Fit to the Stack Loss Data

FIGURE 4-15

Plots of Fitted Stack Loss from
Equation (4-21)

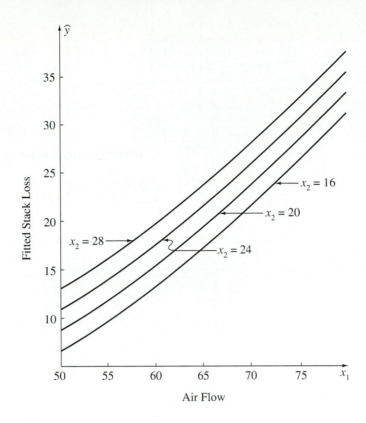

plots against other variables (like machine number or operator) not used in the fitted equation all have the potential to tell an engineer something not previously discovered about a set of data and the process that generated them.

Earlier in this section, there was a discussion of the fact that an "x term" in the equations one fits via least squares can be a known function (e.g., a logarithm) of a basic process variable. In fact, it is frequently helpful to allow an "x term" in equation (4-18) to be a known function of several basic process variables.

EXAMPLE 4-6

Lift/Drag Ratio for a Three-Surface Configuration. P. Burris studied the effects of the positioning (relative to a wing) of a canard (a forward lifting surface) and tail on the lift/drag ratio for a three-surface configuration. Part of his data are given in Table 4-11,

TABLE 4-11

Lift/Drag Ratios for
9 Canard/Tail Position
Combinations

x_1 Canard Position	x_2 Tail Position	y Lift/Drag Ratio
−1.2	−1.2	.858
−1.2	0.0	3.156
−1.2	1.2	3.644
0.0	−1.2	4.281
0.0	0.0	3.481
0.0	1.2	3.918
1.2	−1.2	4.136
1.2	0.0	3.364
1.2	1.2	4.018

PRINTOUT 4-5

```
MTB > read c1-c3
DATA> -1.2 -1.2 .858
DATA> -1.2 0.0 3.156
DATA> -1.2 1.2 3.644
DATA> 0.0 -1.2 4.281
DATA> 0.0 0.0 3.481
DATA> 0.0 1.2 3.918
DATA> 1.2 -1.2 4.136
DATA> 1.2 0.0 3.364
DATA> 1.2 1.2 4.018
DATA> end
     9 ROWS READ
```

Create the x_1x_2 Product

```
MTB > mult c1 by c2 put into c4
MTB > name c1 'x1'  c2 'x2'  c3 'y'  c4 'x1x2'
MTB > print c1 c2 c4 c3

 ROW      x1       x2      x1x2       y

   1    -1.2     -1.2      1.44     0.858
   2    -1.2      0.0      0.00     3.156
   3    -1.2      1.2     -1.44     3.644
   4     0.0     -1.2      0.00     4.281
   5     0.0      0.0      0.00     3.481
   6     0.0      1.2      0.00     3.918
   7     1.2     -1.2     -1.44     4.136
   8     1.2      0.0      0.00     3.364
   9     1.2      1.2      1.44     4.018


MTB > regress c3 3 c1 c2 c4

The regression equation is
y = 3.43 + 0.536 x1 + 0.320 x2 - 0.504 x1x2


Predictor        Coef        Stdev      t-ratio        p
Constant       3.4284   $b_0$   0.2613       13.12      0.000
x1       $b_1$ 0.5361   $b_2$   0.2667        2.01      0.101
x2             0.3201           0.2667        1.20      0.284
x1x2     $b_3$ -0.5042          0.2722       -1.85      0.123

s = 0.7839        R-sq = 64.1%     R-sq(adj) = 42.5%

Analysis of Variance

SOURCE         DF        SS          MS        F        p
Regression      3     5.4771      1.8257     2.97    0.136
Error           5     3.0724      0.6145
Total           8     8.5495

SOURCE         DF     SEQ SS
x1              1     2.4833
x2              1     0.8855
x1x2            1     2.1083
```

where

x_1 = canard placement in inches above the plane defined by the main wing

x_2 = tail placement in inches above the plane defined by the main wing

(The front-to-rear positions of the three surfaces were constant throughout the collection of these particular data.)

A straightforward least squares fitting of the equation

$$y \approx \beta_0 + \beta_1 x_1 + \beta_2 x_2$$

to these data produces R^2 of only .394. Even the addition of squared terms in both x_1 and x_2, i.e., the fitting of

$$y \approx \beta_0 + \beta_1 x_1 + \beta_2 x_2 + \beta_3 x_1^2 + \beta_4 x_2^2$$

produces an increase in R^2 to only .513. However, Printout 4-5 shows that fitting the equation

$$y \approx \beta_0 + \beta_1 x_1 + \beta_2 x_2 + \beta_3 x_1 x_2$$

yields $R^2 = .641$ and the fitted relationship

$$\hat{y} = 3.4284 + .5361 x_1 + .3201 x_2 - .5042 x_1 x_2 \tag{4-22}$$

It is instructive to ponder the implications of relationship (4-22). Figure 4-16 shows the nature of the fitted surface. Clearly, raising the canard (increasing x_1) has markedly different predicted impacts on y, depending on the value of x_2 (the tail position). (It appears that the canard and tail should not be lined up — i.e., one doesn't want x_1 near x_2.) It is the cross product term $x_1 x_2$ in relationship (4-22) that allows the response curves to have different characters for different x_2 values. Without it, the slices of the fitted $(x_1, x_2, \hat{y})$ surface would be parallel for various x_2, much like the situation in Figure 4-15.

FIGURE 4-16

Plots of Fitted Lift/Drag from Equation (4-22)

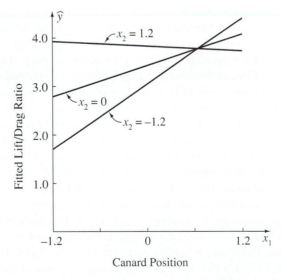

Although the main new point of this example has by now been made, it probably should be mentioned that equation (4-22) is not the last word for fitting the data of Table 4-11. Figure 4-17 gives a plot of the residuals for relationship (4-22) versus canard position x_1, and it shows a strong curvilinear pattern.

FIGURE 4-17

Plot of Residuals from
Equation (4-22) versus x_1

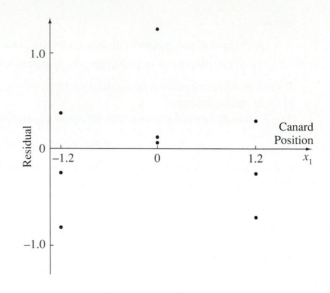

In fact, the fitted equation

$$\hat{y} = 3.9833 + .5361x_1 + .3201x_2 - .4843x_1^2 - .5042x_1x_2 \qquad \textbf{(4-23)}$$

provides $R^2 = .754$ and generally random-looking residuals. It can be verified (by plotting $\hat{y}$ versus x_1 curves for several x_2 values) that the fitted relationship (4-23) yields nonparallel parabolic slices of the fitted $(x_1, x_2, \hat{y})$ surface, instead of the nonparallel linear slices seen in Figure 4-16.

Some Additional Caveats

It should be clear by now that least squares fitting of curves and surfaces is of substantial engineering importance — but that it must be handled with care and thought. Before leaving the subject until Chapter 9, where there is an exposition of methods of formal inference associated with it, a few more warnings must be given.

First, in the same way that Section 4-1 cautioned against extrapolation in the context of fitting a line to (x, y) data, it is necessary to warn of the dangers of extrapolation substantially outside the range of the $(x_1, x_2, \ldots, x_k, y)$ data set in the context of curve or surface fitting. It is sensible to count on a fitted equation to describe the relation of y to a particular set of inputs $x_1, x_2, \ldots, x_k$ only if they are like the x values for the points used to create the equation. The new challenge surface fitting affords is that when several different x variables are involved, it is difficult (or practically impossible) to tell whether a particular $(x_1, x_2, \ldots, x_k)$ vector is in the range of the data used to fit the equation. About all one can do is check to see that it comes close to matching some *single point* in the set *on each coordinate* $x_1, x_2, \ldots, x_k$. It is not sufficient that there be some point with x_1 value near the one of interest, another point with x_2 value near the one of interest, etc. For example, having data with $1 \leq x_1 \leq 5$ and $10 \leq x_2 \leq 20$ doesn't even mean that the (x_1, x_2) pair $(3, 15)$ is necessarily like any of the pairs in the data set. This fact is illustrated in Figure 4-18 for a fictitious set of (x_1, x_2) values.

Another point that needs making (and echoes Section 4-1) is that the fitting of curves and surfaces via least squares can be strongly affected by a few outlying or extreme data points. One can try to identify such points by examining various plots and comparing fits made with and without the suspicious point(s).

FIGURE 4-18

Hypothetical Plot of (x_1, x_2) Pairs

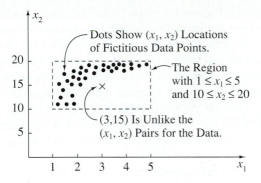

EXAMPLE 4-5
(continued)

Figure 4-14 earlier called attention to the fact that the nitrogen plant data set contains one point with an extreme x_1 value. Figure 4-19 is a scatterplot of (x_1, x_2) pairs for the data in Table 4-9. It shows that by most qualitative standards, observation 1 in Table 4-9 is unusual or outlying.

FIGURE 4-19

Plot of (x_1, x_2) Pairs for the Stack Loss Data

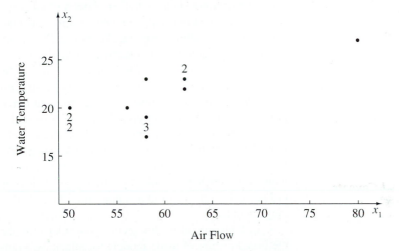

If the fitting of equation (4-21) is redone using only the last 16 data points in Table 4-9, the equation

$$\hat{y} = -56.797 + 1.404x_1 + .601x_2 - .007x_1^2 \qquad \textbf{(4-24)}$$

and $R^2 = .942$ are obtained. One could consider using equation (4-24) as a description of stack loss and limiting attention to x_1 in the range 50 to 62. But it is possible to verify that though some of the coefficients (the b's) in equations (4-21) and (4-24) differ substantially, the two equations produce comparable $\hat{y}$ values for the 16 data points with x_1 between 50 and 62. In fact the largest difference in fitted values is about .4. So, since point 1 in Table 4-9 doesn't radically change predictions made using the fitted equation, it makes sense to leave it in consideration, adopt equation (4-21), and use it to describe stack loss for (x_1, x_2) pairs interior to pattern of scatter on Figure 4-19.

A third caveat has to do with the notion of replication that was first discussed in Section 2-3. It is the fact that the phosphate/compressive strength data of Example 4-3 has several y's for each x that makes it so clear that even the quadratic and cubic curves sketched in Figures 4-9 and 4-10 are inadequate descriptions of the relationship between phosphate and strength. The fitted curves pass clearly outside the range of what look like believable values of y for some values of x. Without such replication, one can't know with real confidence what is permissible (natural) variation about a fitted curve or surface. For example, although the structure of the compressive strength data set of Example 4-3 is strong from the replication point of view, the structure of the lift/drag data set in Example 4-6 is weak from this viewpoint. There is no replication represented in Table 4-11, so one would thus have to have an external value for typical experimental precision in order to identify a fitted value as "obviously" incompatible with an observed one.

The nitrogen plant data set of Example 4-5 was presumably derived from a primarily observational study, where no conscious attempt was probably made to replicate (x_1, x_2, x_3) settings. It is interesting to note, however, that points number 4 and 5 do represent the replication of a single (x_1, x_2, x_3) combination and show a difference in observed stack loss of 1. And this makes the residuals for equation (4-21) (which range from -2.0 to 2.3) seem at least not obviously out of line.

Section 9-2 discusses more formal and precise ways of using data from studies with some replication to judge whether or not a fitted curve or surface "misses," so to speak, some observed y's too badly. For now, simply note that among replication's many virtues is the fact that it allows one to make more reliable judgments about the appropriateness of a fitted equation than are otherwise possible.

The fourth caution regarding curve and surface fitting is that the notion of equation simplicity (*parsimony*) is important for reasons in addition to simplicity of interpretation and reduced expense involved in using the equation. It is also important from the point of view of typically giving smooth interpolation and not *overfitting* a data set.

As a hypothetical example, consider the artificial, generally linear (x, y) data plotted in Figure 4-20. It would be possible to run a (wiggly) $k = 10$ version of the polynomial (4-12) through each of these data points. But in most physical problems, such a curve would do a much worse job of predicting y at values of x not represented by a data point in Figure 4-20 than would a simple fitted line. A 10th-order polynomial would overfit the data in hand.

As a final point in this section, consider how the methods discussed here fit into the broad picture of using models for attacking engineering problems. It must be said that physical theories of physics, chemistry, materials, etc. rarely produce equations of the forms (4-12) or (4-18). Sometimes one can rewrite pertinent equations from those theories in such forms, as was possible with Taylor's equation for tool life earlier in this section. But usually one cannot, and the primary usefulness of the present technique of

FIGURE 4-20

Scatterplot of 11 Pairs (x, y)

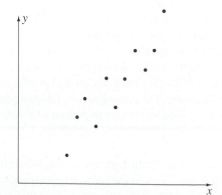

least squares fitting of curves and surfaces is not in the fitting of theoretically generated equations. Instead, the majority of engineering applications of the methods of this section are to the large number of engineering problems where no commonly known physical theory is a description of the real problem faced, and an *empirical* solution must be found. In such cases, the tool of least squares fitting of curves and surfaces can function as a kind of "mathematical French curve," allowing an engineer to develop approximate empirical descriptions of how a response y is related to system inputs $x_1, x_2, \ldots, x_k$.

4-3 Fitted Effects for Factorial Data Structures

In the previous two sections, the discussion has centered on the least squares fitting of equations to data sets where a quantitative response y is presumed to depend on the levels $x_1, x_2, \ldots, x_k$ of quantitative factors. In many engineering applications, at least some of the system "knobs" whose effects must be assessed are basically qualitative rather than quantitative. When a data set has complete factorial structure (review the meaning of this terminology in Section 1-2), it is still possible to describe it in terms of an equation. This equation involves so-called fitted factorial effects. Sometimes, when a few of these fitted effects dominate the rest, a parsimonious version of this equation can adequately describe the data and have intuitively appealing and understandable interpretations. The use of simple plots and the examination of residuals will be discussed in this section, as tools helpful in assessing whether such a simple structure holds.

The discussion begins with the 2-factor case, then considers three (or, by analogy, more) factors. Finally, the special case where each factor has only two levels is discussed. In such 2^p contexts, one can calculate fitted effects efficiently by hand using the Yates algorithm and often generate fitted combination means using the reverse Yates algorithm.

Fitted Effects for 2-Factor Studies

Example 3-9 illustrated how illuminating a plot of sample means versus levels of one of the factors can be in a 2-factor study. Such calculating and plotting is always the place to begin in understanding the story carried by a two-way factorial data set. But in addition, it is helpful to calculate the factor level averages of the sample means and the grand average of the sample means. For factor A having I levels and factor B having J levels, the following notation will be used in this text:

$\bar{y}_{ij}$ = the sample mean of the observed system responses when factor A is at level i and factor B is at level j

$$\bar{y}_{i.} = \frac{1}{J} \sum_{j=1}^{J} \bar{y}_{ij}$$

= the average sample mean when factor A is at level i

$$\bar{y}_{.j} = \frac{1}{I} \sum_{i=1}^{I} \bar{y}_{ij}$$

= the average sample mean when factor B is at level j

$$\bar{y}_{..} = \frac{1}{IJ} \sum_{i,j} \bar{y}_{ij}$$

= the grand average sample mean

These average sample means will prove to be useful descriptive statistics. They can be thought of as row and column averages when one thinks of the $\bar{y}_{ij}$ as laid out in a two-dimensional format, as shown in Figure 4-21.

FIGURE 4-21

Cell Sample Means and Row, Column, and Grand Average Sample Means for a Two-Way Factorial

		Factor B				
		Level 1	Level 2		Level J	
	Level 1	$\bar{y}_{11}$	$\bar{y}_{12}$	• • •	$\bar{y}_{1J}$	$\bar{y}_{1.}$
	Level 2	$\bar{y}_{21}$	$\bar{y}_{22}$	• • •	$\bar{y}_{2J}$	$\bar{y}_{2.}$
Factor A		⋮	⋮		⋮	⋮
	Level I	$\bar{y}_{I1}$	$\bar{y}_{I2}$	• • •	$\bar{y}_{IJ}$	$\bar{y}_{I.}$
		$\bar{y}_{.1}$	$\bar{y}_{.2}$	• • •	$\bar{y}_{.J}$	$\bar{y}_{..}$

EXAMPLE 4-7

Joint Strengths for Three Different Joint Types in Three Different Woods. Kotlers, MacFarland, and Tomlinson studied the tensile strength of three different types of joints made on three different types of wood. Butt, lap, and beveled joints were made in nominal $1'' \times 4'' \times 12''$ pine, oak and walnut specimens using a resin glue. The original intention was to test two specimens of each Joint Type/Wood Type combination. But one operator error and one specimen failure not related to its joint removed two of the original data points from consideration and gave the data in Table 4-12. These data have 3×3 factorial structure. Collecting y's for the nine different combinations into separate samples and calculating means, the $\bar{y}_{ij}$'s are as presented in tabular form in Table 4-13 and plotted in Figure 4-22.

TABLE 4-12

Measured Strengths of 16 Wood Joints

Specimen	Joint	Wood	y Stress at Failure (psi)
1	beveled	oak	1518
2	butt	pine	829
3	beveled	walnut	2571
4	butt	oak	1169
5	beveled	oak	1927
6	beveled	pine	1348
7	lap	walnut	1489
8	beveled	walnut	2443
9	butt	walnut	1263
10	lap	oak	1295
11	lap	oak	1561
12	lap	pine	1000
13	butt	pine	596
14	lap	pine	859
15	butt	walnut	1029
16	beveled	pine	1207

TABLE 4-13

Sample Means for Nine
Wood/Joint Combinations

		WOOD			
		1 (Pine)	2 (Oak)	3 (Walnut)	
	1 (Butt)	$\bar{y}_{11} = 712.5$	$\bar{y}_{12} = 1169.0$	$\bar{y}_{13} = 1146.0$	$\bar{y}_{1.} = 1009.17$
JOINT	2 (Beveled)	$\bar{y}_{21} = 1277.5$	$\bar{y}_{22} = 1722.5$	$\bar{y}_{23} = 2507.0$	$\bar{y}_{2.} = 1835.67$
	3 (Lap)	$\bar{y}_{31} = 929.5$	$\bar{y}_{32} = 1428.0$	$\bar{y}_{33} = 1489.0$	$\bar{y}_{3.} = 1282.17$
		$\bar{y}_{.1} = 973.17$	$\bar{y}_{.2} = 1439.83$	$\bar{y}_{.3} = 1714.00$	$\bar{y}_{..} = 1375.67$

FIGURE 4-22

Plot of Joint Strength Sample
Means versus Wood Type for
Three Joint Types

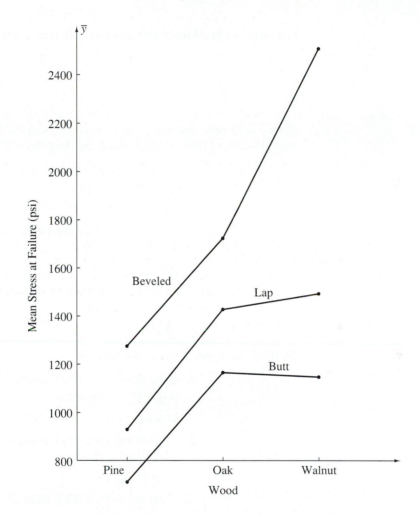

The qualitative messages given by the plot are as follows.

1. Joint types ordered by strength are "Beveled is stronger than lap, which in turn is stronger than butt."

2. Woods ordered by overall strength seem to be "Walnut is stronger than oak, which in turn is stronger than pine."

3. The strength pattern across woods is not consistent from joint type to joint type (or equivalently, the strength pattern across joints is not consistent from wood type to wood type).

The idea of computing *fitted effects* is to invent a way of quantifying such qualitative summaries.

One might take the row and column average means ($\bar{y}_{i\cdot}$'s and $\bar{y}_{\cdot j}$'s, respectively) as measures of average response behavior at different levels of the factors in question. If so, it then makes sense to use the differences between these and the grand average mean $\bar{y}_{\cdot\cdot}$ as measures of the effects of those levels on mean response. This leads to Definition 4-5.

DEFINITION 4-5

In a two-way complete factorial study with factors A and B, the **fitted main effect of factor A at its *i*th level** is

$$a_i = \bar{y}_{i\cdot} - \bar{y}_{\cdot\cdot}$$

Similarly, the **fitted main effect of factor B at its *j*th level** is

$$b_j = \bar{y}_{\cdot j} - \bar{y}_{\cdot\cdot}$$

EXAMPLE 4-7
(continued)

Simple arithmetic and the $\bar{y}$'s in Table 4-13 yield the fitted main effects for the joint strength study of Kotlers, MacFarland, and Tomlinson. First for factor A (the Joint Type),

$$a_1 = \text{the Joint Type fitted main effect for butt joints}$$
$$= 1009.17 - 1375.67$$
$$= -366.5 \text{ psi}$$
$$a_2 = \text{the Joint Type fitted main effect for beveled joints}$$
$$= 1835.67 - 1375.67$$
$$= 460.0 \text{ psi}$$
$$a_3 = \text{the Joint Type fitted main effect for lap joints}$$
$$= 1282.17 - 1375.67$$
$$= -93.5 \text{ psi}$$

Similarly for factor B (the Wood Type),

$$b_1 = \text{the Wood Type fitted main effect for pine}$$
$$= 973.17 - 1375.67$$
$$= -402.5 \text{ psi}$$
$$b_2 = \text{the Wood Type fitted main effect for oak}$$
$$= 1439.83 - 1375.67$$
$$= 64.17 \text{ psi}$$
$$b_3 = \text{the Wood Type fitted main effect for walnut}$$
$$= 1714.00 - 1375.67$$
$$= 338.33 \text{ psi}$$

These fitted main effects quantify the first two qualitative messages carried by the data and listed as 1) and 2) before Definition 4-5. For example,

$$a_2 > a_3 > a_1$$

says that beveled joints are strongest and butt joints the weakest. Further, the fact that the a_i's and b_j's are of roughly the same order of magnitude says that the Joint Type and Wood Type factors are of comparable importance in determining tensile strength.

A difference between fitted main effects for a factor amounts to a difference between corresponding row or column averages and quantifies how different response behavior is for those two levels.

EXAMPLE 4-7
(continued)

For example, comparing pine and oak wood types,

$$b_1 - b_2 = (\bar{y}_{.1} - \bar{y}_{..}) - (\bar{y}_{.2} - \bar{y}_{..})$$
$$= \bar{y}_{.1} - \bar{y}_{.2}$$
$$= 973.17 - 1439.83$$
$$= -466.67 \text{ psi}$$

which indicates that pine joint average strength is about 467 psi less than oak joint average strength.

In some two-factor factorial studies, the fitted main effects as defined in Definition 4-5 pretty much summarize the story told by the combination means $\bar{y}_{ij}$, in the sense that

$$\bar{y}_{ij} \approx \bar{y}_{..} + a_i + b_j \qquad \text{for every } i \text{ and } j \qquad \textbf{(4-25)}$$

Display (4-25) implies, for example, that the pattern of mean responses for level 1 of factor A is the same as for level 2 of A. That is, changing levels of factor B (from say j to j') produces the same change in mean response for level 2 as for level 1 (namely, $b_{j'} - b_j$). In fact, if relation (4-25) holds, there are *parallel traces* of means as one moves across levels of one factor on every level of the other factor.

EXAMPLE 4-7
(continued)

For purposes of illustrating the meaning of expression (4-25), the fitted effects for the Joint Type/Wood Type data have been used to calculate $3 \times 3 = 9$ values of $\bar{y}_{..} + a_i + b_j$ corresponding to the nine experimental combinations. These are given in Table 4-14.

TABLE 4-14

Values of $\bar{y}_{..} + a_i + b_j$ for the Joint Strength Study

		WOOD		
		1 (Pine)	2 (Oak)	3 (Walnut)
	1 (Butt)	$\bar{y}_{..} + a_1 + b_1 =$ 606.67	$\bar{y}_{..} + a_1 + b_2 =$ 1073.33	$\bar{y}_{..} + a_1 + b_3 =$ 1347.50
JOINT	2 (Beveled)	$\bar{y}_{..} + a_2 + b_1 =$ 1433.17	$\bar{y}_{..} + a_2 + b_2 =$ 1899.83	$\bar{y}_{..} + a_2 + b_3 =$ 2174.00
	3 (Lap)	$\bar{y}_{..} + a_3 + b_1 =$ 879.67	$\bar{y}_{..} + a_3 + b_2 =$ 1346.33	$\bar{y}_{..} + a_3 + b_3 =$ 1620.50

For comparison purposes, the $\bar{y}_{ij}$ from Table 4-13 and the $\bar{y}_{..} + a_i + b_j$ from Table 4-14 are plotted on the same sets of axes in Figure 4-23. Notice the parallel traces for the $\bar{y}_{..} + a_i + b_j$ values for the three different joint types. The traces for the $\bar{y}_{ij}$ values for the three different joint types lack parallel character (particularly when walnut is considered), so there are apparently substantial differences between the $\bar{y}_{ij}$'s and the $\bar{y}_{..} + a_i + b_j$'s.

FIGURE 4-23

Plots of $\bar{y}_{ij}$ and $\bar{y}_{..} + a_i + b_j$ versus Wood Type for Three Joint Types

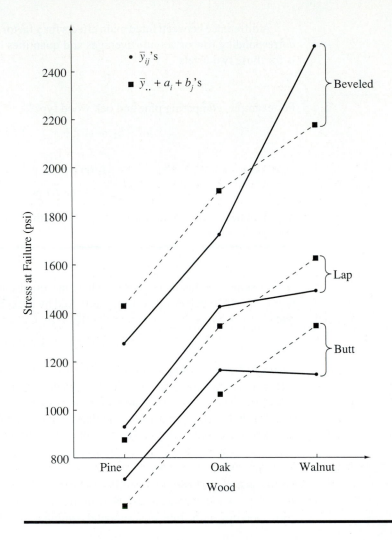

When relationship (4-25) fails to hold, the patterns in mean response across levels of one factor depend on the levels of the second factor. In such cases, the differences between the combination means $\bar{y}_{ij}$ and the values $\bar{y}_{..} + a_i + b_j$ can serve as useful measures of lack of parallelism on the plots of means, and this leads to another definition.

DEFINITION 4-6

In a two-way complete factorial study with factors A and B, the **fitted interaction of factor A at its *i*th level and factor B at its *j*th level** is

$$ab_{ij} = \bar{y}_{ij} - (\bar{y}_{..} + a_i + b_j)$$

The fitted interactions in some sense measure how much pattern the combination means $\bar{y}_{ij}$ carry that is not explainable in terms of the factors A and B separately, but instead must be attributed to their joint effects. Clearly, when relationship (4-25) holds, the fitted interactions ab_{ij} are all small (nearly 0), and system behavior can be thought of as depending separately on level of A and level of B. In such cases, an important practical consequence is that it is possible to develop and make recommendations for levels of the two factors independently of each other. For example, one need not recommend one level of A if B is at its level 1 and another if B is at its level 2.

Consider a study of the effects of factors Tool Type and Turning Speed on the metal removal rate for the use of a lathe. If the fitted interactions are small, one can make turning speed recommendations that remain valid for all tool types. However, if the fitted interactions are important, turning speed recommendations might vary according to tool type.

EXAMPLE 4-7
(continued)

Turning once again to the Joint Type/Wood Type data, consider calculating the fitted interactions. The raw material for these calculations already exists in Tables 4-13 and 4-14. Simply taking differences between entries in these tables cell by cell yields the fitted interactions given in Table 4-15.

TABLE 4-15

Fitted Interactions for the Joint Strength Study

		WOOD		
		1 (Pine)	2 (Oak)	3 (Walnut)
	1 (Butt)	$ab_{11} = 105.83$	$ab_{12} = 95.67$	$ab_{13} = -201.5$
JOINT	2 (Beveled)	$ab_{21} = -155.66$	$ab_{22} = -177.33$	$ab_{23} = 333.0$
	3 (Lap)	$ab_{31} = 49.83$	$ab_{32} = 81.67$	$ab_{33} = -131.5$

It is interesting to compare these fitted interactions to themselves and to the fitted main effects. The largest (in absolute value) fitted interaction corresponds to beveled walnut joints. This is consistent with the visual impression of Figures 4-22 and 4-23: that this Joint Type/Wood Type combination is in some sense most responsible for destroying any nearly parallel structure that might otherwise appear. The fact that (on the whole) the ab_{ij}'s are not as large as the a_i's or b_j's is consistent with the visual impression of Figures 4-23 and 4-24: that the lack of parallelism, while important, is not as important as differences in joint types or wood types.

Simpler Descriptions for Some Two-Way Data Sets

Simply rewriting the equation for ab_{ij} from Definition 4-6, one has

$$\bar{y}_{ij} = \bar{y}_{..} + a_i + b_j + ab_{ij} \qquad \text{(4-26)}$$

That is, $\bar{y}_{..}$, the fitted main effects, and the fitted interactions provide a decomposition or breakdown of the combination sample means into interpretable pieces. These pieces correspond to an overall effect, the effects of factors acting separately, and the effects of factors acting jointly.

Taking a hint from the equation fitting done in the previous two sections, it makes sense to think of (4-26) as a fitted version of an approximate relationship

$$y \approx \mu + \alpha_i + \beta_j + \alpha\beta_{ij} \qquad \text{(4-27)}$$

where μ, α_1, α_2, ..., α_I, β_1, β_2, ..., β_J, $\alpha\beta_{11}$, ..., $\alpha\beta_{1J}$, $\alpha\beta_{21}$, ..., $\alpha\beta_{IJ}$ are some constants and the levels of factors A and B associated with a particular response y pick out which of the α_i's, β_j's and $\alpha\beta_{ij}$'s are appropriate in equation (4-27). By analogy with the previous two sections, consideration should be given to the possibility that a relationship even simpler than equation (4-27) might hold, perhaps not involving $\alpha\beta_{ij}$'s or even α_i's or perhaps β_j's.

It has already been said that when relationship (4-25) is in force, or equivalently

$$ab_{ij} \approx 0 \qquad \text{for every } i \text{ and } j$$

it is possible to understand an observed set of $\bar{y}_{ij}$'s in simplified terms of the factors acting separately. This possibility corresponds to the simplified version of equation (4-27),

$$y \approx \mu + \alpha_i + \beta_j$$

and there are other simplified versions of equation (4-27) that also have appealing interpretations. For example, the simplified version of equation (4-27),

$$y \approx \mu + \alpha_i$$

says that only factor A (not factor B) is important in determining response y. ($\alpha_1, \alpha_2, \ldots, \alpha_I$ still allow for different response behavior for different levels of A.)

Two questions naturally follow on this kind of reasoning. "How does one fit a *reduced* or *simplified* version of equation (4-27) to a data set? And after fitting such an equation, how does one assess the appropriateness of the result?" General answers to these questions are subtle and will be delayed until Section 9-4. But there is one kind of circumstance in which it is possible to give fairly straightforward answers. That is the case where the data are *balanced* — in the sense that all of the samples (leading to the $\bar{y}_{ij}$'s) have the same size. It turns out that when working with balanced data, one can use the fitted effects from Definitions 4-5 and 4-6 to produce fitted responses. And based on such fitted values, the R^2 and residual plotting ideas from the last two sections can be applied here as well.

That is, in general the fitting of a simplified version of equation (4-27) to factorial data via least squares requires the use of a regression program and *dummy variables*. However, when working with balanced data, least squares fitting of such a relationship can also be accomplished by first calculating fitted effects according to Definitions 4-5 and 4-6. Then one adds only those corresponding to terms appearing in the reduced version of equation (4-27) to compute fitted responses, $\hat{y}$. Residuals are then (as before)

$$e = y - \hat{y}$$

(and should look like noise if the simplified equation is an adequate description of the data set). Further, the fraction of raw variation in y accounted for in the fitting process is

$$R^2 = \frac{\sum(y - \bar{y})^2 - \sum(y - \hat{y})^2}{\sum(y - \bar{y})^2} \tag{4-28}$$

where the sums are over all observed y's. (Here, summation notation is being abused even further than usual, by not even subscripting the y's and $\hat{y}$'s.)

EXAMPLE 4-8

(Example 2-13 revisited) **Simplified Description of Two-Way Factorial Golf Ball Flight Data.** G. Gronberg tested drive flight distances for golf balls of several different compressions on several different evenings. Table 4-16 gives a small part of the data that Gronberg collected, representing 80 and 100 compression flight distances (in yards) from two different evenings. Notice that these data are balanced, all four sample sizes being 10.

As discussed in Section 2-4, these data have two-way factorial structure. The factor Evening is not really of primary interest. Rather, it is a blocking factor, its levels creating homogeneous environments in which to compare 80 and 100 compression flight distances. Figure 4-24 gives a graphic using box plots to represent the four samples and emphasizing the factorial structure.

It is straightforward to calculate sample means corresponding to the four cells in Table 4-16 and then find fitted effects. Table 4-17 displays cell, row, column, and grand average means. And based on those values,

$$a_1 = 189.85 - 184.20 = 5.65 \text{ yards}$$
$$a_2 = 178.55 - 184.20 = -5.65 \text{ yards}$$
$$b_1 = 183.00 - 184.20 = -1.20 \text{ yards}$$
$$b_2 = 185.40 - 184.20 = 1.20 \text{ yards}$$

$$ab_{11} = 188.1 - (184.20 + 5.65 + (-1.20)) = -.55 \text{ yards}$$
$$ab_{12} = 191.6 - (184.20 + 5.65 + 1.20) = .55 \text{ yards}$$
$$ab_{21} = 177.9 - (184.20 + (-5.65) + (-1.20)) = .55 \text{ yards}$$
$$ab_{22} = 179.2 - (184.20 + (-5.65) + 1.20) = -.55 \text{ yards}$$

TABLE 4-16

Golf Ball Flight Distances for Four Compression/Evening Combinations

	(FACTOR B) EVENING			
	1		2	
	180	192	196	180
	193	190	192	195
80	197	182	191	197
	189	192	194	192
	187	179	186	193
	180	175	190	185
	185	190	195	167
100	167	185	180	180
	162	180	170	180
	170	185	180	165

(FACTOR A) COMPRESSION

FIGURE 4-24

Golf Ball Flight Distance Boxplots for Four Combinations of Compression and Evening

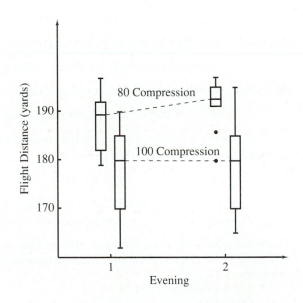

TABLE 4-17

Cell, Row, Column, and Grand Average Means for the Golf Ball Flight Data

		EVENING		
		1	2	
	80	$\bar{y}_{11} = 188.1$	$\bar{y}_{12} = 191.6$	189.85
COMPRESSION	100	$\bar{y}_{21} = 177.9$	$\bar{y}_{22} = 179.2$	178.55
		183.00	185.40	184.20

The fitted effects indicate that most of the differences in the cell means in Table 4-17 are understandable in terms of differences between 80 and 100 compression balls. The effect of differences between evenings appears to be on the order of one-fourth the size of the effect of differences between ball compressions. Further, the pattern of flight distance across the two compressions changed relatively little from evening to evening. These facts are portrayed graphically in Figure 4-25.

FIGURE 4-25

Plots of Sample Means versus Evening for the Golf Ball Flight Data

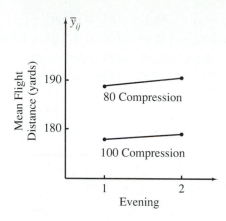

The story told by the fitted effects in this example probably agrees with most readers' intuition. There is little reason *a priori* to expect the relative behaviors of 80 and 100 compression flight distances to change much from evening to evening, but there is slightly more reason to expect the distances to be longer overall on some nights than on others.

It is worth investigating whether the data in Table 4-16 allow the simplest

"Compression effects only"

description, require the somewhat more complicated

"Compression effects and Evening effects but no interactions"

description, or really demand that they be described in terms of

"Compression, Evening and interaction effects"

To do so, one first calculates fitted responses corresponding to the three different possible relationships

$$y \approx \mu + \alpha_i \tag{4-29}$$

$$y \approx \mu + \alpha_i + \beta_j \tag{4-30}$$

$$y \approx \mu + \alpha_i + \beta_j + \alpha\beta_{ij} \tag{4-31}$$

These are generated using the fitted effects; they are collected in Table 4-18. (Not surprisingly, the first and third sets of fitted responses are, respectively, row average and cell means.)

TABLE 4-18

Fitted Responses Corresponding to Equations (4-29), (4-30), and (4-31)

Compression	Evening	For (4-29) $\bar{y}_{..} + a_i = \bar{y}_{i.}$	For (4-30) $\bar{y}_{..} + a_i + b_j$	For (4-31) $\bar{y}_{..} + a_i + b_j + ab_{ij} = \bar{y}_{ij}$
80	1	189.85	188.65	188.10
100	1	178.55	177.35	177.90
80	2	189.85	191.05	191.60
100	2	178.55	179.75	179.20

Residuals $e = y - \hat{y}$ for fitting the three equations (4-29), (4-30), and (4-31) are obtained by subtracting the appropriate entries in, respectively, the 3rd, 4th, or 5th column of Table 4-18 from each of the data values listed in Table 4-16. For example, 40 residuals for the fitting of the "A main effects only" equation (4-29) would be obtained by subtracting 189.85 from every entry in the upper left cell of Table 4-16, subtracting 178.55 from every entry in the lower left cell, 189.85 from every entry in the upper right cell and 178.55 from every entry in the lower right cell.

Figure 4-26 shows normal plots of the residuals from the fitting of the three equations (4-29), (4-30), and (4-31). None of the residual plots is especially linear, but at the same time, none of them is grossly nonlinear either. In particular, the first two, corresponding to simplified versions of relationship (4-27), are not markedly worse than the last one, which corresponds to the use of all fitted effects (both main effects and interactions). From the limited viewpoint of producing residuals with an approximately bell-shaped distribution, the fitting of any of the three equations (4-29), (4-30), and (4-31) would appear approximately equally effective. Of course, as always, the sets of residuals could be plotted in other ways to further investigate the appropriateness of the fitted equations.

The calculation of R^2 values for equations (4-29), (4-30), and (4-31) proceeds as follows. First, since the grand average of all 40 flight distances is $\bar{y} = 184.2$ yards (which in this case also turns out to be $\bar{y}_{..}$),

$$
\begin{aligned}
\sum (y - \bar{y})^2 &= (180 - 184.2)^2 + \cdots + (179 - 184.2)^2 \\
&\quad + (180 - 184.2)^2 + \cdots + (185 - 184.2)^2 \\
&\quad + (196 - 184.2)^2 + \cdots + (193 - 184.2)^2 \\
&\quad + (190 - 184.2)^2 + \cdots + (165 - 184.2)^2 \\
&= 3{,}492.4
\end{aligned}
$$

(This value can easily be obtained on a pocket calculator by using 39 times the sample variance of all 40 flight distances.) Then $\sum (y - \hat{y})^2$ values for the three equations are obtained as the sums of the squared residuals. For example, using Tables 4-16 and 4-18, for equation (4-30),

$$
\begin{aligned}
\sum (y - \hat{y})^2 &= (180 - 188.65)^2 + \cdots + (179 - 188.65)^2 \\
&\quad + (180 - 177.35)^2 + \cdots + (185 - 177.35)^2 \\
&\quad + (196 - 191.05)^2 + \cdots + (193 - 191.05)^2 \\
&\quad + (190 - 179.75)^2 + \cdots + (165 - 179.75)^2 \\
&= 2{,}157.90
\end{aligned}
$$

Finally, equation (4-28) is used. Table 4-19 gives the three values of R^2.

TABLE 4-19

R^2 Values for Fitting Equations (4-29), (4-30), and (4-31) to Gronberg's Data

Equation	R^2
$y \approx \mu + \alpha_i$	.366
$y \approx \mu + \alpha_i + \beta_j$	.382
$y \approx \mu + \alpha_i + \beta_j + \alpha\beta_{ij}$	.386

The story told by the R^2 values is consistent with everything else that's been said in this (by now rather extended) example. None of the values is terribly big, which is consistent with the large within-sample variation in flight distances evident in Figure 4-24. But considering A (Compression) main effects does account for some of

FIGURE 4-26

Normal Plots of Residuals from
Three Different Equations Fitted
to the Golf Data

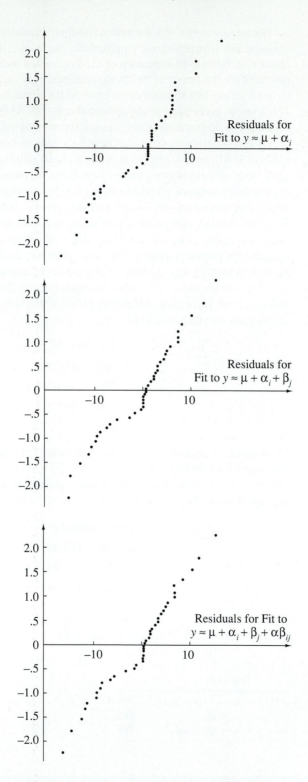

the observed variation in flight distance, and the addition of B (Evening) main effects adds slightly to the variation accounted for. Including interactions into consideration provides essentially negligible additional accounting power.

━━━━ ▬▬▬

Fitted Effects for Three-Way (and Higher) Factorials

The kind of reasoning that has been applied here to the descriptive analysis of two-way factorial data has its natural generalization to complete factorial data structures that are three-way and higher. First, fitted main effects and various kinds of interactions are defined. Then one hopes to discover that a given factorial data set can be adequately described in terms of a few of these that are interpretable when taken as a group and carry the main message in the data. This subsection shows how this program is carried out for 3-factor situations. Once the pattern has been made clear, the reader can carry it out for situations involving more than three factors, working by analogy.

In order to deal with three-way factorial data, yet more notation must be introduced. Unfortunately, this will involve triple subscripts. For factor A having I levels, factor B having J levels, and factor C having K levels, the following notation will be used:

$\bar{y}_{ijk}$ = the sample mean of observed system responses when factor A is at level i, factor B is at level j, and factor C is at level k

$$\bar{y}_{...} = \frac{1}{IJK} \sum_{i,j,k} \bar{y}_{ijk}$$

= the grand average sample mean

$$\bar{y}_{i..} = \frac{1}{JK} \sum_{j,k} \bar{y}_{ijk}$$

= the average sample mean when factor A is at level i

$$\bar{y}_{.j.} = \frac{1}{IK} \sum_{i,k} \bar{y}_{ijk}$$

= the average sample mean when factor B is at level j

$$\bar{y}_{..k} = \frac{1}{IJ} \sum_{i,j} \bar{y}_{ijk}$$

= the average sample mean when factor C is at level k

$$\bar{y}_{ij.} = \frac{1}{K} \sum_{k} \bar{y}_{ijk}$$

= the average sample mean when factor A is at level i and factor B is at level j

$$\bar{y}_{i.k} = \frac{1}{J} \sum_{j} \bar{y}_{ijk}$$

= the average sample mean when factor A is at level i and factor C is at level k

$$\bar{y}_{.jk} = \frac{1}{I} \sum_{i} \bar{y}_{ijk}$$

= the average sample mean when factor B is at level j and factor C is at level k

In these expressions, where a subscript is used as an index of summation, the summation is assumed to extend over all of its I, J, or K possible values.

If one considers the means from a 3-factor study laid out in a geometrical array, it is most natural to think of them laid out in three dimensions. (The rows and columns of a two-dimensional table aren't adequate where three factors are involved.) Figure 4-27 illustrates this situation in general, and the next example will illustrate the use of another three-dimensional display in a 2^3 context.

FIGURE 4-27

IJK Cells in a Three-Dimensional Table

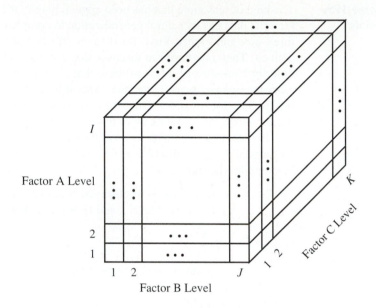

EXAMPLE 4-9

A 2^3 Factorial Experiment on the Strength of a Composite Material. In his article "Application of Two-Cubed Factorial Designs to Process Studies" (ASQC Technical Supplement *Experiments in Industry*, 1985), G. Kinzer discusses two successful 3-factor industrial experiments. In one of them, the strength of a proprietary composite material was thought to be related to three process variables, as indicated in Table 4-20.

TABLE 4-20

Levels of Three Process Variables in a 2^3 Study of Material Strength

Factor	Process Variable	Level 1	Level 2
A	Autoclave Temperature	300°F	330°F
B	Autoclave Time	4 hr	12 hr
C	Time Span (between product formation and autoclaving)	4 hr	12 hr

Five specimens were produced under each of the $2^3 = 8$ combinations of factor levels, and their moduli of rupture were measured (in psi) and averaged to produce the means in Table 4-21. (There were also apparently 10 specimens made with an autoclave temperature of 315°, an autoclave time of 8 hr and a time span of 8 hr, but this point will be ignored for present purposes.)

A helpful display of these means can be made using the corners of a cube as in Figure 4-28. Using this three-dimensional picture, one can think of the average sample means defined before the beginning of this example as averages of $\bar{y}_{ijk}$'s sharing a face or edge of the cube. For example,

$$\bar{y}_{1..} = \frac{1}{2 \cdot 2}(1520 + 2340 + 1670 + 2230) = 1940 \text{ psi}$$

TABLE 4-21

Sample Mean Strengths for 2^3 Treatment Combinations

i Factor A Level	j Factor B Level	k Factor C Level	$\bar{y}_{ijk}$ Sample Mean Strength (psi)
1	1	1	1520
2	1	1	2450
1	2	1	2340
2	2	1	2900
1	1	2	1670
2	1	2	2540
1	2	2	2230
2	2	2	3230

FIGURE 4-28

2^3 Sample Mean Strengths Displayed on the Corners of a Cube

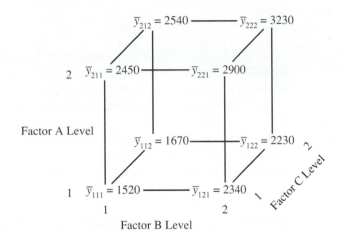

is the average mean on the bottom face, while

$$\bar{y}_{11.} = \frac{1}{2}(1520 + 1670) = 1595 \text{ psi}$$

is the average mean on the lower left edge.

For future reference, all of the average sample means are collected here.

$$\bar{y}_{...} = 2360 \text{ psi} \qquad \bar{y}_{1..} = 1940 \text{ psi}$$
$$\bar{y}_{2..} = 2780 \text{ psi}$$

$$\bar{y}_{.1.} = 2045 \text{ psi} \qquad \bar{y}_{..1} = 2302.5 \text{ psi}$$
$$\bar{y}_{.2.} = 2675 \text{ psi} \qquad \bar{y}_{..2} = 2417.5 \text{ psi}$$

$$\bar{y}_{11.} = 1595 \text{ psi} \qquad \bar{y}_{12.} = 2285 \text{ psi}$$
$$\bar{y}_{21.} = 2495 \text{ psi} \qquad \bar{y}_{22.} = 3065 \text{ psi}$$

$$\bar{y}_{1.1} = 1930 \text{ psi} \qquad \bar{y}_{1.2} = 1950 \text{ psi}$$
$$\bar{y}_{2.1} = 2675 \text{ psi} \qquad \bar{y}_{2.2} = 2885 \text{ psi}$$

$$\bar{y}_{.11} = 1985 \text{ psi} \qquad \bar{y}_{.12} = 2105 \text{ psi}$$
$$\bar{y}_{.21} = 2620 \text{ psi} \qquad \bar{y}_{.22} = 2730 \text{ psi}$$

Analogy with Definition 4-5 provides definitions of fitted main effects in a 3-factor study as the differences between factor-level average means and the grand average mean.

DEFINITION 4-7

In a three-way complete factorial study with factors A, B, and C, the **fitted main effect of factor A at its ith level** is

$$a_i = \bar{y}_{i..} - \bar{y}_{...}$$

The **fitted main effect of factor B at its jth level** is

$$b_j = \bar{y}_{.j.} - \bar{y}_{...}$$

And the **fitted main effect of factor C at its kth level** is

$$c_k = \bar{y}_{..k} - \bar{y}_{...}$$

Using the geometrical representation of factor-level combinations given in Figure 4-27, these fitted effects are averages of $\bar{y}_{ijk}$'s along planes (parallel to one set of faces of the rectangular solid) minus the grand average sample mean.

Next, analogy with Definition 4-6 produces definitions of fitted two-way interactions in a 3-factor study.

DEFINITION 4-8

In a three-way complete factorial study with factors A, B, and C, the **fitted 2-factor interaction of factor A at its ith level** and **factor B at its jth level** is

$$ab_{ij} = \bar{y}_{ij.} - (\bar{y}_{...} + a_i + b_j)$$

the **fitted 2-factor interaction of factor A at its ith level** and **factor C its kth level** is

$$ac_{ik} = \bar{y}_{i.k} - (\bar{y}_{...} + a_i + c_k)$$

and the **fitted 2-factor interaction of factor B at its jth level** and **factor C at its kth level** is

$$bc_{jk} = \bar{y}_{.jk} - (\bar{y}_{...} + b_j + c_k)$$

It may or may not be obvious at this point, but these fitted 2-factor interactions can be thought of in two equivalent ways. They are

> **1.** what one gets as fitted interactions upon averaging across all levels of the factor that is not under consideration to obtain a single two-way table of (average) means and then calculating as per Definition 4-6
> **2.** what one gets as averages, across all levels of the factor not under consideration, of the fitted 2-factor interactions calculated as per Definition 4-6, one level of the excluded factor at a time.

EXAMPLE 4-9
(continued)

To illustrate the meaning of Definitions 4-7 and 4-8, return to the composite material strength study. For example, the fitted A main effects are

$$a_1 = \bar{y}_{1..} - \bar{y}_{...} = 1940 - 2360 = -420 \text{ psi}$$
$$a_2 = \bar{y}_{2..} - \bar{y}_{...} = 2780 - 2360 = 420 \text{ psi}$$

And the fitted AB 2-factor interaction for levels 1 of A and 1 of B is

$$ab_{11} = \bar{y}_{11.} - (\bar{y}_{...} + a_1 + b_1) = 1595 - (2360 + (-420) + (2045 - 2360))$$
$$= -30 \text{ psi}$$

Verify that the entire set of fitted effects for the means of Table 4-21 is as follows.

$a_1 = -420$ psi	$b_1 = -315$ psi	$c_1 = -57.5$ psi
$a_2 = 420$ psi	$b_2 = 315$ psi	$c_2 = 57.5$ psi
$ab_{11} = -30$ psi	$ac_{11} = 47.5$ psi	$bc_{11} = -2.5$ psi
$ab_{12} = 30$ psi	$ac_{12} = -47.5$ psi	$bc_{12} = 2.5$ psi
$ab_{21} = 30$ psi	$ac_{21} = -47.5$ psi	$bc_{21} = 2.5$ psi
$ab_{22} = -30$ psi	$ac_{22} = 47.5$ psi	$bc_{22} = -2.5$ psi

Remember equation (4-26) to see that in 2-factor studies, the fitted grand mean, main effects, and 2-factor interactions completely describe a factorial set of sample means. Such is not the case in 3-factor studies. Instead, a new possibility arises: *3-factor interaction*. Roughly speaking, the fitted 3-factor interactions in a 3-factor study measure how much pattern the combination means carry that is not explainable in terms of the factors A, B, and C acting separately and in pairs. A precise definition is given next.

DEFINITION 4-9

In a three-way complete factorial study with factors A, B, and C, **the fitted 3-factor interaction of A at its *i*th level, B at its *j*th level, and C at its *k*th level** is

$$abc_{ijk} = \bar{y}_{ijk} - (\bar{y}_{...} + a_i + b_j + c_k + ab_{ij} + ac_{ik} + bc_{jk})$$

EXAMPLE 4-9
(continued)

To illustrate the meaning of Definition 4-9, consider again the context of the composite material study. Using the previously calculated fitted main effects and 2-factor interactions, one has (for example)

$$abc_{111} = 1520 - (2360 + (-420) + (-315) + (-57.5) + (-30) + 47.5 + (-2.5))$$
$$= -62.5 \text{psi}$$

Similar calculations can be made to verify that the entire set of 3-factor interactions for the means of Table 4-21 is as follows.

$abc_{111} = -62.5$ psi	$abc_{211} = 62.5$ psi
$abc_{121} = 62.5$ psi	$abc_{221} = -62.5$ psi
$abc_{112} = 62.5$ psi	$abc_{212} = -62.5$ psi
$abc_{122} = -62.5$ psi	$abc_{222} = 62.5$ psi

3-factor interactions are less easily interpreted than main effects or 2-factor interactions. But one insight into their meaning was given immediately before Definition 4-9. Another is the following. If one looks separately at the different levels of (say) factor C, calculates fitted AB interactions, and finds the fitted AB interactions (the pattern of parallelism or nonparallelism) to be essentially the same on all levels of C, then the 3-factor interactions are small (near 0). Otherwise, large 3-factor interactions allow the pattern of AB interaction to change, from one level of C to another.

Simpler Descriptions of Some Three-Way Data Sets

Rewriting the equation in Definition 4-9, one has

$$\bar{y}_{ijk} = \bar{y}_{...} + a_i + b_j + c_k + ab_{ij} + ac_{ik} + bc_{jk} + abc_{ijk} \tag{4-32}$$

This is a breakdown of the combination sample means into somewhat interpretable pieces, corresponding to an overall effect, the factors acting separately, the factors acting in pairs, and the factors acting jointly. Display (4-32) may be regarded as a fitted version of an approximate relationship

$$y \approx \mu + \alpha_i + \beta_j + \gamma_k + \alpha\beta_{ij} + \alpha\gamma_{ik} + \beta\gamma_{jk} + \alpha\beta\gamma_{ijk} \tag{4-33}$$

When approaching the analysis of three-way factorial data, one hopes to discover a simplified version of equation (4-33) that is both interpretable and an adequate description of the data. (Indeed, if it is not possible to do so, it could be argued that little is gained by using the factorial breakdown rather than simply treating the data in question as $I \cdot J \cdot K$ *unstructured* samples.)

As was the case earlier with two-way factorial data, the process of fitting a simplified version of display (4-33) via least squares is, in general, unfortunately somewhat complicated. But when all sample sizes are equal (i.e., the data are balanced), the fitting process can be accomplished by simply adding appropriate fitted effects defined in Definitions 4-7, 4-8, and 4-9. Then the easily generated fitted responses lead to residuals that can be used in residual plotting and the calculation of R^2 for use in investigation of the adequacy of the corresponding simplified version of display (4-33).

EXAMPLE 4-9
(continued)

Looking over the magnitude of the fitted effects for Kinzer's composite material strength study, the A and B main effects clearly dwarf the others, suggesting at least the possibility that the relationship

$$y \approx \mu + \alpha_i + \beta_j \tag{4-34}$$

could be used as a description of the physical system. This relationship doesn't involve factor C at all (either by itself or in combination with A or B) and indicates that responses for a particular AB combination will be comparable for both time spans studied. Further, the fact that display (4-34) doesn't include the $\alpha\beta_{ij}$ term says that factors A and B act on product strength separately, so that their levels can be chosen independently. In geometrical terms corresponding to the kind of plot in Figure 4-28, display (4-34) means that observations from the cube's back face will be comparable to corresponding ones on the front face and that parallelism will prevail on both the front and back faces.

Kinzer's article gives only $\bar{y}_{ijk}$ values, not raw data, so a residual analysis and calculation of R^2 are not possible. But because of the balanced nature of the original data set, fitted values are easily obtained. For example, with factor A at level 1 and B at level 1, using the simplified relationship (4-34) and the fitted main effects found earlier, one has the fitted value

$$\hat{y} = \bar{y}_{...} + a_1 + b_1 = 2360 + (-420) + (-315) = 1625 \text{ psi}$$

All eight fitted values corresponding to equation (4-34) are shown geometrically in Figure 4-29. The fitted values given in the figure might be combined with product requirements and cost information to allow a process engineer to make sound decisions regarding the choice of autoclave temperature, autoclave time, and time span.

FIGURE 4-29

Eight Fitted Responses for Relationship (4-34) and the Composite Strength Study

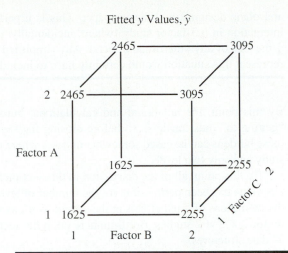

Fitted y Values, $\hat{y}$

In Example 4-9, the simplified version of display (4-33) was especially interpretable because it involved only main effects. But sometimes even versions of relation (4-33) involving interactions can draw attention to what is going on in a data set.

EXAMPLE 4-10

Interactions in a 3-Factor Paper Airplane Experiment. Schmittenberg and Riesterer did a paper airplane study as a class project. They studied the effects of three factors, each at two levels, on flight distance. The factors were Plane Design (A) (design 1 versus design 2), Plane Size (B) (large versus small), and Paper Type (C) (heavy versus light). The means of flight distances they obtained for 15 flights of each of the $8 = 2 \times 2 \times 2$ types of planes are given in Figure 4-30.

FIGURE 4-30

2^3 Sample Mean Flight Distances Displayed on the Corners of a Cube

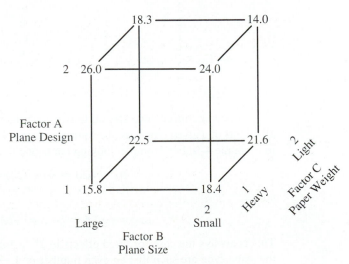

Sample Mean Flight Distances (ft)

Calculate the fitted effects corresponding to the $\bar{y}_{ijk}$'s given in Figure 4-30. Interestingly, when this is done, by far the biggest fitted effects (more than three times larger than any others) are the AC interactions. In retrospect, this makes perfect sense. The strongest message in Figure 4-30 is that plane design 1 should be made with light paper and plane design 2 with heavy paper. This is a perfect example of a strong 2-factor interaction in a 3-factor study (where, incidentally, the fitted 3-factor interactions are $\frac{1}{4}$ the size of any other fitted effects). Any simplified version of display (4-33) used to represent this situation would certainly have to include the $\alpha\gamma_{ik}$ term.

The Yates and Reverse Yates Algorithms

By this point, the notational and calculational burdens of this section are probably "getting to" most readers. So before closing the section, it is important to show that these burdens can be eased somewhat in cases where each factor in a complete factorial study has only two levels.

To this point, all of the discussion in this section has been general, in the sense that any value has been permissible for the number of levels for a factor. In particular, all of the definitions of fitted effects in the section work as well for $3 \times 5 \times 7$ studies as they do for $2 \times 2 \times 2$ studies. But from here on in the section, attention will be restricted to 2^p data structures.

Restricting attention to two-level factors affords several conveniences. One is notational. It is possible to reduce the clutter caused by the multiple subscript "*ijk*" notation, as follows. One first designates one level of each factor as a "high" level and the other as a "low" level. It is then convenient to name the 2^p factorial combinations with names that consist of letters corresponding to those factors which appear in the combination at their high levels. For example, if levels 2 of factors A, B, and C are designated the high levels, shorthand names for the $2^3 = 8$ different ABC combinations are as given in Table 4-22. Using these names, for example, $\bar{y}_a$ can stand for a sample mean where factor A is at its high (or second) level and all other factors are at their low (or first) levels.

TABLE 4-22

Shorthand Names for the 2^3 Factorial Treatment Combinations

Level of Factor A	Level of Factor B	Level of Factor C	Combination Name
1	1	1	(1)
2	1	1	a
1	2	1	b
2	2	1	ab
1	1	2	c
2	1	2	ac
1	2	2	bc
2	2	2	abc

A second convenience special to two-level factorial data structures is the fact that all effects of a given type have the same absolute value. This has already been illustrated in Example 4-9. For example, looking back, one finds that for the data of Table 4-21,

$$a_2 = 420 = -(-420) = -a_1$$

and

$$bc_{22} = -2.5 = bc_{11} = -bc_{12} = -bc_{21}$$

This is always the case for fitted effects in 2^p factorials. In fact, if two fitted effects of the same type are such that an even number of $1 \rightarrow 2$ or $2 \rightarrow 1$ subscript changes are required to get the second from the first, the fitted effects are equal (e.g., $bc_{22} = bc_{11}$). If an odd number are required, then the second fitted effect is -1 times the first (e.g. $bc_{12} = -bc_{22}$). The reason this fact is so useful is that because of it, one needs only to do the arithmetic necessary to find one fitted effect of each type and then choose appropriate signs to get all others of that type.

A famous statistician named Frank Yates is credited with discovering an efficient, mechanical way of generating one fitted effect of each type for a 2^p study. His method produces fitted effects with all "2" subscripts (i.e., corresponding to the "all factors at their high level" combination). The *Yates algorithm* consists of the following steps.

Step 1   Write down the 2^p sample means in a column in what is called *standard order*.

Step 2   Make up another column of numbers by first adding and then subtracting (first from second) the entries in the previous column in pairs.

Step 3   Follow Step 2 a total of p times, and then make up a final column by dividing the entries in the last column by the value 2^p.

The last column (made via Step 3) gives fitted effects (all factors at level 2), again in standard order. Standard order is easily remembered by beginning with (1) and a, then multiplying these 2 names (algebraically) by b to get b and ab, then multiplying these 4 names by c to get c, ac, bc, abc, etc.

EXAMPLE 4-9
(continued)

Table 4-23 shows the use of the Yates algorithm to calculate fitted effects for the 2^3 composite material study.

TABLE 4-23

The Yates Algorithm Applied to the Means of Table 4-21

Combination	$\bar{y}$	Cycle 1	Cycle 2	Cycle 3	Cycle 3 ÷ 8
(1)	1520	3970	9210	18,880	$2360 = \bar{y}_{...}$
a	2450	5240	9670	3,360	$420 = a_2$
b	2340	4210	1490	2,520	$315 = b_2$
ab	2900	5460	1870	−240	$-30 = ab_{22}$
c	1670	930	1270	460	$57.5 = c_2$
ac	2540	560	1250	380	$47.5 = ac_{22}$
bc	2230	870	−370	−20	$-2.5 = bc_{22}$
abc	3230	1000	130	500	$62.5 = abc_{222}$

The entries in the final column of this table are, of course, exactly as listed earlier, and the rest of the fitted effects are easily obtained via appropriate sign changes. This final column is an extremely succinct summary of the fitted effects, which allows one to quickly realize which types of fitted effects are larger than others.

The Yates algorithm is useful not only to find fitted effects. For balanced data sets, it is also possible to modify it slightly to find fitted responses, $\hat{y}$, for a given set of input fitted effects corresponding to a simplified version of a relation like display (4-33). First, one writes down the desired (all factors at their high level) fitted effects (using 0's for those types not considered), in reverse standard order. Then, by applying p cycles of the Yates additions and subtractions, one obtains fitted values, $\hat{y}$, listed in reverse standard order. (Note that no final division is required in this *reverse Yates algorithm*.)

EXAMPLE 4-9
(continued)

Consider fitting the relationship (4-34) to the balanced data set that led to the means of Table 4-21, via the reverse Yates algorithm. Table 4-24 gives the details. The fitted values in the final column are exactly as shown earlier in Figure 4-29.

TABLE 4-24

The Reverse Yates
Algorithm Applied to Fitting
Equation (4-34) to the Data
of Table 4-21

Fitted Effect	Value	Cycle 1	Cycle 2	Cycle 3 ($\hat{y}$)
abc_{222}	0	0	0	$3095 = \hat{y}_{abc}$
bc_{22}	0	0	3095	$2255 = \hat{y}_{bc}$
ac_{22}	0	315	0	$2465 = \hat{y}_{ac}$
c_2	0	2780	2255	$1625 = \hat{y}_c$
ab_{22}	0	0	0	$3095 = \hat{y}_{ab}$
b_2	315	0	2465	$2255 = \hat{y}_b$
a_2	420	315	0	$2465 = \hat{y}_a$
$\bar{y}_{...}$	2360	1940	1625	$1625 = \hat{y}_{(1)}$

The restriction to two-level factors that makes all of these notational and computational devices possible is not as specialized as it may at first seem. When an engineer wishes to study the effects of a large number of factors, even 2^p will be a large number of process conditions to investigate. If more than two levels of factors are considered, the sheer size of a complete factorial study quickly becomes unmanageable. Recognizing this, two-level studies are often used for screening, to identify a few (from many) process variables for subsequent study at more levels, on the basis of their large perceived effects in the screening study. So this 2^p material is in fact quite important to the practice of engineering statistics.

4-4 Transformations and Choice of Measurement Scale

The material in Sections 4-2 and 4-3 can be thought of as an introduction to one of the main themes of engineering statistical analysis: the discovery and use of simply described structure in potentially quite complicated situations. Sometimes this can be facilitated by reexpressing measured process variables and/or responses on some other (nonlinear) scales of measurement than the ones that first come to mind. That is, sometimes no simple description of data set structure may be obvious when regarded on commonly used scales of measurement, but after some or all variables have been transformed, simple structure may become clear.

This section presents several examples of cases where transformations are helpful. In the process, some comments about commonly used types of transformations, as well as more specific reasons for employing transformations, are offered.

**Transformations and
Single Samples**

In Chapter 5, there are a number of standard theoretical distributions that mathematicians regularly use and prove theorems about. When one of these standard models can be used to describe the variation of a system's output, all of those theorems can be brought to bear in making predictions and inferences regarding its behavior. However, when no standard distributional shape can be found to describe a response y, it may nevertheless be possible to so describe $g(y)$ for some function $g(\cdot)$.

EXAMPLE 4-11

Discovery Times at an Auto Shop. Elliot, Kibby, and Meyer studied operations at an auto repair establishment. They collected some data on what they called the "discovery time" associated with diagnosing what repairs the mechanics were going to recommend to the car owners. Thirty such discovery times (in minutes) are given on Figure 4-31, in the form of a stem-and-leaf plot.

The stem-and-leaf plot shows these data to be somewhat skewed to the right. Many of the most common standard methods of statistical inference are based on an assumption

FIGURE 4-31

Stem-and-Leaf Plot of
Discovery Times

```
0 | 4 3
0 | 6 5 5 5 6 6 8 9 8 8
1 | 4 0 3 0
1 | 7 5 9 5 6 9 5
2 | 0
2 | 9 5 7 9
3 | 2
3 | 6
```

that a data-generating mechanism will in the long run produce not skewed, but rather symmetrical, bell-shaped data. Therefore, the use of these methods to draw inferences and make predictions about discovery times at this shop is highly questionable. However, suppose that some transformation could be applied to produce a bell-shaped distribution of transformed discovery times. The standard methods could be used to draw inferences about transformed discovery times, which could then be translated (by undoing the transformation) to inferences about raw discovery times.

One common transformation that has the effect of shortening the right tail of a distribution is the logarithmic transformation. To illustrate its helpfulness in the present context, normal plots of both discovery times and log discovery times are given in Figure 4-32. These plots indicate that Elliot, Kibby, and Meyer could not have reasonably applied standard methods of inference to the discovery times, but they could well have used the methods with log discovery times. The second normal plot is far more linear than the first.

FIGURE 4-32

Normal Plots for
Discovery Times and
Log Discovery Times

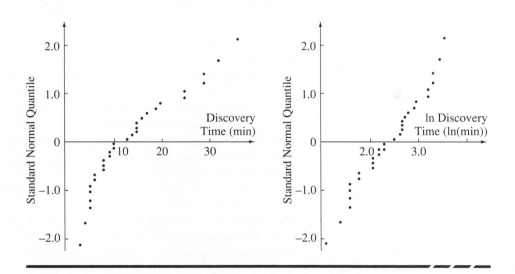

The logarithmic transformation was useful in the preceding example in reducing the skewness of a response distribution. Some other transformations commonly employed to change the shape of a response distribution in statistical engineering studies are the *power transformations*,

$$g(y) = (y - \gamma)^{\alpha} \tag{4-35}$$

In transformation (4-35), the number γ is often taken as a threshold value, corresponding to a minimum possible response. The number α governs the basic shape of a plot of $g(y)$ versus y. For $0 < \alpha < 1$, the transformation (4-35) tends to shorten the right tail of a distribution for y, the shortening becoming more drastic as α approaches 0, but not being as pronounced as that caused by the transformation

$$g(y) = \ln(y - \gamma)$$

For $\alpha > 1$, the transformation (4-35) tends to lengthen the right tail of a distribution for y.

Transformations and Multiple Samples

Comparing several sets of process conditions based on several corresponding samples is one of the archetypal problems of statistical engineering analysis. It is advantageous to do the comparison on a scale where the samples have comparable variabilities, for at least two reasons. The first is the obvious fact that comparisons then reduce simply to comparisons between response locations. Second, standard methods of statistical inference often have well-understood properties only when response variability is comparable for the different sets of conditions.

When response variability is not comparable under different sets of conditions, a transformation can sometimes be applied to all observations to produce transformed responses with nearly comparable variabilities. This possibility of *transforming to stabilize variance* exists when response variance is roughly a function of response mean. Some theoretical calculations (which won't be presented here) suggest the following guidelines as at least a place to begin looking for an appropriate variance-stabilizing transformation.

1. If response standard deviation is approximately proportional to response mean, try a logarithmic transformation.
2. If response standard deviation is approximately proportional to the δ power of the response mean, try transformation (4-35) with $\alpha = 1 - \delta$.

Where several samples (and corresponding $\bar{y}$ and s values) are involved, an empirical way of investigating whether 1) or 2) might be useful is to plot $\ln(s)$ versus $\ln(\bar{y})$ and see if there is approximate linearity. If so, a slope of roughly 1 makes 1) appropriate, while a slope of $\delta \neq 1$ signals what version of 2) might be helpful.

In addition to this empirical way of identifying a potentially variance-stabilizing transformation, theoretical considerations can sometimes provide guidance. The theoretical distributions often used to describe certain types of variables have their own relationships between their (theoretical) means and variances, which can help pick out an appropriate version of 1) or 2). For example, a standard model for data on occurrences per item (see again Section 3-4) has the property that its mean and variance are proportional, suggesting a square root transformation ($\delta = .5$) before analysis of several $\hat{u}$ values. And a standard model for sample variances s^2 (based on responses y that themselves have reasonably bell-shaped distributions) has theoretical standard deviation proportional to its mean, suggesting a logarithmic transformation before analysis of several s^2 values.

Transformations and Simple Structure in Multifactor Studies

In Section 4-2, Taylor's equation for tool life y in terms of cutting speed x was advantageously reexpressed as a linear equation for $\ln(y)$ in terms of $\ln(x)$. This is just one manifestation of the general fact that many approximate laws of physical science and engineering are *power laws*, expressing one quantity as a product of a constant and powers (some possibly negative) of other quantities. That is, they are of the form

$$y \approx \alpha x_1^{\beta_1} x_2^{\beta_2} \cdots x_k^{\beta_k} \tag{4-36}$$

Of course, upon taking logarithms in equation (4-36),

$$\ln(y) \approx \ln(\alpha) + \beta_1 \ln(x_1) + \beta_2 \ln(x_2) + \cdots + \beta_k \ln(x_k) \tag{4-37}$$

which immediately suggests the wide usefulness of the logarithmic transformation for both y and x variables in surface-fitting applications involving power laws.

But there is something else in display (4-37) that bears examination: The k functions of the fundamental x variables enter the equation *additively*. In the language of the previous section, there are *no interactions* between the factors whose levels are specified by the variables $x_1, x_2, \ldots, x_k$. This suggests that even in studies involving only seemingly qualitative factors, if a power law for y is operative and the factors act on different fundamental variables x, a logarithmic transformation will tend to create a simple structure. It will do so by eliminating the need for interactions in describing the response.

EXAMPLE 4-12

Daniel's Drill Advance Rate Study. In his excellent 1976 book *Applications of Statistics to Industrial Experimentation*, Cuthbert Daniel gives an extensive discussion of an unreplicated 2^4 factorial study of the behavior of a new piece of drilling equipment. The response y is a rate of advance of the drill (no units are given), and the experimental factors are Load on the small stone drill (A), Flow Rate through the drill (B), Rotational Speed (C), and Type of Mud used in drilling (D). Daniel's data are given in Table 4-25.

TABLE 4-25

Daniel's 2^4 Drill Advance Rate Data

Combination	y	Combination	y
(1)	1.68	d	2.07
a	1.98	ad	2.44
b	3.28	bd	4.09
ab	3.44	abd	4.53
c	4.98	cd	7.77
ac	5.70	acd	9.43
bc	9.97	bcd	11.75
abc	9.07	abcd	16.30

Application of the Yates algorithm to the data in Table 4-25 (4 cycles are required, as is division of the results of the last cycle by the 4th power of 2) gives the following fitted effects:

$$\bar{y}_{\ldots} = 6.1550$$

$a_2 = .4563 \qquad b_2 = 1.6488 \qquad c_2 = 3.2163 \qquad d_2 = 1.1425$

$ab_{22} = .0750 \qquad ac_{22} = .2975 \qquad ad_{22} = .4213$

$bc_{22} = .7525 \qquad bd_{22} = .2213 \qquad cd_{22} = .7987$

$abc_{222} = .0838 \qquad abd_{222} = .2950 \qquad acd_{222} = .3775 \qquad bcd_{222} = .0900$

$abcd_{2222} = .2688$

Looking at the magnitudes of these fitted effects, the candidate relationships

$$y \approx \mu + \beta_j + \gamma_k + \delta_l \tag{4-38}$$

and

$$y \approx \mu + \beta_j + \gamma_k + \delta_l + \beta\gamma_{jk} + \gamma\delta_{kl} \tag{4-39}$$

are suggested. (The 5 largest fitted effects are, in order of decreasing magnitude, the main effects of C, B, and D, and then the two-factor interactions of C with D and B with C.) Try verifying via the reverse Yates algorithm (4 cycles are required to get $\hat{y}$'s) that fitting equation (4-38) to the balanced data of Table 4-25 produces $R^2 = .875$, and fitting relationship (4-39) produces $R^2 = .948$. But upon closer examination, neither the

FIGURE 4-33

Residual Plots from
Fitting Equation (4-38)
to Daniel's Data

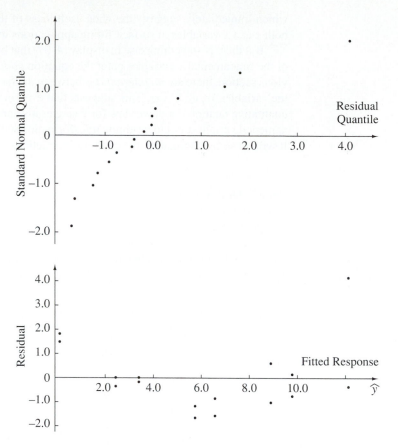

fitted version of equation (4-38) nor the fitted version of relationship (4-39) turns out to be a very good description of these data. Figures 4-33 and 4-34 show normal plots and plots against $\hat{y}$ for residuals from fitted versions of equations (4-38) and (4-39), respectively.

Figure 4-33 shows that the fitted version of equation (4-38) produces several disturbingly large residuals and fitted values that are systematically too small for responses that are small and large, but too large for moderate responses. Such a curved plot of residuals versus $\hat{y}$ in general suggests that a nonlinear transmformation may potentially be effective.

The normal plot for the fitted version of relationship (4-39) in Figure 4-34 looks even worse than the one for equation (4-38). Further, the plot of residuals versus $\hat{y}$ shows that it is the bigger responses that are fitted relatively badly by relationship (4-39). This is an unfortunate circumstance, since presumably one study goal is the optimization of response.

But using $\ln(y)$ as a response variable, the situation is much different. The Yates algorithm produces the following fitted effects.

$\bar{y}_{....} = 1.5977$

$a_2 = .0650$	$b_2 = .2900$	$c_2 = .5772$	$d_2 = .1633$
$ab_{22} = -.0172$	$ac_{22} = .0052$	$ad_{22} = .0334$	
$bc_{22} = -.0251$	$bd_{22} = -.0075$	$cd_{22} = .0491$	
$abc_{222} = .0052$	$abd_{222} = .0261$	$acd_{222} = .0266$	$bcd_{222} = -.0173$
$abcd_{2222} = .0193$			

FIGURE 4-34

Residual Plots from
Fitting Equation (4-39)
to Daniel's Data

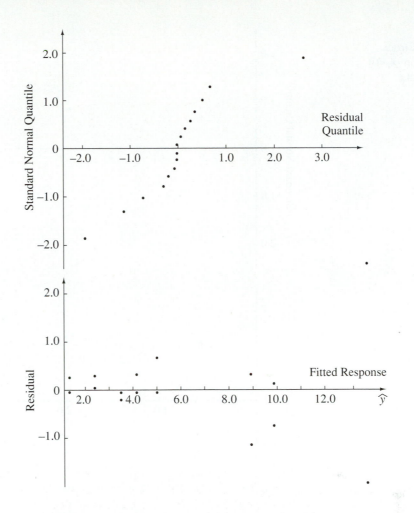

For the logged drill advance rates, the simple relationship

$$\ln(y) \approx \mu + \beta_j + \gamma_k + \delta_l \tag{4-40}$$

yields $R^2 = .976$ and absolutely innocuous residuals. Figure 4-35 shows a normal plot of these and a plot of them against $\hat{y}$.

The point here is that the logarithmic scale appears to be the natural one on which to study drill advance rate. The data can be better described on the log scale without using interaction terms than is possible with interactions on the original scale.

There are sometimes other reasons to consider a logarithmic transformation of a response variable in a multifactor study, besides its potential to produce simple structure. In cases where responses vary over several orders of magnitude, simple curves and surfaces typically don't fit raw y values very well, but they can do a much better job of fitting $\ln(y)$ values (which will usually vary over less than a single order of magnitude). Another potentially helpful property of a log-transformed analysis is that it will never yield physically impossible negative fitted values for a positive variable y. In contrast, an analysis on an original scale of measurement can, rather embarrassingly, do so.

FIGURE 4-35

Residual Plots from
Fitting Equation (4-40)
to Daniel's Data

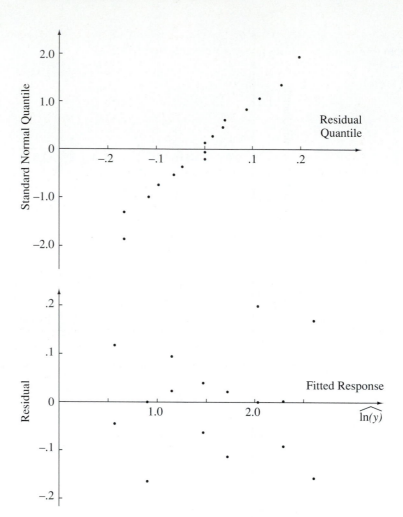

EXAMPLE 4-13

A 3^2 Factorial Chemical Process Experiment. The data in Table 4-26 are from an article by Hill and Demler ("More on Planning Experiments to Increase Research Efficiency," *Industrial and Engineering Chemistry*, 1970). The data concern the running of a chemical process where the objective is to achieve high yield y_1 and low filtration time y_2 by choosing settings for Condensation Temperature, x_1, and the Amount of B employed, x_2.

TABLE 4-26

Yields and Filtration Times
in a 3^2 Factorial Chemical
Process Study

x_1 Condensation Temperature (°C)	x_2 Amount of B (cc)	y_1 Yield (g)	y_2 Filtration Time (sec)
90	24.4	21.1	150
90	29.3	23.7	10
90	34.2	20.7	8
100	24.4	21.1	35
100	29.3	24.1	8
100	34.2	22.2	7
110	24.4	18.4	18
110	29.3	23.4	8
110	34.2	21.9	10

For purposes of this example, consider the second response, filtration time. Fitting the approximate (quadratic) relationship

$$y \approx \beta_0 + \beta_1 x_1 + \beta_2 x_2 + \beta_3 x_1^2 + \beta_4 x_2^2 + \beta_5 x_1 x_2$$

to these data produces the equation

$$\hat{y} = 5179.8 - 56.90 x_1 - 146.0 x_2 + .1733 x_1^2 + 1.222 x_2^2 + .6837 x_1 x_2 \qquad \textbf{(4-41)}$$

and $R^2 = .866$. Equation (4-41) defines a bowl-shaped surface in three dimensions, which has a minimum at about the set of conditions $x_1 = 103.2°C$ and $x_2 = 30.88$ cc. At first glance, it might seem that the development of Equation (4-41) rates as a statistical engineering success story. But there is the embarrassing fact that upon substituting $x_1 = 103.2$ and $x_2 = 30.88$ into Equation (4-41), one gets $\hat{y} = -11.14$ sec, hardly a possible filtration time.

Looking again at the data, it is not hard to see what has gone awry. The largest response is more than 20 times the smallest. So in order to come close to fitting both the extremely large and more moderate responses, the fitted quadratic surface needs to be very steep — so steep that it is forced to dip below the (x_1, x_2)-plane and produce negative $\hat{y}$ values before it can "get turned around" and start to climb again as one moves away from the point of minimum $\hat{y}$ toward larger x_1 and x_2.

One cure for the problem of negative predicted filtration times is to use $\ln(y_2)$ as a response variable. Values of $\ln(y_2)$ are given in Table 4-27 to illustrate the moderating effect the logarithm has on the disparity by a factor of 20 between the largest and smallest filtration times.

TABLE 4-27

Raw Filtration Times and Corresponding Logged Filtration Times

y_2 Filtration Time (sec)	$\ln(y_2)$ Log Filtration Time (ln(sec))
150	5.0106
10	2.3026
8	2.0794
35	3.5553
8	2.0794
7	1.9459
18	2.8904
8	2.0794
10	2.3026

Fitting the approximate quadratic relationship

$$\ln(y_2) \approx \beta_0 + \beta_1 x_1 + \beta_2 x_2 + \beta_3 x_1^2 + \beta_4 x_2^2 + \beta_5 x_1 x_2$$

to the $\ln(y_2)$ values produces the equation

$$\widehat{\ln(y)} = 99.69 - .8869 x_1 - 3.348 x_2 + .002506 x_1^2 + .03375 x_2^2 + .01196 x_1 x_2 \qquad \textbf{(4-42)}$$

and $R^2 = .975$. Equation (4-42) also represents a bowl-shaped surface in three dimensions and has a minimum approximately at the set of conditions $x_1 = 101.5°C$ and $x_2 = 31.6$ cc. The minimum fitted log filtration time is $\widehat{\ln(y_2)} = 1.7582 \ln$ (sec), which translates to a filtration time of 5.8 sec, a far more sensible value than the negative one given earlier.

Notice that the taking of logs in this example had two beneficial effects. The first was to cut the ratio of largest response to smallest down to about 2.5 (from over 20), allowing a good fit (as measured by R^2) for a fitted quadratic in two variables x_1 and x_2. The second was to ensure that minimum predicted filtration time was positive.

Of course, other transformations besides the logarithmic one are also useful in describing the structure of multifactor data sets. Sometimes they are applied to the responses and sometimes to other system variables. As an example of a situation where a transformation like that specified by equation (4-35) is useful in understanding the structure of a sample of bivariate data, consider the following.

EXAMPLE 4-14

Yield Strengths of Copper Deposits and Hall-Petch Theory. In their article "Mechanical Property Testing of Copper Deposits for Printed Circuit Boards" (*Plating and Surface Finishing*, 1988), engineers Lin, Kim, and Weil present some data relating the yield strength of electroless copper deposits to the average grain diameters measured for these deposits. The values in Table 4-28 were deduced from a scatterplot in their paper. These values are plotted in Figure 4-36; they don't seem to promise a simple relationship between grain diameter and yield strength. But in fact, the so called Hall-Petch relationship says that yield strengths of most crystalline materials are proportional to the reciprocal square root of grain diameter. That is, Hall-Petch theory predicts a linear relationship between y and $x^{-.5}$ or between x and y^{-2}. Thus, before trying to further detail the relationship between the two variables, application of transformation (4-35) with $\alpha = -.5$ to x or transformation (4-35) with $\alpha = -2$ to y seems in order. Figure 4-37 shows the partial effectiveness of the reciprocal square root transformation (applied to x) in producing a linear relationship in this context.

TABLE 4-28

Average Grain Diameters and Yield Strengths for Copper Deposits

x Average Grain Diameter (μm)	y Yield Strength (MPa)
.22	330
.27	370
.33	266
.41	270
.48	236
.49	224
.51	236
.90	210

FIGURE 4-36

Scatterplot of Yield Strength versus Average Grain Diameter

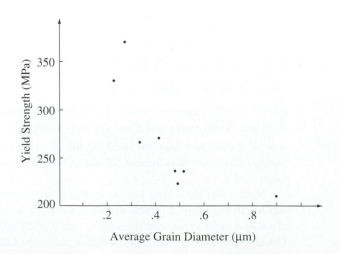

FIGURE 4-37

Scatterplot of Yield Strength versus the Reciprocal Square Root Average Grain Diameter

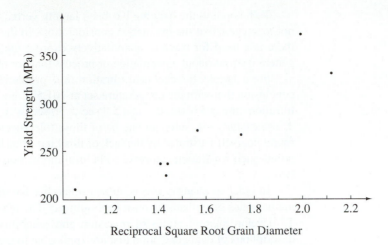

However, it might be pointed out that taking logs of both x and y would also produce linearity if the Hall-Petch (power law) theory is relevant for this material. Further, it would in fact make it possible to discover the relationship empirically, even if it were *a priori* unexpected.

In the preceding example, a directly applicable and well-known physical theory suggests a natural transformation to apply to a set of data. Sometimes physical or mathematical considerations that are related to a problem but do not directly address it may also suggest some things to try in looking for a transformation or transformations to produce simple structure. For example, suppose some other property besides yield strength were of interest and thought to be related to grain size. If a relationship with diameter is not obvious, one might then consider quantifying grain size in terms of cross-sectional area or volume and thus be led to squaring or cubing a measured diameter. To take a different example, if some handling characteristic of an auto is thought to be related to its speed and a relationship with velocity is not obvious, one might remember that kinetic energy is related to velocity squared, thus being led to square the velocity.

To repeat the main point of this section, the search for appropriate transformations is usually a quest for measurement scales on which data set structure is transparent and simple. However, if the original/untransformed scales are the most natural ones on which to report the findings of a study, the data analysis should be done on the transformed scales but then "untransformed" to state the final results.

4-5 *Beyond Descriptive Statistics*

It is hoped that the discussion of engineering statistics so far has made you genuinely ready to accept the need for methods of formal statistical inference. Many real data sets have been examined, and it has been shown that there is often useful structure in them — to some extent obscured by what (for lack of a better term) might be called *background noise*. Recognizing the existence of such variation, one realizes that the data in hand are probably not a perfect representation of the population or process from which they were taken. Thus, any generalization from the sample to a broader sphere will have to be hedged somehow.

To this point, the hedging has been largely verbal, *ad hoc*, and qualitative. This is not to suggest that the qualitative considerations in the examples are unimportant. But there is a need for ways to quantitatively express the precision and reliability of any generalizations about a population or process that are made from data in hand.

For example, the chemical filtration time problem of Example 4-13 produced the conclusion that with the temperature set at 101.5°C and using 31.6 cc of B, a predicted filtration time is 5.8 sec. But is it 5.8 sec $\pm$ 50 sec or $\pm$.05 sec? Then, if one decides on $\pm$ *somevalue*, how sure can one be of those tolerances? (It is a good sign if you have found yourself frustrated by the lack of this kind of quantification in discussions to this point. Such frustration can serve as helpful motivation for study of the next chapter or two.)

In order to quantify precision and reliability for inferences based on samples, the mathematics of probability must be employed. The reasoning behind this is as follows. To develop quantifications of precision and reliability, one must have mathematical descriptions of data generation that are applicable to the circumstances surrounding the original data collection (and any future collection to which one desires to project). Those mathematical models must explicitly allow for (and even predict) the kind of variation that has been faced in the descriptive analyses of data throughout this exposition.

The kinds of models that are most familiar to engineers do not explicitly account for the fact that variation exists. Rather, they are *deterministic*. For example, Newtonian physics predicts that the displacement of a body in free fall in a time t is exactly $\frac{1}{2}gt^2$. In this statement, there is no explicit allowance for, or prediction of, variability. Any observed deviation from the Newtonian predictions is completely unaccounted for. Thus, there is really no logical framework in which to extrapolate from data that don't fit Newtonian predictions exactly to any larger sphere.

Stochastic or *probabilistic* models do explicitly incorporate the feature that even measurements generated under the same set of conditions will exhibit variation. Therefore, they can function as useful descriptions of real-world data collection processes, where many small, unidentifiable causes act to produce the background noise or random variation seen in real data sets. Since variation is accounted for and even predicted by stochastic or probabilistic models, they provide a logical framework in which to quantify precision and reliability and to extrapolate from noisy data to contexts larger than the data set in hand.

In the next chapter, some fundamental concepts of probability will be introduced. Then Chapter 6 begins to use probability as a tool in statistical inference.

EXERCISES

4-1. (§4-1) The following is a small set of artificial data. Show the hand calculations necesssary to do the indicated tasks.

x	y
1	8
2	8
3	6
4	6
5	4

(a) Obtain the least squares line through these data. Make a scatterplot of the data and sketch this line on that scatterplot.
(b) Obtain the sample correlation between x and y for these data.

(c) Obtain the sample correlation between y and $\hat{y}$ for these data, and compare it to your answer to part (b).
(d) Use the formula in Definition 4-3 and compute R^2 for these data. Compare it to the square of your answers to parts (b) and (c).
(e) Find the five residuals from your fit in part (a). How are they portrayed geometrically on the scatterplot for (a)?

4-2. (§4-1) Use a computer package of your own choice and redo the computations and plotting required in Exercise 4-1. Annotate your output, indicating where on the printout you can find the equation of the least squares line, the value of r, the value of R^2, and the residuals.

4-3. (§4-1) Use the method of least squares to fit an equation $y \approx \beta x$ to the small set of artificial data that follows.

x	y
1	2
2	2
3	4
4	4

Cutting Speed, x (sfpm)	Tool Life, y (min)
800	1.00, 0.90, 0.74, 0.66
700	1.00, 1.20, 1.50, 1.60
600	2.35, 2.65, 3.00, 3.60
500	6.40, 7.80, 9.80, 16.50
400	21.50, 24.50, 26.00, 33.00

(This involves more than just adapting a formula for a slope from Section 4-1. Least squares fitting of a line without intercept doesn't generally produce the same slope as when an intercept is included in the fitting.)

4-4. (§4-1) Discuss the purpose of finding and plotting residuals in a statistical analysis.

4-5. (§4-1) The article "Polyglycol Modified Poly (Ethylene Ether Carbonate) Polyols by Molecular Weight Advancement" by R. Harris (*Journal of Applied Polymer Science*, 1990) contains some data on the effect of reaction temperature on the molecular weight of resulting poly polyols. The data for eight experimental runs at temperatures 165°C and above are as follows.

Pot Temperature, x (°C)	Average Molecular Weight, y
165	808
176	940
188	1183
205	1545
220	2012
235	2362
250	2742
260	2935

Use some statistical package to help you complete the following.

(a) What fraction of the observed raw variation in y is accounted for by a linear equation in x?

(b) Fit a linear relationship $y \approx \beta_0 + \beta_1 x$ to these data via least squares. About what change in average molecular weight seems to accompany a 1°C increase in pot temperature (at least over the experimental range of temperatures)?

(c) Compute and plot residuals from the linear relationship fit in (b). Discuss what they suggest about the appropriateness of that fitted equation. (Plot at least residuals versus x, residuals versus $\hat{y}$ and make a normal plot of them.)

(d) These data came from an experiment where the investigator managed the value of x. There is a fairly glaring weakness in the experimenter's data collection efforts. What is it?

(e) Based on your analysis of these data, what average molecular weight would you predict for an additional reaction run at 188°C? At 200°C? Why would or wouldn't you be willing to make a similar prediction of average molecular weight if the reaction is run at 70°C?

4-6. (§4-1 and §4-4) Upon changing measurement scales, nonlinear relationships between two variables can sometimes be made linear. The article "The Effect of Experimental Error on the Determination of the Optimum Metal-Cutting Conditions" by Ermer and Wu (*The Journal of Engineering for Industry*, 1967) contains a data set gathered in a study of tool life in a turning operation.

The data here are part of that larger data set, corresponding to a single feed.

(a) Plot y versus x and calculate R^2 for fitting a linear function of x to y. Does the relationship $y \approx \beta_0 + \beta_1 x$ look like a plausible explanation of tool life in terms of cutting speed?

(b) Take natural logs of both x and y, and repeat part (a) with these log cutting speeds and log tool lives.

(c) Using the logged variables as in (b), fit a linear relationship between the two variables using least squares. Based on this fitted equation, what tool life would you predict for a cutting speed of 550? What approximate relationship between x and y is implied by a linear approximate relationship between $\ln(x)$ and $\ln(y)$? (Give an equation for this relationship.) By the way, Taylor's equation for tool life is $yx^{\alpha} = C$.

4-7. (§4-2) Return to the situation of Exercise 4-5. Fit a quadratic relationship $y \approx \beta_0 + \beta_1 x + \beta_2 x^2$ to the data via least squares. By appropriately plotting residuals and examining R^2 values, determine the advisability of using a quadratic rather than a linear equation to describe the relationship between x and y. If one were to use a quadratic fitted equation, how would the predicted mean molecular weight at 200°C compare to the value that you obtained in part (e) of Exercise 4-5?

4-8. (§4-2) Here are some data taken from the article "Chemithermomechanical Pulp from Mixed High Density Hardwoods" by Miller, Shankar, and Peterson (*Tappi Journal*, 1988). Given are the percent NaOH used as a pretreatment chemical, x_1, the pretreatment time in minutes, x_2, and the the resulting value of a specific surface area variable, y (with units of cm^3/g), for nine batches of pulp produced from a mixture of hardwoods at a treatment temperature of 75° Celsius in mechanical pulping.

% NaOH, x_1	Time, x_2	Specific Surface Area, y
3.0	30	5.95
3.0	60	5.60
3.0	90	5.44
9.0	30	6.22
9.0	60	5.85
9.0	90	5.61
15.0	30	8.36
15.0	60	7.30
15.0	90	6.43

(a) Fit the approximate relationship $y \approx \beta_0 + \beta_1 x_1 + \beta_2 x_2$ to these data via least squares. Interpret the coefficients b_1 and b_2 in the fitted equation. What fraction of the observed raw variation in y is accounted for using this equation?

(b) Compute and plot residuals for your fitted equation from (a). Discuss what these plots indicate about the adequacy of your fitted equation. (At a minimum, you should plot residuals against all of x_1, x_2, and $\hat{y}$, and normal-plot the residuals.)

(c) Make a plot of y versus x_1 for the nine data points and sketch on that plot the three different linear functions of x_1 produced by setting x_2 at first 30, then 60, and then 90 in your fitted equation from (a). How well do fitted responses appear to match observed responses?

(d) What specific surface area would you predict for an additional batch of pulp of this type produced using a 10% NaOH treatment for a time of 70 minutes? Would you be willing to make a similar prediction for 10% NaOH used for 120 minutes? Why or why not?

(e) There are many other possible approximate relationships that one might consider fitting to these data via least squares, one of which is $y \approx \beta_0 + \beta_1 x_1 + \beta_2 x_2 + \beta_3 x_1 x_2$. Fit this equation to the data above and compare the resulting coefficient of determination to the one found in (a). On the basis of these alone, does the use of the more complicated equation seem necessary?

(f) For the equation fit in part (e), repeat the steps of part (c) and compare the plot made here to the one made earlier.

(g) What is an intrinsic weakness of this real published data set?

(h) What kind of structure discussed in Section 1-2 describes this data set? It turns out that since the data set has this special structure and all nine sample sizes are the same (i.e., are all 1), some special relationships hold between the equation fit in (a) and what one gets by separately fitting linear equations in x_1 and then in x_2 to the y data. Fit such one-variable linear equations and compare coefficients and R^2 values to what you obtained in (a). What relationships exist between these?

4-9. (§4-3) Since the data of Exercise 4-8 have complete factorial structure, it is possible (at least temporarily) to ignore the fact that the two experimental factors are basically quantitative and make a factorial analysis of the data.

(a) Compute all fitted factorial main effects and interactions for the data of Exercise 4-8. Interpret the relative sizes of these fitted effects, using a figure like Figure 4-22 to facilitate your discussion.

(b) Compute nine fitted responses for the "main effects only" explanation of y, $y \approx \mu + \alpha_i + \beta_j$. Plot these versus level of the NaOH variable, connecting fitted values having the same level of the Time variable with line segments as in Figure 4-23. Discuss how this plot compares to the two plots of fitted y versus x_1 made in Exercise 4-8.

(c) Use the fitted values computed in (b) and find a value of R^2 appropriate to the "main effects only" representation of y. How does it compare to the values found in Exercise 4-8? Also use the fitted values to compute residuals for this representation. Plot these (versus level of NaOH, level of Time and $\hat{y}$, and in normal plot form). What do they indicate about the "main effects only" explanation of specific area?

4-10. (§4-3 and §4-4) Bachman, Herzberg, and Rich conducted a 2^3 factorial study of fluid flow through thin tubes. They measured the time required for the liquid level in a fluid holding tank to drop from 4 in. to 2 in. for two drain tube diameters and two fluid types. Two different technicians did the measuring; their data are as follows.

Technician	Diameter (in.)	Fluid	Time (sec)
1	.188	water	21.12, 21.11, 20.80
2	.188	water	21.82, 21.87, 21.78
1	.314	water	6.06, 6.04, 5.92
2	.314	water	6.09, 5.91, 6.01
1	.188	ethylene glycol	51.25, 46.03, 46.09
2	.188	ethylene glycol	45.61, 47.00, 50.71
1	.314	ethylene glycol	7.85, 7.91, 7.97
2	.314	ethylene glycol	7.73, 8.01, 8.32

(a) Compute (using the Yates algorithm or otherwise) the values of all the fitted main effects, two-way interactions, and three-way interactions for these data. Do any simple interpretations of these suggest themselves?

(b) The students actually had some physical theory in mind, suggesting that the log of the drain time might be a more convenient response variable than the raw time. Take the logs of the y's and recompute the factorial effects. Does an interpretation of this system in terms of only main effects seem more plausible on the log scale than on the original scale?

(c) Considering the logged drain times as the responses, find fitted values and residuals for a "Diameter and Fluid main effects only" explanation of these data. Compute an R^2 value appropriate to such a view and compare it to an R^2 value that would result from using all factorial effects to describe log drain time. Make and interpret appropriate residual plots.

(d) Based on the analysis from (c), what change in log drain time seems to accompany a change from .188 in. diameter to .314 in. diameter? What does this translate to in terms of raw drain time? Physical theory suggests that raw time is inversely proportional to the fourth power of drain tube radius. Does your answer here seem compatible with that theory?

4-11. (§1-4 and §4-5) Describe in your own words the difference between deterministic and stochastic/probabilistic models. Give an example of a deterministic model that is useful in your field.

4-12. What are some potential benefits that can sometimes be derived from transforming the data before applying standard statistical techniques?

4-13. When analyzing a full factorial data set where the factors involved are quantitative, one can apply either the surface-fitting technology of Section 4-2 or the factorial analysis (in terms of fitted main effects and interactions) material of Section 4-3. What practical engineering advantage does the first offer over the second in such cases?

4-14. Suppose that a response variable, y, obeys an approximate power law in at least two quantitative variables (say, x_1 and x_2). Will there be important interactions? If one analyzes instead the log of y, will there be important interactions? (In order to make this concrete, you may if you wish consider the relationship $y \approx kx_1^2 x_2^{-3}$. Plot, for at least two different values of x_2, y as a function of x_1. Then plot, for at least two different values of x_2, $\ln(y)$

as a function of x_1. What do these plots show in the way of parallelism?)

4-15. Nicholson and Bartle studied the effect of the water/cement ratio on 14-day compressive strength for Portland cement concrete. The water/cement ratios (by volume) and compressive strengths of nine concrete specimens are given below.

Water/Cement Ratio, x	14-Day Compressive Strength, y (psi)
.45	2954, 2913, 2923
.50	2743, 2779, 2739
.55	2652, 2607, 2583

(a) Fit a line to the data here via least squares, showing the hand calculations necessary to do so.
(b) Compute the sample correlation between x and y by hand. Interpret this value.
(c) What fraction of the raw variability in y is accounted for in the fitting of a line to the data?
(d) Compute the residuals from your fitted line and make a normal plot of them. Interpret this plot.
(e) What compressive strength would you predict, based on your calculations from (a), for specimens made using a .48 water/cement ratio?
(f) Use Minitab (or some other statistical package available to you) to find the least squares line, the sample correlation, R^2, and the residuals for this data set.

4-16. Griffith and Tesdall studied the elapsed time in $\frac{1}{4}$ mile runs of a Camaro Z-28 fitted with different sizes of carburetor jetting. Their data from six runs of the car follow.

Jetting Size, x	Elapsed Time, y (sec)
66	14.90
68	14.67
70	14.50
72	14.53
74	14.79
76	15.02

(a) What is an obvious weakness in the students' data collection plan?
(b) Fit both a line and a quadratic equation ($y \approx \beta_0 + \beta_1 x + \beta_2 x^2$) to these data via least squares. Plot both of these equations on a scatterplot of the data.
(c) What fractions of the raw variation in elapsed time are accounted for by the two different fitted equations?
(d) Use your fitted quadratic equation to predict an optimal jetting size (allowing fractional sizes).

4-17. The following are some data taken from "Kinetics of Grain Growth in Powder-formed IN-792: A Nickel-Base Super-alloy" by Huda and Ralph (*Materials Characterization*, September 1990). Three different Temperatures, x_1 (°K), and three different Times, x_2 (min), were used in the heat treating of specimens of a material, and the response

$$y = \text{mean grain diameter } (\mu m)$$

was measured.

Temperature, x_1	Time, x_2	Grain Size, y
1443	20	5
1443	120	6
1443	1320	9
1493	20	14
1493	120	17
1493	1320	25
1543	20	29
1543	120	38
1543	1320	60

(a) What type of data structure did the researchers employ? (Use the terminology of Section 1-2.) What was an obvious weakness in their data collection plan?
(b) Use Minitab or some other regression program to fit the following equations to these data:

$$y \approx \beta_0 + \beta_1 x_1 + \beta_2 x_2$$
$$y \approx \beta_0 + \beta_1 x_1 + \beta_2 \ln(x_2)$$
$$y \approx \beta_0 + \beta_1 x_1 + \beta_2 \ln(x_2) + \beta_3 x_1 \ln(x_2)$$

What are the R^2 values for the three different fitted equations? Compare the three fitted equations in terms of complexity and apparent ability to predict y.
(c) Compute the residuals for the third fitted equation in (b). Plot them against x_1, x_2, and $\hat{y}$. Also normal-plot them. Do any of these plots suggest that the third fitted equation is inadequate as summary of these data? What, if any, possible improvement over the third equation is suggested by these plots?
(d) As a means of understanding the nature of the third fitted equation in (b), make a scatterplot of y vs x_2 using a logarithmic scale for x_2. On this plot, plot three lines representing $\hat{y}$ as a function of x_2 for the three different values of x_1. Qualitatively, how would a similar plot for the second equation differ from this one?
(e) Using the third equation in (b), what mean grain diameter would you predict for $x_1 = 1500$ and $x_2 = 500$?
(f) It is possible to ignore the fact that the Temperature and Time factors are quantitative and make a factorial analysis of these data. Do so. Begin by making a plot similar to Figure 4-22 of the text for these data. Based on that plot, discuss the apparent relative sizes of the Time and Temperature main effects and the Time × Temperature interactions. Then carry out the arithmetic necessary to compute the fitted factorial effects (the fitted main effects and interactions).

4-18. The article "Cyanoacetamide Accelerators for the Epoxide/Isocyanate Reaction" by Eldin and Renner (*Journal of Applied Polymer Science*, 1990) reports the results of a 2^3 factorial experiment. Using cyanoacetamides as catalysts for an epoxy/isocyanate reaction, various mechanical properties of a resulting polymer were studied. One of these was

$$y = \text{impact strength } (kJ/mm^2)$$

The three experimental factors employed and their corresponding experimental levels were as follows.

Factor		Levels
A	Initial Epoxy/Isocyanate Ratio	0.4 (−) vs. 1.2 (+)
B	Flexibilizer Concentration	10 mol % (−) vs. 40 mol % (+)
C	Accelerator Concentration	1/240 mol % (−) vs. 1/30 mol% (+)

(The flexibilizer and accelerator concentrations are relative to the amount of epoxy present initially.) The impact strength data obtained (one observation per combination of levels of the three factors) were as follows.

Combination	y	Combination	y
(1)	6.7	c	6.3
a	11.9	ac	15.1
b	8.5	bc	6.7
ab	16.5	abc	16.4

(a) What is an obvious weakness in the researchers' data collection plan?

(b) Use the Yates algorithm and compute fitted factorial effects corresponding to the "all high" treatment combination (i.e., compute $\bar{y}_{...}$, a_2, b_2, etc.). Interpret these in the context of the original study. (Describe in words which factors and/or combinations of factors appear to have the largest effect(s) on impact strength, and interpret the sign or signs.)

(c) Suppose only factor A is judged to be of importance in determining impact strength. What predicted/fitted impact strengths correspond to this judgment? (Find $\hat{y}$ values using the reverse Yates algorithm or otherwise.) Use these eight values $\hat{y}$ and compute R^2 for the "A main effects only" description of impact strength. (The formula in Definition 4-3 works in this context, as well as in regression.)

(d) Now recognize that the experimental factors here are quantitative, so methods of curve and surface fitting may be applicable. Fit the equation $y \approx \beta_0 + \beta_1$ (epoxy/isocyanate ratio) to the data. What eight values $\hat{y}$ and value of R^2 accompany this fit?

4-19. Timp and M-Sidek studied the strength of mechanical pencil lead. They taped pieces of lead to a desk, with various lengths protruding over the edge of the desk. After fitting a small piece of tape on the free end of a lead piece to act as a stop, they loaded it with paper clips until failure. In one part of their study, they tested leads of two different Diameters, used two different Lengths protruding over the edge of the desk, and tested two different lead Hardnesses. That is, they ran a 2^3 factorial study. Their factors and levels were as follows

Factor A	Diameter	.3 mm (−) vs. .7 mm (+)
Factor B	Length Protruding	3 cm (−) vs. 4.5 cm (+)
Factor C	Hardness	B (−) vs. 2H (+)

and $m = 2$ trials were made at each of the $2^3 = 8$ different sets of conditions. The data the students obtained are given here.

(a) It appears that analysis of these data in terms of the natural logarithms of the numbers of clips first causing failure is more straightforward than the analysis of the raw numbers of clips. So take natural logs and then use the Yates algorithm of Section

Combination	Number of Clips	Combination	Number of Clips
(1)	13, 13	c	16, 15
a	74, 76	ac	89, 88
b	9, 10	bc	10, 12
ab	43, 42	abc	54, 55

4-3 to compute the 2^3 factorial fitted effects. Interpret these fitted effects. In particular, what (in quantitative terms) does the size of the fitted A main effect say about lead strength? Does lead hardness appear to play a dominant role in determining this kind of breaking strength?

(b) Suppose only the main effects of Diameter are judged to be of importance in determining lead strength. Find a predicted log breaking strength for .7 mm, 2H lead when the length protruding is 4.5 cm. Use this to predict the number of clips required to break such a piece of lead.

(c) What, if any, engineering reasons do you have for expecting the analysis of breaking strength to be more straightforward on the log scale than on the original scale?

4-20. Ceramic engineering researchers Leigh and Taylor, in their paper "Computer Generated Experimental Designs" (*Ceramic Bulletin*, 1990), studied the packing properties of crushed T-61 tabular alumina powder. The densities of batches of the material were measured under a total of eight different sets of conditions having a 2^3 factorial structure. That is, the following factors and levels were employed in the study.

Factor A	Mesh Size of Powder Particles	6 mesh (−) vs. 60 mesh (+)
Factor B	Volume of Graduated Cylinder	100 cc (−) vs. 500 cc (+)
Factor C	Vibration of Cylinder	no (−) vs. yes (+)

The mean densities (in g/cc) obtained in 5 determinations for each set of conditions were as follows.

$$\bar{y}_{(1)} = 2.348, \ \bar{y}_a = 2.080, \ \bar{y}_b = 2.298, \ \bar{y}_{ab} = 1.980$$
$$\bar{y}_c = 2.354, \ \bar{y}_{ac} = 2.314, \ \bar{y}_{bc} = 2.404, \ \bar{y}_{abc} = 2.374$$

(a) Compute the fitted 2^3 factorial effects (main effects, 2-factor interactions and 3-factor interaction) corresponding to the following set of conditions: 60 mesh, 500 cc, vibrated cylinder.

(b) If your arithmetic for part (a) is correct, you should have found that the largest of the fitted effects (in absolute value) are (respectively) the C main effect, the A main effect, and then the AC 2-factor interaction. (The next largest fitted effect is only about half of the smallest of these, the AC interaction.) In any case, suppose henceforth that one judges these three fitted effects to summarize the main features of the data set. Interpret this data summary (A and C main effects and AC interactions) in the context of this 3-factor study.

(c) Using your fitted effects from (a) and the data summary from (b) (A and C main effects and AC interactions), what fitted response would you have for these conditions: 60 mesh, 500 cc, vibrated cylinder?

(d) Using your fitted effects from (a), what average change in density would you say accompanies the vibration of the graduated cylinder before density determination?

4-21. The article "An Analysis of Transformations" by Box and Cox (*Journal of the Royal Statistical Society, Series B*, 1964) contains a classical unreplicated 3^3 factorial data set originally taken from an unpublished technical report of Barella and Sust. These researchers studied the behavior of worsted yarns under repeated loading. The response variable was

$$y = \text{the numbers of cycles till failure}$$

for specimens tested with various values of

x_1 = length (250 mm, 300 mm, or 350 mm)

x_2 = amplitude of the loading cycle (8 mm, 9 mm, 10 mm)

x_3 = load (40 g, 45 g, and 50 g)

The researchers' data are given in the accompanying table.

x_1	x_2	x_3	y
250	8	40	674
250	8	45	370
250	8	50	292
250	9	40	338
250	9	45	266
250	9	50	210
250	10	40	170
250	10	45	118
250	10	50	90
300	8	40	1,414
300	8	45	1,198
300	8	50	634
300	9	40	1,022
300	9	45	620
300	9	50	438
300	10	40	442
300	10	45	332
300	10	50	220
350	8	40	3,636
350	8	45	3,184
350	8	50	2,000
350	9	40	1,568
350	9	45	1,070
350	9	50	566
350	10	40	1,140
350	10	45	884
350	10	50	360

(a) In an attempt to find an equation to represent these data, one might first try to fit multivariable polynomials. Use a regression program and fit a full quadratic equation to these data. That is, fit

$$y \approx \beta_0 + \beta_1 x_1 + \beta_2 x_2 + \beta_3 x_3 + \beta_4 x_1^2 + \beta_5 x_2^2 + \beta_6 x_3^2 + \beta_7 x_1 x_2 + \beta_8 x_1 x_3 + \beta_9 x_2 x_3$$

to the data. What fraction of the observed variation in y does it account for? In terms of parsimony (or providing a simple data summary), how does this quadratic equation do as a data summary?

(b) Notice the huge range of values of the response variable. In cases like this, where the response varies over an order of magnitude, taking logarithms of the response often helps produce a simple fitted equation. Here, take (natural) logarithms of all of x_1, x_2, x_3, and y, producing (say) x_1', x_2', x_3', and y', and fit the equation

$$y' \approx \beta_0 + \beta_1 x_1' + \beta_2 x_2' + \beta_3 x_3'$$

to the data. What fraction of the observed variability in $\ln(y)$ does this equation account for? What change in y' seems to accompany a unit (a $1 \ln(g)$) increase in x_3'?

(c) To carry the analysis one step further, note that your fitted coefficients for x_1' and x_2' are nearly the negatives of each other. That suggests that y' depends only on the difference between x_1' and x_2'. To see how this conjecture works, fit the equation

$$y' \approx \beta_0 + \beta_1 (x_1' - x_2') + \beta_2 x_3'$$

to the data. Compute and plot residuals from this relationship (still on the log scale). How does this relationship appear to do as a data summary? What power law for y (on the original scale) in terms of x_1, x_2, and x_3 (on their original scales) is implied by this last fitted equation? How does this equation compare to the one from (a) in terms of parsimony?

(d) Use your equation from (c) to predict the life of an additional specimen of length 300 mm, at an amplitude of 9 mm, under a load of 45 g. Do the same for an additional specimen of length 325 mm, at an amplitude of 9.5 mm, under a load of 47 g. Why would or wouldn't you be willing to make a similar projection for an additional specimen of length 375 mm, at an amplitude of 10.5 mm, under a load of 51 g?

4-22. Bauer, Dirks, Palkovic and Wittmer fired tennis balls out of "Polish cannons" inclined at an angle of 45°, using three different Propellants and two different Charge Sizes of propellant. They observed the distances traveled in the air by the tennis balls. Their data are given in the accompanying table. (Five trials were made for each Propellant/Charge Size combination and the values given are in feet.)

		PROPELLANT		
		Lighter Fluid	Gasoline	Carburetor Fluid
		58	76	90
		50	79	86
	2.5 ml	53	84	79
		49	73	82
		59	71	86
CHARGE SIZE				
		65	96	107
		59	101	102
	5.0 ml	61	94	91
		68	91	95
		67	87	97

Complete a factorial analysis of these data, including a plot of sample means useful for judging the size of Charge Size × Propellant interactions and the computing of fitted main effects and interactions. Write a paragraph summarizing what these data seem to say about how these two variables affect flight distance.

4-23. Below are some data taken from the article "An Analysis of Means for Attribute Data Applied to a 2^4 Factorial Design" by R. Zwickl (*ASQC Electronics Division Technical Supplement*, Fall 1985). They represent numbers of bonds (out of 96) showing evidence of ceramic pull-out on an electronic device called a dual in-line package. (Low numbers are good.) Experimental factors and their levels were as in the accompanying table.

A	Ceramic Surface	unglazed ($-$) vs. glazed ($+$)
B	Metal Film Thickness	normal ($-$) vs. 1.5 times normal ($+$)
C	Annealing Time	normal ($-$) vs. 4 times normal ($+$)
D	Prebond Clean	normal clean ($-$) vs. no clean ($+$)

The resultant numbers of pull-outs for the 2^4 treatment combinations are given next.

Combination	Pull-Outs	Combination	Pull-Outs
(1)	9	d	3
a	70	ad	6
b	8	bd	1
ab	42	abd	7
c	13	cd	5
ac	55	acd	28
bc	7	bcd	3
abc	19	abcd	6

(a) Use the Yates algorithm and identify dominant effects here.

(b) Based on your analysis from (a), postulate a possible "few effects" explanation for these data. Use the reverse Yates algorithm to find fitted responses for such a simplified description of the system. Use the fitted values to compute residuals. Normal-plot these and plot them against levels of each of the four factors, looking for obvious problems with your representation of system behavior.

(c) Based on your "few effects" description of bond strength, make a recommendation for the future making of these devices. (All else being equal, you should choose what appear to be the least expensive levels of factors.)

4-24. Exercise 3-18 concerns a replicated 3^3 factorial data set (weighings of three different masses on three different scales by three different students). Use a full-featured statistical package that will compute fitted effects for such data and write a short summary report stating what those fitted effects reveal about the structure of the weighings data.

4-25. When it is an appropriate description of a two-way factorial data set, what practical engineering advantages does a "main effects only" description offer over a "main effects plus interactions" description?

4-26. The article referred to in Exercise 4-6 actually considers the effects of both cutting speed and feed rate on tool life. The whole data set from the article follows. (The data in Exercise 4-6 are the $x_2 = .01725$ data only.)

(a) Taylor's expanded tool life equation is $yx_1^{\alpha_1} x_2^{\alpha_2} = C$. This relationship suggests that $\ln(y)$ may well be approximately linear in both $\ln(x_1)$ and $\ln(x_2)$. Use a multiple linear regression program

Cutting Speed, x_1 (sfpm)	Feed, x_2 (ipr)	Tool Life, y (min)
800	.01725	1.00, 0.90, 0.74, 0.66
700	.01725	1.00, 1.20, 1.50, 1.60
700	.01570	1.75, 1.85, 2.00, 2.20
600	.02200	1.20, 1.50, 1.60, 1.60
600	.01725	2.35, 2.65, 3.00, 3.60
500	.01725	6.40, 7.80, 9.80, 16.50
500	.01570	8.80, 11.00, 11.75, 19.00
450	.02200	4.00, 4.70, 5.30, 6.00
400	.01725	21.50, 24.50, 26.00, 33.00

to fit the relationship

$$\ln(y) \approx \beta_0 + \beta_1 \ln(x_1) + \beta_2 \ln(x_2)$$

to these data. What fraction of the raw variability in $\ln(y)$ is accounted for in the fitting process? What estimates of the parameters α_1, α_2, and C follow from your fitted equation?

(b) Compute and plot residuals (continuing to work on log scales) for the equation you fit in part (a). Make at least plots of residuals versus fitted $\ln(y)$ and both $\ln(x_1)$ and $\ln(x_2)$, and make a normal plot of these residuals. Do these plots reveal any particular problems with the fitted equation?

(c) Use your fitted equation to predict first a log tool life and then a tool life, if in this machining application a cutting speed of 550 and a feed of .01650 were used.

(d) Plot the ordered pairs appearing in the data set in the (x_1, x_2)-plane. Outline a region in the plane where you would feel reasonably safe using the equation you fit in part (a) to predict tool life.

4-27. K. Casali conducted a gas mileage study on his well-used four-year-old economy car. He drove a 107-mile course a total of eight different times (in comparable weather conditions) at four different speeds, using two different types of gasoline, and ended up with an unreplicated 4×2 factorial study. His data are given in the accompanying table.

Test	Speed (mph)	Gasoline Octane	Gallons Used	Mileage (mpg)
1	65	87	3.2	33.4
2	60	87	3.1	34.5
3	70	87	3.4	31.5
4	55	87	3.0	35.7
5	65	90	3.2	33.4
6	55	90	2.9	36.9
7	70	90	3.3	32.4
8	60	90	3.0	35.7

(a) Make a plot of the mileages that is useful for judging the size of Speed $\times$ Octane interactions. Does it look as if the interactions are large in comparison to the main effects?

(b) Compute the fitted main effects and interactions for the mileages, using the formulas of Section 4-3. Make a plot like Figure 4-23 for comparing the observed mileages to fitted mileages computed supposing that there are no Speed × Octane interactions.

(c) Now fit the equation

$$\text{Mileage} \approx \beta_0 + \beta_1(\text{Speed}) + \beta_2(\text{Octane})$$

to the data and plot lines representing the predicted mileages versus Speed for both the 87 octane and the 90 octane gasolines on the same set of axes.

(d) Now fit the equation Mileage $\approx \beta_0 + \beta_1$(Speed) separately, first to the 87 octane data and then to the 90 octane data. Plot the two different lines on the same set of axes.

(e) Discuss the different appearances of the plots you made in parts (a) through (d) of this exercise, in terms of how well they fit the original data and the different natures of the assumptions involved in producing them.

(f) What was the fundamental weakness in Casali's data collection scheme? A weakness of secondary importance has to do with the fact that tests 1–4 were made 10 days earlier than tests 5–8. Why is this a potential problem?

4-28. The article "Accelerated Testing of Solid Film Lubricants" by Hopkins and Lavik (*Lubrication Engineering*, 1972) contains a nice example of the engineering use of multiple regression. In the study, $m = 3$ sets of journal bearing tests were made on a Mil-L-8937 type film at each combination of three different Loads and three different Speeds. The wear lives of journal bearings, y, in hours, are given next for the tests run by the authors.

Speed, x_1 (rpm)	Load, x_2 (psi)	Wear Life, y (hours)
20	3,000	300.2, 310.8, 333.0
20	6,000	99.6, 136.2, 142.4
20	10,000	20.2, 28.2, 102.7
60	3,000	67.3, 77.9, 93.9
60	6,000	43.0, 44.5, 65.9
60	10,000	10.7, 34.1, 39.1
100	3,000	26.5, 22.3, 34.8
100	6,000	32.8, 25.6, 32.7
100	10,000	2.3, 4.4, 5.8

(a) The authors expected to be able to describe wear life as roughly following the relationship $yx_1x_2 = C$, but they did not find this relationship to be a completely satisfactory model. So instead, they tried using the more general relationship $yx_1^{\alpha_1}x_2^{\alpha_2} = C$. Use a multiple linear regression program to fit the relationship

$$\ln(y) \approx \beta_0 + \beta_1 \ln(x_1) + \beta_2 \ln(x_2)$$

to these data. What fraction of the raw variability in $\ln(y)$ is accounted for in the fitting process? What estimates of the parameters α_1, α_2, and C follow from your fitted equation? Using your estimates of α_1, α_2, and C, plot on the same set of (x_1, y) axes the functional relationships between x_1 and y implied by your fitted equation, for x_2 equal to 3,000, 6,000, and then 10,000 psi, respectively.

(b) Compute and plot residuals (continuing to work on log scales) for the equation you fit in part (a). Make at least plots of residuals

versus fitted $\ln(y)$ and both $\ln(x_1)$ and $\ln(x_2)$, and make a normal plot of these residuals. Do these plots reveal any particular problems with the fitted equation?

(c) Use your fitted equation to predict first a log wear life and then a wear life, if in this application a speed of 20 rpm and a load of 10,000 psi are used.

(d) (Accelerated life testing) As a means of trying to make intelligent data-based predictions of wear life at low stress levels (and correspondingly large lifetimes that would be impractical to observe directly), one might (fully recognizing the inherent dangers of the practice) try to extrapolate using the fitted equation. Use your fitted equation to predict first a log wear life and then a wear life, if a speed of 15 rpm and load of 1,500 psi are used in this application.

4-29. Sari and Al-Ramdhan experimented with an electronic lab kit — in particular, a strobe circuit made using that kit. By varying two capacitances and a resistance, they were able to change the period of oscillation of the circuit in question. They made $m = 3$ counts of the number of cycles completed in 30 seconds for each of 16 different configurations. Their data are shown in the accompanying table.

Resistance, x_1 (Ω)	Capacitance 1, x_2 (μF)	Capacitance 2, x_3 (μF)	Cycles, y
470	200	13.3	37, 36, 37
470	200	3.3	26, 26, 26
100	200	3.3	11, 11, 10
100	200	13.3	13, 14, 13
470	50	13.3	121, 124, 124
470	50	10	127, 126, 129
100	50	3.3	76, 73, 73
100	50	13.3	125, 127, 128
100	100	10	33, 32, 32
470	100	10	57, 54, 55
330	50	10	94, 92, 98
330	200	10	31, 30, 31
330	100	3.3	42, 43, 41
330	100	13.3	41, 42, 41
330	100	10	42, 42, 41
330	100	3.3	76, 72, 77
470	200	13.3	37, 36, 37
470	200	13.3	37, 36, 37

(a) It is possible that if the students had known enough about circuit analysis, they could have come up with an equation based on circuit theory to predict y from x_1, x_2, and x_3. But they were not experts in the subject, so they decided to try and find empirical relationships that might be used to predict the frequency or period of the circuit from the values of x_1, x_2, and x_3. Fit the following four types of equations as descriptions of the relationship of y to x_1, x_2, and x_3.

    (i) An equation for y that is linear in x_1, x_2, and x_3.

    (ii) An equation for y^{-1} that is linear in x_1, x_2, and x_3.

    (iii) A full quadratic equation for y in x_1, x_2, and x_3. (Include linear and quadratic terms in each of x_1, x_2, and x_3, as well as all possible cross-product terms.)

TABLE For Exercise 4-30

Alcohol	Carbon #	x_1	x_2	x_3	x_4	x_5	y
1-pentanol	5	794	138	−78	.811	1.4093	18.2
1-hexanol	6	950	156.5	−52	.814	1.4179	23.3
1-heptanol	7	1105	176	−36	.822	1.4232	29.5
1-octanol	8	1262	196	−15	.827	1.4290	39.1
1-nonanol	9	1422	215	−6	.827	1.4334	46.2
1-decanol	10	1582	231	7	.829	1.4370	50.3
1-undecanol	11	1736	243	11	.830	1.4400	53.2
1-dodecanol	12	1899	260	24	.831		63.6
1-tetradeconol	14	2202	289	38			80.8

(iv) A full quadratic equation for y^{-1} in x_1, x_2, and x_3. (Include linear and quadratic terms in each of x_1, x_2, and x_3, as well as all possible cross-product terms.)

(b) Choose one of the equations from (a) that appears to do an adequate job of fitting the students' data and is as simple as possible. For that equation, compute and plot residuals. (Plot residuals versus fitted values, x_1, x_2, and x_3, and normal-plot the residuals. If your equation is for y^{-1}, compute and plot residuals on the inverse scale.) Do the residual plots indicate any problems with your equation as a description of how y varies with x_1, x_2, and x_3?

(c) Use the fourth fitted equation and predict first y^{-1} and then y for $x_1 = 200 \ \Omega$, $x_2 = 150 \ \mu F$, and $x_3 = 8 \ \mu F$.

4-30. The article "Predicting Cetane Number of *n*-Alcohols and Methyl Esters from Their Physical Properties" by Freedman and Bagby (*Journal of American Oil Chemists' Society*, 1990) considers the problem of potentially avoiding the expensive and time-consuming determination of

> y = Cetane number (a measure of fuel performance determined in a special engine according to ASTM D 613)

for certain alcohols (and esters) by developing equations relating cetane number to more easily determined physical properties. The authors measured y for pure samples of various alcohols and referred to standard handbooks and research articles for values of

> x_1 = Heat of combustion (kg-cal/mole)
>
> x_2 = Boiling point (°C at 760 mm)
>
> x_3 = Melting point (°C)
>
> x_4 = Density
>
> x_5 = Refractive index

for the pure alcohols. (The authors mention the importance of later extending their work to blends, as opposed to the pure alcohols and esters used in the study.) The values reported by the authors are given in the accompanying table.

(a) Fit a linear equation for predicting cetane number from the carbon number of pure alcohols. What is R^2 for this equation?

(b) The researchers found linear equations in each of x_1 and x_5 to be adequate for predicting y. Is either of these approximately linear relationships as strong as the approximately linear relationship between carbon number and cetane number? Explain.

(c) The researchers reported quadratic equations in each of x_2, x_3, and x_4 for predicting y. For each of x_2, x_3, and x_4, fit first a linear equation for y and compute and plot residuals. Then fit quadratic equations. In each case, does consideration of R^2 values and the appearance of the residual plots indicate that the quadratic fit is substantially more effective than the corresponding linear one?

4-31. The article "Statistical Methods for Controlling the Brown Oxide Process in Multilayer Board Processing" by S. Imadi (*Plating and Surface Finishing*, 1988) discusses an experiment conducted to help a circuit board manufacturer measure the concentration of important components in a chemical bath. Various combinations of levels of

> x_1 = % by volume of component A (a proprietary formulation, the major component of which is sodium chlorite)

and

> x_2 = % by volume of component B (a proprietary formulation, the major component of which is sodium hydroxide)

were set in the chemical bath, and the variables

> y_1 = ml of 1N H_2SO_4 used in the first phase of a titration
>
> y_2 = ml of 1N H_2SO_4 used in the second phase of a titration

were measured. Part of the original data collected (corresponding to bath conditions free of Na_2CO_3) follow.

x_1	x_2	y_1	y_2
15	25	3.3	.4
20	25	3.4	.4
20	30	4.1	.4
25	30	4.3	.3
25	35	5.0	.5
30	35	5.0	.3
30	40	5.7	.5
35	40	5.8	.4

(a) Fit equations for both y_1 and y_2 linear in both of the variables x_1 and x_2. Does it appear that the variables y_1 and y_2 are adequately described as linear functions of x_1 and x_2?

(b) Solve your two fitted equations from (a) for x_2 (the concentration of primary interest here) in terms of y_1 and y_2. (Eliminate x_1, by solving the first for x_1 in terms of the other three variables and plugging that expression for x_1 into the second equation.) How does this equation seem to do in terms of, so to speak, predicting x_2 from y_1 and y_2 for the original data? Chemical theory in this situation indicated that $x_2 \approx 8(y_1 - y_2)$. Does your equation seem to do better than the one from chemical theory?

(c) A possible alternative to the calculations in (b) is to simply fit an equation for x_2 in terms of y_1 and y_2 directly via least squares. Fit $x_2 \approx \beta_0 + \beta_1 y_1 + \beta_2 y_2$ to the data, using a regression program. Is this equation the same one as you found in part (b)?

(d) If one were to compare the equations for x_2 derived in (b) and (c) in terms of the sum of squared differences between the predicted and observed values of x_2, which is guaranteed to be the winner? Why?

4-32. The article "Nonbloated Burned Clay Aggregate Concrete" by Martin, Ledbetter, Ahmad, and Britton (*Journal of Materials*, 1972) contains data on both composition and resulting physical property test results for a number of different batches of concrete made using burned clay aggregates. The accompanying data are compressive strength measurements (made according to ASTM C 39 and recorded in psi), y, and splitting tensile strength measurements (made according to ASTM C 496 and recorded in psi), x, for 10 of the batches used in the study.

Batch	1	2	3	4	5
y	1420	1950	2230	3070	3060
x	207	233	254	328	325
Batch	6	7	8	9	10
y	3110	2650	3130	2960	2760
x	302	258	335	315	302

(a) Make a scatterplot of these data and comment on how linear the relation between x and y appears to be for concretes of this type.

(b) Compute the sample correlation between x and y by hand. Interpret this value.

(c) Fit a line to these data using the least squares principle. Show the necessary hand calculations. Sketch this fitted line on your scatterplot from (a).

(d) About what increase in compressive strength appears to accompany an increase of 1 psi in splitting tensile strength?

(e) What fraction of the raw variability in compressive strength is accounted for in the fitting of a line to the data?

(f) Based on your answer to (c), what measured compressive strength would you predict for a batch of concrete of this type, if you were to measure a splitting tensile strength of 245 psi?

(g) Compute the residuals from your fitted line. Plot the residuals against x and against $\hat{y}$. Then make a normal plot of the residuals. What do these plots indicate about the linearity of the relationship between splitting tensile strength and compressive strength?

(h) Use Minitab (or some other statistical package available to you) to find the least squares line, the sample correlation, R^2, and the residuals for these data.

(i) Fit the quadratic relationship $y \approx \beta_0 + \beta_1 x + \beta_2 x^2$ to the data, using Minitab (or some other statistical package available to you). Sketch this fitted parabola on your scatterplot from part (a). Does this fitted quadratic appear to be an important improvement over the line you fit in (c), in terms of describing the relationship of y to x?

(j) How do the R^2 values from parts (h) and (i) compare? Does the increase in R^2 in part (i) speak strongly for the use of the quadratic (as opposed to linear) description of the relationship of y to x for concretes of this type?

(k) If one uses the fitted relationship from part (i) to predict y for $x = 245$, how does the prediction compare to your answer for part (f)?

(l) What do the fitted relationships from parts (c) and (i) give for predicted compressive strengths when $x = 400$ psi? Do these compare to each other as well as your answers to parts (f) and (k)? Why would it be unwise to use either of these predictions without further data collection and analysis?

4-33. In Exercise 4-32, both x and y were really response variables; as such, they were not subject to direct manipulation by the experimenters. That made it difficult to get several (x, y) pairs with a single x value into the data set. In experimental situations where an engineer gets to choose values of an experimental variable x, why is it useful/important to get several y observations for at least some x's?

4-34. Chemical Engineering graduate student S. Osoka studied the effects of an agitator speed, x_1, and a polymer concentration, x_2, on percent recoveries of pyrite, y_1, and kaolin, y_2, from a step of an ore refining process. (High pyrite recovery and low kaolin recovery rates were desirable.) Data from one set of $n = 9$ experimental runs are given here.

x_1 (rpm)	x_2 (ppm)	y_1 (%)	y_2 (%)
1350	80	77	67
950	80	83	54
600	80	91	70
1350	100	80	52
950	100	87	57
600	100	87	66
1350	120	67	54
950	120	80	52
600	120	81	44

(a) What type of data structure did the researcher employ? (Use the terminology of Section 1-2.) What was an obvious weakness in his data collection plan?

(b) Use Minitab or some other regression program to fit the following equations to these data:

$$y_1 \approx \beta_0 + \beta_1 x_1$$
$$y_1 \approx \beta_0 + \beta_2 x_2$$
$$y_1 \approx \beta_0 + \beta_1 x_1 + \beta_2 x_2$$

What are the R^2 values for the three different fitted equations?

Compare the three fitted equations in terms of complexity and apparent ability to predict y_1.

(c) Compute the residuals for the third fitted equation in part (b). Plot them against x_1, x_2, and $\hat{y}_1$. Also normal-plot them. Do any of these plots suggest that the third fitted equation is inadequate as summary of these data?

(d) As a means of understanding the nature of the third fitted equation from part (b), make a scatterplot of y_1 vs x_2. On this plot, plot three lines representing $\hat{y}_1$ as a function of x_2 for the three different values of x_1 represented in the data set.

(e) Using the third equation from part (b), what pyrite recovery rate would you predict for $x_1 = 1000$ rpm and $x_2 = 110$ ppm?

(f) Consider also a multivariable quadratic description of the dependence of y_1 on x_1 and x_2. That is, fit the equation

$$y_1 \approx \beta_0 + \beta_1 x_1 + \beta_2 x_2 + \beta_3 x_1^2 + \beta_4 x_2^2 + \beta_5 x_1 x_2$$

to the data. How does the R^2 value here compare with the ones in part (b)? As a means of understanding this fitted equation, plot on a single set of axes the three different quadratic functions of x_2 obtained by holding x_1 at one of the values of x_1 represented in the data set.

(g) It is possible to ignore the fact that the speed and concentration factors are quantitative and make a factorial analysis of these data. Do so. Begin by making a plot similar to Figure 4-22 for these data. Based on that plot, discuss the apparent relative sizes of the Speed and Concentration main effects and the Speed × Concentration interactions. Then carry out the arithmetic necessary to compute the fitted factorial effects (the fitted main effects and interactions).

(h) If the third equation in part (b) governed y_1, would it lead to Speed × Concentration interactions? What about the equation in part (f)? Explain.

4-35. The data given in Exercise 4-34 concern both responses y_1 and y_2. The analysis suggested in Exercise 4-34 concerned only y_1. Redo all parts of Exercise 4-34, replacing the response y_1 with y_2 throughout.

4-36. K. Fellows conducted a 4-factor experiment, with the response variable the flight distance of a paper airplane when propelled from a launcher fabricated specially for the study. This exercise concerns part of the data he collected, constituting a complete 2^4 factorial. The experimental factors involved and levels used were as given here.

> A   Plane Design   straight wing ($-$) vs tee wing ($+$)
> B   Nose Weight   none ($-$) vs paper clip ($+$)
> C   Paper Type   notebook ($-$) vs construction ($+$)
> D   Wing Tips   straight ($-$) vs bent up ($+$)

The mean flight distances (ft), y, recorded by Fellows for two launches of each plane were as shown in the accompanying table.

(a) Use the Yates algorithm and compute the fitted factorial effects corresponding to the "all high" treatment combination.

(b) Interpret the results of your calculations from (a) in the context of the study. (Describe in words which factors and/or combinations of factors appear to have the largest effect(s) on flight distance. What are the practical implications of these effects?)

Combination	y	Combination	y
(1)	6.25	c	4.75
a	15.50	ac	5.50
b	7.00	bc	4.50
ab	16.50	abc	6.00

Combination	y	Combination	y
d	7.00	cd	4.50
ad	10.00	acd	6.00
bd	10.00	bcd	4.50
abd	16.00	abcd	5.75

(c) Suppose Factors B and D are judged to be inert as far as determining flight distance is concerned. (The main effects of B and D and all interactions involving them are negligible.) What fitted/predicted values correspond to this description of flight distance (A and C main effects and AC interactions only)? Use these 16 values $\hat{y}$ to compute residuals, $y - \hat{y}$. Plot these against $\hat{y}$, levels of A, levels of B, levels of C, and levels of D. Also normal-plot these residuals. Comment on any interpretable patterns in your plots.

(d) Compute an R^2 value corresponding to the description of flight distance used in part (c). (The formula in Definition 4-3 works in this context, as well as in regression. So does the representation of R^2 as the squared sample correlation between y and $\hat{y}$.) Does it seem that the grand mean, A and C main effects, and AC 2-factor interactions provide an effective summary of flight distance?

4-37. The data in the accompanying table appear in the text *Quality Control and Industrial Statistics* by Duncan (taken originally from a paper of L.E. Simon). The data were collected in a study of the effectiveness of armor plate. Armor-piercing bullets were fired at an angle of 40° against armor plate of thickness x_1 (in .001 in.) and Brinell hardness number x_2, and the resulting so-called ballistic limit, y (in ft/sec), was measured.

x_1	x_2	y
253	317	927
258	321	978
259	341	1028
247	350	906
256	352	1159
246	363	1055
257	365	1335
262	375	1392
255	373	1362
258	391	1374
253	407	1393
252	426	1401
246	432	1436
250	469	1327
242	257	950
243	302	998
239	331	1144
242	355	1080
244	385	1276
234	426	1062

(a) Use Minitab or some other regression program to fit the following equations to these data:

$$y \approx \beta_0 + \beta_1 x_1$$
$$y \approx \beta_0 + \beta_2 x_2$$
$$y \approx \beta_0 + \beta_1 x_1 + \beta_2 x_2$$

What are the R^2 values for the three different fitted equations? Compare the three fitted equations in terms of complexity and apparent ability to predict y.

(b) What is the correlation between x_1 and y? The correlation between x_2 and y?

(c) Based on (a) and (b), describe how strongly Thickness and Hardness appear to affect ballistic limit. Review the raw data and speculate as to why the variable with the smaller influence on y seems to be of only minor importance in this data set (although logic says that it must in general have a sizable influence on y).

(d) Compute the residuals for the third fitted equation from (a). Plot them against x_1, x_2, and $\hat{y}$. Also normal-plot them. Do any of these plots suggest that the third fitted equation is seriously deficient as a summary of these data?

(e) Plot the (x_1, x_2) pairs represented in the data set. Why would it be unwise to use any of the fitted equations to predict y for $x_1 = 265$ and $x_2 = 440$?

4-38. Basgall, Dahl, and Warren experimented with smooth and treaded bicycle tires of different widths. Tires were mounted on the same wheel, placed on a bicycle wind trainer, and accelerated to a velocity of 25 miles per hour. Then pedaling was stopped, and the time required for the wheel to cease rolling was recorded. The sample means, y, of five trials for each of six different tires were as follows.

Tire Width	Tread	Time to Stop, y (sec)
700/19c	smooth	7.30
700/25c	smooth	8.44
700/32c	smooth	9.27
700/19c	treaded	6.63
700/25c	treaded	6.87
700/32c	treaded	7.07

(a) Carefully make a plot of times required to stop, useful for investigating the sizes of Width and Tread main effects and Width $\times$ Tread interactions here. Comment briefly on what the plot shows about these effects. Be sure to label the plot very clearly.

(b) Compute the fitted main effects of Width, the fitted main effects of Tread, and the fitted Width $\times$ Tread interactions from the responses above. Discuss how they quantify features that are evident in your plot from (a).

5 Probability: The Mathematics of Randomness

T he theory of probability is the mathematician's description of random variation. A central concept in that theory is the notion of a random (or chance) variable. This chapter introduces that and enough related concepts to provide a minimum background for the application of probability theory in making formal statistical inferences.

The chapter begins with a discussion of discrete random variables, their distributions, and some useful standard discrete models. It next turns to generalities for continuous random variables and some useful standard continuous distributions. Then probability plotting is discussed in general terms, followed by an exploration of its use in assessing the extent to which a data set indicates the appropriateness of a particular continuous probability distribution as a model for a given random variable. Next, the simultaneous modeling of several random variables and the notion of independence are considered. Finally, there is a brief look at random variables that arise as functions of several others, and how randomness of the input variables is translated to randomness of the output variable.

5-1 Discrete Random Variables

The concept of a random variable is introduced in general terms in this section, and then specialization for discrete cases is considered. The notion of a (theoretical) probability distribution for a discrete random variable is discussed. Next are specification of such a distribution via a probability function or cumulative probability function and summarization of discrete distributions in terms of (theoretical) means and variances. Then the so-called binomial, geometric, and Poisson distributions are introduced as examples of useful standard discrete probability models.

Random Variables and Their Distributions

When one collects a piece of data, it is natural to think of the value obtained as subject to chance influences, for a number of reasons. In enumerative contexts, it is common to purposely introduce chance into the data collection process through the use of random sampling techniques. Measurement error is nearly always a factor in statistical engineering studies, and the many small, unnamable causes that work to produce it are conveniently thought of as chance phenomena. In analytical contexts, both transient and reoccurring small changes in system conditions work to make measured responses variable, and this variation is most often conceived of as chance fluctuation. The following terminology will be used for quantities whose values are subject to chance influences.

DEFINITION 5-1

A **random variable** is a quantity that (prior to observation) can be thought of as dependent on chance phenomena. Capital letters near the end of the alphabet are typically used to stand for random variables.

As an example of the use of the notation mentioned in Definition 5-1, consider a situation (like that of Example 3-3) where there is measurement of the torques of bolts holding a component face plate in place. One could think of the next measured value as being subject to chance influences and thus term

$$Z = \text{the next torque recorded}$$

a random variable.

Following Definition 1-9, a distinction was made between discrete and continuous data. That terminology carries over to the present context and inspires two more definitions.

DEFINITION 5-2

A **discrete random variable** is a random variable that has isolated or separated possible values (rather than a continuum of available outcomes).

DEFINITION 5-3

A **continuous random variable** is a random variable that can be idealized as having an entire (continuous) interval of numbers as its set of possible values.

The same discussion that follows Definition 1-9 could be repeated here. Random variables that are basically count variables clearly fall under Definition 5-2 and are discrete. It could be argued that all measurement variables are discrete — on the basis that all measurements are "to the nearest unit." But it is often mathematically convenient, and adequate for practical purposes, to treat them as continuous.

A random variable is, to some extent, *a priori* unpredictable. Therefore, in describing or modeling it, the important thing is to somehow specify its set of potential values and the likelihoods associated with those possible values. That notion leads to the next piece of terminology.

DEFINITION 5-4

To specify a **probability distribution** for a random variable is to give its set of possible values and (in one way or another) consistently assign numbers between 0 and 1 — called **probabilities** — as measures of the likelihood that the various numerical values will occur.

The methods used to specify discrete probability distributions are different from those used to specify continuous probability distributions. So the implications of Definition 5-4 are studied in two steps, beginning in this section with discrete distributions.

Discrete Probability Functions and Cumulative Probability Functions

The basic tool most often used in describing a discrete probability distribution is the probability function.

DEFINITION 5-5

A **probability function** for a discrete random variable X, having possible values $x_1, x_2, \ldots$, is a nonnegative function $f(x)$, with $f(x_i)$ giving the probability that X takes the value x_i.

The remainder of this text uses the fairly standard notational convention that a capital P followed by an expression or phrase enclosed by brackets will be read "the probability" of that expression. In these terms, a probability function for X is a function f such that

$$f(x) = P[X = x]$$

That is, "$f(x)$ is the probability that (the random variable) X takes the value x."

EXAMPLE 5-1

A Torque Requirement Random Variable. Consider again the situation of Example 3-3, where Brenny, Christensen, and Schneider measured bolt torques on the face plates of a heavy equipment component. With

$$Z = \text{the next measured torque for bolt 3}$$

consider treating Z as a discrete random variable and giving a plausible probability function for it.

Finding the relative frequencies for, say, the bolt 3 torque measurements recorded in Table 3-4, one can arrive at the relative frequency distribution given in Table 5-1. This table shows, for example, that over the period the students were collecting data, about 15% of measured torques were 19 ft lb. Suppose one is to project what will happen at the next reading, i.e. talk about the random variable Z. If it is sensible to believe that the same system of causes that produced the data in Table 3-4 will be operating to produce the next bolt 3 torque, then perhaps it also makes sense to base a probability function for Z on the relative frequencies in Table 5-1. That is, specifying a probability function for Z by means of a table, one might employ the values in Table 5-2. (In going from the relative frequencies in Table 5-1 to proposed values for $f(z)$ in Table 5-2, there has been some slightly arbitrary rounding. This has been done so that probability values are expressed to two decimal places. Also, when summed over all possible values z of Z, the probabilities $f(z)$ now total to exactly 1.00.)

TABLE 5-1

Relative Frequency Distribution for Measured Bolt 3 Torques

z Torque (ft lb)	Frequency	Relative Frequency
11	1	$1/34 \approx .02941$
12	1	$1/34 \approx .02941$
13	1	$1/34 \approx .02941$
14	2	$2/34 \approx .05882$
15	9	$9/34 \approx .26471$
16	3	$3/34 \approx .08824$
17	4	$4/34 \approx .11765$
18	7	$7/34 \approx .20588$
19	5	$5/34 \approx .14706$
20	1	$1/34 \approx .02941$
	34	1

TABLE 5-2

A Probability Function for Z

Torque z	Probability $f(z)$
11	.03
12	.03
13	.03
14	.06
15	.26
16	.09
17	.12
18	.20
19	.15
20	.03

Notice several things about Example 5-1. First, although the appropriateness of the probability function in Table 5-2 for describing Z depends essentially on the physical stability of the bolt-tightening process being sampled, there is another way in which relative frequencies can become obvious choices for probabilities. For example, think of treating the 34 torques represented in Table 5-1 as a population, from which $n = 1$ item is to be sampled at random, and

$$Y = \text{the torque value selected}$$

Then the probability function in Table 5-2 is also approximately appropriate for Y. This point is not so important in the specific context of this example as it is in general: In enumerative situations where one item is to be selected at random from a population, an appropriate probability distribution for the value selected is one that is equivalent to the population relative frequency distribution.

Note that this text will usually express probabilities to two decimal places, as in Table 5-2. Although computations will often be carried to several more decimal places, final probabilities will typically be reported only to two places. This is because numbers expressed to more than two places tend to look too impressive and be taken too seriously by the uninitiated. For example, "There is a .097328 probability of booster engine failure" at a certain missile launch. This may represent the results of some very careful mathematical manipulations and be correct to six decimal places in the context of the mathematical model used to obtain the value. But it is doubtful that the model used is a good enough description of physical reality to warrant that much apparent precision. In this author's opinion and experience, apparent precision of two decimal places is about what is warranted in most engineering applications of simple probability.

The probability function shown in Table 5-2 has two properties that are necessary for the mathematical consistency of a discrete probability distribution. They are that the $f(z)$ values are each in the interval $[0, 1]$ and that they total to 1 when summed up. That is, negative probabilities or ones larger than 1 would make no practical sense. A probability of 1 is taken as indicating certainty of occurrence, and a probability of 0 as indicating certainty of nonoccurrence. Thus, according to the model specified in Table 5-2, since the values of $f(z)$ sum to 1, the occurrence of one of the values 11, 12, 13, 14, 15, 16, 17, 18, 19, and 20 ft lb is certain.

A probability function $f(x)$ gives probabilities of occurrence for individual values. Adding appropriate ones of these gives probabilities associated with occurrence of one of a specified type of value for X.

EXAMPLE 5-1
(continued)

As a simple example of the meaning of the preceding statement, consider using $f(z)$, defined in Table 5-2, and finding

$$P[Z > 17] = P[\text{the next torque exceeds 17}]$$

Adding the $f(z)$ entries corresponding to possible values larger than 17 ft lb, one obtains

$$P[Z > 17] = f(18) + f(19) + f(20) = .20 + .15 + .03 = .38$$

One would thus judge the physical likelihood of the next torque being more than 17 ft lb to be about 38%, presuming the model in Table 5-2 is still a sensible description of the next torque value.

If, for example, specifications for torques were 16 ft lb to 21 ft lb, then one could similarly assess the likelihood that the next torque measured will be within specifications as

$$P[16 \leq Z \leq 21] = f(16) + f(17) + f(18) + f(19) + f(20) + f(21)$$
$$= .09 + .12 + .20 + .15 + .03 + .00$$
$$= .59$$

In the torque measurement example, the probability function is most simply specified in tabular form. In many other cases, it is possible to give a formula for $f(x)$.

EXAMPLE 5-2

A Random Tool Serial Number. The last step of the pneumatic tool assembly process studied by Kraber, Rucker, and Williams (see Example 3-11) was to apply a serial number plate to the completed tool. Imagine going to the end of the assembly line at exactly 9:00 A.M. next Monday and observing the number plate whose application begins first after 9:00.

Suppose that

$$W = \text{the last digit of the serial number observed}$$

and consider giving a probability function for W. Suppose further that tool serial numbers begin with some code special to the tool model and end with consecutively assigned numbers reflecting how many tools of the particular model have been produced up through the one being finished. The symmetry of this situation suggests that each possible value of W ($w = 0, 1, \ldots, 9$) is equally likely. That is, a plausible probability function for W is given by the formula

$$f(w) = \begin{cases} .1 & \text{for } w = 0, 1, 2, \ldots, 9 \\ 0 & \text{otherwise} \end{cases}$$

Another way of specifying a discrete probability distribution is sometimes used rather than giving its probability function. That is to specify its *cumulative probability function*. Next, this concept will be defined, and how it is specialized to the case of discrete random variables will be discussed.

DEFINITION 5-6

The **cumulative probability function** for a random variable X is a function $F(x)$ that for each number x gives the probability that X takes that value or a smaller one. In symbols,

$$F(x) = P[X \leq x]$$

Since (for discrete distributions) probabilities are calculated by summing values of $f(x)$, for a discrete distribution,

$$F(x) = \sum_{z \leq x} f(z)$$

(The sum is over possible values less than or equal to x.) In this case, the graph of $F(x)$ will be a stair-step graph with jumps at possible values equal in size to the probabilities associated with those possible values.

EXAMPLE 5-1
(continued)

Values of both the probability function and the cumulative probability function for the torque variable Z are given in Table 5-3. Values of $F(z)$ for other z are also easily obtained. For example,

$$F(10.7) = P[Z \leq 10.7] = 0$$
$$F(16.3) = P[Z \leq 16.3] = P[Z \leq 16] = F(16) = .50$$
$$F(32) = P[Z \leq 32] = 1.00$$

TABLE 5-3

Values of the Probability
Function and Cumulative
Probability Function for Z

z Torque	$f(z) = P[Z = z]$	$F(z) = P[Z \leq z]$
11	.03	.03
12	.03	.06
13	.03	.09
14	.06	.15
15	.26	.41
16	.09	.50
17	.12	.62
18	.20	.82
19	.15	.97
20	.03	1.00

FIGURE 5-1

Graph of the Cumulative
Probability Function for Z

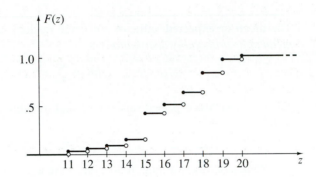

A graph of the cumulative probability function for Z is given in Figure 5-1. It shows the stair-step shape characteristic of cumulative probability functions for discrete distributions.

The information about a discrete distribution carried by a cumulative probability function is equivalent to that carried by the corresponding probability function. The cumulative version is sometimes preferred for table making, because round-off problems are more severe when adding several $f(x)$ terms than when taking the difference of two $F(x)$ values to get a probability associated with a consecutive sequence of possible values.

Summarization of Discrete Probability Distributions

Virtually all of the devices applied to describe relative frequency (empirical) distributions in Chapter 3 have versions that can be applied to describe (theoretical) probability distributions as well.

For a discrete random variable with equally spaced possible values, a *probability histogram* gives a picture of the shape of the variable's distribution. It is made by centering a bar of height $f(x)$ over each possible value x. Probability histograms for the random variables Z and W in Examples 5-1 and 5-2 are given in Figure 5-2. The interpretation of such probability histograms is quite similar to that of relative frequency histograms, except that the areas on them represent theoretical probabilities instead of empirical fractions of data sets. The first graph in Figure 5-2 shows (for example) a torque of 15 ft lb to be mostly likely, while the second shows a uniform probability distribution over the integer values 0 through 9 for the random variable W.

Just as it was convenient to use the arithmetic mean to represent the center of a data set (or its relative frequency distribution), it is convenient to have a notion of mean value for a discrete random variable (or its probability distribution).

FIGURE 5-2

Probability Histograms for
Z and *W*

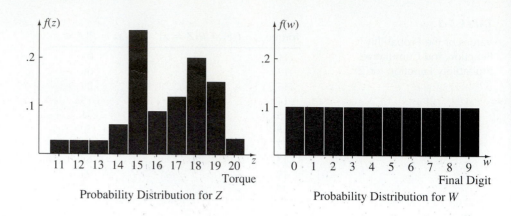

Probability Distribution for *Z*   Probability Distribution for *W*

DEFINITION 5-7

The **mean** or **expected value** of a discrete random variable *X* (sometimes called the *mean of its probability distribution*) is

$$EX = \sum_{x} x f(x) \tag{5-1}$$

EX is read as "the expected value of *X*," and sometimes the notation μ is used in place of *EX*.

Remember the warning in Section 3-3 that μ would stand for both the mean of a population and the mean of a probability distribution. This double use of the symbol makes reasonable sense. Earlier in this section, it was stated that when sampling a single item at random from a finite population, the random value selected has a probability distribution equivalent to the population relative frequency distribution. In such a context, the population mean μ and the expected value of *X*, *EX*, have the same numerical value, so no notational contradiction can occur. (To see that this consistency is built into Definition 5-7, consider calculating the mean of a data set by summing each distinct value times its number of occurrences in the data set and dividing by *N*. This is equivalent to summing each distinct value times its relative frequency of occurrence in the data set — a calculation parallel to that called for in formula (5-1).)

EXAMPLE 5-1
(continued)

Returning to the bolt torque example, consider finding the expected (or theoretical mean) value of the next torque.

$$EZ = \sum_{z} z f(z)$$

$$= 11(.03) + 12(.03) + 13(.03) + 14(.06) + 15(.26)$$
$$+ 16(.09) + 17(.12) + 18(.20) + 19(.15) + 20(.03)$$
$$= 16.35 \text{ ft lb}$$

Verify that this value is essentially the arithmetic mean of the bolt 3 torques listed in Table 3-4. (There is slight disagreement in the third decimal place, but only because the relative frequencies in Table 5-1 were rounded slightly to produce Table 5-2.)

The mean of a discrete probability distribution has a balance point interpretation, much like that associated with the arithmetic mean of a data set. If one thinks of placing point masses of mass $f(x)$ at positions x along a number line, EX is the center of mass of that mass distribution.

EXAMPLE 5-2
(continued)

Considering again the serial number example, and in particular looking at the second part of Figure 5-2, it is clear that if a balance point interpretation of expected value is to hold, EW had better turn out to be 4.5. And indeed,

$$EW = 0(.1) + 1(.1) + 2(.1) + \cdots + 8(.1) + 9(.1) = 45(.1) = 4.5$$

Just as it was convenient to measure the spread of a data set (or its relative frequency distribution) with the variance and standard deviation, it is convenient to have similar notions for a discrete random variable (or its probability distribution).

DEFINITION 5-8

The **variance** of a discrete random variable X (sometimes called the variance of its distribution) is

$$Var\,X = \sum (x - EX)^2 f(x) \qquad \left(= \sum x^2 f(x) - (EX)^2 \right) \qquad \textbf{(5-2)}$$

The **standard deviation** of X is $\sqrt{Var\,X}$. Often the notation σ^2 is used in place of $Var\,X$, and σ is used in place of $\sqrt{Var\,X}$.

The variance of a random variable is its expected or mean squared distance from the center of its distribution — analogous to the moment of inertia from mechanics. The somewhat confusing use of σ^2 to stand for both the variance of a population relative frequency distribution and the variance of a probability distribution is motivated on the same grounds as the double use of μ. (When a random variable X arises as the result of a single random selection from a population, its probability distribution is equivalent to the population relative frequency distribution, and $Var\,X$ is numerically equal to the population variance.)

EXAMPLE 5-1
(continued)

The calculations necessary to produce the bolt torque standard deviation are organized in Table 5-4. So

$$\sigma = \sqrt{Var\,Z} = \sqrt{4.6275} = 2.15 \text{ ft lb}$$

TABLE 5-4

Calculations for $Var\,Z$

z	$f(z)$	$(z - 16.35)^2$	$(z - 16.35)^2 f(z)$
11	.03	28.6225	.8587
12	.03	18.9225	.5677
13	.03	11.2225	.3367
14	.06	5.5225	.3314
15	.26	1.8225	.4739
16	.09	.1225	.0110
17	.12	.4225	.0507
18	.20	2.7225	.5445
19	.15	7.0225	1.0534
20	.03	13.3225	.3997
			$Var\,Z = 4.6275$

Verify that except for a small difference due to round-off, associated with the creation of Table 5-2, this standard deviation of the random variable Z is numerically the same as the population standard deviation associated with the bolt 3 torques in Table 3-4.

EXAMPLE 5-2
(continued)

To illustrate the alternative formula for calculation of a variance given in Definition 5-8, consider finding the variance and standard deviation of the serial number variable W. Table 5-5 shows the calculation of $\sum w^2 f(w)$.

TABLE 5-5

Calculations for $\sum w^2 f(w)$

w	$f(w)$	$w^2 f(w)$
0	.1	0.0
1	.1	.1
2	.1	.4
3	.1	.9
4	.1	1.6
5	.1	2.5
6	.1	3.6
7	.1	4.9
8	.1	6.4
9	.1	8.1
		28.5

Then

$$Var\,W = \sum w^2 f(w) - (EW)^2 = 28.5 - (4.5)^2 = 8.25$$

so that

$$\sqrt{Var\,W} = 2.87$$

Comparing the two probability histograms in Figure 5-2, notice that the distribution of W appears to be more spread out than that of Z. Happily, this is reflected in the fact that

$$\sqrt{Var\,W} = 2.87 > 2.15 = \sqrt{Var\,Z}$$

The Binomial and Geometric Distributions

Discrete probability distributions are sometimes developed from past experience with a particular physical phenomenon (as in Example 5-1). Another way that they are sometimes created is more abstract and less tied to a single particular application. Sometimes a tractable set of mathematical assumptions can be put together, also having the potential for describing a variety of real situations. Those assumptions are manipulated mathematically to derive generic distributions, which are then potentially applicable in describing a number of different random phenomena. One such set of assumptions is that of *independent, identical success-failure trials*.

That is, many engineering situations can be thought of in the following terms. One observes a sequence of repetitions of essentially the same "go–no go" (success-failure) scenario, where the following two conceptions are reasonable.

1. There is a *constant* (possibly known or possibly unknown) *chance of a go/success outcome* on each repetition of the scenario (call this probability p).
2. The repetitions are *independent* in the sense that knowing the outcome of any particular one of them does not change one's assessments of chance related to any others.

Possible examples of this kind of situation are 1) the testing of items manufactured consecutively, where each will be classified as either conforming or nonconforming; 2) observing motorists as they pass a traffic checkpoint and noting whether each is traveling at a legal speed or speeding; 3) measuring the performance of workers in two different workspace configurations and noting whether the performance of each is better in configuration A or configuration B; etc.

In the context of independent, identical success-failure trials, there are two generic kinds of random variables for which it is straightforward to derive appropriate probability distributions. One is the case of a variable that arises as a count of the number of repetitions out of n that yield a go/success result. Such random variables have what are called *binomial* distributions. The other is the case of a variable that arises as the number of repetitions required to first obtain a go/success result. Such variables have *geometric* distributions.

Consider first a variable

> X = the number of go/success results in n independent identical success-
> failure trials

It is easy for a mathematician to show that X has the binomial (n, p) distribution.

DEFINITION 5-9

The **binomial (n, p) distribution** is a discrete probability distribution with probability function

$$f(x) = \begin{cases} \dfrac{n!}{x!\,(n-x)!}\, p^x (1-p)^{n-x} & \text{for } x = 0, 1, \ldots, n \\ 0 & \text{otherwise} \end{cases} \qquad \textbf{(5-3)}$$

for n a positive integer and $0 < p < 1$.

Although there will be no attempt here to "prove" that equation (5-3) is correct, it is at least plausible. In equation (5-3), there is one factor of p for each trial producing a go/success outcome, and one factor of $(1 - p)$ for each trial producing a no go/failure outcome. And the $n!/x!\,(n-x)!$ term is a count of the number of orders in which it would be possible to see x go/success outcomes in n trials.

The name *binomial* distribution derives from the fact that the values $f(0), f(1), f(2), \ldots, f(n)$ are the terms in the expansion of

$$\left(p + (1-p)\right)^n$$

according to the binomial theorem.

EXAMPLE 5-3

The Binomial Distribution and Counts of Reworkable Shafts. Consider again the situation of Example 3-12, where a study was made of the performance of a process for turning steel shafts. Early in that study, percentages of shafts classified as "reworkable" around 20% were common. Suppose one is inspecting such shafts as they are made and that $p = .2$ is indeed a sensible figure for the chance that a given shaft will be judged reworkable. Suppose further that $n = 10$ shafts will be inspected, and one wishes to evaluate the probability that at least two are classified as reworkable.

Adopting a model of independent, identical success-failure trials for shaft conditions,

> U = the number of reworkable shafts in the sample of 10

is a binomial random variable with $n = 10$ and $p = .2$ (the probability that any single shaft is reworkable). So

$$P[\text{at least two reworkable shafts}] = P[U \geqslant 2]$$
$$= f(2) + f(3) + \cdots + f(10)$$
$$= 1 - (f(0) + f(1))$$
$$= 1 - \left(\frac{10!}{0! \, 10!}(.2)^0(.8)^{10} + \frac{10!}{1! \, 9!}(.2)^1(.8)^9 \right)$$
$$= .62$$

Two comments should be made about these calculations. First, the trick employed here, to avoid plugging into the binomial probability function 9 times by recognizing that the $f(u)$'s have to sum up to 1, is a common and useful one. Second, the .62 figure is only as good as the model assumptions that produced it. If an independent, identical success-failure trials description of what to expect fails to accurately portray physical reality, the .62 value is fine mathematics, but possibly a poor description of what will actually happen.

For instance, say that due to tool wear one tends to see 40 shafts in specifications, then 10 reworkable shafts, a tool change, 40 shafts in specifications, and so on. In this case, the binomial distribution would be a very poor description of U, so the .62 figure would be largely irrelevant. (The independence part of the mathematical assumptions leading to the binomial distribution would be inappropriate in this situation.)

———

One kind of circumstance where a model of independent, identical success-failure trials is not exactly appropriate, but a binomial distribution can still be adequate for practical purposes, is in describing the results of simple random sampling from a dichotomous population. That is, if a population of size N contains a fraction p of type A objects, and a fraction $(1 - p)$ of type B objects, and one selects a simple random sample of n of these items,

$$X = \text{the number of type A items in the sample}$$

is, strictly speaking, not a binomial random variable. But if n is a small fraction of N (say, less than 10%), and p is not too extreme (i.e., is not close to either 0 or 1), X is *approximately* binomial (n, p).

EXAMPLE 5-4

Simple Random Sampling from a Lot of Hexamine Pellets. In the pelletizing machine experiment described in Example 3-14, Greiner, Grimm, Larson, and Lukomski found a combination of machine settings that allowed them to produce 66 conforming pellets out of a batch of 100 pellets. Suppose that batch of 100 pellets is treated as a population of interest; consider selecting a simple random sample of size $n = 2$ from that population.

If one defines the random variable

$$V = \text{the number of conforming pellets in the sample of size 2}$$

the most natural probability distribution for V is obtained as follows. Possible values for V are 0, 1, and 2.

$$f(0) = P[V = 0]$$
    = P[first pellet selected is nonconforming and subsequently the second pellet is also nonconforming]
$$f(2) = P[V = 2]$$
    = P[first pellet selected is conforming and subsequently the second pellet selected is conforming]
$$f(1) = 1 - (f(0) + f(2))$$

A plausible way of evaluating $f(0)$ is to then think, "In the long run, the first selection will yield a nonconforming pellet about 34 out of 100 times. Then, considering only cases where this occurs, in the long run the next selection will also yield a nonconforming pellet about 33 out of 99 times." That is, a sensible evaluation of $f(0)$ is

$$f(0) = \frac{34}{100} \cdot \frac{33}{99} = .1133$$

Similarly, one could argue for

$$f(2) = \frac{66}{100} \cdot \frac{65}{99} = .4333$$

and thus

$$f(1) = 1 - (.1133 + .4333) = 1 - .5467 = .4533$$

Now, one cannot think of V as arising from exactly independent trials. For example, knowing that the first pellet selected was conforming would reduce most people's assessment of the chance that the second is also conforming from $\frac{66}{100}$ to $\frac{65}{99}$. Nevertheless, for most practical purposes, V can be thought of as *essentially* binomial with $n = 2$ and $p = .66$. To see this, note that

$$\frac{2!}{0!\,2!}(.34)^2(.66)^0 = .1156 \approx f(0)$$

$$\frac{2!}{1!\,1!}(.34)^1(.66)^1 = .4488 \approx f(1)$$

$$\frac{2!}{2!\,0!}(.34)^0(.66)^2 = .4356 \approx f(2)$$

Here n is a small fraction of N, p is not too extreme, and a binomial distribution is a decent description of a variable arising from simple random sampling.

━━━━━━━━

The reader is encouraged to take time to plot probability histograms for several different binomial distributions. It turns out (not too surprisingly) that for $p < .5$, the resulting shape is right-skewed; for $p > .5$, the resulting shape is left-skewed; and the skewness increases as p moves away from .5, and it decreases as n is increased. Four binomial probability histograms are pictured in Figure 5-3.

Calculation of the mean and variance for binomial random variables is greatly simplified by the fact that when the formulas (5-1) and (5-2) are used with the expression for binomial probabilities in equation (5-3), simple expressions result. That is, for X a binomial (n, p) random variable,

$$EX = \sum_{x=0}^{n} x \frac{n!}{x!\,(n-x)!}\, p^x (1-p)^{n-x} = np \tag{5-4}$$

and

$$Var\,X = \sum_{x=0}^{n} (x - np)^2 \frac{n!}{x!\,(n-x)!}\, p^x (1-p)^{n-x} = np(1-p) \tag{5-5}$$

These formulas may not be at all obvious, but appropriate algebra can show them to be correct.

FIGURE 5-3

Four Binomial
Probability Histograms

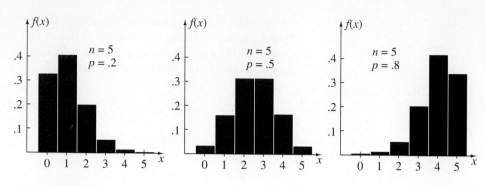

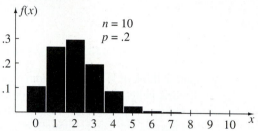

EXAMPLE 5-3
(continued)

Returning to the machining of steel shafts, suppose that in fact a binomial distribution with $n = 10$ and $p = .2$ is appropriate as a model for

$$U = \text{the number of reworkable shafts in the sample of 10}$$

Then, by formulas (5-4) and (5-5),

$$EU = (10)(.2) = 2 \text{ shafts}$$

$$\sqrt{Var\ U} = \sqrt{10(.2)(.8)} = 1.26 \text{ shafts}$$

The form of the so-called geometric distributions is easy to derive from what has been said about the binomial distributions. In the context of a series of independent, identical success-failure trials, let

$$X = \text{the number of trials required to first obtain a go/success result}$$

Then X has the geometric (p) distribution.

DEFINITION 5-10

The **geometric (p) distribution** is a discrete probability distribution with probability function

$$f(x) = \begin{cases} p(1 - p)^{x-1} & \text{for } x = 1, 2, \ldots \\ 0 & \text{otherwise} \end{cases} \tag{5-6}$$

for $0 < p < 1$.

Notice that formula (5-6) makes good intuitive sense. In order for X to take the value x, there must be $x - 1$ consecutive no-go/failure results followed by a go/success. In formula (5-6), there are $x - 1$ terms $(1-p)$ and 1 term p.

Another way to see that formula (5-6) is reasonable is to consider that for X as above and $x = 1, 2, \ldots$

$$1 - F(x) = 1 - P[X \leq x]$$
$$= P[X > x]$$
$$= P[x \text{ no-go/failure results in } x \text{ trials}]$$ **(5-7)**
$$= (1 - p)^x$$

by using the form of the binomial (x, p) probability function given in equation (5-3). Then for $x = 2, 3, \ldots, f(x) = F(x) - F(x - 1) = -(1 - F(x)) + (1 - F(x - 1))$. This, combined with equation (5-7), gives equation (5-6).

The name *geometric* derives from the fact that the values $f(1), f(2), f(3), \ldots$ are terms in the geometric infinite series for

$$p \cdot \frac{1}{1 - (1 - p)}$$

In any case, the geometric distributions are discrete distributions with probability histograms exponentially decaying as x increases. Two different geometric probability histograms are pictured in Figure 5-4.

FIGURE 5-4

Two Geometric
Probability Histograms

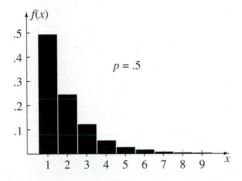

EXAMPLE 5-5

The Geometric Distribution and Shorts in NiCad Batteries. In their article "A Case Study of the Use of an Experimental Design in Preventing Shorts in Nickel-Cadmium Cells" (*Journal of Quality Technology*, 1988), Ophir, El-Gad, and Snyder describe a series of experiments conducted in order to reduce the proportion of cells being scrapped by a battery plant because of internal shorts. The experimental program was successful in reducing the percentage of manufactured cells with internal shorts to around 1%.

Suppose that testing of batteries for shorts begins on a production run in this plant, and let

$$T = \text{the number of the test at which the first short is discovered}$$

A possible model for T (appropriate if the independent, identical success-failure trials description is apt) is geometric with $p = .01$ (the probability that any particular test yields a shorted cell). Then, for example, using equation (5-6),

$$P[\text{the first or second cell tested has a short}] = P[T = 1 \text{ or } 2]$$
$$= f(1) + f(2)$$
$$= (.01) + (.01)(1 - .01)$$
$$= .02$$

Or, for example, using equation (5-7),

$$P[\text{at least 50 cells are tested without finding a short}] = P[T > 50]$$
$$= (1 - .01)^{50}$$
$$= .61$$

Like the binomial distributions, the geometric distributions have means and variances that are simple functions of the identifying parameter p. That is, if X is geometric (p), using equations (5-1), (5-2), and (5-6) and some algebra and calculus, it is possible to show that

$$EX = \sum_{x=1}^{\infty} xp(1 - p)^{x-1} = \frac{1}{p} \tag{5-8}$$

and

$$Var\, X = \sum_{x=1}^{\infty} \left(x - \frac{1}{p} \right)^2 p(1 - p)^{x-1} = \frac{1 - p}{p^2} \tag{5-9}$$

EXAMPLE 5-5
(continued)

In the context of testing batteries for internal shorts, with T as before,

$$ET = \frac{1}{.01} = 100 \text{ batteries}$$

$$\sqrt{Var\, T} = \sqrt{\frac{(1 - .01)}{(.01)^2}} = 99.5 \text{ batteries}$$

Formula (5-8) is an intuitively appealing result. If there is but 1 chance in 100 of encountering a shorted battery at each test, it is sensible to expect to wait through 100 tests on average to encounter the first one.

The Poisson Distributions

As discussed in Section 3-4, it is often of engineering importance to keep track of the total number of occurrences of some relatively rare phenomenon of interest, where the physical or time unit under surveillance has the potential to produce many such occurrences. Cases of floor tiles have potentially many total blemishes on their contents. In a 1-second time interval, there is potentially a large number of messages that can arrive for routing through a switching center in a communications network. And a 1 cc sample of glass potentially contains a large number of imperfections or seeds.

There is thus a need to have appropriate probability distributions to describe random variables that arise as *counts of the number of occurrences of a relatively rare phenomenon across a specified interval of time or space*. By far the most commonly used theoretical distributions in this context are the Poisson distributions (named after the French mathematician Simeon Poisson, who lived in the late 1700s and early 1800s).

DEFINITION 5-11

The **Poisson (λ) distribution** is a discrete probability distribution with probability function

$$f(x) = \begin{cases} \dfrac{e^{-\lambda}\lambda^{x}}{x!} & \text{for } x = 0, 1, 2, \ldots \\ 0 & \text{otherwise} \end{cases} \tag{5-10}$$

for $\lambda > 0$.

The form of equation (5-10) may initially seem unappealing to readers who have no prior exposure to the distribution. But it is one that has both sensible mathematical origins and tractability, and it has also proved itself empirically useful in many different "rare events" circumstances. It would not be consistent with the purposes of this text to expound extensively on the mathematics of the Poisson distributions. However, note that one way to arrive at equation (5-10) is to think of a very large number of trials (opportunities for occurrence), where the probability of success (occurrence) on any one is very small and the product of the number of trials and the success probability is λ. One is then led to the binomial $(n, \frac{\lambda}{n})$ distribution. In fact, for large n, the binomial $(n, \frac{\lambda}{n})$ probability function approximates the one specified in equation (5-10). So one might think of the Poisson distribution arising physically through a mechanism that would present many tiny similar opportunities for independent occurrence or nonoccurrence throughout an interval of time or space.

The Poisson distributions are right-skewed distributions over the values $x = 0, 1, 2, \ldots$, whose probability histograms have peaks near their respective λ's. Two different Poisson probability histograms are shown in Figure 5-5.

λ is, in fact, both the mean and the variance for the Poisson (λ) distribution. That is, if X has the Poisson (λ) distribution, then

$$EX = \sum_{x=0}^{\infty} x \frac{e^{-\lambda}\lambda^{x}}{x!} = \lambda \tag{5-11}$$

FIGURE 5-5

Two Poisson
Probability Histograms

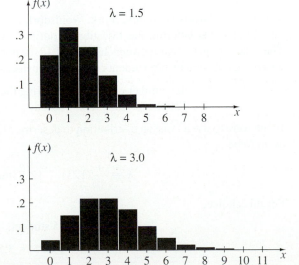

and

$$Var\, X = \sum_{x=0}^{\infty}(x - \lambda)^2 \frac{e^{-\lambda}\lambda^x}{x!} = \lambda \qquad (5\text{-}12)$$

Fact (5-11) is helpful in picking out which Poisson distribution might be useful in describing a particular "rare events" situation.

EXAMPLE 5-6

The Poisson Distribution and Counts of α Particles. A classical data set of Rutherford and Geiger, reported in *Philosophical Magazine* in 1910, concerned the numbers of α particles emitted from a small bar of polonium and colliding with a screen placed near the bar, in 2608 periods of 8 minutes each. The Rutherford and Geiger relative frequency distribution has mean 3.87 and a shape remarkably similar to that of the Poisson probability distribution with mean $\lambda = 3.87$.

Should one attempt to duplicate the Rutherford/Geiger experiment, a reasonable probability function for describing

$S =$ the number of α particles striking the screen in an additional 8-minute period

is then

$$f(s) = \begin{cases} \dfrac{e^{-3.87}(3.87)^s}{s!} & \text{for } s = 0, 1, 2, \ldots \\ 0 & \text{otherwise} \end{cases}$$

Using such a model, one has (for example)

$P[\text{at least 4 particles are recorded}]$
$= P[S \geq 4]$
$= f(4) + f(5) + f(6) + \cdots$
$= 1 - (f(0) + f(1) + f(2) + f(3))$
$= 1 - \left(\dfrac{e^{-3.87}(3.87)^0}{0!} + \dfrac{e^{-3.87}(3.87)^1}{1!} + \dfrac{e^{-3.87}(3.87)^2}{2!} + \dfrac{e^{-3.87}(3.87)^3}{3!}\right)$
$= .54$

EXAMPLE 5-7

Arrivals at a University Library. Stork, Wohlsdorf, and McArthur collected data on numbers of students entering the ISU library during various periods over a week's time. Their data indicate that between 12:00 and 12:10 P.M. on Monday through Wednesday, an average of around 125 students entered. Consider the possibility of modeling

$M =$ the number of students entering the ISU library between
12:00 and 12:01 next Tuesday

If one were to use a Poisson distribution to describe M, the reasonable choice of λ would seem to be

$$\lambda = \frac{1}{10}(125) = 12.5 \text{ students}$$

For this choice,

$$EM = \lambda = 12.5 \text{ students}$$
$$\sqrt{Var\, M} = \sqrt{\lambda} = \sqrt{12.5} = 3.54 \text{ students}$$

and for example, the probability that between 10 and 15 students (inclusive) arrive at the library between 12:00 and 12:01 would be evaluated as

$$P[10 \leq M \leq 15]$$
$$= f(10) + f(11) + f(12) + f(13) + f(14) + f(15)$$
$$= \frac{e^{-12.5}(12.5)^{10}}{10!} + \frac{e^{-12.5}(12.5)^{11}}{11!} + \frac{e^{-12.5}(12.5)^{12}}{12!}$$
$$+ \frac{e^{-12.5}(12.5)^{13}}{13!} + \frac{e^{-12.5}(12.5)^{14}}{14!} + \frac{e^{-12.5}(12.5)^{15}}{15!}$$
$$= .60$$

5-2 *Continuous Random Variables*

It is often a convenient idealization to conceive of a random variable as not discrete but rather continuous in the sense of having a whole continuous interval for its set of possible values. The devices used to describe continuous probability distributions differ somewhat from the tools studied in the last section. So the first tasks here are to introduce the notion of a probability density function, show its relationship to the cumulative probability function for a continuous random variable, and show how it is used to find the mean and variance for a continuous distribution. After this, several standard continuous distributions useful in engineering applications of probability theory will be discussed. That is, the normal (or Gaussian), exponential, Weibull, and beta distributions will be considered.

Probability Density Functions and Cumulative Probability Functions

The methods used to specify and describe probability distributions have almost exact parallels in mechanics. In the last section, it was shown that for discrete distributions, the mean is interpretable as a center of mass or balance point for the distribution. When one begins to consider continuous probability distributions, the analogy to mechanics becomes especially helpful to keep in mind. In mechanics, the properties of a continuous mass distribution are related to the possibly varying density of the mass across its region of location. Amounts of mass in particular regions are obtained from the density by integration. And moments (including the first, which is the center of mass) are also obtained as appropriate integrals.

The concept in probability theory corresponding to that of mass density in mechanics is *probability density*. To specify a continuous probability distribution, one needs to describe "how thick" the probability is in the various parts of the set of possible values. The formal definition that will be employed is the following.

DEFINITION 5-12

A **probability density function** for a continuous random variable X is a nonnegative function $f(x)$ with

$$\int_{-\infty}^{\infty} f(x)\, dx = 1 \tag{5-13}$$

and such that for all $a \leq b$, one is willing to assign $P[a \leq X \leq b]$ according to

$$P[a \leq X \leq b] = \int_{a}^{b} f(x)\, dx \tag{5-14}$$

A generic probability density function is pictured in Figure 5-6. In keeping with equations (5-13) and (5-14), the plot of $f(x)$ does not dip below the x axis, the total area under the curve $y = f(x)$ is 1, and areas under the curve above particular intervals give the probabilities corresponding to those intervals.

FIGURE 5-6

A Generic Probability Density Function

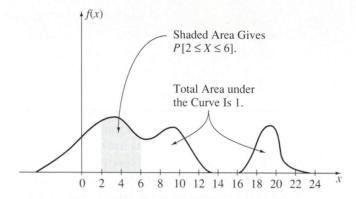

Notice that in direct analogy to what is done in mechanics, if $f(x)$ is indeed the density of probability around x, then the probability in an interval of small length dx around x is approximately $f(x)\,dx$. Then to get a probability between a and b, one needs to sum up such $f(x)\,dx$ values. $\int_a^b f(x)\,dx$ is exactly the limit of $\sum f(x)\,dx$ values as dx gets small, so the expression (5-14) is reasonable.

EXAMPLE 5-8

The Random Time Till a First Arc in the Bob Drop Experiment. Consider once again the bob drop situation first described in Section 1-4 and revisited in Example 4-4. In any use of the apparatus, the bob is almost certainly not released exactly "in sync" with the 60 cycle current that produces the arcs and resulting marks on the paper tape. One could think of a random variable

$Y =$ the time elapsed (in seconds) from the release of the bob until the first arc

as continuous with set of possible values $(0, \frac{1}{60})$.

The first question to pose at this point is "What is a plausible probability density function for Y?" The symmetry of this situation suggests that probability density should be constant over the interval $(0, \frac{1}{60})$ and 0 outside the interval. That is, for any two values y_1, and y_2 in $(0, \frac{1}{60})$, the probability that Y takes a value within a small interval around y_1 of length dy (i.e., $f(y_1)\,dy$ approximately) should be the same as the probability that Y takes a value within a small interval around y_2 of the same length dy (i.e., $f(y_2)\,dy$ approximately). This forces $f(y_1) = f(y_2)$, so there must be a constant probability density on $(0, \frac{1}{60})$.

Now if $f(y)$ is to have the form

$$f(y) = \begin{cases} c & \text{for } 0 < y < \dfrac{1}{60} \\ 0 & \text{otherwise} \end{cases}$$

for some constant c (i.e., is to be as pictured in Figure 5-7), in light of equation (5-13), one must have

$$1 = \int_{-\infty}^{\infty} f(y)\,dy = \int_{-\infty}^{0} 0\,dy + \int_{0}^{1/60} c\,dy + \int_{1/60}^{\infty} 0\,dy = \frac{c}{60}$$

That is, $c = 60$, and thus,

$$f(y) = \begin{cases} 60 & \text{for } 0 < y < \dfrac{1}{60} \\ 0 & \text{otherwise} \end{cases} \tag{5-15}$$

If one does adopt the function specified by equation (5-15) as a probability density describing the distribution of Y, it is then (for example) possible to calculate that

$$P\left[Y \leq \frac{1}{100}\right] = \int_{-\infty}^{1/100} f(y)\,dy = \int_{-\infty}^{0} 0\,dy + \int_{0}^{1/100} 60\,dy = .6$$

FIGURE 5-7

Probability Density
Function for Y

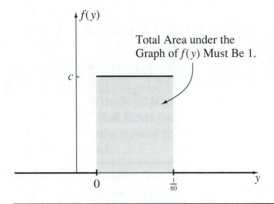

One point about continuous probability distributions that has its parallel in mechanics and is forced by equation (5-14) (but nevertheless may at first seem counterintuitive) involves the probability associated with a continuous random variable assuming a particular prespecified value (say, a). Just as the mass a continuous mass distribution places at a single point is 0, so also is $P[X = a]$ for a continuous random variable X. This follows from equation (5-14), because

$$P[a \leq X \leq a] = \int_{a}^{a} f(x)\,dx = 0$$

One consequence of this mathematical curiosity is that when working with continuous random variables, one typically doesn't need to worry about whether or not inequality signs one writes are strict inequality signs. That is, if X is continuous,

$$P[a \leq X \leq b] = P[a < X \leq b] = P[a \leq X < b] = P[a < X < b]$$

Definition 5-6 gave a perfectly general definition of the cumulative probability function for a random variable (which was specialized in Section 5-1 to the case of a discrete variable). Here equation (5-14) can be used to express the cumulative probability function for a continuous random variable in terms of an integral of its probability density. That is, for X continuous with probability density $f(x)$,

$$F(x) = P[X \leq x] = \int_{-\infty}^{x} f(t)\,dt \tag{5-16}$$

$F(x)$ is obtained from $f(x)$ by integration, and applying the fundamental theorem of calculus to equation (5-16),

$$\frac{d}{dx}F(x) = f(x) \tag{5-17}$$

That is, $f(x)$ is obtained from $F(x)$ by differentiation.

EXAMPLE 5-8
(continued)

The cumulative probability function for Y, the elapsed time from bob release until first arc, is easily obtained from equation (5-15). For $y \leq 0$,

$$F(y) = P[Y \leq y] = \int_{-\infty}^{y} f(t)\,dt = \int_{-\infty}^{y} 0\,dt = 0$$

and for $0 < y \leq \frac{1}{60}$,

$$F(y) = P[Y \leq y] = \int_{-\infty}^{y} f(t)\,dt = \int_{-\infty}^{0} 0\,dt + \int_{0}^{y} 60\,dt = 0 + 60y = 60y$$

and for $y > \frac{1}{60}$,

$$F(y) = P[Y \leq y] = \int_{-\infty}^{y} f(t)\,dt = \int_{-\infty}^{0} 0\,dt + \int_{0}^{1/60} 60\,dt + \int_{1/60}^{y} 0\,dt = 1$$

That is,

$$F(y) = \begin{cases} 0 & \text{if } y \leq 0 \\ 60y & \text{if } 0 < y \leq 1/60 \\ 1 & \text{if } \dfrac{1}{60} < y \end{cases}$$

A plot of $F(y)$ is given in Figure 5-8. Comparing Figure 5-8 to Figure 5-7 shows that indeed the graph of $F(y)$ has slope 0 for $y < 0$ and $y > \frac{1}{60}$ and slope 60 for $0 < y < \frac{1}{60}$. That is, it is clear that $f(y)$ is indeed the derivative of $F(y)$.

FIGURE 5-8

Cumulative Probability
Function for Y

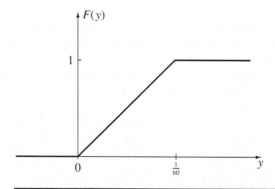

The character of the plot of $F(y)$ in Figure 5-8 is typical of cumulative probability functions for continuous distributions. The graphs of such cumulative probability functions are *continuous* in the sense that their plots are unbroken curves.

**Means and Variances for
Continuous Distributions**

When it comes to summarizing the main features of a continuous distribution, a plot of the probability density $f(x)$ versus x functions as a kind of idealized histogram. It has the same kind of visual interpretations that have already been applied to relative frequency histograms and probability histograms. Further, it is possible to define the mean and variance for a continuous probability distribution and use them as numerical summaries in the same way that means and variances are used to describe data sets and discrete distributions.

DEFINITION 5-13

The **mean** or **expected value** of a continuous random variable X (sometimes called the mean of its probability distribution) is

$$EX = \int_{-\infty}^{\infty} x f(x)\, dx \qquad\qquad \textbf{(5-18)}$$

As is the case for discrete random variables, the notation μ is sometimes used in place of EX.

Formula (5-18) is perfectly plausible, both as a "continuization" of the formula for discrete variables and also in analogy to the center of mass from mechanics. The probability in a small interval around x of length dx is approximately $f(x)\,dx$. So multiplying this by x and summing as in Definition 5-7, one has $\sum x f(x)\,dx$, and formula (5-18) is exactly the limit of such sums as dx gets small. And in mechanics, the center of mass of a continuous mass distribution is of the form given in equation (5-18) except for division by a total mass, which for a probability distribution is 1.

EXAMPLE 5-8
(continued)

Thinking of the probability density in Figure 5-7 as an idealized histogram and thinking of the balance point interpretation of the mean, it is clear that EY had better turn out to be $\frac{1}{120}$ for the elapsed time variable. Happily, equations (5-18) and (5-15) give

$$EY = \int_{-\infty}^{\infty} y f(y)\, dy = \int_{-\infty}^{0} y \cdot 0 \, dy + \int_{0}^{1/60} y \cdot 60 \, dy + \int_{1/60}^{\infty} y \cdot 0 \, dy$$

$$= 30 y^2 \Big|_{0}^{1/60} = \frac{1}{120} \text{ sec}$$

Analogy with the moment of inertia for a continuous mass distribution, or "continuization" of the discrete distribution formula for the variance of a random variable, produces a definition of the variance of a continuous random variable.

DEFINITION 5-14

The **variance** of a continuous random variable X (sometimes called the variance of its probability distribution) is

$$Var\,X = \int_{-\infty}^{\infty} (x - EX)^2 f(x)\, dx \qquad \left(= \int_{-\infty}^{\infty} x^2 f(x)\, dx - (EX)^2 \right) \qquad \textbf{(5-19)}$$

The **standard deviation** of X is $\sqrt{Var\,X}$. Often the notation σ^2 is used in place of $Var\,X$, and σ is used in place of $\sqrt{Var\,X}$.

EXAMPLE 5-8
(continued)

Returning for a final time to the situation of the bob drop and the random variable Y, suppose the standard deviation for Y is needed. Using formula (5-19) and the form of Y's probability density, one has

$$Var\,Y = \int_{-\infty}^{0} \left(y - \frac{1}{120} \right)^2 \cdot 0 \, dy + \int_{0}^{1/60} \left(y - \frac{1}{120} \right)^2 \cdot 60 \, dy + \int_{1/60}^{\infty} \left(y - \frac{1}{120} \right)^2 \cdot 0 \, dy$$

$$= \frac{60 \left(y - \frac{1}{120} \right)^3}{3} \Bigg|_{0}^{1/60}$$

$$= \frac{1}{3} \left(\frac{1}{120} \right)^2$$

So the standard deviation of Y is

$$\sqrt{Var\ Y} = \sqrt{\frac{1}{3}\left(\frac{1}{120}\right)^2} = .0048\ \text{sec}$$

The Normal Probability Distributions

Just as there are a number of standard discrete distributions that are commonly applicable to engineering problems, there are also a number of standard continuous probability distributions that often prove useful in engineering contexts. Indeed, this text has already alluded to the *normal* or *Gaussian distributions* and made use of their properties in producing normal plots. It is now time to introduce them formally.

DEFINITION 5-15

The **normal** or **Gaussian (μ, σ^2) distribution** is a continuous probability distribution with probability density

$$f(x) = \frac{1}{\sqrt{2\pi\sigma^2}}e^{-(x-\mu)^2/2\sigma^2} \qquad \text{for all } x \qquad \textbf{(5-20)}$$

for $\sigma > 0$.

It is not necessarily obvious, but formula (5-20) does yield a legitimate probability density, in that the total area under the curve $y = f(x)$ is 1. Further, it is also the case that

$$\int_{-\infty}^{\infty} x\frac{1}{\sqrt{2\pi\sigma^2}}e^{-(x-\mu)^2/2\sigma^2}\ dx = \mu$$

and

$$\int_{-\infty}^{\infty} (x-\mu)^2\frac{1}{\sqrt{2\pi\sigma^2}}e^{-(x-\mu)^2/2\sigma^2}\ dx = \sigma^2$$

That is, the parameters μ and σ^2 are respectively the mean and variance (as defined in Definitions 5-13 and 5-14) of the distribution.

Figure 5-9 is a graph of the probability density specified by formula (5-20). Calculus can be used to show that the archetypal bell-shaped curve shown there has inflection points at $\mu - \sigma$ and $\mu + \sigma$. The exact form of formula (5-20) is the result of some fairly advanced mathematical manipulations, but it is also a form that turns out to be empirically useful in a great variety of applications.

In theory, probabilities for the normal distributions can be found by integration using formula (5-20). Indeed, readers with pocket calculators that are preprogrammed to do numerical integration may find it instructive to check some of the calculations in the examples that will follow, by straightforward use of formulas (5-14) and (5-20).

FIGURE 5-9

Graph of a Normal Probability Density Function

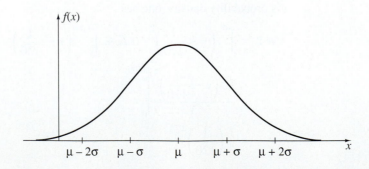

But the usual freshman calculus methods of evaluating integrals via antidifferentiation will fail when it comes to the normal densities, as they do not have antiderivatives that are expressible in terms of elementary functions. Instead, one must use special normal probability tables.

The use of tables for evaluating normal probabilities depends on the following relationship. If X is normally distributed with mean μ and variance σ^2,

$$P[a \leq X \leq b] = \int_a^b \frac{1}{\sqrt{2\pi\sigma^2}} e^{-(x-\mu)^2/2\sigma^2} \, dx = \int_{(a-\mu)/\sigma}^{(b-\mu)/\sigma} \frac{1}{\sqrt{2\pi}} e^{-z^2/2} \, dz \qquad \textbf{(5-21)}$$

where the second inequality follows from the change of variable or substitution $z = \frac{x-\mu}{\sigma}$. But equation (5-21) involves an integral of the normal density with $\mu = 0$ and $\sigma = 1$, so it says that evaluation of all normal probabilities can be reduced to the evaluation of normal probabilities for that special case.

DEFINITION 5-16

The normal distribution with $\mu = 0$ and $\sigma = 1$ is called the **standard normal distribution**.

The relationship between normal (μ, σ^2) and standard normal probabilities is illustrated pictorially in Figure 5-10. Once one realizes that probabilities for all normal distributions can be had by tabulating probabilities for only the standard normal distribution, it is a relatively simple matter to use techniques of numerical integration to produce appropriate tables. The standard normal probability table that will be used in this text (other forms are possible) is given in Appendix D-3. It is in fact a table of the standard normal cumulative probability function. That is, for values z located on the table's margins, the entries in the table body are

$$\Phi(z) = \int_{-\infty}^z \frac{1}{\sqrt{2\pi}} e^{-t^2/2} \, dt$$

(Φ is routinely used to stand for the standard normal cumulative probability function, instead of the more generic F.)

FIGURE 5-10

Illustration of the Relationship between Normal (μ, σ^2) and Standard Normal Probabilities

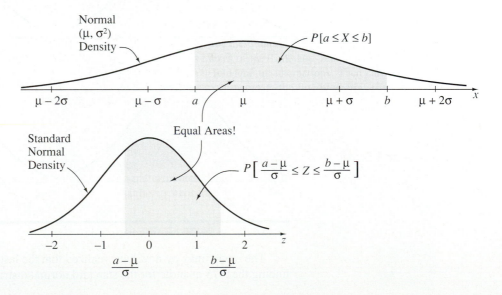

EXAMPLE 5-9

Standard Normal Probabilities. Suppose for sake of illustration that Z is a standard normal random variable. Consider finding some probabilities for Z using the table in Appendix D-3.

By a straight table look-up,

$$P[Z < 1.76] = \Phi(1.76) = .96$$

(The tabled value is .9608, but in keeping with the earlier promise to state final probabilities to only two decimal places, the tabled value was rounded to get .96.)

After two table look-ups and a subtraction,

$$P[.57 < Z < 1.32] = P[Z < 1.32] - P[Z \leq .57]$$
$$= \Phi(1.32) - \Phi(.57)$$
$$= .9066 - .7157$$
$$= .19$$

And a single table look-up and a subtraction yield a right-tail probability like

$$P[Z > -.89] = 1 - P[Z \leq -.89] = 1 - .1867 = .81$$

As the table was used in these examples, probabilities for values z located on the table's margins are found in the table's body. The process can of course be run in reverse. Probabilities located in the table's body can be used to specify values z on the table's margins. For example, consider locating a value z such that

$$P[-z < Z < z] = .95$$

z will then put probability $\frac{1-.95}{2} = .025$ in the right tail of the standard normal distribution — i.e., be such that $\Phi(z) = .975$. Locating .975 in the table body, one sees that $z = 1.96$.

Figure 5-11 illustrates all of the calculations for this example.

FIGURE 5-11

Standard Normal Probabilities for Example 5-9

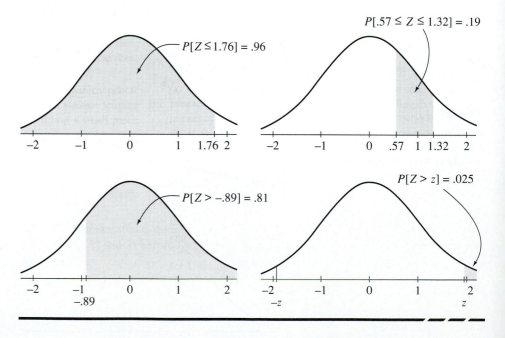

The reader may by now have realized that the last part of Example 5-9 amounts to finding the .975 quantile for the standard normal distribution. In fact, the reader is now in a position to understand the origin of Table 3-10. The standard normal quantiles there were found by looking in the body of Table D-3 for the relevant probabilities and then locating corresponding z's on the margins.

In mathematical symbols, one has that for $\Phi(z)$ the standard normal cumulative probability function and $Q_z(p)$ the standard normal quantile function,

$$\left.\begin{array}{c} \Phi(Q_z(p)) = p \\ Q_z(\Phi(z)) = z \end{array}\right\} \qquad \text{(5-22)}$$

Relationships (5-22) mean that Q_z and Φ are inverse functions. (In fact, further reflection leads to the realization that the relationship $Q = F^{-1}$ is not just a standard normal phenomenon but is in fact true in general for continuous distributions.)

Relationship (5-21) shows how to use the standard normal cumulative probability function to find general normal probabilities. For X normal (μ, σ^2) and a value x associated with X, one converts to units of standard deviations above the mean via

$$z = \frac{x - \mu}{\sigma} \qquad \text{(5-23)}$$

and then consults the standard normal table using z instead of x.

EXAMPLE 5-10

Net Weights of Jars of Baby Food. J. Fisher, in his article "Computer Assisted Net Weight Control" (*Quality Progress*, June 1983), discusses the filling of food containers by weight. In the article, there is a reasonably bell-shaped histogram of individual net weights of jars of strained plums with tapioca. The mean of the values portrayed is about 137.2 g, and the standard deviation is about 1.6 g. The declared (or label) weight on jars of this product is 135.0 g.

Suppose that it is in fact adequate to model

$$W = \text{the next strained plums and tapioca fill weight}$$

with a normal distribution. Further, initially suppose that $\mu = 137.2$ and $\sigma = 1.6$ are appropriate and consider some probabilities associated with W. To begin with, suppose the probability that the next jar filled is below declared weight (i.e., $P[W < 135.0]$) is of interest. Using formula (5-23), one converts $x = 135.0$ to units of standard deviations above μ as

$$z = \frac{135.0 - 137.2}{1.6} = -1.38$$

Then, using Table D-3,

$$P[W < 135.0] = \Phi(-1.38) = .08$$

Thus, this model puts the chance of obtaining a below-nominal fill level at about 8%.

As a second example, consider finding the probability that W is within 1 gram of nominal (i.e., $P[134.0 < W < 136.0]$). Using formula (5-23), one converts both $w_1 = 134.0$ and $w_2 = 136.0$ to z *values* or units of standard deviations above the mean as

$$z_1 = \frac{134.0 - 137.2}{1.6} = -2.00$$

$$z_2 = \frac{136.0 - 137.2}{1.6} = -.75$$

So then

$$P[134.0 < W < 136.0] = \Phi(-.75) - \Phi(-2.00) = .2266 - .0228 = .20$$

The preceding two probabilities and their standard normal counterparts are depicted in Figure 5-12.

FIGURE 5-12

Normal Probabilities for Example 5-10

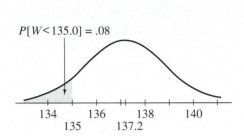

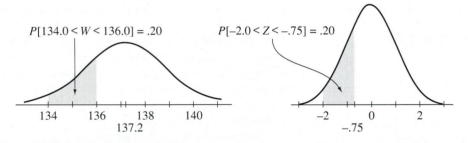

The calculations for this example have thus far consisted of starting with all of the quantities on the right of formula (5-23) and going from the margin of Table D-3 to its body to find probabilities for W. An important variant on this process is to instead go from the body of the table to its margins to obtain z, and then — given only two of the three quantities on the right of formula (5-23) — to solve for the third.

For example, suppose that it is a relatively simple matter to adjust the aim of the filling process (i.e., the mean μ of W). Suppose one then wants to decrease the probability that the next jar is below the declared weight of 135.0 to .01 by increasing μ. What is the minimum μ that will achieve this?

Figure 5-13 shows what to do. μ must be chosen in such a way that $w = 135.0$ becomes the .01 quantile of the normal distribution with mean μ and standard deviation $\sigma = 1.6$. Consulting either Table 3-10 or Table D-3, it is easy to determine that the .01 quantile of the standard normal distribution is

$$z = -2.33$$

So one wants

$$-2.33 = \frac{135.0 - \mu}{1.6}$$

i.e.,

$$\mu = 138.7 \text{ g}$$

FIGURE 5-13

Normal Distribution and
$P[W < 135.0] = .01$

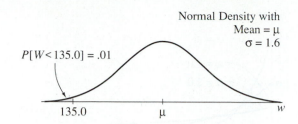

An increase of about $138.7 - 137.2 = 1.5$ g in fill level aim would be required to achieve the desired reduction in probability of a below-nominal fill weight.

In practical terms, the reduction in $P[W < 135.0]$ is bought at the price of increasing the average *give-away cost* associated with filling jars in such a way that on average they contain much more than the nominal contents. In some applications, this type of cost will be prohibitive, but there is another approach open to a process engineer. That is to reduce the variation in fill level through the acquisition of more precise filling equipment (which in itself may not be cheap). In terms of the example and equation (5-23), instead of increasing μ one might consider paying the cost associated with reducing σ. The reader is encouraged to verify that a reduction in σ to about .94 g would also produce $P[W < 135.0] = .01$ without any change in μ.

As Example 5-10 illustrates, equation (5-23) is the fundamental relationship used in textbook exercises involving normal distributions. In such problems, one way or another, three of the four entries in equation (5-23) are specified, and one then is asked to obtain the fourth.

The Exponential Distributions

Section 5-1 discusses the fact that the Poisson distributions are often used as models for the number of occurrences of a relatively rare phenomenon in a specified interval of time. The same mathematical theory that suggests the appropriateness of the Poisson distributions in that context also suggests the usefulness of the *exponential* distributions for describing the waiting time until an occurrence.

DEFINITION 5-17

The **exponential (α) distribution** is a continuous probability distribution with probability density

$$f(x) = \begin{cases} \dfrac{1}{\alpha} e^{-x/\alpha} & \text{for } x > 0 \\ \\ 0 & \text{otherwise} \end{cases} \tag{5-24}$$

for $\alpha > 0$.

Figure 5-14 shows plots of $f(x)$ versus x for three different values of α. Expression (5-24) is extremely tractable, and it is not all difficult to show that α is in fact both the mean and the standard deviation of the exponential (α) distribution. That is,

$$\int_0^\infty x \frac{1}{\alpha} e^{-x/\alpha} \, dx = \alpha$$

and

$$\int_0^\infty (x - \alpha)^2 \frac{1}{\alpha} e^{-x/\alpha} \, dx = \alpha^2$$

FIGURE 5-14

Three Exponential
Probability Densities

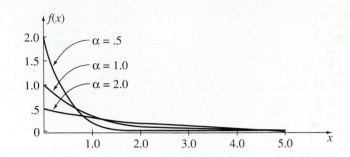

Further, the exponential (α) distribution has a simple cumulative probability function,

$$F(x) = \begin{cases} 0 & \text{if } x \le 0 \\ 1 - e^{-x/\alpha} & \text{if } x > 0 \end{cases} \qquad \text{(5-25)}$$

EXAMPLE 5-11

(Example 5-7 revisited) **The Exponential Distribution and Arrivals at a University Library.** Recall that Stork, Wohlsdorf, and McArthur found the arrival rate of students at the ISU library between 12:00 and 12:10 P.M. early in the week to be about 12.5 students per minute. That translates to a $\frac{1}{12.5} = .08$ min average waiting time between student arrivals.

Consider beginning observation at the ISU library entrance at exactly noon next Tuesday and define the random variable

$T =$ the waiting time (in minutes) until the first student passes through the door

A possible model for T is the exponential distribution with $\alpha = .08$. Using such a description of T, one would for example evaluate the probability of waiting more than 10 seconds ($\frac{1}{6}$ min) for the first arrival as

$$P\left[T > \frac{1}{6}\right] = 1 - F\left(\frac{1}{6}\right) = 1 - \left(1 - e^{-1/6(.08)}\right) = .12$$

This result is pictured in Figure 5-15.

FIGURE 5-15

Exponential Probability for
Example 5-11

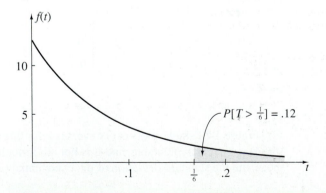

The exponential distribution is the continuous analog of the geometric distribution in several respects. For one thing, both the geometric probability function and the exponential probability density decline exponentially in their arguments x. For another, they are both distributions possessing a kind of *memoryless property*. If the first success in a series of independent identical success failure trials is known not to have occurred through trial t_0, then the additional number of trials (beyond t_0) needed to produce the first success is a geometric (p) random variable (as was the total number of trials required from the beginning). Similarly, if an exponential (α) waiting time is known not to have been completed by time t_0, then the additional waiting time to completion is exponential (α). This memoryless property is related to the *force-of-mortality function* of the distribution being constant. The force-of-mortality function for a distribution is a concept of reliability theory discussed briefly in Appendix B-4.

The Weibull Distributions

The Weibull distributions are a family of probability distributions that generalize the exponential distributions and provide much more flexibility in terms of distributional shape than is available with the single-shape exponential probability densities. The distributions bear the name of Swedish physicist Waloddi Weibull. They have become extremely popular with engineers for describing the strength properties of materials and the life lengths of manufactured devices. The clearest way to specify these distributions is by means of their cumulative probability functions (rather than their probability density functions).

DEFINITION 5-18

The **Weibull (α, β) distribution** is a continuous probability distribution with cumulative probability function

$$F(x) = \begin{cases} 0 & \text{if } x < 0 \\ 1 - e^{-(x/\alpha)^\beta} & \text{if } x \geq 0 \end{cases} \tag{5-26}$$

for parameters $\alpha > 0$ and $\beta > 0$.

Beginning from formula (5-26), it is possible to determine properties of the Weibull distributions. Differentiating formula (5-26), one gets the Weibull (α, β) probability density

$$f(x) = \begin{cases} 0 & \text{if } x < 0 \\ \dfrac{\beta}{\alpha^\beta} x^{\beta-1} e^{-(x/\alpha)^\beta} & \text{if } x > 0 \end{cases} \tag{5-27}$$

which in turn can be shown to yield the mean and variance

$$\mu = \alpha \Gamma\left(1 + \frac{1}{\beta}\right) \tag{5-28}$$

and

$$\sigma^2 = \alpha^2 \left[\Gamma\left(1 + \frac{2}{\beta}\right) - \left(\Gamma\left(1 + \frac{1}{\beta}\right)\right)^2 \right]$$

where $\Gamma(x) = \int_0^\infty t^{x-1} e^{-t}\, dt$ is the *gamma function* of advanced calculus. (For integer values n, $\Gamma(n) = (n-1)!$.) These formulas for $f(x)$, μ, and σ^2 are not particularly illuminating. So it is probably most helpful to simply realize that β controls the shape of the Weibull distribution and that α controls the scale. Look at some plots of Weibull densities to get a feel for these distributions. Figure 5-16 shows plots of $f(x)$ versus x for several (α, β) pairs.

Note that $\beta = 1$ gives the special case of the exponential distributions. For small β, the distributions are decidedly right-skewed, but for β larger than about 3.6, they actually

FIGURE 5-16

Nine Weibull
Probability Densities

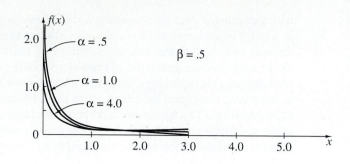

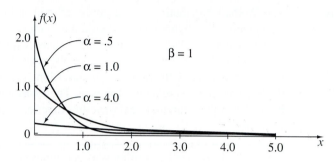

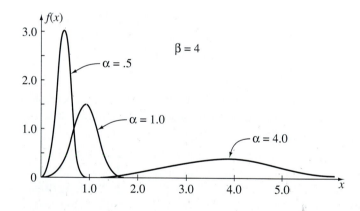

become left-skewed. In terms of determining distributional location, it is probably more revealing than equation (5-28) that upon using equation (5-26), one can find that the median for the Weibull (α, β) distribution is

$$Q(.5) = \alpha e^{-(.3665/\beta)} \tag{5-29}$$

So for large β (for example), the Weibull median is essentially α.

EXAMPLE 5-12

The Weibull Distribution and the Strength of a Ceramic Material. The report "Review of Workshop on Design, Analysis and Reliability Prediction for Ceramics — Part II" by E. Lenoe (*Office of Naval Research Far East Scientific Bulletin,* 1987) suggests that tensile strengths (MPa) of .95 mm rods of HIPped UBE SN-10 with 2.5% yttria material can be described by a Weibull distribution with $\beta = 8.8$ and median 428 MPa. Consider testing the tensile strength of one more rod of this material and define the random variable

$$S = \text{measured tensile strength (MPa)}$$

Under the assumption that S can be modeled using a Weibull distribution with the suggested characteristics, suppose that $P[S \leq 400]$ is needed. Using equation (5-29),

$$428 = \alpha e^{-(.3665/8.8)}$$

Thus, the Weibull scale parameter is

$$\alpha = 446$$

Then, using equation (5-26),

$$P[S \leq 400] = 1 - e^{-(400/446)^{8.8}} = .32$$

Figure 5-17 illustrates this probability calculation.

FIGURE 5-17

Weibull Density and $P[S \leq 400]$

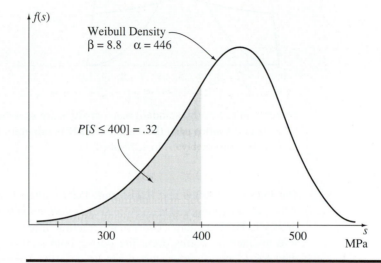

The Beta Distributions

Normal random variables have $(-\infty, \infty)$ as their set of possible values. Exponential and Weibull random variables have $(0, \infty)$ as their set of possible values. In engineering applications of probability theory, it is occasionally helpful to have available a family of distributions whose set of possible values is a finite interval. One such family is the *beta* family of distributions.

DEFINITION 5-19

The **beta (α, β) distribution** over the interval $(a, a + b)$ is a continuous probability distribution with probability density

$$f(x) = \begin{cases} \dfrac{1}{b} \dfrac{\Gamma(\alpha + \beta)}{\Gamma(\alpha)\Gamma(\beta)} \left(\dfrac{x - a}{b}\right)^{\alpha - 1} \left(\dfrac{a + b - x}{b}\right)^{\beta - 1} & \text{for } a < x < a + b \\ 0 & \text{otherwise} \end{cases} \qquad \text{(5-30)}$$

for parameters $\alpha > 0$ and $\beta > 0$.

(Again, as was said following display (5-28), $\Gamma(\cdot)$ is the gamma function, which for integer arguments n has $\Gamma(n) = (n - 1)!$.) It is fairly clear from formula (5-30) that a and b control the interval receiving positive probability, while α and β control the shape of the distribution over that interval. The case $a = 0$ and $b = 1$ is an important one; it yields the *standard beta distributions*. These are distributions over $(0, 1)$ that can be used to describe fractional quantities, such as the fraction of an 8-hour shift that a machine

FIGURE 5-18

Five Standard Beta
Probability Densities

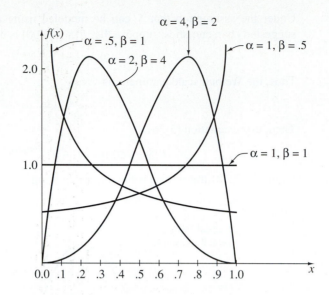

center is idle. Several standard beta densities are sketched in Figure 5-18. Densities for other beta densities with the same α and β parameters have the same basic shapes but are positive on $(a, a + b)$ instead of $(0, 1)$.

EXAMPLE 5-13

The Beta Distribution and Rainstorms. Data gathered by the U.S. Weather Service in Albuquerque (presented in an exercise in *Introduction to Contemporary Statistical Methods* by L. H. Koopmans), concern the fraction of the total rainfall falling during the first five minutes of storms occurring during both summer and nonsummer seasons. The data for 14 nonsummer storms can be described reasonably well by a standard beta distribution with $\alpha = 2.0$ and $\beta = 8.8$.

Consider observing an additional nonsummer storm in Albuquerque and defining the random variable

U = the fraction of the storm's rainfall falling during the first 5 minutes

If one models U as standard beta with $\alpha = 2.0$ and $\beta = 8.8$, it is possible (for example) to calculate the probability that more than 20% of the storm's rain falls during the first 5 minutes as follows. Using formula (5-30) with $a = 0$, $b = 1$, $\alpha = 2.0$, and $\beta = 8.8$

$$P[U > .20] = \frac{\Gamma(2.0 + 8.8)}{\Gamma(2.0)\Gamma(8.8)} \int_{.2}^{1.0} u^{1.0}(1 - u)^{7.8}\,du = (86.24)(.0045) = .39$$

This value is illustrated in Figure 5-19.

In arriving at the value .39, a pocket calculator with a preprogrammed definite integral routine was used to evaluate the integral involved. On that same calculator, values of the Γ function can be obtained using the *factorial function*, $\Gamma(x)$ being evaluated as $(x - 1)!$. This fact can be used to produce the 86.24 factor. Alternatively, the multiplier involving the Γ functions could also be evaluated using the fact that

$$\frac{\Gamma(\alpha)\Gamma(\beta)}{\Gamma(\alpha + \beta)} = \int_0^1 x^{\alpha-1}(1 - x)^{\beta-1}\,dx$$

and employing the numerical integration routine a second time.

FIGURE 5-19

Beta Density and $P[U > .20]$

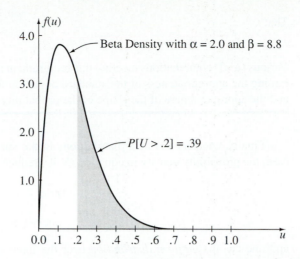

It is possible to express the beta mean and variance in relatively simple terms. Using the form of the probability density specified by equation (5-30) in equations (5-18) and (5-19), one obtains, respectively,

$$\mu = a + b\left(\frac{\alpha}{\alpha + \beta}\right) \tag{5-31}$$

and

$$\sigma^2 = b^2 \frac{\alpha\beta}{(\alpha + \beta)^2(\alpha + \beta + 1)} \tag{5-32}$$

These relationships (and the plotting of candidate densities $f(x)$) can be helpful in picking out potentially useful sets of parameters a, b, α, and β.

EXAMPLE 5-14

The Beta Distribution and PERT Analysis. A common application of nonstandard beta distributions is in Program Evaluation and Review Technique (PERT) analyses. In such analyses, time requirements for component tasks in large projects are modeled as continuous random variables, often with beta distributions. (Entire project time requirements can be deduced from component times and knowledge of any necessary task-sequencing structure.)

Suppose, for example, that in the construction of a large plant for processing crude oil, the installation of temperature-sensing devices is an important component task. One might consider modeling

$V = $ the time required to install the temperature-sensing devices

as a beta random variable. If the engineer in charge considers 1 week the best-case requirement, 5 weeks the worst-case requirement, and 2.5 weeks an average requirement, then plausible beta parameters would be

$$a = 1$$
$$b = 4$$

and via equation (5-31), α and β such that

$$2.5 = 1 + 4\left(\frac{\alpha}{\alpha + \beta}\right)$$

That is,

$$\beta = 1.67\alpha$$

Various (α, β) combinations meeting this requirement might then be compared by determining the appropriateness of the standard deviation they generate via equation (5-32) and the appropriateness of the shape of the probability density they specify.

Finally, consider the beta distributions in the special case of $\alpha = \beta = 1$. In this case, the probability density in equation (5-30) reduces to

$$f(x) = \begin{cases} \dfrac{1}{b} & \text{for } a < x < a + b \\ 0 & \text{otherwise} \end{cases}$$

and one has the *uniform distribution* over the interval $(a, a + b)$. Such a distribution has already been used in this text. The elapsed time variable of Example 5-8, Y, was modeled as uniform on the interval $(0, \frac{1}{60})$.

5-3 *Probability Plotting*

The previous two sections have emphasized that particular probabilities are only as relevant in a given application as are the distributions used to produce them. It would thus be wise to reflect on how one might make a data-based assessment of the appropriateness of a given continuous distribution in a particular real situation. The basic logic that will be employed here was introduced in Section 3-2, where the notion of Q-Q plotting to compare distributional shapes was first introduced. Consider a scenario where one has data consisting of n realizations of a random variable X and is curious whether a probability density with the same shape as $f(x)$ might adequately describe X. To answer this question, one might make and interpret a probability plot consisting of ordered pairs

$$\left(Q_1\left(\frac{i - .5}{n}\right), Q_2\left(\frac{i - .5}{n}\right) \right)$$

where Q_1 is the quantile function for the data set and Q_2 is the quantile function for the probability distribution specified by $f(x)$.

This section will elaborate on the usefulness of this logic. First, some points will be made about probability plotting in the familiar context where $f(x)$ is the standard normal density, (i.e., in the context of normal (probability) plotting). Then the general applicability of the idea will be illustrated by using it in assessing the appropriateness of exponential and Weibull probability models as well. While carrying out this program, the usefulness of probability plotting in the contexts of process capability studies and life data analysis will be briefly indicated.

More on Normal Probability Plots

Definition 5-15 gives the form of the normal or Gaussian probability density with mean μ and variance σ^2. The discussion that follows the definition shows, among other things, that all normal distributions have the same essential shape. Thus, when one makes a theoretical Q-Q plot using standard normal quantiles, one is creating a plot that can be used to judge whether or not there is *any* normal probability distribution that seems a sensible model for the mechanism that generated the data.

EXAMPLE 5-15

Weights of Circulating U.S. Nickels. Ash, Davison, and Miyagawa studied characteristics of U.S. nickels. Among other things, they obtained the weights of 100 nickels to the nearest .01 g. They found those nickels to have a mean weight of 5.002 g and a standard deviation of weights of .055 g. If one then considers the weight of another nickel taken from one's pocket to be a random variable (say, U), it is sensible to think that $EU \approx 5.002$ g and $\sqrt{Var\, U} \approx .055$ g. Further, it would be extremely convenient if a normal distribution could be used to describe U. Then, for example, normal distribution calculations with $\mu = 5.002$ g and $\sigma = .055$ g might be used to assess

$$P[U > 5.05] = P[\text{the nickel weighs over 5.05 g}]$$

A way of determining whether or not the students' data support the use of a normal model for U is to make a normal probability plot. Table 5-6 presents the data collected by Ash, Davison, and Miyagawa; Table 5-7 shows some of the calculations needed to produce a normal probability plot by hand; and Figure 5-20 is a computer-generated normal plot of the nickel weights.

TABLE 5-6

Weights of 100 U.S. Nickels

Weight (g)	Frequency
4.81	1
4.86	1
4.88	1
4.89	1
4.91	2
4.92	2
4.93	3
4.94	2
4.95	6
4.96	4
4.97	5
4.98	4
4.99	7
5.00	12
5.01	10
5.02	7
5.03	7
5.04	5
5.05	4
5.06	4
5.07	3
5.08	2
5.09	3
5.10	2
5.11	1
5.13	1

At least up to the resolution provided by the graphics in Figure 5-20, the plot is pretty linear for weights above, say, 4.90 g. However, there is some indication that the shape of the lower end of the weight distribution differs from that of a Gaussian distribution. Real nickels seem to be somewhat more likely to be light than a normal model would predict. Interestingly enough, it turns out that the four nickels with weights under 4.90 g were all minted in 1970 or before (these data were collected in 1988). This at least suggests the possibility that the nonGaussian shape of the lower end of the weight distribution is related to wear patterns and unusual damage (particularly as regards the extreme lower tail represented by the single 1964 coin with weight 4.81 g).

TABLE 5-7

Example Calculations for a
Normal Plot of Nickel Weights

i	$\dfrac{i - .5}{100}$	$\left(\dfrac{i - .5}{100}\right)$ Weight Quantile	$\left(\dfrac{i - .5}{100}\right)$ Standard Normal Quantile
1	.005	4.81	−2.576
2	.015	4.86	−2.170
3	.025	4.88	−1.960
4	.035	4.89	−1.812
5	.045	4.91	−1.695
6	.055	4.91	−1.598
7	.065	4.92	−1.514
⋮	⋮	⋮	⋮
98	.975	5.10	1.960
99	.985	5.11	2.170
100	.995	5.13	2.576

FIGURE 5-20

Normal Plot of Nickel Weights

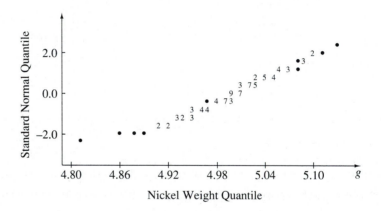

But whatever the origin of the shape in Figure 5-20, its message is clear. For most practical purposes, a normal model for the random variable

$$U = \text{the weight of a nickel taken from one's pocket}$$

will suffice. Bear in mind, however, that such a distribution will tend to slightly overstate probabilities associated with larger weights and understate probabilities associated with smaller weights.

Probability plotting is sometimes used in a slightly indirect manner. That is, if (as in Section 4-4) there is some transformation $g(\cdot)$ so that $Y = g(X)$ can be seen via probability plotting to behave in accord with a convenient probability distribution, a useful model for X itself can often be inferred.

EXAMPLE 5-16

(Example 4-11 revisited) **Normal-Plotting and Logarithms of Auto Shop Discovery Times.** Returning to the situation of Example 4-11, where Elliot, Kibby, and Meyer studied discovery times at an auto repair shop, let

V = the discovery time for the auto repair problem of the next customer to enter the shop

The discovery times (in minutes) collected by Elliot, Kibby, and Meyer are displayed in Figure 5-21. The times themselves did not produce a linear normal plot. But the log discovery times did produce a reasonably linear normal probability plot. (Refer again to Figure 4-32.) This suggests that

$$W = \ln(V)$$

be modeled as normal. Now the natural logarithms of the discovery times have sample mean $\overline{w} \approx 2.46$ and sample standard deviation $s \approx .68$. So a plausible distribution for W is a normal distribution with $\mu = 2.46$ and $\sigma = .68$.

FIGURE 5-21

Stem-and-Leaf Plot of Discovery Times

```
0 | 4 3
0 | 6 5 5 5 6 6 8 9 8 8
1 | 4 0 3 0
1 | 7 5 9 5 6 9 5
2 | 0
2 | 9 5 7 9
3 | 2
3 | 6
```

Since (for example)

$$P[a \leq V \leq b] = P[\ln(a) \leq \ln(V) \leq \ln(b)] = P[\ln(a) \leq W \leq \ln(b)]$$

it is clear that as far as finding probabilities for V goes, one could simply find the corresponding (normal) probabilities for W and never deal directly with a probability density for V. On the other hand, if for some reason it is desirable to have an expression for a probability density for V, it is easy enough to obtain one by writing out an expression for the cumulative probability function for V and differentiating.

For $v > 0$,

$$F(v) = P[V \leq v]$$
$$= P[\ln(V) \leq \ln(v)]$$
$$= P[W \leq \ln(v)]$$
$$= \int_{-\infty}^{\ln(v)} \frac{1}{\sqrt{2\pi(.68)^2}} \exp\left(-\frac{(w - 2.46)^2}{2(.68)^2}\right) dw$$

so for $v > 0$,

$$f(v) = \frac{d}{dv} F(v)$$
$$= \frac{1}{v} \frac{1}{\sqrt{2\pi(.68)^2}} \exp\left(-\frac{(\ln(v) - 2.46)^2}{2(.68)^2}\right) \tag{5-33}$$

A sketch of this probability density is given in Figure 5-22.

The preceding example illustrates a reasonably common line of reasoning, and equation (5-33) is in fact a special case of the *lognormal* probability density.

FIGURE 5-22

Lognormal Probability Density
for $\mu = 2.46$ and $\sigma = .68$

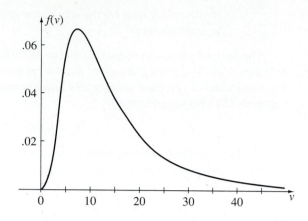

DEFINITION 5-20

The **lognormal (μ, σ^2) distribution** is a continuous probability distribution with probability density

$$f(x) = \begin{cases} \dfrac{1}{x}\dfrac{1}{\sqrt{2\pi\sigma^2}}\exp\left(-\dfrac{(\ln(x)-\mu)^2}{2\sigma^2}\right) & \text{for } x > 0 \\ 0 & \text{otherwise} \end{cases} \qquad \textbf{(5-34)}$$

for $\sigma > 0$.

The most natural way to think about a variable X that is potentially lognormally distributed is to think that $Y = \ln(X)$ is then potentially normally distributed. One may therefore check on the appropriateness of a lognormal model for X by normal-plotting observed values of $\ln(X)$.

Much has been made in this book of the fact that linearity on a Q-Q plot indicates equality of shape. But to this point, no use has been made of the fact that when one has near-linearity on a Q-Q plot, the character of the linear relationship gives information regarding the relative locations and spreads of the two distributions involved. The location of an approximating line on a Q-Q plot provides a comparison of the locations of the two distributions involved. The slope of an approximating line on a Q-Q plot provides a comparison of the spreads (or scales) of the two distributions.

For example, on a normal probability plot, the horizontal coordinate of the point on an approximating line corresponding to a vertical coordinate of 0 (i.e., $Q(.5)$ for the standard normal distribution) provides a mean for a normal distribution fitted to the data set in hand. And the difference between the horizontal coordinates of points on an approximating line corresponding to vertical coordinates of 0 and 1 (i.e., points in the standard normal distribution at the mean and 1 standard deviation away from the mean) provides a standard deviation for a normal distribution fitted to the data set. (That is, the standard deviation is the reciprocal of the slope of the line.) As such, a normal probability plot can be used not only to determine whether some normal distribution might be used to describe a random variable, but also to graphically pick out which normal distribution one might use.

EXAMPLE 5-17

Normal-Plotting and Thread Lengths of U-bolts. Table 5-8 gives thread lengths produced in the manufacture of some U-bolts for auto industry applications. The measurements are in units of .001 in. over nominal, and the particular bolts that gave the measurements

TABLE 5-8

Measured Thread Lengths
for 25 U-Bolts

Thread Length (.001 in. over Nominal)	Tally	Frequency
6	I	1
7		0
8	III	3
9		0
10	IIII	4
11	HHT HHT	10
12		0
13	HHT I	6
14	I	1

in Table 5-8 were sampled from bolts produced by a single machine over a 20-minute period.

Figure 5-23 gives a normal plot of the data in Table 5-8. It indicates that (allowing for the fact that the relatively crude measurement scale employed is responsible for the discrete/rough appearance of the plot) a normal distribution might well have been a sensible probability model for the random variable

L = the actual thread length of an additional U-bolt manufactured in the same time period

FIGURE 5-23

Normal Plot of Thread Lengths
and Eye-Fit Line

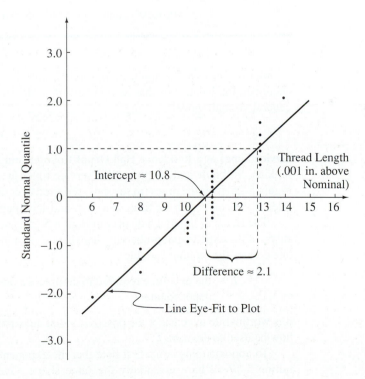

The line eye-fit to the plot further suggests appropriate values for the mean and standard deviation: $\mu \approx 10.8$ and $\sigma \approx 2.1$. It is interesting to note that direct calculation with the data in Table 5-8 gives (respectively) a sample mean and standard deviation of $\bar{l} \approx 10.9$ and $s \approx 1.9$.

In manufacturing contexts like that of the previous example, it is common to make use of the fact that an approximate standard deviation can easily be read from the slope of a normal plot so as to obtain a graphical tool for assessing *process capabilities*. That is, the primary limitation on the performance of an industrial machine or process is typically the basic precision or short-term variation associated with it. For example, suppose the output of such a process or machine over a short period of time has a characteristic of interest that is approximately normally distributed with standard deviation σ. Then, since for any normal random variable X with mean μ and standard deviation σ,

$$P[\mu - 3\sigma < X < \mu + 3\sigma] > .99$$

it makes some sense to use 6σ ($= (\mu + 3\sigma) - (\mu - 3\sigma)$) as a measure of process performance or capability. It is easy to read such a capability figure off a normal plot, and many companies use specially prepared *process capability analysis forms* (which are in essence nothing but pieces of normal probability paper) for this purpose.

EXAMPLE 5-17
(continued)

Figure 5-24 is a plot of the thread length data from Table 5-8, made on a fairly common machine capability analysis sheet. (This author has seen a number of different company logos printed on this type of form.) Notice that using the line on the plot, it is very easy for even someone with limited quantitative background (and perhaps even lacking a basic understanding of the concept of a standard deviation) to arrive at the figure

$$\text{Process capability} \approx 16 - 5 = 11(.001 \text{ in.})$$

Probability Plots for Exponential and Weibull Distributions

In order to illustrate the application of probability plotting to distributions that are not Gaussian, the balance of this section considers its use with first exponential and then general Weibull models.

EXAMPLE 5-18

Service Times at a Residence Hall Depot Counter and Exponential Probability Plotting. Jenkins, Milbrath, and Worth studied service times at a residence hall "depot" counter. Figure 5-25 gives the times (in seconds) required to complete 65 different postage stamp sales at the counter, in the form of a stem-and-leaf diagram.

The shape of the rough display in Figure 5-25 is at least vaguely reminiscent of the shape of the exponential probability densities shown in Figure 5-14. So if, for example, one defines the random variable

$T =$ the next time required to complete a postage stamp sale at the depot counter

it is worthwhile to consider the possibility that an exponential distribution might somehow be used to describe T.

To approach this issue, first note that the exponential distributions defined in Definition 5-17 all have essentially the same shape. One may thus use the exponential distribution with $\alpha = 1$ as a convenient representative of that shape. A plot of $\alpha = 1$ exponential quantiles versus corresponding service time quantiles will give a tool for comparing the empirical shape to the theoretical exponential shape.

For an exponential distribution with mean $\alpha = 1$,

$$F(x) = 1 - e^{-x} \qquad \text{for } x > 0$$

FIGURE 5-24

Thread Length Data Plotted on a Capability Analysis Form (used with permission of Reynolds Metals Company)

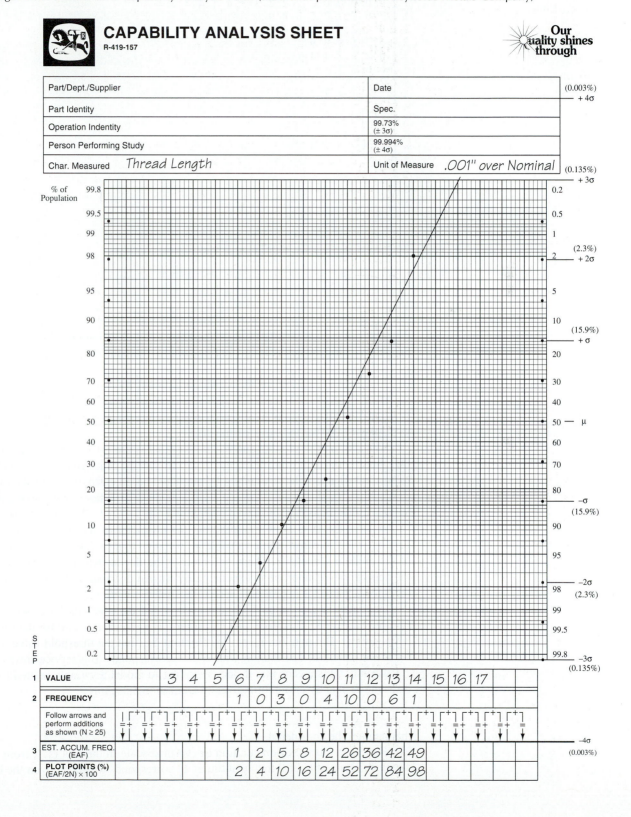

FIGURE 5-25

Stem-and-Leaf Plot of
Service Times

```
0
0   8  8  8  9  9
1   0  0  0  0  0  2  2  2  2  3  3  4  4  4  4  4  4
1   5  6  7  7  7  7  8  8  9  9
2   0  1  2  2  2  2  3  4
2   6  7  8  9  9  9
3   0  2  2  2  4  4
3   6  6  7  7
4   2  3
4   5  6  7  8  8
5
5
6
6
7   0
7
8
8   7
```

So for $0 < p < 1$, setting $F(x) = p$ and solving, one has

$$p = 1 - e^{-x}$$
$$e^{-x} = 1 - p$$
$$x = -\ln(1 - p)$$

That is, $-\ln(1 - p) = Q(p)$, the p quantile of this distribution. Thus, an exponential probability plot can be made by plotting the ordered pairs

$$\left(i\text{th smallest data point}, -\ln\left(1 - \frac{i - .5}{n}\right)\right) \qquad \textbf{(5-35)}$$

Printout 5-1 is from part of a Minitab session used to create a plot of the points indicated in display (5-35) for the service time data.

The plot on the printout shows linearity that is almost remarkable. Except for the fact that the third and fourth largest observed service times (both 48 seconds) appear to be somewhat smaller than might be predicted based on the shape of the exponential distribution, the shape of the empirical service time distribution corresponds quite closely to the exponential shape displayed in Figure 5-14.

As was the case in normal-plotting, the character of the linearity on the plot on Printout 5-1 also carries some valuable information that can be applied to the modeling of the random variable T. The positioning of the line sketched onto the plot indicates the appropriate location of an exponentially shaped distribution for T, and the slope of the line indicates the appropriate spread for that distribution.

As stated in Definition 5-17, the exponential distributions have positive density $f(x)$ for positive x. Therefore, one might term 0 a *threshold value* for the distributions defined there. But note on the plot on Printout 5-1 that the threshold value $(0 = Q(0))$ for the exponential distribution with $\alpha = 1$ corresponds to a service time of roughly 7.5 seconds. This means that if one is looking to model a variable related to T with a distribution exactly of the form given in Definition 5-17, it is

$$S = T - 7.5$$

that ought to be considered.

Further, a change of one unit on the vertical scale in the plot (say, from an $\alpha = 1$ exponential quantile of 0 to a quantile of 1) corresponds to a change on the horizontal

PRINTOUT 5-1

```
MTB > print c1-c4
```

ROW	i	(i-.5)/n	time	Q(p)
1	1	0.007692	8	0.00772
2	2	0.023077	8	0.02335
3	3	0.038462	8	0.03922
4	4	0.053846	9	0.05535
5	5	0.069231	9	0.07174
6	6	0.084615	10	0.08841
7	7	0.100000	10	0.10536
8	8	0.115385	10	0.12260
9	9	0.130769	10	0.14015
10	10	0.146154	10	0.15800
11	11	0.161538	12	0.17619
12	12	0.176923	12	0.19471
13	13	0.192308	12	0.21357
14	14	0.207692	12	0.23281
15	15	0.223077	13	0.25241
16	16	0.238462	13	0.27241
17	17	0.253846	14	0.29282
18	18	0.269231	14	0.31366
19	19	0.284615	14	0.33493
20	20	0.300000	14	0.35667
21	21	0.315385	14	0.37890
22	22	0.330769	14	0.40163
23	23	0.346154	15	0.42488
24	24	0.361538	16	0.44869
25	25	0.376923	17	0.47309
26	26	0.392308	17	0.49809
27	27	0.407692	17	0.52373
28	28	0.423077	17	0.55005
29	29	0.438462	18	0.57708
30	30	0.453846	18	0.60485
31	31	0.469231	19	0.63343
32	32	0.484615	19	0.66284
33	33	0.500000	20	0.69315
34	34	0.515385	21	0.72440
35	35	0.530769	22	0.75666
36	36	0.546154	22	0.79000
37	37	0.561538	22	0.82448
38	38	0.576923	22	0.86020
39	39	0.592308	23	0.89724
40	40	0.607692	24	0.93571
41	41	0.623077	26	0.97571
42	42	0.638462	27	1.01739
43	43	0.653846	28	1.06087
44	44	0.669231	29	1.10633
45	45	0.684615	29	1.15396
46	46	0.700000	29	1.20397
47	47	0.715385	30	1.25662
48	48	0.730769	32	1.31219
49	49	0.746154	32	1.37103
50	50	0.761538	32	1.43355
51	51	0.776923	34	1.50024
52	52	0.792308	34	1.57170
53	53	0.807692	36	1.64866
54	54	0.823077	36	1.73204
55	55	0.838462	37	1.82301

(continued)

PRINTOUT 5-1
(continued)

56	56	0.853846	37	1.92310
57	57	0.869231	42	2.03432
58	58	0.884615	43	2.15948
59	59	0.900000	45	2.30258
60	60	0.915385	46	2.46964
61	61	0.930769	47	2.67031
62	62	0.946154	48	2.92162
63	63	0.961538	48	3.25810
64	64	0.976923	70	3.76892
65	65	0.992308	87	4.86753

```
MTB > plot c4 vs c3
```

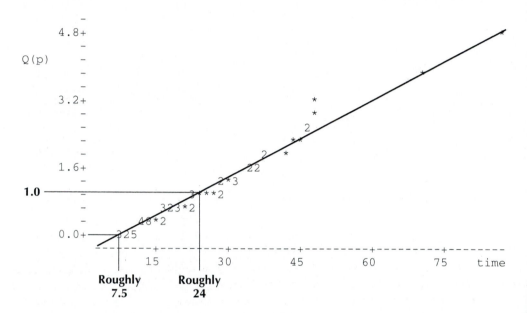

scale of roughly

$$24 - 7.5 = 16.5 \text{ sec}$$

That is, an exponential model for S ought to have an associated spread that is 16.5 times that of the exponential distribution with $\alpha = 1$; S might be modeled as exponential with $\alpha = 16.5$.

So ultimately, the data in Figure 5-25 lead via exponential probability plotting to the suggestion that

$S = T - 7.5$

     = the excess of the next time required to complete a postage stamp sale over the threshold value of 7.5 seconds

be described with the density

$$f(s) = \begin{cases} \dfrac{1}{16.5} e^{-(s/16.5)} & \text{for } s > 0 \\[2mm] 0 & \text{otherwise} \end{cases} \tag{5-36}$$

Of course, probabilities involving T can be computed by first expressing them in terms of S and then using expression (5-36) or the corresponding cumulative probability function.

But if for some reason an expression for a density of T itself is desired, simply shift the density in equation (5-36) to the right 7.5 units to obtain the density

$$f(t) = \begin{cases} \dfrac{1}{16.5}e^{-((t-7.5)/16.5)} & \text{for } t > 7.5 \\ 0 & \text{otherwise} \end{cases}$$

Figure 5-26 shows the probability densities for both S and T.

FIGURE 5-26

Probability Densities for S and T

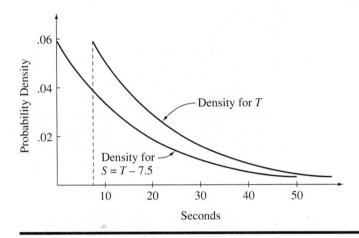

To summarize the main content of the preceding example: Because of the relatively simple form of the exponential $\alpha = 1$ cumulative probability function, it is easy to find quantiles for this distribution. When these are plotted against corresponding quantiles of a data set, one obtains an exponential probability plot. On this plot, linearity indicates exponential shape, the horizontal intercept of a linear plot indicates an appropriate threshold value, and the reciprocal of the slope indicates an appropriate value for the exponential parameter α.

As it turns out, much the same story can be told for the Weibull distributions in general. Recalling Definition 5-18 and the ensuing discussion, one sees that for a given β, the Weibull distributions with parameters α and β all have the same essential shape. Further, considering as a representative of that shape the Weibull distribution with parameters β and $\alpha = 1$, one has, for $x > 0$,

$$F(x) = 1 - e^{-x^{\beta}}$$

Then for $0 < p < 1$, the equation $p = F(x)$ can be easily solved to yield

$$x = (-\ln(1 - p))^{1/\beta}$$

That is, making a plot of the ordered pairs

$$\left(i\text{th smallest data point}, \left(-\ln\left(1 - \frac{i - .5}{n} \right) \right)^{1/\beta} \right) \tag{5-37}$$

is a way of investigating whether a variable might be described with a probability distribution having the Weibull shape for the β in question. On such a plot, linearity indicates equality of shape, the horizontal intercept indicates an appropriate threshold value, and the reciprocal of the slope indicates an appropriate value for the parameter α.

EXAMPLE 5-19

Electrical Insulation Failure Voltages and Weibull Plotting. The data given in the stem-and-leaf plot in Figure 5-27 are failure voltages (in kv/mm) for a type of electrical cable insulation subjected to increasing voltage stress. They were taken from *Statistical Models and Methods for Lifetime Data* by J. F. Lawless.

FIGURE 5-27

Stem-and-Leaf Plot of
Insulation Failure Voltages

```
3 |
3 | 9.4
4 | 5.3
4 | 9.2, 9.4
5 | 1.3, 2.0, 3.2, 3.2, 4.9
5 | 5.5, 7.1, 7.2, 7.5, 9.2
6 | 1.0, 2.4, 3.8, 4.3
6 | 7.3, 7.7
```

Consider making a plot of the ordered pairs indicated by display (5-37), using $\beta = 4$. Figure 5-16 in the previous section shows the shape of a Weibull density with $\beta = 4$, and it is sensible to think that the data pictured in Figure 5-27 might be consistent with the use of such a distribution to describe

R = the voltage at which one additional specimen of this insulation will fail

Some of the calculations needed to make a plot of the points given by equation (5-37) are indicated in Table 5-9, and Figure 5-28 shows the plot that results from the calculations indicated in Table 5-9.

TABLE 5-9

Example Calculations for a
$\beta = 4$ Weibull Probability Plot
of the Failure Voltages

i	ith Smallest Voltage	$p = (i - .5)/20$	$(-\ln(1 - p))^{1/\beta}$
1	39.4	.025	.40
2	45.3	.075	.53
3	49.2	.125	.60
4	49.4	.175	.66
⋮	⋮	⋮	⋮
19	67.3	.925	1.27
20	67.7	.975	1.39

FIGURE 5-28

Weibull Probability Plot
with $\beta = 4$ for Insulation
Failure Voltages

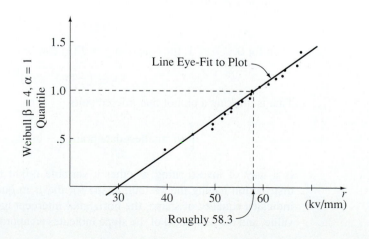

The plot indicates that a Weibull distribution with $\beta = 4$ and $\alpha \approx 58.3 - 30 = 28.3$ is a plausible candidate for describing the random variable

$R - 30 =$ the excess voltage (beyond a threshold of 30) required to cause the next specimen's failure

---

Although the kind of plot indicated by display (5-37) is fairly easy to make and interpret, it is *not* by any means the most common form of probability plotting associated with the Weibull distributions. There are at least two reasons for this. Most importantly, in order to make a plot like the one in Example 5-19, one must input a value of β. Figure 5-16 illustrates clearly that different values of β produce different Weibull shapes, and a candidate shape parameter β is needed when making a plot of the type indicated by display (5-37). But in many applications of the Weibull distributions, one doesn't *a priori* have a value of β in mind. In fact, the first issue to be settled is often determination of what values of β are plausible. In addition, in many applications of the Weibull distributions, there is a natural threshold value: the value 0 that was associated with the introduction of the Weibull distributions in Definition 5-18. But the methodology indicated by display (5-37) doesn't directly constrain the threshold to be 0.

What is typically needed from an initial Weibull-based probability plot (instead of a method that inputs β and can be used to identify a threshold and α) is a method that tacitly inputs a 0 threshold and can be used to identify α and β. This is particularly true in applications to reliability and life data analysis, where the useful life or time to failure of some device is often the variable under study.

Happily enough, it is possible to employ a trick and come up with a probability plotting method that allows identification of values for both α and β in Definition 5-18. The trick is to work on a log scale. That is, if M is a random variable with the Weibull (α, β) distribution, then for $m > 0$,

$$F(m) = 1 - e^{-(m/\alpha)^\beta}$$

so that

$$P[\ln(M) \leq h] = P[M < e^h]$$
$$= 1 - e^{-(e^h/\alpha)^\beta}$$

and for $0 < p < 1$, setting $p = P[\ln(M) < h]$ gives

$$p = 1 - e^{-(e^h/\alpha)^\beta}$$

That is,

$$e^{-(e^h/\alpha)^\beta} = 1 - p$$

$$-\left(\frac{e^h}{\alpha}\right)^\beta = \ln(1 - p)$$

$$\beta \ln\left(\frac{e^h}{\alpha}\right) = \ln(-\ln(1 - p))$$

or finally,

$$\beta h - \beta \ln(\alpha) = \ln(-\ln(1 - p)) \tag{5-38}$$

Now h is (by design) the p quantile of the distribution of $\ln(M)$. So equation (5-38) says that $\ln(-\ln(1 - p))$ is a linear function of $\ln(M)$'s quantile function. The slope of that relationship is β. Further, equation (5-38) shows that when $\ln(-\ln(1-p)) = 0$, the quantile function of $\ln(M)$ has the value $\ln(\alpha)$. So exponentiation of the intercept on the horizontal axis gives α. Thus, one is led to considering a plot of ordered pairs

$$\left(\ln(i\text{th smallest data value}), \ln\left(-\ln\left(1 - \frac{i - .5}{n}\right)\right)\right) \qquad \textbf{(5-39)}$$

If data in hand are consistent with a (0 threshold) Weibull (α, β) model, a reasonably linear plot with slope β and horizontal axis intercept equal to $\ln(\alpha)$ may be expected.

EXAMPLE 5-19
(continued)

Consider again the data given in Figure 5-27 using a plot of the type indicated by display (5-39). Table 5-10 shows some of the calculations that were needed to produce Figure 5-29.

The near-linearity of the plot in Figure 5-29 suggests that a (0 threshold) Weibull distribution might indeed be used to describe the random variable R. A Weibull shape

TABLE 5-10

Example Calculations for a Zero-Threshold Weibull Plot of Failure Voltages

i	ith Smallest Voltage	$\ln(i$th Smallest Voltage$)$	$p = (i - .5)/20$	$\ln(-\ln(1 - p))$
1	39.4	3.67	.025	−3.68
2	45.3	3.81	.075	−2.55
3	49.2	3.90	.125	−2.01
4	49.4	3.90	.175	−1.65
⋮	⋮	⋮	⋮	⋮
19	67.3	4.21	.925	.95
20	67.7	4.22	.975	1.31

FIGURE 5-29

Zero-Threshold Weibull Plot for Insulation Failure Voltages

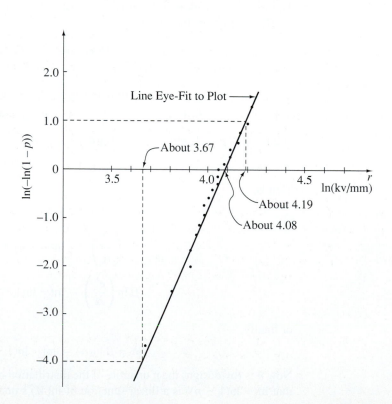

parameter of roughly

$$\beta \approx \text{slope of the fitted line} \approx \frac{1 - (-4)}{4.19 - 3.67} \approx 9.6$$

is indicated. Further, a scale parameter α with

$$\ln(\alpha) \approx \text{horizontal intercept} \approx 4.08$$

and thus

$$\alpha \approx 59$$

appears appropriate.

————

Several comments are in order before closing this section. First, owing to its widespread popularity in life data analysis and strength of materials applications, plotting form (5-39) is what most engineers expect to see when a *Weibull plot* is to be produced. Just as normal-plotting is facilitated by the use of normal probability paper, it is common to see such Weibull plots made on special *Weibull paper*. This is graph paper whose scales are constructed so that instead of using plotting positions given in display (5-39), one can use plotting positions given by

$$\left(i\text{th smallest data value,} \ \frac{i - .5}{n} \right)$$

Figure 5-30 is a reproduction of such a piece of paper. Notice that the determination of β is even facilitated in Figure 5-30 through the inclusion of the protractor in the upper left corner. (The expectation of the paper producers is pretty clearly that most users will encounter Weibull shape parameters of 6 or less.)

Second, some readers may find it disconcerting that in Example 5-19, both a Weibull distribution with threshold 30, $\beta = 4$, and $\alpha = 28.3$ *and* a Weibull distribution with threshold 0, $\beta = 9.6$, and $\alpha = 59$ appeared to be possible models for R. In fact, the situation is even more muddled than this. Normal-plotting will show that both normal and lognormal models are reasonably plausible for R as well. But there is really no logical difficulty here. It is simply the case that for values of their parameters consonant with the particular data in Figure 5-27, these distributions are fairly similar. $n = 20$ observations are not adequate to make possible clear choices between them, and as far as one can tell, any of them could be used to describe R. This type of ambivalence is reasonably common when sample sizes are small. A specialist in the subject matter can often help in identifying which type of distribution is likely to be useful.

Finally, it should be emphasized again that the idea of probability plotting is a quite general one. Its use has been illustrated here only with normal and Weibull distributions. But remember that for any probability density $f(x)$, theoretical Q-Q plotting provides a tool for assessing whether the distributional shape exhibited by $f(x)$ might fruitfully be used in the modeling of a given random variable.

5-4 *Joint Distributions and Independence*

Most applications of probability to engineering statistics involve not one but several random variables. In many cases, the application is distinctly multivariate, and it makes sense to think of more than one process variable as subject to random influences. But even when a situation is univariate, samples larger than size 1 are essentially always

FIGURE 5-30
Weibull Probability Paper

WEIBULL PROBABILITY PAPER

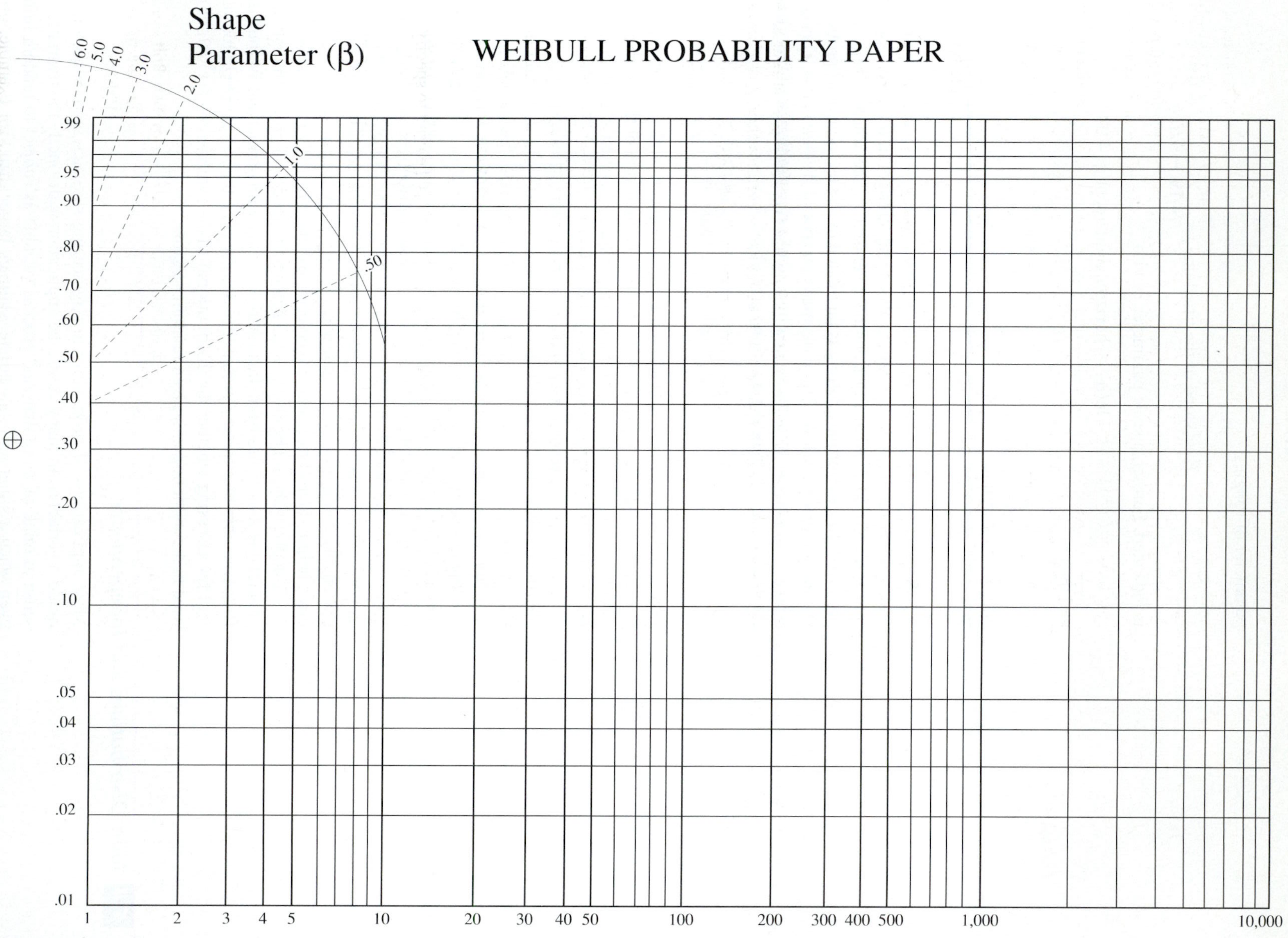

used, and the *n* data values in a sample are usually thought of (and reasonably so) as subject to chance causes. The methods of Sections 5-1 and 5-2 are capable of dealing with only a single random variable at a time. They must be generalized to create methods for modeling several random variables simultaneously if this text is to provide anything approaching a logically complete exposition of the probability needed to support standard methods of formal statistical inference.

Entire books could be written on various aspects of the simultaneous description of many random variables. This section can thus give only a brief introduction to the topic of joint distributions for several random variables. Considering first the comparatively simple case of jointly discrete random variables, the topics of joint and marginal probability functions, conditional distributions and independence will be discussed. The exposition will be done primarily through reference to simple bivariate examples. Next are introduced the analogous concepts of joint and marginal probability density functions, conditional distributions, and independence for jointly continuous random variables. Again, this discussion will be carried out primarily through reference to a bivariate example.

Describing Jointly Discrete Random Variables

When working simultaneously with random variables X and Y, it is important to evaluate not only probabilities associated with the variables individually, but also those associated with them in combination. Take, for example, the assembly of a ring bearing with nominal inside diameter 1.00 inch on a rod with nominal diameter .99 inch. If

$$X = \text{the ring bearing inside diameter}$$
$$Y = \text{the rod diameter}$$

one might be interested in

$$P[X < 1.00] = P[\text{the ring bearing inside diameter is below nominal}]$$
$$P[Y > .99] = P[\text{the rod diameter is above nominal}]$$

However,

$$P[X < Y] = P[\text{there is an interference in assembly}]$$

might also be of critical importance.

For several discrete variables $X, Y, \ldots, Z$, the device used to specify probabilities is a *joint probability function*.

DEFINITION 5-21

A **joint probability function** for discrete random variables $X, Y, \ldots, Z$ is a nonnegative function $f(x, y, \ldots, z)$, giving the probability that (simultaneously) X takes the value x, Y takes the value $y, \ldots$, and Z takes the value z. That is,

$$f(x, y, \ldots, z) = P[X = x \text{ and } Y = y \text{ and} \ldots Z = z]$$

EXAMPLE 5-20

(Example 5-1 revisited) **The Joint Probability Distribution of Two Bolt Torques.** Return again to the situation of Brenny, Christensen, and Schneider and the measuring of bolt torques on the face plates of a heavy equipment component. With

$$X = \text{the next torque recorded for bolt 3}$$
$$Y = \text{the next torque recorded for bolt 4}$$

the data displayed in Table 3-4 and Figure 3-9 suggest, for example, that a sensible value for $P[X = 18 \text{ and } Y = 18]$ might be $\frac{1}{34}$, the relative frequency of this pair in the data set. Similarly, the plausible assignments

$$P[X = 18 \text{ and } Y = 17] = \tfrac{2}{34}$$
$$P[X = 14 \text{ and } Y = 9] = 0$$

also correspond to observed relative frequencies.

If one is willing to accept the whole set of relative frequencies defined by the students' data as defining probabilities for X and Y, these can be collected conveniently in a two-dimensional table specifying a joint probability function for X and Y. This is illustrated in Table 5-11. (To avoid clutter, 0 entries in Table 5-11 have been left blank.)

TABLE 5-11

$f(x, y)$ for the Bolt Torque Problem

y \ x	11	12	13	14	15	16	17	18	19	20
20								$\frac{2}{34}$	$\frac{2}{34}$	$\frac{1}{34}$
19							$\frac{2}{34}$			
18			$\frac{1}{34}$	$\frac{1}{34}$			$\frac{1}{34}$	$\frac{1}{34}$	$\frac{1}{34}$	
17					$\frac{2}{34}$	$\frac{1}{34}$	$\frac{1}{34}$	$\frac{2}{34}$		
16				$\frac{1}{34}$	$\frac{2}{34}$	$\frac{2}{34}$			$\frac{2}{34}$	
15	$\frac{1}{34}$	$\frac{1}{34}$			$\frac{3}{34}$					
14					$\frac{1}{34}$			$\frac{2}{34}$		
13					$\frac{1}{34}$					

The probability function given in tabular form in Table 5-11 has two properties that are necessary for the mathematical consistency of a joint probability function. These are that the $f(x, y, \ldots, z)$ values are each in the interval $[0,1]$ and that they total to 1 when all summed up. By summing up just some of the $f(x, y, \ldots, z)$ values, one obtains probabilities associated with $X, Y, \ldots, Z$ being configured in patterns of interest.

EXAMPLE 5-20
(continued)

For the sake of illustrating how to sum values of a probability function to obtain probabilities related to the joint behavior of the variables being modeled, consider using the joint distribution given in Table 5-11 to evaluate

$$P[X \geq Y]$$
$$P[|X - Y| \leq 1]$$
$$P[X = 17]$$

Take first $P[X \geq Y]$, the probability that the measured bolt 3 torque is at least as big as the measured bolt 4 torque. Figure 5-31 indicates with asterisks which possible combinations of x and y lead to bolt 3 torque at least as large as the bolt 4 torque. Upon referring back to Table 5-11 and adding up those entries corresponding to the cells that contain asterisks, one has

$$P[X \geq Y] = f(15, 13) + f(15, 14) + f(15, 15) + f(16, 16)$$
$$+ f(17, 17) + f(18, 14) + f(18, 17) + f(18, 18)$$
$$+ f(19, 16) + f(19, 18) + f(20, 20)$$
$$= \frac{1}{34} + \frac{1}{34} + \frac{3}{34} + \frac{2}{34} + \cdots + \frac{1}{34}$$
$$= \frac{17}{34}$$

FIGURE 5-31
Combinations of Bolt 3 and Bolt 4 Torques with $x \geqslant y$

y \ x	11	12	13	14	15	16	17	18	19	20
20										*
19									*	*
18								*	*	*
17							*	*	*	*
16						*	*	*	*	*
15					*	*	*	*	*	*
14				*	*	*	*	*	*	*
13			*	*	*	*	*	*	*	*

Similar reasoning allows one to evaluate $P[|X - Y| \leqslant 1]$ — the probability that the bolt 3 and 4 torques are within 1 ft lb of each other. Figure 5-32 shows combinations of x and y with an absolute difference of 0 or 1.

FIGURE 5-32
Combinations of Bolt 3 and Bolt 4 Torques with $|x - y| \leqslant 1$

y \ x	11	12	13	14	15	16	17	18	19	20
20									*	*
19								*	*	*
18							*	*	*	
17						*	*	*		
16					*	*	*			
15				*	*	*				
14			*	*	*					
13			*	*						

Then, adding probabilities corresponding to these combinations,

$$\begin{aligned} P[|X - Y| \leqslant 1] &= f(15,14) + f(15,15) + f(15,16) + f(16,16) \\ &\quad + f(16,17) + f(17,17) + f(17,18) + f(18,17) \\ &\quad + f(18,18) + f(19,18) + f(19,20) + f(20,20) \\ &= \frac{18}{34} \end{aligned}$$

Finally, $P[X = 17]$, the probability that the measured bolt 3 torque is 17 ft lb, is pretty clearly obtained by adding down the $x = 17$ column in Table 5-11. That is,

$$\begin{aligned} P[X = 17] &= f(17,17) + f(17,18) + f(17,19) \\ &= \frac{1}{34} + \frac{1}{34} + \frac{2}{34} \\ &= \frac{4}{34} \end{aligned}$$

The last part of the preceding example shows that starting from a joint probability function for variables $X, Y, \ldots, Z$, it is a simple matter to obtain probabilities for a single one of them by adding $f(x, y, \ldots, z)$ values over all possible combinations of the other variables. Notice that in bivariate examples like the present one, one can add down columns in a two-way table giving $f(x, y)$ to get values for the probability function of X, $f_X(x)$. And one can add across rows in the same table to get values for the probability function of Y, $f_Y(y)$. One can then further think of writing these sums in the margins of the two-way table. So it should not be surprising that probability distributions for individual random variables $X, Y, \ldots, Z$ obtained from their joint distribution are called *marginal distributions*.

DEFINITION 5-22

The individual probability functions for discrete random variables $X, Y, \ldots, Z$ with joint probability function $f(x, y, \ldots, z)$ are called **marginal probability functions**. They are obtained by summing $f(x, y, \ldots, z)$ values over all possible combinations of the other variables. In symbols, the marginal probability function for X, for example, is given by

$$f_X(x) = \sum_{y,\ldots,z} f(x, y, \ldots, z)$$

EXAMPLE 5-20
(continued)

Table 5-12 is a copy of Table 5-11, augmented by the addition of marginal probabilities for X and Y.

TABLE 5-12

Joint and Marginal Probabilities for X and Y

$y \backslash X$	11	12	13	14	15	16	17	18	19	20	$f_Y(y)$
20								$\frac{2}{34}$	$\frac{2}{34}$	$\frac{1}{34}$	$\frac{5}{34}$
19							$\frac{2}{34}$				$\frac{2}{34}$
18			$\frac{1}{34}$	$\frac{1}{34}$			$\frac{1}{34}$	$\frac{1}{34}$	$\frac{1}{34}$		$\frac{5}{34}$
17					$\frac{2}{34}$	$\frac{1}{34}$	$\frac{1}{34}$	$\frac{2}{34}$			$\frac{6}{34}$
16				$\frac{1}{34}$	$\frac{2}{34}$	$\frac{2}{34}$			$\frac{2}{34}$		$\frac{7}{34}$
15	$\frac{1}{34}$	$\frac{1}{34}$			$\frac{3}{34}$						$\frac{5}{34}$
14					$\frac{1}{34}$			$\frac{2}{34}$			$\frac{3}{34}$
13					$\frac{1}{34}$						$\frac{1}{34}$
$f_X(x)$	$\frac{1}{34}$	$\frac{1}{34}$	$\frac{1}{34}$	$\frac{2}{34}$	$\frac{9}{34}$	$\frac{3}{34}$	$\frac{4}{34}$	$\frac{7}{34}$	$\frac{5}{34}$	$\frac{1}{34}$	

Separating off the margins from the two-way table produces tables of marginal probabilities in the familiar format of Section 5-1. For example, the marginal probability function of Y is given separately in Table 5-13.

TABLE 5-13

Marginal Probability Function for Y

y	$f_Y(y)$
13	$\frac{1}{34}$
14	$\frac{3}{34}$
15	$\frac{5}{34}$
16	$\frac{7}{34}$
17	$\frac{6}{34}$
18	$\frac{5}{34}$
19	$\frac{2}{34}$
20	$\frac{5}{34}$

The realization that one can get marginal probability functions from joint probability functions raises the natural question whether the process can be reversed. That is, if one knows $f_X(x)$ and $f_Y(y)$, does one then have exactly one choice for $f(x, y)$? The answer to this question is "No," unless further stipulations are added. Figure 5-33 shows two quite different bivariate joint distributions that nonetheless possess the same marginal distributions. The marked difference between the distributions in Figure 5-33 has to do with the *joint*, rather than individual, behavior of X and Y.

FIGURE 5-33

Two Different Joint Distributions with the Same Marginal Distributions

Distribution 1

y \ x	1	2	3	
3	.4	0	0	.4
2	0	.4	0	.4
1	0	0	.2	.2
	.4	.4	.2	

Distribution 2

y \ x	1	2	3	
3	.16	.16	.08	.4
2	.16	.16	.08	.4
1	.08	.08	.04	.2
	.4	.4	.2	

Conditional Distributions and Independence for Discrete Random Variables

When working with several random variables $X, Y, \dots, Z$, it is often useful to think about what is expected of one of the variables, given the values assumed by all others. For example, in the bolt torque situation, a technician who has just loosened bolt 3 and measured the torque as 15 ft lb ought to have expectations somewhat different from those described by the marginal distribution in Table 5-13 as bolt 4 is loosened. After all, returning to the data in Table 3-4 that led to Table 5-11, it is evident that the relative frequency distribution of bolt 4 torques for those components with bolt 3 torque of 15 ft lb is as in Table 5-14. Somehow, knowing that $X = 15$ ought to make a probability distribution for Y like the relative frequency distribution in Table 5-14 more relevant than the marginal distribution given in Table 5-13.

TABLE 5-14

Relative Frequency Distribution for Bolt 4 Torques When Bolt 3 Torque Is 15 ft lb

y Torque (ft lb)	Relative Frequency
13	$\frac{1}{9}$
14	$\frac{1}{9}$
15	$\frac{3}{9}$
16	$\frac{2}{9}$
17	$\frac{2}{9}$

The theory of probability makes allowance for this notion of "distribution of one variable knowing the values of others" through the concept of conditional distributions.

DEFINITION 5-23

For discrete random variables $X, Y, \dots, Z$ with joint probability function $f(x, y, \dots, z)$, the **conditional probability function of** (for example) X **given** $Y = y, \dots, Z = z$ is the function of x defined by the ratio

$$\frac{f(x, y, \dots, z)}{\sum_x f(x, y, \dots, z)} \tag{5-40}$$

Standard notation for the ratio in expression (5-40) is $f_{X|Y,\dots,Z}(x \mid y, \dots, z)$.

A little reflection on the meaning of expression (5-40) will show that if only two variables X and Y are under consideration, expression (5-40) leads to

$$f_{X|Y}(x \mid y) = \frac{f(x, y)}{f_Y(y)} \tag{5-41}$$

and

$$f_{Y|X}(y \mid x) = \frac{f(x, y)}{f_X(x)} \tag{5-42}$$

And formulas (5-41) and (5-42) are perfectly sensible. Equation (5-41) says that starting from $f(x, y)$ given in a two-way table, if one's attention is restricted to the row specified by $Y = y$, the appropriate (conditional) distribution for X is given by the probabilities in that row (the $f(x, y)$ values) divided by their sum ($f_Y(y) = \sum_x f(x, y)$), so that they are renormalized to total to 1. Similarly, equation (5-42) says that if one's attention is restricted to the column specified by $X = x$, the appropriate conditional distribution for Y is given by the probabilities in that column divided by their sum.

EXAMPLE 5-20
(continued)

To illustrate the use of equations (5-41) and (5-42), consider several of the conditional distributions associated with the joint distribution for the bolt 3 and bolt 4 torques, beginning with the conditional distribution for Y given that $X = 15$.

From equation (5-42),

$$f_{Y|X}(y \mid 15) = \frac{f(15, y)}{f_X(15)}$$

Referring to Table 5-12, the marginal probability associated with $X = 15$ is $\frac{9}{34}$. So dividing values in the $X = 15$ column of that table by $\frac{9}{34}$, one is led to the conditional distribution for Y given in Table 5-15.

TABLE 5-15

The Conditional Probability Function for Y Given $X = 15$

y	$f_{Y|X}(y \mid 15)$
13	$\left(\frac{1}{34}\right) \div \left(\frac{9}{34}\right) = \frac{1}{9}$
14	$\left(\frac{1}{34}\right) \div \left(\frac{9}{34}\right) = \frac{1}{9}$
15	$\left(\frac{3}{34}\right) \div \left(\frac{9}{34}\right) = \frac{3}{9}$
16	$\left(\frac{2}{34}\right) \div \left(\frac{9}{34}\right) = \frac{2}{9}$
17	$\left(\frac{2}{34}\right) \div \left(\frac{9}{34}\right) = \frac{2}{9}$

Comparing Table 5-15 to Table 5-14, one sees that indeed formula (5-42) produces a conditional distribution that agrees with intuition.

For the sake of contrast with $f_{Y|X}(y \mid 15)$, next consider $f_{Y|X}(y \mid 18)$ specified by

$$f_{Y|X}(y \mid 18) = \frac{f(18, y)}{f_X(18)}$$

Consulting Table 5-12 again leads to the conditional distribution for Y given that $X = 18$, shown in Table 5-16.

It is obvious from Tables 5-15 and 5-16 that the conditional distributions of Y given $X = 15$ and given $X = 18$ are quite different. For example, knowing that $X = 18$ would on the whole make one expect Y to be larger than corresponding expectations given that $X = 15$. (The fact that $f_{Y|X}(y \mid x)$ changes with x represents an important enough concept that some formal terminology to describe it will shortly be introduced.)

TABLE 5-16

The Conditional Probability
Function for Y Given $X = 18$

y	$f_{Y|X}(y \mid 18)$
14	$\frac{2}{7}$
17	$\frac{2}{7}$
18	$\frac{1}{7}$
20	$\frac{2}{7}$

To make sure that the meaning of equation (5-41) is also clear, consider the conditional distribution of the bolt 3 torque (X) given that the bolt 4 torque is 20 ($Y = 20$). In this situation, equation (5-41) gives

$$f_{X|Y}(x \mid 20) = \frac{f(x, 20)}{f_Y(20)}$$

(Conditional probabilities for X are the values in the $Y = 20$ row of Table 5-12 divided by the marginal $Y = 20$ value.) Thus, $f_{X|Y}(x \mid 20)$ is given in Table 5-17.

TABLE 5-17

The Conditional Probability
Function for X Given $Y = 20$

x	$f_{X|Y}(x \mid 20)$
18	$\left(\frac{2}{34}\right) \div \left(\frac{5}{34}\right) = \frac{2}{5}$
19	$\left(\frac{2}{34}\right) \div \left(\frac{5}{34}\right) = \frac{2}{5}$
20	$\left(\frac{1}{34}\right) \div \left(\frac{5}{34}\right) = \frac{1}{5}$

The bolt torque example has the feature that (for example) the conditional distributions for Y given various possible values for X differ. Further, these are not generally the same as the marginal distribution for Y. In a sense, X provides some information about Y, in that depending upon its value one has differing probability assessments for Y. Contrast this situation with the following example, which is admittedly somewhat artificial, but nevertheless instructive.

EXAMPLE 5-21

Random Sampling Two Bolt 4 Torques. For the sake of illustration, suppose that the 34 bolt 4 torques obtained by Brenny, Christensen, and Schneider and given in Table 3-4 are written on slips of paper and placed in a hat. Suppose further that the slips are mixed, one is selected, the corresponding torque is noted, and the slip is replaced. Then the slips are again mixed, another is selected, and the second torque is noted. Then define the two random variables

$$U = \text{the value of the first torque selected}$$

$$V = \text{the value of the second torque selected}$$

In this scenario, intuition would dictate that (in contrast to the situation of X and Y in Example 5-20) the variables U and V don't furnish any information about each other. Regardless of what value U takes, the relative frequency distribution of bolt 4 torques in the hat is appropriate as the (conditional) probability distribution for V, and vice versa. That is, not only do U and V share the common marginal distribution given in Table 5-18, but it is also the case that for all u and v, both

$$f_{U|V}(u \mid v) = f_U(u) \tag{5-43}$$

and

$$f_{V|U}(v \mid u) = f_V(v) \tag{5-44}$$

TABLE 5-18

The Common Marginal Probability Function for U and V

u or v	$f_U(u)$ or $f_V(v)$
13	$\frac{1}{34}$
14	$\frac{3}{34}$
15	$\frac{5}{34}$
16	$\frac{7}{34}$
17	$\frac{6}{34}$
18	$\frac{5}{34}$
19	$\frac{2}{34}$
20	$\frac{5}{34}$

Not only do equations (5-43) and (5-44) say that the marginal probabilities in Table 5-18 also serve as conditional probabilities, but rewriting slightly shows that they specify how joint probabilities for U and V must be structured. That is, rewriting the left-hand side of equation (5-43) using expression (5-41), one has

$$\frac{f(u, v)}{f_V(v)} = f_U(u)$$

That is,

$$f(u, v) = f_U(u) f_V(v) \tag{5-45}$$

(The same logic applied to equation (5-44) also leads to equation (5-45).) Expression (5-45) says that joint probability values for U and V are obtained by multiplying corresponding marginal probabilities. So this is an instance in which a joint probability function can be reconstructed from its marginal probability functions. Table 5-19 gives the joint probability function for U and V.

TABLE 5-19

Joint Probabilities for U and V

$v \backslash u$	13	14	15	16	17	18	19	20	$f_V(v)$
20	$\frac{5}{(34)^2}$	$\frac{15}{(34)^2}$	$\frac{25}{(34)^2}$	$\frac{35}{(34)^2}$	$\frac{30}{(34)^2}$	$\frac{25}{(34)^2}$	$\frac{10}{(34)^2}$	$\frac{25}{(34)^2}$	$\frac{5}{34}$
19	$\frac{2}{(34)^2}$	$\frac{6}{(34)^2}$	$\frac{10}{(34)^2}$	$\frac{14}{(34)^2}$	$\frac{12}{(34)^2}$	$\frac{10}{(34)^2}$	$\frac{4}{(34)^2}$	$\frac{10}{(34)^2}$	$\frac{2}{34}$
18	$\frac{5}{(34)^2}$	$\frac{15}{(34)^2}$	$\frac{25}{(34)^2}$	$\frac{35}{(34)^2}$	$\frac{30}{(34)^2}$	$\frac{25}{(34)^2}$	$\frac{10}{(34)^2}$	$\frac{25}{(34)^2}$	$\frac{5}{34}$
17	$\frac{6}{(34)^2}$	$\frac{18}{(34)^2}$	$\frac{30}{(34)^2}$	$\frac{42}{(34)^2}$	$\frac{36}{(34)^2}$	$\frac{30}{(34)^2}$	$\frac{12}{(34)^2}$	$\frac{30}{(34)^2}$	$\frac{6}{34}$
16	$\frac{7}{(34)^2}$	$\frac{21}{(34)^2}$	$\frac{35}{(34)^2}$	$\frac{49}{(34)^2}$	$\frac{42}{(34)^2}$	$\frac{35}{(34)^2}$	$\frac{14}{(34)^2}$	$\frac{35}{(34)^2}$	$\frac{7}{34}$
15	$\frac{5}{(34)^2}$	$\frac{15}{(34)^2}$	$\frac{25}{(34)^2}$	$\frac{35}{(34)^2}$	$\frac{30}{(34)^2}$	$\frac{25}{(34)^2}$	$\frac{10}{(34)^2}$	$\frac{25}{(34)^2}$	$\frac{5}{34}$
14	$\frac{3}{(34)^2}$	$\frac{9}{(34)^2}$	$\frac{15}{(34)^2}$	$\frac{21}{(34)^2}$	$\frac{18}{(34)^2}$	$\frac{15}{(34)^2}$	$\frac{6}{(34)^2}$	$\frac{15}{(34)^2}$	$\frac{3}{34}$
13	$\frac{1}{(34)^2}$	$\frac{3}{(34)^2}$	$\frac{5}{(34)^2}$	$\frac{7}{(34)^2}$	$\frac{6}{(34)^2}$	$\frac{5}{(34)^2}$	$\frac{2}{(34)^2}$	$\frac{5}{(34)^2}$	$\frac{1}{34}$
$f_U(u)$	$\frac{1}{34}$	$\frac{3}{34}$	$\frac{5}{34}$	$\frac{7}{34}$	$\frac{6}{34}$	$\frac{5}{34}$	$\frac{2}{34}$	$\frac{5}{34}$	

Example 5-21 suggests that the intuitive notion that several random variables are unrelated (or furnish no real information about each other) might be formalized in terms of all conditional distributions being equal to their corresponding marginal distributions. Equivalently, it might be phrased in terms of joint probabilities being the products of corresponding marginal probabilities. The formal mathematical terminology is that of *independence* of the random variables.

DEFINITION 5-24

Discrete random variables $X, Y, \ldots, Z$ are called **independent** if their joint probability function $f(x, y, \ldots, z)$ is the product of their respective marginal probability functions. That is, independence means that

$$f(x, y, \ldots, z) = f_X(x) f_Y(y) \cdots f_Z(z) \qquad \text{for all } x, y, \ldots, z \qquad \textbf{(5-46)}$$

If formula (5-46) does not hold, the variables $X, Y, \ldots, Z$ are called **dependent**.

It may not be obvious to the reader, but formula (5-46) does imply that conditional distributions are all equal to their corresponding marginals, so that the definition does fit its unrelatedness motivation.

Looking back through the joint probability tables in this section, it is clear that U and V in Example 5-21 are independent, whereas X and Y in Example 5-20 are dependent. Further, the two joint distributions depicted in Figure 5-33 give an example of a highly dependent joint distribution (the first) and one of independence (the second) that have the same marginals.

The notion of independence is a fundamental one in many engineering applications of probability and statistical inference. When it is plausible to think of random variables as unrelated and therefore relationship (5-46) as appropriate, independence provides tremendous mathematical tractability. Where engineering data is being collected in an analytical context, and care is taken to make sure that all obvious physical causes of *carryover effects* that might influence two or more successive observations are minimal, an assumption of independence between observations is often employed. And in enumerative contexts, relatively small (compared to the population size) simple random samples yield observations that can typically be considered as at least approximately independent.

EXAMPLE 5-21
(continued)

Again for the sake of example, consider putting bolt torques on slips of paper in a hat. The method of torque selection described earlier for producing U and V is not simple random sampling. Simple random sampling as defined in Section 2-2 is *without-replacement* sampling, not the *with-replacement* sampling method used before. Indeed, if the first slip is not replaced before the second is selected, the probabilities in Table 5-19 are not appropriate for describing U and V. For example, if no replacement is done, since only one slip is labeled 13 ft lb, one clearly wants

$$f(13, 13) = P[U = 13 \text{ and } V = 13] = 0$$

not

$$f(13, 13) = \frac{1}{(34)^2}$$

as indicated in Table 5-19. Put differently, if no replacement is done, one clearly wants to use

$$f_{V|U}(13 \mid 13) = 0$$

rather than the value

$$f_{V|U}(13 \mid 13) = f_V(13) = \frac{1}{34}$$

which would be appropriate if sampling is done with replacement. So simple random sampling doesn't lead to exactly independent observations.

But suppose that instead of containing 34 slips labeled with torques, the hat contained 100×34 slips labeled with torques with relative frequencies as in Table 5-18. Then even if sampling were done without replacement, the probabilities developed earlier for U and V would remain at least *approximately* valid. For example, with 3,400 slips and using without-replacement sampling,

$$f_{V|U}(13 \mid 13) = \frac{99}{3,399}$$

is appropriate. Then, using the fact that

$$f_{V|U}(v \mid u) = \frac{f(u, v)}{f_U(u)}$$

so that

$$f(u, v) = f_{V|U}(v \mid u) f_U(u)$$

it is clear that without replacement, the assignment

$$f(13, 13) = \frac{99}{3,399} \cdot \frac{1}{34}$$

is appropriate. But the point here is that

$$\frac{99}{3,399} \approx \frac{1}{34}$$

and

$$\frac{99}{3,399} \cdot \frac{1}{34} \approx \frac{1}{34} \cdot \frac{1}{34}$$

For this hypothetical situation where the population size $N = 3,400$ is much larger than the sample size $n = 2$, independence is a suitable approximate description of observations obtained using simple random sampling.

The discussions in this section have used discrete distributions specified in terms of tables of probability values. Of course, where one has an analytical expression for a joint probability function, convenient analytical expressions for marginal and conditional probability functions can sometimes result as well. But for purposes of exposition, the present approach has been employed to help readers avoid the strong temptation to mistake the forest for the trees, get lost in analytic details, and miss the main ideas involved.

Describing Jointly Continuous Random Variables

All that has been said about joint description of discrete random variables has its analog for continuous variables. Conceptually and computationally, however, the jointly continuous case appears somewhat more challenging than the jointly discrete situation.

Probability density functions replace probability functions, and calculus substitutes for simple arithmetic. But take heart; in the following brief introduction to multivariate continuous distributions, do not get bogged down in details, but rather read for the main ideas.

The counterpart of a joint probability function, the device that is used to specify probabilities for continuous random variables $X, Y, \ldots, Z$, is a *joint probability density*.

DEFINITION 5-25

A **joint probability density** for continuous random variables $X, Y, \ldots, Z$ is a nonnegative function $f(x, y, \ldots, z)$ with

$$\int \cdots \int \int f(x, y, \ldots, z)\, dx\, dy \cdots dz = 1$$

and such that for any region $\mathcal{R}$, one is willing to assign

$$P[(X, Y, \ldots, Z) \in \mathcal{R}] = \int \cdots \int \int_{\mathcal{R}} f(x, y, \ldots, z)\, dx\, dy \cdots dz \qquad \textbf{(5-47)}$$

Instead of summing values of a probability function to find probabilities for a discrete distribution, equation (5-47) tells one (as in Section 5-2) to integrate values of a probability density. Of course, the new complication here is that the integral is multidimensional. But it is still possible to draw on intuition developed in mechanics, remembering that this is exactly the sort of thing that is done to specify mass distributions in several dimensions. (Here, mass is probability, and it is additionally required that the total mass be 1.)

EXAMPLE 5-22

(Example 5-18 revisited) **Residence Hall Depot Counter Service Time and a Continuous Joint Distribution.** As an example that is hopefully both somewhat realistic and reasonably tractable, consider again the depot service time situation of Jenkins, Milbrath, and Worth. As Section 5-3 showed, the students' data suggest an exponential model with $\alpha = 16.5$ for the random variable

S = the excess (over a threshold value of 7.5 sec) time required to complete the next postage stamp sale at the residence hall service counter

Imagine that the true value of S will be measured with a (very imprecise) analog stopwatch, producing the random variable

R = the measured excess service time

For use in this example, consider the function of two variables

$$f(s, r) = \begin{cases} \dfrac{1}{16.5} e^{-s/16.5} \dfrac{1}{\sqrt{2\pi(.25)}} e^{-(r-s)^2/2(.25)} & \text{for } s > 0 \\ 0 & \text{otherwise} \end{cases} \qquad \textbf{(5-48)}$$

as a potential joint probability density for S and R. Figure 5-34 provides a representation of $f(s, r)$, sketched as a surface in three-dimensional space.

The origin of formula (5-48) may appear to be a mystery at this point. There are reasons for suggesting it that may become clear as the discussion proceeds, but regardless of whether the motivation is apparent or not, one can at least entertain the possibility of using it as a joint probability density. $f(s, r)$ defined in equation (5-48) is nonnegative,

FIGURE 5-34

A Joint Probability Density for
S and *R*

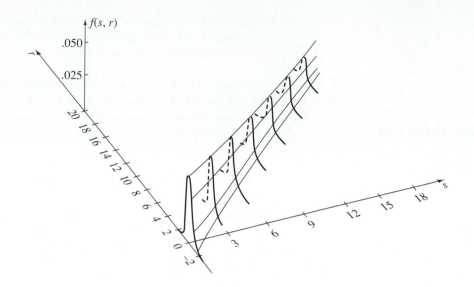

and its integral (the volume underneath the surface sketched in Figure 5-34 over the region in the (s, r)-plane where s is positive) is

$$\int \int f(s, r)\, ds\, dr = \int_0^\infty \int_{-\infty}^\infty \frac{1}{16.5\sqrt{2\pi(.25)}} e^{-(s/16.5)-((r-s)^2/2(.25))}\, dr\, ds$$

$$= \int_0^\infty \frac{1}{16.5} e^{-s/16.5} \left\{ \int_{-\infty}^\infty \frac{1}{\sqrt{2\pi(.25)}} e^{-(r-s)^2/2(.25)}\, dr \right\} ds$$

$$= \int_0^\infty \frac{1}{16.5} e^{-s/16.5}\, ds$$

$$= 1$$

(The integral in braces is 1 because it is the integral of a normal density with mean s and standard deviation .5.) Thus, equation (5-48) specifies a mathematically legitimate joint probability density.

To illustrate the use of a joint probability density in finding probabilities, first consider evaluating $P[R > S]$. Figure 5-35 shows the region in the (s, r)-plane where $f(s, r) > 0$ and $r > s$. It is over this region that one must integrate in order to evaluate $P[R > S]$.

FIGURE 5-35

Region Where $f(s, r) > 0$
and $r > s$

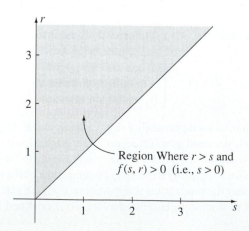

Region Where $r > s$ and
$f(s, r) > 0$ (i.e., $s > 0$)

Then

$$P[R > S] = \int \int_{\mathcal{R}} f(s,r)\,ds\,dr$$

$$= \int_0^\infty \int_s^\infty f(s,r)\,dr\,ds$$

$$= \int_0^\infty \frac{1}{16.5} e^{-s/16.5} \left\{ \int_s^\infty \frac{1}{\sqrt{2\pi(.25)}}\, e^{-(r-s)^2/2(.25)}\,dr \right\} ds$$

$$= \int_0^\infty \frac{1}{16.5} e^{-s/16.5} \left\{ \frac{1}{2} \right\} ds$$

$$= \frac{1}{2}$$

(once again using the fact that the integral in braces is a normal (mean s and standard deviation .5) probability).

As a second example, consider the problem of evaluating $P[S > 20]$ according to the model specified by equation (5-48). Figure 5-36 shows the region over which one must integrate in order to evaluate $P[S > 20]$.

FIGURE 5-36

Region Where $f(s,r) > 0$ and $s > 20$

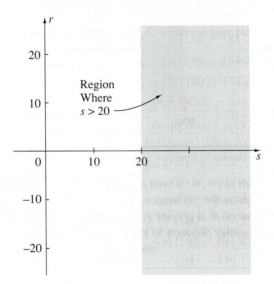

Then

$$P[S > 20] = \int \int_{\mathcal{R}} f(s,r)\,ds\,dr$$

$$= \int_{20}^\infty \int_{-\infty}^\infty f(s,r)\,dr\,ds$$

$$= \int_{20}^\infty \frac{1}{16.5} e^{-s/16.5} \left\{ \int_{-\infty}^\infty \frac{1}{\sqrt{2\pi(.25)}}\, e^{-(r-s)^2/2(.25)}\,dr \right\} ds$$

$$= \int_{20}^\infty \frac{1}{16.5} e^{-s/16.5}\,ds$$

$$= e^{-20/16.5}$$

$$\approx .30$$

The last part of the preceding example illustrates the fact that a joint probability density can be used to find probabilities for individual random variables via *integrating out the others*. In fact, for $X, Y, \ldots, Z$ with joint density $f(x, y, \ldots, z)$,

$$P[X \leq x] = \int_{-\infty}^{x} \int_{-\infty}^{\infty} \cdots \int_{-\infty}^{\infty} f(t, y, \ldots, z) \, dy \cdots dz \, dt$$

This is a statement giving the cumulative probability function for X, and differentiation with respect to x shows that a marginal probability density for X is obtained from the joint density by integrating out the other variables. Putting this in the form of a definition gives the following.

DEFINITION 5-26

The individual probability densities for continuous random variables $X, Y, \ldots, Z$ with joint probability density $f(x, y, \ldots, z)$ are called **marginal probability densities**. They are obtained by integrating $f(x, y, \ldots, z)$ over all possible values of the other variables. In symbols, the marginal probability density function for X, for example, is given by

$$f_X(x) = \int_{-\infty}^{\infty} \cdots \int_{-\infty}^{\infty} f(x, y, \ldots, z) \, dy \cdots dz$$

Compare Definitions 5-22 and 5-26. One does the same kind of thing for jointly continuous variables to find marginal distributions as for jointly discrete variables, except that integration is substituted for summation.

EXAMPLE 5-22
(continued)

Starting with the joint density specified by equation (5-48), it turns out to be possible to arrive at reasonably explicit expressions for the marginal densities for S and R. First considering the density of S, Definition 5-26 declares that for $s > 0$,

$$f_S(s) = \int_{-\infty}^{\infty} \frac{1}{16.5} e^{-s/16.5} \left\{ \frac{1}{\sqrt{2\pi(.25)}} \, e^{-(r-s)^2/2(.25)} \right\} dr$$

$$= \frac{1}{16.5} e^{-s/16.5}$$

Further, since $f(s, r)$ is 0 for negative s, if $s < 0$,

$$f_S(s) = \int_{-\infty}^{\infty} 0 \, dr = 0$$

That is, the form of $f(s, r)$ was chosen so that (as suggested by Example 5-18) S has an exponential distribution with mean $\alpha = 16.5$.

Now, although the determination of $f_R(r)$ is conceptually no different than the determination of $f_S(s)$, the details are a bit more complicated and the result far less clean-cut. Some work (involving completion of a square in the argument of the exponential function and recognition of an integral as a normal probability) will show the interested (and determined) reader that for any r,

$$f_R(r) = \int_{0}^{\infty} \frac{1}{16.5\sqrt{2\pi(.25)}} \, e^{-(s/16.5)-((r-s)^2/2(.25))} \, ds$$

$$= \frac{1}{16.5} \left(1 - \Phi\left(\frac{1}{33} - 2r \right) \right) \exp\left(\frac{1}{2,178} - \frac{r}{16.5} \right)$$

(5-49)

FIGURE 5-37

Marginal Probability
Density for R

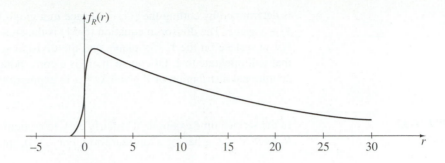

where, as usual, Φ is the standard normal cumulative probability function. A graph of $f_R(r)$ is given in Figure 5-37.

The marginal density for R derived from equation (5-48) does not belong to any standard family of distributions like the normal, Weibull, or beta families. Indeed, there is generally no guarantee that the process of finding marginal densities from a joint density will produce expressions for the densities even as explicit as that in display (5-49). Sometimes numerical integration is required, and the best one can do is to develop a table giving values for a marginal density derived from the integration.

**Conditional Distributions
and Independence
for Continuous
Random Variables**

In order to motivate the definition for conditional distributions derived from a joint probability density, consider again the meaning of a conditional distribution for the discrete case given in Definition 5-23. For jointly discrete variables $X, Y, \ldots, Z$, the conditional distribution for X given $Y = y, \ldots, Z = z$ is specified by holding $y, \ldots, z$ fixed and treating $f(x, y, \ldots, z)$ as a probability function for X after appropriately renormalizing it — i.e., seeing that its values total to 1. The analogous operation for jointly continuous variables is described next.

DEFINITION 5-27

For continuous random variables $X, Y, \ldots, Z$ with joint probability density $f(x, y, \ldots, z)$, the **conditional probability density function of** (for example) **X given $Y = y, \ldots, Z = z$** is the function of x defined by the ratio

$$\frac{f(x, y, \ldots, z)}{\int_{-\infty}^{\infty} f(x, y, \ldots, z)\, dx} \qquad (5\text{-}50)$$

Standard notation for the ratio in expression (5-50) is $f_{X|Y,\ldots,Z}(x \mid y, \ldots, z)$.

In the case that only two variables X and Y are under consideration, expression (5-50) leads to

$$f_{X|Y}(x \mid y) = \frac{f(x, y)}{f_Y(y)} \qquad (5\text{-}51)$$

and

$$f_{Y|X}(y \mid x) = \frac{f(x, y)}{f_X(x)} \qquad (5\text{-}52)$$

Expressions (5-51) and (5-52) are formally identical to the expressions (5-41) and (5-42) relevant for discrete variables. The geometry indicated by equation (5-51) for the two-variable continuous situation is that the shape of $f_{X|Y}(x \mid y)$ as a function of x

is determined by cutting the $f(x, y)$ surface in a graph like that in Figure 5-34 with the $Y = y$ plane. The divisor in equation (5-51) is the area above the (x, y)-plane below the $f(x, y)$ surface on the $Y = y$ plane, and the division serves to produce a function of x that will integrate to 1. Of course, there is a corresponding geometric story told for the conditional distribution of Y given $X = x$ in expression (5-52).

EXAMPLE 5-22
(continued)

In the service time example, it is fairly easy to recognize the conditional distribution of R given $S = s$ as having a familiar form. For $s > 0$, applying expression (5-52), one has

$$f_{R|S}(r \mid s) = \frac{f(s, r)}{f_S(s)} = f(s, r) \div \left(\frac{1}{16.5} e^{-s/16.5} \right)$$

which, using equation (5-48), gives

$$f_{R|S}(r \mid s) = \frac{1}{\sqrt{2\pi(.25)}} e^{-(r-s)^2/2(.25)} \tag{5-53}$$

That is, given that $S = s$, the conditional distribution of R is normal with mean s and standard deviation .5.

This realization is perfectly consistent with the bell-shaped cross sections of $f(s, r)$ shown in Figure 5-34. Basically, the form of $f_{R|S}(r \mid s)$ given in equation (5-53) says that the measured excess service time is the true excess service time plus a normally distributed measurement error that has mean 0 and standard deviation .5. (It is, in fact, this kind of thinking, together with the intention that S itself be exponentially distributed with mean $\alpha = 16.5$, that led to the invention of the form (5-48) in the first place.)

It is evident from expression (5-53) (or for that matter from the way the positions of the bell-shaped contours on Figure 5-34 vary with s) that the variables S and R ought to be called dependent. After all, knowing that $S = s$ gives one the value of R except for a normal error of measurement with mean 0 and standard deviation .5. On the other hand, had it been the case that all conditional distributions of R given $S = s$ were the same (and equal to the marginal distribution of R), one would wish to call S and R independent. The notion of unchanging conditional distributions, all equal to their corresponding marginal, is equivalently and more conveniently expressed in terms of the joint probability density factoring into a product of marginals.

DEFINITION 5-28

Continuous random variables $X, Y, \ldots, Z$ are called **independent** if their joint probability density function $f(x, y, \ldots, z)$ is the product of their respective marginal probability densities. That is, independence means that

$$f(x, y, \ldots, z) = f_X(x) f_Y(y) \cdots f_Z(z) \qquad \text{for all } x, y, \ldots, z \tag{5-54}$$

If expression (5-54) does not hold, the variables $X, Y, \ldots, Z$ are called **dependent**.

Expression (5-54) is formally identical to expression (5-46), which appeared in Definition 5-24 for discrete variables. As was indicated following Definition 5-24, the type of factorization given in expression (5-54) provides great mathematical tractability. It can be appropriate in analytical contexts that are free of carryover effects and in enumerative contexts where relatively small simple random samples are being described.

One final piece of terminology remains to be introduced in this section. It applies equally to jointly discrete and jointly continuous random variables.

DEFINITION 5-29

If random variables $X_1, X_2, \ldots, X_n$ all have the same marginal distribution and are independent, they are termed **iid** or **independent and identically distributed**.

The archetypal examples of iid random variables are successive measurements taken from a stable process and the results of random sampling with replacement from a single population. The question of whether an iid model is appropriate in a given application thus depends on whether or not the data-generating mechanism being studied can be thought of as conceptually equivalent to these.

EXAMPLE 5-23

(Example 5-18 revisited) **Residence Hall Depot Counter Service Times and iid Variables.** Returning once more to the service time example of Jenkins, Milbrath, and Worth, consider the next two excess service times encountered,

S_1 = the next excess (over a threshold of 7.5 sec) time required to complete a postage stamp sale at the residence hall service counter

S_2 = the following excess service time

To the extent that the service process is physically stable, (i.e., excess service times can be thought of in terms of sampling with replacement from a single population), an iid model seems appropriate for S_1 and S_2. Treating excess service times as marginally exponential with mean $\alpha = 16.5$ thus leads to the joint density for S_1 and S_2:

$$f(s_1, s_2) = \begin{cases} \dfrac{1}{(16.5)^2} e^{-(s_1 + s_2)/16.5} & \text{if } s_1 > 0 \text{ and } s_2 > 0 \\ 0 & \text{otherwise} \end{cases}$$

5-5 *Functions of Several Random Variables*

The last section discussed the mathematics used to simultaneously model several random variables, paying particular attention to the notion of independence. An important engineering use of material like that in Section 5-4 is in the analysis and prediction of patterns of variability in quantities that are functions of several other variables, themselves each subject to random variation. For example, a clearance between two mating parts is the difference in one random dimension of the first part and a corresponding random dimension on the second. The sample mean of 34 torques required to loosen a bolt securing a face plate is a function of the 34 random torque values observed. And the measured efficiency of a solar collector is a function of more fundamental variables such as air flow rates, air specific heat, irradiance incident on the plane of the collector, etc., all of which are subject to measurement error.

This section considers methods for studying how the pattern of variation seen in a derived random variable is affected by the pattern of variability associated with variables used to produce it. It begins with a few comments on what is possible using exact methods of mathematical analysis. Then the simple and general tool of simulation is introduced. Next, formulas for means and variances of linear combinations of random variables and the related propagation of error formulas are presented. Last is the pervasive central limit effect, which typically causes variables that can be thought of as averages to have approximately normal distributions.

The Distribution of a Function of Random Variables

The mathematical formalization of the problem to be considered in this section is as follows. Given a joint distribution for the random variables $X, Y, \ldots, Z$ and a function $g(x, y, \ldots, z)$, what can be said about the distribution of the following random variable?

$$U = g(X, Y, \ldots, Z) \tag{5-55}$$

In some special, relatively simple cases, it is possible to figure out exactly what distribution U in equation (5-55) inherits from $X, Y, \ldots, Z$. As an almost trivial example of this, consider the following.

EXAMPLE 5-24

The Distribution of the Clearance between Two Mating Parts with Randomly Determined Dimensions. Suppose that a steel plate with nominal thickness .15 inch is to rest in a groove of nominal width .155 inch, machined on the surface of a steel block. Suppose further that a lot of plates has been made and thicknesses measured, producing the relative frequency distribution in Table 5-20, while a relative frequency distribution for the slot widths measured on a lot of machined blocks is given in Table 5-21.

TABLE 5-20

Relative Frequency Distribution of Plate Thicknesses

Plate Thickness (in.)	Relative Frequency
.148	.4
.149	.3
.150	.3

TABLE 5-21

Relative Frequency Distribution of Slot Widths

Slot Width (in.)	Relative Frequency
.153	.2
.154	.2
.155	.4
.156	.2

If one randomly selects a plate and separately randomly selects a block, a natural joint distribution for the random variables

$$X = \text{the plate thickness}$$
$$Y = \text{the slot width}$$

is one of independence, where the marginal distribution of X is given in Table 5-20 and the marginal distribution of Y is given in Table 5-21. That is, Table 5-22 gives a plausible joint probability function for X and Y.

A variable derived from X and Y that is of substantial potential interest is the clearance involved in the plate/block assembly,

$$U = Y - X$$

Notice that taking the extremes represented in Tables 5-20 and 5-21, U is guaranteed

TABLE 5-22

Marginal and Joint Probabilities for X and Y

$y \backslash x$	.148	.149	.150	$f_Y(y)$
.156	.08	.06	.06	.2
.155	.16	.12	.12	.4
.154	.08	.06	.06	.2
.153	.08	.06	.06	.2
$f_X(x)$	.4	.3	.3	

to be at least $.153 - .150 = .003$ in. but no more than $.156 - .148 = .008$ in. In fact, much more than this can be said. Looking at Table 5-22, one can see that the diagonals of entries lower left to upper right all correspond to the same value of $Y - X$. Adding probabilities on those diagonals produces the distribution of U given in Table 5-23.

TABLE 5-23

The Probability Function for the Clearance $U = Y - X$

u	$f(u)$
.003	.06
.004	$.12 = .06 + .06$
.005	$.26 = .08 + .06 + .12$
.006	$.26 = .08 + .12 + .06$
.007	$.22 = .16 + .06$
.008	.08

Example 5-24 involves a very simple discrete joint distribution and a very simple function g — namely, $g(x, y) = y - x$. Even if $X, Y, \ldots, Z$ are jointly continuous, as long as the joint distribution and $g(x, y, \ldots, z)$ are simple enough, clever use of multivariate calculus can sometimes yield exact analytic expressions for the distribution of U given by expression (5-55). (For example, a brave and interested reader might attempt to verify that the sum of two independent standard normal random variables X and Y is itself normal with mean 0 and variance 2. This can be done by writing out a double integral for $P[U \leq u]$, noticing that the double integral reduces to

$$\int_{-\infty}^{\infty} \Phi(u - y) \frac{1}{\sqrt{2\pi}} e^{-y^2/2} \, dy$$

then differentiating this expression under the integral sign and simplifying to arrive at the normal probability density with mean 0 and variance 2.)

But in general, exact complete solution of the problem of finding the distribution of $U = g(X, Y, \ldots, Z)$ is beyond the realm of what is practically possible. Happily, for most engineering applications of probability where complete solutions are not available, approximate and/or partial solutions suffice to answer the questions of practical interest. The balance of this section studies methods of producing these approximate and/or partial descriptions of the distribution of U, beginning with a brief look at simulation-based methods.

Simulations to Approximate the Distribution of $U = g(X, Y, \ldots, Z)$

There are by now many computer programs and packages that can be used to produce pseudo-random values, intended to behave as if they were realizations of independent random variables following user-chosen marginal distributions. Notice then that if one's model for $X, Y, \ldots, Z$ is one of independence and such a program or package is available, it is a simple matter to generate a simulated value for each of $X, Y, \ldots, Z$ and plug those into g to produce a simulated value for U. If this process is repeated a number of times, a relative frequency distribution for these simulated values of U is developed, and one might reasonably use this relative frequency distribution to approximate an exact distribution for U that might otherwise be intractable.

EXAMPLE 5-25

Uncertainty in the Calculated Efficiency of an Air Solar Collector. The article "Thermal Performance Representation and Testing of Air Solar Collectors" by Bernier and Plett (*Journal of Solar Energy Engineering*, May 1988) considers the testing of air solar collectors. Its analysis of thermal performance based on enthalpy balance leads to the conclusion that under inward leakage conditions, the thermal efficiency of a collector can be expressed as

$$\text{Efficiency} = \frac{M_oC(T_o - T_i) + (M_o - M_i)C(T_i - T_a)}{GA}$$

$$= \frac{C}{GA}\left(M_oT_o - M_iT_i - (M_o - M_i)T_a\right) \qquad (5\text{-}56)$$

where

C = air specific heat (J/kg°C)

G = global irradiance incident on the plane of the collector (W/m^2)

A = collector gross area (m^2)

M_i = inlet mass flowrate (kg/s)

M_o = outlet mass flowrate (kg/s)

T_a = ambient temperature (°C)

T_i = collector inlet temperature (°C)

T_o = collector outlet temperature (°C)

The authors further posit some uncertainty values associated with each of the terms appearing on the right side of equation (5-56) for an example set of measured values of the variables. These are given in Table 5-24.

TABLE 5-24

Reported Uncertainties in the Measured Inputs to Collector Efficiency

Variable	Measured Value	Uncertainty
C	1003.8	1.004 (i.e. $\pm$ .1%)
G	1121.4	33.6 (i.e. $\pm$ 3%)
A	1.58	.005
M_i	.0234	.00035 (i.e. $\pm$ 1.5%)
M_o	.0247	.00037 (i.e. $\pm$ 1.5%)
T_a	−13.22	.5
T_i	−6.08	.5
T_o	24.72	.5*

*This value is not given explicitly in the article. The presumption here is that all temperatures are measured with the same precision.

Plugging the measured values from Table 5-24 into formula (5-56) produces a measured efficiency of about .44. But a question of real engineering importance here is "How good is the .44 value?" That is, how do the uncertainties associated with the measured values affect the reliability of the .44 figure? Should one think of the calculated solar collector efficiency as .44 plus or minus .001, or plus or minus .1, or what?

One way of approaching this question is to ask the related question, "What would be the standard deviation of Efficiency if all of C through T_o were independent random variables with means approximately equal to the measured values and standard deviations related to the uncertainties as, say, half of the uncertainty values?" (This "two sigma" interpretation of uncertainty appears to be at least close to the intention of Bernier and Plett in the original article.)

Printout 5-2 is from a Minitab session in which 100 normally distributed realizations of variables C through T_o were generated (using means equal to measured values and standard deviations equal to half of the corresponding uncertainties) and the resulting efficiencies calculated. Examination of the printout shows that the simulation produced

PRINTOUT 5-2

```
MTB > random 100 c1;
SUBC> normal 1003.8 .502.
MTB > random 100 c2;
SUBC> normal 1121.4 16.8.
MTB > random 100 c3;
SUBC> normal 1.58 .0025.
MTB > random 100 c4;
SUBC> normal .0234 .000175.
MTB > random 100 c5;
SUBC> normal .0247 .000185.
MTB > random 100 c6;
SUBC> normal -13.22 .25.
MTB > random 100 c7;
SUBC> normal -6.08 .25.
MTB > random 100 c8;
SUBC> normal 24.72 .25.
```

Generation of 100 Values Each for C, G, A, M_i, M_o, T_a, T_i, T_o

```
MTB > sub c4 c5 put into c9
MTB > mult c5 c8 put into c10
MTB > mult c4 c7 put into c11
MTB > mult c9 c6 put into c12
MTB > sub c11 c10 put into c10
MTB > sub c12 c10 put into c10
MTB > mult c1 c10 put into c10
MTB > div c10 c2 put into c10
MTB > div c10 c3 put into c10
MTB > name c10 'Effic'
MTB > hist c10
```

Calculation of 100 Efficiency Values

```
Histogram of Effic   N = 100

Midpoint    Count
   0.415       3   ***
   0.420       5   *****
   0.425      10   **********
   0.430      18   ******************
   0.435      15   ***************
   0.440      23   ***********************
   0.445      15   ***************
   0.450       7   *******
   0.455       3   ***
   0.460       1   *
```

Average Simulated Efficiency   **Standard Deviation of Simulated Efficiencies**

```
MTB > describe c10

              N      MEAN    MEDIAN    TRMEAN     STDEV    SEMEAN
Effic       100   0.43649   0.43727   0.43651   0.00949   0.00095

              MIN       MAX        Q1        Q3
Effic     0.41392   0.45915   0.42991   0.44301

MTB > stem c10

Stem-and-leaf of Effic     N  = 100
Leaf Unit = 0.0010

     2    41 34
     4    41 79
    12    42 01122444
    25    42 5566777888999
    44    43 0001111112233344444
   (20)   43 66667777788888888999
    36    44 00001122222333344
    18    44 5566777899
     8    45 00123
     3    45 579
```

a roughly bell-shaped distribution of calculated efficiencies, possessing a mean value of approximately .436 and standard deviation of about .009. Evidently, if one continues with the understanding that uncertainty means something like "2 standard deviations," an uncertainty of .02 is at least of the right order of magnitude to attach to the nominal efficiency figure of .44.

The beauty of Example 5-25 is the ease with which a simulation can be employed to approximate the distribution of a function of several random variables, thus providing a usable answer in a real engineering problem. But the method is so powerful and easy to use that some cautions need to be given about the application of this whole topic (functions of random variables) before going any further.

One must be careful not to expect more than is sensible from a derived probability distribution for

$$U = g(X, Y, \ldots, Z)$$

The output distribution can be no more realistic as a description of U than are the model assumptions used to produce it (i.e., the form of the joint distribution and the form of $g(x, y, \ldots, z)$). It is unfortunately all too common for people to try to apply the methods of this section using a g representing some approximate physical law and U some measurable physical quantity, only to be surprised that the variation in U observed in the real world is substantially larger than that predicted by methods of this section. In such cases, the fault lies not with the methods, but with the naivete of the user. Approximate physical laws are just that, often involving so-called constants that aren't constant, using functional forms that are too simple, and ignoring the influence of variables that aren't obvious or easily measured. Further, although independence of $X, Y, \ldots, Z$ is a very convenient mathematical property, its use is not always justified. When it is inappropriately used as a model assumption, it can produce an inappropriate distribution for U. For these reasons, one should think of the methods of this section (when applied to predict measured variation in a physical quantity subject to an approximate physical law) as useful, but likely to provide only a best-case picture of the variation one should expect to see. Of course, applications of the methods of this section to problems where the model assumptions are known to be more or less exactly correct produce more reliable results.

Means and Variances for Linear Combinations of Random Variables

For engineering purposes, it often suffices to know the mean and variance for U given in formula (5-55) (as opposed to knowing the whole distribution). When this is the case and g is linear, there are explicit formulas that can be used to find the mean and variance.

PROPOSITION 5-1

If $X, Y, \ldots, Z$ are n independent random variables and $a_0, a_1, a_2, \ldots, a_n$ are $n + 1$ constants, then the random variable $U = a_0 + a_1 X + a_2 Y + \cdots + a_n Z$ has mean

$$EU = a_0 + a_1 EX + a_2 EY + \cdots + a_n EZ \tag{5-57}$$

and variance

$$Var\, U = a_1^2\, Var\, X + a_2^2\, Var\, Y + \cdots + a_n^2\, Var\, Z \tag{5-58}$$

As it turns out, formula (5-57) holds regardless of whether or not the variables $X, Y, \ldots, Z$ are independent, and although formula (5-58) does depend upon independence, there is a generalization of it (involving theoretical correlations between the variables) that can be used even if the variables are dependent. However, the form of Proposition 5-1 given here is adequate for present purposes.

One type of application in which Proposition 5-1 is immediately useful is that of tolerancing problems, where it is applied with $a_0 = 0$ and the other a_i's equal to plus and minus 1's.

EXAMPLE 5-24
(continued)

Consider again the situation of the clearance involved in placing a steel plate in a machined slot on a steel block. With X, Y, and U being (respectively) the plate thickness, slot width, and clearance, means and variances for these variables can be calculated from Tables 5-20, 5-21, and 5-23, respectively. The reader is encouraged to do the calculations necessary to verify that

$$EX = .1489 \quad \text{and} \quad Var\, X = 6.9 \times 10^{-7}$$

$$EY = .1546 \quad \text{and} \quad Var\, Y = 1.04 \times 10^{-6}$$

Now, since

$$U = Y - X = (-1)X + 1Y$$

Proposition 5-1 can be applied to conclude that

$$EU = -1EX + 1EY = -.1489 + .1546 = .0057 \text{ in.}$$

$$Var\, U = (-1)^2 6.9 \times 10^{-7} + (1)^2 1.04 \times 10^{-6} = 1.73 \times 10^{-6}$$

so that

$$\sqrt{Var\, U} = .0013 \text{ in.}$$

It is worth the effort to verify that the values of the mean and the standard deviation of the clearance produced above using Proposition 5-1 agree with those obtained using the distribution of U given in Table 5-23 and the formulas for the mean and variance given in Section 5-1. But the advantage of using Proposition 5-1 is that if all that is needed are EU and $Var\, U$, one can avoid going through the intermediate step of somehow deriving the distribution of U. The calculations via Proposition 5-1 use only characteristics of the marginal distributions in producing the required mean and variance.

Another particularly important use of Proposition 5-1 concerns the case where $X, Y, \ldots, Z$ are n iid random variables and each a_i is $\frac{1}{n}$. That is, in cases where random variables $X_1, X_2, \ldots, X_n$ are conceptually equivalent to random selections (*with replacement*) from a single numerical population, Proposition 5-1 tells how the mean and variance of the random variable

$$\bar{X} = \frac{1}{n}X_1 + \frac{1}{n}X_2 + \cdots + \frac{1}{n}X_n$$

are related to the population parameters μ and σ^2. For independent variables $X_1, X_2, \ldots, X_n$ with common mean μ and variance σ^2, Proposition 5-1 shows that

$$E\bar{X} = \frac{1}{n}EX_1 + \frac{1}{n}EX_2 + \cdots + \frac{1}{n}EX_n$$

$$= \frac{1}{n}\mu + \frac{1}{n}\mu + \cdots + \frac{1}{n}\mu$$

$$= \mu \qquad\qquad\qquad\qquad\qquad\qquad\qquad\qquad\qquad\qquad (5\text{-}59)$$

and

$$Var\,\bar{X} = \left(\frac{1}{n}\right)^2 Var\,X_1 + \left(\frac{1}{n}\right)^2 Var\,X_2 + \cdots + \left(\frac{1}{n}\right)^2 Var\,X_n$$

$$= \left(\frac{1}{n}\right)^2 \sigma^2 + \left(\frac{1}{n}\right)^2 \sigma^2 + \cdots + \left(\frac{1}{n}\right)^2 \sigma^2$$

$$= \frac{\sigma^2}{n} \tag{5-60}$$

Notice that since σ^2/n is obviously decreasing in n, equations (5-59) and (5-60) give the reassuring picture of $\bar{X}$ having a distribution centered at the population mean μ, with spread that decreases as the sample size increases.

EXAMPLE 5-26

(Example 5-18 revisited) **The Expected Value and Standard Deviation for a Sample Mean Service Time.** To illustrate the application of formulas (5-59) and (5-60), consider again the situation of Jenkins, Milbrath, and Worth involving stamp sale time requirements at a residence hall depot counter. Suppose that the exponential model with $\alpha = 16.5$ that was derived in Example 5-18 for excess service times continues to be appropriate and that several more postage stamp sales are observed and excess service times noted. With

S_i = the excess (over a 7.5 sec threshold) time required to complete the ith additional stamp sale

consider what means and standard deviations are to be associated with the probability distributions of the sample average, $\bar{S}$, of first the next 4 and then the next 100 excess service times.

First note that $S_1, S_2, \ldots, S_{100}$ are, at least to the extent that the service process is physically stable, reasonably modeled as independent identically distributed exponential random variables with mean $\alpha = 16.5$. Now the exponential distribution with mean $\alpha = 16.5$ has variance equal to $\alpha^2 = (16.5)^2$. So, using formulas (5-59) and (5-60), for the first 4 additional service times,

$$E\bar{S} = \alpha = 16.5 \text{ sec}$$

$$\sqrt{Var\,\bar{S}} = \sqrt{\frac{\alpha^2}{4}} = 8.25 \text{ sec}$$

Then for the first 100 additional service times

$$E\bar{S} = \alpha = 16.5 \text{ sec}$$

$$\sqrt{Var\,\bar{S}} = \sqrt{\frac{\alpha^2}{100}} = 1.65 \text{ sec}$$

Notice that going from a sample size of 4 to a sample size of 100 decreases the standard deviation of $\bar{S}$ by a factor of 5 ($= \sqrt{\frac{100}{4}}$).

It should be evident that relationships (5-59) and (5-60), which perfectly describe the random behavior of $\bar{X}$ under random sampling with replacement, are also approximate descriptions of the behavior of $\bar{X}$ under simple random sampling in enumerative contexts. (Recall Example 5-21 and the discussion about the approximate independence of observations resulting from simple random sampling of large populations.)

The Propagation of Error Formulas

Proposition 5-1 gives exact values for the mean and variance of $U = g(X, Y, \ldots, Z)$ only when g is linear. It doesn't appear to have much to say about situations involving nonlinear functions like the one specified by the right-hand side of expression (5-56) in the example involving solar collector efficiency. But it is often possible to obtain useful approximations to the mean and variance of U by applying Proposition 5-1 to a first-order multivariate Taylor expansion of a not-too-nonlinear g. That is, if g is reasonably well-behaved, then for $x, y, \ldots, z$ (respectively) close to $EX, EY, \ldots, EZ$,

$$\left. \begin{aligned} g(x, y, \ldots, z) &\approx g(EX, EY, \ldots, EZ) + \frac{\partial g}{\partial x} \cdot (x - EX) + \frac{\partial g}{\partial y} \cdot (y - EY) \\ &+ \cdots + \frac{\partial g}{\partial z} \cdot (z - EZ) \end{aligned} \right\} \quad \text{(5-61)}$$

where the partial derivatives are evaluated at $(x, y, \ldots, z) = (EX, EY, \ldots, EZ)$. Now the right side of approximation (5-61) is linear in $x, y, \ldots, z$. Thus, if the variances of $X, Y, \ldots, Z$ are small enough so that with high probability, $X, Y, \ldots, Z$ are such that approximation (5-61) is effective, one might think of plugging $X, Y, \ldots, Z$ into expression (5-61) and applying Proposition 5-1. One then winds up with approximations for the mean and variance of $U = g(X, Y, \ldots, Z)$. The punch line of this reasoning is stated next, as another proposition.

PROPOSITION 5-2

If $X, Y, \ldots, Z$ are independent random variables and g is well-behaved, for small enough variances $Var\, X, Var\, Y, \ldots, Var\, Z$, the random variable $U = g(X, Y, \ldots, Z)$ has approximate mean

$$EU \approx g(EX, EY, \ldots, EZ) \quad \text{(5-62)}$$

and approximate variance

$$Var\, U \approx \left(\frac{\partial g}{\partial x} \right)^2 Var\, X + \left(\frac{\partial g}{\partial y} \right)^2 Var\, Y + \cdots + \left(\frac{\partial g}{\partial z} \right)^2 Var\, Z \quad \text{(5-63)}$$

Formulas (5-62) and (5-63) are often called the *propagation of error* or *transmission of variance* formulas. They describe (in at least an approximate sense) how variability or error is propagated or transmitted through an exact mathematical function.

A little comparison between Propositions 5-1 and 5-2 makes it obvious that in the case that g is exactly linear, expressions (5-62) and (5-63) reduce to expressions (5-57) and (5-58), respectively. (a_1 through a_n are, after all, the partial derivatives of g in the case where $g(x, y, \ldots, z) = a_0 + a_1 x + a_2 y + \cdots + a_n z$.) Proposition 5-2 is purposely vague about when the approximations (5-62) and (5-63) will be adequate for engineering purposes. Mathematically inclined readers will probably not have much trouble constructing examples where the approximations are quite poor. But it has been this author's experience that most often in engineering applications, expressions (5-62) and (5-63) are at least of the right order of magnitude and certainly better than not having any usable approximations.

EXAMPLE 5-27

A Simple Electrical Circuit and the Propagation of Error. Figure 5-38 shows a schematic of an assembly of three resistors. If R_1, R_2, and R_3 are the respective resistances of the three resistors making up the assembly, standard theory says that

$$R = \text{the assembly resistance}$$

FIGURE 5-38

Schematic of a Simple
Assembly of Three Resistors

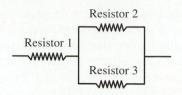

is related to R_1, R_2, and R_3 by

$$R = R_1 + \frac{R_2 R_3}{R_2 + R_3} \tag{5-64}$$

Suppose that a large lot of resistors is manufactured and has a mean resistance of 100 Ω with a standard deviation of resistance of 2 Ω. If three resistors are taken at random from this lot and assembled as in Figure 5-38, consider what formulas (5-62) and (5-63) suggest for an approximate mean and an approximate standard deviation for the resulting assembly resistance.

The g involved here is of course $g(r_1, r_2, r_3) = r_1 + \frac{r_2 r_3}{r_2 + r_3}$, so

$$\frac{\partial g}{\partial r_1} = 1$$

$$\frac{\partial g}{\partial r_2} = \frac{(r_2 + r_3) r_3 - r_2 r_3}{(r_2 + r_3)^2} = \frac{r_3^2}{(r_2 + r_3)^2}$$

$$\frac{\partial g}{\partial r_3} = \frac{(r_2 + r_3) r_2 - r_2 r_3}{(r_2 + r_3)^2} = \frac{r_2^2}{(r_2 + r_3)^2}$$

Also, R_1, R_2, and R_3 are approximately independent with means 100 and standard deviations 2. Then formulas (5-62) and (5-63) suggest that the probability distribution inherited by R has mean

$$ER \approx g(100, 100, 100) = 100 + \frac{(100)(100)}{100 + 100} = 150 \ \Omega$$

and variance

$$Var\,R \approx (1)^2 (2)^2 + \left(\frac{(100)^2}{(100 + 100)^2} \right)^2 (2)^2 + \left(\frac{(100)^2}{(100 + 100)^2} \right)^2 (2)^2 = 4.5$$

so that the standard deviation inherited by R is

$$\sqrt{Var\,R} \approx \sqrt{4.5} = 2.12 \ \Omega$$

As something of a check on how good the 150 Ω and 2.12 Ω values are, 200 sets of normally distributed R_1, R_2, and R_3 values with the specified population mean and standard deviation were simulated and resulting values of R calculated via formula (5-64). These simulated assembly resistances had $\bar{R} = 149.82 \ \Omega$ and a sample standard deviation of 2.25 Ω. A computer-generated histogram of these values is given in Figure 5-39.

Example 5-27 is one in which simulation and the method of Proposition 5-2 produce comparable values for the mean and variance of a derived variable. But it is also one to which the cautions following Example 5-25 apply. Suppose one were to actually

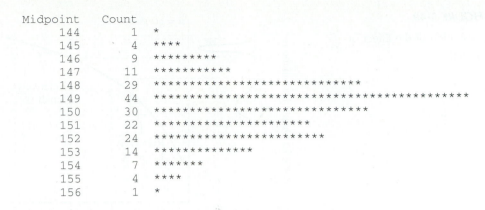

FIGURE 5-39

Histogram of 200 Simulated Values of R

```
Midpoint   Count
     144       1   *
     145       4   ****
     146       9   *********
     147      11   ***********
     148      29   *****************************
     149      44   ********************************************
     150      30   ******************************
     151      22   **********************
     152      24   ************************
     153      14   **************
     154       7   *******
     155       4   ****
     156       1   *
```

take a large batch of resistors possessing a mean resistance of 100 Ω and a standard deviation of resistances of 2 Ω, make up a number of assemblies of the type represented in Figure 5-38, and measure the assembly resistances. One should probably expect the standard deviation figures in Example 5-27 to underpredict the variation observed in the assembly resistances.

The propagation of error and simulation methods may do a good job of approximating the (exact) theoretical mean and standard deviation of assembly resistances. But the extent to which the probability model used for assembly resistances can be expected to represent the physical situation is another matter. Equation (5-64) is highly useful, but of necessity only an approximate description of real assemblies. For example, it ignores small but real temperature, inductance, and other second-order physical effects on measured resistance. In addition, although the probability model allows for variation in the resistances of individual components, it does not account for instrument variation or such vagaries of real-world assembly as the quality of contacts achieved when several parts are connected.

In Example 5-27, because 1) the simulation and propagation of error methods produce comparable results and 2) the simulation method is so easy to use, there arises the question of what the propagation of error formulas have to offer beyond what simulation provides. One important answer to this question concerns intuition that formula (5-63) provides, in terms of the role of derivatives of g in transmitting variation and in terms of partitioning the variance of U into parts derived from essentially different sources.

Considering first the effect that g's partial derivatives have on $Var\ U$, formula (5-63) implies that depending on the size of $\frac{\partial g}{\partial x}$, the variance of X is either inflated or deflated before becoming an ingredient of $Var\ U$. A little reflection on the matter makes it clear that even though formula (5-63) may not be an exact expression, it provides correct intuition. If a given change in x produces a big change in $g(x, y, \ldots, z)$, the impact $Var\ X$ has on $Var\ U$ will be greater than if the change in x produces a small change in $g(x, y, \ldots, z)$. Figure 5-40 is a rough illustration of this point, indicating how in the case that $U = g(X)$, two different approximately normal distributions for X with different means but a common variance produce radically different spreads in the distribution of U, due to differing rates of change of g (different derivatives).

Then, considering the partitioning of the variance of U into interpretable pieces, note that formula (5-63) suggests the possibility of thinking of (for example)

$$\left(\frac{\partial g}{\partial x}\right)^2 Var\ X$$

as the contribution of the variation in X to the variation inherent in U. Comparison

FIGURE 5-40

Illustration of the Effect of $\frac{\partial g}{\partial x}$ on *Var U*

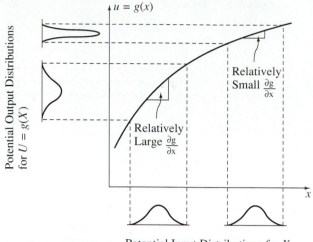

Potential Input Distributions for X

of such individual contributions can make it possible to analyze how various types of potential reductions in input variabilities (possibly available at differing dollar costs) can be expected to affect the output variability in U.

EXAMPLE 5-25
(continued)

As an example of the usefulness of the partitioning of *Var U*, return to the solar collector example. For means of C through T_o taken to be the measured values in Table 5-24, and standard deviations of C through T_o equal to half of the uncertainties listed in the same table, one might well apply the formula (5-63) to the calculated efficiency given in formula (5-56). It is instructive to verify that the squared partial derivatives of Efficiency with respect to each of the inputs, times the variances of those inputs, are as given in Table 5-25.

TABLE 5-25

Contributions to the Output Variation in Collector Efficiency

Variable	Contributions to *Var* Efficiency
C	4.73×10^{-8}
G	4.27×10^{-5}
A	4.76×10^{-7}
M_i	5.01×10^{-7}
M_o	1.58×10^{-5}
T_a	3.39×10^{-8}
T_i	1.10×10^{-5}
T_o	1.22×10^{-5}
Total	8.28×10^{-5}

Thus, the approximate standard deviation for the efficiency variable provided by formula (5-63) is

$$\sqrt{8.28 \times 10^{-5}} \approx .009$$

which agrees quite well with the value obtained earlier via simulation.

What's given in Table 5-25 that doesn't come out of a simulation is some understanding of the biggest contributors to the uncertainty associated with the calculated efficiency. The largest contribution listed in Table 5-25 corresponds to variable G, followed in order by those corresponding to variables M_o, T_o, and T_i. At least for the values of the means used in this example, it is the uncertainties (standard deviations) of those variables that principally produce the uncertainty (standard deviation) of Efficiency. Knowing this gives direction to any efforts one would want to make toward reducing the uncertainty in efficiency, by improvement of measurement methods. Subject to considerations of feasibility and cost, measurement of the variable G deserves first attention, followed by measurement of the variables M_o, T_o, and T_i.

Notice, however, that (a presumably very expensive) reduction of the uncertainty in G alone to essentially 0 would still leave a total in Table 5-25 of about 4.01×10^{-5} and thus an approximate standard deviation for Efficiency of about $\sqrt{4.01 \times 10^{-5}} \approx .006$. Calculations of this kind emphasize the need for reductions in the uncertainties of M_o, T_o, and T_i as well, if dramatic (order of magnitude) improvements in overall uncertainty are to be realized.

The Central Limit Effect

In the engineering application of statistical methods, one of the most frequently used statistics is the sample mean. Earlier in this section formulas (5-59) and (5-60) were seen to relate the mean and variance of the probability distribution of the sample mean to those of a single observation when an iid model is appropriate. One of the most useful facts of applied probability is that if the sample size is reasonably large, it is also possible to predict the approximate *shape* of the probability distribution of $\bar{X}$, independent of the shape of the underlying distribution of individual observations. That is, one has the following fact.

PROPOSITION 5-3

(The Central Limit Theorem) If $X_1, X_2, \ldots, X_n$ are iid random variables (with mean μ and variance σ^2), then for large n, the variable $\bar{X}$ is approximately normally distributed. That is, approximate probabilities for $\bar{X}$ can be calculated using the normal distribution with mean μ and variance $\frac{\sigma^2}{n}$.

Anything approaching a rigorous proof of Proposition 5-3 is both outside the purpose and beyond the mathematical level of this text. But intuition about the effect is fairly easy to develop through the use of an example.

EXAMPLE 5-28

(Example 5-2 revisited) **The Central Limit Effect and the Sample Mean of Tool Serial Numbers.** Consider again the example from Section 5-1, involving the last digit of essentially randomly selected serial numbers of pneumatic tools. Suppose now that

W_1 = the last digit of the serial number observed next Monday at 9 A.M.

W_2 = the last digit of the serial number observed the following Monday at 9 A.M.

A plausible model for the pair of random variables W_1, W_2 is that they are independent, each with the marginal probability function

$$f(w) = \begin{cases} .1 & \text{if } w = 0, 1, 2, \ldots, 9 \\ 0 & \text{otherwise} \end{cases} \tag{5-65}$$

which is pictured in Figure 5-41.

Using such a model, it is a straightforward exercise (along the lines of Example 5-24) to reason that $\bar{W} = \frac{1}{2}(W_1 + W_2)$ has the probability function given in Table 5-26 and pictured in Figure 5-42.

FIGURE 5-41

Probability Histogram for W

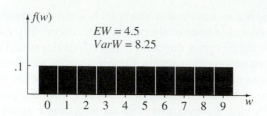

TABLE 5-26

The Probability Function
for $\overline{W}$ for $n = 2$

$\overline{w}$	$f(\overline{w})$
0.0	.01
0.5	.02
1.0	.03
1.5	.04
2.0	.05
2.5	.06
3.0	.07
3.5	.08
4.0	.09
4.5	.10
5.0	.09
5.5	.08
6.0	.07
6.5	.06
7.0	.05
7.5	.04
8.0	.03
8.5	.02
9.0	.01

FIGURE 5-42

Probability Histogram for $\overline{W}$
Based on $n = 2$

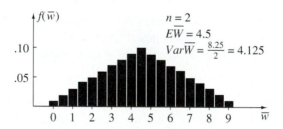

Comparing Figures 5-41 and 5-42, it is clear that even for a completely flat/uniform underlying distribution of W and the small sample size of $n = 2$, the probability distribution of $\overline{W}$ looks far more bell-shaped than the underlying distribution. It should be intuitively clear why this is so. As one moves away from the mean or central value of $\overline{W}$, there are relatively fewer and fewer combinations of w_1 and w_2 that can produce a given value of $\overline{w}$. For example, to observe $\overline{W} = 0$, one must have $W_1 = 0$ and $W_2 = 0$ — that is, one must observe not one but two extreme values. On the other hand, there are 10 different combinations of w_1 and w_2 that lead to $\overline{W} = 4.5$.

In theory, it is possible to use the same kind of logic leading to Table 5-26 to produce exact probability distributions for $\overline{W}$ based on larger sample sizes n. (A truly energetic reader might consider beginning with Table 5-26 and generating the distribution of the average of 2 independent $\overline{W}$'s, each based on $n = 2$ — that is, the probability distribution of $\overline{W}$ based on $n = 4$.) But such work is tedious, and for the purpose of indicating roughly how the central limit effect takes over as n gets larger, it is probably sufficient to approximate the distribution of $\overline{W}$ via simulation for a larger sample size. To this end, 200 sets of values for iid variables $W_1, W_2, \ldots, W_8$ (with marginal distribution given in expression (5-65)) were simulated and each set averaged to produce 200 simulated values of $\overline{W}$ based on $n = 8$. Figure 5-43 is a stem-and-leaf diagram of these 200 values. Notice the bell-shaped character of the plot and the agreement between the simulated mean and variance of $\overline{W}$ and the theoretical values given by formulas (5-59) and (5-60).

FIGURE 5-43

Stem-and-Leaf Plot of 200 Simulated Values of $\overline{W}$ Based on $n = 8$

```
Stem-and-leaf of Wbar       N  = 200
Leaf Unit = 1.0
```

Simulated Mean = 4.49 ≈ 4.50 = EW = $E\overline{W}$

Simulated Variance = 1.12 ≈ $\dfrac{8.25}{8}$ = $\dfrac{Var\,W}{n}$ = $Var\,\overline{W}$

```
    2      1 88
    6      2 0113
   14      2 55778888
   30      3 0111111222333333
   60      3 5555555555667777777888888888888
   95      4 000000011111111222222222222333333333
  (45)     4 5555555566666666666667777777777777788888888888
   60      5 000000011111222222223333333
   32      5 55555556667788
   18      6 00012333
   10      6 5566778
    3      7 12
    1      7 5
```

Proposition 5-3 is somewhat vague, in that what constitutes "large n" isn't obvious. The truth of the matter is that what sample size is required before $\bar{X}$ can be treated as essentially normal does depend to some degree on the shape of the underlying distribution of a single observation. Underlying distributions with decidedly nonnormal shapes require somewhat bigger values of n. But for most engineering purposes, $n \geq 25$ or so is adequate to make $\bar{X}$ essentially normal for the majority of data-generating mechanisms met in practice. (The exceptions are those subject to the occasional production of wildly outlying values.) Indeed, as Example 5-28 suggests, $\bar{X}$ is essentially normal for sample sizes much smaller than 25 in many cases.

The practical usefulness of Proposition 5-3 is that in many circumstances, only a normal table is needed to evaluate probabilities for sample averages.

EXAMPLE 5-26
(continued)

Return one more time to the situation of Jenkins, Milbrath, and Worth, involving stamp sale time requirements, and consider observing and averaging the next $n = 100$ excess service times, to produce

$\bar{S}$ = the sample mean time (over a 7.5 sec threshold) required to complete the next 100 stamp sales

And for the sake of example, consider approximating $P[\bar{S} > 17]$.

As discussed before, an iid model with marginal exponential $\alpha = 16.5$ distribution is plausible for the individual excess service times. Then

$$E\bar{S} = \alpha = 16.5 \text{ sec}$$

and

$$\sqrt{Var\,\bar{S}} = \sqrt{\frac{\alpha^2}{100}} = 1.65 \text{ sec}$$

are appropriate for $\bar{S}$, via formulas (5-59) and (5-60). Further, in view of the fact that $n = 100$ is large, one may use the normal probability table to find approximate probabilities for $\bar{S}$. Figure 5-44 shows an approximate distribution for $\bar{S}$ and the area corresponding to $P[\bar{S} > 17]$.

FIGURE 5-44

Approximate Probability Distribution for $\bar{S}$ and $P[\bar{S} > 17]$

The Approximate Probability Distribution of $\bar{S}$ Is Normal with Mean 16.5 and Standard Deviation 1.65.

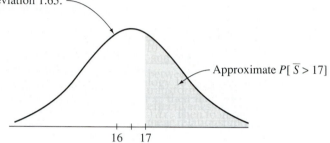

Approximate $P[\bar{S} > 17]$

16   17

As always, one must convert to z values before consulting the standard normal table. In this case, the mean and standard deviation to be used in the conversion are (respectively) 16.5 sec and 1.65 sec. That is, one calculates as

$$z = \frac{\bar{x} - E\bar{X}}{\sqrt{Var\,\bar{X}}} = \frac{\bar{x} - \mu}{\frac{\sigma}{\sqrt{n}}} \tag{5-66}$$

which, when applied in the present context, gives

$$z = \frac{17 - 16.5}{1.65} = .30$$

so

$$P[\bar{S} > 17] \approx P[Z > .30] = 1 - \Phi(.30) = .38$$

Formula (5-66) is boxed in and highlighted in some texts, to stress its particular importance. However, this author prefers to emphasize the logic rather than highlighting the formula. The point is that by Proposition 5-3, $\bar{X}$ is approximately normal for large n. Formulas (5-59) and (5-60) give the appropriate mean and standard deviation. Then the discussion in Section 5-2 shows how an appropriate mean and standard deviation lead to a z value and a corresponding normal probability.

The final example in this section illustrates how the Central Limit Theorem and some idea of a process standard deviation can help guide the choice of sample size in statistical applications.

EXAMPLE 5-29

(Example 5-10 revisited) **Sampling a Jar Filling Process.** The process of filling food containers, discussed by J. Fisher in his 1983 *Quality Progress* article, appears (from a histogram in the paper) to have an inherent standard deviation of measured fill weights on the order of 1.6 g. Suppose that in order to calibrate a fill level adjustment knob on such a process, one will set the knob, fill a run of n jars, and use their sample mean net contents as an indication of the process mean fill level corresponding to that knob setting. Suppose further that one would like to choose a sample size, n, large enough that *a priori* there is an 80% chance the sample mean is within .3 g of the actual process mean.

If the filling process can be thought of as physically reasonably stable, it makes sense to model the n observed net weights as iid random variables with (unknown) marginal mean μ and standard deviation $\sigma = 1.6$ g. Then for large n,

$$\overline{V} = \text{the observed sample average net weight}$$

can be thought of as approximately normal with mean μ and standard deviation $\sigma/\sqrt{n} = 1.6/\sqrt{n}$ (by formulas (5-59) and (5-60) and Proposition 5-3).

Now the requirement that $\overline{V}$ be within .3 g of μ can be written as

$$\mu - .3 < \overline{V} < \mu + .3$$

so the problem at hand is to choose n such that

$$P[\mu - .3 < \overline{V} < \mu + .3] = .80$$

Figure 5-45 pictures the situation. It is a simple matter to verify that the .90 quantile of the standard normal distribution is roughly 1.28 — that is, $P[-1.28 < Z < 1.28] = .8$. So evidently, Figure 5-45 indicates that one wants $\mu + .3$ to have z value 1.28. That is, one wants

$$1.28 = \frac{(\mu + .3) - \mu}{\dfrac{1.6}{\sqrt{n}}}$$

or

$$.3 = 1.28\frac{1.6}{\sqrt{n}}$$

So solving for n, a sample size of $n \approx 47$ would be required to provide the kind of precision of measurement desired.

FIGURE 5-45

Approximate Probability
Distribution for $\overline{V}$

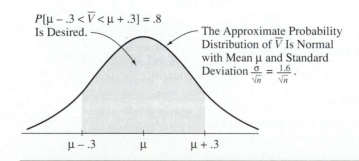

$P[\mu - .3 < \overline{V} < \mu + .3] = .8$
Is Desired.

The Approximate Probability
Distribution of $\overline{V}$ Is Normal
with Mean μ and Standard
Deviation $\frac{\sigma}{\sqrt{n}} = \frac{1.6}{\sqrt{n}}$.

$\mu - .3 \qquad \mu \qquad \mu + .3$

EXERCISES

5-1. (§5-1) Sketch probability histograms for the binomial distributions with $n = 5$ and $p = .1, .3, .5, .7,$ and $.9$. On each histogram, mark the location of the mean and indicate the size of the standard deviation.

5-2. (§5-1) Suppose 90% of all students taking a begining programming course fail to get their first program to run on first submission. Use a binomial distribution and assign probabilities to the possibilities that among a group of six such students,
(a) all fail on their first submissions.
(b) at least four fail on their first submissions.
(c) less than four fail on their first submissions.
Continuing to use this binomial model,
(d) what is the mean number who will fail?
(e) what are the variance and standard deviation of the number who will fail?

5-3. (§5-1) In an experiment to evaluate a new artificial sweetener, ten subjects are all asked to taste cola from three unmarked glasses, two of which contain regular cola while the third contains cola made with the new sweetener. The subjects are asked to identify the glass whose content is different from the other two. If there is no difference between the taste of sugar and the taste of the new sweetener, the subjects would be just guessing.
(a) Make a table for a probability function for

$$X = \text{the number of subjects correctly identifying}$$
$$\text{the artificially sweetened cola}$$

under this hypothesis of no difference in taste.
(b) If in fact seven of the ten subjects correctly identify the artificial sweetener, is this outcome strong evidence of a taste difference? Explain.

5-4. (§5-1) Sketch probability histograms for the Poisson distributions with means $\lambda = .5, 1.0, 2.0,$ and 4.0. On each histogram, mark the location of the mean and indicate the size of the standard deviation.

5-5. (§5-1) A process for making plate glass produces an average of 4 seeds (small bubbles) per 100 square feet. Use Poisson distributions and assess probabilities that
(a) a particular piece of glass 5 ft × 10 ft will contain more than 2 seeds.
(b) a particular piece of glass 5 ft × 5 ft will contain no seeds.

5-6. (§5-1) Transmission line interruptions in a telecommunications network occur at an average rate of 1 per day.
(a) Use a Poisson distribution as a model for

$$X = \text{the number of interruptions in the next 5-day}$$
$$\text{work week}$$

and assess $P[X = 0]$.
(b) Now consider the random variable

$$Y = \text{the number of weeks in the next 4 in which}$$
$$\text{there are no interruptions}$$

What is a reasonable probability model for Y? Assess $P[Y = 2]$.

5-7. (§2-2, §3-1, §3-3, §5-1) Suppose that a small population consists of the $N = 6$ values 2, 3, 4, 4, 5, and 6.
(a) Sketch a relative frequency histogram for this population and compute the population mean μ and standard deviation σ.
(b) Now let $X =$ the value of a single number selected at random from this population. Sketch a probability histogram for this variable X and compute EX and $Var X$.
(c) Now think of drawing a simple random sample of size $n = 2$ from this small population. Make tables giving the probability distributions of the random variables

$$\bar{X} = \text{the sample mean}$$
$$S^2 = \text{the sample variance}$$

(There are 15 different possible unordered samples of 2 out of 6 items. Each of the 15 possible samples is equally likely to be chosen and has its own corresponding $\bar{x}$ and s^2.) Use the tables and make probability histograms for these random variables. Compute $E\bar{X}$ and $Var \bar{X}$. How do these compare to μ and σ^2?

5-8. (§5-2) The random number generator supplied on a certain brand of pocket electronic calculator is not terribly well chosen, in that values it generates are not adequately described by the uniform distribution on the interval $(0, 1)$. Suppose instead that a probability density

$$f(x) = \begin{cases} k(5 - x) & \text{for } 0 < x < 1 \\ 0 & \text{otherwise} \end{cases}$$

is a more appropriate model for $X =$ the next value produced by this random number generator.
(a) Find the value of k.
(b) Sketch the probability density involved here.
(c) Evaluate $P[.25 < X < .75]$.
(d) Compute and graph the cumulative probability function for $X, F(x)$.
(e) Calculate EX and the standard deviation of X.

5-9. (§5-2) Suppose that Z is a standard normal random variable. Evaluate the following probabilities involving Z.
(a) $P[Z < -.62]$
(b) $P[Z > 1.06]$
(c) $P[-.37 < Z < .51]$
(d) $P[|Z| \leq .47]$
(e) $P[|Z| > .93]$
(f) $P[-3.0 < Z < 3.0]$
Find numbers # such that the following statements involving Z are true.
(g) $P[Z \leq \#] = .90$
(h) $P[|Z| < \#] = .90$
(i) $P[|Z| > \#] = .03$

5-10. (§5-2) Suppose that X is a normal random variable with mean 43.0 and standard deviation 3.6. Evaluate the following probabilities involving X.
(a) $P[X < 45.2]$
(b) $P[X \leq 41.7]$

(c) $P[43.8 < X \leqslant 47.0]$

(d) $P[|X - 43.0| \leqslant 2.0]$

(e) $P[|X - 43.0| > 1.7]$

Find numbers # such that the following statements involving X are true.

(f) $P[X < \#] = .95$

(g) $P[X \geqslant \#] = .30$

(h) $P[|X - 43.0| > \#] = .05$

5-11. (§5-2) The diameters of bearing journals that are ground on a certain grinder can be described as normally distributed with mean 2.0005 in. and standard deviation .0004 in.

(a) If engineering specifications on these diameters are 2.0000 in. $\pm$ .0005 in., what fraction of these journals are in specifications?

(b) What adjustment to the grinding process would probably increase the fraction of journal diameters that will be in specifications? What appears to be the best possible fraction of journal diameters inside $\pm$.0005 in. specifications, given the $\sigma = .0004$ in. apparent precision of the grinder being used?

(c) Suppose consideration was being given to purchasing a more expensive/newer grinder, capable of holding tighter tolerances on the parts it produces. What σ would have to be associated with the new machine in order to guarantee that (when perfectly adjusted so that $\mu = 2.0000$) the grinder would produce diameters with at least 95% meeting 2.0000 in. $\pm$ .0005 in. specifications?

5-12. (§5-3) In Exercise 3-6, you were asked to make a normal plot of some data taken from J.S. Hunter's article in the *RCA Engineer*.

(a) Return to that normal plot and use it to derive an approximate mean and standard deviation for percentage yields for the chemical process.

(b) Use a computer package to make a computer-generated version of the normal plot.

5-13. (§5-3) The article "Statistical Investigation of the Fatigue Life of Deep Groove Ball Bearings" by J. Leiblein and M. Zelen (*Journal of Research of the National Bureau of Standards*, 1956) contains the data given below on the lifetimes of 23 ball bearings. The units are 10^6 revolutions before failure.

> 17.88, 28.92, 33.00, 41.52, 42.12, 45.60,
> 48.40, 51.84, 51.96, 54.12, 55.56, 67.80,
> 68.64, 68.64, 68.88, 84.12, 93.12, 98.64,
> 105.12, 105.84, 127.92, 128.04, 173.40

(a) Use a normal plot to assess how well a Gaussian distribution fits these data. Then determine if a lognormal distribution is a better representation of bearing load life. What parameter values would you use in a lognormal description of load life? For these parameters, what are the .05 quantiles of ln(life) and of life?

(b) Use the method of display (5-39) in Section 5-3 and investigate whether the Weibull distribution might be used to describe bearing load life. If a Weibull description is sensible, read appropriate parameter values from the Weibull plot. Then use the form of the Weibull cumulative probability function given in Section 5-2 to find the resulting .05 quantile of the bearing load life distribution.

5-14. (§5-4) Explain in qualitative terms what it means for two random variables X and Y to be independent. What practical advantage is there when X and Y can be described as independent?

5-15. (§5-5) A type of nominal $\frac{3}{4}$ inch plywood is made of 5 layers. Suppose that these layers can be thought of as having thicknesses roughly describable as independent random variables with means and standard deviations as below.

Layer	Mean (in.)	Standard Deviation (in.)
1	.094	.001
2	.156	.002
3	.234	.002
4	.172	.002
5	.094	.001

Find the mean and standard deviation of total thickness associated with the combination of these individual values.

5-16. (§5-5) The coefficient of linear expansion of brass is to be obtained as a laboratory exercise. For a brass bar that is L_1 meters long at $T_1°$C and L_2 meters long at $T_2°$C, this coefficient is

$$\alpha = \frac{L_2 - L_1}{L_1(T_2 - T_1)}$$

Suppose that the equipment to be used in the laboratory is thought to have a standard deviation for repeated length measurements of about .00005 m and a standard deviation for repeated temperature measurements of about .1°C.

(a) If using $T_1 \approx 50°$C and $T_2 \approx 100°$C, $L_1 \approx 1.00000$ m and $L_2 \approx 1.00095$ m are obtained, and it desired to attach an approximate standard deviation to the derived value of α, find such an approximate standard deviation two different ways. First, use simulation as was done in Printout 5-2. Then use the propagation of error formula. How well do your two values agree?

(b) In this particular lab exercise, the precision of which measurements (the lengths or the temperatures) is the primary limiting factor in the precision of the derived coefficient of linear expansion? Explain.

(c) Within limits, the larger $T_2 - T_1$, the better the value for α. What (in qualitative terms) is the physical origin of those limits?

5-17. (§5-5) A pendulum swinging through small angles approximates simple harmonic motion. The period of the pendulum, τ, is (approximately) given by

$$\tau = 2\pi\sqrt{\frac{L}{g}}$$

where L is the length of the pendulum and g is the acceleration due to gravity. This fact can be used to derive an experimental value for g. Suppose that the length L of about 5 ft can be measured with a standard deviation of about .25 in. (about .0208 foot), and the period τ of about 2.48 sec can be measured with standard deviation of about .1 sec. What is a reasonable standard deviation to attach to a value of g derived using this equipment? Is the precision of the length measurement or the precision of the period measurement the principal limitation on the precision of the derived g?

5-18. (§5-5) Consider again the random number generator discussed in Exercise 5-8. Suppose that it is used to generate 25 random numbers, and that these may be reasonably be thought of as independent random variables with common individual (marginal) distribution as given in Exercise 5-8. Let $\bar{X}$ be the sample mean of these 25 values.

(a) What are the mean and standard deviation of the random variable $\bar{X}$?

(b) What is the approximate probability distribution of $\bar{X}$?

(c) Approximate the probability that $\bar{X}$ exceeds .5.

(d) Approximate the probability that $\bar{X}$ takes a value within .02 of its mean.

(e) Redo parts (a) through (d) using a sample size of 100 instead of 25.

5-19. (§5-5) Passing a large production run of piston rings through a grinding operation produces edge widths possessing a standard deviation of .0004 in. A simple random sample of rings is to be taken and their edge widths measured, with the intention of using $\bar{X}$ as an estimate of the population mean thickness μ. Approximate the probabilities that $\bar{X}$ is within .0001 in. of μ for samples of size $n = 25, 100,$ and 400.

5-20. A discrete random variable X can be described using the probability function

x	2	3	4	5	6
$f(x)$	.1	.2	.3	.3	.1

(a) Make a probability histogram for X. Also plot $F(x)$, the cumulative probability function for X.

(b) Find the mean and standard deviation for the random variable X.

5-21. Suppose that for single launches of a space shuttle, there is a constant probability of O-ring failure (say, .15). Consider 10 future launches, and let X be the number of those involving an O-ring failure. Use an appropriate probability model and evaluate all of the following.

(a) $P[X = 2]$

(b) $P[X \geq 1]$

(c) EX

(d) $Var\,X$

(e) the standard deviation of X

5-22. An injection molding process for making auto bumpers leaves an average of 1.3 visual defects per bumper prior to painting of the bumpers. Let Y and Z be the numbers of visual defects on (respectively) the next two bumpers produced. Use an appropriate probability distribution and evaluate the following.

(a) $P[Y = 2]$

(b) $P[Y \geq 1]$

(c) $\sqrt{Var\,Y}$

(d) $P[Y + Z \geq 2]$ (*Hint:* What is a sensible distribution for $Y + Z$, the number of blemishes on two bumpers?)

5-23. Suppose that the random number generator supplied in a pocket calculator actually generates values in such a way that if X is the next value generated, X can be adequately described using

a probability density of the form

$$f(x) = \begin{cases} k((x - .5)^2 + 1) & \text{for } 0 < x < 1 \\ 0 & \text{otherwise} \end{cases}$$

(a) Evaluate k and sketch a graph of $f(x)$.

(b) Evaluate $P[X \geq .5]$, $P[X > .5]$, $P[.75 > X \geq .5]$, and $P[|X - .5| \geq .2]$.

(c) Compute EX and $Var\,X$.

(d) Compute and graph $F(x)$, the cumulative probability function for X. Read from your graph the .8 quantile of the distribution of X.

5-24. Suppose that Z is a standard normal random variable. Evaluate the following probabilities involving Z.

(a) $P[Z \leq 1.13]$

(b) $P[Z > -.54]$

(c) $P[-1.02 < Z < .06]$

(d) $P[|Z| \leq .25]$

(e) $P[|Z| > 1.51]$

(f) $P[-3.0 < Z < 3.0]$

Find numbers # such that the following statements about Z are true.

(g) $P[|Z| < \#] = .80$

(h) $P[Z < \#] = .80$

(i) $P[|Z| > \#] = .04$

5-25. Suppose that X is a normal random variable with mean $\mu = 10.2$ and standard deviation $\sigma = .7$. Evaluate the following probabilities involving X.

(a) $P[X \leq 10.1]$

(b) $P[X > 10.5]$

(c) $P[9.0 < X < 10.3]$

(d) $P[|X - 10.2| \leq .25]$

(e) $P[|X - 10.2| > 1.51]$

Find numbers # such that the following statements about X are true.

(f) $P[|X - 10.2| < \#] = .80$

(g) $P[X < \#] = .80$

(h) $P[|X - 10.2| > \#] = .04$

5-26. In a grinding operation, there is an upper specification of 3.150 in. on a dimension of a certain part after grinding. Suppose that the standard deviation of this dimension for parts of this type ground to any particular mean dimension μ is $\sigma = .002$ in. Suppose further that one desires to have no more than 3% of the parts fail to meet specifications. What is the maximum (minimum machining cost) μ that can be used if this 3% requirement is to be met and this dimension is normally distributed?

5-27. A 10-ft cable is made of 50 strands. Suppose that individually, 10-ft strands have breaking strengths with mean 45 lb and standard deviation 4 lb. Suppose further that the breaking strength of a cable is roughly the sum of the strengths of the strands that make it up.

(a) Find a plausible mean and standard deviation for the breaking strengths of such 10-ft cables.

(b) Evaluate the probability that a 10-ft cable of this type will support a load of 2230 lb. (*Hint:* If $\bar{X}$ is the mean breaking strength of the strands, $\sum(\text{Strengths}) \geqslant 2230$ is the same as $\bar{X} \geqslant (\frac{2230}{50})$. Now use the Central Limit Theorem.)

5-28. The electrical resistivity, ρ, of a piece of wire is a property of the material involved and the temperature at which it is measured. At a given temperature, if a cylindrical piece of wire of length L and cross-sectional area A has resistance R, the material's resistivity is calculated using the formula $\rho = \frac{RA}{L}$. Thus, if a wire's cross section is assumed to be circular with diameter D, the resistivity at a given temperature is

$$\rho = \frac{R\pi D^2}{4L}$$

In a lab exercise to determine the resistivity of copper at 20°C, students measure lengths, diameters, and resistances of wire nominally 1.0 m in length (L), 2.0×10^{-3} m in diameter (D), and of resistance (R) $.54 \times 10^{-2}$ Ω. Suppose that it is sensible to describe the measurement precisions in this laboratory with the standard deviations $\sigma_L \approx 10^{-3}$ m, $\sigma_D \approx 10^{-4}$ m, and $\sigma_R \approx 5 \times 10^{-4}$ Ω.
(a) Find an approximate standard deviation that might be used to describe the expected precision for an experimentally derived value of ρ.
(b) Imprecision in which of the measurements is likely to be the largest contributor to imprecision in measured resistivity? Explain.
(c) One should expect the value derived in (a) to underpredict the kind of variation that would be observed in such laboratory exercises over a period of years. Explain why this is so.

5-29. Suppose that the thickness of sheets of a certain weight of book paper have mean .1 mm and a standard deviation of .003 mm. A particular textbook will be printed on 370 sheets of this paper. Find sensible values for the mean and standard deviation of the thicknesses of copies of this text (excluding, of course, the book's cover).

5-30. Distinguish clearly between the subjects of *probability* and *statistics*. Is one field a subfield of the other?

5-31. Pairs of resistors are to be connected in parallel and a difference in electrical potential applied across the resistor assembly. Ohm's law predicts that in such a situation, the current flowing in the circuit will be

$$I = V\left(\frac{1}{R_1} + \frac{1}{R_2}\right)$$

where R_1 and R_2 are the two resistances and V the potential applied. Suppose that R_1 and R_2 have means $\mu_R = 10$ Ω and standard deviations $\sigma_R = .1$ Ω and that V has mean $\mu_V = 9$ volt and $\sigma_V = .2$ volt.
(a) Find an approximate mean and standard deviation for I, treating R_1, R_2, and V as independent random variables.
(b) Based on your work in (a), would you say that the variation in voltage or the combined variations in R_1 and R_2 are the biggest contributors to variation in current? Explain.

5-32. Students in a materials lab are required to experimentally determine the heat conductivity of aluminum.
(a) If student-derived values are normally distributed about a mean of .5(cal/(cm)(sec)(°C)) with standard deviation of .03, evaluate

the probability that an individual student will obtain a conductivity from .48 to .52.
(b) If student values have the mean and standard deviation given in (a), evaluate the probability that a class of 25 students will produce a sample mean conductivity from .48 to .52.
(c) If student values have the mean and standard deviation given in (a), evaluate the probability that at least 2 of the next 5 values produced by students will be in the range from .48 to .52.

5-33. Suppose that 10-ft lengths of a certain type of cable have breaking strengths with mean $\mu = 450$ lb and standard deviation $\sigma = 50$ lb.
(a) If five of these cables are used to support a single load L, suppose that the cables are loaded in such a way that support fails if any one of the cables has strength below $\frac{L}{5}$. With $L = 2000$ lb, assess the probability that the support fails, if individual cable strength is normally distributed. Do this in two steps. First find the probability that a particular individual cable fails, then use that to evaluate the desired probability.
(b) Approximate the probability that the sample mean strength of 100 of these cables is below 457 lb.

5-34. Find EX and $Var\,X$ for a continuous distribution with probability density

$$f(x) = \begin{cases} .3 & \text{if } 0 < x < 1 \\ .7 & \text{if } 1 < x < 2 \\ 0 & \text{otherwise} \end{cases}$$

5-35. What is the difference between a relative frequency distribution and a probability distribution?

5-36. Explain how an approximate mean μ and standard deviation σ can be read off a plot of standard normal quantiles versus data quantiles made on regular graph paper.

5-37. Suppose that it is adequate to describe the 14-day compressive strengths of test specimens of a certain concrete mixture as normally distributed with mean $\mu = 2930$ psi and standard deviation $\sigma = 20$ psi.
(a) Assess the probability that the next specimen of this type tested for compressive strength will have strength above 2945 psi.
(b) Use your answer to part (a) and assess the probability that in the next four specimens tested, at least one has compressive strength above 2945 psi.
(c) Assess the probability that the next 25 specimens tested have a sample mean compressive strength within 5 psi of $\mu = 2930$ psi.
(d) Suppose that although the particular concrete formula under consideration in this problem is relatively strong, it is difficult to pour in large quantities without the development of serious air pockets (which can have important implications for structural integrity). In fact, suppose that using standard methods of pouring, serious air pockets form at an average rate of 1 per 50 cubic yards of poured concrete. Use an appropriate probability distribution and assess the probability that two or more serious air pockets will appear in a 150-cubic-yard pour to be made tomorrow.

5-38. What is the practical usefulness of the technique of probability plotting?

5-39. For X with a continuous distribution specified by the probability density

$$f(x) = \begin{cases} .5x & \text{for } 0 < x < 2 \\ 0 & \text{otherwise} \end{cases}$$

find $P[X < 1.0]$ and find the mean, EX.

5-40. The viscosity of a liquid may be measured by placing it in a cylindrical container and determining the force needed to turn a cylindrical rotor (of nearly the same diameter as the container) at a given velocity in the liquid. The relationship between the viscosity η, force F, area A of the side of the rotor in contact with the liquid, the size L of the gap between the rotor and the inside of the container, and the velocity v at which the rotor surface moves is

$$\eta = \frac{FL}{vA}$$

Suppose that students are to determine an experimental viscosity for SAE no. 10 oil as a laboratory exercise and that appropriate means and standard deviations for the measured variables F, L, v, and A in this laboratory are as follows.

$$\mu_F = 151 \text{ N} \qquad \sigma_F = .05 \text{ N}$$
$$\mu_A = 1257 \text{ cm}^2 \qquad \sigma_A = .2 \text{ cm}^2$$
$$\mu_L = .5 \text{ cm} \qquad \sigma_L = .05 \text{ cm}$$
$$\mu_v = 30 \text{ cm/sec} \qquad \sigma_v = 1 \text{ cm/sec}$$

(a) Use the propagation of error formulas and find an approximate standard deviation that might serve as a measure of precision for an experimentally derived value of η from this laboratory.

(b) Explain why, if one were to compare experimental values of η obtained for SAE no. 10 oil in similar laboratory exercises conducted over a number of years at a number of different universities, the approximate standard deviation derived in (a) above would be likely to understate the variability actually observed in those values.

5-41. The heat conductivity of a cylindrical bar of diameter D and length L, connected between two constant temperature devices of temperatures T_1 and T_2 (respectively), that conducts Q calories in t seconds is

$$\lambda = \frac{4QL}{\pi(T_1 - T_2)tD^2}$$

In a materials laboratory exercise to determine λ for brass, the following means and standard deviations for the variables D, L, T_1, T_2, Q, and t are appropriate, as are the partial derivatives of λ with respect to the various variables (evaluated at the means of the variables).

	D	L	T_1	T_2	Q	t
μ	1.6 cm	100 cm	100°C	0°C	240 cal	600 sec
σ	.1 cm	.1 cm	1°C	1°C	10 cal	1 sec
partial	−.249	.199	−.00199	.00199	.000825	.000332

(The units of the partial derivatives are the units of λ (cal/(cm)(sec)(°C)) divided by the units of the variable in question.)

(a) Find an approximate standard deviation to associate with an experimentally derived value of λ.

(b) Which of the variables appears to be the biggest contributor to variation in experimentally determined values of λ? Explain your choice.

5-42. Suppose that 15% of all daily oxygen purities delivered by an air-products supplier are below 99.5% purity and that it is plausible to think of daily purities as independent random variables. Evaluate the probability that in the next 5-day workweek, 1 or less delivered purities will fall below 99.5%.

5-43. Suppose that the raw daily oxygen purities delivered by an air-products supplier have a standard deviation $\sigma \approx .1$ (percent), and it is plausible to think of daily purities as independent random variables. Approximate the probability that the sample mean $\bar{X}$ of $n = 25$ delivered purities falls within .03 (percent) of the raw daily purity mean, μ.

5-44. Students are going to measure Young's Modulus for copper by measuring the elongation of a piece of copper wire under a tensile force. For a cylindrical wire of diameter D subjected to a tensile force F, if the initial length (length before applying the force) is L_0 and final length is L_1, Young's Modulus for the material in question is

$$Y = \frac{4FL_0}{\pi D^2(L_1 - L_0)}$$

The test and measuring equipment used in a particular lab are characterized by the standard deviations

$$\sigma_F \approx 10 \text{ lb} \qquad \sigma_D \approx .001 \text{ in.} \qquad \sigma_{L_0} = \sigma_{L_1} = .01 \text{ in.}$$

and in the setup employed, $F \approx 300$ lb, $D \approx .050$ in., $L_0 \approx 10.00$ in., and $L_1 \approx 10.10$ in.

(a) Treating the measured force, diameter, and lengths as independent variables with the preceding means and standard deviations, find an approximate standard deviation to attach to an experimentally derived value of Y. (Partial derivatives of Y at the nominal values of F, D, L_0, and L_1 are approximately $\frac{\partial Y}{\partial F} \approx 5.09 \times 10^4$, $\frac{\partial Y}{\partial D} \approx -6.11 \times 10^8$, $\frac{\partial Y}{\partial L_0} \approx 1.54 \times 10^8$, and $\frac{\partial Y}{\partial L_1} \approx -1.53 \times 10^8$ in the appropriate units.)

(b) Uncertainty in which of the variables is the biggest contributor to uncertainty in Y?

(c) Notice that the equation for Y says that for a particular material (and thus supposedly constant Y), circular wires of constant initial lengths L_0, but of different diameters and subjected to different tensile forces, will undergo elongations $\Delta L = L_1 - L_0$ of approximately

$$\Delta L \approx \kappa \frac{F}{D^2}$$

for κ a constant depending on the material and the initial length. Suppose that one decides to measure ΔL for a factorial arrangement of levels of F and D. Does the above equation predict that F and D *will* or *will not* have important interactions? Explain.

5-45. Exercise 3-19 concerns the lifetimes (in numbers of 24 mm deep holes drilled in 1045 steel before failure) of 12 D952-II (8 mm) drills.

(a) Make a normal plot of the data given in Chapter 3. In what specific way does the shape of the data distribution appear to depart from a Gaussian shape?

(b) The 12 lifetimes have mean $\bar{y} = 117.75$ and standard deviation $s \approx 51.1$. Simply using these in place of μ and σ for the underlying drill life distribution, use the normal table to find an approximate fraction of drill lives below 40 holes.

(c) Based on your answer to (a), if your answer to (b) is seriously different from the real fraction of drill lives below 40, is it most likely high or low? Explain.

5-46. Metal fatigue causes cracks to appear on the skin of older aircraft. Assume that it is reasonable to model the number of cracks appearing on a 1 m^2 surface of planes of a certain model and vintage as Poisson with mean $\lambda = .03$.

(a) If 1 m^2 is inspected, assess the probability that at least one crack is present on that surface.

(b) If 10 m^2 are inspected, assess the probability that at least one crack (total) is present.

(c) If 10 areas, each of size 1 m^2, are inspected, assess the probability that exactly one of these has cracks.

5-47. If a dimension on a mechanical part is normally distributed, how small must the standard deviation be if 95% of such parts are to be within specifications of 2 cm $\pm$ .002 cm when the mean dimension is ideal?

5-48. The fact that the "exact" calculation of normal probabilities requires either numerical integration or the use of tables (ultimately generated using numerical integration) has inspired many people to develop approximations to the standard normal cumulative distribution function. Several of the simpler of these approximations are discussed in the articles "A Simpler Approximation for Areas Under the Standard Normal Curve," by A. Shah (*The American Statistician*, 1985), "Pocket-Calculator Approximation for Areas under the Standard Normal Curve," by R. Norton (*The American Statistician*, 1989), and "Approximations for Hand Calculators Using Small Integer Coefficients," by S. Derenzo (*Mathematics of Computation*, 1977). For $z > 0$, consider the approximations offered in these articles:

$$\Phi(z) \approx g_S(z) = \begin{cases} .5 + \dfrac{z(4.4 - z)}{10} & 0 \leq z \leq 2.2 \\ .99 & 2.2 < z < 2.6 \\ 1.00 & 2.6 \leq z \end{cases}$$

$$\Phi(z) \approx g_N(z) = 1 - \frac{1}{2} \exp\left(-\frac{z^2 + 1.2z^{.8}}{2}\right)$$

$$\Phi(z) \approx g_D(z) = 1 - \frac{1}{2} \exp\left(-\frac{(83z + 351)z + 562}{703/z + 165}\right)$$

Evaluate $g_S(z)$, $g_N(z)$, and $g_D(z)$ for $z = .5, 1.0, 1.5, 2.0,$ and 2.5. How do these values compare to the corresponding entries in Table D-3?

5-49. Exercise 5-48 concerned approximations for normal probabilities. People have also invested a fair amount of effort to find useful formulas approximating standard normal *quantiles*. One such approximation, again taken from the article by S. Derenzo

mentioned in Exercise 5-48, is as follows. For $p > .50$, let $y = -\ln(2(1 - p))$ and

$$Q_z(p) \approx \sqrt{\frac{((4y + 100)y + 205)y^2}{((2y + 56)y + 192)y + 131}}$$

For $p < .50$, let $y = -\ln(2p)$ and

$$Q_z(p) \approx -\sqrt{\frac{((4y + 100)y + 205)y^2}{((2y + 56)y + 192)y + 131}}$$

Use these formulas to approximate $Q_z(p)$ for $p = .01, .05, .1, .3, .7, .9, .95,$ and $.99$. How do the values you obtain compare with the corresponding entries in Table 3-10?

5-50. The data here are from the article "Fiducial Bounds on Reliability for the Two-Parameter Negative Exponential Distribution," by F. Grubbs (*Technometrics*, 1971). They are the mileages at first failure for 19 military personnel carriers.

> 162, 200, 271, 320, 393, 508, 539, 629,
> 706, 777, 884, 1008, 1101, 1182, 1462,
> 1603, 1984, 2355, 2880

(a) Make a histogram of these data. How would you describe its shape?

(b) Make an exponential probability plot for these data. Does it appear that the exponential distribution can be used to model the mileage to failure of this kind of vehicle? In Example 5-18, a threshold service time of 7.5 seconds was suggested by a similar exponential probability plot. Does the plot for these data give a strong indication of the need for a threshold mileage larger than 0 if an exponential distribution is to be used here?

(c) These data have a sample mean of $\bar{x} = 998.1$. Suppose that one takes this fact and the shape of the data set as an indication that an exponential distribution with mean 1,000 is a plausible model for the mileage to failure of this kind of vehicle. Evaluate the probability that a personnel carrier of this type gives less than 500 miles of service before first failure. Evaluate the probability that it gives at least 2,000 miles of service before first failure.

(d) Under the same exponential model assumptions used in (c), find the .05 quantile of the distribution of mileage to first failure. Find the .90 quantile of the distribution.

5-51. The article "Statistical Strength Evaluation of Hot-pressed Si_3N_4," by R. Govila (*Ceramic Bulletin*, 1983), contains summary statistics from an extensive study of the flexural strengths of two high-strength hot-pressed silicon nitrides in $\frac{1}{4}$-point, 4-point bending. The values below are fracture strengths of 30 specimens of one of the materials tested at 20°C. (The units are MPa, and the data were read from a graph in the paper and may therefore individually differ by perhaps as much as 10 MPa from the actual measured values.)

> 514, 533, 543, 547, 584, 619, 653, 684,
> 689, 695, 700, 705, 709, 729, 729, 753,
> 763, 800, 805, 805, 814, 819, 819, 839,
> 839, 849, 879, 900, 919, 979

(a) The materials researcher who collected the original data believed the Weibull distribution to be an adequate model for flexural strength of this material. Make a Weibull probability plot using the method of display (5-39) of Section 5-3 and investigate this possibility. Does a Weibull model fit these data?

(b) Eye-fit a line through your plot from part (a). Use it to help you determine an appropriate shape parameter, β, and an appropriate scale parameter, α, for a Weibull distribution used to describe flexural strength of this material at $20°$C. For a Weibull distribution with your fitted values of α and β, what is the median strength? What is a strength exceeded by 80% of such Si_3N_4 specimens? By 90% of such specimens? By 99% of such specimens?

(c) Make normal plots of the raw data and of the logarithms of the raw data. Comparing the three probability plots made in this exercise, is there strong reason to prefer a Weibull model, a normal model, or a lognormal model over the other two possibilities as a description of the flexural strength?

(d) Eye-fit lines to your plots from part (c). Use them to help you determine appropriate means and standard deviations for normal distributions used to describe flexural strength and the logarithm of flexural strength. Compare the .01, .10, .20, and .50 quantiles of the fitted normal and lognormal distributions for strength to the quantiles you computed in part (b).

5-52. The article "Using Statistical Thinking to Solve Maintenance Problems" by Brick, Michael, and Morganstein (*Quality Progress*, 1989) contains the data below on lifetimes of sinker rollers. Given are the numbers of 8-hour shifts that 17 sinker rollers (at the bottom of a galvanizing pot and used to direct steel sheet through a coating operation) lasted before failing and requiring replacement.

$$10, 12, 15, 17, 18, 18, 20, 20, 21, 21, 23,$$
$$25, 27, 29, 29, 30, 35$$

(a) The authors of the article considered a Weibull distribution to be a likely model for the lifetimes of such rollers. Make a zero-threshold Weibull probability plot for use in assessing the reasonableness of such a description of roller life.

(b) Eye-fit a line to your plot in (a) and use it to estimate parameters for a Weibull distribution for describing roller life.

(c) Use your estimated parameters from (a) and the form of the Weibull cumulative distribution function given in Section 5-2 to estimate the .10 quantile of the roller life distribution.

5-53. The article "Elementary Probability Plotting for Statistical Data Analysis" by J. King (*Quality Progress*, 1988) contains 24 measurements of deviations from nominal, of a distance between two holes drilled in a steel plate. These are reproduced here. The units are mm.

$$-2, -2, 7, -10, 4, -3, 0, 8, -5, 5, -6, 0,$$
$$2, -2, 1, 3, 3, -4, -6, -13, -7, -2, 2, 2$$

(a) Make a dot diagram for these data and compute $\bar{x}$ and s.

(b) Make a normal plot for these data. Eye-fit a line on the plot and use it to find graphical estimates of a process mean and standard deviation for this deviation from nominal. Compare these graphical estimates with the values you calculated in (a).

(c) Engineering specifications on this deviation from nominal were ± 10 mm. Suppose that $\bar{x}$ and s from (a) are adequate approximations of the process mean and standard deviation for this variable. Use the normal distribution with those parameters and compute a fraction of deviations that fall outside specifications. Does it appear from this exercise that the drilling operation is *capable* (i.e., precise) enough to produce essentially all measured deviations in specifications, at least if properly aimed? Explain.

5-54. An engineer is responsible for setting up a monitoring system for a critical diameter on a turned metal part produced in his plant. Engineering specifications for the diameter are 1.180 in. $\pm$ .004 in. For ease of communication, the engineer sets up the following nomenclature for measured diameters on these parts:

Green Zone Diameters: 1.178 in. $\leq$ Diameter $\leq$ 1.182 in.

Red Zone Diameters: Diameter $\leq$ 1.176 in. or
Diameter $\geq$ 1.184 in.

Yellow Zone Diameters: any other Diameter

Suppose that in fact the diameters of parts coming off the lathe in question can be thought of as independent normal random variables with mean $\mu = 1.181$ in. and standard deviation $\sigma = .002$ in.

(a) Find the probabilities that a given diameter falls into each of the three zones.

(b) Suppose that a technician simply begins measuring diameters on consecutive parts and continues until a Red Zone measurement is found. Assess the probability that more than ten parts must be measured. Also, give the expected number of measurements that must be made.

The engineer decides to use the Green/Yellow/Red gaging system in the following way. Every hour, parts coming off the lathe will be checked. First, a single part will be measured. If it is in the Green Zone, no further action is needed that hour. If the initial part is in the Red Zone, the lathe will be stopped and a supervisor alerted. If the first part is in the Yellow Zone, a second part is measured. If this second measurement is in the Green Zone, no further action is required, but if it is in the Yellow or the Red Zone, the lathe is stopped and a supervisor alerted. It is possible to argue that under this scheme (continuing to suppose that measurements are independent normal variables with mean 1.181 in. and standard deviation .002 in.), the probability that the lathe is stopped in any given hour is .1865.

(c) Use the preceding fact and evaluate the probability that the lathe is stopped exactly twice in 8 consecutive hours. Also, what is the expected number of times the lathe will be stopped in 8 time periods?

5-55. Suppose that a pair of random variables have the joint probability density

$$f(x, y) = \begin{cases} 4x(1 - y) & \text{if } 0 \leq x \leq 1 \text{ and } 0 \leq y \leq 1 \\ 0 & \text{otherwise} \end{cases}$$

(a) Find the marginal probability densities for X and Y.

(b) Are X and Y independent? Explain.

(c) Evaluate $P[X + 2Y \geq 1]$.

5-56. A laboratory receives four specimens having identical appearances. However, it is possible that (a single unknown) one of the specimens is contaminated with a toxic material. The lab must test the specimens to find the toxic specimen (if in fact one is contaminated). The testing plan first put forth by the laboratory staff is to test the specimens one at a time, stopping when (and if) a contaminated specimen is found.

Define two random variables

X = the number of contaminated specimens

Y = the number of specimens tested

Let $p = P[X = 0]$ and therefore $P[X = 1] = 1 - p$.
(a) Give the conditional distributions of Y given $X = 0$ and $X = 1$ for the staff's initial testing plan. Then use them to determine the joint probability function of X and Y. (Your joint distribution will involve p, and you may simply fill out tables like the accompanying ones.)

y	$f_{Y\|X}(y \mid 0)$	y	$f_{Y\|X}(y \mid 1)$
1		1	
2		2	
3		3	
4		4	

$f(x, y)$

$y\backslash x$	0	1
1		
2		
3		
4		

(b) Based on your work in (a), find EY, the mean number of specimens tested using the staff's original plan.
(c) A second testing method devised by the staff involves testing composite samples of material taken from possibly more than one of the original specimens. By initially testing a composite of all four specimens, then (if the first test reveals the presence of toxic material) following up with a composite of two, and then an appropriate single specimen, it is possible to do the lab's job in 1 test if $X = 0$, and in 3 tests if $X = 1$. Suppose that because testing is expensive, it is desirable to hold the number of tests to a minimum. For what values of p is this second method preferable to the first? (*Hint:* What is EY for this second method?)

5-57. A random variable X has a cumulative distribution function

$$F(x) = \begin{cases} 0 & \text{for } x \leq 0 \\ \sin(x) & \text{for } 0 < x \leq \pi/2 \\ 1 & \text{for } \pi/2 < x \end{cases}$$

(a) Find $P[X \leq .32]$.
(b) Give the probability density for $X, f(x)$.
(c) Evaluate EX and $Var X$.

5-58. Quality audit records are kept on numbers of major and minor failures of circuit packs during burn-in of large electronic switching devices. They indicate that for a device of this type, the random variables

X = the number of major failures

Y = the number of minor failures

can be described at least approximately by the accompanying joint distribution.

$y\backslash x$	0	1	2
0	.15	.05	.01
1	.10	.08	.01
2	.10	.14	.02
3	.10	.08	.03
4	.05	.05	.03

(a) Find the marginal probability functions for both X and Y — $f_X(x)$ and $f_Y(y)$.
(b) Are X and Y independent? Explain.
(c) Find the mean and variance of X — EX and $Var X$.
(d) Find the mean and variance of Y — EY and $Var Y$.
(e) Find the conditional probability function for Y, given that $X = 0$ — i.e., that there are no major circuit pack failures. (That is, find $f_{Y|X}(y \mid 0)$.) What is the mean of this conditional distribution?

Suppose that demerits are assigned to devices of this type according to the formula $D = 5X + Y$.

(f) Find the mean value of D, ED. (Use your answers to (c) and (d) and formula (5-57) of Section 5-5. Formula (5-57) holds whether or not X and Y are independent.)
(g) Find the probability a device of this type scores 7 or less demerits. That is, find $P[D \leq 7]$.
(h) On average, how many of these devices will have to be inspected in order to find one that scores 7 or less demerits? (Use your answer to (g).)

5-59. Consider jointly continuous random variables X and Y with density

$$f(x, y) = \begin{cases} x + y & \text{for } 0 < x < 1 \text{ and } 0 < y < 1 \\ 0 & \text{otherwise} \end{cases}$$

(a) Find the probability that the product of X and Y is at least $\frac{1}{4}$.
(b) Find the marginal probability density for X. (Notice that Y's is similar.) Use this to find the expected value and standard deviation of X.
(c) Are X and Y independent? Explain.
(d) Compute the mean of $X + Y$. Why can't formula (5-58) of Section 5-5 be used to find the variance of $X + Y$?

5-60. An engineering system consists of two subsystems operating independently of each other. Let

X = the time till failure of the first subsystem

Y = the time till failure of the second subsystem

Suppose that X and Y are independent exponential random variables each with mean $\alpha = 1$ (in appropriate time units).
(a) Write out the joint probability density for X and Y. Be sure to state carefully where the density is positive and where it is 0.

Suppose first that the system is a *series system* (i.e., one that fails when either of the subsystems fail).

(b) The probability that the system is still functioning at time $t > 0$ is then

$$P[X > t \text{ and } Y > t]$$

Find this probability using your answer to (a). (What region in the (x, y)-plane corresponds to the possibility that the system is still functioning at time t?)

(c) If one then defines the random variable

$$T = \text{the time till the system fails}$$

the cumulative probability function for T is

$$F(t) = 1 - P[X > t \text{ and } Y > t]$$

so that your answer to (b) can be used to find the distribution for T. Use your answer to (b) and some differentiation to find the probability density for T. What kind of distribution does T have? What is its mean?

Suppose now that the system is a *parallel system* (i.e., one that fails only when both subsystems fail).

(d) The probability that the system has failed by time t is

$$P[X \le t \text{ and } Y \le t]$$

Find this probability using your answer to part (a).

(e) Now, as before, let T be the time until the system fails. Use your answer to (d) and some differentiation to find the probability density for T. Then calculate the mean of T.

5-61. A machine element is made up of a rod and a ring bearing. The rod must fit through the bearing. If

$$X = \text{the diameter of the rod}$$

$$Y = \text{the inside diameter of the ring bearing}$$

are modeled as independent uniform random variables, X uniform on $(1.97, 2.02)$ and Y uniform on $(2.00, 2.06)$, do the following. (The uniform distributions are mentioned at the very end of Section 5-2.)

(a) Write out the joint probability density for X and Y.

(b) Evaluate $P[Y - X < 0]$, the probability of an interference in assembly.

(c) Find $EX, Var\,X, EY,$ and $Var\,Y$ using the formulas for the mean and variance of a beta distribution given in displays (5-31) and (5-32) of Section 5-2.

(d) Use your answer to (c) and Proposition 5-1 to find the mean and variance of $Y - X$.

5-62. Visual inspection of integrated circuit chips, even under high magnification is often less than perfect. Suppose that an inspector has an 80% chance of detecting any given flaw. We will suppose that the inspector never "cries wolf" — that is, perceives there to be a flaw where none exists. Then consider the random variables

$$X = \text{the true number of flaws on a chip}$$

$$Y = \text{the number of flaws identified by the inspector}$$

(a) What is a sensible conditional distribution for Y given that $X = 5$? Given that $X = 5$, find the (conditional) probability that $Y = 3$.

In general, a sensible conditional probability function for Y given $X = x$ is the binomial probability function with number of trials x and success probability .8. That is, one could use

$$f_{Y|X}(y \mid x) = \begin{cases} \binom{x}{y} .8^y .2^{x-y} & \text{for } y = 0, 1, 2, \dots, x \\ 0 & \text{otherwise} \end{cases}$$

Now suppose that X is modeled as Poisson with mean $\lambda = 3$ — i.e.,

$$f_X(x) = \begin{cases} \dfrac{e^{-3}3^x}{x!} & \text{for } x = 0, 1, 2, 3, \dots \\ 0 & \text{otherwise} \end{cases}$$

Multiplication of the two formulas gives a joint probability function for X and Y.

(b) Find the (marginal) probability that $Y = 0$. (Note that this is obtained by summing $f(x, 0)$ over all possible values of x.)

(c) Find $f_Y(y)$ in general. What (marginal) distribution does Y have?

5-63. Suppose that cans to be filled with a liquid are circular cylinders. The radii of these cans have mean $\mu_r = 1.00$ in. and standard deviation $\sigma_r = .02$ in. The volumes of liquid dispensed into these cans have mean $\mu_v = 15.10$ in.^{3} and standard deviation $\sigma_v = .05$ in.^{3}.

(a) If the volumes dispensed into the cans are approximately normally distributed, about what fraction will exceed 15.07 in.^{3}?

(b) Approximate the probability that the total volume dispensed into the next 100 cans exceeds 1510.5 in.^{3} (If the total exceeds 1510.5, $\bar{X}$ exceeds 15.105.)

(c) Approximate the mean μ_h and standard deviation σ_h of the heights of the liquid in the filled cans. (Recall that the volume of a circular cylinder is $v = \pi r^2 h$, where h is the height of the cylinder.)

(d) Does the variation in bottle radius or the variation in volume of liquid dispensed into the bottles have the biggest impact on the variation in liquid height? Explain.

5-64. Suppose that a pair of random variables have the joint probability density

$$f(x, y) = \begin{cases} \exp(x - y) & \text{if } 0 \le x \le 1 \text{ and } x \le y \\ 0 & \text{otherwise} \end{cases}$$

(a) Evaluate $P[Y \le 1.5]$.

(b) Find the marginal probability densities for X and Y.

(c) Are X and Y independent? Explain.

(d) Find the conditional probability density for Y given $X = .25$, $f_{Y|X}(y \mid .25)$. Given that $X = .25$, what is the mean of Y? (Hint: use $f_{Y|X}(y \mid .25)$.)

5-65. (Defects per Unit Acceptance Sampling) Suppose that in the inspection of an incoming product, nonconformities on an inspection unit are counted. If too many are seen, the incoming lot is rejected and returned to the manufacturer. (For concreteness, you might think of blemishes on rolled paper or wire, where an inspection unit consists of a certain length of material from the roll.) Suppose further that the number of nonconformities on a piece of product of any particular size can be modeled as Poisson with an appropriate mean.

(a) Suppose that this rule is followed: "Accept the lot if on a standard size inspection unit 1 or less nonconformities are seen." The *operating characteristic curve* of this acceptance sampling plan is a plot of the probability that the lot is accepted as a function of λ = the mean defects per inspection unit. (For X = the number of nonconformities seen, X has Poisson distribution with mean λ and $OC(\lambda) = P[X \leq 1]$.) Make a plot of the operating characteristic curve.

(b) Suppose that instead of the rule in (a), this rule is followed: "Accept the lot if on 2 standard size inspection units 2 or less total nonconformities are seen." Make a plot of the operating characteristic curve for this second plan and compare it with the plot from part (a). (Note that here, for X = the total number of nonconformities seen, X has a Poisson distribution with mean 2λ and $OC(\lambda) = P[X \leq 2]$.)

5-66. A data set in the text *Probability Models and Applications* by Olkin, Gleser, and Derman gives the ratios (by weight) of copper content to the sum of copper and lead contents for ore samples taken from the Transvaal vein of level 10 of the Frisco mine. It suggests that a beta distribution with parameters $\alpha = 1.13$ and $\beta = 3.00$ might be useful as a model of such ratios.

(a) Sketch the beta probability density with parameters $\alpha = 1.13$ and $\beta = 3.00$.

Henceforth use the distribution from (a) as a description of

$$X = \text{the next Cu to Cu} + \text{Pb ratio for an ore}$$
$$\text{sample from this mine}$$

(b) What are the mean and standard deviation of X?

(c) Evaluate $P[X \leq .05]$.

(d) Find the median of the distribution for X. How does it compare to EX?

(e) Use your answer to (c) and evaluate the mean number of ore samples that would need to be analyzed in order to find one with a Cu to Cu + Pb ratio of .05 or less.

5-67. A discrete random variable X can be described using the following probability function:

x	1	2	3	4	5
$f(x)$	.61	.24	.10	.04	.01

(a) Make a probability histogram for X. Also plot $F(x)$, the cumulative probability function for X.

(b) Find the mean and standard deviation for the random variable X.

(c) Evaluate $P[X \geq 3]$ and then find $P[X < 3]$.

5-68. A classical data set of Rutherford and Geiger (referred to in Example 5-6) suggests that for a particular experimental setup involving a small bar of polonium, the number of collisions of α particles with a small screen placed near the bar during an 8-minute period can be modeled as a Poisson variable with mean

$\lambda = 3.87$. Consider an experimental setup of this type, and let X and Y be (respectively) the numbers of collisions in the next two 8-minute periods. Evaluate the following.

(a) $P[X \geq 2]$

(b) $\sqrt{Var\,X}$

(c) $P[X + Y = 6]$

(d) $P[X + Y \geq 3]$

(*Hint for parts (c) and (d):* What is a sensible probability distribution for $X + Y$, the number of collisions in a 16-minute period?)

5-69. Suppose that an eddy current nondestructive evaluation technique for identifying cracks in critical metal parts has a probability of around .20 of detecting a single crack of length .003 in. in a certain material. Suppose further that $n = 8$ specimens of this material, each containing a single crack of length .003 in. are submitted for inspection using this technique. Let W be the number of these cracks that are detected. Use an appropriate probability model and evaluate the following.

(a) $P[W = 3]$

(b) $P[W \leq 2]$

(c) EW

(d) $Var\,W$

(e) the standard deviation of W

5-70. In the scenario described in Exercise 5-69, suppose that a series of specimens of the material, each containing a single crack of length .003 in. are submitted for inspection using the eddy current inspection technique. Let Y be the number of specimens inspected in order to obtain the first crack detection. Use an appropriate probability model and evaluate all of the following.

(a) $P[Y = 5]$

(b) $P[Y \leq 4]$

(c) EY

(d) $Var\,Y$

(e) the standard deviation of Y

5-71. Suppose that X is a continuous random variable with probability density of the form

$$f(x) = \begin{cases} k\left(x^2(1-x)\right) & \text{for } 0 < x < 1 \\ 0 & \text{otherwise} \end{cases}$$

(a) Evaluate k and sketch a graph of $f(x)$.

(b) Evaluate $P[X \leq .25]$, $P[X \leq .75]$, $P[.25 < X \leq .75]$, and $P[|X - .5| > .1]$.

(c) Compute EX and $\sqrt{Var\,X}$.

(d) Compute and graph $F(x)$, the cumulative distribution function for X. Read from your graph the .6 quantile of the distribution of X.

5-72. Suppose that engineering specifications on the shelf depth of a certain slug to be turned on a CNC lathe are from .0275 in. to .0278 in. and that values of this dimension produced on the lathe can be described using a normal distribution with mean μ and standard deviation σ.

(a) If $\mu = .0276$ and $\sigma = .0001$, about what fraction of shelf depths are in specifications?

(b) What machine precision (as measured by σ) would be required in order to produce about 98% of shelf depths within engineering specifications (assuming that μ is at the midpoint of the specifications)?

5-73. The resistance of an assembly of several resistors connected in series is the sum of the resistances of the individual resistors. Suppose that a large lot of nominal 10 Ω resistors has mean resistance $\mu = 9.91$ Ω and standard deviation of resistances $\sigma = .08$ Ω. Suppose that 30 resistors are randomly selected from this lot and connected in series.

(a) Find a plausible mean and variance for the resistance of the assembly.

(b) Evaluate the probability that the resistance of the assembly exceeds 298.2 Ω. (*Hint:* If $\bar{X}$ is the mean resistance of the 30 resistors involved, the resistance of the assembly exceeding 298.2 Ω is the same as $\bar{X}$ exceeding 9.94 Ω. Now apply the Central Limit Theorem.)

5-74. At a small metal fabrication company, steel rods of a particular type cut to length have lengths with standard deviation .005 in.

(a) If lengths are normally distributed about a mean μ (which can be changed by altering the setup of a jig) and specifications on this length are 33.69 in. $\pm$ .01 in., what appears to be the best possible fraction of the lengths in specifications? What does μ need to be in order to achieve this fraction?

(b) Suppose now that in an effort to determine the mean length produced using the current setup of the jig, a sample of rods is to be taken and their lengths measured, with the intention of using the value of $\bar{X}$ as an estimate of μ. Approximate the probabilities that $\bar{X}$ is within .0005 in. of μ for samples of sizes $n = 25$, 100, and 400. Do your calculations for this part of the question depend for their validity on the length distribution being normal? Explain.

5-75. Suppose that the measurement of the diameters of #10 machine screws produced on a particular machine yields values that are normally distributed with mean μ and standard deviation $\sigma = .03$ mm.

(a) If $\mu = 4.68$ mm, about what fraction of all measured diameters will fall in the range from 4.65 mm to 4.70 mm?

(b) Use your value from (a) and an appropriate discrete probability distribution to evaluate the probability (assuming $\mu = 4.68$) that among the next five measurements made, exactly four will fall in the range from 4.65 mm to 4.70 mm.

(c) Use your value from (a) and an appropriate discrete probability distribution to evaluate the probability (assuming that $\mu = 4.68$) that if one begins sampling and measuring these screws, the first diameter in the range from 4.65 mm to 4.70 mm will be found on the second, third, or fourth screw measured.

(d) Now suppose that μ is unknown but is to be estimated by $\bar{X}$ obtained from measuring a sample of $n = 25$ screws. Evaluate the probability that the sample mean, $\bar{X}$, takes a value within .01 mm of the long-run (population) mean μ.

(e) What sample size, n, would be required in order to *a priori* be 90% sure that $\bar{X}$ from n measurements will fall within .005 mm of μ?

Introduction to Formal Statistical Inference

6

F ormal statistical inference involves using the tool of probability theory to quantify the reliability of some kinds of data-based conclusions. This chapter introduces the logic involved in several general types of formal statistical inference. Then the most commonly used specific methods that it produces for one- and two-sample statistical studies are discussed.

The chapter begins with an introduction to the topic of confidence interval estimation, using the important case of large-sample inference for a mean as the focus of discussion. Then the topic of significance testing is considered, again using the case of large-sample inference for a mean as the context of discussion. With the general notions in hand, successive sections treat the standard one- and two-sample confidence interval and significance-testing methods for means, then variances, and then proportions. Finally, the important notions of tolerance and prediction intervals are introduced.

6-1 Large-Sample Confidence Intervals for a Mean

Many of the important engineering applications of statistics fit the following archetypal mold. Appropriate values for one or more parameters of a data-generating process are unknown. Based on a set of data, one wishes to identify an interval of values that is likely to contain an unknown parameter (or perhaps a function of one or more parameters) and quantify in some sense "how likely" the interval is to cover the correct value.

For example, a physically stable piece of equipment that dispenses baby food into jars might have an unknown associated mean fill level, μ. Based on some data, one might desire an interval likely to contain μ and an evaluation of the reliability of the interval. Or a machine that puts threads on automotive U-bolts might have an inherent variation in thread lengths, describable in terms of a standard deviation σ. The point of data collection might then be to produce an interval of likely values for σ, together with a statement of how reliable the interval is. Or two different methods of running a pelletizing machine might possess different unknown propensities to produce defective pellets, (say, p_1 and p_2), and a data-based interval for $p_1 - p_2$, together with an associated statement of reliability, might be needed.

The type of formal statistical inference designed to deal with such problems is called *confidence interval estimation*.

DEFINITION 6-1

A **confidence interval** for a parameter (or function of one or more parameters) is a data-based interval of numbers thought likely to contain the parameter (or function of one or more parameters) possessing a stated probability-based *confidence* or reliability.

This section discusses how basic probability facts lead to simple large-sample-size formulas for confidence intervals for a mean μ. The somewhat unusual case where the corresponding standard deviation σ is known is treated first. Then parallel reasoning produces a formula usable in the much more common situation where σ is not known. The section closes with discussions of three issues related to the application of confidence intervals: the modifications necessary to go from *two-sided* confidence intervals to (sometimes more appropriate) *one-sided* intervals; the proper interpretation of the reliability figure attached to a confidence interval; and lastly, some crude sample-size considerations.

A Large-n Confidence
Interval for μ *Involving* σ

The final example in Section 5-5 involved a scenario where a physically stable filling process was known to have an associated standard deviation of $\sigma = 1.6$ g net weight. Since, for large n, the sample mean of iid random variables is approximately normal, it was seen in Example 5-29 that for $n = 47$ and

$$\bar{x} = \text{the sample mean net fill weight of 47 jars filled by the process (g)}$$

there is an approximately 80% chance that $\bar{x}$ is within .3 gram of μ. This fact is pictured again in Figure 6-1.

FIGURE 6-1

Approximate Probability
Distribution for $\bar{x}$ Based on
$n = 47$

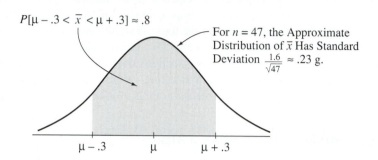

$P[\mu - .3 < \bar{x} < \mu + .3] \approx .8$

For $n = 47$, the Approximate Distribution of $\bar{x}$ Has Standard Deviation $\frac{1.6}{\sqrt{47}} \approx .23$ g.

$\mu - .3 \qquad \mu \qquad \mu + .3$

It is necessary to digress momentarily and discuss a matter of notation. In Chapter 5, capital letters were carefully used as symbols for random variables and corresponding lowercase letters for their possible or observed values. But here a lowercase symbol, $\bar{x}$, has been used for the sample mean *random variable*. This is (for better or for worse) fairly standard statistical usage, and it is in keeping with the kind of convention used in Chapters 3 and 4. At this point, this text is going to abandon strict adherence to the capitalization convention introduced in Chapter 5. Many random variables (like individual values in a sample, sample means, sample variances, fitted factorial effects from samples with factorial structure, fitted regression coefficients, etc.) will be symbolized using lowercase letters (and the same symbols used for their observed values). The Chapter 5 capitalization convention is especially helpful as one is learning the basics of probability. But once those basics are mastered, it is common to abuse notation to some extent and expect a reader to determine from context whether it is a random variable or its observed value that is under discussion. This is the notational route that will be followed here.

The most common way of thinking about a graphic like Figure 6-1 is to conceive of the eventuality that indeed

$$\mu - .3 < \bar{x} < \mu + .3 \tag{6-1}$$

in terms of whether or not $\bar{x}$ falls in an interval of length $2(.3) = .6$ centered about μ. But for present purposes, it is useful to realize that an equivalent way of thinking is to consider whether or not an interval of length .6 centered about $\bar{x}$ falls on top of μ. Algebraically, inequality (6-1) is equivalent to

$$\bar{x} - .3 < \mu < \bar{x} + .3 \tag{6-2}$$

which tends to shift one's attention to this second way of thinking. And the fact that the equivalent expressions (6-1) and (6-2) have about an 80% chance of holding true anytime a sample of 47 fill weights is taken suggests that the random interval

$$(\bar{x} - .3, \bar{x} + .3) \tag{6-3}$$

might be used as an approximate confidence interval for μ, with 80% associated reliability or confidence.

EXAMPLE 6-1

A Confidence Interval for a Process Mean Fill Weight. Suppose that in the present context of determining the process mean fill weight, a sample of $n = 47$ jars produces a sample mean fill weight of $\bar{x} = 138.2$ g. Then expression (6-3) suggests that the interval with endpoints

$$138.2 \text{ g} \pm .3 \text{ g}$$

(i.e., the interval from 137.9 g to 138.5 g) be used as an 80% confidence interval for the process mean fill weight.

It is not hard to see how to generalize the logic that led to expression (6-3) to produce a large-sample confidence interval for a mean μ with any desired reliability for use anytime one knows the corresponding standard deviation σ. That is, anytime an iid model (stable process/random sampling from a single population) is appropriate for the elements of a large sample, the Central Limit Theorem implies that the sample mean $\bar{x}$ is approximately normal with mean μ and standard deviation $\sigma/\sqrt{n}$. Then, if for $p > .5$, z is the p quantile of the standard normal distribution, the probability that

$$\mu - z\frac{\sigma}{\sqrt{n}} < \bar{x} < \mu + z\frac{\sigma}{\sqrt{n}} \tag{6-4}$$

is approximately $1 - 2(1 - p)$. But inequality (6-4) can be rewritten as

$$\bar{x} - z\frac{\sigma}{\sqrt{n}} < \mu < \bar{x} + z\frac{\sigma}{\sqrt{n}} \tag{6-5}$$

and thought of as the eventuality that the random interval with endpoints

$$\bar{x} \pm z\frac{\sigma}{\sqrt{n}} \tag{6-6}$$

brackets the unknown μ. So an interval with the endpoints of expression (6-6) is an approximate confidence interval for μ. The associated confidence level is $(1 - 2(1 - \Phi(z))) \times 100\%$. Or equivalently, for a desired confidence, z should be chosen such that the standard normal probability between $-z$ and z corresponds to that confidence level.

Table 3-10 (of standard normal quantiles) or Table D-3 (of standard normal cumulative probabilities) can be used to verify the appropriateness of the entries in Table 6-1. This table gives values of z for use in expression (6-6), for some commonly used confidence levels.

TABLE 6-1

z's for Use in Two-sided Large-*n* Intervals for μ

Desired Confidence	z
80%	1.28
90%	1.645
95%	1.96
98%	2.33
99%	2.58

EXAMPLE 6-2

A Confidence Interval for the Mean Deviation from Nominal in a Grinding Operation. Dib, Smith, and Thompson studied a grinding process used in the rebuilding of automobile engines. They found that the natural short-term variability associated with the diameters of rod journals on engine crankshafts ground using the process was on the order of

$\sigma = .7 \times 10^{-4}$ in. Suppose that the rod journal grinding process can be thought of as physically reasonably stable over runs of, say, 50 journals or less. Then if 32 consecutive rod journal diameters have mean deviation from nominal of $\bar{x} = -.16 \times 10^{-4}$ in., it is possible to apply expression (6-6) to derive a confidence interval for the current process mean deviation from nominal. For the sake of illustration, consider making a 95% confidence interval. Consulting Table 6-1 (or otherwise, realizing that 1.96 is the .975 quantile of the standard normal distribution), it is clear that $z = 1.96$ is called for in formula (6-6). Thus, a 95% confidence interval for the current process mean deviation from nominal journal diameter has endpoints

$$-.16 \times 10^{-4} \pm (1.96)\frac{.7 \times 10^{-4}}{\sqrt{32}}$$

that is, endpoints

$$-.40 \times 10^{-4} \text{ in. and } .08 \times 10^{-4} \text{ in.}$$

An interval like this one could be of engineering importance in, for example, determining the advisability of making an adjustment to the process aim. The fact that the data-based interval above includes both positive and negative values indicates that although $\bar{x} < 0$, the information in hand doesn't provide enough precision even to tell with any certainty in which direction the grinding process should be adjusted. That is, it isn't clear whether one needs to take off more or to take off less metal in grinding. This, coupled with the fact that potential machine adjustments are probably much coarser than the best-guess misadjustment of $\bar{x} = -.16 \times 10^{-4}$ in., speaks strongly against making a change in the process aim based on the current data.

A Generally Applicable Large-n Confidence Interval for μ

Although expression (6-6) provides a mathematically correct confidence interval formula, the appearance of σ in the formula severely limits its practical usefulness. In engineering contexts, it is fairly rare that one needs to estimate a mean μ but knows (and therefore can plug into a formula) the corresponding σ. Such contexts consist, in this author's experience, primarily of manufacturing situations like those of Examples 6-1 and 6-2, where considerable past experience can be relied on to give a sensible value for σ, but where physical process drifts over time can put the current value of μ in question.

Happily enough, a small modification of the line of reasoning that led to expression (6-6) produces a confidence interval formula for μ that depends only on the characteristics of a sample, not on any characteristic of an underlying distribution. The argument leading to formula (6-6) depends on the fact that for large n, $\bar{x}$ is approximately normal with mean μ and standard deviation $\sigma/\sqrt{n}$ — i.e., that

$$\frac{\bar{x} - \mu}{\dfrac{\sigma}{\sqrt{n}}} \tag{6-7}$$

is approximately standard normal. The appearance of σ in expression (6-7) is what eventually leads to its unfortunate appearance in the confidence interval formula (6-6). As it turns out, however, mathematicians can prove a slight generalization of the Central Limit Theorem, guaranteeing that for large n,

$$\frac{\bar{x} - \mu}{\dfrac{s}{\sqrt{n}}} \tag{6-8}$$

is also approximately standard normal. And expression (6-8) doesn't involve σ.

Beginning with the fact that (when an iid model for observations is appropriate and n is large), expression (6-8) is approximately standard normal, one can reason much as before. For a positive z, the eventuality that

$$-z < \frac{\bar{x} - \mu}{\frac{s}{\sqrt{n}}} < z$$

is equivalent to

$$\mu - z\frac{s}{\sqrt{n}} < \bar{x} < \mu + z\frac{s}{\sqrt{n}}$$

which in turn is equivalent to

$$\bar{x} - z\frac{s}{\sqrt{n}} < \mu < \bar{x} + z\frac{s}{\sqrt{n}}$$

Thus, the interval with random center $\bar{x}$ and random length $2zs/\sqrt{n}$ — i.e., with random endpoints

$$\bar{x} \pm z\frac{s}{\sqrt{n}} \qquad\qquad (6\text{-}9)$$

can be used as an approximate confidence interval for μ that doesn't require σ as input. Of course as before, for a desired confidence, z should be chosen such that the standard normal probability between $-z$ and z corresponds to that confidence level.

EXAMPLE 6-3

Breakaway Torques and Hard Disk Failures. F. Willett, in the article "The Case of the Derailed Disk Drives" (*Mechanical Engineering*, 1988), discusses a statistical engineering study done to isolate the cause of "blink code A failure" in a model of Winchester hard disk drive. Included in that article are the data given in Figure 6-2. These are breakaway torques (units are inch ounces) required to loosen the drive's interrupter flag on the stepper motor shaft, for 26 disk drives returned to the manufacturer for blink code A failure. It is possible to verify that for these data, $\bar{x} = 11.5$ in. oz and $s = 5.1$ in. oz.

FIGURE 6-2

Torques Required to Loosen 26 Interrupter Flags

```
0 | 0 2 3
0 | 7 8 8 9 9
1 | 0 0 0 1 1 2 2 2 3
1 | 5 5 6 6 7 7 7 9
2 | 0
2 |
```

If the disk drives that produced the data in Figure 6-2 are thought of as representing the (much larger) population of drives subject to blink code A failure, it seems reasonable to use an iid model and formula (6-9) in the estimation of the population mean breakaway torque. Choosing, for example, to make a 90% confidence interval for μ, $z = 1.645$ is indicated in Table 6-1. And using formula (6-9), an interval with endpoints

$$11.5 \pm 1.645\frac{5.1}{\sqrt{26}}$$

(i.e., endpoints 9.9 in. oz and 13.1 in. oz) is indicated.

The interval derived here shows clearly that the mean breakaway torque for drives with blink code A failure was substantially below the factory's 33.5 in. oz target value. Interestingly enough, recognition of this phenomenon turned out to be key in finding and eliminating a design flaw in the disk drives.

Some Additional Comments Concerning Confidence Intervals

Formulas (6-6) and (6-9) have to this point been used to make confidence statements of the type "μ is between a and b." But often a statement like "μ is at least c" or "μ is no more than d" would be of more practical engineering use. For example, an automotive engineer might wish to state, "The mean NO emission for this engine is no more than 5 ppm," while a civil engineer might want to make a statement like "The mean compressive strength for specimens of this type of concrete is at least 4188 psi." That is, practical engineering problems are sometimes best addressed using one-sided confidence intervals rather than two-sided ones.

There is no real problem in coming up with formulas for one-sided confidence intervals. If one has a workable two-sided formula, all that must be done is to replace the lower limit with $-\infty$ or the upper limit with $+\infty$ (making the interval bigger and thus more likely to fall on top of the quantity being estimated) and adjust the stated confidence level appropriately upward. This usually means reducing the "unconfidence level" by a factor of 2. This prescription works not only with formulas (6-6) and (6-9) but also with the rest of the two-sided confidence interval formulas introduced in this chapter.

EXAMPLE 6-3
(continued)

In the context of the mean breakaway torque for defective disk drives, consider making a one-sided 90% confidence interval for μ of the form $(-\infty, \#)$, for # an appropriate number. Put slightly differently, consider finding a 90% *upper confidence bound* for μ, (say, #).

According to the discussion preceding this example, beginning with a two-sided 80% confidence interval for μ, one may replace the lower limit with $-\infty$ and arrive at a one-sided 90% confidence interval for μ. That is, using formula (6-9), a 90% upper confidence bound for the mean breakaway torque is

$$\bar{x} + 1.28\frac{s}{\sqrt{n}} = 11.5 + 1.28\frac{5.1}{\sqrt{26}} = 12.8 \text{ in. oz}$$

Equivalently, a 90% one-sided confidence interval for μ is $(-\infty, 12.8)$.

Notice that the 12.8 in. oz figure here is less than (and thus closer to the 11.5 in. oz sample mean than) the 13.1 in. oz upper limit from the 90% two-sided interval found earlier. In the one-sided case, one is essentially declaring $-\infty$ as a lower limit and thus facing no risk of producing an interval containing only numbers larger than the unknown μ. One can thus afford (for a given confidence level) to use an upper limit smaller than that for a corresponding two-sided interval.

As a second issue in the application of the idea of a confidence interval, note that there is little value (and perhaps substantial danger) in creating numerical intervals with the impressive title "confidence intervals" unless one has a correct understanding of the technical meaning of the term *confidence*. And unfortunately, misunderstandings about the meaning abound. Textbooks are full of them, and (when left to their own devices) students inevitably develop their own. So it is worth some ink to carefully lay out what *confidence* does and doesn't mean.

Prior to selecting a sample and plugging sample values into a formula like (6-6) or (6-9), the meaning of a confidence level is reasonably obvious. Choosing, for example, a 90% confidence level and thus $z = 1.645$ for application in formula (6-9), one can think that before the fact of sample selection and calculation, "there is about a 90% chance of winding up with an interval that brackets μ." In symbols, this might be expressed as

$$P\left[\bar{x} - 1.645\frac{s}{\sqrt{n}} < \mu < \bar{x} + 1.645\frac{s}{\sqrt{n}}\right] \approx .90$$

But how should one think about a confidence level *after* the fact of sample selection? This is entirely another matter. Once numbers have been plugged into a formula like (6-6) or (6-9), the die has (so to speak) already been cast, and one has a numerical interval that is either right or wrong. The practical difficulty is that one doesn't know which is the case. But it no longer makes logical sense to attach a probability to the eventuality of correctness of the interval. For example, it would make no sense to look again at the two-sided interval found in Example 6-3 and try to say something like "there is a 90% probability that μ is between 9.9 in. oz and 13.1 in. oz." μ is not a random variable, but rather a fixed (albeit unknown) quantity, that either is or is not between 9.9 and 13.1. There is no probability left in the situation to be discussed.

So where does that leave one? What does it mean that (9.9, 13.1) is a 90% confidence interval for μ? Like it or not, one is left with the realization that the phrase "90% confidence" refers more to the method used to obtain the interval (9.9, 13.1) than to the interval itself. In coming up with the interval, the best methodology available has been used — one that would produce numerical intervals bracketing μ in about 90% of repeated applications. But the effectiveness of the particular interval in this application is unknown, and it is not quantifiable in terms of a probability. A person who (in the course of a lifetime) makes many 90% confidence intervals can expect to have a "lifetime batting average" of about 90%. But the effectiveness of any particular application (at least at the time it is made) will be unknown.

This author has found it useful to give students a short statement summarizing this discussion as "the authorized interpretation of confidence." That statement is included here as a definition.

DEFINITION 6-2	(Interpretation of a Confidence Interval) To say that a numerical interval (a, b) is (e.g.) a 90% confidence interval for a parameter is to say that in obtaining it, one has applied methods of data collection and calculation that would produce intervals bracketing the parameter in about 90% of repeated applications. Whether or not the particular interval (a, b) brackets the parameter is unknown and not describable in terms of a probability.

At this point, the reader may feel somewhat disillusioned, thinking that the statement in Definition 6-2 is a rather weak meaning to attach to the reliability figure associated with a confidence interval. This author has some sympathy with that feeling. Nevertheless, the statement in Definition 6-2 is the correct interpretation and is all that can be rationally expected. Despite the fact that the correct interpretation may be somewhat unappealing, confidence interval methods have proved themselves to be of great practical use.

As a final consideration in this introduction to confidence interval estimation, note that formulas like (6-6) and (6-9) can give some crude quantitative guidance in answering the inevitable question, "How big must n be?" In formula (6-9), for example, if one has in mind 1) a desired confidence level, 2) a worst-case expectation for the sample standard deviation, and 3) a desired precision (in terms of the size of the plus-or-minus part of the formula (6-9)) of estimation of μ, it is a simple matter to solve for a corresponding sample size.

EXAMPLE 6-3
(continued)

Suppose that in the problem of blink code A failure in disk drives, engineers plan to follow up the analysis of the data in Figure 6-2 with the breakaway torque testing of a number of new drives. This will be done after subjecting them to accelerated (high) temperature conditions, in an effort to understand the mechanism behind the creation of low breakaway torques. Further suppose that the mean breakaway torque for temperature-stressed drives is to be estimated with a two-sided 95% confidence interval and that the torque variability expected in the new temperature-stressed drives is no worse than the $s = 5.1$ in. oz figure measured for the drives returned for blink code A failure. Consider the possibility that a ± 1 in. oz precision of estimation is desired. Then using the plus-or-minus part of formula (6-9) and remembering Table 6-1, one is led to set

$$1 = 1.96 \frac{5.1}{\sqrt{n}}$$

which, when solved for n, gives

$$n \approx 100$$

A study involving in the neighborhood of $n = 100$ temperature-stressed new disk drives is indicated by these calculations. If this figure is impractical because of the cost involved, the calculations at least indicate that dropping below this sample size will (unless the variability associated with the stressed new drives is less than that of the returned drives) force a reduction in either the confidence or the precision associated with the final interval.

For two reasons, the kind of calculations in the previous example give somewhat less than an ironclad answer to the question of sample size. The first is that they are only as good as one's prediction of the sample standard deviation, s. If one underpredicts s, an n that is not really large enough will result from this kind of calculations. (And by the same token, if one is excessively conservative and overpredicts s, an unnecessarily large sample size will result.) The second issue is that expression (6-9) remains a large-sample formula. If calculations like those above produce n much smaller than, say, 25 or 30, the value should be increased enough to guarantee that formula (6-9) can be applied. However, in spite of these two caveats, the method of this last example is often extremely helpful in providing at least crude quantitative guidance in determining sample size.

6-2 Large-Sample Significance Tests for a Mean

The last section illustrated how the tool of probability can lead to a type of formal statistical inference called confidence interval estimation. This section makes a parallel introduction of a second type of formal statistical inference: significance testing.

Significance testing can be thought of as using data to quantitatively assess the plausibility of a trial value of a parameter (or function of one or more parameters). This trial value is typically hypothesized to embody a status quo/"pre-data" view of a situation. For example, a process engineer might employ significance testing to assess the plausibility of a target value of 138 g as the current process mean fill level in the dispensing of baby food into jars. In determining whether a machine putting threads on automotive U-bolts is in need of overhaul because of excessive variation in output thread lengths, one might employ significance testing to assess the plausibility of a standard value (say, $\sigma = .5 \times 10^{-4}$ in.) as the standard deviation of thread lengths currently applicable to the machine. Or two different methods of running a pelletizing machine might have unknown propensities to produce defective pellets, (say, p_1 and p_2), and significance testing could be used to assess the plausibility of $p_1 - p_2 = 0$ — i.e., that the two methods are equally effective.

This section describes how basic probability facts lead to simple large-sample-size significance tests for a mean μ. It considers first the somewhat unusual case where the corresponding standard deviation σ is known; in this context, standard significance testing terminology is introduced. Next, a five-step format for summarizing the thought process followed in a significance test is introduced and applied. Then the more common situation of significance testing for μ where σ is not known is considered. The section closes with two discussions about significance-testing logic. One considers its use in decision-making contexts. The second considers the relationship between statistical significance and practical importance, and the comparative importance of significance-testing and confidence interval methods.

Large-n Significance Tests for μ Involving σ

Recalling once more the situation of Example 5-29, consider a scenario where a physically stable filling process is known to have an associated standard deviation of $\sigma = 1.6$ g net weight. Suppose further that with a declared (label) weight of 135 g, process engineers have set a target mean net fill weight at $135 + 3\sigma = 139.8$ g. Finally, suppose that in a routine check of filling process performance, intended to detect any change of the process mean from its target value, a sample of $n = 25$ jars produces a sample mean fill weight of 139.0 g. What does this value of $\bar{x}$ have to say about the plausibility of the current process mean actually being at the target mean of 139.8 g?

The Central Limit Theorem can be called on here. If indeed the current process mean is at the target of 139.8 g, $\bar{x}$ has an approximately normal distribution with mean 139.8 g and standard deviation $\sigma/\sqrt{n} = 1.6/\sqrt{25} = .32$ g, as pictured in Figure 6-3 along with the observed value of $\bar{x} = 139.0$ g.

FIGURE 6-3

Approximate Probability Distribution for $\bar{x}$ if $\mu = 139.8$ and the Observed Value of $\bar{x} = 139.0$

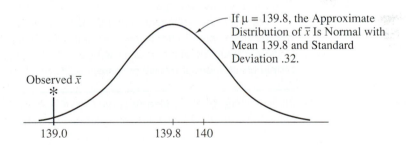

If $\mu = 139.8$, the Approximate Distribution of $\bar{x}$ Is Normal with Mean 139.8 and Standard Deviation .32.

Observed $\bar{x}$

139.0   139.8   140

Figure 6-4 shows the standard normal picture that corresponds to Figure 6-3. It is based on the fact that if the current process mean is on target at 139.8 g, then the fact that $\bar{x}$ is approximately normal with mean μ and standard deviation $\sigma/\sqrt{n} = .32$ g means that

$$Z = \frac{\bar{x} - 139.8}{\dfrac{\sigma}{\sqrt{n}}} = \frac{\bar{x} - 139.8}{.32} \tag{6-10}$$

is approximately standard normal. The observed $\bar{x} = 139.0$ g in Figure 6-3 has corresponding observed $z = -2.5$ in Figure 6-4.

It is obvious from either Figure 6-3 or Figure 6-4 that if indeed the current process mean is on target at 139.8 g and thus that the figures are correct, a fairly extreme/rare

FIGURE 6-4

The Standard Normal Picture
Corresponding to Figure 6-3

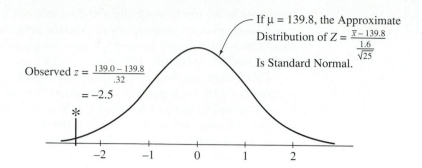

$\bar{x}$, or equivalently z, has been observed. Of course, extreme/rare things occasionally happen, but the nature of the observed $\bar{x}$ (or z) might well instead be considered to make implausible the possibility that the process is on target (the pictures are correct).

In fact, the figures even suggest a way of quantifying their own implausibility: namely, the calculation of a probability associated with values of $\bar{x}$ (or Z) at least as extreme as the one actually observed. Now the phrase "at least as extreme" has to be defined in relation to the original purpose of data collection — to detect either a decrease of μ below target or an increase above target. But with this understanding, not only are values $\bar{x} \leq 139.0$ g ($z \leq -2.5$) as extreme as that observed, but so also are values $\bar{x} \geq 140.6$ g ($z \geq 2.5$). (The first kind of $\bar{x}$ or z suggests a decrease in μ, and the second suggests an increase.) That is, one might quantify the implausibility of the process being on target by noting that if indeed this were so, only a fraction

$$\Phi(-2.5) + \left(1 - \Phi(2.5)\right) = .01$$

of all samples would produce a value of $\bar{x}$ (or Z) as extreme as the one actually observed. Put in those terms, the data seem to speak rather convincingly against the possibility of the process being on target.

The kind of argument that has just been made is an application of typical significance-testing logic. In order to make the pattern of thought obvious, it is useful to isolate some elements of it in definition form. This is done next, beginning with a formal restatement of overall purpose.

DEFINITION 6-3

Statistical **significance testing** is the use of data in the quantitative assessment of the plausibility of some trial value for a parameter (or function of one or more parameters).

Logically, significance testing begins with the specification of some trial or hypothesized value. Special jargon and notation exist for the statement of this value.

DEFINITION 6-4

A **null hypothesis** is a statement of the form

$$\text{Parameter} = \#$$

or

$$\text{Function of parameters} = \#$$

(for some number, #) that forms the basis of investigation in a significance test. A null hypothesis is usually formed to embody a status quo/"pre-data" view of the parameter (or function of the parameter(s)); it is typically symbolized as H_0.

The notion of a null hypothesis is so central to the logic of significance testing that it is common to use the term *hypothesis testing* in place of *significance testing*. The "null" part of the phrase "null hypothesis" refers to the fact that null hypotheses are statements of no difference, or equality. For example, in the context of the filling operation, standard usage would be to write

$$H_0: \mu = 139.8 \tag{6-11}$$

meaning that the null hypothesis is one of no difference between μ and the target value of 139.8 g.

After formulating a null hypothesis, one must specify what kinds of departures from it are going to be of interest.

DEFINITION 6-5

An **alternative hypothesis** is a statement that stands in opposition to the null hypothesis; it specifies what forms of departure from the null hypothesis are of potential concern. An alternative hypothesis is typically symbolized as H_a. It is of the same form as the corresponding null hypothesis, except that the equality sign is replaced by $\neq$, $>$, or $<$.

In many applications, the alternative hypothesis embodies an investigator's suspicions and/or hopes about the true state of affairs, amounting to a kind of *research hypothesis* that the investigator hopes to establish. For example, if an engineer develops what is intended to be a device for improving automotive gas mileage, in testing it, a null hypothesis expressing "no mileage change" and an alternative hypothesis expressing "mileage improvement" would be appropriate.

In the present example of the filling operation, there is no desire on the part of process engineers that μ be other than 139.8 g, but there is a need to detect both the possibility of consistently underfilled ($\mu < 139.8$ g) and the possibility of consistently overfilled ($\mu > 139.8$ g) jars. Thus, an appropriate alternative hypothesis is

$$H_a: \mu \neq 139.8 \tag{6-12}$$

Once null and alternative hypotheses have been established, it is necessary to lay out carefully how the data will be used to evaluate the plausibility of the null hypothesis. This involves specifying a statistic to be calculated, a probability distribution appropriate for it if in fact the null hypothesis is true, and what kinds of observed values will be considered to make the null hypothesis implausible.

DEFINITION 6-6

A **test statistic** is the particular form of numerical data summarization to be used in a significance test. The formula for the test statistic typically involves the number appearing in the null hypothesis.

DEFINITION 6-7

A **reference (or null) distribution** for a test statistic is the probability distribution describing the test statistic, provided the null hypothesis is in fact true.

The type of values of the test statistic considered to cast doubt on the validity of the null hypothesis is specified after looking at the form of the alternative hypothesis. Roughly speaking, one identifies values of the test statistic that are more likely to occur if the alternative hypothesis is true than if the null hypothesis holds.

The discussion of the filling process scenario has vacillated between using $\bar{x}$ and its standardized version Z given in equation (6-10) for a test statistic. Equation (6-10) gives a specialization of the general (large-n, known σ) test statistic for μ,

$$Z = \frac{\bar{x} - \#}{\frac{\sigma}{\sqrt{n}}} \tag{6-13}$$

to the present scenario, where the hypothesized value of μ is 139.8, $n = 25$, and $\sigma = 1.6$. In the long run, it is actually most convenient to think of the test statistic for this kind of problem in the standardized form shown in equation (6-13) rather than as $\bar{x}$ itself. The reason that equation (6-13) is most convenient is that for any problem like the filling example, the reference distribution will be the same — namely, standard normal.

Continuing with the filling example, note that if instead of the null hypothesis (6-11), the alternative hypothesis (6-12) is operating, observed $\bar{x}$'s much larger or much smaller than 139.8 will tend to result. Such $\bar{x}$'s will then, via equation (6-13), translate respectively to large or small (that is, large negative numbers in this case) observed values of Z — i.e., large values of $|z|$. It is such observed values that should be considered to render the null hypothesis implausible.

Having specified how data will be used to judge the plausibility of the null hypothesis, it remains to collect appropriate data, plug them into the formula for the test statistic, and (using the calculated value and the reference distribution), arrive at a quantitative assessment of the plausibility of H_0. There is standard statistical jargon for the form this will take.

DEFINITION 6-8

The **observed level of significance** or **p-value** in a significance test is the probability that the reference distribution assigns to the set of possible values of the test statistic that are at least as extreme as the one actually observed (in terms of casting doubt on the validity of the null hypothesis).

The smaller the observed level of significance, the stronger the evidence against the validity of the null hypothesis. In the context of the filling operation, with an observed value of the test statistic of

$$z = -2.5$$

the p-value or observed level of significance is

$$\Phi(-2.5) + (1 - \Phi(2.5)) = .01$$

which gives fairly strong evidence against the possibility that the process mean is on target.

A Five-Step Format for Summarizing Significance Tests

It is pedagogically helpful to lay down a step-by-step format for organizing write-ups of significance tests, for several reasons. First, it emphasizes the common logical structure shared by all significance tests, making the methods easier to understand as a group than they would be one at a time. For another, the format emphasizes the logical progression of thought in any one application, making it as easy as possible for a user to spot inconsistencies or other errors that might otherwise go undetected.

The significance testing format that will be used in this text includes the following five steps.

Step 1 State the null hypothesis.

Step 2 State the alternative hypothesis.

Step 3 State the test criteria. That is, give the formula for the test statistic (plugging in only a hypothesized value from the null hypothesis, but not any sample information) and the reference distribution. Then state in general terms what observed values of the test statistic will constitute evidence against the validity of the null hypothesis.

Step 4   Show the sample-based calculations.

Step 5   Report an observed level of significance and (to the extent possible) state its implications in the context of the real engineering problem.

EXAMPLE 6-4

A Significance Test Regarding a Process Mean Fill Level. Consider the use of the five-step significance-testing format in summarizing the preceding discussion of the filling process. The essential points are as follows:

1. $H_0: \mu = 139.8$.
2. $H_a: \mu \neq 139.8$.
3. The test statistic is

$$Z = \frac{\bar{x} - 139.8}{\dfrac{\sigma}{\sqrt{n}}}$$

The reference distribution is standard normal, and large observed values $|z|$ will constitute evidence against the validity of H_0.

4. The sample gives

$$z = \frac{139.0 - 139.8}{\dfrac{1.6}{\sqrt{100}}} = -2.5$$

5. The observed level of significance is

$P[\text{a standard normal variable} \leq -2.5] + P[\text{a standard normal variable} \geq 2.5]$

$\qquad = P\left[|\text{a standard normal variable}| \geq 2.5\right]$

$\qquad = .01$

This is reasonably strong evidence that in fact the process mean fill level is not on target.

Generally Applicable Large-n Significance Tests for μ

The significance-testing method used to carry the discussion thus far is easy to discuss and understand, but of limited practical use. The problem with it is that, as introduced, the test statistic (6-13) involves the parameter σ. As remarked in Section 6-1, there are few engineering contexts where one needs to make inferences regarding μ but knows the corresponding σ. They consist primarily of manufacturing situations where considerable past experience provides a value for the process short-term variability, but physical process drifts can put the current process mean in doubt. Happily, because of the same probability fact that made it possible to produce a large-sample confidence interval formula for μ free of σ, it is also possible to do large-n significance testing for μ without having to supply σ.

For observations that are describable as essentially equivalent to random selections with replacement from a single population with mean μ and variance σ^2, if n is large,

$$Z = \frac{\bar{x} - \mu}{\dfrac{s}{\sqrt{n}}}$$

is approximately standard normal. This means that for large n, to test

$$H_0: \mu = \#$$

a widely applicable method will simply be to use the logic already introduced, but with the test statistic

$$Z = \frac{\bar{x} - \#}{\frac{s}{\sqrt{n}}} \qquad\qquad \textbf{(6-14)}$$

in place of statistic (6-13).

EXAMPLE 6-5

(Example 6-3 revisited) **Significance Testing and Hard Disk Failures.** Consider again the problem of blink code A failure in Winchester hard disk drives. Breakaway torques set at the factory on the interrupter flag connection to the the stepper motor shaft averaged 33.5 in. oz, and there was suspicion that blink code A failure was associated with reduced breakaway torque. Recall that a sample of $n = 26$ failed drives had breakaway torques (given in Figure 6-2) producing $\bar{x} = 11.5$ in. oz and $s = 5.1$ in. oz.

Consider the situation of an engineer wishing to judge the extent to which the data in hand debunk the possibility that drives experiencing blink code A failure have mean breakaway torque equal to the factory-set mean value. The five-step significance-testing format can be used.

 1. $H_0: \mu = 33.5$.
 2. $H_a: \mu < 33.5$.

(Here the alternative hypothesis is directional, amounting to a research hypothesis embodying the engineer's suspicions about the relationship between drive failure and breakaway torque.)

 3. The test statistic is

$$Z = \frac{\bar{x} - 33.5}{\frac{s}{\sqrt{n}}}$$

The reference distribution is standard normal, and small observed values z will constitute evidence against the validity of H_0.

(Means less than 33.5 will tend to produce $\bar{x}$'s of the same nature and therefore small — i.e., large negative — z's.)

 4. The sample gives

$$z = \frac{11.5 - 33.5}{\frac{5.1}{\sqrt{26}}} = -22.0$$

 5. The observed level of significance is

$$P[\text{a standard normal variable} < -22.0] \approx 0$$

The sample gives overwhelming evidence that failed drives have mean breakaway torque below the factory-set level.

As a practical matter, it is important not to make too much of a logical jump here, to some kind of premature delusion that this work constitutes the complete solution to the real engineering problem. It's pretty clear that drives returned for blink A code failure have substandard breakaway torques. But in the absence of evidence to the contrary, it is possible that they are no different in that respect from non-failing drives currently in the field. And even if reduced breakaway torque is at fault, a real-world fix of the drive failure problem requires detective work leading to the identification and prevention of the physical mechanism producing reduced torque. This is not to say the significance test lacks importance, but rather to remind the reader that it is but one of many tools an engineer uses to do a job.

Significance Testing and Formal Statistical Decision Making

The basic logic introduced thus far in this section is sometimes applied in a decision-making context, where data are being counted on to provide guidance in choosing between two rival courses of action. In such cases, a decision-making framework is often built into the formal statistical analysis in an explicit way, and some additional terminology and patterns of thought are standard. These are introduced next.

In at least some decision-making contexts, it is possible to conceive of two different possible decisions or courses of action as being related to a null and an alternative hypothesis. For example, in the well-worn filling process scenario, H_0: $\mu = 139.8$ might correspond to the course of action "Leave the process alone," and H_a: $\mu \neq 139.8$ could correspond to the course of action "Adjust the process." When such a correspondence to two hypotheses holds, it is common to identify two different errors that are possible when approaching the decision-making process.

DEFINITION 6-9

When significance testing is used in a decision-making context, the possibility of deciding in favor of H_a when in fact H_0 is true is called a **type I error**.

DEFINITION 6-10

When significance testing is used in a decision-making context, the possibility of deciding in favor of H_0 when in fact H_a is true is called a **type II error**.

The content of these two definitions is conveniently represented in the 2 × 2 table pictured in Figure 6-5. In the filling process context, a type I error would amount to adjusting an on-target process, whereas a type II error would be failing to adjust an off-target process.

FIGURE 6-5

Four Potential Outcomes in a Decision Problem

The True State of Affairs Is Described by:		The Ultimate Decision Is in Favor of: H_0	The Ultimate Decision Is in Favor of: H_a
	H_0		Type I Error
	H_a	Type II Error	

The method by which significance-testing procedures are harnessed and used to actually come to a decision in favor of either H_0 or H_a is as follows. A critical value is chosen, and if the observed level of significance is smaller than the critical value (thus making the validity of the null hypothesis correspondingly implausible), one decides in favor of H_a. Otherwise, the course of action corresponding to H_0 is followed. A little reflection on this plan should make it obvious that the critical value for the observed level of significance ends up being the *a priori* probability the decision maker runs of deciding in favor of H_a, calculated supposing H_0 to be true. There is special terminology for this concept.

DEFINITION 6-11

When significance testing is used in a decision-making context, a critical value separating those large observed levels of significance for which H_0 will be accepted from those small observed levels of significance for which H_0 will be rejected in favor of H_a is called the **type I error probability** or the **significance level**. The symbol α is usually used to stand for the type I error probability.

It is standard practice (although perhaps not always rational) to use small numbers, like .1, .05, or even .01, for α. In a sense, this puts some inertia in favor of H_0 into the decision-making process. (Such a practice guarantees that *type I* errors won't be made very often. But at the same time, it creates an asymmetry in the treatment of H_0 and H_a that is not always justified.)

Definition 6-10 and Figure 6-5 make it clear that type I errors are not the only undesirable possibility to be faced when significance testing is used to support decision making. The possibility of type II errors must also be considered.

DEFINITION 6-12

When significance testing is used in a decision-making context, the probability — calculated supposing a particular parameter value described by H_a holds — that the observed level of significance is bigger than α (i.e., H_0 is not rejected) is called a **type II error probability**. The symbol β is usually used to stand for a type II error probability.

For most of the testing methods studied in this book, calculation of β's is more than the limited introduction to probability given in Chapter 5 will support. But the job can be handled for the simple known-σ situation that was used to introduce the topic of significance testing. And making a few such calculations will provide the reader some intuition consistent with what, qualitatively at least, holds in general.

EXAMPLE 6-4
(continued)

Again consider the filling process, this time supposing that significance testing based on $n = 25$ will be used tomorrow in a decision-making context to decide whether or not to adjust the process. For sake of illustration, type II error probabilities, calculated supposing $\mu = 139.5$ and $\mu = 139.2$ for tests using $\alpha = .05$ and $\alpha = .2$, will be compared.

First considering the situation with $\alpha = .05$, note that one will decide in favor of H_0 if the observed value of Z given in equation (6-10) (generalized in formula (6-13)) is such that

$$|z| < 1.96$$

i.e., if

$$139.8 - 1.96(.32) < \bar{x} < 139.8 + 1.96(.32)$$

i.e., if

$$139.2 < \bar{x} < 140.4 \qquad \text{(6-15)}$$

Now if a μ described by H_a given in display (6-12) is the true process mean, $\bar{x}$ is not approximately normal with mean 139.8 and standard deviation .32, but rather approximately normal with mean μ and standard deviation .32. So for such a μ, expression (6-15) and Definition 6-12 show that the corresponding β will be the probability the corresponding normal distribution assigns to the possibility that $139.2 < \bar{x} < 140.4$. This is pictured in Figure 6-6 for the two means $\mu = 139.5$ and $\mu = 139.2$.

FIGURE 6-6

Approximate Probability Distributions for $\bar{x}$ for Two Different Values of μ Described by H_a and the Corresponding β's, When $\alpha = .05$

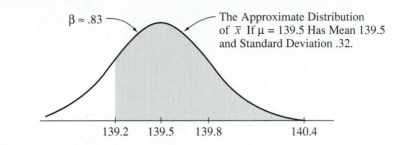

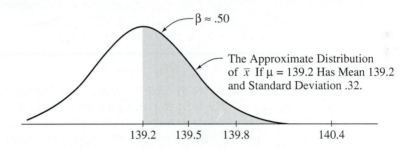

It is an easy matter to calculate z values corresponding to $\bar{x} = 139.2$ and $\bar{x} = 140.4$ using means of 139.5 and 139.2 and a standard deviation of .32 and consult a standard normal table in order to verify the correctness of the two β's marked in Figure 6-6.

Parallel reasoning for the situation with $\alpha = .2$ is as follows. One will decide in favor of H_0 if $|z| < 1.28$ — i.e., if

$$139.4 < \bar{x} < 140.2$$

If a μ described by H_a is the true process mean, $\bar{x}$ is approximately normal with mean μ and standard deviation .32. So the corresponding β will be the probability this normal distribution assigns to the possibility that $139.4 < \bar{x} < 140.2$. This is pictured in Figure 6-7 (p. 294) for the two means $\mu = 139.5$ and $\mu = 139.2$, having corresponding type II error probabilities $\beta = .61$ and $\beta = .27$.

The calculations represented by the two figures are collected in Table 6-2. Notice two features of the table. First, the β values for $\alpha = .05$ are larger than those for $\alpha = .2$. If one wants to run only a 5% chance of (incorrectly) deciding to adjust an on-target process, the price to be paid is a larger probability of failure to recognize an off-target condition. Secondly, the $\mu = 139.2$ β values are smaller than the $\mu = 139.5$ β values. The further the filling process is from being on target, the less likely it is that one will fail to detect the off-target condition.

TABLE 6-2

$n = 25$ Type II Error Probabilities

		μ	
		139.2	139.5
α	.05	.50	.83
	.2	.27	.61

FIGURE 6-7

Approximate Probability
Distributions for $\bar{x}$ for
Two Different Values of μ
Described by H_a and the
Corresponding β's, When
$\alpha = .2$

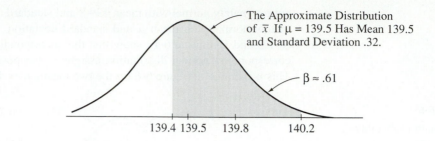

The Approximate Distribution
of $\bar{x}$ If $\mu = 139.5$ Has Mean 139.5
and Standard Deviation .32.

$\beta \approx .61$

139.4 139.5   139.8   140.2

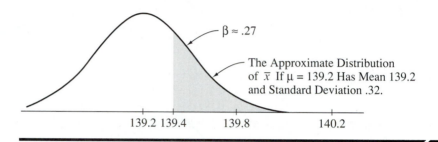

$\beta \approx .27$

The Approximate Distribution
of $\bar{x}$ If $\mu = 139.2$ Has Mean 139.2
and Standard Deviation .32.

139.2 139.4   139.8   140.2

The story told by Table 6-2 applies in qualitative terms to all applications of significance testing in decision-making contexts. The further H_0 is from being true, the smaller the corresponding β. And small α's imply large β's and vice versa.

However, there is one other element of this general picture that plays an important role in the determination of error probabilities. It needs to be mentioned because, to some extent, it reduces the tension between the desire for small α and the corresponding desire for small β's. That is the matter of sample size. If one can afford to increase a sample size, for a given α, the corresponding β's can be reduced.

Redo the calculations of the previous example, this time supposing that $n = 100$ rather than 25. Table 6-3 shows the type II error probabilities that should result, and comparison with Table 6-2 serves to indicate the sample-size effect in the filling process example.

TABLE 6-3

$n = 100$ Type II
Error Probabilities

| | | \multicolumn{2}{c}{μ} |
		139.2	139.5
α	.05	.04	.53
	.2	.01	.28

Sometimes helpful in understanding the standard logic applied when significance testing is employed in a decision-making context is an analogy that involves thinking of the process of coming to a decision as a sort of legal proceeding, like a criminal trial. In a criminal trial, there are two opposing hypotheses, namely

H_0: The defendant is innocent

H_a: The defendant is guilty

Evidence, playing a role similar to the data used in testing, is gathered and used to decide between the two hypotheses. Two types of potential errors exist in a criminal trial: the possibility of convicting an innocent person (parallel to the type I error) and the possibility of acquitting a guilty person (similar to the type II error). A criminal trial is a situation where the two types of errors are definitely thought of as having differing consequences, so the two hypotheses are treated asymmetrically. The *a priori* presumption in a criminal trial is in favor of H_0, the defendant's innocence. In order to keep the chance of a false conviction small (i.e., keep α small), overwhelming evidence is required for conviction, in much the same way that if a small α is used in testing, extreme values of the test statistic are needed in order to indicate rejection of H_0. One consequence of this method of operation in criminal trials is that there is a substantial chance that a guilty individual will be acquitted, in the same way that small α's produce big β's in testing contexts.

This significance testing/criminal trial parallel is useful, but do not make more of it than is justified. One should not jump to the conclusion that all significance-testing applications are properly thought of in this light. In this author's opinion and experience, few engineering scenarios are simple enough to reduce to a "decide between H_0 and H_a" basis. Sensible applications of significance testing are often only steps of "evidence evaluation" in a many-faceted, data-based detective job necessary to solve an engineering problem. (Presently, more will be said about the proper engineering use of significance testing.) And even when a real problem can be reduced to a simple "decide between H_0 and H_a" framework, it need not be the case that the "choose a small α" logic is appropriate. In some engineering contexts, the practical consequences of a type II error are such that rational decision making strikes a balance between the opposing goals of small α and small β's.

Some Additional Comments Concerning Significance Testing

Confidence interval estimation and significance testing are the two most commonly used forms of formal statistical inference. These having been introduced, it is appropriate to offer some comparative comments about their practical usefulness and, in the process, admit to an *estimation orientation* that will be reflected in much of the rest of this book's treatment of formal inference.

As it's been described here, significance testing is essentially a method of quantifying the strength of the evidence against a null hypothesis. If an observed level of significance is small, the evidence is strong. If it is large, the evidence is weak. Confidence interval estimation, on the other hand, is essentially a method of identifying a range of plausible values for a parameter or function of one or more parameters. There are several reasons why, in most engineering problem-solving contexts, the second type of method is more to the point.

In the first place, more often than not in problem-solving contexts, engineers need to know "What is the value of the parameter?" rather than "Is the parameter equal to some hypothesized value?" And it is confidence interval estimation, not significance testing, that is designed to answer the first type of question. A confidence interval for a mean breakaway torque of from 9.9 in. oz to 13.1 in. oz says what values of μ seem plausible. A tiny observed level of significance in testing $H_0: \mu = 33.5$ says only that the data speak clearly against the possibility that $\mu = 33.5$, but it doesn't give any clue to the likely value of μ.

The fact that significance testing doesn't produce any useful indication of what parameter values are plausible is sometimes obscured by careless interpretation of semi-standard testing jargon. For example, it is common in some fields to term p-values less than .05 "statistically significant" and ones less than .01 "highly significant." The danger in this kind of usage is that "significant" can be incorrectly heard to mean "of great practical consequence" and the p-value incorrectly interpreted as a measure of how far a parameter differs from a value stated in a null hypothesis. One reason this interpretation doesn't follow is that the observed level of significance in a test depends not only on how far H_0 appears to be from being correct, but on the sample size as well. Roughly speaking, given a large enough sample size, any departure from H_0, whether of practical importance or not, can be shown to be "highly significant."

EXAMPLE 6-6

Statistical Significance and Practical Importance in a Regulatory Agency Test. One of this author's favorite examples of the above points involves a newspaper article, which appears as Figure 6-8. Apparently the Pass Master manufacturer did enough physical mileage testing (used a large enough n) to produce a p-value less than .05 for testing a null hypothesis of no mileage improvement. That is, a "statistically significant" result was obtained.

FIGURE 6-8

Article from *The Lafayette Journal and Courier*, Page D-3, August 28, 1980. Reprinted by permission of the Associated Press. © 1980 the Associated Press.

EPA Endorses Gasoline Gadget

WASHINGTON (AP) — A gadget that cuts off a car's air conditioner when the vehicle accelerates has become the first product aimed at cutting gasoline consumption to win government endorsement.

The device, marketed under the name "Pass Master," can provide a "small but real fuel economy benefit," the Environmental Protection Agency said Wednesday.

Motorists could realize up to 4 percent fuel reduction while using their air conditioners on cars equipped with the device, the agency said. That would translate into .8-miles-per-gallon improvement for a car that normally gets 20 miles to the gallon with the air conditioner on.

The agency cautioned that the 4 percent figure was a maximum amount and could be less depending on a motorist's driving habits, the type of car and the type of air conditioner.

But still the Pass Master, which sells for less than $15, is the first of 40 products to pass the EPA's tests as making any "statistically significant" improvement in a car's mileage.

But notice that from some perspectives, the size of the actual mileage improvement reported is "small but real," amounting to about .8 mpg. Whether or not the size of this improvement is of *practical importance* is a matter largely separate from the significance-testing result. And an engineer equipped with a confidence interval for the mean mileage improvement is in a better position to judge this than is one who knows only that the p-value was less than .05.

EXAMPLE 6-5
(continued)

To illustrate the effect that sample size has on observed level of significance, return to the breakaway torque problem and consider two hypothetical samples, one based on $n = 25$ and the other on $n = 100$, but both giving $\bar{x} = 32.5$ in. oz and $s = 5.1$ in. oz. For testing H_0: $\mu = 33.5$ with H_a: $\mu < 33.5$, the first hypothetical sample gives

$$z = \frac{32.5 - 33.5}{\frac{5.1}{\sqrt{25}}} = -.98$$

with associated observed level of significance

$$\Phi(-.98) = .16$$

On the other hand, the second hypothetical sample gives

$$z = \frac{32.5 - 33.5}{\frac{5.1}{\sqrt{100}}} = -1.96$$

with corresponding p-value

$$\Phi(-1.96) = .02$$

Because the second sample size is larger, the second sample gives stronger evidence that the mean breakaway torque is below 33.5 in. oz. But the best data-based guess at the difference between μ and 33.5 is $\bar{x} - 33.5 = -1.0$ in. oz in both cases. And it is the size of the difference between μ and 33.5 that is of primary engineering importance.

It is useful to realize that in addition to doing its primary job of providing an interval of plausible values for a parameter, a confidence interval itself also provides some significance-testing information. For example, a 95% confidence interval for a parameter can be thought of as containing all those values of the parameter for which significance tests using the data in hand would produce p-values bigger than 5%. (Those values not covered by the interval would have associated p-values smaller than 5%.)

EXAMPLE 6-5
(continued)

Recall from Section 6-1 that a 90% one-sided confidence interval for the mean breakaway torque for failed drives is $(-\infty, 12.8)$. This means that for any value, #, larger than 12.8 in. oz, a significance test of $H_0: \mu = \#$ with $H_a: \mu < \#$ would produce a p-value less than .1. So clearly, the observed level of significance corresponding to the null hypothesis $H_0: \mu = 33.5$ is less than .1 . (In fact, as was seen earlier in this section, the p-value is 0 to two decimal places.) Put more loosely, the interval $(-\infty, 12.8)$ is a long way from containing 33.5 in. oz and therefore makes such a value of μ quite implausible.

Suppose one accepts the premises that in problem-solving contexts, confidence intervals not only address a more germane issue than significance tests do but also provide some significance-testing information as a by-product. A legitimate question becomes, "How then might an engineer find significance testing, *per se*, worthwhile?" One answer to this question is that in an almost negative way, p-values can help an engineer gauge the extent to which data in hand are inconclusive. When observed levels of significance are large, more information is needed in order to arrive at any definitive judgment. Another answer is that sometimes legal requirements force the use of significance testing in a compliance or effectiveness demonstration. This was the case in Figure 6-8, where before the Pass Master could be marketed, some mileage improvement (no matter how small) had to be legally demonstrated. And finally, there are cases where the use of significance testing in a decision-making framework is necessary and appropriate. An example is acceptance sampling: based on information from a sample of items from a large lot, one must determine whether or not to receive shipment of the lot. So, properly understood and handled, significance testing does have its place in engineering practice. Although, to the extent possible, the rest of this book will feature confidence interval estimation over significance testing, methods of significance testing will not be ignored.

6-3 *One- and Two-Sample Inference for Means*

Sections 6-1 and 6-2 introduced the basic concepts of confidence interval estimation and significance testing. There are quite literally thousands of specific methods of these two types published in the statistical literature. This book can only discuss a small fraction that are particularly well-known and useful to engineers. The next three sections consider the most elementary of these — some of those that are applicable to one- and two-sample studies — beginning in this section with methods of formal inference for means.

Inferences for a single mean, based not on the large samples of Sections 6-1 and 6-2 but instead on small samples, are considered first. In the process, it is necessary to introduce the so-called (Student) t probability distributions. Presented next are methods of formal inference for paired data, based on application of methods of inference for a single mean to paired differences. The section concludes with discussion of both large- and small-n methods for data-based comparison of two means based on independent samples.

Small-Sample Inference for a Single Mean

The most important practical limitation on the use of the methods of the previous two sections is the requirement that the sample size involved must be large. That restriction comes from the fact that without it, there is no probability theorem to appeal to in order to conclude that

$$\frac{\bar{x} - \mu}{\frac{s}{\sqrt{n}}} \tag{6-16}$$

is approximately standard normal. So if, for example, one mechanically uses the large-n confidence interval formula

$$\bar{x} \pm z \frac{s}{\sqrt{n}} \tag{6-17}$$

with a small sample, there is no way of assessing what actual level of confidence should be declared. That is, for small n, using $z = 1.96$ in formula (6-17) generally doesn't produce 95% confidence intervals. And without a further proviso, there is neither any way to tell what confidence might be associated with $z = 1.96$ nor any way to tell how to choose z in order to produce a 95% confidence level.

However, there is one important special circumstance in which it is possible to reason in a way parallel to the work in Sections 6-1 and 6-2 and arrive at confidence interval and significance-testing methods for means based on small sample sizes. That is the situation where it is plausible to model the observations as iid normal random variables (i.e. as conceptually equivalent to random selections (with replacement) from a population with a bell-shaped relative frequency distribution). The reason that the Gaussian case is convenient is that although the variable (6-16) is not approximately standard normal, if the observations are iid normal variables, the quantity (6-16) does have a recognized, tabled distribution, which is called the *Student t distribution*.

DEFINITION 6-13

The (**Student**) t **distribution with degrees of freedom parameter** ν, is a continuous probability distribution with probability density

$$f(t) = \frac{\Gamma\left(\dfrac{\nu + 1}{2}\right)}{\Gamma\left(\dfrac{\nu}{2}\right)\sqrt{\pi\nu}} \left(1 + \frac{t^2}{\nu}\right)^{-(\nu+1)/2} \qquad \text{for all } t \tag{6-18}$$

If a random variable has the probability density given by formula (6-18), it is said to have a t_ν distribution.

The word "Student" in Definition 6-13 was the pen name of the statistician who first came upon formula (6-18).

Expression (6-18) is rather formidable-looking. But the gamma function was introduced in Section 5-2; what's more, no direct computations with expression (6-18) will actually be required in this book. Nevertheless, it is useful to have expression (6-18) available in order to sketch several t probability densities, to get a feel for their shape. Figure 6-9 pictures the t densities for degrees of freedom $\nu = 1, 2, 5$, and 11, along with the standard normal density.

FIGURE 6-9

t Probability Densities for $\nu = 1, 2, 5$, and 11 and the Standard Normal Density

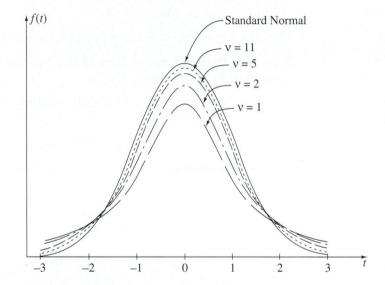

The message carried by Figure 6-9 is that the t probability densities are more or less bell-shaped and symmetric about 0. As compared to the standard normal density, they are flatter, but they seem to be increasingly like the standard normal density as ν gets larger. In fact, for most practical purposes, for ν larger than about 30, the t distribution with ν degrees of freedom and the standard normal distribution are indistinguishable.

Probabilities for the t distributions are not typically found using the density in expression (6-18), as no simple antiderivative for $f(t)$ exists — so numerical integration would be required. (Of course, it is true that these days such work can be done using preprogrammed routines on pocket calculators.) Instead, it is common to use tables (of necessity, fairly sketchy ones) to evaluate t distribution quantiles and to get at least crude bounds on the types of probabilities needed in significance testing. Table D-4 is a typical table of t quantiles. Across the top of the table are several cumulative probabilities. Down the left side are values of the degrees of freedom parameter, ν. In the body of the table are corresponding quantiles. Notice also that the last line of the table is a "$\nu = \infty$" (i.e., standard normal) line.

EXAMPLE 6-7

Use of a Table of t Distribution Quantiles. Suppose that U is a random variable having a t distribution with $\nu = 5$ degrees of freedom. Consider first finding the .95 quantile of U's distribution, then seeing what Table D-4 reveals about $P[U < -1.9]$ and then about $P[|U| > 2.3]$.

First, looking at the $\nu = 5$ row of Table D-4 under the cumulative probability .95, 2.015 is found in the body of the table. That is, $Q(.95) = 2.015$ or (equivalently) $P[U \le 2.015] = .95$.

Then note that by symmetry,

$$P[U < -1.9] = P[U > 1.9] = 1 - P[U \le 1.9]$$

Looking at the $\nu = 5$ row of Table D-4, one can see that 1.9 is between the .90 and .95 quantiles of the t_5 distribution. That is,

$$.90 < P[U \le 1.9] \le .95$$

so finally

$$.05 < P[U < -1.9] < .10$$

Lastly, again by symmetry,

$$P[|U| > 2.3] = P[U < -2.3] + P[U > 2.3] = 2P[U > 2.3] = 2(1 - P[U \le 2.3])$$

Then, from the $\nu = 5$ row of Table D-4, 2.3 is seen to be between the .95 and .975 quantiles of the t_5 distribution. That is,

$$.95 < P[U \le 2.3] < .975$$

so

$$.05 < P[|U| > 2.3] < .10$$

The three calculations of this example are pictured in Figure 6-10.

FIGURE 6-10

Three t_5 Probability
Calculations for Example 6-7

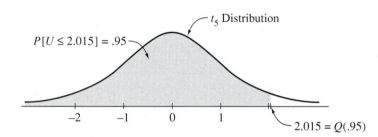

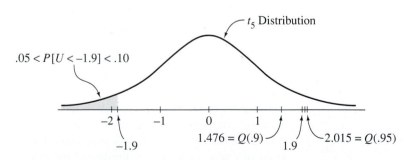

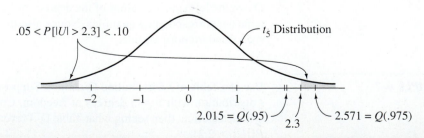

The connection between expressions (6-18) and (6-16) that allows the development of small-n inference methods for a normal underlying distribution is that if an iid normal model is appropriate,

$$\frac{\bar{x} - \mu}{\dfrac{s}{\sqrt{n}}} \tag{6-19}$$

has the t distribution with $\nu = n - 1$ degrees of freedom. Notice that this is consistent with the basic fact used in the previous two sections. That is, for large n, ν is large, so the t_ν distribution is approximately standard normal, and for large n, the variable (6-19) has already been treated as approximately standard normal.

It is common to try to motivate the "degrees of freedom" terminology and $\nu = n - 1$ formula by allusion to the fact that if one constrains a sample of n values to have a particular mean $\bar{x}$, there are really only $n - 1$ unconstrained choices of sample values. This author doesn't personally find that truth particularly illuminating and suggests instead that the reader simply view the fact that ν depends on n as generally plausible and accept the exact relationship $\nu = n - 1$ as just what would result from some further mathematical work.

In any case, having realized that the variable (6-19) can under appropriate circumstances be treated as a t_{n-1} random variable (and having a table of t quantiles available), one is in a position to work in exact analogy to what was done in Sections 6-1 and 6-2 to find methods for confidence interval estimation and significance testing. That is, if a particular data-generating mechanism can be thought of as essentially equivalent to drawing independent observations from a single normal distribution, a two-sided confidence interval for μ has endpoints

$$\bar{x} \pm t \frac{s}{\sqrt{n}} \tag{6-20}$$

where t is chosen such that the t_{n-1} distribution assigns probability corresponding to the desired confidence level to the interval between $-t$ and t. Further, the null hypothesis

$$H_0: \mu = \#$$

can be tested using the statistic

$$T = \frac{\bar{x} - \#}{\dfrac{s}{\sqrt{n}}} \tag{6-21}$$

and a t_{n-1} reference distribution.

Operationally, the only difference between the inference methods indicated here and the large-sample methods of the previous two sections is the exchange of standard normal quantiles and probabilities for ones corresponding to the t_{n-1} distribution. Conceptually, however, the nominal confidence and significance properties here are practically relevant only under the extra proviso of a reasonably normal underlying distribution. Before applying either expression (6-20) or (6-21) in practice, it is advisable to investigate the appropriateness of a Gaussian model assumption.

EXAMPLE 6-8

Small-Sample Confidence Limits for a Mean Spring Lifetime. Part of a data set of W. Armstrong (appearing in *Analysis of Survival Data* by Cox and Oakes) gives numbers of cycles to failure of ten springs of a particular type under a stress of 950 N/mm². These spring-life observations are given in Table 6-4, in units of 1000 cycles.

TABLE 6-4

Cycles to Failure of 10 Springs
under 950 N/mm^2 Stress
(10^3 cycles)

Spring Lifetimes
225, 171, 198, 189, 189
135, 162, 135, 117, 162

An important engineering question in this context might well be "What is the average spring lifetime under these conditions of 950 N/mm^2 stress?" Since only $n = 10$ observations are available, the large-sample confidence interval method of Section 6-1 is not applicable. Instead, only the method indicated by expression (6-20) is a possible option; for it to be appropriate, an underlying normal lifetime distribution must also be appropriate.

Having no relevant base of experience, this author cannot speculate *a priori* about the appropriateness of a normal lifetime model in this context. But at least it is possible to examine the data in Table 6-4 themselves for evidence of strong departure from normality. Figure 6-11 is a normal plot for the data; it shows that in fact no such evidence exists.

FIGURE 6-11

Normal Plot of Spring Lifetimes

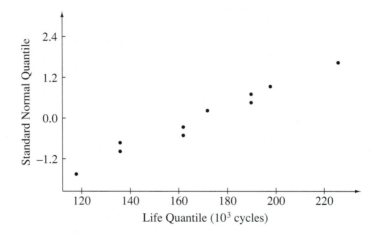

It is easy to verify that for the 10 lifetimes, $\bar{x} = 168.3$ ($\times 10^3$ cycles) and $s = 33.1$ ($\times 10^3$ cycles). So to develop a confidence interval for the mean spring lifetime, one is led to use these values in expression (6-20), along with an appropriately chosen value of t. Using, for example, a 90% confidence level and a two-sided interval, t should be chosen as the .95 quantile of the t distribution with $\nu = n - 1 = 9$ degrees of freedom. That is, one uses the t_9 distribution and chooses $t > 0$ such that

$$P[-t < \text{a } t_9 \text{ random variable} < t] = .90$$

Consulting Table D-4, one finds that the choice $t = 1.833$ is in order. So a two-sided 90% confidence interval for μ has endpoints

$$168.3 \pm 1.833 \frac{33.1}{\sqrt{10}}$$

i.e.,

$$168.3 \pm 19.2$$

i.e.,

$$149.1 \times 10^3 \text{ cycles} \qquad \text{and} \qquad 187.5 \times 10^3 \text{ cycles}$$

As shown in Example 6-8, normal-plotting the data as a rough check on the plausibility of an underlying normal distribution is a sound practice, and one that will be used repeatedly in this text. However, it is important not to expect more than is justified from the method. It is certainly preferable to use it rather than making an unexamined leap to a possibly inappropriate normal assumption. But it is also true that when used with small samples, the method doesn't often provide definitive indications as to whether a Gaussian model can be used. According to conventional statistical wisdom, it is the nature of randomness that small samples from normal distributions (even exactly normal ones) will often have only marginally linear-looking normal plots. At the same time, small samples from even quite nonnormal distributions can often have reasonably linear normal plots. In short, because of sampling variability, small samples don't carry much information about underlying distributional shape. About all that can be counted on from a small-sample preliminary normal plot, like that in Example 6-8, is warning in case of a fairly gross departure from normality associated with an underlying distributional shape that is much heavier in the tails than a Gaussian distribution (i.e., producing more extreme values than a normal shape would).

In applications, it is a good idea to make the effort to (so to speak) calibrate one's normal-plot perceptions, if they are going to be used as a tool for checking a model. One way to do this is to use simulation and generate a number of samples of the size in question from a standard normal distribution and normal-plot these. Then the shape of the normal plot of the data in hand can be compared to the simulations to get some feeling as to whether any nonlinearity it exhibits is more than might be expected from a normal plot of a sample from an underlying normal distribution. To illustrate, Figure 6-12 shows normal plots for several simulated samples of size $n = 10$ from the standard normal distribution. Comparing Figures 6-11 and 6-12, it is clear that indeed the spring-life data carry no strong indication of nonnormality.

Example 6-8 shows the use of the confidence interval formula (6-20), but not the significance testing method (6-21). Since the small-sample method is exactly analogous to the large-sample method of the Section 6-2 (except for the substitution of the t distribution for the standard normal distribution), and the source from which the data were taken doesn't indicate any particular value of μ belonging naturally in a null hypothesis, the use of the method indicated in expression (6-21) by itself will not be illustrated at this point. There is, however, an application of the testing method to paired differences in Example 6-9.

Inference for the Mean of a Paired Difference

An important type of application of the foregoing methods of confidence interval estimation and significance testing is to *paired data*. In many engineering applications, it is natural to make two measurements of essentially the same kind, but differing in timing or physical location, on a single sample of physical objects. The goal in such situations is often to investigate the possibility of consistent differences between the two measurements. (Review the discussion of paired data terminology in Section 1-2.)

EXAMPLE 6-9

Comparing Leading-Edge and Trailing-Edge Measurements on a Shaped Wood Product. Drake, Hones, and Mulholland worked with a company on the monitoring of the operation of an end cut router in the manufacture of a wood product. They measured a critical dimension of a number of pieces of a particular type as they came off the router. Both a leading-edge and a trailing-edge measurement were made on each piece. The design for the piece in question specified that both leading-edge and trailing-edge values were to have target value .172 in. Table 6-5 gives leading- and trailing-edge measurements taken by the students on five consecutive pieces.

FIGURE 6-12
Normal Plots of Samples of Size $n = 10$ from a Standard Normal Distribution

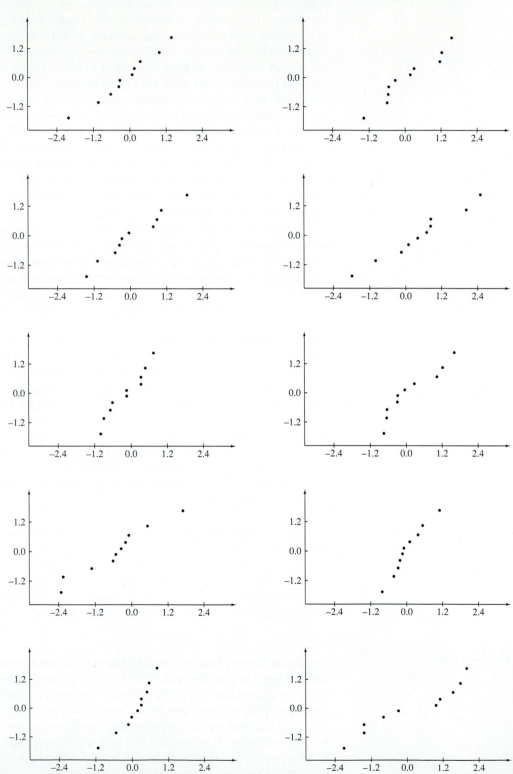

TABLE 6-5

Leading-Edge and
Trailing-Edge Dimensions
for Five Workpieces

Piece	Leading-Edge Measurement (in.)	Training-Edge Measurement (in.)
1	.168	.169
2	.170	.168
3	.165	.168
4	.165	.168
5	.170	.169

In this situation, the correspondence between leading- and trailing-edge dimensions was at least as critical to proper fit in a later assembly operation as was the conformance of the individual dimensions to the nominal value of .172 in. This was thus a paired data situation, where one issue of concern was the possibility of a consistent difference between leading- and trailing-edge dimensions that might be traced to a machine misadjustment or unwise method of router operation.

In situations like Example 6-9, one simple method of investigating the possibility of a consistent difference between paired data is to first reduce the two measurements on each physical object to a single difference between them. One may then apply the methods of confidence interval estimation and significance testing studied thus far to the differences. That is, after reducing paired data to differences $d_1, d_2, \ldots, d_n$, if n (the number of data pairs) is large, endpoints of a confidence interval for the underlying mean difference, μ_d, are

$$\bar{d} \pm z \frac{s_d}{\sqrt{n}} \tag{6-22}$$

where s_d is the sample standard deviation of $d_1, d_2, \ldots, d_n$. Similarly, the null hypothesis

$$H_0: \mu_d = \# \tag{6-23}$$

can be tested using the test statistic

$$Z = \frac{\bar{d} - \#}{\frac{s_d}{\sqrt{n}}} \tag{6-24}$$

and a standard normal reference distribution.

If n is small, in order to come up with methods of formal inference, an underlying normal distribution *of differences* must be plausible. If that is the case, a confidence interval for μ_d has endpoints

$$\bar{d} \pm t \frac{s_d}{\sqrt{n}} \tag{6-25}$$

and the null hypothesis (6-23) can be tested using the test statistic

$$T = \frac{\bar{d} - \#}{\frac{s_d}{\sqrt{n}}} \tag{6-26}$$

and a t_{n-1} reference distribution.

EXAMPLE 6-9
(continued)

To illustrate this method of paired differences, consider testing the null hypothesis H_0: $\mu_d = 0$ and making a 95% confidence interval for any consistent difference between leading- and trailing-edge dimensions, μ_d, based on the data in Table 6-5. One begins

TABLE 6-6

Five Differences in
Leading- and Trailing-Edge
Measurements

Piece	d = Difference in Dimensions (in.)
1	$-.001 \; (= .168 - .169)$
2	$.002 \; (= .170 - .168)$
3	$-.003 \; (= .165 - .168)$
4	$-.003 \; (= .165 - .168)$
5	$.001 \; (= .170 - .169)$

by reducing the $n = 5$ paired observations in Table 6-5 to differences

$$d = \text{leading-edge dimension} - \text{trailing-edge dimension}$$

appearing in Table 6-6.

Figure 6-13 is a normal plot of the $n = 5$ differences in Table 6-6. A little experimenting with normal plots of simulated samples of size $n = 5$ from a Gaussian distribution will convince the interested reader that the lack of linearity in Figure 6-13 would in no way be atypical of Gaussian data. This, together with the fact that normal distributions are very often appropriate for describing machined dimensions of mass-produced parts, suggests the conclusion that the methods represented by expressions (6-25) and (6-26) are in order in this example.

FIGURE 6-13

Normal Plot of
$n = 5$ Differences

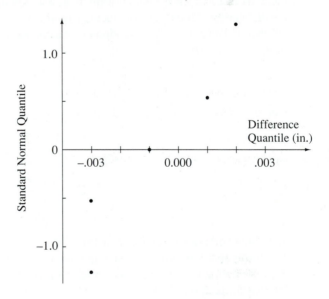

The differences in Table 6-6 have $\bar{d} = -.0008$ in. and $s_d = .0023$ in. So, first investigating the plausibility of a "no consistent difference" hypothesis in a five-step significance testing format, one has the following.

   1. H_0: $\mu_d = 0$.
   2. H_a: $\mu_d \neq 0$.

(There is *a priori* no reason to adopt a one-sided alternative hypothesis.)

   3. The test statistic will be

$$T = \frac{\bar{d} - 0}{\dfrac{s_d}{\sqrt{n}}}$$

The reference distribution will be the t distribution with $\nu = n - 1 = 4$ degrees of freedom. Large observed $|t|$ will count as evidence against H_0 and in favor of H_a.

4. The sample gives

$$t = \frac{-.0008}{\frac{.0023}{\sqrt{5}}} = -.78$$

5. The observed level of significance is $P[|$a t_4 random variable$| \geq .78]$, which can be seen from Table D-4 to be larger than $2(.10) = .2$. The data in hand do not render particularly implausible the hypothesis of no systematic difference between leading- and trailing-edge measurements.

Consulting Table D-4 for the .975 quantile of the t_4 distribution, one sees that $t = 2.776$ is the appropriate multiplier for use in expression (6-25) for 95% confidence. That is, a two-sided 95% confidence interval for the mean difference between the leading- and trailing-edge dimensions for these pieces has endpoints

$$-.0008 \pm 2.776 \frac{.0023}{\sqrt{5}}$$

i.e.,

$$-.0008 \text{ in.} \pm .0029 \text{ in.} \tag{6-27}$$

i.e.,

$$-.0037 \text{ in.} \qquad \text{and} \qquad .0021 \text{ in.}$$

Notice that the confidence interval for μ_d not only gives a range of plausible values but also implicitly says that, since 0 is in the calculated interval, the observed level of significance for testing H_0: $\mu_d = 0$ is more than .05 ($= 1 - .95$). Put slightly differently, it is clear from display (6-27) that the imprecision represented by the plus-or-minus part of the expression is large enough to make it less than certain that the perceived difference, $\bar{d} = -.0008$, is not just a result of sampling variability.

Example 6-9 treats a small-sample problem. No example for large n is included, because after the taking of differences illustrated above, such an example would reduce to rehash of things in Sections 6-1 and 6-2. In fact, since for large n, the t distribution with $\nu = n - 1$ degrees of freedom becomes essentially standard normal, one could even imitate Example 6-9 for large n and get into no logical problems. (Of course, with large n, there would be no need to worry about possible lack of normality of the differences.) So at this point, it makes sense to move on from consideration of the paired difference method.

Large-Sample Comparisons of Two Means (Based on Independent Samples)

One of the principles of effective engineering data collection espoused in Section 2-3 was *comparative study*. The idea of paired differences can be thought of as providing inference methods for a very special kind for comparison, where one sample of items in some sense provides its own basis for comparison. Methods of formal inference that can be used to compare two means, in cases where two different "unrelated" samples form the basis of inference, are studied next, beginning with large-sample methods.

Many scenarios for engineering data collection involve two (or more) samples of data collected under different conditions, where primary interest centers on comparing the two conditions in terms of their impact on the mean of some response variable.

EXAMPLE 6-10

Comparing the Packing Properties of Molded and Crushed Pieces of a Solid. A company research effort involved finding a workable geometry for packing molded pieces of a crystalline solid. One comparison made by the company was between the weight of pieces molded into a particular geometry, which could be poured into a standard container, and the weight of irregularly shaped pieces (obtained through a crushing process), which could be poured into the same container. A series of 24 attempts to pack both molded and crushed pieces of the solid produced the data (in grams) that are given in Figure 6-14 in the form of a back-to-back stem-and-leaf diagram.

FIGURE 6-14

Back-to-Back Stem-and-Leaf
Plots of Packing Weights for
Molded and Crushed Pieces

Molded		Crushed
	11	
7.9	11	
4.5, 3.6, 1.2	12	
9.8, 8.9, 7.9, 7.1, 6.1, 5.7, 5.1	12	
2.3, 1.3, 0.0	13	
8.0, 7.0, 6.5, 6.3, 6.2	13	
2.2, 0.1	14	
	14	
2.1, 1.2, 0.2	15	
	15	
	16	1.8
	16	5.8, 9.6
	17	1.3, 2.0, 2.4, 3.3, 3.4, 3.7
	17	6.6, 9.8
	18	0.2, 0.9, 3.3, 3.8, 4.9
	18	5.5, 6.5, 7.1, 7.3, 9.1, 9.8
	19	0.0, 1.0
	19	

Notice that although the same number of molded and crushed weights are represented in the figure, there are two distinctly different samples represented. This is in no way comparable to the paired difference situation treated in Example 6-9, and a different method of formal inference is appropriate.

In situations like Example 6-10, it is useful to adopt subscript notation for both the parameters and the statistics — for example, letting μ_1 and μ_2 stand for underlying distributional means corresponding to the first and second conditions and $\bar{x}_1$ and $\bar{x}_2$ stand for corresponding sample means. Now if the two data-generating mechanisms are conceptually essentially equivalent to sampling with replacement from (respectively) distribution 1 and distribution 2, the exposition in Section 5-5 says that $\bar{x}_1$ has mean μ_1 and variance σ_1^2/n_1, and $\bar{x}_2$ has mean μ_2 and variance σ_2^2/n_2.

The difference in sample means $\bar{x}_1 - \bar{x}_2$ is a natural statistic to use in comparing μ_1 and μ_2. Proposition 5-1 implies that if it is reasonable to think of the two samples as separately chosen/independent, the random variable has

$$E(\bar{x}_1 - \bar{x}_2) = \mu_1 - \mu_2$$

and

$$Var(\bar{x}_1 - \bar{x}_2) = \frac{\sigma_1^2}{n_1} + \frac{\sigma_2^2}{n_2}$$

If in addition, n_1 and n_2 are large (so that $\bar{x}_1$ and $\bar{x}_2$ are each approximately normal), it turns out that $\bar{x}_1 - \bar{x}_2$ is approximately normal — i.e., that

$$\frac{\bar{x}_1 - \bar{x}_2 - (\mu_1 - \mu_2)}{\sqrt{\dfrac{\sigma_1^2}{n_1} + \dfrac{\sigma_2^2}{n_2}}} \qquad (6\text{-}28)$$

has an approximately standard normal probability distribution.

Now it is possible to begin with the fact that the variable (6-28) is approximately standard normal and end up with confidence interval and significance-testing methods for $\mu_1 - \mu_2$ by using logic exactly parallel to that in the "known-σ" parts of Sections 6-1 and 6-2. But in terms of practical utility, it is far more useful to begin instead with an expression that is free of the typically unknown parameters σ_1 and σ_2. Happily, probabilists, beginning with an "independent, iid samples from two different distributions" model, can prove the more useful fact that for large n_1 and n_2, not only is the variable (6-28) approximately standard normal, but so is

$$\frac{\bar{x}_1 - \bar{x}_2 - (\mu_1 - \mu_2)}{\sqrt{\dfrac{s_1^2}{n_1} + \dfrac{s_2^2}{n_2}}} \qquad (6\text{-}29)$$

Then the standard logic of Section 6-1 shows that a two-sided large-sample confidence interval for the difference $\mu_1 - \mu_2$ based on two independent samples has endpoints

$$\bar{x}_1 - \bar{x}_2 \pm z\sqrt{\dfrac{s_1^2}{n_1} + \dfrac{s_2^2}{n_2}} \qquad (6\text{-}30)$$

where z is chosen such that the probability that the standard normal distribution assigns to the interval between $-z$ and z corresponds to the desired confidence. And the logic of Section 6-2 shows that under the same conditions,

$$H_0\!: \mu_1 - \mu_2 = \#$$

can be tested using the statistic

$$Z = \frac{\bar{x}_1 - \bar{x}_2 - \#}{\sqrt{\dfrac{s_1^2}{n_1} + \dfrac{s_2^2}{n_2}}} \qquad (6\text{-}31)$$

and a standard normal reference distribution.

EXAMPLE 6-10
(continued)

In the packing geometry problem, the crushed pieces were *a priori* expected to pack better than the (for other purposes more convenient) molded pieces. Consider using the preceding methods to test the statistical significance of the difference in mean weights and also to make a 95% one-sided confidence interval for the difference (declaring that the crushed mean weight minus the molded mean weight is at least some number).

First, it must be admitted that the sample sizes here ($n_1 = n_2 = 24$) are borderline for being termed large. It would be preferable to have a few more observations of each type, but lacking them, this author would personally go ahead and use the methods of expressions (6-30) and (6-31) but remain properly wary of the results if they in any way produce a "close call" in engineering or business terms.

Proceeding to significance testing, arbitrarily labeling "crushed" condition 1 and "molded" condition 2 and calculating from the data in Figure 6-14 that $\bar{x}_1 = 179.55$ g, $s_1 = 8.34$ g, $\bar{x}_2 = 132.97$ g, and $s_2 = 9.31$ g, the five-step testing format produces the following summary.

1. H_0: $\mu_1 - \mu_2 = 0$.
2. H_a: $\mu_1 - \mu_2 > 0$.

(The research hypothesis here is that the crushed mean exceeds the molded mean so that the difference, taken in this order, is positive.)

3. The test statistic is

$$Z = \frac{\bar{x}_1 - \bar{x}_2 - 0}{\sqrt{\dfrac{s_1^2}{n_1} + \dfrac{s_2^2}{n_2}}}$$

The reference distribution is standard normal, and large observed values z will constitute evidence against H_0 and in favor of H_a.
4. The samples give

$$z = \frac{179.55 - 132.97 - 0}{\sqrt{\dfrac{(8.34)^2}{24} + \dfrac{(9.31)^2}{24}}} = 18.3$$

5. The observed level of significance is P[a standard normal variable ≥ 18.3] ≈ 0. The data present rather overwhelming evidence that $\mu_1 - \mu_2 > 0$ — i.e., that the mean packed weight of crushed pieces exceeds that of the molded pieces.

Then turning to a confidence interval for $\mu_1 - \mu_2$, note that only the lower endpoint given in display (6-30) will be used, so $z = 1.645$ will be appropriate. That is, with 95% confidence, one concludes that the difference (crushed minus molded) in means exceeds

$$(179.55 - 132.97) - 1.645\sqrt{\frac{(8.34)^2}{24} + \frac{(9.31)^2}{24}}$$

i.e., exceeds

$$46.58 - 4.20 = 42.38 \text{ g}$$

Or differently put, a 95% one-sided confidence interval for $\mu_1 - \mu_2$ is

$$(42.38, \infty)$$

Students are sometimes uneasy about the arbitrary choice involved in designating one condition in a two-sample study as number 1 and the other as number 2 and the consequently arbitrary order in which the differences in both sample and population means are taken. The fact is that either order can be used. As long as one follows through consistently with a given order, the real-world conclusions reached will be completely unaffected by the arbitrary choice. For example, in Example 6-10, if one had labeled the molded condition number 1 and the crushed condition number 2, an appropriate one-sided confidence for the molded mean minus the crushed mean would have turned out to be

$$(-\infty, -42.38)$$

which in practical terms has the same meaning as the one in the example.

It is also all too common for students, when faced with the description of an engineering scenario involving a comparison, to be confused about whether to apply a method of formal inference for paired differences or one for two independent samples. Again, just remember that paired data typically arise from two measurements on each element of a single sample of objects. This stands in contrast to single measurements made on each element of two samples. In the woodworking scenario of Example 6-9, the data were paired because both leading-edge and trailing-edge measurements were made on each piece. If (for some unknown reason) one decided to take leading-edge measurements from one group of items and trailing-edge measurements from another, a two-sample (not a paired difference) analysis would be in order.

Small-Sample Comparisons of Two Means (Based on Independent Samples from Normal Distributions with a Common Variance)

The final inference methods to be considered in this section are those for the difference in two means in cases where at least one of n_1 and n_2 is small. The discussion at the beginning of this section showed that in order to produce confidence intervals and a significance-testing method for a single mean based on a small sample, it is necessary to restrict attention to cases where the underlying distribution is reasonably Gaussian. It turns out that when comparing two means based on independent samples, the two associated standard deviations must be comparable as well. In fact, the exact probability model supporting the methods about to be introduced is one of independent samples from two normal distributions with $\sigma_1 = \sigma_2$. A way of making at least a rough check on the plausibility of these model assumptions in an application is to normal-plot two samples on the same set of axes, checking not only for approximate linearity but also for approximate equality of slope.

EXAMPLE 6-8
(continued)

The data of W. Armstrong on spring lifetimes (appearing in the book by Cox and Oakes) not only concern spring longevity at a 950 N/mm^2 stress level, but also longevity at a 900 N/mm^2 stress level (and several others for that matter). Table 6-7 repeats the 950 N/mm^2 data from before and gives the lifetimes of ten springs at the 900 N/mm^2 stress level as well.

TABLE 6-7

Spring Lifetimes under Two Different Levels of Stress (10^3 cycles)

950 N/mm^2 Stress	900 N/mm^2 Stress
225, 171, 198, 189, 189	216, 162, 153, 216, 225
135, 162, 135, 117, 162	216, 306, 225, 243, 189

Figure 6-15 consists of normal plots for the two samples made on a single set of axes. In light of the kind of variation in linearity and slope exhibited in Figure 6-12 by the normal plots for samples of this size ($n = 10$) from a single normal distribution, there is certainly no strong evidence on Figure 6-15 against the appropriateness of using an "equal variances, normal distributions" model in an analysis of the spring data.

If one is going to use a probability model that includes the assumption that $\sigma_1 = \sigma_2$, then calling the common value σ, it makes sense that both s_1 and s_2 will approximate σ. That suggests that they should somehow be combined into a single estimate of the basic, baseline variation present. As it turns out, mathematical tractability dictates a particular method of combining or *pooling* the individual s's to arrive at a single estimate of σ.

FIGURE 6-15

Normal Plots of Spring
Lifetimes under Two Different
Levels of Stress

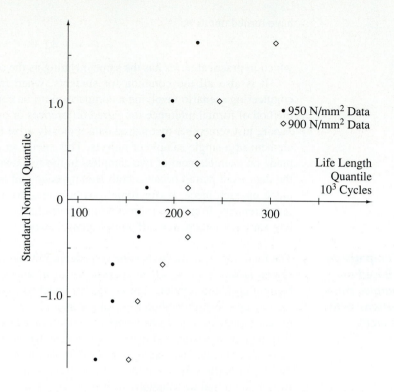

DEFINITION 6-14

If two numerical samples of respective sizes n_1 and n_2 produce respective sample variances s_1^2 and s_2^2, the **pooled sample variance**, s_P^2, is the weighted average of s_1^2 and s_2^2 where the weights are the sample sizes minus 1. That is,

$$s_P^2 = \frac{(n_1 - 1)s_1^2 + (n_2 - 1)s_2^2}{(n_1 - 1) + (n_2 - 1)} = \frac{(n_1 - 1)s_1^2 + (n_2 - 1)s_2^2}{n_1 + n_2 - 2} \qquad \textbf{(6-32)}$$

The **pooled sample standard deviation**, s_P, is the square root of s_P^2.

EXAMPLE 6-8
(continued)

In the spring-life context, making the arbitrary choice to call the 900 N/mm^2 stress level condition 1 and the 950 N/mm^2 stress level condition 2, one has $s_1 = 42.9$ (10^3 cycles) and $s_2 = 33.1$ (10^3 cycles). So pooling the two sample variances via formula (6-32) produces

$$s_P^2 = \frac{(10 - 1)(42.9)^2 + (10 - 1)(33.1)^2}{(10 - 1) + (10 - 1)} = 1,468(10^3 \text{ cycles})^2$$

Then, taking the square root,

$$s_P = \sqrt{1,468} = 38.3(10^3 \text{ cycles})$$

s_P is a kind of average of s_1 and s_2. Its exact form is dictated more by considerations of mathematical convenience (which are off the main track of this presentation) than by obvious intuition. But take some comfort in the fact that s_P is at least guaranteed to fall between the two values s_1 and s_2.

In the argument leading to large-sample inference methods for $\mu_1 - \mu_2$, the quantity given in expression (6-28),

$$\frac{\bar{x}_1 - \bar{x}_2 - (\mu_1 - \mu_2)}{\sqrt{\dfrac{\sigma_1^2}{n_1} + \dfrac{\sigma_2^2}{n_2}}}$$

was briefly considered. In the present $\sigma_1 = \sigma_2$ context, this can be rewritten as

$$\frac{\bar{x}_1 - \bar{x}_2 - (\mu_1 - \mu_2)}{\sigma\sqrt{\dfrac{1}{n_1} + \dfrac{1}{n_2}}} \tag{6-33}$$

and the fact is that (continuing to restrict attention to the case of underlying Gaussian distributions) the variable (6-33) is standard normal. One could use this fact to produce methods for confidence interval estimation and significance testing. But for use, these would require the input of the parameter σ, which is typically unknown. So instead of beginning with expression (6-28) or (6-33), it is standard to replace σ in expression (6-33) with s_P and begin with the quantity

$$\frac{(\bar{x}_1 - \bar{x}_2) - (\mu_1 - \mu_2)}{s_P\sqrt{\dfrac{1}{n_1} + \dfrac{1}{n_2}}} \tag{6-34}$$

It turns out that the form of expression (6-34) is crafted exactly so that under the present model assumptions, the variable (6-34) has a well-known, tabled probability distribution: the t distribution with $\nu = (n_1 - 1) + (n_2 - 1) = n_1 + n_2 - 2$ degrees of freedom. (Notice that the $n_1 - 1$ degrees of freedom associated with the first sample add together with the $n_2 - 1$ degrees of freedom associated with the second to produce $n_1 + n_2 - 2$ overall.) This probability fact, again via the kind of reasoning developed in Sections 6-1 and 6-2, produces inference methods for $\mu_1 - \mu_2$ that can be used even if one or both of n_1 and n_2 are small.

That is, a two-sided confidence interval for the difference $\mu_1 - \mu_2$, based on independent samples from underlying normal distributions with a common variance, has endpoints

$$\bar{x}_1 - \bar{x}_2 \pm t s_P \sqrt{\dfrac{1}{n_1} + \dfrac{1}{n_2}} \tag{6-35}$$

where t is chosen such that the probability that the $t_{n_1+n_2-2}$ distribution assigns to the interval between $-t$ and t corresponds to the desired confidence. And under the same conditions,

$$\mathrm{H}_0 \colon \mu_1 - \mu_2 = \#$$

can be tested using the statistic

$$T = \frac{\bar{x}_1 - \bar{x}_2 - \#}{s_P\sqrt{\dfrac{1}{n_1} + \dfrac{1}{n_2}}} \tag{6-36}$$

and a $t_{n_1+n_2-2}$ reference distribution.

EXAMPLE 6-8
(continued)

Return to the spring-life scenario, to illustrate small-sample formal inference for two means. First consider testing the hypothesis of equal mean lifetimes with an alternative of increased lifetime accompanying a reduction in stress level. Then consider making a two-sided 95% confidence interval for the difference in mean lifetimes.

Continuing to call the 900 N/mm^2 stress level condition 1 and the 950 N/mm^2 stress level condition 2, one has (from Table 6-7) that $\bar{x}_1 = 215.1$ and $\bar{x}_2 = 168.3$, while (from before) $s_P = 38.3$. The five-step significance-testing format then gives the following.

1. $H_0: \mu_1 - \mu_2 = 0$.
2. $H_a: \mu_1 - \mu_2 > 0$.

(The engineering expectation is that condition 1 produces the larger lifetimes.)

3. The test statistic is

$$T = \frac{\bar{x}_1 - \bar{x}_2 - 0}{s_P\sqrt{\dfrac{1}{n_1} + \dfrac{1}{n_2}}}$$

The reference distribution is t with $10 + 10 - 2 = 18$ degrees of freedom, and large observed t will count as evidence against H_0.

4. The samples give

$$t = \frac{215.1 - 168.3 - 0}{38.3\sqrt{\dfrac{1}{10} + \dfrac{1}{10}}} = 2.7$$

5. The observed level of significance is $P[$ a t_{18} random variable $\geq 2.7]$, which (according to Table D-4) is between .01 and .005. This is strong evidence that the lower stress level is associated with larger mean spring lifetimes.

Then, if one is to use expression (6-35) to produce a two-sided 95% confidence interval, the choice of t as the .975 quantile of the t_{18} distribution is in order. Endpoints of the confidence interval for $\mu_1 - \mu_2$ are

$$(215.1 - 168.3) \pm 2.101(38.3)\sqrt{\frac{1}{10} + \frac{1}{10}}$$

i.e.,

$$46.8 \pm 36.0$$

i.e.,

$$10.8 \times 10^3 \text{ cycles} \quad \text{and} \quad 82.8 \times 10^3 \text{ cycles}$$

The story told by these calculations is that the data in Table 6-7 provide enough information to establish convincingly that increased stress is associated with reduced mean spring life. But although the apparent magnitude of that reduction when moving from the 900 N/mm^2 level to the 950 N/mm^2 level is 46.8×10^3 cycles, the variability present in the data is large enough (and the sample sizes small enough) that only a precision of $\pm 36.0 \times 10^3$ cycles can be attached to the figure 46.8×10^3 cycles.

The inference methods represented by displays (6-35) and (6-36) are the last of the standard one- and two-sample methods for means. In the next two sections, parallel methods for variances and proportions are considered. But before leaving this section to consider those methods, a final comment is appropriate about the small-sample methods presented in this section.

This discussion has emphasized that strictly speaking, the nominal properties (in terms of *a priori* coverage probabilities for confidence intervals and relevant *p*-value declarations for significance tests) of the small-sample methods depend on the appropriateness of exactly Gaussian underlying distributions and (in these last cases) exactly equal variances. On the other hand, when actually applying the methods, rather crude probability-plotting checks have been used for verifying (only) that the models are roughly plausible. According to conventional statistical wisdom (based on a number of theoretical studies of the properties of the methods of this section), the small-sample methods presented here are remarkably robust to all but gross departures from the assumptions of Gaussian distributions and equal variance. That is, as long as the model assumptions are roughly a description of reality, the nominal confidence levels and *p*-values will not be ridiculously incorrect. (For example, a nominally 90% confidence interval method might in reality be only an 80% method, but it will not be only a 20% confidence interval method.) So the kind of plotting that has been illustrated here is often taken as adequate precaution against unjustified application of the small-sample inference methods for means.

6-4 One- and Two-Sample Inference for Variances

By now, at a number of different times and in various ways, this text has indicated that engineers must often pay close attention to the measurement, the prediction, and sometimes the physical reduction of the variability associated with a system response. Accordingly, it makes sense to consider the problems of formal inference for a single variance and formal inference for comparing two variances. In the process of doing so, two more standard families of probability distributions — the χ^2 distributions and the F distributions — will be introduced.

Inference for the Variance of a Normal Distribution

The key step in developing each type of formal inference discussed in this chapter has been to find a random quantity involving both the parameter (or function of parameters) of interest and sample-based quantities that can be shown to have some well-known distribution under appropriate circumstances. Repeatedly, some variable involving population means, sample means, and sample variances has been identified as approximately standard normal or as having a t distribution. Accordingly, it has been possible to reason out methods for confidence interval estimation and significance testing for the mean(s) involved. It turns out that existing inference methods for a single variance rely on a type of continuous probability distribution that has not yet been discussed in this book: the χ^2 distributions.

DEFINITION 6-15

The χ^2 (**Chi-squared**) **distribution with degrees of freedom parameter** ν, is a continuous probability distribution with probability density

$$f(x) = \begin{cases} \dfrac{1}{2^{\nu/2}\Gamma\left(\dfrac{\nu}{2}\right)} x^{(\nu/2)-1} e^{-x/2} & \text{for } x > 0 \\[2ex] 0 & \text{otherwise} \end{cases} \tag{6-37}$$

If a random variable has the probability density given by formula (6-37), it is said to have the χ^2_ν distribution.

Form (6-37) is not terribly inviting, but neither is it intractable. For instance, it is easy enough to use it to make the kind of sketch in Figure 6-16 for comparing the shapes of the χ_ν^2 distributions for various choices of ν.

FIGURE 6-16

χ^2 Probability Densities for $\nu = 1, 2, 3, 5,$ and 8

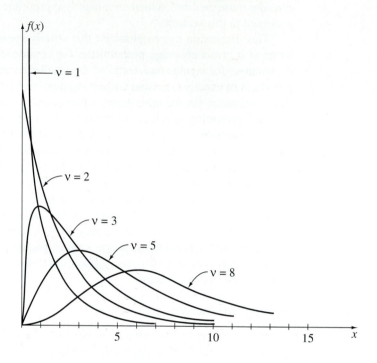

The χ_ν^2 distribution has mean ν and variance 2ν. For $\nu = 2$, it is exactly the exponential distribution with mean 2. For large ν, the χ_ν^2 distributions look increasingly bell-shaped (and can in fact be approximated by normal distributions with matching means and variances). Rather than using form (6-37) to find χ^2 probabilities, it is more common to use (necessarily abbreviated) tables of χ^2 quantiles. Table D-5 is one such table. Across the top of the table are several cumulative probabilities. Down the left side of the table are values of the degrees of freedom parameter, ν. In the body of the table are corresponding quantiles.

EXAMPLE 6-11

Use of a Table of χ^2 Distribution Quantiles. Suppose that U is a random variable with a χ_3^2 distribution. Consider first finding the .95 quantile of U's distribution and then seeing what Table D-5 says about $P[U < .4]$ and $P[U > 10.0]$.

First, looking at the $\nu = 3$ row of Table D-5 under the cumulative probability .95, one finds 7.815 in the body of the table. That is, $Q(.95) = 7.815$, or (equivalently) $P[U \leq 7.815] = .95$.

Then note that again using the $\nu = 3$ line of Table D-5, .4 lies between the .05 and .10 quantiles of the χ_3^2 distribution. Thus,

$$.05 < P[U < .4] < .10$$

Finally, since 10.0 lies between the ($\nu = 3$ line) entries of the table corresponding to cumulative probabilities .975 and .99 (i.e., the .975 and .99 quantiles of the χ_3^2 distribution), one may reason that

$$.01 < P[U > 10.0] < .025$$

The reason that the χ^2 distributions are of interest here is a probability fact concerning the behavior of the random variable s^2 if the observations from which it is calculated are iid normal random variables. Under such assumptions,

$$\frac{(n-1)s^2}{\sigma^2} \tag{6-38}$$

has a χ^2_{n-1} distribution. This fact is what is needed to identify inference methods for σ.

That is (subject to the limitations imposed by the rather abbreviated form of Table D-5), given a desired confidence level concerning σ, one can choose χ^2 quantiles (say, L and U) such that the probability that a χ^2_{n-1} random variable will take a value between L and U corresponds to that confidence level. (Typically, L and U are chosen to "split the 'unconfidence' between the upper and lower χ^2_{n-1} tails" — for example, using the .05 and .95 χ^2_{n-1} quantiles for L and U, respectively, if 90% confidence is of interest.) Then, because with an underlying normal distribution model, the variable (6-38) has a χ^2_{n-1} distribution, the probability that

$$L < \frac{(n-1)s^2}{\sigma^2} < U \tag{6-39}$$

corresponds to the desired confidence level. But expression (6-39) is algebraically equivalent to the eventuality that

$$\frac{(n-1)s^2}{U} < \sigma^2 < \frac{(n-1)s^2}{L}$$

But this then means that when an engineering data-generating mechanism can be thought of as essentially equivalent to random sampling from a normal distribution, a two-sided confidence interval for σ^2 has endpoints

$$\frac{(n-1)s^2}{U} \quad \text{and} \quad \frac{(n-1)s^2}{L} \tag{6-40}$$

where L and U are such that the χ^2_{n-1} probability assigned to the interval (L, U) corresponds to the desired confidence.

Further, there is an obvious significance-testing method for σ^2 to be derived from the notion that the variable (6-38) has a χ^2_{n-1} distribution. That is, subject to the same limitations needed to validate the confidence interval method,

$$H_0: \sigma^2 = \#$$

can be tested using the statistic

$$X^2 = \frac{(n-1)s^2}{\#} \tag{6-41}$$

and a χ^2_{n-1} reference distribution.

One feature of the testing methodology that may need comment concerns the computing of p-values in the case that the alternative hypothesis is of the form H_a: $\sigma^2 \neq \#$. (p-values for the one-sided alternative hypotheses H_a: $\sigma^2 < \#$ and H_a: $\sigma^2 > \#$ are, respectively, the left and right χ^2_{n-1} tail areas beyond the observed value of X^2.) The fact that the χ^2 distributions have no point of symmetry leaves some doubt for two-sided significance testing as to what types of observed values ought to be considered more extreme than the one provided by a sample. That is, how should one translate an observed value of X^2 into a (two-sided) p-value? The convention that will be used here is as follows. If the observed value is larger than the χ^2_{n-1} median, the (two-sided) p-value will be twice the χ^2_{n-1} probability to the right of the observed value. If the observed value of X^2 is smaller than the χ^2_{n-1} median, the (two-sided) p-value will be twice the χ^2_{n-1} probability to the left of the observed value.

Knowing that display (6-40) gives endpoints for a confidence interval for σ^2 also leads to knowing how to get a confidence interval for a function of σ^2. The square roots of the values in display (6-40) give endpoints for a confidence interval for the standard deviation, σ. And 6 times the square roots of the values in display (6-40) could be used as endpoints of a confidence interval for the "6σ" *capability of a process.*

EXAMPLE 6-12

Inference for the Capability of a CNC Lathe. Cowan, Renk, Vander Leest, and Yakes worked with a manufacturer of high-precision metal parts on a project involving a computer numerically controlled (CNC) lathe. A critical dimension of one particular part produced on the lathe had engineering specifications of the form

Nominal dimension $\pm$ .0020 in.

An important practical issue in such situations is whether or not the machine in question is capable of meeting specifications of this type. One way of addressing this concern is to collect data and perform inference for the intrinsic machine short-term variability, represented as a standard deviation. Table 6-8 gives values of the critical dimension measured on 20 parts machined on the lathe in question over a three-hour period. The units are .0001 in. over nominal.

TABLE 6-8

Measurements of a Dimension on 20 Parts Machined on a CNC Lathe

Measured Dimension (.0001 in. over nominal)	Frequency
8	1
9	1
10	10
11	4
12	3
13	1

For purposes of illustration, suppose one takes the $\pm$.0020 in. engineering specification as a statement of worst acceptable "$\pm 3\sigma$" machine capability, accordingly uses the data in Table 6-8, and (since $\frac{.0020}{3} \approx .0007$) tests H_0: $\sigma = .0007$. This significance test will then be followed by the making of a one-sided 99% confidence interval of the form $(0, \#)$ for 3σ.

The relevance of the methods represented by displays (6-40) and (6-41) depends on the appropriateness of an underlying normal distribution as a description of the critical dimension as machined in the three-hour period in question. In this regard, note that (at least after allowing for the fact of the obvious discreteness of measurement introduced by gaging read to .0001 in.) the normal plot of the data from Table 6-8 shown in Figure 6-17 is not distressing in its departure from linearity.

Further, at least over periods where manufacturing processes like the one in question are physically stable, normal distributions often prove to be quite adequate models for measured dimensions of mass-produced parts. Evidence available on the machining process (in the form of so-called control charts), which is not presented here, indicated that for practical purposes, the machining process was stable over the three-hour period in question. So one may proceed to use the normal-based methods, with no strong reason to doubt their relevance.

Direct calculation with the data of Table 6-8 shows that $s = 1.1 \times 10^{-4}$ in. So using the five-step significance-testing format, one has the following.

FIGURE 6-17

Normal Plot of Measurements on 20 Parts Machined on a CNC Lathe

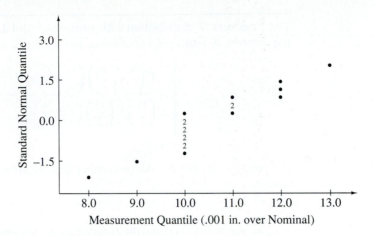

1. $H_0: \sigma = .0007$.
2. $H_a: \sigma > .0007$.

(The possibility that the machine is not capable of holding to the stated tolerances is the one of most practical concern, and this is described in terms of σ larger than standard.)

3. The test statistic is

$$X^2 = \frac{(n - 1)s^2}{(.0007)^2}$$

The reference distribution is χ^2 with $\nu = (20 - 1) = 19$ degrees of freedom, and large observed values of X^2 (resulting from large values of s^2) will constitute evidence against the validity of H_0.

4. The sample gives

$$x^2 = \frac{(20 - 1)(.00011)^2}{(.0007)^2} = .5$$

5. The observed level of significance is $P[\text{a } \chi^2_{19} \text{ random variable} \geq .5]$. Now .5 is smaller than the .005 quantile of the χ^2_{19} distribution, so the p-value exceeds .995. There is nothing in the data in hand to indicate that the machine is incapable of holding to the given tolerances.

According to Table D-5, the .01 quantile of the χ^2_{19} distribution is 7.633. So using expression (6-40), a 99% upper confidence bound for 3σ is

$$3\sqrt{\frac{(20 - 1)(1.1 \times 10^{-4} \text{ in.})^2}{7.633}} = 5.0 \times 10^{-4} \text{ in.}$$

When this is compared to the $\pm 20 \times 10^{-4}$ in. engineering requirement, it shows that the lathe in question is clearly capable of producing the kind of precision specified for the given dimension.

Inference for the Ratio of Two Variances (Based on Independent Samples from Normal Distributions)

To move from inference for a single variance to inference for comparing two variances requires the introduction of yet another new family of probability distributions: (Snedecor's) F distributions.

DEFINITION 6-16

The **(Snedecor) F distribution with numerator and denominator degrees of freedom parameters ν_1 and ν_2** is a continuous probability distribution with probability density

$$f(x) = \begin{cases} \dfrac{\Gamma\left(\dfrac{\nu_1 + \nu_2}{2}\right)\left(\dfrac{\nu_1}{\nu_2}\right)^{\nu_1/2} x^{(\nu_1/2)-1}}{\Gamma\left(\dfrac{\nu_1}{2}\right)\Gamma\left(\dfrac{\nu_2}{2}\right)\left(1 + \dfrac{\nu_1 x}{\nu_2}\right)^{(\nu_1+\nu_2)/2}} & \text{for } x > 0 \\[2em] 0 & \text{otherwise} \end{cases} \qquad (6\text{-}42)$$

If a random variable has the probability density given by formula (6-42), it is said to have the F_{ν_1,ν_2} distribution.

As Figure 6-18 reveals, the F distributions are strongly right-skewed distributions, whose densities achieve their maximum values at arguments somewhat less than 1. Roughly speaking, the smaller the values ν_1 and ν_2, the more asymmetric and spread out is the corresponding F distribution.

FIGURE 6-18

Four Different F
Probability Densities

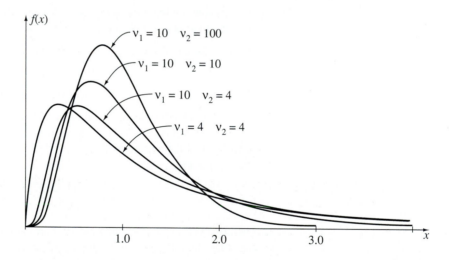

Direct use of formula (6-42) to find probabilities for the F distributions requires numerical integration methods. For purposes of applying the F distributions in statistical inference, the path typically taken is to make use of some fairly abbreviated tables of F distribution quantiles instead. In Appendix Tables D-6, a set of such tables of F quantiles is presented. Notice that the body of a particular one of these tables, for a single p, gives the F distribution p quantiles, for various combinations of ν_1 (the numerator degrees of freedom) and ν_2 (the denominator degrees of freedom). The values of ν_1 are given across the top margin of the table and the values of ν_2 down the left margin.

Notice also in Tables D-6 that the tables of p quantiles are only for p larger than .5. It turns out that one needs F distribution quantiles for p smaller than .5 as well. Rather than making up tables of such values, it is standard practice to make use of a computational trick: One can get quantiles for small p by using a relationship that holds between F_{ν_1,ν_2} and F_{ν_2,ν_1} quantiles. The fact is that if one lets Q_{ν_1,ν_2} stand for the F_{ν_1,ν_2} quantile function and Q_{ν_2,ν_1} stand for the quantile function for the F_{ν_2,ν_1} distribution,

$$Q_{\nu_1,\nu_2}(p) = \frac{1}{Q_{\nu_2,\nu_1}(1-p)} \qquad (6\text{-}43)$$

Expression (6-43) means that a small lower percentage point of an F distribution may be obtained by taking the reciprocal of a corresponding small upper percentage point of the F distribution with degrees of freedom reversed.

EXAMPLE 6-13

Use of Tables of F Distribution Quantiles. Suppose that V is an $F_{3,5}$ random variable. Consider finding the .95 and .01 quantiles of V's distribution and then seeing what Tables D-6 reveal about $P[V > 4.0]$ and $P[V < .3]$.

First, a direct look-up in the $p = .95$ table of quantiles, in the $\nu_1 = 3$ column and $\nu_2 = 5$ row, produces the number 5.41. That is, $Q(.95) = 5.41$, or (equivalently) $P[V < 5.41] = .95$.

As far as finding the $p = .01$ quantile of the $F_{3,5}$ distribution goes, expression (6-43) must be used. That is,

$$Q_{3,5}(.01) = \frac{1}{Q_{5,3}(.99)}$$

so that using the $\nu_1 = 5$ column and $\nu_2 = 3$ row of the table of F .99 quantiles, one has

$$Q_{3,5}(.01) = \frac{1}{28.24} = .04$$

Next considering $P[V > 4.0]$, one finds (using the $\nu_1 = 3$ columns and $\nu_2 = 5$ rows of Tables D-6) that 4.0 lies between the .90 and .95 quantiles of the $F_{3,5}$ distribution. That is,

$$.90 < P[V \leq 4.0] < .95$$

so that

$$.05 < P[V > 4.0] < .10$$

Finally, considering $P[V < .3]$, note that none of the entries of Tables D-6 is less than 1.00. So to place the value .3 in the $F_{3,5}$ distribution, one must locate its reciprocal, 3.33, in the $F_{5,3}$ distribution and then make use of expression (6-43). Using the $\nu_1 = 5$ columns and $\nu_2 = 3$ rows of Tables D-6, one finds that 3.33 is between the .75 and .90 quantiles of the $F_{5,3}$ distribution. So by expression (6-43), .3 is between the .1 and .25 quantiles of the $F_{3,5}$ distribution, and

$$.10 < P[V < .3] < .25$$

The extra gyrations one is forced to go through to find small F distribution quantiles are an artifact of standard table-making practice, rather than being any intrinsic extra difficulty associated with the F distributions.

The reason that the F distributions are of use at this point is a probability fact that ties the behavior of ratios of independent sample variances based on samples from underlying normal distributions to the variances σ_1^2 and σ_2^2 of those underlying distributions. That is, when s_1^2 and s_2^2 come from independent samples from normal distributions, the variable

$$\frac{s_1^2}{\sigma_1^2} \cdot \frac{\sigma_2^2}{s_2^2} \tag{6-44}$$

has an F_{n_1-1,n_2-1} distribution. (Notice that s_1^2 has $n_1 - 1$ associated degrees of freedom and is in the numerator of this expression, while s_2^2 has $n_2 - 1$ associated degrees of freedom and is in the denominator, providing motivation for the language introduced in Definition 6-16.)

Having an appropriate probability distribution for the quantity (6-44) is exactly what is needed to produce formal inference methods for the ratio σ_1^2/σ_2^2. For example, providing the necessary probability model assumptions are satisfied, it is possible to pick appropriate F quantiles L and U such that the probability that the variable (6-44) falls between L and U corresponds to a desired confidence level. (Typically, L and U are chosen to "split the 'unconfidence'" between the upper and lower F_{n_1-1,n_2-1} tails.) But the eventuality that

$$L < \frac{s_1^2}{\sigma_1^2} \cdot \frac{\sigma_2^2}{s_2^2} < U$$

is algebraically equivalent to

$$\frac{1}{U} \cdot \frac{s_1^2}{s_2^2} < \frac{\sigma_1^2}{\sigma_2^2} < \frac{1}{L} \cdot \frac{s_1^2}{s_2^2}$$

That is, when a data-generating mechanism can be thought of as essentially equivalent to independent random sampling from two normal distributions, a two-sided confidence interval for σ_1^2/σ_2^2 has endpoints

$$\frac{s_1^2}{U \cdot s_2^2} \qquad \text{and} \qquad \frac{s_1^2}{L \cdot s_2^2} \tag{6-45}$$

where L and U are (F_{n_1-1,n_2-1} quantiles) such that the F_{n_1-1,n_2-1} probability assigned to the interval (L, U) corresponds to the desired confidence.

In addition, there is an obvious significance-testing method for σ_1^2/σ_2^2 to be derived from the notion that the variable (6-44) has an F_{n_1-1,n_2-1} distribution. That is, subject to the same modeling limitations as needed to validate the confidence interval method,

$$\text{H}_0: \frac{\sigma_1^2}{\sigma_2^2} = \# \tag{6-46}$$

can be tested using the statistic

$$F = \frac{s_1^2/s_2^2}{\#} \tag{6-47}$$

and an F_{n_1-1,n_2-1} reference distribution. (It is worth noting that the choice of $\# = 1$ in displays (6-46) and (6-47), so that the null hypothesis is one of equality of variances, is the only one commonly used in practice.) p-values for the one-sided alternative hypotheses $\text{H}_a: \sigma_1^2/\sigma_2^2 < \#$ and $\text{H}_a: \sigma_1^2/\sigma_2^2 > \#$ are (respectively) the left and right F_{n_1-1,n_2-1} tail areas beyond the observed values of the test statistic. For the two-sided alternative hypothesis $\text{H}_a: \sigma_1^2/\sigma_2^2 \neq \#$, the standard convention is to report twice the F_{n_1-1,n_2-1} probability to the right of the observed f if $f > 1$ and to report twice the F_{n_1-1,n_2-1} probability to the left of the observed f if $f < 1$.

EXAMPLE 6-14

Comparing Uniformity of Hardness Test Results for Two Types of Steel. Condon, Smith, and Woodford did some hardness testing on specimens of 4% carbon steel. Part of their data are given in Table 6-9, where Rockwell hardness measurements for ten specimens from a lot of heat treated steel specimens and five specimens from a lot of cold rolled steel specimens are represented.

TABLE 6-9	Heat Treated (Rockwell Hardness)	Cold Rolled (Rockwell Hardness)
Rockwell Hardness Measurements for Steel Specimens of Two Types	32.8, 44.9, 34.4, 37.0, 23.6, 29.1, 39.5, 30.1, 29.2, 19.2	21.0, 24.5, 19.9, 14.8, 18.8

Consider comparing measured hardness *uniformity* for these two steel types (rather than mean hardness, as might have been done in Section 6-3). Figure 6-19 shows side-by-side dot diagrams for the two samples and suggests that there is a larger variability associated with the heat treated specimens than with the cold rolled specimens.

FIGURE 6-19

Dot Diagrams of Hardness for Heat Treated and Cold Rolled Steels

Heat Treated

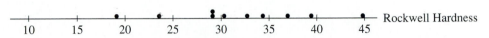

Cold Rolled

The two normal plots in Figure 6-20 indicate no obvious problems with a model assumption of normal underlying distributions.

Then, arbitrarily choosing to call the heat treated condition number 1 and cold rolled condition 2, one has $s_1 = 7.52$ and $s_2 = 3.52$, and a five-step significance test of equality of variances based on the variable (6-47) proceeds as follows.

1. $H_0: \dfrac{\sigma_1^2}{\sigma_2^2} = 1$.

2. $H_a: \dfrac{\sigma_1^2}{\sigma_2^2} \neq 1$.

(If there is any reason to pick a one-sided alternative hypothesis here, the author (who is not a materials expert) doesn't know it.)

3. The test statistic is

$$F = \frac{s_1^2}{s_2^2}$$

The reference distribution is the $F_{9,4}$ distribution, and both large observed f and small observed f will constitute evidence against H_0.

4. The samples give

$$f = \frac{(7.52)^2}{(3.52)^2} = 4.6$$

5. Since the observed f is larger than 1, for the two-sided alternative, the p-value is

$$2P[\text{an } F_{9,4} \text{ random variable} \geq 4.6]$$

FIGURE 6-20

Normal Plots of Hardness
for Heat Treated and Cold
Rolled Steels

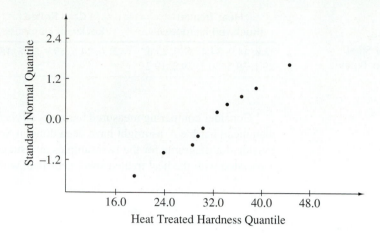

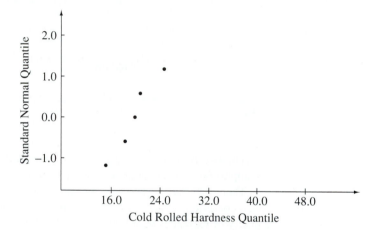

From Tables D-6, 4.6 is between the $F_{9,4}$ distribution .9 and .95 quantiles, so the observed level of significance is between .1 and .2. This makes moderately (but not completely) implausible the possibility that the heat treated and cold rolled variabilities are comparable.

In an effort to pin down the relative sizes of the heat treated and cold rolled hardness variabilities, one may use the square roots of the expressions in display (6-45) and also give a 90% two-sided confidence interval for σ_1/σ_2. Now the .95 quantile of the $F_{9,4}$ distribution is 6.0, while the .95 quantile of the $F_{4,9}$ distribution is 3.63, implying that the .05 quantile of the $F_{9,4}$ distribution is $\frac{1}{3.63}$. Thus, a 90% confidence interval for the ratio of standard deviations σ_1/σ_2 has endpoints

$$\sqrt{\frac{(7.52)^2}{6.0(3.52)^2}} \quad \text{and} \quad \sqrt{\frac{(7.52)^2}{(1/3.63)(3.52)^2}}$$

That is,

$$.87 \quad \text{and} \quad 4.07$$

Notice that the fact that the interval (.87, 4.07) covers values both smaller and larger than 1 indicates that the data in hand do not provide definitive evidence even as to which of the two variabilities in material hardness is larger.

One of the most important engineering applications of the inference methods represented by displays (6-45) through (6-47) is in the comparison of inherent precisions for different pieces of equipment and for different methods of operating a single piece of equipment.

EXAMPLE 6-15

Comparing Uniformities of Operation of Two Ream Cutters. Abassi, Afinson, Shezad, and Yeo worked with a company that cuts rolls of paper into sheets. The uniformity of the sheet lengths is important, because the better the uniformity, the closer the average sheet length can be set to the nominal value without producing undersized sheets, thereby reducing the company's giveaway costs. The students compared the uniformity of sheets cut on a ream cutter having a manual brake to the uniformity of sheets cut on a ream cutter that had an automatic brake. The basis of that comparison was estimated standard deviations of sheet lengths cut by the two machines — just the kind of information used to frame formal inferences in this section. The students estimated $\sigma_{\text{manual}}/\sigma_{\text{automatic}}$ to be on the order of 1.5 and predicted a period of two years or less for the recovery of the capital improvement cost of equipping all the company's ream cutters with automatic brakes.

The methods of this section are, strictly speaking, normal distribution methods, and it is worthwhile to ask, "How essential is this Gaussian restriction to the predictable behavior of these inference methods for one and two variances?" There is a remark at the end of Section 6-3 to the effect that the methods presented there for *means* are fairly robust to moderate violation of the section's model assumptions. Unfortunately, conventional statistical wisdom is that such is *not* the case for the methods for *variances* presented here.

These are methods whose nominal confidence levels and *p*-values can be fairly badly misleading unless the corresponding model is a good one. This makes the kind of careful data scrutiny that has been implemented in the examples (in the form of normal-plotting) essential to the responsible use of the methods of this section. And it suggests that since normal-plotting itself isn't typically terribly revealing unless the sample size involved is moderate to large, formal inferences for variances will be most safely made on the basis of moderate to large normal-looking samples.

The importance of the "normal distribution(s)" restriction to the predictable operation of the methods of this section is not the only reason to prefer large sample sizes for inferences on variances. A little experience with the formulas in this section will convince the reader that (even granting the appropriateness of Gaussian models) small samples often do not prove adequate to answer practical questions about variances. χ^2 and F confidence intervals for variances and variance ratios based on small samples can be so big as to be of little practical value, and the engineer will typically be driven to large sample sizes in order to solve variance-related real-world problems. Please do not take this as in any way a failing of the present methods. It is simply a warning and quantification of the fact that learning about variances requires more data than (for example) learning about means.

6-5   One- and Two-Sample Inference for Proportions

The methods of formal statistical inference that have been introduced to this point are useful in the analysis of quantitative data. Occasionally, however, engineering studies produce only qualitative data, and one is faced with the problem of making properly hedged inferences from such data. This section considers how the sample fraction $\hat{p}$, defined in Section 3-4 as a summary measure for qualitative data, can be used as the basis for formal statistical inferences. It begins with the use of $\hat{p}$ from a single sample to make formal inferences about a single system or population. The section then considers the use of sample proportions from two samples to make formal inferences comparing two systems or populations.

Inference for a Single Proportion

Recall from Section 3-4 that the notation $\hat{p}$ is used for the fraction of a sample that possesses a characteristic of engineering interest. A sample of pellets produced by a pelletizing machine might individually prove conforming or nonconforming, and $\hat{p}$ could be the sample fraction conforming. Or in another context, a sample of turned steel shafts might individually prove acceptable, reworkable, or scrap; $\hat{p}$ could be the sample fraction reworkable.

If formal statistical inferences are to be based on $\hat{p}$, it is obvious that one must be able to think of the physical situation involved in such a way that $\hat{p}$ is related to some germane parameter characterizing it. This is in direct analogy to what has already been done here for means and variances. For example, when one can think of a data-generating mechanism as essentially iid (for example, derived from a physically stable process), the sample mean of quantitative data, $\bar{x}$, gives a basis for inferences on the underlying parameter μ.

Accordingly, this section considers scenarios where $\hat{p}$ is derived from a data-generating mechanism that is essentially iid in the sense that 1) all sampled items possess the same *a priori* likelihood of having the characteristic of interest and 2) whether or not a given item in the sample has the characteristic is unrelated to whether or not any other item in the sample has the characteristic. The underlying likelihood of an item having the characteristic will be denoted as p, and inferences will concern this parameter. Applications will include inferences about physically stable processes, where p is a system's propensity to produce an item with the characteristic of interest. And they will include inferences drawn about population proportions p, in enumerative contexts involving large populations. For example, the methods that are about to be introduced are used both to make inferences about the routine operation of a physically stable pelletizing machine and also to make inferences about the fraction of nonconforming machine parts contained in a specific lot of 10,000 such parts.

The realization that the discussion here will center on scenarios of the stable process/random sampling type should remind the reader of that part of Section 5-1 that concerned independent, identical, success-failure trials models — and specifically the binomial distributions. In fact, a little review of that material should convince the reader that the probabilistic connection that will allow the use of $\hat{p}$ as a basis of inference for p is that under the present assumptions, the variable

$n\hat{p}$ = the number of items in the sample with the characteristic of interest

has the binomial (n, p) distribution.

Now the sample fraction $\hat{p}$ is just a scale change away from $n\hat{p}$, so facts about the distribution of $n\hat{p}$ have immediate counterparts regarding the distribution of $\hat{p}$. For example, Section 5-1 stated that the mean and variance for the binomial (n, p) distribution are (respectively) np and $np(1 - p)$. This (together with Proposition 5-1) implies that $\hat{p}$ has

$$E\hat{p} = E\left(\frac{1}{n} \cdot n\hat{p}\right) = \frac{1}{n}E(n\hat{p}) = \frac{1}{n} \cdot np = p \tag{6-48}$$

and

$$Var\,\hat{p} = Var\left(\frac{1}{n} \cdot n\hat{p}\right) = \left(\frac{1}{n}\right)^2 Var(n\hat{p}) = \frac{np(1 - p)}{n^2} = \frac{p(1 - p)}{n} \tag{6-49}$$

Equations (6-48) and (6-49) provide a crude but reassuring picture of the behavior of the statistic $\hat{p}$. They show that the probability distribution of $\hat{p}$ is centered at the underlying parameter p, with a variability that decreases as n increases.

EXAMPLE 6-16

(Example 5-3 revisited) **Means and Standard Deviations of Sample Fractions of Reworkable Shafts.** Consider again the situation of the *Quality Progress* study of H. Rowe, concerning the performance of a process for turning steel shafts. Assume for the time being that the process is physically stable and that the likelihood that a given shaft is reworkable is $p = .20$. Then consider $\hat{p}$, the sample fraction of reworkable shafts in samples of first $n = 4$ and then $n = 100$ shafts.

Expressions (6-48) and (6-49) show that for the $n = 4$ sample size,

$$E\hat{p} = p = .2$$

$$\sqrt{Var\,\hat{p}} = \sqrt{\frac{p(1-p)}{n}} = \sqrt{\frac{(.2)(.8)}{4}} = .2$$

Similarly, for the $n = 100$ sample size,

$$E\hat{p} = p = .2$$

$$\sqrt{Var\,\hat{p}} = \sqrt{\frac{(.2)(.8)}{100}} = .04$$

Comparing the two standard deviations, it is clear that the effect of a change in sample size from $n = 4$ to $n = 100$ is to produce a factor of 5 ($= \sqrt{100/4}$) decrease in the standard deviation of $\hat{p}$, while the distribution of $\hat{p}$ is centered at p for both sample sizes.

It is possible to work directly with the fact that $n\hat{p}$ has a binomial distribution and come up with exact methods for confidence interval estimation and significance testing for p, based on any sample size n. However, this section restricts attention to large-n situations and provides approximate methods for confidence interval estimation and significance testing. This is for two reasons. First, the methods presented here are much easier to explain and understand than are the exact methods. And second, unless n is large, $\hat{p}$ doesn't really carry much information about p. So for small n, the exact methods may be mathematically correct, but they are of limited practical value; while for large n, they give inferences nearly equivalent to those of the approximate methods presented here.

The basic new insight needed to provide inference methods based on $\hat{p}$ is the fact that for large n, the binomial (n, p) distribution (and therefore also the distribution of $\hat{p}$) is approximately normal. That is, for large n, approximate probabilities for $n\hat{p}$ (or $\hat{p}$) can be found using the normal distribution with mean $\mu = np$ (or $\mu = p$) and variance $\sigma^2 = np(1 - p)$ (or $\sigma^2 = \frac{p(1-p)}{n}$).

EXAMPLE 6-16
(continued)

In the shaft turning example, consider evaluating the probability that for a sample of $n = 100$ shafts, $\hat{p} \geq .25$. Notice that $\hat{p} \geq .25$ is equivalent here to the eventuality that $n\hat{p} \geq 25$. So one could in theory use the form of the binomial probability function given in Definition 5-9 and evaluate the desired probability exactly as

$$P[\hat{p} \geq .25] = P[n\hat{p} \geq 25] = f(25) + f(26) + \cdots + f(99) + f(100)$$

But direct use of the binomial probability function for large n and even moderate numbers of successes is unpleasant (at least unless appropriate computer software is available). Instead of making such laborious calculations, it is common (and typically adequate for practical purposes) to settle instead for a Gaussian approximation to probabilities such as the one desired here.

Figure 6-21 shows the normal distribution with mean $\mu = p = .2$ and standard deviation $\sigma = \sqrt{p(1-p)/n} = .04$ and the corresponding probability assigned to the interval $[.25, \infty)$. Conversion of .25 to a z value and then an approximate probability proceeds as follows.

$$z = \frac{.25 - E\hat{p}}{\sqrt{Var\,\hat{p}}} = \frac{.25 - .2}{.04} = 1.25$$

so

$$P[\hat{p} \geqslant .25] \approx 1 - \Phi(1.25) = .1056 \approx .11$$

As it turns out, the exact value of $P[\hat{p} \geqslant .25]$ (calculated to four decimal places using the binomial probability function) is .1314. It is possible to improve somewhat on normal approximations to large-n binomial probabilities like this one, by making appropriate .5 unit corrections to values of $n\hat{p}$ before calculating z values. That is, when one uses the binomial probability function directly, the probability given by the area under a binomial probability histogram corresponding to the "$X = 25$," "$X = 26$,"... bars (extending from 24.5 to the right) is being calculated. It thus makes sense to begin the normal approximation with $n\hat{p} = 24.5$ (rather than 25), so $\hat{p} = .245$ (rather than $\hat{p} = .25$).

FIGURE 6-21

Approximate Probability
Distribution for $\hat{p}$

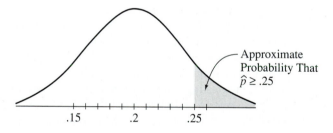

For $n = 100$, the Approximate
Distribution of $\hat{p}$ Is Normal with
Mean .2 and Standard Deviation .04.

Approximate
Probability That
$\hat{p} \geq .25$

.15   .2   .25

Figure 6-22 illustrates this kind of thinking (which incidentally produces a normal approximation agreeing with the exact probability to two decimal places in the present example).

FIGURE 6-22

Schematic of an Exact
Binomial Probability
Histogram for $n\hat{p}$
and an Approximating
Normal Density

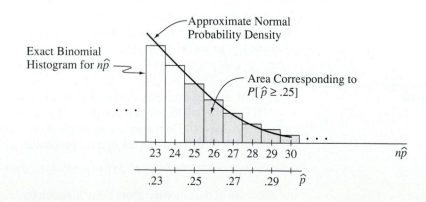

Approximate Normal
Probability Density

Exact Binomial
Histogram for $n\hat{p}$

Area Corresponding to
$P[\hat{p} \geq .25]$

23 24 25 26 27 28 29 30 $n\hat{p}$

.23   .25   .27   .29 $\hat{p}$

An adjustment of this kind is sometimes called a *continuity correction*, since it becomes relevant because a continuous distribution is being used to approximate a discrete one. No extensive use of the continuity correction will be made in this book. This brief discussion has been included because the reader is likely to encounter the terminology elsewhere, and some familiarity with it is in order.

The statement that for large n, the random variable $\hat{p}$ is approximately normal is actually a version of the Central Limit Theorem. For a given n, the approximation is best for moderate p (i.e., p near .5), and a common rule of thumb is to require that both the expected number of successes and the expected numer of failures be at least 5 before making use of a normal approximation to the binomial (n, p) distribution. This is a requirement that

$$np \geqslant 5 \qquad \text{and} \qquad n(1 - p) \geqslant 5$$

which amounts to a requirement that

$$5 \leqslant np \leqslant n - 5 \tag{6-50}$$

(Notice that in Example 6-16, $np = 100(.2) = 20$ and $5 \leqslant 20 \leqslant 95$.)

An alternative, and typically somewhat more stringent rule of thumb (which comes from a requirement that the mean of the binomial distribution be at least 3 standard deviations from both 0 and n) is to require that

$$9 \leqslant (n + 9)p \leqslant n \tag{6-51}$$

before using the normal approximation. (Again in Example 6-16, $(n + 9)p = (100 + 9)(.2) = 21.8$ and $9 \leqslant 21.8 \leqslant 100$.)

The approximate normality of $\hat{p}$ for large n implies that for large n,

$$\frac{\hat{p} - p}{\sqrt{\dfrac{p(1 - p)}{n}}} \tag{6-52}$$

is approximately standard normal. This and the reasoning of Section 6-2 then imply that the null hypothesis

$$H_0\colon p = \#$$

can be tested using the statistic

$$Z = \frac{\hat{p} - \#}{\sqrt{\dfrac{\#(1 - \#)}{n}}} \tag{6-53}$$

and a standard normal reference distribution. Further, reasoning parallel to that in Section 6-1 (beginning with the fact that the variable (6-52) is approximately standard normal), leads to the conclusion that an interval with endpoints

$$\hat{p} \pm z\sqrt{\frac{p(1 - p)}{n}} \tag{6-54}$$

(where z is chosen such that the standard normal probability between $-z$ and z corresponds to a desired confidence) is a mathematically valid two-sided confidence interval for p.

However, the endpoints indicated by expression (6-54) are of no practical use as they stand, since they involve the unknown parameter p. There are two standard ways of remedying this situation. One draws its motivation from the simple plot of $p(1 - p)$ as a function of p that is shown in Figure 6-23. That is, from Figure 6-23 it is easy to see that $p(1 - p) \leqslant (.5)^2 = .25$, so the plus-or-minus part of formula (6-54) has (for $z > 0$)

$$z\sqrt{\frac{p(1 - p)}{n}} \leqslant z\frac{1}{2\sqrt{n}}$$

Thus, modifying the endpoints in formula (6-54) by replacing the plus-or-minus part with $\pm z/2\sqrt{n}$ produces an interval that is guaranteed to be as wide as necessary to give the desired approximate confidence level. That is, the interval with endpoints

$$\hat{p} \pm z\frac{1}{2\sqrt{n}} \tag{6-55}$$

where z is chosen such that the standard normal probability between $-z$ and z corresponds to a desired confidence, is a practically usable large-n two-sided *conservative* confidence interval for p. (Appropriate use of only one of the endpoints in display (6-55), as usual, gives a one-sided confidence interval.)

FIGURE 6-23

Plot of $p(1 - p)$ versus p

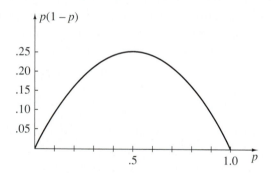

The other common method of dealing with the fact that the endpoints in formula (6-54) are of no practical use is to begin one's search for a formula from a point other than the approximate standard normal distribution of the variable (6-52). That is, it turns out that for large n, not only is the variable (6-52) approximately standard normal, but so is

$$\frac{\hat{p} - p}{\sqrt{\dfrac{\hat{p}(1 - \hat{p})}{n}}} \tag{6-56}$$

And the denominator of the quantity (6-56) (which amounts to an estimated standard deviation for $\hat{p}$) is free of the parameter p. So when manipulations parallel to those in Section 6-1 are applied to expression (6-56), one ends up with the conclusion that the interval with endpoints

$$\hat{p} \pm z\sqrt{\frac{\hat{p}(1 - \hat{p})}{n}} \tag{6-57}$$

can be used as a two-sided large-n confidence interval for p with confidence level corresponding to the standard normal probability assigned to the interval between $-z$ and z. (One-sided confidence limits are obtained in the usual way, using only one of the endpoints in display (6-57) and appropriately adjusting the confidence level.)

EXAMPLE 6-17

Inference for the Fraction of Dry Cells with Internal Shorts. The article "A Case Study of the Use of an Experimental Design in Preventing Shorts in Nickel-Cadmium Cells" by Ophir, El-Gad, and Snyder (*Journal of Quality Technology*, 1988) describes a series of experiments conducted in order to reduce the proportion of cells being scrapped by a battery plant because of internal shorts. Although the paper is not completely explicit on this point, it seems from several comments in the paper that at the beginning of the study about 6% of the cells produced were being scrapped because of internal shorts.

At one point in the work described in the paper, in a sample of 235 cells produced under a particular trial set of plant operating conditions, there were 9 cells with shorts. Consider what formal inferences can be drawn about the set of operating conditions based on such data. To begin with, note that $\hat{p} = \frac{9}{235} = .038$ and that if one desires, for example, a two-sided 95% confidence interval for p, expression (6-57) produces endpoints

$$.038 \pm 1.96 \sqrt{\frac{(.038)(1 - .038)}{235}}$$

i.e.,

$$.038 \pm .025$$

i.e.,

$$.013 \quad \text{and} \quad .063 \qquad\qquad \textbf{(6-58)}$$

Notice that according to display (6-58), although $\hat{p} = .038 < .06$ (and there is thus indication that the trial conditions furnish an improvement over the standard ones), the case for this is not airtight. The data in hand seem to allow some possibility that p for the trial conditions even exceeds .06. And the ambiguity is further emphasized if the conservative formula (6-55) is used in place of expression (6-57). Verify that instead of 95% confidence endpoints of $.038 \pm .025$, formula (6-55) gives endpoints $.038 \pm .064$.

To illustrate the significance-testing method represented by expression (6-53), consider testing with an alternative hypothesis that the trial conditions are an improvement over the standard ones. One then has the following summary.

1. $H_0: p = .06$.
2. $H_a: p < .06$.
3. The test statistic is

$$Z = \frac{\hat{p} - .06}{\sqrt{\dfrac{(.06)(1 - .06)}{n}}}$$

The reference distribution is standard normal, and small observed values z will count as evidence against H_0.
4. The sample gives

$$z = \frac{.038 - .06}{\sqrt{\dfrac{(.06)(1 - .06)}{235}}} = -1.42$$

5. The observed level of significance is then

$$\Phi(-1.42) = .08$$

This is strong but not overwhelming evidence that the trial conditions are an improvement on the standard ones.

It needs to be emphasized again that these inferences depend for their practical relevance on the appropriateness of the "stable process/independent, identical trials" model for the battery-making process and extend only as far as that description continues to make sense. It is important that the experience reported in the article was gained under (presumably physically stable) regular production, so there is reason to hope that a single "independent, identical trials" model can describe both experimental and future process behavior.

Section 6-1 illustrated the fact that that the form of the large-n confidence interval for a mean can be used to guide sample size choices for estimating μ. The same is true regarding the estimation of p. That is, if one has in mind a desired confidence level and a worst-case (largest) expectation for $\hat{p}(1 - \hat{p})$ in expression (6-57) (or if one plans to use expression (6-55)) and has a desired precision of estimation of p, it is a simple matter to solve for a corresponding sample size.

EXAMPLE 6-17
(continued)

Return to the nicad battery example and suppose that for some reason a better fix on the implications of the trial set of operating conditions (leading to the 9 shorted cells in sample of 235) was desired. In fact, suppose that p is to be estimated with a two-sided conservative 95% confidence interval, and a $\pm.01$ (fraction defective) precision of estimation is desired. Then, using the plus-or-minus part of expression (6-55) (or equivalently, the plus-or-minus part of expression (6-57) under the worst-case scenario that $\hat{p} = .5$), one is led to set

$$.01 = 1.96 \frac{1}{2\sqrt{n}}$$

From this, one finds that a sample size of

$$n \approx 9{,}604$$

is required.

In most engineering contexts, probably including that of the battery factory, a sample size on this order is ridiculously large from a practical point of view. Rethinking the calculation by planning the use of expression (6-57) and adopting the point of view that, say, 10% is a worst-case expectation for $\hat{p}$ (and thus $.1(1-.1) = .09$ is a worst-case expectation for $\hat{p}(1 - \hat{p})$), one might be led instead to set

$$.01 = 1.96 \sqrt{\frac{(.1)(1 - .1)}{n}}$$

However, solving for n, one has

$$n \approx 3{,}458$$

which is still probably out of the realm of what is practically workable.

The moral of these calculations should be clear. Something has to give. The kind of large confidence and somewhat precise estimation requirements set at the beginning here cannot typically be simultaneously satisfied using a realistic sample size. One or the other of the requirements is going to have to be relaxed.

The sample-size conclusions just illustrated are typical, and they provide quantitative justification for two points already made in this book in qualitative terms. First, loosely speaking, qualitative data carry less information than corresponding numbers of quantitative data (and therefore usually require very large samples to produce definitive inferences). This makes measurements generally preferable to qualitative observations in engineering applications. Second, if inferences about p based on even large values of n are often disappointing in their precision or reliability, there is little practical motivation to consider small-sample inference for p.

Inference for the Difference between Two Proportions (Based on Independent Samples)

Two separately derived sample proportions $\hat{p}_1$ and $\hat{p}_2$, representing different processes or populations, can form the basis of simple formal inference methods for comparing those processes or populations. The logic behind those methods concerns the difference $\hat{p}_1 - \hat{p}_2$. If 1) the "independent, identical success-failure trials" description applies separately to the mechanisms that generate two samples, 2) the two samples are reasonably described as separately derived/independent, and 3) both n_1 and n_2 are large, a very simple approximate description of the distribution of $\hat{p}_1 - \hat{p}_2$ results.

To begin with, notice that (assuming the samples are independent) Proposition 5-1 and the discussion in this section concerning the mean and variance of a single sample proportion imply that $\hat{p}_1 - \hat{p}_2$ has

$$E(\hat{p}_1 - \hat{p}_2) = E\hat{p}_1 + (-1)E\hat{p}_2 = p_1 - p_2 \tag{6-59}$$

and

$$Var(\hat{p}_1 - \hat{p}_2) = (1)^2\, Var\,\hat{p}_1 + (-1)^2\, Var\,\hat{p}_2 = \frac{p_1(1 - p_1)}{n_1} + \frac{p_2(1 - p_2)}{n_2} \tag{6-60}$$

Then the approximate normality of the independent variables $\hat{p}_1$ and $\hat{p}_2$ for large sample sizes turns out to imply the approximate normality of the difference $\hat{p}_1 - \hat{p}_2$. That is, $\hat{p}_1 - \hat{p}_2$ has an approximately Gaussian distribution with mean $p_1 - p_2$ and variance given by equation (6-60).

EXAMPLE 6-16
(continued)

Consider again the context of the turning of steel shafts, and imagine that two different, physically stable lathes produce reworkable shafts at respective rates of 20 and 25%. Then suppose that samples of (respectively) $n_1 = 50$ and $n_2 = 50$ shafts produced by the machines are taken, and the sample fractions reworkable $\hat{p}_1$ and $\hat{p}_2$ are found. Consider approximating the probability that $\hat{p}_1 \geqslant \hat{p}_2$ (i.e., that $\hat{p}_1 - \hat{p}_2 \geqslant 0$).

Using expressions (6-59) and (6-60), the variable $\hat{p}_1 - \hat{p}_2$ has

$$E(\hat{p}_1 - \hat{p}_2) = .20 - .25 = -.05$$

and

$$\sqrt{Var(\hat{p}_1 - \hat{p}_2)} = \sqrt{\frac{(.20)(1 - .20)}{50} + \frac{(.25)(1 - .25)}{50}} = \sqrt{.00695} = .083$$

Figure 6-24 shows the approximate normal distribution of $\hat{p}_1 - \hat{p}_2$ and the area corresponding to $P[\hat{p}_1 - \hat{p}_2 \geqslant 0]$. The z value corresponding to $\hat{p}_1 - \hat{p}_2 = 0$ is computed as

$$z = \frac{0 - E(\hat{p}_1 - \hat{p}_2)}{\sqrt{Var(\hat{p}_1 - \hat{p}_2)}} = \frac{0 - (-.05)}{.083} = .60$$

so that

$$P[\hat{p}_1 - \hat{p}_2 \geqslant 0] = 1 - \Phi(.60) = .27$$

FIGURE 6-24

Approximate Probability
Distribution for $\hat{p}_1 - \hat{p}_2$

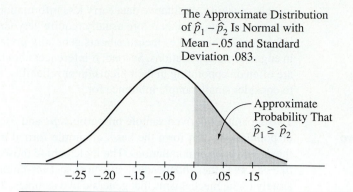

The Approximate Distribution
of $\hat{p}_1 - \hat{p}_2$ Is Normal with
Mean $-.05$ and Standard
Deviation $.083$.

The large-n_1, n_2 approximate normality of $\hat{p}_1 - \hat{p}_2$ translates to the realization that

$$\frac{\hat{p}_1 - \hat{p}_2 - (p_1 - p_2)}{\sqrt{\dfrac{p_1(1 - p_1)}{n_1} + \dfrac{p_2(1 - p_2)}{n_2}}} \tag{6-61}$$

is approximately standard normal, and this observation forms the basis for inference concerning $p_1 - p_2$. Consider first the issue of confidence interval estimation for $p_1 - p_2$. The familiar argument of Section 6-1 (beginning with the quantity (6-61)) shows

$$\hat{p}_1 - \hat{p}_2 \pm z\sqrt{\frac{p_1(1 - p_1)}{n_1} + \frac{p_2(1 - p_2)}{n_2}} \tag{6-62}$$

to be a mathematically correct but practically unusable formula for endpoints of a confidence interval for $p_1 - p_2$. Conservative modification of expression (6-62), via replacement of both $p_1(1 - p_1)$ and $p_2(1 - p_2)$ with $.25$, shows that the two-sided interval with endpoints

$$\hat{p}_1 - \hat{p}_2 \pm z \cdot \frac{1}{2}\sqrt{\frac{1}{n_1} + \frac{1}{n_2}} \tag{6-63}$$

is a large-sample two-sided conservative confidence interval for $p_1 - p_2$ with confidence at least that corresponding to the standard normal probability between $-z$ and z. (One-sided intervals are obtained from expression (6-63) in the usual way.)

In addition, in by now familiar fashion, one can also begin with the fact that for large sample sizes, the modification of the variable (6-61),

$$\frac{\hat{p}_1 - \hat{p}_2 - (p_1 - p_2)}{\sqrt{\dfrac{\hat{p}_1(1 - \hat{p}_1)}{n_1} + \dfrac{\hat{p}_2(1 - \hat{p}_2)}{n_2}}} \tag{6-64}$$

is approximately standard normal and be led to the conclusion that the interval with endpoints

$$\hat{p}_1 - \hat{p}_2 \pm z\sqrt{\frac{\hat{p}_1(1 - \hat{p}_1)}{n_1} + \frac{\hat{p}_2(1 - \hat{p}_2)}{n_2}} \tag{6-65}$$

is a large-sample two-sided confidence interval for $p_1 - p_2$ with confidence corresponding to the standard normal probability assigned to the interval between $-z$ and z. (Again, use of only one of the endpoints in display (6-65) gives a one-sided confidence interval.)

EXAMPLE 6-18

(Example 3-14 revisited) **Comparing Fractions Conforming for Two Methods of Operating a Pelletizing Process.** Greiner, Grim, Larson, and Lukomski studied a number of different methods of running a pelletizing process. Two of these involved using a mix with 20% reground powder with respectively small (condition 1) and large (condition 2) shot sizes. Of $n_1 = n_2 = 100$ pellets produced under these two sets of conditions, sample fractions $\hat{p}_1 = .38$ and $\hat{p}_2 = .29$ of the pellets proved to conform to specifications. Consider making a 90% confidence interval for comparing the two methods of process operation (i.e., an interval for $p_1 - p_2$).

Use of expression (6-65) shows that the interval with endpoints

$$.38 - .29 \pm 1.645 \sqrt{\frac{(.38)(1 - .38)}{100} + \frac{(.29)(1 - .29)}{100}}$$

i.e.,

$$.09 \pm .109$$

i.e.,

$$-.019 \quad \text{and} \quad .199 \qquad \textbf{(6-66)}$$

is a 90% confidence interval for $p_1 - p_2$, the difference in long-run fractions of conforming pellets that would be produced under the two sets of conditions. Notice that although appearances are that condition 1, involving a small shot size, has the higher associated likelihood of producing a conforming pellet, the case for this made by the data in hand is again not airtight. The interval (6-66) seems to allow some possibility that $p_1 - p_2 < 0$—i.e., that p_2 actually exceeds p_1. (The reader might well verify that here the conservative interval indicated by expression (6-63) has endpoints of the form $.09 \pm .116$ and thus tells a similar story.)

The usual significance-testing method related to the fact that the variable (6-61) is approximately standard normal applies to the null hypothesis

$$H_0: p_1 - p_2 = 0 \qquad \textbf{(6-67)}$$

i.e., the hypothesis that the parameters p_1 and p_2 are equal. Notice that if $p_1 = p_2$ and the common value is denoted as p, expression (6-61) can be rewritten as

$$\frac{\hat{p}_1 - \hat{p}_2}{\sqrt{p(1 - p)}\sqrt{\dfrac{1}{n_1} + \dfrac{1}{n_2}}} \qquad \textbf{(6-68)}$$

Now the variable (6-68) cannot serve as a test statistic for the null hypothesis (6-67), since it involves the presumably unknown hypothesized common value of p_1 and p_2. What is done to modify the variable (6-68), in order to arrive at a usable test statistic, is to replace p with a sample-based estimate, obtained by pooling together the two samples. That is, with

$$\hat{p} = \frac{n_1 \hat{p}_1 + n_2 \hat{p}_2}{n_1 + n_2} \qquad \textbf{(6-69)}$$

(the total number of items in the two samples with the characteristic of interest, divided by the total number of items in the two samples), a significance test of hypothesis (6-67) can be carried out using the test statistic

$$Z = \frac{\hat{p}_1 - \hat{p}_2}{\sqrt{\hat{p}(1 - \hat{p})}\sqrt{\dfrac{1}{n_1} + \dfrac{1}{n_2}}} \tag{6-70}$$

It turns out that if $H_0: p_1 - p_2 = 0$ is true, Z in equation (6-70) is approximately standard normal, so a standard normal reference distribution is in order.

EXAMPLE 6-18
(continued)

As further confirmation of the fact that in the pelletizing problem, sample fractions of $\hat{p}_1 = .38$ and $\hat{p}_2 = .29$ based on samples of size $n_1 = n_2 = 100$ are not completely convincing evidence of a real difference in process performance for small and large shot sizes, consider testing $H_0: p_1 - p_2 = 0$ with $H_a: p_1 - p_2 \neq 0$. As a preliminary step, note that from expression (6-69),

$$\hat{p} = \frac{100(.38) + 100(.29)}{100 + 100} = \frac{67}{200} = .335$$

Then the five-step summary gives the following.

1. $H_0: p_1 - p_2 = 0$.
2. $H_a: p_1 - p_2 \neq 0$.
3. The test statistic is

$$Z = \frac{\hat{p}_1 - \hat{p}_2}{\sqrt{\hat{p}(1 - \hat{p})}\sqrt{\dfrac{1}{n_1} + \dfrac{1}{n_2}}}$$

The reference distribution is standard normal, and large observed values $|z|$ will constitute evidence against H_0.
4. The samples give

$$z = \frac{.38 - .29}{\sqrt{(.335)(1 - .335)}\sqrt{\dfrac{1}{100} + \dfrac{1}{100}}} = 1.35$$

5. The p-value is $P[|\text{a standard normal variable}| \geq 1.35]$. That is, the p-value is

$$\Phi(-1.35) + (1 - \Phi(1.35)) = .18$$

The data furnish only fairly weak evidence of a real difference in long-run fractions of conforming pellets for the two shot sizes.

The kind of results seen in Example 6-18 may take some getting used to. The fact that even with sample sizes as large as 100, sample fractions differing by nearly .1 are still not necessarily conclusive evidence of a difference in p_1 and p_2, is just another manifestation of the earlier point that individual qualitative observations carry disappointingly little information.

A final reminder of the large-sample nature of the methods presented here is in order before closing this section. The methods here all rely (for the agreement of nominal and actual confidence levels or the validity of their p-values) on the adequacy of normal approximations to binomial distributions. The approximations, as has been said, are often workable provided expression (6-50) or (6-51) holds. When testing $H_0: p = \#$, it is easy to plug both n and $\#$ into expression (6-50) or (6-51) and verify that the inequalities hold before putting great stock in normal-based p-values. But since when estimating p or $p_1 - p_2$ or testing $H_0: p_1 - p_2 = 0$, no parallel check is obvious, it is not completely clear how to screen potential application scenarios for ones where the nominal confidence levels or p-values are possibly misleading. What is often done in these last three types of applications is to plug both n and $\hat{p}$ (or both n_1 and $\hat{p}_1$ and n_2 and $\hat{p}_2$) into expression (6-50) or (6-51) and verify that the inequalities hold before setting much stock in nominal (normal-based) confidence levels and p-values. Since these random quantities are only approximations to the corresponding nonrandom quantities, one will occasionally be misled regarding the appropriateness of the normal approximations by such empirical checks. But they are better than automatic application, protected by no check at all.

6-6 Prediction and Tolerance Intervals

The methods of confidence interval estimation and significance testing introduced thus far in this chapter have concerned the problem of reasoning from sample information to statements about underlying *parameters* of the data generation, such as μ, σ, and p. These are extremely important engineering tools, but they often fail to directly address the question of real interest in a statistical engineering study. Sometimes what is really needed as the ultimate product of a statistical analysis is not a statement about a parameter but rather an indication of reasonable bounds on other *individual values* generated by the process under study. For example, assume one is about to purchase a new automobile. For some purposes, knowing that "the mean EPA mileage for this model is likely in the range 25 mpg $\pm$ .5 mpg" is not nearly as useful as knowing that "the EPA mileage figure for the particular auto you are ordering is likely in the range 25 mpg $\pm$ 3 mpg." Both of these statements may be quite accurate, but they serve different purposes. The first statement is one about a mean mileage and the second is about an individual mileage. And it is only statements of the first type that have been directly treated thus far.

This section indicates what is possible in the way of formal statistical inferences, not for parameters but rather for individual values generated by a stable data-generating mechanism. There are two types of formal inference methods aimed in this general direction — *statistical prediction interval* methods and *statistical tolerance interval* methods — and methods of both types will be discussed. The section begins with prediction intervals for a normal distribution. Then tolerance intervals for a normal distribution are considered. Finally, there is a discussion of how it is possible to use minimum and/or maximum values in a sample to create prediction and tolerance intervals for even nonnormal underlying distributions.

Prediction Intervals for a Normal Distribution

One fruitful way to phrase the question of inference for additional individual values produced by a process is the following. If $x_1, x_2, \ldots, x_n$ are data in hand, generated by a single stable process, and x_{n+1} is one additional value to be generated by the same mechanism, how might the data in hand be used to create a numerical interval likely to bracket (the as yet unrevealed) x_{n+1}? How, for example, might mileage tests on ten autos of a particular model be used to predict the results of the same test applied to an eleventh auto?

If the underlying distribution is adequately described as Gaussian with mean μ and variance σ^2, there is a simple line of reasoning based on the random variable

$$\bar{x} - x_{n+1}$$

(6-71)

that leads to an answer to this question. That is, *a priori* the random variable in expression (6-71) has, by the methods of Section 5-5 (Proposition 5-1 in particular),

$$E(\bar{x} - x_{n+1}) = E\bar{x} + (-1)Ex_{n+1} = \mu - \mu = 0 \tag{6-72}$$

and

$$Var(\bar{x} - x_{n+1}) = Var\,\bar{x} + (-1)^2\,Var\,x_{n+1} = \frac{\sigma^2}{n} + \sigma^2 = \left(1 + \frac{1}{n}\right)\sigma^2 \tag{6-73}$$

Further, it turns out that the difference (6-71) is normally distributed, so the variable

$$\frac{(\bar{x} - x_{n+1}) - 0}{\sigma\sqrt{1 + \frac{1}{n}}} \tag{6-74}$$

is standard normal. And taking one more step, if s^2 is the usual sample variance of $x_1, x_2, \ldots, x_n$, it turns out that substituting s for σ in expression (6-74) produces a variable

$$\frac{(\bar{x} - x_{n+1}) - 0}{s\sqrt{1 + \frac{1}{n}}} \tag{6-75}$$

which a probabilist can show to have a t distribution with $\nu = n - 1$ degrees of freedom.

Now the useful thing about the realization that the variable (6-75) has a t distribution is that (upon identifying x_{n+1} with μ and $\sqrt{1 + (1/n)}$ with $\sqrt{1/n}$), the variable (6-75) is formally similar to the t-distributed variable used to derive a small-sample confidence interval for μ. In fact, algebraic steps parallel to those used in the first part of Section 6-3 show that if $t > 0$ is such that the t_{n-1} distribution assigns, say, .95 probability to the interval between $-t$ and t, there is then .95 probability that

$$\bar{x} - ts\sqrt{1 + \frac{1}{n}} < x_{n+1} < \bar{x} + ts\sqrt{1 + \frac{1}{n}}$$

But this reasoning suggests in general that the interval with endpoints

$$\bar{x} \pm ts\sqrt{1 + \frac{1}{n}} \tag{6-76}$$

can be used as a two-sided interval to predict x_{n+1}, and that the probability-based reliability figure attached to the interval should be the t_{n-1} probability assigned to the interval from $-t$ to t. The interval (6-76) is a called a *prediction interval* for a single additional normal observation (with associated confidence corresponding to the t_{n-1} probability assigned to the interval from $-t$ to t). In general, the language indicated in Definition 6-17 will be used.

DEFINITION 6-17

A **prediction interval** for a single additional observation is a data-based interval of numbers thought likely to contain the observation, possessing a stated probability-based confidence or reliability.

It is the fact that a finite sample gives only a somewhat clouded picture of an underlying process or population distribution that prevents the making of a normal distribution prediction interval from being a trivial matter of straightforward probability calculations like those in Section 5-2. That is, suppose one had enough data to "know" the mean, μ, and variance, σ^2, of an underlying normal distribution. Then, since 1.96 is the .975 standard normal quantile, the interval with endpoints

$$\mu - 1.96\sigma \quad \text{and} \quad \mu + 1.96\sigma \quad \quad \textbf{(6-77)}$$

has a 95% chance of bracketing the next value generated by the underlying distribution. The fact that (when based only on small samples), one's perception of μ and σ is noisy forces one to abandon expression (6-77) for an interval like (6-76). It is thus comforting that for large n and 95% confidence, formula (6-76) can be expected to produce an interval with endpoints approximating those in display (6-77). That is, for large n and 95% confidence, $t \approx 1.96$, $\sqrt{1 + (1/n)} \approx 1$, and one expects that typically $\bar{x} \approx \mu$ and $s \approx \sigma$, so that expressions (6-76) and (6-77) will essentially agree. The beauty of expression (6-76) is that it allows in a rational fashion for the uncertainties involved in the $\mu \approx \bar{x}$ and $\sigma \approx s$ approximations.

EXAMPLE 6-19

(Example 6-8 revisited) **Predicting a Spring Lifetime.** Recall from Section 6-3 that $n = 10$ spring lifetimes under 950 N/mm^2 stress conditions given in Table 6-4 produced a fairly linear normal plot, and $\bar{x} = 168.3$ ($\times 10^3$ cycles) and $s = 33.1$ ($\times 10^3$ cycles). Consider now the problem of predicting the lifetime of an additional spring of this type (under the same test conditions) with 90% confidence.

Using $\nu = 10 - 1 = 9$ degrees of freedom, the .95 quantile of the t distribution is (from Table D-4) 1.833. So employing expression (6-76), one has that two-sided 90% prediction limits for an additional spring lifetime are

$$168.3 \pm 1.833(33.1)\sqrt{1 + \frac{1}{10}}$$

i.e.,

$$104.7 \times 10^3 \text{ cycles} \quad \text{and} \quad 231.9 \times 10^3 \text{ cycles} \quad \quad \textbf{(6-78)}$$

Note that the interval indicated by display (6-78) is not at all the same as the confidence interval for μ found in Example 6-8. The limits of

$$149.1 \times 10^3 \text{ cycles} \quad \text{and} \quad 187.5 \times 10^3 \text{ cycles}$$

found there apply to the mean spring lifetime, μ, not to an additional observation x_{11} as the ones in display (6-78) do.

EXAMPLE 6-20

Predicting the Weight of a Newly Minted Penny. The delightful book *Experimentation and Measurement* by W. J. Youden (published as NBS Special Publication 672 by the U.S. Department of Commerce) contains a data set giving the weights of $n = 100$ newly minted U.S. pennies measured to 10^{-4} g but reported only to the nearest .02 g. These data are reproduced in Table 6-10.

Printout 6-1 represents part of a Minitab session (involving the Minitab "INVCDF" command) used to generate a normal probability plot of the penny weight data. It shows that a Gaussian distribution is a plausible model for weights of newly minted pennies.

Further, calculation with the values in Table 6-10 shows that for the penny weights, $\bar{x} = 3.108$ g and $s = .043$ g. Then interpolation in Table D-4 shows the .9 quantile of the t_{99} distribution to be about 1.290, so that using only the "plus" part of expression (6-76), one finds that a one-sided 90% prediction interval of the form $(-\infty, \#)$ for the

TABLE 6-10

Weights of 100 Newly Minted U.S. Pennies

Penny Weight (g)	Frequency
2.99	1
3.01	4
3.03	4
3.05	4
3.07	7
3.09	17
3.11	24
3.13	17
3.15	13
3.17	6
3.19	2
3.21	1

PRINTOUT 6-1

```
MTB > hist c1

Histogram of C1   N = 100

Midpoint   Count
    3.00      1   *
    3.02      4   ****
    3.04      4   ****
    3.06      4   ****
    3.08      7   *******
    3.10     17   *****************
    3.12     24   ************************
    3.14     17   *****************
    3.16     13   *************
    3.18      6   ******
    3.20      2   **
    3.22      1   *

MTB > set c2
DATA> 1:100
DATA> end
MTB > sub .5 c2 c2
MTB > div c2 100 c2
MTB > invcdf c2 c2;
SUBC> norm 0 1.
MTB > plot c2 c1
```

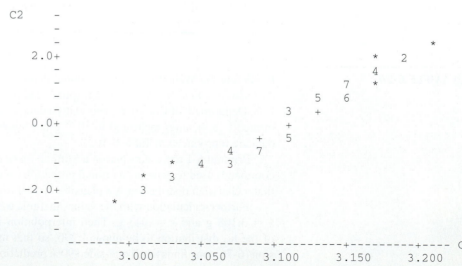

weight of an additional penny has upper endpoint

$$3.108 + 1.290(.043)\sqrt{1 + \frac{1}{100}}$$

i.e.,

$$3.164 \text{ g} \qquad\qquad (6\text{-}79)$$

Two points should be obvious to the reader from this example. First, one can modify the two-sided prediction limits for a single observation in expression (6-76) to get a one-sided limit exactly as one modifies two-sided confidence limits for a parameter to get a one-sided limit. Second, the calculation represented by the result (6-79) is, because $n = 100$ is a fairly large sample size, only marginally different from what one would get assuming $\mu = 3.108$ g exactly and $\sigma = .043$ g exactly, and using the .9 normal quantile 1.282 to calculate

$$\mu + 1.282\sigma = 3.108 + (1.282)(.043) = 3.163 \text{ g} \qquad (6\text{-}80)$$

But the fact that the result (6-79) is slightly larger than the final result in display (6-80) is meant to reflect the small uncertainty involved in the use of $\bar{x}$ in place of μ and s in place of σ.

The name "prediction interval" probably carries with it connotations that should be dispelled before going any further. That is, the word *prediction* tends to make one think of future and thus potentially different conditions. But no such meaning should be associated with statistical prediction intervals. That is, it is important to remember that the assumption behind formula (6-76) is that $x_1, x_2, \ldots, x_n$ and x_{n+1} are *all* generated according to the *same* underlying distribution. If (for example, because of potential physical changes in a system during a time lapse between the generation of $x_1, x_2, \ldots, x_n$ and the generation of x_{n+1}) no single stable process model for the generation of all $n + 1$ observations is appropriate, then neither is the use of formula (6-76). Statistical inference is not a crystal ball for foretelling an erratic and patternless future, but rather a methodology for quantifying the extent of one's knowledge about a pattern of variation existing in a consistent present. It has implications in other times and at other places only if that same pattern of variation can be expected to repeat itself in those conditions.

While reflecting on the applicability of prediction interval methods, it is also appropriate to make a comment or two regarding the meaning of the confidence or reliability figure attached to a prediction interval. Since a prediction interval is doing a different job than the confidence intervals of previous sections, the meaning of confidence given in Definition 6-2 isn't quite germane here.

Clearly, prior to the generation of any of $x_1, x_2, \ldots, x_n, x_{n+1}$, planned use of expression (6-76) gives a guaranteed probability of success in bracketing x_{n+1}. And after all of $x_1, x_2, \ldots, x_n, x_{n+1}$ have been generated, one has either been completely successful or completely unsuccessful in bracketing x_{n+1}. But it is not altogether obvious how to think about "confidence" of prediction after $x_1, x_2, \ldots, x_n$ are in hand, but prior to the generation of x_{n+1}. For example, in the context of Example 6-19, having used sample data to arrive at the prediction limits in display (6-78) — i.e.,

$$104.7 \times 10^3 \text{ cycles} \qquad \text{to} \qquad 231.9 \times 10^3 \text{ cycles}$$

since x_{11} is a random variable, it would make sense to contemplate

$$P[104.7 \times 10^3 \leq x_{11} \leq 231.9 \times 10^3]$$

However, there is no guarantee of the value for the probability or any way to determine it. In particular, it is *not* necessarily .9 (the confidence level associated with the prediction interval). That is, there is no practically usable way to employ probability to describe the likely effectiveness of a numerical prediction interval. One is thus left with the interpretation of confidence of prediction given in Definition 6-18.

DEFINITION 6-18

(Interpretation of a Prediction Interval)   To say that a numerical interval (a, b) is (for example) a 90% prediction interval for an additional observation x_{n+1} is to say that in obtaining it, one has applied methods of data collection and calculation that would produce intervals bracketing an $(n + 1)$th observation in about 90% of repeated applications of the entire process of 1) selecting the sample $x_1, \ldots, x_n$, 2) calculating an interval, and 3) generating a single additional observation x_{n+1}. Whether or not x_{n+1} will fall into the numerical interval (a, b) is not known, and although there is some probability associated with that eventuality, it is not possible to evaluate it. And in particular, it need not be 90%.

In regard to the last two sentences in Definition 6-18, it should be said that when using a 90% prediction interval method, although some samples $x_1, \ldots, x_n$ produce numerical intervals with probability less than .9 of bracketing x_{n+1} and others produce numerical intervals with probability more than .9, the average for all samples $x_1, \ldots, x_n$ does turn out to be .9. The practical problem is simply that with data $x_1, \ldots, x_n$ in hand, one doesn't know whether one is above, below, or at the .9 figure.

Tolerance Intervals for a Normal Distribution

The emphasis, when making a prediction interval of the type just discussed, is on a *single* additional observation beyond those *n* already in hand. But in practical engineering problems, many more items beyond those making up a sample may be of interest. In such cases, one may wish to declare a data-based interval likely to encompass *most* measurements from the rest of these items. (It should be intuitively plausible that if many more items are involved and measurements are not *a priori* restricted to some particular finite interval, trying to declare a data-based interval likely to encompass *all* measurements from the rest of these items would be too ambitious a goal.)

It should make intuitive sense that not only are prediction intervals not designed for the purpose of being likely to encompass most of the measurements from the additional items of interest; by most standards, they are also not adequate to do so. The discussion following Definition 6-18 shows that, for example, only on average is the fraction of an underlying normal distribution bracketed by a 90% prediction interval equal to 90%. So a crude analysis (identifying the mean fraction bracketed with the median fraction bracketed) then suggests that the probability that the actual fraction bracketed is at least 90% is on the order of only .5. That is, a 90% prediction interval is not constructed to be big enough for the present purpose. What is needed instead is a statistical tolerance interval.

DEFINITION 6-19

A **statistical tolerance interval for a fraction p of an underlying distribution** is a data-based interval thought likely to contain at least a fraction p and possessing a stated (usually large) probability-based confidence or reliability.

The derivation of appropriate tolerance interval formulas (even for an underlying normal distribution) requires probability background well beyond what has been developed in this text. But results of that work look and behave about as one would expect.

That is, it is possible, for a given confidence and fraction p of an underlying normal distribution, for a probabilist to find a corresponding constant τ_2 such that the two-sided interval with endpoints

$$\bar{x} \pm \tau_2 s \tag{6-81}$$

is a tolerance interval for a fraction p of the underlying distribution. The τ_2 appearing in expression (6-81) is, for common (large) confidence levels, larger than the multiplier $t\sqrt{1 + (1/n)}$ appearing in expression (6-76) for two-sided confidence of prediction corresponding to the fraction p. On the other hand, as n gets large, both τ_2 from expression (6-81) and $t\sqrt{1 + (1/n)}$ from expression (6-76) tend to the $(1 - \frac{p}{2})$ standard normal quantile. Table D-7-A gives some values of τ_2 for 95% and 99% confidence and $p = .9, .95,$ and $.99$. (The use of this table will be demonstrated momentarily.)

In a way that is perhaps slightly unexpected, given the situation for prediction intervals and confidence intervals for parameters, the factors τ_2 are not used to make one-sided tolerance intervals. Instead, another set of constants that will here be called τ_1 values have been developed. They are such that for a given confidence and fraction p of an underlying normal distribution, both of the one-sided intervals

$$(-\infty, \bar{x} + \tau_1 s) \tag{6-82}$$

and

$$(\bar{x} - \tau_1 s, \infty) \tag{6-83}$$

are tolerance intervals for a fraction p of the underlying distribution. Although the values τ_1 are not obtained by simply changing a confidence level or value of p and then consulting a table of values of τ_2, they do in some respects behave like values of τ_2. τ_1 appearing in intervals (6-82) and (6-83) is, for common confidence levels, larger than the multiplier $t\sqrt{1 + (1/n)}$ appearing in expression (6-76) for one-sided confidence of prediction corresponding to the fraction p. And as n gets large, both τ_1 from expression (6-82) or (6-83) and $t\sqrt{1 + (1/n)}$ from expression (6-76) tend to the $(1 - p)$ standard normal quantile. Table D-7-B gives some values of τ_1 for 95% and 99% confidence and $p = .9, .95,$ and $.99$.

EXAMPLE 6-19
(continued)

Consider making a two-sided 95% tolerance interval for 90% of additional spring life-times based on the data of Table 6-4. As earlier, for these data, $\bar{x} = 168.3 \, (\times 10^3 \text{ cycles})$ and $s = 33.1 \, (\times 10^3 \text{ cycles})$. Then consulting Table D-7-A, one finds that since $n = 10$, $\tau_2 = 2.856$ is appropriate for use in expression (6-81). That is, two-sided 95% tolerance limits for 90% of additional spring lifetimes are

$$168.3 \pm 2.856 \, (33.1)$$

i.e.,

$$73.8 \times 10^3 \text{ cycles} \quad \text{and} \quad 262.8 \times 10^3 \text{ cycles} \tag{6-84}$$

It is obvious from comparing displays (6-78) and (6-84) that the effect of moving from the prediction of a single additional spring lifetime to attempting to bracket most of a large number of additional lifetimes is to increase the size of the declared interval.

EXAMPLE 6-20
(continued)

Consider again the new penny weights given in Table 6-10 and the problem of making a one-sided 95% tolerance interval of the form $(-\infty, \#)$ for the weights of 90% of additional pennies minted under the same conditions as those represented in the table.

Remembering that for the penny weights, $\bar{x} = 3.108$ g and $s = .043$ g, and using Table D-7-B for $n = 100$, one sees that the desired upper tolerance bound for 90% of the penny weights is

$$3.108 + 1.527(.043) = 3.174 \text{ g}$$

As expected, this is larger (more conservative) than the value of 3.164 g given in display (6-79) as a one-sided 90% prediction limit for a single additional penny weight.

Having reasoned out fairly carefully what the interpretation of confidence should be for intervals meant to bracket parameters, and what confidence of prediction means, the correct interpretation of the confidence level for a tolerance interval should be fairly easy to grasp. Clearly, prior to the generation of $x_1, x_2, \ldots, x_n$, planned use of expression (6-81), (6-82), or (6-83) gives a guaranteed probability of success in bracketing a fraction of at least p of the underlying distribution. But after observing $x_1, \ldots, x_n$ and making a numerical interval, it is impossible to know whether one's attempt has or has not been successful. That is, one is left with the following interpretation of confidence for a tolerance interval.

DEFINITION 6-20

(Interpretation of a Tolerance Interval) To say that a numerical interval (a, b) is (for example) a 90% tolerance interval for a fraction p of an underlying distribution is to say that in obtaining it, one has applied methods of data collection and calculation that would produce intervals bracketing a fraction of at least p of the underlying distribution in about 90% of repeated applications (of generation of $x_1, \ldots, x_n$ and subsequent calculation). Whether or not the numerical interval (a, b) actually contains at least a fraction p is unknown and not describable in terms of a probability.

Prediction and Tolerance Intervals Based on Minimum and/or Maximum Values in a Sample

Formulas (6-76), (6-81), (6-82), and (6-83) for prediction and tolerance limits are definitely normal distribution formulas. The question then arises, "What if an engineering data-generation process is stable but does not produce normally distributed observations? How, if at all, can one make prediction or tolerance limits?" Two kinds of answers to this question will be illustrated in this text. The first employs the transformation idea presented in Section 4-4, and the second involves the use of minimum and/or maximum sample values to establish prediction and/or tolerance bounds.

In the first place, it was first observed in Section 4-4 that if a response variable y fails to be normally distributed, it may still be possible to find some transformation g (essentially specifying a revised scale of measurement) such that $g(y)$ is normal. Then normal-based methods might be applied to $g(y)$, and answers of interest translated back into statements about y instead of $g(y)$.

EXAMPLE 6-21

(Example 4-11 revisited) **Prediction and Tolerance Intervals for Discovery Times Obtained Using a Transformation.** Section 5-3 argued that the auto service discovery time data of Elliot, Kibby, and Meyer given in Figure 4-31 are not themselves Gaussian-looking, but that their natural logarithms are. This, together with the facts that the $n = 30$ natural logarithms have $\bar{x} = 2.46$ and $s = .68$, can be used to make prediction or tolerance intervals for log discovery times.

For example, using expression (6-76) to make a two-sided 99% prediction interval for an additional log discovery time, one has endpoints

$$2.46 \pm 2.756(.68)\sqrt{1 + \frac{1}{30}}$$

i.e.,

$$.55 \ln \min \quad \text{and} \quad 4.37 \ln \min \tag{6-85}$$

And using expression (6-81) to make, for example, a 95% tolerance interval for 99% of additional log discovery times, one is led to the interval with endpoints

$$2.46 \pm 3.355(.68)$$

i.e.,

$$.18 \ln \min \quad \text{and} \quad 4.74 \ln \min \tag{6-86}$$

The point here is that intervals specified in displays (6-85) and (6-86) for log discovery times have, via exponentiation, their counterparts for raw discovery times. That is, exponentiation of the values in display (6-85) gives a 99% prediction interval for another discovery time of from

$$1.7 \min \quad \text{to} \quad 79.0 \min$$

And exponentiation of the values in display (6-86) gives a 95% tolerance interval for 99% of additional discovery times of from

$$1.2 \min \quad \text{to} \quad 114.4 \min$$

When it is not possible to find a transformation that will allow the use of normal-based formulas, but some other standard family of distributions (e.g., the Weibull family) can be used to model the response, a hunt through the statistical literature can sometimes turn up prediction and tolerance limits designed specifically for that family. (The book *Statistical Intervals: A Guide for Practitioners*, by Hahn and Meeker, is a good place to begin such a search.) This text can't begin to give methods corresponding to formulas (6-76), (6-81), (6-82), and (6-83) for every type of probability distribution that has been discussed. What will be done is to point out that intervals from the smallest observation and/or to the largest value in a sample can be used as prediction and/or tolerance intervals for *any* underlying distribution.

That is, if $x_1, x_2, \ldots, x_n$ are values in a sample and $\min(x_1, \ldots, x_n)$ and $\max(x_1, \ldots, x_n)$ are (respectively) the smallest and largest values among $x_1, x_2, \ldots, x_n$, consider the potential use of the intervals

$$(-\infty, \max(x_1, \ldots, x_n)) \tag{6-87}$$

$$(\min(x_1, \ldots, x_n), \infty) \tag{6-88}$$

and

$$(\min(x_1, \ldots, x_n), \max(x_1, \ldots, x_n)) \tag{6-89}$$

as prediction or tolerance intervals. It turns out that independent of exactly what underlying continuous distribution is operating, if the generation of $x_1, x_2, \ldots, x_n$ (and if relevant, x_{n+1}) can be described as a stable process, it is possible to evaluate the confidence associated with intervals (6-87), (6-88), and (6-89).

Consider first the use of interval (6-87) or (6-88) as a one-sided prediction interval for a single additional observation x_{n+1}. The associated confidence level is

$$\text{One-sided prediction confidence level} = \frac{n}{n+1} \tag{6-90}$$

Then, considering the use of interval (6-89) as a two-sided prediction interval for a single additional observation x_{n+1}, the associated confidence level is

$$\text{Two-sided prediction confidence level} = \frac{n-1}{n+1} \qquad \textbf{(6-91)}$$

The corresponding statements for use of intervals (6-87), (6-88), and (6-89) as tolerance intervals must of necessity involve p, the fraction of the underlying distribution one hopes to bracket. The fact is that using interval (6-87) or (6-88) as a one-sided tolerance interval for a fraction p of an underlying distribution, the associated confidence level is

$$\text{One-sided confidence level} = 1 - p^n \qquad \textbf{(6-92)}$$

And when interval (6-89) is used as a tolerance interval for a fraction p of an underlying distribution, the appropriate associated confidence is

$$\text{Two-sided confidence level} = 1 - p^n - n(1-p)p^{n-1} \qquad \textbf{(6-93)}$$

EXAMPLE 6-19
(continued)

Return one more time to the spring-life scenario, and consider the use of interval (6-89) as first a prediction interval and then a tolerance interval for 90% of additional spring lifetimes. Notice in Table 6-4 that the smallest and largest of the observed spring lifetimes are, respectively,

$$\min(x_1, \ldots, x_{10}) = 117 \times 10^3 \text{ cycles}$$

and

$$\max(x_1, \ldots, x_{10}) = 225 \times 10^3 \text{ cycles}$$

so the numerical interval under consideration is the one with endpoints $117\,(\times 10^3 \text{ cycles})$ and $225\,(\times 10^3 \text{ cycles})$.

Then expression (6-91) means that this interval can be used as a prediction interval with

$$\text{Prediction confidence} = \frac{10-1}{10+1} = \frac{9}{11} = 82\%$$

And expression (6-93) says that as a tolerance interval for a fraction $p = .9$ of many additional spring lifetimes, the interval can be used with associated confidence

$$\text{Confidence} = 1 - (.9)^{10} - 10(1 - .9)(.9)^9 = 26\%$$

EXAMPLE 6-20
(continued)

Looking for a final time at the penny weight data in Table 6-10, consider the use of interval (6-87) as first a prediction interval and then a tolerance interval for 99% of additional penny weights. Notice that in Table 6-10, the largest of the $n = 100$ weights is 3.21 g, so

$$\max(x_1, \ldots, x_{100}) = 3.21 \text{ g}$$

Then expression (6-90) says that when used as an upper prediction limit for a single additional penny weight, the prediction confidence associated with 3.21 g is

$$\text{Prediction confidence} = \frac{100}{100+1} = 99\%$$

And expression (6-92) shows that as a tolerance interval for 99% of many additional penny weights, the interval $(-\infty, 3.21)$ has associated confidence

$$\text{Confidence} = 1 - (.99)^{100} = 63\%$$

A little experience with formulas (6-90), (6-91), (6-92), and (6-93) and real data sets that include an extreme data point or two will convince the reader that the intervals (6-87), (6-88), and (6-89) often carry disappointingly small confidence coefficients and/or are disappointingly large when used as prediction or tolerance intervals. Usually (but not always), one can do better in terms of high confidence and short intervals if (possibly after transformation) the normal distribution methods discussed earlier can be applied. Or, to some extent, the present intervals' susceptibility to the influence of a few extreme values can be remedied (at the expense of some complication) by replacing the intervals (6-87) through (6-89) with ones involving observations not quite at the extremes of the sample, and finding corresponding formulas for confidence levels. But the beauty of intervals (6-87), (6-88), and (6-89) is that they are both widely applicable (in even nonGaussian contexts) and extremely simple.

As this section closes, it is worth emphasizing that prediction and tolerance interval methods are very useful engineering tools. Historically, they probably haven't been used as much as they should be, for lack of accessible textbook material on the methods. This section may have served to make the reader aware of the existence of the methods as the appropriate form of formal inference when the focus of a statistical engineering study is on individual values generated by a process, rather than on process parameters. When the few particular methods discussed here don't prove adequate for practical purposes, the reader should look into the topic further, beginning with the book by Hahn and Meeker mentioned earlier.

EXERCISES

6-1. Consider the breaking strength data of Example 3-5. Notice that the normal plot of these data given as Figure 3-18 is reasonably linear. It may thus be sensible to suppose that breaking strengths for generic towels of this type (as measured by the students) are adequately modeled as normal. Under this assumption:

(a) (§6-3) Make and interpret 95% two-sided and one-sided confidence intervals for the mean breaking strength of generic towels. (For the one-sided interval, make an interval of the form $(\#, \infty)$.)

(b) (§6-6) Make and interpret 95% two-sided and one-sided prediction intervals for a single additional generic towel breaking strength. (For the one-sided interval, give the lower prediction bound.)

(c) (§6-6) Make and interpret 95% two-sided and one-sided tolerance intervals for 99% of generic towel breaking strengths. (For the one-sided interval, give the lower tolerance bound.)

(d) (§6-4) Make and interpret 95% two-sided and one-sided confidence intervals for σ, the standard deviation of generic towel breaking strengths.

(e) (§6-3) Put yourself in the position of a quality control inspector, concerned that the mean breaking strength not fall under 9,500 g. Assess the strength of the evidence in the data that the mean generic towel strength is in fact below the 9,500 g target. (Show the whole five-step significance-testing format.)

(f) (§6-4) Now put yourself in the place of a quality control inspector concerned that the breaking strength be reasonably consistent — i.e., that σ be small. Suppose in fact that it is desirable that σ be no more than 400 g. Use the significance-testing format and assess the strength of the evidence given in the data that in fact σ exceeds the target standard deviation.

6-2. Consider the situation of Example 1-1.

(a) (§6-3) Use the five-step significance-testing format to assess the strength of the evidence collected by the engineer in charge of this study to the effect that the laying method is superior to the hanging method in terms of mean runouts produced.

(b) (§6-3) Make and interpret 90% two-sided and one-sided confidence intervals for the improvement in mean runout produced by the laying method over the hanging method. (For the one-sided interval, give a lower bound for $\mu_{\text{hung}} - \mu_{\text{laid}}$.)

(c) (§6-1) Make and interpret a 90% confidence interval for the mean runout for laid gears.

(d) (§3-2, §5-3, and §6-6) What is it about Figure 1-1 that makes it questionable whether "normal distribution" prediction and tolerance interval formulas ought to be used to describe runouts for laid gears? Suppose instead that one were to use the methods of the last part of Section 6-6 to make prediction and tolerance intervals for laid gear runouts. What confidence could be associated with the largest observed laid runout as a prediction bound for a single additional laid runout? What confidence could be associated with with the largest observed laid runout as a tolerance bound for 95% of additional laid gear runouts?

6-3. Consider the situation of Example 4-1. In particular, limit attention to those densities obtained under the 2,000 and 4,000 psi pressures. (One can then view the six corresponding densities as two samples of size $n_1 = n_2 = 3$.)

(a) (§6-3) Assess the strength of the evidence that increasing pressure increases the mean density of the resulting cylinders. Use the five-step significance-testing format.

(b) (§6-3) Give a 99% lower confidence bound for the increase in mean density associated with the change from 2,000 to 4,000 psi conditions.

(c) (§6-4) Assess the strength of the evidence (in the six density values) that the variability in density differs for the 2,000 and 4,000 psi conditions (i.e., that $\sigma_{2,000} \neq \sigma_{4,000}$).

(d) (§6-4) Give a 90% two-sided confidence interval for the ratio of density standard deviations for the two pressures.

(e) (§6-3 and §6-4) What model assumptions stand behind the formal inferences you made in parts (a) through (d) above?

6-4. Consider again the situation of Exercise 3-3: investigating the torques required to loosen two particular bolts holding an assembly on a piece of machinery.

(a) (§6-3) What model assumptions are needed in order to compare top-bolt and bottom-bolt torques here? Make a plot for investigating the necessary distributional assumption.

(b) (§6-3) Assess the strength of the evidence that there is a mean increase in required torque as one moves from the top to the bottom bolts.

(c) (§6-3) Give a 98% two-sided confidence interval for the mean difference in torques between the top and bottom bolts.

6-5. Consider the situation of Example 3-14, and in particular the results for the 50% reground mixture.

(a) (§6-5) Make and interpret 95% one-sided and two-sided confidence intervals for the fraction of conforming pellets that would be produced using the 50% mixture and the small shot size. (For the one-sided interval, give a lower confidence bound.) Use both methods of dealing with the fact that $\sigma_{\hat{p}}$ is not known, and compare the resulting pairs of intervals.

(b) (§6-5) If records show that past pelletizing performance was such that 55% of the pellets produced were conforming, does the value in Table 3-20 constitute strong evidence that the conditions of 50% reground mixture and small shot size provide an improvement in yield? Show the five-step format.

(c) (§6-5) Compare the small and large shot size conditions using a 95% two-sided confidence interval for the difference in fractions conforming. Interpret the interval in the context of the example.

(d) (§6-5) Assess the strength of the evidence given in Table 3-20 that the shot size affects the fraction of pellets conforming (when the 50% reground mixture is used).

6-6. In Exercise 3-15, there is a data set consisting of the aluminum contents of 26 bihourly samples of recycled PET plastic from a recycling facility. Those 26 measurements have $\bar{y} = 142.7$

ppm and $s \approx 98.2$ ppm. Use these facts to respond to the following.

(a) Make a 90% two-sided confidence interval for the mean aluminum content of such specimens over the 52-hour study period.

(b) Make a 95% two-sided confidence interval for the mean aluminum content of such specimens over the 52-hour study period. How does this compare to your answer to part (a)?

(c) Make a 90% one-sided upper confidence bound for the mean aluminum content of such samples over the 52-hour study period. (Find # such that $(-\infty, \#)$ is a 90% confidence interval.) How does this value compare to the upper endpoint of your interval from part (a)?

(d) Make a 95% one-sided upper confidence bound for the mean aluminum content of such samples over the 52-hour study period. How does this value compare to your answer to part (c)?

(e) Interpret your interval from (a) for someone with little statistical background. (Speak in the context of the recycling study, and use Definition 6-2 as your guide.)

6-7. In the context of the aluminum contamination study discussed in Exercise 6-6 and in Exercise 3-15, it was in fact desirable to have mean aluminum content for samples of recycled plastic below 200 ppm. Use the five-step significance-testing format and determine the strength of the evidence in the data that in fact this contamination goal has been violated. (You will want to begin with $H_0: \mu = 200$ ppm and use $H_a: \mu > 200$ ppm.)

6-8. Again in the context of Exercise 3-15, if one uses the interval from 30 ppm to 511 ppm as a prediction interval for a single additional aluminum content measurement from the study period, what associated prediction confidence level can be stated? What confidence can be associated with this interval as a tolerance interval for 90% of all such aluminum content measurements?

6-9. The natural logarithms of the aluminum contents discussed in Exercise 3-15 have a reasonably bell-shaped relative frequency distribution. Further, these 26 log aluminum contents have sample mean 4.9 and sample standard deviation .59. Use this information to respond to the following.

(a) Give a two-sided 99% tolerance interval for 90% of additional log aluminum contents at the Rutgers recycling facility. Then translate this interval into a 99% tolerance interval for 90% of additional raw aluminum contents.

(b) Make a 90% prediction interval for one additional log aluminum content and translate it into a prediction interval for a single additional aluminum content.

(c) How do the intervals from (a) and (b) compare?

6-10. Simple counting with the data of Exercise 3-15 shows that 18 out of the 26 PET samples had aluminum contents above 100 ppm. Give a two-sided approximate 95% confidence interval for the fraction of all such samples with aluminum contents above 100 ppm.

6-11. Heyde, Kuebrick, and Swanson measured the heights of 405 steel punches of a particular type. These were all from a single manufacturer and were supposed to have heights of .500 in. (The stamping machine in which these are used is designed to use .500 in. punches.) The students' measurements had $\bar{x} = .5002$ in. and $s = .0026$ in. (The raw data are given in Exercise 3-22.)

(a) Make a 98% two-sided confidence interval for the mean height of such punches produced by this manufacturer under conditions similar to those existing when the students' punches were manufactured.

(b) Use the five-step format and test the hypothesis that the mean height of such punches is "on spec" (i.e., is .500 in.).

(c) In the students' application, the mean height of the punches did not tell the whole story about how they worked in the stamping machine. Several of these punches had to be placed side by side and used to stamp the same piece of material. In this context, what other feature of the height distribution is almost certainly of practical importance?

6-12. Discuss in the context of Exercise 6-11, part (b), the potential difference between statistical significance and practical importance.

6-13. Losen, Cahoy, and Lewis measured the lengths of some spanner bushings of a particular type purchased from a local machine supply shop. The lengths obtained by one of the students were as follows. (The units are inches.)

1.1375, 1.1390, 1.1420, 1.1430, 1.1410, 1.1360, 1.1395, 1.1380,

1.1350, 1.1370, 1.1345, 1.1340, 1.1405, 1.1340, 1.1380, 1.1355

(a) If one is to, e.g., make a confidence interval for the population mean measured length of these bushings via the formulas in Section 6-3, what model assumption must be employed? Make a probability plot to assess the reasonableness of the assumption.

(b) Make a 90% two-sided confidence interval for the mean measured length for bushings of this type measured by this student.

(c) Give an upper bound for the mean length with 90% associated confidence.

(d) Make a 90% two-sided prediction interval for a single additional measured bushing length.

(e) Make a 95% two-sided tolerance interval for 99% of additional measured bushing lengths.

(f) Consider the statistical interval derived from the minimum and maximum sample values — namely,

$$(1.1340, 1.1430)$$

What confidence level should be associated with this interval as a prediction interval for a single additional bushing length? What confidence level should be associated with this interval as a tolerance interval for 99% of additional bushing lengths?

6-14. B. Choi tested the stopping properties of various bike tires on various surfaces. For one thing, he tested both treaded and smooth tires on dry concrete. The lengths of skid marks produced in his study under these two conditions were as follows (in cm).

Treaded	365, 374, 376, 391, 401, 402
Smooth	341, 348, 349, 355, 375, 391

(a) In order to make formal inferences about $\mu_{\text{Treaded}} - \mu_{\text{Smooth}}$ based on these data, what must one be willing to use for model assumptions? Make a plot useful for investigating the reasonableness of those assumptions.

(b) Proceed under the necessary model assumptions to assess the strength of Choi's evidence of a difference in mean skid lengths.

(c) Make the necessary model assumptions and find a 95% two-sided confidence interval for $\mu_{\text{Treaded}} - \mu_{\text{Smooth}}$.

6-15. The study mentioned in Exercise 6-13 also included measurement of the outside diameters of the 16 bushings. Two of the students measured each of the bushings, with the results given here.

Bushing	1	2	3	4
Student A	.3690	.3690	.3690	.3700
Student B	.3690	.3695	.3695	.3695
Bushing	5	6	7	8
Student A	.3695	.3700	.3695	.3690
Student B	.3695	.3700	.3700	.3690
Bushing	9	10	11	12
Student A	.3690	.3695	.3690	.3690
Student B	.3700	.3690	.3695	.3695
Bushing	13	14	15	16
Student A	.3695	.3700	.3690	.3690
Student B	.3690	.3695	.3690	.3690

(a) If one wants to compare the two students' average measurements, the formulas used in Exercise 6-14 are not appropriate. Why?

(b) Make a 95% two-sided confidence interval for the mean difference in outside diameter measurements for the two students.

6-16. Return to data on Choi's bicycle stopping distance, given in Exercise 6-14.

(a) Operating under the assumption that treaded tires produce normally distributed stopping distances, give a two-sided 95% confidence interval for the standard deviation of treaded tire stopping distances.

(b) Operating under the assumption that smooth tires produce normally distributed stopping distances, give a 99% upper confidence bound for the standard deviation of smooth tire stopping distances.

(c) Operating under the assumption that both treaded and smooth tires produce normally distributed stopping distances, assess the strength of Choi's evidence that treaded and smooth stopping distances differ in their variability. (Use H_0: $\sigma_{\text{Treaded}} = \sigma_{\text{Smooth}}$ and H_a: $\sigma_{\text{Treaded}} \neq \sigma_{\text{Smooth}}$ and show the whole five-step format.)

(d) Operating under the assumption that both treaded and smooth tires produce normally distributed stopping distances, give a 90% two-sided confidence interval for the ratio $\sigma_{\text{Treaded}}/\sigma_{\text{Smooth}}$.

6-17. Find the following quantiles using the tables of Appendix D:
(a) the .90 quantile of the t_5 distribution
(b) the .10 quantile of the t_5 distribution
(c) the .95 quantile of the χ_7^2 distribution
(d) the .05 quantile of the χ_7^2 distribution
(e) the .95 quantile of the F distribution with numerator degrees of freedom 8 and denominator degrees of freedom 4
(f) the .05 quantile of the F distribution with numerator degrees of freedom 8 and denominator degrees of freedom 4

6-18. Find the following quantiles using the tables of Appendix D:
(a) the .99 quantile of the t_{13} distribution
(b) the .01 quantile of the t_{13} distribution
(c) the .975 quantile of the χ_3^2 distribution

(d) the .025 quantile of the χ^2_3 distribution

(e) the .75 quantile of the F distribution with numerator degrees of freedom 6 and denominator degrees of freedom 12

(f) the .25 quantile of the F distribution with numerator degrees of freedom 6 and denominator degrees of freedom 12

6-19. Specifications on the punch heights referred to in Exercise 6-11 were .500 in. to .505 in. In the sample of 405 punches measured by Hyde, Kuebrick, and Swanson, there were only 290 punches meeting these specifications. Suppose that the 405 punches can be thought of as a random sample of all such punches manufactured by the supplier under standard manufacturing conditions. Give an approximate 99% two-sided confidence interval for the standard fraction of nonconforming punches of this type produced by the punch supplier.

6-20. Consider two hypothetical machines producing a particular wigdet. If samples of $n_1 = 25$ and $n_2 = 25$ widgets produced by the respective machines have fractions nonconforming $\hat{p}_1 = .2$ and $\hat{p}_2 = .32$, is this strong evidence of a difference in machine nonconforming rates? What does this suggest about the kind of sample sizes one typically needs in order to reach definitive conclusions based on attributes or qualitative data?

6-21. Ho, Lewer, Peterson, and Riegel worked with the lack of flatness in a particular kind of manufactured steel disk. Fifty different parts of this type were measured for what the students called "wobble," with the results that the 50 (positive) values obtained had mean $\bar{x} = .0287$ in. and standard deviation $s = .0119$ in.

(a) Give a 95% two-sided confidence interval for the mean wobble of all such disks.

(b) Give a lower bound for the mean wobble possessing a 95% confidence level.

(c) Suppose that these disks are ordered with the requirement that the mean wobble not exceed .025 in. Assess the strength of the evidence in the students' data to the effect that the requirement is being violated. Show the whole five-step format.

(d) Is the requirement of part (c) the same as an upper specification of .025 in. on individual wobbles? Explain. Is it possible for a lot with many individual wobbles exceeding .025 in. to meet the requirement of part (c)?

(e) Of the measured wobbles, 19 were .030 in. or more. Use this fact and make an approximate 90% two-sided confidence interval for the fraction of all such disks with wobbles of at least .030 in.

6-22. What is the practical consequence of using a "normal distribution" confidence interval formula when in fact the underlying data-generating mechanism can not be adequately described using a Gaussian distribution? Say something more specific/informative than "An error might be made," or "The interval might not be valid." (What, e.g., can be said about the real confidence level that ought to be associated with a nominally 90% confidence interval in such a situation?)

6-23. T. Johnson tested properties of several brands of 10 lb test monofilament fishing line. Part of his study involved measuring the stretch of a fixed length of line under a 3.5 kg load. Test results for three pieces of two of the brands follow. The units are cm.

Brand B	.86, .88, .88
Brand D	1.06, 1.02, 1.04

(a) Considering first only Brand B, use "normal distribution" model assumptions and give a 90% upper prediction bound for the stretch of an additional piece of Brand B line.

(b) Again considering only Brand B, use "normal distribution" model assumptions and give a 95% upper tolerance bound for stretch measurements of 90% of such pieces of Brand B line.

(c) Again considering only Brand B, use "normal distribution" model assumptions and give 90% two-sided confidence intervals for the mean and for the standard deviation of the Brand B stretch distribution.

(d) Compare the Brand B and Brand D standard deviations of stretch using an appropriate 90% two-sided confidence interval.

(e) Compare the Brand B and Brand D mean stretch values using an appropriate 90% two-sided confidence interval. Does this interval give clear indication of a difference in mean stretch values for the two brands?

(f) Carry out a formal significance test of the hypothesis that the two brands have the same mean stretch values. (Use a two-sided alternative hypothesis.) Does the conclusion you reach here agree with your answer to part (e)?

6-24. The accompanying data are $n = 10$ daily measurements of the purity (in percent) of oxygen being delivered by a certain industrial air products supplier. (These data are similar to some given in a November 1990 article in *Chemical Engineering Progress* and used in Exercise 3-23.)

$$99.77 \quad 99.66 \quad 99.61 \quad 99.59 \quad 99.55$$
$$99.64 \quad 99.53 \quad 99.68 \quad 99.49 \quad 99.58$$

(a) Make a normal plot of these data. What does the normal plot reveal about the shape of the purity distribution? ("It is not bell-shaped" is not an adequate answer. Say how its shape departs from the Gaussian shape.)

(b) What statistical "problems" are caused by lack of a normal distribution shape for data such as these?

As a way to deal with problems like those from part (b), one might try transforming the original data. Next are values of $y' = \ln(y - 99.3)$ corresponding to each of the original data values y, and some summary statistics for the transformed values.

$-.76$	-1.02	-1.17	-1.24	-1.39	$\bar{y}' = -1.203$
-1.08	-1.47	$-.97$	-1.66	-1.27	$s_{y'} = .263$

(c) Make a normal plot of the transformed values and verify that it is very linear.

(d) Make a 95% two-sided prediction interval for the next transformed purity delivered by this supplier. What does this "untransform" to in terms of raw purity?

(e) Make a 99% two-sided tolerance interval for 95% of additional transformed purities from this supplier. What does this "untransform" to in terms of raw purity?

(f) Suppose that the air products supplier advertises a median purity of at least 99.5%. This corresponds to a median (and therefore mean) transformed value of at least -1.61. Test the supplier's claim (H_0: $\mu_{y'} = -1.61$) against the possibility that the purity is substandard. Show and carefully label all five steps.

6-25. Exercise 3-19 contains a data set on the lifetimes (in numbers of 24 mm deep holes drilled in 1045 steel before tool failure) of 12 D952-II (8 mm) drills. The data there have mean $\bar{y} = 117.75$ and $s = 51.1$ holes drilled. Suppose that a normal distribution can be used to roughly describe drill lifetimes.

(a) Give a 90% lower confidence bound for the mean lifetime of drills of this type in this kind of industrial application.

(b) Based on your answer to (a), do you think a hypothesis test of H_0: $\mu = 100$ versus H_a: $\mu > 100$ would have a large p-value or a small p-value? Explain.

(c) Give a 90% lower prediction bound for the next life length of a drill of this type in this kind of industrial application.

(d) Give two-sided tolerance limits with 95% confidence for 90% of all life lengths for drills of this type in this kind of industrial application.

(e) Give two-sided 90% confidence limits for the standard deviation of life lengths for drills of this type in this kind of industrial application.

6-26. In estimating a proportion p, one will use a two-sided interval $\hat{p} \pm \Delta$. Suppose that one wishes to make a 95% confidence interval and have $\Delta \leq .01$. About what sample size will be needed to guarantee this?

6-27. Interpret the statement, "The interval from 6.3 to 7.9 is a 95% confidence interval for the mean μ."

6-28. Confidence, prediction, and tolerance intervals are all intended to do different jobs. What are these jobs? Consider the differing situations of an official of the EPA, a consumer about to purchase a single auto, and a design engineer trying to equip a certain model of auto with a gas tank large enough that most autos produced will have highway cruising ranges of at least 350 miles. Argue that depending on the point of view adopted, a lower confidence bound for a mean mileage, a lower prediciton bound for an individual mileage, or a lower tolerance bound for most mileages would be of interest.

6-29. M. Murphy recorded the mileages he obtained while commuting to school in his 9-year-old economy car. He kept track of the mileage for ten different tankfuls of fuel, involving gasoline of two different octanes. His data follow.

87 Octane	90 Octane
26.43, 27.61, 28.71, 28.94, 29.30	30.57, 30.91, 31.21, 31.77, 32.86

(a) Make normal plots for these two samples of size 5 on the same set of axes. Does the "equal variances, normal distributions" model appear reasonable for describing this situation?

(b) Find s_P for these data. What is this quantity measuring?

(c) Give a 95% two-sided confidence interval for the difference in mean mileages obtainable under these circumstances using the fuel of the two different octanes. From the nature of this confidence interval, would you expect to find a large p-value or a small p-value when testing H_0: $\mu_{87} = \mu_{90}$ versus H_a: $\mu_{87} \neq \mu_{90}$?

(d) Conduct a significance test of H_0: $\mu_{87} = \mu_{90}$ against the alternative that the higher-octane gasoline provides a higher mean mileage.

(e) Give 95% lower prediction bounds for the next mileages experienced, using first 87 octane fuel and then 90 octane fuel.

(f) Give 95% lower tolerance bounds for 95% of additional mileages experienced, using first 87 octane fuel and then 90 octane fuel.

6-30. Eastman, Frye, and Schnepf worked with a company that mass-produces plastic bags. Their work at the company eventually focused on startup problems of a particular machine that could be operated at either a high speed or a low speed. One part of the data they collected consisted of counts of faulty bags produced in the first 250 manufactured after changing a roll of plastic feedstock. The counts they obtained for both low- and high-speed operation of the machine were 147 faulty ($\hat{p}_H = \frac{147}{250}$) under high-speed operation and 12 faulty under low-speed operation ($\hat{p}_L = \frac{12}{250}$). In what follows, suppose that it is sensible to think of the machine as operating in a physically stable fashion during the production of the first 250 bags after changing a roll of plastic, with a constant probability (p_H or p_L) of any particular bag produced being faulty.

(a) Give a 95% upper confidence bound for p_H.

(b) Give a 95% upper confidence bound for p_L.

(c) Compare p_H and p_L using an appropriate two-sided 95% confidence interval. Does this interval provide a clear indication of a difference in the effectiveness of the machine at startup when run at the two speeds? What kind of a p-value (big or small) would you expect to find in a test of H_0: $p_H = p_L$ versus H_a: $p_H \neq p_L$?

(d) Use the five-step format and test H_0: $p_H = p_L$ versus H_a: $p_H \neq p_L$.

6-31. Hamilton, Seavey, and Stucker measured resistances, diameters, and lengths for seven copper wires at two different temperatures and used these to compute experimental resistivities for copper at these two temperatures. Their resulting resistivities follow. The units are 10^{-8} Ωm.

Wire	0.0°C	21.8°C
1	1.52	1.72
2	1.44	1.56
3	1.52	1.68
4	1.52	1.64
5	1.56	1.69
6	1.49	1.71
7	1.56	1.72

(a) Suppose that primary interest here centers on the difference between resistivities at the two different temperatures. Make a normal plot of the seven observed differences here. Does it appear that a normal distribution description of the observed difference in resistivities at these two temperatures is plausible?

(b) Give a 90% two-sided confidence interval for the mean difference in resistivity measurements for copper wire of this type at 21.8°C and 0.0°C.

(c) Give a 90% two-sided prediction interval for an additional difference in resistivity measurements for copper wire of this type at 21.8°C and 0.0°C.

6-32. The students referred to in Exercise 6-31 also measured the resistivities for seven aluminum wires at the same temperatures. The 21.8°C measurements that they obtained follow.

2.65, 2.83, 2.69, 2.73, 2.53, 2.65, 2.69

(a) Give a 99% two-sided confidence interval for the mean resistivity value derived from such experimental determinations.

(b) Give a 95% two-sided prediction interval for the next resistivity value that would be derived from such an experimental determination.

(c) Give a 95% two-sided tolerance interval for 99% of resistivity values derived from such experimental determinations.

(d) Give a 95% two-sided confidence interval for the standard deviation of resistivity values derived from such experimental determinations.

(e) How strong is the evidence that there is a real difference in the precisions with which the aluminum resistivities and the copper resistivities can be measured at 21.8°C? (Carry out a significance test of H_0: $\sigma_{copper} = \sigma_{aluminum}$ versus H_a: $\sigma_{copper} \neq \sigma_{aluminum}$ using the data of this problem and the 21.8°C data of Exercise 6-31.)

(f) Again using the data of this exercise and Exercise 6-31, give a 90% two-sided confidence interval for the ratio $\sigma_{copper}/\sigma_{aluminum}$.

6-33. (The Stein Two-Stage Estimation Procedure) One of the most common of all questions faced by engineers planning a data-based study is how much data to collect. The last part of Example 6-3 illustrates a rather crude method of producing an answer to the sample-size question when estimation of a single mean is involved. In fact, in such circumstances, a more careful two-stage procedure due to Charles Stein can sometimes be used to find appropriate sample sizes.

Suppose that one wishes to use an interval of the form $\bar{x} \pm \Delta$, with a particular confidence coefficient to estimate the mean μ of a normal distribution. If it is desirable to have $\Delta \leq \#$ for some number # and one can collect data in two stages, it is possible to choose an overall sample size to satisfy these criteria as follows. After taking a small or moderate initial sample of size n_1 (n_1 must be at least 2 and is typically at least 4 or 5), one computes the sample standard deviation of the initial data — say, s_1. Then if t is the appropriate t_{n_1-1} distribution quantile for producing the desired (one- or two-sided) confidence, it is necessary to find the smallest integer n such that

$$n \geq \left(\frac{ts_1}{\#}\right)^2$$

If this integer is larger than n_1, then $n_2 = n - n_1$ additional observations are taken. (Otherwise, $n_2 = 0$.) Finally, with $\bar{x}$ the sample mean of all the observations (from both the initial and any subsequent sample), the formula $\bar{x} \pm ts_1/\sqrt{n_1 + n_2}$ (with t still based on $n_1 - 1$ degrees of freedom) is used to estimate μ.

Suppose that in estimating the mean resistance of a production run of resistors, it is desirable to have two-sided confidence level 95% and the "± part" of the interval be no longer than .5 Ω.

(a) If an initial sample of $n_1 = 5$ resistors produces a sample standard deviation of 1.27 Ω, how many (if any) additional resistors should be sampled in order to meet the stated goals?

(b) If all of the $n_1 + n_2$ resistors taken together produce the sample mean $\bar{x} = 102.8$ Ω, what confidence interval for μ should be declared?

6-34. Example 5-18 concerns some data on service times at a residence hall depot counter. The data portrayed in Figure 5-25 are decidedly nonnormal-looking, so prediction and tolerance interval formulas based on normal distributions are not appropriate for use with these data. However, the largest of the $n = 65$ observed service times in that figure is 87 sec.

(a) What prediction confidence level can be associated with 87 sec as an upper prediction bound for a single additional service time?

(b) What confidence level can be associated with 87 sec as an upper tolerance bound for 95% of service times?

6-35. Caliste, Duffie, and Rodriguez studied the process of key-making using a manual machine at a local lumber yard. The records of two different employees who made keys during the study period were as follows. Employee 1 made a total of 54 different keys, 5 of which were returned as not fitting their locks. Employee 2 made a total of 73 different keys, 22 of which were returned as not fitting their locks.

(a) Give approximate 95% two-sided confidence intervals for the long-run fractions of faulty keys produced by these two different employees.

(b) Give an approximate 95% two-sided confidence interval for the difference in long-run fractions of faulty keys produced by these two different employees.

(c) Assess the strength of the evidence provided in these two samples of a real difference in the keymaking proficiencies of these two employees. (Test H_0: $p_1 = p_2$ using a two-sided alternative hypothesis.)

6-36. The article "Optimizing Heat Treatment with Factorial Design" by T. Lim (*JOM*, 1989) discusses the improvement of a heat treating process for gears through the use of factorial experimentation. To compare the performance of the heat treating process under the original settings of process variables to that

OPTIMIZED SETTINGS		
Gear	y_1 (mm)	y_2 (mm)
1A	.036	.050
2A	.040	.054
3A	.026	.043
4A	.051	.071
5A	.034	.043
6A	.050	.058
7A	.059	.061
8A	.055	.048
9A	.051	.060
10A	.050	.033

ORIGINAL SETTINGS		
Gear	y_1 (mm)	y_2 (mm)
1B	.056	.070
2B	.064	.062
3B	.070	.075
4B	.037	.060
5B	.054	.071
6B	.060	.070
7B	.065	.060
8B	.060	.060
9B	.051	.070
10B	.062	.070

using the "improved" settings (identified through factorial experimentation), $n_1 = n_2 = 10$ gears were treated under both sets of conditions. Then measures of flatness, y_1 (in mm of distortion), and concentricity, y_2 (again in mm of distortion) were made on each of the gears. The data shown were read from graphs in the article (and may in some cases differ by perhaps $\pm.002$ mm from the original measurements).

(a) What assumptions are necessary in order to make inferences regarding the parameters of the y_1 (or y_2) distribution for the optimized settings of the process variables?

(b) Make a normal plot for the optimized settings' y_1 values. Does it appear that it is reasonable to treat the optimized settings' flatness distribution as Gaussian? Explain.

(c) Suppose that the optimized settings' flatness distribution is normal, and do the following.

   (i) Give a 90% two-sided confidence interval for the mean flatness distortion value for gears of this type.

   (ii) Give a 90% two-sided prediction interval for an additional flatness distortion value.

   (iii) Give a 95% two-sided tolerance interval for 90% of additional flatness distortion values.

   (iv) Give a 90% two-sided confidence interval for the standard deviation of flatness distortion values for gears of this type.

(d) Repeat parts (b) and (c) using the optimized settings' y_2 values and concentricity instead of flatness.

(e) Explain why it is not possible to base formal inferences (tests and confidence intervals), for comparing the standard deviations of the y_1 and y_2 distributions for the optimized process settings, on the sample standard deviations of the y_1 and y_2 measurements from gears 1A through 10A.

(f) What assumptions are necessary in order to make comparisons between parameters of the y_1 (or y_2) distributions for the original and optimized settings of the process variables?

(g) Make normal plots of the y_1 data for the original settings and for the optimized settings on the same set of axes. Does an "equal variances, normal distributions" model appear tenable here? Explain.

(h) Supposing that the flatness distortion distributions for the original and optimized process settings are adequately described as Gaussian with a common standard deviation, do the following.

   (i) Use an appropriate significance test to assess the strength of the evidence in the data to the effect that the optimized settings produce a reduction in mean flatness distortion.

   (ii) Give a 90% lower confidence bound on the reduction in mean flatness distortion provided by the optimized process settings.

(i) Repeat parts (g) and (h) using the y_2 values and concentricity instead of flatness.

6-37. R. Behne measured air pressure in car tires in a student parking lot. Shown here is one summary of the data he reported. Any tire with pressure reading more than 3 psi below its recommended value is considered underinflated for the purposes of the summary below, while any tire with pressure reading more than 3 psi above its recommended value is considered overinflated. The counts in the accompanying table are the numbers of cars (out of 25 checked) falling into the four possible categories.

		UNDERINFLATED TIRES	
		None	At Least One Tire
OVERINFLATED TIRES	None	6	5
	At Least One Tire	10	4

(a) Behne's sample was in all likelihood a convenience sample (as opposed to a genuinely simple random sample) of the cars in the large lot. Does it make sense to argue in this case that the data can be treated as if the sample were a simple random sample? On what basis? Explain.

(b) Give a two-sided 90% confidence interval for the fraction of all cars in the lot with at least one underinflated tire.

(c) Give a two-sided 90% confidence interval for the fraction of the cars in the lot with at least one overinflated tire.

(d) Give a 90% lower confidence bound on the fraction of cars in the lot with at least one misinflated tire.

(e) Why can't the data here be used with formula (6-65) of Section 6-5 to make a confidence interval for the difference in the fraction of cars with at least one underinflated tire and the fraction with at least one overinflated tire?

6-38. The article "A Recursive Partitioning Method for the Selection of Quality Assurance Tests" by Raz and Bousum (*Quality Engineering*, 1990) contains some data on the fractions of torque converters manufactured in a particular facility failing a final inspection (and thus requiring some rework). For a particular family of four-element converters, about 39% of 442 converters tested were out of specifications on a high-speed operation inlet flow test.

(a) If plant conditions tomorrow are like those under which the 442 converters were manufactured, give a two-sided 98% confidence interval for the probability that a given converter manufactured will fail the high-speed inlet flow test.

(b) Suppose that a process change is instituted in an effort to reduce the fraction of converters failing the high-speed inlet flow test. If only 32 out of the first 100 converters manufactured fail the high-speed inlet flow test, is this convincing evidence that a real process improvement has been accomplished? (Give and interpret a 90% two-sided confidence interval for the change in test failure probability.)

6-39. Return to the situation of Exercise 3-14 and the measured gains of 120 amplifiers. The nominal/design value of the gain was 10.0 dB; 16 of the 120 amplifiers measured had gains above nominal. Give a 95% two-sided confidence interval for the fraction of all such amplifiers with above-nominal gains.

6-40. The article "Multi-functional Pneumatic Gripper Operating under Constant Input Actuation Air Pressure" by J. Przybyl (*Journal of Engineering Technology*, 1988) discusses the performance of a 6-digit pneumatic robotic gripper. One part of the article concerns the gripping pressure (measured by manometers) delivered to objects of different shapes for fixed input air pressures. The data given here are the measurements (in psi) reported for an actuation pressure of 40 psi for (respectively) a 1.7 in. $\times$ 1.5 in. $\times$ 3.5 in. rectangular bar and a circular bar of radius 1.0 in. and length 3.5 in.

(a) Compare the variabilities of the gripping pressures delivered to the two different objects using an appropriate 98% two-sided

Rectangular Bar	Circular Bar
76	84
82	87
85	94
88	80
82	92

confidence interval. Does there appear to be much evidence in the data of a difference between these? Explain.

(b) Supposing that the variabilities of gripping pressure delivered by the gripper to the two different objects are comparable, give a 95% two-sided confidence interval for the difference in mean gripping pressures delivered.

(c) The data here came from the operation of a single prototype gripper. Why would you expect to see more variation in measured gripping pressures than that represented here if each measurement in a sample were made on a different gripper? Strictly speaking, to what do the inferences in (a) and (b) apply? To the single prototype gripper or to all grippers of this design? Discuss this issue.

6-41. DuToit, Hansen, and Osborne measured the diameters of some no. 10 machine screws with two different calipers (digital and vernier scale). Part of their data are recorded here. Given in the small frequency table are the measurements obtained on 50 screws by one of the students using the digital calipers.

Diameter (mm)	Frequency
4.52	1
4.66	4
4.67	7
4.68	7
4.69	14
4.70	9
4.71	4
4.72	4

(a) Compute the sample mean and standard deviation for these data.

(b) Use your sample values from (a) and make a 98% two-sided confidence interval for the mean diameter of such screws as measured by this student with these calipers.

(c) Repeat part (b) using 99% confidence. How does this interval compare with the one from (b)?

(d) Use your values from (a) and find a 98% lower confidence bound for the mean diameter. (Find a number # such that (#, ∞) is a 98% confidence interval.) How does this value compare to the lower endpoint of your interval from (b)?

(e) Repeat (d) using 99% confidence. How does the value computed here compare to your answer to (d)?

(f) Interpret your interval from (b) for someone with little statistical background. (Speak in the context of the diameter measurement study, and use Definition 6-2 as your guide.)

6-42. In the context of the machine screw diameter study of Exercise 6-41, suppose that the nominal diameter of such screws is 4.70 mm. Use the five-step significance-testing format of Section 6-2 and assess the strength of the evidence provided by the data that the long-run mean measured diameter differs from

nominal. (You will want to begin with $H_0: \mu = 4.70$ mm and use $H_a: \mu \neq 4.70$ mm.)

6-43. Discuss, in the context of Exercise 6-42, the potential difference between statistical significance and practical importance.

6-44. A sample of 95 U-bolts produced by a small company has thread lengths with a mean of $\bar{x} = 10.1$ (.001 in. above nominal) and $s = 3.2$ (.001 in.).

(a) Give a 95% two-sided confidence interval for the mean thread length (measured in .001 in. above nominal). Judging from this interval, would you expect a small or a large p-value when testing $H_0: \mu = 0$ versus $H_a: \mu \neq 0$? Explain.

(b) Use the five-step format of Section 6-2 and assess the strength of the evidence provided by the data to the effect that the population mean thread length exceeds nominal.

6-45. D. Kim did some crude tensile strength testing on pieces of some nominally .012 in. diameter wire of various lengths. Below are Kim's measured strengths (kg) for pieces of wire of lengths 25 cm and 30 cm.

25 cm Lengths	30 cm Lengths
4.00, 4.65, 4.70, 4.50	4.10, 4.50, 3.80, 4.60
4.40, 4.50, 4.50, 4.20	4.20, 4.60, 4.60, 3.90

(a) If one is to make a confidence interval for the mean measured strength of 25-cm pieces of this wire using the methods of Section 6-3, what model assumption must be employed? Make a probability plot useful in assessing the reasonableness of the assumption.

(b) Make a 95% two-sided confidence interval for the mean measured strength of 25-cm pieces of this wire.

(c) Give a 95% lower confidence bound for the mean measured strength of 25-cm pieces.

(d) Make a 95% two-sided prediction interval for a single additional measured strength for a 25-cm piece of wire.

(e) Make a 99% two-sided tolerance interval for 95% of additional measured strengths of 25-cm pieces of this wire.

(f) Consider the statistical interval derived from the minimum and maximum sample values for the 25-cm lengths — namely, (4.00, 4.70). What confidence should be associated with this interval as a prediction interval for a single additional measured strength? What confidence should be associated with this interval as a tolerance interval for 95% of additional measured strengths for 25-cm pieces of this wire?

(g) In order to make formal inferences about $\mu_{25} - \mu_{30}$ based on these data, what must one be willing to use for model assumptions? Make a plot useful for investigating the reasonableness of those assumptions.

(h) Proceed under the assumptions discussed in part (g) and assess the strength of the evidence provided by Kim's data to the effect that an increase in specimen length produces a decrease in measured strength.

(i) Proceed under the necessary model assumptions to give a 98% two-sided confidence interval for $\mu_{25} - \mu_{30}$.

6-46. The machine screw measurement study of DuToit, Hansen, and Osborne referred to in Exercise 6-41 involved measurement of diameters of each of 50 screws with both digital and vernier scale calipers. For the student referred to in Exercise 6-41, the differences in measured diameters (digital − vernier, with units of mm) had the following frequency distribution.

Difference	−.03	−.02	−.01	.00	.01	.02	.03
Frequency	1	3	11	19	10	6	0

(a) Make a 90% two-sided confidence interval for the mean difference in digital and vernier readings for this student.

(b) Assess the strength of the evidence provided by these differences to the effect that there is a systematic difference in the readings produced by the two calipers (at least when employed by this student).

(c) Briefly discuss why your answers to parts (a) and (b) of this exercise are compatible. (Discuss how the outcome of part (b) could have easily been anticipated from the outcome of part (a).)

6-47. The article "Influence of Final Recrystallization Heat Treatment on Zircaloy-4 Strip Corrosion" by Foster, Dougherty, Burke, Bates, and Worcester (*Journal of Nuclear Materials*, 1990) reported some summary statistics from the measurement of the diameters of 821 particles observed in a bright field TEM micrograph of a Zircaloy-4 specimen. The sample mean diameter was $\bar{x} = .055$ μm, and the sample standard deviation of the diameters was $s = .028$ μm.

(a) The engineering researchers wished to establish from their observation of this single specimen the impact of a certain combination of specimen lot and heat treating regimen on particle size. Briefly discuss why data such as the ones summarized above have serious limitations for this purpose. (*Hints:* The apparent "sample size" here is huge. But of what does one have a sample? How widely do the researchers want their results to apply? In the light of this desire, is the "real" sample size really so large?)

(b) Use the sample information and give a 98% two-sided confidence interval for the mean diameter of particles in this particular Zircaloy-4 specimen.

(c) Suppose that a standard method of heat treating for such specimens is believed to produce a mean particle diameter of .057 μm. Assess the strength of the evidence contained in the sample of diameter measurements to the effect that the specimen's mean

particle diameter is different from the standard. Show the whole five-step format.

(d) Discuss, in the context of part (c), the potential difference between the mean diameter being statistically different from .057 μm and there being a difference between μ and .057 that is of practical importance.

6-48. Return to Kim's tensile strength data given in Exercise 6-45.

(a) Operating under the assumption that measured tensile strengths of 25-cm lengths of the wire studied are normally distributed, give a two-sided 98% confidence interval for the standard deviation of measured strengths.

(b) Operating under the assumption that measured tensile strengths of 30-cm lengths of the wire studied are normally distributed, give a 95% upper confidence bound for the standard deviation of measured strengths.

(c) Operating under the assumption that both 25- and 30-cm lengths of the wire have normally distributed measured tensile strengths, assess the strength of Kim's evidence that 25- and 30-cm lengths differ in variability of their measured tensile strengths. (Use H_0: $\sigma_{25} = \sigma_{30}$ and H_a: $\sigma_{25} \neq \sigma_{30}$ and show the whole five-step format.)

(d) Operating under the assumption that both 25- and 30-cm lengths produce normally distributed tensile strengths, give a 98% two-sided confidence interval for the ratio σ_{25}/σ_{30}.

6-49. Find the following quantiles:

(a) the .99 quantile of the χ_4^2 distribution

(b) the .025 quantile of the χ_4^2 distribution

(c) the .99 quantile of the F distribution with numerator degrees of freedom 3 and denominator degrees of freedom 15

(d) the .25 quantile of the F distribution with numerator degrees of freedom 3 and denominator degrees of freedom 15

6-50. The digital and vernier caliper measurements of no. 10 machine screw diameters summarized in Exercise 6-46 are such that for 19 out of 50 of the screws, there was no difference in the measurements. Based on these results, give a 95% confidence interval for the long-run fraction of such measurements by the student technician that would produce agreement between the digital and vernier caliper measurements.

7   Inference for Unstructured Multisample Studies

hapter 6 introduced the basics of formal statistical inference and applied them to the analysis of data from one- and two-sample statistical engineering studies. This chapter begins to consider formal inference methods for the analysis of multisample studies, with a look at methods that make no explicit use of structure relating the samples (except possibly the order in which samples are collected). That is, the study of inference methods specifically crafted for use in factorial, fractional factorial, and hierarchical studies and in curve- and surface-fitting analyses will be delayed until subsequent chapters. Therefore, the methods of this chapter have the widest potential applicability of the multisample inference methods studied in this text. They are relevant when there is no obvious standard structure relating several samples, but they also have potential application in any of the structured contexts of Chapters 8 through 10, if one temporarily ignores the structure.

The chapter begins with a discussion of the standard one-way model typically used in the analysis of measurement data from unstructured multisample statistical engineering studies, and the role of residuals in judging the appropriateness of that model. The making of statistical intervals in multisample contexts is then considered, including individual and simultaneous confidence interval methods for linear combinations of mean responses and also prediction and tolerance intervals. Next are introduced the one-way analysis of variance (ANOVA) and analysis of means (ANOM) tests for the hypothesis of equality of several means, along with a method of estimating *variance components* related to the first of these. The chapter then covers the use of Shewhart control charts for the monitoring of the performance of processes via the periodic examination of samples taken on them. The $\bar{x}$, R, and s control charts for measurement data are studied, and their relation to the ANOVA and (particularly) the ANOM methods are noted. The chapter then closes with a section on p charts and u charts, tools that are useful when only attributes data are available for use in process-monitoring situations.

7-1   The One-Way Normal Model

One of the messages that is at least implicit in Chapters 1 through 4 of this book is that typical statistical engineering studies produce samples taken under not one or two, but rather many different sets of conditions. It is more common for a civil engineer to collect a small-to-moderate amount of data on the strength properties of each of many different concrete formulas than to do an exhaustive investigation of the properties of one or two formulas. Aeronautical engineers usually want to compare the aerodynamic properties of more than two lifting bodies. A chemical engineer is typically faced with many different possible sets of conditions under which to run a production process and thus often faces the problem of comparing responses from not just two, but a number of these. And so on. The point is that although the inference methods of Chapter 6 are a start, they are not a complete statistical tool kit for engineering problem solving. Methods of formal inference appropriate to multisample studies are also needed.

This section begins to provide such methods. First the reader is reminded of the usefulness of some of the simple graphical tools of Chapter 3 for making qualitative comparisons in multisample studies. Next is introduced the "equal variances, underlying normal distributions" model that forms the basis of standard formal inference methods for unstructured multisample measurement data. The role of residuals in evaluating the reasonableness of that model in a particular application is explained and emphasized. The section then proceeds to introduce the notion of combining sample variances from a number of samples to produce a single pooled estimate of baseline variation. Finally, there is a discussion of how the computation and use of *standardized residuals* can be helpful when sample sizes vary considerably in a multisample study.

Graphical Comparison of Several Samples of Measurement Data

Any thoughtful analysis of several samples of engineering measurement data ought to begin with the making of one or more graphical representations of those data and the visual inspection of the plots. Where samples are small, side-by-side dot diagrams are the most natural graphical tool. Where sample sizes are moderate to large (say, at least six or so data points per sample), side-by-side boxplots are effective at giving a succinct visual description of the samples.

EXAMPLE 7-1

Comparing Compressive Strengths for Eight Different Concrete Formulas. Armstrong, Babb, and Campen did compressive strength testing on 16 different concrete formulas. Part of their data are given in Table 7-1, where eight different formulas are represented. (Actually, the only differences between the formulas 1 through 8 included here are their water/cement ratios. Formula 1 had the lowest water/cement ratio, and the ratio increased with formula number in the progression .40, .44, .49, .53, .58, .62, .66, .71. Of course, knowing these actual water/cement ratios suggests that a curve-fitting analysis might be useful with these data, but for the time being this possibility will be ignored.)

Notice that making side-by-side dot diagrams for these eight samples of sizes $n_1 = n_2 = n_3 = n_4 = n_5 = n_6 = n_7 = n_8 = 3$ amounts to making a scatterplot of compressive strength versus formula number. Such a plot is shown in Figure 7-1.

TABLE 7-1

Compressive Strengths for 24 Concrete Specimens

Specimen	Concrete Formula	28-Day Compressive Strength (psi)
1	1	5,800
2	1	4,598
3	1	6,508
4	2	5,659
5	2	6,225
6	2	5,376
7	3	5,093
8	3	4,386
9	3	4,103
10	4	3,395
11	4	3,820
12	4	3,112
13	5	3,820
14	5	2,829
15	5	2,122
16	6	2,971
17	6	3,678
18	6	3,325
19	7	2,122
20	7	1,372
21	7	1,160
22	8	2,051
23	8	2,631
24	8	2,490

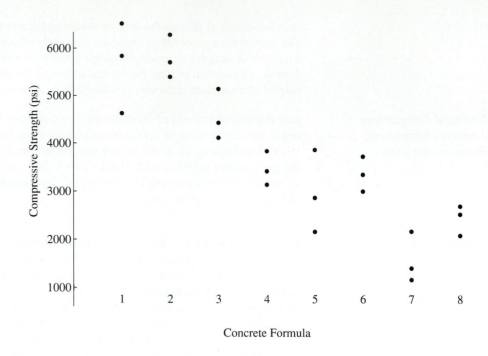

The general message conveyed by Figure 7-1 is that there are some pretty clear differences in mean compressive strengths between the formulas, but that the variabilities in compressive strengths are roughly comparable for the eight different formulas represented.

EXAMPLE 7-2

Comparing Empirical Spring Constants for Three Different Types of Springs. Hunwardsen, Springer, and Wattonville did some testing of three different types of steel springs. They made experimental determinations of spring constants for $n_1 = 7$ springs of type 1 (a 4 in. design with a theoretical spring constant of 1.86), $n_2 = 6$ springs of type 2 (a 6 in. design with a theoretical spring constant of 2.63), and $n_3 = 6$ springs of type 3 (a 4 in. design with a theoretical spring constant of 2.12), using an 8.8 lb actual load. The students' experimental values are given in Table 7-2.

TABLE 7-2

Empirical Spring Constants

Type 1 Springs	Type 2 Springs	Type 3 Springs
1.99, 2.06, 1.99	2.85, 2.74, 2.74	2.10, 2.01, 1.93
1.94, 2.05, 1.88	2.63, 2.74, 2.80	2.02, 2.10, 2.05
2.30		

These samples are just barely large enough to produce meaningful boxplots. Figure 7-2 gives a side-by-side boxplot representation of these data.

The primary qualitative message carried by Figure 7-2 is that there is a substantial difference in empirical spring constants between the 6 in. spring type and the two 4 in. spring types, but that no such difference between the two 4 in. spring types is obvious (in spite of the difference in theoretical constants for the two types). Of course, the material in Table 7-2 could also be presented in side-by-side dot diagram form, as in Figure 7-3.

FIGURE 7-2

Side-by-Side Boxplots of
Empirical Spring Constants for
Springs of Three Types

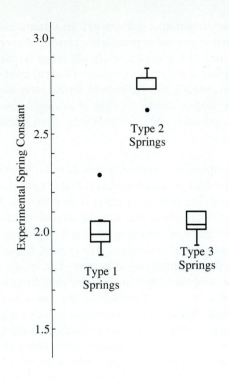

FIGURE 7-3

Side-by-Side Dot Diagrams for
Three Samples of Empirical
Spring Constants

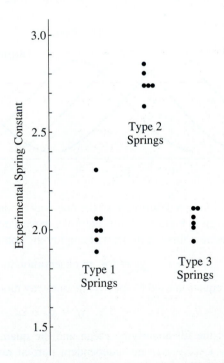

A primary function of methods of formal statistical inference is to sharpen and quantify somewhat vague and imprecise impressions that one gets when making a descriptive analysis of data. But it should practically never be the case that an intelligent graphical look at data and a correct application of formal inference methods tell completely different stories. Indeed, the methods of formal inference offered for simple unstructured multisample studies can be thought of as *confirmatory* — in cases like Examples 7-1 and 7-2, they should confirm what is clear from a descriptive or *exploratory* look at the data.

The One-Way (Normal) Multisample Model, Fitted Values, and Residuals

Chapter 6 emphasized repeatedly that to make formal inferences, one must develop a model for the data generation process that is both tractable and a plausible description of the process. The present situation is no different. As it turns out, standard inference methods for unstructured multisample studies are derived from the natural extension of the model that was used in Section 6-3 to provide small-sample inference methods for the comparison of two means. The present discussion will be carried out under the assumption that r samples of respective sizes $n_1, n_2, \ldots, n_r$ are roughly describable as independent samples from normal underlying distributions with a common variance — say, σ^2. Just as the $r = 2$ version of this *one-way* (as opposed, e.g., to several-way factorial) model led to useful inference methods for $\mu_1 - \mu_2$, this general version will lead to a variety of useful inference methods for r-sample studies. Figure 7-4 shows a number of different normal distributions with a common standard deviation. It can be thought of as representing essentially what must be generating the data behind the scenes if the methods of this chapter are to be applied.

FIGURE 7-4

r Normal Distributions with a Common Standard Deviation

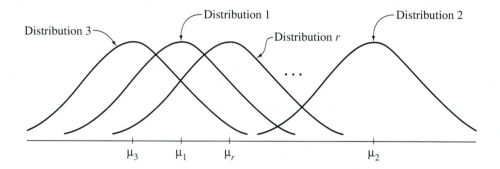

In addition to a description of the one-way model in words and the pictorial representation given in Figure 7-4, it is helpful to have a description of the model in symbols. This and the next three sections will employ the notation

$$y_{ij} = \text{the } j\text{th observation in sample } i$$

The model equation used to specify the one-way model is then

$$y_{ij} = \mu_i + \epsilon_{ij} \tag{7-1}$$

where μ_i is the ith underlying mean and the quantities $\epsilon_{11}, \epsilon_{12}, \ldots, \epsilon_{1n_1}, \epsilon_{21}, \epsilon_{22}, \ldots, \epsilon_{2n_2}, \ldots, \epsilon_{r1}, \epsilon_{r2}, \ldots, \epsilon_{rn_r}$ are independent normal random variables with mean 0 and variance σ^2. (In this statement, the means $\mu_1, \mu_2, \ldots, \mu_r$ and the variance σ^2 are typically unknown parameters.)

Equation (7-1) says exactly what is conveyed by Figure 7-4 and the statement in words (that the data are describable as independent samples from normal distributions with a common variance). But it says it in a way that is suggestive of another useful pattern of thinking, reminiscent of the "residual" notion that was used extensively in

Sections 4-1, 4-2, and 4-3. That is, equation (7-1) can be interpreted as saying that an observation in sample i is made up of the corresponding underlying mean plus some random noise, namely

$$\epsilon_{ij} = y_{ij} - \mu_i$$

This is a theoretical counterpart of an empirical notion met in Chapter 4. There, it was useful to decompose data into fitted values (corresponding to some equation found empirically to describe a data set) and the resultant residuals (that ought to appear patternless or random if in fact the equation was effective as a description of the data set).

In the present situation, since any possible structure relating the r different samples is specifically being ignored, it may not be immediately obvious how to apply the notions of fitted values and residuals. But a little reflection on the matter and contemplation of equation (7-1) ought to convince most readers that a plausible meaning for

$$\hat{y}_{ij} = \text{the fitted value corresponding to } y_{ij}$$

in the present context is the ith sample mean

$$\bar{y}_i = \frac{1}{n_i} \sum_{j=1}^{n_i} y_{ij}$$

That is,

$$\hat{y}_{ij} = \bar{y}_i \tag{7-2}$$

(This is not only intuitively plausible, but it is also consistent with what was done in Sections 4-1 and 4-2. If one fits the approximate relationship $y_{ij} \approx \mu_i$ to the data via least squares — i.e., by minimizing $\sum_{ij}(y_{ij} - \mu_i)^2$ over choices of $\mu_1, \mu_2, \ldots, \mu_r$ — each minimizing value of μ_i is easily shown to be $\bar{y}_i$.)

Taking equation (7-2) to specify fitted values for an r-sample study, the pattern established in Chapter 4 (specifically, Definition 4-4) then says that residuals ought to be defined as the differences between observed values and sample means. That is, with

$$e_{ij} = \text{the residual corresponding to } y_{ij}$$

one has

$$e_{ij} = y_{ij} - \hat{y}_{ij} = y_{ij} - \bar{y}_i \tag{7-3}$$

Rearranging display (7-3), one has the relationship

$$y_{ij} = \hat{y}_{ij} + e_{ij} = \bar{y}_i + e_{ij} \tag{7-4}$$

which is an empirical counterpart of the theoretical statement (7-1). In fact, combining equations (7-1) and (7-4) into a single statement, one obtains

$$y_{ij} = \mu_i + \epsilon_{ij} = \bar{y}_i + e_{ij} \tag{7-5}$$

which is itself a specific instance of a pattern of thinking that runs through all of the common normal-distribution-based methods of analysis for multisample studies. That is, in words, equation (7-5) says

$$\text{Observation} = \text{deterministic response} + \text{noise} = \text{fitted value} + \text{residual} \tag{7-6}$$

and display (7-6) is a paradigm that provides a unified way of approaching the majority of the analysis methods presented in the rest of this book.

The decompositions (7-5) and (7-6) provide theoretical motivation for the kind of normal-plotting of residuals that was done (in a somewhat *ad hoc* manner) in Chapter 4.

In the process, they suggest methods that are useful for investigating the reasonableness of the present model in a given application. That is, the decompositions (7-5) and (7-6) suggest that fitted values ($\hat{y}_{ij} = \bar{y}_i$) are meant to approximate the deterministic part of a system response (μ_i), and residuals (e_{ij}) are therefore meant to approximate the corresponding noise in the response (ϵ_{ij}). The fact that the ϵ_{ij} in equation (7-1) are assumed to be iid normal ($0, \sigma^2$) random variables then suggests that the e_{ij} ought to look at least approximately like a random sample from a normal distribution. So one is led to the normal-plotting of an entire set of residuals (as in Chapter 4) as a way of checking on the reasonableness of equation (7-1). Also, the plotting of residuals in display (7-3) against fitted values of equation (7-2), time order of observation, or any other potentially relevant variable — hoping (as in Chapter 4) to see only random scatter — are other ways of investigating the appropriateness of the present model assumptions in equation (7-1).

These kinds of plotting, which combine together residuals from all r samples, are particularly useful in practice, because when r is large at all, budget constraints on total data collection costs often force the individual sample sizes $n_1, n_2, \ldots, n_r$ to be fairly small. This makes it a fruitless proposition to investigate "single variance, normal distributions" model assumptions using (for example) sample-by-sample normal plots. Of course, where all of $n_1, n_2, \ldots, n_r$ are of a decent size, a sample-by-sample approach ought to be used as well.

EXAMPLE 7-1
(continued)

Returning again to the concrete strength study of Armstrong, Babb, and Campen, consider the problem of investigating the reasonableness of using the model (7-1) for analyzing these data. Figure 7-1 should be considered a first step in this investigation. As remarked earlier, it conveys the visual impression that at least the "equal variances" part of the one-way model assumptions is plausible. Next, it makes sense to compute some summary statistics and examine them, particularly the sample standard deviations. Table 7-3 gives sample sizes, sample means, and sample standard deviations for the data in Table 7-1.

TABLE 7-3

Summary Statistics for the Concrete Strength Study

i Concrete Formula	n_i Sample Size	$\bar{y}_i$ Sample Mean (psi)	s_i Sample Standard Deviation (psi)
1	$n_1 = 3$	$\bar{y}_1 = 5,635.3$	$s_1 = 965.6$
2	$n_2 = 3$	$\bar{y}_2 = 5,753.3$	$s_2 = 432.3$
3	$n_3 = 3$	$\bar{y}_3 = 4,527.3$	$s_3 = 509.9$
4	$n_4 = 3$	$\bar{y}_4 = 3,442.3$	$s_4 = 356.4$
5	$n_5 = 3$	$\bar{y}_5 = 2,923.7$	$s_5 = 852.9$
6	$n_6 = 3$	$\bar{y}_6 = 3,324.7$	$s_6 = 353.5$
7	$n_7 = 3$	$\bar{y}_7 = 1,551.3$	$s_7 = 505.5$
8	$n_8 = 3$	$\bar{y}_8 = 2,390.7$	$s_8 = 302.5$

At first glance, it might seem worrisome that in this table, s_1 is more than three times the size of s_8. But a little reflection on the matter leads to the conclusion that the sample sizes here are so small that a largest ratio of sample standard deviations on the order of 3.2 is hardly unusual (for $r = 8$ samples of size 3 from a normal distribution). Note from the F tables in Appendix D-6 that for samples of size 3, even if only 2 (rather than 8) sample standard deviations were involved, a ratio of sample variances of $(965.6/302.5)^2 \approx 10.2$ would yield a p-value between .10 and .20 for testing the null hypothesis of equal variances with a two-sided alternative. The sample standard deviations in Table 7-3 really carry no strong indication that the one-way model is inappropriate.

Since the individual sample sizes are so small, trying to see anything useful in eight separate normal plots of the samples is essentially hopeless. But one can hope to gain some insight by calculating and plotting all $8 \times 3 = 24$ residuals. Some of the calculations necessary to compute residuals for the data in Table 7-1 (using the fitted values appearing as sample means in Table 7-3) are shown in Table 7-4.

Figures 7-5 and 7-6 are, respectively, a plot of residuals versus fitted y (e_{ij} versus $\bar{y}_{ij}$) and a normal plot of all 24 residuals.

TABLE 7-4

Example Computations of
Residuals for the Concrete
Strength Study

Specimen	i Concrete Formula	y_{ij} Compressive Strength (psi)	$\hat{y}_{ij} = \bar{y}_i$ Fitted Value	e_{ij} Residual
1	1	5,800	5,635.3	164.7
2	1	4,598	5,635.3	−1,037.3
3	1	6,508	5,635.3	872.7
4	2	5,659	5,753.3	−94.3
5	2	6,225	5,753.3	471.7
⋮	⋮	⋮	⋮	⋮
22	8	2,051	2,390.7	−339.7
23	8	2,631	2,390.7	240.3
24	8	2,490	2,390.7	99.3

FIGURE 7-5

Plot of Residuals versus
Fitted Responses for the
Compressive Strengths

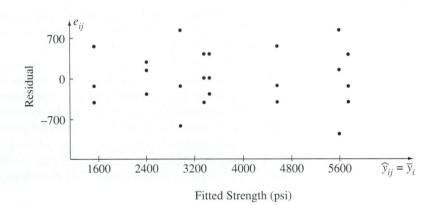

Fitted Strength (psi)

FIGURE 7-6

Normal Plot of the
Compressive Strength Residuals

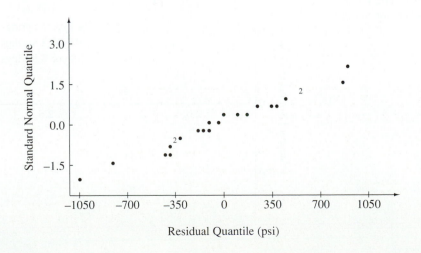

Residual Quantile (psi)

Figure 7-5 gives no indication of any kind of strong dependence of σ on μ (which would violate the one-way model's "constant variance" restriction). And the plot in Figure 7-6 is reasonably linear, thus providing no obvious difficulty with the one-way model's assumption of Gaussian distributions. In all, it seems from examination of both the raw data and the residuals that analysis of the data in Table 7-1 on the basis of model (7-1) is perfectly sensible.

EXAMPLE 7-2
(continued)

The spring testing data of Hunwardsen, Springer, and Wattonville can also be examined with the potential use of the one-way model (7-1) in mind. Figures 7-2 and 7-3 indicate that fairly comparable variabilities are associated with experimental spring constants for the $r = 3$ different spring types, although the single very large value (for spring type 1) causes some doubt both in terms of this judgment and also (by virtue of its position on its boxplot as an outlying value) regarding a "normal distribution" description of type 1 experimental constants. Summary statistics for these samples are given in Table 7-5.

TABLE 7-5

Summary Statistics for the Empirical Spring Constants

i Spring Type	n_i	$\bar{y}_i$	s_i
1	7	2.030	.134
2	6	2.750	.074
3	6	2.035	.064

It is interesting to observe that without the single extreme value of 2.30, the first sample standard deviation would be .068, remarkably in line with those of the second and third samples. But even the ratio of largest to smallest sample variance of $(.134/.064)^2 = 4.38$ is not an absolutely compelling reason to abandon a standard one-way model description of the spring constant scenario. A look at the F tables with $\nu_1 = 6$ and $\nu_2 = 5$ shows that 4.38 is between the $F_{6,5}$ distribution .9 and .95 quantiles. So even if there were only two rather than three samples involved, a variance ratio of 4.38 would yield a p-value between .1 and .2 for (two-sided) testing of equality of variances. This author would personally investigate further, making and examining some more plots, before letting the single type 1 empirical spring constant of 2.30 cause him to abandon the use of the highly tractable model (7-1) in analysis of these data.

FIGURE 7-7

Normal Plots of Empirical Spring Constants for Springs of Three Types

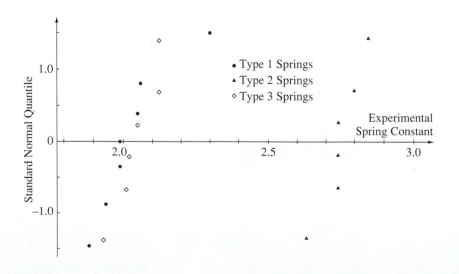

• Type 1 Springs
▲ Type 2 Springs
◇ Type 3 Springs

Sample sizes $n_1 = 7$ and $n_2 = n_3 = 6$ are large enough that it makes sense to look at sample-by-sample normal plots of the spring constant data. Such plots, drawn on the same set of axes, are shown in Figure 7-7.

Further, use of the fitted values ($\bar{y}_i$) listed in Table 7-5 with the original data given in Table 7-2 produces 19 residuals, as partially illustrated in Table 7-6. Then Figures 7-8 and 7-9, respectively, show a plot of residuals versus fitted responses and a normal plot of all 19 residuals.

TABLE 7-6

Example Computations of
Residuals for the Spring
Constant Study

i Spring Type	j Observation Number	y_{ij} Spring Constant	$\hat{y}_{ij} = \bar{y}_i$ Sample Mean	e_{ij} Residual
1	1	1.99	2.030	−.040
⋮	⋮	⋮	⋮	⋮
1	7	2.30	2.030	.270
2	1	2.85	2.750	.100
⋮	⋮	⋮	⋮	⋮
2	6	2.80	2.750	.050
3	1	2.10	2.035	.065
⋮	⋮	⋮	⋮	⋮
3	6	2.05	2.035	.015

FIGURE 7-8

Plot of Residuals versus Fitted
Responses for the Empirical
Spring Constants

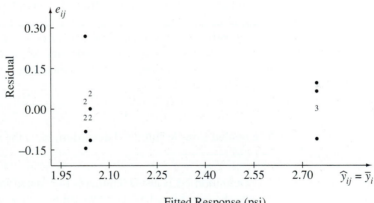

Fitted Response (psi)

FIGURE 7-9

Normal Plot of the Spring
Constant Residuals

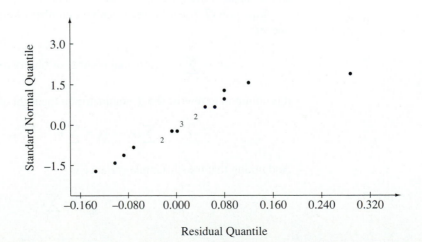

Residual Quantile

But Figures 7-8 and 7-9 again draw attention to the largest type 1 empirical spring constant. Compared to the other measured values, 2.30 is simply too large (and thus produces a residual that is too large compared to all the rest) to permit serious use of model (7-1) with the spring constant data. Barring the possibility that checking of original data sheets would show the 2.30 value to be an arithmetic blunder or gross error of measurement (which could be corrected or legitimately force elimination of the 2.30 value from consideration), it appears that the use of model (7-1) in the making of formal inferences about the $r = 3$ spring types could produce inferences with true (and unknown) properties quite different from their nominal properties.

One might, of course, limit attention to spring types 2 and 3, as there appears to be nothing in the second or third samples to render the "equal variances, normal distributions" model untenable for the two corresponding spring types. But the pattern of variation for springs of type 1 appears to be detectably different from that for springs of types 2 and 3, and the standard one-way model is not appropriate when all three types are considered.

A Pooled Estimate of Variance for Multisample Studies

The "equal variances, normal distributions" model (7-1) has as a fundamental parameter σ, the standard deviation associated with responses from any of conditions $1, 2, 3, \ldots, r$. In a manner similar to what was done in the $r = 2$ situation of Section 6-3, it is typical in multisample studies (provided, of course, that the kind of checks implemented in the previous examples indicate that it makes sense) to pool the r sample variances to arrive at a single estimate of σ derived from all r samples.

DEFINITION 7-1

If r numerical samples of respective sizes $n_1, n_2, \ldots, n_r$ produce sample variances $s_1^2, s_2^2, \ldots, s_r^2$, the **pooled sample variance**, s_P^2, is the weighted average of the sample variances, where the weights are the sample sizes minus 1. That is,

$$s_P^2 = \frac{(n_1 - 1)s_1^2 + (n_2 - 1)s_2^2 + \cdots + (n_r - 1)s_r^2}{(n_1 - 1) + (n_2 - 1) + \cdots + (n_r - 1)} \tag{7-7}$$

The **pooled sample standard deviation**, s_P, is the square root of s_P^2.

Definition 7-1 is just Definition 6-14 restated for the case of more than two samples. As was the case for s_P based on two samples, s_P is guaranteed to lie between the largest and smallest of the s_i and is a particularly convenient form of compromise value.

Equation (7-7) can be rewritten in a number of equivalent forms. For one thing, if one lets

$$n = \sum_{i=1}^{r} n_i = \text{the total number of observations in all } r \text{ samples}$$

it is common to rewrite the denominator on the right of equation (7-7) as

$$\sum_{i=1}^{r}(n_i - 1) = \sum_{i=1}^{r} n_i - \sum_{i=1}^{r} 1 = n - r$$

And noting that the ith sample variance is

$$s_i^2 = \frac{1}{n_i - 1} \sum_{j=1}^{n_i} (y_{ij} - \bar{y}_i)^2$$

the numerator on the right of equation (7-7) is

$$\sum_{i=1}^{r}(n_i - 1)\left(\frac{1}{(n_i - 1)}\sum_{j=1}^{n_i}(y_{ij} - \bar{y}_i)^2\right) = \sum_{i=1}^{r}\sum_{j=1}^{n_i}(y_{ij} - \bar{y}_i)^2 \tag{7-8}$$

$$= \sum_{i=1}^{r}\sum_{j=1}^{n_i}e_{ij}^2 \tag{7-9}$$

It is common to see s_P^2 defined in terms of the right-hand side of equation (7-8) or (7-9) divided by $n - r$. This text has chosen to introduce s_P^2 in terms of equation (7-7), however, because form (7-7) makes it easiest to see the motivation behind the quantity.

EXAMPLE 7-1
(continued)

For the compressive strength data of Armstrong, Babb, and Campen, each of $n_1, n_2, \ldots, n_8$ are 3, and s_1 through s_8 are given in Table 7-3. So using equation (7-7), one obtains

$$\begin{aligned} s_P^2 &= \frac{(3 - 1)(965.6)^2 + (3 - 1)(432.3)^2 + \cdots + (3 - 1)(302.5)^2}{(3 - 1) + (3 - 1) + \cdots + (3 - 1)} \\ &= \frac{2[(965.6)^2 + (432.3)^2 + \cdots + (302.5)^2]}{16} \\ &= \frac{2{,}705{,}705}{8} \\ &= 338{,}213(\text{psi})^2 \end{aligned}$$

and thus

$$s_P = \sqrt{338{,}213} = 581.6 \text{ psi}$$

The meaning of this value is that one estimates that if a large number of specimens of any one of formulas 1 through 8 were tested, a standard deviation of compressive strengths on the order of 582 psi would be obtained.

---

s_P is an estimate of the intrinsic or baseline variation present in a response variable, calculated supposing that the baseline variation is constant across the conditions under which the samples were collected. When that supposition is reasonable, the pooling idea allows one to combine a number of individually unreliable small-sample estimates into a single, relatively more reliable combined estimate. It is a fundamental measure that figures prominently in a variety of useful methods of formal inference.

On occasion, it is helpful to have not only a single number as a data-based best guess at σ^2, but a confidence interval as well. Happily enough, the one-way model assumptions lead to such an interval based on s_P^2. That is, it turns out that under model restrictions (7-1), a probabilist can show that the variable

$$\frac{(n - r)s_P^2}{\sigma^2}$$

has a χ_{n-r}^2 distribution. Thus, in a manner exactly parallel to the derivation in Section 6-4, a two-sided confidence interval for σ^2 has endpoints

$$\frac{(n - r)s_P^2}{U} \quad \text{and} \quad \frac{(n - r)s_P^2}{L} \tag{7-10}$$

where L and U are such that the χ^2_{n-r} probability assigned to the interval (L, U) corresponds to the desired confidence level. And, of course, a one-sided interval is available by using only one of the endpoints (7-10) and choosing U or L such that the χ^2_{n-r} probability assigned to the interval $(0, U)$ or (L, ∞) corresponds to the desired confidence.

EXAMPLE 7-1
(continued)

In the scenario of concrete compressive strength, consider the use of display (7-10) in making a two-sided 90% confidence interval for σ. Since $n - r = 16$ degrees of freedom are associated with s_p^2, one consults Table D-5 for the .05 and .95 quantiles of the χ^2_{16} distribution. These are 7.962 and 26.296, respectively. Thus, from display (7-10), a confidence interval for σ^2 has endpoints

$$\frac{16(581.6)^2}{26.296} \quad \text{and} \quad \frac{16(581.6)^2}{7.962}$$

So a two-sided 90% confidence interval for σ has endpoints

$$\sqrt{\frac{16(581.6)^2}{26.296}} \quad \text{and} \quad \sqrt{\frac{16(581.6)^2}{7.962}}$$

that is,

$$453.7 \text{ psi} \quad \text{and} \quad 824.5 \text{ psi}$$

Standardized Residuals

In discussing the use of residuals in checking of the appropriateness of the one-way model (7-1), the reasoning has been that the residuals e_{ij} are meant to approximate the corresponding random errors ϵ_{ij}. Since the model assumptions are that the ϵ_{ij} are iid normal variables, the e_{ij} ought to look approximately like iid normal variables. This is sensible rough-and-ready reasoning, adequate for many circumstances. But strictly speaking, the e_{ij} are neither independent nor identically distributed, and it can be important to recognize this.

As an extreme example of the lack of independence of the residuals for a given sample i, consider a case where $n_i = 2$. Since

$$e_{ij} = y_{ij} - \bar{y}_i$$

one immediately knows that $e_{i1} = -e_{i2}$. And one can further apply Proposition 5-1 to show that if the sample sizes n_i are varied, the residuals don't have the same variance (and therefore can't be identically distributed). That is, since

$$e_{ij} = y_{ij} - \bar{y}_i = \left(\frac{n_i - 1}{n_i}\right) y_{ij} - \frac{1}{n_i} \sum_{j' \neq j} y_{ij'}$$

one has

$$Var\, e_{ij} = \left(\frac{n_i - 1}{n_i}\right)^2 \sigma^2 + \left(-\frac{1}{n_i}\right)^2 (n_i - 1)\sigma^2 = \frac{n_i - 1}{n_i}\sigma^2 \tag{7-11}$$

So, for example, residuals from a sample of size $n_i = 2$ have variance $\sigma^2/2$, while those from a sample of size $n_i = 100$ have variance $99\sigma^2/100$, and one ought to expect residuals from larger samples to be somewhat bigger than those from small samples.

A way of addressing at least the issue that residuals need not have a common variance is through the use of standardized residuals.

DEFINITION 7-2

If a residual e has variance $a \cdot \sigma^2$ for some positive constant a, and s is some estimate of σ, the **standardized residual** corresponding to e is defined by

$$e^* = \frac{e}{s\sqrt{a}} \tag{7-12}$$

The division by $s\sqrt{a}$ in equation (7-12) is a division by an estimated standard deviation of e and serves to, so to speak, put all of the residuals on the same scale, giving them a common standard deviation (of 1.0).

Plotting with standardized residuals

$$e_{ij}^* = \frac{e_{ij}}{s_P\sqrt{\dfrac{n_i - 1}{n_i}}} \tag{7-13}$$

is a somewhat more refined way of judging the adequacy of the one-way model than the plotting of raw residuals e_{ij} illustrated in Examples 7-1 and 7-2. Of course, when all n_i are the same, as in Example 7-1, the plotting of the standardized residuals in equation (7-13) is completely equivalent to plotting with the raw residuals. And as a practical matter, unless some n_i are very small and others are very large, the standardization used in equation (7-13) typically doesn't have much effect on the appearance of residual plots.

EXAMPLE 7-2
(continued)

In the context of the spring constant study, allowing for the fact that sample 1 is larger than the other two (and thus according to the model (7-1) should produce larger residuals) doesn't materially change the outcome of the residual analysis. To see this, note that for the data of Table 7-2, using the summary statistics in Table 7-5,

$$s_P^2 = \frac{(7 - 1)(.134)^2 + (6 - 1)(.074)^2 + (6 - 1)(.064)^2}{(7 - 1) + (6 - 1) + (6 - 1)} = .0097$$

so that

$$s_P = \sqrt{.0097} = .099$$

Then using equation (7-13), each residual from sample 1 should be divided by

$$.099\sqrt{\frac{7 - 1}{7}} = .0913$$

in order to get standardized residuals, while each residual from the second and third samples should be divided by

$$.099\sqrt{\frac{6 - 1}{6}} = .0900$$

in order to get standardized residuals. Clearly, .0913 and .0900 are not much different, and dividing the sample 1 residuals by .0913 and the sample 2 and 3 residuals by .0900 before plotting has little effect on the appearance of residual plots. By way of example, a normal plot of all 19 standardized residuals is given in Figure 7-10. Verify its similarity to the normal plot of all 19 raw residuals given in Figure 7-9.

FIGURE 7-10

Normal Plot of the
Spring Constant
Standardized Residuals

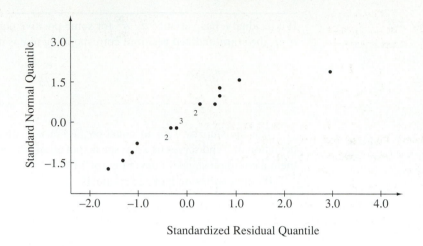

The notion of standardized residuals is often introduced only in the context of curve- and surface-fitting analyses, where the variances of residuals $e = (y - \hat{y})$ depend not only on the sizes of the samples involved but also on the associated values of the independent or predictor variables $(x_1, x_2, \ldots, \text{etc.})$. The notion has been introduced here, not only because it can be of importance in the present situation if the sample sizes vary widely, but also because it is particularly easy to motivate the idea in the unstructured multisample context. This is due, of course, to the particular form taken by the residuals and the resulting simple form of equation (7-11) and hence of equation (7-13).

7-2 *Simple Statistical Intervals in Multisample Studies*

Sections 6-3 and 6-6 illustrate how useful confidence intervals for means and differences in means and prediction and tolerance intervals can be in one- and two-sample statistical engineering studies. The enterprises of estimating an individual mean, comparing a pair of means, predicting an additional observation and locating the bulk of an underlying distribution are every bit as important when there are r samples involved as they are in one- and two-sample studies. Of course, the methods of Chapter 6 can be applied in r-sample studies by simply limiting attention to one or two of the samples at a time. But since individual sample sizes in multisample studies are often small, such a strategy of inference often turns out to be relatively uninformative. Happily enough, under the one-way model assumptions discussed in the previous section, it is possible to base inference methods parallel to those in Sections 6-3 and 6-6 on the pooled standard deviation, s_P. In the process, one produces statistical intervals that are relatively more informative than those that result from the direct application of the formulas from Sections 6-3 and 6-6 in the present context.

This section first considers the confidence interval estimation of a single mean and of the difference between two means under the "equal variances, normal distributions" model. Next is discussed an extension of some of the prediction and tolerance interval ideas of Section 6-6 to the multisample context. There follows an investigation of the possibility of developing a confidence interval formula for any linear combination of underlying means. Finally, the section closes with some comments concerning the notions of individual and simultaneous confidence levels, in situations where several confidence intervals are to be made.

Intervals for Means and for Comparing Means

The primary drawback to applying the small-sample confidence interval formulas from Section 6-3 in a multisample context is that typical small sample sizes lead to small degrees of freedom, large t multipliers in the plus-or-minus parts of the interval formulas, and thus intervals that are on average relatively long. One of the reasons that the one-way model assumptions are so useful is that based on them, one can develop formulas essentially identical to those in Section 6-3, except that s_P replaces estimates of σ based on only one or two samples. The associated degrees of freedom for the t multipliers are accordingly adjusted (upward) to $n - r$ in light of the replacement.

That is, in a development parallel to that in Section 6-3, a mathematician can show that under the "equal variances, normal distributions" model, the variable

$$\frac{\bar{y}_i - \mu_i}{\dfrac{s_P}{\sqrt{n_i}}}$$

has a t_{n-r} distribution. Hence, a two-sided confidence interval for the ith mean, μ_i, has endpoints

$$\bar{y}_i \pm t\frac{s_P}{\sqrt{n_i}} \tag{7-14}$$

where the associated confidence is the probability assigned to the interval from $-t$ to t by the t_{n-r} distribution. The reader should verify that this is exactly formula (6-20) from Section 6-3, except that s_P has replaced s_i and the degrees of freedom have been adjusted from $n_i - 1$ to $n - r$.

In the same way, it is possible for a mathematician to establish that under the one-way normal model, for conditions i and i', the variable

$$\frac{\bar{y}_i - \bar{y}_{i'} - (\mu_i - \mu_{i'})}{s_P\sqrt{\dfrac{1}{n_i} + \dfrac{1}{n_{i'}}}}$$

has a t_{n-r} distribution. Hence, a two-sided confidence interval for $\mu_i - \mu_{i'}$ has endpoints

$$\bar{y}_i - \bar{y}_{i'} \pm ts_P\sqrt{\frac{1}{n_i} + \frac{1}{n_{i'}}} \tag{7-15}$$

where the associated confidence is the probability assigned to the interval from $-t$ to t by the t_{n-r} distribution. Display (7-15) is essentially formula (6-35) of Section 6-3, except that s_P is calculated based on r samples instead of two and the degrees of freedom are $n - r$ instead of $n_i + n_{i'} - 2$.

Of course, use of only one endpoint from formula (7-14) or (7-15) produces a one-sided confidence interval with associated confidence corresponding to the t_{n-r} probability assigned to the interval $(-\infty, t)$ (for $t > 0$).

The virtues of formulas (7-14) and (7-15) (in comparison to the corresponding formulas from Section 6-3) are that (when they are appropriate) for a given confidence, they will tend to produce shorter (and thus more informative) intervals than their Chapter 6 counterparts.

EXAMPLE 7-3

(Example 7-1 revisited) **Confidence Intervals for Individual, and Differences of, Mean Concrete Compressive Strengths.** In the situation of the concrete strength study of Armstrong, Babb, and Campen, consider making a 90% two-sided confidence interval for the mean compressive strength of an individual concrete formula. Then consider making a 90% two-sided confidence interval for the difference in mean compressive strengths for two different formulas. Since $n = 24$ and $r = 8$, there are $n - r = 16$ degrees of freedom associated with $s_P = 581.6$. Then the .95 quantile of the t_{16} distribution is 1.746, which is appropriate for use in both formulas (7-14) and (7-15).

Turning first to the estimation of a single mean compressive strength, since each n_i is 3, the plus-or-minus part of formula (7-14) gives

$$t\frac{s_P}{\sqrt{n_i}} = 1.746\frac{581.6}{\sqrt{3}} = 586.3 \text{ psi}$$

Thus a ± 586.3 psi precision could be attached to any one of the sample means in Table 7-7 as an estimate of the corresponding formula's mean strength.

TABLE 7-7

Concrete Formula Sample
Mean Strengths

Concrete Formula	Sample Mean Strength (psi)
1	5,635.3
2	5,753.3
3	4,527.3
4	3,442.3
5	2,923.7
6	3,324.7
7	1,551.3
8	2,390.7

For example, since $\bar{y}_3 = 4{,}527.3$ psi, a 90% two-sided confidence interval for μ_3 has endpoints

$$4{,}527.3 \pm 586.3$$

that is,

$$3{,}941.0 \text{ psi} \quad \text{and} \quad 5{,}113.6 \text{ psi}$$

In parallel fashion, consider the estimation of the difference in two mean compressive strengths. Again, since each n_i is 3, the plus-or-minus part of formula (7-15) gives

$$ts_P\sqrt{\frac{1}{n_i} + \frac{1}{n_{i'}}} = 1.746(581.6)\sqrt{\frac{1}{3} + \frac{1}{3}} = 829.1 \text{ psi}$$

Thus, a ± 829.1 psi precision could be attached to any difference between sample means in Table 7-7 as an estimate of the corresponding difference in formula mean strengths. For instance, since $\bar{y}_3 = 4{,}527.3$ psi and $\bar{y}_7 = 1{,}551.3$ psi, a 90% two-sided confidence interval for $\mu_3 - \mu_7$ has endpoints

$$(4{,}527.3 - 1{,}551.3) \pm 829.1$$

That is,

$$2{,}146.9 \text{ psi} \quad \text{and} \quad 3{,}805.1 \text{ psi}$$

The use of $n - r = 16$ degrees of freedom in Example 7-3 instead of $n_i - 1 = 2$ and $n_i + n_{i'} - 2 = 4$ reflects the reduction in uncertainty associated with s_P as an estimate of σ as compared to that of s_i and of s_P based on only two samples. That reduction is, of course, bought at the price of restriction to problems where the "equal variances" model is tenable.

Prediction and Tolerance Intervals

The same kind of substitution of s_P for s_i that produces formula (7-14) in parallel to formula (6-20) from Section 6-3 gives a prediction interval for a single additional observation from a particular one of the r underlying normal distributions in the one-way model. That is, if y_{in_i+1} is a single additional observation from distribution i, the standard model assumptions can be shown to imply that

$$\frac{\bar{y}_i - y_{in_i+1}}{s_P\sqrt{1 + \dfrac{1}{n_i}}}$$

has a t_{n-r} distribution. This fact in turn leads to the conclusion that a two-sided prediction interval for a single additional observation from the ith underlying distribution has endpoints

$$\bar{y}_i \pm t s_P\sqrt{1 + \frac{1}{n_i}} \tag{7-16}$$

where the associated prediction confidence is the probability assigned to the interval from $-t$ to t by the t_{n-r} distribution. The reader should check that formula (7-16) is essentially formula (6-76) of Section 6-6 , except that s_P has replaced s_i and the degrees of freedom have been adjusted from $n_i - 1$ to $n - r$. As was the case with confidence interval formulas (7-14) and (7-15), it is possible to use only one of the endpoints from display (7-16), reduce the "unconfidence" by half, and arrive at a one-sided prediction interval for a single additional observation from the ith underlying distribution.

EXAMPLE 7-3
(continued)

Consider the problem of producing a one-sided 90% prediction interval of the form $(\#, \infty)$ for, say, a single additional specimen strength for concrete formula number 3. Using $n - r = 16$ degrees of freedom, the .9 quantile of the t_{16} distribution for use in formula (7-16) is 1.337. Then, since $\bar{y}_3 = 4{,}527.3$ and $s_P = 581.6$, a 90% lower prediction bound for an additional formula 3 compressive strength is

$$4{,}527.3 - 1.337(581.6)\sqrt{1 + \frac{1}{3}} = 3{,}629.4 \text{ psi}$$

In a manner reminiscent of the discussion in Section 6-6, the prediction interval formula (7-16) is meant not to bracket a fixed percentage of an underlying distribution, but instead to locate a single additional observation. The job of locating most of a large number of additional compressive strengths for concrete formula 3, for example, is one for a tolerance interval. In theory, it is no harder for a probabilist to create one- and two-sided tolerance interval formulas parallel to (6-81), (6-82), and (6-83) of Section 6-6 for the present multisample situation than it was to create the one-sample formulas. But it turns out that the tabling of appropriate constants τ gets out of hand, at least for a book like this one. (The appropriate constants depend not only on p, the desired confidence level, and n_i, but also on $n - r$.)

What can be done in a way that is consistent with the level and goals of this text is to provide *one-sided* tolerance intervals for a single underlying distribution, giving both a fairly exact method of determining appropriate constants for 95% confidence and a simple but somewhat crude method of approximating the constants in general. That is, intervals

$$(\bar{y}_i - \tau s_P, \infty) \tag{7-17}$$

and

$$(-\infty, \bar{y}_i + \tau s_P) \tag{7-18}$$

can be used as one-sided tolerance intervals for a fraction p of the ith underlying distribution, provided τ is appropriately chosen (depending upon p, n_i, $n - r$, and the desired confidence level).

As it turns out, τ in formulas (7-17) and (7-18) must be chosen based on a particular quantile of a particular *noncentral t* distribution. The exact form of the noncentral t distributions is not particularly illuminating. Rather than discussing them, what will be done here is to present both a method of using the small table appearing in Appendix D-8 to find τ values for 95% confidence and also a general approximation for τ values.

The method by which Table D-8 can be used to produce values of τ for 95% confidence is as follows. One first computes a value of a parameter ξ according to

$$\xi = \frac{\sqrt{n_i} \, Q_z(p)}{\sqrt{2(n - r) + n_i Q_z^2(p)}} \tag{7-19}$$

($Q_z(p)$ is the standard normal p quantile.) Next, Table D-8 is entered using this value and the degrees of freedom parameter $\nu = n - r$ to produce a value λ in the body of the table (via interpolation as needed). Finally, using λ taken from the table, τ appropriate for use in formulas (7-17) and (7-18) can be found using the relationship

$$\tau = \frac{\sqrt{n_i} Q_z(p) + \lambda \sqrt{1 + \dfrac{n_i Q_z^2(p) - \lambda^2}{2(n - r)}}}{\sqrt{n_i} \left(1 - \dfrac{\lambda^2}{2(n - r)}\right)} \tag{7-20}$$

A common method of approximating τ for confidence levels other than 95% follows from approximating the distribution of the variables $\bar{y}_i \pm \tau s_P$ with certain normal distributions. For a one-sided tolerance interval with confidence level γ for a fraction p of the ith underlying distribution, one may use either formula (7-17) or (7-18), with τ determined as in equation (7-20), by employing the approximation

$$\lambda \approx Q_z(\gamma) \tag{7-21}$$

EXAMPLE 7-3
(continued)

Consider finding a one-sided 95% tolerance interval of the form $(\#, \infty)$ for 90% of many additional formula 3 concrete specimen compressive strengths. Since $n_3 = 3$, $n - r = 16$, and an interval for a fraction $p = .9$ of the compressive strengths is desired, formula (7-19) shows that one wants

$$\xi = \frac{\sqrt{3}(1.282)}{\sqrt{2(16) + 3(1.282)^2}} = .37$$

for use in Table D-8. Then, interpolation in the $\nu = 16$ column of the table gives

$$\lambda \approx 1.6812$$

Finally, formula applying (7-20), it is clear that

$$\tau = \frac{\sqrt{3}(1.282) + 1.6812\sqrt{1 + \dfrac{(3(1.282)^2 - (1.6812)^2)}{2(16)}}}{\sqrt{3}\left(\dfrac{1 - (1.6812)^2}{2(16)}\right)} = 2.5053$$

is appropriate for use in formula (7-17). That is, a 95% lower tolerance bound for a fraction $p = .9$ of many additional formula 3 concrete specimen compressive strengths is, from formula (7-17),

$$\bar{y}_3 - 2.5053 s_P = 4{,}527.3 - 2.5053(581.6) = 3{,}070.2 \text{ psi}$$

It is interesting to note that in this particular problem, since

$$Q_z(.95) = 1.645$$

the approximation of λ by $Q_z(\gamma)$ suggested in equation (7-21) is a fairly good one, and the approximate value of τ derived from use of equations (7-21) and (7-20), namely $\tau \approx 2.4734$, is close to the more exact value 2.5053.

As expected, the tolerance bound of 3,070.2 psi is looser than the corresponding prediction bound found earlier. Roughly speaking, the tolerance interval is meant to accomplish a more ambitious task than the prediction interval and must therefore be smaller.

The virtue of formulas (7-16), (7-17), and (7-18) in comparison to the corresponding formulas from Section 6-6 is that if the "equal variances" restriction is appropriate in a multisample problem, the present formulas lead to intervals that are on average "tighter" and therefore more informative than the use of the formulas from Section 6-6 one sample at a time. The proper interpretation of intervals constructed via formulas (7-16), (7-17), and (7-18) is, however, exactly that given in Definitions 6-18 and 6-20.

As indicated earlier, tabulation problems will prevent presentation of the extension of the one-sample two-sided tolerance interval methods of Section 6-6. However, it may be pointed out that one could (in a fairly crude manner) use both the lower tolerance bound indicated by formula (7-17) and the upper tolerance bound indicated by formula (7-18) and come up with a two-sided interval. As a tolerance interval for a fraction $1 - 2(1 - p) = 2p - 1$ of the ith underlying distribution, such an interval has associated confidence *at least* $1 - 2(1 - \gamma) = 2\gamma - 1$. (The associated actual confidence is not calculable by elementary methods.) For instance, if (as in Example 7-3)

$$(4{,}527.3 - 2.5053(581.6), \infty)$$

and

$$(-\infty, 4{,}527.3 + 2.5053(581.6))$$

are both 95% tolerance intervals for 90% of additional formula 3 specimen compressive strengths, then the interval

$$(4{,}527.3 - 2.5053(581.6), 4{,}527.3 + 2.5053(581.6))$$

can be used as a tolerance interval for 80% of additional formula 3 strengths with associated confidence of at least 90%.

Intervals for General Linear Combinations of Means

There is an important and simple generalization of the formulas (7-14) and (7-15) that is easy to state and motivate at this point. Its most common applications are in the context of factorial studies, but it is pedagogically most sound to introduce the method in the unstructured r-sample context, so that the logic behind it is clear and seen not to be limited to factorial analyses. The basic notion behind the generalization is that μ_i and $\mu_i - \mu_{i'}$ are particular linear combinations of the r means $\mu_1, \mu_2, \ldots, \mu_r$, and the same logic that produces confidence intervals for μ_i and $\mu_i - \mu_{i'}$ will produce a confidence interval for any linear combination of the r means.

That is, suppose that for constants $c_1, c_2, \ldots, c_r$, the quantity

$$L = c_1\mu_1 + c_2\mu_2 + \cdots + c_r\mu_r \tag{7-22}$$

is of engineering interest. (Note that, for example, if all c_i's except c_3 are 0 and $c_3 = 1$, $L = \mu_3$, the mean response from condition 3. Similarly, if $c_3 = 1$, $c_5 = -1$, and all other c_i's are 0, $L = \mu_3 - \mu_5$, the difference in mean responses from conditions 3 and 5.) A natural data-based way to approximate L is to replace the theoretical or underlying means, μ_i, with empirical or sample means, $\bar{y}_i$. That is, define an estimate of L (say, $\hat{L}$), by

$$\hat{L} = c_1\bar{y}_1 + c_2\bar{y}_2 + \cdots + c_r\bar{y}_r \tag{7-23}$$

(Clearly, if $L = \mu_3$, then $\hat{L} = \bar{y}_3$, while if $L = \mu_3 - \mu_5$, then $\hat{L} = \bar{y}_3 - \bar{y}_5$.)

The one-way model assumptions make it very easy to describe the distribution of $\hat{L}$ given in equation (7-23). Since $E\bar{y}_i = \mu_i$ and $Var\,\bar{y}_i = \sigma^2/n_i$, one can appeal again to Proposition 5-1 and conclude that

$$E\hat{L} = c_1 E\bar{y}_1 + c_2 E\bar{y}_2 + \cdots + c_r E\bar{y}_r$$
$$= c_1\mu_1 + c_2\mu_2 + \cdots + c_r\mu_r$$
$$= L$$

and

$$Var\,\hat{L} = c_1^2\,Var\,\bar{y}_1 + c_2^2\,Var\,\bar{y}_2 + \cdots + c_r^2\,Var\,\bar{y}_r$$
$$= c_1^2\frac{\sigma^2}{n_1} + c_2^2\frac{\sigma^2}{n_2} + \cdots + c_r^2\frac{\sigma^2}{n_r}$$
$$= \sigma^2\left(\frac{c_1^2}{n_1} + \frac{c_2^2}{n_2} + \cdots + \frac{c_r^2}{n_r}\right)$$

Further, the fact that the one-way model restrictions imply that the $\bar{y}_i$ are independent normal variables in turn implies that $\hat{L}$ is normally distributed. So the standardized version of $\hat{L}$,

$$\frac{\hat{L} - E\hat{L}}{\sqrt{Var\,\hat{L}}} = \frac{\hat{L} - L}{\sigma\sqrt{\dfrac{c_1^2}{n_1} + \dfrac{c_2^2}{n_2} + \cdots + \dfrac{c_r^2}{n_r}}} \tag{7-24}$$

is standard normal, and the usual manipulations beginning with this fact would produce an unusable confidence interval for L involving the unknown parameter σ. The obvious way to try to make this type of reasoning lead to a result of practical importance is to begin not with the variable (7-24), but with

$$\frac{\hat{L} - L}{s_P\sqrt{\dfrac{c_1^2}{n_1} + \dfrac{c_2^2}{n_2} + \cdots + \dfrac{c_r^2}{n_r}}} \tag{7-25}$$

instead. The fact is that the "equal variances, normal distributions" assumptions imply that the variable (7-25) has a t_{n-r} distribution. And this leads in the standard way to the fact that the interval with endpoints

$$\hat{L} \pm t s_{\text{P}} \sqrt{\frac{c_1^2}{n_1} + \frac{c_2^2}{n_2} + \cdots + \frac{c_r^2}{n_r}} \tag{7-26}$$

can be used as a two-sided confidence interval for L with associated confidence corresponding to the t_{n-r} probability assigned to the interval between $-t$ and t. Further, a one-sided confidence interval for L can be obtained by using only one of the endpoints in display (7-26) and appropriately adjusting the confidence level upward by reducing the unconfidence in half.

It is worthwhile to verify that the general formula (7-26) reduces to the formula (7-14) if a single c_i is 1 and all others are 0. And if one c_i is 1, one other is -1, and all others are 0, the general formula (7-26) reduces to the formula (7-15).

EXAMPLE 7-4

Comparing Absorbency Properties for Three Brands of Paper Towels. D. Speltz did some absorbency testing for several brands of paper towels. His study included (among others) a generic brand and two national brands from the middle price range. $n_1 = n_2 = n_3 = 5$ tests were made on towels of each of these $r = 3$ brands, and the numbers of milliliters of water (out of a possible 100) not absorbed out of a graduated cylinder were recorded. Some summary statistics for the tests on these brands are given in Table 7-8.

TABLE 7-8

Summary Statistics for Absorbencies of Three Brands of Paper Towels

Brand	i	n_i	$\bar{y}_i$	s_i
Generic	1	5	93.2 ml	.8 ml
National B	2	5	81.0 ml	.7 ml
National V	3	5	83.8 ml	.8 ml

Plots (not shown here) of the raw absorbency values and residuals, after the fashion of the previous section, indicate no problems with the use of the one-way model in the analysis of the absorbency data for these three towel brands.

One question of practical interest that can be phrased in a way that makes the formula (7-26) useful is "Do the national brands absorb more than the generic brand?" A way of quantifying this question is to ask for a two-sided 95% confidence interval estimate for

$$L = \mu_1 - \frac{1}{2}(\mu_2 + \mu_3) \tag{7-27}$$

the difference between the average liquid left by the generic brand and the arithmetic mean of the national brand averages.

With L as in equation (7-27), formula (7-23) shows that

$$\hat{L} = 93.2 - \frac{1}{2}(81.0) - \frac{1}{2}(83.8) = 10.8 \text{ ml}$$

is an estimate of the increased absorbency offered by the national brands. Using the standard deviations given in Table 7-8 gives

$$s_{\text{P}}^2 = \frac{(5-1)(.8)^2 + (5-1)(.7)^2 + (5-1)(.8)^2}{(5-1) + (5-1) + (5-1)} = .59$$

and thus

$$s_P = \sqrt{.59} = .77 \text{ mL}$$

Now $n - r = 15 - 3 = 12$ degrees of freedom are associated with s_P, and the .975 quantile of the t_{12} distribution for use in (7-26) is 2.179. In addition, since $c_1 = 1$, $c_2 = -\frac{1}{2}$, and $c_3 = -\frac{1}{2}$ and all three sample sizes are 3,

$$\sqrt{\frac{c_1^2}{n_1} + \frac{c_2^2}{n_2} + \frac{c_3^2}{n_3}} = \sqrt{\frac{1}{3} + \frac{\left(-\frac{1}{2}\right)^2}{3} + \frac{\left(-\frac{1}{2}\right)^2}{3}} = .71$$

So finally, endpoints for a two-sided 95% confidence interval for L given in equation (7-27) are

$$10.8 \pm 2.179(.77)(.71)$$

that is,

$$10.8 \pm 1.2$$

i.e.,

$$9.6 \text{ ml} \quad \text{and} \quad 12.0 \text{ ml} \tag{7-28}$$

The interval indicated in display (7-28) shows definitively the substantial advantage in absorbency held by the national brands over the generic, particularly in view of the fact that the amount actually absorbed by the generic brand appears to average only about 6.8 ml (= 100 ml − 93.2 ml).

EXAMPLE 7-5

A Confidence Interval for a Main Effect in a 2^2 Factorial Brick Fracture Strength Study. Graves, Lundeen, and Micheli studied the fracture strength properties of brick bars. They included several experimental variables in their study, including both bar composition and heat treating regimen. Part of their data are given in Table 7-9. Modulus of rupture values under a bending load are given in psi for $n_1 = n_2 = n_3 = n_4 = 3$ bars of $r = 4$ types.

TABLE 7-9

Modulus of Rupture Measurements for Brick Bars in a 2^2 Factorial Study

i Bar Type	% Water in Mix	Heat Treating Regimen	MOR (psi)
1	17	slow cool	4911, 5998, 5676
2	19	slow cool	4387, 5388, 5007
3	17	fast cool	3824, 3140, 3502
4	19	fast cool	4768, 3672, 3242

Notice that the data represented in Table 7-9 have a 2×2 complete factorial structure. Indeed, if one returns to Section 4-3 (in particular, to Definition 4-5), it becomes clear that in the present notation, the fitted main effect of the factor Heat Treating Regimen at its slow cool level is

$$\frac{1}{2}(\bar{y}_1 + \bar{y}_2) - \frac{1}{4}(\bar{y}_1 + \bar{y}_2 + \bar{y}_3 + \bar{y}_4) \tag{7-29}$$

But the variable (7-29) is the $\hat{L}$ for the linear combination of mean strengths $\mu_1, \mu_2, \mu_3,$

and μ_4 given by

$$L = \frac{1}{4}\mu_1 + \frac{1}{4}\mu_2 - \frac{1}{4}\mu_3 - \frac{1}{4}\mu_4 \qquad (7\text{-}30)$$

So subject to the relevance of the "equal variances, normal distributions" description of modulus of rupture for fired brick clay bodies of the four types represented in Table 7-9 (which the reader is invited to investigate) one can apply formula (7-26) in order to develop a precision figure to attach to the fitted effect represented by the variable (7-29).

Table 7-10 gives summary statistics for the data of Table 7-9.

TABLE 7-10

Summary Statistics
for the Modulus of
Rupture Measurements

i Bar Type	$\bar{y}_i$	s_i
1	5,528.3	558.3
2	4,927.3	505.2
3	3,488.7	342.2
4	3,894.0	786.8

Using the values in Table 7-10, one is led to

$$\hat{L} = \frac{1}{2}(\bar{y}_1 + \bar{y}_2) - \frac{1}{4}(\bar{y}_1 + \bar{y}_2 + \bar{y}_3 + \bar{y}_4)$$

$$= \frac{1}{4}\bar{y}_1 + \frac{1}{4}\bar{y}_2 - \frac{1}{4}\bar{y}_3 - \frac{1}{4}\bar{y}_4$$

$$= \frac{1}{4}(5,528.3 + 4,927.3 - 3,488.7 - 3,894.0)$$

$$= 768.2 \text{ psi}$$

and

$$s_P = \sqrt{\frac{(3-1)(558.3)^2 + (3-1)(505.2)^2 + (3-1)(342.2)^2 + (3-1)(786.8)^2}{(3-1) + (3-1) + (3-1) + (3-1)}}$$

$$= 570.8 \text{ psi}$$

Further, $n - r = 12 - 4 = 8$ degrees of freedom are associated with s_P. Therefore, if one wanted (for example) a two-sided 98% confidence interval for L given in equation (7-30), the necessary .99 quantile of the t_8 distribution is 2.896. Then, since

$$\sqrt{\frac{\left(\frac{1}{4}\right)^2}{3} + \frac{\left(\frac{1}{4}\right)^2}{3} + \frac{\left(-\frac{1}{4}\right)^2}{3} + \frac{\left(-\frac{1}{4}\right)^2}{3}} = .2887$$

a two-sided 98% confidence interval for L has endpoints

$$768.2 \pm 2.896(570.8)(.2887)$$

that is,

$$291.1 \text{ psi} \qquad \text{and} \qquad 1,245.4 \text{ psi} \qquad (7\text{-}31)$$

Display (7-31) establishes convincingly the effectiveness of a slow cool regimen in increasing MOR. It says that the differences in sample mean MOR values for slow and fast cooled bricks are not simply reflecting background variation. In fact, multiplying the endpoints in display (7-31) each by 2 in order to get a confidence interval for

$$2L = \frac{1}{2}(\mu_1 + \mu_2) - \frac{1}{2}(\mu_3 + \mu_4)$$

one sees that (when averaged over 17% and 19% water mixtures) the slow cool regimen seems to offer an increase in MOR in the range from

$$582.2 \text{ psi} \quad \text{to} \quad 2,490.8 \text{ psi}$$

Individual and Simultaneous Confidence Levels

This section has introduced a variety of statistical intervals for use in multisample statistical engineering studies. In a particular application, one might wish to use several of these methods, perhaps several times each. For example, even in the relatively simple context of Example 7-4 (the paper towel absorbency study of D. Speltz), considering just confidence intervals involving mean absorbencies (i.e., leaving aside the use of prediction and/or tolerance intervals), it would be reasonable to desire confidence intervals for each of

$$\mu_1, \mu_2, \mu_3, \mu_1 - \mu_2, \mu_1 - \mu_3, \mu_2 - \mu_3, \mu_1 - \frac{1}{2}(\mu_2 + \mu_3)$$

Since it is a fact of life that one often makes many confidence statements in multisample studies, it is important to reflect again briefly on the meaning of a confidence level and remember that it is attached to one interval at a time. The point is that if, in a multisample study, one makes several 90% confidence intervals, the 90% figure applies to the intervals individually. One is "90% sure" of the first interval, separately "90% sure" of the second, separately "90% sure" of the third, and so on. If one were interested in announcing a reliability figure for the intervals jointly (i.e., an *a priori* probability that all the intervals are effective), it is not clear how to arrive at such a figure, but it is fairly obvious that it must be less than 90%. That is, the *simultaneous* confidence to be associated with a group of several confidence intervals is generally not easy to determine, but it is typically less (and sometimes much less) than the *individual* confidence level(s) associated with the intervals one at a time.

There are at least three different approaches to be taken once one recognizes that there is a difference between simultaneous and individual confidence and that the confidence levels attached to the intervals studied thus far are meant only as one-at-a-time guarantees of effectiveness. The most obvious option is to make individual confidence intervals and be careful to interpret them as such. One then must be careful to recognize that as the number of intervals one makes with a given nominal confidence increases, so does the likelihood that among them are one or more intervals that fail to cover the quantities they are meant to locate.

A second way of handling the issue of simultaneous versus individual confidence is to use very large individual confidence levels for the separate intervals and then use a somewhat crude inequality to find at least a minimum value for the simultaneous confidence associated with an entire group of intervals. That is, if k confidence intervals have associated confidences $\gamma_1, \gamma_2, \ldots, \gamma_k$, the *Bonferroni Inequality* says that the simultaneous or joint confidence that all k intervals are effective (say, γ) satisfies

$$\gamma \geq 1 - \big((1 - \gamma_1) + (1 - \gamma_2) + \cdots + (1 - \gamma_k)\big) \qquad (7\text{-}32)$$

(Basically, this statement says that the joint unconfidence associated with k intervals is no worse than the sum of the k individual unconfidences. For example, five intervals with individual 99% confidence levels have a joint or simultaneous confidence level of at least 95%.)

The third way of approaching the issue of simultaneous confidence is to develop and employ methods that for some specific, useful set of unknown quantities provide intervals with a known level of simultaneous confidence. There are whole books full of such simultaneous inference methods. In the next section, a few of the better known and simplest of these are discussed.

7-3 *Some Simultaneous Confidence Interval Methods*

Section 7-2 illustrated the fact that that there are several kinds of confidence intervals for means and linear combinations of means that one might wish to make in a multisample statistical engineering study. The issue of individual versus simultaneous confidence was also raised, but only the use of the Bonferroni Inequality was given as a means of controlling a simultaneous confidence level.

This section presents three methods for making a number of confidence intervals and in the process maintaining a desired simultaneous confidence. The first of these is due to Pillai and Ramachandran; it provides a guaranteed simultaneous confidence in the estimation of all r individual underlying means. The second is Tukey's method for the simultaneous confidence interval estimation of all differences in pairs of underlying means. Last is a method (due to Dunnett) for the simultaneous comparison of each of the last $(r - 1)$ underlying means to the first mean in an r-sample study. This method is particularly appropriate where a study involves one standard or control condition and $r - 1$ experimental or treatment conditions.

The Pillai-Ramachandran Method

One of the things that is typically of interest in an r-sample statistical engineering study is the estimation of all r individual mean responses $\mu_1, \mu_2, \ldots, \mu_r$. If one applies the individual confidence interval formula of Section 7-2,

$$\bar{y}_i \pm t \frac{s_P}{\sqrt{n_i}} \tag{7-33}$$

r times to estimate these means, the only handle one has on the corresponding simultaneous confidence is given by the Bonferroni Inequality, formula (7-32) of Section 7-2. This fairly crude tool says that if, for example, $r = 8$ and one wants 95% simultaneous confidence, individual "unconfidences" of $\frac{.05}{8} = .00625$ (i.e., individual confidences of 99.375%) for the eight different intervals will suffice to produce the desired simultaneous confidence.

Another approach to the setting of simultaneous confidence limits on all of $\mu_1, \mu_2, \ldots, \mu_r$ is to replace t in formula (7-33) with a multiplier derived specifically for the purpose of providing an exact, stated *simultaneous* confidence in the estimation of all the means. Such multipliers have been derived by Pillai and Ramachandran for the cases of simultaneous 95% confidence, where either all of the intervals for the r means are two-sided or all are one-sided. That is, Table D-9-A gives values of constants k_2^* such that the r two-sided intervals with respective endpoints

$$\bar{y}_i \pm k_2^* \frac{s_P}{\sqrt{n_i}} \tag{7-34}$$

have simultaneous 95% confidence as intervals for the means $\mu_1, \mu_2, \ldots, \mu_r$. (These values k_2^* are in fact .95 quantiles of the *Studentized maximum modulus distributions* and are taken from a paper by Pillai and Ramachandran. Tables for other quantiles of this family of distributions could be constructed using their methods if a simultaneous confidence level other than 95% were desired.)

Table D-9-B gives values of some other constants k_1^* such that if for each i from 1 to r, an interval of the form

$$\left(-\infty, \bar{y}_i + k_1^* \frac{s_P}{\sqrt{n_i}}\right)$$ (7-35)

or of the form

$$\left(\bar{y}_i - k_1^* \frac{s_P}{\sqrt{n_i}}, \infty\right)$$ (7-36)

is made as a confidence interval for μ_i, the simultaneous confidence associated with the collection of r one-sided intervals is 95%. (These k_1^* values are in fact .95 quantiles of the *Studentized extreme deviate distributions*. As in the two-sided case above, it would be possible to construct tables of other quantiles if a simultaneous confidence other than 95% were desired.)

In this book, the use of r intervals of the form (7-34) or of forms (7-35) or (7-36) will be called the P-R method of simultaneous confidence intervals. In order to enter tables D-9 and apply the P-R method, one must find (using interpolation as needed) the table entry in the column corresponding to the number of samples, r, and the row corresponding to the degrees of freedom associated with s_P, namely $\nu = n - r$.

EXAMPLE 7-6

(Example 7-3 revisited) **Simultaneous Confidence Intervals for Eight Mean Concrete Compressive Strengths.** Consider again the concrete strength study of Armstrong, Babb, and Campen. Recall that tests on $n_i = 3$ specimens of each of $r = 8$ different concrete formulas gave $s_P = 581.6$ psi. Using formula (7-33) and remembering that there are $n - r = 16$ degrees of freedom associated with s_P, one has that if two-sided intervals for the formula mean compressive strengths with individual 95% confidence are desired, endpoints of

$$\bar{y}_i \pm 2.120 \frac{581.6}{\sqrt{3}}$$

that is,

$$\bar{y}_i \pm 711.9 \text{ psi}$$ (7-37)

should be used.

In contrast to intervals (7-37), consider the use of (7-34) to produce $r = 8$ two-sided intervals for the formula mean strengths with simultaneous 95% confidence. Interpolation in Table D-9-A shows that $k_2^* \approx 3.10$ is appropriate in this application. So each concrete formula mean compressive strength, μ_i, should be estimated using

$$\bar{y}_i \pm 3.10 \frac{581.6}{\sqrt{3}}$$

that is,

$$\bar{y}_i \pm 1{,}040.9 \text{ psi}$$ (7-38)

Expressions (7-37) and (7-38) provide two-sided intervals for the eight mean compressive strengths. If one-sided intervals of the form (#, ∞) were desired instead, consultation of the t table for the .95 quantile of the t_{16} distribution and use of formula (7-33) shows that the values

$$\bar{y}_i - 1.746\frac{581.6}{\sqrt{3}}$$

that is,

$$\bar{y}_i - 586.3 \text{ psi} \qquad\qquad (7\text{-}39)$$

are individual 95% lower confidence bounds for the formula mean compressive strengths, μ_i. At the same time, consultation of Table D-9-B shows that for simultaneous 95% confidence, use of $k_1^* = 2.78$ in formula (7-36) is appropriate, and the values

$$\bar{y}_i - 2.78\frac{581.6}{\sqrt{3}}$$

that is,

$$\bar{y}_i - 933.5 \text{ psi} \qquad\qquad (7\text{-}40)$$

are simultaneous 95% lower confidence bounds for the formula mean compressive strengths, μ_i.

━━━━ ▬▬ ▬▬

Comparison of intervals (7-37) with intervals (7-38) and intervals (7-39) with intervals (7-40) shows clearly the impact of requiring simultaneous rather than individual confidence. For a given nominal confidence level, the simultaneous intervals must be bigger (more conservative) than the corresponding individual intervals.

It is fairly common practice to summarize the information about mean responses gained in a multisample study in terms of a plot of sample means versus sample numbers, enhanced with "error bars" around the sample means to indicate the uncertainty associated with locating the underlying means. Various conventions (including some very poor ones) are in use for the making of these bars. When looking at such a plot, one typically forms an overall visual impression. Therefore, it is this author's opinion that error bars derived from the simultaneous confidence limits of expression (7-34) are the most sensible representation of what is known about a group of r means. For example, Figure 7-11 is a graphical representation of the eight formula sample mean strengths given in Table 7-7 with $\pm 1{,}040.9$ psi error bars, as indicated by expression (7-38).

When looking at a display like Figure 7-11, it is important to remember that what is represented is the precision of one's knowledge about the *mean* strengths, rather than any kind of predictions for individual compressive strengths. In this regard, the similarity of the spread of the samples on the side-by-side dot diagram given as Figure 7-1 and the size of the error bars here should not be thought to hold in general. As sample sizes increase, spreads on displays of individual measurements like Figure 7-1 will tend to stabilize (representing the spreads of the underlying distributions), while the lengths of error bars associated with means will shrink to 0 as increased information gives sharper and sharper evidence about the underlying means.

In any case, Figure 7-11 shows clearly that the information resident in the data is quite adequate to establish the existence of differences in formula mean compressive strengths.

Tukey's Method

A second set of confidence intervals that is often of interest in an r-sample statistical engineering study consists of intervals for the differences in all $\frac{r(r-1)}{2}$ pairs of mean responses μ_i and $\mu_{i'}$. Section 7-2 argued that a single difference in mean responses,

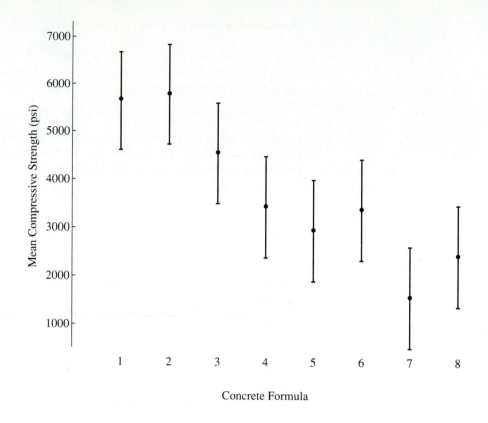

$\mu_i - \mu_{i'}$, can be estimated using an interval with endpoints

$$\bar{y}_i - \bar{y}_{i'} \pm t s_\mathrm{P} \sqrt{\frac{1}{n_i} + \frac{1}{n_{i'}}} \qquad \text{(7-41)}$$

where, of course, the associated confidence level is an individual one. But if, for example, $r = 8$, there are 28 different two-at-a-time comparisons of underlying means to be considered (μ_1 versus μ_2, μ_1 versus $\mu_3, \ldots, \mu_1$ versus μ_8, μ_2 versus $\mu_3, \ldots,$ and μ_7 versus μ_8). If one wanted to try to guarantee a reasonable simultaneous confidence level for all these comparisons via the crude Bonferroni idea, a huge individual confidence level would be required for the intervals (7-41). (For example, the Bonferroni Inequality would require 99.82% individual confidence for 28 intervals in order to guarantee simultaneous 95% confidence.)

A better approach to the setting of simultaneous confidence limits on all of the differences $\mu_i - \mu_{i'}$ is to replace t in formula (7-41) with a multiplier derived specifically for the purpose of providing exact, stated simultaneous confidence in the estimation of all such differences. J. Tukey first pointed out that it is possible to provide such multipliers using quantiles of the *Studentized range distributions*. Tables D-10-A and D-10-B give values of constants q^* such that the set of two-sided intervals with respective endpoints

$$\bar{y}_i - \bar{y}_{i'} \pm \frac{q^*}{\sqrt{2}} s_\mathrm{P} \sqrt{\frac{1}{n_i} + \frac{1}{n_{i'}}} \qquad \text{(7-42)}$$

has simultaneous confidence at least 95% or 99%, depending on whether $Q(.95)$ is read from Table D-10-A or $Q(.99)$ is read from Table D-10-B, in the estimation of all differences $\mu_i - \mu_{i'}$. (If all the sample sizes $n_1, n_2, \ldots, n_r$ are equal, the 95% or 99%

nominal simultaneous confidence figure is exact, while if the sample sizes are not all equal, the true value is at least as big as the nominal value.)

In order to enter Tables D-10 and apply Tukey's method, one must find (using interpolation as needed) the column corresponding to the number of samples/means to be compared and the row corresponding to the degrees of freedom associated with s_P, (namely, $\nu = n - r$).

EXAMPLE 7-6
(continued)

Consider the making of confidence intervals for differences in formula mean compressive strengths. If a 95% two-sided individual confidence interval is desired for a specific difference $\mu_i - \mu_{i'}$, formula (7-41) shows that appropriate endpoints are

$$\bar{y}_i - \bar{y}_{i'} \pm 2.120(581.6)\sqrt{\frac{1}{3} + \frac{1}{3}}$$

that is,

$$\bar{y}_i - \bar{y}_{i'} \pm 1{,}006.7 \text{ psi} \tag{7-43}$$

On the other hand, if one plans to estimate *all* differences in mean compressive strengths and desires *simultaneous* 95% confidence, by formula (7-42) two-sided intervals with endpoints

$$\bar{y}_i - \bar{y}_{i'} \pm \frac{4.90}{\sqrt{2}}(581.6)\sqrt{\frac{1}{3} + \frac{1}{3}}$$

that is,

$$\bar{y}_i - \bar{y}_{i'} \pm 1{,}645.4 \text{ psi} \tag{7-44}$$

are in order (4.90 is the value in the $r = 8$ column and $\nu = 16$ row of Table D-10-A.) In keeping with the fact that the confidence level associated with the intervals (7-44) is a simultaneous one, the Tukey intervals are wider than those indicated in formula (7-43).

It is worth noting that the plus-or-minus part of display (7-44) is not as big as twice the plus-or-minus part of expression (7-38). Thus, when looking at a plot like Figure 7-11, it is not necessary that the error bars around two means fail to overlap before it is safe to judge the corresponding underlying means to be detectably different. Rather, it is only necessary that the two sample means differ by the plus-or-minus part of formula (7-42) — 1,645.4 psi in the present situation.

Dunnett's Method

It is sometimes the case in an r-sample statistical engineering study that one of the r samples (say, sample 1) represents a standard or control set of conditions, while the other samples represent $r - 1$ possible alternatives to the standard or control. In such cases, rather than being interested in simultaneous intervals for all $\frac{r(r-1)}{2}$ differences $\mu_i - \mu_{i'}$, one might well be primarily interested in only simultaneous intervals for the $(r - 1)$ differences between the alternative means and the control mean, $\mu_i - \mu_1$. In such cases, use of Tukey's method amounts to overkill, producing excessively long intervals. Instead, what is needed is a method specifically crafted to produce a given simultaneous confidence in the estimation of only the $(r - 1)$ differences $\mu_i - \mu_1$.

C. Dunnett first developed a method for the simultaneous estimation of the $\mu_i - \mu_1$. His method, applicable to cases where $n_2 = n_3 = \cdots = n_r$, is to replace t in formula (7-41) with an appropriate multiplier t^* so that a given level of simultaneous confidence is obtained. Tables D-11 give values that can be used to produce constants t^* such that the set of $(r - 1)$ two-sided intervals with respective endpoints

$$\bar{y}_i - \bar{y}_1 \pm t^* s_{\mathrm{P}} \sqrt{\frac{1}{n_1} + \frac{1}{n_i}} \tag{7-45}$$

have simultaneous confidence 95% or 99% (depending upon whether Table D-11-A or D-11-B is used) in the estimation of the differences $\mu_2 - \mu_1, \mu_3 - \mu_1, \ldots, \mu_r - \mu_1$. When all samples sizes are equal, (i.e., $n_1 = n_2 = \cdots = n_r$), t^* is read directly from the body of the table by finding the large entry in the column corresponding to the number of alternatives to the standard or control (i.e., $r - 1$) and in the row corresponding to $\nu = n - r$ degrees of freedom. When n_1 is larger than $n_2 = n_3 = \cdots = n_r$, both the large entry, found as above and the small number appearing as its superscript come into play. Letting m stand for $n_2 = n_3 = \cdots = n_r$, l stand for the large entry, and s stand for the superscript, one employs

$$t^* = \left(1 + \frac{s}{100}\left(1 - \frac{m}{n_1}\right)\right)l \tag{7-46}$$

in formula (7-45).

EXAMPLE 7-7

(Example 7-4 revisited) **Simultaneous Confidence Intervals for Comparing Mean Absorbencies of Two National Brands of Paper Towels to That of a Generic Brand.** Consider again D. Speltz's paper towel absorbency testing, involving a generic brand and two national brands. Suppose that the performance of the generic brand is treated as a control or standard, and simultaneous 95% confidence intervals for the two differences in absorbency between the national brands and the generic are desired. That is, in this case where $r = 3$ and $n_1 = n_2 = n_3 = 5$, intervals for $\mu_2 - \mu_1$ and $\mu_3 - \mu_1$ possessing 95% simultaneous confidence are of interest.

Summary statistics from Speltz's study are given again in Table 7-11. They can be used once again to show that $s_{\mathrm{P}} = .77$ ml, with $\nu = n - r = 15 - 3 = 12$ associated degrees of freedom.

TABLE 7-11

Summary Statistics for Absorbencies of Three Brands of Paper Towels

Brand	i	$\bar{y}_i$	s_i
Generic	1	93.2 ml	.8 ml
National B	2	81.0 ml	.7 ml
National V	3	83.8 ml	.8 ml

Notice that using formula (7-41), two-sided individual 95% confidence limits for $\mu_2 - \mu_1$ or $\mu_3 - \mu_1$ are

$$(\bar{y}_i - 93.2) \pm 2.179(.77)\sqrt{\frac{1}{5} + \frac{1}{5}}$$

that is,

$$(\bar{y}_i - 93.2) \pm 1.06 \text{ ml} \tag{7-47}$$

Using formula (7-42) to produce two-sided simultaneous 95% confidence limits for all of $\mu_2 - \mu_1$, $\mu_3 - \mu_1$, and $\mu_3 - \mu_2$, limits of

$$\bar{y}_i - \bar{y}_{i'} \pm \frac{3.77}{\sqrt{2}}(.77)\sqrt{\frac{1}{5} + \frac{1}{5}}$$

that is,

$$(\bar{y}_i - \bar{y}_{i'}) \pm 1.30 \text{ ml} \qquad \text{(7-48)}$$

are appropriate. But using formula (7-45) to produce two-sided simultaneous 95% confidence limits for only $\mu_2 - \mu_1$ and $\mu_3 - \mu_1$, appropriate limits are

$$(\bar{y}_i - 93.2) \pm 2.50(.77)\sqrt{\frac{1}{5} + \frac{1}{5}}$$

that is,

$$(\bar{y}_i - 93.2) \pm 1.22 \text{ ml} \qquad \text{(7-49)}$$

Expressions (7-47), (7-48), and (7-49) show the effect (on required interval length) of holding a given confidence level simultaneously for an increasing number of quantities. The Dunnett intervals are intermediate in length between the individual and Tukey intervals. Since the Dunnett confidence guarantee covers only comparisons with the control, not comparisons among the other means, the Dunnett intervals can be less conservative than the more numerous Tukey intervals.

EXAMPLE 7-8

Use of Dunnett's Method Where the Control Sample Size Is Larger Than the Treatment Sample Size. As a synthetic example of the use of formula (7-46), suppose that a statistical engineering study is run to compare $r - 1 = 4$ alternative methods of operating a chemical process to the standard one presently in use. Suppose further that $n_1 = 6$ yield measurements are taken on production runs using the standard method, that $n_2 = n_3 = n_4 = n_5 = 3$ yield measurements are obtained for each of the alternative methods, and that simultaneous 95% intervals are desired for the four differences between alternative and standard mean yields.

In this situation, consultation of Table D-11-A with $r - 1 = 4$ and $\nu = 18 - 5 = 13$ produces a tabled value of 2.78, with a superscript of 3.4 . Formula (7-46) then gives

$$t^* = \left(1 + \frac{3.4}{100}\left(1 - \frac{3}{6}\right)\right)2.78 = 2.83$$

for use in expression (7-45), producing interval endpoints

$$\bar{y}_i - \bar{y}_1 \pm 2.83s_P\sqrt{\frac{1}{6} + \frac{1}{3}}$$

This section has mentioned only three of many existing methods of simultaneous confidence interval estimation for multisample studies. These three should serve to indicate the general character of such methods and illustrate the implications of a simultaneous (as opposed to individual) confidence requirement.

One final word of caution has to do with the theoretical justification of all of the methods found in this section. It is the "equal variances, normal distributions" model that supports these engineering tools. If one is to put any real faith in the nominal confidence levels attached to the P-R, Tukey, and Dunnett methods presented here, that faith should be based on evidence (typically gathered, at least to some extent, as illustrated in Section 7-1) that the standard one-way normal model is a sensible description of the physical situation involved.

7-4 *The Analysis of Means (ANOM) and Analysis of Variance (ANOVA) Methods*

This book's approach to the formal analysis of data from multisample statistical engineering studies has to this point been completely "interval-oriented." But there are also significance-testing methods that are appropriate to the multiple-sample context, and this section considers some of these and related issues raised by their introduction. It begins with some brief introductory comments regarding significance testing in the r-sample situation. Then the analysis of means (ANOM) test and the one-way analysis of variance (ANOVA) test for the equality of r means are discussed. The section next considers the notion of an ANOVA table and the organization and intuition that it provides. Finally, there is a brief look at the *one-way random effects model* and ANOVA-based estimates of its parameters.

Significance Testing and Multisample Studies

Just as there are many quantities for which one might want to produce statistical intervals based on data from a multisample study, there are potentially many issues of statistical significance to be judged. For instance, one might desire p-values for hypotheses like

$$H_0: \mu_3 = 7 \tag{7-50}$$

$$H_0: \mu_3 - \mu_7 = 0 \tag{7-51}$$

$$H_0: \mu_1 - \frac{1}{2}(\mu_2 + \mu_3) = 0 \tag{7-52}$$

Logically, the confidence interval methods discussed in detail in Section 7-2 must have their significance-testing analogs for treating hypotheses that, like all three of these, involve linear combinations of the means $\mu_1, \mu_2, \ldots, \mu_r$.

In general, (for the same probability reasons that gave confidence intervals for linear combinations of means), under the standard one-way model, if

$$L = c_1\mu_1 + c_2\mu_2 + \cdots + c_r\mu_r$$

the hypothesis

$$H_0: L = \# \tag{7-53}$$

can be tested using the test statistic

$$T = \frac{\hat{L} - \#}{s_P\sqrt{\dfrac{c_1^2}{n_1} + \dfrac{c_2^2}{n_2} + \cdots + \dfrac{c_r^2}{n_r}}} \tag{7-54}$$

and a t_{n-r} reference distribution. This fact specializes to cover hypotheses of types (7-50) to (7-52), by appropriate choice of the c_i and #.

Significance tests of the general type indicated by displays (7-53) and (7-54) are often useful in r-sample studies and are to some extent implicit in the individual confidence intervals that are made for linear combinations of means. (Those values not covered by a confidence interval for L have corresponding p-values smaller than 1 minus the interval's confidence level. So some idea as to the p-value associated with a particular number # can be inferred.) But the significance-testing methods most often associated with the one-way normal model are not for hypotheses of the type (7-53). Instead, the most common methods concern the hypothesis that all r underlying means have the same value. In symbols, this is

$$H_0: \mu_1 = \mu_2 = \cdots = \mu_r \tag{7-55}$$

Given that one is working under the assumptions of the "constant variance, normal distributions" model to begin with, hypothesis (7-55) amounts to a statement that all r underlying distributions are essentially the same — or in common parlance, "There are no differences between treatments."

Hypothesis (7-55) can be thought of in terms of the simultaneous equality of $\frac{r(r-1)}{2}$ pairs of means — that is, as equivalent to the statement that simultaneously

$$\mu_1 - \mu_2 = 0, \quad \mu_1 - \mu_3 = 0, \quad \dots, \quad \mu_1 - \mu_r = 0,$$
$$\mu_2 - \mu_3 = 0, \quad \dots, \quad \text{and} \quad \mu_{r-1} - \mu_r = 0$$

And this fact should remind the reader of the ideas about simultaneous confidence intervals from the previous section (specifically, Tukey's method). In fact, one way of judging the statistical significance of an r-sample data set in reference to hypothesis (7-55) is to apply Tukey's method of simultaneous interval estimation and note whether or not all the intervals for differences in means cover 0. If they all do, the associated p-value is larger than 1 minus the simultaneous confidence level. If not all of the intervals include 0, the associated p-value is smaller than 1 minus the simultaneous confidence level. (If simultaneous 95% intervals all include 0, no differences between means are definitively established, and the corresponding p-value exceeds .05.)

This author admits a bias toward estimation over testing per se. A logical consequence of this bias is a fondness for the notion of deriving a rough idea of a p-value for hypothesis (7-55) as a byproduct of the use of Tukey's method. But two other famous significance-testing methods for hypothesis (7-55) also deserve discussion: the analysis of means test (due to Ott) and the one-way analysis of variance test.

Ott's ANOM Test

Ellis Ott introduced a method for testing hypothesis (7-55) that has enjoyed great popularity among some working engineers and industrial statisticians. It is typically implemented in an appealing graphical form and treated as a decision-making tool, where the two possible decisions corresponding to H_0 given in display (7-55) and

$$H_a: \text{not } H_0 \tag{7-56}$$

are, respectively, "The data in hand are adequate to establish definitively that there are differences among the means" and "More data are needed to establish the nature of any differences between the means."

Ott's ANOM test is most often applied in contexts where the sample sizes are all equal, (i.e., where $n_1 = n_2 = \cdots = n_r$). Here, m will stand for this common value, and the discussion of the ANOM test will be restricted to the equal sample size (or *balanced data*) situation. The ANOVA test that is presented next will be discussed without any equality restriction on the n_i; it can thus be used when the data are *unbalanced*, in the sense that the sample sizes are varied. Or, if the reader is intent on using ANOM in an analysis involving unequal sample sizes, appropriate modifications for that more general situation can be found in the January 1983 issue of the *Journal of Quality Technology* (which was dedicated to Ott and devoted to ANOM).

In notation like that used in Section 4-3, let

$$\bar{y}_{\cdot} = \frac{1}{r} \sum_{i=1}^{r} \bar{y}_i \tag{7-57}$$

If in fact hypothesis (7-55) were true in an application, $\bar{y}_{\cdot}$ would (at least in the situation of equal sample sizes) be the most obvious data-based approximation to the common value of $\mu_1, \mu_2, \dots, \mu_r$. Further, the differences

$$\bar{y}_i - \bar{y}_{\cdot} \tag{7-58}$$

would seem to be representing only random variation. It is a simple matter to use Proposition 5-1 (in a way very similar to that leading to equation (7-11) of Section 7-1) to reason that under the one-way model, if hypothesis (7-55) is true, the differences (7-58) are normal with

$$E(\bar{y}_i - \bar{y}_.) = 0$$

and

$$Var(\bar{y}_i - \bar{y}_.) = \frac{\sigma^2}{m}\left(\frac{r-1}{r}\right)$$

These facts should provide some motivation for using the statistic

$$H = \frac{\max_i |\bar{y}_i - \bar{y}_.|}{s_P\sqrt{\dfrac{r-1}{mr}}} \tag{7-59}$$

as a test statistic for hypothesis (7-55), where large observed values of H will constitute evidence against H_0. And the variable (7-59) is a version of Ott's ANOM test statistic. Tables D-12 contain some (large-p) quantiles of an appropriate reference distribution for use with statistic (7-59) in testing hypothesis (7-55). (As was the case with the F tables, each individual table contains p quantiles for a single p.) One simply locates r across the top of the table, finds $\nu = n - r = mr - r = r(m - 1)$ down the left side, and reads the corresponding quantile out of the body of the table.

EXAMPLE 7-9

(Example 7-1 revisited) **An ANOM Test of the Equality of Eight Concrete Formula Mean Strengths.** The concrete strength study of Armstrong, Babb, and Campen involved $r = 8$ samples of constant size $m = n_1 = n_2 = \cdots = n_8 = 3$. The eight sample means $\bar{y}_i$ give $\bar{y}_. = 3{,}693.6$ and absolute differences $|\bar{y}_i - \bar{y}_.|$ in Table 7-12.

TABLE 7-12

Sample Means and Their Deviations from $\bar{y}_.$ in the Concrete Strength Study

i Formula	$\bar{y}_i$	$|\bar{y}_i - \bar{y}_.|$
1	5,635.3	1,941.7
2	5,753.3	2,059.7
3	4,527.3	833.7
4	3,442.3	251.3
5	2,923.7	769.9
6	3,324.7	368.9
7	1,551.3	2,142.3
8	2,390.7	1,302.9

The largest of the $|\bar{y}_i - \bar{y}_.|$ values in Table 7-12 is 2,142.3, corresponding to formula 7. Then, using display (7-59), the observed value of the ANOM test statistic is

$$h = \frac{2{,}142.3}{581.6\sqrt{\dfrac{8-1}{3(8)}}} = 6.82$$

since (from before) $s_P = 581.6$ psi. Then, entering Tables D-12 with $r = 8$ and $\nu = n - r = r(m - 1) = 16$, it is clear that the p-value here is less than .001 (i.e., that 6.82

exceeds the .999 quantile of the reference distribution). As has been indicated a number of times by now, the data present overwhelming evidence that the mean compressive strengths for the formulas differ. As was remarked in Example 7-1 when these data were introduced, the difference between the formulas is only one of water content. So this small p-value can be taken as strong confirmation of the importance of water content in determining the strength of concrete of the type the students studied.

The graphical form in which the results of ANOM tests are usually presented is as follows. Sample means $\bar{y}_i$ are plotted against i on axes that have a *center line* drawn at $\bar{y}_.$ and upper and lower *decision limits* drawn at

$$\bar{y}_. \pm h s_P \sqrt{\frac{r-1}{mr}} \qquad (7\text{-}60)$$

where h is a value taken from Tables D-12 corresponding to a desired type I error probability. If all $\bar{y}_i$'s plot inside the decision limits, the corresponding p-value is larger than the chosen type I error probability; otherwise, it is smaller. Although (strictly speaking) it is not completely logical to do so, it is also common practice to identify those $\bar{y}_i$ that plot outside the decision limits as significantly different from those that don't (all of which are judged not significantly different). While not being completely happy with this particular practice, this author does find the ANOM graphic to be a helpful addition to the tools that have already been used here to represent the information carried by r samples of measurement data.

EXAMPLE 7-12
(continued)

If, for example, one opts for $\alpha = .01$ decision limits for the concrete strength study, Table D-12-C shows $h = 3.87$ to be appropriate for use in formula (7-60). So upper and lower decision limits should be drawn at

$$3{,}693.6 \pm 3.87(581.6)\sqrt{\frac{8-1}{3(8)}}$$

that is,

$$2{,}478.0 \text{ psi} \qquad \text{and} \qquad 4{,}909.2 \text{ psi}$$

Figure 7-12 is then a graphical representation of the ANOM treatment of the concrete strength data. The plot in Figure 7-12 can be thought of as showing the size of differences in the sample means, relative to the size of the variation between them that would be required to produce statistical significance. Of course, it is meant to convey a somewhat different message than Figure 7-1 (which is meant to depict individual strengths) or Figure 7-11 (which is meant to convey the notion of precision of simultaneous estimation of the means). (Of the two, Figure 7-12 is more closely related to Figure 7-11 than to Figure 7-1.)

ANOM is sometimes applied in a form slightly different from that indicated in formulas (7-59) or (7-60), actually closer to Ott's original suggestion. That is, s_P in formulas (7-59) and (7-60) is sometimes replaced by an estimate of σ based not on pooling r sample variances, but rather based on averaging r sample ranges. There are good reasons for using the version presented here (having to do with theoretical results to the effect that s_P is superior to a multiple of the average sample range as an

FIGURE 7-12

Graphical Representation of the ANOM Test for Differences in Mean Concrete Strengths

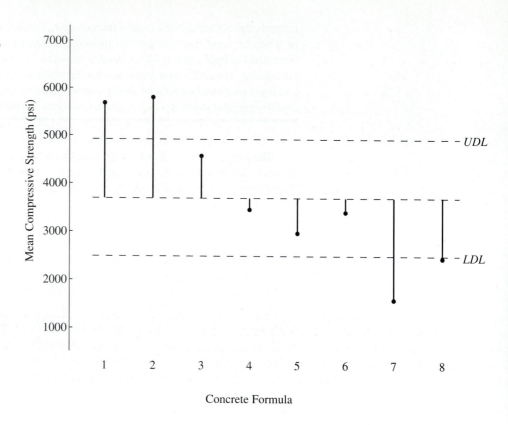

estimate of σ, and the fact that when s_P is used it is possible to work out the entries of Table D-12 in more or less exact fashion). But it is worth mentioning the possibility of using a multiple of an average sample range, $\bar{R}$, in place of s_P (and finding a modified reference distribution for statistic (7-59) and multiple h for decision limits (7-60)). This is because of the similarity between the resulting variant of ANOM and what will be seen in Section 7-5 to be standard practice in Shewhart control charting.

The One-Way ANOVA F Test

There is a second well-known method of testing hypothesis (7-55) of equality of r means, which has come to be known as the *one-way analysis of variance F test*. (At this point it may seem rather strange that a test about means is called by a name apparently emphasizing variance. The motivation for this jargon should become be clearer by the end of this section.) This test lacks the kind of appealing graphical representation possessed by the ANOM test just discussed. On the other hand, it has some theoretical advantages over the ANOM test (having to do with its superior ability to detect differences in the r means). And it also turns out to be associated with a very helpful way of thinking about partitioning the overall variability that one encounters in a response variable.

One way of looking at the ANOM test statistic in formula (7-59) is that its numerator, $\max |\bar{y}_i - \bar{y}_\cdot|$, is a measure of the variability existing in the data *between* the r samples, while the denominator $s_P\sqrt{\frac{r-1}{mr}}$ provides a way of standardizing or judging the numerator based on s_P, which is itself derived from variation *within* the samples. Similar thinking, based however on a different representation of the variation between the samples, can be used to motivate the ANOVA F statistic. But one small matter of notation must be dealt with before introducing that test statistic.

Repeatedly in the balance of this book, it will be convenient to have symbols for the summary measures of Section 3-3 (sample means and variances) applied to the data from multisample studies, *ignoring the fact that there are r different samples involved* (i.e., as if all the responses in hand came from a single set of process conditions). Already the unsubscripted letter n has been used to stand for $n_1 + n_2 + \cdots + n_r$, the number of observations one would have ignoring the fact that r samples are involved. This kind of convention will now be formally extended to include statistics calculated from the n responses. For emphasis, the convention will be stated in definition form.

DEFINITION 7-3

(A Notational Convention for Multisample Studies). In multisample studies, symbols for sample sizes and sample statistics appearing without subscript indices or dots will be understood to be calculated from all responses in hand, obtained by combining all samples.

So n will stand for the total number of data points (even in an r-sample study), $\bar{y}$ for the grand sample average of response y, and s^2 for a grand sample variance calculated completely ignoring sample boundaries.

For present purposes (of writing down a test statistic for testing hypothesis (7-55)), one needs to make use of $\bar{y}$, the grand sample average. It is important to recognize that $\bar{y}$ and $\bar{y}_{\cdot}$ from formula (7-57) are not necessarily the same unless all sample sizes are equal. That is, when the sample sizes are varied, $\bar{y}$ is the (unweighted) arithmetic average of the raw data values y_{ij} but is a weighted average of the sample means $\bar{y}_i$. On the other hand, $\bar{y}_{\cdot}$ is the (unweighted) arithmetic mean of the sample means $\bar{y}_i$ but is a weighted average of the raw data values y_{ij}. For example, in the simple case that $r = 2$, $n_1 = 2$, and $n_2 = 3$,

$$\bar{y} = \frac{1}{5}(y_{11} + y_{12} + y_{21} + y_{22} + y_{23}) = \frac{2}{5}\bar{y}_1 + \frac{3}{5}\bar{y}_2$$

while

$$\bar{y}_{\cdot} = \frac{1}{2}(\bar{y}_1 + \bar{y}_2) = \frac{1}{4}y_{11} + \frac{1}{4}y_{12} + \frac{1}{6}y_{21} + \frac{1}{6}y_{22} + \frac{1}{6}y_{23}$$

and, in general, $\bar{y}$ and $\bar{y}_{\cdot}$ will not be the same.

Now, under hypothesis (7-55), that $\mu_1 = \mu_2 = \cdots = \mu_r$, $\bar{y}$ is a natural estimate of the common mean. (After all, the hypothesis implies that all underlying distributions are the same, so the data in hand are reasonably thought of not as r different samples, but rather as a single sample of size n.) Then (in a manner similar to the development of the ANOM test statistic) the differences $\bar{y}_i - \bar{y}$ are indicators of possible differences between the μ_i. It turns out to be convenient to summarize the size of the $\bar{y}_i - \bar{y}$ values in terms of a kind of total of squares — namely,

$$\sum_{i=1}^{r} n_i(\bar{y}_i - \bar{y})^2 \tag{7-61}$$

One can think of statistic (7-61) either in terms of a weighted sum of the quantities $(\bar{y}_i - \bar{y})^2$ or in terms of an unweighted sum, where there is a term in the sum for each raw data point, and therefore n_i of the type $(\bar{y}_i - \bar{y})^2$. The quantity (7-61) is an alternative to the ANOM max $|\bar{y}_i - \bar{y}_{\cdot}|$ as a measure of the between-sample variation existing in the data. For a given set of sample sizes, the larger it is, the more variation there is between the sample means $\bar{y}_i$.

The way that the quantity (7-61) is standardized in order to produce a test statistic for hypothesis (7-55) is as follows. One simply divides it by $(r - 1)s_P^2$, giving

$$F = \frac{\dfrac{1}{r - 1}\displaystyle\sum_{i=1}^{r} n_i(\bar{y}_i - \bar{y})^2}{s_P^2} \tag{7-62}$$

as a test statistic. The fact is that if $H_0: \mu_1 = \mu_2 = \cdots = \mu_r$ is true, the one-way model assumptions imply that F in equation (7-62) has an $F_{r-1,n-r}$ distribution. So the hypothesis of equality of r means can be tested using the statistic in equation (7-62) with an $F_{r-1,n-r}$ reference distribution, where large observed values of F are taken as evidence against H_0 in favor of H_a: not H_0, the possibility that not all of $\mu_1, \mu_2, \ldots, \mu_r$ are the same.

EXAMPLE 7-9
(continued)

Returning again to the context of the concrete compressive strength study, it is obvious from Table 7-12 that since each $n_i = 3$, in this situation,

$$\sum_{i=1}^{r} n_i(\bar{y}_i - \bar{y})^2 = 3(1{,}941.7)^2 + 3(2{,}059.7)^2 + \cdots + 3(2{,}142.3)^2 + 3(1{,}302.9)^2$$

$$= 47{,}360{,}780 \text{ (psi)}^2$$

In order to use this to judge statistical significance, one standardizes via equation (7-62) to arrive at the observed value of the test statistic

$$f = \frac{\dfrac{1}{8 - 1}(47{,}360{,}780)}{(581.6)^2} = 20.0$$

But it is an easy matter to verify from Tables D-6 that 20.0 is larger than the .999 quantile of the $F_{7,16}$ distribution. So using this F test to judge statistical significance,

$$p\text{-value} = P[\text{an } F_{7,16} \text{ random variable} \geq 20.0] < .001$$

and one has essentially the same bottom line as provided by the ANOM test. That is, the data provide overwhelming evidence that not all of $\mu_1, \mu_2, \ldots, \mu_8$ are equal.

For pedagogical (and other) reasons, the ANOM and ANOVA tests have been presented after discussing interval-oriented methods of inference for r-sample studies. But it is worth noting that if they are to be used in applications, these testing methods typically belong chronologically before the use of the estimation methods. That is, the ANOM and ANOVA methods can serve as screening devices, letting one know whether the data in hand are adequate to differentiate conclusively between the means, or whether more data are needed.

The One-Way ANOVA Identity and Table

Associated with the ANOVA test statistic is some strong intuition related to the partitioning of observed variability. This is related to an algebraic identity that is stated here in the form of a proposition.

PROPOSITION 7-1

For any n numbers y_{ij}

$$(n - 1)s^2 = \sum_{i=1}^{r} n_i(\bar{y}_i - \bar{y})^2 + (n - r)s_P^2 \tag{7-63}$$

or in other symbols,

$$\sum_{i,j}(y_{ij} - \bar{y})^2 = \sum_{i=1}^{r} n_i(\bar{y}_i - \bar{y})^2 + \sum_{i=1}^{r}\sum_{j=1}^{n_i}(y_{ij} - \bar{y}_i)^2 \qquad (7\text{-}64)$$

Proposition 7-1 states what is commonly known as the *one-way analysis of variance identity*; it should begin to shed some light on the somewhat mysterious phrase "analysis of variance." It says that an overall measure of variability in the response y, namely,

$$(n-1)s^2 = \sum_{i,j}(y_{ij} - \bar{y})^2$$

can be partitioned or decomposed algebraically into two parts. One,

$$\sum_{i=1}^{r} n_i(\bar{y}_i - \bar{y})^2$$

can be thought of as being associated with variation between the samples or "treatments," and the other,

$$(n-r)s_{\text{P}}^2 = \sum_{i=1}^{r}\sum_{j=1}^{n_i}(y_{ij} - \bar{y}_i)^2$$

is associated with variation within the samples (and in fact consists of the sum of the squared residuals). The F statistic (7-62), developed to use in testing $\text{H}_0\colon \mu_1 = \mu_2 = \cdots = \mu_r$, has a numerator related to the first of these measures (the one representing variation between samples) and a denominator related to the second (the one representing variation within samples). So the ANOVA F statistic does amount to a kind of <u>analyzing</u> of the raw <u>variability</u> in y.

In recognition of their prominence in the calculation of the one-way ANOVA F statistic and their usefulness as descriptive statistics in their own right, the three sums (of squares) appearing in formula (7-63) or (7-64) are usually given special names and shorthand. These are stated here in definition form.

DEFINITION 7-4

In a multisample study, $(n-1)s^2$, the sum of squared differences between the raw data values and the grand sample mean, will be called the **total sum of squares** and symbolized as $SSTot$.

DEFINITION 7-5

In an unstructured multisample study, $\sum n_i(\bar{y}_i - \bar{y})^2$ will be called the **treatment sum of squares** and symbolized as $SSTr$.

DEFINITION 7-6

In a multisample study, the sum of squared residuals, $\sum(y - \hat{y})^2$ ($(n-r)s_{\text{P}}^2$ in the unstructured situation), will be called the **error sum of squares** and symbolized as SSE.

In the new notation introduced in these definitions, Proposition 7-1 states that in an unstructured multisample context,

$$SSTot = SSTr + SSE \qquad (7\text{-}65)$$

Partially as a means of organizing calculation of the F statistic given in formula (7-62) and partially because it is an effective tool for reinforcing and extending the variance partitioning insight provided by formulas (7-63), (7-64), and (7-65), it is useful to make an ANOVA table. There are, in fact, many forms of ANOVA tables corresponding to various multisample analyses. The form most germane to the present situation is given in symbolic form as Table 7-13.

TABLE 7-13

General Form of the One-Way ANOVA Table

ANOVA TABLE (FOR TESTING $H_0: \mu_1 = \mu_2 = \cdots = \mu_r$)				
Source	SS	df	MS	F
Treatments	$SSTr$	$r-1$	$SSTr/(r-1)$	$MSTr/MSE$
Error	SSE	$n-r$	$SSE/(n-r)$	
Total	$SSTot$	$n-1$		

The column headings in Table 7-13 are Source (of variation), Sum of Squares (corresponding to the source), degrees of freedom (corresponding to the source), Mean Square (corresponding to the source), and F (for testing the significance of the source in contributing to the overall observed variability). The entries in the Source column of the table are shown here as being Treatments, Error, and Total, but the name Treatments is sometimes replaced by Between (Samples), and the name Error is sometimes replaced by Within (Samples) or Residual. The first two entries in the SS column must sum to the third, as is the case for the df column. Notice that the entries in the df column are those attached to the numerator and denominator, respectively, of the test statistic in equation (7-62). The ratios of sums of squares to degrees of freedom are called mean squares, here the mean square for treatments ($MSTr$) and the mean square for error (MSE). Verify that in the present context, $MSE = s_P^2$ and $MSTr$ is the numerator of the F statistic given in equation (7-62), so that indeed the single ratio appearing in the F column does give the observed value of F for testing $H_0: \mu_1 = \mu_2 = \cdots = \mu_r$.

EXAMPLE 7-9
(continued)

Consider once more the concrete strength study scenario. It is possible to return to the raw data given in Table 7-1 and find that $\bar{y} = 3{,}693.6$, so

$$SSTot = (n-1)s^2$$

$$= (5{,}800 - 3{,}693.6)^2 + (4{,}598 - 3{,}693.6)^2 + (6{,}508 - 3{,}693.6)^2$$

$$+ \cdots + (2{,}631 - 3{,}693.6)^2 + (2{,}490 - 3{,}693.6)^2$$

$$= 52{,}772{,}190 \ (\text{psi})^2$$

(Actually, the easiest way to carry out this computation (short of using a full-blown statistical package) is by using a preprogrammed pocket calculator to find s^2 and then multiplying the result by $n - 1 = 23$.)

Further, as in Section 7-1, $s_P^2 = 338{,}213.1 \ (\text{psi})^2$ and $n - r = 16$, so

$$SSE = (n-r)s_P^2 = 5{,}411{,}410 \ (\text{psi})^2$$

And from earlier in this section,

$$SSTr = \sum_{i=1}^{r} n_i(\bar{y}_i - \bar{y})^2 = 47{,}360{,}780$$

Then, plugging these and appropriate degrees of freedom values into the general form of the one-way ANOVA table, one obtains the table that is appropriate in the concrete compressive strength study, presented here as Table 7-14.

TABLE 7-14

One-Way ANOVA Table for the
Concrete Strength Study

ANOVA TABLE (FOR TESTING $H_0: \mu_1 = \mu_2 = \cdots = \mu_8$)

Source	SS	df	MS	F
Treatments	47,360,780	7	6,765,826	20.0
Error	5,411,410	16	338,213	
Total	52,772,190	23		

Notice that, as promised by the one-way ANOVA identity, the sum of the treatment and error sums of squares is the total sum of squares. Also, Table 7-14 serves as a helpful summary of the testing process, showing at a glance the observed value of F, the appropriate degrees of freedom, and $s_P^2 = MSE$.

The especially alert reader may recall having used some kind of breakdown of a "raw variation in the data" figure earlier in this text (namely, in Chapter 4). In fact, there is a direct connection between the present discussion and the discussion and use of R^2 in Sections 4-1, 4-2, and 4-3. (See Definition 4-3 and its use throughout those three sections.) In the present notation, the coefficient of determination defined as a descriptive measure in Section 4-1 is

$$R^2 = \frac{SSTot - SSE}{SSTot} \tag{7-66}$$

(since fitted values for the present situation are the sample means and SSE is the sum of squared residuals here, just as it was earlier). Expression (7-66) is a perfectly general recasting of the definition of R^2 into "SS" notation. In the present one-way context, the one-way identity (7-65) makes it possible to rewrite the numerator of the right-hand side of formula (7-66) as $SSTr$. So in an unstructured r-sample study (where fitted values are sample means),

$$R^2 = \frac{SSTr}{SSTot} \tag{7-67}$$

That is, the first entry in the SS column of the ANOVA table divided by the total entry of that column can be thought of as "the fraction of the raw variability in y accounted for in the process of fitting the equation $y_{ij} \approx \mu_i$ to the data."

EXAMPLE 7-9
(continued)

In the concrete compressive strength study, a look at Table 7-14 shows that

$$R^2 = \frac{47,360,780}{52,772,190} = .897$$

That is, another way to describe these data is to say that differences between concrete formulas account for nearly 90% of the raw variability observed in compressive strength.

So the ANOVA breakdown of variability not only facilitates the testing of H_0: $\mu_1 = \mu_2 = \cdots = \mu_r$, but it also makes direct connection with the earlier descriptive analyses of what part of the raw variability is accounted for in fitting a model equation.

**Random Effects Models
and Analyses**

On occasion, the r particular conditions leading to the r samples in a multisample statistical engineering study are not so much of interest in and of themselves, as they are of interest as representative of a wider set of conditions. For example, suppose $r = 5$

production workers are observed taking readings with a new type of gage, and for each, $n_i = 4$ times required to make readings are recorded for a study of the advisability of purchasing more such gages. The five workers involved are really of interest primarily in that they shed light on the much larger group of workers who will potentially use the gages. Or when a batch chemical process is run many times a day, and $n_i = 2$ impurity analyses are run on $r = 3$ of these batches each day as a quality monitoring measure, the three batches involved are of interest primarily because they represent the much larger number of batches made that day. Or in the nondestructive testing of critical metal parts, if $n_i = 3$ mechanical wave travel time measurements are made on each of $r = 6$ parts selected from a large lot of such parts, the six particular parts involved are of interest primarily as they provide information on the whole lot.

In such situations, rather than making the particular r means actually represented in the data (i.e., $\mu_1, \mu_2, \ldots, \mu_r$) the focus of efforts at formal inference, it is more natural to think in terms of making inferences about *the mechanism that generates the means* μ_i. Interestingly enough, it is possible, under appropriate model assumptions, to use the ANOVA ideas introduced in this section in this enterprise, at least when all sample sizes are equal. The balance of this section is concerned with how this is done.

The most commonly used probability model for the analysis of r-sample data, where the r conditions actually studied are thought of as representing a much wider set of conditions of interest, is a variation on the one-way model of this chapter, called the *one-way random effects model*. It is built on the usual one-way normal assumptions that

$$y_{ij} = \mu_i + \epsilon_{ij} \tag{7-68}$$

where the ϵ_{ij} are iid normal $(0, \sigma^2)$ random variables. But it doesn't treat the means μ_i as parameters/unknown constants. Instead, the means $\mu_1, \mu_2, \ldots, \mu_r$ are treated as (unobservable) random variables independent of the ϵ_{ij}'s and themselves iid according to some normal distribution with an unknown mean μ and unknown variance (say, σ_τ^2). The random variables μ_i are now called *random* (treatment) *effects*, and the variances σ^2 and σ_τ^2 are called *variance components*. The objects of formal inference then become μ (the mean of the random effects) and the two variance components σ^2 and σ_τ^2.

EXAMPLE 7-10

Magnesium Contents at Different Locations on an Alloy Rod and the Random Effects Model. Youden's *Experimentation and Measurement* contains an interesting data set concerned with the magnesium contents of different parts of a long rod of magnesium alloy. A single ingot of magnesium alloy had been drawn into a rod of about 100 m length, with a square cross section about 4.5 cm on a side. $r = 5$ flat test pieces about 1.2 cm thick were cut from the rod (after it had been cut into 100 bars and five of these randomly selected to represent the rod), and multiple magnesium determinations were made on the five specimens. $n_i = 10$ of the resulting measurements for each specimen are given in Table 7-15. (There were actually twice this number made, representing duplicate determinations for the 50 values given in Table 7-15. Some additional structure in Youden's data will also be ignored for present purposes; the measurements were made at the same ten geometrically laid out locations on the five test pieces.) The units of measurement in Table 7-15 are .001% magnesium.

Notice that in this example, on the order of

$$\frac{100 \times 100 \text{ cm}}{1.2 \text{ cm}} \approx 8{,}300$$

test specimens could be cut from the 100 m rod. The purpose of creating the rod was to provide secondary standards for field calibration of chemical analysis instruments.

TABLE 7-15

Measured Magnesium Contents
for Five Alloy Specimens

Specimen 1	Specimen 2	Specimen 3	Specimen 4	Specimen 5
76	69	73	73	70
71	71	69	75	66
70	68	68	69	68
67	71	69	72	68
71	66	70	69	64
65	68	70	69	70
67	71	65	72	69
71	69	67	63	67
66	70	67	69	69
68	68	64	69	67
$\bar{y}_1 = 69.2$	$\bar{y}_2 = 69.1$	$\bar{y}_3 = 68.2$	$\bar{y}_4 = 70.0$	$\bar{y}_5 = 67.8$
$s_1 = 3.3$	$s_2 = 1.7$	$s_3 = 2.6$	$s_4 = 3.3$	$s_5 = 1.9$

That is, laboratories purchasing pieces of this rod could use them as being of "known" magnesium content to calibrate their instruments. As such, the practical issues at stake here are not primarily how the $r = 5$ particular test specimens analyzed compare. Rather, the issues are what the overall magnesium content is and whether or not the rod is consistent enough in content along its length to be of any use as a calibration tool. That is, a random effects model and inference for the mean effect μ and the variance components are quite natural in this situation. Here, σ_τ^2 represents the variation in magnesium content among the potentially 8,300 different test specimens, and σ^2 represents measurement error plus variation in magnesium content within the 1.2 cm thick specimens, test location to test location.

When all of the r sample sizes n_i are the same, (say, equal to m), it turns out to be quite easy to do some diagnostic checking of the aptness of the normal random effects model (7-68) and make subsequent inferences about μ, σ^2, and σ_τ^2. So this discussion will be limited to cases of equal sample sizes.

As far as investigation of the reasonableness of the model restrictions on the distribution of the μ_i and inference for μ are concerned, a key observation is that

$$\bar{y}_i = \frac{1}{m} \sum_{j=1}^{m} (\mu_i + \epsilon_{ij}) = \mu_i + \bar{\epsilon}_i$$

(where, of course, $\bar{\epsilon}_i$ is the sample mean of $\epsilon_{i1}, \ldots, \epsilon_{im}$). Note also that under the random effects model (7-68), these $\bar{y}_i = \mu_i + \bar{\epsilon}_i$ are iid Gaussian variables with mean μ and variance $\sigma_\tau^2 + \sigma^2/m$. So to begin with, normal-plotting the $\bar{y}_i$ is a sensible method of at least indirectly investigating the appropriateness of the normal distribution assumption for the μ_i. In addition, the fact that the $\bar{y}_i$ are (under the assumptions of the normal random effects model) independent normal variables with mean μ and a common variance suggests that the small sample inference methods from Section 6-3 should simply be applied to the sample means $\bar{y}_i$ in order to do inference for μ. Of course, in doing so, the "sample size" involved is the number of $\bar{y}_i$'s — namely, r.

EXAMPLE 7-10
(continued)

For the magnesium alloy rod, the $r = 5$ sample means are in Table 7-15. Figure 7-13 gives a normal plot of those five values, showing no obvious problems with a normal random effects model for magnesium contents.

FIGURE 7-13

Normal Plot of Five Specimen
Mean Magnesium Contents

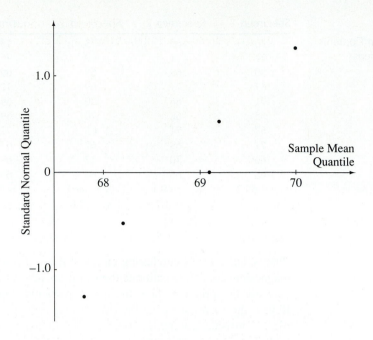

Then calculation with the $\bar{y}_i$ values gives

$$\bar{y}_{.} = \frac{1}{5}\sum_{i=1}^{5}\bar{y}_i = 68.86$$

$$\frac{1}{5-1}\sum_{i=1}^{5}(\bar{y}_i - \bar{y}_{.})^2 = .76$$

so that

$$\sqrt{\frac{1}{5-1}\sum_{i=1}^{5}(\bar{y}_i - \bar{y}_{.})^2} = .87$$

So, applying the small-sample confidence interval formula for a single mean from Section 6-3, it is clear that (since $r - 1 = 4$ degrees of freedom are appropriate) a two-sided 95% confidence for μ has endpoints

$$68.86 \pm 2.776\frac{.87}{\sqrt{5}}$$

that is,

$$67.78 \times 10^{-3}\% \qquad \text{and} \qquad 69.94 \times 10^{-3}\% \tag{7-69}$$

The endpoints (7-69) provide a notion of precision appropriate for the number $68.86 \times 10^{-3}\%$ as an estimate of the rod's mean magnesium content.

━━━━━━━ ▬ ▬ ▬

It is useful to write out in symbols what was just done to get a confidence interval for μ. That is, a sample variance of $\bar{y}_i$'s was used. This is

$$\frac{1}{r-1}\sum_{i=1}^{r}(\bar{y}_i - \bar{y}_{.})^2 = \frac{1}{m(r-1)}\sum_{i=1}^{r}m(\bar{y}_i - \bar{y})^2 = \frac{1}{m(r-1)}SSTr = \frac{1}{m}MSTr$$

because all n_i are m, and $\bar{y}_{\cdot} = \bar{y}$ in this case. But this means that under the assumptions of the one-way normal random effects model, a two-sided confidence interval for μ has endpoints

$$\bar{y}_{\cdot} \pm t\sqrt{\frac{MSTr}{mr}} \tag{7-70}$$

where t is such that the probability the t_{r-1} distribution assigns to the interval between $-t$ and t corresponds to the desired confidence. One-sided intervals are obtained in the usual way, by employing only one of the endpoints in display (7-70).

ANOVA-Based Inference for Variance Components

Turning attention to the variance components in the random effects model (7-68), first note that as far as diagnostic checking of the assumption that the ϵ_{ij} are iid normal variables and inference for $\sigma^2 = Var\,\epsilon_{ij}$ are concerned, all of the methods of Section 7-1 remain in force. If one thinks of holding the μ_i fixed in formula (7-68), it is clear that (conditional on the μ_i) the random effects model treats the r samples as random samples from normal distributions with a common variance. So before doing inference for σ^2 (or σ_{τ}^2 for that matter) via usual normal theory formulas, it is advisable to do the kind of sample-by-sample normal-plotting and plotting of residuals illustrated in Section 7-1. And if it is then plausible that the ϵ_{ij} are iid normal $(0, \sigma^2)$ variables, formula (7-10) of Section 7-1 can be used to produce a confidence interval for σ^2, and significance testing for σ^2 can be done based on the fact that $r(m-1)s_P^2/\sigma^2$ has a $\chi^2_{r(m-1)}$ distribution.

Inference for σ_{τ}^2 borrows from things already discussed but also provides a new wrinkle or two of its own. First, significance testing for

$$H_0: \sigma_{\tau}^2 = 0 \tag{7-71}$$

is made possible by the observation that if H_0 is true, then (just as when $H_0: \mu_1 = \mu_2 = \cdots = \mu_r$ in the case where the μ_i are not random effects but fixed parameters) the $n = mr$ observations are all coming from a single normal distribution. So if

$$F = \frac{MSTr}{MSE} \tag{7-72}$$

has an $F_{r-1, n-r}$ distribution when the μ_i are fixed parameters and $H_0: \mu_1 = \mu_2 = \cdots = \mu_r$ holds, the same must be true under the assumptions of the random effects model (7-68) when the null hypothesis (7-71) holds. Thus, the same one-way ANOVA F test used to test $H_0: \mu_1 = \mu_2 = \cdots = \mu_r$ when the means μ_i are considered fixed parameters can also be used to test $H_0: \sigma_{\tau}^2 = 0$ under the assumptions of the random effects model.

As far as estimation goes, it doesn't turn out to be possible to give a simple confidence interval formula for σ_{τ}^2 directly. But what can be done in a straightforward fashion is to give both a natural ANOVA-based single-number estimate of σ_{τ}^2 and a confidence interval for the ratio σ_{τ}^2/σ^2. To accomplish the first of these, consider the mean values of random variables $MSTr$ and $MSE \,(= s_P^2)$ under the assumptions of the random effects model. Not too surprisingly,

$$E(MSE) = Es_P^2 = \sigma^2 \tag{7-73}$$

(After all, s_P^2 has been used to approximate σ^2. That the "center" of the probability distribution of s_P^2 is σ^2 should therefore seem only reassuring.) And it is further possible for a mathematician to show that

$$E(MSTr) = \sigma^2 + m\sigma_{\tau}^2 \tag{7-74}$$

Then, from equations (7-73) and (7-74),

$$\frac{1}{m}\big(E(MSTr) - E(MSE)\big) = \sigma_\tau^2$$

or

$$E\frac{1}{m}(MSTr - MSE) = \sigma_\tau^2 \qquad \textbf{(7-75)}$$

So equation (7-75) suggests that the random variable

$$\frac{1}{m}(MSTr - MSE) \qquad \textbf{(7-76)}$$

is one whose distribution is centered about the variance component σ_τ^2 and thus is a natural ANOVA-based estimate of σ_τ^2. The variable in formula (7-76) is potentially negative; when that occurs, common practice is to estimate σ_τ^2 by 0. So the formula actually used to estimate σ_τ^2 is

$$\max\left(0, \frac{1}{m}(MSTr - MSE)\right)$$

Facts (7-73) and (7-75), which motivate this method of estimating σ_τ^2, are important enough that they are often included as entries in an <u>E</u>xpected <u>M</u>ean <u>S</u>quare column added to the one-way ANOVA table when testing $H_0 \colon \sigma_\tau^2 = 0$.

The basic probability fact that enables confidence interval estimation of the ratio of variance components σ_τ^2/σ^2 is that under the assumptions of the random effects model (7-68),

$$\frac{\dfrac{MSTr}{\sigma^2 + m\sigma_\tau^2}}{\dfrac{MSE}{\sigma^2}}$$

has an $F_{r-1,n-r}$ distribution. Some algebraic manipulations beginning from this fact show that the interval with endpoints

$$\frac{1}{m}\left(\frac{MSTr}{U \cdot MSE} - 1\right) \qquad \text{and} \qquad \frac{1}{m}\left(\frac{MSTr}{L \cdot MSE} - 1\right) \qquad \textbf{(7-77)}$$

can be used as a two-sided confidence interval for σ_τ^2/σ^2, where the associated confidence is the probability the $F_{r-1,n-r}$ distribution assigns to the interval (L, U). One-sided intervals for σ_τ^2/σ^2 can be had by using only one of the endpoints and choosing L or U such that the probability assigned by the $F_{r-1,n-r}$ distribution to (L, ∞) or $(0, U)$ corresponds to the desired confidence.

EXAMPLE 7-10
(continued)

Consider again the situation of the measured magnesium contents for specimens cut from the 100 m alloy rod. Some normal plotting shows the "single variance Gaussian ϵ_{ij}" part of the model assumptions (7-68) to be at least not obviously flawed. Sample-by-sample normal plots show fair linearity (at least after allowing for the discreteness introduced in the data by the measurement scale used), except perhaps for sample 4, with its five identical values. The five sample standard deviations are roughly of the same order of magnitude, and the normal plot of residuals in Figure 7-14 is pleasantly linear. So it is sensible to consider formal inference for σ^2 and σ_τ^2 based on the normal theory model.

FIGURE 7-14

Normal Plot of Residuals for the Magnesium Content Study

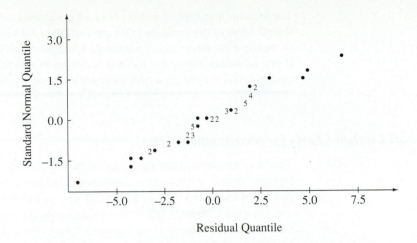

Table 7-16 is an ANOVA table for the data of Table 7-15. From Table 7-16, the p-value for testing $H_0: \sigma_\tau^2 = 0$ is the $F_{4,45}$ probability to the right of 1.10. According to Tables D-6, this is larger than .25, giving very weak evidence of detectable variation between specimen mean magnesium contents.

TABLE 7-16

ANOVA Table for the Magnesium Content Study

ANOVA TABLE (FOR TESTING $H_0: \sigma_\tau^2 = 0$)

Source	SS	df	MS	EMS	F
Treatments	30.32	4	7.58	$\sigma^2 + 10\sigma_\tau^2$	1.10
Error	309.70	45	6.88	σ^2	
Total	340.02	49			

The *EMS* column in Table 7-16 reminds one first that $MSE = s_P^2 = 6.88$ serves as an estimate of σ^2. So one would estimate that multiple magnesium determinations on a given specimen would have a typical standard deviation on the order of $\sqrt{6.88} = 2.6 \times 10^{-3}\%$. Then the expected mean squares further suggest that σ_τ^2 be approximated by

$$\frac{1}{10}(MSTr - MSE) = \frac{1}{10}(7.58 - 6.88) = .07$$

so an estimate of σ_τ is then

$$\sqrt{.07} = .26 \times 10^{-3}\%$$

That is, the standard deviation of specimen mean magnesium contents is estimated to be on the order of $\frac{1}{10}$ of the standard deviation associated with multiple measurements on a single specimen.

A confidence interval for σ^2 could be made using formula (7-10) of Section 7-1. That will not be done here, but formula (7-77) will be used to make a one-sided 90% confidence interval of the form (0,#) for σ_τ/σ. The .90 quantile of the $F_{45,4}$ distribution is about 3.80, so the .10 quantile of the $F_{4,45}$ distribution is about $\frac{1}{3.80}$. Then taking the root of the second endpoint given in display (7-77), a 90% upper confidence bound for σ_τ/σ is

$$\sqrt{\frac{1}{10}\left(\frac{7.58}{\left(\frac{1}{3.80}\right)6.88} - 1\right)} = .56$$

The bottom line here is that σ_τ is small compared to σ and is not even clearly other than 0. Most of the variation in the data of Table 7-15 is thus associated with the making of multiple measurements on a single specimen. Of course, this would be good news if one wished to cut up the rod and distribute pieces of it as having known magnesium contents, and thus being useful tools for instrument calibration.

7-5 *Shewhart Control Charts for Measurement Data*

This text has repeatedly made use of the phrase "stable process" and alluded to the fact that unless a data-generation process has associated with it a single, repeatable pattern of variation, there is no logical way to move from data in hand to predictions and inferences about the physical mechanism producing the data. Also stressed has been the notion that "baseline" or "inherent" variation evident in the output of a process is a principal limitation on system performance and the quality of any product of the process. But so far no tools have been given for empirically evaluating the extent to which a data-generation mechanism can be thought of as stable or determining what the baseline variation of a process is (or, equivalently, separating baseline or inherent variation from that which is possibly avoidable).

W. Shewhart, working in the late 1920s and early 1930s at Bell Laboratories, developed an extremely simple yet effective tool for displaying and analyzing the results of a series of samples periodically drawn from a process, with the related goals of determining the stability (or lack of stability) of the process and separating baseline or inherent process variation from that which may be avoidable. This tool has become known as the *Shewhart control chart*. (Actually, this author much prefers the nonstandard name *Shewhart monitoring chart*, to avoid the connotations of automatic/feedback process adjustment that the word *control* may carry for readers familiar with the field of engineering control.)

This section and the next present a brief introduction to the topic of Shewhart control charts, beginning here with charts for measurement data. It begins with some generalities, discussing Shewhart's conceptualization of process variability, reviewing his charting idea in general terms, considering the notions of using Shewhart charts both "with standards given" and also "retrospectively," and discussing the methods in relation to the other methods of Chapters 6 and 7. Then the specific instances of Shewhart control charts for means, ranges, and standard deviations are considered in turn. Finally, the section closes with a few comments about the importance of control charts in the improvement of modern industrial processes.

Generalities about Shewhart Control Charts

As the word has been used in this book, *stability* of an engineering data-generating process has referred to a consistency or repeatability over time. When one thinks of empirically assessing the stability of a process, it is therefore clear that samples of data taken from the process at different points in time will be needed.

EXAMPLE 7-11

Monitoring the Lengths of Sheets Cut on a Ream Cutter. Shervheim and Snider worked with a company on the cutting of a rolled material into sheets using a ream cutter. In one part of their study of the behavior of the cutter, every two minutes they sampled five consecutive sheets and measured their lengths. Part of the students' length data are given in Table 7-17, in units of $\frac{1}{64}$ inch over a certain reference length.

TABLE 7-17

Lengths of 22 Samples of Five
Sheets Cut on a Ream Cutter

Sample	Time	Excess Length
1	12:40	9, 10, 7, 8, 10
2	12:42	6, 10, 8, 8, 10
3	12:44	11, 10, 9, 5, 11
4	12:46	10, 9, 9, 8, 7
5	12:48	7, 5, 11, 9, 5
6	12:50	9, 9, 10, 7, 9
7	12:52	10, 8, 6, 11, 8
8	12:54	7, 10, 8, 8, 9
9	12:56	10, 9, 9, 5, 12
10	12:58	8, 10, 6, 8, 10
11	1:00	8, 10, 4, 7, 8
12	1:02	8, 10, 10, 6, 9
13	1:04	10, 8, 6, 7, 10
14	1:06	8, 6, 10, 8, 8
15	1:08	13, 5, 8, 8, 13
16	1:10	10, 4, 9, 10, 8
17	1:12	7, 7, 9, 7, 8
18	1:14	9, 7, 7, 9, 6
19	1:16	5, 10, 5, 8, 10
20	1:18	9, 6, 8, 9, 11
21	1:20	6, 10, 11, 5, 6
22	1:22	15, 3, 7, 9, 11

One of the goals of the study was to investigate the stability of the cutting process over time. The kind of multisample data the students collected, where the samples were separated and ordered in time, are ideal for that purpose.

As a parenthetical remark, notice that data (like those in Table 7-17) collected for purposes of assessing process stability will often be naturally thought of as r samples of some fixed sample size m, lacking any standard structure except for that related to the fact that they were taken in a particular time order. That is why Shewhart control charting material has been placed in this chapter with inference methods for unstructured multisample studies.

Shewhart's fundamental qualitative insight regarding variation in a process over time (seen in data taken on the process, or on a product of the process) is that

$$\begin{array}{c}\text{Overall process} \\ \text{variation}\end{array} = \text{baseline variation} + \begin{array}{c}\text{variation that} \\ \text{can be eliminated}\end{array} \tag{7-78}$$

Shewhart conceived of the baseline process variation as that which will remain even under the most careful process monitoring and appropriate physical interventions — an inherent property of a particular system configuration, which cannot be reduced without basic changes in the physical process or how it is run. This variation is considered to be due to *common* (universal) *causes* or *system causes*. Other terms used for it are *random variation* and *short-term variation*. In the context of the cutting operation of Example 7-11, this kind of variation might be seen in four or five consecutive sheet lengths cut on a single ream cutter, from a single roll of material, without any intervening operator adjustments, following a particular plant standard method of machine operation, etc. It is that variation that comes from hundreds of small unnamable, unidentifiable physical causes. When only this kind of variation is acting, it is reasonable to term a process "stable."

The second component of overall process variation on the right of equation (7-78) is that which can potentially be eliminated by appropriate physical intervention (provided it can be identified). This kind of variation has variously been called variation due to *special* or *assignable causes*, *nonrandom variation*, and *long-term variation*. In the sheet-cutting example, this might be variation in sheet length brought about by undesirable changes in tension on the material being cut, roller slippage on the cutter, unwarranted operator adjustments to the machine, eccentricities associated with how a particular incoming roll of material was wound, etc. Shewhart reasoned that a first step in producing good process performance (and therefore quality products of the process) is to be able to separate the two kinds of variation. One would then have a basis for knowing when to intervene and find and eliminate the cause of any assignable variation, thereby producing process stability.

Shewhart's graphical method for separating the two components of overall variation in equation (7-78) is based on the following logic. First, periodically taken samples on process behavior are gathered and reduced to appropriate summary statistics, and the summary statistics are plotted against time order of observation. To this simple time plotting of summary statistics (which was already mentioned in Section 3-3), Shewhart added the notion that lines be drawn on the chart to separate the values of the plotted statistic that are consistent with a "baseline variation only" picture of process performance from those that are not consistent with such a view (and are therefore presumably attributable to the operation of nonrandom or special causes). Shewhart called these lines of demarcation *control limits*. (The terms *action limits* and *decision limits* are also sometimes used.) When all plotted points fall within the control limits, one judges the process to be stable, subject only to chance causes. But when a point falls outside the limits, physical investigation and intervention is called for, to eliminate any assignable cause of variation. Figure 7-15 is a hypothetical plot of a generic control chart for a summary statistic, (say, w), showing upper and lower control limits (*UCL* and *LCL*), some plotted values, and one "out of control" point.

There are any number of potential types of chart that fit the general pattern of Figure 7-15. For example, common possibilities in the sheet-cutting scenario of Example 7-11 include control charts for the sample mean (sheet length) and charts for the sample

FIGURE 7-15

Generic Shewhart Control Chart for a Statistic w

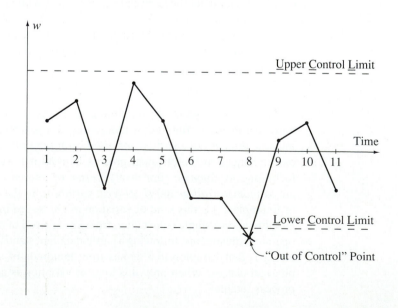

range or sample standard deviation. These three possibilities will presently be discussed in detail. But first, some generalities still need to be considered.

For one thing, there remains the matter of how to set the position of the control limits. Shewhart's idea was as follows. He argued that one should apply probability theory and develop appropriate stable process/iid observations distributions for the plotted statistics. Then small upper and lower percentage points for these could be used to establish control limits. As an example, the central limit material in Section 5-5 should have conditioned the reader to think of sample means as at least approximately Gaussian with expected value equal to an underlying mean μ, and standard deviation $\sigma/\sqrt{m}$, where σ is an underlying standard deviation and m is the sample size. So for plotting sample means, one might set the upper and lower control limits at (respectively) small upper and lower percentage points of the normal distribution with mean μ and standard deviation $\sigma/\sqrt{m}$, where μ and σ are some kind of process mean and process short-term standard deviation, respectively.

The way that Shewhart's idea on the setting of control limits is implemented in practice is the following. One tries to identify a standard probability model for describing process behavior that 1) is specified by a few parameters and 2) implies the relevance of some standard type of distribution for the plotted statistic. The distribution of the plotted statistic is then a function of the process parameters. After identifying sensible values for those parameters, the control limits are developed as functions of them. For example, in the case of control charts for sample means, the emphasis in developing control limits is on finding appropriate values for the underlying process mean and standard deviation. These then imply appropriate values for control limits for sample means via the use of the Gaussian distribution with mean μ and standard deviation $\sigma/\sqrt{m}$ for $\bar{x}$.

In this discussion of Shewhart control charts, two different possible circumstances will be distinguised, having to do with the origin of the values of the process parameters used to derive control limits. In some applications, values of process parameters (and therefore, by implication, parameters for the "stable process, iid observations" distribution of the plotted statistic) and thus control limits are provided from outside the data being used to produce the plotted values. Such circumstances will be called *standards given* situations; for emphasis, the meaning of this term is stated here in definition form.

DEFINITION 7-7

When control limits are derived from data, requirements, or knowledge of the behavior of a process that are outside the information contained in the samples whose summary statistics are to be plotted, the charting is said to be done with **standards given**.

For example, suppose that in the sheet-cutting context of Example 7-11, past experience with the ream cutter indicates that a process short-term standard deviation on the order of $\sigma = 1.9$ ($\frac{1}{64}$ in.) is appropriate when the cutter is operating as it should. Further, suppose that legal and other considerations have led to the establishment of a target process mean of $\mu = 10.0$ ($\frac{1}{64}$ in. above the reference length). Then control limits based on these values and applied to data collected tomorrow would be "standards given" control limits.

One way to think about a "standards given" control chart (which relates it to the material in Chapter 6) is as a graphical means of repeatedly testing the hypothesis

$$H_0: \text{Process parameters are at their standard values.} \qquad (7\text{-}79)$$

When a plotted point lies inside control limits, one is directed to a decision in favor of hypothesis (7-79) for the time period in question. A point plotting outside limits makes hypothesis (7-79) untenable at the time represented by the sample.

In contrast to "standards given" applications, there are situations in which no externally derived values for process parameters are used. Instead, a single set of samples taken from the process over a period of time is used to both develop a plausible set of parameters for the process and also to judge the stability of the process over the period represented by the data (on the basis of control limits derived from the parameter values estimated from the data). The terms *retrospective* or "*as past data*" will be used in this text for such control charting applications.

DEFINITION 7-8

When control limits are derived from the same samples whose summary statistics are plotted, the charting is said to be done **retrospectively** or "**as past data.**"

In the context of Example 7-11, control limits derived from the data in Table 7-17 and applied to summary statistics for those same data would be "as past data" control limits for assessing the cutting process stability over the period from 12:40 through 1:22 on the day the data were taken.

A way of thinking about a retrospective control chart (which relates it to the material of Section 7-4) is as a graphical means of testing the hypothesis

H_0: A single set of process parameters was acting throughout the   **(7-80)**
time period studied.

When a point or points plot outside of control limits derived from the whole data set, the hypothesis (7-80) of process stability over the period represented by the data becomes untenable.

"Standards Given" $\bar{x}$ Control Charts

The single most famous and frequently used Shewhart control chart is one in which the plotted statistic is the sample mean of measurements taken on a process or its product. Types of control charts are typically named according to the symbols usually employed for the plotted statistics; the following discussion concerns *Shewhart $\bar{x}$ charts*. In using this terminology (and other notation from the statistical quality control field), this text must choose a path through existing notational conflicts between the most common usages in control charting contexts and those for other multisample statistical engineering studies. Before going on, the options that will be exercised here must be explained.

In the first place, to this point in Chapter 7 (also in Chapter 4, for that matter) the symbol y has been used for the basic response variable in a multisample statistical engineering study, $\bar{y}_i$ for a sample mean, and $\bar{y}.$ and $\bar{y}$ for unweighted and weighted averages of the $\bar{y}_i$, respectively. In contrast, in Chapters 3 and 6, where the discussion centered primarily on one- and two-sample studies, x was used as the basic response variable and $\bar{x}$ (or $\bar{x}_i$ in the case of two-sample studies) to stand for a sample mean. Standard usage in Shewhart control charting is to use the x and $\bar{x}$ ($\bar{x}_i$) convention, and the precedent is so strong that this section will adopt it as well. This is despite the fact that many samples are involved, and it would be perfectly logical (but nonstandard) to use the notation already established in previous sections of this chapter. In addition, historical momentum in control charting applications dictates that rather than using $\bar{x}.$ notation,

$$\bar{\bar{x}} = \frac{1}{r} \sum_{i=1}^{r} \bar{x}_i$$ **(7-81)**

is used for the average of sample means. Because of its pervasive use in applications, this convention will also be adopted in this section. But this "bar bar" or "double bar" notation is used in this book *only* in this section.

Something should also be said about the notation used for sample sizes in this section. In both control charting applications and the other kinds of multisample contexts discussed in the chapter, it is common to use the notation n_i for an individual sample size. But there is some conflict between what has been done in this chapter and typical control charting notation when all sample sizes n_i have a common value. The convention here has been to use m for such a common value and n for $\sum n_i$. More or less standard quality control notation is to instead use n for a common sample size. In this matter, this book will break with the quality control convention and continue to use the conventions established thus far in Chapter 7 regarding sample sizes, believing that to do otherwise invites too much confusion. But the reader is hereby alerted to the fact that the m used here is usually going to appear as n in other treatments of control charting.

Having dealt with the notational problems, the main discussion may resume. Consider first the making of a "standards given" Shewhart $\bar{x}$ chart based on samples of size m. An iid model for observations from a process with mean μ and standard deviation σ produces

$$E\bar{x} = \mu \tag{7-82}$$

and

$$\sqrt{Var\,\bar{x}} = \frac{\sigma}{\sqrt{m}} \tag{7-83}$$

and (if the underlying process distribution is reasonably normal or m is of moderate size) an approximately normal distribution for $\bar{x}$. The fact that essentially all of the probability of a normal distribution is within 3 standard deviations of its mean led Shewhart to suggest that for given process standards μ and σ, $\bar{x}$ chart control limits could be set at

$$LCL_{\bar{x}} = \mu - 3\frac{\sigma}{\sqrt{m}} \qquad \text{and} \qquad UCL_{\bar{x}} = \mu + 3\frac{\sigma}{\sqrt{m}} \tag{7-84}$$

Additionally, he suggested drawing a *center line* on an $\bar{x}$ chart at the standard mean μ.

As a practical matter, the kind of limits in formula (7-84) have proved themselves of great utility even in cases where m is fairly small and there is no reason to expect a normal distribution for observations in a sampling period. Formulas (7-82) and (7-83) hold regardless of whether a process short-term distribution is normal, and the 3-sigma (of the plotted statistic, $\bar{x}$) control limits in display (7-84) tend to bracket most of the distribution of $\bar{x}$ under nearly any circumstances. (Indeed, a crude but universal analysis, based on a probability theory version of the Chebyschev Theorem stated in Section 3-3 for relative frequency distributions, guarantees that limits (7-84) will bracket at least $\frac{8}{9}$ of the distribution of $\bar{x}$ in any stable process context.)

EXAMPLE 7-11
(continued)

Consider the use of process standards $\mu = 10$ and $\sigma = 1.9$ in $\bar{x}$ charting based on the data given in Table 7-17. With these standard values for μ and σ, since the $r = 22$ samples are all of size $m = 5$, formulas (7-84) indicate control limits

$$UCL_{\bar{x}} = 10 + 3\frac{1.9}{\sqrt{5}} = 12.55$$

$$LCL_{\bar{x}} = 10 - 3\frac{1.9}{\sqrt{5}} = 7.45$$

for use, along with a center line drawn at $\mu = 10$. Table 7-18 gives some sample-by-sample summary statistics for the data of Table 7-17, including the sample means $\bar{x}_i$. Figure 7-16 is a "standards given" Shewhart $\bar{x}$ chart for the same data.

TABLE 7-18

Sample-by-Sample Summary
Statistics for 22 Samples of
Sheet Lengths

i Sample	$\bar{x}_i$	s_i	R_i
1	8.8	1.30	3
2	8.4	1.67	4
3	9.2	2.49	6
4	8.6	1.14	3
5	7.4	2.61	6
6	8.8	1.10	3
7	8.6	1.95	5
8	8.4	1.14	3
9	9.0	2.55	7
10	8.4	1.67	4
11	7.4	2.19	6
12	8.6	1.67	4
13	8.2	1.79	4
14	8.0	1.41	4
15	9.4	3.51	8
16	8.2	2.49	6
17	7.6	.89	2
18	7.6	1.34	3
19	7.6	2.51	5
20	8.6	1.82	5
21	7.6	2.70	6
22	9.0	4.47	12
	$\sum \bar{x} = 183.4$	$\sum s = 44.41$	$\sum R = 109$

FIGURE 7-16

"Standards Given" Shewhart
$\bar{x}$ Control Chart for Cut
Sheet Lengths

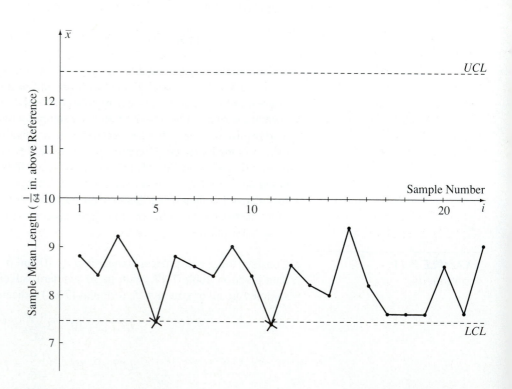

Figure 7-16 shows two points plotting below the lower control limit: the sample means for samples 5 and 11. But it is perfectly obvious from the plot what was going on in the data of Table 7-17 to produce the "out of control" points and corresponding debunking of hypothesis (7-79). Not one of the $r = 22$ plotted sample means lies at or above the target value of 10. If an average sheet length of $\mu = 10$ was truly desired, a simple adjustment was needed, to increase sheet lengths roughly

$$10 - \bar{\bar{x}} = 10 - 8.3 = 1.7 \left(\frac{1}{64} \text{ in.} \right)$$

The true process mean operating to produce the data was clearly below the standard mean.

Retrospective $\bar{x}$ Control Charts

Retrospective (or "as past data") control limits for $\bar{x}$ come about by replacing μ and σ in formulas (7-84) with estimates made from the data in hand, under the supposition that the process was stable over the period represented by the data. That is, in calculating such estimates, one supposes that (as stated in hypothesis (7-80)) a single set of parameters is adequate to describe process behavior during the study period. Notice that the present situation supposing process stability is exactly that met in the ANOM and ANOVA material of Section 7-4 under the hypothesis of equality of r means. So one way to think about a retrospective $\bar{x}$ chart is as a graphical test of the constancy of the process mean over time. Further, the analogy with the material of Section 7-4 suggests natural estimates of μ and σ for use in formulas (7-84).

In Section 7-4, $\bar{y}.$ and $\bar{\bar{y}}$ were used to approximate a hypothesized common value of $\mu_1, \mu_2, \ldots, \mu_r$. In the present notation, this suggests replacing μ in formulas (7-84) with $\bar{\bar{x}}$.

Regarding the development of an estimate of σ for use in formulas (7-84), analogy with all that has gone before in this chapter suggests s_P. And indeed, s_P is a perfectly rational choice of estimate for σ. But it is not one that is commonly used. Historical precedent/accident in the quality control field has made other estimates much more widely used. These must therefore be discussed, not so much because they are better than s_P, but because they represent standard statistical engineering practice.

The most common way of approximating a supposedly constant σ in control charting contexts is based on probability facts about the range, R, of a sample of m observations from a Gaussian distribution. It is possible for a probabilist to derive the probability density for R defined in Definition 3-9, supposing m iid Gaussian variables with mean μ and standard deviation σ are involved. That density will not be given in this book, but it is useful to know that the mean of that distribution is (for a given sample size m) proportional to σ. The constant of proportionality is typically called d_2, and in symbols,

$$ER = d_2 \sigma \tag{7-85}$$

or equivalently,

$$\sigma = \frac{ER}{d_2} \tag{7-86}$$

Values of d_2 for various m are given in Table D-2. (Return to the comments preceding display (3-3) in Section 3-3 and recognize that what was cryptic there should now make sense.)

Statements (7-85) and (7-86) are theoretical. The way they find practical relevance is to think that under the hypothesis that the process standard deviation is constant, the sample mean of sample ranges

$$\bar{R} = \frac{1}{r} \sum_{i=1}^{r} R_i \tag{7-87}$$

can be expected to approximate the theoretical mean range, ER. That is, from statement (7-86), it seems that

$$\frac{\bar{R}}{d_2} \tag{7-88}$$

is a plausible way to estimate σ. On theoretical grounds, $\bar{R}/d_2$ is inferior to s_P, but it has the weight of historical precedent behind it, and it is relatively simple to calculate (a virtue that was of primary importance before the advent of widespread computing power).

A second estimate of σ with quality control origins comes about by making the same kind of argument that led to statistic (7-88), beginning not with R but instead with s. That is, the fact that it is possible for a probabilist to derive a χ^2_{m-1} probability density for $(m - 1)s^2/\sigma^2$ if s^2 is based on m iid normal (μ, σ^2) random variables has been used extensively (beginning in Section 6-4) in this text. That density can in turn be used to find a density for s and thus the theoretical mean of s. As it turns out, although $Es^2 = \sigma^2$, the theoretical mean of s is not quite σ, but rather a multiple of σ (for a given sample size m). The constant of proportionality is typically called c_4, and in symbols,

$$Es = c_4\sigma \tag{7-89}$$

or equivalently,

$$\sigma = \frac{Es}{c_4} \tag{7-90}$$

It is possible to write out an explicit expression for c_4:

$$c_4 = \sqrt{\frac{2}{m-1}} \left(\frac{\Gamma(m/2)}{\Gamma\left(\frac{m-1}{2}\right)} \right)$$

and values of c_4 for various m are given in Table D-2. From that table, it is easy to see that as a function of m, c_4 increases from about .8 when $m = 2$ to essentially 1 for large m.

The practical use made of the theoretical statements (7-89) and (7-90) is to think that the sample average of the sample standard deviations

$$\bar{s} = \frac{1}{r} \sum_{i=1}^{r} s_i \tag{7-91}$$

can be expected to approximate the theoretical mean (sample) standard deviation Es, so that (from statement (7-90)) a plausible estimate of σ becomes

$$\frac{\bar{s}}{c_4} \tag{7-92}$$

It is worth remarking that $\bar{s}$ is not the same as s_P, even when all sample sizes are the same. s_P is derived by averaging sample variances and then taking a square root. $\bar{s}$ comes from taking the square roots of the sample variances and then averaging. In general, these two orders of operation do not produce the same results.

In any case, commonly used retrospective control limits for $\bar{x}$ are obtained by substituting $\bar{\bar{x}}$ given in formula (7-81) for μ and either of the estimates of σ given in displays (7-88) or (7-92) for σ in the formulas (7-84). Further, an "as past data" center line for an $\bar{x}$ chart is typically set at $\bar{\bar{x}}$.

EXAMPLE 7-11
(continued)

Consider the retrospective $\bar{x}$ control charting of the ream cutter data. Using the column totals given in Table 7-18, one finds from formulas (7-81), (7-87), and (7-91) that here

$$\bar{\bar{x}} = \frac{183.4}{22} = 8.3$$

$$\bar{R} = \frac{109}{22} = 4.95$$

$$\bar{s} = \frac{44.41}{22} = 2.019$$

Then, consulting Table D-2 with a sample size of $m = 5$, one finds that $d_2 = 2.326$, so an estimate of σ based on $\bar{R}$ is (from expression (7-88))

$$\frac{\bar{R}}{d_2} = \frac{4.95}{2.326} = 2.13$$

Also, Table D-2 shows that for a sample size of $m = 5$, $c_4 = .9400$, so an estimate of σ based on $\bar{s}$ is (from expression (7-92))

$$\frac{\bar{s}}{c_4} = \frac{2.019}{.94} = 2.15$$

(The reader might find it interesting at this point to calculate s_P and compare it to these values. Beginning from the standard deviations in Table 7-18, this author got $s_P = 2.19$.)

Using (for example) statistic (7-88), one is thus led to substitute 8.3 for μ and 2.13 for σ in "standards given" formulas (7-84) to obtain the retrospective limits

$$LCL_{\bar{x}} = 8.3 - 3\frac{2.13}{\sqrt{5}} = 5.44 \quad \text{and} \quad UCL_{\bar{x}} = 8.3 + 3\frac{2.13}{\sqrt{5}} = 11.16$$

Figure 7-17 shows an "as past data" Shewhart $\bar{x}$ control chart for the ream cutter data, using limits based on $\bar{R}$.

Notice the contrast between the pictures of the ream cutter performance given by Figures 7-16 and 7-17. Figure 7-16 shows clearly that process parameters are not at their standard values, but Figure 7-17 shows that it is perhaps plausible to think of the data in Table 7-17 as coming from *some* stable data-generating mechanism. The observed $\bar{x}$'s hover nicely (indeed — as will be argued at the end of the next section — perhaps too nicely) about a central value, showing no "out of control" points or obvious trends. That hypothesis (7-80) is at least approximately true is believable on the basis of Figure 7-17.

Several comments should be made before turning to a discussion of other Shewhart control charts for measurements. First, note the similarity in both spirit and graphical display between the retrospective Shewhart $\bar{x}$ chart and the ANOM test presented in the previous section. There are differences in exactly how the decision limits are set, and the time order between samples inherent in Shewhart charting is not necessarily

FIGURE 7-17

Retrospective Shewhart
$\bar{x}$ Control Chart for Cut
Sheet Lengths

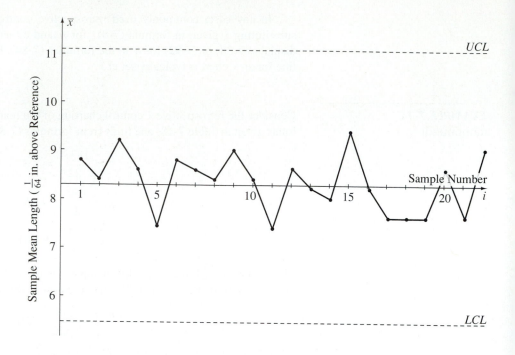

present when using Ott's ANOM test. But the notion of looking at plotted differences between sample means and a grand mean, as compared to some limits of plausible random variation, is common to the methods. And the similarity between the pervasive Shewhart $\bar{x}$ chart and ANOM is often cited as a strong argument for the use of the ANOM in place of ANOVA methods. Once one is familiar with $\bar{x}$ charts (as many working engineers are), it is a relatively small step to understand ANOM.

A second comment concerning the $\bar{x}$ charts is the importance of remembering that what is represented on an $\bar{x}$ chart is behavior (both expected and observed) of sample means, *not* individual measurements. It is unfortunately all too common to see engineering specifications (which refer to individual measurements) marked on $\bar{x}$ control charts either in place of, or in addition to, proper control limits. But how sample means compare to specifications for individual measurements tells nothing about either the stability of the process as represented in the means or the acceptability of individual measurements according to the stated engineering requirements. It is simply bad practice to mix (or mix up) control limits and specifications.

A third comment has to do with the fairly arbitrary choice of 3-sigma control limits in formulas (7-84). A legitimate question is, "Why not 2-sigma or 2.5-sigma or 3.09-sigma limits?" There is no completely convincing theoretical answer to this question. Indeed, arguments in favor of other multiples than 3 for use in formulas (7-84) are heard from time to time. But the forces of historical precedent and many years of successful application combine to make the use of 3-sigma limits nearly universal.

As a final comment regarding $\bar{x}$ charts, the basic "standards given" formulas for control limits (7-84) are sometimes combined with formula (7-88) or (7-92) for estimating σ, and $\bar{\bar{x}}$ is put in place of μ to obtain formulas for retrospective control limits for $\bar{x}$. For example, using the estimate of σ in display (7-88), one obtains the formulas

$$LCL_{\bar{x}} = \bar{\bar{x}} - 3\frac{\bar{R}}{d_2\sqrt{m}} \qquad \text{and} \qquad UCL_{\bar{x}} = \bar{\bar{x}} + 3\frac{\bar{R}}{d_2\sqrt{m}} \qquad \textbf{(7-93)}$$

In fact, it is standard practice to use the abbreviation

$$A_2 = \frac{3}{d_2 \sqrt{m}}$$

and rewrite the limits in formulas (7-93) as

$$LCL_{\bar{x}} = \bar{\bar{x}} - A_2 \bar{R} \qquad \text{and} \qquad UCL_{\bar{x}} = \bar{\bar{x}} + A_2 \bar{R} \qquad \qquad \textbf{(7-94)}$$

Values of A_2 are given along with the other control chart constants in Table D-2. It is worthwhile to verify that the use of formulas (7-94) in the context of Example 7-11 produces exactly the numerical retrospective control limits for $\bar{x}$ found earlier.

The version of retrospective $\bar{x}$ chart limits related to the estimate of σ in display (7-92) is

$$LCL_{\bar{x}} = \bar{\bar{x}} - 3\frac{\bar{s}}{c_4 \sqrt{m}} \qquad \text{and} \qquad UCL_{\bar{x}} = \bar{\bar{x}} + 3\frac{\bar{s}}{c_4 \sqrt{m}} \qquad \qquad \textbf{(7-95)}$$

It is also standard practice to use the abbreviation

$$A_3 = \frac{3}{c_4 \sqrt{m}}$$

and rewrite the limits in display (7-95) as

$$LCL_{\bar{x}} = \bar{\bar{x}} - A_3 \bar{s} \qquad \text{and} \qquad UCL_{\bar{x}} = \bar{\bar{x}} + A_3 \bar{s} \qquad \qquad \textbf{(7-96)}$$

Values of A_3 are given in Table D-2.

Control Charts for Ranges The $\bar{x}$ control chart is aimed primarily at monitoring the constancy of the average process response, μ, over time. It deals only indirectly with the process short-term variation σ. (If σ increases beyond a standard value, it will produce $\bar{x}_i$ more variable than expected and eventually trigger an "out of control" point. But such a possible change in σ is detected most effectively by directly monitoring the spread of samples taken from the process.) Thus, in applications, $\bar{x}$ charts are almost always accompanied by companion charts intended to monitor σ.

The conceptually simplest and most common Shewhart control charts for monitoring the process standard deviation are the R charts, the charts for sample ranges. In their "standards given" version, they are based again on the fact that it is possible for a probabilist to find a probability density for R based on m iid normal (μ, σ^2) random variables. Using this density, not only is it possible to show that $ER = d_2\sigma$, but the standard deviation of the probability distribution can be found as well, and it turns out (for a given m) to be proportional to σ. The constant of proportionality is called d_3 and is tabled for various m in Table D-2. That is, for R based on m iid Gaussian observations,

$$\sqrt{\text{Var}\,R} = d_3 \sigma \qquad \qquad \textbf{(7-97)}$$

Although the information about the theoretical distribution of R provided by formulas (7-85) and (7-97) is admittedly somewhat scanty, it is enough to suggest possible "standards given" 3-sigma (of R) control limits for R. A plausible "center" line for a "standards given" R chart is at $ER = d_2\sigma$, and (using formula (7-97)) control limits are

$$LCL_R = ER - 3\sqrt{\text{Var}\,R} = d_2\sigma - 3d_3\sigma = (d_2 - 3d_3)\sigma \qquad \qquad \textbf{(7-98)}$$

$$UCL_R = ER + 3\sqrt{\text{Var}\,R} = (d_2 + 3d_3)\sigma \qquad \qquad \textbf{(7-99)}$$

The limit indicated in formula (7-98) turns out to be negative for $m \leq 6$; for those sample sizes, since ranges are nonnegative, no lower control limit is used. Formulas (7-98) and

(7-99) are typically simplified by the introduction of yet more notation. That is, standard quality control usage is to let

$$D_1 = (d_2 - 3d_3) \qquad \text{and} \qquad D_2 = (d_2 + 3d_3)$$

and rewrite formulas (7-98) and (7-99) as

$$LCL_R = D_1\sigma \qquad \text{and} \qquad UCL_R = D_2\sigma \qquad \text{(7-100)}$$

Like the other control chart constants, D_1 and D_2 appear in Table D-2 for various sample sizes m. Note that for $m \leq 6$, there is no tabled value for D_1, as no lower limit is in order.

EXAMPLE 7-11
(continued)

Consider a "standards given" control chart analysis for the sheet length ranges given in Table 7-18, using a standard $\sigma = 1.9$ ($\frac{1}{64}$ in.). Since samples of size $m = 5$ are involved, Table D-2 shows that $d_2 = 2.326$ and $D_2 = 4.918$ are appropriate for use in establishing a "standards given" control chart for R. The center line should be drawn at

$$d_2\sigma = 2.326(1.9) = 4.4$$

and the upper control limit should be set at

$$D_2\sigma = 4.918(1.9) = 9.3$$

(Since $m \leq 6$, no lower control limit will be used.) Figure 7-18 shows a "standards given" control chart for ranges of the sheet lengths. It is clear from the figure that for the most part, a constant process standard deviation of $\sigma = 1.9$ is plausible, except for the clear indication to the contrary at sample 22. The 22nd observed range, $R = 12$, is simply larger than expected based on a sample of size $m = 5$ from a normal distribution with $\sigma = 1.9$. In practice, it would be in order to undertake a physical search for the cause of the apparent increase in process variability associated with the last sample taken.

FIGURE 7-18

"Standards Given" Shewhart R Chart for Cut Sheet Lengths

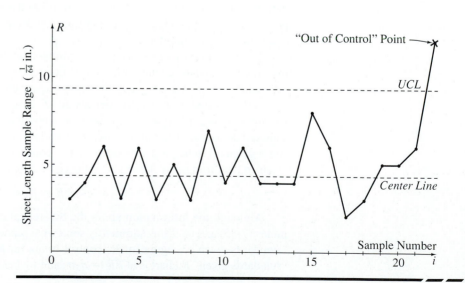

As was the case for $\bar{x}$ charts, combination of formulas for the estimation of (supposedly constant) process parameters with the "standards given" limits (7-100) produces retrospective control limits for R charts. For example, basing an estimate of σ on $\bar{R}$ as in display (7-88), one is led (not too surprisingly) to a retrospective center line for R at

$d_2(\bar{R}/d_2) = \bar{R}$ and retrospective control limits

$$LCL_R = \frac{D_1\bar{R}}{d_2} \quad \text{and} \quad UCL_R = \frac{D_2\bar{R}}{d_2} \tag{7-101}$$

The abbreviations

$$D_3 = \frac{D_1}{d_2} \quad \text{and} \quad D_4 = \frac{D_2}{d_2}$$

are commonly used, and limits (7-101) are written as

$$LCL_R = D_3\bar{R} \quad \text{and} \quad UCL_R = D_4\bar{R} \tag{7-102}$$

Values of the constants D_3 and D_4 are found in Table D-2 for various m.

EXAMPLE 7-11
(continued)

For the ream cutter data, $\bar{R} = \frac{109}{22}$, so retrospective control limits for ranges of the type (7-102) put a center line at

$$\bar{R} = 4.95$$

and since for $m = 5, D_4 = 2.115$,

$$UCL_R = 2.115\left(\frac{109}{22}\right) = 10.5$$

Look again at Figure 7-18 and note that the use of these retrospective limits (instead of the $\sigma = 1.9$ standards given limits of Figure 7-18) does not materially alter the appearance of the plot. The range for sample 22 still plots above the upper control limit. It is not plausible that a single σ stands behind all of the 22 plotted ranges (not even $\sigma \approx \bar{R}/d_2 = 2.13$). It is pretty clear that a different physical mechanism must have been acting at sample 22 than was operative earlier.

For pedagogical reasons, $\bar{x}$ charts were considered here before turning to charts aimed at monitoring σ. In terms of order of attention in an application, however, R (or s) charts are traditionally (and correctly) given first priority. They deal directly with the baseline component of process variation. Thus (so conventional wisdom goes), if they show lack of stability, there is little reason to go on to considering the behavior of means (which deals primarily with the long-term component of process variation) until appropriate physical changes bring the ranges (or standard deviations) to the place of repeatability.

Control Charts for Standard Deviations

Less common but nevertheless important alternatives to range charts are control charts for standard deviations, s. In their "standards given" version, they are based on the fact that it is possible to find a probability density for s calculated from m iid normal (μ, σ^2) random variables. Using this density, not only can one show that $Es = c_4\sigma$, but it is also possible to find the standard deviation of s. It turns out that

$$\sqrt{Var\, s} = \sqrt{1 - c_4^2}\, \sigma \tag{7-103}$$

Then formulas (7-89) and (7-103) taken together yield "standards given" 3-sigma control limits for s. That is, with a center line at $c_4\sigma$, one employs the limits

$$LCL_s = c_4\sigma - 3\sqrt{1 - c_4^2}\, \sigma = \left(c_4 - 3\sqrt{1 - c_4^2}\right)\sigma$$

$$UCL_s = c_4\sigma + 3\sqrt{1 - c_4^2}\, \sigma = \left(c_4 + 3\sqrt{1 - c_4^2}\right)\sigma$$

Standard notation is to let

$$B_5 = \left(c_4 - 3\sqrt{1 - c_4^2}\right) \qquad \text{and} \qquad B_6 = \left(c_4 + 3\sqrt{1 - c_4^2}\right)$$

so, ultimately, "standards given" control limits for s become

$$LCL_s = B_5\sigma \qquad \text{and} \qquad UCL_s = B_6\sigma \qquad\qquad \text{(7-104)}$$

As expected, the constants B_5 and B_6 are tabled in Table D-2. For $m \leqslant 5, c_4 - 3\sqrt{1 - c_4^2}$ turns out to be negative, so no value is shown in Table D-2 for B_5, and no lower control limit for s is typically used for such sample sizes.

EXAMPLE 7-11
(continued)

Returning once more to the ream cutter example of Shervheim and Snider, consider the monitoring of σ through the use of sample standard deviations rather than ranges, based on a standard of $\sigma = 1.9$ ($\frac{1}{64}$ in.). Table D-2 with sample size $m = 5$ once again gives $c_4 = .9400$ and also shows that $B_6 = 1.964$. So an s chart for the data of Table 7-17 has a center line at

$$c_4\sigma = (.94)(1.9) = 1.79$$

and an upper control limit at

$$UCL_s = B_6\sigma = 1.964(1.9) = 3.73$$

and, since the sample size is only 5, no lower control limit.

Figure 7-19 is a "standards given" Shewhart s chart for the s values given in Table 7-18. The story told by Figure 7-19 is essentially identical to that conveyed by the range chart in Figure 7-18. Only at sample 22 does the hypothesis that $\sigma = 1.9$ become untenable, and the need for physical intervention is indicated there.

FIGURE 7-19

"Standards Given" s Chart for Cut Sheet Lengths

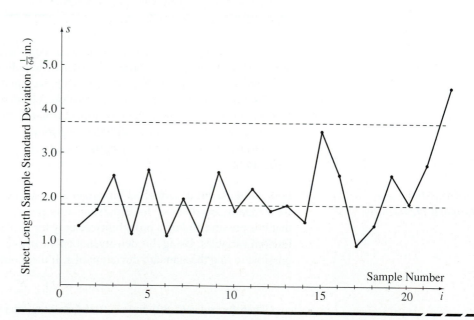

As was the case for $\bar{x}$ and R charts, retrospective control limits for s can be had by replacing the parameter σ in the "standards given" limits (7-104) with any appropriate estimate. The most common way of proceeding in the present situation is to employ the estimate $\bar{s}/c_4$ and thus end up with a retrospective center line for an s chart at $c_4(\bar{s}/c_4) = \bar{s}$ and retrospective control limits

$$LCL_s = \frac{B_5\bar{s}}{c_4} \quad \text{and} \quad UCL_s = \frac{B_6\bar{s}}{c_4} \tag{7-105}$$

And using the abbreviations

$$B_3 = \frac{B_5}{c_4} \quad \text{and} \quad B_4 = \frac{B_6}{c_4}$$

the retrospective limits (7-105) are written as

$$LCL_s = B_3\bar{s} \quad \text{and} \quad UCL_s = B_4\bar{s} \tag{7-106}$$

Values of the constants B_3 and B_4 are given in Table D-2.

EXAMPLE 7-11
(continued)

For the ream cutter data, $\bar{s} = \frac{44.41}{22} = 2.02$, so retrospective control limits for standard deviations of the type (7-106) put a center line at

$$\bar{s} = 2.02$$

and, since $B_4 = 2.089$ for $m = 5$,

$$UCL_s = 2.089 \left(\frac{44.41}{22} \right) = 4.22$$

Look again at Figure 7-19 and verify that the use of these retrospective limits (instead of the $\sigma = 1.9$ "standards given" limits) wouldn't much change the appearance of the plot. As was the case for the retrospective R chart analysis, these retrospective s chart limits still put sample 22 in a class by itself, suggesting that a different physical mechanism produced it than that which led to the other 21 samples.

Ranges are easier to calculate than standard deviations and are certainly easier to explain as well. As a result, R charts are more popular than s charts. In fact, R charts are so common that the phrase "$\bar{x}$ and R charts" is typically spoken in quality control circles in such a way that the $\bar{x}/R$ pair is almost implied to be a single inseparable entity. However, statistical theory supports the view that when computational problems and conceptual understanding are not issues, s charts are preferable to R charts because of their superior sensitivity to changes in σ. So even though the R and s charts give nearly identical pictures of cutting process behavior in the example here, and notwithstanding that the R charts are so popular, this author considers the s chart notion easily worth including in this short discussion of Shewhart control charts.

A useful final observation about the s chart idea is that for r-sample statistical engineering studies where all sample sizes are the same, the "as past data" control limits in display (7-106) can provide some rough help in the model-checking activities of Section 7-1 as regards the "single variance" assumption of the one-way model. $B_3\bar{s}$ and $B_4\bar{s}$ can be treated as rough limits on the variation in sample standard deviations deemed to be consistent with the one-way model's assumption of a single variance.

EXAMPLE 7-12

(Example 7-1 revisited) **s Chart Control Limits and the "Equal Variances" Assumption in the Concrete Strength Study.** In the concrete compressive strength study of Armstrong,

Babb, and Campen, the $r = 8$ sample standard deviations based on samples of size $m = 3$ given in Table 7-3 have $\bar{s} = 534.8$ psi. Then for $m = 3$, $B_4 = 2.568$, and so

$$B_4\bar{s} = 2.568(534.8) = 1,373 \text{ psi}$$

The largest of the eight values s_i in Table 7-3 is 965.6, and there are thus no "out of control" standard deviations. So as in Section 7-1, no strong evidence against the relevance of the "single variance" model assumption is discovered here.

Control Charts for Measurements and Industrial Process Improvement

The $\bar{x}$ and R (or $\bar{x}$ and s) control chart combination is an important engineering tool for the improvement of manufacturing processes. U.S. companies have trained literally hundreds of thousands of workers in the making of Shewhart $\bar{x}$ and R charts over the past few years, hoping for help in meeting the challenge of international competition. The record of success produced by this training effort is mixed. It is thus worth pausing briefly to reflect on what aid the tools of this section can and cannot rationally be expected to provide in the effort to improve industrial processes.

In the first place, it is clear that warnings of assignable variation provided by Shewhart control charts are helpful in reducing the variation of an industrial process only to the extent that they are acted on in a timely and competent fashion. If "out of control" signals don't lead to appropriate physical investigation and action to eliminate assignable causes, they contribute nothing toward improved process behavior. If workers collect data to be archived away on $\bar{x}$ and R chart forms and do not have the authority, skills, or motivation to intervene intelligently when excess process variation is indicated, they are engaged in a futile activity.

Control charts can signal the need for process intervention. But perhaps nearly as important is the fact that they also tell a user when not to be alarmed at observed variation and give in to the temptation to adjust a stable process. This is the other side of the intervention coin. Inadvisably adjusting an industrial process that is subject only to common or random causes degrades its behavior rather than improving it. Rational use of Shewhart control charts can help prevent this possibility.

It is also important to say that even when properly made and acted on, Shewhart control charts can do only so much towards the improvement of industrial processes. They can be a tool for helping to reduce variation to the minimum possible for a given system configuration (in terms of equipment, methods of operation, etc.). But once that minimum has been reached, all that Shewhart charting does is to help maintain that configuration's best performance — to maintain the "baseline variation only" situation corresponding to the *status quo* way of doing things.

In a modern world economy, however, it is increasingly obvious that companies cannot hope to be leaders in their industries by being content simply to maintain stable, *status quo* methods of operation. Instead, ways must be found for improving beyond today's methods for tomorrow. This requires thought and, often, engineering experimentation. The philosophies and methods of experimental design and engineering data collection and analysis discussed in this book have an important role in that quest for improvement beyond today's best industrial methodology. But the particular role of control charting in such efforts is only indirect. By using control charts and bringing a current process to stability, one provides a basis or foundation for improvement through experimentation and reconfiguration. Indeed, it can be argued fairly convincingly that unless an existing process is repeatable, there is no sensible way of evaluating the impact of experimental changes one might make in it, trying to find tomorrow's improved version of the process. It is important to realize, however, that the Shewhart control charts provide only the foundation rather than the necessary subject matter expertise or statistical tools needed to guide the experimental search for improved ways of doing things.

7-6 Shewhart Control Charts for Qualitative and Count Data

The previous section discussed Shewhart $\bar{x}$, R, and s control charts, treating them as tools for studying the stability of a system over time. This section focuses on how the Shewhart control charting idea can be applied to attributes data.

The discussion begins with p charts. Next are introduced u charts and their specialization to the case of a constant-size inspection unit, the c charts. Finally, consideration is given to a number of common nonrandom patterns that can appear on both variables control charts and attributes control charts. Possible physical causes for them and some formal rules that are often recommended for automating their recognition are discussed.

p Charts

This text has consistently indicated that when they are available, measurements are generally preferable to attributes data. But in some situations, the only available information on the stability of a process takes the form of qualitative or count data. Consideration of the topic of control charting in such situations will begin here, with cases where what is available for plotting are sample fractions, $\hat{p}_i$, of items produced by a process having some attribute of interest. The most common use of this is where $\hat{p}_i$ is the fraction of a sample of n_i items that is nonconforming according to some engineering standard or specification. So this section will use the "fraction nonconforming" language, in spite of the fact that $\hat{p}_i$ can be the sample fraction having any attribute of interest (desirable, undesirable, or indifferent).

The basic probability facts supporting control charting for the fraction nonconforming are exactly those that were used in Section 6-5 to develop methods of confidence interval estimation and hypothesis testing based on $\hat{p}$. That is, if a process being monitored is stable over time, each $n_i\hat{p}_i$ is usefully modeled as binomial (n_i, p), where p is a constant likelihood that any sampled item is nonconforming. (In contrast to the treatment of the control charts for measurements, this section will explicitly allow for sample sizes n_i varying in time. Charts for measurements are almost always based on fairly small but constant sample sizes. But charts for attributes data typically involve larger sample sizes that sometimes vary.)

As in Section 6-5, a binomial model for $n_i\hat{p}_i$ leads immediately to a mean and standard deviation for $\hat{p}_i$:

$$E\hat{p}_i = p \tag{7-107}$$

and

$$\sqrt{Var\,\hat{p}_i} = \sqrt{\frac{p(1-p)}{n_i}} \tag{7-108}$$

But then formulas (7-107) and (7-108) suggest obvious "standards given" 3-sigma control limits for the sample "fraction nonconforming" $\hat{p}_i$. That is, if p is a standard likelihood that any single item in a sample is nonconforming, then a "standards given" p chart has a center line at p and control limits

$$LCL_{\hat{p}_i} = p - 3\sqrt{\frac{p(1-p)}{n_i}} \tag{7-109}$$

$$UCL_{\hat{p}_i} = p + 3\sqrt{\frac{p(1-p)}{n_i}} \tag{7-110}$$

In the event that formula (7-109) produces a negative value, no lower control limit is used.

EXAMPLE 7-13

p **Chart Monitoring of a Pelletizing Process.** Kaminski, Rasavahn, Smith, and Weitekamper worked on the same pelletizing process already used as an example several times in this book. (See Examples 1-2, 3-14, 5-4, and 6-18.) Extensive data collection on two different days led the students to establish $p = .61$ as a standard rate of nonconforming tablets produced by the process, when run under a shop standard operating regimen. On a third day, the students took $r = 25$ samples of $n_1 = n_2 = \cdots = n_{25} = m = 30$ consecutive pellets at intervals as they came off the machine and plotted sample "fractions nonconforming" $\hat{p}_i$ on a "standards given" p chart made with $p = .61$. The resulting data are given in Table 7-19.

TABLE 7-19

Numbers and Fractions of Nonconforming Pellets in 25 Samples of Size 30

i Sample	$n_i\hat{p}_i$ Number Nonconforming	$\hat{p}_i$
1	13	.43
2	12	.40
3	9	.30
4	15	.50
5	17	.57
6	13	.43
7	20	.67
8	18	.60
9	18	.60
10	16	.53
11	15	.50
12	17	.57
13	15	.50
14	20	.67
15	10	.33
16	12	.40
17	17	.57
18	14	.47
19	16	.53
20	10	.33
21	14	.47
22	13	.43
23	17	.57
24	10	.33
25	12	.40
$\sum n_i\hat{p}_i = 363$		

For samples of size $n_i = m = 30$, 3-sigma "standards given" p chart control limits are, from formulas (7-109) and (7-110),

$$LCL_{\hat{p}_i} = .61 - 3\sqrt{\frac{(.61)(1 - .61)}{30}} = .34$$

$$UCL_{\hat{p}_i} = .61 + 3\sqrt{\frac{(.61)(1 - .61)}{30}} = .88$$

and a center line at .61 is appropriate. Figure 7-20 is such a "standards given" p chart for the data of Table 7-19.

FIGURE 7-20

"Standards Given" p Chart for Nonconforming Pellets

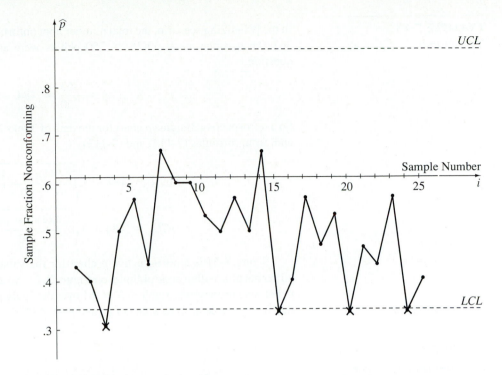

The fact that four $\hat{p}_i$ values plot below the lower control limit in Figure 7-20, and the fact that the $\hat{p}_i$ values run consistently below the chart's center line, make untenable the hypothesis that the pelletizing process was stable at the standard value of 61% nonconforming on the day these data were gathered. By the way, notice that in this example, points plotting "out of control" on the low side are an indication of process *improvement*. They nevertheless represent a circumstance warranting physical attention, to determine the physical cause for the reduced fraction defective and hopefully to learn how to make the improvement permanent.

To make retrospective limits for a p chart, one must settle on a method of estimating the (supposedly constant) process parameter p. Here the pooling idea introduced in the two-sample context of Section 6-5 will be used. That is, as a direct extension of formula (6-69) of Section 6-5, let

$$\hat{p} = \frac{n_1 \hat{p}_1 + n_2 \hat{p}_2 + \cdots + n_r \hat{p}_r}{n_1 + n_2 + \cdots + n_r} \qquad \textbf{(7-111)}$$

($\hat{p}$ is the total number nonconforming divided by the total number inspected; when sample sizes vary, it is a weighted average of the $\hat{p}_i$.)

With $\hat{p}$ as in formula (7-111), an "as past data" Shewhart p chart has a center line at $\hat{p}$ and

$$LCL_{\hat{p}_i} = \hat{p} - 3\sqrt{\frac{\hat{p}(1-\hat{p})}{n_i}} \qquad \textbf{(7-112)}$$

$$UCL_{\hat{p}_i} = \hat{p} + 3\sqrt{\frac{\hat{p}(1-\hat{p})}{n_i}} \qquad \textbf{(7-113)}$$

As in the "standards given" context, when formula (7-112) produces a negative value, no lower control limit is used for $\hat{p}_i$.

EXAMPLE 7-13
(continued)

In the pelletizing scenario, the total number nonconforming in the samples was $\sum n_i \hat{p}_i = 363$. Then, since $mr = 30(25) = 750$ pellets were actually inspected on the day in question,

$$\hat{p} = \frac{363}{750} = .484$$

So a retrospective 3-sigma p chart for the data of Table 7-19 has a center line at $\hat{p} = .484$ and, from formulas (7-112) and (7-113),

$$LCL_{\hat{p}_i} = .484 - 3\sqrt{\frac{(.484)(1-.484)}{30}} = .21$$

$$UCL_{\hat{p}_i} = .484 + 3\sqrt{\frac{(.484)(1-.484)}{30}} = .76$$

Figure 7-21 is a retrospective p chart for the situation of Kaminski et al. Clearly, all points plot within control limits on Figure 7-21. So although it is not tenable that the pelletizing process was stable at $p = .61$ over the study period, it is completely plausible that it was stable at some value of p (and $\hat{p} = .484$ is a sensible guess for that value).

FIGURE 7-21

Retrospective p Chart for
Nonconforming Pellets

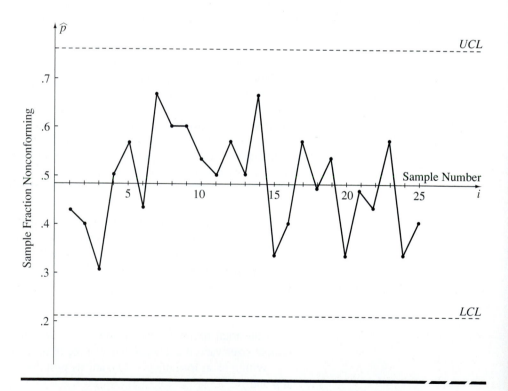

Because of the inherent limitations of categorical data in engineering contexts, little more will be said in this book about formal inference based on sample fractions beyond that which was already said in Section 6-5. For example, formal significance tests of equality of r proportions, parallel to the tests of equality of r means presented in Section 7-4, won't be discussed. However, note that in rough terms, just as retrospective $\bar{x}$ charts serve as a kind of graphical test of $H_0: \mu_1 = \mu_2 = \cdots = \mu_r$, and retrospective R

and s charts can be thought of as tools for judging the reasonableness of H_0: $\sigma_1 = \sigma_2 = \cdots = \sigma_r$, the retrospective p chart can be interpreted as a graphical tool for judging how sensible the hypothesis H_0: $p_1 = p_2 = \cdots = p_r$ appears.

u Charts

Section 3-4 introduced the notation $\hat{u}$ for the ratio of the number of occurrences of a phenomenon of interest to the total number of inspection units or items sampled, in contexts where there may be multiple occurrences on a given item or inspection unit. The most common kind of control charting application involving such ratios is the case where the phenomenon of interest is nonconformance to some engineering standard or specification. This section will use the terminology of nonconformances per unit, in spite of the fact that $\hat{u}$ can be the sample occurrence rate for any type of phenomenon (desirable, undesirable, or indifferent).

The theoretical basis for control charting based on nonconformances per unit is found in the Poisson distributions of Section 5-1. That is, if for some specified inspection unit or unit of process output of a given size, a physically stable process has an associated mean nonconformances per unit of λ, and

X_i = the number of nonconformances observed on k_i units inspected at time i

then a reasonable model for X_i is often the Poisson distribution with mean $k_i\lambda$. Then the material in Section 5-1 produces the fact that both $EX_i = k_i\lambda$ and $Var\, X_i = k_i\lambda$.

But notice that if $\hat{u}_i$ is the sample nonconformances per unit observed at period i,

$$\hat{u}_i = \frac{X_i}{k_i}$$

so Proposition 5-1 can be applied to produce a sensible mean and standard deviation for $\hat{u}_i$. That is,

$$E\hat{u}_i = E\frac{X_i}{k_i} = \frac{1}{k_i}EX_i = \frac{1}{k_i}(k_i\lambda) = \lambda \tag{7-114}$$

$$Var\,\hat{u}_i = Var\,\frac{X_i}{k_i} = \frac{1}{k_i^2}Var\,X_i = \frac{1}{k_i^2}(k_i\lambda) = \frac{\lambda}{k_i}$$

so

$$\sqrt{Var\,\hat{u}_i} = \sqrt{\frac{\lambda}{k_i}} \tag{7-115}$$

The relationships (7-114) and (7-115) then form the basis for "standards given" 3-sigma control limits for $\hat{u}_i$. That is, if λ is a standard mean nonconformances per unit, then a "standards given" u chart has a center line at λ and

$$LCL_{\hat{u}_i} = \lambda - 3\sqrt{\frac{\lambda}{k_i}} \tag{7-116}$$

$$UCL_{\hat{u}_i} = \lambda + 3\sqrt{\frac{\lambda}{k_i}} \tag{7-117}$$

The difference in formula (7-116) can turn out negative; when it does, no lower control limit is used.

Another matter of notation must be discussed at this point. It has to do with the use of the symbol λ in the present context. λ is the symbol commonly used (as in Section 5-1) for a Poisson mean, and this fact is the basis for the usage here. However, it is more common in statistical quality control circles to use c or even c' for a standard mean nonconformances per unit. In fact, the case of the u chart where all k_i are the same (and thus one may as well define a standard inspection unit to be k_i times the original unit and λ to be k_i times the original λ, making all "k's" equal to 1) is usually referred to as a c chart. This author believes the λ notation used here represents the path of least confusion through this notational conflict and thus will not use c or c'. However, be aware that at least in the quality control world, there is a more popular alternative to the present λ convention.

When the limits (7-116) and (7-117) are used with nonconformances per unit data, one is essentially checking whether the prespecified λ is a plausible description of a physical process at each time period covered by the data. Often, however, there is no obvious standard occurrence rate λ, and the question to be addressed by the (retrospective) u charting is whether the process appears stable. That is, the question is whether or not it seems plausible that some (single) λ describes the process over all time periods covered by the data. What is needed, in order to modify formulas (7-116) and (7-117) to produce retrospective control limits for such cases, is a way to use the $\hat{u}_i$ to make a single estimate of a supposedly constant λ. This text's approach to this problem is to make an estimate exactly analogous to the pooled estimate of p in formula (7-111). That is, let

$$\hat{u} = \frac{k_1\hat{u}_1 + k_2\hat{u}_2 + \cdots + k_r\hat{u}_r}{k_1 + k_2 + \cdots + k_r} \tag{7-118}$$

that is, $\hat{u}$ is the total number of nonconformances observed divided by the total number of units inspected. Then combining formula (7-118) with limits (7-116) and (7-117), a retrospective 3-sigma u chart has a center line at $\hat{u}$ and

$$LCL_{\hat{u}_i} = \hat{u} - 3\sqrt{\frac{\hat{u}}{k_i}} \tag{7-119}$$

$$UCL_{\hat{u}_i} = \hat{u} + 3\sqrt{\frac{\hat{u}}{k_i}} \tag{7-120}$$

As the reader might by now expect, when formula (7-119) gives a negative value, no lower control limit is employed.

EXAMPLE 7-14

(Example 3-13 revisited) **u Chart Monitoring of the Defects per Truck Found at Final Assembly.** In his book *Statistical Quality Control Methods*, I. W. Burr discusses the use of u charts to monitor the performance of an assembly process at a station in a truck assembly plant. Part of Burr's data were given earlier in Table 3-19. Table 7-20 gives a (partially overlapping) $r = 30$ production days' worth of Burr's data. (The values were extrapolated from Burr's figures and the fact that truck production through November 20 was 95 trucks/day and was 130 trucks/day thereafter. Burr gives only $\hat{u}_i$ values, production rates, and the fact that all trucks produced were inspected.)

Consider the problem of control charting for these data. Since Burr gave no figure λ for the plant's standard errors per truck, this problem will be approached as one of making a retrospective u chart. Using formula (7-118), and the column totals from Table 7-20, one first has

$$\hat{u} = \frac{\sum k_i\hat{u}_i}{\sum k_i} = \frac{6,078}{3,445} = 1.764$$

So an "as past data" u chart will have a center line at 1.764 errors/truck. From formulas

TABLE 7-20

Numbers and Rates of
Nonconformances for a
Truck Assembly Process

i Sample	Date	k_i Trucks Produced	$k_i \hat{u}_i$ Errors Found	$\hat{u}_i$ Errors/Truck
1	11/4	95	114	1.20
2	11/5	95	142	1.50
3	11/6	95	146	1.54
4	11/7	95	257	2.70
5	11/8	95	185	1.95
6	11/11	95	228	2.40
7	11/12	95	327	3.44
8	11/13	95	269	2.83
9	11/14	95	167	1.76
10	11/15	95	190	2.00
11	11/18	95	199	2.09
12	11/19	95	180	1.89
13	11/20	95	171	1.80
14	11/21	130	163	1.25
15	11/22	130	205	1.58
16	11/25	130	292	2.25
17	11/26	130	325	2.50
18	11/27	130	267	2.05
19	11/29	130	190	1.46
20	12/2	130	200	1.54
21	12/3	130	185	1.42
22	12/4	130	204	1.57
23	12/5	130	182	1.40
24	12/6	130	196	1.51
25	12/9	130	140	1.08
26	12/10	130	165	1.27
27	12/11	130	153	1.18
28	12/12	130	181	1.39
29	12/13	130	185	1.42
30	12/16	130	270	2.08
		$\sum k_i = 3{,}445$	$\sum k_i \hat{u}_i = 6{,}078$	

(7-119) and (7-120), for the first 13 days (where each k_i was 95),

$$LCL_{\hat{u}_i} = 1.764 - 3\sqrt{\frac{1.764}{95}} = 1.355 \text{ errors/truck}$$

$$UCL_{\hat{u}_i} = 1.764 + 3\sqrt{\frac{1.764}{95}} = 2.173 \text{ errors/truck}$$

On the other hand, for the last 17 days (during which 130 trucks were produced each day),

$$LCL_{\hat{u}_i} = 1.764 - 3\sqrt{\frac{1.764}{130}} = 1.415 \text{ errors/truck}$$

$$UCL_{\hat{u}_i} = 1.764 + 3\sqrt{\frac{1.764}{130}} = 2.113 \text{ errors/truck}$$

Notice that since k_i appears in the denominator of the plus-or-minus part of control limit formulas (7-116), (7-117), (7-119), and (7-120), the larger the inspection effort at a given time period, the tighter the corresponding control limits. This is perfectly logical. A bigger "sample size" at a given period ought to make the corresponding $\hat{u}_i$ a more reliable indicator of λ, so less variation about a standard or estimated common value is tolerated.

Figure 7-22 is a retrospective u chart for the data of Table 7-20. The figure shows that the data-generating process behind Burr's data can in no way be thought of as stable or subject to only random causes. There is too much variation in the $\hat{u}_i$ to be explainable as due only to small unidentifiable causes. Some of the variation can probably be thought of in terms of a general downward trend, perhaps associated with workers gaining job skills. But even accounting for that, there is substantial erratic fluctuation of the $\hat{u}_i$ — which couldn't fit between control limits, no matter where they might be centered. These data simply represent a real engineering process that, according to accepted standards, is not repeatable enough to allow (without appropriate sleuthing and elimination of large causes of variation) anything but "one day at a time" inferences about its behavior.

FIGURE 7-22

Retrospective u Chart for Truck Assembly Errors

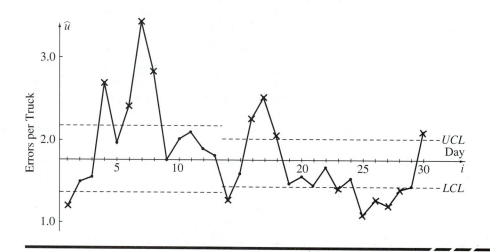

This book has had little to say about formal inference from data with an underlying Poisson distribution. But it should be obvious that retrospective u charts like the one in Example 7-14 can be thought of as rough graphical tests of the hypothesis H_0: $\lambda_1 = \lambda_2 = \cdots = \lambda_r$ for Poisson-distributed $X_i = k_i \hat{u}_i$.

Common Control Chart Patterns and Special Checks

Shewhart control charts (both those for measurements and those for attributes data) are useful not only because they supply semiformal information of a hypothesis-testing type; much important qualitative information is also carried by patterns that can sometimes be seen in the charts' simple plots. Section 3-3 included some comments about engineering information carried in plots of summary statistics against time. Shewhart charts are, of course, such plots — augmented with control limits. It is thus appropriate to amplify and extend those comments somewhat, in light of the extra element of information provided by the control limits.

In the first place, it should probably be explicitly stated what one expects to see on a 3-sigma control chart, if the "stable process" probability model supporting its use is appropriate, before discussing interesting possible departures from the expected. If a physically stable process lies behind the statistics plotted on a Shewhart control chart, it is expected (tacitly assuming the distribution of the plotted values to be mound-shaped) that 1) most plotted points will lie in the middle, (say, the middle $\frac{2}{3}$) of the region delineated by the control limits around the center line, 2) a few, (say, on the order of 1 in 20) points will lie outside this region but inside the control limits, 3) essentially no points will lie outside the control limits, and 4) there will be no obvious trends in time for any sizable part of the chart. That is, one expects to see a random scatter/white noise plot that fills, but essentially remains within, the region bounded by the control limits. When something other is seen, even if no points plot outside the control limits, there is reason to consider the possibility that something in addition to chance causes is active in the data-generating mechanism.

Repeated cyclical (often "up, then back down again") patterns sometimes show up on Shewhart control charts. Occasionally, these even remain contained between the control limits. But such behavior is not characteristic of plots resulting from underlying iid/stable process data-generating mechanisms. When it occurs, the alert engineer will look for identifiable physical causes of variation whose effects would come and go on about the same schedule as the ups and downs seen on the control chart. Sometimes cyclical patterns end up being associated with variables like daily temperature effects, which may be largely beyond a user's control. But at other times, they have to do with things like different (rotating) operators' slightly different methods of machine operation, which can be mostly eliminated via standardization, training, and awareness.

Again, the expectation is that points plotted on a Shewhart control chart should (over time) pretty much fill up but rarely plot outside the region delineated by control limits. This can be violated in two different ways, both of which suggest the need for engineering attention. In the first place, more variation than expected (like that evident on Figure 7-22), which produces multiple points outside the control limits, is often termed *instability*. And (after eliminating the possibility of a blunder in calculations) it is nearly airtight evidence of one or more unregulated and perhaps even unrecognized process variables having effects so large that they must be regulated. Such erratic behavior can sometimes be traced to material or components from several different suppliers having somewhat different physical properties and entering a production line in a mixed or haphazard order. Also, ill-advised operators may overadjust equipment (without any basis in control charting); this can take a fairly stable process and make it unstable.

Less variation than expected on a Shewhart chart presents an interesting quandary. Look again at Figure 7-17 and reflect on the fact that the plotted $\bar{x}$'s on that chart probably hug the center line more than they should if they in fact were independent, normal $\bar{x}$'s. They don't come close to filling up the region between the control limits. The reader's first reaction to this state of affairs might well be, "So what? Isn't small variation good?" Small variation is indeed a virtue, but when points on a control chart hug the center line, what one has is *unbelievably* small variation, which may conceal a blunder in calculation or (almost paradoxically) unnecessarily large but nonrandom variation.

In the first place, the simplest possible explanation of a plot like Figure 7-17 is that the process short-term variation, σ, has been overestimated — either because a standard σ is not applicable or because of some blunder in calculation or logic. Notice that using a value for σ that is bigger than what is really called for when making the limits

$$LCL_{\bar{x}} = \mu - 3\frac{\sigma}{\sqrt{m}} \qquad \text{and} \qquad UCL_{\bar{x}} = \mu + 3\frac{\sigma}{\sqrt{m}}$$

will spread the control limits too wide and produce an $\bar{x}$ chart that is insensitive to changes in μ — that is, insensitive to process instability. So this possibility is not one that ought to be taken lightly.

A more subtle possible source of unbelievably small variation on a Shewhart chart has to do with the (usually unwitting) mixing of several consistently different types of observations in the calculation of a single statistic that is naively thought to be representing only one kind of observation. This often happens when data are being taken from a production stream where multiple heads or cavities on a machine (or various channels of another type of multiple-channel process) are represented in a regular order in the stream. For example, items machined on heads 1, 2, and 3 of a machine might show up downstream in a production process in the order 1, 2, 3, 1, 2, 3, 1, 2, 3, etc. Then, if there is more difference between the different types of observations than there is within a given type, values of a single statistic calculated using observations of several types can be remarkably (and excessively) consistent.

Consider, for example, the possibility that a five-headed machine has heads that are detectably/consistently different. Suppose four of the five are perfectly adjusted and always produce conforming items, and the fifth is severely misadjusted and always produces nonconforming items. Although 20% of the items produced are nonconforming, a binomial distribution model with $p = .2$ will typically overpredict the variation that will be seen in $n_i \hat{p}_i$ for samples of items from this process. Indeed, samples of size $m = 5$ of consecutive items coming off this machine would have $\hat{p}_i = .2$, always. Clearly, no $\hat{p}_i$'s would approach p chart control limits.

Or in a measurement data context, with the same hypothetical five-headed machine, consider the possibility that four of the five heads always produce a part dimension at the target of 8 in., (plus or minus, say, .01 in.), whereas the fifth head is grossly misadjusted, always producing the dimension at 9 in. (plus or minus .01 in.). Then, in this exaggerated example, naive mixing together of the output of all five heads will produce ranges unbelievably stable at about 1 in. and sample means (of five consecutive pieces) unbelievably stable at about 8.2 in. But the super-stability is not a cause for rejoicing, but rather a cause for thought and investigation that could well lead to the physical elimination of the differences between the various mechanisms producing the data — in this case, the fixing of the faulty head.

The possibility of unnatural consistency on a Shewhart chart, brought on by more or less systematic sampling of detectably different data streams, is often called *stratification* in quality control circles. (The term *stratification* is used in a slightly different way by survey samplers, but that is another story and off the track of this discussion.) Although there is no way of verifying this suspicion, this author suspects that some form of stratification was indeed at work in the production of the ream cutter data of Shervheim and Snider and the $\bar{x}$ chart in Figure 7-17. For example, multiple blades set at not quite equal angles on a roller that cuts sheets (as sketched in Figure 7-23) could produce consistently different consecutive sheet lengths and unbelievably stable $\bar{x}$'s. Or even with only a single blade on the cutter roller, regular patterns in material tension, brought on by slight eccentricities of feeder rollers, could also produce consistent patterns in consecutive sheet lengths, and thus too much stability on the $\bar{x}$ chart.

FIGURE 7-23

Schematic of a Roller Cutter

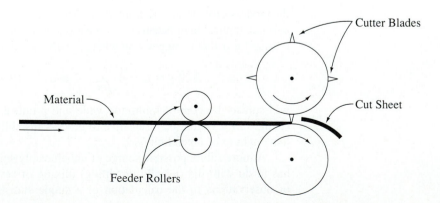

Other nonrandom patterns sometimes appearing on control charts include both *gradual* and more *sudden changes in level* and unabated *trends* up or down. Gradual changes in level can sometimes be traced to machine warmup phenomena, slow changeovers in raw material source, or introduction of operator training. And phenomena like tool wear and machine degradation over time will typically produce patterns of plotted points moving in a single direction until there is some sort of human intervention.

The terms *grouping* and *bunching* are used to describe irregular patterns on control charts where plotted points tend to come in sets of similar values, but where the pattern is neither regular/repeatable enough to be termed cyclical nor consistent enough in one direction to merit the use of the term *trend*. Such grouping can be brought about (for example) by calibration changes in a measuring instrument and, in machining processes, by fixture changes.

Finally, *runs* of many consecutive points on one side of a center line are sometimes seen on control charts. Figure 7-16, the "standards given" $\bar{x}$ chart for the sheet-length data of the last section, is an extreme example of a chart exhibiting a run. On "standards given" charts, runs (even when not accompanied by points plotting outside control limits) tend to discredit the chart's center line value as a plausible median for the distribution of the plotted statistic. On $\bar{x}$ charts, that translates to a discrediting of the target process mean as the value of the true process mean, thus indicating that the process is misaimed. (In the sheet-length situation of Figure 7-16, average sheet length is clearly below the target length.) And on a *p* or *u* chart, it indicates the inappropriateness of the supposedly standard rate of nonconforming items or nonconformances. On retrospective control charts, runs on one side of the center line are usually matched by runs on the other side, and one of the earlier terminologies (of cycles, trends or grouping) can typically be applied in addition to the term *runs*.

In recognition of the fact that the elementary "wait for a point to plot outside of control limits" mode of using control charts is blind to the various interpretable patterns discussed here, a variety of (relatively *ad hoc*) *special checks* have been developed, and applying them to Shewhart control charts is sometimes advocated. To give the reader the flavor of these checks for unnatural patterns, two of the most famous sets are shown in Tables 7-21 and 7-22. Besides many other different sets appearing in standard quality control books, companies making serious use of control charts often develop their own collections of such rules. The two sets given here are included more to show what is

TABLE 7-21

Western Electric Alarm Rules (from the *AT&T Quality Control Handbook*)

- A single point outside 3-sigma limits
- 2 out of any 3 successive points outside 2-sigma limits on one side of the center line
- 4 out of any 5 successive points outside 1-sigma limits on one side of the center line
- 8 consecutive points on one side of the center line

TABLE 7-22

Alarm Rules of L.S. Nelson (from the *Journal of Quality Technology*)

- a single point outside 3-sigma limits
- 9 points in a row on one side of the center line
- 6 points in a row increasing or decreasing
- 14 points in a row alternating up and down
- 2 out of any 3 successive points outside 2-sigma limits on one side of the center line
- 4 out of any 5 successive points outside 1-sigma limits on one side of the center line
- 15 points in a row inside 1-sigma limits
- 8 points in a row with none inside 1-sigma limits

possible than to advocate them in particular. The real bottom line of this discussion is simply that when used judiciously (overinterpretation of control chart patterns is a real temptation that also must be avoided), the qualitative information carried by patterns on Shewhart control charts can be an important engineering tool.

■ EXERCISES

7-1. Return again to the data of Example 4-1. These may be viewed as simply $r = 5$ samples of $m = 3$ densities. (For the time being, ignore the fact that the pressure variable is quantitative and that curve fitting seems a most natural method of analysis to apply to these data.) Use the material of Chapter 7 and do the following.
(a) (§7-1) Compute and make a normal plot for the residuals for the one-way model. What does the plot indicate about the appropriateness of the one-way model assumptions here?
(b) (§7-1) Using the five samples, find s_P, the pooled estimate of σ. What does this value measure? Give a two-sided 90% confidence interval for σ based on s_P.
(c) (§7-2) Individual two-sided confidence intervals for the five different means here would be of the form $\bar{y}_i \pm \Delta$ for a number Δ. If 95% individual confidence is desired, what value of Δ should be used here?
(d) (§7-2) Individual two-sided confidence intervals for the differences in the five different means would be of the form $\bar{y}_i - \bar{y}_{i'} \pm \Delta$ for a number Δ. If 95% individual confidence is desired, what value of Δ should be used here?
(e) (§7-2) Give a 95% two-sided individual confidence interval for the difference between the average density for the 10,000, 8,000, and 6,000 psi conditions and the average density for the 4,000 and 2,000 psi conditions.
(f) (§7-2) Give two-sided 95% prediction intervals based on s_P for the next densities produced under first the 4,000 psi and then the 8,000 psi conditions.
(g) (§7-2) Give 95% lower tolerance bounds based on s_P for 99% of additional densities produced under first the 4,000 psi and then the 8,000 psi conditions.
(h) (§7-2) If one uses the Δ from (c) and makes the five corresponding intervals, what minimum simultaneous confidence level is guaranteed by the Bonferroni Inequality of Section 7-2?
(i) (§7-3) Using the P-R method, what Δ would be used to make two-sided intervals of the form $\bar{y}_i \pm \Delta$ for all five mean densities, possessing simultaneous 95% confidence?
(j) (§7-3) Using the Tukey method, what Δ would be used to make two-sided intervals of the form $\bar{y}_i - \bar{y}_{i'} \pm \Delta$ for all differences in the five mean densities, possessing simultaneous 95% confidence?
(k) (§7-3) Suppose that one thinks of the 2,000 psi condition as being a standard or control condition. Using the Dunnett method, what Δ would be used to make two-sided intervals of the form $\bar{y}_i - \bar{y}_1 \pm \Delta$ for all differences between the other four mean densities and the control mean density, possessing simultaneous 95% confidence?
(l) (§7-4) Carry out an ANOM test of the hypothesis that all five conditions produce the same mean density. Show the appropriate five-step significance-testing summary, and also make a graphical display like that in Figure 7-12.

(m) (§7-4) Make an ANOVA table for these data. You should do the calculations by hand first and then check your arithmetic using a statistical computer package. Then use the calculations to find both R^2 for the one-way model and also the observed level of significance for an F test of the null hypothesis that all five pressures produce the same mean density.

7-2. The following data are taken from the paper "Zero-Force Travel-Time Parameters for Ultrasonic Head-Waves in Railroad Rail" by Bray and Leon-Salamanca (*Materials Evaluation*, 1985). Given are measurements in nanoseconds of the travel time (in excess of 36.1 μsec) of a certain type of mechanical wave induced by mechanical stress in railroad rails. Three measurements were made on each of six different rails.

Rail	Travel Time (nanoseconds above 36.1 μsec)
1	55, 53, 54
2	26, 37, 32
3	78, 91, 85
4	92, 100, 96
5	49, 51, 50
6	80, 85, 83

(a) (§7-4) Make appropriate plots for checking the appropriateness of a one-way random effects analysis of these data. What do these suggest?
(b) (§7-4) Ignoring any possible problems with the standard assumptions of the random effects model revealed in (a), make an ANOVA table for these data and find estimates of σ and σ_τ. What, in the context of this situation, do these two estimates measure?
(c) (§7-4) Find and interpret a two-sided 90% confidence interval for the ratio σ_τ / σ.

7-3. (§7-5) The following are some data taken from a larger set in *Statistical Quality Control* by Grant and Leavenworth, giving the drained weights (in ounces) of contents of size No. $2\frac{1}{2}$ cans of standard grade tomatoes in puree. Twenty samples of three cans taken from a canning process at regular intervals are represented.

Sample	x_1	x_2	x_3	Sample	x_1	x_2	x_3
1	22.0	22.5	22.5	11	20.0	19.5	21.0
2	20.5	22.5	22.5	12	19.0	21.0	21.0
3	20.0	20.5	23.0	13	19.5	20.5	21.0
4	21.0	22.0	22.0	14	20.0	21.5	24.0
5	22.5	19.5	22.5	15	22.5	19.5	21.0
6	23.0	23.5	21.0	16	21.5	20.5	22.0
7	19.0	20.0	22.0	17	19.0	21.5	23.0
8	21.5	20.5	19.0	18	21.0	20.5	19.5
9	21.0	22.5	20.0	19	20.0	23.5	24.0
10	21.5	23.0	22.0	20	22.0	20.5	21.0

(a) Suppose that standard values for the process mean and standard deviation of drained weights (μ and σ) in this canning plant are 21.0 oz and 1.0 oz, respectively. Make and interpret "standards given" $\bar{x}$ and R charts based on these samples. What do these charts indicate about the behavior of the filling process over the time period represented by these data?

(b) As an alternative to the "standards given" range chart made in part (a), make a "standards given" s chart based on the 20 samples. How does its appearance compare to that of the R chart?

Now suppose that no standard values for μ and σ have been provided.

(c) Find one estimate of σ for the filling process based on the average of the 20 sample ranges, $\bar{R}$, and another based on the average of 20 sample standard deviations, $\bar{s}$. How do these compare to the pooled sample standard deviation (of Section 7-1), s_P, here?

(d) Use $\bar{\bar{x}}$ and your estimate of σ based on $\bar{R}$ and make retrospective control charts for $\bar{x}$ and R. What do these indicate about the stability of the filling process over the time period represented by these data?

(e) Use $\bar{\bar{x}}$ and your estimate of σ based on $\bar{s}$ and make retrospective control charts for $\bar{x}$ and s. How do these compare in appearance to the retrospective charts for process mean and variability made in part (d)?

7-4. (§7-6) The accompanying data are some taken from *Statistical Quality Control Methods* by I. W. Burr, giving the numbers of beer cans found to be defective in periodic samples of 312 cans at a brewery.

Sample	Defectives	Sample	Defectives
1	6	11	7
2	7	12	7
3	5	13	6
4	7	14	6
5	5	15	6
6	5	16	6
7	4	17	23
8	5	18	10
9	12	19	8
10	6	20	5

(a) Suppose that the company standard for the fraction of cans defective is that $p = .02$ of the cans be defective on average. Use this value and make a "standards given" p chart based on these data. Does it appear that the process fraction defective was stable at the $p = .02$ value over the period represented by these data?

(b) Make a retrospective p chart for these data. What does this chart indicate about the stability of the canning process?

7-5. (§7-6) The accompanying table lists some data on outlet leaks found in the first assembling of two radiator parts, again taken from Burr's *Statistical Quality Control Methods*. Each radiator may have several leaks.

Make a retrospective u chart based on these data. What does it indicate about the stability of the assembly process?

7-6. Hoffman, Jabaay, and Leuer did a study of pencil lead strength. They loaded pieces of lead of the same diameter (supported on two ends) in their centers and recorded the forces at

TABLE For Exercise 7-5

Date	Tested	Leaks	Date	Tested	Leaks
6/3	39	14	6/14	50	10
6/4	45	4	6/17	32	3
6/5	46	5	6/18	50	11
6/6	48	13	6/19	33	1
6/7	40	6	6/20	50	3
6/10	58	2	6/24	50	6
6/11	50	4	6/25	50	8
6/12	50	11	6/26	50	5
6/13	50	8	6/27	50	2

Total tested = 841   Total leaks = 116

which they failed. Part of their data are given here (in grams of load applied at failure).

4H lead	56.7, 63.8, 56.7, 63.8, 49.6
H lead	99.2, 99.2, 92.1, 106.0, 99.2
B lead	56.7, 63.8, 70.9, 63.8, 70.9

(a) If one is to apply the methods of Chapter 7 in the analysis of these data, what model assumptions must be made? Make three normal plots of these samples on the same set of axes, and also make a normal plot of residuals for the one-way model, as means of investigating the reasonableness of these assumptions. Comment on the plots.

(b) Compute a pooled estimate of variance based on these three samples. What is the corresponding value of s_P?

(c) Use the value of s_P that you calculated in (b) and make (individual) 95% two-sided confidence intervals for each of the three mean lead strengths μ_{4H}, μ_H, and μ_B.

(d) Use s_P and make (individual) 95% two-sided confidence intervals for each of the three differences in mean lead strengths, $\mu_{4H} - \mu_H$, $\mu_{4H} - \mu_B$, and $\mu_H - \mu_B$.

(e) Suppose that for some reason it is desirable to compare the mean strength of B lead to the average of the mean strengths of 4H and H leads. Give a 95% two-sided confidence interval for the quantity $\frac{1}{2}(\mu_{4H} + \mu_H) - \mu_B$.

(f) Use the P-R method of simultaneous confidence intervals and make simultaneous 95% two-sided confidence intervals for the three mean strengths μ_{4H}, μ_H, and μ_B. How do the lengths of these intervals compare to the lengths of the intervals you found in part (c)? Why is it sensible that the lengths should be related in this way?

(g) Use the Tukey method of simultaneous confidence intervals and make simultaneous 95% two-sided confidence intervals for the three differences in mean lead strengths, $\mu_{4H} - \mu_H$, $\mu_{4H} - \mu_B$, and $\mu_H - \mu_B$. How do the lengths of these intervals compare to the lengths of the intervals you found in part (d)?

(h) Use the ANOM test statistic and assess the strength of the evidence against $H_0: \mu_{4H} = \mu_H = \mu_B$ in favor of $H_a:$ not H_0. Show the whole five-step format.

(i) As an alternative to what you did in part (h), use the one-way ANOVA test statistic and assess the strength of the evidence against $H_0: \mu_{4H} = \mu_H = \mu_B$ in favor of $H_a:$ not H_0. Show the whole five-step format.

(j) Make the ANOVA table corresponding to the significance test you carried out in part (i).

(k) As a means of checking your work for parts (i) and (j) of this problem, use Minitab or some other statistical package to produce the required ANOVA table, F statistic, and p-value.

7-7. Allan, Robbins, and Wyckoff worked with a machine shop that employs a CNC (computer numerically controlled) lathe in the manufacture of a part for a heavy equipment maker. Some summary statistics for measurements of a particular diameter on the part for 20 hourly samples of $m = 4$ parts turned on the lathe are given here. (The means are in 10^{-4} in. above 1.1800 in. and the ranges are in 10^{-4} in.)

Sample	1	2	3	4	5
$\bar{x}$	9.25	8.5	9.5	6.25	5.25
R	1	2	2	8	7
Sample	6	7	8	9	10
$\bar{x}$	5.25	5.75	19.5	10.0	9.5
R	5	5	1	3	1
Sample	11	12	13	14	15
$\bar{x}$	9.5	9.75	12.25	12.75	14.5
R	6	1	9	2	7
Sample	16	17	18	19	20
$\bar{x}$	8.0	10.0	10.25	8.75	10.0
R	3	0	1	3	0

(a) The mid-specification for the diameter in question was 1.1809 in. Suppose that a standard σ for diameters turned on this machine is 2.5×10^{-4} in. Use these two values and find "standards given" control limits for $\bar{x}$ and R. Make both $\bar{x}$ and R charts using these, and comment on what the charts indicate about the turning process.

(b) In contrast to part (a) where standards were furnished, compute retrospective or "as past data" control limits for both $\bar{x}$ and R. Make both $\bar{x}$ and R charts using these and comment on what the charts indicate about the turning process.

(c) If one were to judge the sample ranges to be stable, it would then make sense to use $\bar{R}$ to develop an estimate of the turning process short-term standard deviation σ. Find such an estimate.

(d) The engineering specifications for the turned diameter are (still in .0001 in. above 1.1800 in.) from 4 to 14. Supposing that the average diameter could be kept on target (at the mid-specification), does your estimate of σ from part (c) suggest that the turning process would then be *capable* of producing most diameters in these specifications? Explain.

7-8. Becker, Francis, and Nazarudin conducted a study of the effectiveness of commercial clothes dryers in removing water from different types of fabric. The following are some summary statistics from a part of their study, where a garment made of one of $r = 3$ different blends was wetted and dried for 10 minutes in a particular dryer and the (water) weight loss (in grams) measured. Each of the three different garments was tested three times.

100% Cotton	Cotton/Polyester	Cotton/Acrylic
$n_1 = 3$	$n_2 = 3$	$n_3 = 3$
$\bar{y}_1 = 85.0$ g	$\bar{y}_2 = 348.3$ g	$\bar{y}_3 = 258.3$ g
$s_1 = 25.0$ g	$s_2 = 88.1$ g	$s_3 = 63.3$ g

(a) What restrictions/model assumptions are required in order to do formal inference based on the data summarized here (if one wishes to pool information on the baseline variability involved and use the formulas of this chapter)? Henceforth assume that those model assumptions are a sensible description of this situation.

(b) Find s_P and the associated degrees of freedom.

(c) What does s_P measure?

(d) Give a 90% lower confidence bound for the mean amount of water that can be removed from the cotton garment by this dryer in a 10-minute period.

(e) Give a 90% two-sided prediction interval for the next amount of water that would be removed from the cotton garment by this dryer in a 10-minute period.

(f) Give a 95% lower tolerance bound for 90% of the measured amounts of water that would be removed from the cotton garment in many future repetitions of this test.

(g) Give a 90% two-sided confidence interval for comparing the means for the two blended garments.

(h) Suppose that all pairs of fabrics are to be compared using intervals of the form $\bar{y}_i - \bar{y}_{i'} \pm \Delta$ and that simultaneous 95% confidence is desired. Find Δ.

(i) A partially completed ANOVA table for testing $H_0: \mu_1 = \mu_2 = \mu_3$ follows. Finish filling in the table, then find a p-value for a significance test of this hypothesis.

ANOVA TABLE

Source	SS	df	MS	F
	24,787			
	132,247			

7-9. The article "Behavior of Rubber-Based Elastomeric Construction Adhesive in Wood Joints" by P. Pellicane (*Journal of Testing and Evaluation*, 1990) compared the performance of $r = 8$ different commercially available construction adhesives. $m = 8$ joints glued with each glue were tested for strength, giving results summarized as follows. (The units are kN.)

Glue (i)	1	2	3	4
$\bar{y}_i$	1821	1968	1439	616
s_i	214	435	243	205
Glue (i)	5	6	7	8
$\bar{y}_i$	1354	1424	1694	1669
s_i	135	191	225	551

(a) Temporarily considering only the test results for glue 1, give a 95% lower tolerance bound for the strengths of 99% of joints made with glue 1.

(b) Still considering only the test results for glue 1, give a 95% lower confidence bound for the mean strength of joints made with glue 1.

(c) Now considering only the test results for glues 1 and 2, assess the strength of the evidence against the possibility that glues 1 and 2 produce joints with the same mean strength. Show the whole five-step significance-testing format.

(d) What model assumptions stand behind the formulas you used in parts (a) and (b)? In part (c)?

Henceforth, consider test results from all eight glues when making your analyses.

(e) Find a pooled sample standard deviation and give its degrees of freedom.

(f) Repeat parts (a) and (b) using the pooled standard deviation instead of only s_1. What extra model assumption is required to do this (beyond what was used in parts (a) and (b))?

(g) Find the value of an F statistic for testing $H_0: \mu_1 = \mu_2 = \cdots = \mu_8$ and give its degrees of freedom. (*Hint:* These data are balanced. You ought to be able to use the $\bar{y}$'s and the sample variance routine on your calculator to help get the numerator for this statistic.)

(h) Find the value of the ANOM statistic for testing $H_0: \mu_1 = \mu_2 = \cdots = \mu_8$. What do Tables D-12 indicate about the associated p-value?

(i) Simultaneous 95% two-sided confidence limits for the mean strengths for the eight glues are of the form $\bar{y}_i \pm \Delta$ for an appropriate number Δ. Find Δ.

(j) Simultaneous 95% two-sided confidence limits for all differences in mean strengths for the eight glues are of the form $\bar{y}_i - \bar{y}_{i'} \pm \Delta$ for a number Δ. Find Δ.

7-10. The following are some general questions about the random effects analyses discussed at the end of Section 7-4.

(a) Explain in general terms when a random effects analysis is appropriate for use with multisample data.

(b) Consider a scenario where $r = 5$ different technicians employed by a company each make $m = 2$ measurements of the diameter of a particular widget using a particular gage, in a study of how technician differences show up in diameter data the company collects. Under what circumstances would a random effects analysis of the resulting data be appropriate?

(c) Suppose that the following ANOVA table was made in a random effects analysis of data like those described in part (b). Give estimates of the standard deviation associated with repeat diameter measurements for a given technician and the standard deviation of long-run mean measurements for various technicians. The sums of squares are in units of square inches.

ANOVA TABLE

Source	SS	df	MS	F
Technician	.0000136	4	.0000034	1.42
Error	.0000120	5	.0000024	
Total	.0000256	9		

7-11. Explain the difference between several intervals having associated 95% individual confidences and having associated 95% simultaneous confidence.

7-12. Example 4-7 treats some data collected by Kotlers, MacFarland, and Tomlinson while studying strength properties of wood joints. Part of those data (stress at failure values in units of psi for four out of the original nine wood/joint type combinations) are reproduced here, along with $\bar{y}$ and s for each of the four samples represented.

	WOOD TYPE	
	Pine	Oak
Butt	829 596 $\bar{y} = 712.5$ $s = 164.8$	1169 $\bar{y} = 1169$
Lap	1000 859 $\bar{y} = 929.5$ $s = 99.7$	1295 1561 $\bar{y} = 1428.0$ $s = 188.1$

JOINT TYPE (row label at left)

(a) Treating pine/butt joints alone, give a 95% two-sided confidence interval for mean strength for such joints. (Here, base your interval on only the pine/butt data.)

(b) Treating only lap joints, how strong is the evidence shown here of a difference in mean joint strength between pine and oak woods? (Here use only the pine/lap and oak/lap data.) Use the five-step format.

(c) Give a 90% two-sided confidence interval for comparing the strength standard deviations for pine/lap and oak/lap joints.

Henceforth, consider all four samples.

(d) Assuming that all four wood type/joint type conditions are thought to have approximately the same associated variability in joint strength, give an estimate of this supposedly common standard deviation.

(e) It is possible to compute simultaneous 95% lower (one-sided) confidence limits for mean joint strengths for all four wood type/joint type combinations. Give these.

(f) Give a 95% lower (one-sided) prediction limit for the strength of an additional pine/butt joint, using your answer to (d).

(g) Suppose that one desires to compare butt joint strength to lap joint strength, and in fact wants a 95% two-sided confidence interval for

$$\frac{1}{2}(\mu_{\text{pine/butt}} + \mu_{\text{oak/butt}}) - \frac{1}{2}(\mu_{\text{pine/lap}} + \mu_{\text{oak/lap}})$$

Give such a confidence interval, again making use of your answer to (d).

7-13. In an industrial application of Shewhart $\bar{x}$ and R control charts, 20 successive hourly samples of $m = 2$ high-precision metal parts were taken, and a particular diameter on the parts was measured. $\bar{x}$ and R values were calculated for each of the 20 samples, and these had

$$\bar{\bar{x}} = .35080 \text{ in.} \quad \text{and} \quad \bar{R} = .00019 \text{ in.}$$

(a) Give retrospective control limits that you would use in an analysis of the $\bar{x}$ and R values.

(b) The engineering specifications for the diameter being measured were .3500 in. $\pm$ .0020 in. Unfortunately, even practicing engineers sometimes have difficulty distinguishing in their thinking and speech between specifications and control limits. Briefly (but carefully) discuss the difference in meaning between the control limits for $\bar{x}$ found in part (a) and these engineering specifications. (To what quantities do the two apply? What are the different purposes for the two? Where do the two come from? And so on.)

7-14. Here are some summary statistics produced by Davies and Sehili for ten samples of $m = 4$ pin head diameters formed on a certain type of electrical component. The sampled components were groups of consecutive items taken from the output of a machine approximately once every ten minutes. The units are .001 in.

Sample	$\bar{x}$	R	s	Sample	$\bar{x}$	R	s
1	31.5	3	1.29	6	33.0	3	1.41
2	30.75	2	.96	7	33.0	2	.82
3	29.75	3	1.26	8	33.0	4	1.63
4	30.50	3	1.29	9	34.0	2	.82
5	32.0	0	0	10	26.0	0	0

$$\sum \bar{x} = 313.5 \qquad \sum R = 22 \qquad \sum s = 9.48$$

(a) Assuming that the basic short-term variability of the mechanism producing pin head diameters is constant, it makes sense to try to quantify it in terms of a standard deviation σ. Various estimates of that σ are possible. Give three such possible estimates, based on $\bar{R}$, $\bar{s}$, and s_P.
(b) Using each of your estimates from (a), give retrospective control limits for both $\bar{x}$ and R.
(c) Compare the $\bar{x}$'s given above to your control limits from (b) based on $\bar{R}$. Are there any points that would plot outside control limits on a Shewhart $\bar{x}$ chart? On a Shewhart R chart?
(d) For the company manufacturing these parts, what are the practical implications of your analysis in parts (b) and (c)?

7-15. Dunnwald, Post, and Kilcoin studied the viscosities of various weights of various brands of motor oil. Some summary statistics for part of their data are given here. Summarized are $m = 10$ measurements of the viscosities of each of $r = 4$ different weights of Brand M motor oil at room temperature. Units are seconds required for a ball to drop a particular distance through the oil.

10W30	SAE 30	10W40	20W50
$\bar{y}_1 = 1.385$	$\bar{y}_2 = 2.066$	$\bar{y}_3 = 1.414$	$\bar{y}_4 = 4.498$
$s_1 = .091$	$s_2 = .097$	$s_3 = .150$	$s_4 = .204$

(a) Find the pooled sample standard deviation here. What are the associated degrees of freedom?
(b) If the P-R method is used to find simultaneous 95% two-sided confidence intervals for all four mean viscosities, the intervals produced are of the form $\bar{y}_i \pm \Delta$, for Δ an appropriate number. Find Δ.
(c) If the Tukey method is used to find simultaneous 95% two-sided confidence intervals for all differences in mean viscosities, the intervals produced are of the form $\bar{y}_i - \bar{y}_{i'} \pm \Delta$, for Δ an appropriate number. Find Δ.
(d) Carry out an ANOM test of the hypothesis that the four oil weights have the same mean viscosity.
(e) Carry out an ANOVA test of the hypothesis that the four oil weights have the same mean viscosity.

7-16. Explain briefly how a Shewhart $\bar{x}$ chart can help reduce variation in, say, a widget diameter, first by signaling the need for process intervention/adjustment and then also by preventing adjustments when no "out of control" signal is given.

7-17. Why is it essential to have an operational definition of a nonconformance if one is to make effective practical use of a Shewhart c chart?

7-18. Why might it well be argued that the name *control* chart invites confusion?

7-19. What must an engineering application of control charting involve beyond the simple naming of points plotting out of control, if it is to be practically effective?

7-20. Because of modern business pressures, it is not uncommon for standards for fractions nonconforming to be in the range of 10^{-4} to 10^{-6}.
(a) What are "standards given" 3-sigma control limits for a p chart with standard fraction nonconforming 10^{-4} and sample size 100?
(b) If p becomes twice the standard value (of 10^{-4}), what is the probability that the scheme from (a) detects this state of affairs at the first subsequent sample? (Use your answer to (a) and the binomial distribution for $n = 100$ and $p = 2 \times 10^{-4}$.)
(c) What does (b) suggest about the feasibility of doing process monitoring for very small fractions defective based on attributes data?

7-21. Suppose that a company standard for the mean number of visual imperfections on a square foot of plastic sheet is $\lambda = .04$.
(a) Give upper control limits for the number of imperfections found on pieces of material .5 ft $\times$ .5 ft and then 5 ft $\times$ 5 ft.
(b) What would you tell a worker who, instead of inspecting a 10 ft $\times$ 10 ft specimen of the plastic (counting total imperfections on the whole), wants to inspect only a 1 ft $\times$ 1 ft specimen and multiply the observed count of imperfections by 100?

7-22. Bailey, Goodman, and Scott worked on a process for attaching metal connectors to the ends of hydraulic hoses. One part of that process involved grinding rubber off the ends of the hoses. The amount of rubber removed is termed the skive length. The values in the accompanying table are skive length means and standard deviations for 20 samples of five consecutive hoses ground on one grinder. Skive length is expressed in .001 in. above the target length.
(a) What do these values indicate about the stability of the skiving process? Show appropriate work and explain fully.
(b) Give an estimate of the process short-term standard deviation based on the given values.
(c) If specifications on the skive length are $\pm .006$ in., and over short periods, skive length can be thought of as normally distributed, what does your answer to (b) indicate about the best possible fraction (for perfectly adjusted grinders) of skives in specifications? Give a number.
(d) Based on your answer to (b), give control limits for future control of skive length means and ranges for samples of size $m = 3$.
(e) Suppose that hoses from all grinders used during a given shift are all dumped into a common bin. If upon sampling, say, 20 hoses from this bin at the end of a shift, the 20 measured skive lengths

Sample	$\bar{x}$	s
1	$-.4$	5.27
2	0.0	4.47
3	-1.4	3.29
4	1.8	2.28
5	1.4	1.14
6	0.0	4.24
7	$-.4$	4.39
8	1.4	4.51
9	.2	4.32
10	-3.2	2.05
11	-2.2	5.50
12	-5.2	2.86
13	$-.8$	1.30
14	.8	2.68
15	-2.0	2.92
16	$-.2$	1.30
17	-6.6	2.30
18	-1.0	4.21
19	-3.2	5.76
20	-2.4	4.28
	-23.4	69.07

have a standard deviation twice the size of your answer to (b), what possible explanations come to mind for this?

(f) Suppose current policy is to sample five consecutive hoses once an hour for each grinder. An alternative possibility is to sample one hose every 12 minutes for each grinder.

    (i) Briefly discuss practical tradeoffs that you see between the two possible sampling methods.

    (ii) If in fact the new sampling scheme were adopted, would you recommend treating the five hoses from each hour as a sample of size 5 and doing $\bar{x}$ and R charting with $m = 5$? Explain.

7-23. Two different types of nonconformance can appear on widgets manufactured by Company V. Counts of these on ten widgets produced one per hour are below.

Widget	1	2	3	4	5	6	7	8	9	10
Type A Defects	4	2	1	2	2	2	0	2	1	0
Type B Defects	0	2	2	4	2	4	3	3	7	2
Total Defects	4	4	3	6	4	6	3	5	8	2

(a) Considering first total nonconformances, is there evidence here of process instability? Show appropriate work.

(b) What statistical indicators might you expect to observe in data like these if in fact type A and B defects have a common cause mechanism?

(c) (**Charts for Demerits**) For the sake of example, suppose that type A defects are judged twice as important as type B defects. One might then consider charting

    $X =$ Demerits

       $= 2(\text{number of A defects}) + (\text{number of B defects})$

If one can model (number of A defects) and (number of B defects) as independent Poisson random variables, it is relatively easy to come up with sensible control limits. (Remember that the vari-

ance of a sum of independent random variables is the sum of the variances.)

    (i) If the mean number of A defects per widget is λ_1 and the mean number of B defects per widget is λ_2, what are the mean and variance for X? Use your answers to give "standards given" control limits for X.

    (ii) In light of your answer to (i), what numerical limits for X would you use to analyze the above values "as past data"?

7-24. A manufacturer of U-bolts collects data on the thread lengths of the bolts that it produces. 19 samples of five consecutive bolts gave the thread lengths indicated below (in .001 in. above nominal).

Sample	Thread Lengths	$\bar{x}$	R	s	
1	11, 14, 14, 10, 8	11.4	6	2.61	
2	14, 10, 11, 10, 11	11.2	4	1.64	
3	8, 13, 14, 13, 10	11.6	6	2.51	
4	11, 8, 13, 11, 13	11.2	5	2.05	
5	13, 10, 11, 11, 11	11.2	3	1.10	
6	11, 10, 10, 11, 13	11.0	3	1.22	$\sum \bar{x} = 212.4$
7	8, 6, 11, 11, 11	9.4	5	2.30	
8	10, 11, 10, 14, 10	11.0	4	1.73	$\sum R = 77$
9	11, 8, 11, 8, 10	9.6	3	1.52	
10	6, 6, 11, 13, 11	9.4	7	3.21	$\sum s = 32.92$
11	11, 14, 13, 8, 11	11.4	6	2.30	
12	8, 11, 10, 11, 14	10.8	6	2.17	
13	11, 11, 13, 8, 13	11.2	5	2.05	
14	11, 8, 11, 11, 11	10.4	3	1.34	
15	11, 11, 13, 11, 11	11.4	2	.89	
16	14, 13, 13, 13, 14	13.4	1	.55	
17	14, 13, 14, 13, 11	13.0	3	1.22	
18	13, 11, 11, 11, 13	11.8	2	1.10	
19	14, 11, 11, 11, 13	12.0	3	1.41	

(a) Compute two different estimates of the process short-term standard deviation of thread length, one based on the sample ranges and one based on the sample standard deviations.

(b) Use your estimate from (a) based on sample standard deviations and compute control limits for the sample ranges R, and then compute control limits for the sample standard deviations s. Applying these to the R and s values, what is suggested about the threading process?

(c) Using a center line at $\bar{\bar{x}}$, and your estimate of σ based on the sample standard deviations, compute control limits for the sample means $\bar{x}$. Applying these to the $\bar{x}$ values here, what is suggested about the threading process?

(d) A check of the control chart form from which these data were taken shows that the coil of the heavy wire from which these bolts are made was changed just before samples 1, 9, and 16 were taken. What insight, if any, does this information provide into the possible origins of any patterns you see in the data?

(e) Suppose that a customer will purchase bolts of the type represented in the data above only if essentially all bolts received can be guaranteed to have thread lengths within .01 in. of nominal. Does it appear that with proper process monitoring and adjustment, the equipment and manufacturing practices in use at this company will be able to produce only bolts meeting these standards? Explain in quantitative terms. If one's equipment were not adequate to meet such requirements, name two options that might be taken and their practical pros and cons.

7-25. (**Variables versus Attributes Control Charting**) Suppose that a dimension of parts produced on a certain machine over a short period can be thought of as normally distributed with some mean μ and standard deviation $\sigma = .005$ in. Suppose further that values of this dimension more than .0098 in. from the 1.000 in. nominal value are considered nonconforming. Finally, suppose that hourly samples of ten of these parts are to be taken.
(a) If μ is exactly on target (i.e., $\mu = 1.000$ in.), about what fraction of parts will be nonconforming? Is it possible for the fraction nonconforming ever to be any less than this figure?
(b) One could use a p chart based on $m = 10$ to monitor process performance in this situation. What would be "standards given" 3-sigma control limits for the p chart, using your answer from part (a) as the standard value of p?
(c) What is the probability that a particular sample of $m = 10$ parts will produce an "out of control" signal on the chart from (b) if μ remains at its standard value of $\mu = 1.000$ in.? How does this compare to the same probability for a 3-sigma $\bar{x}$ chart for $m = 10$ set up with a center line at 1.000? (For the p chart, use a binomial probability calculation. For the $\bar{x}$ chart, use the facts that $\mu_{\bar{x}} = \mu$ and $\sigma_{\bar{x}} = \sigma/\sqrt{m}$.)
(d) Compare the probability that a particular sample of $m = 10$ parts will produce an "out of control" signal on the p chart from (b) to the probability that the sample will produce an "out of control" signal on the ($m = 10$) 3-sigma $\bar{x}$ chart first mentioned in (c), supposing that in fact $\mu = 1.005$ in. What moral is told by your calculations here and in part (c)?

7-26. In a particular defects/unit context, the number of standard size units inspected at a given opportunity varies. With

$$X_i = \text{the number of defects found on sample } i$$
$$k_i = \text{the number of units inspected at time } i$$
$$\hat{u}_i = X_i/k_i$$

the following were obtained at eight consecutive periods.

i	1	2	3	4	5	6	7	8
k_i	1	2	1	3	2	1	1	3
$\hat{u}_i$	0	1.5	0	.67	2	0	0	.33

(a) What do these values suggest about the stability of the process under study?
(b) Suppose that henceforth, k_i is going to be held constant, and that standard quality will be defined as a mean of 1.2 defects per unit. Compare 3-sigma Shewhart c charts based on $k_i = 1$ and on $k_i = 2$, in terms of the probabilities that a given sample produces an "out of control" signal if
　　(i) the actual defect rate is standard.
　　(ii) the actual defect rate is twice standard.

7-27. Successive samples of carriage bolts are checked for length using "a go–no go" gage. The results from ten successive samples are as follows.

Sample	1	2	3	4	5	6	7	8	9	10
Sample Size	30	20	40	30	20	20	30	20	20	20
Nonconforming	2	1	5	1	2	1	3	0	1	2

What do these values indicate about the stability of the bolt cutting process?

7-28. State briefly the practical goals of control charting and action on "out of control" signals produced by the charts.

7-29. Explain why too little variation appearing on a Shewhart control chart need not be a good sign.

7-30. The article "How to Use Statistics Effectively in a Pseudo-Job Shop" by G. Fellers (*Quality Engineering*, 1990) discusses some applications of statistical methods in the manufacture of corrugated cardboard boxes. One part of the article concerns the analysis of a variable called box "skew" which quantifies how far from being perfectly square boxes are. This response variable, which will here be called y, is measured in units of $\frac{1}{32}$ in. $r = 24$ customer orders (each requiring a different machine setup) were studied, and from each, the skews, y, of five randomly selected boxes were measured. A partial ANOVA table made in summary of the data follows.

	ANOVA TABLE			
Source	SS	df	MS	F
Order (setup)	1052.39			
Error				
Total	1405.59	119		

(a) Complete the ANOVA table.
(b) In a given eight-hour day, around 20 different orders are run in this plant on each machine. This situation is one in which a random effects analysis is most natural. Explain why.
(c) Find estimates of σ and σ_τ. What, in the context of this situation, do these two estimates measure?
(d) Find and interpret a two-sided 90% confidence interval for the ratio σ_τ/σ.
(e) If there is variability in skew, customers must continually adjust automatic folding and packaging equipment in order to prevent machine jam-ups. Such variability is therefore highly undesirable for the box manufacturer, who wishes to please customers. What does your analysis from (c) and (d) indicate about how the manufacturer should proceed in any attempts to reduce variability in skew? (What is the big component of variance, and what kind of actions might be taken to reduce it? For example, is a need for the immediate purchase of new high-precision manufacturing equipment indicated?)

7-31. The article "High Tech, High Touch" by J. Ryan (*Quality Progress*, 1987) discusses the quality enhancement processes used by Martin Marietta in the production of the space shuttle external (liquid oxygen) fuel tanks. It includes a graph giving counts of major hardware nonconformances for each of 41 tanks produced. The accompanying data are approximate counts read from that graph for the last 35 tanks. (The first six tanks were of a different design than the others and are therefore not included here.)

Tank	Nonconformances	Tank	Nonconformances
1	537	19	157
2	463	20	120
3	417	21	148
4	370	22	65
5	333	23	130
6	241	24	111
7	194	25	65
8	185	26	74
9	204	27	65
10	185	28	148
11	167	29	74
12	157	30	65
13	139	31	139
14	130	32	213
15	130	33	222
16	267	34	93
17	102	35	194
18	130		

(a) Make a retrospective c chart for these data. Is there evidence of real quality improvement in this series of counts of nonconformances? Explain.

(b) Consider only the last 17 tanks. Does it appear that quality was stable over the production period represented by these tanks? (Make another retrospective c chart.)

(c) It is possible that some of the figures read from the graph in the original article may differ from the the real figures by as much as, say, 15 nonconformances. Would this measurement error account for the apparent lack of stability you found in (a) or (b) above? Explain.

7-32. Kaminski, Rasavahn, Smith, and Weitekamper worked with the same pelletizing machine referred to in Examples 1-2, 3-14, and 6-18. They collected process monitoring data on several different days of operation. The accompanying table shows counts of nonconforming pellets in periodic samples of size $m = 30$ from two different days. (The pelletizing on day 1 was done with 100% fresh material, and on the second day, a mixture of fresh and reground materials was used.)

(a) Make a retrospective p chart for the day 1 data. Is there evidence of process instability in the day 1 data? Explain.

(b) Treating the day 1 data as a single sample of size 750 from the day's production of pellets, give a 90% two-sided confidence interval for the fraction nonconforming produced on the day in question.

(c) In light of your answers to parts (a) and (b), explain why a process being in control or stable does not necessarily mean that it is producing a satisfactory fraction of conforming product.

(d) Repeat parts (a) and (b) for the day 2 data.

(e) Try making a single retrospective control chart for the two days taken together. Do points plot out of control on this single chart? Explain why this does or does not contradict the results of parts (a), (b), and (d).

(f) Treating the data from days 1 and 2 as two samples of size 750 from the respective days' production of pellets, give a two-sided 98% confidence interval for the difference in fractions of nonconforming pellets produced on the two days.

TABLE For Exercise 7-32

DAY 1		DAY 2	
Sample	Nonconforming	Sample	Nonconforming
1	16	1	14
2	18	2	20
3	17	3	17
4	18	4	13
5	22	5	12
6	14	6	12
7	16	7	14
8	18	8	15
9	18	9	19
10	19	10	21
11	20	11	18
12	25	12	14
13	14	13	13
14	13	14	9
15	23	15	16
16	13	16	16
17	23	17	15
18	15	18	11
19	14	19	17
20	23	20	8
21	17	21	16
22	20	22	13
23	16	23	16
24	19	24	15
25	22	25	13

7-33. Eastman, Frye, and Schnepf counted defective plastic bags in 15 consecutive groups of 250 coming off a converting machine immediately after a changeover to a new roll of plastic. Their counts are as follows.

Sample	Nonconforming
1	147
2	93
3	41
4	0
5	18
6	0
7	31
8	22
9	0
10	0
11	0
12	0
13	0
14	0
15	0

Is it plausible that these data came from a physically stable process, or is it clear that there is some kind of startup phenomenon involved here? Make and interpret an appropriate control chart to support your answer.

7-34. Sinnott, Thomas, and White compared several properties of five different brands of 10W30 motor oil. In one part of their study, they measured the boiling points of the oils. $m = 3$ measurements for each of the $r = 5$ oils follow. (Units are degrees F.)

Brand C	Brand H	Brand W	Brand Q	Brand P
378	357	321	353	390
386	365	303	349	378
388	361	306	353	381

(a) Compute and make a normal plot for the residuals for the one-way model. What does the plot indicate about the appropriateness of the one-way model assumptions?

(b) Using the five samples, find s_P, the pooled estimate of σ. What does this value measure? Give a two-sided 90% confidence interval for σ based on s_P.

(c) Individual two-sided confidence intervals for the five different means here would be of the form $\bar{y}_i \pm \Delta$ for an appropriate number Δ. If 90% individual confidence is desired, what value of Δ should be used?

(d) Individual two-sided confidence intervals for the differences in the five different means would be of the form $\bar{y}_i - \bar{y}_{i'} \pm \Delta$ for a number Δ. If 90% individual confidence is desired, what value of Δ should be used here?

(e) Give 95% lower prediction bounds based on s_P for the next boiling points for Brand C and Brand P oils.

(f) Give 95% lower tolerance bounds based on s_P for 99% of additional boiling points for Brand C and Brand P oils.

(g) Using the P-R method, what Δ would be used to make two-sided intervals of the form $\bar{y}_i \pm \Delta$ for all five mean boiling points, possessing simultaneous 95% confidence?

(h) Using the Tukey method, what Δ would be used to make two-sided intervals of the form $\bar{y}_i - \bar{y}_{i'} \pm \Delta$ for all differences in the five mean boiling points, possessing simultaneous 99% confidence?

(i) Carry out an ANOM test of the hypothesis that all five oils have the same mean boiling point. Show the appropriate five-step significance-testing summary, and also make a graphical display like that in Figure 7-12.

(j) Make an ANOVA table for these data. Then use the calculations to find both R^2 for the one-way model and also the observed level of significance for an F test of the null hypothesis that all five oils have the same mean boiling point.

(k) It is likely that the measurements represented here were all made on a single can of each brand of oil. (The students' report was not explicit about this point.) If so, the formal inferences made here are really most honestly thought of as applying to the five particular cans used in the study. Discuss why the inferences would not necessarily extend to all cans of the brands included in the study, and describe the conditions under which you might be willing to make such an extension. Is the situation different if, for example, each of the measurements comes from a different can of oil, taken from different shipping lots? Explain.

7-35. Baik, Johnson, and Umthun worked with a small metal fabrication company on monitoring the performance of a process for cutting metal rods. Specifications for the lengths of these rods were 33.69 in. $\pm$ .03 in. Measured lengths of rods in 15 samples of $m = 4$ rods, made over a period of two days, were as shown in the accompanying table. (The data are recorded in inches above

Sample	Rod Lengths	$\bar{x}$	R	s
1	.0075, .0100, .0135, .0135	.01113	.0060	.00293
2	−.0085, .0035, −.0180, .0010	−.00550	.0215	.00981
3	.0085, .0000, .0100, .0020	.00513	.0100	.00487
4	.0005, −.0005, .0145, .0170	.00788	.0175	.00916
5	.0130, .0035, .0120, .0070	.00888	.0095	.00444
6	−.0115, −.0110, −.0085, −.0105	−.01038	.0030	.00131
7	−.0080, −.0070, −.0060, −.0045	−.00638	.0035	.00149
8	−.0095, −.0100, −.0130, −.0165	−.01225	.0070	.00323
9	.0090, .0125, .0125, .0080	.01050	.0045	.00235
10	−.0105, −.0100, −.0150, −.0075	−.01075	.0075	.00312
11	.0115, .0150, .0175, .0180	.01550	.0065	.00297
12	.0020, .0005, .0010, .0010	.00113	.0015	.00063
13	−.0010, −.0025, −.0020, −.0030	−.00213	.0020	.00085
14	−.0020, .0015, .0025, .0025	.00113	.0045	.00214
15	−.0010, −.0015, −.0020, −.0045	−.00225	.0035	.00155

$$\bar{\bar{x}} = .00078 \qquad \bar{R} = .0072 \qquad \bar{s} = .00339$$

the target value of 33.69, and the first five samples were made on day 1, while the remainder were made on day 2.)

(a) Find a retrospective center line and control limits for all 15 sample ranges. Apply them to the ranges and say what is indicated about the rod cutting process.

(b) Repeat part (a) for the sample standard deviations rather than ranges.

The initial five samples were taken while the operators were first learning to cut these particular rods. Suppose that it therefore makes sense to look separately at the last ten samples. These samples have $\bar{\bar{x}} = -.00159$, $\bar{R} = .00435$, and $\bar{s} = .001964$.

(c) Both the ranges and standard deviations of the last ten samples look reasonably stable. What about the last ten $\bar{x}$'s? (Compute control limits for the last ten $\bar{x}$'s, based on either $\bar{R}$ or $\bar{s}$, and say what is indicated about the rod cutting process.)

As a matter of fact, the cutting process worked as follows. Rods were welded together at one end in bundles of 80, and the whole bundle cut at once. The four measurements in each sample came from a single bundle. (There are 15 bundles represented.)

(d) How does this explanation help you understand the origin of patterns discovered in the data in parts (a) through (c)?

(e) Find an estimate of the "process short-term σ" for the last ten samples. What is it really measuring in the present context?

(f) Use your estimate from (e) and, assuming that lengths of rods from a single bundle are approximately normally distributed, compute an estimate of the fraction of lengths in a bundle that are in specifications, if in fact $\mu = 33.69$ in.

(g) If one simply pools together the last ten samples (making a single sample of size 40) and computes the sample standard deviation, one gets the value $s = .00898$. This is much larger than any s recorded for one of the samples, and should be much larger than your value from (e). What is the origin of this difference in magnitude?

7-36. Consider the last ten samples from Exercise 7-35. Upon considering the physical circumstances that produced the data, it becomes sensible to replace the control chart analysis done there with a random effects analysis simply meant to quantify the within- and between-bundle variance components.

(a) Make an ANOVA table for these ten samples of size 4. Based on the mean squares, find estimates of σ, the standard deviation of lengths for a given bundle, and σ_τ, the standard deviation of bundle mean lengths.

(b) Find and interpret a two-sided 90% confidence interval for the ratio σ_τ/σ.

(c) What is the principal origin of variability in the lengths of rods produced by this cutting method? (Is it variability of lengths within bundles or differences between bundles?)

7-37. The following data appear in the text *Quality Control and Industrial Statistics* by A. J. Duncan. They represent the numbers of disabling injuries suffered and millions of man hours worked at a large corporation in 12 consecutive months.

Month	1	2	3	4	5	6
Injuries	11	4	5	8	4	4
10^6 man hr	.175	.178	.175	.180	.183	.198

Month	7	8	9	10	11	12
Injuries	9	12	2	6	6	7
10^6 man hr	.210	.212	.210	.211	.195	.200

(a) Temporarily assuming the injury rate per man hour to be stable over the period studied, find a sensible estimate of the mean injuries per 10^6 man hours.

(b) Based on your figure from (a), find "control limits" for the observed rates in each of the 12 months. Do these data appear to be consistent with a "stable system" view of the corporation's injury production mechanisms? Or are there months that are clearly distinguishable from the others in terms of accident rates?

7-38. Eder, Williams, and Bruster studied the force (applied to the cutting arm handle) required to cut various types of paper in a standard paper trimmer. The students used stacks of five sheets of four different types of paper and recorded the forces needed to move the cutter arm (and thus cut the paper). The data that follow (the units are ounces) are for $m = 3$ trials with each of the four paper types and also for a "baseline" condition where no paper was loaded into the trimmer.

No Paper	Newsprint	Construction	Computer	Magazine
24, 25, 31	61, 51, 52	72, 70, 77	59, 59, 70	54, 59, 61

(a) If one is to apply the methods of this chapter in the analysis of these data, what model assumptions must be made? With small sample sizes such as those here, only fairly crude checks on the appropriateness of the assumptions are possible. One possibility is to compute residuals and normal-plot them. Do this and comment on the appearance of the plot.

(b) Compute a pooled estimate of the standard deviation based on these five samples. What is s_P supposed to be measuring in the present situation?

(c) Use the value of s_P and make (individual) 95% two-sided confidence intervals for each of the five mean force requirements $\mu_{\text{No paper}}$, $\mu_{\text{Newsprint}}$, $\mu_{\text{Construction}}$, μ_{Computer}, and μ_{Magazine}

(d) Individual confidence intervals for the differences between particular pairs of mean force requirements are of the form $\bar{y}_i - \bar{y}_{i'} \pm \Delta$ for an appropriate value of Δ. Use s_P and find Δ if individual 95% two-sided intervals are desired.

(e) Suppose that it is desirable to compare the "no paper" force requirement to the average of the force requirements for the various papers. Give a 95% two-sided confidence interval for the quantity

$$\mu_{\text{No paper}} - \frac{1}{4}\left(\mu_{\text{Newsprint}} + \mu_{\text{Construction}} + \mu_{\text{Computer}} + \mu_{\text{Magazine}}\right)$$

(f) Use the P-R method of simultaneous confidence intervals and make simultaneous 95% two-sided confidence intervals for the five mean force requirements. How do the lengths of these intervals compare to the lengths of the intervals you found in part (c)? Why is it sensible that the lengths should be related in this way?

(g) Simultaneous confidence intervals for the differences between all pairs of mean force requirements are of the form $\bar{y}_i - \bar{y}_{i'} \pm \Delta$ for an appropriate value of Δ. Use s_P and find Δ if simultaneous 95% two-sided intervals are desired. How does this value compare to the value you found in part (d)?

(h) In the context of this paper cutting study, it is plausible that one might want to make simultaneous two-sided 95% confidence intervals for only the four differences between the "no paper" mean and each of the different paper mean force requirements. Which method discussed in Section 7-3 would be appropriate for use in doing this? If such intervals are to be of the form $\bar{y}_i - \bar{y}_1 \pm \Delta$, find an appropriate value of Δ.

(i) Use the ANOM test statistic and assess the strength of the students' evidence against H_0: $\mu_{\text{No paper}} = \mu_{\text{Newsprint}} = \mu_{\text{Construction}} = \mu_{\text{Computer}} = \mu_{\text{Magazine}}$ in favor of H_a: not H_0. Show the whole five-step format.

(j) As an alternative to what you did in part (i), use the one-way ANOVA test statistic and assess the strength of the students' evidence against H_0: $\mu_{\text{No paper}} = \mu_{\text{Newsprint}} = \mu_{\text{Construction}} = \mu_{\text{Computer}} = \mu_{\text{Magazine}}$ in favor of H_a: not H_0. Show the whole five-step format.

(k) Make the ANOVA table corresponding to the significance test you carried out in part (j).

7-39. Duffy, Marks, and O'Keefe did some testing of the 28-day compressive strengths of 3 in. $\times$ 6 in. concrete cylinders. In part of their study, concrete specimens made with a .50 water/cement ratio and different percentages of entrained air were cured in a moisture room and subsequently strength tested. $m = 4$ specimens of each type produced the measured strengths (in 10^3 psi) summarized as follows.

3% Air	6% Air	10% Air
$\bar{y}_1 = 5.3675$	$\bar{y}_2 = 4.9900$	$\bar{y}_3 = 2.9250$
$s_1 = .1638$	$s_2 = .1203$	$s_3 = .2626$

(a) Find the pooled sample standard deviation and its associated degrees of freedom.

Use your answer to part (a) throughout the rest of this problem.

(b) Give a 99% lower confidence bound for the mean strength of 3% air specimens.

(c) Give a 99% lower prediction bound for the strength of a single additional 3% air specimen.

(d) Give a 95% lower tolerance bound for 99% of strengths of additional 3% entrained air specimens.

(e) Give a 99% two-sided confidence interval for comparing the mean strengths of 3% air and 10% air specimens.

(f) Suppose that mean strengths of specimens for all pairs of levels of entrained air are to be compared using intervals of the form $\bar{y}_i - \bar{y}_{i'} \pm \Delta$. Find Δ for simultaneous 95% two-sided confidence limits.

(g) A partially completed ANOVA table for testing $H_0: \mu_1 = \mu_2 = \mu_3$ follows. Finish filling in the table, then find a p-value for an F test of this hypothesis.

ANOVA TABLE

Source	SS	df	MS	F
Total	14.1608			

(h) The ANOM test of $H_0: \mu_1 = \mu_2 = \mu_3$ is an alternative method to the one used in part (g). Give the value of the test statistic that would be used in the ANOM test and find the associated p-value.

8

Inference for Multisample Studies with Full Factorial Structure

*I*n Chapter 7, this book's exposition of inference methods that are useful in multisample statistical engineering studies began with the most generic of those methods: ones that neither require nor make use of any special structure relating the samples. Those are both widely applicable and practically informative. But Chapter 4 illustrated on a descriptive level (rather than inferential level) the discovery, interpretation, and physical exploitation of simple patterns in means from *structured* multisample statistical engineering studies. This is often the way engineers make progress in the designing, building, operation, and improvement of physical systems. So it is natural to consider methods of inference crafted for the purpose of helping in the identification and interpretation of such patterns.

This chapter begins an exposition of inference methods for structured multisample studies, starting with methods appropriate in the analysis of data from complete factorial studies. It builds on the descriptive statistics material of Section 4-3 and the tools of inference discussed in Chapter 7 to provide methods for studies where samples of a response variable are available from all combinations of each of several levels of several factors. First are considered basic inference methods for 2-factor situations where there is some replication (i.e., not all sample sizes are 1). Then there is a discussion of some methods that are applicable under an assumption of negligible interactions in a balanced two-way factorial. Next, methods of inference for *p*-way factorial studies are introduced, with particular emphasis on those where each factor is represented at only two levels. The chapter closes with an exploration of the insights provided for full factorial studies with balanced data by extensions of the ANOVA methods introduced in Section 7-4.

8-1 Basic Inference in Two-Way Factorials with Some Replication

This section considers basic methods of inference from data possessing complete two-way factorial structure, in cases where there is replication somewhere in the data set — i.e., at least one of the sample sizes is larger than 1. It begins by pointing out that the material in Sections 7-1 through 7-4 is immediately applicable and can often be useful in sharpening the kind of preliminary graphical analysis of two-way data suggested in Section 4-3. Then there is a discussion of how rewriting the one-way model from Section 7-1 in a form that recognizes 2-factor structure in a set of means leads to natural definitions of (theoretical) factorial effects that are completely analogous to the (empirical) fitted factorial effects introduced in Chapter 4. The section concludes by developing both individual and simultaneous confidence interval methods for two-way factorial effects, based on the fact that they are linear combinations of underlying means.

One-Way Methods in Two-Way Factorials

The intention in this section is to provide basic inference methods for two-way factorial studies involving some replication. Example 8-1 is of this type.

EXAMPLE 8-1

(Example 4-7 revisited) **Joint Strengths for Three Different Joint Types in Three Different Woods.** Consider again the wood joint strength study of Kotlers, MacFarland, and Tomlinson, discussed in descriptive terms in Section 4-3. Table 8-1 reorganizes the data given earlier in Table 4-12 into a (two-way) 3×3 table showing the nine different

TABLE 8-1

Joint Strengths for 3^2 Combinations of Joint Type and Wood

		WOOD		
		1 (Pine)	2 (Oak)	3 (Walnut)
	1 (Butt)	829, 596	1169	1263, 1029
JOINT	2 (Beveled)	1348, 1207	1518, 1927	2571, 2443
	3 (Lap)	1003, 859	1295, 1561	1489

samples of one or two joint strengths, for all combinations of three different woods and three different joint types.

The data in Table 8-1 have complete two-way factorial structure, and seven of the nine combinations represented in the table provide some replication.

Unfortunately, as more and more complicated inference methods are considered, students sometimes lose track of a fundamental observation regarding data sets like that in Example 8-1: All of the material in Sections 7-1 through 7-4 remains applicable when thinking about such structured data sets. That is, the data in Table 8-1 constitute $r = 9$ samples of sizes 1 and 2. Provided the graphical and numerical checks of Section 7-1 reveal no obvious problems with the one-way model for joint strengths, all of the interval-oriented methods and testing methods of Sections 7-2 through 7-4 can be brought to bear on the task of understanding the information carried by the data.

One way in which this observation is particularly helpful is in indicating the precision of estimated means on plots of sample means from multifactor studies. Section 4-3 discussed how near-parallelism on plots of sample means versus levels of one factor leads to simple interpretations of two-way factorials. By marking either individual or simultaneous confidence limits as *error bars* around the sample means on such a plot, it is possible to get a rough idea of the detectability or statistical significance of any empirical lack of parallelism.

EXAMPLE 8-1
(continued)

The place to begin a formal analysis of the wood joint strength data is with consideration of the appropriateness of the "equal variances, normal distributions" model for joint strength. Table 8-2 gives some summary statistics for the data of Table 8-1. (The means were given earlier in Table 4-13.)

TABLE 8-2

Sample Means and Standard Deviations for Nine Joint/Wood Combinations

		WOOD		
		1 (Pine)	2 (Oak)	3 (Walnut)
	1 (Butt)	$\bar{y}_{11} = 712.5$ $s_{11} = 164.8$	$\bar{y}_{12} = 1,169$	$\bar{y}_{13} = 1,146$ $s_{13} = 165.5$
JOINT	2 (Beveled)	$\bar{y}_{21} = 1,277.5$ $s_{21} = 99.7$	$\bar{y}_{22} = 1,722.5$ $s_{22} = 289.2$	$\bar{y}_{23} = 2,507$ $s_{23} = 90.5$
	3 (Lap)	$\bar{y}_{31} = 929.5$ $s_{31} = 99.7$	$\bar{y}_{32} = 1,428$ $s_{32} = 188.1$	$\bar{y}_{33} = 1489$

Residuals for the joint strength data are obtained by subtracting the sample means in Table 8-2 from the corresponding observations in Table 8-1. Notice that in this data set, the sample sizes are so small that the residuals will obviously be highly dependent. Those from samples of size 2 will be plus-and-minus a single number corresponding to that sample. Those from samples of size 1 will be zero. So there is reason to expect residual plots to show some effects of this dependence, and this should be kept in mind when using them. Figure 8-1 is a normal plot of the 16 residuals, and its complete symmetry (with respect to the positive and negative residuals) is caused by this dependence. (One could in fact argue for some kind of plotting of only 7 points — one residual or absolute residual from each of the combinations where there is replication — on the basis that a plot of 16 points appears to carry more information than it really does.) But at least one can say that Figure 8-1 indicates no clear problems with the one-way model's normal-distribution assumption.

FIGURE 8-1

Normal Plot of 16 Residuals for the Wood Joint Strength Study

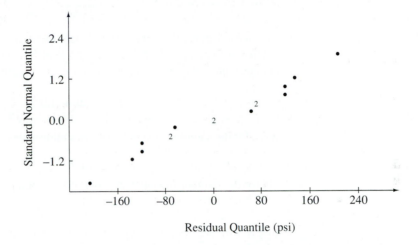

Of course, the sample standard deviations in Table 8-2 vary somewhat, but the ratio between the largest and smallest (a factor of about 3) is in no way unbelievable based on these sample sizes of 2. (Even if only 2 rather than 7 sample variances were involved, since 9 is between the .75 and .9 quantiles of the $F_{1,1}$ distribution, the observed level of significance for testing the equality of the two underlying variances would exceed $.2 = 2(1 - .9)$.) And Figure 8-2, which is a plot of residuals versus sample means, suggests no trend in σ as a function of mean response.

FIGURE 8-2

Plot of Residuals versus Sample Means for the Joint Strength Study

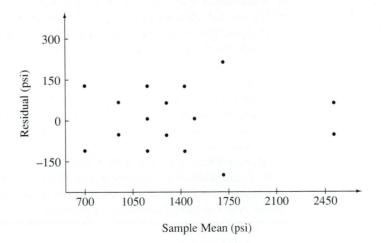

In sum, the very small sample sizes represented in Table 8-1 make definitive investigation of the appropriateness of the one-way normal model assumptions impossible. But the limited checks that are possible provide no indication of serious problems with those restrictions in reference to the joint strength data.

Notice that for these data,

$$s_P^2 = \frac{(2-1)s_{11}^2 + (2-1)s_{13}^2 + (2-1)s_{21}^2 + \cdots + (2-1)s_{32}^2}{(2-1) + (2-1) + (2-1) + \cdots + (2-1)}$$

$$= \frac{1}{7}\left((164.8)^2 + (165.5)^2 + \cdots + (188.1)^2\right)$$

$$= 28,805 \ (\text{psi})^2$$

So

$$s_P = \sqrt{28,805} = 169.7 \ \text{psi}$$

where s_P has 7 degrees of freedom associated with it.

Then, for example, from formula (7-14) of Section 7-2, individual two-sided 99% confidence intervals for the combination mean strengths would have endpoints

$$\bar{y}_{ij} \pm 3.499(169.7)\frac{1}{\sqrt{n_{ij}}}$$

That is, for the samples of size 1, one uses endpoints

$$\bar{y}_{ij} \pm 593.9 \tag{8-1}$$

while for the samples of size 2, appropriate end points are

$$\bar{y}_{ij} \pm 419.9 \tag{8-2}$$

Figure 8-3 shows a plot of sample mean strengths (like that in Figure 4-22), which has been enhanced with error bars around the means constructed using expressions (8-1) and (8-2). Notice, by the way, that the Bonferroni Inequality puts the simultaneous confidence associated with all nine of the indicated intervals at a minimum of 91% ($.91 = 1 - 9(1 - .99)$).

The important message carried by Figure 8-3, not present in the unenhanced plot of Figure 4-22, is the relatively large imprecision associated with the sample means as estimates of underlying mean strengths. And that imprecision has implications with regard to the statistical detectability of factorial effects. For example, by moving near the extremes on some error bars in Figure 8-3, one might imagine finding nine means within the indicated intervals such that their connecting line segments would exhibit parallelism. That is, the plot with confidence intervals for means marked on it already suggests that the empirical interactions between Wood Type and Joint Type seen in these data may not be large enough to distinguish from background noise. Or if they are detectable, they may be only barely so.

The issues of whether the empirical differences between woods and between joint types are distinguishable from experimental variation are perhaps somewhat easier to call. There is consistency in the patterns "Walnut is stronger than oak is stronger than pine" and "Beveled is stronger than lap is stronger than butt." This, combined with differences at least approaching the size of indicated imprecisions, suggests that although it isn't clear whether the empirical lack of parallelism is large enough to support firm conclusions about the nature of interactions, firm statements about the main effects of Wood Type and Joint Type are likely possible.

This author considers the kind of analysis made thus far on the joint strength data extremely important and illuminating. This discussion will proceed to more complicated statistical methods for such problems. But these often amount primarily to a further

FIGURE 8-3

Plot of Joint Strength Sample
Means versus Wood Types
Enhanced with Error Bars
Based on Individual 99%
Confidence Intervals

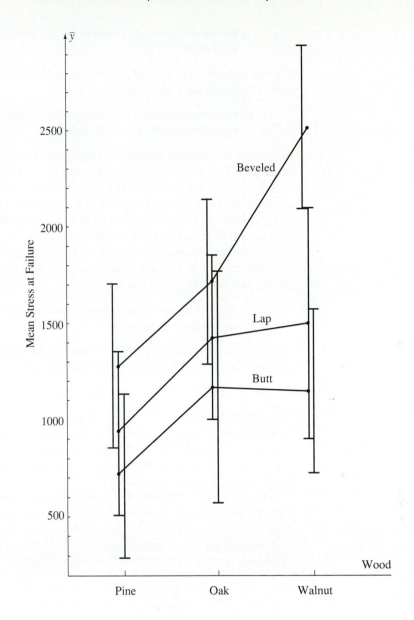

refinement and quantification of the two-way factorial story already told graphically by
a plot like Figure 8-3.

**Two-Way Factorial Notation
and Definitions of Effects**

In order to efficiently discuss formal inference in two-way factorial studies, it is useful
to modify the generic multisample notation used in Chapter 7 along the lines hinted at
in the discussion of descriptive statistics for such situations in Section 4-3. Consider
combinations of factor A having I levels and factor B having J levels, and use the triple
subscript notation (sad, but unavoidable):

y_{ijk} = the kth observation in the sample from the ith level of A and jth
level of B

Then for $I \cdot J$ different samples corresponding to the possible combinations of a level of
A with a level of B, let

n_{ij} = the number of observations in the sample from the ith level of A and jth level of B

Use the notations $\bar{y}_{ij}$, $\bar{y}_{i.}$, and $\bar{y}_{.j}$ introduced in Section 4-3, and in the obvious way (actually already used in Example 8-1), let

s_{ij} = the sample standard deviation of the n_{ij} observations in the sample from the ith level of A and the jth level of B

Notice that this notation amounts to adding another subscript to the notation introduced in Chapter 7 in order to acknowledge the two-way structure in the samples (which wasn't assumed in Chapter 7). In Chapter 7, it was most natural to think of r samples as numbered $i = 1$ to r and laid out in a single row. Here it is appropriate to think of $r = I \cdot J$ samples laid out in the cells of a two-way table like Table 8-1 and named by their row number i and column number j.

In addition to using this notation for empirical quantities, it is also useful to modify the notation used in Chapter 7 for theoretical quantities (model parameters) to fit the present situation. That is, if one lets

μ_{ij} = the underlying mean response corresponding to the ith level of factor A and jth level of factor B

the model assumptions that the $I \cdot J$ samples are roughly describable as independent samples from normal distributions with a common variance σ^2 can be written as

$$y_{ijk} = \mu_{ij} + \epsilon_{ijk} \tag{8-3}$$

where the quantities $\epsilon_{111}, \ldots, \epsilon_{11n_{11}}, \epsilon_{121}, \ldots, \epsilon_{12n_{12}}, \ldots, \epsilon_{IJ1}, \ldots, \epsilon_{IJn_{IJ}}$ are independent normal $(0, \sigma^2)$ random variables. Equation (8-3) is sometimes called the *two-way (normal) model equation*, but is nothing but a rewrite of the basic one-way model equation of Chapter 7 in a notation that recognizes the special organization of $r = I \cdot J$ samples into rows and columns, as in Table 8-1.

The descriptive analysis of two-way factorials in Section 4-3 relied on computing from the sample means $\bar{y}_{ij}$ their row averages $\bar{y}_{i.}$ and column averages $\bar{y}_{.j}$ and then using these to define fitted factorial effects. Analogous operations performed on the underlying or theoretical means μ_{ij} lead to appropriate definitions for corresponding underlying or theoretical factorial effects. That is, to begin with, let

$$\mu_{i.} = \frac{1}{J} \sum_{j=1}^{J} \mu_{ij}$$

= the average underlying mean when factor A is at level i

$$\mu_{.j} = \frac{1}{I} \sum_{i=1}^{I} \mu_{ij}$$

= the average underlying mean when factor B is at level j

$$\mu_{..} = \frac{1}{IJ} \sum_{i,j} \mu_{ij}$$

= the grand average underlying mean

Figure 8-4 shows these as row, column, and grand averages of the μ_{ij}; this is the theoretical counterpart of Figure 4-21.

Then, following the pattern established in Definitions 4-5 and 4-6 for sample quantities, one has the following two definitions for theoretical quantities.

DEFINITION 8-1

In a two-way complete factorial study with factors A and B, the **main effect of factor A at its *i*th level** is

$$\alpha_i = \mu_{i.} - \mu_{..}$$

Similarly, the **main effect of factor B at its *j*th level** is

$$\beta_j = \mu_{.j} - \mu_{..}$$

FIGURE 8-4

Underlying Cell Mean Responses and Their Row, Column, and Grand Averages

Factor B

	Level 1	Level 2		Level *J*	
Level 1	μ_{11}	μ_{12}	$\cdots$	μ_{1J}	$\mu_{1.}$
Level 2	μ_{21}	μ_{22}	$\cdots$	μ_{2J}	$\mu_{2.}$
Level *I*	μ_{I1}	μ_{I2}	$\cdots$	μ_{IJ}	$\mu_{I.}$
	$\mu_{.1}$	$\mu_{.2}$	$\cdots$	$\mu_{.J}$	$\mu_{..}$

Factor A

These main effects are measures of how (population or underlying) mean responses change from row to row or from column to column in Figure 8-4. The fitted main effects of Section 4-3 can be thought of as empirical approximations to them. It is a consequence of the form of Definition 8-1 that (like their empirical counterparts) the main effects of a given factor sum to 0 over levels of that factor. That is, simple algebra shows that

$$\sum_{i=1}^{I} \alpha_i = 0$$

and

$$\sum_{j=1}^{J} \beta_j = 0$$

Next is a definition of theoretical interactions.

DEFINITION 8-2

In a two-way complete factorial study with factors A and B, the **interaction of factor A at its *i*th level and factor B at its *j*th level** is

$$\alpha\beta_{ij} = \mu_{ij} - (\mu_{..} + \alpha_i + \beta_j)$$

The interactions in a two-way set of underlying means μ_{ij} measure lack of parallelism that would exist on a plot of the (typically unknown) parameters μ_{ij} against levels of one of the factors. They measure how much pattern there is in the underlying means μ_{ij} that is not explainable in terms of the factors A and B separately, but must instead be attributed to their joint effects. The fitted interactions of Section 4-3 are meant as empirical approximations of these underlying quantities. Small fitted interactions ab_{ij} indicate small underlying interactions $\alpha\beta_{ij}$ and thus make it justifiable to think of the two factors A and B as operating separately on the response variable when one is called on to make engineering judgments or decisions.

The form of Definition 8-2 has several simple algebraic consequences that are occasionally useful to know. One is that (like fitted interactions) interactions $\alpha\beta_{ij}$ sum to 0 over levels of either factor. That is, as defined,

$$\sum_{i=1}^{I} \alpha\beta_{ij} = \sum_{j=1}^{J} \alpha\beta_{ij} = 0$$

Another simple consequence of the definition of interactions is that upon adding $(\mu_{..} + \alpha_i + \beta_j)$ to both sides of the equation defining $\alpha\beta_{ij}$, one obtains a decomposition of each μ_{ij} into a grand mean plus an A main effect plus a B main effect plus an AB interaction:

$$\mu_{ij} = \mu_{..} + \alpha_i + \beta_j + \alpha\beta_{ij} \tag{8-4}$$

The identity (8-4) is sometimes combined with the two-way model equation (8-3) to obtain the equivalent model equation

$$y_{ijk} = \mu_{..} + \alpha_i + \beta_j + \alpha\beta_{ij} + \epsilon_{ijk} \tag{8-5}$$

Here the factorial effects appear explicitly as going into the makeup of the observations. Although there are circumstances where representation (8-5) is essential, in most respects this author finds it most useful to think of the two-way model assumptions in form (8-3) and just remember that the α_i, β_j, and $\alpha\beta_{ij}$ are simple functions of the $I \cdot J$ means μ_{ij}.

Individual Confidence Intervals for Interactions and Concerning Main Effects

The primary new wrinkle in two-way factorial inference is the drawing of inferences concerning the interactions and main effects, with the possibility of finding A, B, or A and B "main effects only" models adequate to describe responses, and subsequently using such simplified descriptions in making predictions about system behavior. The basis of inference for the α_i, β_j, and $\alpha\beta_{ij}$ is that they are linear combinations of the underlying means μ_{ij}. (That is, for properly chosen "c's," the factorial effects are "L's" from Section 7-2.) And the fitted effects defined in Definitions 4-5 and 4-6 are the corresponding linear combinations of the sample means $\bar{y}_{ij}$. (That is, the fitted factorial effects are the corresponding "$\hat{L}$'s.")

EXAMPLE 8-1
(continued)

To illustrate the fact that the effects defined in Definitions 8-1 and 8-2 are linear combinations of the underlying means μ_{ij}, consider α_1, β_3, and $\alpha\beta_{23}$ in the context of the wood joint strength study. First,

$$\alpha_1 = \mu_{1.} - \mu_{..}$$

$$= \frac{1}{3}(\mu_{11} + \mu_{12} + \mu_{13}) - \frac{1}{9}(\mu_{11} + \mu_{12} + \cdots + \mu_{32} + \mu_{33})$$

$$= \frac{2}{9}\mu_{11} + \frac{2}{9}\mu_{12} + \frac{2}{9}\mu_{13} - \frac{1}{9}\mu_{21} - \frac{1}{9}\mu_{22} - \frac{1}{9}\mu_{23} - \frac{1}{9}\mu_{31} - \frac{1}{9}\mu_{32} - \frac{1}{9}\mu_{33}$$

and a_1 is the corresponding linear combination of the $\bar{y}_{ij}$. Similarly,

$$\beta_3 = \mu_{.3} - \mu_{..}$$
$$= \frac{1}{3}(\mu_{13} + \mu_{23} + \mu_{33}) - \frac{1}{9}(\mu_{11} + \mu_{12} + \cdots + \mu_{32} + \mu_{33})$$
$$= \frac{2}{9}\mu_{13} + \frac{2}{9}\mu_{23} + \frac{2}{9}\mu_{33} - \frac{1}{9}\mu_{11} - \frac{1}{9}\mu_{12} - \frac{1}{9}\mu_{21} - \frac{1}{9}\mu_{22} - \frac{1}{9}\mu_{31} - \frac{1}{9}\mu_{32}$$

and b_3 is the corresponding linear combination of the $\bar{y}_{ij}$. And finally,

$$\alpha\beta_{23} = \mu_{23} - (\mu_{..} + \alpha_2 + \beta_3)$$
$$= \mu_{23} - (\mu_{..} + (\mu_{2.} - \mu_{..}) + (\mu_{.3} - \mu_{..}))$$
$$= \mu_{23} - \mu_{2.} - \mu_{.3} + \mu_{..}$$
$$= \mu_{23} - \frac{1}{3}(\mu_{21} + \mu_{22} + \mu_{23}) - \frac{1}{3}(\mu_{13} + \mu_{23} + \mu_{33})$$
$$+ \frac{1}{9}(\mu_{11} + \mu_{12} + \cdots + \mu_{33})$$
$$= \frac{4}{9}\mu_{23} - \frac{2}{9}\mu_{21} - \frac{2}{9}\mu_{22} - \frac{2}{9}\mu_{13} - \frac{2}{9}\mu_{33} + \frac{1}{9}\mu_{11} + \frac{1}{9}\mu_{12}$$
$$+ \frac{1}{9}\mu_{31} + \frac{1}{9}\mu_{32}$$

and ab_{23} is the corresponding linear combination of the $\bar{y}_{ij}$. ▬ ▬ ▬

Once one realizes that the factorial effects are simple linear combinations of the μ_{ij}, it is a small step to recognize that formula (7-26) of Section 7-2 (giving endpoints for a confidence interval for a general linear combination of means, L) can be applied to do formal inference for the effects. For example, the question of whether the lack of parallelism evident in Figure 8-3 is large enough to be statistically detectable can be approached quantitatively by looking at confidence intervals for the $\alpha\beta_{ij}$. And quantitative comparisons between joint types can be based on confidence intervals for differences between the A main effects, $\alpha_i - \alpha_{i'} = \mu_{i.} - \mu_{i'.}$, while quantitative comparisons between woods can be based on differences between the B main effects, $\beta_j - \beta_{j'} = \mu_{.j} - \mu_{.j'}$.

The biggest impediment to applying formula (7-26) of Section 7-2 to do inference for factorial effects is determining how the "$\sum c_i^2/n_i$" term appearing in the formula should look. In the preceding 3×3 factorial example, a number of rather odd-looking coefficients c_{ij} appeared when writing out expressions for the α_i, β_j, and $\alpha\beta_{ij}$ in terms of the basic means μ_{ij}. However, it is possible to discover and write down general formulas. Although the derivations won't be given here, many readers will be able (if interested) to do the algebra necessary to verify that the formulas that follow are correct.

Considering first the matter of inference for the interactions in a two-way factorial, with $L = \alpha\beta_{ij}$ it is possible to verify that

$$\sum_{i,j} \frac{c_{ij}^2}{n_{ij}} = \left(\frac{1}{IJ}\right)^2 \left(\frac{(I-1)^2(J-1)^2}{n_{ij}} + (I-1)^2 \sum_{j' \neq j} \frac{1}{n_{ij'}} + (J-1)^2 \sum_{i' \neq i} \frac{1}{n_{i'j}} + \sum_{i' \neq i, j' \neq j} \frac{1}{n_{i'j'}} \right)$$

$$(8\text{-}6)$$

Expression (8-6) is admittedly fairly ugly. But in the case of balanced data (say, where all $n_{ij} = m$), expression (8-6) simplifies considerably, to

$$\sum_{i,j} \frac{c_{ij}^2}{m} = \frac{(I-1)(J-1)}{mIJ} \tag{8-7}$$

As far as inference for individual main effects in a two-way factorial is concerned, with $L = \alpha_i$ it is possible to show that

$$\sum_{i,j} \frac{c_{ij}^2}{n_{ij}} = \left(\frac{1}{IJ}\right)^2 \left((I-1)^2 \sum_j \frac{1}{n_{ij}} + \sum_{i' \neq i,j} \frac{1}{n_{i'j}}\right) \tag{8-8}$$

while with $L = \beta_j$ the corresponding formula is

$$\sum_{i,j} \frac{c_{ij}^2}{n_{ij}} = \left(\frac{1}{IJ}\right)^2 \left((J-1)^2 \sum_i \frac{1}{n_{ij}} + \sum_{i,j' \neq j} \frac{1}{n_{ij'}}\right) \tag{8-9}$$

As was the case for the interactions, formulas (8-8) and (8-9) simplify considerably when all $n_{ij} = m$. That is, with balanced data and $L = \alpha_i$,

$$\sum_{i,j} \frac{c_{ij}^2}{n_{ij}} = \frac{I-1}{mIJ} \tag{8-10}$$

and with $L = \beta_j$,

$$\sum_{i,j} \frac{c_{ij}^2}{n_{ij}} = \frac{J-1}{mIJ} \tag{8-11}$$

Often, differences between main effects are of more interest than the individual main effects themselves. And it is easy to find formulas for the "$\sum c_i^2 / n_i$" term in expression (7-26) of Section 7-2 for differences in main effects. That is, with $L = \alpha_i - \alpha_{i'} = \mu_{i.} - \mu_{i'.}$,

$$\sum_{i,j} \frac{c_{ij}^2}{n_{ij}} = \frac{1}{J^2} \left(\sum_j \frac{1}{n_{ij}} + \sum_j \frac{1}{n_{i'j}}\right) \tag{8-12}$$

while with $L = \beta_j - \beta_{j'} = \mu_{.j} - \mu_{.j'}$, the corresponding formula is

$$\sum_{i,j} \frac{c_{ij}^2}{n_{ij}} = \frac{1}{I^2} \left(\sum_i \frac{1}{n_{ij}} + \sum_i \frac{1}{n_{ij'}}\right) \tag{8-13}$$

Once again, equal samples sizes lead to attractive simplifications of formulas (8-12) and (8-13). That is, when all $n_{ij} = m$, $L = \alpha_i - \alpha_{i'}$ implies that

$$\sum_{i,j} \frac{c_{ij}^2}{n_{ij}} = \frac{2}{mJ} \tag{8-14}$$

and $L = \beta_j - \beta_{j'}$ implies that

$$\sum_{i,j} \frac{c_{ij}^2}{n_{ij}} = \frac{2}{mI} \tag{8-15}$$

Having written down the series of formulas (8-6) through (8-15), the form of individual confidence intervals for any of the quantities $L = \alpha\beta_{ij}$, α_i, β_j, $\alpha_i - \alpha_{i'}$, or $\beta_j - \beta_{j'}$ is obvious. In the formula for confidence interval endpoints

$$\hat{L} \pm t s_P \sqrt{\sum_{i,j} \frac{c_{ij}^2}{n_{ij}}} \tag{8-16}$$

one computes s_P by pooling the $I \cdot J$ sample variances in the usual way (arriving at an estimate with $n - r = n - IJ$ associated degrees of freedom), uses the fitted effects from Section 4-3 to find $\hat{L}$, chooses an appropriate one of formulas (8-6) through (8-15) to give the quantity under the radical, and chooses t from Table D-4 according to a desired confidence and degrees of freedom $\nu = n - IJ$.

EXAMPLE 8-1
(continued)

Consider making formal inferences for the factorial effects in the wood joint strength study of Kotlers, MacFarland, and Tomlinson. Suppose that inferences are to be phrased in terms of two-sided 99% individual confidence intervals and one begins by considering the interactions $\alpha\beta_{ij}$.

Unfortunately, despite the students' best efforts to the contrary, the sample sizes represented in Table 8-1 are not all the same. So one is forced to use formula (8-6) rather than the simpler formula (8-7) when making confidence intervals for the $I \cdot J = 3 \cdot 3 = 9$ interactions $\alpha\beta_{ij}$. Table 8-3 collects the sums of reciprocal sample sizes appearing in formula (8-6) for each of the nine combinations of $i = 1, 2, 3$ and $j = 1, 2, 3$.

TABLE 8-3

Sums of Reciprocal Sample Sizes Needed in Making Confidence Intervals for Joint/Wood Interactions

i	j	$\dfrac{1}{n_{ij}}$	$\sum_{j' \neq j} \dfrac{1}{n_{ij'}}$	$\sum_{i' \neq i} \dfrac{1}{n_{i'j}}$	$\sum_{i' \neq i,\, j' \neq j} \dfrac{1}{n_{i'j'}}$
1	1	.5	1.5	1.0	2.5
1	2	1.0	1.0	1.0	2.5
1	3	.5	1.5	1.5	2.0
2	1	.5	1.0	1.0	3.0
2	2	.5	1.0	1.5	2.5
2	3	.5	1.0	1.5	2.5
3	1	.5	1.5	1.0	2.5
3	2	.5	1.5	1.5	2.0
3	3	1.0	1.0	1.0	2.5

For example, for the combination $i = 1$ and $j = 1$,

$$\frac{1}{n_{11}} = \frac{1}{2} = .5$$

$$\frac{1}{n_{12}} + \frac{1}{n_{13}} = \frac{1}{1} + \frac{1}{2} = 1.5$$

$$\frac{1}{n_{21}} + \frac{1}{n_{22}} = \frac{1}{2} + \frac{1}{2} = 1.0$$

$$\frac{1}{n_{22}} + \frac{1}{n_{23}} + \frac{1}{n_{32}} + \frac{1}{n_{33}} = \frac{1}{2} + \frac{1}{2} + \frac{1}{2} + \frac{1}{1} = 2.5$$

The entries in Table 8-3 lead to values for $\sum c_{ij}^2/n_{ij}$ via formula (8-6). Then, since (from before) $s_P = 169.7$ psi with 7 associated degrees of freedom, and since the .995 quantile of the t_7 distribution is 3.499, it is clear how to calculate the plus-or-minus part of formula (8-16) in order to get two-sided 99% confidence intervals for the $\alpha\beta_{ij}$. In addition, remember that all nine fitted interactions were calculated in Section 4-3 and collected in Table 4-15. Table 8-4 gives the $\sqrt{\sum c_{ij}^2/n_{ij}}$ values, the fitted interactions ab_{ij}, and the plus-or-minus part of two-sided 99% individual confidence intervals for the interactions $\alpha\beta_{ij}$.

To illustrate the calculations represented in the third column of Table 8-4, consider the combination with $i = 1$ (butt joints) and $j = 1$ (pine wood). Since $I = 3$ and $J = 3$, formula (8-6) shows that for $L = \alpha\beta_{11}$

$$\sum \frac{c_{ij}^2}{n_{ij}} = \left(\frac{1}{3 \cdot 3}\right)^2 \left(\frac{2^2 \cdot 2^2}{2} + 2^2(1.5) + 2^2(1.0) + 2.5\right) = .2531$$

from which

$$\sqrt{\sum \frac{c_{ij}^2}{n_{ij}}} = \sqrt{.2531} = .5031$$

TABLE 8-4

99% Individual Two-Sided
Confidence Intervals for Joint
Type/Wood Type Interactions

i	j	$\sqrt{\sum \dfrac{c_{ij}^2}{n_{ij}}}$	ab_{ij} (psi)	$ts_P\sqrt{\sum \dfrac{c_{ij}^2}{n_{ij}}}$ (psi)
1	1	.5031	105.83	298.7
1	2	.5720	95.67	339.6
1	3	.5212	−201.5	309.5
2	1	.4843	−155.67	287.6
2	2	.5031	−177.33	298.7
2	3	.5031	333.0	298.7
3	1	.5031	49.83	298.7
3	2	.5212	81.67	309.5
3	3	.5720	−131.5	339.6

Notice the practical implications of the calculations summarized in Table 8-4. All but one of the intervals centered at an ab_{ij} with a half width given in the last column of the table would cover 0. Only for $i = 2$ (beveled joints) and $j = 3$ (walnut wood) is the magnitude of the fitted interaction big enough to put its associated confidence interval entirely to one side of 0. That is, most of the lack of parallelism seen in Figure 8-3 is potentially attributable to experimental variation, but that associated with beveled joints and walnut wood can be definitely differentiated from background noise. (Admittedly, however, if it were only somewhat less pronounced, it too would be potentially attributable to random variation.) This suggests that if mean joint strength differences on the order of 333 ± 299 psi are of engineering importance, it is not adequate to think of the factors Joint Type and Wood Type as operating separately on joint strength across all three levels of each factor. On the other hand, if one were to restrict attention to either butt and lap joints or to pine and oak woods, a "negligible interactions" conceptualization of joint strength would perhaps be tenable.

To illustrate the use of formula (8-16) in making inferences about main effects on joint strength, consider comparing joint strengths for pine and oak woods. The rather extended analysis of interactions here and the character of Figure 8-3 suggest that the strength profiles of pine and oak across the three joint types are comparable, at least up to the precision provided by the students' data. So estimation of $\beta_1 - \beta_2 = \mu_{.1} - \mu_{.2}$ can be thought of as amounting not only to the estimation of the difference in average (across joint types) mean strengths of pine and oak joints, but also to the difference in mean strengths of pine and oak joints for any of the three joint types individually. $\beta_1 - \beta_2$ is thus a quantity of real interest.

Once again, since the data in Table 8-1 are not balanced, it will be necessary to use the more complicated expression (8-13) rather than formula (8-15) in making a confidence interval for $\beta_1 - \beta_2$. For $L = \beta_1 - \beta_2$, formula (8-13) gives

$$\sum_{i,j} \frac{c_{ij}^2}{n_{ij}} = \frac{1}{3^2}\left[\left(\frac{1}{2} + \frac{1}{2} + \frac{1}{2}\right) + \left(\frac{1}{1} + \frac{1}{2} + \frac{1}{2}\right)\right] = .3889$$

So, since from the fitted effects in Section 4-3

$$b_1 = -402.5 \text{ psi} \qquad \text{and} \qquad b_2 = 64.17 \text{ psi}$$

formula (8-16) shows that endpoints of a two-sided 99% confidence interval for $L = \beta_1 - \beta_2$ are

$$(-402.5 - 64.17) \pm 3.499(169.7)\sqrt{.3889}$$

that is,

$$-466.67 \pm 370.29$$

that is,

$$-836.96 \text{ psi} \qquad \text{and} \qquad -96.38 \text{ psi}$$

This analysis establishes that the oak joints are on average from 96 psi to 837 psi stronger than comparable pine joints. This may seem to be a rather weak conclusion, given the apparent strong increase in sample strengths as one moves from pine to oak in Figure 8-3. But it is as strong a statement as is justified in the light of the large confidence requirement (99%) and the substantial imprecision in the students' data (related to the small sample sizes and a fairly large pooled standard deviation, $s_P = 169.7$ psi). If ±370 psi precision for comparing pine and oak joint strength is not adequate for engineering purposes and large confidence is still desired, these calculations point to the need for more data in order to sharpen that comparison.

Much of the computational complexity of the previous discussion results from the fact that although Kotlers, MacFarland, and Tomlinson intended to obtain samples of size $m = 2$ from all nine combinations of Joint Type and Wood Type, circumstances largely beyond their control conspired to give them two sample sizes of $n_{ij} = 1$ and seven of $n_{ij} = 2$. Note that if everything had gone as planned, for the inference for interactions, formula (8-7) could have been used instead of formula (8-6), and one would have had

$$\sum_{i,j} \frac{c_{ij}^2}{n_{ij}} = \frac{2 \cdot 2}{2 \cdot 3 \cdot 3} = \frac{2}{9}$$

and

$$\sqrt{\sum_{i,j} \frac{c_{ij}^2}{n_{ij}}} = .4714$$

Much of the unpleasantness associated with calculations for Tables 8-3 and 8-4 would have been avoided. (And for that matter, for the estimation of $\beta_1 - \beta_2$, formula (8-15) could have been used instead of formula (8-13).)

But the example is instructive as it stands. Things don't always go as planned, and even though it is beginning to become apparent that balanced data provide great computational advantages in factorial inference, one doesn't always end up with balanced data. Some texts treat factorial analyses for balanced data only. But despite the unavoidable notational and computational annoyances involved, it is essential to present the general (possibly unbalanced) story if one wants to provide practical engineering tools. It would seem silly to adopt the point of view, "If you lose a data point, you're out of luck."

Tukey's Method for Comparing Main Effects

Formula (8-16) is meant to guarantee *individual* confidence levels for intervals made using it. (Of course, the crude Bonferroni idea can be used to find minimum simultaneous confidence levels to associate with several intervals made up using formula (8-16) to estimate and/or compare factorial effects.) When interactions in a two-way factorial study are judged to be negligible, questions of practical engineering importance can usually be phrased in terms of comparing the various A or B main effects. It is then useful to have a method designed specifically to produce a *simultaneous* confidence level for the comparison of all pairs of A or B main effects. Happily enough, Tukey's method (discussed in Section 7-3) can be used not only for comparing all pairs of means μ_{ij}; it can also be easily modified to produce simultaneous confidence intervals for all differences in α_i's or in β_j's. That is, two-sided simultaneous confidence intervals for all possible differences in A main effects $\alpha_i - \alpha_{i'} = \mu_{i.} - \mu_{i'.}$ can be made using respective endpoints

$$\bar{y}_{i.} - \bar{y}_{i'.} \pm \frac{q^*}{\sqrt{2}} s_P \frac{1}{J} \sqrt{\sum_j \frac{1}{n_{ij}} + \sum_j \frac{1}{n_{i'j}}} \tag{8-17}$$

where q^* is taken from Tables D-10 using $\nu = n - IJ$ degrees of freedom, number of means to be compared I, and the .95 or .99 quantile figure (depending whether 95% or 99% simultaneous confidence is desired). Expression (8-17) amounts to the specialization of formula (8-16) to $L = \alpha_i - \alpha_{i'}$ with t replaced by $q^*/\sqrt{2}$. When all $n_{ij} = m$, one can replace the use of expression (8-12) in formula (8-17) with the use of expression (8-14) and obtain the less unwieldy formula

$$\bar{y}_{i.} - \bar{y}_{i'.} \pm \frac{q^* s_P}{\sqrt{Jm}} \tag{8-18}$$

Corresponding to formulas (8-17) and (8-18) are, of course, formulas for endpoints of simultaneous two-sided confidence intervals for all possible differences in B main effects $\beta_j - \beta_{j'} = \mu_{.j} - \mu_{.j'}$ — namely,

$$\bar{y}_{.j} - \bar{y}_{.j'} \pm \frac{q^*}{\sqrt{2}} s_P \frac{1}{I} \sqrt{\sum_i \frac{1}{n_{ij}} + \sum_i \frac{1}{n_{ij'}}} \tag{8-19}$$

and

$$\bar{y}_{.j} - \bar{y}_{.j'} \pm \frac{q^* s_P}{\sqrt{Im}} \tag{8-20}$$

where q^* is taken from Tables D-10 using $\nu = n - IJ$ degrees of freedom and number of means to be compared J.

EXAMPLE 8-2

A 3 × 2 Factorial Study of Ultimate Tensile Strength for Drilled Aluminum Strips. Clubb and Goedken studied the effects on tensile strength of holes drilled in 6 in.-by-2 in. 2024–T3 aluminum strips .0525 in. thick. A hole of diameter .149 in., .185 in., or .221 in. was centered either .5 in. or 1.0 in. from the edge and 3.0 in. from each end of each of 18 strips. Ultimate axial stress was then measured for each on an MTS machine. The tensile strengths (in pounds) obtained in this 3 × 2 complete factorial study are given in Table 8-5.

TABLE 8-5

Tensile Strengths of
Aluminum Strips for
3 × 2 Combinations
of Drilled Hole Size
and Placement

		PLACEMENT	
		1 (.5 in. from Edge)	2 (1.0 in. from Edge)
	1 (.149 in.)	5728, 5664, 5514	5689, 5891, 5611
SIZE 2 (.185 in.)		5508, 5620, 5375	5579, 5557, 5778
	3 (.221 in.)	5469, 5440, 5460	5621, 5664, 5523

It can be verified via the usual plots that, except possibly for the fact that the sample for hole size 3 (.221 in.) and placement 1 (.5 in. from edge) has a spread that is surprisingly smaller than the others, there is little in these data to cast doubt on the relevance of the usual one-way normal model assumptions. Some figuring not illustrated in this book shows that it is not really that unusual to have a ratio of largest sample variance to smallest sample variance (among 6 based on samples of size $m = 3$) of the kind produced by Table 8-5 (namely, $(144.5)^2/(14.8)^2 \approx 96$), assuming the correctness of the one-way model. Thus, in view of the great tractability of the usual assumptions, this author would proceed to use them, keeping in mind that no close calls should be based on them.

Note since the two factors Size and Placement are quantitative, one could consider use of surface fitting methods in the analysis of the data of Table 8-5. This methodology is discussed on an inferential level in Chapter 9, but for the time being, the discussion will be confined to a factorial analysis of the data.

The data of Table 8-5 have the pleasant feature of being balanced, with a common sample size of $m = 3$. Table 8-6 gives sample means and standard deviations for the tensile strengths.

TABLE 8-6

Sample Means and
Standard Deviations for
3 × 2 Size/Placement
Combinations

		PLACEMENT	
		1 (.5 in. from Edge)	2 (1.0 in. from Edge)
	1 (.149 in.)	$\bar{y}_{11} = 5635.3$ lb $s_{11} = 109.8$ lb	$\bar{y}_{12} = 5730.3$ lb $s_{12} = 144.5$ lb
SIZE 2 (.185 in.)		$\bar{y}_{21} = 5501.0$ lb $s_{21} = 122.6$ lb	$\bar{y}_{22} = 5638.0$ lb $s_{22} = 121.7$ lb
	3 (.221 in.)	$\bar{y}_{31} = 5456.3$ lb $s_{31} = 14.8$ lb	$\bar{y}_{32} = 5602.7$ lb $s_{32} = 72.3$ lb

Pooling the $3 \cdot 2 = 6$ sample variances gives

$$s_P = \sqrt{\frac{(3-1)(109.8)^2 + (3-1)(144.5)^2 + \cdots + (3-1)(72.3)^2}{(3-1) + (3-1) + \cdots + (3-1)}} = 106.7 \text{ lb}$$

with $\nu = mIJ - IJ = 3 \cdot 3 \cdot 2 - 3 \cdot 2 = 12$ associated degrees of freedom. Then consider illustrating the experimental results graphically. Notice that the P-R method for making simultaneous two-sided 95% confidence intervals for $r = 6$ means based on $\nu = 12$ degrees of freedom is (from formula (7-34) of Section 7-3) to use endpoints on the order of

$$\bar{y}_{ij} \pm 3.1 \frac{106.7}{\sqrt{3}}$$

for estimating each μ_{ij}. ($k_2^* \approx 3.1$ was obtained via rough interpolation in Table D-9-A.)

FIGURE 8-5

Plot of Aluminum Strip
Sample Means, Enhanced
with Error Bars Based
on 95% Simultaneous
Confidence Intervals

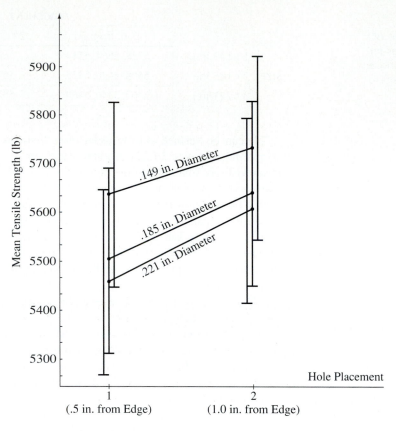

This is approximately

$$\bar{y}_{ij} \pm 191$$

Figure 8-5 is a plot of the $3 \times 2 = 6$ sample mean tensile strengths enhanced with ± 191 lb error bars.

The lack of parallelism in Figure 8-5 is fairly small, both compared to the absolute size of the strengths being measured and also relative to the kind of uncertainty about the individual mean strengths indicated by the 95% simultaneous confidence error bars. It is straightforward to use the methods of Section 4-3, arriving at the following values for the fitted effects in this study.

$$a_1 = 88.9 \qquad b_1 = -63.1$$
$$a_2 = -24.4 \qquad b_2 = 63.1$$
$$a_3 = -64.4$$

$$ab_{11} = 15.6 \qquad ab_{12} = -15.6$$
$$ab_{21} = -5.4 \qquad ab_{22} = 5.4$$
$$ab_{31} = -10.1 \qquad ab_{32} = 10.1$$

Then, since the data of Table 8-5 are balanced, one may use formula (8-7) together with formula (8-16). So individual confidence intervals for the interactions $\alpha\beta_{ij}$ are of the form

$$ab_{ij} \pm t(106.7) \sqrt{\frac{(3-1)(2-1)}{3 \cdot 3 \cdot 2}}$$

that is,

$$ab_{ij} \pm t(35.6)$$

Clearly, for any sensible confidence level (producing *t* of at least 1), such intervals would all cover 0. One therefore has an even more quantitative way of phrasing the lack of statistical detectability of the interactions already represented in Figure 8-5.

It thus seems sensible to proceed to consideration of the main effects in this tensile strength study. To illustrate the application of Tukey's method to factorial main effects, consider first simultaneous 95% two-sided confidence intervals for the three differences $\alpha_1 - \alpha_2$, $\alpha_1 - \alpha_3$, and $\alpha_2 - \alpha_3$. Applying formula (8-18) with $\nu = 12$ degrees of freedom and $I = 3$ means to be compared, one has from Table D-10-A that intervals with endpoints

$$\bar{y}_{i.} - \bar{y}_{i'.} \pm \frac{(3.77)(106.7)}{\sqrt{2 \cdot 3}}$$

that is, roughly

$$\bar{y}_{i.} - \bar{y}_{i'.} \pm 164 \text{ lb}$$

are in order. Note that no difference between the a_i's exceeds 164 lb. That is, if simultaneous 95% confidence is desired in the comparison of the hole sizes, one must judge the students' data to be interesting — perhaps even suggestive of a decrease in strength with increased diameter — but nevertheless statistically inconclusive. To really pin down the impact of hole size on tensile strength, larger samples are needed.

To see that the Clubb and Goedken data do tell at least some story in a reasonably conclusive manner, consider finally the use of formulas (8-15) and (8-16) to make a two-sided 95% confidence interval for $\beta_2 - \beta_1$, the difference in mean strengths for strips with centered holes as compared to ones with holes .5 in. from the strip edge. The desired interval has endpoints

$$b_2 - b_1 \pm t s_{\text{P}} \sqrt{\frac{2}{mI}}$$

that is,

$$63.1 - (-63.1) \pm 2.179(106.7)\sqrt{\frac{2}{3(3)}}$$

that is,

$$126.2 \pm 109.6$$

that is,

$$16.6 \text{ lb} \quad \text{and} \quad 235.8 \text{ lb}$$

Thus, although the students' data don't provide much precision, they *are* adequate to establish clearly the existence of some decrease in tensile strength as a hole is moved from the center of the strip towards its edge.

Formulas (8-17) through (8-20) are, mathematically speaking, perfectly valid providing only that the basic "equal variances, underlying normal distributions" model is a reasonable description of an engineering application. (Under the basic model (8-3), formulas (8-17) and (8-19) provide an actual simultaneous confidence at least as big as the nominal one, and when all $n_{ij} = m$, formulas (8-18) and (8-20) provide actual simultaneous confidence equal to the nominal one.) But note that in practical terms, the inferences they provide (and indeed the ones provided by formula (8-16) for differences in main effects) are not of much interest unless the interactions $\alpha\beta_{ij}$ have been judged to be negligible. That is, when the $\alpha\beta_{ij}$ are negligible, the $\alpha_i - \alpha_{i'}$ represent not only average differences in factor A level i and level i' mean responses, $\mu_{i.} - \mu_{i'.}$, but all *individual* differences $\mu_{ij} - \mu_{i'j}$ as well. When plots of combination means versus (say) levels of factor B essentially exhibit parallelism, the average difference between the A level i profile and the A level i' profile is equal to the difference between each (i, j) and (i', j) pair of means. Then $\alpha_i - \alpha_{i'}$ can truly be thought of as *the* change in mean response accompanying a change in level of factor A from i to i'.

When there is statistically detectable and physically important lack of parallelism (nonnegligible interactions) in a two-way factorial, this author does not recommend the mechanical use of formal inference methods like (8-16) through (8-20) for main effects. Nonnegligible interactions constitute a warning that the patterns of change in mean response, as one moves between levels of one factor, (say, B) are different for various levels of the other factor (say, A). That is, the pattern in the μ_{ij} is not a simple one generally describable in terms of the two factors separately. Rather than trying to understand the pattern in terms of main effects, something else must be done.

As discussed in Section 4-4, sometimes a transformation can produce a response variable that can be understood in terms of main effects only. At other times, restriction of attention to part of an original factorial produces a study (of reduced scope) where it makes sense to think in terms of main effects only. (In Example 8-1, for example, consideration of only butt and lap joints with the three different woods gives an arena where "negligible interactions" may be a sensible conceptualization of joint strength.) Or it may be most natural to mentally separate an $I \times J$ factorial into $I(J)$ different $J(I)$ level studies on the effects of factor B(A) at different levels of A(B). (One might think of the 3×3 wood joint strength study in Example 8-1 as three different studies, one for each joint type, of the effects of wood type on strength.) Or if none of these approaches to analyzing a two-way factorial data set possessing nonnegligible interactions is attractive in a particular application, it is always possible to ignore the two-way structure completely and treat the $I \cdot J$ samples as arising from simply $I \cdot J$ unstructured different conditions.

8-2 No-Interaction Analyses in Balanced Two-Way Factorials

The previous section introduced some basic methods of inference that are applicable in any complete two-way factorial study, provided it contains some replication. The aims were primarily to aid judgments about the statistical and practical importance of interactions and then to measure the main effects of the factors in cases where the interactions are negligible. This section will explore further the implications (in terms of formal inference methods) of a *no-interaction* view of a two-way factorial study.

First there is a brief discussion of the relationship between a no-interaction model for a two-way factorial and the usual "equal variances, normal distributions" model, and a review of the diagnostics that are available to help judge the appropriateness of the no-interaction description. For the case of balanced data, a no-interaction estimate of response variance σ^2 is introduced. Then some statistical interval methods based specifically on the no-interaction model are considered for the case of balanced data. Confidence intervals for mean responses and for comparing main effects and then prediction and tolerance intervals are discussed. Then the observation is made that when the *a priori* expectation is that interactions in a two-factor study are negligible, this belief/model can make possible formal inference based on the no-interaction estimate of σ^2 even when the study includes no replication. A common application of this notion is considered — namely, to the analysis of randomized complete block studies where the block size is the same as the number of treatment levels. The section closes with a few additional comments intended to help put the material of the section into proper perspective.

The No-Interaction 2-Factor Model and Balanced-Data Fitted Values, Residuals, and Variance Estimate

Section 8-1 argued that the basic "equal variances, underlying normal distributions" model for two-way factorial data sets

$$y_{ijk} = \mu_{ij} + \epsilon_{ijk} \tag{8-21}$$

can be rewritten in the equivalent form

$$y_{ijk} = \mu_{..} + \alpha_i + \beta_j + \alpha\beta_{ij} + \epsilon_{ijk} \tag{8-22}$$

where the α_i, β_j, and $\alpha\beta_{ij}$ are defined in terms of the mean responses μ_{ij} in Definitions 8-1 and 8-2. Models (8-21) and (8-22) place no restrictions on the means μ_{ij}, but Sections 8-1 and 4-3 emphasize the great utility/simplification provided in 2-factor contexts when the μ_{ij} are judged to have essentially a "parallel profiles" pattern. In algebraic terms, this is the possibility that

$$\alpha\beta_{ij} \approx 0 \qquad \text{for all } i, j \tag{8-23}$$

and equation (8-23) represents a substantial restriction on the sets of values that $I \cdot J$ means μ_{ij} can assume.

It turns out to be both possible and practically useful to develop methods of statistical inference based on the assumption that interactions are negligible. There are at least two different ways of expressing the complete set of mathematical assumptions under which such methods are applicable. The first is to combine equations (8-21) and (8-23) and write

$$y_{ijk} = \mu_{ij} + \epsilon_{ijk} \qquad \text{and all } \alpha\beta_{ij} = 0 \tag{8-24}$$

(where the ϵ_{ijk} are iid normal $(0, \sigma^2)$ random variables), and the second is to combine equations (8-22) and (8-23) and write

$$y_{ijk} = \mu_{..} + \alpha_i + \beta_j + \epsilon_{ijk} \tag{8-25}$$

(where again the ϵ_{ijk} are iid normal $(0, \sigma^2)$ random variables). Both displays (8-24) and (8-25) are statements in symbols of the assumptions that observations are normally distributed with a common variance and that the (theoretical) means of those observations, the μ_{ij}, possess the parallelism/no-interaction property.

From display (8-24) (or the word statement of the model assumptions), it should be clear that statistical methods based on the model specified by display (8-24) or (8-25) will not be as widely applicable as those based on the assumptions represented by equation (8-21) or (8-22). Statements (8-24) and (8-25) are restricted versions of statements (8-21) and (8-22). However, when the no-interaction assumptions are appropriate, statistical methods based on them will typically yield more pointed real-world engineering implications than those based on the more general assumptions (8-21) and (8-22). The "trick," of course, is determining when the more specialized no-interaction model is appropriate to use.

Much of what has already been said about two-way factorial analyses has bearing on the question of when the no-interaction model appears adequate to describe the results of a statistical engineering study. The plots of sample means against levels of one of the factors (introduced in Section 4-3 and enhanced with error bars in Section 8-1) and the confidence intervals for interactions (discussed in Section 8-1) are generally applicable methods that are relevant to this question. And (as discussed in Section 4-3) when factorial data are balanced, it is a relatively simple matter to find (least squares) fitted values for a simplified relationship like

$$y \approx \mu_{..} + \alpha_i + \beta_j$$

(by simply adding appropriate fitted effects) and to use the fitted values in computing residuals. These in turn lead to residual plots (both in normal plot form, as in Example 4-8, and in plots versus $\hat{y}$ or other variables, like those in Example 4-12) and R^2 comparisons (like those made in Example 4-8).

EXAMPLE 8-3

(Example 8-2 revisited) **The No-Interaction Two-Way Model and the Drilled Aluminum Strip Tensile Strength Study.** For the drilled aluminum strip tensile strength study of Clubb and Goedken, the plot of sample means enhanced with error bars in Figure 8-5 and confidence intervals for the interactions presented in Example 8-2 made it tenable to adopt a no-interaction description of tensile strength. To further investigate the appropriateness of displays (8-24) and (8-25) for describing how tensile strength depends on hole size and placement, consider finding R^2 values for both the relationships

$$y \approx \mu_{..} + \alpha_i + \beta_j + \alpha\beta_{ij}$$

(corresponding to model (8-21) or (8-22)) and

$$y \approx \mu_{..} + \alpha_i + \beta_j$$

(corresponding to model (8-24) or (8-25)) and making residual plots for the second of these.

Fitted values corresponding to model (8-21) or (8-22) are the sample means. (Model (8-21) is after all nothing but the one-way model of Chapter 7 written in fancy notation.) So first considering the R^2 value related to model (8-21) or (8-22),

$$\sum (y_{ijk} - \hat{y}_{ijk})^2 = \sum (y_{ijk} - \bar{y}_{ij})^2$$
$$= (n - IJ)s_P^2$$
$$= (18 - 6)(11{,}375.58)$$
$$= 136{,}507$$

Since it is possible to check that for the data of Table 8-5,

$$\sum (y_{ijk} - \bar{y})^2 = 286{,}229$$

the R^2 value corresponding to model (8-21) is

$$R^2 = \frac{286{,}229 - 136{,}507}{286{,}229} = .523$$

Because the data in Table 8-5 are balanced, least squares fitted values corresponding to the no-interaction model can be found by using Definition 4-5 to compute fitted main effects and then employing the relationship

$$\hat{y} = \bar{y}_{..} + a_i + b_j = \bar{y}_{..} + (\bar{y}_{i.} - \bar{y}_{..}) + (\bar{y}_{.j} - \bar{y}_{..}) = \bar{y}_{i.} + \bar{y}_{.j} - \bar{y}_{..} \qquad \text{(8-26)}$$

Table 8-7 gives observations, fitted values from equation (8-26), and residuals corresponding to the no-interaction model.

TABLE 8-7

Observations, Fitted Values, and Residuals for the No-Interaction Model and the Drilled Aluminum Strip Strength Study

(i, j)	y Tensile Strength (psi)	$\hat{y} = \bar{y}_{..} + a_i + b_j$	$e = y - \hat{y}$
(1,1)	5728	5,619.78	108.22
(1,1)	5664	5,619.78	44.22
(1,1)	5514	5,619.78	−105.78
(1,2)	5689	5,745.88	−56.88
(1,2)	5891	5,745.88	145.12
(1,2)	5611	5,745.88	−134.88
(2,1)	5508	5,506.44	1.56
(2,1)	5620	5,506.44	113.56
(2,1)	5375	5,506.44	−131.44
(2,2)	5579	5,632.56	−53.56
(2,2)	5557	5,632.56	−75.56
(2,2)	5778	5,632.56	145.44
(3,1)	5469	5,466.44	2.56
(3,1)	5440	5,466.44	−26.44
(3,1)	5460	5,466.44	−6.44
(3,2)	5621	5,592.56	28.44
(3,2)	5664	5,592.56	71.44
(3,2)	5523	5,592.56	−69.56

Then corresponding to model (8-24) or (8-25) is

$$\sum(y_{ijk} - \hat{y}_{ijk})^2 = (108.22)^2 + \cdots + (71.44)^2 + (-69.56)^2 = 138{,}750$$

so the R^2 value corresponding to the no-interaction model is

$$R^2 = \frac{286{,}229 - 138{,}750}{286{,}229} = .515$$

Notice that this value is very little less than the value of R^2 corresponding to the "no restrictions on the μ_{ij}" model (8-21). This is yet another way of expressing the fact that relative to the within-sample variability present in the students' data, the lack of parallelism evident in Figure 8-5 is rather small.

Figures 8-6 and 8-7 are (respectively) a plot of residuals versus $\hat{y}$ and a normal plot of residuals for the no-interaction model. Neither is remarkable, and in the end one finds little evidence against the appropriateness of a no-interaction description of tensile strength.

FIGURE 8-6

Plot of Residuals versus Predicted Responses for the No-Interaction Model in the Aluminum Strip Strength Study

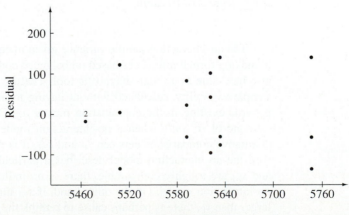

FIGURE 8-7

Normal Plot of Residuals
for the No-Interaction Model
in the Aluminum Strip
Strength Study

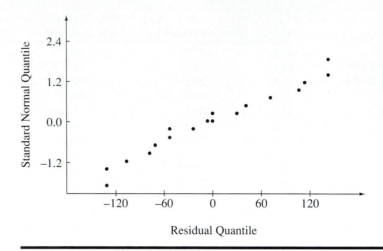

Section 7-4 introduced the notion that the basic estimate of σ^2 in a multisample study, (namely, s_P^2) can be thought of as the sum of squared residuals for the one-way model divided by an appropriate number of degrees of freedom. A recurring theme in the rest of this book concerns alternative estimates of σ^2. In cases where some restriction of the general one-way model holds (in terms of some special relationship between the means involved, such as lack of interaction in a two-way factorial), it is possible to derive a corresponding estimate of σ^2 as an alternative to s_P^2. Such estimates of σ^2 will have the structure of a sum of squared residuals for the model in question, divided by an appropriate number of degrees of freedom. In the present situation, model (8-24) or (8-25) suggests a "no-interaction s^2."

DEFINITION 8-3

In a balanced complete two-way factorial study with fitted grand mean $\bar{y}_{..}$, fitted A main effects a_i, fitted B main effects b_j, and common sample size m,

$$s_{NI}^2 = \frac{1}{mIJ - I - J + 1} \sum_{i,j,k} \left(y_{ijk} - (\bar{y}_{..} + a_i + b_j) \right)^2$$

$$= \frac{1}{mIJ - I - J + 1} \sum_{i,j,k} (y_{ijk} - \bar{y}_{i.} - \bar{y}_{.j} + \bar{y}_{..})^2 \qquad \textbf{(8-27)}$$

will be called a **no-interaction sample variance**. Associated with it are $\nu = mIJ - I - J + 1$ degrees of freedom.

The no-interaction sample variance given in equation (8-27) will presently be used in making formal inferences based on balanced complete two-way factorial data. But it also has value as a rough diagnostic tool. That is, s_P^2 is the basic measure of observed sample variability, calculated without having to make any special assumptions about patterns existing in the combination means μ_{ij}. On the other hand, s_{NI}^2 is a sensible estimate of σ^2 *only* when a no-interaction model is appropriate. By making rough qualitative comparisons between s_P and s_{NI}, it is possible to get another view of how well the no-interaction model describes a particular engineering situation. Where s_P and s_{NI} are roughly comparable, there is no indication of statistical inadequacy of a no-interaction model. On the other hand, if s_{NI} turns out to be an order of magnitude larger than s_P, there is perhaps cause to rethink the use of a model like that specified in display (8-24) or (8-25).

EXAMPLE 8-3
(continued)

In the aluminum strip tensile strength study, the sum of the squares of the residuals (for a no-interaction model) found in the last column of Table 8-7 is about 138,750. Then, from equation (8-27),

$$s_{NI}^2 = \frac{1}{3 \cdot 3 \cdot 2 - 3 - 2 + 1}(138{,}750) = 9{,}910.7 \text{ lb}^2$$

so

$$s_{NI} = \sqrt{9{,}910.7} = 99.6 \text{ lb}$$

In Example 8-2, it was shown that for this data set, $s_P = 106.7$ lb. Comparing the values of s_{NI} and s_P, one sees that s_{NI} is in line with the more generally applicable s_P. This is yet another way of phrasing the fact that there is no strong evidence of inadequacy of a no-interaction view of ultimate tensile strength.

s_{NI}^2 defined in equation (8-27) and the statistical interval methods based on it (which are about to be discussed) are balanced-data entities. The reader may well be asking, "Why the restriction to balanced data? And what do I do if not all n_{ij} are the same?" The answers to these questions were alluded to already in Section 4-3. The problem with trying to use a formula like (8-27) with unbalanced two-way factorial data is that the process of least squares fitting of a no-interaction description to such data doesn't typically yield $\bar{y}_{..} + a_i + b_j$ as a formula for fitted values. So while one can imagine formally calculating $\bar{y}_{..} + a_i + b_j$ values for unbalanced data, there are typically better fitted values available consistent with the no-interaction model. The values $y_{ijk} - (\bar{y}_{..} + a_i + b_j)$ aren't appropriate as residuals, and a formula like (8-27) won't generally give a sensible no-interaction estimate of σ^2. What must be done to develop inference methods analogous to the ones that follow is to make use of formal inference tools for regression analysis and the technique of "dummy variables." A discussion of that methodology must wait until the end of Chapter 9.

Balanced-Data Confidence Intervals under the No-Interaction Model

The statistical interval methods in Section 8-1 are general ones, in the sense that they are valid for any sample sizes and regardless of whether or not interactions are present. The formulas in the rest of this section are not so general, being mathematically defensible only under the no-interaction model and applicable only to balanced data. Nevertheless, when appropriate, they tend to produce shorter (and therefore more informative) intervals than their more general counterparts. That is, where applicable, they tend to provide sharper real-world engineering insights than the methods of Section 8-1.

As a first instance of what is possible under model (8-24) or (8-25) with balanced data, consider the confidence interval estimation of a single mean response μ_{ij}. A no-interaction assumption implies that

$$\mu_{ij} = \mu_{..} + \alpha_i + \beta_j$$

and then (if the data are balanced) a sensible empirical approximation of μ_{ij} is given in relationship (8-26) — that is, by

$$\hat{y} = \bar{y}_{..} + a_i + b_j = \bar{y}_{i.} + \bar{y}_{.j} - \bar{y}_{..} \tag{8-28}$$

It is further possible to verify that Proposition 5-1 and the model assumption of equal variances in equation (8-24) or (8-25) gives

$$Var\,\hat{y} = \frac{\sigma^2}{m} \cdot \frac{I + J - 1}{IJ} \tag{8-29}$$

Then (accepting that under the no-interaction model, s_{NI}^2 is a balanced-data estimate of the response variance having $\nu = mIJ - I - J + 1$ associated degrees of freedom) it is completely plausible that under model (8-24) or (8-25), a two-sided individual confidence interval for the mean response at level i of factor A and level j of factor B can be made using endpoints

$$(\bar{y}_{..} + a_i + b_j) \pm ts_{NI}\sqrt{\frac{I + J - 1}{mIJ}} \tag{8-30}$$

where t is chosen such that the probability the $t_{mIJ-I-J+1}$ distribution assigns to the interval between $-t$ and t corresponds to the desired confidence. (One-sided individual confidence intervals are made from formula (8-30) in the usual way.)

EXAMPLE 8-3
(continued)

Consider making individual two-sided 99% confidence intervals for mean ultimate tensile strengths of aluminum strips for the hole sizes and placements studied by Clubb and Goedken. $s_{NI} = 99.6$ lb has associated with it

$$\nu = 3 \cdot 3 \cdot 2 - 3 - 2 + 1 = 14$$

degrees of freedom, and the .995 quantile of the t_{14} distribution is 2.977. So the plus-or-minus part of formula (8-30) gives

$$2.977(99.6)\sqrt{\frac{3 + 2 - 1}{3 \cdot 3 \cdot 2}} = 139.8 \text{ lb}$$

Then individual 99% confidence intervals for mean tensile strengths for the $3 \cdot 2 = 6$ different hole size/location combinations can be found by attaching an uncertainty of ± 139.8 lb to the six different fitted mean responses appearing in the third column of Table 8-7. For example, an interval for the mean strength of strips with .149 in. diameter holes placed .5 in. from their edges would have endpoints

$$5,619.8 \pm 139.8$$

that is,

$$5,480.0 \text{ lb} \quad \text{and} \quad 5,759.6 \text{ lb}$$

It is worth pointing out what the more general formula (7-14) of Section 7-2 would give for two-sided individual 99% confidence limits for mean responses in this situation. Since $s_P = 106.7$ lb has associated with it

$$\nu = n - r = 3 \cdot 3 \cdot 2 - 6 = 12$$

degrees of freedom, and the .995 quantile of the t_{12} distribution is 3.055, the plus-or-minus part of the general (pooled variance) confidence interval for an individual mean gives

$$3.055\frac{106.7}{\sqrt{3}} = 188.2 \text{ lb}$$

Thus, individual 99% confidence intervals for the mean tensile strengths can be found by attaching an uncertainty of ± 188.2 lb to the six sample means appearing in Table 8-6. For example, the mean strength of strips with .149 in. diameter holes placed .5 in. from their edges can be estimated using an interval with endpoints

$$5,635.3 \pm 188.2$$

that is,

$$5{,}447.1 \text{ lb} \qquad \text{and} \qquad 5{,}823.5 \text{ lb}$$

Notice that not only is the second interval in the preceding example centered at a somewhat different value than the first (5,635.3 lb versus the earlier 5,619.8 lb), but the second plus-or-minus figure attached to the fitted value is larger (188.2 lb as opposed to 139.8 lb). These are both typical of what happens when one compares intervals made using formula (8-30) of this section and formula (7-14) of Section 7-2 in situations where the no-interaction model adequately describes balanced two-way factorial data. $\bar{y}_{..} + a_i + b_j$ employed in formula (8-30) will not typically be exactly the same as $\bar{y}_{ij}$, as it employs data not only from level i of A and level j of B, but from all $I \cdot J$ samples in the estimation of μ_{ij} (through the no-interaction assumption).

Further, although when the no-interaction model is appropriate, one can expect s_{NI} and s_P to be comparable, the respective degrees of freedom associated with them are related by

$$mIJ - I - J + 1 > (m - 1)IJ$$

So for a given confidence level, the t value appearing in formula (8-30) will be smaller than the one in formula (7-14) of Section 7-2. Further, the radicals appearing in the two formulas are related by

$$\sqrt{\frac{I + J - 1}{mIJ}} < \sqrt{\frac{1}{m}}$$

The facts that the respective t values and radicals are related in this way then both work to make the intervals in formula (8-30) shorter (and thus more informative) on average than those made without employing the no-interaction model assumption.

This is another clear example of a pervasive qualitative principle of statistical inference: The more specialized a model one finds to describe a physical situation, the sharper one's data-based inferences can be made. The practical tension is to avoid going too far in the direction of specialization and attempting to use models that are not general enough to describe the particular engineering application.

A second example of the kind of formal inference methods for balanced data possible under model (8-24) or (8-25) concerns the comparison of main effects. Under a no-interaction model assumption, the individual and simultaneous confidence interval methods for $\alpha_i - \alpha_{i'}$ ($\beta_j - \beta_{j'}$) discussed in Section 8-1 can be sharpened by exchange of s_{NI} for s_P and appropriate adjustment of the degrees of freedom. That is, under the no-interaction model assumptions, a two-sided balanced-data individual confidence interval for the difference in A main effects at levels i and i' ($\alpha_i - \alpha_{i'} = \mu_{i.} - \mu_{i'.}$) can be made using endpoints

$$\bar{y}_{i.} - \bar{y}_{i'.} \pm t s_{NI} \sqrt{\frac{2}{Jm}} \tag{8-31}$$

where t is chosen such that the probability the $t_{mIJ-I-J+1}$ distribution assigns to the interval between $-t$ and t corresponds to the desired confidence. One-sided individual confidence intervals can be made from formula (8-31) in the usual way, and there is also a Tukey version of the intervals (8-31). That is, under the assumptions of model (8-24) or (8-25), two-sided simultaneous confidence intervals for all possible differences in

A main effects ($\alpha_i - \alpha_{i'} = \mu_{i.} - \mu_{i'.}$) can be made using respective endpoints

$$\bar{y}_{i.} - \bar{y}_{i'.} \pm \frac{q^* s_{\text{NI}}}{\sqrt{Jm}} \tag{8-32}$$

where q^* is taken from Tables D-10 using $\nu = mIJ - I - J + 1$ degrees of freedom and number of means to be compared I. The .95 or .99 quantile figure is used, depending on whether 95% or 99% simultaneous confidence is desired.

There are, of course, formulas for comparing B main effects that are parallel to formulas (8-31) and (8-32). Endpoints for individual confidence intervals for a difference in B main effects ($\beta_j - \beta_{j'} = \mu_{.j} - \mu_{.j'}$) are

$$\bar{y}_{.j} - \bar{y}_{.j'} \pm t s_{\text{NI}} \sqrt{\frac{2}{Im}} \tag{8-33}$$

where t has $\nu = mIJ - I - J + 1$ associated degrees of freedom. And endpoints for simultaneous two-sided confidence intervals for all differences in B main effects are

$$\bar{y}_{.j} - \bar{y}_{.j'} \pm \frac{q^* s_{\text{NI}}}{\sqrt{Im}} \tag{8-34}$$

where q^* has $\nu = mIJ - I - J + 1$ associated degrees of freedom and is chosen for comparing J means.

EXAMPLE 8-3
(continued)

In Section 8-1, there is a comparison of hole size main effects for the Clubb and Goedken data using simultaneous 95% confidence intervals based on s_P. To illustrate the use of the preceding formulas based on the no-interaction model and s_{NI}, note that for $\nu = 3 \cdot 3 \cdot 2 - 3 - 2 + 1 = 14$ degrees of freedom and $I = 3$ means to be compared, the .95 Studentized range quantile is, from Table D-10-A, 3.70. So for simultaneous 95% confidence, each difference in hole size main effects $\alpha_i - \alpha_{i'} = \mu_{i.} - \mu_{i'.}$ can be estimated via formula (8-32) using a two-sided interval with endpoints

$$\bar{y}_{i.} - \bar{y}_{i'.} \pm \frac{3.70(99.6)}{\sqrt{2 \cdot 3}}$$

that is,

$$\bar{y}_{i.} - \bar{y}_{i'.} \pm 150.4$$

Recall from Section 8-1 that for the tensile strength data,

$$a_1 = 88.9, \qquad a_2 = -24.4, \qquad \text{and} \qquad a_3 = -64.4$$

and therefore that hole size 1 (.149 in.) mean strengths appear to exceed hole size 3 (.221 in.) mean strengths by about

$$\bar{y}_{1.} - \bar{y}_{3.} = a_1 - a_3 = 88.9 + 64.4 = 153.3 \text{ lb}$$

Then, since 153.3 exceeds 150.4 in absolute value, this analysis establishes a detectably nonzero difference in mean strengths, depending upon hole size.

It is important, however, to note that a similar "Tukey" analysis in Section 8-1 based on s_P rather than s_{NI} did not identify any differences in hole size main effects as detectably nonzero. (Review the calculations in this example and Example 8-2. The principal contribution to this phenomenon in this particular case is the fact that here $s_{\text{NI}} < s_P$. The difference between the 14 degrees of freedom associated with s_{NI} and the 12 associated with s_P doesn't have much impact on the calculations.) Three facts are relevant to the practical interpretation of this result. 1) 153.3 is close to 150.4. 2) The earlier analysis of hole size effects was inconclusive. 3) At the beginning of Example 8-2, there was some unease with the model of assumption of equal variances, so "close calls" should not be based on standard inference methods and these data. Considering these facts, in some real engineering situations one might well want to reserve judgment on the importance of hole size in determining tensile strength. If there were potentially severe practical consequences associated with incorrectly judging the importance of changes in hole diameter of the magnitude .149 in. to .221 in., this author would probably lobby for the collection of more data in a case like this before being willing to base engineering decisions on the experimental results.

Balanced-Data Prediction and Tolerance Intervals under the No-Interaction Model

Prediction and tolerance intervals have the same roles in multisample statistical engineering studies with factorial structure as they do in one-sample or unstructured multisample studies. They are meant to locate one potential additional observation or some specified portion of all potential additional observations at some particular combination of a level of factor A and one of factor B. It is perfectly legitimate to essentially ignore two-way factorial structure and apply formula (7-16), (7-17), or (7-18) of Section 7-2 to produce prediction or tolerance intervals in a two-way factorial study. Indeed, when important interactions are evident in the data, this is the only sensible way to proceed. But, when the two-way no-interaction model is appropriate, one can derive formulas from it that will (on average) produce shorter and therefore more informative intervals than those produced simply using the formulas from Section 7-2.

Consider first the problem of making a prediction interval for a single additional observation from the response distribution at level i of factor A and level j of factor B, based on balanced data. A combination of the kinds of reasoning used to produce formula (8-30) and formula (7-16) from Section 7-2 shows that if $y_{ijn_{ij}+1}$ is a single additional observation from the ith level of A and jth level of B, the no-interaction model assumptions imply that

$$\frac{y_{ijn_{ij}+1} - (\bar{y}_{..} + a_i + b_j)}{s_{\mathrm{NI}}\sqrt{1 + \dfrac{I + J - 1}{mIJ}}}$$

has a $t_{mIJ-I-J+1}$ distribution. This in turn implies that a two-sided prediction interval for a single additional observation has endpoints

$$(\bar{y}_{i.} + \bar{y}_{.j} - \bar{y}_{..}) \pm ts_{\mathrm{NI}}\sqrt{1 + \frac{I + J - 1}{mIJ}} \tag{8-35}$$

where the associated prediction confidence is the probability assigned to the interval from $-t$ to t by the $t_{mIJ-I-J+1}$ distribution. And as usual, one of the two endpoints indicated by formula (8-35) can be used to make a one-sided prediction interval with "unconfidence" only half that of the corresponding two-sided interval.

EXAMPLE 8-3
(continued)

Consider the problem of producing a one-sided 90% prediction interval of the form $(\#, \infty)$ for, say, an additional tensile strength for an aluminum strip of the type tested by Clubb and Goedken having a .149 in. diameter hole placed .5 in. from its edge. Since the .90 quantile of the t_{14} distribution is 1.345 and (from before) $\bar{y}_{..} + a_1 + b_1 = \bar{y}_{1.} + \bar{y}_{.1} - \bar{y}_{..} = 5{,}619.8$, formula (8-35) yields the 90% lower prediction bound

$$5{,}619.8 - 1.345(99.6)\sqrt{1 + \frac{3 + 2 - 1}{3 \cdot 3 \cdot 2}} = 5{,}471.7 \text{ lb}$$

As far as producing tolerance intervals under the no-interaction model assumptions goes, the situation is much like that in Section 7-2. It turns out that what can be sensibly done here is to provide balanced-data one-sided tolerance intervals for a single underlying distribution, giving a fairly exact method of determining necessary constants for 95% confidence and a more crude method of approximating the constants in general. That is, intervals

$$(\bar{y}_{i.} + \bar{y}_{.j} - \bar{y}_{..} - \tau s_{\text{NI}}, \infty) \tag{8-36}$$

and

$$(-\infty, \bar{y}_{i.} + \bar{y}_{.j} - \bar{y}_{..} + \tau s_{\text{NI}}) \tag{8-37}$$

can be used as balanced-data one-sided tolerance intervals for a fraction p of the ijth underlying distribution, provided τ is appropriately chosen (depending on p, I, J, m, and the desired confidence level).

Table D-8 can be used to produce values of τ for 95% confidence. One first computes the value of a parameter ξ according to

$$\xi = \frac{\sqrt{\dfrac{mIJ}{I + J - 1}} Q_z(p)}{\sqrt{2(mIJ - I - J + 1) + \dfrac{mIJ Q_z^2(p)}{I + J - 1}}} \tag{8-38}$$

where, as usual, $Q_z(p)$ is the standard normal p quantile. Next Table D-8 is entered using this value, and the degrees of freedom parameter $\nu = mIJ - I - J + 1$ to produce a value λ in the body of the table. Finally, using λ, τ appropriate for use in formula (8-36) or (8-37) can be found using the relationship

$$\tau = \frac{Q_z(p) + \lambda \sqrt{\left(\dfrac{I + J - 1}{mIJ}\right)\left(1 + \dfrac{\dfrac{mIJ Q_z^2(p)}{I + J - 1} - \lambda^2}{2(mIJ - I - J + 1)}\right)}}{1 - \dfrac{\lambda^2}{2(mIJ - I - J + 1)}} \tag{8-39}$$

For confidence levels other than 95%, only an approximate method of finding τ for use in formulas (8-36) and (8-37) can be given here. That is, for a one-sided tolerance interval with confidence level γ for a fraction p of the ijth underlying distribution, one may use either formula (8-36) or (8-37) with τ determined as in equation (8-39) by employing the approximation

$$\lambda \approx Q_z(\gamma) \tag{8-40}$$

EXAMPLE 8-3
(continued)

For purposes of comparing the result to the prediction interval found earlier, consider finding a 95% tolerance interval of the form $(\#, \infty)$ for 90% of many additional tensile strengths of aluminum strips having .149 in. diameter holes placed .5 in. from their edges. For this level of confidence, one may use Table D-8 and begin by calculating an appropriate value of ξ via formula (8-38). In the present situation,

$$\xi = \frac{\sqrt{\dfrac{3 \cdot 3 \cdot 2}{3 + 2 - 1}}(1.282)}{\sqrt{2(3 \cdot 3 \cdot 2 - 3 - 2 + 1) + \dfrac{3 \cdot 3 \cdot 2(1.282)^2}{3 + 2 - 1}}} = .46$$

So consulting Table D-8, for this value of ξ and $\nu = mIJ - I - J + 1 = 14$, one finds via rough interpolation that $\lambda \approx 1.683$ is indicated for use in formula (8-39). Plugging this value of λ into formula (8-39), one obtains

$$\tau = \frac{1.282 + 1.683\sqrt{\left(\dfrac{3+2-1}{3\cdot 3\cdot 2}\right)\left(1 + \dfrac{\dfrac{3\cdot 3\cdot 2(1.282)^2}{3+2-1} - (1.683)^2}{2(3\cdot 3\cdot 2 - 3 - 2 + 1)}\right)}}{1 - \dfrac{(1.683)^2}{2(3\cdot 3\cdot 2 - 3 - 2 + 1)}} = 2.378$$

Finally, a 95% lower tolerance bound for 90% of additional tensile strengths with .149 in. diameter holes placed .5 in. from the strip edges is (from formula (8-36))

$$5,619.8 - 2.378(99.6) = 5,382.9 \text{ lb}$$

As expected, this value is smaller than the 90% lower prediction bound found earlier. (As always, the tolerance bound is doing a more ambitious job than the corresponding prediction bound and must therefore be looser.)

Notice also that use of the approximation (8-40) would give $\lambda \approx Q_z(.95) = 1.645$. The reader can check that use of this value of λ in formula (8-39) then produces $\tau = 2.347$ and thus the value 5,383.1 lb, which is somewhat larger than the more exact bound. Of course, for 95% confidence there is no reason to resort to approximation (8-40). But the comparison here between $\lambda \approx 1.683$ and $Q_z(.95) = 1.645$ illustrates the fact that approximation (8-40) generally serves to get one into the right ballpark.

Complete Two-Way Factorials with No Replication

One of the important recurring qualitative themes of this book is the importance of replication in statistical engineering studies. In spite of this fact, it can be noted that in all of the formulas of this section, there is nothing to preclude the possibility that $m = 1$ (i.e., that there is no replication whatsoever in the data set under consideration). At first sight, it might thus appear that these formulas with $m = 1$ somehow represent a clever sidestepping of the basic need to ascertain the size of the baseline variation present in an engineering context. This is, of course, not a completely accurate analysis of the situation. And it is important to consider the practical limitations on the use of the methods of this section with $m = 1$.

It is essential to bear in mind that the mathematical basis of the methods in this section is the no-interaction model. s_{NI} that appears in the formulas of this section (and can be computed from two-way factorial data even when $m = 1$) is an estimate of σ, the basic measure of spread for each underlying distribution, only when use of the no-interaction model makes sense. When there is no replication, s_{NI} is derived only by taking the observed "lack of parallelism" in the data as an indication of the size of σ. (To see this, return to Definition 8-3 and note that when $m = 1$,

$$s_{\text{NI}}^2 = \frac{1}{(I-1)(J-1)} \sum_{i,j} \left(\bar{y}_{ij} - (\bar{y}_{..} + a_i + b_j)\right)^2 \tag{8-41}$$

That is, s_{NI}^2 is a kind of average of the squared fitted interactions.)

It is a basic failing of unreplicated data that extremely limited tools are available for investigating the reasonableness of assumptions like those of model (8-24) or (8-25). With $m = 1$, there is no s_P to gage s_{NI} against, or to use in making inferences for interactions as in Section 8-1. About all one can do toward assessing how reasonable a no-interaction model looks based on $I \cdot J$ samples of $m = 1$ observation is to compute and plot residuals. (And because of linear dependencies existing among the residuals, this is not of much help unless I and J are big enough that the degrees of freedom $(I - 1)(J - 1)$ is of reasonable size.) So although the formulas of this section are in theory applicable with $m = 1$ under the no-interaction model, it is often essentially impossible to empirically validate those model assumptions based on data lacking replication.

If the formulas of this section are to be used in the analysis of two-way factorial data with $m = 1$, the use of a no-interaction model will typically have to be supported by *a priori* (in addition to data-based) considerations. Sometimes past experience with the effects of two variables A and B on a response y may make an engineer "sure" that they act "separately" (i.e., that there are no interactions). (Of course, how often it is sensible to say that one "knows" enough about a system to be sure that interactions are negligible, and yet needs data to detail the main effects, is open to debate.)

There is one kind of circumstance where it is quite common and perhaps logically defensible to assume *a priori* that model (8-24) or (8-25) is appropriate and base inferences on unreplicated two-way factorial data and the methods of this section. That is where one of the factors involved in an experiment is a *blocking factor*. In such cases, it may make good intuitive sense to think, "Factor B is not a fundamental process variable but was created only to produce homogeneous environments for the comparison of levels of A. I expect that except for baseline variation and possible differences in block average, the pattern of responses (to changes in level of A) will be the same in each block." When this is the case, the no-interaction model becomes potentially relevant, and $m = 1$ simply means that each block is exactly large enough to accommodate each level of the primary experimental variable once.

EXAMPLE 8-4

An Unreplicated Randomized Complete Block Experiment for Comparing Six Different Measurement Devices. Chapter 13 of the NBS (now NIST) Handbook 91, *Experimental Statistics* by M.G. Natrella, contains a discussion of (an unreplicated) randomized complete block experiment, performed primarily to compare the operation of six different measurement apparatuses in determining the so-called conversion gain of resistors. The conversion gains (ratios of available current-noise power to applied direct-current power, expressed in decibel units) of four different resistors were measured on each of the six test sets. The resulting values are given in Table 8-8.

TABLE 8-8

Conversion Gains for Four Resistors Measured Once on Each of Six Test Sets

		RESISTOR (BLOCK)			
		1	2	3	4
	1	138.0	152.2	153.6	141.4
	2	141.6	152.2	154.0	141.5
TEST SET	3	137.5	152.1	153.8	142.6
	4	141.8	152.2	153.6	142.2
	5	138.6	152.0	153.2	141.1
	6	139.6	152.8	153.6	141.9

Notice that in the context of this example, it could make sense to expect *a priori* that model (8-24) or (8-25) will describe how factors A(Test Set) and B(Resistor) affect measured conversion gain. If the testing apparatuses are doing their job, it is natural to expect patterns of measured gain across instruments (reflecting calibration differences) to be fairly consistent resistor to resistor, with differences between the physical properties of the resistors showing up in the measurements only as differences in resistor mean responses. That is, *a priori* the no-interaction feature of model (8-24) or (8-25) seems quite plausible. The iid Gaussian $(0, \sigma^2)$ ϵ's part of the model assumptions also has appeal, in that a normal distribution is often an adequate model for errors of physical measurement. And the single-σ feature of the assumptions (8-24) or (8-25) can be interpreted in the present context as a statement of comparable precisions for the test apparatuses.

So this example is one in which considerations outside the data themselves suggest the possibility of using the methods of this section even though $m = 1$. (This is not to say that those in charge of the study that produced the data of Table 8-8 should have been easily satisfied that no replication was necessary or possible. Indeed, it appears that getting at least some replication somewhere in the two-way factorial represented in Table 8-8 would have been relatively easy and extremely advantageous.) With $m = 1$, one uses the data values in Table 8-8 as combination sample means $\bar{y}_{ij}$ and applies the formulas of Section 4-3 to arrive at fitted effects for use in the formulas of this section. The following are row and column average means for the $\bar{y}_{ij}$ appearing in Table 8-8.

$$\bar{y}_{1.} = 146.30 \qquad \bar{y}_{.1} = 139.52 \qquad \bar{y}_{..} = 146.80$$
$$\bar{y}_{2.} = 147.33 \qquad \bar{y}_{.2} = 152.25$$
$$\bar{y}_{3.} = 146.50 \qquad \bar{y}_{.3} = 153.63$$
$$\bar{y}_{4.} = 147.45 \qquad \bar{y}_{.4} = 141.78$$
$$\bar{y}_{5.} = 146.23$$
$$\bar{y}_{6.} = 146.98$$

These in turn lead to the fitted values $\hat{y} = \bar{y}_{..} + a_i + b_j = \bar{y}_{i.} + \bar{y}_{.j} - \bar{y}_{..}$ and residuals given in Table 8-9.

TABLE 8-9

Gains, Fitted Gains, and Residuals for the No-Interaction Model

i Test Set	j Resistor	y Conversion Gain	$\hat{y}$ Fitted Gain	$e = y - \hat{y}$ Residual
1	1	138.0	139.02	−1.02
1	2	152.2	151.75	.45
1	3	153.6	153.13	.47
1	4	141.4	141.28	.12
2	1	141.6	140.05	1.55
2	2	152.2	152.78	−.58
2	3	154.0	154.16	−.16
2	4	141.5	142.31	−.81
3	1	137.5	139.22	−1.72
3	2	152.1	151.95	.15
3	3	153.8	153.33	.47
3	4	142.6	141.48	1.12
4	1	141.8	140.17	1.63
4	2	152.2	152.90	−.70
4	3	153.6	154.28	−.68
4	4	142.2	142.43	−.23
5	1	138.6	138.95	−.35
5	2	152.0	151.68	.32
5	3	153.2	153.06	.14
5	4	141.1	141.21	−.11
6	1	139.6	139.70	−.10
6	2	152.8	152.43	.37
6	3	153.6	153.81	−.21
6	4	141.9	141.96	−.06

Even though there is no replication in this situation, it is the case that with $I = 6$ and $J = 4$, $(I - 1)(J - 1) = 15$ is large enough that examination of residuals is potentially fruitful. Figure 8-8 is a normal plot of the residuals found in the last column of Table 8-9; the plot is not particularly remarkable.

FIGURE 8-8

Normal Plot of Residuals for the No-Interaction Model for Measured Gains

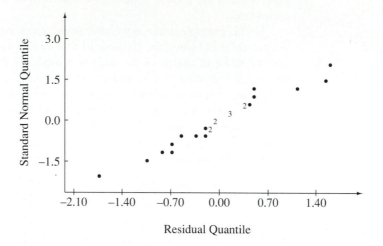

However, Figure 8-9 is more interesting. It is a plot of residuals against fitted response, showing clearly that there are two distinctly different kinds of fitted responses (big ones and small ones) and that the residuals are smaller for the larger fitted gains.

FIGURE 8-9

Plot of Residuals versus Fitted Response for the No-Interaction Model for Measured Gains

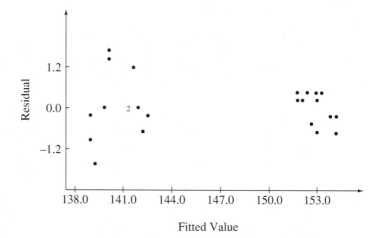

Looking back in Table 8-8, it is obvious (and could be seen from of a plot of residuals versus resistor number as well) that resistors 1 and 4 have low measured conversion gains and big residuals, while resistors 2 and 3 have high measured conversion gains and small residuals. Figure 8-10, which is a plot of observed values against test set number, makes quite clear what is going on in the data of Table 8-8. High gains are simply measured more consistently than low ones. So although it was initially argued that model (8-24) or (8-25) might be useful in the analysis of these data, the data themselves indicate that this may be the case only when attention is restricted to either high or low gains. So, for purposes of finally illustrating the use of some of the methods of this section when $m = 1$, confine attention to the behavior of the test sets for conversion gains around 150. That is, for the rest of this example, only the data for resistors 2 and 3 will be used.

FIGURE 8-10

Plot of Observed Gains versus
Test Set Numbers

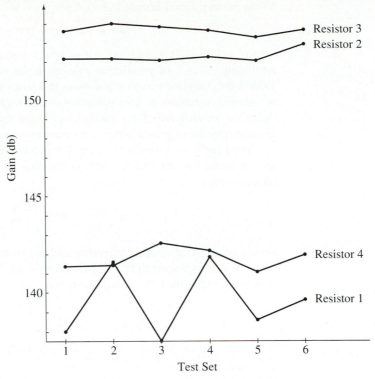

TABLE 8-10

Gains, Fitted Gains, and
Residuals for Resistors 2 and 3
under the No-Interaction Model

i Test Set	j Resistor	y Conversion Gain	$\hat{y}$ Fitted Gain	$e = y - \hat{y}$ Residual
1	2	152.2	152.21	−.01
1	3	153.6	153.59	.01
2	2	152.2	152.41	−.21
2	3	154.0	153.79	.21
3	2	152.1	152.26	−.16
3	3	153.8	153.64	.16
4	2	152.2	152.21	−.01
4	3	153.6	153.59	.01
5	2	152.0	151.91	.09
5	3	153.2	153.29	−.09
6	2	152.8	152.51	.29
6	3	153.6	153.89	−.29

Restricting attention to the second and third columns of Table 8-8 requires the recomputation of fitted values and residuals. The results of this recomputation (involving now only half of the data of Table 8-8 in finding $\bar{y}_{i.}$ and $\bar{y}_{..}$ values) are given in Table 8-10.

No plots of the residuals will be included here, but the reader is invited to verify that they carry no strong evidence against the usefulness of model (8-24) or (8-25) in the analysis of the high-gain data. Therefore, it makes sense to compute s_{NI} and use the Tukey method to compare the six different test sets. First, since restricting attention to resistors 2 and 3 gives $I = 6$ and $J = 2$, from formula (8-27) or (8-41) and the last column of Table 8-10,

$$s_{NI}^2 = \frac{1}{(6-1)(2-1)}((-.01)^2 + (.01)^2 + (-.21)^2 + \cdots + (.29)^2 + (-.29)^2)$$

$$= .0647$$

So the no-interaction sample standard deviation is

$$s_{NI} = .25$$

with $(I - 1)(J - 1) = mIJ - I - J + 1 = 5$ associated degrees of freedom. That is, attributing all lack of parallelism evident in the data in the two middle columns of Table 8-8 to baseline variation and using that lack of parallelism to produce an estimate of inherent variation in gain measurement at high gains, one ends up with the .25 figure for an estimate of the standard deviation that would be associated with many remeasurements of gain of a particular resistor on a particular test set.

Then applying formula (8-32) to the present situation, it is clear that to compare test set means via simultaneous 95% confidence intervals for differences, a precision of plus-or-minus

$$q^* \frac{s_{NI}}{\sqrt{Jm}} = 6.03 \frac{(.25)}{\sqrt{2 \cdot 1}} = 1.07$$

should be attached to each difference in pairs of test set averages. ($\nu = 5$ degrees of freedom and six means to be compared were used in looking up q^*.) Considering only resistors 2 and 3, the test set mean responses are

$$\bar{y}_{1.} = 152.9 \qquad \bar{y}_{3.} = 152.95 \qquad \bar{y}_{5.} = 152.6$$
$$\bar{y}_{2.} = 153.1 \qquad \bar{y}_{4.} = 152.9 \qquad \bar{y}_{6.} = 153.2$$

from which it is clear that the present data are not adequate to definitively identify calibration differences in the six test sets. (No pair of means differs by as much as 1.07.) Substantially more data would be needed to identify with any certainty the nature of differences between the measurement apparatuses.

Some Additional Comments

Before closing this section, it seems wise to try to put the methods into clearer perspective relative to those of the previous section and those of Chapter 7. For one thing, some readers may be wondering when the formulas of Section 8-1 (based on s_P) should be used in estimating main effects and when those of the present section (based on s_{NI}) are more appropriate. This author's view of the matter is as follows. At the early stages of analysis of two-way factorial data, while one is in the process of deciding whether some simplified form of the general two-way model (8-21) or (8-22) might be adequate, the more general estimation methods of Section 8-1 are appropriate. However, if one does settle on the no-interaction model (8-24) or (8-25) as a workable description of an engineering system, then for balanced data, the more specialized (and thus typically more informative) methods of this section should be used to make the sharpest inferences possible.

Another related point is that although this section has concentrated on the model equation

$$y_{ijk} = \mu_{..} + \alpha_i + \beta_j + \epsilon_{ijk}$$

other simplifications are possible as well, the two most easily understood being

$$y_{ijk} = \mu_{..} + \alpha_i + \epsilon_{ijk} \tag{8-42}$$

and

$$y_{ijk} = \mu_{..} + \beta_j + \epsilon_{ijk} \tag{8-43}$$

(Other less interpretable ones could be obtained from the general form by deleting one or both of the "main effect" terms, but not the interaction term.) Note that for the simplified models (8-42) and (8-43), it should be obvious how to proceed with formal inference. Model (8-42) expresses the possibility that all observations from a given level of factor A can be thought of as composing a single sample from an underlying normal distribution with mean $\mu_{..} + \alpha_i$ and variance σ^2. One is thus back in the situation of Chapter 7 with $r = I$ unstructured samples, and all of the methods of that chapter can be applied if one determines that model (8-42) is relevant. Of course, parallel statements can be made regarding model (8-43). And outside of the model (8-24) or (8-25) discussed extensively in this section, only simplifications (8-42) and (8-43) are commonly used. Thus, taken together, Chapter 7 and these first two sections of Chapter 8 pretty much summarize interval-oriented two-way factorial inference.

A final point of clarification regards the typical difference between analyses of randomized complete block studies based on model (8-24) or (8-25) (as illustrated in Example 8-4) and alternative analyses based on model (8-42). For such studies, a naive method of statistical analysis would be to essentially ignore the existence of the blocking variable and operate as if the data were simply I unrelated data samples from the various different levels of the experimental factor A. In the context of Example 8-4, this would amount to ignoring the fact that in the part of the data ultimately used here, two and only two different resistors were used, and thinking of the measured gains as samples of size two of "measurements for resistors of the type studied." In qualitative terms, the effect of failing to explicitly recognize the existence of the blocking variable in such cases is to attribute to baseline or unavoidable experimental variation that variability that is legitimately assignable to differences block-to-block (e.g., resistor-to-resistor). One thereby inflates one's perception of the size of experimental error. For example, considering the resistors 2 and 3 part of the data in Table 8-8, earlier calculations produced $s_{NI} = .25$. But if one were to naively treat the two gain measurements for each test set as composing a sample of size 2, the pooled sample standard deviation from the six different "samples" of size 2 turns out to be $s_P = 1.01$. Ignoring the existence of the blocking variable (resistor) quadruples one's perception of the size of experimental variation.

This notion, that in the analysis of randomized complete block studies the use of model (8-24) or (8-25) is a way to explicitly account for block-to-block variation separately from variation assignable to differences between levels of the primary experimental variable, is so common that the no-interaction model is sometimes termed "the" *randomized complete block model*. This author is not particularly fond of this terminology. Its use tends to cause students to lose track of the fact that blocking has to do with the physical conduct of an experiment, not with how the resulting data are subsequently manipulated. Besides, the use of model (8-24) or (8-25) may or may not be appropriate in the analysis of a given randomized complete block experiment. (*A priori* expectations of lack of treatment by block interactions may not turn out to be supported by the data.) On the other hand, a no-interaction analysis can be appropriate in circumstances other than those where one of the factors is a blocking factor. For these reasons, this author much prefers to use the "no-interaction model" terminology employed in this section instead of "randomized complete block model" jargon.

8-3 *p*-Factor Studies with Two Levels for Each Factor

The two previous sections have looked at formal inference methods for statistical studies possessing two-way factorial structure. This section presents methods of formal inference for complete *p*-way factorials, paying primary attention to those cases where each of *p* factors is represented at only two levels.

The section begins by again pointing out the relevance of the one-way methods of Chapter 7 to structured (in this case, p-way factorial) situations. Next are considered the p-way factorial normal model, definitions of effects in that model, and basic confidence interval methods for the effects, noting the extremely simple forms these take in the 2^p case. Then attention is completely restricted to 2^p studies, and a further method is presented for identifying simplified versions of the full p-way model that are potentially adequate to describe particular applications. That is, the use of the principle of effect sparsity and normal plotting of the fitted effects is considered. Then, for balanced 2^p studies, there follows a review of the fitting of reduced models via the reverse Yates algorithm and the role of residuals in checking the efficacy of such simplifications of the full factorial model. Finally, interval-oriented inference methods based on simplified models in balanced 2^p studies are discussed.

One-Way Methods in p-way Factorials

The place to begin in the analysis of data from a p-way factorial statistical engineering study is to recognize that fundamentally one is just working with several samples of data. Subject to the relevance of the "equal variances, normal distributions" model assumptions of Chapter 7, the inference methods of that chapter are then available for use in analyzing the data.

EXAMPLE 8-5

A 2^3 Factorial Study of Power Requirements in Metal Cutting. In *Fundamental Concepts in the Design of Experiments*, C.R. Hicks describes a study conducted by Purdue University engineering graduate student L.D. Miller on power requirements for cutting malleable iron using ceramic tooling. Miller studied the effects of the three factors

- A   Tool Type (type 1 or type 2)
- B   Tool Bevel Angle (15° or 30°)
- C   Type of Cut (continuous or interrupted)

on the power required to make a cut on a lathe at a particular depth of cut, feed rate, and spindle speed. The response variable used was the vertical deflection (in mm) of the indicator needle on a dynamometer (a measurement proportional to the horsepower required to make the particular cut). Miller's data (as reported by Hicks) are given in Table 8-11.

TABLE 8-11

Dynamometer Readings for 2^3 Treatment Combinations in a Metal Cutting Study

Tool Type	Bevel Angle	Type of Cut	*y* Dynamometer Reading (mm)
1	15°	continuous	29.0, 26.5, 30.5, 27.0
2	15°	continuous	28.0, 28.5, 28.0, 25.0
1	30°	continuous	28.5, 28.5, 30.0, 32.5
2	30°	continuous	29.5, 32.0, 29.0, 28.0
1	15°	interrupted	28.0, 25.0, 26.5, 26.5
2	15°	interrupted	24.5, 25.0, 28.0, 26.0
1	30°	interrupted	27.0, 29.0, 27.5, 27.5
2	30°	interrupted	27.5, 28.0, 27.0, 26.0

The most elementary view possible of the power requirement data in Table 8-11 is as $r = 8$ samples of size $m = 4$. Simple summary statistics for these $2^3 = 8$ samples are given in Table 8-12.

To the extent that the one-way normal model is an adequate description of this power requirement study, the methods of Chapter 7 are available for use in analyzing the data

TABLE 8-12

Summary Statistics for
2^3 Samples of Dynamometer
Readings in a Metal
Cutting Study

Tool Type	Bevel Angle	Type of Cut	$\bar{y}$	s
1	15°	continuous	28.250	1.848
2	15°	continuous	27.375	1.601
1	30°	continuous	29.875	1.887
2	30°	continuous	29.625	1.702
1	15°	interrupted	26.500	1.225
2	15°	interrupted	25.875	1.548
1	30°	interrupted	27.750	0.866
2	30°	interrupted	27.125	0.854

of Table 8-11. The reader is encouraged to verify that plotting of residuals (obtained by subtracting the $\bar{y}$ values in Table 8-12 from the corresponding raw data values of Table 8-11) reveals only one slightly unpleasant feature of the power requirement data relative to the potential use of standard methods of inference. That is that when plotted against levels of the Type of Cut variable, the residuals for interrupted cuts are seen to be on the whole somewhat smaller than those for continuous cuts. (This phenomenon is also obvious in retrospect from the sample standard deviations in Table 8-12. These are smaller for the second four samples than for the first four.) But the disparity in the size of the residuals is not huge (nothing like an order of magnitude). So although there may be some empirical basis for suspecting some improvement in power requirement consistency for interrupted cuts as opposed to continuous ones, the tractability of the "equal variances, normal distributions" model and the kind of robustness arguments put forth at the end of Section 6-3 once again suggest that the standard model and methods be used. This is sensible, provided one then treats resulting inferences as approximate and avoids basing real-world close calls on them.

The pooled sample variance here is

$$s_P^2 = \frac{(4-1)(1.848)^2 + (4-1)(1.601)^2 + \cdots + (4-1)(.854)^2}{(4-1) + (4-1) + \cdots + (4-1)} = 2.226$$

so

$$s_P = 1.492 \text{ mm}$$

with $\nu = n - r = 32 - 8 = 24$ associated degrees of freedom. Then, for example, the P-R method of simultaneous inference from Section 7-3 would produce two-sided simultaneous 95% confidence intervals for mean dynamometer readings with endpoints of the form

$$\bar{y} \pm 2.96 \frac{1.492}{\sqrt{4}}$$

that is, of the form

$$\bar{y} \pm 2.21 \text{ mm}$$

(There is enough precision provided by the data to think of the sample means in Table 8-12 as roughly "all good to within 2.21 mm.") And the other methods of Sections 7-1 through 7-4 based on s_P might be used as well.

The bases of *p*-way factorial inference are exactly analogous to those for two-way factorial inference discussed in the previous two sections. Methods of *p*-way inference are particular applications and extensions of one-way methods. As such, their relevance in given engineering applications should be investigated by the kind of one-way model checks discussed in Section 7-1 and alluded to in the previous example. If use of the one-way methods is badly inappropriate in a p-factor study, then none of the p-way methods that follow here are appropriate either.

p-way Factorial Notation, Definitions of Effects, and Related Confidence Interval Methods

Section 8-1 illustrated that standard notation in two-way factorials requires triple subscripts for naming observations. In a general p-way factorial, $(p + 1)$ subscript notation is required. It is clear that such notation, though necessary, quickly gets out of hand. As in Section 4-3 (on a descriptive level) the exposition here will explicitly develop only the general factorial notation for $p = 3$, leaving the reader to infer by analogy how things would have to go for $p = 4, 5$, etc. (Of course, when specializing to the 2^p situation a bit later in this section, the accompanying special notation introduced in Section 4-3 makes it possible to treat even large-p situations fairly explicitly.)

Then for $p = 3$ factors A, B, and C having (respectively) I, J, and K levels, let

y_{ijkl} = the lth observation in the sample from the ith level of A, jth level of B and kth level of C

For the $I \cdot J \cdot K$ different samples corresponding to the possible combinations of a level of A with one of B and one of C, let

n_{ijk} = the number of observations in the sample from the ith level of A, jth level of B, and kth level of C

$\bar{y}_{ijk}$ = the sample mean of the n_{ijk} observations in the sample from the ith level of A, jth level of B, and kth level of C

s_{ijk} = the sample standard deviation of the n_{ijk} observations in the sample from the ith level of A, jth level of B, and kth level of C

and further continue the dot notations used in Section 4-3 for unweighted averages of the $\bar{y}_{ijk}$ of various kinds. In comparison to the notation of Chapter 7, this amounts to adding two subscripts in order to acknowledge the three-way structure in the samples.

The use of additional subscripts is helpful not only for naming empirical quantities like those above, but also for naming theoretical quantities. That is, with

μ_{ijk} = the underlying mean response corresponding to the ith level of A, jth level of B, and kth level of C

the standard one-way normal model assumptions can be rewritten as

$$y_{ijkl} = \mu_{ijk} + \epsilon_{ijkl} \tag{8-44}$$

where the ϵ_{ijkl} terms are iid normal random variables with mean 0 and variance σ^2. Formula (8-44) could be called the *three-way (normal) model equation*, because it recognizes the special organization of the $I \cdot J \cdot K$ samples according to combinations of levels of the three factors. But beyond this, it says no more or less than the one-way model equation from Section 7-1.

The initial objects of inference in three-way factorial analyses are usually linear combinations of theoretical (underlying) means μ_{ijk}, analogous to the linear combinations of sample means $\bar{y}_{ijk}$ called fitted effects in Section 4-3. Thus, it is necessary to carefully define the theoretical or underlying main effects, 2-factor interactions, and 3-factor interactions for a three-way factorial study. In the definitions that follow, a dot appearing as a subscript will (as usual) be understood to indicate that an average has been taken over all levels of the factor corresponding to the dotted subscript. Consider first main effects. Parallel to Definition 4-7 for fitted main effects is a definition of theoretical main effects.

DEFINITION 8-4

In a three-way complete factorial study with factors A, B, and C, the **main effect of factor A at its *i*th level** is

$$\alpha_i = \mu_{i..} - \mu_{...}$$

the **main effect of factor B at its *j*th level** is

$$\beta_j = \mu_{.j.} - \mu_{...}$$

and the **main effect of factor C at its *k*th level** is

$$\gamma_k = \mu_{..k} - \mu_{...}$$

These main effects measure how (when averaged over all combinations of levels of the other factors) underlying mean responses change from level to level of the factor in question. Definition 8-4 has the algebraic consequences that

$$\sum_{i=1}^{I} \alpha_i = 0, \qquad \sum_{j=1}^{J} \beta_j = 0, \qquad \text{and} \qquad \sum_{k=1}^{K} \gamma_k = 0$$

The theoretical counterpart of Definition 4-8 is a definition of theoretical 2-factor interactions.

DEFINITION 8-5

In a three-way complete factorial study with factors A, B, and C, the **2-factor interaction of factor A at its *i*th level and factor B at its *j*th level** is

$$\alpha\beta_{ij} = \mu_{ij.} - (\mu_{...} + \alpha_i + \beta_j)$$

the **2-factor interaction of A at its *i*th level and C at its *k*th level** is

$$\alpha\gamma_{ik} = \mu_{i.k} - (\mu_{...} + \alpha_i + \gamma_k)$$

and the **2-factor interaction of B at its *j*th level and C at its *k*th level** is

$$\beta\gamma_{jk} = \mu_{.jk} - (\mu_{...} + \beta_j + \gamma_k)$$

Like their empirical counterparts defined in Section 4-3, the 2-factor interactions in a three-way study are measures of lack of parallelism on two-way plots of (underlying) means obtained by averaging out over levels of the "other" factor. And it is an algebraic consequence of the form of Definition 8-5 that

$$\sum_{i=1}^{I} \alpha\beta_{ij} = \sum_{j=1}^{J} \alpha\beta_{ij} = 0, \quad \sum_{i=1}^{I} \alpha\gamma_{ik} = \sum_{k=1}^{K} \alpha\gamma_{ik} = 0, \quad \text{and} \quad \sum_{j=1}^{J} \beta\gamma_{jk} = \sum_{k=1}^{K} \beta\gamma_{jk} = 0$$

Finally, there is the matter of three-way interactions in a three-way factorial study. Direct analogy with the definition of fitted three-way interactions given as Definition 4-9 gives the following.

DEFINITION 8-6

In a three-way complete factorial study with factors A, B, and C, the **3-factor interaction of factor A at its *i*th level, factor B at its *j*th level, and factor C at its *k*th level** is

$$\alpha\beta\gamma_{ijk} = \mu_{ijk} - (\mu_{...} + \alpha_i + \beta_j + \gamma_k + \alpha\beta_{ij} + \alpha\gamma_{ik} + \beta\gamma_{jk})$$

Like their fitted counterparts, the (theoretical) 3-factor interactions are measures of patterns in the μ_{ijk} not describable in terms of the factors separately or in pairs. Or differently put, they measure how much what one would call the AB interactions, upon confining attention to a single level of C, change from level to level of C. And, like the fitted 3-factor interactions defined in Section 4-3, the theoretical 3-factor interactions defined here sum to 0 over levels of any one of the factors. That is,

$$\sum_{i=1}^{I} \alpha\beta\gamma_{ijk} = \sum_{j=1}^{J} \alpha\beta\gamma_{ijk} = \sum_{k=1}^{K} \alpha\beta\gamma_{ijk} = 0$$

The fundamental observation that makes inference for the factorial effects defined in Definitions 8-4, 8-5, and 8-6 possible is that they are particular linear combinations of the means μ_{ijk} (*L*'s from Section 7-2), and the corresponding fitted effects from Section 4-3 are the corresponding linear combinations of the sample means $\bar{y}_{ijk}$ ($\hat{L}$'s from Section 7-2). So at least in theory, to make confidence intervals for the factorial effects, one needs only to figure out exactly what coefficients are applied to each of the means and use formula (7-26) of Section 7-2. The practical difficulties with this approach (at least if it were to be implemented by hand) are that for general I, J, and K there can be a huge number of these effects to deal with, and figuring out exactly what coefficients (*c*'s from Section 7-2) are applied to each μ_{ijk} to arrive at a given factorial effect can be a laborious process.

EXAMPLE 8-6

Finding Coefficients on Means for a Factorial Effect in a Three-Way Factorial. Consider a hypothetical example in which A appears at $I = 2$ levels, B at $J = 2$ levels, and C at $K = 3$ levels. For the sake of illustration, consider how one would make a confidence interval for $\alpha\gamma_{23}$. By Definitions 8-4 and 8-5,

$$\alpha\gamma_{23} = \mu_{2\cdot3} - (\mu_{\cdots} + \alpha_2 + \gamma_3)$$

$$= \mu_{2\cdot3} - (\mu_{2\cdots} + \mu_{\cdot\cdot3} - \mu_{\cdots})$$

$$= \frac{1}{2}(\mu_{213} + \mu_{223}) - \frac{1}{6}(\mu_{211} + \mu_{221} + \mu_{212} + \mu_{222} + \mu_{213} + \mu_{223})$$

$$- \frac{1}{4}(\mu_{113} + \mu_{213} + \mu_{123} + \mu_{223}) + \frac{1}{12}(\mu_{111} + \mu_{211} + \cdots + \mu_{223})$$

$$= \frac{1}{6}\mu_{213} + \frac{1}{6}\mu_{223} - \frac{1}{12}\mu_{211} - \frac{1}{12}\mu_{221} - \frac{1}{12}\mu_{212} - \frac{1}{12}\mu_{222}$$

$$- \frac{1}{6}\mu_{113} - \frac{1}{6}\mu_{123} + \frac{1}{12}\mu_{111} + \frac{1}{12}\mu_{121} + \frac{1}{12}\mu_{112} + \frac{1}{12}\mu_{122}$$

so the "$\sum c_i^2/n_i$" applicable to estimating $\alpha\gamma_{23}$ via formula (7-26) of Section 7-2 would be

$$\sum \frac{c_{ijk}^2}{n_{ijk}} = \left(\frac{1}{6}\right)^2 \left(\frac{1}{n_{213}} + \frac{1}{n_{223}} + \frac{1}{n_{113}} + \frac{1}{n_{123}}\right)$$

$$+ \left(\frac{1}{12}\right)^2 \left(\frac{1}{n_{211}} + \frac{1}{n_{221}} + \frac{1}{n_{212}} + \frac{1}{n_{222}} + \frac{1}{n_{111}} + \frac{1}{n_{121}} + \frac{1}{n_{112}} + \frac{1}{n_{122}}\right)$$

and using this expression, endpoints for a confidence interval for $\alpha\gamma_{23}$ would be

$$ac_{23} \pm ts_P\sqrt{\sum \frac{c_{ijk}^2}{n_{ijk}}}$$

The preceding algebra is not difficult, but is hardly pleasant either. It is possible to work out general formulas for the "$\sum c_i^2/n_i$" terms for factorial effects in arbitrary p-way factorials and implement them in computer software. But it would not be consistent with the purposes of this book to try to lay out those general formulas here. If a factorial is large at all (and is not one where all factors have two levels), in practice a computer-implemented statistical analysis package must be used to calculate even all of the fitted effects themselves (let alone corresponding plus-or-minus figures for their estimation).

However, in the special case of p-way factorials where each factor has only two levels there is no difficulty in describing explicitly how to make confidence intervals for the factorial effects or in carrying out a fairly complete analysis of all of these by hand for p as large as even 4 or 5. This is because the 2^p case of the general p-way factorial structure allows three important simplifications. First, it turns out that for any factorial effect in a 2^p factorial, the coefficients applied to the underlying means to produce the effect are all $\pm\frac{1}{2^p}$. So the "$\sum c_i^2/n_i$" term needed to make a confidence interval for any effect in a 2^p factorial is

$$\left(\pm\frac{1}{2^p}\right)^2 \left(\frac{1}{n_{(1)}} + \frac{1}{n_a} + \frac{1}{n_b} + \frac{1}{n_{ab}} + \cdots\right)$$

where the subscripts (1), a, b, ab, etc. refer to the combination-naming convention for 2^p factorials introduced in Section 4-3. (Recall that a particular combination of levels of p factors is named by a string of letters identifying those effects appearing in the combination at their high levels.)

So if one lets E stand for a generic effect in a 2^p factorial (a particular kind of L from Section 7-2) and $\hat{E}$ be the corresponding fitted effect (the corresponding $\hat{L}$ from Section 7-2), then endpoints of an individual two-sided confidence interval for E are

$$\hat{E} \pm ts_P\frac{1}{2^p}\sqrt{\frac{1}{n_{(1)}} + \frac{1}{n_a} + \frac{1}{n_b} + \frac{1}{n_{ab}} + \cdots} \qquad \textbf{(8-45)}$$

where the associated confidence is the probability that the t distribution with $\nu = n - r = n - 2^p$ degrees of freedom assigns to the interval between $-t$ and t. The usual device of using only one endpoint from formula (8-45) and halving the unconfidence produces a corresponding one-sided confidence interval for the effect. And it is worth noting that in balanced-data situations where all sample sizes are equal to m, formula (8-45) can be written even more simply as

$$\hat{E} \pm t\frac{s_P}{\sqrt{m2^p}} \qquad \textbf{(8-46)}$$

There is a second simplification of the general p-way factorial situation afforded in the 2^p case. Because of the way factorial effects sum to 0 over levels of any factor involved, estimating *one* effect of each type is sufficient to completely describe a 2^p factorial. For example, since in a 2^p factorial,

$$\alpha\beta_{11} = -\alpha\beta_{21} = -\alpha\beta_{12} = \alpha\beta_{22}$$

it is necessary to estimate only one AB interaction to have detailed what is known about 2-factor interactions of A and B. There is no need to labor in finding separate estimates of $\alpha\beta_{11}$, $\alpha\beta_{12}$, $\alpha\beta_{21}$, and $\alpha\beta_{22}$. Appropriate sign changes on an estimate of $\alpha\beta_{22}$ suffice to cover the problem.

The third important fact making analysis of 2^p factorial effects so tractable is the existence of the Yates algorithm. As demonstrated in Example 4-9, it is really quite simple to use the algorithm to mechanically generate one fitted effect of each type for a given 2^p data set: those effects corresponding to the high levels of all factors. Together, these three facts make it not only possible but in fact quite easy to do basic confidence interval estimation of the effects in a 2^p study possessing at least some replication.

EXAMPLE 8-5
(continued)

Consider again the metal working power requirement study introduced earlier in this section. Agreeing to (arbitrarily) name tool type 2, the 30° tool bevel angle, and the interrupted cut type as the high levels of (respectively) factors A, B, and C, the eight combinations of the three factors are listed in Table 8-12 in Yates standard order. Then, taking the sample means from that table in the order listed, one can apply the Yates algorithm to produce the fitted effects for the high levels of all factors, as in Table 8-13.

TABLE 8-13

The Yates Algorithm Applied to the Means in Table 8-12

Combination	$\bar{y}$	Cycle 1	Cycle 2	Cycle 3	Cycle 3 ÷ 8
(1)	28.250	55.625	115.125	222.375	$27.7969 = \bar{y}_{...}$
a	27.375	59.500	107.250	−2.375	$-.2969 = a_2$
b	29.875	52.375	−1.125	6.375	$.7969 = b_2$
ab	29.625	54.875	−1.250	.625	$.0781 = ab_{22}$
c	26.500	−.875	3.875	−7.875	$-.9844 = c_2$
ac	25.875	−.250	2.500	−.125	$-.0156 = ac_{22}$
bc	27.750	−.625	.625	−1.375	$-.1719 = bc_{22}$
abc	27.125	−.625	0.000	−.625	$-.0781 = abc_{222}$

Recalling that for the data of Table 8-11, $s_P = 1.492$ mm with 24 associated degrees of freedom, one has (from formula (8-46)) that for (say) individual 90% confidence, the factorial effects in this example can be estimated with two-sided intervals having endpoints

$$\hat{E} \pm 1.711 \frac{1.492}{\sqrt{4 \cdot 2^3}}$$

that is, approximately

$$\hat{E} \pm .45$$

Then, comparing the fitted effects in the last column of Table 8-13 to the ±.45 value, note that only the main effects of Tool Bevel Angle (factor B) and Type of Cut (factor C) are statistically detectable. And for example, it appears that running the machining process at the high level of factor B (the 30° bevel angle) produces a dynamometer reading that is on average between approximately

$$2(.80 - .45) \text{ mm} \quad \text{and} \quad 2(.80 + .45) \text{ mm}$$

that is, between

$$.7 \text{ mm} \quad \text{and} \quad 2.5 \text{ mm}$$

higher than when the process is run at the low level of factor B (the 15° bevel angle). (The difference between B main effects at the high and low levels of B is $\beta_2 - \beta_1 = \beta_2 - (-\beta_2) = 2\beta_2$, hence the multiplication by 2 of the end points of the confidence interval for β_2.)

2^p Studies without Replication and the Normal-Plotting of Fitted Effects

The use of formula (8-45) or (8-46) in judging the detectability of 2^p factorial effects is an extremely practical and effective method. But it depends for its applicability on there being replication somewhere in the data set, so that one can arrive at a pooled sample standard deviation s_P. Unfortunately, it is not uncommon that poorly informed people fail to realize the importance of this and therefore routinely do unreplicated 2^p factorial studies. Although such studies should be avoided whenever possible, various methods of analysis have been suggested for them. This author's favorite one follows from a very clever line of reasoning due originally to an engineering statistician named Cuthbert Daniel (already mentioned in this book in Example 4-12).

Daniel's idea was to invoke a principle of *effect sparsity*. He reasoned that in many real engineering systems, the effects of only a relatively few factors are the primary determiners of system mean response. Thus, in terms of the type of 2^p factorial effects used here, a relatively few of $\alpha_2, \beta_2, \alpha\beta_{22}, \gamma_2, \alpha\gamma_{22}, \ldots$, etc., often dominate the rest (are much larger in absolute value than the majority). In turn, this would imply that often amongst the fitted effects $a_2, b_2, ab_{22}, c_2, ac_{22}, \ldots$, etc., there are a few with sizable means, and the others have means that are (relatively speaking) near 0. Daniel's idea for identifying those situations where a few effects dominate the rest was to normal-plot the fitted effects for the "all high treatment" combination (obtained, for example, by use of the Yates algorithm). When a few plot in positions much more extreme than would be predicted from putting a line through the majority of the points, one identifies them as the likely principal determiners of system behavior. (Actually, Daniel originally proposed making a *half normal plot* of the absolute values of the fitted effects, in order to eliminate any visual effect of the somewhat arbitrary naming of one level of each factor as the high level. For several reasons, among them simplicity of exposition, this treatment will use the full normal plot modification of Daniel's method. The idea of half normal plotting is considered further in Exercise 8-18.)

EXAMPLE 8-7

(Example 4-12 revisited) **Identifying Detectable Effects in an Unreplicated 2^4 Factorial Drill Advance Rate Study Via Normal Plotting of Fitted Effects.** Section 4-4 discussed an example of an unreplicated 2^4 factorial experiment taken from Daniel's *Applications of Statistics to Industrial Experimentation*. There the effects of the four factors

    A   Load
    B   Flow Rate
    C   Rotational Speed
    D   Type of Mud

on the logarithm of an advance rate of a small stone drill were considered. The Yates algorithm applied to the $16 = 2^4$ observed log advance rates produced the following fitted effects.

$$\bar{y}_{....} = 1.5977$$

$$a_2 = .0650 \qquad b_2 = .2900 \qquad c_2 = .5772 \qquad d_2 = .1633$$

$$ab_{22} = -.0172 \qquad ac_{22} = .0052 \qquad ad_{22} = .0334$$

$$bc_{22} = -.0251 \qquad bd_{22} = -.0075 \qquad cd_{22} = .0491$$

$$abc_{222} = .0052 \qquad abd_{222} = .0261 \qquad acd_{222} = .0266$$

$$bcd_{222} = -.0173 \qquad abcd_{2222} = .0193$$

Figure 8-11 is a normal plot of the 15 fitted effects a_2 through $abcd_{2222}$.

Applying Daniel's reasoning, it is obvious that the points corresponding to the C, B, and D main effects plot off any sensible line established through the bulk of the plotted points. So it becomes natural to think that these main effects are detectably larger than the other effects, and therefore distinguishable from experimental error even if the other effects are not. Thus, it seems that drill behavior is potentially describable in terms of the (separate) action of the factors Rotational Speed, Flow Rate, and Mud Type.

FIGURE 8-11

Normal Plot of the Fitted Effects for Daniel's Drill Advance Rate Study

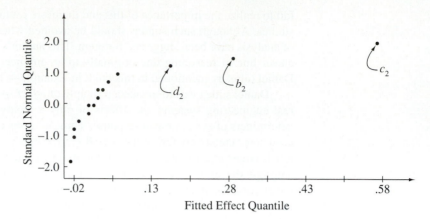

By the way, note that since the fitted effects plotted regard the natural logarithm of advance rate, the fact that $c_2 = .5772$ says that changing from the low level of rotational speed to the high level produces roughly an increase of $2(.5772) \approx 1.15$ in the natural log of the advance rate — i.e., increases the advance rate by a factor of $e^{1.15} \approx 3.2$.

Example 8-7 is one in which the normal plotting clearly identifies a few effects as larger than the others. However, such a plot sometimes has a fairly straight-line appearance. When this happens, the message to be heard is that the fitted effects are potentially explainable as the result of background variation. And it is risky at best to make real-world engineering decisions based on fitted effects that haven't been definitively established as representing consistent system reactions to changes in level of the corresponding factors. A linear normal plot of fitted effects from an unreplicated 2^p study says that one hasn't learned enough from the data in hand to go forward with any degree of certainty about the correctness of inferences drawn from the existing data. More data are needed.

This normal-plotting device has been introduced primarily as a tool for analyzing data lacking any replication. However, it should be said that the method is useful even in cases where there is some replication and s_P can therefore be calculated and formula (8-45) or (8-46) used to judge the detectability of the various factorial effects. Some practice making and using such plots will teach the reader that the process of doing so often amounts to a helpful kind of "data fondling." Many times, a bit of thought makes it possible to trace an unusual pattern on such a plot back to a previously unnoticed peculiarity in the data.

As an example of this, consider what a normal plot of fitted effects would point out about the following eight hypothetical sample means.

$$\bar{y}_{(1)} = 95 \qquad \bar{y}_c = 145$$
$$\bar{y}_a = 101 \qquad \bar{y}_{ac} = 103$$
$$\bar{y}_b = 106 \qquad \bar{y}_{bc} = 107$$
$$\bar{y}_{ab} = 106 \qquad \bar{y}_{abc} = 97$$

This is an exaggerated example of a phenomenon that sometimes occurs less blatantly in practice: that $2^p - 1$ of the sample means are more or less comparable, while one of the means is clearly different. When this occurs (unless the unusual mean corresponds to the "all high treatment" combination), a normal plot of fitted effects roughly like the one in Figure 8-12 will follow. About half the fitted effects will be large positively and the other half large negatively. (When the unusual mean is the one corresponding to the "all high" combination, the fitted effects will all have the same sign.)

FIGURE 8-12

Normal Plot of Fitted Effects for Eight Hypothetical Means

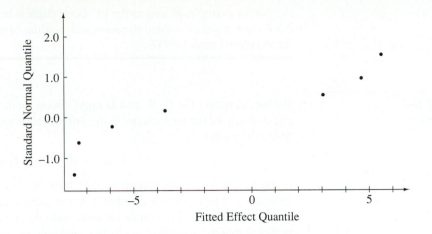

Fitting and Checking Simplified Models in Balanced 2^p Factorial Studies and a Corresponding Variance Estimate

When beginning the analysis of 2^p factorial data, one hopes that a simplified *p*-way model involving only a few main effects and/or low order interactions will be adequate to describe it. Analyses based on formulas (8-45) or (8-46) or normal-plotting are ways of identifying such potential descriptions of special *p*-way structure. Once one has identified such potential simplifications of the 2^p analog of model (8-44), it is often of interest to go beyond that identification to the fitting and checking (residual analysis) of an appropriate simplified model, and on even to the making of formal inferences under the restricted/simplified model assumptions. As has been indicated (particularly in the last section), in general these further steps require the use of dummy variables and multiple linear regression methods — the discussion of which will wait until the end of Chapter 9. But when a 2^p factorial data set is *balanced*, it is possible to handle the model fitting, checking, and subsequent interval-oriented inference quite neatly.

With balanced 2^p factorial data, producing least squares fitted values is no more difficult than adding together (with appropriate signs) desired fitted effects and the grand sample mean. Or equivalently and more efficiently, the reverse Yates algorithm can be used.

EXAMPLE 8-5
(continued)

In the context of the metal working power requirement study and the data of Table 8-11, only the B and C main effects seem detectably nonzero. So it is reasonable to contemplate the simplified version of model (8-44),

$$y_{ijkl} = \mu_{\cdots} + \beta_j + \gamma_k + \epsilon_{ijkl} \tag{8-47}$$

for possible use in describing dynamometer readings. From Table 8-13, the fitted version of $\mu_{\cdots}$ is $\bar{y}_{\cdots} = 27.7969$, the fitted version of β_2 is $b_2 = .7969$, and the fitted version of γ_2 is $c_2 = -.9844$. Then, simply adding together appropriate signed versions of the fitted effects, for the four possible combinations of j and k, one gets the corresponding fitted responses in Table 8-14.

TABLE 8-14

Fitted Responses for a "B and C Main Effects Only" Description of Power Requirement

j	k	b_j	c_k	$\hat{y} = \bar{y}_{\cdots} + b_j + c_k$
1	1	−.7969	.9844	27.9844
2	1	.7969	.9844	29.5782
1	2	−.7969	−.9844	26.0156
2	2	.7969	−.9844	27.6094

So for example, as long as the 15° bevel angle is being considered and a continuous cut is contemplated, a fitted dynamometer reading of about 27.98 is appropriate under the simplified model (8-47).

EXAMPLE 8-7
(continued)

Having identified the C, B, and D main effects as detectably larger than the A main effect or any of the interactions in the drill advance rate study, it is natural to consider fitting the model

$$y_{ijkl} = \mu_{....} + \beta_j + \gamma_k + \delta_l + \epsilon_{ijkl} \qquad (8\text{-}48)$$

to the logarithms of the unreplicated 2^4 factorial data of Table 4-25. (Note that even though $p = 4$ factors are involved here, five subscripts are not required, since one doesn't need a subscript to differentiate between multiple members of the 2^4 different samples in this unreplicated context. y_{ijkl} is the single observation at the ith level of A, jth level of B, kth level of C, and lth level of D.) Since the drill advance rate data are balanced (all sample sizes are $m = 1$), the fitted effects given earlier (calculated without reference to the simplified model) serve as fitted effects under model (8-48), and fitted responses under model (8-48) are obtainable by simple addition and subtraction using those.

Since there are eight different combinations of j, k, and l, eight different linear combinations of $\bar{y}_{....}$, b_2, c_2, and d_2 are required. While these could be treated one at time, it is probably more efficient to generate them all at once using the reverse Yates algorithm. And in this particular example, there are at least two different ways in which the reverse Yates algorithm might be used. The first and most straightforward is to carry through four cycles of the algorithm, setting only fitted effects $\bar{y}_{....}$, b_2, c_2, and d_2 to values other than 0 (among a total of 16 effects) for generating the full set of 2^4 fitted values, as in Table 8-15.

TABLE 8-15

The Reverse Yates Algorithm Used to Fit the "B, C, and D Main Effects" Model to Daniel's Data

Fitted Effect	Value	Cycle 1	Cycle 2	Cycle 3	Cycle 4 ($\hat{y}$)
$abcd_{2222}$	0	0	0	.1633	2.6282
bcd_{222}	0	0	.1633	2.4649	2.6282
acd_{222}	0	0	.5772	.1633	2.0482
cd_{22}	0	.1633	1.8877	2.4649	2.0482
abd_{222}	0	0	0	.1633	1.4738
bd_{22}	0	.5772	.1633	1.8849	1.4738
ad_{22}	0	.2900	.5772	.1633	.8938
d_2	.1633	1.5977	1.8877	1.8849	.8938
abc_{222}	0	0	0	.1633	2.3016
bc_{22}	0	0	.1633	1.3105	2.3016
ac_{22}	0	0	.5772	.1633	1.7216
c_2	.5772	.1633	1.3077	1.3105	1.7216
ab_{22}	0	0	0	.1633	1.1472
b_2	.2900	.5772	.1633	.7305	1.1472
a_2	0	.2900	.5772	.1633	.5672
$\bar{y}_{....}$	1.5977	1.5977	1.3077	.7305	.5672

From Table 8-15 it is evident, for example, that the fitted mean responses for combinations bcd and abcd ($\hat{y}_{bcd}$ and $\hat{y}_{abcd}$) are both 2.6282.

A second, slightly more subtle way to employ the reverse Yates algorithm in this example would be to recognize that since factor A and its interactions do not appear in

TABLE 8-16

An Alternative Use of the Reverse Yates Algorithm to Produce the Fitted Values of Table 8-15

Fitted Effect	Value	Cycle 1	Cycle 2	Cycle 3
bcd_{222}	0	0	.1633	2.6282
cd_{22}	0	.1633	2.4649	2.0482
bd_{22}	0	.5772	.1633	1.4738
d_2	.1633	1.8877	1.8849	.8938
bc_{22}	0	0	.1633	2.3016
c_2	.5772	.1633	1.3105	1.7216
b_2	.2900	.5772	.1633	1.1472
$\bar{y}_{\ldots}$	1.5977	1.3077	.7305	.5672

model (8-48), it would suffice to carry through three cycles of the algorithm, considering only the three factors B, C, and D. One generates $2^3 = 8$ fitted means from the nonzero fitted effects $\bar{y}_{\ldots}$, b_2, c_2, and d_2 among a total of only eight effects, as in Table 8-16.

The fitted values in the last column of Table 8-16 then serve as fitted values in the original 2^4 study for combinations of factors having levels of B, C, and D as indicated by the row names and either level of factor A.

Fitted means derived as in these examples lead in the usual way to residuals, R^2 values, and plots for checking on the reasonableness of simplified versions of the general 2^p version of model (8-44). In addition, corresponding to simplified or *reduced* models like (8-47) or (8-48), there are what will here be called *few-effects* s^2 values. When $m > 1$, these can be compared to s_P^2, as another means of investigating the reasonableness of the corresponding models.

DEFINITION 8-7

In a balanced complete 2^p factorial study, if a reduced model involving u different effects (including the grand mean) has corresponding fitted values $\hat{y}$ and thus residuals $y - \hat{y}$, the quantity

$$s_{\mathrm{FE}}^2 = \frac{1}{m2^p - u} \sum (y - \hat{y})^2 \qquad (8\text{-}49)$$

will be called a **few-effects sample variance**. Associated with it are $\nu = m2^p - u$ degrees of freedom.

As was the case with s_{NI}^2 in the previous section, the quantity (8-49) represents an estimate of the basic background variance, whenever the corresponding simplified/reduced/few-effects model is an adequate description of the study. When it is not, s_{FE} will tend to overestimate σ, so comparing s_{FE} to s_P is a way of investigating the appropriateness of that description.

It is not obvious at this point, but there is a helpful alternative way to calculate the value of s_{FE}^2 given in formula (8-49). It turns out that

$$s_{\mathrm{FE}}^2 = \frac{1}{m2^p - u} \left[SSTot - m2^p \sum \hat{E}^2 \right] \qquad (8\text{-}50)$$

where the sum is over the squares of the $u - 1$ fitted effects corresponding to those main effects and interactions appearing in the reduced model equation, and (as always) $SSTot = \sum (y - \bar{y})^2 = (n - 1)s^2$.

EXAMPLE 8-5
(continued)

It was remarked at the beginning of this section that residuals for the power requirement data based on model (8-44) are obtained by subtracting sample means in Table 8-12 from observations in Table 8-11. Under model (8-47), however, the fitted values in Table 8-14 are appropriate for producing residuals. The fitted means and residuals for a "B and C main effects only" description of this 2^3 data set are given in Table 8-17.

TABLE 8-17

Residuals for the "B and C Main Effects Only" Model of Power Requirements

Combination	$\hat{y}$	Residuals $(y - \hat{y})$
(1)	27.9844	1.0156, −1.4844, 2.5156, −.9844
a	27.9844	.0156, .5156, .0156, −2.9844
b	29.5782	−1.0782, −1.0782, .4218, 2.9218
ab	29.5782	−.0782, 2.4218, −.5782, −1.5782
c	26.0156	1.9844, −1.0156, .4844, .4844
ac	26.0156	−1.5156, −1.0156, 1.9844, −.0156
bc	27.6094	−.6094, 1.3906, −.1094, −.1094
abc	27.6094	−.1094, .3906, −.6094, −1.6094

FIGURE 8-13

Normal Plot of Residuals for the Power Requirements Study

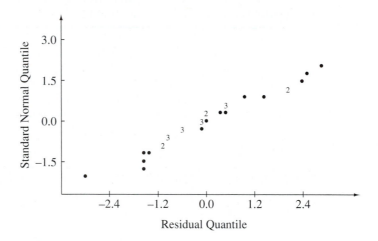

FIGURE 8-14

Plot of Residuals versus Fitted Power Requirements

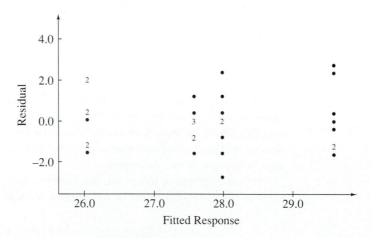

Figure 8-13 is a normal plot of these residuals, and Figure 8-14 is a plot of the residuals against the fitted values.

If there is anything remarkable in these plots, it is that Figure 8-14 contains a hint that smaller mean response has associated with it smaller response variability. In fact, looking back at Table 8-17, it is easy to see that the two smallest fitted means correspond to the high level of C (i.e., interrupted cuts). That is, the hint of change in response variation shown in Figure 8-14 is the same phenomenon related to cut type that was discussed when these data were first introduced. It appears that power requirements for interrupted cuts may be slightly more consistent than for continuous cuts. But on the whole, there is little in the two figures to invalidate model (8-47) as at least a rough-and-ready description of the mechanism behind the data of Table 8-11.

Note that for the power requirement data,

$$SSTot = (n - 1)s^2 = 108.93$$

Then, since $s_p^2 = 2.226$, the one-way ANOVA identity (7-63) (of Section 7-4) says that

$$SSTr = 108.93 - 24(2.226) = 55.51$$

so R^2 corresponding to the general model (8-44) is

$$R^2 = \frac{55.51}{108.93} = .51$$

On the other hand, it is possible to verify that for the simplified model (8-47), squaring and summing the residuals in Table 8-17 gives

$$\sum (y - \hat{y})^2 = 57.60$$

So for the "B and C main effects only" description of dynamometer readings,

$$R^2 = \frac{108.93 - 57.60}{108.93} = .47$$

Thus, although one will at best account for only about 51% of the raw variation in dynamometer readings, fitting the simple model (8-47) will account for nearly all of that potentially assignable variation. So from this point of view as well, model (8-47) seems attractive as a description of power requirement.

Note that formulas (8-49) and (8-50) imply that for balanced 2^p factorial data, fitting reduced models gives

$$\sum (y - \hat{y})^2 = SSTot - m2^p \sum \hat{E}^2$$

So it is not surprising that using the $b_2 = .7969$ and $c_2 = -.9844$ figures from before, one finds that

$$\begin{aligned} SSTot - m2^p \sum \hat{E}^2 &= 108.93 - 4 \cdot 2^3 \cdot \left((.7969)^2 + (-.9844)^2 \right) \\ &= 108.93 - 51.33 \\ &= 57.60 \end{aligned}$$

which is the value of $\sum (y - \hat{y})^2$ just used in finding R^2 for the reduced model. At any rate, from formula (8-49) or (8-50), it is then clear that (corresponding to reduced model (8-47))

$$s_{FE}^2 = \frac{1}{4 \cdot 2^3 - 3}(57.60) = 1.986$$

so

$$s_{FE} = \sqrt{1.986} = 1.409 \text{ mm}$$

which agrees closely with the value of s_p. Once again on this account, description (8-47) seems quite workable.

EXAMPLE 8-7
(continued)

Table 8-18 contains the log advance rates, fitted values, and residuals for Daniel's unreplicated 2^4 example. (The raw data were given in Table 4-25, and it is the few-effects model (8-48) that is under consideration.)

The reader can verify by plotting that the residuals in Table 8-18 are not in any way remarkable. Further, it is possible to check that

$$SSTot = \sum (y - \bar{y})^2 = 7.2774$$

and

$$\sum (y - \hat{y})^2 = .1736$$

So (as indicated earlier in Example 4-12) for the use of model (8-48),

$$R^2 = \frac{7.2774 - .1736}{7.2774} = .976$$

Since there is no replication in this data set, fitting the 4-factor version of the general model (8-44) would give a perfect fit, R^2 equal to 1.000, all residuals equal to 0, and no value of s_P^2. Thus, there is really nothing to judge $R^2 = .976$ against in relative terms. But even in absolute terms it appears that the "B, C, and D main effects only" model for log advance rate fits the data well.

TABLE 8-18

Responses, Fitted Values, and Residuals for the "B, C, and D Main Effects" Model and Daniel's Drill Advance Rate Data

Combination	y ln(advance rate)	$\hat{y}$	$e = y - \hat{y}$
(1)	.5188	.5672	−.0484
a	.6831	.5672	.1159
b	1.1878	1.1472	.0406
ab	1.2355	1.1472	.0883
c	1.6054	1.7216	−.1162
ac	1.7405	1.7216	.0189
bc	2.2996	2.3016	−.0020
abc	2.2050	2.3016	−.0966
d	.7275	.8938	−.1663
ad	.8920	.8938	−.0018
bd	1.4085	1.4738	−.0653
abd	1.5107	1.4738	.0369
cd	2.0503	2.0482	.0021
acd	2.2439	2.0482	.1957
bcd	2.4639	2.6282	−.1643
abcd	2.7912	2.6282	.1630

An estimate of the variability of log advance rates for a fixed combination of factor levels derived under the assumptions of model (8-48), is (from formula (8-49))

$$s_{FE} = \sqrt{\frac{1}{1 \cdot 2^4 - 4}(.1736)} = .120$$

As noted, there's no s_P to compare this value to, but it is at least consistent with the kind of variation in y seen in Table 8-18 when one compares responses for pairs of combinations that (like combinations b and ab) differ only in level of the factor A.

Statistical Intervals for Balanced 2^p Studies under Few-Effects Models

This section has already made heavy use of the fact that since the basic p-way factorial model is just a rewritten version of the one-way normal model from Chapter 7, the confidence interval methods of that chapter can all see application in p-way factorial studies. The same is also true for the other methods of Sections 7-1 through 7-4, including the prediction and tolerance interval methods. But the fact of the matter is that when a simplified/few-effects model is appropriate, sharper real-world engineering conclusions can usually be had by using methods based on the simplified model than by applying the general methods of Chapter 7. And when attention is restricted to balanced 2^p studies, it is possible to write down simple, explicit formulas for several useful forms of interval-oriented inference.

As a first example of what is possible under a few-effects model in a balanced 2^p factorial study, consider the confidence interval estimation of a particular mean response. It turns out that for balanced data, the 2^p fitted effects (including the grand mean) that come out of the Yates algorithm are independent normal variables with means equal to the corresponding underlying effects and variances $\sigma^2/m2^p$. So, if a simplified version of model (8-44) involving u effects (including the overall mean) is appropriate, a fitted response $\hat{y}$ has mean equal to the corresponding underlying mean, and

$$Var\,\hat{y} = u\frac{\sigma^2}{m2^p}$$

It should then be plausible that under a simplified/restricted/few-effects model in a balanced 2^p factorial study, a two-sided interval with endpoints

$$\hat{y} \pm ts_{FE}\sqrt{\frac{u}{m2^p}} \tag{8-51}$$

may be used as an individual confidence interval for the corresponding mean response. The associated confidence is the probability that the t distribution with $\nu = m2^p - u$ degrees of freedom assigns to the interval between $-t$ and t. And a one-sided confidence interval for the mean response can be obtained in the usual way, by employing only one of the endpoints indicated in formula (8-51) and appropriately adjusting the confidence level.

EXAMPLE 8-5
(continued)

Consider estimating the mean dynamometer reading corresponding to a 15° bevel angle and interrupted cut, using the "B and C main effects only" description of Miller's power requirement study. (These are the conditions that appear to produce the smallest mean power requirement.) Using (for example) 95% confidence, a fitted value of 26.02 from Table 8-14 and $s_{FE} = 1.409$ mm possessing $\nu = 4 \cdot 2^3 - 3 = 29$ associated degrees of freedom in formula (8-51), one is led to a two-sided interval with endpoints

$$26.02 \pm 2.045(1.409)\sqrt{\frac{3}{4 \cdot 2^3}}$$

That is, one is led to an interval with endpoints of approximately

$$26.02 \text{ mm} \pm .88 \text{ mm} \tag{8-52}$$

that is,

$$25.14 \text{ mm} \qquad \text{and} \qquad 26.90 \text{ mm}$$

In contrast to this interval, one might consider what the method of Section 7-2 provides for a 95% confidence interval for the mean reading for tool type 1, a 15° bevel angle, and interrupted cuts. Since $s_P = 1.492$ with $\nu = 24$ associated degrees of freedom, and (from Table 8-12) $\bar{y}_c = 26.50$, formula (7-14) of Section 7-2 produces a two-sided confidence interval for μ_c with endpoints

$$26.50 \pm 2.064(1.492)\frac{1}{\sqrt{4}}$$

that is,

$$26.50\text{mm} \pm 1.54 \text{ mm} \tag{8-53}$$

A major practical difference between intervals (8-52) and (8-53) is the apparent increase in precision provided by interval (8-52), due in numerical terms primarily to the "extra" $\sqrt{3/8}$ factor present in the first plus-or-minus calculation but not in the second. However, it must be remembered that the extra precision is bought at the price of the use of model (8-47) and the consequent use of all observations in the generation of $\hat{y}_c$ (rather than only the observations from the single sample corresponding to combination c).

A second instance of a balanced-data confidence interval method based on a few-effects simplification of the general 2^p model is that for estimating the effects included in the model. It comes about by replacing s_P in formula (8-46) with s_{FE} and appropriately adjusting the degrees of freedom associated with the t quantile. That is, under a few-effects model in a 2^p study with balanced data, a two-sided individual confidence interval for an effect included in the model is

$$\hat{E} \pm t\frac{s_{FE}}{\sqrt{m2^p}} \tag{8-54}$$

where $\hat{E}$ is the corresponding fitted effect and the confidence associated with the interval is the probability that the t distribution with $\nu = m2^p - u$ degrees of freedom assigns to the interval between $-t$ and t. One-sided intervals are made from formula (8-54) in the usual way.

It it is worth highlighting one possibility afforded by formula (8-54). Unlike formula (8-46), formula (8-54) can be used in studies where $m = 1$. This makes it possible to attach precision figures to estimated effects in the same way that in Section 8-2 it was possible to attach precision figures to differences in estimated main effects in unreplicated two-way factorials under a no-interaction assumption.

EXAMPLE 8-7
(continued)

Consider again Daniel's drill advance rate study and, for example, the effect of high level of rotational speed on the natural logarithm of advance rate. Under the "B, C, and D main effects only" description of log advance rate, $s_{FE} = .120$ with $\nu = 1 \cdot 2^4 - 4 = 12$ associated degrees of freedom. Also, $c_2 = .5772$. Then (for example) using a 95% confidence level, from formula (8-54), a two-sided interval for γ_2 under the simplified model has endpoints

$$.5772 \pm 2.179\frac{.120}{\sqrt{1 \cdot 2^4}}$$

that is,

$$.5772 \pm .0654$$

that is,

$$.5118 \quad \text{and} \quad .6426$$

This in turn translates (via multiplication by 2) to an increase of between

$$1.0236 \quad \text{and} \quad 1.2852$$

in average log advance rate as one moves from the low level of rotational speed to the high level. And upon exponentiation, a multiplication of median advance rate by a factor between

$$2.78 \quad \text{and} \quad 3.62$$

is indicated as one moves between levels of rotational speed. (Notice that since a normal mean is also the distribution's median, and under a transformation the median of the transformed values is the transformation applied to the median, the inference about the mean logged rate can be translated to one about the median rate. However, since the mean of transformed values is not in general the transformed mean, the interval obtained by exponentiation unfortunately does not apply to the mean advance rate.)

As additional instances of sharpened methods of inference that are available under few-effects models in balanced 2^p factorial studies, the discussion in this section will end with a look at prediction and tolerance intervals. The increased precision associated with the fitted values corresponding to a simplified model, (which in general makes formula (8-51) more informative than formula (7-14) of Section 7-2) also has its influence when making prediction and tolerance intervals.

To begin with, the same kind of reasoning that has been used repeatedly to produce prediction intervals can be applied in the present context. The result is that under a simplified model including u terms (including the overall mean) and based on balanced 2^p factorial data, a two-sided prediction interval for a single additional observation at some particular combination of factor levels has endpoints

$$\hat{y} \pm t s_{\text{FE}} \sqrt{1 + \frac{u}{m2^p}} \tag{8-55}$$

where the associated prediction confidence is the probability assigned to the interval from $-t$ to t by the t distribution with $\nu = m2^p - u$ degrees of freedom. One-sided prediction intervals can be made using only one of the endpoints indicated in formula (8-55) and adjusting the prediction confidence level accordingly.

As in Sections 7-2 and 8-2, it is practical to give only one-sided tolerance interval methods in the present situation. That is, under a few-effects model and using balanced 2^p factorial data, for a given combination of factor levels, the intervals

$$(\hat{y} - \tau s_{\text{FE}}, \infty) \tag{8-56}$$

$$(-\infty, \hat{y} + \tau s_{\text{FE}}) \tag{8-57}$$

can be used as one-sided tolerance intervals for a fraction p of the underlying distribution, providing τ is appropriately chosen. And as usual, a fairly exact method of finding appropriate τ values for 95% confidence can be given along with a general approximation that can be employed for other confidence levels. (In doing so, one must, however, avoid a potential ambiguity of notation brought about by the fact that the letter p is being used to stand both for the number of two-level factors involved in a study and for the fraction of the underlying distribution. The method of avoiding ambiguity employed here will be to write Q_z in place of $Q_z(p)$, this book's usual symbolism for the standard normal p quantile, in the next two formulas. All p's appearing in formulas (8-58) and (8-59) thus refer to the number of factors being studied.)

A method by which Table D-8 can be used to produce values of τ for 95% confidence is as follows. One first computes a value of a parameter ξ according to

$$\xi = \frac{\sqrt{\dfrac{m2^p}{u}}\, Q_z}{\sqrt{2\,(m2^p - u) + \dfrac{m2^p Q_z^2}{u}}} \qquad (8\text{-}58)$$

Then Table D-8 is entered using this value and the degrees of freedom parameter $\nu = m2^p - u$ to produce a value λ in the body of the table (via interpolation as needed). Finally, using λ taken from the table, τ appropriate for use in formulas (8-56) and (8-57) can be found using the relationship

$$\tau = \frac{\sqrt{\dfrac{m2^p}{u}}\, Q_z + \lambda \sqrt{1 + \dfrac{\dfrac{m2^p Q_z^2}{u} - \lambda^2}{2(m2^p - u)}}}{\sqrt{\dfrac{m2^p}{u}}\left(1 - \dfrac{\lambda^2}{2(m2^p - u)}\right)} \qquad (8\text{-}59)$$

A method of approximating τ is as follows. For a one-sided tolerance interval with confidence level γ for a fraction p of the underlying distribution corresponding to a given set of levels of the factors, one may use formula (8-56) or (8-57) with τ determined as in formula (8-59) by employing the approximation

$$\lambda \approx Q_z(\gamma) \qquad (8\text{-}60)$$

EXAMPLE 8-7
(continued)

First consider the problem of making a one-sided 90% prediction interval of the form $(\#, \infty)$ for a single additional log advance rate when all factors are set at their high level. (From Table 8-15, the advance rate seems highest when all of factors B, C, and D are at their high levels.) Since for model (8-48), $\hat{y}_{abcd} = 2.6282$ and $s_{FE} = .120$ with $\nu = 12$ associated degrees of freedom, formula (8-55) produces a 90% lower prediction bound for an additional log advance rate of

$$2.6282 - 1.356(.120)\sqrt{1 + \frac{4}{1 \cdot 2^4}} = 2.4463$$

This translates to a 90% lower prediction bound for an additional raw advance rate of

$$e^{2.4463} = 11.55$$

Then, as a contrast to the 90% lower prediction bound, consider computing a 95% lower tolerance bound for 90% of additional log advance rates using formula (8-56). Since 95% confidence is desired, one may employ Table D-8, and it is first necessary to evaluate ξ. From formula (8-58),

$$\xi = \frac{\sqrt{\dfrac{1 \cdot 2^4}{4}}\,(1.282)}{\sqrt{2(1 \cdot 2^4 - 4) + \dfrac{1 \cdot 2^4(1.282)^2}{4}}} = .46$$

So, consulting Table D-8, one finds that $\lambda \approx 1.685$ is in order for use in formula (8-59). (Note that approximation (8-60) would lead to use of $\lambda \approx 1.645$.) Then formula (8-59) says that

$$\tau = \frac{\sqrt{\dfrac{1 \cdot 2^4}{4}}\,(1.282) + 1.685\sqrt{1 + \dfrac{\dfrac{1 \cdot 2^4 (1.282)^2}{4} - (1.685)^2}{2(1 \cdot 2^4 - 4)}}}{\sqrt{\dfrac{1 \cdot 2^4}{4}\left(1 - \dfrac{(1.685)^2}{2(1 \cdot 2^4 - 4)}\right)}} = 2.481$$

So the desired 95% lower tolerance bound for 90% of additional log advance rates is

$$2.6282 - 2.481(.120) = 2.3305$$

And this figure translates to a 95% lower tolerance bound for 90% of raw advance rates of

$$e^{2.3305} = 10.28$$

8-4   ANOVA Methods for Balanced Factorial Studies

The first three sections of this chapter have discussed interval-oriented inference for full factorial statistical engineering studies. It is also possible to develop significance-testing methods for such situations. The most famous of these are extensions of the analysis of variance methods introduced in Section 7-4, applicable to balanced full factorial studies.

This section discusses some of the ANOVA methods for balanced factorial data, beginning with the two-way ANOVA methods. Then the p-way ANOVA is considered, paying primary attention to the special case where all p factors have only two levels. Finally, random effects models in two- and p-way studies, and the making of ANOVA-based estimates of their variance components, are introduced.

Two-Way ANOVA

Section 7-4 shows how a measure of overall observed variability in a general multisample statistical study can be partitioned into pieces that can be thought of as due (respectively) to treatments and error and can form the basis for an F test of the null hypothesis of no differences between treatments. As it turns out, when a multisample study has balanced two-way factorial structure, an even more extensive partitioning of the overall observed variability is possible. This partitioning allows one to think of pieces of the treatment variability as due to A main effects, B main effects, and AB interactions and to develop F tests of the null hypotheses of no A main effects, of no B main effects, and of no AB interactions.

Written in two-way factorial notation, the basic one-way ANOVA identity, $SSTot = SSTr + SSE$, for balanced data is

$$\sum_{i,j,k} (y_{ijk} - \bar{y})^2 = m \sum_{i,j} (\bar{y}_{ij} - \bar{y})^2 + \sum_{i,j,k} (y_{ijk} - \bar{y}_{ij})^2 \tag{8-61}$$

But it is also an algebraic fact that for balanced two-way factorial data,

$$\left. \begin{aligned}
m \sum_{i,j} (\bar{y}_{ij} - \bar{y})^2 &= mJ \sum_{i=1}^{I} a_i^2 + mI \sum_{j=1}^{J} b_j^2 + m \sum_{i,j} ab_{ij}^2 \\
&= mJ \sum_{i=1}^{I} (\bar{y}_{i.} - \bar{y}_{..})^2 + mI \sum_{j=1}^{J} (\bar{y}_{.j} - \bar{y}_{..})^2 \\
&\quad + m \sum_{i,j} (\bar{y}_{ij} - \bar{y}_{i.} - \bar{y}_{.j} + \bar{y}_{..})^2
\end{aligned} \right\} \tag{8-62}$$

And the three terms on the right of formula (8-62) (which represent, respectively, the sums over all data points of the squares of corresponding fitted A main effects, fitted B main effects, and fitted AB interactions) provide a breakdown of $SSTr$ into interpretable pieces. One can think of them as representing variation in response assignable to (respectively) A main effects, B main effects, and AB interactions. They are prominent enough in the practice of engineering statistics that they deserve special names and shorthand, and these are given next in definition form.

DEFINITION 8-8

In a balanced two-way factorial study with a common sample size m,

$$mJ \sum_{i=1}^{I} a_i^2 = mJ \sum_{i=1}^{I} (\bar{y}_{i.} - \bar{y}_{..})^2$$

will be called the **sum of squares due to factor A** and symbolized as SSA.

DEFINITION 8-9

In a balanced two-way factorial study with a common sample size m,

$$mI \sum_{j=1}^{J} b_j^2 = mI \sum_{j=1}^{J} (\bar{y}_{.j} - \bar{y}_{..})^2$$

will be called the **sum of squares due to factor B** and symbolized as SSB.

DEFINITION 8-10

In a balanced two-way factorial study with a common sample size m,

$$m \sum_{i,j} ab_{ij}^2 = m \sum_{i,j} (\bar{y}_{ij} - \bar{y}_{i.} - \bar{y}_{.j} + \bar{y}_{..})^2$$

will be called the **sum of squares due to AB interactions** and symbolized as $SSAB$.

In the notation of Definitions 8-8, 8-9, and 8-10, the identity (8-62) can be written as

$$SSTr = SSA + SSB + SSAB \qquad \textbf{(8-63)}$$

or going a step further and combining formula (8-63) and the basic one-way ANOVA identity, one has, for balanced two-way factorial data,

$$SSTot = SSA + SSB + SSAB + SSE \qquad \textbf{(8-64)}$$

Formula (8-64) is often called the *two-way ANOVA identity*.

EXAMPLE 8-8

(Example 8-2 revisited) **Two-Way ANOVA in the Drilled Aluminum Strip Tensile Strength Study.** Consider again the aluminum strip tensile strength study of Clubb and Goedken. Recall that Table 8-5 gives $m = 3$ ultimate tensile strength measurements for (respectively) $I = 3$ times $J = 2$ combinations of hole size (factor A) and hole placement (factor B). Straightforward calculation with the data of Table 8-5 shows that

$$SSTot = \sum_{i,j,k} (y_{ijk} - \bar{y})^2 = 286{,}229$$

Referring to Example 8-2, one has that, for the tensile strength data,

$$SSE = (n - r)s_P^2 = 12(106.7)^2 = 136{,}507$$

The one-way ANOVA identity then shows that

$$SSTr = SSTot - SSE = 286{,}229 - 136{,}507 = 149{,}722$$

The discussion in Example 8-2 also includes the values of the fitted factorial effects for the tensile strength data. These can be used to find SSA, SSB, and $SSAB$ as follows.

$$
\begin{aligned}
SSA &= mJ \sum_{i=1}^{I} a_i^2 \\
&= 3(2)((88.88)^2 + (-24.44)^2 + (-64.44)^2) \\
&= 75{,}911
\end{aligned}
$$

Similarly,

$$
\begin{aligned}
SSB &= mI \sum_{j=1}^{J} b_j^2 \\
&= 3(3)((-63.06)^2 + (63.06)^2) \\
&= 71{,}568
\end{aligned}
$$

And finally,

$$
\begin{aligned}
SSAB &= m \sum_{i,j} ab_{ij}^2 \\
&= 3((15.55)^2 + (-15.55)^2 + (-5.44)^2 + (5.44)^2 + (-10.11)^2 + (10.11)^2) \\
&= 2{,}243
\end{aligned}
$$

Notice that

$$75{,}911 + 71{,}568 + 2{,}243 = 149{,}722$$

so that — exactly as promised in identity (8-63) — SSA, SSB, and $SSAB$ total to $SSTr$, thereby providing a partition of the variability assignable to differences between aluminum strip types into interpretable pieces.

Although it is perfectly possible to produce sums of squares like those in Example 8-8 using only a pocket calculator, it is most common these days to use commercial statistical packages to do such calculations. Printout 8-1 is from a Minitab session intended to produce values for SSA, SSB, and $SSAB$ from the tensile strength data.

The form of Printout 8-1 already suggests that it is common to display the sums of squares defined in Definitions 8-8, 8-9, and 8-10 as entries in an ANOVA table somewhat more complicated than that introduced in Section 7-4. The form of that (two-way ANOVA) table will be discussed presently, but it is first necessary to discuss the significance tests typically associated with the partition of $SSTr$ into SSA, SSB, and $SSAB$.

It is the case that provided $m > 1$, under the basic "equal variances, normal distributions" model

$$y_{ijk} = \mu_{ij} + \epsilon_{ijk} \tag{8-65}$$

used throughout Section 8-1, it is possible to base significance tests for the hypotheses

$$H_0: \alpha_i = 0 \qquad \text{for all } i \ (\mu_{1.} = \mu_{2.} = \cdots = \mu_{I.}) \tag{8-66}$$

$$H_0: \beta_j = 0 \qquad \text{for all } j \ (\mu_{.1} = \mu_{.2} = \cdots = \mu_{.J}) \tag{8-67}$$

$$H_0: \alpha\beta_{ij} = 0 \qquad \text{for all } i, j \tag{8-68}$$

PRINTOUT 8-1

```
MTB > print c1 c2 c3

ROW    size   placemnt   strength

  1      1        1         5728
  2      1        1         5664
  3      1        1         5514
  4      1        2         5689
  5      1        2         5891
  6      1        2         5611
  7      2        1         5508
  8      2        1         5620
  9      2        1         5375
 10      2        2         5579
 11      2        2         5557
 12      2        2         5778
 13      3        1         5469
 14      3        1         5440
 15      3        1         5460
 16      3        2         5621
 17      3        2         5664
 18      3        2         5523

MTB > twoway c3 c1 c2

ANALYSIS OF VARIANCE    strength

SOURCE           DF          SS          MS
size              2       75911       37956
placemnt          1       71568       71568
INTERACTION       2        2243        1122
ERROR            12      136507       11376
TOTAL            17      286229
```

on *SSE* and the corresponding sums of squares, as follows. The hypothesis (8-66) of no A main effects can be tested using the statistic

$$F = \frac{SSA/(I-1)}{SSE/IJ(m-1)} \tag{8-69}$$

and an $F_{I-1,IJ(m-1)}$ reference distribution, where large observed values of F are taken as evidence against H_0 in favor of H_a: not H_0, the possibility that there are nonzero A main effects, (i.e., that not all row average underlying means are the same). Similarly, the hypothesis (8-67) of no B main effects can be tested using the statistic

$$F = \frac{SSB/(J-1)}{SSE/IJ(m-1)} \tag{8-70}$$

and an $F_{J-1,IJ(m-1)}$ reference distribution, where large observed values of F are taken as evidence against H_0 in favor of H_a: not H_0, the possibility that there are nonzero B main effects, (i.e., that not all column average underlying means are the same). Finally, the hypothesis (8-68) of no AB interactions can be tested using the statistic

$$F = \frac{SSAB/(I-1)(J-1)}{SSE/IJ(m-1)} \tag{8-71}$$

and an $F_{(I-1)(J-1),IJ(m-1)}$ reference distribution, where large observed values of F are taken as evidence against H_0 in favor of H_a: not H_0, the possibility that there are nonzero

AB interactions, (i.e., that there is lack of the parallelism property for the underlying means).

The denominator of each of the statistics (8-69), (8-70), and (8-71) is s_P^2, the basic estimate of σ^2, the underlying variance. It has been written in displays (8-69) through (8-71) as a ratio of SSE to $\nu = n - r = mIJ - IJ = IJ(m - 1)$, simply to emphasize once again its structure as a sum of squared residuals divided by a corresponding degrees of freedom value. Similarly, the numerators in statistics (8-69) through (8-71) each have the form of a sum of squares divided by a corresponding numerator degrees of freedom value. It is interesting (and not accidental) that

$$(I - 1) + (J - 1) + (I - 1)(J - 1) = IJ - 1 \qquad \textbf{(8-72)}$$

that is, that degrees of freedom for A plus degrees of freedom for B plus degrees of freedom for AB interaction total to degrees of freedom for treatments. ($r - 1 = IJ - 1$, since there are $I \cdot J$ different samples in a complete two-way factorial.)

The reader should by now be prepared to accept a table like that appearing on Printout 8-1 as a convenient way to organize the calculation of the test statistics (8-69) through (8-71) and display the partition of $SSTr$ according to formula (8-63) and the corresponding degrees of freedom according to formula (8-72). The general format that will be used for making a two-way ANOVA table in this text is shown as Table 8-19.

TABLE 8-19

General Form of the Two-Way Balanced-Data ANOVA Table

ANOVA TABLE

Source	SS	df	MS	F
Treatments	$SSTr$	$IJ - 1$	$SSTr/(IJ - 1)$	$MSTr/MSE$
A	SSA	$I - 1$	$SSA/(I - 1)$	MSA/MSE
B	SSB	$J - 1$	$SSB/(J - 1)$	MSB/MSE
AB interactions	$SSAB$	$(I - 1)(J - 1)$	$SSAB/(I - 1)(J - 1)$	$MSAB/MSE$
Error	SSE	$IJ(m - 1)$	$SSE/IJ(m - 1)$	
Total	$SSTot$	$mIJ - 1$		

The A, B, and AB rows of Table 8-19 are in some ways a refinement or amplification of the Treatments row. And the F statistics in the last column are exactly statistic (7-62) of Section 7-4 (for testing whether there are any differences at all among the IJ combination means), followed by statistics (8-69), (8-70), and (8-71) from this section.

EXAMPLE 8-8
(continued)

The sums of squares calculated earlier for the Clubb and Goedken tensile strength data lead to the two-way ANOVA table for the data of Table 8-5 that is given here as Table 8-20.

TABLE 8-20

ANOVA Table for the Aluminum Strip Tensile Strength Study

ANOVA TABLE

Source	SS	df	MS	F
Treatments	149,722	5	29,944	2.6
Hole Size	75,911	2	37,956	3.3
Hole Placement	71,568	1	71,568	6.3
Size × Placement Interaction	2,243	2	1,122	.1
Error	136,507	12	11,376	
Total	286,229	17		

The p-values associated with the observed F values in Table 8-20 are determined using F tables with numerator degrees of freedom parameters given in the corresponding rows and 12 denominator degrees of freedom. Table 8-21 summarizes these p-values.

TABLE 8-21

p-values for the Significance Tests Indicated in Table 8-20

Hypothesis	p-value
$H_0: \mu_{11} = \mu_{12} = \cdots = \mu_{32}$	$.05 < P[\text{an } F_{5,12} \text{ variable} > 2.6] < .10$
$H_0: \alpha_i = 0 \text{ all } i$	$.05 < P[\text{an } F_{2,12} \text{ variable} > 3.3] < .10$
$H_0: \beta_j = 0 \text{ all } j$	$.01 < P[\text{an } F_{1,12} \text{ variable} > 6.3] < .05$
$H_0: \alpha\beta_{ij} = 0 \text{ all } i, j$	$.25 < P[\text{an } F_{2,12} \text{ variable} > .1]$

The story told by the p-values in Table 8-21 is, of course, consistent with the picture of the students' results that emerged from the graphical and interval-oriented analysis in Section 8-1. For example, the strongest message in the data is the one against the hypothesis of no hole placement effects. And there is clearly no definitive evidence of size by placement interactions in the data.

The partition of treatment variability into three interpretable pieces (as in Table 8-20) has important intuitive appeal. So also does the fact that the three F statistics (8-69) through (8-71) each provide a single observed level of significance for *all effects of a given type at once*. (For example, all $3 \times 2 = 6$ interactions were considered at once in the calculation of the statistic (8-71) for the tensile strength data.) In fact, an energetic reader who consults other texts regarding the analysis of two-way factorial data will probably find the two-way ANOVA to be the main (and often only) topic treated.

Nevertheless, it is the author's hope that the discussion in this chapter has convinced the reader of the primary importance of graphical and interval-oriented methods of analysis for 2-factor studies. ANOVA calculations like those in Example 8-8 alone do not come close to giving adequate practical guidance or insight in 2-factor engineering applications. Knowing, for example, that there are statistically significant B main effects gives an engineer no clue as to the absolute size of those effects, no indication of which level of B is best, nor any sound means of predicting system performance if a given AB combination is to be used. Insight into those matters comes instead from the material presented in Sections 8-1 and 8-2. This author grants the usefulness of two-way ANOVA but sees it as typically an adjunct methodology — not the only or even the primary tool of formal statistical inference that ought to be used in the analysis of 2-factor statistical engineering studies.

p-Way ANOVA

Just as it is possible in balanced data sets to decompose $SSTr$ in a way that recognizes two-way complete factorial structure, it is similarly possible to break $SSTr$ down in a way that recognizes complete three-way (or higher) factorial structure in balanced data sets. The basic fact that makes this possible is that in a balanced complete factorial situation, $SSTr$ is the sum over all data points of the squares of the corresponding fitted factorial main effects and interactions. For example, in the $p = 3$ context, it turns out to be the case that

$$
\begin{aligned}
SSTr &= \sum_{i,j,k,l} (a_i^2 + b_j^2 + c_k^2 + ab_{ij}^2 + ac_{ik}^2 + bc_{jk}^2 + abc_{ijk}^2) \\
&= mJK \sum_{i=1}^{I} a_i^2 + mIK \sum_{j=1}^{J} b_j^2 + mIJ \sum_{k=1}^{K} c_k^2 + mK \sum_{i,j} ab_{ij}^2 \\
&\quad + mJ \sum_{i,k} ac_{ik}^2 + mI \sum_{j,k} bc_{jk}^2 + m \sum_{i,j,k} abc_{ijk}^2
\end{aligned}
\right\} \quad \text{(8-73)}
$$

Terms like those on the right of identity (8-73) are important enough to merit their own terminology and symbols. These are given next in general form, applicable to any balanced complete factorial data set.

DEFINITION 8-11

In a balanced p-way factorial study, the sum over all data points of the squared fitted effects of a given type will be called the **sum of squares due to the corresponding factor or interaction between the corresponding factors**. Such a sum of squares will be symbolized as SS followed by the letter designation(s) associated with the factor (or factors). For example, if $p \geq 3$, the sum over all data points of the quantities abc_{ijk}^2 will be called the sum of squares for ABC 3-factor interaction and symbolized as $SSABC$.

A little thought should make clear that computational formulas are possible for p-way factorial sums of squares. For one thing, sums can be written as extending only over those subscripts appearing on fitted effects in question, if one compensates by multiplying the sum by the product of m (the common sample size) and the numbers of levels of factors not involved in the fitted effects in question. For example, as indicated in identity (8-73), in a three-way factorial,

$$SSAB = mK \sum_{i,j} ab_{ij}^2 \tag{8-74}$$

Further, formulas from Section 4-3 for the fitted effects in terms of sample means are often combined with formulas like (8-74) to give formulas for sums of squares written in terms of sample means.

This text is not going to write out such general computational formulas for factorial sums of squares, believing that three-way and higher ANOVA calculations will typically be done only with the aid of commercial statistical packages. In the rare case that hand calculation is necessary, the verbal description in Definition 8-11 should be adequate to enable most readers to decompose $SSTr$ into its parts.

However, in one specific circumstance it is quite easy and useful to write out a computational means of obtaining the sums of squares defined in Definition 8-11. That is the case where all p factors have exactly two levels. For a balanced 2^p factorial study with common sample size m, if $\hat{E}$ is a fitted effect (obtained through the use of the Yates algorithm or otherwise), then the sum of squares corresponding to the type of effect estimated by $\hat{E}$ is

$$SS = m2^p(\hat{E})^2 \tag{8-75}$$

EXAMPLE 8-9

(Example 8-5 revisited) **Three-Way ANOVA in the Metal Cutting Power Requirement Study.** Consider again the power requirement study of L.D. Miller, which was used as an example in the previous section. Miller's data, given in Table 8-11, represent samples of size $m = 4$ taken in a complete 2^3 factorial fashion in a study of the effects on power requirements on a lathe of Tool Type (factor A), Tool Bevel Angle (factor B), and Type of Cut (factor C).

Example 8-5 showed that for the power requirement data,

$$SSTr = \sum_{i,j,k} m(\bar{y}_{ijk} - \bar{y})^2 = 55.51$$

In addition, the application of the Yates algorithm to the power requirement data gives the following fitted effects.

$$a_2 = -.2969 \qquad b_2 = .7969 \qquad c_2 = -.9844$$
$$ab_{22} = .0781 \qquad ac_{22} = -.0156 \qquad bc_{22} = -.1719$$
$$abc_{222} = -.0781$$

Then, by relationship (8-75),

$$SSA = 4 \cdot 2^3(-.2969)^2 = 2.82$$
$$SSB = 4 \cdot 2^3(.7969)^2 = 20.32$$
$$SSC = 4 \cdot 2^3(-.9844)^2 = 31.01$$
$$SSAB = 4 \cdot 2^3(.0781)^2 = .20$$
$$SSAC = 4 \cdot 2^3(-.0156)^2 = .01$$
$$SSBC = 4 \cdot 2^3(-.1719)^2 = .95$$
$$SSABC = 4 \cdot 2^3(-.0781)^2 = .20$$

Verify that the seven component sums of squares given here total to $SSTr = 55.51$, as promised in the discussion preceding this example.

As was the case with two-way factorials, the factorial sums of squares from a balanced p-way study not only partition $SSTr$ in an intuitively appealing way, but they can also be used to develop significance tests for the null hypotheses that all effects of a given type are 0. The general situation is that in a balanced p-way factorial, a hypothesis of the form

$$\text{H}_0: \text{all effects of type *** are 0} \tag{8-76}$$

can be tested using a test statistic

$$F = \frac{SS***/df***}{SSE/df\ error} \tag{8-77}$$

where large observed values of F count as evidence against H_0 in favor of H_a: not H_0, the possibility that some effects of type *** are nonzero. The appropriate reference distribution is F with numerator degrees of freedom $\nu_1 = df***$ equal to a product of terms, each of which is a number of levels minus 1 corresponding to a factor involved in the naming of the effect type ***. The appropriate denominator degrees of freedom, $\nu_2 = df\ error$, are those associated with s_P^2, which in the case of balanced factorial data amount to $(m - 1)$ times the product of the numbers of levels of all factors included in the study. For example, in a three-way factorial study, this means that

$$\text{H}_0: \alpha\beta_{ij} = 0 \qquad \text{for all } i, j$$

can be tested using the statistic

$$F = \frac{SSAB/(I - 1)(J - 1)}{SSE/(m - 1)IJK}$$

and an $F_{(I-1)(J-1),(m-1)IJK}$ reference distribution.

Recognizing the intuitive appeal of the ANOVA partition of $SSTr$ for balanced p-way factorials and the fact that tests of the form (8-76) and (8-77) are possible, the kind of expanded ANOVA table shown for the $p = 2$-way case (as Table 8-19) can also be made for larger values of p. To illustrate the pattern to be followed, Table 8-22 shows the form appropriate if $p = 3$ factors are involved.

TABLE 8-22

General Form of the Three-Way Balanced-Data ANOVA Table

ANOVA TABLE

Source	SS	df	MS	F
Treatments	$SSTr$	$IJK - 1$	$SSTr/(IJK - 1)$	$MSTr/MSE$
A	SSA	$I - 1$	$SSA/(I - 1)$	MSA/MSE
B	SSB	$J - 1$	$SSB/(J - 1)$	MSB/MSE
C	SSC	$K - 1$	$SSC/(K - 1)$	MSC/MSE
AB	$SSAB$	$(I - 1)(J - 1)$	$SSAB/(I - 1)(J - 1)$	$MSAB/MSE$
AC	$SSAC$	$(I - 1)(K - 1)$	$SSAC/(I - 1)(K - 1)$	$MSAC/MSE$
BC	$SSBC$	$(J - 1)(K - 1)$	$SSBC/(J - 1)(K - 1)$	$MSBC/MSE$
ABC	$SSABC$	$(I - 1)(J - 1)(K - 1)$	$\dfrac{SSABC}{(I - 1)(J - 1)(K - 1)}$	$MSABC/MSE$
Error	SSE	$(m - 1)IJK$	$SSE/(m - 1)IJK$	
Total	$SSTot$	$mIJK - 1$		

EXAMPLE 8-9
(continued)

Returning again to the situation of Example 8-5, recall that for the power requirement data, $SSTot = 108.93$ and $SSE = SSTot - SSTr = 108.93 - 55.51 = 53.42$. Then use of the factorial sums of squares derived from relationship (8-75) produces the version of Table 8-22 appropriate in the power requirement study — namely, Table 8-23.

TABLE 8-23

ANOVA Table for the Power Requirement Study

ANOVA TABLE

Source	SS	df	MS	F
Treatments	55.51	7	7.93	3.6
Tool	2.82	1	2.82	1.3
Bevel	20.32	1	20.32	9.1
Cut	31.01	1	31.01	13.9
Tool × Bevel	.20	1	.20	.1
Tool × Cut	.01	1	.01	.004
Bevel × Cut	.95	1	.95	.4
Tool × Bevel × Cut	.20	1	.20	.1
Error	53.42	24	2.226	
Total	108.93	31		

p-values for the Tool through Tool × Bevel × Cut observed F values listed in Table 8-23 would be based on 1 and 24 degrees of freedom, and it is straightforward to check that those for the Bevel and Cut main effects lie between .01 and .001, while the others are quite large. That is, the results of the F tests are consistent with the discussions in Section 8-3 leading to the conclusion that a "B and C main effects only" description of the power requirement appears adequate on the basis of the data in Table 8-11.

It should be noted that the amount of extra insight provided by an ANOVA table for a 2^p study is probably not as great as for other p-way factorials, where some factor or factors have more than two levels. That is because the F tests for the various types of effects are exactly equivalent to two-sided t tests based on the statistics

$$\frac{\hat{E} - 0}{s_P\sqrt{\dfrac{1}{m2^p}}}$$

That is, the *p*-values corresponding to the *F* statistics could be roughly inferred from the making of confidence intervals for the (2^p factorial) effects as discussed in Section 8-3. When each factor has only two levels, the fact that the effects of a given type sum to 0 over all indices means that the *F* tests represented by displays (8-76) and (8-77) are really tests that a single linear combination of means (and therefore its negative) is 0, not tests that simultaneously several different linear combinations are 0. So the reader should probably view Table 8-23 and the observed *F* values in it as of interest primarily as illustrating the meaning of the entries in the generic Table 8-22, not because it adds much at this point to one's understanding of the power requirement study.

Two-Way Random Effects Models and Analyses

In one kind of circumstance, this author finds two-way (and to a lesser extent, *p*-way) ANOVA methods to be of primary usefulness. This is where the levels of the factors included in a statistical engineering study are not so much of interest in and of themselves, as they are of interest as representative of a wider set of possible levels. For example, suppose that in a study at a large company, $I = 3$ machinists each use $J = 3$ different lathes to machine a critical dimension of $m = 2$ parts of a given type. The particular machinists and lathes used to gather the balanced two-way factorial data might well be of interest primarily in so far as they shed light on the variation inherent in how company machinists and lathes in general produce the critical dimension. In such cases, it would seem that formal inference should not focus on the effects of the particular factor levels included in the study. Rather, it would seem that statistical methods should deal with the variability that stands behind and generates any differences between the effects of the particular factor levels included in the study. And it turns out to be possible to use the balanced-data multi-way ANOVA calculations introduced in this section in the quantification of that underlying variability.

EXAMPLE 8-10

Two-Way ANOVA in a Gage Repeatability and Reproducibility Study. P. Tsai, in the article "Variable Gauge Repeatability and Reproducibility Study Using the Analysis of Variance Method" (*Quality Engineering*, 1988), discusses a so-called gage R and R study conducted at GMC Inland Division. A gage fixture built in a division tool shop was intended for use in checking an injection-molded part during production. A particular critical part dimension had specifications of 685 ± .5 mm.

Two operators (from the large number who would normally use the fixture in question) each measured the critical part dimension on ten different parts, two times apiece. (The operators went through the whole process of loading the parts into the fixture and measuring and removing them for each reading. The actual measuring was done automatically, and the operators weren't given any readings until all 20 of their measurements had been completed.) The 40 measurements obtained in this process are reproduced here in Table 8-24. Measurements are in .001 mm above nominal (i.e., above 685 mm).

The primary questions of interest in Tsai's gage/fixture qualification study were the following.

1. What is the *repeatability* of the measurement system? That is, what is a measure of variation associated with repeat readings for a single operator with a single part?
2. What is the *reproducibility* of the measurement system, in terms of a measure of variation associated with different operators collecting readings on a given part?

The issue of part-to-part variation in measurements (which would be of central importance in subsequent use of the gaging to monitor manufacturing performance) was not a main focus of attention in this particular study. The fact that multiple parts were

TABLE 8-24

40 Measurements Obtained in
a Gage R and R Study

	Operator 1	Operator 2
Part 1	−289 −273	−324 −309
Part 2	−311 −327	−340 −333
Part 3	−295 −318	−335 −326
Part 4	−301 −303	−304 −333
Part 5	−265 −288	−289 −279
Part 6	−298 −304	−305 −299
Part 7	−273 −293	−287 −250
Part 8	−276 −301	−275 −305
Part 9	−328 −341	−316 −314
Part 10	−293 −282	−300 −297

used can be thought of in the present context as an effort to ensure that the results of the study would be generally applicable to this type of part — that is, that the fixture was subjected to the range of conditions one would expect to meet in practice.

A two-way ANOVA table for this example is given as Table 8-25. The point of the present discussion will be that the calculations summarized in the ANOVA table can form the basis of inferences regarding not only the repeatability and reproducibility questions of interest in the original study, but also regarding the variability of parts represented by those in the study.

TABLE 8-25

ANOVA Table for Tsai's
Gage R and R Study

ANOVA TABLE

Source	SS	df	MS
Treatments	14,936	19	
Parts	11,750	9	1306
Operators	648	1	648
Parts × Operators	2,538	9	282
Error	3,270	20	163
Total	18,206	39	

The most commonly used probability model for the analysis of balanced two-way factorial data where the $I \cdot J$ conditions actually studied are thought of as particular combinations of potentially much larger sets of levels of factors A and B is a variation on the two-way model used in this chapter, called the *two-way random effects model*. It is built on the usual two-way normal assumptions that

$$y_{ijk} = \mu_{..} + \alpha_i + \beta_j + \alpha\beta_{ij} + \epsilon_{ijk} \qquad \text{(8-78)}$$

where the ϵ_{ijk} are iid normal $(0, \sigma^2)$ random variables. But it doesn't treat the effects α_i, β_j, and $\alpha\beta_{ij}$ as parameters/unknown constants. Rather, the variables $\alpha_1, \alpha_2, \ldots, \alpha_I$, $\beta_1, \beta_2, \ldots, \beta_J, \alpha\beta_{11}, \alpha\beta_{12}, \ldots, \alpha\beta_{IJ}$ are treated as (unobservable) random variables independent of the ϵ_{ijk}'s. These *random effects* are themselves assumed to be independent and normally distributed with 0 means and

$$Var\, \alpha_i = \sigma_\alpha^2$$
$$Var\, \beta_j = \sigma_\beta^2$$
$$Var\, \alpha\beta_{ij} = \sigma_{\alpha\beta}^2$$

where the variances $\sigma^2, \sigma_\alpha^2, \sigma_\beta^2$, and $\sigma_{\alpha\beta}^2$ are unknown parameters, typically termed *variance components*. The variance components σ_α^2 and σ_β^2 are thought of as quantifying the underlying variation in (respectively) factor A and factor B main effects, while $\sigma_{\alpha\beta}^2$ quantifies the overall lack of parallelism in mean responses for particular combinations of levels of A and B. The primary objects of formal inference then become these variance components.

It is an interesting and important conceptual point that the standard two-way model discussed in Section 8-1 is just a rewrite of the standard one-way model introduced in Section 7-1, but the two-way random effects model is *not* simply a rewrite of the one-way random effects model introduced toward the end of Section 7-4. Instead, it is inherently more complicated. It is possible to check that the one-way random effects model (7-68) (of Section 7-4) declares observations from any two different samples to be independent/uncorrelated. On the other hand, the two-way random effects assumptions imply that observations from different samples sharing the same level of A (or B) are dependent/correlated, basically because they share a common α_i (or β_j) term. The rather complicated nature of the two-way random effects model (and correspondingly more complicated nature of the p-way random effects models for larger p) turns out to result in some complications and restrictions on what inferences can be drawn regarding variance components using elementary methods. So be alert to the fact that in comparison to what was done in Section 7-4, the discussion that follows concerning p-way random effects analyses may seem somewhat incomplete.

The complex nature of the two-way random effects model is already obvious when one contemplates trying to assess the reasonableness of the model in a particular application, prior to using it in the drawing of inferences. The usual residual plotting (of residuals $e_{ijk} = y_{ijk} - \bar{y}_{ij}$, in normal plots, or versus $\bar{y}_{ij}$ or time order of observation, etc.) is a sensible way of investigating the appropriateness of the assumption that the ϵ_{ijk} are iid normal variables. But there is no simple, widely accepted method of checking the distributional assumptions on the random effects α_i, β_j, and $\alpha\beta_{ij}$. (The obvious possibility of doing normal-plotting for the a_i's and b_j's is not completely satisfactory, because fitted effects turn out to approximate not simply the corresponding underlying random effects, but rather the corresponding underlying random effects plus some other random variables. The interested reader can, for example, probably do the algebra necessary to verify that a normal plot of a_i's would provide a check only on the normality of the α_i's plus corresponding averages over j and k of $\alpha\beta_{ij}$'s and ϵ_{ijk}'s.) This lack of a complete set of simple diagnostic tools for the two-way random effects model is unfortunate. The investigator who is planning to use the methods that follow has a serious responsibility to consider intuitively how much sense the model assumptions make in a particular application. It is also prudent not to make real-world close calls on the basis of inferences in the two-way random effects model.

Perhaps the most revealing elementary consequence of the two-way random effects model assumptions is the set of formulas for expected mean squares implied by the model. It turns out that for balanced data under the two-way random effects model,

$$
\left.
\begin{aligned}
E(s_P^2) &= E(MSE) = \sigma^2 \\
E(MSAB) &= m\sigma_{\alpha\beta}^2 + \sigma^2 \\
E(MSA) &= mJ\sigma_{\alpha}^2 + m\sigma_{\alpha\beta}^2 + \sigma^2 \\
E(MSB) &= mI\sigma_{\beta}^2 + m\sigma_{\alpha\beta}^2 + \sigma^2
\end{aligned}
\right\}
\tag{8-79}
$$

That is, under the two-way random effects model, the (random) mean squares have average values related to the variance components in the relatively simple ways detailed by formulas (8-79). Display (8-79) says that (as always) s_P^2 reflects the basic baseline variation, $MSAB$ reflects variation in the $\alpha\beta_{ij}$'s plus baseline variation, and the mean squares for the main effects reflect variation in the corresponding main effects plus baseline variation plus variation associated with the interactions.

The formulas (8-79) are revealing enough that they are usually included in an *EMS* column of a two-way ANOVA table when doing a random effects analysis. They turn out to be correct indicators of how to proceed with both hypothesis testing and estimation for the variance components $\sigma_{\alpha\beta}^2$, σ_{α}^2, and σ_{β}^2. For example, consider first making inferences about $\sigma_{\alpha\beta}^2$, the interaction variance component. Formulas (8-79) suggest that on average, the difference between $MSAB$ and MSE is $m\sigma_{\alpha\beta}^2$. So a sensible estimate of $\sigma_{\alpha\beta}^2$ would seem to be

$$
\frac{1}{m}(MSAB - MSE)
$$

or (recognizing that this can on occasion turn out negative)

$$
\max\left(0, \frac{1}{m}(MSAB - MSE)\right)
\tag{8-80}
$$

Further, the form of formulas (8-79) suggests that a test of $H_0: \sigma_{\alpha\beta}^2 = 0$ might somehow be based on a comparison of $MSAB$ and MSE. And in fact, it is possible for a mathematician to show that $H_0: \sigma_{\alpha\beta}^2 = 0$ can be tested using the statistic

$$
F = \frac{MSAB}{MSE}
\tag{8-81}
$$

an $F_{(I-1)(J-1),(m-1)IJ}$ reference distribution and counting large observed values of the test statistic as evidence against H_0 in favor of $H_a: \sigma_{\alpha\beta}^2 > 0$.

Similar reasoning applied to the question of inference for the variance components σ_{α}^2 and σ_{β}^2 also suggests theoretically correct estimation and testing methods. However, these methods have what may at first glance seem to be a somewhat surprising twist to them. Formulas (8-79) suggest that when it comes to inference for the variance of main effects, it is $MSAB$ rather than MSE that provides the proper background against which to evaluate MSA or MSB. (A large difference between, e.g., MSA and MSE is potentially attributable to a large value of σ_{α}^2 *or* a large value of $\sigma_{\alpha\beta}^2$.) At any rate, it is the case that a standard estimate of σ_{α}^2 is

$$
\max\left(0, \frac{1}{mJ}(MSA - MSAB)\right)
\tag{8-82}
$$

and a corresponding estimate of σ_{β}^2 is

$$
\max\left(0, \frac{1}{mI}(MSB - MSAB)\right)
\tag{8-83}
$$

Further, $H_0: \sigma_\alpha^2 = 0$ can be tested using

$$F = \frac{MSA}{MSAB} \tag{8-84}$$

and an $F_{(I-1),(I-1)(J-1)}$ reference distribution, where large observed values of F count as evidence against H_0 in favor of $H_a: \sigma_\alpha^2 > 0$. Similarly, $H_0: \sigma_\beta^2 = 0$ can be tested using

$$F = \frac{MSB}{MSAB} \tag{8-85}$$

and an $F_{(J-1),(I-1)(J-1)}$ reference distribution, where large observed values of F count as evidence against H_0 in favor of $H_a: \sigma_\beta^2 > 0$.

Since the F ratios indicated by equations (8-84) and (8-85) are different from those appropriate under the standard (fixed effects) two-way model of Section 8-1, it is worthwhile to write out completely the (modified) form taken by a two-way ANOVA table when a random effects analysis is being made. This general form is given in Table 8-26.

TABLE 8-26

General Form of the Two-Way Balanced-Data Random Effects ANOVA Table

ANOVA TABLE

Source	SS	df	MS	EMS	F
Treatments	SSTr	$IJ - 1$			
A	SSA	$I - 1$	$SSA/(I - 1)$	$mJ\sigma_\alpha^2 + m\sigma_{\alpha\beta}^2 + \sigma^2$	MSA/MSAB
B	SSB	$J - 1$	$SSB/(J - 1)$	$mI\sigma_\beta^2 + m\sigma_{\alpha\beta}^2 + \sigma^2$	MSB/MSAB
AB interaction	SSAB	$(I - 1)(J - 1)$	$SSAB/(I - 1)(J - 1)$	$m\sigma_{\alpha\beta}^2 + \sigma^2$	MSAB/MSE
Error	SSE	$(m - 1)IJ$	$SSE/(m - 1)IJ$	σ^2	
Total	SSTot	$mIJ - 1$			

EXAMPLE 8-10
(continued)

Returning again to the gage/fixture qualification study of Tsai, the complete ANOVA table for a two-way random effects analysis of the data in Table 8-24 is given in Table 8-27.

TABLE 8-27

Random Effects ANOVA Table for Tsai's Gage R and R Study

ANOVA TABLE

Source	SS	df	MS	EMS	F
Treatments	14,936	19			
Parts	11,750	9	1306	$4\sigma_\alpha^2 + 2\sigma_{\alpha\beta}^2 + \sigma^2$	4.6
Operators	648	1	648	$20\sigma_\beta^2 + 2\sigma_{\alpha\beta}^2 + \sigma^2$	2.3
Parts × Operators	2,538	9	282	$2\sigma_{\alpha\beta}^2 + \sigma^2$	1.7
Error	3,270	20	163	σ^2	
Total	18,206	39			

So, for instance, the p-value associated with a significance test of $H_0: \sigma_\beta^2 = 0$ is $P[\text{an } F_{1,9} \text{ random variable} > 2.3]$, which is between .25 and .10 according to Tables D-6. The reader should verify that in sum, the three observed F values in the last column of Table 8-27 show reasonably strong evidence of nonzero part-to-part variation in the critical dimension, but only weaker evidence of statistically detectable operator-to-operator variation and part-by-operator interaction.

What are probably of more practical use in the context of gage/fixture qualification than the significance values associated with the observed F ratios in Table 8-27 are the estimates of the variance components derived using $MSE = s_P^2$ and the formulas (8-80), (8-82), and (8-83). That is, the natural estimate of the baseline variability (σ^2) associated with repeat measurements of a single part by a single operator is

$$s_P^2 = MSE = 163$$

Further, from formula (8-83), an estimate of σ_β^2 (the operator variance) is

$$\frac{1}{2(10)}(648 - 282) = 18.3$$

and, from formula (8-80), an estimate of $\sigma_{\alpha\beta}^2$ (the interaction variance) is

$$\frac{1}{2}(282 - 163) = 59.5$$

It is instructive to translate the variance component language more directly into the repeatability and reproducibility language of the real statistical engineering study. Notice first that σ is the theoretical standard deviation that would be associated with repeat measurements of a single part by a single operator, so

$$\sqrt{s_P^2} = \sqrt{163} = 13 \times 10^{-3} \text{ mm}$$

or perhaps some company standard multiple of it (it appears from Tsai's article that GMC uses a 5.15 multiplier) might well be termed the estimated gage/fixture *repeatability*. Next notice that the two-way random effects model implies that for a single part (fixed i) even if gage repeatability were 0 ($\sigma^2 = 0$, so that all ϵ_{ijk} were 0), the variance associated with multiple operators producing measurements would be the variance produced by variables $\beta_j + \alpha\beta_{ij}$ as j varies over operators — that is,

$$\sigma_\beta^2 + \sigma_{\alpha\beta}^2$$

This suggests (on a standard deviation rather than a variance scale) that

$$\sqrt{\sigma_\beta^2 + \sigma_{\alpha\beta}^2}$$

or some standard multiple thereof would be a sensible theoretical measure of measurement system *reproducibility*. This can, of course, be estimated by plugging the ANOVA-based estimates of variance components into the theoretical expression. In the present case, this thinking leads to

$$\sqrt{18.3 + 59.5} = 9 \times 10^{-3} \text{ mm}$$

as an estimate of reproducibility based on Tsai's data.

Finally, recall from the original description of this scenario that specifications on the dimension being gaged were $685 \pm .5$ mm. The $\pm$ part of this specification extends 1 mm, i.e., 1000×10^{-3} mm. It is thus comforting to note that the measurement imprecisions conveyed by the repeatability and reproducibility figures above (even when multiplied by the GMC 5.15 factor) are an order of magnitude smaller than the spread in tolerances that the gaging in question was being developed to guarantee. The gaging tested seems adequate to do the process monitoring job it was made for.

The tests and estimates (8-80) through (8-85) pretty much summarize what is easily done in the way of inference for variance components in the two-way random effects model. It is also possible to make a confidence interval for σ^2 based on s_P^2, exactly as in Section 7-4. But elementary methods of confidence interval estimation for the other variance components do not seem possible. In spite of their somewhat limited nature, this author still believes that the existing methods often provide important engineering insights in those balanced two-way factorial studies where a random effects viewpoint makes sense.

p-Way Random Effects Models and Analyses

The pattern of reasoning established in the preceding discussion of two-way random effects analyses can to some degree be extended to the (less frequently occurring) case of balanced p-way factorials where the studied levels of all p factors are of interest primarily as they are representative of wider sets of possible levels of those factors. For example, a critical dimension of the same $I = 10$ parts might be measured by $J = 2$ operators on $K = 3$ different gages $m = 2$ times each in a study of the integrity of a company's gaging system. In such a study, the particular parts, operators, and gages involved are of interest primarily as they represent the performance of all company operators using all company gages to measure all parts produced by the company. And in such a case, it would seem natural to hope for $p = 3$-way analogs of the two-way random effects analysis methods just discussed.

To begin with, a p-way factorial random effects model has the same form as the standard p-way normal model alluded to in Section 8-3, except that the effects $\alpha_i, \beta_j, \alpha\beta_{ij}, \gamma_k, \ldots$, are understood to be unobservable (independent, mean 0, normal) random variables, each effect type having its own variance component ($\sigma_\alpha^2, \sigma_\beta^2, \sigma_{\alpha\beta}^2, \sigma_\gamma^2, \ldots$, etc.). The variance components are the usual objects of inference in a p-way random effects analysis, and the easiest way to understand how ANOVA-based inference must proceed is through seeing how expected values of ANOVA mean squares are related to the variance components.

It is always the case that $Es_P^2 = \sigma^2$, and it turns out to be possible to give an easy algorithm for writing out the other expected mean squares under balanced-data p-way random effects models. The expected mean square for a given factorial source is a linear combination of variance components structured as follows. First, σ^2 is a summand in all *EMS*'s. Then, for a given source, there are terms for all effects types whose names share all letters appearing in the name of the source in question. Those terms are made up of the product of the effect type variance component and m times the product of the numbers of levels of all factors whose names don't appear in the name of the effect. For example, for $p = 3$ factor studies, Table 8-28 gives degrees of freedom and expected mean squares for the various ANOVA sources.

TABLE 8-28

Degrees of Freedom and Expected Mean Squares in a Three-Way Random Effects Analysis

Source	df	MS
A	$(I-1)$	$mJK\sigma_\alpha^2 + mK\sigma_{\alpha\beta}^2 + mJ\sigma_{\alpha\gamma}^2 + m\sigma_{\alpha\beta\gamma}^2 + \sigma^2$
B	$(J-1)$	$mIK\sigma_\beta^2 + mK\sigma_{\alpha\beta}^2 + mI\sigma_{\beta\gamma}^2 + m\sigma_{\alpha\beta\gamma}^2 + \sigma^2$
C	$(K-1)$	$mIJ\sigma_\gamma^2 + mJ\sigma_{\alpha\gamma}^2 + mI\sigma_{\beta\gamma}^2 + m\sigma_{\alpha\beta\gamma}^2 + \sigma^2$
AB	$(I-1)(J-1)$	$mK\sigma_{\alpha\beta}^2 + m\sigma_{\alpha\beta\gamma}^2 + \sigma^2$
AC	$(I-1)(K-1)$	$mJ\sigma_{\alpha\gamma}^2 + m\sigma_{\alpha\beta\gamma}^2 + \sigma^2$
BC	$(J-1)(K-1)$	$mI\sigma_{\beta\gamma}^2 + m\sigma_{\alpha\beta\gamma}^2 + \sigma^2$
ABC	$(I-1)(J-1)(K-1)$	$m\sigma_{\alpha\beta\gamma}^2 + \sigma^2$
Error	$(m-1)IJK$	σ^2

For *some* variance components in p-way random effects models, it is possible to find F tests of the hypotheses that they are 0. If two expected mean squares differ only by a single term, the ratio of the corresponding mean squares can be used (exactly as discussed in the $p = 2$ case) to make an F ratio for testing the hypothesis that the variance component is 0. For example, when $p = 3$, Table 8-28 shows that H_0: $\sigma^2_{\alpha\beta\gamma} = 0$ can be tested using the statistic $F = MSABC/MSE$. Similarly, H_0: $\sigma^2_{\alpha\beta} = 0$ can be tested using the statistic $F = MSAB/MSABC$. But (for $p > 2$) not all such hypotheses have obvious F tests. For example, perusal of Table 8-28 suggests that for $p = 3$, none of the hypotheses H_0: $\sigma^2_{\alpha} = 0$, H_0: $\sigma^2_{\beta} = 0$, and H_0: $\sigma^2_{\gamma} = 0$ have natural corresponding F ratios. Various more complicated approximate testing methods have been suggested for judging the statistical significance of such variance components, but for several reasons (including what this author sees as a lack of broad utility), this testing issue will not be pursued further in this book. The reader who has occasion to really need such a significance testing method should see (for example) Chapters 21 and 22 of *Applied Linear Statistical Models* (3rd ed.) by Neter, Wasserman, and Kutner.

Simple estimation of variance components in balanced p-way factorial studies goes forward without problem, provided one is moderately resourceful in finding linear combinations of expected mean squares that are equal to variance components of interest. One then simply uses the corresponding linear combination of mean squares (or 0, should the linear combination turn out negative) as a natural ANOVA-based estimate of any variance component of interest. For example, with $p = 3$ factors, Table 8-28 shows $E(MSBC)$ and $E(MSABC)$ to differ only by $mI\sigma^2_{\beta\gamma}$, suggesting the estimate of $\sigma^2_{\beta\gamma}$

$$\max\left(0, \frac{1}{mI}(MSBC - MSABC)\right)$$

Or as a further example, notice that

$$E(MSA) - E(MSAB) - E(MSAC) + E(MSABC) = mJK\sigma^2_{\alpha}$$

suggesting the estimate of σ^2_{α}

$$\max\left(0, \frac{1}{mJK}(MSA - MSAB - MSAC + MSABC)\right)$$

And (in contrast to the somewhat incomplete hypothesis testing situation) it is possible to find such elementary estimates for all variance components in balanced p-way factorials.

Before closing this discussion of factorial random effects analyses, a final comment regarding the methods must be made. That is that there is actually another possibility in addition to the *fixed effects* analyses (which treat all factorial effects as fixed unknown parameters) and the *random effects* analyses just discussed (which treat all factorial effects as random variables generated by underlying distributions whose variances are the factorial variance components). That possibility is to treat some effects as fixed and some as random.

For example, in the context of the gage/fixture study of Example 8-10, a small company having only two operators might well want to treat the ten parts used in the study as representative of many other parts, but at the same time be specifically interested in the gaging performance of the two particular operators in the study. In such a case, analysis methods beyond those presented here are appropriate. Those methods are based on *mixed effects* models, which are yet more complicated than either the basic p-way (fixed effects) models of the first three sections of this chapter or the factorial random effects models met in this section. Some reasonably elementary inference methods for mixed effects models are discussed, for example, in the book by Neter, Wasserman, and Kutner mentioned earlier. But the matter will not be pursued here, because the topic is not as essential for engineers as ones yet to be discussed. However, if the need arises to look further into the topic, the material given in this section on ANOVA-based methods is adequate background for further inquiry into the subject.

EXERCISES

8-1. (§8-1) The accompanying table shows part of the data of Dimond and Dix, referred to in Examples 1-7 and 3-9. The values are the shear strengths (in lb) for $m = 3$ tests on joints of various combinations of Wood Type and Glue Type.

Wood	Glue	Joint Shear Strengths
pine	white	130, 127, 138
pine	carpenter's	195, 194, 189
pine	cascamite	195, 202, 207
fir	white	95, 119, 62
fir	carpenter's	137, 157, 145
fir	cascamite	152, 163, 155

(a) Make a plot of the six combination means like the one in Figure 3-22 and enhance it with error bars derived using the P-R method of making 95% simultaneous two-sided confidence intervals.
(b) Compute the fitted main effects and interactions from the six combination sample means. Use these to make individual 95% confidence intervals for all of the main effects and interactions in this 2×3 factorial study. What do these indicate about the detectability of the various effects?
(c) Use Tukey's method for simultaneous comparison of main effects and give simultaneous 95% confidence intervals for all differences in Glue Type main effects.

8-2. (§8-2) Consider again the situation of Exercise 8-1 and suppose that a no-interaction model is judged to be adequate here.
(a) Compute an estimate of σ based on the no-interaction assumption, and compare it to the pooled sample standard deviation.
(b) Compare an individual two-sided 95% confidence interval for the mean strength of pine/white joints made using the no-interaction assumption to one based on the methods of Section 7-2.
(c) Compare an individual two-sided 95% prediction interval for the next strength of a pine/white joint made using the no-interaction assumption to one based on the methods of Section 7-2.
(d) Compare the results of part (c) of Exercise 8-1 with the results of applying the Tukey method to the problem of comparing glue types under the no-interaction assumption.

8-3. (§8-4) Consider once again the situation of Exercise 8-1. These complete factorial data are balanced, so the two-way ANOVA methods of Section 8-4 are applicable.
(a) Find SSA, SSB, and $SSAB$ and verify that they total to $SSTr$. Do the calculations first by hand, and then check yourself using a statistical computer package.
(b) Make an ANOVA table for these data and use it in tests of the three hypotheses that there are no interactions, that there are no Wood Type main effects, and that there are no Glue Type main effects.

8-4. (§8-3) Consider again the situation of Exercise 4-10.
(a) For the logged responses, make individual 95% confidence intervals for the effects corresponding to the high levels of all three factors. Which effects are statistically detectable?

(b) Fit an appropriate few-effects model suggested by your work in (a) to these data. Compare the corresponding value of s_{FE} to the value of s_P.
(c) Compare a two-sided individual 95% confidence interval for the mean (logged) response for combination (1) made using the fitted few-effects model to one based on the methods of Section 7-2.
(d) Compare a two-sided 95% prediction interval for the next (logged) response for combination (1) made using the the fitted few-effects model to one based on the methods of Section 7-2. What do these intervals give in terms of raw (unlogged) times?

8-5. (§8-3) Exercise 4-23 concerned some data on the making of Dual In-line Packages and the number of pullouts produced on such devices under 2^4 different combinations of manufacturing conditions. Return to that exercise, and if you have not already done so, use the Yates algorithm and compute fitted 2^4 factorial effects for the data set.
(a) Use normal-plotting to identify detectable effects here.
(b) Based on your analysis from (a), postulate a possible few-effects model for this situation. Use the reverse Yates algorithm to fit such a model to these data. Use the fitted values to compute residuals. Normal-plot these and plot them against levels of each of the four factors, looking for obvious problems with the model.
(c) Based on your few-effects model, make a recommendation for the future making of these devices. Give a 95% two-sided confidence interval (based on the few-effects model) for the mean pullouts you expect to experience if your advice is followed.
(d) Give a 95% prediction interval for the number of pullouts experienced on the next DIP made under the conditions you recommend.
(e) Finally, give a 90% upper tolerance bound for 95% of all additional pullout counts for DIPs made under the conditions you recommend.

8-6. B. Choi conducted a replicated full factorial study of the stopping properties of various types of bicycle tires on various riding surfaces. Three different Types of Tires were used on the bike, and three different Pavement Conditions were used. For each Tire Type/Pavement Condition combination, $m = 6$ skid mark lengths were measured. The accompanying table shows some summary statistics for the study. (The units are cm.)

	Dry Concrete	Wet Concrete	Dirt
Smooth Tires	$\bar{y}_{11} = 359.8$ $s_{11} = 19.2$	$\bar{y}_{12} = 366.5$ $s_{12} = 26.4$	$\bar{y}_{13} = 393.0$ $s_{13} = 25.4$
Reverse Tread	$\bar{y}_{21} = 343.0$ $s_{21} = 15.5$	$\bar{y}_{22} = 356.7$ $s_{22} = 37.4$	$\bar{y}_{23} = 375.7$ $s_{23} = 39.9$
Treaded Tires	$\bar{y}_{31} = 384.8$ $s_{31} = 15.4$	$\bar{y}_{32} = 400.8$ $s_{32} = 60.8$	$\bar{y}_{33} = 402.5$ $s_{33} = 32.8$

(a) Compute s_P for Choi's data set. What is this supposed to be measuring?
(b) Make a plot of the sample means similar to Figure 8-3. Use error bars for the means calculated from 95% two-sided individual

confidence limits for the means. (Make use of your value of s_P from (a).)

(c) Based on your plot from (b), which factorial effects appear to be distinguishable from background noise? (Tire Type main effects? Pavement Condition main effects? Tire × Pavement interactions?)

(d) Compute all of the fitted factorial effects for Choi's data. (Find the a_i's, b_j's, and ab_{ij}'s defined in Section 4-3.)

(e) If one wishes to find individual 95% two-sided confidence intervals for the interactions $\alpha\beta_{ij}$, intervals of the form $ab_{ij} \pm \Delta$ are appropriate. Find Δ. Based on this value, are there statistically detectable interactions here? How does this conclusion compare with your more qualitative answer to part (c)?

(f) If one wishes to compare Tire Type main effects, confidence intervals for the differences $\alpha_i - \alpha_{i'}$ are in order. Find individual 95% two-sided confidence intervals for $\alpha_1 - \alpha_2$, $\alpha_1 - \alpha_3$, and $\alpha_2 - \alpha_3$. Based on these, are there statistically detectable Tire Type main effects here? How does this conclusion compare with your answer to part (c)?

(g) Redo part (f), this time using simultaneous 95% two-sided confidence intervals.

8-7. Return to the situation of Exercise 4-18. That exercise concerned some unreplicated 2^3 factorial data taken from a study of the mechanical properties of a polymer. If you have not already done so, use the Yates algorithm to compute fitted 2^3 factorial effects for the data given in that exercise. Then make a normal plot of the seven fitted effects $a_2, b_2, \ldots, abc_{222}$ as a means of judging the statistical detectability of the various effects on impact strength. Interpret this plot.

8-8. Exercise 4-19 concerns a 2^3 study of mechanical pencil lead strength done by Timp and M-Sidek. Return to that exercise, and if you have not already done so, use the Yates algorithm to compute fitted 2^3 effects for the logged data.

(a) Compute s_P for the logged data. Individual confidence intervals for the theoretical 2^3 effects are of the form (fitted effect) $\pm \Delta$. Find Δ if 95% individual two-sided intervals are of interest.

(b) Based on your value from part (a), which of the factorial effects are statistically detectable? Considering only those effects that are both statistically detectable and large enough to have a material impact on the breaking strength, interpret the results of the students' experiment. (For example, if the A main effect is judged to be both detectable and of practical importance, what does moving from the .3 diameter to the .7 diameter do to the breaking strength? Remember to translate back from the log scale when making these interpretations.)

(c) Use the reverse Yates algorithm to produce fitted $\ln(y)$ values for a few-effects model corresponding to your answer to (b). Use the fitted values to compute residuals (still on the log scale). Normal-plot these and plot them against levels of each of the three factors and against the fitted values, looking for obvious problems with the few-effects model.

(d) Based on your few-effects model, give a 95% two-sided prediction interval for the next $\ln(y)$ that would be produced by the abc treatment combination. By exponentiating the endpoints of this interval, give a 95% two-sided prediction interval for the next number of clips required to break a piece of lead under this set of conditions.

(e) Finally, based on your few-effects model, give a 95% lower tolerance bound for 90% of $\ln(y)$ values that would be produced by the abc treatment combination. By exponentiating this, give a 95% lower tolerance bound for 90% of the numbers of clips required to break pieces of lead under this set of conditions.

8-9. The following are the weights recorded by $I = 3$ different students when weighing the same nominally 5 g mass with $J = 2$ different scales $m = 2$ times apiece. (They are part of the much larger data set given in Exercise 3-18.)

	Scale 1	Scale 2
Student 1	5.03, 5.02	5.07, 5.09
Student 2	5.03, 5.01	5.02, 5.07
Student 3	5.06, 5.00	5.10, 5.08

Corresponding fitted factorial effects are: $a_1 = .00417$, $a_2 = -.01583$, $a_3 = .01167$, $b_1 = -.02333$, $b_2 = .02333$, $ab_{11} = -.00417$, $ab_{12} = .00417$, $ab_{21} = .01083$, $ab_{22} = -.01083$, $ab_{31} = -.00667$, and $ab_{32} = .00667$. Further, a pooled standard deviation is $s_P = .02483$.

(a) If one wishes to enhance a plot of sample means with error bars derived from 95% two-sided individual confidence limits for the mean weights, what plus-or-minus value would be used to make those error bars? Make such a plot and discuss the likely statistical detectability of the interactions.

(b) Individual 95% two-sided confidence limits for the interactions $\alpha\beta_{ij}$ are of the form $ab_{ij} \pm \Delta$. Find Δ here. Based on this, are the interactions statistically detectable?

(c) Compare the Student main effects using individual 95% two-sided confidence intervals.

(d) Compare the Student main effects using simultaneous 95% two-sided confidence intervals.

8-10. Heyde, Kuebrich, and Swanson conducted a gage repeatability and reproducibility study similar to the one discussed in Example 8-10. Each of the three students used a certain 1 in. micrometer caliper to measure the heights of ten steel punches $m = 3$ times. The data they collected are shown in the accompanying table. The measurements are in 10^{-3} in.

(a) Make an analysis of variance table for this set of two-way factorial data. (You will probably want to make use of a computer program if one is available. The calculations would be quite tedious by hand.)

	Student 1	Student 2	Student 3
Punch 1	496, 496, 499	497, 499, 497	497, 498, 496
Punch 2	498, 497, 499	498, 496, 499	497, 499, 500
Punch 3	498, 498, 498	497, 498, 497	496, 498, 497
Punch 4	497, 497, 498	496, 496, 499	498, 497, 497
Punch 5	499, 501, 500	499, 499, 499	499, 499, 500
Punch 6	499, 498, 499	500, 499, 497	498, 498, 498
Punch 7	503, 499, 502	498, 499, 499	500, 499, 502
Punch 8	500, 499, 499	501, 498, 499	500, 501, 499
Punch 9	499, 500, 499	500, 500, 498	500, 499, 500
Punch 10	497, 496, 496	500, 494, 496	496, 498, 496

Treat these data from a two-way random effects perspective that would be appropriate if one thinks of the students as representing a larger pool of potential technicians who might use this caliper and the punches as representing a much larger pool of punches that will be measured using this device.

(b) Give expressions for the expected mean squares in this context. You will want to use formulas (8-79) of Section 8-4 in this endeavor. Then use the observed mean squares and find estimates of the variance components σ^2, σ_α^2, σ_β^2, and $\sigma_{\alpha\beta}^2$.

(c) Recognizing that σ is the theoretical standard deviation that would be associated with repeat measurements of a single punch by a single technician, find an estimate of this quantity. (Recall that this or some multiple of it might well be called the gage repeatability.) In light of the fact that specifications on these punch heights were .500 in. to .505 in., what does the repeatability figure indicate about the feasibility of using the caliper in question to check punches for correct heights?

(d) Compute an estimate of

$$\sqrt{\sigma_\beta^2 + \sigma_{\alpha\beta}^2}$$

as a measure of gage reproducibility (variation in measurements that would be experienced even if there were no repeatability problem, as a result of operator-to-operator variability). How does this figure compare to the value in (c)?

8-11. The students who conducted the study used in Exercise 8-10 came to the conclusion that the micrometer caliper was not going to provide sufficient precision for their purposes. They therefore found a more precise gage (a depth micrometer) to use in subsequent work. But before proceeding to the main part of their study, they conducted a gage repeatability and reproducibility study on the new gage, exactly parallel to the one described in Exercise 8-10. An ANOVA table summarizing the results of the second study is given below. The units of the sums of squares and mean squares are 10^{-6} in.^{2}.

Use this analysis of variance table and do parts (b), (c), and (d) from Exercise 8-10 for this second, more precise gage.

ANOVA TABLE

Source	SS	df	MS
Punches	114.46	9	12.717
Students	.02	2	.011
Punches × Students	1.31	18	.073
Error	7.33	60	.122
Total	123.12	89	

8-12. The oil viscosity study of Dunnwald, Post, and Kilcoin (referred to in Exercise 7-15) was actually a 3×4 full factorial study. Some summary statistics for the entire data set are recorded in the accompanying table. Summarized are $m = 10$ measurements of the viscosities of each of four different weights of three different brands of motor oil at room temperature. Units are seconds required for a ball to drop a particular distance through the oil.

(a) Find the pooled sample standard deviation here. What are the associated degrees of freedom?

(b) Make a plot of sample means useful for investigating the size of Brand × Weight interactions. Enhance this plot by adding error bars derived from 99% individual confidence intervals for the cell means. Does it appear that there are important and statistically detectable interactions here?

(c) If the Tukey method is used to find simultaneous 95% two-sided confidence intervals for all differences in Brand main effects, the intervals produced are of the form $\bar{y}_{i.} - \bar{y}_{i'.} \pm \Delta$, for Δ an appropriate number. Find Δ.

(d) If the Tukey method is used to find simultaneous 95% two-sided confidence intervals for all differences in Weight main effects, the intervals produced are of the form $\bar{y}_{.j} - \bar{y}_{.j'} \pm \Delta$, for Δ an appropriate number. Find Δ.

(e) Based on your answers to (c) and (d), would you say that there are statistically detectable Brand and/or Weight main effects on viscosity?

(f) This author strongly suspects that the "$m = 10$" viscosity measurements made for each of the 12 Brand/Weight combinations were made on oil from a single quart of that type of oil. If this is the case, s_P, one's baseline measure of variability, is measuring only the variability associated with experimental technique (not, e.g., from quart to quart of a given type of oil). One might thus argue that the real-world inferences to be made, properly speaking, extend only to the particular quarts used in the study. Discuss how these interpretations (of s_P and the extent of statistically based inferences) would be different if in fact the students used different quarts of oil in producing the "$m = 10$" different viscosity measurements in each cell.

8-13. The article "Effect of Temperature on the Early-Age Properties of Type I, Type III and Type I/Fly Ash Concretes" by N. Gardner (*ACI Materials Journal*, 1990) contains summary statistics for a very large study of the properties of several concretes under a variety of curing conditions. The accompanying table presents some of the statistics from that paper. Given here are the sample means and standard deviations of 14-day compressive

TABLE For Exercise 8-12

	10W30	SAE 30	10W40	20W50
Brand M	$\bar{y}_{11} = 1.385$	$\bar{y}_{12} = 2.066$	$\bar{y}_{13} = 1.414$	$\bar{y}_{14} = 4.498$
	$s_{11} = .091$	$s_{12} = .097$	$s_{13} = .150$	$s_{14} = .204$
Brand C	$\bar{y}_{21} = 1.319$	$\bar{y}_{22} = 2.002$	$\bar{y}_{23} = 1.415$	$\bar{y}_{24} = 4.662$
	$s_{21} = .088$	$s_{22} = .089$	$s_{23} = .115$	$s_{24} = .151$
Brand H	$\bar{y}_{31} = 1.344$	$\bar{y}_{32} = 2.049$	$\bar{y}_{33} = 1.544$	$\bar{y}_{34} = 4.549$
	$s_{31} = .066$	$s_{32} = .089$	$s_{33} = .068$	$s_{34} = .171$

TABLE For Exercise 8-13

	0°C	10°C	20°C	30°C
.55 Water/Cement Ratio	$\bar{y}_{11} = 28.99$ $s_{11} = .91$	$\bar{y}_{12} = 30.24$ $s_{12} = 1.26$	$\bar{y}_{13} = 33.99$ $s_{13} = 1.85$	$\bar{y}_{14} = 36.02$ $s_{14} = .93$
.35 Water/Cement Ratio	$\bar{y}_{21} = 38.70$ $s_{21} = .77$	$\bar{y}_{22} = 36.16$ $s_{22} = 1.92$	$\bar{y}_{23} = 40.18$ $s_{23} = 2.86$	$\bar{y}_{24} = 42.36$ $s_{24} = 1.35$

TABLE For Exercise 8-14

	0°C	10°C	20°C	30°C
.55 Water/Cement Ratio	$\bar{y}_{11} = 47.82$ $s_{11} = 4.03$	$\bar{y}_{12} = 42.75$ $s_{12} = 2.96$	$\bar{y}_{13} = 42.38$ $s_{13} = 2.62$	$\bar{y}_{14} = 43.45$ $s_{14} = 1.80$
.35 Water/Cement Ratio	$\bar{y}_{21} = 42.14$ $s_{21} = 2.64$	$\bar{y}_{22} = 36.72$ $s_{22} = 3.03$	$\bar{y}_{23} = 36.72$ $s_{23} = 1.51$	$\bar{y}_{24} = 37.70$ $s_{24} = .89$

strengths for $m = 5$ specimens of Type I cement/fly ash concrete for all possible combinations of $I = 2$ water-cement ratios and $J = 4$ curing temperatures. The units are MPa.

(a) Find the pooled sample standard deviation here. What are the associated degrees of freedom?

(b) Make a plot of sample means useful for investigating the size of Ratio × Temperature interactions. Enhance this plot by adding error bars derived from simultaneous 95% confidence intervals for the cell means. Does it appear that there are important and statistically detectable interactions here? What practical implications would this have for a cold-climate civil engineer?

(c) Compute the fitted factorial effects from the eight sample means.

(d) If one wished to make individual 95% confidence intervals for the Ratio × Temperature interactions $\alpha\beta_{ij}$, these would be of the form $ab_{ij} \pm \Delta$, for an appropriate value of Δ. Find this Δ. Based on this value, do you judge any of the interactions to be statistically detectable?

8-14. The same article referred to in Exercise 8-13 reported summary statistics (similar to the ones for Type I cement/fly ash concrete used above) for the 14-day compressive strengths of Type III cement concrete. These are shown in the accompanying table.

(a) Find the pooled sample standard deviation here. What are the associated degrees of freedom? ($m = 5$, as in Exercise 8-13.)

(b) Make a plot of sample means useful for investigating the size of Ratio × Temperature interactions. Enhance this plot by adding error bars derived from simultaneous 95% confidence intervals for the cell means. Does it appear that there are important and statistically detectable interactions here? What practical implications would this have for a cold-climate civil engineer?

(c) Compute the fitted factorial effects from the eight sample means.

(d) If one wished to make individual 95% confidence intervals for the Ratio × Temperature interactions $\alpha\beta_{ij}$, these would be of the form $ab_{ij} \pm \Delta$, for an appropriate value of Δ. Find this Δ. Based on this value, do you judge any of the interactions to be statistically detectable?

(e) Give and interpret a 90% confidence interval for the difference in water/cement ratio main effects, $\alpha_2 - \alpha_1$. How would this be of practical use to a cold-climate civil engineer?

8-15. Suppose that in the context of Exercises 8-13 and 8-14, one judges that for the Type I cement/fly ash concrete there are important Ratio × Temperature interactions, but that for the Type III cement concrete there are not important Ratio × Temperature interactions. Taking the whole data set from both exercises together (both concrete types), would there be important (3-factor) Type × Ratio × Temperature interactions? Explain.

8-16. Consider again the situation of Exercise 8-14 and suppose that one judges a no-interaction model to be adequate here.

(a) Compute an estimate of σ based on the no-interaction assumption, and compare it to the pooled sample standard deviation, s_P. (It turns out that for balanced complete two-way factorial data,

$$\sum_{i,j,k} \left(y_{ijk} - \bar{y}_{i.} - \bar{y}_{.j} + \bar{y}_{..} \right)^2$$
$$= (m - 1)IJs_P^2 + m \sum_{i,j} \left(\bar{y}_{ij} - \bar{y}_{i.} - \bar{y}_{.j} + \bar{y}_{..} \right)^2$$

so the information given in Exercise 8-14 is adequate to do this problem.)

(b) Compare an individual two-sided 95% confidence interval for the mean strength of .35 water/cement ratio concrete specimens cured at 0°C, made using the no-interaction assumption, to one based on the methods of Section 7-2.

(c) Compare a two-sided 95% prediction interval for the strength of an additional .35 water/cement ratio concrete specimen cured at 0°C, made using the no-interaction assumption, to one based on the methods of Section 7-2.

(d) Compare a 95% lower tolerance bound for the strengths of 90% of .35 water/cement ratio concrete specimens cured at 0°C, made using the no-interaction assumption, to one based on the methods of Section 7-2.

8-17. The ISU M.S. thesis "An Accelerated Engine Test for Crankshaft and Bearing Compatibility," by P. Honan, discusses an industrial experiment run to investigate the effects of three factors on the wear of engine bearings. Two levels of each of the three factors

A Crankshaft Material cast nodular iron ($-$) vs. forged steel ($+$)
B Bearing Material aluminum ($-$) vs. copper/lead ($+$)
C Debris Added to Oil none ($-$) vs. 5.5 g SAE fine dust every 25 hours ($+$)

were used in a 100-hour, 20-step engine probe test. Two response variables were measured:

$$y_1 = \text{rod journal wear } (\mu m)$$

$$y_2 = \text{main journal wear } (\mu m)$$

The values of y_1 and y_2 reported by Honan are as follows.

Combination	y_1	y_2	Combination	y_1	y_2
(1)	2.7	5.6	c	3.1	3.2
a	.9	1.4	ac	18.6	27.3
b	3.0	7.1	bc	2.5	6.0
ab	1.1	1.6	abc	60.3	99.7

(a) Use the Yates algorithm and compute the fitted effects of the three experimental factors on both the rod and main bearing wear figures.

(b) Because there was no replication in this relatively expensive industrial experiment, there is no real option for judging the statistical significance of the 2^3 factorial effects except the use of normal-plotting. Make normal plots of the seven fitted effects, $a_2, b_2, \ldots, abc_{222}$ for both response variables. Do these identify one or two of the 2^3 factorial effects as clearly larger than the others? How hopeful are you that there is a simple, intuitively appealing few-effects description of the effects of factors A, B, and C on y_1 and y_2?

(c) Your normal plots from (b) ought to each have an interesting gap in the middle of the plot. Explain the origin of both that gap and the fact that all of your fitted effects should be positive, in terms of the relative magnitudes of the responses listed above. (How, for example, does the response for combination abc enter into the calculation of the various fitted effects?)

(d) One simple way to describe the outcomes obtained in this study is as having *one very big response and one moderately big response*. Is there much chance that this pattern in y_1 and y_2 is in fact due only to random variation (i.e., that none of the factors have any effect here)? Make a normal plot of the raw y_1 values and one for the raw y_2 values to support your answer.

8-18. There is a certain degree of arbitrariness in the choice to use signs on the fitted effects corresponding to the "all high treatment" combination, when normal-plotting fitted 2^p factorial effects. This can be eliminated by probability plotting the absolute values of the fitted effects, and using not standard normal quantiles but rather quantiles for the distribution of the absolute value of a standard normal random variable. This notion is called *half normal-plotting the absolute fitted effects*, since the probability density of the absolute value of a standard normal variable looks like the right half of the standard normal density (multiplied by 2). The half

normal quantiles are related to the standard normal quantiles via

$$Q(p) = Q_z\left(\frac{1+p}{2}\right)$$

and one interprets a half normal plot in essentially the same way that a normal plot is interpreted. That is, one thinks of the smaller plotted values as establishing a pattern of random-looking variation and identifies any of the larger values plotting off a line on the plot established by the small values as detectably larger than the others.

(a) Redo part (a) of Exercise 8-5, using a half normal plot of the absolute values of the fitted effects. (Your ith plotted point will have horizontal coordinate equal to the ith smallest absolute fitted effect and vertical coordinate equal to the $\frac{i-.5}{15}$ half normal quantile.) Are the conclusions about the statistical detectability of effects here the same as those you reached in Exercise 8-5?

(b) Redo Exercise 8-7, using a half normal plot of the absolute values of the fitted effects. (Your ith plotted point will have horizontal coordinate equal to the ith smallest absolute fitted effect and vertical coordinate equal to the $\frac{i-.5}{7}$ half normal quantile.) Are the conclusions about the statistical detectability of effects here the same as those you reached in Exercise 8-7?

8-19. The text *Engineering Statistics* by Hogg and Ledolter contains an account (due originally to R. Snee) of a partially replicated 2^3 factorial industrial experiment. Under investigation were the effects of the following factors and levels on the percentage impurity, y, in a chemical product.

A Polymer Type standard ($-$) vs. new (but expensive) ($+$)
B Polymer Concentration .01% ($-$) vs. .04% ($+$)
C Amount of an Additive 2 lbs ($-$) vs. 12 lbs ($+$)

The data that were obtained are as follows.

Combination	y (%)	Combination	y (%)
(1)	1.0	c	.9, .7
a	1.0, 1.2	ac	1.1
b	.2	bc	.2, .3
ab	.5	abc	.5

(a) Compute the fitted 2^3 factorial effects corresponding to the "all high treatment" combination.

(b) Compute the pooled sample standard deviation, s_P.

(c) Use your value of s_P from (b) and find the plus-or-minus part of 90% individual confidence intervals for the 2^3 factorial effects.

(d) Based on your calculation in (c), which of the effects do you judge to be detectable in this 2^3 study?

(e) Write a paragraph or two for your engineering manager, summarizing the results of this experiment and making recommendations for the future running of this process. (Remember that you want low y and, all else being equal, low production cost.)

8-20. The article "Use of Factorial Designs in the Development of Lighting Products" by J. Scheesley (*Experiments in Industry: Design, Analysis and Interpretation of Results*, American Society for Quality Control, 1985) discusses a large industrial experiment intended to compare the use of two different types of lead wire in the manufacture of incandescent light bulbs under a variety of plant circumstances. The primary response variable in the study was

y = average number of leads missed per hour (because of misfeeds into automatic assembly equipment)

which was measured and recorded on the basis of eight-hour shifts. Consider here only part of the original data, which may be thought of as having replicated 2^4 factorial structure. That is, consider the following factors and levels.

A   Lead Type   standard $(-)$ vs. new $(+)$
B   Plant   1 $(-)$ vs. 2 $(+)$
C   Machine Type   standard $(-)$ vs. high speed $(+)$
D   Shift   1st $(-)$ vs. 2nd $(+)$

$m = 4$ values of y (each requiring an eight-hour shift to produce) for each combination of levels of factors A, B, C, and D gave the accompanying $\bar{y}$ and s^2 values.

Combination	$\bar{y}$	s^2	Combination	$\bar{y}$	s^2
(1)	28.4	97.6	d	36.8	146.4
a	21.9	15.1	ad	19.2	24.8
b	20.2	5.1	bd	19.9	5.7
ab	14.3	61.1	abd	22.5	22.5
c	30.4	43.5	cd	25.5	53.4
ac	25.1	96.2	acd	21.5	56.6
bc	38.2	100.8	bcd	22.0	10.4
abc	12.8	23.6	abcd	22.5	123.8

(a) Compute the pooled sample standard deviation. What does it measure in the present context? (Variability in hour-to-hour missed lead counts? Variability in shift-to-shift missed lead per hour figures?)

(b) Use the Yates algorithm and compute the fitted 2^4 factorial effects.

(c) Which of the effects are statistically detectable here? (Use individual two-sided 98% confidence limits for the effects to make this determination.) Is there a simple interpretation of this set of effects?

(d) Would you be willing to say, on the basis of your analysis in (a) through (c), that the new lead type will provide an overall reduction in the number of missed leads? Explain.

(e) Would you be willing to say, on the basis of your analysis in (a) through (c), that a switch to the new lead type will provide a reduction in missed leads for every set of plant circumstances? Explain.

8-21. Maxwell, Taylor, and Thelen conducted a gage repeatability and reproducibility study similar to the one in Example 8-10. Each of the three students measured the lengths of six size AA batteries from a single manufacturer $m = 3$ times with a micrometer. The measurements they reported are shown in the accompanying table, expressed in 10^{-4} in. above 1.9000 in.

(a) Make an analysis of variance table for this set of two-way factorial data.

Treat these data from a two-way random effects perspective that would be appropriate if one thinks of the students as repre-

Battery	Student	Length, y
1	1	630, 610, 605
1	2	628, 615, 620
1	3	619, 611, 611
2	1	517, 500, 500
2	2	524, 528, 515
2	3	519, 520, 524
3	1	529, 517, 518
3	2	517, 538, 539
3	3	536, 510, 540
4	1	542, 519, 531
4	2	532, 540, 546
4	3	548, 528, 530
5	1	627, 634, 630
5	2	631, 630, 635
5	3	613, 613, 612
6	1	616, 611, 612
6	2	610, 625, 618
6	3	615, 620, 615

senting a larger pool of potential technicians who might use this micrometer and the batteries as representing a much larger pool of batteries that will be measured using this device.

(b) Give expressions for the expected mean squares in this context. You will want to use formulas (8-79) of Section 8-4 in this endeavor. Then use the observed mean squares and find estimates of the variance components σ^2, σ_α^2, σ_β^2, and $\sigma_{\alpha\beta}^2$.

(c) Recognizing that σ is the theoretical standard deviation that would be associated with repeat measurements of a single battery by a single technician, find an estimate of this quantity. (Recall that this or some multiple of it might well be called the gage repeatability.)

(d) Compute an estimate of

$$\sqrt{\sigma_\beta^2 + \sigma_{\alpha\beta}^2}$$

as a measure of gage reproducibility (variation in measurements that would be experienced even if there were no repeatability problem, as a result of operator-to-operator variability). How does this figure compare to the value in (c)?

8-22. DeBlieck, Rohach, Topf, and Wilcox conducted a replicated 3×3 factorial study of the uniaxial force required to buckle household cans. A single brand of cola cans, a single brand of beer cans, and a single brand of soup cans were used in the study. The cans were prepared by bringing them to $0°C$, $22°C$, or $200°C$ before testing. The forces required to buckle each of $m = 3$ cans for the nine different Can Type/Temperature combinations follow.

Can Type	Temperature	Force Required, y (lb)
cola	$0°C$	174, 306, 192
cola	$22°C$	150, 188, 125
cola	$200°C$	200, 198, 204
beer	$0°C$	234, 246, 300
beer	$22°C$	204, 339, 254
beer	$200°C$	414, 200, 286
soup	$0°C$	570, 704, 632
soup	$22°C$	667, 593, 647
soup	$200°C$	600, 620, 596

(a) Make a plot of the nine combination sample means like the one in Figure 3-22, but enhance it with error bars derived using 98% individual two-sided confidence intervals.

(b) Compute the fitted main effects and interactions from the nine combination sample means. Use these to make individual 98% confidence intervals for all of the main effects and interactions in this 3×3 factorial study. What do these indicate about the detectability of the various effects?

(c) Use Tukey's method for simultaneous comparison of main effects and give simultaneous 99% confidence intervals for all differences in Can Type main effects. Then use the same method and give simultaneous 99% confidence intervals for all differences in Temperature main effects.

Suppose that one judges a no-interaction model to be adequate here.

(d) Compute an estimate of σ based on the no-interaction assumption and compare it to the pooled sample standard deviation, s_P, used in (a) through (c).

(e) Compare an individual two-sided 98% confidence interval for the mean force required to buckle soup cans at 22°C, made using the no-interaction assumption, to one based on the methods of Section 7-2.

(f) Compare an individual two-sided 98% prediction interval for the next force required to buckle a soup can at 22°C, made using the no-interaction assumption, to one based on the methods of Section 7-2.

(g) Compare the results in part (c) with the results of applying the Tukey method to the problem of comparing can types under the no-interaction assumption.

(h) These complete factorial data are balanced, so the two-way ANOVA methods of Section 8-4 are applicable. Find SSA, SSB, and $SSAB$ and verify that they total to $SSTr$. Do the calculations first by hand, and then check yourself using a statistical computer package.

(i) Make an ANOVA table for these data and use it in tests of the three hypotheses that there are no interactions, that there are no Can Type main effects, and that there are no Temperature main effects.

8-23. Consider again the 2^4 factorial data set in Exercise 4-36. (Paper airplane flight distances collected by K. Fellows were studied there.) As a means of making the evaluation of which of the fitted effects produced by the Yates algorithm appear to be detectable, normal-plot the fitted effects for the 2^4 data set. Interpret the plot.

8-24. The full paper airplane flight study of K. Fellows, referred to in part in Exercise 8-23 and Exercise 4-36, was in fact a $3 \times 2 \times 3 \times 2$ factorial study. The factors and levels included in the study were as follows.

	Factor	Level 1	Level 2	Level 3
A	Plane Design	Straight Wing	Tee Wing	Delta Wing
B	Nose Weight	None	Paper Clip	
C	Paper Type	Typing	Notebook	Construction
D	Wing Tips	Straight	Bent Up	

The following is the full set of flight length data collected by Fellows. (y is in feet.)

A	B	C	D	y	A	B	C	D	y
1	1	1	1	5.00	2	2	1	1	10.00
1	1	1	2	6.00	2	2	1	2	14.75
1	1	2	1	6.25	2	2	2	1	16.50
1	1	2	2	7.00	2	2	2	2	16.00
1	1	3	1	4.75	2	2	3	1	6.00
1	1	3	2	4.50	2	2	3	2	5.75
1	2	1	1	6.75	3	1	1	1	4.50
1	2	1	2	7.25	3	1	1	2	4.50
1	2	2	1	7.00	3	1	2	1	5.75
1	2	2	2	10.00	3	1	2	2	4.50
1	2	3	1	4.50	3	1	3	1	4.50
1	2	3	2	4.50	3	1	3	2	4.25
2	1	1	1	10.00	3	2	1	1	4.50
2	1	1	2	8.50	3	2	1	2	5.00
2	1	2	1	15.50	3	2	2	1	5.25
2	1	2	2	10.00	3	2	2	2	4.50
2	1	3	1	5.50	3	2	3	1	5.75
2	1	3	2	6.00	3	2	3	2	4.25

An ANOVA table for these data is presented next.

ANOVA TABLE

Source	SS	df	MS
A	205.21	2	102.61
B	12.54	1	12.54
C	96.35	2	48.17
D	.02	1	.02
A × B	6.30	2	3.15
A × C	74.16	4	18.54
A × D	3.47	2	1.73
B × C	4.26	2	2.13
B × D	4.17	1	4.17
C × D	4.04	2	2.02
A × B × C	5.23	4	1.31
A × B × D	5.36	2	2.68
A × C × D	10.05	4	2.51
B × C × D	4.35	2	2.17
A × B × C × D	4.14	4	1.04
Total	439.64	35	

(a) Compute the fitted A main effects a_1, a_2, and a_3. Show the calculations needed to verify that the sum of squares for the A main effects shown in the ANOVA table is correct.

(b) There was no real replication in this study, so there is no s_P to use as a denominator for F tests of hypotheses that various kinds of effects are zero. Nevertheless, one can look at the mean squares as useful summary measures for comparing the apparent sizes of the effects of various types. If one were to go to the trouble of computing all of the fitted effects from these data, which type would be the largest? The second largest? The third largest? What few-effects description of airplane flight distance is suggested for this four-way factorial situation?

(c) How do your conclusions in (b) compare to the conclusions reached in Exercise 8-23 on the basis of a 2^4 factorial subset of the full $3 \times 2 \times 3 \times 2$ factorial data set?

TABLE For Exercise 8-25

	FACTOR B	WIDTH	
	1 narrow (< 2 mm)	2 medium (3.5 mm)	3 wide (5.5 mm)
FACTOR A BRAND 1	$\bar{y}_{11} = 2.811$ kg $s_{11} = .0453$ kg	$\bar{y}_{12} = 4.164$ kg $s_{12} = .2490$ kg	$\bar{y}_{13} = 8.001$ kg $s_{13} = .8556$ kg
2	$\bar{y}_{21} = 2.459$ kg $s_{21} = .4697$ kg	$\bar{y}_{22} = 4.111$ kg $s_{22} = .1030$ kg	$\bar{y}_{23} = 6.346$ kg $s_{23} = .1924$ kg

8-25. Boston, Franzen, and Hoefer conducted a 2×3 factorial study of the strengths of rubber bands. Two different brands of bands were studied; from both companies, bands of three different widths were used. For each Brand/Width combination, the strengths of $m = 5$ bands of length 18–20 cm were determined by loading the bands till failure. Some summary statistics from the study are presented in the accompanying table.

(a) Compute s_P for the rubber band strength data. What is this supposed to measure?

(b) Make a plot of sample means similar to Figure 8-3. Use error bars for the means, calculated from 95% two-sided individual confidence limits for the means. (Make use of your value of s_P.)

(c) Based on your plot from (b), which factorial effects appear to be distinguishable from background noise? (Brand main effects? Width main effects? Brand $\times$ Width interactions?)

(d) Compute all of the fitted factorial effects for the strength data. (Find the a_i's, the b_j's, and the ab_{ij}'s defined in Section 4-3.)

(e) If one wishes to find individual 95% confidence intervals for the interactions $\alpha\beta_{ij}$, intervals of the form $ab_{ij} \pm \Delta$ are appropriate. Find Δ. Based on this value, are there statistically detectable interactions here? How does this conclusion compare with your more qualitative answer to part (c)?

(f) If one wishes to compare Width main effects, confidence intervals for the differences $\beta_j - \beta_{j'}$ are in order. Find individual 95% two-sided confidence intervals for $\beta_1 - \beta_2$, $\beta_1 - \beta_3$, and $\beta_2 - \beta_3$. Based on these, are there statistically detectable Width main effects here? How does this compare with your answer to part (c)?

(g) Redo part (f), this time using simultaneous 95% two-sided confidence intervals.

8-26. A classic unreplicated 2^4 factorial study, used as an example in *Experimental Statistics* (NBS Handbook #91) by M. G. Natrella, concerns flame tests of fire-retardant treatments for cloth.

The factors and levels used in the study were

A	Fabric Tested	sateen ($-$) vs. monk's cloth ($+$)
B	Treatment	X ($-$) vs. Y ($+$)
C	Laundering Condition	before laundering ($-$) vs. after laundering ($+$)
D	Direction of Test	warp ($-$) vs. fill ($+$)

The response variable, y, is the inches burned on a standard-size sample in the flame test. The data reported by Natrella follow.

Combination	y	Combination	y
(1)	4.2	d	4.0
a	3.1	ad	3.0
b	4.5	bd	5.0
ab	2.9	abd	2.5
c	3.9	cd	4.0
ac	2.8	acd	2.5
bc	4.6	bcd	5.0
abc	3.2	abcd	2.3

(a) Use the (four-cycle) Yates algorithm and compute the fitted 2^4 factorial effects for the study.

(b) Make either a normal plot or a half normal plot using the fitted effects from part (a). What subject matter interpretation of the data is suggested by the plot? (See Exercise 8-18 regarding half normal-plotting.)

(c) Natrella's original analysis of these data produced the conclusion that both the A main effects and the AB two-factor interactions are statistically detectable and of practical importance. This author (based on a plot like the one asked for in (b)) is inclined to doubt that the data are really adequate to detect the AB interaction. But for the sake of example, temporarily accept the conclusion of Natrella's analysis. What does it say in practical terms about the fire-retardant treating of cloth? (How would you explain the results to a clothing manufacturer?)

9

Inference for Curve- and Surface-Fitting Analyses of Multisample Studies

T he two previous chapters began a study of statistical inference methods for multi-sample studies, by considering first those which make no explicit use of structure relating several samples, and then discussing some directed at the analysis and exploitation of full factorial structure. In this chapter, the exposition will take a somewhat different turn. The discussion here will primarily consider inference methods for multisample studies (full factorial and otherwise) where the factors involved are inherently quantitative and it is reasonable to believe that some approximate functional relationship holds between the values of the system/input/independent variables and observed system responses. That is, this chapter introduces and applies inference methods for the curve- and surface-fitting contexts discussed in Sections 4-1 and 4-2.

The chapter begins with a discussion of the simplest possible situation of this type — namely, where a response variable y is approximately linearly related to a single quantitative input variable x. In this specific context, it is possible to give explicit formulas and illustrate what is possible in the way of inference methods for surface-fitting analyses. The second section then treats the general problem of statistical inferences in multiple regression (curve- and surface-fitting) analyses. In the general situation, it is not expedient to produce many computational formulas. So the exposition relies instead on summary measures commonly appearing on multiple regression printouts from commercially available statistical packages. (Matrix formulas for the inference methods can, however, be found in Appendix A.) Then application of the general methodologies is made in two directions. First are considered the fitting and interpretation of quadratic response surfaces and the related question of experimental strategies in response optimization studies. The final section in the chapter then explores a "trick" by which inference methods for multiple regression can be used to provide a full range of inference tools for factorial analyses based on even unbalanced data sets.

9-1 Inference Methods Related to the Least Squares Fitting of a Line

This section considers formal inference methods that are applicable in statistical engineering studies where a response variable y is approximately linearly related to an input/system variable x. It begins by introducing the (normal) simple linear regression model and discussing how the general notion of basing an estimate of response variance on residuals can be applied in this context. Next there is a look at the standardized residuals. Then inference for the rate of change ($\Delta y/\Delta x$) parameter in the model is considered, along with inference for the average response for a given setting of x. There follows a discussion of prediction and tolerance intervals for responses at a given setting of x. Next is a brief exposition of the relevance of ANOVA ideas to the present situation. The section then closes with an illustration of how commercially available statistical software can be used to expedite the calculations necessary to apply the inference methods introduced in the section.

The Simple Linear Regression Model, Corresponding Variance Estimate, and Standardized Residuals

Chapter 7 introduced the one-way (equal variances, normal distributions) model as the most common probability basis of inference methods for multisample studies. It was represented in symbols as

$$y_{ij} = \mu_i + \epsilon_{ij} \tag{9-1}$$

where the means $\mu_1, \mu_2, \ldots, \mu_r$ were treated as r unrestricted parameters. In Chapter 8, it was convenient (for example) to rewrite equation (9-1) in two-way contexts as

$$y_{ijk} = \mu_{ij} + \epsilon_{ijk} \qquad (= \mu_{..} + \alpha_i + \beta_j + \alpha\beta_{ij} + \epsilon_{ijk}) \tag{9-2}$$

where the μ_{ij} are still unrestricted, and to consider restrictions/simplifications of model (9-2) such as

$$y_{ijk} = \mu_{..} + \alpha_i + \beta_j + \epsilon_{ijk} \tag{9-3}$$

Model (9-3) really differs from model (9-2) or (9-1) only in the fact that it postulates a special form or restriction for the means μ_{ij}. Expression (9-3) says that the means must satisfy a parallelism relationship.

Turning now to the matter of inference based on data pairs $(x_1, y_1), (x_2, y_2), \ldots, (x_n, y_n)$ exhibiting an approximately linear scatterplot, one once again proceeds by imposing a restriction on the one-way model (9-1) in order to produce an appropriate probability model to support the enterprise. In words, the model assumptions will be that there are underlying normal distributions for the response y with a common variance σ^2, but means $\mu_{y|x}$ that change linearly in x. In symbols, it is typical to write that for $i = 1, 2, \ldots, n$,

$$y_i = \beta_0 + \beta_1 x_i + \epsilon_i \tag{9-4}$$

where the ϵ_i are (unobservable) iid normal $(0, \sigma^2)$ random variables, the x_i are known constants, and β_0, β_1, and σ^2 are unknown model parameters (fixed constants). Model (9-4) is commonly known as *the (normal) simple linear regression model*. If one thinks of the different values of x in an (x, y) data set as separating it into various samples of y's, expression (9-4) can be thought of as the specialization of model (9-1) where the (previously unrestricted) means $\mu_{y|x}$ satisfy the linear relationship $\mu_{y|x} = \beta_0 + \beta_1 x$. Figure 9-1 is a pictorial representation of the "constant variance, normal, linear (in x) mean" model.

FIGURE 9-1

Graphical Representation
of the Simple Linear
Regression Model

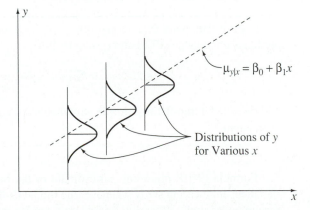

A consequence of the fact that the simple linear regression model (9-4) can be thought of as a restriction of the general one-way model (9-1) is that inferences about quantities involving those x values represented in the data (like the mean response at a single x or the difference between mean responses at two different values of x) will typically be sharper when methods based on model (9-4) can be used in place of the general methods of Chapter 7. And to the extent that one dares to assume that description (9-4) models system behavior for values of x not included in the data, a model like (9-4) provides the basis for inferences involving limited interpolation and extrapolation on the system variable x.

Section 4-1 contains an extensive discussion of the use of least squares in the fitting of the approximately linear relation

$$y \approx \beta_0 + \beta_1 x \tag{9-5}$$

to a set of (x, y) data. Rather than redoing that discussion here in Chapter 9, it is most sensible simply to observe that the discussion in Section 4-1 can be thought of as a discussion of fitting and the use of residuals in model checking for the simple linear regression model (9-4). In particular, associated with the simple linear regression model are the (least squares) estimates of β_1 and β_0 given by

$$b_1 = \frac{\sum xy - \dfrac{(\sum x)(\sum y)}{n}}{\sum x^2 - \dfrac{(\sum x)^2}{n}} \tag{9-6}$$

and

$$b_0 = \bar{y} - b_1 \bar{x} \tag{9-7}$$

and corresponding fitted values

$$\hat{y}_i = b_0 + b_1 x_i \tag{9-8}$$

and residuals

$$e_i = y_i - \hat{y}_i \tag{9-9}$$

Further, in a manner consistent with all that was said in Chapters 7 and 8 about estimating σ^2 under model (9-1) and its various specializations, the residuals (9-9) can be used to make up an estimate of σ^2. As always, one divides a sum of squared residuals by an appropriate number of degrees of freedom. That is, one can make the following definition of a simple linear regression or line-fitting sample variance.

DEFINITION 9-1

For a set of data pairs $(x_1, y_1), (x_2, y_2), \ldots, (x_n, y_n)$ where least squares fitting of a line produces fitted values (9-8) and residuals (9-9),

$$s_{\text{LF}}^2 = \frac{1}{n-2} \sum (y - \hat{y})^2 = \frac{1}{n-2} \sum e^2 \tag{9-10}$$

will be called a **line-fitting sample variance**. Associated with it are $\nu = n - 2$ degrees of freedom.

Formula (9-10) represents an estimate of the basic background variation σ^2 in a statistical engineering study whenever the model (9-4) is an adequate description of the system under study. When it is not, s_{LF} will tend to overestimate σ, so comparing s_{LF} to s_{P} is another way of investigating the appropriateness of that description.

A number of computational formulas are available for use as alternatives to the straightforward application of formula (9-10). One such formula is

$$s_{\text{LF}}^2 = \frac{1}{n-2} \left(\sum y^2 - \frac{(\sum y)^2}{n} - \frac{\left(\sum xy - \dfrac{\sum x \sum y}{n} \right)^2}{\sum x^2 - \dfrac{(\sum x)^2}{n}} \right) \tag{9-11}$$

EXAMPLE 9-1

(Example 4-1 revisited) **Inference in the Ceramic Powder Pressing Study.** The main example in this section will be the pressure/density study of Benson, Locher, and Watkins (used extensively in Section 4-1 to illustrate the possibilities in the descriptive analysis of (x, y) data). Figure 9-2 shows again the $n = 15$ data pairs (x, y) from Table 4-1, representing

$$x = \text{the pressure setting used (psi)}$$
$$y = \text{the density obtained (g/cc)}$$

in the dry pressing of a ceramic compound into cylinders.

FIGURE 9-2

Scatterplot of Density versus
Pressing Pressure

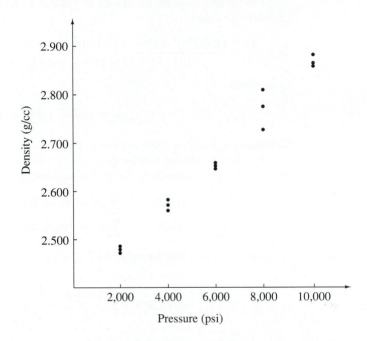

Recall further from the calculation of R^2 in Example 4-1 that the data of Table 4-1 produce fitted values in Table 4-2 and then

$$\sum (y - \hat{y})^2 = .005153$$

So for the pressure/density data, one has (via formula (9-10)) that

$$s_{\text{LF}}^2 = \frac{1}{15 - 2}(.005153) = .000396 \ (\text{g/cc})^2$$

so

$$s_{\text{LF}} = \sqrt{.000396} = .0199 \ \text{g/cc}$$

The physical interpretation is that if one accepts the appropriateness of model (9-4) in this powder pressing example, for any fixed pressure the standard deviation of densities associated with many cylinders made at that pressure would be approximately .02 g/cc.

The original data in this example can be thought of as organized into $r = 5$ separate samples of size $m = 3$, one for each of the pressures 2,000 psi, 4,000 psi, 6,000 psi, 8,000 psi, and 10,000 psi. It is instructive to consider what this thinking leads to for an alternative estimate of σ — namely, s_{P}. Table 9-1 gives $\bar{y}$ and s values for the five samples.

TABLE 9-1

Sample Means and Standard
Deviations of Densities for Five
Different Pressing Pressures

x Pressure (psi)	$\bar{y}$ Sample Mean	s Sample Standard Deviation
2,000	2.479	.0070
4,000	2.569	.0110
6,000	2.652	.0056
8,000	2.769	.0423
10,000	2.866	.0114

The sample standard deviations in Table 9-1 can be employed in the usual way to calculate s_P. That is,

$$s_P^2 = \frac{(3-1)(.0070)^2 + (3-1)(.0110)^2 + \cdots + (3-1)(.0114)^2}{(3-1) + (3-1) + \cdots + (3-1)} = .000424 \ (\text{g/cc})^2$$

from which

$$s_P = \sqrt{s_P^2} = .0206 \ \text{g/cc}$$

Comparing s_{LF} and s_P, there is clearly no indication of poor fit carried by these values.

It is worth verifying that formula (9-11) does indeed produce (except for round-off) the same value as formula (9-10) in this example. From Example 4-1,

$$\sum x = 90,000 \qquad \sum x^2 = 660,000,000 \qquad \sum xy = 245,870$$
$$\sum y = 40.005 \qquad \sum y^2 = 106.982701$$

and plugging these into formula (9-11) gives

$$s_{LF}^2 = \frac{1}{13}\left(106.982701 - \frac{(40.005)^2}{15} - \frac{\left(245,870 - \frac{(90,000)(40.005)}{15}\right)^2}{660,000,000 - \frac{(90,000)^2}{15}}\right)$$

$$= .000396 \ (\text{g/cc})^2$$

as before.

▬▬▬ ▬ ▬

Section 4-1 includes some plotting of the residuals (9-9) for the pressure/density data (in particular, a normal plot that appears as Figure 4-7) for the purpose of assessing how sensible a linear (except for background noise) description of the x/y relationship appears. Although the (raw) residuals (9-9) are most easily calculated, most commercially available regression programs provide standardized residuals as well as, or even in preference to, the raw residuals. (At this point, the reader should review the discussion concerning standardized residuals surrounding Definition 7-2.) The fact is that in curve- and surface-fitting analyses, the variances of the residuals depend on where the corresponding x's fit in the overall pattern of sets of values of system variables. Standardizing before plotting is a way to prevent mistaking a pattern on a residual plot that is explainable on the basis of these theoretically different variances of residuals, for one that is indicative of problems with the basic model. In the line-fitting context, for a given x with corresponding response y, theory says that under model (9-4),

$$Var(y - \hat{y}) = \sigma^2\left(1 - \frac{1}{n} - \frac{(x - \bar{x})^2}{\sum(x - \bar{x})^2}\right) \qquad \textbf{(9-12)}$$

So using formula (9-12) and Definition 7-2, one sees that corresponding to the data pair (x_i, y_i) is the standardized residual for simple linear regression

$$e_i^* = \frac{e_i}{s_{\text{LF}}\sqrt{1 - \dfrac{1}{n} - \dfrac{(x_i - \bar{x})^2}{\sum(x - \bar{x})^2}}} \tag{9-13}$$

The more sophisticated method of examining residuals under model (9-4) is thus to make various plots of the values (9-13) instead of plotting the raw residuals (9-9).

EXAMPLE 9-1
(continued)

Consider how the standardized residuals for the pressure/density data set are related to the raw residuals. Recalling that

$$\sum(x - \bar{x})^2 = \sum x^2 - \frac{(\sum x)^2}{n}$$

$$= 660,000,000 - \frac{(90,000)^2}{15}$$

$$= 120,000,000$$

and that the x_i values in the original data included only the pressures 2,000 psi, 4,000 psi, 6,000 psi, 8,000 psi, and 10,000 psi, it is easy to obtain the necessary values of the radical in the denominator of expression (9-13). These are collected in Table 9-2.

TABLE 9-2

Calculations for Standardized Residuals for the Cylinder Densities

x	$\sqrt{1 - \dfrac{1}{15} - \dfrac{(x - 6,000)^2}{120,000,000}}$
2,000	.894
4,000	.949
6,000	.966
8,000	.949
10,000	.894

The entries in Table 9-2 show, for example, that one should expect residuals corresponding to $x = 6,000$ psi to be (on average) about $.966/.894 = 1.08$ times as large as residuals corresponding to $x = 10,000$ psi. Division of raw residuals by s_{LF} times the appropriate entry of the second column of Table 9-2 then puts them all on equal footing, so to speak. Table 9-3 shows both the raw residuals for the pressure/density study (taken from Table 4-5) and their standardized counterparts.

TABLE 9-3

Residuals and Standardized Residuals for the Pressure/Density Study

x	e	Standardized Residual
2,000	.0137, .0067, −.0003	.77, .38, −.02
4,000	−.0117, .0003, .0103	−.62, .02, .55
6,000	−.0210, −.0100, −.0140	−1.09, −.52, −.73
8,000	−.0403, .0097, .0437	−2.13, .51, 2.31
10,000	−.0007, .0173, −.0037	−.04, .97, −.21

In the present situation, since the values .894, .949, and .966 are roughly comparable, the standardization via formula (9-13) doesn't materially effect conclusions about model adequacy. For example, Figures 9-3 and 9-4 are normal plots of (respectively) raw

FIGURE 9-3

Normal Plot of Residuals
for a Linear Fit to the
Pressure/Density Data

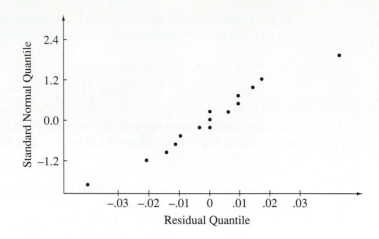

FIGURE 9-4

Normal Plot of Standardized
Residuals for a Linear Fit to
the Pressure/Density Data

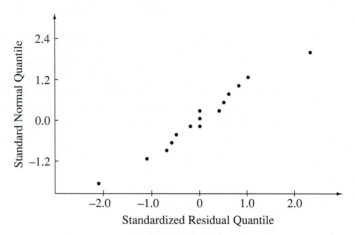

residuals and standardized residuals; for all intents and purposes, they are identical. So any conclusions (like those made in Section 4-1 based on Figure 4-7) about model adequacy supported by Figure 9-3 are equally supported by Figure 9-4, and vice versa.

In other situations, however (especially those where a data set contains a few very extreme x values), the standardization process can involve more widely varying denominators for formula (9-13) than those implied by Table 9-2 and thereby materially affect the results of a residual analysis.

**Inference for the
Slope Parameter**

Especially in engineering applications of the simple linear regression model (9-4) where x represents a variable that can be physically manipulated by the engineer, the parameter β_1 is of fundamental interest. It is the rate of change of average response y with respect to x, and it governs the impact of a change in x on the system output. Inference for β_1 turns out to be fairly simple, because of the distributional properties that b_1 (the slope of the least squares line) inherits from the model. That is, under model (9-4), b_1 can be shown to have a normal distribution with

$$Eb_1 = \beta_1$$

and

$$Var\, b_1 = \frac{\sigma^2}{\sum(x - \bar{x})^2} \tag{9-14}$$

which in turn imply that

$$\frac{b_1 - \beta_1}{\dfrac{\sigma}{\sqrt{\sum(x - \bar{x})^2}}}$$

is standard normal. In a manner similar to many of the arguments in Chapters 6 and 7, this motivates the fact that the quantity

$$\frac{b_1 - \beta_1}{\dfrac{s_{\text{LF}}}{\sqrt{\sum(x - \bar{x})^2}}} \tag{9-15}$$

has a t_{n-2} distribution. The standard arguments of Chapter 6 applied to expression (9-15) then show that

$$H_0: \beta_1 = \# \tag{9-16}$$

can be tested using the test statistic

$$T = \frac{b_1 - \#}{\dfrac{s_{\text{LF}}}{\sqrt{\sum(x - \bar{x})^2}}} \tag{9-17}$$

and a t_{n-2} reference distribution. Perhaps more importantly, under the simple linear regression model (9-4), a two-sided confidence interval for β_1 can be obtained using endpoints

$$b_1 \pm t \frac{s_{\text{LF}}}{\sqrt{\sum(x - \bar{x})^2}} \tag{9-18}$$

where the associated confidence is the probability assigned to the interval between $-t$ and t by the t_{n-2} distribution. A one-sided interval is made in the usual way, based on one endpoint from formula (9-18).

EXAMPLE 9-1
(continued)

In the context of the powder pressing study, Section 4-1 showed that the slope of the least squares line through the pressure/density data is

$$b_1 = .000048\bar{6} \text{ (g/cc)/psi}$$

Then, for example, a 95% confidence interval for β_1 can be made using the .975 quantile of the t_{13} distribution in formula (9-18). That is, one can use endpoints

$$.000048\bar{6} \pm 2.160 \frac{.0199}{\sqrt{120,000,000}}$$

that is,

$$.000048\bar{6} \pm .0000039$$

that is,

$$.0000448 \text{ (g/cc)/psi} \quad \text{and} \quad .0000526 \text{ (g/cc)/psi}$$

It is worth noting that a confidence interval like this one for β_1 can be translated into a confidence interval for a difference in mean responses for two different values of x. That is because according to model (9-4), two different values of x differing by

Δx have mean responses differing by $\beta_1 \Delta x$. One then simply multiplies endpoints of a confidence interval for β_1 by Δx to obtain a confidence interval for the difference in mean responses. For example, in the present context, since $8,000 - 6,000 = 2,000$, the difference between mean densities at 8,000 psi and 6,000 psi levels has a 95% confidence interval with endpoints

$$2,000(.0000448) \text{ g/cc} \quad \text{and} \quad 2,000(.0000526) \text{ g/cc}$$

that is,

$$.0896 \text{ g/cc} \quad \text{and} \quad .1052 \text{ g/cc}$$

The practical importance of formula (9-18) is that it allows one to attach a kind of precision to the slope of the least squares line that was first found to be so helpful in Chapter 4. It is useful to consider how that precision is related to study characteristics that are potentially under an investigator's control. Notice that both formulas (9-14) and (9-18) indicate that the larger $\sum(x - \bar{x})^2$ is (i.e., the more spread out the x_i values are), the more precision b_1 offers as an estimate of the underlying slope β_1. Thus, as far as the estimation of β_1 is concerned, in statistical engineering studies where x represents the value of a system variable under the control of an experimenter, the experimenter should choose settings of x with the largest possible sample variance. (In fact, if one has n observations to spend and can choose values of x anywhere in some interval $[a, b]$, taking $\frac{n}{2}$ of them at $x = a$ and $\frac{n}{2}$ at $x = b$ produces the best possible precision for estimating the slope β_1.)

However, this advice (to spread the x_i's out) must be taken with a grain of salt. Remember that in practical terms, the approximately linear relationship (9-4) may hold over only a limited range of possible x values. Choosing experimental values of x beyond the limits where it is reasonable to expect formula (9-4) to hold, somehow thinking that it will make it possible to estimate slope precisely, is of course nonsensical. And it is also important to recognize that precise estimation of β_1 under the assumptions of model (9-4) is not the only consideration when planning to collect (x, y) data. For example, it is usually also important in applications to be in a position to tell when the linear form of (9-4) is inappropriate. That dictates that data be collected at a number of different settings of x, not simply at the smallest and largest values possible.

Inference for the Mean System Response for a Particular Value of x

Chapters 7 and 8 repeatedly considered the problem of estimating the mean of the response variable y under a particular one (or combination) of the levels of the factor (or factors) of interest. In the present context, the analog is the problem of estimating the mean response for a fixed value of the system variable x,

$$\mu_{y|x} = \beta_0 + \beta_1 x \tag{9-19}$$

The natural data-based approximation of the mean in formula (9-19) is the corresponding y value taken from the least squares line. The notation

$$\hat{y} = b_0 + b_1 x \tag{9-20}$$

will be used for this value on the least squares lines, in spite of the fact that the value in formula (9-20) may not be a fitted value in the sense that the phrase has most often been used to this point. (x need not be equal to any of $x_1, x_2, \ldots, x_n$ for both expressions (9-19) and (9-20) to make sense.) As it turns out, the simple linear regression model (9-4) leads to tractable distributional properties for $\hat{y}$ that then produce inference methods for $\mu_{y|x}$.

Under model (9-4), $\hat{y}$ can be shown to have a normal distribution with

$$E\hat{y} = \mu_{y|x} = \beta_0 + \beta_1 x$$

and

$$Var\,\hat{y} = \sigma^2\left(\frac{1}{n} + \frac{(x - \bar{x})^2}{\sum(x - \bar{x})^2}\right) \tag{9-21}$$

In expression (9-21), please note that notation is being abused somewhat. The i subscripts and indices of summation in $\sum(x - \bar{x})^2$ have been suppressed. This summation runs over the n values x_i included in the original data set. On the other hand, in the $(x - \bar{x})^2$ term appearing as a numerator in expression (9-21), the x involved is not necessarily equal to any of $x_1, x_2, \ldots, x_n$, but rather is simply the value of the system variable at which the mean response is to be estimated. In any case, one then has the implication that

$$\frac{\hat{y} - \mu_{y|x}}{\sigma\sqrt{\dfrac{1}{n} + \dfrac{(x - \bar{x})^2}{\sum(x - \bar{x})^2}}}$$

has a standard normal distribution. This in turn motivates the fact that

$$\frac{\hat{y} - \mu_{y|x}}{s_{LF}\sqrt{\dfrac{1}{n} + \dfrac{(x - \bar{x})^2}{\sum(x - \bar{x})^2}}} \tag{9-22}$$

has a t_{n-2} distribution. The standard arguments of Chapter 6 applied to expression (9-22) then show that

$$H_0: \mu_{y|x} = \# \tag{9-23}$$

can be tested using the test statistic

$$T = \frac{\hat{y} - \#}{s_{LF}\sqrt{\dfrac{1}{n} + \dfrac{(x - \bar{x})^2}{\sum(x - \bar{x})^2}}} \tag{9-24}$$

and a t_{n-2} reference distribution. Further, under the simple linear regression model (9-4), a two-sided individual confidence interval for $\mu_{y|x}$ can be obtained using endpoints

$$\hat{y} \pm t s_{LF}\sqrt{\frac{1}{n} + \frac{(x - \bar{x})^2}{\sum(x - \bar{x})^2}} \tag{9-25}$$

where the associated confidence is the probability assigned to the interval between $-t$ and t by the t_{n-2} distribution. A one-sided interval is made in the usual way based on one endpoint from formula (9-25).

EXAMPLE 9-1
(continued)

Returning again to the pressure/density study of Benson, Locher, and Watkins, consider making individual 95% confidence intervals for the mean densities of cylinders produced first at 4,000 psi and then at 5,000 psi.

Treating first the 4,000 psi condition, note that the corresponding estimate of mean density is

$$\hat{y} = 2.375 + .0000486(4,000) = 2.5697 \text{ g/cc}$$

Further, from formula (9-25) and the fact that the .975 quantile of the t_{13} distribution is 2.160, the implication is that a precision of plus-or-minus

$$2.160(.0199)\sqrt{\frac{1}{15} + \frac{(4,000 - 6,000)^2}{120,000,000}} = .0136 \text{ g/cc}$$

ought to be attached to the 2.5697 g/cc figure. That is, the endpoints of a two-sided 95% confidence interval for the mean density under the 4,000 psi condition are

$$2.5561 \text{ g/cc} \quad \text{and} \quad 2.5833 \text{ g/cc}$$

Then, under the $x = 5,000$ psi condition, the corresponding estimate of mean density is

$$\hat{y} = 2.375 + .0000486(5,000) = 2.6183 \text{ g/cc}$$

Using formula (9-25), it is then clear that a precision of plus-or-minus

$$2.160(.0199)\sqrt{\frac{1}{15} + \frac{(5,000 - 6,000)^2}{120,000,000}} = .0118 \text{ g/cc}$$

should be attached to the 2.6183 g/cc figure. That is, the endpoints of a two-sided 95% confidence interval for the mean density under the 5,000 psi condition are

$$2.6065 \text{ g/cc} \quad \text{and} \quad 2.6301 \text{ g/cc}$$

The reader should compare the plus-or-minus parts of the two confidence intervals found here. The interval for $x = 5,000$ psi is shorter and therefore more informative than the interval for $x = 4,000$ psi. The origin of this discrepancy should be clear, at least upon scrutiny of either formula (9-21) or (9-25). For the students' data, $\bar{x} = 6,000$ psi, and $x = 5,000$ psi is closer to $\bar{x}$ than is $x = 4,000$ psi, so the $(x - \bar{x})^2$ term (and thus the interval length) is smaller for $x = 5,000$ psi than for $x = 4,000$ psi.

The phenomenon noted in the preceding example — that the length of a confidence interval for $\mu_{y|x}$ increases as one moves away from $\bar{x}$ — has an intuitively plausible implication for the planning of experiments where an approximately linear relationship between y and x is expected, and x is under the investigator's control. If there is an interval of values of x over which one wishes to have good precision in estimating mean responses, it is only sensible to center one's data collection efforts in that interval.

It is worth noting that proper use of displays (9-23), (9-24), and (9-25) gives inference methods for the parameter β_0 in model (9-4). Since β_0 is the y intercept of the linear relationship (9-19), by setting $x = 0$ in displays (9-23), (9-24), and (9-25) one arrives at tests and confidence intervals for β_0. In practical terms, however, unless $x = 0$ is a feasible value for the input variable and the region where the linear relationship (9-19) is a sensible description of physical reality includes $x = 0$, inference for β_0 alone is rarely of practical interest.

The confidence intervals represented by formula (9-25) carry individual associated confidence levels. Section 7-3 showed that it is possible (using the P-R method) to give simultaneous confidence intervals for r possibly different means in the unstructured multisample situation. This comes about essentially by appropriately increasing the t multiplier used in the plus-or-minus part of the formula for individual confidence intervals. Here it turns out to be possible, by replacing t in formula (9-25) with a larger value, to give simultaneous confidence intervals for *all* means $\mu_{y|x}$. That is, under model

(9-4), simultaneous two-sided confidence intervals for all mean responses $\mu_{y|x}$ can be made using respective endpoints

$$(b_0 + b_1x) \pm \sqrt{2f}\, s_{\mathrm{LF}} \sqrt{\frac{1}{n} + \frac{(x - \bar{x})^2}{\sum(x - \bar{x})^2}} \qquad \textbf{(9-26)}$$

where for positive f, the associated simultaneous confidence is the $F_{2,n-2}$ probability assigned to the interval $(0, f)$.

Of course, the practical meaning of the phrase "for all means $\mu_{y|x}$" is more like "for all mean responses in an interval where the simple linear regression model (9-4) is a workable description of the relationship between x and y." As is always the case in curve- and surface-fitting situations, extrapolation outside of the range of x values where one has data (and even to some extent interpolation inside that range) is risky business. When it is done, it should be supported by subject matter expertise to the effect that it is justifiable.

It may be somewhat difficult to grasp the meaning of a simultaneous confidence figure applicable to *all* possible intervals of the form (9-26). To this point, the confidence levels considered have been finite sets of intervals. Probably the best way to understand the theoretically infinite set of intervals given by formula (9-26) is as defining a region in the (x, y)-plane thought likely to contain the line $y = \beta_0 + \beta_1 x$. Figure 9-5 is a sketch of a typical confidence region represented by formula (9-26). About the least squares line there is a region indicated, whose vertical extent increases with distance from $\bar{x}$ and which has the stated confidence in covering the line describing the relationship between x and the corresponding mean system response, $\mu_{y|x}$.

FIGURE 9-5

Region in the (x, y)-Plane Defined by Simultaneous Confidence Intervals for All Values of $\mu_{y|x}$

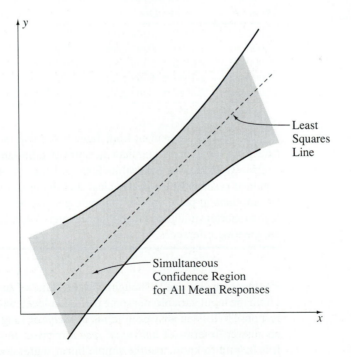

EXAMPLE 9-1
(continued)

It is instructive to compare what the P-R method of Section 7-3 and formula (9-26) give for simultaneous 95% confidence intervals for mean cylinder densities produced under the five conditions actually used by the students in their study.

First, formula (7-34) of Section 7-3 shows that with $n - r = 15 - 5 = 10$ degrees of freedom for s_P and $r = 5$ conditions under study, 95% simultaneous two-sided confidence intervals for all five mean densities are of the form

$$\bar{y}_i \pm 3.10 \frac{s_P}{\sqrt{n_i}}$$

which in the present context is

$$\bar{y}_i \pm 3.10 \frac{.0206}{\sqrt{3}}$$

that is,

$$\bar{y}_i \pm .0369 \text{ g/cc}$$

Then, since $\nu_1 = 2$ and $\nu_2 = 13$ degrees of freedom are involved in the use of formula (9-26), simultaneous intervals of the form

$$\hat{y} \pm \sqrt{2(3.81)}\, s_{\text{LF}} \sqrt{\frac{1}{15} + \frac{(x - 6,000)^2}{120,000,000}}$$

are indicated. Table 9-4 shows the five intervals that result from the use of each of the two simultaneous confidence methods.

TABLE 9-4

Simultaneous 95% Confidence Intervals for Mean Cylinder Densities

x Pressure	$\mu_{y\|x}$ (P-R Method) Mean Density	$\mu_{y\|x}$ (from formula (9-26)) Mean Density
2,000 psi	$2.4790 \pm .0369$ g/cc	$2.4723 \pm .0246$ g/cc
4,000 psi	$2.5693 \pm .0369$ g/cc	$2.5697 \pm .0174$ g/cc
6,000 psi	$2.6520 \pm .0369$ g/cc	$2.6670 \pm .0142$ g/cc
8,000 psi	$2.7687 \pm .0369$ g/cc	$2.7643 \pm .0174$ g/cc
10,000 psi	$2.8660 \pm .0369$ g/cc	$2.8617 \pm .0246$ g/cc

Two points are evident from Table 9-4. First, the intervals that result from formula (9-26) are somewhat wider than the corresponding individual intervals given by formula (9-25). (Compare the interval in Table 9-4 for $x = 4,000$ psi to the individual interval obtained earlier.) But it is also clear that the use of the simple linear regression model assumptions in preference to the more general one-way assumptions of Chapter 7 can lead to shorter simultaneous confidence intervals and correspondingly sharper real-world engineering inferences.

Prediction and Tolerance Intervals

Inference for $\mu_{y\|x}$ can be thought of as one kind of answer to the qualitative question, "If I hold the input variable x at some particular level, what can I expect in terms of a system response?" It is an answer in terms of *mean* or long-run average response. Sometimes an answer in terms of *individual responses* is of more practical use, and in such cases it is helpful to know that the simple linear regression model assumptions (9-4) lead to their own specialized formulas for prediction and tolerance intervals.

The basic fact that makes possible prediction intervals under the assumptions (9-4) is that if y_{n+1} is one additional observation, coming from the distribution of responses corresponding to a particular x, and $\hat{y}$ is the corresponding fitted value at that x (based on the original n data pairs), then

$$\frac{y_{n+1} - \hat{y}}{s_{LF}\sqrt{1 + \frac{1}{n} + \frac{(x - \bar{x})^2}{\sum(x - \bar{x})^2}}}$$

has a t_{n-2} distribution. This fact leads in the usual way to the conclusion that under model (9-4) the two-sided interval with endpoints

$$\hat{y} \pm ts_{LF}\sqrt{1 + \frac{1}{n} + \frac{(x - \bar{x})^2}{\sum(x - \bar{x})^2}} \qquad (9\text{-}27)$$

can be used as a prediction interval for an additional observation y at a particular value of the input variable x. The associated prediction confidence is the probability that the t_{n-2} distribution assigns to the interval between $-t$ and t. One-sided intervals are made in the usual way, by employing only one of the endpoints (9-27) and adjusting the confidence level appropriately.

It is possible not only to derive prediction interval formulas from the simple linear regression model assumptions, but also to develop relatively simple formulas for one-sided tolerance bounds. That is, the intervals

$$(\hat{y} - \tau s_{LF}, \infty) \qquad (9\text{-}28)$$

$$(-\infty, \hat{y} + \tau s_{LF}) \qquad (9\text{-}29)$$

can be used as one-sided tolerance intervals for a fraction p of the underlying distribution of responses corresponding to a particular value of the system variable x, provided τ is appropriately chosen (depending upon p, n, x, and the desired confidence level).

In a manner analogous to what has been done in this book in a number of contexts by now, one can use the table in Appendix D-8 to produce reasonably exact values of τ for 95% confidence and can approximate values of τ for other confidence levels. In order to write down reasonably clean formulas for τ, the notation

$$A = \sqrt{\frac{1}{n} + \frac{(x - \bar{x})^2}{\sum(x - \bar{x})^2}} \qquad (9\text{-}30)$$

will be adopted for the multiplier that is used (e.g., in formula (9-25)) to go from an estimate of σ to an estimate of the standard deviation of $\hat{y}$. Then, for purposes of entering Table D-8, one first computes a value of a parameter ξ according to

$$\xi = \frac{Q_z(p)}{\sqrt{2A^2(n - 2) + Q_z^2(p)}} \qquad (9\text{-}31)$$

Next, Table D-8 is entered using ξ and the degrees of freedom parameter $\nu = n - 2$ to produce a value λ in the body of the table (via interpolation as needed). Finally, using λ taken from the table, τ appropriate for use in interval (9-28) or (9-29) can be found using the relationship

$$\tau = \frac{Q_z(p) + A\lambda\sqrt{1 + \frac{1}{2(n - 2)}\left(\frac{Q_z^2(p)}{A^2} - \lambda^2\right)}}{1 - \frac{\lambda^2}{2(n - 2)}} \qquad (9\text{-}32)$$

When a confidence level other than 95% is desired, rather than using formula (9-31) and Table D-8 to produce a value of λ, one can use the fact that

$$\lambda \approx Q_z(\gamma) \qquad (9\text{-}33)$$

used in formula (9-32) and then formula (9-28) or (9-29) produces a one-sided tolerance interval with approximate confidence level γ for a fraction p of the underlying distribution corresponding to a given value x of the input variable.

EXAMPLE 9-1
(continued)

To illustrate the use of prediction and tolerance interval formulas in the simple linear regression context, consider calculating a 90% lower prediction bound for a single additional density in powder pressing, if a pressure of 4,000 psi is employed. Then, additionally consider finding a 95% lower tolerance bound for 90% of many additional cylinder densities if that pressure is used.

Treating first the prediction problem, formula (9-27) shows that an appropriate prediction bound is

$$2.5697 - 1.350(.0199)\sqrt{1 + \frac{1}{15} + \frac{(4,000 - 6,000)^2}{120,000,000}} = 2.5796 - .0282$$

that is,

$$2.5514 \text{ g/cc}$$

If, rather than predicting a single additional density for $x = 4,000$ psi, one wishes to locate 90% of additional densities corresponding to a 4,000 psi pressure, a tolerance bound is in order. For 95% confidence, one first uses formula (9-30) and finds that

$$A = \sqrt{\frac{1}{15} + \frac{(4,000 - 6,000)^2}{120,000,000}} = .3162$$

Then, using formula (9-31), one wants to enter Table D-8 with

$$\xi = \frac{1.282}{\sqrt{2(.3162)^2(15 - 2) + (1.282)^2}} = .62$$

Interpolation in Table D-8 using $\nu = 13$ degrees of freedom in turn produces

$$\lambda \approx 1.681$$

Next, applying formula (9-32), one has

$$\tau = \frac{1.282 + (.3162)(1.681)\sqrt{1 + \frac{1}{2(15 - 2)}\left(\frac{(1.282)^2}{(.3162)^2} - (1.681)^2\right)}}{1 - \frac{(1.681)^2}{2(15 - 2)}} = 2.174$$

So finally, a 95% lower tolerance bound for 90% of densities produced using a pressure of 4,000 psi is (via formula (9-28))

$$2.5697 - 2.174(.0199) = 2.5697 - .0433$$

that is,

$$2.5264 \text{ g/cc}$$

The fact that curve-fitting facilitates and even invites interpolation and extrapolation makes it imperative that care be taken in the interpretation of prediction and tolerance intervals. All of the caveats regarding the interpretation of prediction and tolerance intervals raised in Section 6-6 apply equally to the present situation. But the new element here (that formally, the intervals can be made for system conditions (values of x) where one has absolutely no data) requires additional caution. If one is to use formulas (9-27), (9-28), and (9-29) at a value of x not represented among $x_1, x_2, \ldots, x_n$, it must be plausible to assume that model (9-4) is adequate not only to describe system behavior at those x values where one has data, but at the additional value of x as well. And even when this is "plausible" on intuitive grounds, this author would generally treat the application of formulas (9-27), (9-28), and (9-29) to new values of x with a good dose of care. Should one's (unverified) judgment prove wrong, the nominal confidence level has unknown practical relevance.

Simple Linear Regression and ANOVA

Sections 7-4 and 8-4 illustrate how, for unstructured and balanced full factorial studies, partition of the total sum of squares into interpretable pieces provides not only intuition and quantification regarding the origin of observed variation, but also the bases for F tests of related hypotheses that are potentially of interest. It turns out that something similar is possible in simple linear regression contexts.

In the unstructured context of Section 7-4, it was useful to name the difference between $SSTot$ and SSE. The corresponding convention for curve- and surface-fitting situations is stated next in definition form.

DEFINITION 9-2

In curve- and surface-fitting analyses of multisample studies, the difference $SSTot - SSE$ will be called the **regression sum of squares** and symbolized as SSR.

It is not obvious, but the difference referred to in Definition 9-2 can be shown in general to have the form of a sum over all data points of the squares of appropriate quantities. In the present context of fitting a line by least squares, it turns out that

$$SSR = \sum_{i=1}^{n} (\hat{y}_i - \bar{y})^2$$

So the sum of squares terminology in Definition 9-2 is justified, at least in this simplest situation, where the curve being fit is a straight line.

Without using the particular terminology of Definition 9-2, this text has already made fairly extensive use of $SSR = SSTot - SSE$. A review of Definitions 4-3, 7-4, and 7-6 will show that in curve- and surface-fitting contexts, one can write

$$R^2 = \frac{SSR}{SSTot} \tag{9-34}$$

that is, SSR is the numerator of the coefficient of determination defined first in Definition 4-3. It is commonly thought of as the part of the raw variability in y that is accounted for in the curve- or surface-fitting process.

SSR and SSE not only provide an appealing partition of $SSTot$, but also form the raw material for an F test of

$$H_0: \beta_1 = 0 \tag{9-35}$$

versus

$$H_a: \beta_1 \neq 0 \tag{9-36}$$

The fact is that under model (9-4), hypothesis (9-35) can be tested using the statistic

$$F = \frac{SSR/1}{s_{\text{LF}}^2} = \frac{SSR/1}{SSE/(n-2)} \tag{9-37}$$

and an $F_{1,n-2}$ reference distribution, where large observed values of the test statistic constitute evidence against H_0 in favor of H_a.

Earlier in this section, the general null hypothesis H_0: $\beta_1 = \#$ was tested using the t statistic (9-17). It is thus reasonable to consider the relationship of the F test indicated in displays (9-35), (9-36), and (9-37) to the earlier t test. In the first place, the null hypothesis (9-35) is clearly a special form of hypothesis (9-16), H_0: $\beta_1 = \#$. It is, in fact, the most frequently tested version of hypothesis (9-16), because it can (within limits) be interpreted as the null hypothesis that mean response doesn't depend on x. This is because when hypothesis (9-35) is true within the simple linear regression model (9-4), one is left with the relationship $\mu_{y|x} = \beta_0 + 0 \cdot x = \beta_0$, which obviously doesn't depend on x. (Actually, a better interpretation of a test of hypothesis (9-35) is as a test of whether a linear term in x adds significantly to one's ability to model the response y after accounting for an overall mean response.)

If one then considers testing hypotheses (9-35) and (9-36), it would appear that the $\# = 0$ versions of formula (9-17) and formula (9-37) represent two different testing methods. But they are in fact equivalent. That is, the statistic (9-37) turns out to be the square of the $\# = 0$ version of statistic (9-17), and (two-sided) observed significance levels based on statistic (9-17) and the t_{n-2} distribution turn out to be the same as observed significance levels based on statistic (9-37) and the $F_{1,n-2}$ distribution. So, from one point of view, the F test specified in displays (9-35), (9-36), and (9-37) is redundant, given the earlier discussion. But it is introduced here because of its relationship to ANOVA ideas and the fact that its natural generalization to more complex curve- and surface-fitting contexts (discussed in the next section) cannot be made equivalent to a t test.

The partition of $SSTot$ into its parts SSR and SSE and the calculation of the statistic (9-37) can be displayed in ANOVA table format. Table 9-5 shows the general format that this book will use in the simple linear regression context.

TABLE 9-5

General Form of the
ANOVA Table for Simple
Linear Regression

ANOVA TABLE				
Source	SS	df	MS	F
Regression	SSR	1	$SSR/1$	MSR/MSE
Error	SSE	$n-2$	$SSE/(n-2)$	
Total	$SSTot$	$n-1$		

EXAMPLE 9-1
(continued)

Recall again from the discussion of the pressure/density example in Section 4-1 that

$$SSTot = \sum(y - \bar{y})^2 = .289366$$

(Alternatively, one could use the values of $\sum y^2$ and $\sum y$ given earlier in this section and the fact that $SSTot = \sum y^2 - (\sum y)^2/n$.) Also, from earlier in this section,

$$SSE = \sum(y - \hat{y})^2 = .005153$$

Thus,

$$SSR = .289366 - .005153 = .284213$$

TABLE 9-6

ANOVA Table for the
Pressure/Density Data

ANOVA TABLE				
Source	SS	df	MS	F
Regression	.284213	1	.284213	717
Error	.005153	13	.000396	
Total	.289366	14		

and the specific version of Table 9-5 applicable in the present example is given as Table 9-6.

Notice then that the observed level of significance for testing $H_0: \beta_1 = 0$ is

$$P[\text{an } F_{1,13} \text{ random variable } > 717] < .001$$

and one has very strong evidence against the possibility that $\beta_1 = 0$. A linear term in Pressure is an important contributor to one's ability to describe the behavior of Cylinder Density. This is, of course, completely consistent with the earlier interval-oriented analysis that produced 95% confidence limits for β_1 of

$$.0000448 \text{ (g/cc)/psi} \quad \text{and} \quad .0000526 \text{ (g/cc)/psi}$$

that do not bracket 0.

The value of $R^2 = .9822$ (found first in Section 4-1) can also be easily derived, using the entries of Table 9-6 and the relationship (9-34).

Simple Linear Regression and Statistical Software

Many of the calculations needed for the methods of this section are facilitated by linear regression programs, which are available for computers of all sizes. None of the methods of this section are so computationally intensive that they really require the use of such software, but it is worthwhile to consider its use in the simple linear regression context. Knowing where on a typical printout to find the various summary statistics corresponding to calculations made in this section, one learns where to look for important summary statistics for the more complicated curve- and surface-fitting analyses of subsequent sections. Printout 9-1 is from a Minitab analysis of the pressure/density data.

PRINTOUT 9-1

```
MTB > print c1 c2

  ROW   pressure   density

    1       2000      2.486
    2       2000      2.479
    3       2000      2.472
    4       4000      2.558
    5       4000      2.570
    6       4000      2.580
    7       6000      2.646
    8       6000      2.657
    9       6000      2.653
   10       8000      2.724
   11       8000      2.774
   12       8000      2.808
   13      10000      2.861
   14      10000      2.879
   15      10000      2.858

MTB > regress c2 1 c1;
SUBC> predict 5000.
```

(continued)

PRINTOUT 9-1
(continued)

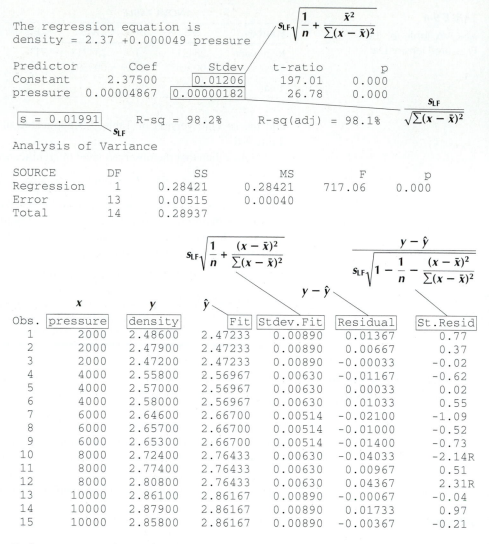

The regression equation is
density = 2.37 +0.000049 pressure

$$s_{LF}\sqrt{\frac{1}{n} + \frac{\bar{x}^2}{\sum(x-\bar{x})^2}}$$

Predictor	Coef	Stdev	t-ratio	p
Constant	2.37500	0.01206	197.01	0.000
pressure	0.00004867	0.00000182	26.78	0.000

$$\frac{s_{LF}}{\sqrt{\sum(x-\bar{x})^2}}$$

s = 0.01991   R-sq = 98.2%   R-sq(adj) = 98.1%

s_{LF}

Analysis of Variance

SOURCE	DF	SS	MS	F	p
Regression	1	0.28421	0.28421	717.06	0.000
Error	13	0.00515	0.00040		
Total	14	0.28937			

$$s_{LF}\sqrt{\frac{1}{n} + \frac{(x-\bar{x})^2}{\sum(x-\bar{x})^2}}$$

$$\frac{y-\hat{y}}{s_{LF}\sqrt{1 - \frac{1}{n} - \frac{(x-\bar{x})^2}{\sum(x-\bar{x})^2}}}$$

$y - \hat{y}$

Obs.	pressure	density	Fit	Stdev.Fit	Residual	St.Resid
1	2000	2.48600	2.47233	0.00890	0.01367	0.77
2	2000	2.47900	2.47233	0.00890	0.00667	0.37
3	2000	2.47200	2.47233	0.00890	-0.00033	-0.02
4	4000	2.55800	2.56967	0.00630	-0.01167	-0.62
5	4000	2.57000	2.56967	0.00630	0.00033	0.02
6	4000	2.58000	2.56967	0.00630	0.01033	0.55
7	6000	2.64600	2.66700	0.00514	-0.02100	-1.09
8	6000	2.65700	2.66700	0.00514	-0.01000	-0.52
9	6000	2.65300	2.66700	0.00514	-0.01400	-0.73
10	8000	2.72400	2.76433	0.00630	-0.04033	-2.14R
11	8000	2.77400	2.76433	0.00630	0.00967	0.51
12	8000	2.80800	2.76433	0.00630	0.04367	2.31R
13	10000	2.86100	2.86167	0.00890	-0.00067	-0.04
14	10000	2.87900	2.86167	0.00890	0.01733	0.97
15	10000	2.85800	2.86167	0.00890	-0.00367	-0.21

R denotes an obs. with a large st. resid.

Fit	Stdev.Fit	95% C.I.	95% P.I.
2.61833	0.00545	(2.60655,2.63012)	(2.57373,2.66294)

Printout 9-1 is reasonably typical of summaries of regression analyses printed by commercially available statistical packages. The most basic piece of information on the printout is, of course, the fitted equation. Immediately below it is a table giving (to more significant digits) the estimated coefficients (b_0 and b_1), their estimated standard deviations, and the *t* ratios that are made up as the quotients. The printout includes the values of s_{LF} and R^2 and an ANOVA table much like Table 9-6. For the several observed values of test statistics printed out (including the observed value of *F* from formula (9-37)), Minitab gives observed levels of significance. The ANOVA table is followed by a table of values of *y*, fitted *y*,

$$s_{LF}\sqrt{\frac{1}{n} + \frac{(x-\bar{x})^2}{\sum(x-\bar{x})^2}}$$

and residual, and standardized residual corresponding to the n data points. Minitab's regression program has an option that allows one to request fitted values, confidence intervals for $\mu_{y|x}$, and prediction intervals for x values of interest, and Printout 9-1 finishes with this information for the value $x = 5,000$.

The reader is encouraged to compare the information on Printout 9-1 with the various results obtained in Example 9-1 and verify that everything on the printout (except the "adjusted R^2" value) is indeed familiar.

9-2 *Inference Methods for General Least Squares Curve- and Surface-Fitting*

The previous section explored the range of formal inference methods available in contexts where a system response y is approximately linearly related to a single input or system variable x. Confidence interval estimation, hypothesis testing, prediction and tolerance intervals, and ANOVA were all seen to have their simple linear regression versions. This section makes a parallel study of the situation in more general curve- and surface-fitting contexts, where a system response is approximately linearly related to k input/system variables $x_1, x_2, \ldots, x_k$. First, the multiple linear regression model and its corresponding variance estimate and standardized residuals are introduced. Then, in turn, there are discussions of how multiple linear regression computer programs can 1) facilitate inference for rate of change parameters in the model, 2) make possible inference for the mean system response at a given combination of values for $x_1, x_2, \ldots, x_k$ and the making of prediction and tolerance intervals, and 3) allow the use of ANOVA methods in multiple regression contexts.

The Multiple Linear Regression Model, Corresponding Variance Estimate, and Standardized Residuals

This section considers situations like those treated on a descriptive level in Section 4-2, where for k system variables $x_1, x_2, \ldots, x_k$ and a response variable y, an approximate relationship like

$$y \approx \beta_0 + \beta_1 x_1 + \beta_2 x_2 + \cdots + \beta_k x_k \tag{9-38}$$

holds. As in Section 4-2, the form (9-38) covers not only those circumstances where $x_1, x_2, \ldots, x_k$ all represent physically different variables, but also describes contexts where some of the variables are functions of others. For example, the relationship

$$y \approx \beta_0 + \beta_1 x_1 + \beta_2 x_1^2$$

can be thought of as a $k = 2$ version of formula (9-38) where x_2 is a deterministic function of x_1, $x_2 = x_1^2$.

As in Section 4-2, a double subscript notation will be used here for the values of the input variables. Thus, the problem considered here is that of inference based on the data vectors $(x_{11}, x_{21}, \ldots, x_{k1}, y_1), (x_{12}, x_{22}, \ldots, x_{k2}, y_2), \ldots, (x_{1n}, x_{2n}, \ldots, x_{kn}, y_n)$. As always, a probability model is needed to support formal inferences for such data, and the one considered here is an appropriate specialization of the general one-way normal model of Section 7-1. That is, the standard assumptions of the multiple linear regression model are that there are underlying normal distributions for the response y with a common variance σ^2, but means $\mu_{y|x_1, x_2, \ldots, x_k}$ that change linearly with each of $x_1, x_2, \ldots, x_k$. In symbols, it is typical to write that for $i = 1, 2, \ldots, n$,

$$y_i = \beta_0 + \beta_1 x_{1i} + \beta_2 x_{2i} + \cdots + \beta_k x_{ki} + \epsilon_i \tag{9-39}$$

where the ϵ_i are (unobservable) iid normal $(0, \sigma^2)$ random variables, the $x_{1i}, x_{2i}, \ldots, x_{ki}$ are known constants, and $\beta_0, \beta_1, \beta_2, \ldots, \beta_k$ and σ^2 are unknown model parameters (fixed constants). This is the specialization of the general one-way model

$$y_{ij} = \mu_i + \epsilon_{ij}$$

to the situation where the means $\mu_{y|x_1,x_2,...,x_k}$ satisfy the relationship

$$\mu_{y|x_1,x_2,...,x_k} = \beta_0 + \beta_1 x_1 + \beta_2 x_2 + \cdots + \beta_k x_k \qquad \textbf{(9-40)}$$

If one thinks of formula (9-40) as defining a surface in $(k + 1)$-dimensional space, then the model equation (9-39) simply says that responses y differ from corresponding values on that surface by mean 0, variance σ^2 random noise. Figure 9-6 illustrates this point for the simple $k = 2$ case (where x_1 and x_2 are not functionally related).

FIGURE 9-6

Graphical Representation of the Multiple Linear Regression Model
$y = \beta_0 + \beta_1 x_1 + \beta_2 x_2 + \epsilon$

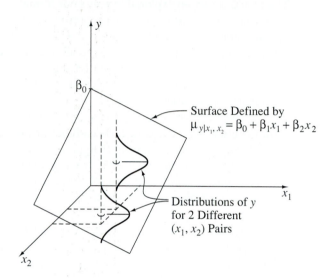

Since the multiple linear regression model (9-39) is a restriction of the general one-way model, inferences about quantities involving those $(x_1, x_2, \ldots, x_k)$ combinations represented in the data, like the mean response at a single $(x_1, x_2, \ldots, x_k)$ combination or the difference between two such mean responses, will typically be sharper when methods based on model (9-39) can be used in place of the general methods of Chapter 7. And as was true for simple linear regression, to the extent that one dares to assume that description (9-39) models system behavior for values of $x_1, x_2, \ldots, x_k$ not included in the data, a model like (9-39) provides the basis for inferences involving limited interpolation and extrapolation on the system variables $x_1, x_2, \ldots, x_k$.

Section 4-2 contains a discussion of using computer programs in the least squares fitting of the approximate relationship (9-38) to a set of $(x_1, x_2, \ldots, x_k, y)$ data. That discussion can be thought of as covering the fitting and use of residuals in model checking for the multiple linear regression model (9-39). Section 4-2 did not produce explicit formulas for $b_0, b_1, b_2, \ldots, b_k$, the (least squares) estimates of $\beta_0, \beta_1, \beta_2, \ldots, \beta_k$. Instead it relied on the computer programs to produce those estimates. Of course, once one has least squares estimates of the β's, corresponding fitted values immediately become

$$\hat{y}_i = b_0 + b_1 x_{1i} + b_2 x_{2i} + \cdots + b_k x_{ki} \qquad \textbf{(9-41)}$$

with residuals

$$e_i = y_i - \hat{y}_i \qquad \textbf{(9-42)}$$

In a manner consistent with all that has been said in this book about estimating σ^2 under the basic one-way model and its various specializations, the residuals (9-42) can be used to make up an estimate of σ^2. One divides a sum of squared residuals by an appropriate number of degrees of freedom. That is, one can make the following definition of a multiple linear regression or surface-fitting sample variance.

DEFINITION 9-3

For a set of n data vectors $(x_{11}, x_{21}, \ldots, x_{k1}, y_1), (x_{12}, x_{22}, \ldots, x_{k2}, y_2), \ldots, (x_{1n}, x_{2n}, \ldots, x_{kn}, y_n)$ where least squares fitting produces fitted values given by formula (9-41) and residuals (9-42),

$$s_{SF}^2 = \frac{1}{n-k-1} \sum (y - \hat{y})^2 = \frac{1}{n-k-1} \sum e^2 \qquad \text{(9-43)}$$

will be called a **surface-fitting sample variance**. Associated with it are $\nu = n - k - 1$ degrees of freedom.

Compare Definitions 9-1 and 9-3 and notice that the $k = 1$ version of s_{SF}^2 is just s_{LF}^2 from simple linear regression. For general k, formula (9-43) represents an estimate of the basic background variation σ^2, provided model (9-39) is an adequate description of the system under study. When it is not, s_{SF} will tend to overestimate σ, so comparing s_{SF} to s_P is another way of investigating the appropriateness of that description.

EXAMPLE 9-2

(Example 4-5 revisited) **Inference in the Nitrogen Plant Study.** The main example in this section will be the nitrogen plant data set given in Table 4-9. Recall that in the discussion of Example 4-5, with

$$x_1 = \text{a measure of air flow}$$
$$x_2 = \text{the cooling water inlet temperature}$$
$$y = \text{a measure of stack loss}$$

the fitted equation

$$\hat{y} = -15.409 - .069x_1 + .528x_2 + .007x_1^2$$

appeared to be a sensible data summary. Accordingly, consider the making of inferences based on the $k = 3$ version of model (9-39),

$$y_i = \beta_0 + \beta_1 x_{1i} + \beta_2 x_{2i} + \beta_3 x_{1i}^2 + \epsilon_i \qquad \text{(9-44)}$$

Printout 9-2 is from a Minitab analysis of the data of Table 4-9. Among many other things, it gives the values of the residuals from the fitted version of formula (9-44) for all $n = 17$ data points. It is then possible to apply Definition 9-3 and produce a surface-fitting estimate of the parameter σ^2 in the model (9-44). That is,

$$s_{SF}^2 = \frac{1}{17-3-1}\left((.320)^2 + (-.680)^2 + \cdots + (.819)^2 + (.053)^2\right)$$
$$= 1.26$$

so a corresponding estimate of σ is

$$s_{SF} = \sqrt{1.26}$$
$$= 1.125$$

(The units of y—and therefore s_{SF}—are .1% of incoming ammonia escaping unabsorbed.)

It may seem a waste to do even these calculations, since (like most multiple regression programs) Minitab's regression program outputs s_{SF} as part of its analysis. The reader should take time to locate the value $s_{SF} = 1.125$ on Printout 9-2. The physical interpretation of this value is that if one accepts the relevance of model (9-44) in this example, for fixed values of air flow and inlet temperature (and therefore air flow squared),

PRINTOUT 9-2

```
MTB > print c1 c4 c2 c3

  ROW    x1   x1**2    x2    y

   1     50    2500    18     8
   2     50    2500    18     7
   3     50    2500    19     8
   4     50    2500    19     8
   5     50    2500    20     9
   6     56    3136    20    15
   7     58    3364    17    13
   8     58    3364    18    14
   9     58    3364    18    14
  10     58    3364    18    11
  11     58    3364    19    12
  12     58    3364    23    15
  13     62    3844    22    18
  14     62    3844    23    18
  15     62    3844    24    19
  16     62    3844    24    20
  17     80    6400    27    37

MTB > regress c3 3 c1 c2 c4
* NOTE *      x1 is highly correlated with other   predictor
variables
* NOTE *    x1**2 is highly correlated with other   predictor
variables

The regression equation is
y = - 15.4 - 0.069 x1 + 0.528 x2 + 0.00682 x1**2

Predictor        Coef        Stdev     t-ratio          p
Constant       -15.41        12.60       -1.22      0.243
x1            -0.0691       0.3984       -0.17      0.865
x2            0.5278       0.1501        3.52      0.004
x1**2        0.006818     0.003178        2.15      0.051


 s = 1.125       R-sq = 98.0%    R-sq(adj) = 97.5%
       └─ s_SF
Analysis of Variance

SOURCE        DF          SS          MS          F          p
Regression     3       799.80      266.60     210.81      0.000
Error         13        16.44        1.26
Total         16       816.24

SOURCE        DF      SEQ SS
x1             1       775.48
x2             1        18.49
x1**2          1         5.82
```

(continued)

PRINTOUT 9-2
(continued)

$$y - \hat{y}$$

Obs.	x1	y	Fit	Stdev.Fit	Residual	St.Resid
1	50.0	8.000	7.680	0.493	0.320	0.32
2	50.0	7.000	7.680	0.493	-0.680	-0.67
3	50.0	8.000	8.208	0.499	-0.208	-0.21
4	50.0	8.000	8.208	0.499	-0.208	-0.21
5	50.0	9.000	8.735	0.548	0.265	0.27
6	56.0	15.000	12.657	0.298	2.343	2.16R
7	58.0	13.000	12.490	0.595	0.510	0.53
8	58.0	14.000	13.018	0.475	0.982	0.96
9	58.0	14.000	13.018	0.475	0.982	0.96
10	58.0	11.000	13.018	0.475	-2.018	-1.98
11	58.0	12.000	13.546	0.378	-1.546	-1.46
12	58.0	15.000	15.657	0.513	-0.657	-0.66
13	62.0	18.000	18.125	0.407	-0.125	-0.12
14	62.0	18.000	18.653	0.462	-0.653	-0.64
15	62.0	19.000	19.181	0.553	-0.181	-0.18
16	62.0	20.000	19.181	0.553	0.819	0.84
17	80.0	37.000	36.947	1.121	0.053	0.57 X

```
R denotes an obs. with a large st. resid.
X denotes an obs. whose X value gives it large influence.
```

the standard deviation associated with many days' stack losses produced under those conditions would be expected to be approximately .1125%.

It is worth noticing that among the 17 data points in Table 4-9, there are only 12 different airflow/inlet temperature combinations (and therefore (x_1, x_2, x_1^2) vectors). One can thus think of the original data as organized into $r = 12$ separate samples, one for each different (x_1, x_2, x_1^2) vector, and find an estimate of σ that doesn't depend for its validity on the appropriateness of the model assumption that $\mu_{y|x_1,x_2} = \beta_0 + \beta_1 x_1 + \beta_2 x_2 + \beta_3 x_1^2$. That is, one can compute a value for s_P and compare it to s_{SF} as a check on the appropriateness of model (9-44). Table 9-7 organizes the calculation of that pooled estimate of σ.

Then

$$s_P^2 = \frac{1}{17 - 12}((2 - 1)(.5) + (2 - 1)(0.0) + (3 - 1)(3.0) + (2 - 1)(.5))$$

$$= 1.40$$

TABLE 9-7

Twelve Sample Means and Four Sample Variances for the Stack Loss Data

x_1 Air Flow	x_2 Inlet Temperature	y Stack Loss	$\bar{y}$	s^2
50	18	8, 7	7.5	.5
50	19	8, 8	8.0	0.0
50	20	9	9.0	—
56	20	15	15.0	—
58	17	13	13.0	—
58	18	14, 14, 11	13.0	3.0
58	19	12	12.0	—
58	23	15	15.0	—
62	22	18	18.0	—
62	23	18	18.0	—
62	24	19, 20	19.5	.5
80	27	37	37.0	—

so

$$s_P = \sqrt{s_P^2} = \sqrt{1.40} = 1.183$$

The fact that $s_{SF} = 1.125$ and $s_P = 1.183$ are in substantial agreement is consistent with the work in Example 4-5, which found the equation

$$\hat{y} = -15.409 - .069x_1 + .528x_2 + .007x_1^2$$

to be a good summarization of the nitrogen plant data.

s_{SF} is basic to all of formal statistical inference based on model (9-39). But before using it to make statistical intervals and do significance testing, note also that it is useful for producing standardized residuals for the multiple linear regression model. That is, it turns out to be possible to find positive constants $a_1, a_2, \ldots, a_n$ (which are each complicated functions of all of $x_{11}, x_{21}, \ldots, x_{k1}, x_{12}, x_{22}, \ldots, x_{k2}, \ldots, x_{1n}, x_{2n}, \ldots, x_{kn}$) such that the ith residual $e_i = y_i - \hat{y}_i$ has

$$Var(y_i - \hat{y}_i) = a_i\sigma^2$$

Then, recalling Definition 7-2, corresponding to the data point $(x_{1i}, x_{2i}, \ldots, x_{ki}, y_i)$ is the standardized residual for multiple linear regression

$$e_i^* = \frac{e_i}{s_{SF}\sqrt{a_i}} \tag{9-45}$$

It is not possible to include here a simple formula for the a_i that are needed to compute standardized residuals. (They are of interest only as building blocks in formula (9-45) anyway.) But just as for simple linear regression, it is easy to read the values of the standardized residuals (9-45) off a typical multiple regression printout and to plot them in the usual ways as means of checking the apparent appropriateness of a candidate version of model (9-39) fit to a set of n data points $(x_1, x_2, \ldots, x_k, y)$.

EXAMPLE 9-2
(continued)

As an illustration of the use of standardized residuals, consider again Printout 9-2. The annotations on that printout locate the columns of residuals and standardized residuals for model (9-44) and the nitrogen plant data. Figure 9-7 depicts normal probability plots, first of the raw residuals and then of the standardized residuals.

There are only the most minor differences between the appearances of the two plots in Figure 9-7, suggesting that in the present context, decisions concerning the appropriateness of model (9-44) based on raw residuals will not be much altered by the more sophisticated consideration of standardized residuals instead.

Inference for the Parameters
$\beta_0, \beta_1, \beta_2, \ldots, \beta_k$

Section 9-1 considered inference for the slope parameter β_1 in simple linear regression, treating it as a rate of change (of average y as a function of x). In the multiple regression context, if $x_1, x_2, \ldots, x_k$ are all physically different system variables, the coefficients $\beta_1, \beta_2, \ldots, \beta_k$ can again be thought of as rates of change of average response with respect to $x_1, x_2, \ldots, x_k$, respectively. (They are partial derivatives of $\mu_{y|x_1,x_2,\ldots,x_k}$ with respect to the x's.) On the other hand, when some x's are functionally related to others (for instance, if one is considering a $k = 2$ version of relationship (9-40) with $\mu_{y|x} = \beta_0 + \beta_1x + \beta_2x^2$), individual interpretation of the β's can be less straightforward. In any case, the β's do

FIGURE 9-7

Normal Plots of Residuals and
Standardized Residuals for the
Stack Loss Data

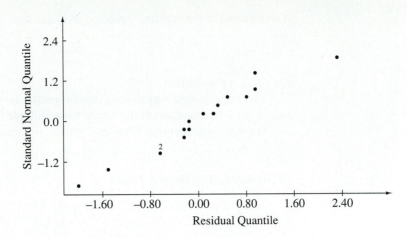

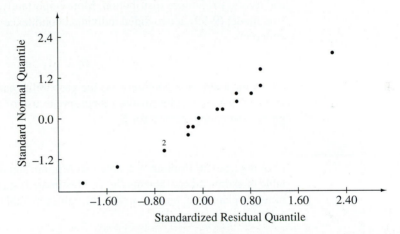

determine the nature of the surface represented by

$$\mu_{y|x_1,x_2,\ldots,x_k} = \beta_0 + \beta_1 x_1 + \beta_2 x_2 + \cdots + \beta_k x_k$$

and it is possible to do formal inference for $\beta_0, \beta_1, \ldots, \beta_k$ individually in much the same way as inference was done for β_1 in the last section. In many instances, important physical interpretations can be found for such inferences. (For example, if one begins with $\mu_{y|x} = \beta_0 + \beta_1 x + \beta_2 x^2$, an inference that β_2 is positive would say that the mean response is concave up as a function of x, and thus might potentially be shown to have a minimum value.)

The key to formal inference for the β's is that based on model (9-39), a mathematician can show that there are positive constants $d_0, d_1, d_2, \ldots, d_k$ (which are each complicated functions of all of $x_{11}, \ldots, x_{k1}, x_{12}, \ldots, x_{k2} \ldots, x_{1n}, \ldots, x_{kn}$) such that the least squares coefficients $b_0, b_1, \ldots, b_k$ are normally distributed with

$$Eb_l = \beta_l$$

and

$$Var\, b_l = d_l \sigma^2$$

This in turn makes it plausible that for $l = 0, 1, 2, \ldots, k$, the quantity

$$s_{SF}\sqrt{d_l} \qquad\qquad (9\text{-}46)$$

is an estimate of the standard deviation of b_l and that

$$\frac{b_l - \beta_l}{s_{SF}\sqrt{d_l}} \tag{9-47}$$

has a t_{n-k-1} distribution.

There is no simple way to write down formulas for the constants d_l, but the products $s_{SF}\sqrt{d_l}$ are a standard part of the output from multiple linear regression programs.

The usual arguments of Chapter 6 applied to expression (9-47) then show that

$$H_0: \beta_l = \# \tag{9-48}$$

can be tested using the test statistic

$$T = \frac{b_l - \#}{s_{SF}\sqrt{d_l}} \tag{9-49}$$

and a t_{n-k-1} reference distribution. More importantly, under the multiple linear regression model (9-39), a two-sided individual confidence interval for β_l can be made using endpoints

$$b_l \pm t s_{SF}\sqrt{d_l} \tag{9-50}$$

where the associated confidence is the probability assigned to the interval between $-t$ and t by the t_{n-k-1} distribution. Appropriate use of only one of the endpoints (9-50) gives a one-sided interval for β_l.

EXAMPLE 9-2
(continued)

Looking again at Printout 9-2, note that Minitab's multiple regression output includes a table of estimated coefficients (b_l) and (estimated) standard deviations ($s_{SF}\sqrt{d_l}$). For the nitrogen plant data, one has in fact the values in Table 9-8.

TABLE 9-8

Fitted Coefficients and Estimates of Their Standard Deviations for the Stack Loss Data

Estimated Coefficient	(Estimated) Standard Deviation of the Estimate
$b_0 = -15.41$	$s_{SF}\sqrt{d_0} = 12.60$
$b_1 = -.0691$	$s_{SF}\sqrt{d_1} = .3984$
$b_2 = .5278$	$s_{SF}\sqrt{d_2} = .1501$
$b_3 = .006818$	$s_{SF}\sqrt{d_3} = .003178$

Then (for example) since the upper .05 point of the t_{13} distribution is 1.771, from formula (9-50) a two-sided 90% confidence interval for β_2 in model (9-44) has endpoints

$$.5278 \pm 1.771(.1501)$$

that is,

.2620 (.1% nitrogen loss/degree)   and   .7936 (.1% nitrogen loss/degree)

This interval establishes that there is an increase in mean stack loss y with increased inlet temperature x_2 (the interval contains only positive values) and gives a way of assessing the likely impact on y of various changes in x_2. For example, if x_1 (and therefore $x_3 = x_1^2$) is held constant but x_2 is increased by $2°$, one can anticipate an increase in mean stack loss of between

.5240 (.1% nitrogen loss)   and   1.5873 (.1% nitrogen loss)

As a second example of the use of formula (9-50), note that a 90% two-sided confidence interval for β_3 has endpoints

$$.006818 \pm 1.771(.003178)$$

that is,

$$.0012 \quad \text{and} \quad .0124$$

β_3 controls the amount and direction of curvature (in the variable x_1) possessed by the surface specified by $\mu_{y|x_1,x_2} = \beta_0 + \beta_1 x_1 + \beta_2 x_2 + \beta_3 x_1^2$. Since the interval contains only positive values, it shows that at the 90% confidence level, there is some important concave-up curvature in the air flow variable needed to describe the stack loss variable. This is consistent with the picture of fitted mean response given previously in Figure 4-15.

However, check that if 95% confidence is used in the calculation of the two-sided interval for β_3, the resulting confidence interval contains values on both sides of 0. The message carried by this fact is that if this higher level of confidence is needed, the data in hand are not adequate to establish definitively the nature of any curvature in mean stack loss as a function of inlet temperature. Any real curvature must be judged weak enough in comparison to the basic background variation, so more data are needed to decide whether the surface is concave up, linear, or concave down in the variable x_1.

Very often multiple regression programs output not only the estimated standard deviations of fitted coefficients (9-46) but also the ratios

$$\frac{b_l}{s_{SF}\sqrt{d_l}}$$

and associated two-sided p-values for testing

$$H_0: \beta_l = 0$$

Review Printout 9-2 and note that, for example, the two-sided p-value for testing H_0: $\beta_3 = 0$ in model (9-44) is slightly larger than .05. This is completely consistent with the preceding discussion regarding the interpretation of interval estimates of β_3.

Inference for the Mean System Response for a Particular Set of Values for $x_1, x_2, \ldots, x_k$

Inference methods for the parameters $\beta_0, \beta_1, \ldots, \beta_k$ provide insight into the nature of the relationships between $x_1, x_2, \ldots, x_k$ and the mean response y. But other methods are needed to answer the important engineering question, "What can be expected in terms of system response if I use a particular combination of levels of the system variables $x_1, x_2, \ldots, x_k$?" An answer to this question will first be phrased in terms of inference methods for the mean system response (9-40).

In a manner similar to what was done in Section 9-1, the notation

$$\hat{y} = b_0 + b_1 x_1 + b_2 x_2 + \cdots + b_k x_k \tag{9-51}$$

will here be used for the value produced by the least squares equation when a particular set of numbers $x_1, x_2, \ldots, x_k$ is plugged into it. This is in spite of the fact that $\hat{y}$ may not be a fitted value in the strict sense of the phrase. (The vector $(x_1, x_2, \ldots, x_k)$ may not match any data vector $(x_{1i}, x_{2i}, \ldots, x_{ki})$ used to produce the least squares coefficients $b_0, b_1, \ldots, b_k$.) As it turns out, the multiple linear regression model (9-39) leads to tractable distributional properties for $\hat{y}$, which then produce inference methods for $\mu_{y|x_1,x_2,\ldots,x_k}$.

Under model (9-39), it turns out to be possible to find a positive constant A depending in a complicated way upon $x_1, x_2, \ldots, x_k$ and all of $x_{11}, \ldots, x_{k1}, x_{12}, \ldots, x_{k2}, \ldots,$ $x_{1n}, \ldots, x_{kn}$ (the locations at which inference is desired and the original data points were collected) so that $\hat{y}$ has a normal distribution with

$$E\hat{y} = \mu_{y|x_1, x_2, \ldots, x_k} = \beta_0 + \beta_1 x_1 + \cdots + \beta_k x_k$$

and

$$\text{Var}\, \hat{y} = \sigma^2 A^2 \tag{9-52}$$

In view of formula (9-52), it is thus plausible that

$$s_{\text{SF}} \cdot A \tag{9-53}$$

can be used as an estimated standard deviation for $\hat{y}$ and that inference methods for the mean system response can be based on the fact that

$$\frac{\hat{y} - \mu_{y|x_1, x_2, \ldots, x_k}}{s_{\text{SF}} \cdot A}$$

has a t_{n-k-1} distribution. That is,

$$H_0 : \mu_{y|x_1, x_2, \ldots, x_k} = \# \tag{9-54}$$

can be tested using the test statistic

$$T = \frac{\hat{y} - \#}{s_{\text{SF}} \cdot A} \tag{9-55}$$

and a t_{n-k-1} reference distribution. Further, under the multiple linear regression model (9-39), a two-sided confidence interval for $\mu_{y|x_1, x_2, \ldots, x_k}$ can be made using endpoints

$$\hat{y} \pm t s_{\text{SF}} \cdot A \tag{9-56}$$

where the associated confidence is the probability assigned to the interval between $-t$ and t by the t_{n-k-1} distribution. One-sided intervals based on formula (9-56) are made in the usual way.

The practical obstacle to be overcome in the use of these methods is the computation of the quantity A. Once again, although it is not possible here to give a simple formula for A, most multiple regression programs allow one to obtain A for $(x_1, x_2, \ldots, x_k)$ vectors of interest. Minitab, for example, will automatically (for the least abbreviated choice of its output format) produce values of statistic (9-53) corresponding to each data point $(x_{1i}, x_{2i}, \ldots, x_{ki}, y_i)$, labeled as (the estimated) *standard deviation* (of the) *fit*. And an option available through the use of a subcommand of the "REGRESS" command makes it possible to obtain similar information for *any* choice of $(x_1, x_2, \ldots, x_k)$.

EXAMPLE 9-2
(continued)

For the sake of illustration, consider the problem of estimating the mean stack loss if the nitrogen plant of Example 4-5 is operated consistently with $x_1 = 58$ and $x_2 = 19$. (Notice that this means that $x_3 = x_1^2 = 3364$ is involved.) Now the conditions $x_1 = 58$, $x_2 = 19$ and $x_3 = 3364$ match perfectly those of data point number 11 on Printout 9-2. Thus, $\hat{y}$ and $s_{\text{SF}} \cdot A$ for these conditions may be read directly from the printout as 13.546 and .378, respectively. Then, for example, from formula (9-56), a 90% two-sided confidence interval for the mean stack loss corresponding to an air flow of 58 and water inlet temperature of 19 would have endpoints

$$13.546 \pm 1.771(.378)$$

that is,

$$12.88 \text{ (.1\% nitrogen loss)} \quad \text{and} \quad 14.22 \text{ (.1\% nitrogen loss)}$$

As a second illustration of the use of formula (9-56), suppose that one is contemplating setting plant standard operating conditions at an air flow of $x_1 = 60$ and a water inlet temperature of $x_2 = 20$ and wants an interval estimate for the mean stack loss implied by those conditions. Notice that the $x_1 = 60$, $x_2 = 20$, and $x_3 = x_1^2 = 3600$ vector does not exactly match that of any of the $n = 17$ data points available. Therefore, conceptually one must be willing to do some interpolation/extrapolation to make the desired interval. And operationally it will not be possible to simply read appropriate values of $\hat{y}$ and $s_{SF} \cdot A$ off Printout 9-2.

Location of the point with coordinates $x_1 = 60$ and $x_2 = 20$ on a scatterplot of (x_1, x_2) values for the original $n = 17$ data points (like Figure 4-19) reveals that the candidate operating conditions do not appear to be wildly different from those used to develop the fitted equation. So there is hope that the use of formula (9-56) will provide an inference of some practical relevance. Accordingly, a version of the regression program used to produce Printout 9-2 was rerun with the intention of finding values of $\hat{y}$ and $s_{SF} \cdot A$ for the candidate operating conditions via the use of the Minitab "PREDICT" subcommand. The output is reproduced here as Printout 9-3.

Reading from the final section of the printout, one has $\hat{y} = 15.544$ and $s_{SF} \cdot A = .383$, so a 90% two-sided confidence interval for the mean stack loss has endpoints

$$15.544 \pm 1.771(.383)$$

PRINTOUT 9-3

Abbreviated Output Requested

```
MTB > brief 1            x₁  x₂  x₁²
MTB > regress c3 3 c1 c2 c4;
SUBC> predict 60 20 3600.        x₁, x₂, x₁² Conditions of Interest
* NOTE *        x1 is highly correlated with other  predictor
variables
* NOTE *     x1**2 is highly correlated with other  predictor
variables


The regression equation is
y = - 15.4 - 0.069 x1 + 0.528 x2 + 0.00682 x1**2

Predictor        Coef        Stdev      t-ratio          p
Constant       -15.41        12.60        -1.22      0.243
x1            -0.0691       0.3984        -0.17      0.865
x2            0.5278        0.1501         3.52      0.004
x1**2        0.006818      0.003178        2.15      0.051

s = 1.125        R-sq = 98.0%      R-sq(adj) = 97.5%

Analysis of Variance

SOURCE          DF           SS           MS          F          p
Regression       3        799.80       266.60     210.81     0.000
Error           13         16.44         1.26
Total           16        816.24

      Fit   Stdev.Fit          95% C.I.             95% P.I.
   15.544       0.383    ( 14.717, 16.372)    ( 12.977, 18.111)
      ŷ       s_SF · A
```

In the printout above: $x_1 \ x_2 \ x_1^2$ label the predict command values. The Fit value 15.544 is labeled $\hat{y}$ and the Stdev.Fit value 0.383 is labeled $s_{SF} \cdot A$.

that is,

$$14.86 \ (.1\% \text{ nitrogen loss}) \quad \text{and} \quad 16.22 \ (.1\% \text{ nitrogen loss})$$

(Of course, endpoints of a similar 95% interval could be read directly from the printout.)

It is impossible to overemphasize the fact that the preceding two intervals are dependent for their practical relevance on that of model (9-44) for not only those (x_1, x_2) pairs in the original data, but (in the second case) also for the $x_1 = 60$ $x_2 = 20$ set of conditions. Formulas like (9-56) always allow for imprecision due to statistical fluctuations/background noise in the data. They *do not*, however, allow for discrepancies related to the application of a model in a regime over which it is not appropriate. Formula (9-56) is an important and useful formula. But it should be used thoughtfully, with no expectation that it will magically do more than help quantify the precision provided by the data in the context of a particular set of model assumptions.

Lacking a simple explicit formula for A, it is difficult to be very concrete about how this quantity varies. In qualitative terms, it does change with the $(x_1, x_2, \ldots, x_k)$ vector under consideration. It is smallest when this vector is near the center of the cloud of points $(x_{1i}, x_{2i}, \ldots, x_{ki})$ in k-dimensional space corresponding to the n data points used to fit model (9-39). The fact that it can vary substantially is obvious from Printout 9-2. There one can see that for the nitrogen plant illustration, the estimated standard deviation of $\hat{y}$ given in display (9-53) varies from .298 to 1.121, indicating that A for data point 17 is about 3.8 times the size of A for data point 6. ($\frac{1.121}{.298} \approx 3.8$.) That is, the precision with which a mean response is determined can vary widely over the region where it is sensible to use a fitted equation.

Formula (9-56) provides individual confidence intervals for mean responses. Simultaneous intervals are also easily obtained by a modification of formula (9-56) similar to the one provided in the previous section for the simple linear regression context. That is, under model (9-39), simultaneous two-sided confidence intervals for all mean responses $\mu_{y|x_1, x_2, \ldots, x_k}$ can be made using respective endpoints

$$\hat{y} \pm \sqrt{(k+1)f} s_{\text{SF}} \cdot A \tag{9-57}$$

where for positive f, the associated confidence is the $F_{k+1, n-k-1}$ probability assigned to the interval $(0, f)$. Formula (9-57) is clearly related to formula (9-56) through the replacement of the multiplier t by the (larger for a given nominal confidence) multiplier $\sqrt{(k+1)f}$. When it is applied only to $(x_1, x_2, \ldots, x_k)$ vectors found in the original n data points, formula (9-57) can be thought of as an alternative to the P-R method of simultaneous intervals for means, appropriate to surface fitting problems. When model (9-39) is indeed appropriate, formula (9-57) will usually give shorter simultaneous intervals than the P-R method.

EXAMPLE 9-2
(continued)

For making simultaneous 90% confidence intervals for the mean stack losses at the 12 different sets of plant conditions represented in the original nitrogen plant data set, one could use formula (9-57) with $k = 3$, $f = 2.43$ (the .9 quantile of the $F_{4,13}$ distribution), and the $\hat{y}$ and corresponding $s_{\text{SF}} \cdot A$ values appearing on Printout 9-2.

**Prediction and
Tolerance Intervals**

The second kind of answer that statistical theory can provide to the engineering question "What is to be expected in terms of system response if one uses a particular combination of levels of the system variables $x_1, x_2, \ldots, x_k$?" has to do with individual responses rather than mean responses. That is, the same factor A referred to in making confidence intervals for mean responses can be used to develop prediction and tolerance intervals for surface-fitting situations.

In the first place, reasoning exactly parallel to that used to produce prediction intervals for simple linear regression leads to the conclusion that under model (9-39), the two-sided interval with endpoints

$$\hat{y} \pm t s_{SF}\sqrt{1 + A^2} \tag{9-58}$$

can be used as a prediction interval for an additional observation at a particular combination of levels of the variables $x_1, x_2, \ldots, x_k$. The associated prediction confidence is the probability that the t_{n-k-1} distribution assigns to the interval between $-t$ and t. One-sided intervals are made in the usual way, by employing only one of the endpoints (9-58) and adjusting the confidence level appropriately.

In order to use formula (9-58), one can either take $s_{SF} \cdot A$ and s_{SF} from a multiple regression printout, and obtain A via division, or equivalently, use a small amount of algebra to rewrite formula (9-58) as

$$\hat{y} \pm t\sqrt{s_{SF}^2 + (s_{SF} \cdot A)^2} \tag{9-59}$$

and substitute s_{SF} and $s_{SF} \cdot A$ directly into formula (9-59).

In order to find one-sided tolerance bounds in the surface-fitting context, one begins with the value of A corresponding to a particular combination of the levels of the variables $x_1, x_2, \ldots, x_k$. For a 95% tolerance bound for a fraction p of the underlying distribution of responses, one first computes a value of a parameter ξ according to

$$\xi = \frac{Q_z(p)}{\sqrt{2A^2(n - k - 1) + Q_z^2(p)}} \tag{9-60}$$

and enters Table D-8 using ξ and $\nu = n - k - 1$ degrees of freedom to find a value of λ. As is by now usual, for other confidence levels besides 95%, one employs the approximation

$$\lambda \approx Q_z(\gamma) \tag{9-61}$$

for confidence level γ. Then, from either Table D-8 or approximation (9-61), a value of λ is in turn used to produce a value of τ via

$$\tau = \frac{Q_z(p) + A\lambda\sqrt{1 + \dfrac{1}{2(n - k - 1)}\left(\dfrac{Q_z^2(p)}{A^2} - \lambda^2\right)}}{1 - \dfrac{\lambda^2}{2(n - k - 1)}} \tag{9-62}$$

Finally, the interval

$$(\hat{y} - \tau s_{SF}, \infty) \tag{9-63}$$

or

$$(-\infty, \hat{y} + \tau s_{SF}) \tag{9-64}$$

can be used as a one-sided tolerance interval for a fraction p of the underlying distribution of responses corresponding to a particular combination of levels of the variables $x_1, x_2, \ldots, x_k$. The associated confidence level is 95% if formula (9-60) and Table D-8 are used or approximately γ if formula (9-61) is used to produce λ.

EXAMPLE 9-2
(continued)

Returning to the nitrogen plant example, consider first the calculation of a 90% lower prediction bound for a single additional stack loss y, if an air flow of $x_1 = 58$ and water inlet temperature of $x_2 = 19$ are used. Then consider also a 95% lower tolerance bound for 90% of many additional stack loss values if the plant is run under those conditions.

Treating the prediction interval problem, recall that for $x_1 = 58$ and $x_2 = 19$, $\hat{y} = 13.546$ and $s_{SF} \cdot A = .378$. Since $s_{SF} = 1.125$ and the .9 quantile of the t_{13} distribution is 1.350, formula (9-59) shows that the desired 90% lower prediction bound for an additional stack loss under such plant operating conditions is

$$13.546 - 1.350\sqrt{(1.125)^2 + (.378)^2}$$

that is, approximately

$$11.94 \ (.1\% \text{ nitrogen loss})$$

If one wishes not to predict a single additional stack loss, but rather to locate 90% of many additional stack losses with 95% confidence, expression (9-60) is the place to begin. Note that for $x_1 = 58$ and $x_2 = 19$,

$$A = .378/1.125 = .336$$

so, using expression (9-60),

$$\xi = \frac{1.282}{\sqrt{2(.336)^2(17 - 3 - 1) + (1.282)^2}}$$
$$= .599$$

and $\nu = 17 - 3 - 1 = 13$ degrees of freedom are called for when using Table D-8. Interpolating in that table, one obtains roughly

$$\lambda \approx 1.6815$$

Next, applying formula (9-62), one has

$$\tau = \frac{1.282 + (.378)(1.6815)\sqrt{1 + \dfrac{1}{2(17 - 3 - 1)}\left(\dfrac{(1.282)^2}{(.378)^2} - (1.6815)^2\right)}}{1 - \dfrac{(1.6815)^2}{2(17 - 3 - 1)}} = 2.262$$

So finally, a 95% lower tolerance bound for 90% of stack losses produced under operating conditions of $x_1 = 58$ and $x_2 = 19$ is, via formula (9-63),

$$13.546 - 2.262(1.125) = 13.546 - 2.545$$

that is,

$$11.001 \ (.1\% \text{ nitrogen loss})$$

Of course, the caveats raised in the previous section concerning prediction and tolerance intervals in simple regression all apply to the present case of multiple regression as well. So do points similar to those made earlier in this section in reference to confidence intervals for the mean system response. Although they are extremely useful engineering tools, statistical intervals are never any better than the models on which they are based.

Multiple Regression and ANOVA

It is this author's view that formal inference in curve- and surface-fitting contexts can (and typically should) be carried out primarily using interval-oriented methods. Nevertheless, testing and ANOVA methods do have their place, and the discussion now turns to the matter of what ANOVA ideas provide in the way of inference methods and additional qualitative insight in multiple regression.

As always, $SSTot$ will stand for $\sum(y - \bar{y})^2$ and SSE for $\sum(y - \hat{y})^2$. Remember also that Definition 9-2 introduced the notation SSR for the difference $SSTot - SSE$. As remarked following Definition 9-2, the coefficient of determination can be written in terms of SSR and $SSTot$ as

$$R^2 = \frac{SSR}{SSTot}$$

Further, under model (9-39), these sums of squares ($SSTot$, SSE, and SSR) form the basis of an F test of the hypothesis

$$H_0: \beta_1 = \beta_2 = \cdots = \beta_k = 0 \tag{9-65}$$

versus

$$H_a: \text{not } H_0 \tag{9-66}$$

The fact is that hypothesis (9-65) can be tested using the statistic

$$F = \frac{SSR/k}{SSE/(n - k - 1)} \tag{9-67}$$

and an $F_{k,n-k-1}$ reference distribution, where large observed values of the test statistic constitute evidence against H_0 in favor of H_a. (The denominator of statistic (9-67) is another way of writing s_{SF}^2.)

Notice that hypothesis (9-65) in the context of model (9-39) implies that the mean response doesn't depend on any of the process variables $x_1, x_2, \ldots, x_k$. That is, if all of β_1 through β_k are 0, model statement (9-39) reduces to

$$y_i = \beta_0 + \epsilon_i$$

So (within limits) a test of hypothesis (9-65) can be interpreted as a test of whether the mean response is related to any of the input variables under consideration. The calculations leading to statistic (9-67) are most often organized in a table quite similar to the one discussed in Section 9-1 for testing $H_0: \beta_1 = 0$ in simple linear regression. The general form of that table is given as Table 9-9.

TABLE 9-9

General Form of the ANOVA Table for Testing $H_0: \beta_1 = \beta_2 = \cdots = \beta_k = 0$ in Multiple Regression

ANOVA TABLE				
Source	SS	df	MS	F
Regression	SSR	k	SSR/k	MSR/MSE
Error	SSE	$n - k - 1$	$SSE/(n - k - 1)$	
Total	$SSTot$	$n - 1$		

EXAMPLE 9-2
(continued)

Once again turning to the analysis of the nitrogen plant data under model (9-44), consider testing the hypothesis $H_0: \beta_1 = \beta_2 = \beta_3 = 0$ — that is, mean stack loss doesn't depend on air flow (or its square) or water inlet temperature. Printout 9-2 includes an ANOVA table for testing this hypothesis, which is essentially reproduced here as Table 9-10.

From Table 9-10, it is clear that the observed value of the F statistic (9-67) is 210.81, which is to be compared to $F_{3,13}$ quantiles in order to produce an observed level of significance. As indicated in Printout 9-2, the $F_{3,13}$ probability to the right of the value

	ANOVA TABLE			
Source	SS	df	MS	F
Regression (on x_1, x_2, x_1^2)	799.80	3	266.60	210.81
Error	16.44	13	1.26	
Total	816.24	16		

210.81 is 0 (to three decimal places), which is definitive evidence that not all of β_1, β_2, and β_3 can be 0. Taken as a group, the variables x_1, x_2, and $x_3 = x_1^2$ definitely enhance one's ability to predict stack loss.

As a sidelight, note also that the value of the coefficient of determination here can be calculated using sums of squares given in Table 9-10 as

$$R^2 = \frac{SSR}{SSTot} = \frac{799.80}{816.24} = .980$$

which is the value for R^2 advertised long ago in Example 4-5. Also, the error mean square, $MSE = 1.26$, is (as expected) exactly the value of s_{SF}^2 calculated earlier in this example.

It is a matter of simple algebra to verify that R^2 and the F statistic (9-67) are equivalent in the sense that

$$F = \frac{\dfrac{R^2}{k}}{\dfrac{1 - R^2}{n - k - 1}} \tag{9-68}$$

so the F test of hypothesis (9-65) can be thought of in terms of attaching a (quantitative) p-value to the statistic R^2, which has heretofore been used in a rather qualitative fashion to judge the usefulness of a particular equation in summarizing patterns in a multivariate data set. This is a valuable development, but it should be remembered that it is R^2 rather than F that has the more direct interpretation as a measure of what fraction of raw variability the fitted equation accounts for. F and its associated p-value take account of the sample size n in a way that R^2 doesn't; they really measure statistical detectability rather than variation accounted for. This means that an equation that accounts for a fraction of observed variation that is relatively small by most standards can produce a very impressive p-value. If this point is not clear, try using formula (9-68) to find the p-value for a situation where $n = 1000$, $k = 4$, and $R^2 = .1$.

From Section 4-2 on, R^2 values have been used in this book for informal comparisons of various potential summary equations for a data set. It turns out that it is sometimes possible to attach p-values to such comparisons through the use of the corresponding regression sums of squares and another F test.

Suppose that one has in mind two different regression models for describing a data set — the first of form (9-39) for k input variables $x_1, x_2, \ldots, x_k$, and the second being a specialization of the first where some p of the coefficients β (say, $\beta_{l_1}, \beta_{l_2}, \ldots, \beta_{l_p}$) are all 0 (i.e., a specialization not involving input variables $x_{l_1}, x_{l_2}, \ldots, x_{l_p}$). The first of these models will be called the *full* regression model and the second a *reduced* regression model. When one informally compares R^2 values for two such models, the comparison is essentially between SSR values, since the two R^2 values share the same denominator

$SSTot$. As it turns out, the two SSR values can be used to produce an observed level of significance for the comparison.

Under model (9-39) (the full model), the hypothesis

$$H_0: \beta_{l_1} = \beta_{l_2} = \cdots = \beta_{l_p} = 0 \tag{9-69}$$

(that the reduced model holds) can be tested against

$$H_a: \text{not } H_0 \tag{9-70}$$

using the test statistic

$$F = \frac{\dfrac{SSR_f - SSR_r}{p}}{\dfrac{SSE_f}{(n - k - 1)}} \tag{9-71}$$

and an $F_{p,n-k-1}$ reference distribution, where large observed values of the test statistic constitute evidence against H_0 in favor of H_a. In expression (9-71), the "f" and "r" subscripts of course refer to the full and reduced regressions. The calculation of statistic (9-71) can be facilitated by expanding the basic ANOVA table for the full model (Table 9-9). Table 9-11 shows one possible form this expansion can take.

TABLE 9-11

Expanded ANOVA Table for Testing $H_0: \beta_{l_1} = \beta_{l_2} = \cdots = \beta_{l_p} = 0$ in Multiple Regression

ANOVA TABLE

Source	SS	df	MS	F
Regression (full)	SSR_f	k		
Regression (reduced)	SSR_r	$k - p$		
Regression (full | reduced)	$SSR_f - SSR_r$	p	$(SSR_f - SSR_r)/p$	$MSR_{f|r}/MSE_f$
Error	SSE_f	$n - k - 1$	$SSE_f/(n - k - 1)$	
Total	$SSTot$	$n - 1$		

EXAMPLE 9-2
(continued)

In the nitrogen plant example, consider the comparison of the two possible descriptions of stack loss

$$y \approx \beta_0 + \beta_1 x_1 \tag{9-72}$$

(stack loss is approximately a linear function of air flow only) and

$$y \approx \beta_0 + \beta_1 x_1 + \beta_2 x_2 + \beta_3 x_1^2 \tag{9-73}$$

(the description of stack loss that has been used throughout this section). Although a printout won't be included here to show it, it is a simple matter to verify that the fitting of expression (9-72) to the nitrogen plant data produces $SSR = 775.48$ and therefore $R^2 = .950$. Fitting expression (9-73), on the other hand, gives $SSR = 799.80$ and $R^2 = .980$. Since expression (9-72) can be thought of as the specialization/reduction of expression (9-73) obtained by dropping the last $p = 2$ terms, one can formalize the comparison of these two SSR (or R^2) values with a p-value. That is, one can test

$$H_0: \beta_2 = \beta_3 = 0$$

in the (full) model (9-73). Table 9-12 organizes the calculation of the observed value of the statistic (9-71) for this problem.

TABLE 9-12

ANOVA Table for Testing H_0: $\beta_2 = \beta_3 = 0$ in Model (9-73) for the Stack Loss Data

ANOVA TABLE

Source	SS	df	MS	F
Regression (x_1, x_2, x_1^2)	799.80	3		
Regression (x_1)	775.48	1		
Regression $(x_2, x_1^2 \mid x_1)$	24.32	2	12.16	9.7
Error (x_1, x_2, x_1^2)	16.44	13	1.26	
Total	816.24	16		

When compared with tabled $F_{2,13}$ percentage points, the observed value of 9.7 is seen to produce a *p*-value between .01 and .001. There is strong evidence in the nitrogen plant data that an explanation of mean response in terms of expression (9-73) (pictured, for example, in Figure 4-15) is superior to one in terms of expression (9-72) (which could be pictured as a single linear response in x_1 for all x_2).

It is worth noting that the F statistic (9-71) can be written in terms of R^2 values as

$$F = \frac{\dfrac{R_f^2 - R_r^2}{p}}{\dfrac{1 - R_f^2}{n - k - 1}} \tag{9-74}$$

so that the test of hypothesis (9-69) is indeed a way of attaching a *p*-value to the comparison of two R^2's. However, just as was remarked earlier concerning the test of hypothesis (9-65), it is the R^2's themselves (not necessarily the value of F in formula (9-74)) that indicate how much additional variation a full model accounts for over a reduced model. The observed F value or associated *p*-value measures the extent to which that increase is distinguishable from background noise.

To conclude this section, something needs to be said about the relationship between the tests of hypotheses (9-48) (with # = 0), mentioned earlier, and the tests of hypothesis (9-69) based on the F statistic (9-71). When $p = 1$ (the full model contains only one more term than the reduced model), observed levels of significance based on statistic (9-71) are in fact equal to two-sided observed levels of significance based on # = 0 versions of statistic (9-49). But for cases where $p \geq 2$, the tests of the hypotheses that individual β's are 0 (one at a time) are not an adequate substitute for the tests of hypothesis (9-69). For example, in the full model

$$y = \beta_0 + \beta_1 x_1 + \beta_2 x_2 + \beta_3 x_3 + \epsilon \tag{9-75}$$

testing

$$H_0: \beta_2 = 0 \tag{9-76}$$

and then testing

$$H_0: \beta_3 = 0 \tag{9-77}$$

using the earlier methods need not be at all equivalent to making a single test of

$$H_0: \beta_2 = \beta_3 = 0 \tag{9-78}$$

This fact may at first seem paradoxical. But what can happen is the following. Should the variables x_2 and x_3 be reasonably highly correlated in the data set, it is possible to get large *p*-values for tests of both hypothesis (9-76) and (9-77) and yet a tiny *p*-value for a test of hypothesis (9-78). The message carried by such an outcome is that (due to the fact that the variables x_2 and x_3 appear in the data set to be more or less equivalent) in the presence of x_1 and x_2, x_3 is not needed to model y; and in the presence of x_1 and x_3, x_2 is not needed to model y; but one or the other of the two variables x_2 and x_3 is needed to help model y even in the presence of x_1. So, the F test of hypothesis (9-69) is more than just a fancy version of several tests of hypotheses H_0: $\beta_l = 0$; it is an important addition to an engineer's curve- and surface-fitting tool kit.

9-3 *The Use of Quadratic Response Surfaces*

The study of least squares surface-fitting methods in Chapter 4 and thus far in this chapter has indicated that in engineering applications, the particular forms of equations fit to data are often not motivated by any physical theory but are rather only meant to serve as simple, convenient empirical descriptions of system behavior. (The β's, for example, are often not expected to have any more fundamental physical significance than their usual interpretation as partial derivatives of the mean response function.) It is thus useful to have available a relatively simple and well-understood yet reasonably flexible general form to use in such empirical surface-fitting contexts. The quadratic or *second-order* functions of k physically independent system variables $x_1, x_2, \ldots, x_k$ provide one such general form.

This section considers several aspects of the use of quadratic response surfaces. It begins with a discussion of their form and the data requirements for their fitting and considers two data structures that make it possible to use them. Next are considered some devices for interpreting them, beginning with general graphical methods and then including an analytical approach that is special to the case of quadratic surfaces. Finally, the section closes with a general discussion of search strategies in engineering process optimization studies.

Quadratic Surfaces and Their Fitting

This discussion will continue the notational conventions of multiple regression used in Section 9-2 and primarily confine itself to situations where k system input variables $x_1, x_2, \ldots, x_k$ are not functionally related. In the spirit of the linear regression material in this book, the simplest possible approximate relationship between $x_1, x_2, \ldots, x_k$ and a response variable y is, of course,

$$y \approx \beta_0 + \beta_1 x_1 + \beta_2 x_2 + \cdots + \beta_k x_k \tag{9-79}$$

Thought of as defining a surface in $k + 1$ dimensions, equation (9-79) is the description of a line/plane/multidimensional generalization of a plane (depending on the size of k). For many engineering purposes, the relationship (9-79) is too simple to be of great use, because it fails to allow for any *curvature* in the surface it defines. For example, in the simple $k = 1$ case, if one is trying to optimize some system response y as a function of the input x_1 and believes that such an optimal value exists somewhere in the range $x_1 = 5$ to $x_1 = 10$, use of the linear equation in x_1

$$y \approx \beta_0 + \beta_1 x_1 \tag{9-80}$$

to describe the relationship between x_1 and y automatically indicates a belief that the optimum value of x_1 is either 5 or 10 (depending on the sign of β_1 and whether one's objective is to maximize or to minimize y). Geometrically, the fact that equation (9-80) allows for no curvature in the shape of the response function means that no value of x_1 strictly between 5 and 10 can possibly produce a predicted optimum y.

Of course, the simplest generalization of equation (9-80) that allows some curvature in the shape of a plot of y versus x_1 is the quadratic equation in x_1

$$y \approx \beta_0 + \beta_1 x_1 + \beta_2 x_1^2 \tag{9-81}$$

It has already been shown (for example, in Example 4-3) that relationship (9-81) can be a marked improvement over equation (9-80) as a description of (x_1, y) engineering data. It is the multivariate generalization of relationship (9-81) that is considered in this section.

The full k-variable quadratic generalization of relationships (9-79) and (9-81) includes not only all of the terms $x_1, x_2, \ldots, x_k$ and their squares $x_1^2, x_2^2, \ldots, x_k^2$ but also all products of pairs of the variables. For example, the quadratic generalization of relationship (9-81) for $k = 3$ process variables is

$$\left. \begin{aligned} y &\approx \beta_0 + \beta_1 x_1 + \beta_2 x_2 + \beta_3 x_3 + \beta_4 x_1^2 + \beta_5 x_2^2 + \beta_6 x_3^2 + \beta_7 x_1 x_2 \\ &+ \beta_8 x_1 x_3 + \beta_9 x_2 x_3 \end{aligned} \right\} \quad \textbf{(9-82)}$$

Without using the quadratic response surface terminology, expressions like (9-82) have already been found very useful in this text. The nitrogen plant example in both Chapter 4 and the previous section (Examples 4-5 and 9-2), the canard and tail position example of Burris (Example 4-6), and the chemical process filtration time example in Section 4-4 (Example 4-13) all involved the fitting and use of second-order response functions.

Although it may not be obvious *a priori*, the possibility of using a full quadratic expression like (9-82) effectively to represent a set of $(x_1, x_2, \ldots, x_k, y)$ data depends on there being enough different kinds of $(x_1, x_2, \ldots, x_k)$ vectors represented in the data that various versions of the full quadratic can be distinguished. For example, in the simple $k = 1$ case, it should be obvious that having data only for $x_1 = 0$ and $x_1 = 1$ would not be adequate to allow the least squares fitting of the quadratic equation (9-81). There are, as an arbitrary example, many different versions of equation (9-81) with $y = 5$ for $x_1 = 0$ and $y = 7$ for $x_1 = 1$, including

$$y \approx 5 + 2x_1 + 0x_1^2$$
$$y \approx 5 - 8x_1 + 10x_1^2$$
$$y \approx 5 + 10x_1 - 8x_1^2$$

which have plots with quite different shapes. The first is linear, the second is concave up with a minimum at $x_1 = .4$, and the third is concave down with a maximum at $x_1 = .625$. This is illustrated in Figure 9-8. The point is that data from at least three different x_1 values are needed in order to fit a one-variable quadratic equation like relationship (9-81).

FIGURE 9-8

Plots of Three Different Quadratic Functions Passing through the Points $(x_1, y) = (0, 5)$ and $(x_1, y) = (1, 7)$

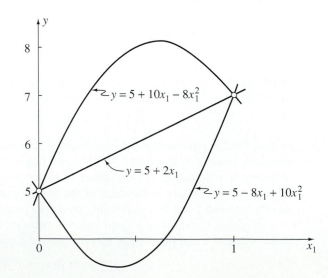

What would happen operationally if one blindly tried to use a regression program to fit equation (9-81) to a set of (x_1, y) data having only two different x_1 values in it? The program would either "bomb" or politely refuse the user's request, perhaps fitting instead the simpler equation (9-80).

For $k > 1$, it is not so easy to describe exactly what is needed in order to be able to fit a full quadratic response surface. One must have at least as many different $(x_1, x_2, \ldots, x_k)$ vectors as there are coefficients β in the quadratic function of $(x_1, x_2, \ldots, x_k)$, and there must be at least three different values for each of the variables $x_1, x_2, \ldots, x_k$ represented in the data. But even this may not be sufficient, as the following example shows.

EXAMPLE 9-3

Data Requirements for the Fitting of a Quadratic Function of Two Variables. Consider the fitting of the full quadratic relationship in x_1 and x_2

$$y \approx \beta_0 + \beta_1 x_1 + \beta_2 x_2 + \beta_3 x_1^2 + \beta_4 x_2^2 + \beta_5 x_1 x_2 \qquad \textbf{(9-83)}$$

to a set of (x_1, x_2, y) data. Note that there are six coefficients β in expression (9-83). But note also that even (x_1, x_2, y) data from each of the nine different (x_1, x_2) combinations listed in Table 9-13 would fail to allow the fitting of equation (9-83).

Figure 9-9 locates the (x_1, x_2) vectors of Table 9-13 in the plane.

TABLE 9-13

Nine Combinations (x_1, x_2)

x_1	x_2
-2	0
-1	0
0	0
1	0
2	0
0	-1
0	-2
0	1
0	2

FIGURE 9-9

Scatterplot of Nine Points (x_1, x_2)

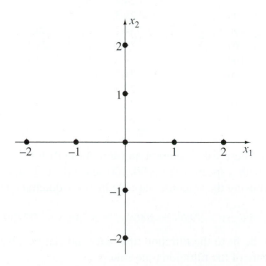

The point here is that although there are nine different combinations and at least three different values of each of x_1 and x_2 represented in Table 9-13, the pattern evident in Figure 9-9 makes it impossible to distinguish various versions of expression (9-83) based on only y values at the combinations listed in Table 9-13. For example, verify that the two equations

$$y \approx 5 + x_1 - x_2 + 3x_1^2 + 4x_2^2 - 10x_1x_2$$
$$y \approx 5 + x_1 - x_2 + 3x_1^2 + 4x_2^2 + 10x_1x_2$$

produce exactly the same values of y for those (x_1, x_2) in Table 9-13, but radically different y values for (x_1, x_2) vectors off both the x_1 and x_2 axes (i.e., with both $x_1 \neq 0$ and $x_2 \neq 0$). In order to apply fully quadratic response surfaces successfully, one needs more than just three different values of each input variable and a number of x combinations at least equal to the number of β's fitted.

This text will not attempt to describe exactly what conditions must be met in order to use $(x_1, x_2, \ldots, x_k, y)$ data to fit a full quadratic response surface in k variables. Instead, this discussion will simply 1) raise the possibility that haphazardly collected data are not necessarily sufficient to allow the use of a full quadratic response surface and 2) provide the reader with two standard types of experimental plans that are guaranteed to provide enough information for the fitting of the k-variable version of expression (9-82).

3^k and Central Composite Designs

One type of data structure that is guaranteed to make possible the fitting of a quadratic response surface is a full 3^k factorial structure in the variables $x_1, x_2, \ldots, x_k$.

EXAMPLE 9-4

(Example 4-13 revisited) **A 3^2 Factorial Chemical Process Experiment and Quadratic Surface Fitting.** The Hill and Demler data given first in Table 4-26 are repeated here in Table 9-14. They represent the yields y_1 and filtration times y_2 associated with the running of a chemical process at a particular condensation temperature x_1 and amount of B x_2.

TABLE 9-14

Yields and Filtration Times in a 3^2 Factorial Chemical Process Study

x_1 Condensation Temperature (°C)	x_2 Amount of B (cc)	y_1 Yield (g)	y_2 Filtration Time (sec)
90	24.4	21.1	150
90	29.3	23.7	10
90	34.2	20.7	8
100	24.4	21.1	35
100	29.3	24.1	8
100	34.2	22.2	7
110	24.4	18.4	18
110	29.3	23.4	8
110	34.2	21.9	10

These data represent an (unreplicated) full 3^2 factorial study in the factors x_1 and x_2, with respective levels 90, 100, and 110, and 24.4, 29.3, and 34.2. Section 4-4 showed that using the response variable $\ln(y_2)$, a quadratic response surface of the form

$$\widehat{\ln(y_2)} = 99.69 - .8869x_1 - 3.348x_2 + .002506x_1^2 + .03375x_2^2 + .01196x_1x_2$$

can be fit to the filtration time data and can be advantageously used to summarize the nature of the filtration time data.

For k larger than 2 or 3, the number of different combinations of three levels of each of k variables grows at a rate that makes the collection of full 3^k factorial data impractical in many statistical engineering contexts. (For example, a complete 3^4 factorial study already requires a minimum of 81 data points.) And as it turns out, a full 3^k factorial is to some degree overkill in terms of the amount of data needed to fit a k-variable quadratic response function. One can make do with less. Statisticians have developed a number of different smaller data structures, with the specific end in mind of efficient (both in terms of relatively small data requirements and relatively precise resultant b's) fitting of second-order response surfaces. One of the most popular of these is the *central composite* data structure.

DEFINITION 9-4

For k system variables $x_1, x_2, \ldots, x_k$ a **central composite design** in the variables is a collection of $(x_1, x_2, \ldots, x_k)$ vectors consisting of

1. a full (or for large k, possibly fractional) 2^k factorial
2. a (usually replicated) center point
3. two star points on each axis of the design

After properly coding the $x_1, x_2, \ldots, x_k$ variables, these points are expressible as

1. all possible combinations (or a fraction thereof) of $x_1 = \pm 1, x_2 = \pm 1, \ldots, x_k = \pm 1$
2. a (usually repeated) point with $x_1 = x_2 = x_3 = \cdots = x_k = 0$
3. for some $\alpha > 1$ all vectors of the form $(\pm \alpha, 0, 0, \ldots, 0), (0, \pm \alpha, 0, \ldots, 0) \ldots$ and $(0, 0, 0, \ldots, \pm \alpha)$

Figures 9-10 and 9-11 give geometric representations of the $k = 2$ and $k = 3$ central composite designs.

FIGURE 9-10

A $k = 2$ Central Composite Design

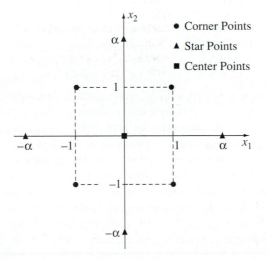

EXAMPLE 9-5

A $k = 4$ Central Composite Burnishing Process Experiment. Pande and Patil, in "Investigations on Vibratory Burnishing Process" (*International Journal of Machine Tool Design Research*, 1984) studied the effects of four vibratory ball burnishing process variables x_1, x_2, x_3, and x_4 on measures of surface roughness y_1 and microhardness y_2 for burnished 70mm diameter ring-shaped specimens of M.S. (C-20) BHN-170 material. Their four experimental variables and two responses were as listed in Table 9-15.

FIGURE 9-11

A $k = 3$ Central
Composite Design

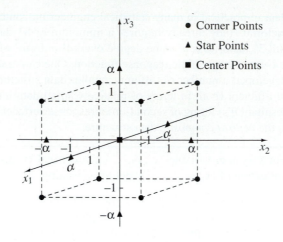

TABLE 9-15

Experimental Variables
and Responses in a
Burnishing Study

Factor	Variable
Burnishing Speed	$x_1 = \dfrac{\ln(V) - 4.61}{1.56}$, where V is rpm
Burnishing Feed	$x_2 = \dfrac{\ln(s) + 2.04}{2.68}$, where s is mm/rev
Burnishing Force	$x_3 = \dfrac{f - 15}{5}$, where f is ball force in kgf
Burnishing Frequency	$x_4 = \dfrac{\nu - 30}{10}$, where ν is in Hz
Surface Roughness	y_1 in μm
Microhardness	y_2 in kgf/mm^2

The investigators hoped ultimately to optimize y_1 and y_2 and therefore used an experimental plan sufficient to allow the fitting of quadratic response surfaces. Notice that since $k = 4$, a minimum of 81 different specimens would have to have been made in order to complete a full 3^4 factorial study. The authors actually used a central composite plan with $\alpha = 2$ in their data collection. They tested 31 (instead of 81) specimens and arrived at the data given in Table 9-16. (The data points are listed in the order "16 2^k factorial points, then 7 center points, then 8 star points.")

Using the data from their central composite design, Pande and Patil were able to fit second-order response surfaces for both y_1 and y_2, which for each response involved a total of 15 coefficients β (an intercept, four for the x terms, four for the x^2 terms, and six for the cross product terms). Unfortunately, as it turned out, not even the second-order descriptions of y_1 and y_2 were particularly apt or helpful. Not everything that even enlightened engineers do will always turn out to be simply understood.

One of the virtues of the central composite designs is that they lend themselves naturally to *sequential experimental strategy*. That is, it is possible to begin a program of engineering experimentation by completing first the 2^k factorial part of the plan plus perhaps a few center points, then to add the star points only as it becomes obvious that the response truly involves important curvature over the range of values being studied. The 2^k part of the design is adequate to allow analyses of the sort discussed in Section 8-3 and to support the fitting of a first-order model (9-79). As will be seen in Section 9-4,

TABLE 9-16

Surface Roughness and
Microhardness of 31 Specimens

x_1 Speed	x_2 Feed	x_3 Force	x_4 Frequency	y_1 Surface Roughness	y_2 Microhardness
−1	−1	−1	−1	.85	309
1	−1	−1	−1	.66	268
−1	1	−1	−1	.93	309
1	1	−1	−1	1.30	277
−1	−1	1	−1	1.05	405
1	−1	1	−1	1.08	373
−1	1	1	−1	1.04	287
1	1	1	−1	1.05	359
−1	−1	−1	1	.63	389
1	−1	−1	1	.46	389
−1	1	−1	1	.27	405
1	1	−1	1	.75	320
−1	−1	1	1	.70	423
1	−1	1	1	1.08	345
−1	1	1	1	.69	389
1	1	1	1	1.14	359
0	0	0	0	.95	373
0	0	0	0	.79	405
0	0	0	0	.87	389
0	0	0	0	.88	373
0	0	0	0	.93	359
0	0	0	0	.80	359
0	0	0	0	.93	345
−2	0	0	0	.79	359
2	0	0	0	.86	320
0	−2	0	0	.56	309
0	2	0	0	1.15	277
0	0	−2	0	1.07	259
0	0	2	0	.97	389
0	0	0	−2	.93	309
0	0	0	2	.41	389

important interactions in 2^k factorial analyses are indicative of some kinds of curvature in the corresponding response, and poor R^2 values for a first-order model (9-79) also suggest that a second-order description of the response may be needed. Repeated center point observations further facilitate the detection of curvature, for two reasons. In the first place, there are types of curvature that would not be revealed by 2^k data but that can be seen when a center point is added to a 2^k design in k quantitative variables. Secondly, replication produces a value of s_P for comparison with s_{SF} in judging how well a first-order model describes a situation.

A popular quick check, useful for detecting some kinds of curvature (in a response y based on unreplicated 2^k factorial data augmented by a repeated center point), is based on the fact that under a first-order multiple regression model, the arithmetic mean of the average responses at the 2^k corner points is the same as the mean response at the center point. So (with the obvious meanings for $\bar{y}_{center}$ and $\bar{y}_{corners}$) the difference

$$\bar{y}_{center} - \bar{y}_{corners} \tag{9-84}$$

is a measure of (some kinds of) curvature and can be used as an additional tool for assessing the fit of a model (9-79) to data from a first phase of a central composite experimental design. In fact, it is even easy to apply formula (7-26) of Section 7-2 to make a confidence interval for the corresponding linear combination of underlying mean responses based on measure (9-84). That is, with n_{center} standing for the number of center points used, one might attach a precision of

$$\pm t s_{\text{P}} \sqrt{\frac{1}{n_{\text{center}}} + \frac{1}{2^k}} \qquad (9\text{-}85)$$

to the measure (9-84), where the associated two-sided confidence is the probability that the t distribution with $\nu = n_{\text{center}} - 1$ degrees of freedom assigns to the interval between $-t$ and t.

When the methods of Section 8-3 reveal important interactions, and/or diagnostic tools for checking the adequacy of a first-order regression model suggest important departures from linearity, then the star points can be used to augment the 2^k design plus center points, and a full quadratic response function can be fitted.

EXAMPLE 9-6

A Central Composite Study for Optimizing Bread Wrapper Seal Strength. The article "Sealing Strength of Wax-Polyethylene Blends" by Brown, Turner, and Smith (*Tappi*, 1958), which is also employed as an example in the useful book *Response Surface Methodology* by R. Myers, contains an interesting central composite data set. The effects of the three process variables Seal Temperature, Cooling Bar Temperature, and % Polyethylene Additive on the seal strength y of a bread wrapper stock were studied using a central composite design. With the coding of the process variables indicated in Table 9-17, the data in Table 9-18 were obtained.

TABLE 9-17

Coding of Three Process Variables in a Seal Strength Study

	Factor	Variable	
A	Seal Temperature	$x_1 = \dfrac{t_1 - 255}{30}$	where t_1 is in °F
B	Cooling Bar Temperature	$x_2 = \dfrac{t_2 - 55}{9}$	where t_2 is in °F
C	Polyethylene Content	$x_3 = \dfrac{c - 1.1}{.6}$	where c is in %

The reader can verify that if one fits a first-order model

$$y = \beta_0 + \beta_1 x_1 + \beta_2 x_2 + \beta_3 x_3 + \epsilon \qquad (9\text{-}86)$$

to the first 14 data points listed in Table 9-18 (the 2^3 factorial corner points and six center points), a coefficient of determination of only $R^2 = .27$ is obtained, along with an estimate of σ equal to $s_{\text{SF}} = 1.75$. The pooled sample standard deviation (coming from the six center points) is quite a bit smaller than s_{SF} — namely, $s_{\text{P}} = 1.00$. Between the small value of R^2 and the moderate difference between s_{SF} and s_{P} here, there is already some indication that model (9-86) may be a poor description of the data in Table 9-18.

Using the Yates algorithm to compute 2^3 fitted effects from the eight corner points, one arrives at the values

$$
\begin{array}{lll}
a_2 = -.875 & ab_{22} = -.350 & abc_{222} = .175 \\
b_2 = .275 & ac_{22} = -.500 & \\
c_2 = .775 & bc_{22} = .150 &
\end{array}
$$

To evaluate the statistical detectability of the underlying 2^3 factorial effects, one can base confidence intervals on these fitted values using the $m = 1, p = 3, \nu = 5$ version

TABLE 9-18

Seal Strengths Produced
under 15 Different Sets
of Process Conditions

x_1	x_2	x_3	Seal Strength y (g/in.)
-1	-1	-1	6.6
1	-1	-1	6.9
-1	1	-1	7.9
1	1	-1	6.1
-1	-1	1	9.2
1	-1	1	6.8
-1	1	1	10.4
1	1	1	7.3
0	0	0	10.1
0	0	0	9.9
0	0	0	12.2
0	0	0	9.7
0	0	0	9.7
0	0	0	9.6
-1.682	0	0	9.8
1.682	0	0	5.0
0	-1.682	0	6.9
0	1.682	0	6.3
0	0	-1.682	4.0
0	0	1.682	8.6

of formula (8-46) from Section 8-3, with s_P taken from the six center points. (There is no logical problem with using a pooled estimate of σ derived from a point or points other than those involved in computing the fitted effects, provided that the usual model assumption of constant variance is appropriate for all sets of conditions from which one uses data.) That is, a precision of

$$\pm t(1.00)\frac{1}{\sqrt{1 \cdot 2^3}} = \pm.35t$$

could be attached to each fitted effect. It is clear from this that at typical confidence levels, none of the fitted interactions are large enough to be clearly distinguishable from background noise. That is, this is an example where the fitted interactions from the 2^k corner points alone do not clearly indicate curvature in the response.

However, consider the curvature check provided by expressions (9-84) and (9-85). It is possible to verify that the first eight responses listed in Table 9-18 give $\bar{y}_{\text{corners}} = 7.65$ and the responses from the six center points give $\bar{y}_{\text{center}} = 10.2$. Thus,

$$\bar{y}_{\text{center}} - \bar{y}_{\text{corners}} = 10.2 - 7.65 = 2.55 \tag{9-87}$$

The measure of precision that expression (9-85) indicates should be attached to this 2.55 figure is

$$\pm t(1.00)\sqrt{\frac{1}{6} + \frac{1}{8}} = \pm.54t \tag{9-88}$$

where t has 5 associated degrees of freedom. It is thus clear that the difference between $\bar{y}_{\text{center}}$ and $\bar{y}_{\text{corners}}$ is definitive evidence of curvature in the response function.

If the seal strength study were being done in stages, the poor value of R^2 for model (9-86), the moderate-to-large difference between s_P and s_{SF}, and now the fairly clear indication of curvature provided by calculations (9-87) and (9-88) would be reason to augment the 2^3 factorial plus center points with the star points and try a second-order description of seal strength. As it turns out, fitting the expression (9-82) to the entire set of data in Table 9-18 produces the equation

$$\left. \begin{aligned} \hat{y} = 10.165 &- 1.104x_1 + .0872x_2 + 1.020x_3 - .7596x_1^2 - 1.042x_2^2 \\ &- 1.148x_3^2 - .3500x_1x_2 - .5000x_1x_3 + .1500x_2x_3 \end{aligned} \right\} \quad \textbf{(9-89)}$$

with a coefficient of determination of $R^2 = .86$ and $s_{\text{SF}} = 1.09$. At least on the basis of the two measures R^2 and s_{SF}, this second-order description of seal strength seems much superior to a first-order description.

The 3^k and central composite experimental designs are by no means the only *response surface designs* that have been proposed. But they are among the most popular and serve well to illustrate the issues that arise in the collection of data adequate to fit full quadratic response functions in k process variables.

Graphical Interpretation of Fitted Quadratic Surfaces

Once one recognizes that quadratic expressions like (9-82) are among the simplest and most tractable functions of k system variables allowing for curvature in a response, collects data adequate to fit such an expression, and then uses a multiple regression program to produce a fitted equation, the matter of useful interpretation of that equation remains. In one sense, the interpretation of a fitted equation like (9-89) is "just" a matter of analytic geometry/freshman calculus. But in practical terms, most of us (including this author) have only the vaguest recollection of that material. Therefore, it will be useful here to discuss some methods for trying to see what an equation like (9-89) means in the engineering context where it was derived.

When k is fairly small (say, no more than 3) one possible approach to understanding the nature of a fitted quadratic response surface is through plots of $\hat{y}$ versus a particular system variable for the value(s) of the other system variable(s) held fixed. This is the method that was used in Figure 4-15 for the nitrogen plant data and in Figure 4-16 for the tail and canard positioning data of Burris. Note that in light of the inference material presented in Section 9-2, it is perfectly possible to enhance such plots by indicating the precision with which the mean responses are known, through error bars based on confidence intervals.

EXAMPLE 9-4
(continued)

Printout 9-4 is from the fitting of a full quadratic response surface for $\ln(y_2)$ to the data of Table 9-14. It is included here for purposes of providing the estimated standard deviations of fitted values $\hat{y}$ that are needed to use formulas (9-56) or (9-57) from Section 9-2.

Using the fitted equation from Printout 9-4 along with error bars based on 90% simultaneous confidence intervals (via formula (9-57) of the previous section), one might produce a display like that in Figure 9-12 for representing what has been learned about the log filtration time from the data in Table 9-14.

Figure 9-12 at least suggests that minimum mean $\ln(y_2)$ may be achievable using x_1 around 100 and x_2 between 29.3 and 34.2.

In Figure 9-12, there is one "extra" variable, x_2. For representing $k = 3$ quadratic surfaces, one of course needs two extra variables and thus traces of fitted response for fixed pairs of values of the extra variables.

A second kind of graphical representation that can be helpful in understanding the nature of fitted quadratic functions for fairly small k (say, not more than 4) is the *contour*

PRINTOUT 9-4

```
MTB > print c1-c7

   ROW     x1     x2    time    logtime     x1**2      x2**2    x1*x2

     1     90   24.4     150    5.01064       8100     595.36     2196
     2     90   29.3      10    2.30259       8100     858.49     2637
     3     90   34.2       8    2.07944       8100    1169.64     3078
     4    100   24.4      35    3.55535      10000     595.36     2440
     5    100   29.3       8    2.07944      10000     858.49     2930
     6    100   34.2       7    1.94591      10000    1169.64     3420
     7    110   24.4      18    2.89037      12100     595.36     2684
     8    110   29.3       8    2.07944      12100     858.49     3223
     9    110   34.2      10    2.30259      12100    1169.64     3762

MTB > brief 3
MTB > regress c4 5 c1 c2 c5 c6 c7
* NOTE *       x1 is highly correlated with other predictor variables
* NOTE *       x2 is highly correlated with other predictor variables
* NOTE *    x1**2 is highly correlated with other predictor variables
* NOTE *    x2**2 is highly correlated with other predictor variables
* NOTE *    x1*x2 is highly correlated with other predictor variables

The regression equation is
logtime = 99.7 - 0.887 x1 - 3.35 x2 + 0.00251 x1**2 + 0.0337 x2**2
          + 0.0120 x1*x2

Predictor        Coef        Stdev     t-ratio         p
Constant        99.69        20.88        4.78     0.017
x1            -0.8869       0.3755       -2.36     0.099
x2            -3.3475       0.5210       -6.43     0.008
x1**2        0.002506     0.001836        1.37     0.266
x2**2        0.033745     0.007647        4.41     0.022
x1*x2        0.011956     0.002649        4.51     0.020

s = 0.2596     R-sq = 97.5%     R-sq(adj) = 93.4%

Analysis of Variance

SOURCE        DF          SS          MS         F          p
Regression     5      7.9441      1.5888     23.57      0.013
Error          3      0.2022      0.0674
Total          8      8.1464

SOURCE        DF      SEQ SS
x1             1      0.7493
x2             1      4.3834
x1**2          1      0.1256
x2**2          1      1.3129
x1*x2          1      1.3729

Obs.     x1    logtime       Fit  Stdev.Fit    Residual    St.Resid
  1      90    5.0106     4.8416     0.2330      0.1691        1.48
  2      90    2.3026     2.5907     0.1935     -0.2882       -1.66
  3      90    2.0794     1.9604     0.2330      0.1191        1.04
  4     100    3.5553     3.6517     0.1935     -0.0964       -0.56
  5     100    2.0794     1.9867     0.1935      0.0927        0.54
  6     100    1.9459     1.9422     0.1935      0.0037        0.02
  7     110    2.8904     2.9631     0.2330     -0.0727       -0.64
  8     110    2.0794     1.8840     0.1935      0.1955        1.13
  9     110    2.3026     2.4253     0.2330     -0.1227       -1.07
```

FIGURE 9-12

Plots of Fitted Log Filtration
Time Enhanced with Error Bars

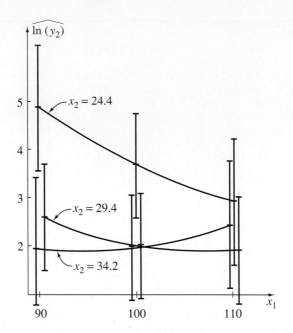

plot. A contour plot is based on exactly the same idea as a topographic map. For a given pair of system variables (say, for the sake of example, x_1 and x_2) one can, for fixed values of all other input variables, sketch out the loci of points in the (x_1, x_2)-plane that produce several particular values of $\hat{y}$. It is reasonably easy to do this by hand, using the quadratic formula to pick out (x_1, x_2) pairs, and nowadays many statistical analysis computer programs also provide contour plots for quadratic surfaces fairly automatically.

EXAMPLE 9-4
(continued)

Using the fitted expression

$$\widehat{\ln(y_2)} = 99.69 - .8869x_1 - 3.348x_2 + .002506x_1^2 + .03375x_2^2 + .01196x_1x_2$$

for the Hill and Demler data and the quadratic formula, it is clear that if for a particular value of x_1, a value v of $\widehat{\ln(y_2)}$ is taken on, the x_2 values at which this happens must satisfy

$$x_2 = \frac{(3.348 - .01196x_1) \pm \sqrt{\begin{array}{c}(3.348 - .01196x_1)^2 \\ - 4(.03375)(.002506x_1^2 - .8869x_1 + 99.69 - v)\end{array}}}{2(.03375)}$$

Substituting v's of interest into this expression allows one to find (x_1, x_2) pairs on given $\widehat{\ln(y_2)} = v$ contours. Such a method was used to produce Figure 9-13.

Alternatively, Printout 9-5 illustrates a Minitab run producing a similar representation of the fitted $\ln(y_2)$.

These contour plots are somewhat simpler to interpret than Figure 9-12. But they do not naturally carry the information about precision of estimation that can be placed on a figure like Figure 9-12. However, the location of the predicted optimum filtration time is most clearly seen on the contour plots.

The use of contour plots for cases where more than two system variables are involved requires the making of a different plot for each value (or set of values) for the extra variable(s). Nevertheless, such sets of contour plots can often be illuminating.

FIGURE 9-13

Contour Plot of Fitted Log
Filtration Time

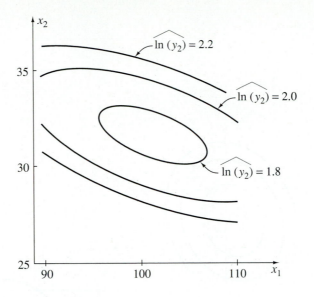

PRINTOUT 9-5

```
MTB > grid c2=25:35, c1=90:110
MTB > name c1 'x1'
MTB > name c2 'x2'
MTB > let c3=99.69-.8869*c1-
3.348*c2+.002506*c1**2+.03375*c2**2+.01196*c1*c2
MTB > contour c3 c2 c1

A = 1.75, 2.00     B = 2.25, 2.50      C = 2.75, 3.00
D = 3.25, 3.50     E = 3.75, 4.00      F = 4.25, 4.50
     35.00+ ,AAAAAAAAAAAAAAAAA,,,,,,,,,,,,,,,,,,BBBBBBBBBB.....
         - AAAAAAAAAAAAAAAAAAAAAAAAAAAAAAAAA,,,,,,,,,,,,,BBBBBBB
   x2    - AAAAAAAAAAAAAAAAAAAAAAAAAAAAAAAAAAAAAAAAAA,,,,,,,,,,,
         - AAAAAAAAAAAAAAAAAAAAAAAAAAAAAAAAAAAAAAAAAAAAAAAAAA,,,,,
         - ,AAAAAAAAAAAAAAAAAAAAAAAAAAAAAAAAAAAAAAAAAAAAAAAAAAAAAA
     31.50+ ,,,,,AAAAAAAAAAAAAAAAAAAAAAAAAAAAAAAAAAAAAAAAAAAAAAAAAA
         - ,,,,,,,,,,AAAAAAAAAAAAAAAAAAAAAAAAAAAAAAAAAAAAAAAAAAAAA
         - BBBBB,,,,,,,,,,,AAAAAAAAAAAAAAAAAAAAAAAAAAAAAAAAAAAAAA
         - ..BBBBBBBBB,,,,,,,,,,,,AAAAAAAAAAAAAAAAAAAAAAAAAAAAAAA
         - C........BBBBBBBBB,,,,,,,,,,,,,AAAAAAAAAAAAAAAAAAAAA
     28.00+ ,CCCCCCC........BBBBBBBBBB,,,,,,,,,,,,,,,,,,,,,,,
         - DD,,,,,,CCCCCCC.........BBBBBBBBBBBBBB,,,,,,,,,,,,,,,
         - ...DDDDDD,,,,,,,CCCCCCCCC..........BBBBBBBBBBBBBBBBBB
         - EEEEE......DDDDDD,,,,,,,CCCCCCCCC..............BBB
         - FF,,,,,EEEEE......DDDDDDD,,,,,,,,CCCCCCCCCCC.....
         -----+---------+---------+---------+---------+----x1
            92.00     96.00    100.00    104.00    108.00
```

EXAMPLE 9-6
(continued)

Figure 9-14 shows a series of five contour plots made for the fitted equation (9-89) for bread wrapper seal strength, corresponding to $x_3 = -2, -1, 0, 1,$ and 2. The figure suggests that optimum predicted bread wrapper seal strength may be achievable for x_3 between 0 and 1, with x_1 between -2 and -1, and x_2 between 0 and 1.

When an engineering problem involves more than one response variable, one of the especially helpful things that can be done is to overlay contour plots for the

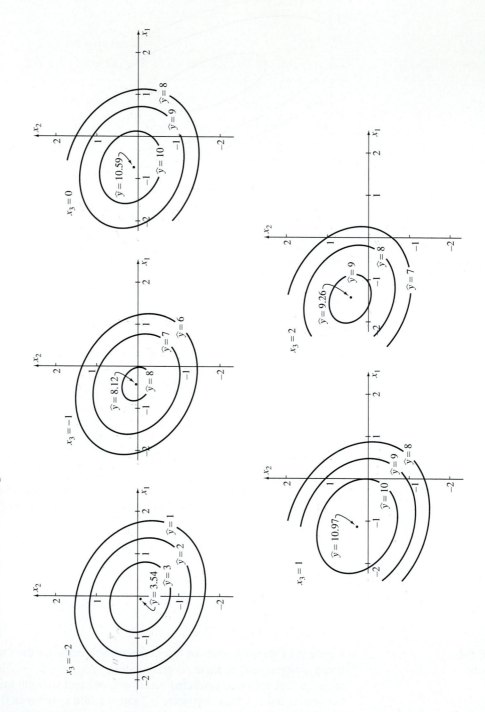

FIGURE 9-14

A Series of Contour Plots for Seal Strength

different response variables and therefore consider the implications of various choices of $x_1, x_2, \ldots, x_k$ settings for the different responses simultaneously. This is particularly helpful when the responses being considered are more or less competing responses, as when measures of both *cost* and *benefit* are considered.

EXAMPLE 9-4
(continued)

The fitting of a full quadratic response surface to the yield data in Table 9-14 can be shown to produce the equation

$$\hat{y}_1 = -113.2 + 1.254x_1 + 5.068x_2 - .009333x_1^2 - .1180x_2^2 + .01990x_1x_2$$

In this chemical example, notice that it is natural to expect the engineering desires for large yield y_1 and small filtration time y_2 to pull one in opposite directions. Figure 9-15 is a representation of this situation, showing both yield and filtration time contour plots. It could be used to help choose values of the process variables x_1 and x_2 to strike a balance between somewhat competing demands on y_1 and y_2.

FIGURE 9-15

Overlaid Contour Plots for
Yield and Filtration Time

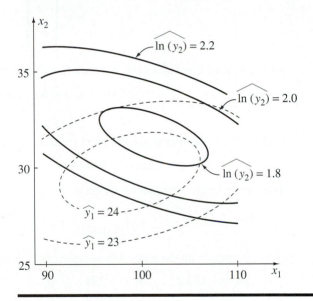

Analytical Interpretation of Fitted Quadratic Surfaces

The effective use of graphical methods for the analysis of fitted second-order response surfaces is limited primarily to small values of k. It is thus useful (particularly for k bigger than 2) to have some analytic tools for determining the nature of a quadratic response function. It ought to be at least plausible to most readers that it is the coefficients of the squared and cross product terms that determine the nature of a quadratic surface. After all, in the simplest case of $k = 1$, it is b_2 that determines whether the relationship

$$\hat{y} = b_0 + b_1x_1 + b_2x_1^2$$

describes a parabola opening up (the $b_2 > 0$ case), a parabola opening down (the $b_2 < 0$ case), or a line (the $b_2 = 0$ case).

In order to describe how the coefficients in a fitted quadratic response function determine the shape of the fitted surface, it is necessary to use some matrix notation. Temporarily, the way the fitted coefficients b are subscripted will be modified as follows. The meaning of b_0 will remain unchanged. b_1 through b_k will be the coefficients for the k system variables x_1 through x_k. b_{11} through b_{kk} will be the coefficients for the k squares

x_1^2 through x_k^2. And for each $i \neq j$, b_{ij} will be the coefficient of the $x_i x_j$ cross product. One can then define a vector b and matrix B as

$$b = \begin{bmatrix} b_1 \\ b_2 \\ \vdots \\ b_k \end{bmatrix}$$

$$B = \begin{bmatrix} b_{11} & \frac{1}{2}b_{12} & \cdots & \frac{1}{2}b_{1k} \\ \frac{1}{2}b_{12} & b_{22} & \cdots & \frac{1}{2}b_{2k} \\ \vdots & \vdots & & \vdots \\ \frac{1}{2}b_{1k} & \frac{1}{2}b_{2k} & \cdots & b_{kk} \end{bmatrix}$$

and with

$$x = \begin{bmatrix} x_1 \\ x_2 \\ \vdots \\ x_k \end{bmatrix}$$

write a general fitted quadratic response function in matrix notation as

$$\hat{y} = b_0 + x'b + x'Bx \tag{9-90}$$

EXAMPLE 9-4
(continued)

In the Hill and Demler study, the fitted yield equation was

$$\hat{y}_1 = -113.2 + 1.254x_1 + 5.068x_2 - .009333x_1^2 - .1180x_2^2 + .01990x_1x_2$$

This can be written in matrix notation as

$$\hat{y}_1 = -113.2 + (x_1, x_2)\begin{pmatrix} 1.254 \\ 5.068 \end{pmatrix} + (x_1, x_2)\begin{bmatrix} -.009333 & \frac{1}{2}(.01990) \\ \frac{1}{2}(.01990) & -.1180 \end{bmatrix}\begin{pmatrix} x_1 \\ x_2 \end{pmatrix}$$

and in keeping with the preceding notation, one could write the coefficients b as

$$b_0 = -113.2 \qquad b_1 = 1.254 \qquad b_{11} = -.009333 \qquad b_{12} = .01990$$
$$b_2 = 5.068 \qquad b_{22} = -.1180$$

In the $k = 1$ case of equation (9-81), provided b_2 is not 0, $\hat{y}$ has critical point (a minimum or maximum) at

$$x_1 = -\frac{b_1}{2b_2}$$

The corresponding fact for the general k situation is that provided the matrix B is nonsingular, the expression (9-90) has a stationary point (i.e., a point at which first partial derivatives with respect to $x_1, x_2, \ldots, x_k$ are all 0) where

$$x = -\frac{1}{2}B^{-1}b \tag{9-91}$$

And depending upon the nature of B, the stationary point will be either a minimum, a maximum, or a *saddle point* of the fitted response. (Moving away from a saddle point in some directions produces an increase in $\hat{y}$, while moving away in other directions produces a decrease.)

It turns out that it is the *eigenvalues* of $\boldsymbol{B}$ that are critical in determining what shape equation (9-90) represents. The eigenvalues of $\boldsymbol{B}$ are the k solutions of the equation (in λ)

$$\det(\boldsymbol{B} - \lambda\boldsymbol{I}) = 0 \qquad\qquad \textbf{(9-92)}$$

where $\boldsymbol{I}$ is the identity matrix.

When all solutions to equation (9-92) are positive, expression (9-90) represents a surface that is bowl-shaped up and has a minimum at the point (9-91). When all solutions to equation (9-92) are negative, expression (9-90) represents a surface that is bowl-shaped down and has a maximum at the point (9-91). When some solutions to equation (9-92) are positive and some are negative, the surface (9-90) has neither a maximum nor minimum (unless one restricts attention to some bounded region of $\boldsymbol{x}$ vectors).

EXAMPLE 9-4
(continued)

Figure 9-15 indicates that the fitted yield equation in the study of Hill and Demler is bowl-shaped down. To verify this analytically, note that for fitted yield y_1,

$$\det(\boldsymbol{B} - \lambda\boldsymbol{I}) = \det\begin{bmatrix} -.009333 - \lambda & \frac{1}{2}(.01990) \\ \frac{1}{2}(.01990) & -.1180 - \lambda \end{bmatrix}$$

$$= (.009333 + \lambda)(.1180 + \lambda) - \frac{1}{4}(.01990)^2$$

$$= \lambda^2 + .1273\lambda + .001002$$

Then setting $\det(\boldsymbol{B} - \lambda\boldsymbol{I}) = 0$, the quadratic formula provides roots

$$\lambda = \frac{-.1273 \pm \sqrt{(.1273)^2 - 4(.001002)}}{2}$$

that is,

$$\lambda = -.0086 \qquad \text{and} \qquad \lambda = -.1187$$

providing analytical confirmation of the inverted bowl picture provided by Figure 9-15. The exact location of maximum fitted yield can be found using expression (9-91).

$$-\frac{1}{2}\boldsymbol{B}^{-1}\boldsymbol{b} = -\frac{1}{2}\begin{bmatrix} -.009333 & \frac{1}{2}(.01990) \\ \frac{1}{2}(.01990) & -.1180 \end{bmatrix}^{-1} \begin{pmatrix} 1.254 \\ 5.068 \end{pmatrix}$$

which after some work simplifies to

$$-\frac{1}{2}\boldsymbol{B}^{-1}\boldsymbol{b} = \begin{pmatrix} 99.09 \\ 29.79 \end{pmatrix}$$

indicating that maximum fitted yield occurs when $x_1 = 99.09$ and $x_2 = 29.79$.

EXAMPLE 9-7

(Example 4-6 revisited) **Analysis of the Fitted Lift/Drag Ratio for a Three-Surface Configuration.** In Chapter 4, in Burris' lifting body study, for y a lift/drag ratio, x_1 a canard placement (in inches above the plane of a main wing), and x_2 a tail placement (in inches above a main wing), the fitted equation

$$\hat{y} = 3.983 + .5361x_1 + .3021x_2 - .4843x_1^2 - .5042x_1x_2 \tag{9-93}$$

was a reasonable data summary. (The inclusion of an x_2^2 term in the fitted equation provides a fitted surface that is only slightly different, and little appreciable increase in R^2.) Figure 9-16 is a contour plot of the fitted response surface (9-93); it shows it to be saddle-shaped.

FIGURE 9-16

Contour Plot of Fitted Lift/Drag Ratio for a Three-Surface Configuration

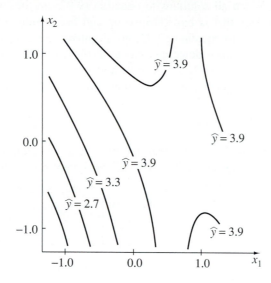

Analytic confirmation of this saddle shape can be had via consideration of the eigenvalues of the B matrix corresponding to equation (9-93). For fitted response function (9-93),

$$B = \begin{bmatrix} -.4843 & \frac{1}{2}(-.5042) \\ \frac{1}{2}(-.5042) & 0 \end{bmatrix}$$

so

$$\det(B - \lambda I) = (-.4843 - \lambda)(-\lambda) - \frac{1}{4}(-.5042)^2$$

$$= \lambda^2 + .4843\lambda - .0636$$

Thus, $\det(B - \lambda I) = 0$ implies that

$$\lambda = \frac{-.4843 \pm \sqrt{(.4843)^2 + 4(.0636)}}{2}$$

that is, that

$$\lambda = -.5918 \text{ or } \lambda = .1075$$

That is, B has both a positive and a negative eigenvalue, confirming the saddle shape of the surface indicated in Figure 9-16.

For $k > 2$, the determination of the eigenvalues of B is typically handled using some kind of numerical analysis software.

PRINTOUT 9-6

```
MTB > read 3 by 3 matrix m1
DATA> -.7596 -.175 -.250
DATA> -.175 -1.042 .075
DATA> -.250 .075 -1.148
      3 ROWS READ
MTB > read 3 by 1 matrix m2
DATA> -1.104
DATA> .0872
DATA> 1.020
      3 ROWS READ
MTB > eigen for m1 put into c1
MTB > print c1

C1
 -1.27090  -1.11680  -0.56190

MTB > invert m1 m3
MTB > mult m3 m2 m4
MTB > mult m4 -.5 m5
MTB > print m5
 MATRIX M5

 -1.01104
  0.26069
  0.68146
```

EXAMPLE 9-6
(continued)

Printout 9-6 illustrates the use of Minitab in the analytic investigation of the nature of the fitted surface (9-89) in the bread wrapper seal strength study of Brown, Turner, and Smith. The printout shows the three eigenvalues of $\boldsymbol{B}$ in this study to be negative, and the fitted seal strength therefore has a maximum. This maximum is predicted to occur at the combination of values of the system variables $x_1 = -1.01$, $x_2 = .26$, and $x_3 = .68$.

It can happen that one or more of the eigenvalues of $\boldsymbol{B}$ for a quadratic surface are 0. When this happens, $\boldsymbol{B}$ is singular, and one has a surface possessing a *ridge* geometry instead of a bowl or saddle geometry. Such surfaces can fail to have any stationary points or have whole infinite loci of stationary points. In applications, one will rarely if ever encounter a situation where $\boldsymbol{B}$ has an eigenvalue of exactly 0, but where one or more of the eigenvalues are of much smaller order of magnitude than the others, *approximately* ridge geometries will be encountered.

EXAMPLE 9-8

Two Simple Quadratic Surfaces with Ridge Geometries. Two simple $k = 2$ quadratic surfaces that illustrate the notion of ridge geometry are specified by

$$y = -x_2^2 \tag{9-94}$$

and

$$y = x_1 - x_2^2 \tag{9-95}$$

Figure 9-17 shows contour plots for equations (9-94) and (9-95). The figure reveals that equation (9-94) represents a stationary (flat) ridge geometry and equation (9-95) represents a rising ridge geometry.

FIGURE 9-17

Contour Plots for Two
Ridge Geometries

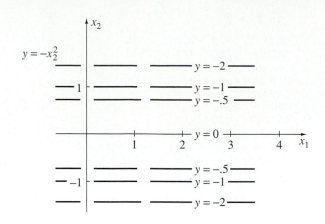

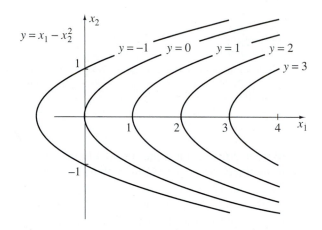

It is a simple exercise (left to the reader) to show that indeed

$$B = \begin{bmatrix} 0 & 0 \\ 0 & -1 \end{bmatrix}$$

corresponding to both equations (9-94) and (9-95) has one 0 eigenvalue, and that equation (9-94) has the entire x_1 axis as its set of stationary points, while equation (9-95) has no stationary points.

Because the engineering purpose for fitting and interpreting second-order surfaces is often to guide the search for optimum process operating conditions, it becomes especially important to caution against the possibility of overinterpreting such a surface. That is, before making engineering decisions or process changes based on the nature of a fitted surface, it is important to be sure that the nature of the surface is well enough determined to justify its use in guiding those decisions or changes.

One useful rule of thumb (suggested by Box, Hunter, and Hunter in their book *Statistics for Experimenters*) is to check that for a fitted surface involving a total of l coefficients b (including b_0),

$$\max \hat{y} - \min \hat{y} > 4\sqrt{\frac{l \cdot s_{\text{SF}}^2}{n}} \tag{9-96}$$

before trying to make decisions based on its nature (bowl shape up or down, saddle, ridge, etc.) and even limited interpolation or extrapolation. Criterion (9-96) is a comparison of the movement of the fitted surface across those n data points in hand, to 4 times an estimate of the root of the average variance associated with the n fitted values $\hat{y}$ in hand. If criterion (9-96) is not satisfied, the interpretation to be made is that the fitted surface is so flat (relative to the precision with which it is determined) as to make it impossible to tell with any certainty the true nature of how mean response varies as a function of the system variables.

Response Optimization Search Strategies

Second-order response surfaces are important tactical tools in engineering studies that are aimed at optimizing one or more responses summarizing system performance. The discussion of central composite designs and their natural two-stage implementation opened the question of what experimental strategy should be associated specifically with the use of quadratic surfaces. Before closing this section, it seems appropriate to consider the somewhat broader question of general experimental strategy in engineering process optimization studies (part of whose answer can include the wise use of second-order response surfaces).

One famous kind of optimization strategy for industrial processes should be mentioned here (although it doesn't make use of second-order surfaces). *Evolutionary Operation* (EVOP) was originally suggested by Box and Wilson and fully exposited in a 1969 book entitled *Evolutionary Operation*, by Box and Draper. Since its original introduction, many variations have been suggested, including a "simplex" version due to Spendley, Hext, and Himsworth. Only the original factorial version will be discussed here.

EVOP is basically a program of habitual/ongoing, cautious experimentation to be used on full-scale (operating) industrial processes. The suggested experimentation involves only small changes in factor levels. These typically give only weak indications of possible modifications for process improvement, but they are also unlikely to produce a production disaster by the running of an unexpectedly bad combination of process parameters. The idea is to make good product, but also (over time) to accumulate information on a direction in which process parameters might be moved so as to enhance performance.

If process performance can be characterized by l response variables, $y_1, y_2, \ldots, y_l$, and is thought to be determined by k quantitative variables, $x_1, x_2, \ldots, x_k$, whose settings can be chosen by those running the process, a factorial EVOP program can be run according to the following steps.

Step 1   If current operating conditions are $x_1^*, x_2^*, \ldots, x_k^*$, an EVOP committee picks two or three of the k variables that seem most promising in terms of improving all of $y_1, y_2, \ldots, y_l$. (For the sake of exposition, suppose these are x_1 and x_2.)

Step 2   Levels of x_1 and x_2 that bracket x_1^* and x_2^* but are still considered safe are chosen, and x_1 and x_2 are varied regularly in a 2^2 factorial pattern plus center point (x_1^*, x_2^*), until the means of all of $y_1, y_2, \ldots, y_l$ can be estimated with enough precision that all five combinations of the variables x_1 and x_2 can be clearly distinguished.

Step 3   If, taking account of Step 2 and expert engineering judgment, there appears to be a corner point of the 2^2 design that is superior to (x_1^*, x_2^*), a new center of operation is established by moving from (x_1^*, x_2^*) toward the superior corner point, without moving the new center of operation outside the current 2^2 design. The program then returns to Step 2.

Step 4   If, taking account of Step 2 and expert engineering judgment, there doesn't appear to be a good direction to move, the program returns to Step 1, and attention is turned to some other set of x variables.

This type of program is relatively conservative in that only two or three factors are varied at once, small moves are recommended, and no moves are made without a committee decision. One reason for such an approach is the possibly highly multidimensional nature of good process performance. What might improve one response might degrade another. An overriding constraint is that experimentation on an ongoing production process must often be calculated to avoid major disruptions or even temporary significant degradations of that process.

In other engineering contexts besides that of ongoing production processes (such as limited-duration studies in full-scale facilities and laboratory or pilot plant studies), more aggressive experimentation is often appropriate. Experimental search strategies for the optimization of a single response in these situations typically involve the sequential use of both first-order and second-order response surfaces, fitted *locally* to describe the impact of changes in process variables $x_1, x_2, \ldots, x_k$ on the response y. That is, variations on the following steps are commonly used.

Step 1   Beginning with a best guess for optimal settings of process variables $x_1^*, x_2^*, \ldots, x_k^*$, one collects 2^k factorial data, plus data from the conditions $x_1^*, x_2^*, \ldots, x_k^*$ used as a center point. A first-order equation like equation (9-79) is fitted to the 2^k plus center point data, and appropriate diagnostics are employed — plots of residuals and the kinds of checks used with the 2^k plus center point part of the bread wrapper seal strength data in Example 9-6 — to determine the importance of any response curvature.

Step 2   If Step 1 reveals no important curvature, a search is made (data are collected) along a ray from the point $(x_1^*, x_2^*, \ldots, x_k^*)$ in the direction of steepest ascent or descent, depending on whether large y or small y is desired. That is, if the fitted equation has coefficients $b_0, b_1, \ldots, b_k$, data are collected at $(x_1, x_2, \ldots, x_k)$ points of the form

$$(x_1^* + ub_1, x_2^* + ub_2, \ldots, x_k^* + ub_k)$$

for various positive (negative) u if a maximum (minimum) y is sought. u is increased (decreased) until the response ceases to increase (decrease). At such a point, one returns to Step 1, using the point of maximum response as a new center point.

(It should be noted here that the notion of *direction of steepest ascent/descent* employed in this step is somewhat unsatisfactory, in that it depends upon the scaling or units used in expresssing the x variables. Physically different search patterns end up being recommended when different units are employed. The most common method of dealing with this unpleasant ambiguity is to choose units for this step where the x values in the 2^k factorial of Step 1 are all ± 1 and the center point is at $(0, 0, \ldots, 0)$.)

Step 3   If Step 1 reveals important curvature, the 2^k plus center point design is augmented with star points, and a full second-order response surface is fitted. If an adequately fitting model is obtained and $\boldsymbol{B}$ has only negative (positive) eigenvalues, a search is made along the line from the point $(x_1^*, x_2^*, \ldots, x_k^*)$ to the transpose of $-\frac{1}{2}\boldsymbol{B}^{-1}\boldsymbol{b}$. (Points on this line segment are various weighted averages of $(x_1^*, x_2^*, \ldots, x_k^*)$ and this transpose.) Where maximum (minimum) response is observed, a new starting point is established, and one returns to Step 1.

This kind of process must of course terminate (if experimentation is to be of finite duration) when the investigator is reasonably sure that all possible important improvements in response have been realized.

9-4 *Regression and Factorial Inference*

The applicability of a fair portion of the factorial inference methods discussed in this book is limited to balanced-data situations. In some instances, formulas don't even make sense unless data sets are balanced, and in other cases, balanced-data formulas look generally applicable but in fact are not. In particular, remember that all of the no-interaction methods discussed in Section 8-2 for two-way factorials, the use of the reverse Yates algorithm to fit few-effects 2^p factorial models, the methods of interval-oriented inference for balanced 2^p studies under few-effects models discussed in Section 8-3, and the factorial ANOVA methods discussed in Section 8-4 are all limited to balanced-data applications.

In statistical engineering studies (by accident if not by design), one will occasionally be faced with the analysis of unbalanced factorial data and want to make formal inferences based on few-effects models. Happily enough, this can be accomplished through clever use of the multiple regression formulas provided in Section 9-2. This section discusses how factorial analyses can be thought of in multiple regression terms and how standard regression analysis software can be used to fit few-effects models to (even unbalanced) factorial data and to support the making of some important types of statistical intervals.

The section begins with a discussion of two-way factorial cases and shows how the device of dummy variables can be used in two-way contexts. Then three-way (and higher) situations are considered, paying particular attention to the 2^p factorials. Finally, the discussion concludes with a few comments about ANOVA for unbalanced factorial data.

Regression with Dummy Variables and Two-Way Factorials

The basic multiple regression model equation used in Section 9-2,

$$y_i = \beta_0 + \beta_1 x_{1i} + \beta_2 x_{2i} + \cdots + \beta_k x_{ki} + \epsilon_i \tag{9-97}$$

looks deceptively simple. Versions of it can in fact be used in a wide variety of different contexts. It has already been shown, for example, that by choosing some of the x variables to represent squares and products of other x's, model (9-97) can be used as the basis of fitting and inference for quadratic response surfaces. With proper choice of x variables, this model can also be used as the basis of fitting and inference for few-effects models in factorial analyses (even for unbalanced data).

For purposes of illustration, consider the case of a complete two-way factorial study with $I = 3$ levels of factor A and $J = 3$ levels of factor B. Verify that (in the usual two-way factorial notation introduced in Definitions 8-1 and 8-2) the basic constraints on the main effects and two-factor interactions, $\sum_i \alpha_i = 0$, $\sum_j \beta_j = 0$ and $\sum_i \alpha\beta_{ij} = \sum_j \alpha\beta_{ij} = 0$, imply that the $I \cdot J = 3 \cdot 3 = 9$ different mean responses in such a study,

$$\mu_{ij} = \mu_{..} + \alpha_i + \beta_j + \alpha\beta_{ij} \tag{9-98}$$

can be written as displayed in Table 9-19.

At first glance, the advantage of writing out these mean responses in terms of only effects corresponding to the first 2 ($= I - 1$) levels of A and first 2 ($= J - 1$) levels of B may not be so obvious. But the fact is that doing so expresses the 9 ($= I \cdot J$) different means in terms of only as many different parameters as there are means, and allows one to look for a regression-type analog of expression (9-98).

To this end, notice first that $\mu_{..}$ appears in each mean response listed and therefore plays a role much like that of the intercept term β_0 in a regression model. Further, the two A main effects, α_1 and α_2, appear with positive signs when (respectively) $i = 1$ or 2, but with negative signs when $i = 3$ ($= I$). In a similar manner, the first two B main

TABLE 9-19

Mean Responses in a 3^2 Factorial Study

i Level of A	j Level of B	Mean Response
1	1	$\mu_{..} + \alpha_1 + \beta_1 + \alpha\beta_{11}$
1	2	$\mu_{..} + \alpha_1 + \beta_2 + \alpha\beta_{12}$
1	3	$\mu_{..} + \alpha_1 - \beta_1 - \beta_2 - \alpha\beta_{11} - \alpha\beta_{12}$
2	1	$\mu_{..} + \alpha_2 + \beta_1 + \alpha\beta_{21}$
2	2	$\mu_{..} + \alpha_2 + \beta_2 + \alpha\beta_{22}$
2	3	$\mu_{..} + \alpha_2 - \beta_1 - \beta_2 - \alpha\beta_{21} - \alpha\beta_{22}$
3	1	$\mu_{..} - \alpha_1 - \alpha_2 + \beta_1 - \alpha\beta_{11} - \alpha\beta_{21}$
3	2	$\mu_{..} - \alpha_1 - \alpha_2 + \beta_2 - \alpha\beta_{12} - \alpha\beta_{22}$
3	3	$\mu_{..} - \alpha_1 - \alpha_2 - \beta_1 - \beta_2 + \alpha\beta_{11} + \alpha\beta_{12} + \alpha\beta_{21} + \alpha\beta_{22}$

effects, β_1 and β_2, appear with positive signs when (respectively) $j = 1$ or 2, but with negative signs when $j = 3$ ($= J$). If one tries to think of the four A and B main effects used in Table 9-19 in terms of coefficients β in a regression model, it soon becomes clear how to invent "system variables" x to make the regression coefficients β appear with correct signs in the expressions for means μ_{ij}.

That is, one defines four *dummy variables* x_1^A, x_2^A, x_1^B, and x_2^B as follows. Let

$$x_1^A = \begin{cases} 1 & \text{if the response } y \text{ is from level 1 of A} \\ -1 & \text{if the response } y \text{ is from level 3 of A} \\ 0 & \text{otherwise} \end{cases}$$

$$x_2^A = \begin{cases} 1 & \text{if the response } y \text{ is from level 2 of A} \\ -1 & \text{if the response } y \text{ is from level 3 of A} \\ 0 & \text{otherwise} \end{cases}$$

$$x_1^B = \begin{cases} 1 & \text{if the response } y \text{ is from level 1 of B} \\ -1 & \text{if the response } y \text{ is from level 3 of B} \\ 0 & \text{otherwise} \end{cases}$$

$$x_2^B = \begin{cases} 1 & \text{if the response } y \text{ is from level 2 of B} \\ -1 & \text{if the response } y \text{ is from level 3 of B} \\ 0 & \text{otherwise} \end{cases}$$

Then, making the correspondences indicated in Table 9-20, the $\mu_{..} + \alpha_i + \beta_j$ part of expression (9-98) can be written in regression notation as

$$\mu_{y|x_1^A, x_2^A, x_1^B, x_2^B} = \beta_0 + \beta_1 x_1^A + \beta_2 x_2^A + \beta_3 x_1^B + \beta_4 x_2^B$$

TABLE 9-20

Correspondences between Regression Coefficients and the Grand Mean and Main Effects in a 3^2 Factorial Study

Regression Coefficient	Corresponding 3×3 Factorial Effect
β_0	$\mu_{..}$
β_1	α_1
β_2	α_2
β_3	β_1
β_4	β_2

What is even more clever is that since the x's used here take only the values -1, 0, and 1, so also do their products. It turns out that taken in pairs (one x^A variable with one x^B variable), the products produce the correct (-1, 0, or 1) multipliers for the 2-factor interactions $\alpha\beta_{11}, \alpha\beta_{12}, \alpha\beta_{21}$, and $\alpha\beta_{22}$ appearing in Table 9-19. That is, if one thinks of the interactions $\alpha\beta_{ij}$ in terms of regression coefficients β, with the additional

TABLE 9-21

Correspondence between
Regression Coefficients
and Interactions in a
3^2 Factorial Study

Regression Coefficient	Corresponding 3×3 Factorial Effect
β_5	$\alpha\beta_{11}$
β_6	$\alpha\beta_{12}$
β_7	$\alpha\beta_{21}$
β_8	$\alpha\beta_{22}$

correspondences listed in Table 9-21, the entire expression (9-98) can be written in regression notation as

$$\mu_{y|x_1^A, x_2^A, x_1^B, x_2^B} = \beta_0 + \beta_1 x_1^A + \beta_2 x_2^A + \beta_3 x_1^B + \beta_4 x_2^B + \beta_5 x_1^A x_1^B$$
$$+ \beta_6 x_1^A x_2^B + \beta_7 x_2^A x_1^B + \beta_8 x_2^A x_2^B \tag{9-99}$$

The advantage of rewriting the factorial-type expression (9-98) as a regression-type expression (9-99) is that it then becomes obvious how to fit few-effects models and do inference under those models even for unbalanced factorial data. Nowhere in Section 9-2 was there any requirement that the data set be balanced. So the methods there can be used (employing properly constructed x variables and properly interpreting a corresponding regression printout) to fit reduced versions of model (9-99) and make confidence, prediction, and tolerance intervals under those reduced models.

The general $I \times J$ two-way factorial version of this story is similar. One defines $I - 1$ factor A dummy variables $x_1^A, x_2^A, \ldots, x_{I-1}^A$ according to

$$x_i^A = \begin{cases} 1 & \text{if the response } y \text{ is from level } i \text{ of A} \\ -1 & \text{if the response } y \text{ is from level } I \text{ of A} \\ 0 & \text{otherwise} \end{cases} \tag{9-100}$$

and $J - 1$ factor B dummy variables $x_1^B, x_2^B, \ldots, x_{J-1}^B$ according to

$$x_j^B = \begin{cases} 1 & \text{if the response } y \text{ is from level } j \text{ of B} \\ -1 & \text{if the response } y \text{ is from level } J \text{ of B} \\ 0 & \text{otherwise} \end{cases} \tag{9-101}$$

and uses a regression program to do the computations. Estimated regression coefficients of x_i^A or x_j^B variables alone are interpreted as estimated main effects, while those for $x_i^A x_j^B$ cross products are interpreted as estimated 2-factor interactions.

EXAMPLE 9-9

(Examples 4-7 and 8-1 revisited) **A Factorial Analysis of Unbalanced Wood Joint Strength Data Using a Regression Program.** Consider again the wood joint strength study of Kotlers, MacFarland, and Tomlinson. The discussion in Section 8-1 showed that if only the wood types pine and oak were considered, a no-interaction description of joint strength for butt, beveled, and lap joints might be appropriate. The corresponding part of the (originally 3×3 factorial) data of Kotlers, MacFarland, and Tomlinson is given again in Table 9-22.

TABLE 9-22

Strengths of 11 Wood Joints

		WOOD TYPE	
		1 (Pine)	2 (Oak)
	1 (Butt)	829, 596	1169
JOINT TYPE	2 (Beveled)	1348, 1207	1518, 1927
	3 (Lap)	1000, 859	1295, 1561

Notice that because these data are unbalanced (due to the unfortunate loss of one butt/oak response), it is not possible either to fit a no-interaction model to these data by simply adding together fitted effects (defined in Section 4-3) or to use methods of Section 8-2 to produce confidence intervals for mean responses or prediction or tolerance intervals. But it *is* possible to use the dummy variable regression approach based on formulas (9-100) and (9-101) to do these things. Consider the regression data set version of Table 9-22 given in Table 9-23.

TABLE 9-23

Joint Strength Data Prepared for a Factorial Analysis Using a Regression Program

i Joint Type	*j* Wood Type	x_1^A	x_2^A	x_1^B	*y*
1	1	1	0	1	829, 596
1	2	1	0	−1	1169
2	1	0	1	1	1348, 1207
2	2	0	1	−1	1518, 1927
3	1	−1	−1	1	1000, 859
3	2	−1	−1	−1	1295, 1561

Printout 9-7 shows the result of fitting the two regression models

$$y = \beta_0 + \beta_1 x_1^A + \beta_2 x_2^A + \beta_3 x_1^B + \beta_4 x_1^A x_1^B + \beta_5 x_2^A x_1^B + \epsilon \qquad \textbf{(9-102)}$$
$$y = \beta_0 + \beta_1 x_1^A + \beta_2 x_2^A + \beta_3 x_1^B + \epsilon \qquad \textbf{(9-103)}$$

to the data of Table 9-23.

The first regression run, corresponding to model (9-102), is the full model or $\mu_{ij} = \mu_{..} + \alpha_i + \beta_j + \alpha\beta_{ij}$ description of the data. For that run, the reader should verify the applicability of the correspondences between fitted regression coefficients *b* and fitted effects (defined in Section 4-3), listed in Table 9-24. (For example, $\bar{y}_{..} = 1206.5$ and $\bar{y}_{1.} = 940.75$, so $a_1 = 940.75 − 1206.5 = −265.75$, which is the value of the fitted regression coefficient b_1.)

TABLE 9-24

Correspondence between Fitted Regression Coefficients and Fitted Factorial Effects for the Wood Joint Strength Data

Fitted Regression Coefficient	Value	Corresponding Fitted Effect
b_0	1206.5	$\bar{y}_{..}$
b_1	−265.75	a_1
b_2	293.50	a_2
b_3	−233.33	b_1
b_4	5.08	ab_{11}
b_5	10.83	ab_{21}

Model (9-102), like the two-way model (8-5) of Section 8-1, represents no restriction or simplification of the basic one-way model, so least squares estimates of parameters that are linear combinations of underlying means are simply the same linear combinations of sample means. Further, the fitted *y* values are (as expected) simply the sample means $\bar{y}_{ij}$.

On the other hand, the printout for the second regression run, the one corresponding to model (9-103), is the $\mu_{ij} = \mu_{..} + \alpha_i + \beta_j$ description of the data and has a somewhat different flavor. For one thing, the fitted regression coefficients *b* for model (9-103) are not equal to the (full model) fitted factorial effects defined in Section 4-3. (The *b*'s are least squares estimates of the underlying effects for the no-interaction model. But when factorial data are unbalanced, these do not necessarily turn out to be equal to the quantities defined in Section 4-3. For example, b_1 from the second fitted equation is

-264.48, which is the least squares estimate of α_1 in a no-interaction model but differs from $a_1 = -264.75$.) In a similar vein, the fitted responses are neither sample means nor sums of $\bar{y}_{..}$ plus the full-model fitted main effects defined in Section 4-3. Of course, since the x variables take only values -1, 0, and 1, the fitted responses *are* sums and differences of the least squares estimates of the underlying parameters $\mu_{..}, \alpha_1, \alpha_2, \beta_1$ in the no-interaction model.

Inference under model (9-102) is simply inference under the usual one-way normal model, and all of Sections 7-1 through 7-4 and 8-1 can be used. It is then reassuring that (for example) for the first regression run, $s_{SF} = s_P = 182.2$ and that (for example) for butt joints and pine wood (levels 1 of both A and B), the estimated standard deviation for $\hat{y} = \bar{y}_{11}$ is

$$128.9 = s_{SF} \cdot A = \frac{s_P}{\sqrt{n_{11}}} = \frac{182.2}{\sqrt{2}}$$

PRINTOUT 9-7

```
MTB > print c1-c4

ROW    xa1    xa2    xb1      y

  1      1      0      1     829
  2      1      0      1     596
  3      1      0     -1    1169
  4      0      1      1    1348
  5      0      1      1    1207
  6      0      1     -1    1518
  7      0      1     -1    1927
  8     -1     -1      1    1000
  9     -1     -1      1     859
 10     -1     -1     -1    1295
 11     -1     -1     -1    1561

MTB > mult c1 c3 c5
MTB > name c5 'xa1*xb1'
MTB > mult c2 c3 c6
MTB > name c6 'xa2*xb1'
MTB > regress c4 5 c1 c2 c3 c5 c6

The regression equation is
y = 1207 - 266 xa1 + 293 xa2 - 233 xb1 + 5.1 xa1*xb1 + 10.8 xa2*xb1

Predictor        Coef       Stdev     t-ratio        p
Constant      1206.50       56.82       21.23    0.000
xa1           -265.75       85.91       -3.09    0.027
xa2            293.50       77.43        3.79    0.013
xb1           -233.33       56.82       -4.11    0.009
xa1*xb1          5.08       85.91        0.06    0.955
xa2*xb1         10.83       77.43        0.14    0.894

s = 182.2      R-sq = 88.5%      R-sq(adj) = 77.1%

Analysis of Variance

SOURCE        DF          SS          MS        F       p
Regression     5     1283527      256705     7.73   0.021
Error          5      166044       33209
Total         10     1449571
```

(continued)

PRINTOUT 9-7
(continued)

```
SOURCE          DF      SEQ SS
xa1             1       120144
xa2             1       577927
xb1             1       583908
xa1*xb1         1          897
xa2*xb1         1          650

Obs.     xa1         y        Fit  Stdev.Fit  Residual  St.Resid
  1     1.00      829.0     712.5     128.9     116.5      0.90
  2     1.00      596.0     712.5     128.9    -116.5     -0.90
  3     1.00     1169.0    1169.0     182.2       0.0       * X
  4     0.00     1348.0    1277.5     128.9      70.5      0.55
  5     0.00     1207.0    1277.5     128.9     -70.5     -0.55
  6     0.00     1518.0    1722.5     128.9    -204.5     -1.59
  7     0.00     1927.0    1722.5     128.9     204.5      1.59
  8    -1.00     1000.0     929.5     128.9      70.5      0.55
  9    -1.00      859.0     929.5     128.9     -70.5     -0.55
 10    -1.00     1295.0    1428.0     128.9    -133.0     -1.03
 11    -1.00     1561.0    1428.0     128.9     133.0      1.03

X denotes an obs. whose X value gives it large influence.

MTB > regress c4 3 c1 c2 c3

The regression equation is
y = 1207 - 264 xa1 + 293 xa2 - 234 xb1

Predictor       Coef       Stdev     t-ratio         p
Constant     1207.14       47.38       25.48     0.000
xa1          -264.48       70.62       -3.74     0.007
xa2           292.86       65.11        4.50     0.000
xb1          -233.97       47.38       -4.94     0.000

s = 154.7       R-sq = 88.4%      R-sq(adj) = 83.5%

Analysis of Variance

SOURCE          DF          SS          MS        F        p
Regression       3     1281979      427327    17.85    0.001
Error            7      167591       23942
Total           10     1449571

SOURCE          DF      SEQ SS
xa1             1       120144
xa2             1       577927
xb1             1       583908

Obs.     xa1         y        Fit  Stdev.Fit  Residual  St.Resid
  1     1.00      829.0     708.7      94.8     120.3      0.98
  2     1.00      596.0     708.7      94.8    -112.7     -0.92
  3     1.00     1169.0    1176.6     109.4      -7.6     -0.07
  4     0.00     1348.0    1266.0      90.7      82.0      0.65
  5     0.00     1207.0    1266.0      90.7     -59.0     -0.47
  6     0.00     1518.0    1734.0      90.7    -216.0     -1.72
  7     0.00     1927.0    1734.0      90.7     193.0      1.54
  8    -1.00     1000.0     944.8      90.7      55.2      0.44
  9    -1.00      859.0     944.8      90.7     -85.8     -0.68
 10    -1.00     1295.0    1412.7      90.7    -117.7     -0.94
 11    -1.00     1561.0    1412.7      90.7     148.3      1.18
```

To illustrate how inference under a no-interaction model would proceed for the unbalanced 3×2 factorial joint strength data, consider making a 95% two-sided confidence interval for the mean strength of butt/pine joints and then a 90% lower prediction bound for the strength of a single joint of the same kind. Note that for data point 1 (a butt/pine observation) on the second regression run, $\hat{y} = 708.7$ and $s_{SF} \cdot A = 94.8$ where $s_{SF} = 154.7$ has 7 associated degrees of freedom. So from formula (9-56) of Section 9-2, two-sided 95% confidence limits for mean butt/pine joint strength are

$$708.7 \pm 2.365(94.8)$$

that is,

$$484.5 \text{ psi} \quad \text{and} \quad 932.9 \text{ psi}$$

Similarly, using formula (9-59) of Section 9-2, a 90% lower prediction limit for a single additional butt/pine joint strength is

$$708.7 - 1.415\sqrt{(154.7)^2 + (94.8)^2} = 452.0 \text{ psi}$$

From these two calculations, it should be clear that other methods from Section 9-2 could be used here as well. Although the details won't be worked out here, the reader should have no trouble, for example, finding and using residuals and standardized residuals for the no-interaction model based on formulas (9-42) and (9-45) of Section 9-2, giving simultaneous confidence intervals for all six mean responses under the no-interaction model using formula (9-57) of Section 9-2, or giving one-sided tolerance bounds for certain joint/wood combinations under the no-interaction model using formula (9-63) or (9-64) from that section.

Regression with Dummy Variables and Higher-Way Factorials

The pattern of analysis set out in the preceding discussion and example for two-way factorials carries over quite naturally to three-way and higher factorials. In order to use a multiple regression program to fit and make possible inference for simplified versions of the p-way factorial (equal variances, normal distributions) model, proceed as follows. $I - 1$ dummy variables $x_1^A, x_2^A, \ldots, x_{I-1}^A$ are defined (as before) to carry information about I levels of factor A, $J - 1$ dummy variables $x_1^B, x_2^B, \ldots, x_{J-1}^B$ are defined (as before) to carry information about J levels of factor B, $K - 1$ dummy variables $x_1^C, x_2^C, \ldots, x_{K-1}^C$ are defined (as before) to carry information about K levels of factor C, $\ldots$, etc. Products of pairs of these, one each from the groups representing two different factors, carry information about 2-factor interactions of the factors. Products of triples of these, one each from the groups representing three different factors, carry information about 3-factor interactions of the factors. And so on.

When something short of the largest possible regression model is fitted to an unbalanced factorial data set, the estimated coefficients b that result are the least squares estimates of the underlying factorial effects in the few-effects model. Usually, these differ somewhat from the (full-model) fitted effects defined in Section 4-3. All of the machinery of Section 9-2 can be applied to create fitted values, residuals, and standardized residuals; to plot these to do model checking; to make confidence intervals for mean responses; and to create prediction and tolerance intervals.

It should be noted that (as in Example 9-9) when the dummy variable approach is used, the fitted coefficients b correspond to fitted effects for the levels 1 through $I - 1$, $J - 1$, $K - 1$, etc. of the factors. For two-level factorials, this would mean that the fitted coefficients are estimated factorial effects for the "all low" treatment combination. However, because of extensive use of the Yates algorithm in this text, the alert reader will think first in terms of the 2^p factorial effects for the "all high" treatment combination.

Two sensible courses of action then suggest themselves for the analysis of unbalanced 2^p factorial data. One might proceed exactly as just indicated, using dummy variables x_1^A, x_1^B, x_1^C, etc. and various products of the same, taking care to remember to interpret b's as low-level fitted effects and subsequently to switch signs as appropriate in order to get high-level fitted effects. The other possibility is to depart slightly from the program laid out for general p-way factorials in 2^p cases: Instead of using the variables x_1^A, x_1^B, x_1^C, etc. and their products when doing regression, one may use the variables

$$x_2^A = -x_1^A = \begin{cases} 1 & \text{if the response } y \text{ is from the high level of A} \\ -1 & \text{if the response } y \text{ is from the low level of A} \end{cases}$$

$$x_2^B = -x_1^B = \begin{cases} 1 & \text{if the response } y \text{ is from the high level of B} \\ -1 & \text{if the response } y \text{ is from the low level of B} \end{cases}$$

$$x_2^C = -x_1^C = \begin{cases} 1 & \text{if response } y \text{ is from the high level of C} \\ -1 & \text{if response } y \text{ is from the low level of C} \end{cases}$$

etc. and their products when doing regression. When the variables x_2^A, x_2^B, x_2^C, etc. are used, the fitted b's are the estimated "all high" 2^p factorial effects.

EXAMPLE 9-10

(Example 8-5 revisited) **A Factorial Analysis of Unbalanced 2^3 Power Requirement Data Using Regression.** Return to the situation of the 2^3 metalworking power requirement study of Miller. The original data set (given in Table 8-11) is a balanced data set, with the common sample size being $m = 4$. For the sake of illustrating how regression with dummy variables can be used in the analysis of unbalanced higher-way factorial data, consider artificially unbalancing Miller's data by supposing that somehow the first data point appearing in Table 8-11 has gotten lost. The portion of Miller's data that will be used here is then given in Table 9-25.

TABLE 9-25

Dynamometer Readings for 2^3 Treatment Combinations

Tool Type	Bevel Angle	Type of Cut	y Dynamometer Reading (mm)
1	15°	continuous	26.5, 30.5, 27.0
2	15°	continuous	28.0, 28.5, 28.0, 25.0
1	30°	continuous	28.5, 28.5, 30.0, 32.5
2	30°	continuous	29.5, 32.0, 29.0, 28.0
1	15°	interrupted	28.0, 25.0, 26.5, 26.5
2	15°	interrupted	24.5, 25.0, 28.0, 26.0
1	30°	interrupted	27.0, 29.0, 27.5, 27.5
2	30°	interrupted	27.5, 28.0, 27.0, 26.0

Check that for this slightly altered data set, the Yates algorithm produces the fitted effects

$$a_2 = -.2656 \qquad ab_{22} = .0469 \qquad abc_{222} = -.0469$$
$$b_2 = .8281 \qquad ac_{22} = -.0469$$
$$c_2 = -.9531 \qquad bc_{22} = -.2031$$

and $s_P = 1.51$ with $\nu = 23$ associated degrees of freedom. Formula (8-45) of Section 8-3 then shows that, (say) two-sided 90% confidence intervals based on these fitted effects would have plus-and-minus parts of the form

$$\pm 1.714(1.51)\frac{1}{2^3}\sqrt{\frac{7}{4} + \frac{1}{3}} = \pm.47$$

Just as in Example 8-5, where all $n = 32$ data points were used, one might thus judge only the B and C main effects to be clearly larger than background noise.

The first regression run shown in Printout 9-8 supports exactly these same conclusions. This first regression run was made using all seven of the terms

$$x_2^A, \ x_2^B, \ x_2^C, \ x_2^A x_2^B, \ x_2^A x_2^C, \ x_2^B x_2^C, \ \text{and } x_2^A x_2^B x_2^C$$

(i.e., using the *full model* in regression terminology and the *unrestricted 2^3 factorial model* in the terminology of Section 8-3). On the first run in Printout 9-8, one can identify the fitted regression coefficients b with the fitted factorial effects in the pairs indicated in Table 9-26.

TABLE 9-26

Correspondence between Fitted Regression Coefficients and Fitted Factorial Effects for the First Regression Run of Printout 9-8

Fitted Regression Coefficient	Fitted Factorial Effect
b_0	$\bar{y}...$
b_1	a_2
b_2	b_2
b_3	c_2
b_4	ab_{22}
b_5	ac_{22}
b_6	bc_{22}
b_7	abc_{222}

Analysis of the data of Table 9-25 based on a full factorial model

$$y_{ijkl} = \mu_{...} + \alpha_i + \beta_j + \gamma_k + \alpha\beta_{ij} + \alpha\gamma_{ik} + \beta\gamma_{jk} + \alpha\beta\gamma_{ijk} + \epsilon_{ijkl}$$

that is,

$$y_i = \beta_0 + \beta_1 x_{2i}^A + \beta_2 x_{2i}^B + \beta_3 x_{2i}^C + \beta_4 x_{2i}^A x_{2i}^B + \beta_5 x_{2i}^A x_{2i}^C + \beta_6 x_{2i}^B x_{2i}^C$$
$$+ \ \beta_7 x_{2i}^A x_{2i}^B x_{2i}^C + \epsilon_i$$

is a logical first step. Based on that step, it seems desirable to fit and draw inferences based on a "B and C main effects only" description of y. Since the data in Table 9-25 are unbalanced, the naive use of the reverse Yates algorithm with the (full factorial) fitted effects will not produce appropriate fitted values. The estimates $\bar{y}_{...}, b_2$, and c_2 are simply *not* the least squares estimates of $\mu_{...}, \beta_2$, and γ_2 for the "B and C main effects only" model in this unbalanced data situation.

However, what can be done is to fit the reduced regression model

$$y_i = \beta_0 + \beta_2 x_{2i}^B + \beta_3 x_{2i}^C + \epsilon_i$$

to the data. The second regression run in Printout 9-8 represents the use of this technique. Locate on that printout the (reduced-model) estimates of the factorial effects $\mu_{...}, \beta_2$, and γ_2 and note that they differ somewhat from $\bar{y}_{...}, b_2$, and c_2 as defined in Section 4-3 and calculated via the Yates algorithm. Note also that the four different possible fitted mean responses, along with their estimated standard deviations, are as given in Table 9-27.

TABLE 9-27

Fitted Values and Their Estimated Standard Deviations for a "B and C Main Effects Only" Analysis of the Unbalanced Power Requirement Data

Bevel Angle	x_2^B	Type of Cut	x_2^C	$\hat{y}$	$s_{SF} \cdot A$
15°	-1	continuous	-1	27.88	.46
30°	1	continuous	-1	29.54	.44
15°	-1	interrupted	1	25.98	.44
30°	1	interrupted	1	27.64	.44

PRINTOUT 9-8

```
MTB > print c1-c4

ROW    xa2    xb2    xc2       y

  1     -1     -1     -1    26.5
  2     -1     -1     -1    30.5
  3     -1     -1     -1    27.0
  4      1     -1     -1    28.0
  5      1     -1     -1    28.5
  6      1     -1     -1    28.0
  7      1     -1     -1    25.0
  8     -1      1     -1    28.5
  9     -1      1     -1    28.5
 10     -1      1     -1    30.0
 11     -1      1     -1    32.5
 12      1      1     -1    29.5
 13      1      1     -1    32.0
 14      1      1     -1    29.0
 15      1      1     -1    28.0
 16     -1     -1      1    28.0
 17     -1     -1      1    25.0
 18     -1     -1      1    26.5
 19     -1     -1      1    26.5
 20      1     -1      1    24.5
 21      1     -1      1    25.0
 22      1     -1      1    28.0
 23      1     -1      1    26.0
 24     -1      1      1    27.0
 25     -1      1      1    29.0
 26     -1      1      1    27.5
 27     -1      1      1    27.5
 28      1      1      1    27.5
 29      1      1      1    28.0
 30      1      1      1    27.0
 31      1      1      1    26.0

MTB > mult c1 c2 c5
MTB > name c5 'xa*xb'
MTB > mult c1 c3 c6
MTB > name c6 'xa*xc'
MTB > mult c2 c3 c7
MTB > name c7 'xb*xc'
MTB > mult c1 c7 c8
MTB > name c8 'xa*xb*xc'
MTB > regress c4 7 c1 c2 c3 c5 c6 c7 c8

The regression equation is
y = 27.8 - 0.266 xa2 + 0.828 xb2 - 0.953 xc2 + 0.047 xa*xb
    - 0.047 xa*xc - 0.203 xb*xc - 0.047 xa*xb*xc

Predictor        Coef        Stdev     t-ratio        p
Constant      27.7656       0.2731      101.68    0.000
xa2           -0.2656       0.2731       -0.97    0.341
xb2            0.8281       0.2731        3.03    0.006
xc2           -0.9531       0.2731       -3.49    0.002
xa*xb          0.0469       0.2731        0.17    0.865
xa*xc         -0.0469       0.2731       -0.17    0.865
xb*xc         -0.2031       0.2731       -0.74    0.465
xa*xb*xc      -0.0469       0.2731       -0.17    0.865
```

(continued)

PRINTOUT 9-8
(continued)

```
s = 1.514        R-sq = 51.0%      R-sq(adj) = 36.0%

Analysis of Variance

SOURCE         DF          SS          MS        F        p
Regression      7      54.748       7.821     3.41    0.012
Error          23      52.688       2.291
Total          30     107.435

SOURCE         DF      SEQ SS
xa2             1       2.202
xb2             1      22.645
xc2             1      28.398
xa*xb           1       0.091
xa*xc           1       0.051
xb*xc           1       1.293
xa*xb*xc        1       0.068

Obs.     xa2         y       Fit  Stdev.Fit  Residual  St.Resid
  1    -1.00     26.500    28.000     0.874    -1.500     -1.21
  2    -1.00     30.500    28.000     0.874     2.500      2.02R
  3    -1.00     27.000    28.000     0.874    -1.000     -0.81
  4     1.00     28.000    27.375     0.757     0.625      0.48
  5     1.00     28.500    27.375     0.757     1.125      0.86
  6     1.00     28.000    27.375     0.757     0.625      0.48
  7     1.00     25.000    27.375     0.757    -2.375     -1.81
  8    -1.00     28.500    29.875     0.757    -1.375     -1.05
  9    -1.00     28.500    29.875     0.757    -1.375     -1.05
 10    -1.00     30.000    29.875     0.757     0.125      0.10
 11    -1.00     32.500    29.875     0.757     2.625      2.00R
 12     1.00     29.500    29.625     0.757    -0.125     -0.10
 13     1.00     32.000    29.625     0.757     2.375      1.81
 14     1.00     29.000    29.625     0.757    -0.625     -0.48
 15     1.00     28.000    29.625     0.757    -1.625     -1.24
 16    -1.00     28.000    26.500     0.757     1.500      1.14
 17    -1.00     25.000    26.500     0.757    -1.500     -1.14
 18    -1.00     26.500    26.500     0.757     0.000      0.00
 19    -1.00     26.500    26.500     0.757     0.000      0.00
 20     1.00     24.500    25.875     0.757    -1.375     -1.05
 21     1.00     25.000    25.875     0.757    -0.875     -0.67
 22     1.00     28.000    25.875     0.757     2.125      1.62
 23     1.00     26.000    25.875     0.757     0.125      0.10
 24    -1.00     27.000    27.750     0.757    -0.750     -0.57
 25    -1.00     29.000    27.750     0.757     1.250      0.95
 26    -1.00     27.500    27.750     0.757    -0.250     -0.19
 27    -1.00     27.500    27.750     0.757    -0.250     -0.19
 28     1.00     27.500    27.125     0.757     0.375      0.29
 29     1.00     28.000    27.125     0.757     0.875      0.67
 30     1.00     27.000    27.125     0.757    -0.125     -0.10
 31     1.00     26.000    27.125     0.757    -1.125     -0.86

R denotes an obs. with a large st. resid.

MTB > regress c4 2 c2 c3

The regression equation is
y = 27.8 + 0.832 xb2 - 0.949 xc2
```

(continued)

PRINTOUT 9-8
(continued)

```
Predictor        Coef       Stdev      t-ratio         p
Constant      27.7619      0.2553       108.73     0.000
xb2            0.8319      0.2553         3.26     0.003
xc2           -0.9494      0.2553        -3.72     0.001

s = 1.420        R-sq = 47.4%      R-sq(adj) = 43.7%

Analysis of Variance

SOURCE        DF          SS          MS          F          p
Regression     2      50.972      25.486      12.64     0.000
Error         28      56.463       2.017
Total         30     107.435

SOURCE        DF      SEQ SS
xb2            1      23.093
xc2            1      27.879

Obs.      xb2         y        Fit  Stdev.Fit   Residual    St.Resid
  1     -1.00    26.500    27.879      0.457     -1.379       -1.03
  2     -1.00    30.500    27.879      0.457      2.621        1.95
  3     -1.00    27.000    27.879      0.457     -0.879       -0.65
  4     -1.00    28.000    27.879      0.457      0.121        0.09
  5     -1.00    28.500    27.879      0.457      0.621        0.46
  6     -1.00    28.000    27.879      0.457      0.121        0.09
  7     -1.00    25.000    27.879      0.457     -2.879       -2.14R
  8      1.00    28.500    29.543      0.437     -1.043       -0.77
  9      1.00    28.500    29.543      0.437     -1.043       -0.77
 10      1.00    30.000    29.543      0.437      0.457        0.34
 11      1.00    32.500    29.543      0.437      2.957        2.19R
 12      1.00    29.500    29.543      0.437     -0.043       -0.03
 13      1.00    32.000    29.543      0.437      2.457        1.82
 14      1.00    29.000    29.543      0.437     -0.543       -0.40
 15      1.00    28.000    29.543      0.437     -1.543       -1.14
 16     -1.00    28.000    25.981      0.437      2.019        1.49
 17     -1.00    25.000    25.981      0.437     -0.981       -0.73
 18     -1.00    26.500    25.981      0.437      0.519        0.38
 19     -1.00    26.500    25.981      0.437      0.519        0.38
 20     -1.00    24.500    25.981      0.437     -1.481       -1.10
 21     -1.00    25.000    25.981      0.437     -0.981       -0.73
 22     -1.00    28.000    25.981      0.437      2.019        1.49
 23     -1.00    26.000    25.981      0.437      0.019        0.01
 24      1.00    27.000    27.644      0.437     -0.644       -0.48
 25      1.00    29.000    27.644      0.437      1.356        1.00
 26      1.00    27.500    27.644      0.437     -0.144       -0.11
 27      1.00    27.500    27.644      0.437     -0.144       -0.11
 28      1.00    27.500    27.644      0.437     -0.144       -0.11
 29      1.00    28.000    27.644      0.437      0.356        0.26
 30      1.00    27.000    27.644      0.437     -0.644       -0.48
 31      1.00    26.000    27.644      0.437     -1.644       -1.22

R denotes an obs. with a large st. resid.
```

The fitted *y* values and estimated standard deviations in Table 9-27 can be used in the formulas of Section 9-2 to produce confidence intervals for the four mean responses, prediction intervals, tolerance intervals, and so on based on the "B and C main effects only" model. All of this can be done despite the fact that the data of Table 9-25 are unbalanced.

Example 9-10 has been treated as if the lack of balance in the data came about by misfortune. And the lack of balance in Example 9-9 did come about in such a fashion. It should thus be noted that lack of balance in p-way factorial data can also be the result of careful planning. Consider, for example, a 2^4 factorial situation where one's budget can support collection of 20 observations but not as many as 32. In such a case, complete replication of the 16 combinations of two levels of four factors in order to achieve balance is not possible. But it makes far more sense to replicate only four of the 16 combinations (and thus be able to calculate s_P and honestly assess the size of background variation) than to achieve balance by using no replication. By now it should be obvious how one would subsequently go about the analysis of the resulting partially replicated factorial data.

Unbalanced Factorial Data and ANOVA

Section 8-4 considered the use of ANOVA methods for balanced full factorial data. There it was pointed out that sums of squared fitted effects of various types 1) partition the one-way $SSTr$ into interpretable pieces that can be thought of as "due to" effects of various types and 2) form the bases of F tests of the hypotheses that all effects of a given type are 0 in the full factorial model. (Although it wasn't discussed in Section 8-4, it is also true that those same sums of squares will appear in F tests of the hypotheses that all effects of a given type are 0 in any few-effects model that includes them.)

When factorial data are unbalanced, the sums of squared fitted effects do not turn out to possess these properties; indeed, there is no universal or even wide agreement as to what one ought to mean by symbols like SSA, SSB, $SSAB$, etc. for unbalanced data. Nevertheless, that does not prevent people who write statistical software from creating programs that output numbers for unbalanced factorial data and calling them SSA, SSB, $SSAB$, etc. Such numbers are usually developed by taking the difference between the regression sum of squares for a dummy variable regression model including the corresponding x (or products of x) variables (on the one hand) and the regression sum of squares for that same model with those variables deleted (on the other hand).

This is a reasonably sensible approach, in that one can try to think of sums of squares for effects in terms of additional raw variability accounted for when the effects of a given type are included in the model. The only problem is that for unbalanced data, the additional variation accounted for typically depends on what other effects have already been taken into account. For instance, in Example 9-10, the difference between the regression sums of squares for the models

$$y_i = \beta_0 + \beta_1 x_{2i}^A + \beta_2 x_{2i}^B + \beta_3 x_{2i}^C + \beta_4 x_{2i}^A x_{2i}^B + \beta_5 x_{2i}^A x_{2i}^C + \beta_6 x_{2i}^B x_{2i}^C$$
$$+ \beta_7 x_{2i}^A x_{2i}^B x_{2i}^C + \epsilon_i$$

and

$$y_i = \beta_0 + \beta_1 x_{2i}^A + \beta_3 x_{2i}^C + \beta_4 x_{2i}^A x_{2i}^B + \beta_5 x_{2i}^A x_{2i}^C + \beta_6 x_{2i}^B x_{2i}^C$$
$$+ \beta_7 x_{2i}^A x_{2i}^B x_{2i}^C + \epsilon_i$$

(which would be used for testing the null hypothesis of no B main effects in the full factorial model) does not turn out to be the same as the difference between the regression sums of squares for models

$$y_i = \beta_0 + \beta_2 x_{2i}^B + \beta_3 x_{2i}^C + \epsilon_i$$

and

$$y_i = \beta_0 + \beta_3 x_{2i}^C + \epsilon_i$$

(which would be used for testing the null hypothesis of no B main effects in the "B and C main effects only" model).

Unfortunately, the software documentation of statistical packages often neglects to make clear what the full and reduced models are for the production of unbalanced data factorial sums of squares. There has been a fair amount of haggling among statisticians about what choices make the most sense, but this author has little interest in advocating any particular choice here. For unbalanced factorial data, there is no single most natural partitioning of $SSTr$. Rather than trying to advocate one of many possible partitionings tied to a particular series of progressively larger regression models, the situation will be left here as it stands. One must simply admit that some of the attractive intuition provided by factorial ANOVA fails to carry over to unbalanced data, but note that for a particular full or few-effects factorial model, it *is* possible to use a regression-type ANOVA F test from Section 9-2 to test whether all effects of a given type in that model are 0.

EXERCISES

9-1. (§9-1) Return to the situation of Exercise 4-5 and the polymer molecular weight study of R. Harris.

(a) Find s_{LF} for these data. What does this intend to measure in the context of the engineering problem?

(b) Plot both residuals versus x and the standardized residuals versus x. How much difference is there in the appearance of these two plots?

(c) Give a 90% two-sided confidence interval for the increase in mean average molecular weight that accompanies a 1°C increase in temperature here.

(d) Give individual 90% two-sided confidence intervals for the mean average molecular weight at 212°C and also at 250°C.

(e) Give simultaneous 90% two-sided confidence intervals for the two means indicated in part (d).

(f) Give 90% lower prediction bounds for the next average molecular weight, first at 212°C and then at 250°C.

(g) Give 95% lower tolerance bounds for 90% of average molecular weights, first at 212°C and then at 250°C.

(h) Make an ANOVA table for testing $H_0: \beta_1 = 0$ in the simple linear regression model. What is the p-value here for a two-sided test of this hypothesis?

9-2. (§9-2) Return to the situation of Exercise 4-16 and the carburetion study of Griffith and Tesdall. Consider an analysis of these data based on the model $y = \beta_0 + \beta_1 x + \beta_2 x^2 + \epsilon$.

(a) Find s_{SF} for these data. What does this intend to measure in the context of the engineering problem?

(b) Plot both residuals versus x and the standardized residuals versus x. How much difference is there in the appearance of these two plots?

(c) Give 90% individual two-sided confidence intervals for all of β_0, β_1, and β_2.

(d) Give individual 90% two-sided confidence intervals for the mean elapsed time with a carburetor jetting size of 70 and then with a jetting size of 76.

(e) Give simultaneous 90% two-sided confidence intervals for the two means indicated in part (d).

(f) Give 90% lower prediction bounds for an additional elapsed time with a carburetor jetting size of 70 and also with a jetting size of 76.

(g) Give 95% lower tolerance bounds for 90% of additional elapsed times, first with a carburetor jetting size of 70 and then with a jetting size of 76.

(h) Make an ANOVA table for testing $H_0: \beta_1 = \beta_2 = 0$ in the model $y = \beta_0 + \beta_1 x + \beta_2 x^2 + \epsilon$. What is the meaning of this hypothesis in the context of the study and the quadratic model? What is the p-value?

(i) Use a t statistic and test the null hypothesis $H_0: \beta_2 = 0$. What is the meaning of this hypothesis in the context of the study and the quadratic model?

9-3. (§9-2) Return to the situation of Exercise 4-8, and the chemithermomechanical pulp study of Miller, Shankar, and Peterson. Consider an analysis of the data there based on the model $y = \beta_0 + \beta_1 x_1 + \beta_2 x_2 + \epsilon$.

(a) Find s_{SF}. What does this intend to measure in the context of the engineering problem?

(b) Plot both residuals and standardized residuals versus x_1, x_2, and $\hat{y}$. How much difference is there in the appearance of these pairs of plots?

(c) Give 90% individual two-sided confidence intervals for all of β_0, β_1, and β_2.

(d) Give individual 90% two-sided confidence intervals for the mean specific surface area, first when $x_1 = 9.0$ and $x_2 = 60$ and then when $x_1 = 10.0$ and $x_2 = 70$. (If you are using Minitab, you will need to use the Minitab "PREDICT" subcommand.)

(e) Give simultaneous 90% two-sided confidence intervals for the two means indicated in part (d).

(f) Give 90% lower prediction bounds for the next specific surface area, first when $x_1 = 9.0$ and $x_2 = 60$ and then when $x_1 = 10.0$ and $x_2 = 70$.

(g) Give 95% lower tolerance bounds for 90% of specific surface areas, first when $x_1 = 9.0$ and $x_2 = 60$ and then when $x_1 = 10.0$ and $x_2 = 70$.

(h) Make an ANOVA table for testing $H_0: \beta_1 = \beta_2 = 0$ in the model $y = \beta_0 + \beta_1 x_1 + \beta_2 x_2 + \epsilon$. What is the p-value?

9-4. Return to the situation of Exercise 4-15 and the concrete strength study of Nicholson and Bartle.

(a) Find estimates of the parameters β_0, β_1, and σ in the simple linear regression model $y = \beta_0 + \beta_1 x + \epsilon$. How does your estimate of σ based on the simple linear regression model compare with the pooled sample standard deviation, s_P?

(b) Compute residuals and standardized residuals. Plot both against x and $\hat{y}$ and normal-plot them. How much do the appearances of the plots of the standardized residuals differ from those of the raw residuals?

(c) Make a 90% two-sided confidence interval for the increase in mean compressive strength that accompanies a .1 increase in water/cement ratio. (Note: This is $.1\beta_1$).

(d) Test the hypothesis that the mean compressive strength doesn't depend on the water/cement ratio. What is the p-value?

(e) Make a 95% two-sided confidence interval for the mean strength of specimens with water/cement ratio .5 (based on the simple linear regression model).

(f) Make a 95% two-sided prediction interval for the strength of an additional specimen with water/cement ratio .5 (based on the simple linear regression model).

(g) Make a 95% lower tolerance bound for the strengths of 90% of additional specimens with water/cement ratio .5 (based on the simple linear regression model).

9-5. Return to the situation of Exercise 4-17 and the grain growth study of Huda and Ralph. Consider an analysis of the researchers' data based on the model

$$y = \beta_0 + \beta_1 x_1 + + \beta_2 \ln(x_2) + \beta_3 x_1 \ln(x_2) + \epsilon$$

(a) Use a multiple regression program to fit this model to the data given in Chapter 4. Based on this fit, what is your estimate of the standard deviation of grain size, y, associated with different specimens treated using a fixed temperature and time?

(b) Make a plot of the observed y's versus the corresponding $\ln(x_2)$'s. On this plot, sketch the linear fitted response functions ($\hat{y}$ versus $\ln(x_2)$) for $x_1 = 1443$, 1493, and 1543. Notice that the fit to the researchers' data is excellent. However, notice also that the model has four β's and was fit based on only nine data points. What possibility therefore needs to be kept in mind when making predictions based on this model?

(c) Make a 95% two-sided confidence interval for the mean y when a temperature of $x_1 = 1493°$K and a time of $x_2 = 120$ minutes are used.

(d) Make 95% two-sided prediction interval for an additional grain size, y, when a temperature of $x_1 = 1493°$K and a time of $x_2 = 120$ minutes are used.

(e) Find a 95% two-sided confidence interval for the mean y when a temperature of $x_1 = 1500°$K and a time of $x_2 = 500$ minutes are used. (If you are using Minitab, you will need to use the subcommand "PREDICT" with your basic regression command in order to do this.)

(f) What does the hypothesis H$_0$: $\beta_1 = \beta_2 = \beta_3 = 0$ mean in the context of this study and the model being used in this exercise? Find the p-value associated with an F test of this hypothesis.

(g) What does the hypothesis H$_0$: $\beta_3 = 0$ mean in the context of this study and the model being used in this exercise? Find the p-value associated with a two-sided t test of this hypothesis.

9-6. (§9-3) Flood and Shankwitz reported the results of a metallurgical engineering design project that involved the study of

Time x_1 (min)	Temperature x_2 (°F)	Increase in Hardness, y
5	800	0, 0, −1
5	900	−3, −2, 1
5	1000	−1, −1, 0
5	1100	−4, 1, 3
50	800	3, 4, −1
50	900	−3, −1, 1
50	1000	−4, −1, −3
50	1100	−4, −4, −2
150	800	4, 2, −2
150	900	−1, −1, −2
150	1000	−4, −5, −7
150	1100	−7, −5, −8
500	800	1, −3, 0
500	900	−2, −8, −2
500	1000	−8, −7, −7
500	1100	−11, −9, −5

tempering response of a certain grade of stainless steel. Slugs of this steel were preprocessed to reasonably uniform hardnesses, which were measured and recorded. The slugs were then tempered at various temperatures for various lengths of time. The hardnesses were then remeasured and the change in hardness computed. The data in the accompanying table were obtained in this replicated 4×4 factorial study.

(a) Fit the quadratic model

$$y = \beta_0 + \beta_1 \ln(x_1) + \beta_2 x_2 + \beta_3 (\ln(x_1))^2 + \beta_4 x_2^2$$
$$+ \beta_5 x_2 \ln(x_1) + \epsilon$$

to these data. What fraction of the observed variability in hardness increase is accounted for in the fitting of the quadratic response surface? What is your estimate of the standard deviation of hardness changes that would be experienced at any fixed combination of time and temperature? How does this estimate compare with s_P? Does there appear to be enough difference between the two values to cast serious doubt on the appropriateness of the regression model?

(b) There was some concern on the project group's part that the 5-minute time was completely unlike the other times and should not be considered in the same analysis as the longer times. Temporarily delete the 12 slugs treated only 5 minutes from consideration, refit the quadratic model, and compare fitted values for the 36 slugs tempered longer than 5 minutes for this regression to those from part (a). How different are these two sets of values?

Henceforth consider the quadratic model fitted to all 48 data points.

(c) Make a contour plot showing how y varies with $\ln(x_1)$ and x_2. In particular, use it to identify the region of $\ln(x_1)$ and x_2 values where the tempering seems to provide an increase in hardness. Sketch the corresponding region in the (x_1, x_2)-plane.

(d) For the $x_1 = 50$ and $x_2 = 800$ set of conditions,

    (i) give a 95% two-sided confidence interval for the mean increase in hardness provided by tempering.

    (ii) give a 95% two-sided prediction interval for the increase in hardness produced by tempering an additional slug.

(iii) give a 95% lower tolerance bound for the hardness increases of 90% of such slugs undergoing tempering.

9-7. The article "Orthogonal Design for Process Optimization and its Application in Plasma Etching" by Yin and Jillie (*Solid State Technology*, 1987) discusses a 4-factor experiment intended to guide optimization of a nitride etch process on a single wafer plasma etcher. Data were collected at only nine out of $3^4 = 81$ possible combinations of three levels of each of the four factors (actually making up an *orthogonal array*, a concept discussed in some detail in Section 10-3). The factors involved in the experimentation were the Power applied to the cathode x_1, the Pressure in the reaction chamber x_2, the spacing or Gap between the anode and the cathode x_3, and the Flow of the reactant gas C_2F_6, x_4. Three different responses were measured, an etch rate for SiN y_1, a uniformity for SiN y_2, and a selectivity of the process (for silicon nitride) between silicon nitride and polysilicon y_3. Eight of the nine different combinations were run once, while one combination was run three times. The researchers reported the data given in the accompanying table.

x_1 (W)	x_2 (mTorr)	x_3 (cm)	x_4 (sccm)	y_1 (Å/min)	y_2 (%)	y_3 (SiN/poly)
275	450	0.8	125	1075	2.7	1.63
275	500	1.0	160	633	4.9	1.37
275	550	1.2	200	406	4.6	1.10
300	450	1.0	200	860	3.4	1.58
300	500	1.2	125	561	4.6	1.26
275	450	0.8	125	1052	1.7	1.72
300	550	0.8	160	868	4.6	1.65
325	450	1.2	160	669	5.0	1.42
325	500	0.8	200	1138	2.9	1.69
325	550	1.0	125	749	5.6	1.54
275	450	0.8	125	1037	2.6	1.72

The data are listed in the order in which they were actually collected. Notice that the conditions under which the first, sixth, and eleventh data points were collected are the same — that is, there is some replication in this fractional factorial data set.

(a) The fact that the first, sixth, and last data points were collected under the same set of process conditions provides some check on the consistency of experimental results across time in this study. What else might (should) have been done in this study to try to make sure that time trends in an extraneous variable don't get confused with the effects of the experimental variables (in particular, the effect of x_1, as the experiment was run)? (Consider again the ideas of Section 2-3.)

(b) Fit a model linear in all of x_1, x_2, x_3, and x_4 to each of the three response variables. Notice that although such a model appears to provide a good fit to the y_3 data, the situations for y_1 and y_2 are not quite so appealing. (Compare s_{SF} to s_P for y_1 and note that R^2 for the second variable is relatively low, at least compared to what one can achieve for y_3.)

(c) In search of better-fitting equations for the y_2 (or y_1) data, one might consider fitting a full quadratic equation in x_1, x_2, x_3, and x_4 to the data. What happens when you attempt to do this using a

regression package? (The problem is that the data given here are not adequate to distinguish between various possible quadratic response surfaces in four variables.)

(d) In light of the difficulty experienced in (c), a natural thing to do might be to try to fit quadratic surfaces involving only some of all possible second-order terms. Fit the two models for y_2 including
   (i) $x_1, x_2, x_3, x_4, x_1^2, x_2^2, x_3^2$, and x_4^2 terms
   (ii) $x_1, x_2, x_4, x_1^2, x_2^2, x_4^2, x_1x_2$, and x_2x_4 terms
How do these two fitted equations compare in terms of $\hat{y}_2$ values for (x_1, x_2, x_3, x_4) combinations in the data set? How do $\hat{y}_2$ values compare for the two fitted equations when $x_1 = 325$, $x_2 = 550$, $x_3 = 1.2$, and $x_4 = 200$? (Notice that although this last combination is not in the data set, there are values of the individual variables in the data set matching these.) What is the practical engineering difficulty faced in a situation like this, where there is not enough data available to fit a full quadratic model but it doesn't seem that a model linear in the variables is an adequate description of the response?

Henceforth, confine attention to y_3, and consider an analysis based on a model linear in all of x_1, x_2, x_3, and x_4.

(e) Give a 90% two-sided individual confidence interval for the increase in mean selectivity ratio that accompanies a 1 watt increase in power.

(f) What appear to be the optimal settings of the variables x_1, x_2, x_3, and x_4 (within their respective ranges of experimentation)? Refer to the coefficients of your fitted equation from (b).

(g) Give a 90% two-sided confidence interval for the mean selectivity ratio at the combination of settings that you identified in (f). What cautions would you include in a report in which this interval is to appear? (Under what conditions is your calculated interval going to have real-world meaning?)

9-8. The article "How to Optimize and Control the Wire Bonding Process: Part II" by Scheaffer and Levine (*Solid State Technology*, 1991) discusses the use of a $k = 4$ factor central composite design in the improvement of the operation of the K&S 1484XQ bonder. The effects of the variables Force, Ultrasonic Power, Temperature, and Time on the final ball bond shear strength were studied. The accompanying table gives data like those collected by the authors. (The original data were not given in the paper, but enough information was given to produce these simulated values that have structure like the original data.)

(a) Fit both the full quadratic response surface and the simpler linear response surface to these data. On the basis of simple examination of the the R^2 values, does it appear that the quadratic surface is enough better as a data summary to make it worthwhile to suffer the increased complexity that it brings with it? How do the s_{SF} values for the two fitted models compare to s_P computed from final six data points listed here (repeated center points)?

(b) Conduct a formal test (in the full quadratic model) of the hypothesis that the linear model $y = \beta_0 + \beta_1 x_1 + \beta_2 x_2 + \beta_3 x_3 + \beta_4 x_4 + \epsilon$ is an adequate description of the response. Does your p-value support your qualitative judgment from part (a)?

(c) In the linear model $y = \beta_0 + \beta_1 x_1 + \beta_2 x_2 + \beta_3 x_3 + \beta_4 x_4 + \epsilon$, give a 90% confidence interval for β_2. Interpret this interval in the context of the original engineering problem. (What is β_2 supposed

TABLE For Exercise 9-8

Force x_1 (gm)	Power x_2 (mw)	Temp. x_3 (°C)	Time x_4 (ms)	Strength y (gm)
30	60	175	15	26.2
40	60	175	15	26.3
30	90	175	15	39.8
40	90	175	15	39.7
30	60	225	15	38.6
40	60	225	15	35.5
30	90	225	15	48.8
40	90	225	15	37.8
30	60	175	25	26.6
40	60	175	25	23.4
30	90	175	25	38.6
40	90	175	25	52.1
30	60	225	25	39.5
40	60	225	25	32.3
30	90	225	25	43.0
40	90	225	25	56.0
25	75	200	20	35.2
45	75	200	20	46.9
35	45	200	20	22.7
35	105	200	20	58.7
35	75	150	20	34.5
35	75	250	20	44.0
35	75	200	10	35.7
35	75	200	30	41.8
35	75	200	20	36.5
35	75	200	20	37.6
35	75	200	20	40.3
35	75	200	20	46.0
35	75	200	20	27.8
35	75	200	20	40.3

to measure?) Would you expect the p-value from a test of H_0: $\beta_2 = 0$ to be large or to be small?

(d) Use the linear model and find a 95% lower tolerance bound for 98% of bond shear strengths at the center point $x_1 = 35, x_2 = 75$, $x_3 = 200$, and $x_4 = 20$.

9-9. (Testing for "Lack of Fit" to a Regression Model) In curve- and surface-fitting problems where there is some replication, this text has used the informal comparison of s_{SF} (or s_{LF}) to s_P as a means of detecting poor fit of a regression model. It is actually possible to use these values to conduct a formal significance test for lack of fit. That is, under the one-way normal model of Chapter 7, it is possible to test

$$H_0: \mu_{y|x_1, x_2, ..., x_k} = \beta_0 + \beta_1 x_1 + \beta_2 x_2 + \cdots + \beta_k x_k$$

using the test statistic

$$F = \frac{\dfrac{(n-k-1)s_{SF}^2 - (n-r)s_P^2}{r-k-1}}{s_P^2}$$

and an $F_{r-k-1, n-r}$ reference distribution, where large values of F count as evidence against H_0. (If s_{SF} is much larger than s_P, the difference in the numerator of F will be large, producing a large sample value and a small observed level of significance.)

(a) It is not possible to use the lack of fit test described above in any of Exercises 9-1, 9-2, 9-3, or 9-5. Why?
(b) For the situation of Exercise 9-4, conduct a formal test of lack of fit of the linear relationship $\mu_{y|x} = \beta_0 + \beta_1 x$ to the concrete strength data.
(c) For the situation of Exercise 9-6, conduct a formal test of lack of fit of the full quadratic relationship

$$\mu_{y|x_1, x_2} = \beta_0 + \beta_1 \ln(x_1) + \beta_2 x_2 + \beta_3 (\ln(x_1))^2 + \beta_4 x_2^2 + \beta_5 x_2 \ln(x_1)$$

to the hardness increase data.
(d) For the situation of Exercise 9-8, conduct a formal test of lack of fit of the linear relationship

$$\mu_{y|x_1, x_2, x_3, x_4} = \beta_0 + \beta_1 x_1 + \beta_2 x_2 + \beta_3 x_3 + \beta_4 x_4$$

to the ball bond shear strength data.

9-10. (§9-4) Return to the situation of Exercise 8-19 and the chemical product impurity study. The analysis suggested in that exercise leads to the conclusion that only the A and B main effects are detectably nonzero. The data are unbalanced, so it is not possible to use the reverse Yates algorithm to fit the "A and B main effects only" model to the data.
(a) Use the regression techniques of Section 9-4 to fit the "A and B main effects only" model. (You should be able to pattern what you do after Example 9-10.) How do A and B main effects estimated on the basis of this few-effects/simplified description of the pattern of response compare with what you obtained for fitted effects using the Yates algorithm?
(b) Compute and plot residuals for the few-effects model. (Plot against levels of A, B, and C, against $\hat{y}$, and normal-plot them.) Do any of these plots indicate any problems with the few-effects model?
(c) How does s_{FE} (which you can read directly off your printout as s_{SF}) compare with s_P in this situation? Do the two values carry any strong suggestion of lack of fit?

9-11. Return to the situation of Exercises 4-34 and 4-35 and the ore refining study of S. Osoka. In that study, the object was to discover settings of the process variables x_1 and x_2 that would simultaneously maximize y_1 and minimize y_2.
(a) Fit full quadratic response functions for y_1 and y_2 to the data given in Chapter 4. Compute and plot residuals for these two fitted equations. Comment on the appearance of these plots and what they indicate about the appropriateness of the fitted response surfaces.
(b) Judge the usefulness of the surfaces fitted in part (a) against criterion (9-96) of Section 9-3. Do the response surfaces appear to be determined adequately to support further analysis (involving optimization, for example)?
(c) Use the analytic method discussed in Section 9-3 to investigate the nature of the response surfaces fitted in part (a). According to the signs of the eigenvalues, what kinds of surfaces were fitted to y_1 and y_2, respectively?
(d) Make contour plots of the fitted y_1 and y_2 response surfaces from (a) on a single set of x_1/x_2 axes. Use these to help locate (at

least approximately) a point (x_1, x_2) with maximum predicted y_1, subject to a constraint that predicted y_2 be no larger than 55.

(e) For the point identified in part (d), give 90% two-sided prediction intervals for the next values of y_1 and y_2 that would be produced by this refining process. Also give a 95% lower tolerance bound for 90% of additional pyrite recoveries and a 95% upper tolerance bound for 90% of additional kaolin recoveries at this combination of x_1 and x_2 settings.

9-12. Return to the concrete strength testing situation of Exercise 4-32.

(a) Find estimates of the parameters β_0, β_1, and σ in the simple linear regression model $y = \beta_0 + \beta_1 x + \epsilon$.

(b) Compute standardized residuals and plot them in the same ways that you are asked to plot the ordinary residuals in part (g) of the problem in Chapter 4. How much do the appearances of the new plots differ from the earlier ones?

(c) Make a 95% two-sided confidence interval for the increase in mean compressive strength that accompanies a 5 psi increase in splitting tensile strength. (Note: This is $5\beta_1$.)

(d) Make a 90% two-sided confidence interval for the mean strength of specimens with splitting tensile strength 300 psi (based on the simple linear regression model).

(e) Make a 90% two-sided prediction interval for the strength of an additional specimen with splitting tensile strength 300 psi (based on the simple linear regression model).

(f) Find a 95% lower tolerance bound for the strengths of 90% of additional specimens with splitting tensile strength 300 psi (based on the simple linear regression model).

9-13. Wiltse, Blandin, and Schiesel experimented with a grain thresher built for an agricultural engineering design project. They ran efficiency tests on the cleaning chamber of the machine. This part of the machine sucks air through threshed material, drawing light (non-seed) material out an exhaust port, while the heavier seeds fall into a collection tray. Air flow is governed by the spacing of an air relief door. The following are the weights, y (in grams), of the portions of 14 gram samples of pure oat seeds run through the cleaning chamber that ended up in the collection tray. Four different door spacings x were used, and 20 trials were made at each door spacing.

.500 in. Spacing

12.00, 12.30, 12.45, 12.45, 12.50, 12.50, 12.50, 12.60, 12.65, 12.70, 12.70, 12.80, 12.90, 12.90, 13.00, 13.00, 13.00, 13.10, 13.20, 13.20

.875 in. Spacing

12.40, 12.80, 12.80, 12.90, 12.90, 12.90, 12.90, 13.00, 13.00, 13.00, 13.00, 13.20, 13.20, 13.20, 13.30, 13.40, 13.40, 13.45, 13.45, 13.70

1.000 in. Spacing

12.00, 12.80, 12.80, 12.90, 12.90, 13.00, 13.00, 13.00, 13.15, 13.20, 13.20, 13.30, 13.40, 13.40, 13.45, 13.50, 13.60, 13.60, 13.60, 13.70

1.250 in. Spacing

12.10, 12.20, 12.25, 12.25, 12.30, 12.30, 12.30, 12.40, 12.50, 12.50, 12.50, 12.60, 12.60, 12.85, 12.90, 12.90, 13.00, 13.10, 13.15, 13.25

Use the quadratic model $y = \beta_0 + \beta_1 x + \beta_2 x^2 + \epsilon$ and do the following.

(a) Find an estimate of σ in the model above. What is this supposed to measure? How does your estimate compare to s_P here? What does this comparison suggest to you?

(b) Use an F statistic and test the null hypothesis $H_0: \beta_1 = \beta_2 = 0$. (You may take values off a printout to do this, but show the whole five-step significance-testing format.) What is the meaning of this hypothesis in the present context?

(c) Use a t statistic and test the null hypothesis $H_0: \beta_2 = 0$. (Again, you may take values off a printout to do this, but show the whole five-step significance-testing format.) What is the meaning of this hypothesis in the present context?

(d) Give a 90% lower confidence bound for the mean weight of the part of such samples that would wind up in the collection tray using a 1.000 in. door spacing.

(e) Give a 90% lower prediction bound for the next weight of the part of such a sample that would wind up in the collection tray using a 1.000 in. door spacing.

(f) Give a 95% lower tolerance for 90% of the weights of all such samples that would wind up in the collection tray using a 1.000 in. door spacing.

9-14. Return to the armor testing context of Exercise 4-37. In what follows, base your answers on the model $y = \beta_0 + \beta_1 x_1 + \beta_2 x_2 + \epsilon$.

(a) Use a multiple regression program to fit this model to the data given in Chapter 4. Based on this fit, what is your estimate of the standard deviation of ballistic limit, y, associated with different specimens of a given thickness and Brinell hardness?

(b) Find and plot the standardized residuals. (Plot them versus x_1, versus x_2, and versus $\hat{y}$ and normal-plot them.) Comment on the appearance of your plots.

(c) Make 90% two-sided confidence intervals for β_1 and for β_2. Based on the second of these, what increase in mean ballistic limit would you expect to accompany a 20-unit increase in Brinell hardness number?

(d) Make a 95% two-sided confidence interval for the mean ballistic limit when a thickness of $x_1 = 258$ (.001 in.) and a Brinell hardness of $x_2 = 391$ are involved.

(e) Make a 95% two-sided prediction interval for an additional ballistic limit when a thickness of $x_1 = 258$ (.001 in.) and a Brinell hardness of $x_2 = 391$ are involved.

(f) Find a 95% lower tolerance bound for 98% of additional ballistic limits when a thickness of $x_1 = 258$ (.001 in.) and a Brinell hardness of $x_2 = 391$ are involved.

(g) Find a 95% two-sided confidence interval for the mean ballistic limit when a thickness of $x_1 = 260$ (.001 in.) and a Brinell hardness of $x_2 = 380$ are involved.

(h) What does the hypothesis $H_0: \beta_1 = \beta_2 = 0$ mean in the context of this study and the model being used in this exercise? Find the p-value associated with an F test of this hypothesis.

(i) What does the hypothesis $H_0: \beta_1 = 0$ mean in the context of this study and the model being used in this exercise? Find the p-value associated with a two-sided t test of this hypothesis.

10 Inference for Multisample Studies with Specialized Structure

T he last two chapters presented methods of statistical inference associated with 1) the factorial effects that can be used to describe full factorial data sets and 2) the curves and surfaces that are sometimes useful for summarizing patterns in system responses when all factors of interest are quantitative. These methods are widely applicable in multifactor statistical engineering studies. This chapter introduces several additional methods of statistical analysis that are often extremely useful in engineering problems, but that are somewhat more specialized than the methods of Chapters 8 and 9.

The chapter begins with three sections treating the analysis of data from fractional factorial studies. The first section treats the problems of choosing $\frac{1}{2}$ fractions of 2^p factorial studies and then analyzing the data generated. A follow-up section treats these same issues for general $\frac{1}{2}, \frac{1}{4}, \frac{1}{8}, \ldots$, etc. fractions of 2^p studies. Following the coverage of these 2^{p-q} fractional factorials is a presentation of the analysis of data from some other (orthogonal array) fractional factorial data structures, with particular attention to the Latin squares. Next, ANOVA-based estimation of variance components in balanced hierarchical studies is considered. The chapter closes with a discussion of data collection and analysis methods that are appropriate where system behavior is thought to be determined by the proportions in which several different components appear in a mixture.

10-1 Standard Fractions of Two-Level Factorials, Part I: $\frac{1}{2}$ Fractions

The notion of a fractional factorial data structure was first introduced in Section 1-2. As yet, this text has done little to indicate either how such a structure might be chosen or how analysis of fractional factorial data might proceed. The delay in getting to these topics is a reflection of their subtle nature rather than any lack of importance. Indeed, the technology of fractional factorial experimentation and analysis is one of the most important tools in the modern engineer's kit. This is especially true where many factors potentially affect a response variable and there is little *a priori* knowledge about the relative impacts of these factors.

This section and the next treat the (standard) 2^{p-q} fractional factorials — the class of fractional factorial structures for which advantageous methods of data collection and analysis can be presented most easily and completely. These structures, involving $\frac{1}{2^q}$ of all possible combinations of levels of p two-level factors, are among the most useful fractional factorial designs for application in engineering experimentation. In addition, they clearly illustrate the general issues that arise anytime only a fraction of a complete factorial set of factor-level combinations can be included in a multifactor study.

This section begins with some general qualitative remarks about fractional factorial experimentation. The standard $\frac{1}{2}$ fractions of 2^p studies (the 2^{p-1} fractional factorials) are then discussed in detail. The section covers, in turn, the proper choice of such fractions, the resultant aliasing or confounding patterns, and corresponding methods of data analysis. The section closes with a few remarks about qualitative issues, addressed to the practical use of 2^{p-1} designs.

General Observations about Fractional Factorial Studies

In many of the physical systems engineers work on, there are many factors potentially affecting a response y. In such cases, even when the number of levels considered for each factor is only two, there can be an impossible number of different combinations of levels of the factors to consider. For instance, if $p = 10$ factors are to be considered, even when limiting attention to only two levels of each factor, one would be faced with collecting at least $2^{10} = 1024$ data points in order to complete a full factorial study. In most engineering contexts, restrictions on time and other resources would make a study of this size infeasible. One could try to guess which few factors are most important in determining the response y and do a smaller complete factorial study on those factors (holding the levels of the remaining factors fixed). But there would obviously be a risk of guessing wrong as to which factors are most important and therefore failing to discover the real pattern of how factors affect the response.

A superior alternative is to conduct one's investigation in at least two stages. A relatively small screening study (or several of them), intended to identify those factors most likely influencing the response, can be done first. This can be followed up with a more detailed study (or studies) in those variables, intended to make clear the pattern of their influence on the response. It is in the initial screening phase of such a program of engineering experimentation that small fractions of 2^p studies are most appropriate. Tools such as larger fractions, full factorials, and response surface designs like the central composite designs, are appropriate for the later stage (or stages) of study.

Once the reality of resource limitations has made an investigator consider fractional factorial experimentation, several qualitative points should become clear. For one thing, it should be obvious that one cannot hope to learn as much from a fraction of a full factorial study as from the full factorial itself. (There is no Santa Claus, who for the price of eight observations will give one as much information as can be obtained from 16.) What in fact happens is that fractional factorial experiments inevitably leave some ambiguity in the interpretation of their results. Through careful planning of exactly which fraction of a full factorial to use, one tries to hold the ambiguity to a minimum and to make sure it is of a type that is most tolerable, recognizing that some ambiguity is unavoidable. Not all fractions of a given size from a particular full factorial study have the same potential for producing useful information.

EXAMPLE 10-1

Choosing Half of a 2^2 Factorial Study. As a completely artificial but nevertheless instructive example of the preceding points, suppose that two factors A and B each have two levels (low and high) and that instead of conducting a full 2^2 factorial study, one contemplates collecting a response or responses at only $\frac{1}{2}$ of the four possible combinations

$$(1), \text{ a, b, and ab}$$

That is, only two of the four combinations are to be represented in a fractional factorial study.

If (1) is chosen as one of the two combinations to be studied, two of the three possible choices of the other combination can fairly obviously be eliminated from consideration as the second combination to include in the study. The possibility of studying the combinations

$$(1) \text{ and a}$$

is no good, since in both cases the factor B is held at its low level. Therefore, no information at all would be obtained on B's impact on the response variable. Similarly, the possibility of studying the combinations

$$(1) \text{ and b}$$

can be eliminated, since no information would be obtained on factor A's impact on the response. So that leaves only the set of combinations

$$(1) \text{ and } ab$$

as a $\frac{1}{2}$ fraction of the full 2^2 factorial that is at all sensible (if combination (1) is to be included). Similar reasoning eliminates all other pairs of combinations from potential use except the pair

$$a \text{ and } b$$

But now notice that any experiment that includes only combinations

$$(1) \text{ and } ab$$

or combinations

$$a \text{ and } b$$

must inevitably produce somewhat ambiguous results. Since one moves from combination (1) to combination ab (or from a to b) by changing levels of *both* factors, if a large difference in response is observed, it will not be clear whether the difference is due to A, due to B, or even due to AB interaction.

At least in qualitative terms, such is the nature of all fractional factorial studies. Although very poor choices of experimental combinations may be avoided, one must be prepared to accept some level of ambiguity as the price for not conducting a full factorial study.

EXAMPLE 10-2

Half of a Hypothetical 2^2 Factorial. As a second hypothetical but instructive example of the issues that must be dealt with in fractional factorial experimentation, consider an imaginary engineering system whose behavior is governed principally by the levels of three factors: A, B, and C. (For the sake of concreteness, suppose that A is a temperature, B is a pressure, and C is a catalyst type, and that the effects of these on the yield y of a chemical process are under consideration.) Suppose further that in a 2^3 study of this system, the factorial effects on an underlying mean response μ are given by

$$\mu_{...} = 10, \qquad \alpha_2 = 3, \qquad \beta_2 = 1, \qquad \gamma_2 = 2,$$
$$\alpha\beta_{22} = 2, \qquad \alpha\gamma_{22} = 0, \qquad \beta\gamma_{22} = 0, \qquad \alpha\beta\gamma_{222} = 0$$

Either through the use of the reverse Yates algorithm or otherwise, it is possible to verify that corresponding to these effects are then the eight combination means

$$\mu_{(1)} = 6, \qquad \mu_a = 8, \qquad \mu_b = 4, \qquad \mu_{ab} = 14,$$
$$\mu_c = 10, \qquad \mu_{ac} = 12, \qquad \mu_{bc} = 8, \qquad \mu_{abc} = 18$$

Now imagine that for some reason, only four of the eight combinations of levels of A, B, and C will be included in an engineering study of this system, and that in fact one decides (by means to be discussed later) to include the combinations

$$a, \ b, \ c, \ \text{and } abc$$

in the study. Suppose further that the background noise is negligible, so that observations for a given treatment combination are essentially equal to the corresponding underlying mean. Then one essentially knows the values of

$$\mu_a = 8, \qquad \mu_b = 4, \qquad \mu_c = 10, \qquad \mu_{abc} = 18$$

FIGURE 10-1

2^3 Hypothetical Means, with Four Known Means Circled

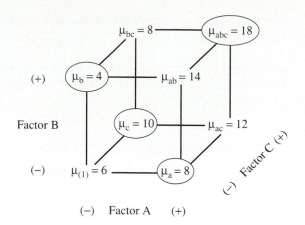

Figure 10-1 shows the complete set of eight combination means laid out in the usual way on the corners of a cube, with the four observed means circled.

As a sidelight, note the admirable symmetry possessed by the four circled corners on Figure 10-1. Each face of the cube has two circled corners (both levels of all factors appear twice in the choice of treatment combinations). Each edge has one circled corner (each combination of all pairs of factors appears once). And collapsing the cube in any one of the three possible directions (left to right, top to bottom, or front to back) would give a full factorial set of four combinations. (Ignoring the level of any one of A, B, or C in the four combinations a, b, c, and abc gives a full factorial in the other two factors.)

But now consider what an engineer possessing only the values of μ_a, μ_b, μ_c, and μ_{abc} might be led to conclude about the system. In particular, begin with the matter of evaluating an A main effect. Recall from Definition 8-4 that

$$\alpha_2 = \mu_{2..} - \mu_{...}$$

$$= \left(\begin{array}{l}\text{the average of all four mean}\\\text{responses where A is at its}\\\text{second or high level}\end{array}\right) - \left(\begin{array}{l}\text{the grand average of all}\\\text{eight mean responses}\end{array}\right)$$

which can be thought of as the right face average minus the grand average for the cube in Figure 10-1. Armed only with the four means μ_a, μ_b, μ_c, and μ_{abc} (the four circled corners on Figure 10-1), it is of course not possible to compute α_2. But what one might try to do is to make a calculation similar to the one that produces α_2 using only the available means. That is, one might try calculating

$$\alpha_2^* = \text{a "}\tfrac{1}{2} \text{ fraction A main effect"}$$

$$= \left(\begin{array}{l}\text{the average of the available}\\\text{two means where A is at its}\\\text{high level}\end{array}\right) - \left(\begin{array}{l}\text{the grand average of the}\\\text{available four means}\end{array}\right)$$

$$= \frac{1}{2}(\mu_a + \mu_{abc}) - \frac{1}{4}(\mu_a + \mu_b + \mu_c + \mu_{abc})$$

$$= \frac{1}{2}(8 + 18) - \frac{1}{4}(8 + 4 + 10 + 18)$$

$$= 13 - 10$$

$$= 3$$

And, amazingly enough, $\alpha_2^* = \alpha_2$ here.

It appears that, using only four combinations, one can learn as much about the A main effect as if all eight combination means were in hand! This, of course, is too good to be true in general, as can be illustrated if one tries to make a parallel calculation for a C main effect:

$$\gamma_2^* = \text{a "}\tfrac{1}{2}\text{ fraction C main effect"}$$

$$= \left(\begin{array}{c}\text{the average of the two}\\ \text{available means where}\\ \text{C is at its high level}\end{array}\right) - \left(\begin{array}{c}\text{the grand average of the}\\ \text{four available means}\end{array}\right)$$

$$= \frac{1}{2}(\mu_c + \mu_{abc}) - \frac{1}{4}(\mu_a + \mu_b + \mu_c + \mu_{abc})$$

$$= 4$$

Unfortunately, this hypothetical example began with $\gamma_2 = 2$. Here, the $\frac{1}{2}$ fraction calculation gives something quite different from the full factorial calculation.

The key to understanding how one can apparently get something for nothing in the case of the A main effects in this example, but cannot do so in the case of the C main effects, is to know that (in general) for this $\frac{1}{2}$ fraction,

$$\alpha_2^* = \alpha_2 + \beta\gamma_{22}$$

and

$$\gamma_2^* = \gamma_2 + \alpha\beta_{22}$$

Since this numerical example began with $\beta\gamma_{22} = 0$, one is "fortunate" — it turns out numerically that $\alpha_2^* = \alpha_2$. On the other hand, since $\alpha\beta_{22} = 2 \neq 0$, one is "unfortunate" — it turns out numerically that $\gamma_2^* = \gamma_2 + 2 \neq \gamma_2$.

As it happens, relationships like these for α_2^* and γ_2^* hold for all $\frac{1}{2}$ fraction versions of the full factorial effects. These relationships can be thought of as detailing the nature of the ambiguity inherent in the use of the $\frac{1}{2}$ fraction of the full 2^3 factorial set of combinations. Essentially, based on data from four out of eight possible combinations, one will be unable to distinguish between certain pairs of effects, such as the A main effect and BC 2-factor interaction pair here.

━━━━━━━━━━━━━━━━━━   ▬ ▬ ▬

Choice of Standard $\frac{1}{2}$
Fractions of 2^p Studies

The use of standard 2^{p-q} fractional factorial data structures depends on having answers for the following three basic questions.

1. How does one rationally choose $\frac{1}{2^q}$ of 2^p possible combinations of factor levels to include in a study?

2. How does one determine the pattern of ambiguities implied by a given choice of 2^{p-q} combinations?

3. How does one do data analysis for a particular choice of 2^{p-q} combinations?

These questions will be answered in this section for the case of $\frac{1}{2}$ fractions (2^{p-1} fractional factorials) and for general q in the next section.

In order to arrive at what is in some sense a best possible choice of $\frac{1}{2}$ of 2^p combinations of levels of p factors, one may do the following. For the first $p - 1$ factors, write out (in a kind of truth table fashion) all 2^{p-1} possible combinations of these factors. By multiplying plus and minus signs (thinking of multiplying plus and minus 1's) corresponding to levels of the first factors, one then arrives at a set of plus and minus signs that can be used to prescribe how to choose levels for the last factor (to be used in combination with the indicated levels of the first $p - 1$ factors).

EXAMPLE 10-3

A 2^{5-1} Chemical Process Experiment. In his article "Experimenting with a Large Number of Variables" (ASQC Technical Supplement *Experiments in Industry*, 1985), R. Snee discusses a successful 2^{5-1} experiment on a chemical process, where the response of interest, y, was a coded color index of the product. The factors studied and their levels were as in Table 10-1.

TABLE 10-1

Five Chemical Process Variables and Their Experimental Levels

Factor	Process Variable	Factor Levels
A	Solvent/Reactant	low $(-)$ vs. high $(+)$
B	Catalyst/Reactant	.025 $(-)$ vs. .035 $(+)$
C	Temperature	150°C $(-)$ vs. 160°C $(+)$
D	Reactant Purity	92% $(-)$ vs. 96% $(+)$
E	pH of Reactant	8.0 $(-)$ vs. 8.7 $(+)$

The standard recommendation for choice of a $\frac{1}{2}$ fraction was followed in Snee's study. Table 10-2 shows an appropriate set of 16 lines of plus and minus signs for generating the $\frac{1}{2} \cdot 32 = 16$ combinations included in Snee's study. The first four columns of this table can be thought of as specifying levels of factors A, B, C, and D for the $16 = 2^4$ possible combinations of levels of these factors (written down in Yates standard order). (The first line, for example, indicates the low level of all of these first four factors.) The last column of this table is obtained by multiplying the first four plus or minus signs (plus or minus 1's) in a given row. It is this last column that can be used to determine how to choose a level of factor E for use when the factors A through D are at the levels indicated in the first four columns.

TABLE 10-2

Signs for Specifying a Standard 2^{5-1} Fractional Factorial

A	B	C	D	ABCD Product
−	−	−	−	+
+	−	−	−	−
−	+	−	−	−
+	+	−	−	+
−	−	+	−	−
+	−	+	−	+
−	+	+	−	+
+	+	+	−	−
−	−	−	+	−
+	−	−	+	+
−	+	−	+	+
+	+	−	+	−
−	−	+	+	+
+	−	+	+	−
−	+	+	+	−
+	+	+	+	+

In Snee's study, the signs in the ABCD Product column were used without modification to specify levels of E. The corresponding treatment combination names (written in the same order as in Table 10-2) and the data reported by Snee are given in Table 10-3.

Notice that the 16 combinations listed in Table 10-3 are $\frac{1}{2}$ of the $2^5 = 32$ possible combinations of levels of these five factors. (They are in fact those 16 that have an odd number of factors appearing at their high levels).

TABLE 10-3

16 Combinations and Observed Color Indices in Snee's 2^{5-1} Study

Combination	Color Index, y
e	−.63
a	2.51
b	−2.68
abe	−1.66
c	2.06
ace	1.22
bce	−2.09
abc	1.93
d	6.79
ade	6.47
bde	3.45
abd	5.68
cde	5.22
acd	9.38
bcd	4.30
abcde	4.05

EXAMPLE 10-4

A 2^{5-1} Agricultural Engineering Study. The article "An Application of Fractional Factorial Experimental Designs" by Mary Kilgo (*Quality Engineering*, 1988) provides an interesting complement to the previous example. In one part of an agricultural engineering study concerned with the use of carbon dioxide at very high pressures to extract oil from peanuts, the effects of five factors on a percent yield variable y were studied in a 2^{5-1} fractional factorial experiment. The five factors and their levels (as named in Kilgo's article) are given in Table 10-4.

TABLE 10-4

Five Peanut Processing Variables and Their Experimental Levels

Factor	Process Variable	Factor Levels
A	Pressure	415 bars (−) vs. 550 bars (+)
B	Temperature	25°C (−) vs. 95°C (+)
C	Peanut Moisture	5% (−) vs. 15% (+)
D	Flow Rate	40 l/min (−) vs. 60 l/min (+)
E	Average Particle Size	1.28 mm (−) vs. 4.05 mm (+)

Interestingly enough, rather than studying the 16 combinations obtainable using the final column of Table 10-2 directly, Kilgo *switched* all of the signs in the ABCD product column before assigning levels of E. This leads to the use of "the other" 16 out of 32 possible combinations (in fact, those having an even number of factors appearing at their high levels). The 16 combinations studied and corresponding responses reported by Kilgo are given in Table 10-5 in the same order for factors A through D as in Table 10-2.

The difference between the combinations listed in Tables 10-3 and 10-5 deserves some thought. As Kilgo named the factor levels, the two lists of combinations are quite different. But verify that if she had made the slightly less natural but nevertheless permissible choice to call the 4.05 mm level of factor E low (−) level and the 1.28 mm level the high (+) level, the names of the physical combinations actually studied would be exactly those in Table 10-3 rather than those in Table 10-5.

The point here is that due to the rather arbitrary nature of how one chooses to name high and low levels of two factors, the names of different physical combinations are themselves to some extent arbitrary. In choosing fractional factorials, one chooses some particular naming convention and then has the freedom to choose levels of the

TABLE 10-5

16 Combinations and Observed Yields in Kilgo's 2^{5-1} Study

Combination	Yield, y (%)
(1)	63
ae	21
be	36
ab	99
ce	24
ac	66
bc	71
abce	54
de	23
ad	74
bd	80
abde	33
cd	63
acde	21
bcde	44
abcd	96

last factor (or factors for $q > 1$ cases) by either using the product column(s) directly or after switching signs. The decision whether or not to switch signs does affect exactly which physical combinations will be run and thus how the data should be interpreted in the subject matter context. But generally, the different possible choices (to switch or not switch signs) are *a priori* equally attractive. For systems that happen to have relatively simple structure, all possible results of these arbitrary choices typically lead to similar engineering conclusions. When systems turn out to have complicated structures, the whole notion of fractional factorial experimentation loses its appeal. Different arbitrary choices lead to different perceptions of system behavior, none of which (usually) correctly portrays the complicated real situation.

Aliasing in the Standard $\frac{1}{2}$ Fractions

Once one has chosen a $\frac{1}{2}$ fraction of a 2^p study, the next problem to be faced is determining the nature of the ambiguities that must arise from its use. It turns out that for 2^{p-1} data structures of the type described here, one can begin with a kind of statement of how the fractional factorial plan was derived, and through a system of *formal multiplication* arrive at an understanding of which (full) factorial effects cannot be separated on the basis of the fractional factorial data. Some terminology is given next, in the form of a formal definition.

DEFINITION 10-1

When it is only possible to estimate the sum (or difference) of two or more (full) factorial effects on the basis of data from a fractional factorial, those effects are said to be **aliased** or **confounded** and are sometimes called **aliases**. In this text, the phrase **alias structure** of a fractional factorial plan will be used to mean a complete specification of all sets of aliased effects.

As an example of the use of this terminology, return to Example 10-2. There, since it is possible only to estimate $\alpha_2 + \beta\gamma_{22}$, not either of α_2 or $\beta\gamma_{22}$ individually, the A main effect is confounded with (or aliased with) the BC 2-factor interaction. The A main effect and the BC 2-factor interaction are aliases.

The way the system of formal multiplication works for detailing the alias structure of one of the recommended 2^{p-1} factorials is as follows. One begins by writing

$$\begin{pmatrix} \text{the name of the} \\ \text{last factor} \end{pmatrix} \leftrightarrow \pm \begin{pmatrix} \text{the product of names of} \\ \text{the first } p-1 \text{ factors} \end{pmatrix} \qquad \textbf{(10-1)}$$

where the plus or minus sign is determined by whether the signs were left alone or switched in the specification of levels of the last factor. The double arrow in expression (10-1) will be read as "is aliased with." And since expression (10-1) really says how the fractional factorial under consideration was chosen, expression (10-1) will be called the plan's *generator*. The generator (10-1) for a 2^{p-1} plan says that the (high level) main effect of the last factor will be aliased with plus or minus the (all factors at their high levels) $p-1$ factor interaction of the first $p-1$ factors.

EXAMPLE 10-3
(continued)

In Snee's 2^{5-1} study, the generator

$$\text{E} \leftrightarrow \text{ABCD}$$

was used. Therefore the (high level) E main effect is aliased with the (all high levels) ABCD 4-factor interaction. That is, only $\epsilon_2 + \alpha\beta\gamma\delta_{2222}$ can be estimated based on the $\frac{1}{2}$ fraction data, not either of its summands individually.

EXAMPLE 10-4
(continued)

In Kilgo's 2^{5-1} study, the generator

$$\text{E} \leftrightarrow -\text{ABCD}$$

was used. The (high level) E main effect is aliased with minus the (all high levels) ABCD 4-factor interaction. That is, only $\epsilon_2 - \alpha\beta\gamma\delta_{2222}$ can be estimated based on the $\frac{1}{2}$ fraction data, not either of the terms individually.

The entire alias structure for a $\frac{1}{2}$ fraction follows from the generator (10-1) by multiplying both sides of the expression by various factor names, using two special conventions. These are that any letter multiplied by itself produces the symbol "I" and that any letter multiplied by "I" is that letter again. Applying the first of these conventions to expression (10-1), one might multiply both sides of the expression by the name of the last factor and have the relation

$$\text{I} \leftrightarrow \pm \text{ the product of names of all } p \text{ factors} \qquad \textbf{(10-2)}$$

Expression (10-2) is interpreted as meaning that the grand mean is aliased with plus or minus the (all factors at their high level) p-factor interaction. There is further special terminology for an expression like that in display (10-2).

DEFINITION 10-2

The list of all aliases of the grand mean for a 2^{p-q} fractional factorial is called the **defining relation** for the design.

By first translating a generator (or generators in the case of $q > 1$) into a defining relation and then multiplying through the defining relation by a product of letters corresponding to an effect of interest, one can immediately identify all aliases of that effect.

EXAMPLE 10-3
(continued)

In Snee's 2^{5-1} experiment, the generator was

$$E \leftrightarrow ABCD$$

When multiplied through by E, this gives the experiment's defining relation

$$I \leftrightarrow ABCDE \qquad \text{(10-3)}$$

which indicates that the grand mean $\mu_{....}$ is aliased with the 5-factor interaction $\alpha\beta\gamma\delta\epsilon_{22222}$. Then, for example, multiplying through defining relation (10-3) by the product AC produces the relationship

$$AC \leftrightarrow BDE$$

Thus, the AC 2-factor interaction is aliased with the BDE 3-factor interaction. In fact, the entire alias structure for the Snee study can be summarized in terms of the aliasing of 16 different pairs of effects. These are indicated in Table 10-6, which was developed by using the defining relation (10-3) to find successively (in Yates order) the aliases of all effects involving only factors A, B, C, and D.

TABLE 10-6

The Complete Alias Structure for Snee's 2^{5-1} Study

I ↔ ABCDE
A ↔ BCDE
B ↔ ACDE
AB ↔ CDE
C ↔ ABDE
AC ↔ BDE
BC ↔ ADE
ABC ↔ DE
D ↔ ABCE
AD ↔ BCE
BD ↔ ACE
ABD ↔ CE
CD ↔ ABE
ACD ↔ BE
BCD ↔ AE
ABCD ↔ E

From Table 10-6, one sees that for Snee's study, main effects are confounded with 4-factor interactions and 2-factor interactions with 3-factor interactions. This degree of ambiguity is in fact as mild as is possible in a 2^{5-1} study.

EXAMPLE 10-4
(continued)

In Kilgo's peanut oil extraction study, since the generator is $E \leftrightarrow -ABCD$, the defining relation is $I \leftrightarrow -ABCDE$, and the alias structure is that given in Table 10-6, except that a minus sign should be inserted on one side or the other of every row of the table. So, for example, one may estimate $\alpha\beta_{22} - \gamma\delta\epsilon_{222}$ but neither $\alpha\beta_{22}$ nor $\gamma\delta\epsilon_{222}$ separately, based on Kilgo's data.

Data Analysis for 2^{p-1} Fractional Factorials

Once one understands the alias structure of a 2^{p-1} fractional factorial, the question of how to analyze data from such a study has a simple answer. One temporarily ignores the last factor and (using the Yates algorithm or otherwise) computes and somehow judges the statistical significance and apparent real importance of the "effects" computed for the complete factorial in $p - 1$ two-level factors. Then one seeks a plausible simple interpretation of these, recognizing that they are estimates not of the effects in the first $p - 1$ factors alone, but of those effects plus their aliases. Where some replication is available, the judging of statistical significance can be done through the use of confidence intervals. Where all 2^{p-1} samples are of size 1, the device of normal-plotting fitted (sums of) effects is usual.

EXAMPLE 10-3
(continued)

Consider the analysis of Snee's data given in Table 10-3. The data in Table 10-3 are listed in Yates standard order for factors A, B, C, and D (if ones ignores the existence of factor E). Then, according to the prescription for analysis just given, the first step to be taken is to use the Yates algorithm (for four factors) on the data. These calculations are summarized in Table 10-7.

TABLE 10-7

The Yates Algorithm for a 2^4 Factorial Applied to Snee's 2^{5-1} Data

y	Cycle 1	Cycle 2	Cycle 3	Cycle 4	Cycle 4 ÷ 16	Sum Estimated
$-.63$	1.88	-2.46	$.66$	46.00	2.875	$\mu + \alpha\beta\gamma\delta\epsilon_{22222}$
2.51	-4.34	3.12	45.34	13.16	$.823$	$\alpha_2 + \beta\gamma\delta\epsilon_{2222}$
-2.68	3.28	22.39	7.34	-20.04	-1.253	$\beta_2 + \alpha\gamma\delta\epsilon_{2222}$
-1.66	$-.16$	22.95	5.82	$.88$	$.055$	$\alpha\beta_{22} + \gamma\delta\epsilon_{222}$
2.06	13.26	4.16	-9.66	6.14	$.384$	$\gamma_2 + \alpha\beta\delta\epsilon_{2222}$
1.22	9.13	3.18	-10.38	1.02	$.064$	$\alpha\gamma_{22} + \beta\delta\epsilon_{222}$
-2.09	14.60	1.91	2.74	$.66$	$.041$	$\beta\gamma_{22} + \alpha\delta\epsilon_{222}$
1.93	8.35	3.91	-1.86	$.02$	$.001$	$\alpha\beta\gamma_{222} + \delta\epsilon_{22}$
6.79	3.14	-6.22	5.58	44.68	2.793	$\delta_2 + \alpha\beta\gamma\epsilon_{2222}$
6.47	1.02	-3.44	$.56$	-1.52	$-.095$	$\alpha\delta_{22} + \beta\gamma\epsilon_{222}$
3.45	$-.84$	-4.13	$-.98$	$-.72$	$-.045$	$\beta\delta_{22} + \alpha\gamma\epsilon_{222}$
5.68	4.02	-6.25	2.00	-4.60	$-.288$	$\alpha\beta\delta_{222} + \gamma\epsilon_{22}$
5.22	$-.32$	-2.12	2.78	-5.02	$-.314$	$\gamma\delta_{22} + \alpha\beta\epsilon_{222}$
9.38	2.23	4.86	-2.12	2.98	$.186$	$\alpha\gamma\delta_{222} + \beta\epsilon_{22}$
4.30	4.16	2.55	6.98	-4.90	$-.306$	$\beta\gamma\delta_{222} + \alpha\epsilon_{22}$
4.05	$-.25$	-4.41	-6.96	-13.94	$-.871$	$\alpha\beta\gamma\delta_{2222} + \epsilon_2$

Each entry in the final column of Table 10-7 gives the name of the effect that the corresponding numerical value in the "Cycle 4 ÷ 16" column would be estimating if factor E weren't present, *plus* the alias of that effect. The numbers in the next-to-last column must be interpreted in light of the fact that they are estimating sums of 2^5 factorial effects.

Since there is no replication indicated in Table 10-3, one can only turn to the technique of normal-plotting fitted (sums of) effects to try to identify those that are distinguishable from noise. Figure 10-2 is a normal plot of the last 15 entries of the Cycle 4 ÷ 16 column of Table 10-7. (Since in most contexts one is *a priori* willing to grant that the overall mean response is other than 0, the estimate of it plus its alias(es) is rarely included in such a plot.)

Depending upon how one draws one's line through the small estimated (sums of) effects in Figure 10-2, the estimate corresponding to D + ABCE, and possibly B + ACDE, E + ABCD, and A + BCDE as well, are seen to be distinguishable in

FIGURE 10-2

Normal Plot of Estimated Sums of Effects in Snee's 2^{5-1} Study

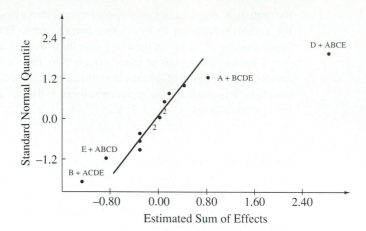

magnitude from the others. (The line in Figure 10-2 has been drawn in keeping with the view that there are four statistically detectable sums of effects, primarily because a half-normal plot of the absolute values of the estimates — not included here — supports that view.) If one adopts the view that there are indeed four detectable (sums of) effects indicated by Figure 10-2, it is clear that the simplest possible interpretation of this outcome is that in fact the four large estimates are each reflecting primarily the corresponding main effects (and not the aliased 4-factor interactions). That is, a tentative (because of the incomplete nature of fractional factorial information) description of the chemical process is that D (reactant purity), B (catalyst/reactant), A (solvent/reactant), and E (pH of reactant) main effects are (in that order) the principal determinants of product color.

Depending on the engineering objectives for product color index y, this tentative description of the system could have several possible interpretations. If large y were desirable, the high levels of A and D and low levels of B and E appear most attractive. If small y were desirable, the situation would be reversed. But in fact, Snee's study was done not to figure out how to maximize or minimize y, but rather to determine how to reduce variation in y. The engineering implications of the "D, B, A, and E main effects only" tentative system description are thus to focus attention on the need to control variation first in level of factor D (reactant purity), then in level of factor B (catalyst/reactant), then in level of factor A (solvent/reactant), and finally in level of factor E (pH of reactant).

EXAMPLE 10-4
(continued)

Verify that for Kilgo's data in Table 10-5, use of the (four-cycle) Yates algorithm on the data as listed (in standard order for factors A, B, C, and D, ignoring factor E) produces the estimated (differences of) effects given in Table 10-8.

TABLE 10-8

Estimated Differences of 2^5 Factorial Effects from Kilgo's 2^{5-1} Study

Value	Difference Estimated	Value	Difference Estimated
54.3	$\mu_{\ldots} - \alpha\beta\gamma\delta\epsilon_{22222}$	0.0	$\delta_2 - \alpha\beta\gamma\epsilon_{2222}$
3.8	$\alpha_2 - \beta\gamma\delta\epsilon_{2222}$	−2.0	$\alpha\delta_{22} - \beta\gamma\epsilon_{222}$
9.9	$\beta_2 - \alpha\gamma\delta\epsilon_{2222}$	−.9	$\beta\delta_{22} - \alpha\gamma\epsilon_{222}$
2.6	$\alpha\beta_{22} - \gamma\delta\epsilon_{222}$	−3.1	$\alpha\beta\delta_{222} - \gamma\epsilon_{22}$
.6	$\gamma_2 - \alpha\beta\delta\epsilon_{2222}$	1.1	$\gamma\delta_{22} - \alpha\beta\epsilon_{222}$
.6	$\alpha\gamma_{22} - \beta\delta\epsilon_{222}$	.1	$\alpha\gamma\delta_{222} - \beta\epsilon_{22}$
1.5	$\beta\gamma_{22} - \alpha\delta\epsilon_{222}$	3.5	$\beta\gamma\delta_{222} - \alpha\epsilon_{22}$
1.8	$\alpha\beta\gamma_{222} - \delta\epsilon_{22}$	22.3	$\alpha\beta\gamma\delta_{2222} - \epsilon_2$

The last 15 of these estimated differences are normal-plotted in Figure 10-3. It is evident from the figure that the two estimated (differences of) effects corresponding to

$$\beta_2 - \alpha\gamma\delta\epsilon_{2222} \quad \text{and} \quad \alpha\beta\gamma\delta_{2222} - \epsilon_2$$

are significantly larger than the other 13 estimates. The simplest possible interpretation of this outcome is that the two large estimates are each reflecting primarily the corresponding main effects (not the aliased 4-factor interactions). That is, a tentative description of the oil extraction process is that average particle size (factor E) and temperature (factor B), acting more or less separately, are the principle determinants of yield. Since this is an example where the ultimate engineering objective is to maximize response, it is further clear that since the two large estimates are both positive, for best yield one would prefer the high level of B (95°C temperature) and low level of E (1.28mm particle size). ($-\epsilon_2$ is apparently positive, and since $\epsilon_1 = -\epsilon_2$, the superiority of the low level of E is indicated. Differently put, ϵ_2 appears to be negative, so the high level of E is to be avoided.)

FIGURE 10-3

Normal Plot of Estimated Differences of Effects in Kilgo's 2^{5-1} Study

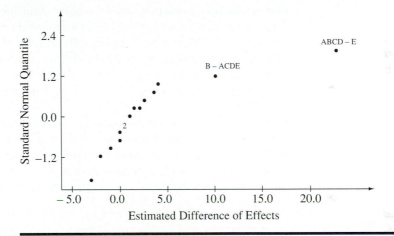

Some Additional Comments

The next section treats general $\frac{1}{2^q}$ fractions of 2^p factorials. But before closing this discussion of the special case of $q = 1$, several comments should be interjected. The first concerns the range of statistical methods that will be provided here for use with fractional factorials.

By now it ought to seem quite standard that this book presents not only methods for the estimation and testing of means for the various structured multisample situations introduced, but also prediction, tolerance, and simultaneous interval methods for those contexts. However, these other methods will not be provided for the 2^{p-1} or other fractional factorial structures.

As was indicated at the beginning of this section, fractional factorial (2^{p-1} and otherwise) experimentation is, in this author's opinion, best thought of as a step along the way in a sequential program of experimentation. It provides tentative insights that can be pursued in more detail in successive experimental stages. Even when an engineer thinks that a fractional factorial has really "nailed things down," only subsequent confirmatory experience with the system, under those conditions predicted to be optimal, really eliminates the possibility of misperception that is inherent in fractional factorial study. In light of this, it doesn't seem appropriate here to do more than provide ways of developing the tentative insights that are properly the product of fractional factorial studies.

It is of course true that one could take fractional factorial data and arrive at prediction and tolerance intervals (either through the use of a regression approach like that in Section 9-4 or by ignoring some factors thought to be unimportant after an analysis like the ones in this section and the next two, and thus considering the data to constitute a full factorial in the other important factors). For example, in Kilgo's study in Example 10-4, if the suspicion that only the E and B main effects are important is in fact correct, then the 16 data points given in Table 10-5 constitute essentially a 2^2 factorial (in B and E) replicated $m = 4$ times and the methods of Section 8-3 could be used accordingly. But it shouldn't be inferred that such a practice is either routinely justifiable or the primary way that fractional designs ought to be used. So this exposition will not go beyond simple inference for (sums and differences of) effects.

A second comment regards the sense in which the $\frac{1}{2}$ fractions recommended here are the best ones possible. Other possible $\frac{1}{2}$ fractions could be developed (essentially by using a product column of signs derived from levels of fewer than all $p - 1$ of the first factors to assign levels of the last one). But the alias structures associated with those alternatives are less attractive than the ones encountered in this section. That is, here effects have been found to be aliased in pairs: main effects with $p - 1$ factor interactions, 2-factor interactions with $p - 2$-factor interactions, and so on. Any other $\frac{1}{2}$ fractions fundamentally different from the ones discussed here would have several main effects aliased with interactions of $p - 2$ or less factors, so they would be more likely to produce data incapable of separating important effects. The "l order effects aliased with $p - l$ order effects" structure of this section is simply the best one can do with a 2^{p-1} fractional factorial.

The last matter for consideration in this discussion of 2^{p-1} studies is what directions a follow-up investigation might take in order to resolve ambiguities left after one completes a 2^{p-1} study. Sometimes several different simple descriptions of system structure remain equally plausible after analysis of an initial $\frac{1}{2}$ fraction of a full factorial study. One approach to resolving these is to complete the factorial and "run the other $\frac{1}{2}$ fraction."

EXAMPLE 10-5

A 2^{4-1} Fabric Tenacity Study Followed Up by a Second 2^{4-1} Study. Researchers Johnson, Clapp, and Baqai, in "Understanding the Effect of Confounding in Design of Experiments: A Case Study in High Speed Weaving" (*Quality Engineering*, 1989), discuss a study done to evaluate the effects of four two-level factors on a measure of woven fabric tenacity. The factors that were studied are indicated in Table 10-9.

TABLE 10-9

Four Weaving Process Variables and Their Experimental Levels

Factor	Weaving Process Variable	Factor Levels
A	Side of Cloth (l. to r.)	nozzle side ($-$) vs. opposite side ($+$)
B	Yarn Type	air spun ($-$) vs. ring spun ($+$)
C	Pick Density	35 ppi ($-$) vs. 50 ppi ($+$)
D	Air Pressure	30 psi ($-$) vs. 45 psi ($+$)

Factor A reflects the left-to-right location on the fabric width from which a tested sample is taken. Factor C reflects a count of yarns per inch inserted in the cloth, top to bottom, during weaving. Factor D reflects the air pressure used to propel the yarn across the fabric width during weaving.

Initially, a replicated 2^{4-1} study was done using the generator D $\leftrightarrow$ ABC. $m = 5$ pieces of cloth were tested for each of the eight different factor-level combinations studied. The resulting mean fabric tenacities $\bar{y}$, expressed in terms of strength per unit linear density, are given in Table 10-10.

TABLE 10-10

Eight Sample Means from a 2^{4-1} Fabric Tenacity Experiment

Combination	$\bar{y}$ (g/den.)	Combination	$\bar{y}$ (g/den.)
(1)	24.50	cd	25.68
ad	22.05	ac	24.51
bd	24.52	bc	24.68
ab	25.00	abcd	24.23

Although it is not absolutely clear in the original article, it also appears that pooling the eight s^2 values from the original $\frac{1}{2}$ fraction gave $s_P \approx 1.16$. (The authors report $\pm.375$ 95% confidence limits for the estimated sums of effects, but they don't say exactly how those were derived.)

Apply the (three-cycle) Yates algorithm to the means listed in Table 10-10 (in the order given) and verify that the estimated sums of effects corresponding to the means in Table 10-10 are those given in Table 10-11.

TABLE 10-11

Estimated Sums of 2^4 Effects in a 2^{4-1} Fabric Weaving Experiment

Estimate	Sum of Effects Estimated
24.396	$\mu_{....} + \alpha\beta\gamma\delta_{2222}$
$-.449$	$\alpha_2 + \beta\gamma\delta_{222}$
.211	$\beta_2 + \alpha\gamma\delta_{222}$
.456	$\alpha\beta_{22} + \gamma\delta_{22}$
.379	$\gamma_2 + \alpha\beta\delta_{222}$
.044	$\alpha\gamma_{22} + \beta\delta_{22}$
$-.531$	$\beta\gamma_{22} + \alpha\delta_{22}$
$-.276$	$\alpha\beta\gamma_{222} + \delta_2$

Temporarily ignoring the existence of factor D, one would derive confidence intervals based on these estimates using the $m = 5$ and $p = 3$ version of formula (8-46) from Section 8-3. That is, using 95% two-sided individual confidence intervals, since $\nu = 8(5 - 1) = 32$ degrees of freedom are associated with s_P, a precision of roughly

$$\pm 2.04 \frac{1.16}{\sqrt{5 \cdot 8}} = \pm.375$$

should be associated with each of the estimates in Table 10-11. By this standard, the estimates corresponding to the A + BCD, AB + CD, C + ABD, and BC + AD sums are statistically significant. Two reasonably plausible and equally simple tentative interpretations of this outcome are that

1. There are detectable A and C main effects and detectable 2-factor interactions of A with B and D.
2. There are detectable A and C main effects and detectable 2-factor interactions of C with B and D.

(For that matter, there are others that the reader may well find as plausible as these two.)

In any case, the ambiguities left by the collection of the data summarized in Table 10-10 were unacceptable to Johnson, Clapp, and Baqai. To remedy the situation, they subsequently completed the 2^4 factorial study by collecting data from the other eight combinations defined by the generator D $\leftrightarrow$ $-$ABC. The means they obtained are given in Table 10-12.

One should honestly consider (and hopefully eliminate) the possibility that there is a systematic difference between the values in Table 10-10 and in Table 10-12 as a result of some unknown factor or factors that changed in the time lapse between the collection of the first block of observations and the second block. If that possibility can

TABLE 10-12

Eight More Sample Means from a Second 2^{4-1} Fabric Tenacity Study

Combination	$\bar{y}$	Combination	$\bar{y}$
d	23.73	c	24.63
a	23.55	acd	25.78
b	25.98	bcd	24.10
abd	23.64	abc	23.93

be eliminated, it would make sense to put together the two data sets, treat them as a single full 2^4 factorial data set, and employ the methods of Section 8-3 in their analysis. (Some repetition of a combination or combinations included in the first study phase — e.g., the center point of the design — would have been advisable to allow at least a cursory check on the possibility of a systematic block effect.)

Johnson, Clapp, and Baqai don't say explicitly what sample sizes were used to produce the $\bar{y}$'s in Table 10-12. (Presumably, $m = 5$ was used.) Nor do they give a value for s_P based on all 2^4 samples, so it is not possible to give a complete analysis of the full factorial data à la Section 8-3. But it is possible to note what results from the use of the Yates algorithm with the full factorial set of $\bar{y}$'s. This is summarized in Table 10-13.

TABLE 10-13

Fitted Effects from the Full 2^4 Factorial Fabric Tenacity Study

Effect	Estimate	Effect	Estimate
$\mu_{\ldots}$	$\bar{y}_{\ldots} = 24.407$	δ_2	$d_2 = -.191$
α_2	$a_2 = -.321$	$\alpha\delta_{22}$	$ad_{22} = .029$
β_2	$b_2 = .103$	$\beta\delta_{22}$	$bd_{22} = -.197$
$\alpha\beta_{22}$	$ab_{22} = .011$	$\alpha\beta\delta_{222}$	$abd_{222} = .093$
γ_2	$c_2 = .286$	$\gamma\delta_{22}$	$cd_{22} = .446$
$\alpha\gamma_{22}$	$ac_{22} = .241$	$\alpha\gamma\delta_{222}$	$acd_{222} = .108$
$\beta\gamma_{22}$	$bc_{22} = -.561$	$\beta\gamma\delta_{222}$	$bcd_{222} = -.128$
$\alpha\beta\gamma_{222}$	$abc_{222} = -.086$	$\alpha\beta\gamma\delta_{2222}$	$abcd_{2222} = -.011$

The statistical significance of the entries of Table 10-13 will not be judged here. But note that the picture of fabric tenacity given by the fitted effects in this table is somewhat more complicated than either of the tentative descriptions derived from the original 2^{4-1} study. The fitted effects, listed in order of decreasing absolute value, are

$$\text{BC, CD, A, C, AC, BD, D}, \ldots, \text{etc.}$$

Although tentative description 2) accounts for the first four of these, the A and C main effects indicated in Table 10-13 are not really as large as one might have guessed looking only at Table 10-11. Further, the AC 2-factor interaction appears from Table 10-13 to be nearly as large as the C main effect. This is obscured in the original 2^{4-1} fractional factorial, because the AC 2-factor interaction is aliased with an apparently fairly large BD 2-factor interaction *of opposite sign*.

Ultimately, this example is one of a fairly complicated system of effects. It admirably illustrates the difficulties and even errors of interpretation that can arise when only fractional factorial data are available for use in studying such systems.

▬▬ ▬▬

In conclusion, it should be said that when a 2^{p-1} fractional factorial seems to leave only very mild ambiguities of interpretation, it can be possible to resolve those with the use of only a *few* additional data points (rather than requiring the addition of the entire other $\frac{1}{2}$ fraction of combinations). But this is a more advanced topic than is sensibly discussed here. The interested reader can refer to Chapter 14 of Daniel's *Applications of Statistics to Industrial Experimentation* for an illuminating discussion of this matter.

10-2 Standard Fractions of Two-Level Factorials; Part II: General 2^{p-q} Studies

Section 10-1 began the study of fractional factorials with the $\frac{1}{2}$ fractions of 2^p factorials, considering in turn the issues of choice, determination of the corresponding alias structure, and data analysis. It happens that the approaches used to treat 2^{p-1} studies extend very easily and naturally to the smaller $\frac{1}{2^q}$ fractions of 2^p factorials for $q > 1$.

This section first shows how the ideas of Section 10-1 are generalized to cover the general 2^{p-q} situation; then considers the notion of design resolution and its implications for comparing alternative possible 2^{p-q} plans. Next an introduction is given to how the 2^{p-q} ideas can be employed where a blocking variable (or variables) dictate the use of a number of blocks equal to a power of 2. The section concludes with some comments regarding wise use of this 2^{p-q} material.

Using 2^{p-q} Fractional Factorials

The method of choosing a $\frac{1}{2}$ fraction of a 2^p factorial recommended in the previous section used a column of signs developed as products of plus and minus signs for *all* of the first $p-1$ factors. The key to understanding how the ideas of the previous section generalize to $\frac{1}{4}, \frac{1}{8}, \frac{1}{16}$, etc. fractions of 2^p studies is to realize that that there are several possible similar columns that could be developed using only *some* of the first $p-1$ factors. When one moves from $\frac{1}{2}$ fractions to $\frac{1}{2^q}$ fractions of 2^p factorials, one makes use of such columns in assigning levels of the last q factors and then develops and uses an alias structure consistent with the choice of columns.

For example, first consider the situation for cases where $p - q = 3$ — that is, where $2^3 = 8$ different combinations of levels of p two-level factors are going to be included in a study. A table of signs specifying all eight possible combinations of levels of the first three factors A, B, and C, with four additional columns made up as the possible products of the first three columns, is given in Table 10-14.

TABLE 10-14

Signs for Specifying all Eight Combinations of Three Two-Level Factors and Four Sets of Products of Those Signs

A	B	C	AB Product	AC Product	BC Product	ABC Product
−	−	−	+	+	+	−
+	−	−	−	−	+	+
−	+	−	−	+	−	+
+	+	−	+	−	−	−
−	−	+	+	−	−	+
+	−	+	−	+	−	−
−	+	+	−	−	+	−
+	+	+	+	+	+	+

The final column of Table 10-14 can be used to choose levels of factor D for a best possible 2^{4-1} fractional factorial study. But it is also true that two or more of the product columns in Table 10-14 can be used to choose levels of several additional factors (beyond the first three). If this is done, one winds up with a fractional factorial that can be understood in the same ways it is possible to make sense of the standard 2^{p-1} data structures discussed in Section 10-1.

EXAMPLE 10-6

A 2^{6-3} **Propellant Slurry Study.** The text *Probability and Statistics for Engineers and Scientists*, by Walpole and Myers, contains an interesting 2^{6-3} fractional factorial data set taken originally from the *Proceedings of the 10th Conference on the Design of Experiments in Army Research, Development and Testing* (ARO-D Report 65-3). The study investigated the effects of six two-level factors on X-ray intensity ratios associated with a particular component of propellant mixtures in X-ray fluorescent analyses of propellant slurry. Factors A, B, C, and D represent the concentrations (at low and high levels) of four propellant components. Factors E and F represent the weights (also at low and high levels) of fine and coarse particles present.

Eight different combinations of levels of factors A, B, C, D, E, and F were each tested twice for X-ray intensity ratio, y. The eight combinations actually included in the study can be thought of as follows. Using the columns of Table 10-14, levels of factor D were chosen using the signs in the ABC product column directly; levels of factor E were chosen by reversing the signs in the BC product column; and levels of factor F were chosen by reversing the signs of the AC product column. Verify that such a prescription implies that the eight combinations included in the study (written down in Yates order for factors A, B, and C) were as displayed in Table 10-15.

TABLE 10-15

Combinations Included in the 2^{6-3} Propellant Slurry Study

A	B	C	F	E	D	Combination Name
−	−	−	−	−	−	(1)
+	−	−	+	−	+	adf
−	+	−	−	+	+	bde
+	+	−	+	+	−	abef
−	−	+	+	+	+	cdef
+	−	+	−	+	−	ace
−	+	+	+	−	−	bcf
+	+	+	−	−	+	abcd

The eight combinations indicated in Table 10-15 are, of course, $\frac{1}{8}$ of the 64 different possible combinations of levels of the six factors.

━━━━━━━━━━━━━━━━━━━━━━━━━━ ━ ━ ━

The development of 2^{p-q} fractional factorials has been illustrated with eight combination (i.e. $p - q = 3$) plans. But it should be obvious that there are 16-row, 32-row, 64-row, . . . , etc. versions of Table 10-14 and that using any of these, one can assign levels for the last q factors according to signs in product columns and end up with a $\frac{1}{2^q}$ fraction of a full 2^p factorial plan. When this is done, it turns out that the 2^p factorial effects are aliased in 2^{p-q} groups of 2^q effects each. The determination of this alias structure can be made by using q generators to develop a defining relation for the fractional factorial. A general definition of the notion of generators for a 2^{p-q} fractional factorial is next.

DEFINITION 10-3

When a 2^{p-q} fractional factorial comes about by assigning levels of each of the last q factors based on a different column of products of signs for the first $p - q$ factors, the q different relationships

$$\begin{pmatrix}\text{the name of an}\\\text{additional factor}\end{pmatrix} \leftrightarrow \pm \begin{pmatrix}\text{a product of names of some}\\\text{of the first } p - q \text{ factors}\end{pmatrix}$$

(corresponding to how the combinations are chosen) are called **generators** of the plan.

Each generator can be translated into a statement with I on the left side and then taken individually, multiplied in pairs, multiplied in triples, and so on until the whole *defining relation* is developed. (In doing so, one may make use of the convention that minus any letter times minus that letter is I.)

EXAMPLE 10-6
(continued)

In the Army propellant example, the three generators that led to the combinations in Table 10-15 were

$$D \leftrightarrow ABC$$
$$E \leftrightarrow -BC$$
$$F \leftrightarrow -AC$$

Multiplying through by the left sides of these, one obtains the three relationships

$$I \leftrightarrow ABCD \qquad\qquad\qquad \textbf{(10-4)}$$
$$I \leftrightarrow -BCE \qquad\qquad\qquad \textbf{(10-5)}$$
$$I \leftrightarrow -ACF \qquad\qquad\qquad \textbf{(10-6)}$$

But in light of the conventions of formal multiplication, it would seem (for example) that if $I \leftrightarrow ABCD$ and $I \leftrightarrow -BCE$, it should also be the case that

$$I \leftrightarrow (ABCD) \cdot (-BCE)$$

that is,

$$I \leftrightarrow -ADE$$

Similarly, using relationships (10-4) and (10-6), one obtains

$$I \leftrightarrow -BDF$$

using relationships (10-5) and (10-6), one obtains

$$I \leftrightarrow ABEF$$

and finally, using all three relationships (10-4), (10-5), and (10-6), one has

$$I \leftrightarrow CDEF$$

Combining all of this, the complete defining relation for this 2^{6-3} study is

$$I \leftrightarrow ABCD \leftrightarrow -BCE \leftrightarrow -ACF \leftrightarrow -ADE \leftrightarrow -BDF \leftrightarrow ABEF \leftrightarrow CDEF \qquad \textbf{(10-7)}$$

Defining relation (10-7) is rather formidable looking, but it tells the whole truth about what can be learned based on the $\frac{1}{8}$ of 64 possible combinations of six two-level factors. Relation (10-7) specifies all effects that will be aliased with the grand mean. Appropriately multiplying through expression (10-7) gives one all aliases of any effect of interest. For example, multiplying through relation (10-7) by A gives

$$A \leftrightarrow BCD \leftrightarrow -ABCE \leftrightarrow -CF \leftrightarrow -DE \leftrightarrow -ABDF \leftrightarrow BEF \leftrightarrow ACDEF$$

and one sees that for example, the (high level) A main effect will be indistinguishable from minus the (all high levels) CF 2-factor interaction.

With a 2^{p-q} fractional factorial's defining relation in hand, the analysis of data proceeds exactly as indicated earlier for $\frac{1}{2}$ fractions. One computes estimates of (sums and differences of) effects via the Yates algorithm, judges their statistical detectability using confidence-interval or normal-potting methods, and then seeks a plausible tentative interpretation of the important estimates in light of the alias structure.

EXAMPLE 10-6
(continued)

In the Army propellant study, $m = 2$ trials for each of the 2^{6-3} combinations listed in Table 10-15 gave $s_{\text{P}}^2 = .02005$ and the sample averages listed in Table 10-16.

TABLE 10-16

Eight Sample Means from the 2^{6-3} Propellant Slurry Study

Combination	$\bar{y}$	Combination	$\bar{y}$
(1)	1.1214	cdef	.9285
adf	1.0712	ace	1.1635
bde	.9415	bcf	.9561
abef	1.1240	abcd	.9039

Temporarily ignoring all but the three factors A, B, and C, the (three-cycle) Yates algorithm can be used on the sample means, as shown in Table 10-17.

TABLE 10-17

The Yates Algorithm for a 2^3 Factorial Applied to the 2^{6-3} Propellant Data

$\bar{y}$	Cycle 1	Cycle 2	Cycle 3	Cycle 3 $\div$ 8	Sum Estimated
1.1214	2.1926	4.2581	8.2101	1.0263	$\mu_{.....}$ + aliases
1.0712	2.0655	3.9520	.3151	.0394	α_2 + aliases
.9415	2.0920	.1323	−.3591	−.0449	β_2 + aliases
1.1240	1.8600	.1828	−.0545	−.0068	$\alpha\beta_{22}$ + aliases
.9285	−.0502	−.1271	−.3061	−.0383	γ_2 + aliases
1.1635	.1825	−.2320	.0505	.0063	$\alpha\gamma_{22}$ + aliases
.9561	.2350	.2327	−.1049	−.0131	$\beta\gamma_{22}$ + aliases
.9039	−.0522	−.2872	−.5199	−.0650	$\alpha\beta\gamma_{222}$ + aliases

Remember that the estimates in the next-to-last column of Table 10-17 must be interpreted in the light of the alias structure for the original experimental plan. So for example, since (both from the original generators and from relation (10-7)) one knows that D $\leftrightarrow$ ABC, the −.0650 value on the last line of Table 10-17 is estimating

$$\alpha\beta\gamma_{222} + \delta_2 \pm \text{ (six other effects)}$$

So if one were expecting a large main effect of factor D, one would expect it to be evident in the −.0650 value.

Since a value of s_P is available here, there is no need to resort to normal-plotting to judge the statistical detectability of the values coming out of the Yates algorithm. Instead (still temporarily calculating as if only the first three factors were present) one can make confidence intervals based on the estimates, by employing the $\nu = 8 = 16 - 8$, $m = 2$, and $p = 3$ version of formula (8-46) from Section 8-3. That is, using 95% two-sided individual confidence intervals, a precision of

$$\pm 2.306 \frac{\sqrt{.02005}}{\sqrt{2 \cdot 2^3}} = \pm.0817$$

should be attached to each of the estimates in Table 10-17. By this standard, none of the estimates from the propellant study are clearly different from 0. For engineering purposes, the bottom line is that more data are needed before even the most tentative conclusions about system behavior should be made.

EXAMPLE 10-7

A 2^{5-2} Catalyst Development Experiment. Hansen and Best, in their paper "How to Pick a Winner" (presented at the 1986 annual meeting of the American Statistical Association), described several industrial experiments conducted in a research program aimed at the development of an effective catalyst for producing ethyleneamines by the amination of monoethanolamine. One of these was a partially replicated 2^{5-2} fractional factorial study in which the response variable y was percent water produced during the reaction period. The five two-level experimental factors were as in Table 10-18. (The T-372 support was an alpha-alumina support and the T-869 support was a silica alumina support.)

TABLE 10-18

Five Catalysis Variables and Their Experimental Levels

Factor	Process Variable	Levels
A	Ni/Re Ratio	2/1 $(-)$ vs. 20/1 $(+)$
B	Precipitant	$(NH_4)_2CO_3$ $(-)$ vs. none $(+)$
C	Calcining Temperature	300°C $(-)$ vs. 500°C $(+)$
D	Reduction Temperature	300° C $(-)$ vs. 500°C $(+)$
E	Support Used	T-372 $(-)$ vs. T-869 $(+)$

The fractional factorial described by Hansen and Best has generators $D \leftrightarrow ABC$ and $E \leftrightarrow BC$. The resulting defining relation is then

$$I \leftrightarrow ABCD \leftrightarrow BCE \leftrightarrow ADE$$

where the fact that the ADE 3-factor interaction is aliased with the grand mean can be seen by multiplying together ABCD and BCE, which (from the generators) themselves represent effects aliased with the grand mean. So here one sees that effects will be aliased together in eight groups of four.

The data reported by Hansen and Best, and some corresponding summary statistics, are given in Table 10-19.

TABLE 10-19

Data from a 2^{5-2} Catalyst Study and Corresponding Sample Means and Variances

Combination	% Water Produced, y	$\bar{y}$	s^2
e	8.70, 11.60, 9.00	9.767	2.543
ade	26.80	26.80	—
bd	24.88	24.88	—
ab	33.15	33.15	—
cd	28.90, 30.98	29.940	2.163
ac	30.20	30.20	—
bce	8.00, 8.69	8.345	.238
abcde	29.30	29.30	—

The pooled sample variance derived from the values in Table 10-19 is

$$s_P^2 = \frac{(3-1)(2.543) + (2-1)(2.163) + (2-1)(.238)}{(3-1) + (2-1) + (2-1)}$$

$$= 1.872$$

with $\nu = (3-1) + (2-1) + (2-1) = 4$ associated degrees of freedom. The corresponding pooled sample standard deviation is

$$\sqrt{s_P^2} = \sqrt{1.872} = 1.368$$

So temporarily ignoring the existence of factors D and E, one can return to Section 8-3 and use the $p = 3$ version of formula (8-45) to derive precisions to attach to the estimates (of sums of 2^5 factorial effects) that would result from the use of the Yates algorithm on the sample means in Table 10-19. That is, for 95% two-sided individual confidence intervals, one would attach precisions of

$$\pm 2.776(1.368)\frac{1}{2^3}\sqrt{\frac{1}{3} + \frac{1}{1} + \frac{1}{1} + \frac{1}{1} + \frac{1}{2} + \frac{1}{1} + \frac{1}{2} + \frac{1}{1}}$$

that is,

$$\pm 1.195\% \text{ water}$$

to the estimates.

The reader can verify that in fact the (three-cycle) Yates algorithm applied to the means in Table 10-19 gives the estimates in Table 10-20.

TABLE 10-20

Estimates of Sums of Effects for the Catalyst Study

Sum of Effects Estimated	Estimate
grand mean + aliases	24.048
A + aliases	5.815
B + aliases	−.129
AB + aliases	1.492
C + aliases	.399
AC + aliases	−.511
BC + aliases	−5.495
ABC + aliases	3.682

Examining those estimates in Table 10-20 whose magnitudes make them statistically detectable according to a ± 1.195 criterion, one has (in order of decreasing magnitude)

$$\alpha_2 + \beta\gamma\delta_{222} + \alpha\beta\gamma\epsilon_{2222} + \delta\epsilon_{22} \text{ estimated as } 5.815$$
$$\beta\gamma_{22} + \alpha\delta_{22} + \epsilon_2 + \alpha\beta\gamma\delta\epsilon_{22222} \text{ estimated as } -5.495$$
$$\alpha\beta\gamma_{222} + \delta_2 + \alpha\epsilon_{22} + \beta\gamma\delta\epsilon_{2222} \text{ estimated as } 3.682$$
$$\alpha\beta_{22} + \gamma\delta_{22} + \alpha\gamma\epsilon_{222} + \beta\delta\epsilon_{222} \text{ estimated as } 1.492$$

The simplest possible tentative interpretation of the first two of these results is that the A and E main effects are large enough to see above the background variation. What to make of the third, given the first two, is not so clear. The large 3.682 estimate can equally simply be tentatively attributed to a D main effect or to an AE 2-factor interaction. (Interestingly, Hansen and Best reported that subsequent experimentation was done with the purpose of determining the importance of the D main effect, and indeed, the importance of this factor in determining y was established.)

Exactly what to make of the fourth statistically significant estimate is even less clear. Because of that, it is comforting that, although big enough to be detectable, it is less than half the size of the third largest estimate. In the particular real situation, the authors seem to have found an "A, E, and D main effects only" description of y useful in subsequent work with the chemical system.

Design Resolution

The results of five different real applications of 2^{p-q} plans have been discussed in Examples 10-3, 10-4, 10-5, 10-6, and 10-7. Accordingly, it should be clear how important it is to have the simplest alias structure possible when it comes time to interpret the results of a fractional factorial statistical engineering study. One hopes to have low-order effects (like main effects and 2-factor interactions) aliased not with other low-order effects, but rather only with high-order effects (many-factor interactions). It is the defining relation that governs how the 2^p factorial effects are divided up into groups of aliases. And a little reflection should make clear that if there are only long products of factor names appearing in the defining relation, low-order effects are aliased only with high-order effects. On the other hand, if there are short products of factor names appearing, there will be low-order effects aliased with other low-order effects. As a kind of measure of quality of a 2^{p-q} plan, it is thus common to adopt the following notion of *design resolution*.

DEFINITION 10-4

The **resolution** of a 2^{p-q} fractional factorial plan is the number of letters in the shortest product appearing in its defining relation.

In general, when contemplating the use of a 2^{p-q} design, one wants the largest resolution possible for a given investment in 2^{p-q} combinations. For a given number of combinations, not all choices of generators give the same resolution. In Section 10-1, the prescription given for the $\frac{1}{2}$ fractions was intended to give 2^{p-1} fractional factorials of resolution p (the largest resolution possible). For general 2^{p-q} studies, one must be a bit careful in choosing generators. What seems like the most obvious choice of 2^{p-q} combinations need not be the best in terms of resolution.

EXAMPLE 10-8

Resolution 4 in a 2^{6-2} Study. Suppose, for sake of example, that one is planning a 2^{6-2} study — that is, intending to examine 16 out of 64 possible combinations of levels of factors A, B, C, D, E, and F. A rather natural choice of two generators for such a study is

$$E \leftrightarrow ABCD$$
$$F \leftrightarrow ABC$$

Verify that the corresponding defining relation is

$$I \leftrightarrow ABCDE \leftrightarrow ABCF \leftrightarrow DEF$$

The resulting design is of resolution 3, and there are some main effects aliased with (only) 2-factor interactions.

On the other hand, the perhaps slightly less natural choice of generators

$$E \leftrightarrow BCD$$
$$F \leftrightarrow ABC$$

has defining relation

$$I \leftrightarrow BCDE \leftrightarrow ABCF \leftrightarrow ADEF$$

and is of resolution 4. No main effect is aliased with any interaction of order less than 3. This second choice is better than the first in terms of resolution.

Table 10-21 indicates what is possible in terms of resolution for various numbers of factors and combinations for a 2^{p-q} fractional factorial. The table was derived from a more detailed one on page 410 of *Statistics for Experimenters* by Box, Hunter, and Hunter, which gives not only the best resolutions possible, but also generators for designs achieving those resolutions. The more limited information in Table 10-21 is sufficient for most purposes. Once one is sure what is possible, it is usually relatively painless to do the trial-and-error work needed to produce a plan of highest possible resolution. And it is probably worth doing as an exercise, to help one consider the pros and cons of various choices of generators for a given set of real factors.

Table 10-21 has no entries in the "8 combinations" row for more than 7 factors. If the table were extended beyond 11 factors, there would be no entries in the "16

TABLE 10-21

Best Resolutions Possible for Various Numbers of Combinations in a 2^{p-q} Study

		NUMBER OF FACTORS (p)							
		4	5	6	7	8	9	10	11
	8	4	3	3	3	—	—	—	—
NUMBER OF	16		5	4	4	4	3	3	3
COMBINATIONS (2^{p-q})	32			6	4	4	4	4	4
	64				7	5	4	4	4
	128					8	6	5	5

samples" row beyond 15 factors, no entries in the "32 samples" row beyond 31 factors, etc. The reason for this should be fairly obvious. For 8 combinations, there are only seven columns total to use in Table 10-14. Corresponding tables for 16 combinations would have only 15 columns total, for 32 combinations only 31 columns total, etc.

As they have been described here, one can use 2^{p-q} fractional factorials to study at most $2^t - 1$ factors in 2^t samples. The cases of 7 factors in 8 combinations, 15 factors in 16 combinations, 31 factors in 32 combinations, etc. represent a kind of extreme situation where a maximum number of factors is studied (at the price of creating a worst possible alias structure) in a given number of combinations. For the case of 7 factors in 8 combinations, effects are aliased in 8 groups of 16; for the case of 15 factors in 16 combinations, the effects are aliased in 16 groups of 2048; etc. These extreme cases of $2^t - 1$ factors in 2^t combinations are sometimes called *saturated* fractional factorials. They obviously have very complicated alias structures and can be of much help only when an engineer is somehow *a priori* sure that a few main effects will completely swamp background noise and all interactions in a statistical engineering study.

EXAMPLE 10-9

A 16-Run 15-Factor Process Development Study. The article "What Every Technologist Should Know About Experimental Design" by C. Hendrix (*Chemtech*, 1979) includes the results from an unreplicated 16-run (saturated) 15-factor experiment. The response, y, was a measure of cold crack resistance for an industrial product. Experimental factors and levels were as listed in Table 10-22.

The experimental plan used was defined by the generators

$$E \leftrightarrow ABCD, \ F \leftrightarrow BCD, \ G \leftrightarrow ACD, \ H \leftrightarrow ABC, \ J \leftrightarrow ABD, \ K \leftrightarrow CD,$$

$$L \leftrightarrow BD, \ M \leftrightarrow AD, \ N \leftrightarrow BC, \ O \leftrightarrow AC, \text{ and } P \leftrightarrow AB$$

TABLE 10-22

15 Process Variables and Their Experimental Levels

Factor	Process Variable	Levels
A	Coating Roll Temperature	115° ($-$) vs. 125° ($+$)
B	Solvent	Recycled ($-$) vs. Refined ($+$)
C	Polymer X-12 Preheat	No ($-$) vs. Yes ($+$)
D	Web Type	LX-14 ($-$) vs. LB-17 ($+$)
E	Coating Roll Tension	30 ($-$) vs. 40 ($+$)
F	Number of Chill Rolls	1 ($-$) vs. 2 ($+$)
G	Drying Roll Temperature	75° ($-$) vs. 80° ($+$)
H	Humidity of Air Feed to Dryer	75% ($-$) vs. 90% ($+$)
J	Feed Air to Dryer Preheat	No ($-$) vs. Yes ($+$)
K	Dibutylfutile in Formula	12% ($-$) vs. 15% ($+$)
L	Surfactant in Formula	.5% ($-$) vs. 1% ($+$)
M	Dispersant in Formula	.1% ($-$) vs. 2% ($+$)
N	Wetting Agent in Formula	1.5% ($-$) vs. 2.5% ($+$)
O	Time Lapse Before Coating Web	10 min ($-$) vs. 30 min ($+$)
P	Mixer Agitation Speed	100 rpm ($-$) vs. 250 rpm ($+$)

TABLE 10-23

16 Experimental Combinations and Measured Cold Crack Resistances

Combination	y	Combination	y
eklmnop	14.8	dfgjnop	17.8
aghjkln	16.3	adefhmn	18.9
bfhjkmo	23.5	bdeghlo	23.1
abefgkp	23.9	abdjlmp	21.8
cfghlmp	19.6	cdehjkp	16.6
acefjlo	18.6	acdgkmo	16.7
bcegjmn	22.3	bcdfkln	23.5
abchnop	22.2	abcdefghjklmnop	24.9

The combinations actually run and the cold crack resistances observed are given in Table 10-23.

Ignoring all factors but A, B, C, and D, the combinations listed in Table 10-23 are in Yates standard order and are therefore ready for use in finding estimates of sums of effects. Table 10-24 shows the results of using the (four-cycle) Yates algorithm on the 16 observations listed in Table 10-23. A normal plot of the last 15 of these estimates is shown in Figure 10-4. It is clear from the figure that the two estimates corresponding to B + aliases and F + aliases are detectably larger than the rest.

TABLE 10-24

Estimates of Sums of Effects for the 2^{15-11} Process Development Study

Sum of Effects Represented	Estimate
grand mean + aliases	20.28
A + aliases	.13
B + aliases	2.87
P + aliases (including AB)	−.08
C + aliases	.27
O + aliases (including AC)	−.08
N + aliases (including BC)	−.19
H + aliases (including ABC)	.36
D + aliases	.13
M + aliases (including AD)	.03
L + aliases (including BD)	.04
J + aliases (including ABD)	−.06
K + aliases (including CD)	−.26
G + aliases (including ACD)	.29
F + aliases (including BCD)	1.06
E + aliases (including ABCD)	.11

FIGURE 10-4

Normal Plot of Estimated Sums of Effects in the 2^{15-11} Process Development Study

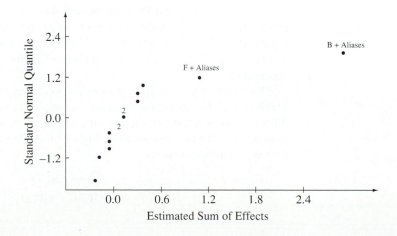

It is not feasible to write out the whole defining relation for this 2^{15-11} study. But one must remember that effects are aliased in 16 groups of $2^{11} = 2048$. In particular (though it would certainly be convenient if the 2.87 estimate in Table 10-24 could be thought of as essentially representing β_2), β_2 has 2047 aliases, some of them as simple as 2-factor interactions. By the same token, it would certainly be convenient if the small estimates in Table 10-24 were indicating that all summands of the sum of effects they represent were small. But the possibility of cancellation in the summation must not be overlooked.

The point is that only the most tentative description of this system should be drawn from even this very simple "two large estimates" outcome. The data in Table 10-23 hint at the primary importance of factors B and F in determining cold crack resistance, but the case is hardly airtight. There is a suggestion of a direction for further experimentation and discussion with process experts, but certainly no detailed map of the countryside where one is going.

Two-Level Factorials and Fractional Factorials in Blocks

A somewhat specialized but occasionally useful adaptation of the 2^{p-q} material presented here has to do with the analysis of full or fractional two-level factorial studies that are run in complete or incomplete blocks. When the number of blocks under consideration is itself a power of 2, clever use of the methods developed in this chapter can guide the choice of which combinations to place in incomplete blocks, as well as the analysis of data from both incomplete and complete block studies.

The basic idea used is to formally represent a 2^t level factor Blocks as t "extra" two-level factors. One lets combinations of levels of these extra factors define the blocks into which combinations of levels of the factors of interest are placed. In data analysis, one recognizes effects involving only the extra factors as Block main effects and effects involving both the extra factors and the factors of interest as Block by Treatment interactions. In carrying out this program, it is fairly common (though not necessarily safe) to operate as if the Block by Treatment interactions were all negligible. How such choice and analysis of blocked 2^{p-q} studies proceeds will be illustrated with a series of three examples that are variations on Example 10-5.

EXAMPLE 10-10

(Example 10-5 revisited) **A 2^4 Fabric Tenacity Study Run in Two Blocks.** In the weaving study of Johnson, Clapp, and Baqai, four factors A, B, C, and D were studied. The discussion in Section 10-1 described how the authors originally ran a replicated 2^{4-1} fractional factorial with defining relation I $\leftrightarrow$ ABCD and followed up later with a second 2^{4-1} fractional factorial having defining relation I $\leftrightarrow$ −ABCD, thus completing the full 2^4 factorial. However, since the study of the two $\frac{1}{2}$ fractions was separated in time, it is sensible to think of the two parts of the study as different *blocks* — that is, to think of a fifth two-level factor (say, E) representing the time of observation.

How then might one think of this situation in terms of the formal multiplication used in this chapter to understand fractional factorial alias structures? Notice that there are 16 different samples and five factors for consideration, which suggests that somehow (at least in formal terms) this situation might be thought of as a 2^{5-1} data structure. Further, the two formal expressions

$$I \leftrightarrow ABCD \tag{10-8}$$

$$I \leftrightarrow -ABCD \tag{10-9}$$

which define the two sets of 8 out of 16 ABCD combinations, might lead one to imagine that they result from a formal expression like

$$I \leftrightarrow ABCDE \qquad (10\text{-}10)$$

where E can be thought of as contributing either the plus or the minus signs in expressions (10-8) and (10-9). In fact, if one calls block 1 (the first set of 8 samples) the high level of E, expression (10-10) leads to exactly the $I \leftrightarrow ABCD \frac{1}{2}$ fraction of 2^4 combinations of A, B, C, and D for use as block 1 and the $I \leftrightarrow -ABCD \frac{1}{2}$ fraction for use as block 2. This can be seen in Table 10-25.

TABLE 10-25

A 2^{5-1} Fractional Factorial or a 2^4 Factorial in Two Blocks

Block 1	Block 2
e	a
abe	b
ace	c
bce	abc
ade	d
bde	abc
cde	acd
abcde	bcd

With factor E designating block number, the two columns of Table 10-25 taken together designate the $I \leftrightarrow ABCDE \frac{1}{2}$ fraction of 2^5 A, B, C, D, and E combinations. And (ignoring the e) the first column of Table 10-25 designates the $I \leftrightarrow ABCD \frac{1}{2}$ fraction of 2^4 A, B, C, and D combinations, while the second designates the $I \leftrightarrow -ABCD \frac{1}{2}$ fraction of 2^4 A, B, C, and D combinations.

Once it is clear that the Johnson, Clapp, and Baqai study can be thought of in terms of expression (10-10) with the two-level blocking factor E, it is also clear how any *block effects* will show up during data analysis. If one temporarily ignores the blocks and uses the Yates algorithm to compute fitted 2^4 factorial effects, it will then be necessary to remember, for example, that the fitted ABCD 4-factor interaction reflects not only $\alpha\beta\gamma\delta_{2222}$, but any block main effects as well. And for example, any 2-factor interaction of A and blocks will be reflected in the fitted BCD 3-factor interaction. Of course, if one for some reason believes that all interactions with blocks are negligible, it would then be expected that all fitted effects except that for the ABCD 4-factor interaction would indeed represent the appropriate 2^4 factorial effects.

EXAMPLE 10-11

A 2^4 Factorial Run in Four Blocks. For the sake of example, suppose that Johnson, Clapp, and Baqai had *a priori* planned to conduct a full 2^4 factorial set of ABCD combinations in four incomplete blocks (of four combinations each). Consider how those blocks might have been chosen and how subsequent data analysis would have proceeded.

The four-level factor Blocks can here be thought of in terms of the combinations of two extra two-level factors, which one might designate as E and F. In order to accommodate the original four factors and these two additional ones in 16 ABCDEF combinations, one must choose a 2^{6-2} design by specifying two generators. The choices

$$E \leftrightarrow BCD \qquad (10\text{-}11)$$

$$F \leftrightarrow ABC \qquad (10\text{-}12)$$

leading to the defining relation

$$I \leftrightarrow BCDE \leftrightarrow ABCF \leftrightarrow ADEF \tag{10-13}$$

will be used here.

The four different combinations of levels of E and F ((1), e, f, and ef) can be thought as designating in which block a given ABCD combination should appear. So generators (10-11) and (10-12) prescribe the division of the full 2^4 factorial into the blocks indicated in Table 10-26.

TABLE 10-26

A 2^{6-2} Fractional Factorial or a 2^4 Factorial in Four Blocks

Block 1	Block 2	Block 3	Block 4
(1)	abe	af	bef
bc	ace	abcf	cef
abd	de	bdf	adef
acd	bcde	cdf	abcdef

The reader should verify that (ignoring factors E and F) the 16 combinations in Table 10-26 are indeed a full 2^4 factorial in A, B, C, and D. Further, the combinations are indeed the $\frac{1}{4}$ fraction of a full 2^6 factorial in A, B, C, D, E, and F prescribed by generators (10-11) and (10-12).

As always, the defining relation (given here in display (10-13)) describes how effects are aliased. Table 10-27 indicates the aliases of each of the 2^4 factorial effects, obtained by multiplying through relation (10-13) by the various combinations of the letters A, B, C, and D.

TABLE 10-27

Aliases of the 2^4 Factorial Effects When Run in Four Blocks Prescribed by Generators (10-11) and (10-12)

$I \leftrightarrow BCDE \leftrightarrow ABCF \leftrightarrow ADEF$
$A \leftrightarrow ABCDE \leftrightarrow BCF \leftrightarrow DEF$
$B \leftrightarrow CDE \leftrightarrow ACF \leftrightarrow ABDEF$
$AB \leftrightarrow ACDE \leftrightarrow CF \leftrightarrow BDEF$
$C \leftrightarrow BDE \leftrightarrow ABF \leftrightarrow ACDEF$
$AC \leftrightarrow ABDE \leftrightarrow BF \leftrightarrow CDEF$
$BC \leftrightarrow DE \leftrightarrow AF \leftrightarrow ABCDEF$
$ABC \leftrightarrow ADE \leftrightarrow F \leftrightarrow BCDEF$
$D \leftrightarrow BCE \leftrightarrow ABCDF \leftrightarrow AEF$
$AD \leftrightarrow ABCE \leftrightarrow BCDF \leftrightarrow EF$
$BD \leftrightarrow CE \leftrightarrow ACDF \leftrightarrow ABEF$
$ABD \leftrightarrow ACE \leftrightarrow CDF \leftrightarrow BEF$
$CD \leftrightarrow BE \leftrightarrow ABDF \leftrightarrow ACEF$
$ACD \leftrightarrow ABE \leftrightarrow BDF \leftrightarrow CEF$
$BCD \leftrightarrow E \leftrightarrow ADF \leftrightarrow ABCEF$
$ABCD \leftrightarrow AE \leftrightarrow DF \leftrightarrow BCEF$

Notice from Table 10-27 that the BCD and ABC 3-factor interactions are aliased with block main effects. So is the AD 2-factor interaction, since one of its aliases is EF, which involves only the two-level extra factors E and F used to represent the four-level factor Blocks. On the other hand, if interactions with Blocks are negligible, it is *only* these three of the 2^4 factorial effects that are aliased with other possibly nonnegligible effects. (For any other of the 2^4 factorial effects, each alias involves both letters from the group A, B, C, and D and also from the group E and F — and is therefore some kind of Block by Treatment interaction.)

Analysis of data from a plan like that in Table 10-26 would proceed as indicated repeatedly in this chapter. The Yates algorithm applied to sample means listed in Yates standard order for factors A, B, C, and D produces estimates that are interpreted in light of the alias structure laid out in Table 10-27.

EXAMPLE 10-12

A 2^{4-1} Fractional Factorial Run in Four Blocks. As a final variant on the 4-factor weaving example, consider how the original $\frac{1}{2}$ fraction of the 2^4 factorial might itself have been run in four incomplete blocks of two combinations. (Imagine that for some reason, only two combinations could be prepared on any single day and that there was some fear of Day effects related to environmental changes, instrument drift, etc.)

Only eight combinations are to be chosen. In doing so, one needs to account for the four experimental factors A, B, C, and D and two extras E and F, which can be used to represent the four-level factor Blocks. Starting with the first three experimental factors A, B, and C (three of them because $2^3 = 8$), one needs to choose three generators. The original 2^{4-1} study had generator

$$D \leftrightarrow ABC$$

so it is natural to begin there. For the sake of example, consider also the generators

$$E \leftrightarrow BC$$
$$F \leftrightarrow AC$$

These give the defining relation

$$I \leftrightarrow ABCD \leftrightarrow BCE \leftrightarrow ACF \leftrightarrow ADE \leftrightarrow BDF \leftrightarrow ABEF \leftrightarrow CDEF \qquad \textbf{(10-14)}$$

and the prescribed set of combinations listed in Table 10-28.

TABLE 10-28

A 2^{6-3} Fractional Factorial or a 2^{4-1} Fractional Factorial in Four Blocks

Block 1	Block 2	Block 3	Block 4
ab	ade	bdf	ef
cd	bce	acf	abcdef

Some experimenting with relation (10-14) will show that *all* 2-factor interactions of the four original experimental factors A, B, C, and D are aliased not only with other 2-factor interactions of experimental factors, but also with block main effects. Thus, any systematic block-to-block changes would further confuse one's perception of 2-factor interactions of the experimental factors. But at least the main effects of A, B, C, and D are not aliased with Block main effects.

Examples 10-10 through 10-12 all treat situations where blocks are *incomplete* — in the sense that they don't each contain every combination of the experimental factors studied. *Complete block* plans with 2^t blocks can also be developed and analyzed through the use of t "extra" two-level factors to represent the single (2^t level) factor Blocks. The path to be followed is by now worn enough through use in this chapter that further examples will not be included. But the reader should have no trouble figuring out, for example, how to analyze a full 2^4 factorial that is run completely once in each of two blocks, or even how to analyze a standard 2^{4-1} fractional factorial that is run completely once in each of four blocks.

Some Additional Comments

This 2^{p-q} fractional factorial material is fascinating, and extremely useful when used with a proper understanding of both its power and its limitations. However, an engineer who tries to use it in a cookbook fashion will usually wind up frustrated and disillusioned. The implications of aliasing must be thoroughly understood for successful use of the material. And a clear understanding of these implications will work to keep the engineer from routinely trying to study many factors based on very small amounts of data in a one-shot experimentation mode.

This author has unfortunately seen a number of engineers newly introduced to fractional factorial experimentation (by others, of course) who have tried to routinely draw final engineering conclusions about multifactor systems based on as few as eight data points. The folly of such a method of operation should be apparent. Economy of experimental effort involves not just collecting a small amount of data on a multifactor system, but rather collecting the minimum amount sufficient for a practically useful and reliable understanding of system behavior. Just a few expensive engineering errors, traceable to naive and overzealous use of fractional factorial experimentation, will easily negate any supposed savings generated by overly frugal data collection.

Although several 8-combination plans have been used as examples in this section, it should be noted that such designs are often too small to provide much useful information on the behavior of real engineering systems. Typically, 2^{p-q} studies with $p - q \geq 4$ are recommended as far more likely to lead to a satisfactory understanding of system behavior.

It has been said several times that when intelligently used as factor-screening tools, 2^{p-q} fractional factorial studies will usually be followed up with more complete experimentation, such as a larger fraction, a response surface design, or a complete factorial (often in a reduced set of factors). It is also true that techniques exist for choosing a relatively small second fraction in such a way as to resolve certain particular types of ambiguities of interpretation that can remain after the analysis of an initial fractional factorial. The interested reader can refer to Section 12.5 of *Statistics for Experimenters* by Box, Hunter, and Hunter for discussions of how to choose an additional fraction to "dealias" a particular main effect and all its associated interactions or to "dealias" all main effects.

10-3   *Latin Squares and Other Orthogonal Fractional Factorial Designs*

As the discussion of the 2^{p-q} fractional factorial designs in the previous two sections suggests, it is a fairly subtle business to choose p-factor fractional factorial designs and subsequently to perform data analysis for them. Unless substantial caution is exercised, the potential for drawing erroneous engineering conclusions from fractional factorial studies even multiplies as one leaves the realm of 2^{p-q} studies and begins to consider the possibility of more than two levels for some or all of p factors. Except in those (rare) situations where it is somehow *a priori* clear that there are no important interactions whatsoever, the safest advice that can be given in (large) p-factor engineering experimentation contexts is either to confine oneself to the surface-fitting approaches of design and analysis in Section 9-3 and the 2^{p-q} approaches of Sections 10-1 and 10-2, or to seek the expert advice of a statistician who specializes in the area of experimental design.

In spite of the soundness of this advice, both some classical textbooks and many current engineering experimentation short courses promote the use of p-factor experimental plans unlike any discussed in Sections 9-3, 10-1, or 10-2. So, primarily for the sake of completeness, this section will try to point out what is true about the use of one other currently popular kind of fractional factorial design: *orthogonal array* plans.

The Latin squares introduced in Section 2-4 are a well-known type of orthogonal array design; they will serve as examples in the present discussion.

The section begins by defining the property of orthogonality and giving several examples of orthogonal array fractional factorial plans. Then there is discussion of some of the weaknesses and difficulties of interpretation that arise when one attempts to use such data collection plans in the study of general p-factor engineering systems. The section concludes with some inference methods that can be used with data from orthogonal array plans under the (very restrictive/specialized) assumption that there are no important interaction effects (in the corresponding complete factorial).

Orthogonal Arrays

Consider now statistical studies in p-variable engineering systems where factors A, B, C, etc. have (respectively) I, J, K, etc. levels, so that a full factorial study would involve $I \cdot J \cdot K \cdot \cdots$ different samples. In such contexts, there is a trend these days to recommend fractional factorial experimental plans with a kind of symmetry or balance property, described in the following definition.

DEFINITION 10-5

In a p-factor study, if for every pair of factors, each level of the first appears in the n data points in combination with each level of the second the same number of times, the study is called an **orthogonal array plan**.

One attractive quality of the orthogonality idea in Definition 10-5 is the symmetry that it provides. It will also be seen later in this section that under no-interaction restrictions, the orthogonality idea provides some very simple inference formulas. Several examples of plans with the orthogonality property follow.

EXAMPLE 10-13

An Unreplicated 2^{3-1} Fractional Factorial as an Orthogonal Array Plan. Balanced 2^{p-q} fractional factorial plans of the type discussed in Sections 10-1 and 10-2 provide important instances of orthogonal arrays. For example, consider the unreplicated ($m = 1$) 2^{3-1} plan specified in Table 10-29.

TABLE 10-29

A Standard 2^{3-1} Fractional Factorial

Combination	Level of A	Level of B	Level of C
c	−	−	+
a	+	−	−
b	−	+	−
abc	+	+	+

It is a simple exercise to verify that in Table 10-29, each level of A is paired once with each level of B, each level A is paired once with each level of C, and each level of B is paired once with each level of C. The unreplicated standard $\frac{1}{2}$ fraction of the 2^3 design specified in Table 10-29 is in fact an orthogonal array plan for $p = 3$ factors and $n = 4$ data points. (For what it's worth, it is also a 2×2 Latin square.)

EXAMPLE 10-14

A 3 × 3 Latin Square as an Orthogonal Array Plan. Figure 2-8 and Table 2-8 describe a 3 × 3 Latin square. Consider an unreplicated fractional factorial study in the three (three-level) factors A, B, and C that is specified by this Latin square. Table 10-30 (a reproduction of Table 2-8) gives prescriptions for the $n = 9$ of the $3 \cdot 3 \cdot 3 = 27$ different possible combinations of levels of these three factors that would make up such a study.

TABLE 10-30

A 3 × 3 Latin Square

Combination	Level of A	Level of B	Level of C
1	1	1	1
2	1	2	3
3	1	3	2
4	2	1	2
5	2	2	1
6	2	3	3
7	3	1	3
8	3	2	2
9	3	3	1

Notice that in Table 10-30, each level of A is paired once with each level of B, each level of A is paired once with each level of C, and each level of B is paired once with each level of C. The unreplicated Latin square design specified in Table 10-30 is an orthogonal array plan for $p = 3$ factors and $n = 9$ data points. (In general, an unreplicated $k \times k$ Latin square is an orthogonal array plan for $p = 3$ factors and k^2 data points.)

EXAMPLE 10-15

A Graeco-Latin Square as an Orthogonal Array Plan. Table 10-31 specifies an orthogonal array plan for $p = 4$ factors in $n = 16$ data points. (The reader can verify that for every pair of factors, each level of the first appears once with each level of the second.)

TABLE 10-31

A Graeco-Latin Square for Factors at Four Levels Each

Data Point	Level of A	Level of B	Level of C	Level of D
1	1	1	1	1
2	1	2	2	2
3	1	3	3	3
4	1	4	4	4
5	2	1	2	3
6	2	2	1	4
7	2	3	4	1
8	2	4	3	2
9	3	1	3	4
10	3	2	4	3
11	3	3	1	2
12	3	4	2	1
13	4	1	4	2
14	4	2	3	1
15	4	3	2	4
16	4	4	1	3

The pattern of factor-level combinations in Table 10-31 is very specialized/symmetric. Considering any three of the four factors, one has a 4×4 Latin square. Table 10-31 in fact specifies what is known as a 4×4 *Graeco-Latin square* design. The origin of this nomenclature will not be pursued in this text, but the jargon is sometimes seen in more advanced texts on (specialized) experimental design.

EXAMPLE 10-16

An "L_{18}" Orthogonal Array Experiment in Integrated Circuit Fabrication. As an additional example of what is possible in the way of orthogonal arrays, consider a situation involving $p = 8$ factors, A at two levels and B, C, D, E, F, G, and H each at three levels. Phadke, Kackar, Speeney, and Grieco, in "Off-Line Quality Control in Integrated Circuit Fabrication Using Experimental Design" (*Bell System Technical Journal*, 1982), describe an orthogonal array experiment using samples of size $m = 2$ for each of the 18 different combinations of levels of these factors given in Table 10-32.

TABLE 10-32

18 Combinations in the Integrated Circuit Fabrication Experiment

Combination	A	B	C	D	E	F	G	H
1	1	1	1	1	1	1	1	1
2	1	1	2	2	2	2	2	2
3	1	1	3	3	3	3	3	3
4	1	2	1	1	2	2	3	3
5	1	2	2	2	3	3	1	1
6	1	2	3	3	1	1	2	2
7	1	3	1	2	1	3	2	3
8	1	3	2	3	2	1	3	1
9	1	3	3	1	3	2	1	2
10	2	1	1	3	3	2	2	1
11	2	1	2	1	1	3	3	2
12	2	1	3	2	2	1	1	3
13	2	2	1	2	3	1	3	2
14	2	2	2	3	1	2	1	3
15	2	2	3	1	2	3	2	1
16	2	3	1	3	2	3	1	2
17	2	3	2	1	3	1	2	3
18	2	3	3	2	1	2	3	1

The replication of each of the combinations indicated in Table 10-32 gave an orthogonal array study for $p = 8$ factors in $n = 36$ observations. The authors attribute the pattern of combinations listed in Table 10-32 to Dr. Genichi Taguchi, who has called the pattern in Table 10-32 the "L_{18}" array (presumably because 18 combinations are involved).

The orthogonality property of Definition 10-5 provides a kind of symmetry in a p-factor fractional factorial plan. But the orthogonality property in and of itself is not enough to guarantee the usefulness of a plan in engineering applications. Next, consider some of the limitations/deficiencies that will typically characterize (even orthogonal) fractional factorials in p factors when more than two levels of some factors are involved.

Practical Limitations

As a first step in understanding the limitations of (even orthogonal) fractional factorial designs when some or all of the factors have more than two levels, consider their use in response surface fitting contexts. The full 3^p factorial designs and central composite designs discussed in Section 9-3 (and other such *response surface designs*) have the virtue that they provide enough data for the fitting of a full quadratic response surface in p variables. Typical (small) orthogonal array fractional factorials do *not* provide enough data for the fitting of such surfaces. Unfortunately, they are nowadays often recommended to "assess both linear and quadratic effects" of factors. This advice is poor; their use as response surface designs is full of pitfalls. There follows an example of the kind of thing that can easily happen.

EXAMPLE 10-17

Deficiencies of an Orthogonal Array Plan As a Response Surface Design. Table 10-33 contains hypothetical data from a $p = 4$ factor study of the effects of variables x_1, x_2, x_3, and x_4 on a response y.

TABLE 10-33

Hypothetical Data from a 4-Factor Orthogonal Array Study

x_1	x_2	x_3	x_4	y
-1	-1	-1	-1	11
-1	0	1	0	10
-1	1	0	1	9
0	-1	0	0	10
0	0	-1	1	10
0	1	1	-1	10
1	-1	1	1	9
1	0	0	-1	10
1	1	-1	0	11

The $n = 9$ combinations of x_1, x_2, x_3, and x_4 in Table 10-33 have the orthogonality property of Definition 10-5. (Indeed, the design is a 3×3 Graeco-Latin square, popularly called "the L_9 array" these days.) But they are simply not adequate to allow a sensible second-order response surface analysis. A full quadratic response function

$$y \approx \beta_0 + \beta_1 x_1 + \beta_2 x_2 + \beta_3 x_3 + \beta_4 x_4 + \beta_5 x_1^2 + \beta_6 x_2^2 + \beta_7 x_3^2 + \beta_8 x_4^2$$
$$+ \beta_9 x_1 x_2 + \beta_{10} x_1 x_3 + \beta_{11} x_1 x_4 + \beta_{12} x_2 x_3 + \beta_{13} x_2 x_4 + \beta_{14} x_3 x_4 \quad \textbf{(10-15)}$$

has 15 coefficients that need to be estimated, and $n = 9$ data points are not sufficient to do the job. One ends up with many quite different versions of expression (10-15) as equally plausible descriptions of the data in Table 10-33. For example, the reader should take the time to verify that the two equations

$$y = 10 + x_1 x_2 \quad \textbf{(10-16)}$$
$$y = 10 - .5x_3 - .5x_4 + .5x_3^2 - .5x_4^2 \quad \textbf{(10-17)}$$

both describe the data in Table 10-33 exactly. But they would have quite different engineering interpretations and can produce quite different predicted responses for x_1, x_2, x_3, x_4, combinations not among the nine listed in Table 10-33.

Expression (10-16) says that x_3 and x_4 are without effect on y, while x_1 and x_2 interact, either one (for a fixed value of the other) affecting y in a linear fashion. On the other hand, expression (10-17) says that x_1 and x_2 are without effect on y, while x_3 and x_4 act separately on y (don't interact), both in a quadratic fashion. Further, for the set of conditions $x_1 = 1, x_2 = 1, x_3 = 1$, and $x_4 = 1$, equation (10-16) predicts $y = 11$, while equation (10-17) predicts $y = 9$. The problem with the design in Table 10-33 as a response surface design is that it can provide no data-based reason to prefer one or the other of equations (10-16) or (10-17) (and their corresponding different interpretations and predictions). (And of course, there are other possibilities besides equations (10-16) and (10-17) for quadratic response surfaces that describe the data in Table 10-33 exactly.)

Sections 10-1 and 10-2 lay out a coherent method for selection and interpretation of two-level fractional factorial data in terms of 2^p factorial effects. Unfortunately, no similar simple general story can be told for (even orthogonal) fractional factorials outside the realm of exclusively two-level factors. There remains the problem of confounding of effects (main effects, 2-factor interactions, 3-factor interactions, etc.). In fact, it is exacerbated by the fact that (for example) the presence of $I = 3$ levels of factor A would mean that there are now three A main effects α_1, α_2, and α_3 (with $\alpha_1 + \alpha_2 +$

$\alpha_3 = 0$) to be dealt with instead of just two (with $\alpha_1 + \alpha_2 = 0$) as in Sections 10-1 and 10-2. Unfortunately, there exists no simple device, like the formal multiplication technique used for 2^{p-q} designs, for figuring out the aliasing implications of a general fractional factorial design. Without such a device, one is left with no rational program for interpreting fractional factorial data in terms of the general p-way factorial effects introduced in Section 8-3.

All possible effect structures leading to a given fractional set of means cannot be identified and weighed against one another for engineering plausibility. So, if one allows the possibility of important interactions in an engineering system, typical orthogonal array fractional factorial plans are of little help in understanding the structure of mean responses or predicting responses for combinations of factor-levels not included in the study.

EXAMPLE 10-14
(continued)

As a hypothetical but instructive example of the quandary one faces when considering different plausible interpretations of general (even orthogonal) fractional factorial data in terms of factorial effects, recall the 3×3 Latin square of Table 10-30. For the sake of illustration, imagine that mean responses for the nine out of $3 \cdot 3 \cdot 3 = 27$ different possible combinations of levels of the qualitative factors A, B, and C are as in Table 10-34.

TABLE 10-34

Hypothetical Data from a
3×3 Latin Square

Level of A	Level of B	Level of C	Mean Response
1	1	1	13
1	2	3	11
1	3	2	12
2	1	2	9
2	2	1	13
2	3	3	11
3	1	3	5
3	2	2	6
3	3	1	10

When one realizes that there are potentially three A main effects α_1, α_2, α_3; three B main effects β_1, β_2, β_3; three C main effects γ_1, γ_2, γ_3; nine AB interactions $\alpha\beta_{ij}$; nine AC interactions $\alpha\gamma_{ik}$; nine BC interactions $\beta\gamma_{jk}$; and 27 ABC interactions $\alpha\beta\gamma_{ijk}$ impacting the response in a situation like this, it is clear that $n = 9$ data points are inadequate to completely sort out the system structure. The reader is encouraged to verify that the two different particular sets of 3^3 factorial effects

$$\mu_{...} = 10 \qquad \alpha_1 = 2, \ \alpha_2 = 1, \ \alpha_3 = -3 \qquad \beta_1 = -1, \ \beta_2 = 0, \ \beta_3 = 1$$
$$\gamma_1 = 2, \ \gamma_2 = -1, \ \gamma_3 = -1$$
with all other effects 0

and

$$\mu_{...} = 10 \qquad \alpha_1 = 2, \ \alpha_2 = 1, \ \alpha_3 = -3 \qquad \beta_1 = -1, \ \beta_2 = 0, \ \beta_3 = 1$$
$$\alpha\beta_{11} = 2, \ \alpha\beta_{12} = -1, \ \alpha\beta_{13} = -1$$
$$\alpha\beta_{21} = -1, \ \alpha\beta_{22} = 2, \ \alpha\beta_{23} = -1$$
$$\alpha\beta_{31} = -1, \ \alpha\beta_{32} = -1, \ \alpha\beta_{33} = 2$$
with all other effects 0

both lead to the nine means in Table 10-34. That is, on the basis of data from the 3×3 Latin square, there is no way to distinguish between a situation where factors A, B, and

C all have main effects but there are no interactions, and a situation where two factors have main effects and 2-factor interactions and the third is completely inert. In fact, one cannot even simply determine all of the possible patterns of effects that might plausibly produce the nine responses in Table 10-34.

A fairly bleak picture has been painted thus far in this section with regard to the engineering usefulness of small orthogonal array plans. But it is necessary to understand their limitations before going on to their possible use, if predictable practical disasters are to be avoided. For a further discussion of these matters, refer to the lively article "Let's All Beware the Latin Square" by J.S. Hunter (*Quality Engineering*, 1989).

No-Interaction Fitted Effects, Fitted Values, Residuals, and Variance Estimate for Orthogonal Array Designs

There are, perhaps, some p-factor engineering situations in which it is possible to "know" *a priori* that no interactions whatsoever will be important in determining a response y. For instance, one might have a validated physical theory saying that a response y is related to variables x_1, x_2, and x_3 through a power law like

$$y \approx \beta_0 x_1{}^{\beta_1} x_2{}^{\beta_2} x_3{}^{\beta_3}$$

The relationship

$$\ln(y) \approx \ln(\beta_0) + \beta_1 \ln(x_1) + \beta_2 \ln(x_2) + \beta_3 \ln(x_3)$$

follows from such a power law, and there are thus no interactions of the variables x_1, x_2, and x_3 in determining $\ln(y)$. Or, in a situation where factor A is a primary experimental factor, but B and C are blocking factors that in similar studies in the past have shown no interactions with similar experimental variables or with each other, it may be safe to plan a statistical engineering study presuming that no interactions will be of importance. In such cases, under a no-interaction restriction on the usual p-way factorial model (with constant variance and normal distributions), data from orthogonal array plans are often sufficient to provide fitted responses, residuals, and estimates of and comparisons between the main effects associated with different levels of the p various factors.

For definiteness here, formulas will be written primarily for $p = 3$ orthogonal array plans, but it should be clear how to make up similar formulas for p of any size. The keys to inference under a no-interaction $p = 3$-way model

$$y_{ijkl} = \mu_{...} + \alpha_i + \beta_j + \gamma_k + \epsilon_{ijkl} \tag{10-18}$$

using orthogonal array data, are that sample means of all observations at a given level of a particular factor can be used to estimate $\mu_{...}$ plus the main effect of that factor at the given level, and that the grand mean of all observations can be used to estimate $\mu_{...}$.

That is, suppose that one has m observations at each of an integer number, $\frac{n}{m}$, of different combinations of levels of A, B, and C. Then for $\frac{n}{m}$ triples (i, j, k), sample means $\bar{y}_{ijk}$ are available. (If a fractional factorial orthogonal array plan is under consideration, $\frac{n}{m} < I \cdot J \cdot K$.) The abbreviations

$$\bar{y}^*_{i..} = \frac{mI}{n} \sum_{j,k} \bar{y}_{ijk} \tag{10-19}$$

$$\bar{y}^*_{.j.} = \frac{mJ}{n} \sum_{i,k} \bar{y}_{ijk} \tag{10-20}$$

$$\bar{y}^*_{..k} = \frac{mK}{n} \sum_{i,j} \bar{y}_{ijk} \tag{10-21}$$

$$\bar{y}^*_{...} = \frac{m}{n} \sum_{i,j,k} \bar{y}_{ijk} \tag{10-22}$$

will be used here, where the first sum is over all combinations with A at level i, the second is over all combinations with B at level j, the third is over all combinations with C at level k, and the fourth is over all combinations in the orthogonal array plan. Using expressions (10-19) through (10-22), estimates of the A, B, and C main effects in model (10-18) can be made from orthogonal array data as follows:

$$a_i^* = \bar{y}_{i..}^* - \bar{y}_{...}^* \qquad \text{(10-23)}$$

$$b_j^* = \bar{y}_{.j.}^* - \bar{y}_{...}^* \qquad \text{(10-24)}$$

$$c_k^* = \bar{y}_{..k}^* - \bar{y}_{...}^* \qquad \text{(10-25)}$$

The stars are used here as in Section 10-1 to remind the reader that unlike the fitted effects defined in Section 4-3, the quantities defined in expressions (10-23) through (10-25) are computed based on samples from *a fraction* of all $I \cdot J \cdot K$ possible combinations of factor levels. Expressions (10-23) through (10-25) then lead to fitted/predicted responses for the no-interaction model (10-18)

$$\hat{y}_{ijk} = \bar{y}_{...}^* + a_i^* + b_j^* + c_k^* \qquad \text{(10-26)}$$

by straightforward addition.

EXAMPLE 10-18

A 4×4 Latin Square Chemical Cell Temperature Study. Chapter 13 of the NBS (now NIST) Handbook 91, *Experimental Statistics*, by M.G. Natrella, contains a discussion of a 4×4 unreplicated Latin square study involving the use of chemical cells as a means of setting up a reference temperature. (Apparently, physical considerations in the context of the study made a no-interaction description of the system plausible, so a Latin square study was potentially adequate to determine the sizes of various effects.)

On each of four different days, each of four different thermometers was placed in one of four different chemical cells and a temperature recorded. The pattern of data collection and the resulting measured temperatures (in $.0001\degree$C above a common value not reported in the handbook) were as given in Table 10-35.

It is then possible to verify that for the data in Table 10-35,

$$\bar{y}_{1..}^* = 35.00, \qquad \bar{y}_{2..}^* = 19.50, \qquad \bar{y}_{3..}^* = 36.00, \qquad \bar{y}_{4..}^* = 36.50$$

$$\bar{y}_{.1.}^* = 28.25, \qquad \bar{y}_{.2.}^* = 35.00, \qquad \bar{y}_{.3.}^* = 35.25, \qquad \bar{y}_{.4.}^* = 28.50$$

$$\bar{y}_{..1}^* = 35.25, \qquad \bar{y}_{..2}^* = 30.75, \qquad \bar{y}_{..3}^* = 31.25, \qquad \bar{y}_{..4}^* = 29.75$$

$$\bar{y}_{...}^* = 31.75$$

TABLE 10-35

Measured Temperatures in a 4×4 Latin Square Chemical Cell Temperature Study

Cell	Thermometer	Day	Temperature
1	1	1	36
1	2	2	38
1	3	3	36
1	4	4	30
2	1	3	17
2	2	4	18
2	3	1	26
2	4	2	17
3	1	2	30
3	2	3	39
3	3	4	41
3	4	1	34
4	1	4	30
4	2	1	45
4	3	2	38
4	4	3	33

From these means and formulas (10-23) through (10-25), estimated Cell (A), Thermometer (B), and Day (C) main effects are

$$a_1^* = 3.25, \qquad a_2^* = -12.25, \qquad a_3^* = 4.25, \qquad a_4^* = 4.75$$
$$b_1^* = -3.50, \qquad b_2^* = 3.25, \qquad b_3^* = 3.50, \qquad b_4^* = -3.25$$
$$c_1^* = 3.50, \qquad c_2^* = -1.00, \qquad c_3^* = -.50, \qquad c_4^* = -2.00$$

And, for example, the fitted value for cell 1, thermometer 1, and day 1 is

$$\hat{y}_{111} = 31.75 + 3.25 + (-3.50) + 3.50 = 35.00$$

The possibility of generating no-interaction fitted values (10-26), means that (except in those $m = 1$ cases where n is so small that fitted values coincide with observed values) one can compute and plot residuals for model (10-18) as a minimal data-based kind of check on the appropriateness of that model. Further, in what is by now standard fashion, appropriately averaged squared residuals can be used to provide a no-interaction sample variance.

DEFINITION 10-6

For an orthogonal array plan in p factors and n observations with fitted values $\hat{y}$,

$$s_{\mathrm{NI}}^2 = \frac{1}{n + p - 1 - (I + J + \cdots)} \sum (y - \hat{y})^2 \tag{10-27}$$

(where the sum in expression (10-27) is over all n data points y and the subscripts have been suppressed) will be called a **no-interaction sample variance**. Associated with it are $\nu = n + p - 1 - (I + J + \cdots)$ degrees of freedom.

Where sample sizes for given ABC (etc.) combinations, m, are larger than 1, rough comparison of s_{NI} to s_P provides the usual check of fit for the (specialized) no-interaction model. Further, s_{NI} can be used to make formal inferences about the various main effects under model (10-18).

EXAMPLE 10-18
(continued)

Table 10-36 is an expansion of Table 10-35 to include all 16 fitted values and the corresponding residuals.

TABLE 10-36

Observations, Fitted Values, and Residuals for a No-Interaction Analysis of the Chemical Cell Temperature Data

Cell	Thermometer	Day	Temperature, y	$\hat{y}$	$e = y - \hat{y}$
1	1	1	36	35.00	1.00
1	2	2	38	37.25	.75
1	3	3	36	38.00	−2.00
1	4	4	30	29.75	.25
2	1	3	17	15.50	1.50
2	2	4	18	20.75	−2.75
2	3	1	26	26.50	−.50
2	4	2	17	15.25	1.75
3	1	2	30	31.50	−1.50
3	2	3	39	38.75	.25
3	3	4	41	37.50	3.50
3	4	1	34	36.25	−2.25
4	1	4	30	31.00	−1.00
4	2	1	45	43.25	1.75
4	3	2	38	39.00	−1.00
4	4	3	33	32.75	.25

The ambitious reader is invited to plot the residuals in Table 10-36 against chemical cell number, against thermometer number, against day number, and against $\hat{y}$ and to normal-plot them. None of these plots turns out to have a very unusual appearance, and so the residuals do not obviously discredit the no-interaction description (10-18) of this 3-factor study.

Proceeding then to apply formula (10-27) to produce a no-interaction sample variance, one has

$$s_{\mathrm{NI}}^2 = \frac{((1.00)^2 + (.75)^2 + (-2.00)^2 + \cdots + (-1.00)^2 + (.25)^2)}{16 + 3 - 1 - (4 + 4 + 4)}$$

$$= \frac{43.5}{6}$$

$$= 7.25$$

so

$$s_{\mathrm{NI}} = 2.69(.0001°\mathrm{C})$$

with $\nu = 6$ associated degrees of freedom. There was no replication in this study and thus no value of s_P against which to compare s_{NI} in an effort to further investigate the plausibility of the no-interaction model (10-18).

━━━ ▰▰▰▰

It is worth noting that for orthogonal array plans, one may employ the dummy variable idea of Section 9-4 and a multiple regression program to arrive at fitted values (10-26) and the no-interaction sample variance s_{NI}^2. (Whether or not a design is orthogonal, the analysis route of using the regression approach of Section 9-4 is available. But the point here is that when the orthogonality condition in Definition 10-5 is satisfied, the regression calculations produce exactly the quantities discussed here in terms of sample means.) As an example of this, Printout 10-1 shows a Minitab multiple regression run using $I - 1 = 3$ factor A (Cell) dummy variables, $J - 1 = 3$ factor B (Thermometer) dummy variables, and $K - 1 = 3$ factor C (Day) dummy variables.

The reader should verify that in Printout 10-1, the estimated effects $\bar{y}_{...}^*$, a_1^*, a_2^*, a_3^*, b_1^*, b_2^*, b_3^*, c_1^*, c_2^*, and c_3^* from formulas (10-22) through (10-25) appear as estimated regression coefficients b_0 through b_{10}, respectively; fitted values (10-26) are displayed as regression $\hat{y}$'s; and s_{NI} from formula (10-27) appears as s_{SF}.

There is a further useful possibility raised by the realization that the idea of regression with dummy variables can be used in the analysis of orthogonal array data. The estimated standard deviations of fitted values $\hat{y}$ typically provided by multiple regression programs enable one to make confidence intervals for mean responses, prediction intervals for additional responses, and even tolerance intervals under the no-interaction model (10-18), by means of the formulas in Section 9-2. Actually, explicit formulas could be given for such inference methods, but that will be avoided here, on the grounds that to write out such formulas invites abuse. The grossly incomplete nature of many orthogonal array designs that are popular these days makes it impossible to evaluate (in a serious, data-based fashion) the practical applicability of the no-interaction model (10-18). All inferences based on model (10-18) should therefore be treated as quite tenuous. In particular, it seems quite risky to make inferences concerning responses at various ABC (etc.) combinations (particularly ones not included in the design). So, in a manner consistent with what was done in Sections 10-1 and 10-2, only a limited spectrum of formal inference methods for orthogonal array designs will be considered here, essentially confined to methods for comparing main effects for different levels of a given factor.

PRINTOUT 10-1

```
MTB > print c1-c10

   ROW   xa1   xa2   xa3   xb1   xb2   xb3   xc1   xc2   xc3     y

     1     1     0     0     1     0     0     1     0     0    36
     2     1     0     0     0     1     0     0     1     0    38
     3     1     0     0     0     0     1     0     0     1    36
     4     1     0     0    -1    -1    -1    -1    -1    -1    30
     5     0     1     0     1     0     0     0     0     1    17
     6     0     1     0     0     1     0    -1    -1    -1    18
     7     0     1     0     0     0     1     1     0     0    26
     8     0     1     0    -1    -1    -1     0     1     0    17
     9     0     0     1     1     0     0     0     1     0    30
    10     0     0     1     0     1     0     0     0     1    39
    11     0     0     1     0     0     1    -1    -1    -1    41
    12     0     0     1    -1    -1    -1     1     0     0    34
    13    -1    -1    -1     1     0     0    -1    -1    -1    30
    14    -1    -1    -1     0     1     0     1     0     0    45
    15    -1    -1    -1     0     0     1     0     1     0    38
    16    -1    -1    -1    -1    -1    -1     0     0     1    33

MTB > regress c10 9 c1-c9

The regression equation is
y = 31.8 + 3.25 xa1 - 12.2 xa2 + 4.25 xa3 - 3.50 xb1 + 3.25 xb2
         + 3.50 xb3 + 3.50 xc1 - 1.00 xc2 - 0.50 xc3

Predictor        Coef        Stdev      t-ratio          p
Constant      31.7500       0.6731        47.17      0.000
xa1             3.250        1.166         2.79      0.032
xa2           -12.250        1.166       -10.51      0.000
xa3             4.250        1.166         3.65      0.011
xb1            -3.500        1.166        -3.00      0.024
xb2             3.250        1.166         2.79      0.032
xb3             3.500        1.166         3.00      0.024
xc1             3.500        1.166         3.00      0.024
xc2            -1.000        1.166        -0.86      0.424
xc3            -0.500        1.166        -0.43      0.683

s = 2.693       R-sq = 96.0%      R-sq(adj) = 90.1%

Analysis of Variance

SOURCE        DF           SS           MS          F          p
Regression     9      1057.50       117.50      16.21      0.001
Error          6        43.50         7.25
Total         15      1101.00

SOURCE        DF       SEQ SS
xa1            1         4.50
xa2            1       704.17
xa3            1        96.33
xb1            1         0.13
xb2            1       117.04
xb3            1        65.33
xc1            1        60.50
xc2            1         8.17
xc3            1         1.33
```

(continued)

PRINTOUT 10-1
(continued)

Obs.	xa1	y	Fit	Stdev.Fit	Residual	St.Resid
1	1.00	36.000	35.000	2.129	1.000	0.61
2	1.00	38.000	37.250	2.129	0.750	0.45
3	1.00	36.000	38.000	2.129	-2.000	-1.21
4	1.00	30.000	29.750	2.129	0.250	0.15
5	0.00	17.000	15.500	2.129	1.500	0.91
6	0.00	18.000	20.750	2.129	-2.750	-1.67
7	0.00	26.000	26.500	2.129	-0.500	-0.30
8	0.00	17.000	15.250	2.129	1.750	1.06
9	0.00	30.000	31.500	2.129	-1.500	-0.91
10	0.00	39.000	38.750	2.129	0.250	0.15
11	0.00	41.000	37.500	2.129	3.500	2.12R
12	0.00	34.000	36.250	2.129	-2.250	-1.36
13	-1.00	30.000	31.000	2.129	-1.000	-0.61
14	-1.00	45.000	43.250	2.129	1.750	1.06
15	-1.00	38.000	39.000	2.129	-1.000	-0.61
16	-1.00	33.000	32.750	2.129	0.250	0.15

R denotes an obs. with a large st. resid.

Comparison of Main Effects for Orthogonal Array Designs

The basic probability fact that makes possible interval-based inference for comparing main effects of a given factor based on orthogonal array data is that model (10-18) implies that differences of fitted effects a_i^* (or b_j^*, or c_k^*, etc.) have very simple distributions. That is, under the no-interaction restriction of model (10-18),

$$a_i^* - a_{i'}^* = \bar{y}_{i..}^* - \bar{y}_{i'..}^*$$

is normal with

$$E(a_i^* - a_{i'}^*) = \alpha_i - \alpha_{i'}$$
$$Var(a_i^* - a_{i'}^*) = Var(\bar{y}_{i..}^* - \bar{y}_{i'..}^*)$$
$$= \frac{I\sigma^2}{n} + \frac{I\sigma^2}{n}$$
$$= \frac{2I}{n}\sigma^2$$

It is then perfectly plausible that under assumptions (10-18), two-sided (orthogonal array) individual confidence limits for the difference in factor A main effects at levels i and i' ($\alpha_i - \alpha_{i'}$) can be made using the endpoints

$$\bar{y}_{i..}^* - \bar{y}_{i'..}^* \pm ts_{\text{NI}}\sqrt{\frac{2I}{n}} \tag{10-28}$$

where t is chosen so that the probability the $t_{n+p-1-(I+J+\cdots)}$ distribution assigns to the interval between $-t$ and t corresponds to the desired confidence. As usual, a one-sided limit can be made from formula (10-28) by appropriately adjusting the stated confidence, and there is also a Tukey version of formula (10-28). That is, under the model restrictions (10-18), two-sided simultaneous confidence intervals for all possible differences in A main effects ($\alpha_i - \alpha_{i'}$) can be made using respective endpoints

$$\bar{y}_{i..}^* - \bar{y}_{i'..}^* \pm q^* s_{\text{NI}}\sqrt{\frac{I}{n}} \tag{10-29}$$

where q^* is taken from Tables D-10 using $\nu = n + p - 1 - (I + J + \cdots)$ degrees of freedom and number of means to be compared I. The .95 or .99 quantile figure is used, depending on whether 95% or 99% simultaneous confidence is desired.

There are, of course, formulas parallel to formulas (10-28) and (10-29) for comparing the main effects of other factors besides A in a p-factor orthogonal array study. One simply replaces $\bar{y}_{i..}^* - \bar{y}_{i'..}^*$ by appropriate differences in sample means, and the number I by the number of levels of the factor in question.

EXAMPLE 10-18
(continued)

Consider the problem of comparing factor A (Chemical Cell) main effects in the reference temperature study of Natrella. There are $\nu = 6$ degrees of freedom associated with $s_{\text{NI}} = 2.69$. So if, for example, one desires a 95% individual confidence interval for comparing the effects of two particular chemical cells, formula (10-28) indicates that a precision of

$$\pm 2.447(2.69)\sqrt{\frac{2 \cdot 4}{16}}$$

that is,

$$\pm 4.65(.0001°C)$$

should be attached to a corresponding difference $\bar{y}_{i..}^* - \bar{y}_{i'..}^* (= a_i^* - a_{i'}^*)$. For instance, cells 1 and 2 would be judged to differ in mean temperature by an amount

$$(35.0 - 19.5) \pm 4.65$$

that is, by

$$\text{from } 10.85 \, (.0001°C) \text{ to } 20.15 \, (.0001°C)$$

for any given day and thermometer. Cell 1 is clearly somewhat warmer than cell 2.

If one is to make *simultaneous* 95% intervals for comparing all pairs of chemical cells, $\nu = 6$ and $I = 4$ are used to locate an appropriate multiplier q^* in Tables D-10. Then from formula (10-29), for 95% simultaneous confidence, each difference in cell mean temperatures can be stated with a precision of

$$\pm 4.90(2.69)\sqrt{\frac{4}{16}}$$

that is,

$$\pm 6.59(.0001°C)$$

In this particular example, since 4 is not only the number of chemical cells (I), but also the number of thermometers (J) and number of days (K) involved, the 4.65 and 6.59 figures found here apply not only to comparisons between cells, but to comparisons between thermometers and between days as well. And reviewing the sample means $\bar{y}_{i..}^*$, $\bar{y}_{.j.}^*$, and $\bar{y}_{..k}^*$ found earlier, one can see (for example) that by the Tukey 95% simultaneous confidence criterion, there are statistically detectable differences between chemical cells and between thermometers, but no statistically detectable differences between days.

It cannot be emphasized strongly enough that formulas (10-28) and (10-29) provide inferences for differences in main effects based on orthogonal array data *only* when the no-interaction model (10-18) is appropriate. The problem is not only that main effects lose much of their practical interpretability in the presence of important interactions. An additional pitfall is that when only fractional factorial data are used to try to compare main effects, one ends up estimating not simply a difference in main effects, but that difference plus a sum of interactions peculiar to the particular design. This has already been studied in detail in the relatively tame situation of 2^{p-q} studies, but it is true in some (typically much less manageable) form of all fractional factorial plans. The reader might find it instructive to verify, for example, that in the 4×4 Latin square situation of Example 10-18, in general (i.e., lacking the no-interaction restriction (10-18)), $\bar{y}_{1..}^{*} - \bar{y}_{2..}^{*}$ has mean value (and thus actually estimates)

$$\alpha_1 - \alpha_2 + \left(\frac{1}{4} (\beta\gamma_{11} + \beta\gamma_{22} + \beta\gamma_{33} + \beta\gamma_{44}) \right.$$

$$+ \frac{1}{4} (\alpha\beta\gamma_{111} + \alpha\beta\gamma_{122} + \alpha\beta\gamma_{133} + \alpha\beta\gamma_{144}) - \frac{1}{4} (\beta\gamma_{13} + \beta\gamma_{24} + \beta\gamma_{31} + \beta\gamma_{42})$$

$$\left. - \frac{1}{4} (\alpha\beta\gamma_{213} + \alpha\beta\gamma_{224} + \alpha\beta\gamma_{231} + \alpha\beta\gamma_{242}) \right)$$

An engineer who naively hopes to "economize" on data collection in a complicated many-factor system that is likely to have important interaction effects, by using a Latin square or other multilevel fractional factorial orthogonal array design of unknown aliasing structure, is practically guaranteed to end up misled, disillusioned, and quite possibly exceedingly embarrassed.

As a final comment in this section, note that little to nothing has been said here about the making up or choice of orthogonal array fractional factorial plans. This is, of course, no accident. Given the rather pessimistic posture adopted here regarding the overall practical usefulness of most (other than two-level) orthogonal array fractional factorial plans, serious consideration of the matter hardly seems justified. In those large-p cases where the design material of Sections 9-3, 10-1, and 10-2 is not adequate for practical purposes and a no-interaction model like (10-18) is almost certainly appropriate, the reader is referred to the existing statistical literature on experimental design for methods of choosing orthogonal array designs. Consult experimental design texts such as *Practical Experimental Designs* by W. J. Diamond, *Fundamental Concepts in the Design of Experiments* by C. R. Hicks, *Statistics for Experimenters* by Box, Hunter, and Hunter, and *Statistical Design and Analysis of Experiments* by Mason, Gunst, and Hess. The early definitive work on the topic is the paper "Orthogonal Main Effect Plans for Asymmetrical Factorial Experiments" by S. Addleman (*Technometrics*, 1962) and the Iowa State University technical reports of Addleman and Kempthorne that preceded the paper.

10-4 *Balanced Hierarchical Studies*

Section 1-2 (specifically, Definition 1-15) introduced the notion of a hierarchical data structure and remarked that hierarchical studies are often useful tools in situations where engineers need to identify sources of unwanted variation in some measured quantity, y. This section is finally going to provide some guidance in the analysis of data from such studies. The analytical tools to be discussed can be thought of as essentially extensions of the one-way ANOVA random effects methods introduced at the end of Section 7-4.

The discussion here begins with the introduction of a commonly used random effects probability model for balanced hierarchical studies. Then some graphical methods that can be useful in getting the feel of a hierarchical data set and in investigating the relevance of the random effects model are considered. The section concludes by presenting the form of an ANOVA table corresponding to that model and attendant estimates of variance components, normal theory significance tests of various hypotheses that particular variance components are 0, and some confidence interval methods.

***A Random Effects
Model for Balanced
Hierarchical Studies***

In keeping with Definition 1-15 and Figure 1-3 (reproduced here as Figure 10-5), consider multisample studies possessing a balanced, tree-type structure. That is, this discussion will treat situations involving a series of factors A, B, C, etc., where levels of B actually represent J sublevels of each of I levels of factor A; levels of C actually represent K sublevels of each of the $J \cdot I$ levels of B within A, etc. The discussion here will be limited to the balanced-data situation where each level of A has the same number of (sub)levels of B, each level of B within A has the same number of (sub)levels of C, etc. Unbalanced cases are notoriously far more difficult to analyze, although some statistical methods have been developed for handling particular special patterns of "unbalance." (In this regard, refer to "The Use and Analysis of Staggered Nested Designs" by Smith and Beverly (*Journal of Quality Technology*, 1981).) Further, for clarity of exposition, the formulas and discussion will center on the 3-factor case where only factors A, B, and C are involved. Formulas for more than three factors (more than two levels of hierarchy) are exactly analogous to the ones presented here. And once the story told here for three factors is understood, the reader ought to be able to work by analogy, if applications involving more extensive nesting should be encountered.

FIGURE 10-5

Schematic of a Generic
Balanced Hierarchical Data
Structure

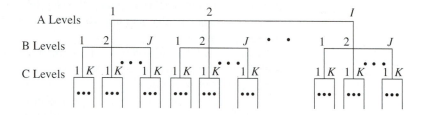

EXAMPLE 10-19

A Hierarchical Copper Content Study. The situation described by G. Wernimont in "Statistical Quality Control in the Chemical Laboratory" (*Quality Engineering*, 1989) will be used as an example of a balanced hierarchical study. (The article is actually reprinted from the older publication, *Industrial Quality Control*.) Wernimont's study concerned the copper content of bronze castings. $I = 11$ castings (factor A) were selected for study; $J = 2$ physical specimens (factor B) were taken from each casting; and for each of the $J \cdot I = 22$ different physical specimens, $K = 2$ laboratory determinations (factor C) of copper content, y, were made. Wernimont's data are given in Table 10-37. The $n = I \cdot J \cdot K = 44$ data points in the table are $r = I \cdot J = 22$ samples of size $m = K = 2$, which are naturally aggregated into $I = 11$ groups of size $J = 2$.

In most applications of hierarchical data structures, the particular levels of the series of factors actually studied are of interest only to the extent that they provide information on a wider universe of possible levels. For example, in Example 10-19, the particular castings studied are probably of interest only to the extent that they provide information on all such castings. The specimens for a given casting are of interest only to the extent that they provide information on all such physical samples that could be taken from a given casting. The laboratory measurements made for a given sample are of interest only to the extent that they represent all similar determinations that could be made for a given physical specimen. In recognition of this character of typical hierarchical studies, the statistical methods usually applied are random effects methods, centering on inference for variance components. This section primarily concerns such methods. But it is worth noting in passing that in those fairly unusual circumstances that the $(I \cdot J)$ basic samples in a (3-factor) hierarchical study represent levels of the factors that are of interest in their

Casting	Specimen	Determination	Copper Content, y (%)
1	1	1	85.54
1	1	2	85.56
1	2	1	85.51
1	2	2	85.54
2	1	1	85.54
2	1	2	85.60
2	2	1	85.25
2	2	2	85.25
3	1	1	85.72
3	1	2	85.77
3	2	1	84.94
3	2	2	84.95
4	1	1	85.48
4	1	2	85.50
4	2	1	84.98
4	2	2	85.02
5	1	1	85.54
5	1	2	85.57
5	2	1	85.84
5	2	2	85.84
6	1	1	85.72
6	1	2	85.86
6	2	1	85.81
6	2	2	85.91
7	1	1	85.72
7	1	2	85.76
7	2	1	85.81
7	2	2	85.84
8	1	1	86.12
8	1	2	86.12
8	2	1	86.12
8	2	2	86.20
9	1	1	85.47
9	1	2	85.49
9	2	1	85.75
9	2	2	85.77
10	1	1	84.98
10	1	2	85.10
10	2	1	85.90
10	2	2	85.90
11	1	1	85.12
11	1	2	85.17
11	2	1	85.18
11	2	2	85.24

TABLE 10-37

Measured Copper Contents for 11 Castings

own right, the methods of Chapter 7 (in particular that dealing with the estimation of linear combinations of r means) are available for use. For example, the first four samples of size 2 in Table 10-37 could be used to make a confidence interval for the difference in averages of the first two and second two means, as a way of comparing casting 1 to casting 2.

In order to describe the probability model typically used to support inferences from (3-factor) hierarchical studies, a triple-subscript notation will be used, which formally looks much like that employed in Section 4-3 and in Chapter 8 when two-way factorial situations were considered. (The reader will simply have to remember from context that the data structure under discussion in this section is hierarchical, not factorial.) That is, let

$$y_{ijk} = \text{the } k\text{th observation at the } j\text{th level of factor} \qquad \textbf{(10-30)}$$
$$\text{B within the } i\text{th level of factor A}$$

The model equation commonly used to describe the observations y_{ijk} in a hierarchical study is then of the form

$$y_{ijk} = \mu + \alpha_i + \beta_{ij} + \epsilon_{ijk} \qquad \textbf{(10-31)}$$

which has both formal similarities and dissimilarities to other model equations encountered in this book. But it is, of course, the meaning of the terms in equation (10-31) that makes the hierarchical random effects model what it is. In model (10-31), the overall mean response μ is an (unknown) fixed parameter, while all of $\alpha_1, \alpha_2, \dots, \alpha_I$, $\beta_{11}, \beta_{12}, \dots, \beta_{1J}, \beta_{21}, \dots, \beta_{IJ}, \epsilon_{111}, \epsilon_{112}, \dots, \epsilon_{11K}, \epsilon_{121}, \dots, \epsilon_{IJK}$ are treated as independent normal random variables. The α_i's are assumed to have mean 0 and variance σ_α^2, the β_{ij}'s to have mean 0 and variance σ_β^2, and the ϵ_{ijk}'s to have mean 0 and variance σ^2, where σ_α^2, σ_β^2, and σ^2 are themselves (unknown) fixed parameters. The parameters σ_α^2, σ_β^2 and σ^2 are the variance components for model (10-31) and are typically the quantities of primary interest in a hierarchical study. In the context of Example 10-19, these components would be measures of (respectively) how much castings differ, how much specimens within castings differ, and how much laboratory copper determinations differ for a given specimen.

It is possible to rewrite equation (10-31) in order to emphasize its similarity to equation (7-68) of Section 7-4, by letting $\mu_i = \mu + \alpha_i$ and writing

$$y_{ijk} = \mu_i + \beta_{ij} + \epsilon_{ijk}$$

where now the μ_i are iid normal random variables with mean μ and variance σ_α^2. However, in this section form (10-31) of the model equation will be most useful.

In order to describe inference methods under model (10-31), notation for certain averages of the basic observations y_{ijk} is needed. So, in the obvious ways, let

$$\bar{y}_{ij} = \frac{1}{K} \sum_k y_{ijk} \qquad \textbf{(10-32)}$$

$$\bar{y}_{i.} = \frac{1}{J} \sum_j \bar{y}_{ij} = \frac{1}{IJ} \sum_{j,k} y_{ijk} \qquad \textbf{(10-33)}$$

$$\bar{y}_{..} = \frac{1}{I} \sum_i \bar{y}_{i.} = \frac{1}{IJK} \sum_{i,j,k} y_{ijk} \qquad \textbf{(10-34)}$$

Again as in equation (10-30), though the formal notation in formulas (10-32) through (10-34) looks like the notation used for factorial analyses, keep in mind that the quite different, hierarchical structure is at work producing the averages in formulas (10-32) through (10-34).

EXAMPLE 10-19
(continued)

The reader can verify that the data in Table 10-37 produce the various sample means collected in Table 10-38.

Further, the grand mean for the data is

$$\bar{y}_{..} = \frac{941.2500}{11} = 85.568$$

TABLE 10-38

Specimen and Casting Mean
Measured Copper Contents

i Casting	j Specimen	$\bar{y}_{ij}$ Specimen Mean	$\bar{y}_{i.}$ Casting Mean
1	1	85.550	85.5375
1	2	85.525	
2	1	85.570	85.4100
2	2	85.250	
3	1	85.745	85.3450
3	2	84.945	
4	1	85.490	85.2450
4	2	85.000	
5	1	85.555	85.6975
5	2	85.840	
6	1	85.790	85.8250
6	2	85.860	
7	1	85.740	85.7825
7	2	85.825	
8	1	86.120	86.1400
8	2	86.160	
9	1	85.480	85.6200
9	2	85.760	
10	1	85.040	85.4700
10	2	85.900	
11	1	85.145	85.1775
11	2	85.210	

(The four decimal places displayed in Table 10-38 for the $\bar{y}_{i.}$ values are recorded to prevent round-off problems later in this example, not because they are significant in any usual sense of the word.)

In order to anticipate how the analysis of hierarchical data will proceed, it is worthwhile to consider what kinds of variation affect the individual data values and the various sample means. For example, in the copper content context, for given i and j (for a given physical specimen), variability between the various y_{ijk}'s is attributable to the analysis-to-analysis variation represented in model (10-31) by the ϵ_{ijk}'s, whose distribution is governed by σ^2. For a given i (for a given casting), variability between the various $\bar{y}_{ij}$'s is attributable to both the analysis-to-analysis variation and also the specimen-to-specimen variation represented in model (10-31) by the β_{ij}'s, whose distribution is governed by σ_β^2. Finally, variability between the various $\bar{y}_{i.}$'s is traceable not only to analysis-to-analysis variation and specimen-to-specimen variation, but also to variation between castings, represented in model (10-31) by the α_i's, whose distribution is governed by σ_α^2.

Graphical Tools for Balanced Hierarchical Studies

One helpful way to graphically summarize the results of a statistical engineering study with balanced hierarchical structure is through the use of the Shewhart control charts of Section 7-5. This is true whether or not there is a relevant time dimension involved in the data structure. One can treat the original data values y_{ijk} as $I \cdot J$ samples of size K, with means $\bar{y}_{ij}$ (using $\bar{y}_{i.}$'s or $\bar{y}_{..}$ as center lines or line on an $\bar{x}$ chart), then treat the $\bar{y}_{ij}$'s as I samples of size J with means $\bar{y}_{i.}$ (using $\bar{y}_{..}$ as the center line on an $\bar{x}$ chart), and so on (if more than two levels of hierarchy are involved). At a given level of data aggregation, the appearance of the $\bar{x}$ chart indicates the extent to which variation in the means of a given type exceeds a baseline level established by factors "further down in the hierarchy." At that same level of aggregation, the appearance of an s or R chart indicates the degree to which that baseline is consistent (as required if one is to make formal inferences based on model (10-31)).

EXAMPLE 10-19
(continued)

Table 10-39 provides the 22 sample variances and standard deviations for the $K = 2$ copper determinations made on each specimen and 11 sample variances and standard deviations calculated from the $J = 2$ values $\bar{y}_{ij}$ for a given casting i.

TABLE 10-39

Sample Variances and Standard Deviations of Individual Copper Measurements and Specimen Mean Measurements

i Casting	j Sample	s_{ij}^2	s_{ij}	Sample Variance of $\bar{y}_{i1}$ and $\bar{y}_{i2}$	Standard Deviation of $\bar{y}_{i1}$ and $\bar{y}_{i2}$
1	1	.00020	.01414	.00031	.01768
1	2	.00045	.02121		
2	1	.00180	.04243	.05120	.22627
2	2	0	0		
3	1	.00125	.03536	.32000	.56569
3	2	.00005	.00707		
4	1	.00020	.01414	.12005	.34648
4	2	.00080	.02828		
5	1	.00045	.02121	.04061	.20153
5	2	0	0		
6	1	.00980	.09899	.00245	.04950
6	2	.00500	.07071		
7	1	.00080	.02828	.00361	.06010
7	2	.00045	.02121		
8	1	0	0	.00080	.02828
8	2	.00320	.05657		
9	1	.00020	.01414	.03920	.19799
9	2	.00020	.01414		
10	1	.00720	.08485	.36980	.60811
10	2	0	0		
11	1	.00125	.03536	.00211	.04596
11	2	.00180	.04243		

Consider first the construction of an "$\bar{x}$ and s" chart pair based on the 22 means $\bar{y}_{ij}$ and standard deviations s_{ij}. From Table 10-39,

$$\bar{s} = \frac{1}{22}(.01414 + .02121 + .04243 + \cdots + .04243) = .02957$$

Then, from Table D-2, one has that $B_4 = 3.267$ for samples of size 2, so a retrospective upper "control limit" for sample standard deviations of two copper determinations based on $\bar{s}$ is

$$B_4\bar{s} = 3.267(.02957) = .09661$$

Further, since for samples of size 2, $A_3 = 2.659$ retrospective "control limits" for averages of two copper determinations for a given specimen will be some central value plus or minus

$$A_3\bar{s} = 2.659(.02957) = .07863$$

Figure 10-6 shows an $\bar{x}$ and s chart pair for the 22 samples of two measurements on each specimen with $\bar{x}$ control limits based on a center line at $\bar{y}_{..}$ and also ones at $\bar{y}_{i.}$ for each i.

FIGURE 10-6

$\bar{x}$ and s Charts for Samples
of Two Measurements on
22 Specimens

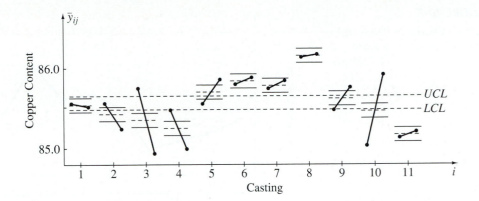

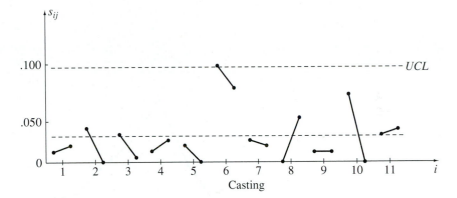

Due to the (presumably) arbitrary nature of the numbering of castings in Table 10-37, no practical conclusions are justifiable on the basis of any trends one perceives in Figure 10-6. But the overall messages of "clear lack of control for means" and "some hint of instability of standard deviations" conveyed by Figure 10-6 do have practical implications. In the first place, the possible instability of standard deviations is a warning that any formal inferences based on model (10-31) (and therefore a constant σ^2 restriction) need to be treated somewhat tenuously. And the "lack of control" exhibited by $\bar{y}_{ij}$'s hints strongly that formal inference methods will identify fairly large variance component(s) σ_β^2 (and probably σ_α^2).

Moving up one level of hierarchy, next consider the making of an "$\bar{x}$ and s" chart pair based on the 11 means $\bar{y}_{i.}$, using 11 "sample standard deviations" calculated from the 2 values $\bar{y}_{i1}$ and $\bar{y}_{i2}$ for each i. The average of the 11 standard deviations given in the last column of Table 10-39 is

$$\frac{1}{11}(.01768 + .22627 + \cdots + .60811 + .04596) = .21342$$

Then, since 2 means $\bar{y}_{ij}$ are involved in the computation of each sample standard deviation in the last column of Table 10-39, $B_4 = 3.267$ should be used to compute a retrospective upper "control limit" for those values — namely,

$$3.267(.21342) = .69724$$

Further, treating the variability in $\bar{y}_{ij}$'s for a given i as the baseline against which to judge variability in $\bar{y}_{i.}$'s, "control limits" for casting means $\bar{y}_{i.}$ will be some central value plus or minus

$$A_3(.21342) = 2.659(.21342) = .56748$$

FIGURE 10-7

$\bar{x}$ and s Charts for "Samples" of Two Specimen Average Measurements on 11 Castings

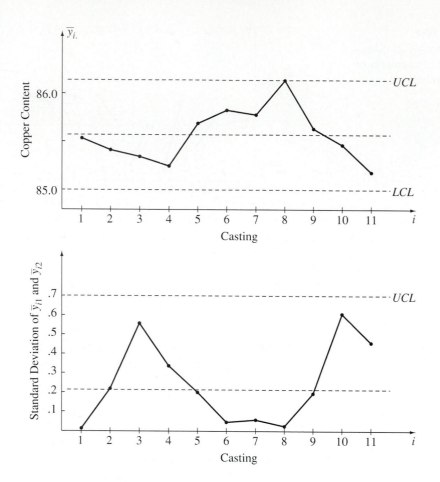

where (as in the case of B_4 just employed) A_3 for sample size $J = 2$ is used. Figure 10-7 shows an $\bar{x}$ and s chart pair for the 11 samples of two $\bar{y}_{ij}$'s for given castings, with $\bar{x}$ chart control limits based on a center line at $\bar{y}_{..}$.

The fairly tame appearance of the charts in Figure 10-7 suggests first that determination-to-determination plus specimen-to-specimen variation is reasonably consistent from casting to casting. Second, it suggests that casting-to-casting variation is not large in comparison to the baseline variability provided by the determination-to-determination plus specimen-to-specimen sources.

The picture of copper content provided by Figures 10-6 and 10-7 pretty much tells the story contained in the data. More formal analytical methods will really provide only quantification of the qualitative conclusions that can be reached based on the figures.

Control charting gives one some feel for the general message carried by balanced hierarchical data. In addition, the s or R charts of samples of observations or means provide some rough diagnostic help in checking the reasonableness of model (10-31)'s assumption of constant variance components. However, there is also the "normal distributions" feature of model (10-31) to be considered. Some help in contemplating the appropriateness of that feature of the model, and also in locating outlying observations and/or means, can be obtained by normal-plotting "residuals" of observations or means at each level of aggregation of the data. That is, one can normal-plot the values

$$y_{ijk} - \bar{y}_{ij}$$

and the values

$$\bar{y}_{ij} - \bar{y}_{i.}$$

and the values

$$\bar{y}_{i.} - \bar{y}_{..} \qquad \text{(or equivalently, } \bar{y}_{i.})$$

as at least a rough check on the assumptions that the various distributions are normal, and as a way of identifying y_{ijk}'s, $\bar{y}_{ij}$'s, or $\bar{y}_{i.}$'s that (so to speak) don't fit with the rest of the data. (It should be noted that even this normal-plotting is not a completely satisfactory check on the normality of the underlying distributions of α_i's, β_{ij}'s, and ϵ_{ijk}'s. That is, these plots should indeed appear normal, but the situation is similar to that for two-way factorial random effects models met in Section 8-4. For instance, $\bar{y}_{i.}$ values represent not simply α_i's, but α_i's plus corresponding averages over j and k of β_{ij}'s and ϵ_{ijk}'s. So normal-plotting them only indirectly addresses the question of normality for the α_i's themselves.)

EXAMPLE 10-19
(continued)

Table 10-40 indicates 44 values $y_{ijk} - \bar{y}_{ij}$ and 22 values $\bar{y}_{ij} - \bar{y}_{i.}$ derived using the data of Table 10-37 and the means of Table 10-38.

TABLE 10-40

Copper Measurements Minus Specimen Means and Specimen Means Minus Casting Means

i Casting	j Specimen	$y_{ijk} - \bar{y}_{ij}$	$\bar{y}_{ij} - \bar{y}_{i.}$
1	1	$-.010, .010$	$-.0125, .0125$
1	2	$-.015, .015$	
2	1	$-.030, .030$	$-.1600, .1600$
2	2	$0, 0$	
3	1	$-.025, .025$	$-.4000, .4000$
3	2	$-.005, .005$	
4	1	$-.010, .010$	$-.2450, .2450$
4	2	$-.020, .020$	
5	1	$-.015, .015$	$-.1425, .1425$
5	2	$0, 0$	
6	1	$-.070, .070$	$-.0350, .0350$
6	2	$-.050, .050$	
7	1	$-.020, .020$	$-.0425, .0425$
7	2	$-.015, .015$	
8	1	$0, 0$	$-.0200, .0200$
8	2	$-.040, .040$	
9	1	$-.010, .010$	$-.1400, .1400$
9	2	$-.010, .010$	
10	1	$-.060, .060$	$-.4300, .4300$
10	2	$0, 0$	
11	1	$-.025, .025$	$-.0325, .0325$
11	2	$-.030, .030$	

Figures 10-8, 10-9, and 10-10 are normal plots of (respectively) the $y_{ijk} - \bar{y}_{ij}$ values, the $\bar{y}_{ij} - \bar{y}_{i.}$ values, and the casting means $\bar{y}_{i.}$. They are remarkably linear, indicating no obvious difficulties with the "normal distributions" part of the model restrictions (10-31) and serving to flag no individual determinations, no particular specimens, and no particular castings as out of line with the rest. It is worth remarking in this regard that although the s chart in Figure 10-6 suggests that s_{61} (the sample standard deviation from the first specimen of casting 6) is unusually large, the "residuals" $y_{611} - \bar{y}_{61}$ and $y_{612} - \bar{y}_{61}$ are not distinguishable as outliers in Figure 10-8.

The device of normal-plotting the means $\bar{y}_{i.}$ has one more use. It makes perfectly good sense in a hierarchical study to think of the means $\bar{y}_{i.}$ as individual "observations"

FIGURE 10-8

Normal Plot of Individual Measured Copper Contents Minus Specimen Means

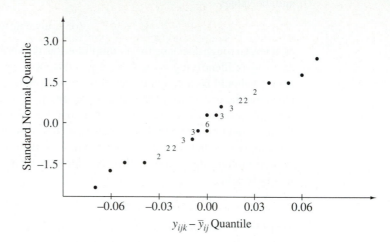

FIGURE 10-9

Normal Plot of Specimen Means Minus Casting Means

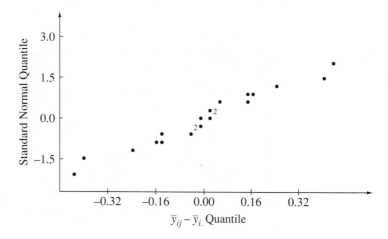

FIGURE 10-10

Normal Plot of Casting Means

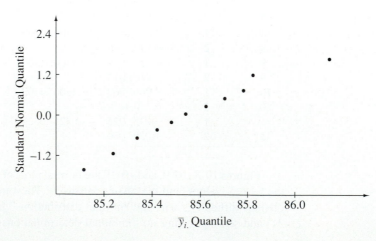

and to consider using them in inference for μ and in the prediction of future "observations" of the same type. Normal-plotting the $\bar{y}_{i.}$'s gives one an idea as to whether the small sample inference methods of Sections 6-3 and 6-6 are appropriate for use in such work.

EXAMPLE 10-19
(continued)

The normal plot of the 11 casting means $\bar{y}_{i.}$ in Figure 10-10 is reasonably linear, suggesting the appropriateness of the normal theory formulas of Chapter 6 in making inferences about mean copper content. The reader can further verify that the $\bar{y}_{i.}$'s have mean

$$\bar{y}_{..} = 85.568$$

and sample standard deviation

$$\sqrt{\frac{1}{11-1} \sum_{i=1}^{11} (\bar{y}_{i.} - \bar{y}_{..})^2} = .283$$

So, for example, two-sided 95% confidence limits for the mean copper content of such castings, μ, based on the t_{10} distribution are

$$85.568 \pm 2.228 \frac{.283}{\sqrt{11}}$$

that is,

$$85.378\% \quad \text{and} \quad 85.758\%$$

Or two-sided 95% prediction limits for one additional measured copper content of a casting (derived from the mean of two determinations on each of two physical samples taken from the casting) are

$$85.568 \pm 2.228(.283)\sqrt{1 + \frac{1}{11}}$$

that is,

$$84.909\% \quad \text{and} \quad 86.227\%$$

ANOVA and Inference for Variance Components

First in Section 7-4 and then again in Section 8-4, breakdowns of the total sum of squares have led to inference methods for variance components in random effects models. Something quite similar is possible in the present context of balanced hierarchical studies. The fundamental algebraic identity that is used in the present context is stated next.

PROPOSITION 10-1

For $I \cdot J \cdot K$ numbers y_{ijk},

$$\sum_{i,j,k} (y_{ijk} - \bar{y}_{..})^2 = JK \sum_{i} (\bar{y}_{i.} - \bar{y}_{..})^2 + K \sum_{i,j} (\bar{y}_{ij} - \bar{y}_{i.})^2 + \sum_{i,j,k} (y_{ijk} - \bar{y}_{ij})^2 \qquad \textbf{(10-35)}$$

The sum of squares on the left of equation (10-35) is of course $SSTot$. The last sum of squares on the right of equation (10-35) would in other contexts be called SSE; it is a measure of variation between the K levels of C within all of the $J \cdot I$ levels of B within

A. The first two sums of squares on the right of equation (10-35) are (respectively) measures of variation between the I levels of A and between the J levels of B within all of the I levels of A. It is common in balanced hierarchical contexts to adopt special nomenclature for these sums of squares; that is now introduced in the form of definitions.

DEFINITION 10-7

In a (3-factor) balanced hierarchical study, the quantity

$$\sum_{i,j,k} (y_{ijk} - \bar{y}_{ij})^2$$

will be called the **sum of squares for C within B and A** and symbolized as $SSC(B(A))$.

DEFINITION 10-8

In a (3-factor) balanced hierarchical study, the quantity

$$K \sum_{i,j} (\bar{y}_{ij} - \bar{y}_{i.})^2$$

will be called the **sum of squares for B within A** and symbolized as $SSB(A)$.

DEFINITION 10-9

In a (3-factor) balanced hierarchical study, the quantity

$$JK \sum_{i} (\bar{y}_{i.} - \bar{y}_{..})^2$$

will be called the **sum of squares for A** and symbolized as SSA.

In terms of the symbolism introduced in Definitions 10-7 through 10-9, the identity (10-35) can be rewritten as

$$SSTot = SSA + SSB(A) + SSC(B(A)) \tag{10-36}$$

Equations (10-35) and (10-36) represent an ANOVA partition of the overall raw variability in y into pieces interpretable in the hierarchical framework.

For purposes of calculation on a hand-held electronic calculator, it is worth noting that the various sums of squares in equations (10-35) or (10-36) are essentially sample variances or sums thereof. As is always the case,

$$SSTot = (n - 1)s^2 \tag{10-37}$$

That is, $SSTot$ can be obtained by almost direct use of a calculator's sample variance function on the raw data points. Next, careful consideration of the form of SSA shows that

$$SSA = JK(I - 1)(\text{sample variance of the } \bar{y}_{i.}\text{'s}) \tag{10-38}$$

so once one has calculated $\bar{y}_{i.}$'s, a simple calculator's sample variance function can be easily used to find SSA. Then a little reflection also shows that

$$SSB(A) = K(J - 1) \sum_{i} (\text{sample variance of the } J \text{ means } \bar{y}_{ij}) \tag{10-39}$$

that is, that by computing and summing I sample variances for groups of J means $\bar{y}_{ij}$, one can obtain $SSB(A)$. Finally,

$$SSC(B(A)) = (K - 1) \sum_{i,j} s_{ij}^2 \tag{10-40}$$

where (as earlier in Example 10-19) s_{ij}^2 stands for the sample variance of the K observations y_{ijk} at the jth level of B within the ith level of A.

EXAMPLE 10-19
(continued)

Most of the raw materials for the computation of sums of squares for the copper content study have already been assembled in the various tables of this section. To begin with, the reader may verify that the grand sample variance of all 44 observations in Table 10-37 is approximately .1195, so from formula (10-37),

$$SSTot = (n - 1)s^2 = 43(.1195) = 5.1385$$

Next, the 11 values $\bar{y}_{i.}$ in Table 10-38 have sample variance of approximately .0801, so from formula (10-38),

$$SSA = JK(I - 1)(\text{sample variance of the } \bar{y}_{i.}\text{'s}) = 2 \cdot 2 \cdot (11 - 1)(.0801) = 3.2031$$

Also, the values in the next-to-last column of Table 10-39 total to .95013, so from formula (10-39),

$$SSB(A) = K(J - 1) \sum_i (\text{sample variance of the } J \text{ means } \bar{y}_{ij}) = 2(2 - 1)(.95014) = 1.9003$$

Finally, the total of the 22 sample variances s_{ij}^2 in Table 10-39 is .0351, so from formula (10-40),

$$SSC(B(A)) = (K - 1) \sum_{i,j} s_{ij}^2 = (2 - 1)(.0351) = .0351$$

The numerical values computed here for sums of squares do sum up as indicated in formulas (10-35) and (10-37). That is,

$$SSTot = 5.1385 = 3.2031 + 1.9003 + .0351 = SSA + SSB(A) + SSC(B(A))$$

One of the virtues of the random effects model (10-31) is the relatively simple and appealing formulas that it yields for expected values of mean squares related to the sums of squares introduced in Definitions 10-7 through 10-9. That is, associated with $SSC(B(A))$ are degrees of freedom $\nu = IJ(K - 1)$, and

$$MSC(B(A)) = \frac{SSC(B(A))}{IJ(K - 1)} \qquad (10\text{-}41)$$

is the mean square for C (within B and A). (This mean square is nothing more than the familiar s_P^2 from Chapter 7, based on $I \cdot J$ samples of size K.) Next, associated with $SSB(A)$ are degrees of freedom $\nu = I(J - 1)$, and

$$MSB(A) = \frac{SSB(A)}{I(J - 1)} \qquad (10\text{-}42)$$

is the mean square for B (within A). Associated with SSA are degrees of freedom $\nu = I - 1$, and

$$MSA = \frac{SSA}{I - 1} \qquad (10\text{-}43)$$

is the mean square for A. The mean squares (10-41) through (10-43) turn out, under the assumptions of model (10-31), to have simple expected values.

Under model (10-31) (actually even without the normality part of the assumptions),

$$E(MSC(B(A))) = \sigma^2 \qquad (10\text{-}44)$$

$$E(MSB(A)) = \sigma^2 + K\sigma_\beta^2 \qquad (10\text{-}45)$$

$$E(MSA) = \sigma^2 + K\sigma_\beta^2 + JK\sigma_\alpha^2 \qquad (10\text{-}46)$$

Much as for the one-way and factorial random effects situations seen earlier in Sections 7-4 and 8-4, the simple expected mean square expressions (10-44) through (10-46) suggest simple ANOVA-based estimates of the variance components.

That is, a standard estimate of σ^2 is

$$MSC(B(A)) = s_P^2 \qquad (10\text{-}47)$$

a standard estimate of σ_β^2 is

$$\max\left(0, \frac{1}{K}(MSB(A) - MSC(B(A)))\right) \qquad (10\text{-}48)$$

and a corresponding estimate of σ_α^2 is

$$\max\left(0, \frac{1}{JK}(MSA - MSB(A))\right) \qquad (10\text{-}49)$$

EXAMPLE 10-19
(continued)

As a means of quantifying the contributions of determination-to-determination variability, specimen-to-specimen variability, and casting-to-casting variability to the variation observed in measured copper contents, consider computing the estimates (10-47) through (10-49) and their square roots. First, applying formulas (10-47) and (10-41), one has an estimated determination-to-determination variance of

$$MSC(B(A)) = \frac{.0351}{11 \cdot 2 \cdot (2-1)} = .0016$$

or on a standard deviation scale, an estimate of σ is

$$\sqrt{.0016} = .040(\%)$$

Next, applying formulas (10-48), (10-41), and (10-42), one has an estimated specimen-to-specimen variance within a casting of

$$\max\left(0, \frac{1}{2}\left(\frac{1.9003}{11 \cdot (2-1)} - .0016\right)\right) = .0856$$

or on a standard deviation scale, σ_β is estimated as

$$\sqrt{.0856} = .293(\%)$$

Finally, applying formulas (10-49), (10-42), and (10-43), one has an estimated casting-to-casting variance of

$$\max\left(0, \frac{1}{2 \cdot 2}\left(\frac{3.2031}{(11-1)} - \frac{1.9003}{11 \cdot (2-1)}\right)\right) = .0369$$

or on a standard deviation scale, σ_α is estimated as

$$\sqrt{.0369} = .192(\%)$$

These estimated variance components are consistent with the more qualitative impressions drawn earlier about the data of Table 10-37 based on the control charts. The sample-to-sample variation is large compared to the determination-to-determination variation, and the casting-to-casting variation smaller than the sample-to-sample variation as well.

In trying to interpret estimated variance components, it is useful to think of a given component as describing the variation in response that would be seen if only the corresponding kind of variation existed. For example, if measurement were perfect and all castings were perfectly homogeneous, the estimate of σ_α (namely, .192%) might describe the variation in response that would be experienced. In this regard, it is also common to report the estimated variance components as fractions of their sum. For example, here casting-to-casting variation would be assigned only a fraction

$$\frac{.0369}{.0016 + .0856 + .0369} = .30 \tag{10-50}$$

of the variation in measured copper content. The rationale for making such ratios is that the sum in the denominator on the left of equation (10-50) is an estimate of

$$\sigma_\alpha^2 + \alpha_\beta^2 + \sigma^2$$

According to model (10-31), this is the variance of measured copper content that would be experienced if (under the present physical conditions) a single determination were made on a single physical sample from each casting.

The forms of expected mean squares (10-44), (10-45), and (10-46) suggest sensible estimates of variance components. They also suggest that ratios of mean squares might provide test statistics for significance tests concerning the variance components. This is in fact the case. That is, under the full set of model assumptions (10-31), the hypothesis $H_0: \sigma_\beta^2 = 0$ can be tested using the statistic

$$F = \frac{MSB(A)}{MSC(B(A))} \tag{10-51}$$

and an $F_{I(J-1),IJ(K-1)}$ reference distribution, where large observed values of F count as evidence against H_0 in favor of the hypothesis $H_a: \sigma_\beta^2 > 0$. In a similar fashion, under the assumptions (10-31), the hypothesis $H_0: \sigma_\alpha^2 = 0$ can be tested using the statistic

$$F = \frac{MSA}{MSB(A)} \tag{10-52}$$

and an $F_{I-1,I(J-1)}$ reference distribution, where large observed values of F count as evidence against H_0 in favor of the hypothesis $H_a: \sigma_\alpha^2 > 0$.

The whole set of ANOVA calculations introduced here is often summarized in tabular form. Table 10-41 gives the general form of an ANOVA table for balanced hierarchical studies with three factors (i.e., two levels) of hierarchy.

TABLE 10-41

General Form of the ANOVA Table for Three-Level Balanced Hierarchical Analyses

ANOVA TABLE

Source	SS	df	MS	EMS	F
A	SSA	$I-1$	MSA	$\sigma^2 + K\sigma_\beta^2 + JK\sigma_\alpha^2$	$MSA/MSB(A)$
B(A)	$SSB(A)$	$I(J-1)$	$MSB(A)$	$\sigma^2 + K\sigma_\beta^2$	$MSB(A)/MSC(B(A))$
C(B(A))	$SSC(B(A))$	$IJ(K-1)$	$MSC(B(A))$	σ^2	
Total	$SSTot$	$IJK-1$			

EXAMPLE 10-19
(continued)

The particular ANOVA table appropriate for summarizing many of the calculations in this example is given here as Table 10-42.

TABLE 10-42

ANOVA Table for the Copper Content Analysis

ANOVA TABLE

Source	SS	df	MS	EMS	F
Castings	3.2031	10	.3203	$\sigma^2 + 2\sigma_\beta^2 + 4\sigma_\alpha^2$	1.85
Samples	1.9003	11	.1728	$\sigma^2 + 2\sigma_\beta^2$	108
Determinations	.0351	22	.0016	σ^2	
Total	5.1385	43			

Comparing the observed f's in Table 10-42 to respectively tabled $F_{10,11}$ and $F_{11,22}$ quantiles, it is clear that the data in hand provide only very weak evidence against H_0: $\sigma_\alpha^2 = 0$, but overwhelming evidence against H_0: $\sigma_\beta^2 = 0$. There is again clear indication of specimen-to-specimen variation, but only relatively weak indication of casting-to-casting variation.

Only partially satisfactory simple methods are available to make confidence intervals for the variance components σ^2, σ_β^2, and σ_α^2 in model (10-31). In a manner similar to the situation described at the end of Section 7-4, it *is* possible to make a two-sided confidence interval for σ^2 using endpoints

$$\frac{IJ(K-1)MSC(B(A))}{U} \quad \text{and} \quad \frac{IJ(K-1)MSC(B(A))}{L} \qquad \textbf{(10-53)}$$

where L and U are such that the $\chi^2_{IJ(K-1)}$ probability assigned to the interval (L, U) corresponds to the desired confidence. (As usual, use of only one of the endpoints (10-53) can produce a one-sided interval for σ^2.) But confidence intervals aimed at σ_β^2 and σ_α^2 directly are not available.

What can be done simply is to give intervals for the ratios σ_β^2/σ^2 and $\sigma_\alpha^2/(\sigma_\beta^2 + \sigma^2/K)$. That is, under model (10-31), endpoints

$$\frac{1}{K}\left(\frac{MSB(A)}{U \cdot MSC(B(A))} - 1\right) \quad \text{and} \quad \frac{1}{K}\left(\frac{MSB(A)}{L \cdot MSC(B(A))} - 1\right) \qquad \textbf{(10-54)}$$

can be used to give a two-sided confidence interval for the ratio σ_β^2/σ^2, where the associated confidence is the probability the $F_{I(J-1),IJ(K-1)}$ distribution assigns to the interval (L, U). And again under model (10-31), endpoints

$$\frac{1}{J}\left(\frac{MSA}{U \cdot MSB(A)} - 1\right) \quad \text{and} \quad \frac{1}{J}\left(\frac{MSA}{L \cdot MSB(A)} - 1\right) \qquad \textbf{(10-55)}$$

can be used to give a two-sided confidence interval for the ratio $\sigma_\alpha^2/(\sigma_\beta^2 + \sigma^2/K)$, where the associated confidence is the probability the $F_{I-1,I(J-1)}$ distribution assigns to the interval (L, U). Although formulas (10-54) and (10-55) address the sizes of σ_β^2 and σ_α^2 only indirectly, they can nevertheless provide additional insight into how balanced hierarchical data ought to be interpreted.

EXAMPLE 10-19
(continued)

Using formula (10-53) and .05 and .95 quantiles of the χ^2_{22} distribution, one has that endpoints of a two-sided 90% confidence interval for σ (the determination-to-determination standard deviation) are

$$\sqrt{\frac{11 \cdot 2 \cdot (2-1)(.0016)}{33.924}} \quad \text{and} \quad \sqrt{\frac{11 \cdot 2 \cdot (2-1)(.0016)}{12.338}}$$

that is,

$$.032 \ (\%) \quad \text{and} \quad .053 \ (\%)$$

Then using formula (10-54) and .05 and .95 quantiles of the $F_{11,22}$ distribution, a two-sided 90% confidence interval for σ_β/σ has endpoints

$$\sqrt{\frac{1}{2}\left(\frac{.1728}{(2.26).0016} - 1\right)} \quad \text{and} \quad \sqrt{\frac{1}{2}\left(\frac{.1728}{\left(\frac{1}{2.63}\right).0016} - 1\right)}$$

that is,

$$4.84 \quad \text{and} \quad 11.90$$

Finally, a two-sided 90% confidence interval for $\sigma_\alpha^2/(\sigma_\beta^2 + \sigma^2/2)$ can be made from formula (10-55) using the .05 and .95 quantiles of the $F_{10,11}$ distribution; it has endpoints

$$\frac{1}{2}\left(\frac{.3203}{(2.85).1728} - 1\right) \quad \text{and} \quad \frac{1}{2}\left(\frac{.3203}{\left(\frac{1}{2.94}\right).1728} - 1\right)$$

that is, (after accounting for the fact that the ratio can't be negative),

$$0 \quad \text{and} \quad 2.22$$

The last interval is for an admittedly somewhat obscure ratio. But, for example, taken together with an inference that

$$\frac{\sigma_\beta}{\sigma} \geqslant 4.84$$

it would indicate that

$$2.22 \geqslant \frac{\sigma_\alpha^2}{\sigma_\beta^2 + \frac{1}{2}\sigma^2} \geqslant \frac{\sigma_\alpha^2}{\sigma_\beta^2 + \frac{1}{2}\left(\frac{\sigma_\beta}{4.84}\right)^2}$$

that is,

$$\frac{\sigma_\alpha^2}{\sigma_\beta^2} \leqslant 2.22\left(1 + \frac{1}{2(4.84)^2}\right)$$

that is,

$$\frac{\sigma_\alpha}{\sigma_\beta} \leqslant 1.51$$

(The exact confidence to associate with this kind of inference is not so obvious, but at least the Bonferroni approach of Section 7-2 could be applied to say that since it depends on two separate 90% individual inferences, this bound is at least an 80% bound. One could possibly even argue that since only one endpoint of each of the two two-sided 90% intervals was involved, the appropriate Bonferroni figure is 90% instead of 80%.)

The calculations in this section have proceeded largely by hand. It is thus worth noting that most commercially available statistical packages can be used to automate at least the ANOVA portion of the calculations. Printout 10-2 is a Minitab printout created to essentially reproduce Table 10-42.

PRINTOUT 10-2

```
MTB > print c1-c4

ROW    C1   C2   C3      C4

  1     1    1    1    85.54
  2     1    1    2    85.56
  3     1    2    1    85.51
  4     1    2    2    85.54
  5     2    1    1    85.54
  6     2    1    2    85.60
  7     2    2    1    85.25
  8     2    2    2    85.25
  9     3    1    1    85.72
 10     3    1    2    85.77
 11     3    2    1    84.94
 12     3    2    2    84.95
 13     4    1    1    85.48
 14     4    1    2    85.50
 15     4    2    1    84.98
 16     4    2    2    85.02
 17     5    1    1    85.54
 18     5    1    2    85.57
 19     5    2    1    85.84
 20     5    2    2    85.84
 21     6    1    1    85.72
 22     6    1    2    85.86
 23     6    2    1    85.81
 24     6    2    2    85.91
 25     7    1    1    85.72
 26     7    1    2    85.76
 27     7    2    1    85.81
 28     7    2    2    85.84
 29     8    1    1    86.12
 30     8    1    2    86.12
 31     8    2    1    86.12
 32     8    2    2    86.20
 33     9    1    1    85.47
 34     9    1    2    85.49
 35     9    2    1    85.75
 36     9    2    2    85.77
 37    10    1    1    84.98
 38    10    1    2    85.10
 39    10    2    1    85.90
 40    10    2    2    85.90
 41    11    1    1    85.12
 42    11    1    2    85.17
 43    11    2    1    85.18
 44    11    2    2    85.24

MTB > anova c4=c1 c2(c1);
SUBC> random c1.
```

(continued)

PRINTOUT 10-2
(continued)

```
Factor   Type Levels Values
C1       random   11    1    2    3    4    5    6    7    8    9
                        10   11
C2(C1) random     2    1    2

Analysis of Variance for C4

Source      DF        SS         MS        F        P
C1          10    3.20306    0.32031    1.85    0.163
C2(C1)      11    1.90030    0.17275  108.29    0.000
Error       22    0.03510    0.00160
Total       43    5.13845
```

10-5 *Mixture Studies*

An important class of chemical and materials engineering problems is that where a substance of interest (for example, a gasoline blend, steel of a certain type, or a compound used to make plastic pipe) is made up of a mixture of k "pure" components. In such cases, the physical properties of the substance typically depend on the proportions of the various pure components appearing in the mixture. A whole subfield of statistical methodology has developed around the problem of identifying appropriate data collection and interpretation methods for describing the dependence of physical properties of a mixture on the proportions of its pure ingredients. This section provides a very brief, limited introduction to the large field of statistical methods for mixture studies.

The discussion begins by considering two special features of mixture studies that set them apart from conventional surface-fitting problems (treated, e.g., in Sections 4-2, 9-2, and 9-3 of this book). Then a graphical device that is useful in mixture studies involving a fairly small number of components is discussed. Next is an introductory discussion of some of the issues involved and some simple tools of experimental design for mixture studies. Finally, the section closes with a few comments on the kinds of other tools for planning and analyzing mixture studies that can be found in the engineering statistics literature, beyond those discussed here.

A Special Feature: Linear Dependence of Proportions

In this section, $x_1, x_2, \ldots, x_k$ will stand for the proportions of k different pure ingredients making up a mixture, and y will be some physical property of the mixture that is of engineering interest. In one sense, the planning and analysis of a mixture study can be thought of as simply an exercise in surface fitting for y as a function of the k system variables $x_1, x_2, \ldots, x_k$. But there are features of the mixture scenario that make some specialized ways of thinking certainly expedient, if perhaps not always absolutely necessary. These are related to implications of the following facts.

1. $x_1, x_2, \ldots, x_k$ must sum to 1.
2. In practical situations, the proportions $x_1, x_2, \ldots, x_k$ are often constrained far beyond the limitation that they sum to 1 and the obvious restriction that each proportion x_i must be between 0 and 1.

Consider first the implications of the fact that in a mixture study the system variables $x_1, x_2, \ldots, x_k$ are linearly dependent, satisfying the equation

$$x_1 + x_2 + \cdots + x_k = 1 \tag{10-56}$$

for the process of surface fitting by least squares and related inference. The simplest nontrivial type of surface one might consider fitting to mixture data is one that is linear in each of the proportions $x_1, x_2, \ldots, x_k$. But the form of approximate relationship

$$y \approx \beta_0 + \beta_1 x_2 + \beta_2 x_2 + \cdots + \beta_k x_k \tag{10-57}$$

used extensively in Sections 4-2, 9-2, and 9-3 doesn't turn out to be suited for use in the mixture context. The problem with form (10-57) is that because of relationship (10-56), there will be many sets of coefficients $b_0, b_1, \ldots, b_k$ producing the same $\hat{y}$'s for all physically possible sets of values for $x_1, x_2, \ldots, x_k$.

EXAMPLE 10-20

Two Equivalent Sets of Parameters in a Linear $k = 3$ Mixture System. As a hypothetical example, consider a $k = 3$ mixture system and the two sets of coefficients β_0, β_1, β_2, and β_3 for use in relationship (10-57)

$$\beta_0 = 0, \qquad \beta_1 = 1, \qquad \beta_2 = 2, \qquad \beta_3 = 2$$

and

$$\beta_0 = 1, \qquad \beta_1 = 0, \qquad \beta_2 = 1, \qquad \beta_3 = 1$$

The two corresponding equations

$$y \approx x_1 + 2x_2 + 2x_3 \tag{10-58}$$

and

$$y \approx 1 + x_2 + x_3 \tag{10-59}$$

appear to be quite different. But for (x_1, x_2, x_3) vectors with $x_1 + x_2 + x_3 = 1$, they produce the same values of y. (If one replaces x_1 in equation (10-58) with $1 - x_2 - x_3$, equation (10-59) results.) Of course, there are many other equations of form (10-57) besides equations (10-58) and (10-59) that also produce the same values of y for (x_1, x_2, x_3) points satisfying relationship (10-56).

An equation of form (10-57) in a mixture context is simply *overparameterized*. To arrive at a single best-fitting approximate expression for y that is linear in the system variables, one must take some step to impose a further restriction on relationship (10-57) — that is, to reduce the number of parameters β in the expression. Two sensible routes for doing so are as follows. In the first place, one might well argue that there are only $k - 1$ free component proportions in the first place (any one of the k of them being derivable as 1 minus the sum of the others) and determine to fit a general $k - 1$-variable linear expression

$$y \approx \beta_0 + \beta_1 x_1 + \beta_2 x_2 + \cdots + \beta_{k-1} x_{k-1} \tag{10-60}$$

in preference to equation (10-57). This thinking works, and it has the appealing feature that after mentally reducing the number of components to $k - 1$ from k, all of the surface-fitting and inference material of Sections 4-2, 9-2, and 9-3 applies to mixture contexts verbatim.

But the device of mentally dropping a component out of explicit consideration during the fitting process has several drawbacks. For one thing, the process seems asymmetric, involving an arbitrary choice of which component to suppress. For another, the coefficients β end up having fairly strange physical interpretations. In expression (10-60), for example, β_1 is the rate of change in y as a function of x_1, when $x_2, x_3, \ldots, x_{k-1}$ are held fixed, but a Δ change in x_1 is accompanied by a $-\Delta$ change in x_k. So although the use of expression (10-60) provides a straightforward route to applying standard regression material in a mixture context, it is not always directly discussed in treatments of the mixture problem.

Instead, the restriction of formula (10-57) that is often adopted is one where the constant term is dropped. That is, the form of linear approximate relationship most often discussed in mixture contexts is

$$y \approx \beta_1 x_1 + \beta_2 x_2 + \cdots + \beta_k x_k \tag{10-61}$$

In expression (10-61), the coefficients β have relatively simple interpretations. β_1 through β_k are the responses predicted to accompany the k pure components. But since there is no intercept term in expression (10-61), it does not have exactly the form of any regression equation yet given serious treatment in this book. So at first glance, it may not be completely obvious either how one ought to fit an expression like (10-61) via least squares or how to approach related formal inference making.

There are at least two ways to think about fitting and related inference for an equation like (10-61) in a mixture context. The first is to realize that completely standard methods, based on conceptualization (10-60), must produce results that are either equivalent to or translatable to methods based on equation (10-61). For example, if one considers minimizing

$$S'(\beta_0', \beta_1', \ldots, \beta_{k-1}') = \sum_{i=1}^{n} \left(y_i - (\beta_0' + \beta_1' x_{1i} + \beta_2' x_{2i} + \cdots + \beta_{k-1}' x_{k-1\,i})\right)^2$$

by choice of $\beta_0', \beta_1', \ldots, \beta_{k-1}'$ and

$$S(\beta_1, \beta_2, \ldots, \beta_k) = \sum_{i=1}^{n} \left(y_i - (\beta_1 x_{1i} + \beta_2 x_{2i} + \cdots + \beta_k x_{ki})\right)^2$$

by choice of $\beta_1, \beta_2, \ldots, \beta_k$, it is easy to see that for mixture data (satisfying formula (10-56)), the minimizing coefficients for the two problems satisfy the k equations

$$\left.\begin{array}{c} b_1 = b_0' + b_1' \\ b_2 = b_0' + b_2' \\ \vdots \\ b_{k-1} = b_0' + b_{k-1}' \\ b_k = b_0' \end{array}\right\} \tag{10-62}$$

so regular regression based on equation (10-60) can produce the coefficients for equation (10-61) via some simple additions.

Further, formal inferences regarding the mean system response and future responses and the making of tolerance intervals under the related (but different-appearing) models

$$y = \beta_0 + \beta_1 x_1 + \beta_2 x_2 + \cdots + \beta_{k-1} x_{k-1} + \epsilon \tag{10-63}$$

$$y = \beta_1 x_1 + \beta_2 x_2 + \cdots + \beta_k x_k + \epsilon \tag{10-64}$$

are exactly the same. And an F test of H_0: $\beta_1 = \beta_2 = \cdots = \beta_{k-1} = 0$ under model (10-63) is eqivalent to an F test of H_0: $\beta_1 = \beta_2 = \cdots = \beta_k$ under model (10-64). In both cases, the full model is one of linear response of y in $x_1, x_2, \ldots, x_k$, while the reduced model is one of constant response.

About the only real difficulty with the advice "Do computations using conceptualizations (10-60) and (10-63) and then translate results if you wish to apply conceptualizations (10-61) and (10-64)" is that, following this advice, the route to formal inferences for the individual coefficients $\beta_1, \beta_2, \ldots, \beta_k$ in equation (10-64) is rather circuitous. So, as a second method of doing fitting and inference related to equations

(10-61) and (10-64), most multiple linear regression programs have a "no intercept" option that fits the coefficients β in equation (10-61) directly and also provides the raw material necessary for doing inference under model (10-64) in mixture contexts.

EXAMPLE 10-21

A $k = 5$ Component Gasoline Octane Study. Table 10-43 contains some $k = 5$ mixture data from a gasoline blending study used as an example in "Developing Blending Models for Gasoline and Other Mixtures" by R. Snee (*Technometrics*, 1981).

TABLE 10-43

Research Octane Measurements for 16 Different Gasoline Blends

x_1 Butane	x_2 Alkylate	x_3 Light Straight Run	x_4 Reformate	x_5 Cat Cracked	y Research Octane
.10	.10	.05	.20	.55	95.1
.10	.00	.15	.20	.55	93.4
.00	.10	.15	.20	.55	93.3
.10	.10	.15	.20	.45	94.1
.00	.00	.05	.40	.55	91.8
.10	.00	.05	.40	.45	91.8
.00	.10	.05	.40	.45	92.5
.00	.00	.15	.40	.45	90.5
.00	.00	.05	.35	.60	92.7
.10	.10	.15	.25	.40	93.5
.10	.00	.05	.25	.60	94.8
.00	.10	.05	.25	.60	93.7
.00	.00	.15	.25	.60	92.5
.10	.10	.05	.35	.40	93.1
.10	.00	.15	.35	.40	91.8
.00	.10	.15	.35	.40	91.6

The response variable in the study, y, was the research octane of the mixture at 2.0 g Pb/gal. The reader should verify that as expected, for each of the $n = 16$ data points listed in Table 10-43, the proportions of the components x_1, x_2, x_3, x_4, and x_5 sum to 1.00. Printout 10-3 shows the results of two Minitab regression runs, first for the straightforward fitting of

$$y \approx \beta_0 + \beta_1 x_1 + \beta_2 x_2 + \beta_3 x_3 + \beta_4 x_4$$

(à la equation (10-60)) and then the no-intercept fitting of

$$y \approx \beta_1 x_1 + \beta_2 x_2 + \beta_3 x_3 + \beta_4 x_4 + \beta_5 x_5$$

(à la equation (10-61)).

Notice that, as indicated in display (10-62), the intercept in the first regression run corresponds to the coefficient of x_5 in the second, and the fitted coefficients of x_1 through x_4 in the second run can be obtained as the sums of the corresponding coefficient from the first run plus the intercept from that run. Also as expected, s_{SF} values for the two runs agree. And the fitted values, estimated standard deviations of fitted values, residuals, and standardized residuals are also the same for the two runs.

It is worth noting, however, that no R^2 value is printed for the second run. This is because in general, R^2 as defined in Definition 4-3 needs to make sense only in the context of fitting an equation that includes an intercept term. Unless an equation that is to be fitted includes a constant term, the formula in Definition 4-3 can yield negative values. For mixture data, equation (10-64) *implicitly* includes a constant term, and thus a meaningful R^2 value can always be announced. But since the no-intercept regression

PRINTOUT 10-3

```
MTB > print c1-c6

ROW      x1      x2      x3      x4      x5      y

  1     0.1     0.1    0.05    0.20    0.55    95.1
  2     0.1     0.0    0.15    0.20    0.55    93.4
  3     0.0     0.1    0.15    0.20    0.55    93.3
  4     0.1     0.1    0.15    0.20    0.45    94.1
  5     0.0     0.0    0.05    0.40    0.55    91.8
  6     0.1     0.0    0.05    0.40    0.45    91.8
  7     0.0     0.1    0.05    0.40    0.45    92.5
  8     0.0     0.0    0.15    0.40    0.45    90.5
  9     0.0     0.0    0.05    0.35    0.60    92.7
 10     0.1     0.1    0.15    0.25    0.40    93.5
 11     0.1     0.0    0.05    0.25    0.60    94.8
 12     0.0     0.1    0.05    0.25    0.60    93.7
 13     0.0     0.0    0.15    0.25    0.60    92.5
 14     0.1     0.1    0.05    0.35    0.40    93.1
 15     0.1     0.0    0.15    0.35    0.40    91.8
 16     0.0     0.1    0.15    0.35    0.40    91.6

MTB > regress c6 4 c1-c4

The regression equation is
y = 97.6 + 4.82 x1 + 3.07 x2 - 12.4 x3 - 12.9 x4

Predictor        Coef       Stdev     t-ratio         p
Constant      97.5929      0.5203      187.56     0.000
x1              4.821       1.685        2.86     0.015
x2              3.071       1.685        1.82     0.096
x3            -12.429       1.685       -7.37     0.000
x4            -12.857       1.192      -10.79     0.000

s = 0.3153      R-sq = 95.2%      R-sq(adj) = 93.5%

Analysis of Variance

SOURCE        DF          SS          MS          F         p
Regression     4     21.6839      5.4210      54.53     0.000
Error         11      1.0936      0.0994
Total         15     22.7775

SOURCE        DF      SEQ SS
x1             1      5.0625
x2             1      3.6100
x3             1      1.4400
x4             1     11.5714
```

(continued)

program is meant to handle data that may not happen to satisfy the restriction (10-56), no R^2 value is routinely printed. In a similar vein, the "total sum of squares" printed for the second run is $\sum y_i^2$, not the familiar $SSTot = \sum(y_i - \bar{y})^2$, and the printed value for the F statistic is for testing $H_0: \beta_1 = \beta_2 = \cdots = \beta_k = 0$ in model (10-64) — hardly a hypothesis of real interest.

Some relevant information is provided by the second/no-intercept regression run that is not available (without extra complication) from the first. That is the set of estimated standard deviations of the estimates of coefficients β_1, β_2, β_3, β_4, and β_5 in the $k = 5$ version of model (10-64). These allow (in the obvious way) for the making of confidence

PRINTOUT 10-3
(continued)

Obs.	x1	y	Fit	Stdev.Fit	Residual	St.Resid
1	0.100	95.1000	95.1893	0.1812	-0.0893	-0.35
2	0.100	93.4000	93.6393	0.1812	-0.2393	-0.93
3	0.000	93.3000	93.4643	0.1812	-0.1643	-0.64
4	0.100	94.1000	93.9464	0.1604	0.1536	0.57
5	0.000	91.8000	91.8286	0.1604	-0.0286	-0.11
6	0.100	91.8000	92.3107	0.1812	-0.5107	-1.98
7	0.000	92.5000	92.1357	0.1812	0.3643	1.41
8	0.000	90.5000	90.5857	0.1812	-0.0857	-0.33
9	0.000	92.7000	92.4714	0.1604	0.2286	0.84
10	0.100	93.5000	93.3036	0.1604	0.1964	0.72
11	0.100	94.8000	94.2393	0.1812	0.5607	2.17R
12	0.000	93.7000	94.0643	0.1812	-0.3643	-1.41
13	0.000	92.5000	92.5143	0.1812	-0.0143	-0.06
14	0.100	93.1000	93.2607	0.1812	-0.1607	-0.62
15	0.100	91.8000	91.7107	0.1812	0.0893	0.35
16	0.000	91.6000	91.5357	0.1812	0.0643	0.25

```
R denotes an obs. with a large st. resid.

MTB > noconstant
MTB > regress c6 5 c1-c5

The regression equation is
y = 102 x1 + 101 x2 + 85.2 x3 + 84.7 x4 + 97.6 x5
```

Predictor	Coef	Stdev	t-ratio	p
Noconstant				
x1	102.414	1.515	67.61	0.000
x2	100.664	1.515	66.45	0.000
x3	85.164	1.430	59.54	0.000
x4	84.7357	0.7448	113.76	0.000
x5	97.5928	0.5203	187.56	0.000

```
s = 0.3153
Analysis of Variance
```

SOURCE	DF	SS	MS	F	p
Regression	5	138071	27614	277765.62	0.000
Error	11	1	0		
Total	16	138072			

SOURCE	DF	SEQ SS
x1	1	69863
x2	1	23201
x3	1	25460
x4	1	16050
x5	1	3497

(continued)

intervals for the β's. For example, since it is clear from the perspective of model (10-63), with $k = 5$, that $11 = 16 - 4 - 1$ degrees of freedom should be associated with s_{SF}, using the .975 quantile of the t_{11} distribution, two-sided 95% confidence limits for β_1 are

$$102.41 \pm 2.201(1.515)$$

that is,

$$99.08 \text{ octane} \quad \text{and} \quad 105.74 \text{ octane}$$

PRINTOUT 10-3
(continued)

Obs.	x1	y	Fit	Stdev.Fit	Residual	St.Resid
1	0.100	95.1000	95.1893	0.1812	-0.0893	-0.35
2	0.100	93.4000	93.6393	0.1812	-0.2393	-0.93
3	0.000	93.3000	93.4643	0.1812	-0.1643	-0.64
4	0.100	94.1000	93.9464	0.1604	0.1536	0.57
5	0.000	91.8000	91.8286	0.1604	-0.0286	-0.11
6	0.100	91.8000	92.3107	0.1812	-0.5107	-1.98
7	0.000	92.5000	92.1357	0.1812	0.3643	1.41
8	0.000	90.5000	90.5857	0.1812	-0.0857	-0.33
9	0.000	92.7000	92.4714	0.1604	0.2286	0.84
10	0.100	93.5000	93.3036	0.1604	0.1964	0.72
11	0.100	94.8000	94.2393	0.1812	0.5607	2.17R
12	0.000	93.7000	94.0643	0.1812	-0.3643	-1.41
13	0.000	92.5000	92.5143	0.1812	-0.0143	-0.06
14	0.100	93.1000	93.2607	0.1812	-0.1607	-0.62
15	0.100	91.8000	91.7107	0.1812	0.0893	0.35
16	0.000	91.6000	91.5357	0.1812	0.0643	0.25

R denotes an obs. with a large st. resid.

(However, a "pure butane octane rating" interpretation of this confidence interval should not be attempted (or at least should be taken with a large grain of salt). The conditions $x_1 = 1.00$, $x_2 = 0$, $x_3 = 0$, $x_4 = 0$, $x_5 = 0$ represent a clear extrapolation from the data of Table 10-43, so any such physical interpretation of the coefficients is highly tenuous at best. The fitted equation may predict octane quite poorly away from the x_1, x_2, x_3, x_4, x_5 region over which it was developed.)

━━ ━━ ━━

The equations (10-60) and (10-61) provide for description of only the simplest possible mixture response systems. More complicated systems require the use of more complicated equations. Many possible forms for such equations have been suggested, but only one additional kind of equation will be discussed here: the mixture version of the fully quadratic response functions considered in Section 9-3.

For sake of illustration, consider the three-component mixture situation — the $k = 3$ case. Reasoning as before for the linear case, one might expect that since $x_3 = 1 - x_1 - x_2$, the quadratic function in two variables x_1 and x_2,

$$y \approx \beta_0 + \beta_1 x_1 + \beta_2 x_2 + \beta_3 x_1^2 + \beta_4 x_2^2 + \beta_5 x_1 x_2 \tag{10-65}$$

is the most general possible quadratic relationship between y and x_1, x_2, and (implicitly) x_3. In fact, in a three-component mixture problem, one can advantageously use equation (10-65) and material directly from Section 9-3 in the description of many $k = 3$ component mixtures with curvature in mean response. But equation (10-65), like equation (10-60), has an asymmetric appearance. This asymmetry can be remedied in the following way. Replacing β_0 with $\beta_0(x_1 + x_2 + x_3)$, $\beta_3 x_1^2$ with $\beta_3 x_1(1 - x_2 - x_3)$, and $\beta_4 x_2^2$ with $\beta_4 x_2(1 - x_1 - x_3)$ in equation (10-65) and collecting terms, equation (10-65) may be rewritten as

$$y \approx (\beta_0 + \beta_1 + \beta_3)x_1 + (\beta_0 + \beta_2 + \beta_4)x_2 + \beta_0 x_3$$
$$+ (\beta_5 - \beta_3 - \beta_4)x_1 x_2 - \beta_3 x_1 x_3 - \beta_4 x_2 x_3$$

This in turn (correctly) suggests that the general quadratic response equation typically used in $k = 3$ variable mixture studies is

$$y \approx \beta_1 x_1 + \beta_2 x_2 + \beta_3 x_3 + \beta_4 x_1 x_2 + \beta_5 x_1 x_3 + \beta_6 x_2 x_3 \tag{10-66}$$

Fitting and inference for fully quadratic response functions in $k = 3$ components can thus be based on the quadratic model with intercept and $k - 1 = 2$ components corresponding to equation (10-65),

$$y = \beta_0 + \beta_1 x_1 + \beta_2 x_2 + \beta_3 x_1^2 + \beta_4 x_2^2 + \beta_5 x_1 x_2 + \epsilon \tag{10-67}$$

directly using the tools of Sections 4-2, 9-2, and 9-3. Or they may be based on the model without intercept or squared terms in $k = 3$ components corresponding to equation (10-66),

$$y = \beta_1 x_1 + \beta_2 x_2 + \beta_3 x_3 + \beta_4 x_1 x_2 + \beta_5 x_1 x_3 + \beta_6 x_2 x_3 + \epsilon \tag{10-68}$$

where a no-intercept regression program option must be used. As was the case for the linear models (10-63) and (10-64), the only real advantage this author sees in using representation (10-68) directly instead of representation (10-67), is that a no-intercept regression run using model (10-68) will produce estimated standard deviations of the individual estimated coefficients for representation (10-68). This advantage seems less important in the quadratic response case than in the linear situation, however, since the β's in the quadratic expression (10-68) have little individual interpretability.

This brief discussion of quadratic response modeling for $k = 3$ component mixtures extends without essential complication to mixtures of larger numbers of components.

Another Special Feature: Constraints on Proportions

There is a second feature of many real mixture studies that sets them apart from the conventional surface-fitting contexts encountered in Sections 4-2, 9-2, and 9-3. That is the fact that real mixture experiments are often conducted under a number of constraints dictated by physical property considerations and/or cost considerations on the mixture proportions $x_1, x_2, \ldots, x_k$. (A standard example is the mixture of sand, cement, and water that makes up concrete. In order to produce a mixture useful for building purposes, none of the proportions of the components can be made too extreme.) Constraints can produce very irregularly shaped experimental regions (in terms of the variables $x_1, x_2, \ldots, x_k$) that are, of course, impossible to visualize effectively when $k > 3$. The practical problem an experimenter then faces is figuring out how to spread out experimental combinations $(x_1, x_2, \ldots, x_k)$ in the experimental region so as to adequately probe the shape of the response over the whole region of k-dimensional space of interest, and to allow the effective fitting of response surfaces of interest. The major part of the large statistical literature on mixture experiments deals with this question. Later, this section discusses some simple methods of design that have been suggested for studies involving only the simplest kinds of constraints on $x_1, x_2, \ldots, x_k$. But at this point, several examples can be used to give the reader a feel for the complexity of the problem.

EXAMPLE 10-21
(continued)

It was noted in passing that the data of Table 10-43 do not cover the entire space of vectors $(x_1, x_2, \ldots, x_5)$ with nonnegative entries summing to 1.00. In fact, the $n = 16$ combinations in Table 10-43 were chosen both to allow fitting of the linear response function (*a priori* expected to adequately describe gasoline octane for blends of interest) and to adequately cover the region of five-dimensional $(x_1, x_2, \ldots, x_5)$ space defined by equation (10-56) and the five inequalities (themselves reflecting practical engineering considerations on physical properties and costs of the blend),

$$\left. \begin{array}{c} 0 \leq x_1 \leq .10 \\ 0 \leq x_2 \leq .10 \\ .05 \leq x_3 \leq .15 \\ .20 \leq x_4 \leq .40 \\ .40 \leq x_5 \leq .60 \end{array} \right\} \tag{10-69}$$

The exact details of the considerations that led to the particular choice of 16 mixtures in Table 10-43 are given in the original article by Snee, but for present purposes it suffices to note that even with a fairly simple set of constraints like (10-69) the question of sensible experimental design is nontrivial. It is impossible to visualize the part of five-dimensional space specified by conditions (10-56) and (10-69) in geometric terms, and any attempts simply to use the factorial, fractional factorial, or response surface design ideas already discussed in this book fail because of the restriction that the component proportions must sum to 1. (Note, for example, that using the two extreme values for each proportion specified by conditions (10-69) and trying to create a 2^5 design or fraction thereof, or even considering only four of the variables and trying to create a 2^4 design, fails because the points so constructed do not necessarily satisfy all of conditions (10-69) and (10-56).)

EXAMPLE 10-22

A Second $k = 5$ Component Gasoline Blending Study. As an example of a set of constraints more complicated than those in display (10-69), consider a second gasoline blending study used as an example in Snee's 1981 *Technometrics* article "Developing Blending Models for Gasoline and Other Mixtures." Snee discussed a $k = 5$ component blending problem with components, economically determined constraints on the proportions, and approximate pure component responses y (research octane) as in Table 10-44.

TABLE 10-44

Components, Constraints, and Pure Component Octanes in a Gasoline Blending Study

Component	Constraint	Approximate Octane
Butane	$0 \leq x_1 \leq .15$	101.8
Isopentane	$0 \leq x_2 \leq .30$	99.6
Reformate	$0 \leq x_3 \leq .35$	112.4
Cat Cracked	$0 \leq x_4 \leq .60$	94.2
Alkylate	$0 \leq x_5 \leq .60$	99.8

In this situation, engineers expected to need the complexity of at least a quadratic response function to adequately model y. That is, data adequate to fit the equation

$$\begin{aligned} y \approx \beta_1 x_1 + \beta_2 x_2 + \beta_3 x_3 + \beta_4 x_4 + \beta_5 x_5 + \beta_6 x_1 x_2 + \beta_7 x_1 x_3 + \beta_8 x_1 x_4 \\ + \beta_9 x_1 x_5 + \beta_{10} x_2 x_3 + \beta_{11} x_2 x_4 + \beta_{12} x_2 x_5 + \beta_{13} x_3 x_4 + \beta_{14} x_3 x_5 \\ + \beta_{15} x_4 x_5 \end{aligned} \qquad \textbf{(10-70)}$$

were needed. To make matters conceptually worse, not only were the economic constraints in Table 10-44 operative, but further economic considerations forced the use of the constraint

$$x_4 + x_5 \leq .70$$

and volatility considerations on the blend forced use of the constraint

$$x_1 + x_2 \leq .30$$

and final octane requirements suggested that (at least in approximate terms) the proportions had to satisfy

$$97 \leq 101.8 x_1 + 99.6 x_2 + 112.4 x_3 + 94.2 x_4 + 99.8 x_5 \leq 101$$

Again, it is a distinctly nontrivial problem to choose experimental blends to meet all of these constraints and provide information adequate to fit equation (10-70) while thoroughly exploring the experimental region.

EXAMPLE 10-23

Constraints in an Artificial $k = 3$ Component Mixture Study. As a means of visualizing the difficulties involved in choosing experimental plans for constrained mixture studies, consider an artificial $k = 3$ example. Suppose that $k = 3$ mixture components are constrained to satisfy the inequalities

$$\left.\begin{array}{r} .2 \leqslant x_1 \\ 0 \leqslant 2x_1 - x_2 \\ .2 \leqslant x_3 \leqslant .6 \end{array}\right\} \qquad \text{(10-71)}$$

The third of these constraints can be written in terms of x_1 and x_2 as

$$.2 \leqslant 1 - x_1 - x_2 \leqslant .6$$

that is,

$$.4 \leqslant x_1 + x_2 \leqslant .8$$

Figure 10-11 is a depiction of the set of (x_1, x_2, x_3) points satisfying these constraints in terms of the first two coordinates, x_1 and x_2 (the third being determined by the first two via equation (10-56)).

FIGURE 10-11

Three-Component Mixtures
Satisfying Constraints (10-71)

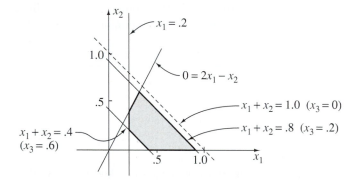

The shaded region in Figure 10-11 has what would be called a fairly odd shape, by most standards. The mixture study experimental design problem under the set of constraints (10-71) amounts to choosing points to cover the shaded region and allow sensible statistical analysis. Precisely how to accomplish this is not a simple question.

Practical experimental design for multicomponent constrained mixture studies is an interesting art these days, incorporating elements of both computer-aided optimization and more qualitative considerations. Various algorithms for searching for "optimal" placements of n mixture vectors $(x_1, x_2, \dots, x_k)$ in constrained regions have been suggested and implemented in computer software. The optimality criteria used for comparing candidate experimental designs usually amount to particular quantifications of the criteria "small variances for estimated coefficients in a particular mixture response model" or "small prediction variances across the experimental region for a given response model." Armed with candidate plans produced by such an algorithm, practitioners then typically consider *ad hoc* modifications to provide for important replication, the ability to fit higher-order models, good properties according to mathematical optimality criteria not used to produce the design, and so on. For a more thorough introduction to the practice of experimental planning of mixture studies than is possible here, refer to the book *Experiments with Mixtures: Designs, Models and the Analysis of Mixture Data* by J.A. Cornell, or to the pamphlet *How to Run Mixture Experiments for Product Quality* by the same author, available as part of the series "ASQC Basic References in Quality Control: Statistical Techniques," published by the American Society for Quality Control.

A Useful Graphical Device

Three-component mixture studies are an important special case of the general mixture problem, where it is possible to graphically display both the sets of proportions making up an experimental design and the nature of response surfaces fitted to a set of data. One somewhat asymmetric way of accomplishing these tasks is to picture only, say, the first two mixture proportions x_1 and x_2 when depicting the experimental region, and to make contour plots of $\hat{y}$ versus only x_1 and x_2. This was the route taken in Figure 10-11.

One way of removing the distressing asymmetry inherent in suppressing one proportion when depicting features of a $k = 3$ component mixture study will be discussed here. It is related to consideration of the nature of the region in three-dimensional space occupied by all three component mixture vectors. This region turns out to amount to a triangular portion of the plane defined by the relationship $x_1 + x_2 + x_3 = 1$ residing in the first octant — that is, in the region where all of the proportions x_1, x_2, and x_3 are nonnegative. Figure 10-12 depicts this region, as well as a line of sight defined perpendicular to the plane $x_1 + x_2 + x_3 = 1$.

FIGURE 10-12

Three-Component Mixtures

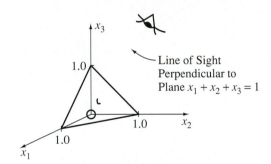

"Elimination" of $x_3 = 1 - x_1 - x_2$ for purposes of plotting features of a three-component mixture study amounts to doing plotting on the (x_1, x_2)-plane and viewing the plot in a direction perpendicular to the (x_1, x_2)-plane. A more symmetric possibility suggested by Figure 10-12 is to project any plotted features from a graphic in the (x_1, x_2)-plane straight up to the triangular region defined in Figure 10-12 and then view along a line of sight perpendicular to the plane of the triangular region. This process is equivalent to doing plotting in the kind of equilateral-triangular coordinate system portrayed in Figure 10-13.

In the coordinate system of Figure 10-13, a mixture with proportions x_1, x_2, and x_3 is located by moving a fraction x_1 of the distance from the $x_1 = 0$ edge perpendicularly toward the $x_1 = 1.0$ corner, a fraction x_2 of the distance from the $x_2 = 0$ edge perpendicularly toward the $x_2 = 1.0$ corner, and (because of the geometry involved and the fact that $x_3 = 1 - x_1 - x_2$) a fraction x_3 of the distance from the $x_3 = 0$ edge perpendicularly toward the $x_3 = 1.0$ corner.

EXAMPLE 10-23

(continued)

Figure 10-11 was a representation of the set of mixture proportions x_1, x_2, x_3 satisfying constraints (10-71), plotted in regular coordinates on the (x_1, x_2)-plane. Projecting that set up onto the triangular region in Figure 10-12 and then viewing along a line of sight perpendicular to the $x_1 + x_2 + x_3 = 1$ plane would produce a graphic like Figure 10-14 (which, of course, could also be produced directly from the use of the coordinate system in Figure 10-13).

FIGURE 10-13

A Triangular Coordinate
System for Portraying Three-
Component Mixtures

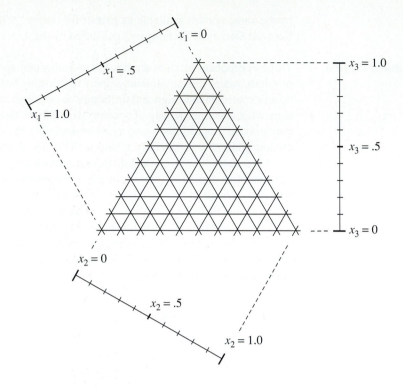

FIGURE 10-14

Three-Component Mixtures
Satisfying Constraints (10-71)

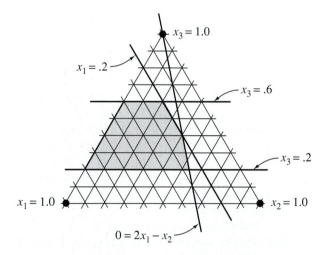

The triangular coordinate system of Figure 10-13 is useful not only for visualizing experimental regions, constraints, and potential sets of proportions (x_1, x_2, x_3) for use as design points in a mixture study, but also as a symmetric coordinate system in which to produce contour plots for summarizing response surfaces fitted to three-component mixture data.

EXAMPLE 10-24

Using Contour Plots on Triangular Coordinate Systems in a Pipe Compound Mixture Experiment. In the article "Design and Analysis of an ABS Pipe Compound Experiment" by Koons and Wilt (1985 ASQC Technical Supplement, *Experiments in Industry*), the authors describe a $k = 3$-component mixture experiment where three physical properties of acrylonitrile-butadiene styrene (ABS) material were studied. The three response variables were

$$y_1 = \text{Izod impact strength (ft lb/in.)}$$
$$y_2 = \text{deflection temperature under load (°F)}$$
$$y_3 = \text{yield strength (psi)}$$

and the mixture components were

$$x_1 = \text{proportion of grafted polybutadiene}$$
$$x_2 = \text{proportion of styrene-acrylonitrile copolymer}$$
$$x_3 = \text{proportion of coal tar pitch}$$

The practical question of interest was whether or not the low cost and readily available coal tar pitch could be used in making the ABS material without significant degradation of resultant physical properties of the mixture. An experimental region defined by the constraints

$$\left.\begin{array}{c} .45 \le x_1 \le .75 \\ .30 \le x_2 \le .55 \\ x_3 \le .25 \end{array}\right\} \tag{10-72}$$

was employed. This triangular region is pictured as part of the entire mathematically conceivable set of mixture proportions in Figure 10-15. The $n = 10$ data points collected in the study are listed in Table 10-45.

FIGURE 10-15

Experimental Region in the Pipe Compound Experiment

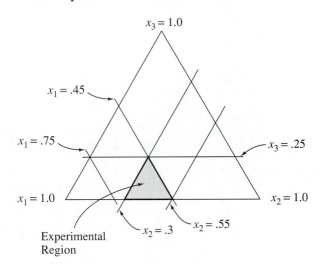

TABLE 10-45

Data From a Three-Component Pipe Compound Mixture Study

x_1	x_2	x_3	y_1	y_2	y_3
.45	.55	0	6.1	213	6080
.45	.425	.125	2.1	191	5740
.45	.30	.25	.8	183	5065
.50	.45	.05	4.9	202	5615
.50	.35	.15	2.5	186	5385
.535	.38	.085	4.4	197	5410
.575	.425	0	7.3	204	5280
.575	.30	.125	3.9	188	5080
.60	.35	.05	6.1	192	5035
.70	.30	0	7.4	205	4350

The reader is encouraged to plot the ten points (x_1, x_2, x_3) of Table 10-45 in the triangular coordinates of Figure 10-15 and note that they have the virtue of covering the experimental region. They are also adequate to allow fitting of the fully quadratic response surface (10-65) or (10-66) to each of the response variables. Printout 10-4 shows Minitab runs for fitting both forms of the quadratic relationship to the first of the three response variables. (Other regression runs not shown on the printout were made to fit both forms of the quadratic relationship to the other two response variables.)

PRINTOUT 10-4

```
MTB > print c1-c11

ROW      x1      x2      x3      y1      y2      y3      x1sq      x2sq

  1    0.450   0.550   0.000    6.1    213    6080   0.202500   0.302500
  2    0.450   0.425   0.125    2.1    191    5740   0.202500   0.180625
  3    0.450   0.300   0.250    0.8    183    5065   0.202500   0.090000
  4    0.500   0.450   0.050    4.9    202    5615   0.250000   0.202500
  5    0.500   0.350   0.150    2.5    186    5385   0.250000   0.122500
  6    0.535   0.380   0.085    4.4    197    5410   0.286225   0.144400
  7    0.575   0.425   0.000    7.3    204    5280   0.330625   0.180625
  8    0.575   0.300   0.125    3.9    188    5080   0.330625   0.090000
  9    0.600   0.350   0.050    6.1    192    5035   0.360000   0.122500
 10    0.700   0.300   0.000    7.4    205    4350   0.490000   0.090000


ROW       x1x2        x1x3        x2x3

  1     0.247500    0.000000    0.000000
  2     0.191250    0.056250    0.053125
  3     0.135000    0.112500    0.075000
  4     0.225000    0.025000    0.022500
  5     0.175000    0.075000    0.052500
  6     0.203300    0.045475    0.032300
  7     0.244375    0.000000    0.000000
  8     0.172500    0.071875    0.037500
  9     0.210000    0.030000    0.017500
 10     0.210000    0.000000    0.000000

MTB > regress c4 5 c1 c2 c7 c8 c9
 * NOTE *       x1 is highly correlated with other   predictor variables
 * NOTE *       x2 is highly correlated with other   predictor variables
 * NOTE *     x1sq is highly correlated with other   predictor variables
 * NOTE *     x2sq is highly correlated with other   predictor variables
 * NOTE *     x1x2 is highly correlated with other   predictor variables

The regression equation is
y1 = 16.8 - 22.1 x1 - 111 x2 + 7.67 x1sq + 84.7 x2sq + 133 x1x2

Predictor         Coef        Stdev      t-ratio          p
Constant        16.757        5.774         2.90      0.044
x1             -22.07        15.00        -1.47      0.215
x2            -110.71        14.49        -7.64      0.002
x1sq             7.666        9.834         0.78      0.479
x2sq            84.739        9.874         8.58      0.001
x1x2           133.15        17.18         7.75      0.001

s = 0.1411      R-sq = 99.8%      R-sq(adj) = 99.6%
```

(continued)

PRINTOUT 10-4
(continued)

```
Analysis of Variance

SOURCE        DF            SS           MS          F          p
Regression     5       45.2453       9.0491     454.26      0.000
Error          4        0.0797       0.0199
Total          9       45.3250

SOURCE        DF        SEQ SS
x1             1       20.9616
x2             1       22.4859
x1sq           1        0.0921
x2sq           1        0.5088
x1x2           1        1.1969

MTB > noconstant
MTB > regress c4 6 c1-c3 c9-c11
* NOTE *        x1 is highly correlated with other  predictor variables
* NOTE *        x2 is highly correlated with other  predictor variables
* NOTE *        x3 is highly correlated with other  predictor variables
* NOTE *      x1x2 is highly correlated with other  predictor variables
* NOTE *      x1x3 is highly correlated with other  predictor variables

The regression equation is
y1 = 2.35 x1 - 9.22 x2 + 16.8 x3 + 40.7 x1x2 - 7.67 x1x3 - 84.7 x2x3

Predictor       Coef        Stdev      t-ratio         p
Noconstant
x1             2.353        1.724        1.36       0.244
x2            -9.217        3.204       -2.88       0.045
x3            16.757        5.774        2.90       0.044
x1x2          40.745        9.874        4.13       0.015
x1x3          -7.666        9.834       -0.78       0.479
x2x3         -84.739        9.874       -8.58       0.001

s = 0.1411
Analysis of Variance

SOURCE        DF            SS           MS          F          p
Regression     6      252.270        42.045    2110.64      0.000
Error          4        0.080         0.020
Total         10      252.350

SOURCE        DF        SEQ SS
x1             1      221.808
x2             1        0.559
x3             1       28.105
x1x2           1        0.312
x1x3           1        0.018
x2x3           1        1.467
```

Table 10-46 gives the coefficients b obtained in the process of fitting the relationship (10-65) — that is,

$$y \approx \beta_0 + \beta_1 x_1 + \beta_2 x_2 + \beta_3 x_1^2 + \beta_4 x_2^2 + \beta_5 x_1 x_2$$

to the three responses, and Table 10-47 gives the same information for the fitting of the relationship (10-66),

$$y \approx \beta_1 x_1 + \beta_2 x_2 + \beta_3 x_3 + \beta_4 x_1 x_2 + \beta_5 x_1 x_3 + \beta_6 x_2 x_3$$

TABLE 10-46

Fitted Coefficients for Quadratic (10-65) and the Pipe Compound Data

Response	b_0	b_1	b_2	b_3	b_4	b_5
y_1	16.76	-22.07	-110.71	7.67	84.74	133.15
y_2	339.0	-462.5	-376.8	359.1	350.2	449.1
y_3	-7643	32519	25381	-22989	-9334	-29893

TABLE 10-47

Fitted Coefficients for Quadratic (10-66) and the Pipe Compound Data

Response	b_1	b_2	b_3	b_4	b_5	b_6
y_1	2.35	-9.22	16.76	40.75	-7.67	-84.74
y_2	235.5	312.4	339.0	-260.2	-359.1	-350.2
y_3	1887	8404	-7644	2430	22989	9334

Contour plotting with the relationships summarized in Table 10-46 or Table 10-47 is an effective way of visualizing the messages contained in the data of Table 10-45. (All three R^2 values for these quadratic fits exceed .95.) Computer programs exist for contour plotting with mixture models like (10-66) in the triangular coordinate system of Figure 10-13. For example, one such program can be found in the 1981 *Journal of Quality Technology* paper, "Response Surface Contour Plots for Mixture Problems," by Koons and Heasley. Alternatively, one can begin with the more conventional-looking quadratic response function (10-65), apply either the manual or computer methods discussed in Section 9-3 to produce contours, and then simply plot them in triangular coordinates if so desired. Figure 10-16 is a contour plot of the fitted quadratic response surface (10-65) for y_1, drawn in regular (x_1, x_2) rectangular coordinates. Figure 10-17 is the same plot, simply redrawn in the triangular coordinate system. In both plots, only the mixture proportions satisfying constraints (10-72) are depicted in the plotting.

FIGURE 10-16

Contour Plot of Fitted Impact Strength in Rectangular Coordinates

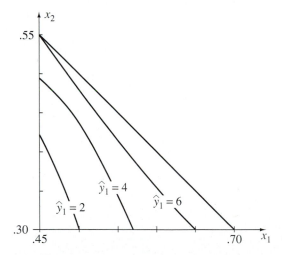

The overlaying of contour plots for different responses is just as helpful in the mixture context as it proved to be in the more conventional setting of Section 9-3. In the ABS pipe compound experiment, the engineers decided to consider mixtures with

$$
\left.
\begin{aligned}
\hat{y}_1 &> 1.2 \\
\hat{y}_2 &> 195 \\
\hat{y}_3 &> 5050
\end{aligned}
\right\}
\tag{10-73}
$$

for possible production. (The 1.2, 195, and 5050 values are lower ASTM specifications for these properties of a certain type and grade of pipe, plus heuristic "safety factors"

FIGURE 10-17

Contour Plot of Fitted
Impact Strength in
Triangular Coordinates

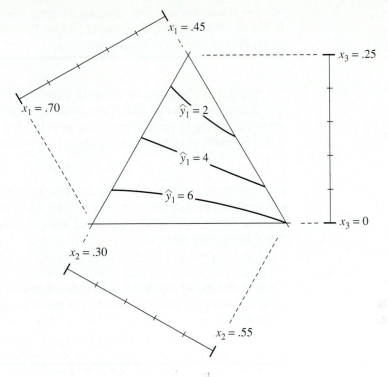

introduced by the engineers to allow for a certain amount of uncertainty associated with the fitted equations.) Figure 10-18 shows the $\hat{y}_1 = 1.2$, $\hat{y}_2 = 195$, and $\hat{y}_3 = 5050$ contours based on the fitted quadratic equations; it thus locates a region of mixture proportions satisfying the requirements (10-73).

FIGURE 10-18

Three-Component Mixtures
Predicted to Satisfy
Constraints (10-73)

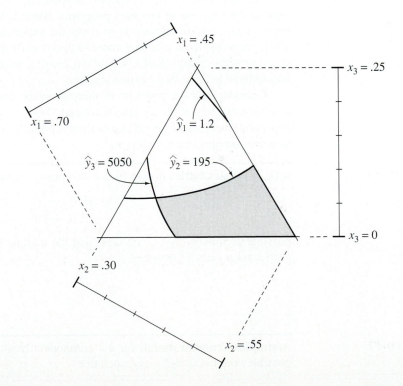

For production purposes, the choice among mixture formulas in the shaded region of Figure 10-18 might (for example) be based on economic considerations and/or the availability of the various components.

━━━━━━━━━━━━━━━━━━━━━━━━ ▬ ▬ ▬

The device of plotting in triangular coordinates is, fairly obviously, most suited to $k = 3$ component studies. But it can sometimes also be used to advantage in mixture studies involving larger numbers of components. For example, in $k = 4$ contexts, for a series of values of x_4, the sets of (x_1, x_2, x_3) vectors satisfying the constraint

$$x_1 + x_2 + x_3 = 1 - x_4$$

can be pictured as a series of triangular regions (of size decreasing with increasing x_4), and corresponding contour plots can then be made, as a way of studying the shape of a fitted surface. (A similar strategy, fixing pairs of values of x_4 and x_5, can be adopted in $k = 5$ component studies.)

EXAMPLE 10-25

Contour Plotting for a Simple $k = 4$ Mixture Response Surface. Figure 10-19, for $x_4 = 0$, .3, and .6 shows a series of contour plots representing the fairly simple linear response surface

$$\hat{y} = 3x_1 + 6x_2 - 3x_3 - 2x_4$$

━━━━━━━━━━━━━━━━━━━━━━━━ ▬ ▬ ▬

Some Issues and Methods in Experimental Design

This section has already indicated that in general, the problem of experimental design for mixture studies for moderate-to-large k is usually attacked with the aid of specialized computer programs. These attempt to optimize measures of design quality over choices of n design points $(x_1, x_2, \ldots, x_k)$. It is beyond the scope of this section to consider the logic or even the use of any such programs. But it is feasible to consider several well-known heuristics for attempting to cover the experimental region in the simple cases where the proportions are either unconstrained or effectively only constrained by bounds stated in terms of individual x_i's. And it is possible to discuss some of the general issues encountered in designing mixture studies.

Consider first the problem of mixture study design when the proportions of k components are completely unconstrained. Two *ad hoc* (but nevertheless intuitively appealing) methods of spreading out design points in such cases are to use the simplex lattice or simplex centroid designs.

DEFINITION 10-10

A (k, t) **simplex lattice design** for a k-component mixture study consists of all

$$\frac{(k + t - 1)!}{t!\,(k - 1)!}$$

possible vectors $(x_1, x_2, \ldots, x_k)$ satisfying the mixture restriction (10-56) and such that each x_i takes one of the values

$$0, \frac{1}{t}, \frac{2}{t}, \ldots, \frac{t-1}{t}, 1$$

DEFINITION 10-11

A **simplex centroid design** for a k-component mixture study consists of all $2^k - 1$ possible vectors $(x_1, x_2, \ldots, x_k)$ such that

1. 1 x_i is 1.0 while all other x_i's are 0, or
2. 2 x_i's are $\frac{1}{2}$ while all other x_i's are 0, or
3. 3 x_i's are $\frac{1}{3}$ while all other x_i's are 0, or
    $\vdots$
4. $k - 1$ x_i's are $\frac{1}{k-1}$ while the other x_i is 0, or
5. all k x_i's are $\frac{1}{k}$.

FIGURE 10-19

Contour Plots for a
Simple $k = 4$ Mixture
Response Surface

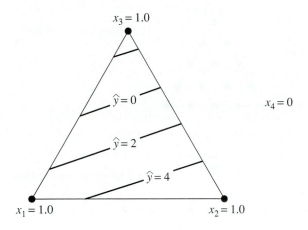

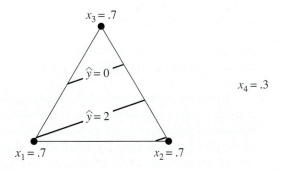

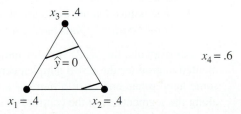

Figure 10-20 depicts two different simplex lattice designs and the simplex centroid design for $k = 3$ component mixtures, in the triangular coordinate system.

Several facts should be obvious from studying Definitions 10-10 and 10-11 and Figure 10-20. For one thing, the simplex lattice designs for $t < k$ will completely fail to place observations in the interior of the experimental region (i.e., at least one and usually several components i will have $x_i = 0$ and thus fail to appear in the chosen mixtures). In fact, one must have $t > k$ in order for there to be any more than one $(x_1, x_2, \ldots, x_k)$

FIGURE 10-20

Two $k = 3$ Simplex Lattice Designs and a Simplex Centroid Design

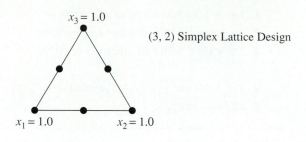

$x_3 = 1.0$

(3, 2) Simplex Lattice Design

$x_1 = 1.0 \qquad x_2 = 1.0$

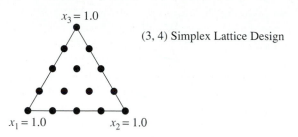

$x_3 = 1.0$

(3, 4) Simplex Lattice Design

$x_1 = 1.0 \qquad x_2 = 1.0$

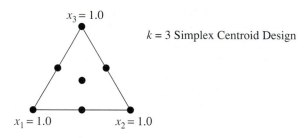

$x_3 = 1.0$

$k = 3$ Simplex Centroid Design

$x_1 = 1.0 \qquad x_2 = 1.0$

design point in the interior of the experimental region. And in a similar fashion, the simplex centroid designs place only a single point in the interior of the experimental region.

Intuitively, such potential lack of coverage of the interior of an experimental region is quite unappealing, leaving one essentially in the position of extrapolating interior behavior of the response(s) from boundary behavior. A useful kind of *ad hoc* modification of simplex lattice designs for small t and simplex centroid designs (which can help remedy the problem of lack of interior coverage) is to add

1. the centroid point $x_1 = \frac{1}{k}, x_2 = \frac{1}{k}, \ldots, x_k = \frac{1}{k}$, if it is not already included, and
2. some points that are averages of the centroid and existing design points (particularly the "corner" points having a single $x_i = 1.0$).

For example, the (3, 2) simplex lattice design of Figure 10-20 could be first augmented to produce the $k = 3$ simplex centroid design shown in the figure, and then for some appropriate choice of a number u, $(0 < u < 1)$, points a fraction u of the way along the segments from the centroid of the region to the corners could be added. That is, one could add three points to the third design pictured in Figure 10-20, of the form

$$u(1, 0, 0) + (1 - u)\left(\frac{1}{3}, \frac{1}{3}, \frac{1}{3}\right)$$

$$u(0, 1, 0) + (1 - u)\left(\frac{1}{3}, \frac{1}{3}, \frac{1}{3}\right)$$

$$u(0, 0, 1) + (1 - u)\left(\frac{1}{3}, \frac{1}{3}, \frac{1}{3}\right)$$

The addition of such extra points upsets some simple hand-calculation formulas for the fitting of low-order models to simplex lattice or simplex centroid data. But in this time of abundant computing resources, this fact seems of no consequence in light of the fuller picture of response behavior provided by the addition of such points.

Another matter raised by consideration of Definitions 10-10 and 10-11 and Figure 10-20 is that at least as typically presented, these elementary designs do not call for any replication. (For that matter, most optimal design algorithms also produce prescriptions for experiments employing no replication.) But as a practical matter, replication is every bit as important in mixture studies as it is in any other statistical engineering context. As always, without replication one has no really firm basis on which to judge the size of background variation or to evaluate whether a fitted response surface is tracking a real pattern in response or simply following a random ghost. So as a practical matter, initial candidates for mixture designs often must be augmented not only to improve their coverage of the interior of an experimental region, but to include some replication as well. When only a few sample sizes can be taken to be larger than 1, symmetry considerations suggest that the first priority for replication should, in fact, be points in the interior of the experimental region.

Definitions 10-10 and 10-11 give possible experimental designs for unconstrained mixture studies. But it ought to be obvious that essentially the same kind of geometrical patterns of experimental points could be used anytime constraints produce an experimental region of the same shape as the unconstrained region. For example, Figure 10-15 shows that natural constraints in the ABS pipe compound study led to a triangular experimental region, geometrically similar to the full unconstrained set of (x_1, x_2, x_3) vectors satisfying equation (10-56). So the simplex lattice and simplex centroid prescriptions ought to have versions that are applicable under the constraints of the study.

The fact is that if an experimental region in a k-component mixture study is to be geometrically similar to the full unconstrained region, it is describable in terms of only lower bounds on the individual proportions. If $c_1, c_2, \ldots, c_k$ are k (possibly 0) nonnegative numbers such that $\sum c_i < 1$ and the inequalities $x_1 \geq c_1, x_2 \geq c_2, \ldots, x_k \geq c_k$ describe the experimental region, the following device can be used to translate an experimental design for an unconstrained mixture study into a corresponding design under the lower bound constraints. Let

$$C = \sum_{i=1}^{k} c_i \tag{10-74}$$

and for $i = 1, 2, \ldots, k$, define variables z_i by

$$z_i = \frac{x_i - c_i}{1 - C} \tag{10-75}$$

The z_i's are sometimes called *pseudocomponents*. Equation (10-75) can be solved for x_i, giving

$$x_i = c_i + (1 - C)z_i \tag{10-76}$$

and an unconstrained design for $z_1, z_2, \ldots, z_k$ can be translated via equation (10-76) to a constrained design for $x_1, x_2, \ldots, x_k$.

EXAMPLE 10-24
(continued)

As it turns out, the constraints (10-72) from the ABS pipe compound study are equivalent to the $k = 3$ lower bound constraints

$$.45 \leq x_1$$
$$.30 \leq x_2$$
$$0 \leq x_3$$

So from equations (10-74) and (10-75), pseudocomponents z_1, z_2, and z_3 can be found using $c_1 = .45$, $c_2 = .30$, $c_3 = 0$, and $C = .75$ — that is,

$$z_1 = \frac{x_1 - .45}{.25}$$

$$z_2 = \frac{x_2 - .30}{.25}$$

$$z_3 = \frac{x_3 - 0}{.25}$$

Table 10-48 shows again the $n = 10$ design points reported by Koons and Wilt, both in their original proportion forms and in their pseudocomponent forms.

TABLE 10-48

Proportional and Pseudocomponent Forms of the Pipe Compound Study Design Points

Data Point	x_1	x_2	x_3	z_1	z_2	z_3
1	.45	.55	0	0	1	0
2	.45	.425	.125	0	.5	.5
3	.45	.30	.25	0	0	1
4	.50	.45	.05	.2	.6	.2
5	.50	.35	.15	.2	.2	.6
6	.535	.38	.085	.34	.32	.34
7	.575	.425	0	.5	.5	0
8	.575	.30	.125	.5	0	.5
9	.60	.35	.05	.6	.2	.2
10	.70	.30	0	1	0	0

By looking at the pseudocomponent columns in Table 10-48, the reader should be able to verify that data points 1, 2, 3, 6, 7, 8, and 10 constitute (at least approximately) a $k = 3$ simplex centroid design. The points 4, 5, and 9 lie (at least approximately) a fraction $u = .39$ of the way along the line segments from the centroid of the region (point 6) toward the corners (points 1, 3, and 10, respectively).

It might also be remarked that it is probably less than ideal that all 10 sample sizes were 1 (i.e., there was no replication) in this study.

The geometric nature of the points making up a simplex centroid design suggests a way of generalizing its basic features beyond cases where the experimental region in a mixture study is describable in terms of only lower bounds to cases where the experimental region is described by lower and/or upper bounds on each x_i. That is, a way to think of the prescription given by Definition 10-11 is to begin with a set of corner points, then to add points in the middle of edges, then to add points in the middle of two-dimensional faces, . . . , and finally to add a point in the middle of the experimental region. And it is fairly straightforward (though often tedious) to identify such corner (or vertex) points and middle (or centroid) points in a mixture problem, provided there are $2k$ numbers $c_1, c_2, \ldots, c_k$ and $d_1, d_2, \ldots, d_k$ with $\sum c_i < 1$ and $0 \leq c_i < d_i \leq 1$ for all i and such that the inequalities $c_1 \leq x_1 \leq d_1, c_2 \leq x_2 \leq d_2, \ldots$ and $c_k \leq x_k \leq d_k$ describe the experimental region.

To find all corner points under such constraints, for each set of $k - 1$ variables x_i, one may write out the set of 2^{k-1} factorial combinations of extreme permissible levels of the $k - 1$ proportions and check to see if the implied value of the excluded proportion satisfies its constraints. If so, the point is a vertex. If not, it lies outside the experimental region.

Once one has identified all vertices of a region describable entirely in terms of upper and lower bounds on individual proportions, centroids of the region of various types can be found. The overall centroid (the center of the essentially $(k - 1)$-dimensional experimental region) can be found by averaging all vertices. Centroids of $(k - 2)$-dimensional faces can be found by — one variable at a time — averaging the vertices at represented levels of that variable. Centroids of $(k - 3)$-dimensional faces can be found by (two) variables at a time) averaging the vertices at represented pairs of levels of those variables. And so on.

Of course, exactly what to do with a set of vertices and centroids after finding them is another question entirely. It has already been noted that centroid designs largely neglect the interior of experimental regions. Besides, for large k, the total number of centroids and vertices will greatly exceed the minimum number of design points needed to fit first- or second-order response surfaces. But the set of centroids does at least constitute a set of sensible candidate design points for examination by design optimization search programs, and for enlightened tinkering by experimenters taking into account a variety of criteria (including the need for replication and coverage of the entire experimental region) when choosing a mixture design.

Notice as well that this discussion has not even touched the issue of locating vertices and centroids of experimental regions that (like the shaded area in Figure 10-14) can't be described by constraints on proportions one at a time.

Additional Comments

The intention here has been to provide a reasonably short but practically usable look at the statistical design and analysis of mixture studies. Much more could be said; a fairly regular stream of papers in the area appears in journals like *Technometrics* and *Journal of Quality Technology*. So before leaving the subject of mixture studies, it seems reasonable to list a few of the additional kinds of tools that are available to the reader who wishes to make a more extensive search of the literature on mixture experiments.

For one thing, a wider variety of models for describing mixture data have been developed than the simple linear and quadratic ones used in this section. Higher-order polynomials in the proportions x_i of course allow for fitting of more complicated patterns in response variables. In some nearly unconstrained mixture contexts, extreme response behavior at very low levels of some proportions can be successfully described using equations involving x_i^{-1} terms. And in some situations, models built using ratio terms like $x_i/x_{i'}$ are useful.

In the way of graphical tools, various displays beyond those used here have been suggested. For example, *response traces* are sometimes made, essentially by plotting for a given proportion x_i, the value of $\hat{y}$ as a function of x_i as one moves along a line segment connecting the design centroid with the vertex of the experimental region corresponding to $x_i = 1.0$.

Computationally, some combinations of simple mixture designs and particular models produce most attractive manual formulas for estimating coefficients and mean responses.

Finally, there are numerous optimization/search algorithms and software available for aiding the mixture experimental design process. It seems clear that engineers planning to make serious use of the topics of this section in any more than, say, $k = 4$ component studies will find themselves driven to study and gain access to such software.

EXERCISES

10-1. (§10-2) In a $p = 7$-factor study, only 32 different combinations of levels of (two-level factors) A, B, C, D, E, F, and G will included, at least initially. The generators F $\leftrightarrow$ ABCD and G $\leftrightarrow$ ABCE will be used to choose the 32 combinations to include in the study.

(a) Write out the whole defining relation for the experiment that is contemplated here.

(b) Based on your answer to part (a), what effects will be aliased with the C main effect in the experiment that is being planned?

(c) When running the experiment, what levels of factors F and G are used when all of A, B, C, D, and E are at their low levels? What levels of factors F and G are used when A, B, and C are at their high levels and D and E are at their low levels?

(d) Suppose that after listing the data (observed y's) in Yates standard order as regards factors A, B, C, D, and E, one uses the Yates algorithm to compute 32 fitted sums of effects. Suppose further that the fitted values appearing on the A + aliases, ABCD + aliases, and BCD + aliases rows of the Yates algorithm computations are the only ones judged to be of both statistical significance and practical importance. What is the simplest possible interpretation of this result?

10-2. (§10-1) In a 2^{5-1} study with defining relation I $\leftrightarrow$ ABCDE, it is possible for both the A main effect and the BCDE 4-factor interaction to be of large magnitude, but for both of them to go undetected. How might this quite easily happen?

10-3. (§10-2) In a 2^{5-2} study, where four sample sizes are 1 and four sample sizes are 2, $s_P = 5$. If 90% two-sided confidence limits are going to be used to judge the statistical detectability of sums of effects, what plus-or-minus value will be used?

10-4. (§10-2) Consider planning, executing, and analyzing the results of a 2^{6-2} fractional factorial experiment based on the two generators E $\leftrightarrow$ ABC and F $\leftrightarrow$ BCD.

(a) Write out the defining relation (i.e., the whole list of aliases of the grand mean) for such a plan.

(b) When running the experiment, what levels of factors E and F are used when all of A, B, C, and D are at their low levels? When A is at its high level, but B, C, and D are at their low levels?

(c) Suppose that $m = 3$ data points from each of the 16 combinations of levels of factors (specified by the generators) give a value of $s_P \approx 2.00$. If 90% two-sided confidence individual intervals are to be made to judge the statistical significance of the estimated (sums of) effects produced by the Yates algorithm here, what is the value of the plus-or-minus part of each of those intervals?

10-5. The paper "How to Optimize and Control the Wire Binding Process: Part I" by Scheaffer and Levine (*Solid State Technology*, November 1990) contains the results of a 2^{5-1} fractional factorial experiment with additional repeated center point, run in an effort to determine how to improve the operation of a K&S Model 1484 XQ wire bonder. The generator E $\leftrightarrow$ ABCD was used in setting up the 2^{5-1} part of the experiment involving the factors and levels indicated in the accompanying table.

The response variable, y, was a force (in grams) required to pull wire bonds made on the machine under a particular combi-

Factor A	Constant Velocity	.6 in./sec ($-$) vs. 1.2 in./sec ($+$)
Factor B	Temperature	150°C ($-$) vs. 200°C ($+$)
Factor C	Bond Force	80 g ($-$) vs. 120 g ($+$)
Factor D	Ultrasonic Power	120 mW ($-$) vs. 200 mW ($+$)
Factor E	Bond Time	10 ms ($-$) vs. 20 ms ($+$)

nation of levels of the factors. (Each y was actually an average of the pull forces required on a 30 lead test sample.) The responses from the 2^{5-1} part of the study were as below.

Combination	y
e	8.5
a	7.9
b	7.7
abe	8.7
c	9.0
ace	9.2
bce	8.6
abc	9.5
d	5.8
ade	8.0
bde	7.8
abd	8.7
cde	6.9
acd	8.5
bcd	8.6
abcde	8.3

In addition, three runs were made at a constant velocity of .9 in./sec, a temperature of 175°C, a bond force of 100 g, a power of 160 mW and a bond time of 15 ms. These produced y values of 8.1, 8.6, and 8.1.

(a) Place the 16 observations from the 2^{5-1} part of the experiment in Yates standard order as regards levels of factors A through D. Use the four-cycle Yates algorithm to compute fitted sums of 2^5 effects. Identify what sum of effects each of these estimates. (For example, the first estimates $\mu_{.....} + \alpha\beta\gamma\delta\epsilon_{22222}$.)

(b) The three center points can be thought of as providing a pooled sample variance here. You may verify that $s_P = .29$. If one then wishes to make confidence intervals for the sums of effects, it is possible to use the $m = 1$, $p = 4$, and $\nu = 2$ version of formula (8-46) of Section 8-3. What is the plus-or-minus value that comes from this program, for individual 95% two-sided confidence intervals? Using this value, which of the fitted sums of effects would you judge to be statistically detectable? Does this list suggest to you any particularly simple/intuitive description of how bond strength depends on the levels of the five factors?

(c) Based on your analysis from (b), if you had to guess what levels of the factors A, C, and D should be used for high bond strength, what would you recommend? If the CE + ABD fitted sum reflects primarily the CE 2-factor interaction, what level of E then seems best? Which of the combinations actually observed had these levels of factors A, C, D, and E? How does its response compare to the others?

10-6. An engineer who is planning a 2^{6-2} fractional factorial experiment has in mind two different possible sets of generators:

> Set 1   E $\leftrightarrow$ ABCD and F $\leftrightarrow$ ABC
>
> Set 2   E $\leftrightarrow$ BCD and F $\leftrightarrow$ ABC

Write out the whole defining relation for both of the choices above. Which one is most appealing and why?

10-7. (§§10-1, 10-2, 10-3) What are the advantages and disadvantages of fractional factorial experimentation in comparison to factorial experimentation?

10-8. (§10-2) Under what circumstances can one hope to be successful experimenting with (say) 12 factors in (say) 16 experimental runs (i.e., based on 16 data points)?

10-9. (§§10-1, 10-2) What is the principle of "sparsity of effects" and how can it be used in the analysis of unreplicated 2^p and 2^{p-q} experiments?

10-10. In Section 10-1, you were advised to choose $\frac{1}{2}$ fractions of 2^p factorials by using the generator

> last factor $\leftrightarrow$ product of all other factors

For example, this means that in choosing $\frac{1}{2}$ of 2^4 possible combinations of levels of factors A, B, C, and D, you were advised to use the generator D $\leftrightarrow$ ABC. There are other possibilities. For example, one could use the generator D $\leftrightarrow$ AB.

(a) Using this alternative plan (specified by D $\leftrightarrow$ AB), what eight different combinations of factor levels would be run? (Use the standard naming convention, listing, for each of the eight sets of experimental conditions to be run, those factors appearing at their high levels.)

(b) For the alternative plan specified by D $\leftrightarrow$ AB, list all eight pairs of effects of factors A, B, C, and D that would be aliased. (You may, if you wish, list eight sums of the effects $\mu_{....}, \alpha_2, \beta_2, \alpha\beta_{22}, \gamma_2, \ldots$ etc. that can be estimated.)

(c) Suppose that in an analysis of data from an experiment run according to the alternative plan (with D $\leftrightarrow$ AB), the Yates algorithm is used with $\bar{y}$'s listed according to Yates standard order for factors A, B, and C. Give four equally plausible interpretations of the eventuality that the first four lines of the Yates calculations produce large estimated sums of effects (in comparison to the other four, for example).

(d) Why might one well argue that the choice D $\leftrightarrow$ ABC is superior to the choice D $\leftrightarrow$ AB?

10-11. $p = 5$ factors A, B, C, D, and E are to be studied in a 2^{5-2} fractional factorial study. The two generators D $\leftrightarrow$ AB and E $\leftrightarrow$ AC are to be used in choosing the eight ABCDE combinations to be included in the study.

(a) Give the list of eight different combinations of levels of the factors that will be included in the study. (Use the convention of naming, for each sample, those factors that should be set at their high levels.)

(b) Give the list of all effects aliased with the A main effect if this experimental plan is adopted.

10-12. The following are eight sample means listed in Yates standard order, considering levels of three two-level factors A, B, and C.

$$70, 61, 72, 59, 68, 64, 69, 69$$

(a) Use the Yates algorithm here to compute eight estimates of effects from the sample means.

(b) Temporarily suppose that no value for s_P is available. Make a plot appropriate to identifying those estimates from (a) that are likely to represent something more than background noise. Based on the appearance of your plot, which if any of the estimated effects are clearly representing something more than background noise?

(c) As it turned out, $s_P = .9$, based on $m = 2$ observations at each of the eight different sets of conditions. Based on 95% individual two-sided confidence intervals for the underlying effects estimated from the eight $\bar{y}$'s, which estimated effects are clearly representing something other than background noise? (If confidence intervals $\hat{E} \pm \Delta$ were to be made, show the calculation of Δ and state which estimated effects are clearly representing more than noise.)

Still considering the eight sample means, henceforth suppose that by some criteria, one considers only the estimates ending up on the first, second, and sixth lines of the Yates calculations to be both statistically detectable and of practical importance.

(d) If in fact the eight $\bar{y}$'s came from a (4-factor) 2^{4-1} experiment with generator D $\leftrightarrow$ ABC, how would one typically interpret the result that the first, second, and sixth lines of the Yates calculations (for means in standard order for factors A, B, and C) give statistically detectable and practically important values?

(e) If in fact the eight $\bar{y}$'s came from a (5-factor) 2^{5-2} experiment with generators D $\leftrightarrow$ ABC and E $\leftrightarrow$ AC, how would one typically interpret the result that the first, second, and sixth lines of the Yates calculations (for means in standard order for factors A, B, and C) give statistically detectable and practically important values?

10-13. A production engineer who wishes to study six two-level factors in eight experimental runs decides to use the generators D $\leftrightarrow$ AB, E $\leftrightarrow$ AC, and F $\leftrightarrow$ BC in planning a 2^{6-3} fractional factorial experiment.

(a) What eight combinations of levels of the six factors will be run? (Name them using the usual convention of prescribing for each run which of the factors will appear at their high levels.)

(b) What seven other effects will be aliased with the A main effect in the engineer's study?

10-14. The article "Going Beyond Main-Effect Plots" by Kenett and Vogel (*Quality Progress*, 1991) outlines the results of a 2^{5-1} fractional factorial industrial experiment concerned with the improvement of the operation of a wave soldering machine. The effects of the five factors Conveyor Speed (A), Preheat Temperature (B), Solder Temperature (C), Conveyor Angle (D), and Flux Concentration (E) on the variable

> y = the number of faults per 100 solder joints (computed from inspection of 12 circuit boards)

were studied. (The actual levels of the factors employed were not given in the article.) The combinations studied and the values of y that resulted are given next.

Combination	y
(1)	.037
a	.040
b	.014
ab	.042
ce	.063
ace	.100
bce	.067
abce	.026
de	.351
ade	.360
bde	.329
abde	.173
cd	.372
acd	.184
bcd	.158
abcd	.131

For reasons that are not clear, Kenett and Vogel used a nonstandard $\frac{1}{2}$ fraction of the full 2^5 factorial. That is, the recommendations of Section 10-1 were not followed in choosing which 16 of the 32 possible combinations of levels of factors A through E to include in the wave soldering study. In fact, the generator $E \leftrightarrow -CD$ was apparently employed.

(a) Verify that the combinations listed above are in fact those prescribed by the relationship $E \leftrightarrow -CD$. (For example, with all of A through D at their low levels, note that the low level of E is indicated by multiplying minus signs for C and D by another minus sign. Thus, combination (1) is one of the 16 prescribed by the generator.)

(b) Write the defining relation for the experiment. What is the resolution of the design chosen by the authors? What resolution does the standard choice of $\frac{1}{2}$ fraction provide? Unless there were some unspecified extenuating circumstances that dictated the authors' choice of $\frac{1}{2}$ fraction, why does it seem to be an unwise one?

(c) Write out the 16 different differences of effects that can be estimated based on the data above. (For example, one of these is $\mu_{.....} - \gamma\delta\epsilon_{222}$, another is $\alpha_2 - \alpha\gamma\delta\epsilon_{2222}$, etc.)

(d) Notice that the combinations listed here are in Yates standard order as regards levels of factors A through D. Use the four-cycle Yates algorithm and find the fitted differences of effects. Normal-plot these and identify any statistically detectable differences. Notice that by virtue of the choice of $\frac{1}{2}$ fraction made by the authors, the most obviously statistically significant difference is that of a main effect and a 2-factor interaction.

10-15. The article "Robust Design: A Cost-Effective Method for Improving Manufacturing Processes" by Kacker and Shoemaker (*AT&T Technical Journal*, 1986) discusses the use of a 2^{8-4} fractional factorial experiment in the improvement of the performance of a step in an integrated circuit fabrication process. The initial step in fabricating silicon wafers for IC devices is to grow an epitaxial layer of sufficient (and hopefully adequately uniform) thickness on polished wafers. The engineers involved in running

Factor A	Arsenic Flow Rate	55% (−) vs. 59% (+)
Factor B	Deposition Temperature	1210°C (−) vs. 1220°C (+)
Factor C	Code of Wafers	668G4 (−) vs. 678G4 (+)
Factor D	Susceptor Rotation	continuous (−) vs. oscillating (+)
Factor E	Deposition Time	high (−) vs. low (+)
Factor F	HCl Etch Temperature	1180°C (−) vs. 1215°C (+)
Factor G	HCl Flow Rate	10% (−) vs. 14 % (+)
Factor H	Nozzle Position	2 (−) vs. 6 (+)

this part of the production process considered the effects of eight factors (listed in the accompanying table) on the properties of the deposited epitaxial layer.

A batch of 14 wafers is processed at one time, and the experimenters measured thickness at five locations on each of the wafers processed during one experimental run. These $14 \times 5 = 70$ measurements from each run of the process were then reduced to two response variables:

$y_1 = $ the mean of the 70 thickness measurements

$y_2 = $ the logarithm of the variance the 70 thickness measurements

y_2 is a measure of uniformity of the epitaxial thickness, and y_1 is (clearly) a measure of the magnitude of the thickness. The authors reported results from the experiment as shown in the accompanying table.

Combination	y_1 (μm)	y_2
(1)	14.821	−.4425
afgh	14.888	−1.1989
begh	14.037	−1.4307
abef	13.880	−.6505
cefh	14.165	−1.4230
aceg	13.860	−.4969
bcfg	14.757	−.3267
abch	14.921	−.6270
defg	13.972	−.3467
adeh	14.032	−.8563
bdfh	14.843	−.4369
abdg	14.415	−.3131
cdgh	14.878	−.6154
acdf	14.932	−.2292
bcde	13.907	−.1190
abcdefgh	13.914	−.8625

It is possible to verify that the combinations listed here come from the use of the four generators $E \leftrightarrow BCD$, $F \leftrightarrow ACD$, $G \leftrightarrow ABD$, and $H \leftrightarrow ABC$.

(a) Write out the whole defining relation for this experiment. (The grand mean will have 15 aliases.) What is the resolution of the design?

(b) Consider first the response y_2, the measure of uniformity of the epitaxial layer. Use the Yates algorithm and normal- and/or half normal-plotting (see Exercise 8-18) to identify statistically detectable fitted sums of effects. Suppose that only the two largest (in magnitude) of these are judged to be both statistically significant and of practical importance. What is suggested about how levels of the factors might henceforth be set in order to minimize y_2? From the limited description of the process above, does it appear that these settings require any extra manufacturing expense? **(c)** Turn now to the response y_1. Again use the Yates algorithm and normal- and/or half normal-plotting to identify statistically detectable sums of effects. Which of the factors seems to be most important in determining the average epitaxial thickness? In fact, the target thickness for this deposition process was 14.5 μm. Does it appear that by appropriately choosing a level of this variable it may be possible to get the mean thickness on target? Explain. (As it turns out, the thought process outlined here allowed the engineers to significantly reduce the variability in epitaxial thickness while getting the mean on target, improving on previously standard process operating methods.)

10-16. Arndt, Cahill, and Hovey worked with a plastics manufacturer and experimented on an extrusion process. They conducted a 2^{6-2} fractional factorial study with some partial "replication" (the reason for the quote marks will be discussed later). The experimental factors in their study were as follows.

Factor A	Bulk Density, a measure of the weight per unit volume of the raw material used
Factor B	Moisture, the amount of water added to the raw material mix
Factor C	Crammer Current, the amperage supplied to the crammer-auger
Factor D	Extruder Screw Speed
Factor E	Front End Temperature, a temperature controlled by heaters on the front end of the extruder
Factor F	Back End Temperature, a temperature controlled by heaters on the back end of the extruder

Physically low and high levels of these factors were identified. Using the two generators E $\leftrightarrow$ AB and F $\leftrightarrow$ AC, 16 different combinations of levels of the factors were chosen for inclusion in a plant experiment, where the response of primary interest was the output of the extrusion process in terms of pounds of useful product per hour. A coded version of the data the students obtained is given in the accompanying table. (The data have been rescaled

Combination	y
ef	13.99
a	6.76
bf	20.71
abe	11.11, 11.13
ce	19.61
acf	15.73
bc	23.45
abcef	20.00
def	24.94
ad	24.03, 25.03
bdf	24.97
abde	24.29
cde	24.94, 25.21
acdf	24.32, 24.48
bcd	30.00
abcdef	33.08

by subtracting a particular value and dividing by another, so as to disguise the original responses without destroying their basic structure. You may think of these values as output measured in numbers of some undisclosed units above an undisclosed baseline value.)

(a) The students who planned this experiment hadn't been exposed to the concept of design resolution. What does Table 10-21 indicate is the best possible resolution for a 2^{6-2} fractional factorial experiment? What is the resolution of the one that the students planned? Why would they have been better off with a different plan than the one specified by the generators E $\leftrightarrow$ AB and F $\leftrightarrow$ AC?

(b) Find a choice of generators E $\leftrightarrow$ (some product of letters A through D) and F $\leftrightarrow$ (some other product of letters A through D) that provides maximum resolution for a 2^{6-2} experiment.

(c) The combinations here are listed in Yates standard order as regards factors A through D. Compute $\bar{y}$'s and then use the (four-cycle) Yates algorithm and compute 16 estimated sums of 2^6 factorial effects.

(d) When the extrusion process is operating, many pieces of product can be produced in an hour, but the entire data collection process leading to the data here took over eight hours. (Note, for example, that changing temperatures on industrial equipment requires time for parts to heat up or cool down, changing formulas of raw material means that one must let one batch clear the system, etc.) The repeat observations above were obtained from *two consecutive pieces of product*, made minutes apart, without any change in the extruder setup in between their manufacture. With this in mind, discuss why a pooled standard deviation based on these four "samples of size 2" is quite likely to underrepresent the level of "baseline" variability in the output of this process under a fixed combination of levels of factors A through F. Argue that it would have been extremely valuable to have (for example) rerun one or more of the combinations tested early in the study again late in the study.

(e) Use the pooled sample standard deviation from the repeat observations and compute (using the $p = 4$ version of formula (8-45) in Section 8-3) the plus-or-minus part of 90% two-sided confidence limits for the 16 sums of effects estimated in part (c), acting as if the value of s_P were a legitimate estimate of background variability. Which sums of effects are statistically detectable by this standard? How do you interpret this in light of the information in part (d)?

(f) As an alternative to the analysis in part (e), make a normal plot of the last 15 of the 16 estimated sums of effects you computed in part (c). Which sums of effects appear to be statistically detectable? What is the simplest interpretation of your findings in the context of the industrial problem? (What has been learned about how to run the extruding process?)

(g) Briefly discuss where to go from here, if it is your job to optimize the extrusion process (maximize y). You may want to refer again to Section 9-3 as well as to Section 10-2. What data would you collect next, and what would you be planning to do with them? (Chapter 11 might give you some good ideas as well.)

10-17. (§10-4) The article "Variability Estimates of Sliver Weights at Different Carding Stages and a Suggested Sampling Plan for Jute Processing" by A. Lahiri (*Journal of the Textile*

Institute, 1990) concerns the partitioning of variability in sliver weight in textile manufacture. (A sliver is a continuous strand or band of loose, untwisted wool, cotton, etc., produced along the way to making yarn.) For a particular mill, $I = 3$ (of many) machines were studied, using $K = 5$ (10 m) pieces of sliver cut from each of $J = 5$ rolls produced on each of the machines. The weights of each of the 75 pieces of sliver were determined and produced the following ANOVA table. (The original units of weight are not given in the article.)

ANOVA TABLE

Source	SS	df	MS
Machines	1966	2	983
Rolls	644	12	53.7
Pieces	280	60	4.7
Total	2890	74	

(a) Find estimates of the three variance components σ^2, σ_β^2, and σ_α^2 in this balanced hierarchical study. State carefully what each of these variance components means in the context of the problem.

(b) Use your estimates from (a) and compute fractions of the variance in weights of such pieces of sliver in this mill that could be attributed to each of Machines, Rolls (within a machine), and Pieces (within a roll). If it was your job to reduce the variability in sliver weight in the mill in question, where would you start looking for physical means to increase uniformity? (What has been learned about the sources of variation here?)

(c) Use formulas (10-53) and (10-54) of Section 10-4 to give 90% two-sided confidence intervals for σ^2 and the ratio σ_β^2/σ^2.

10-18. The article "The Successful Use of the Taguchi Method to Increase Manufacturing Process Capability" by S. Shina (*Quality Engineering*, 1991) discusses the use of a 2^{8-3} fractional factorial experiment to improve the operation of a wave soldering process for through-hole printed circuit boards. The experimental factors and levels studied were as shown in the accompanying table.

Factor A	Preheat Temperature	180° ($-$) vs. 220° ($+$)
Factor B	Solder Wave Height	.250 ($-$) vs. .400 ($+$)
Factor C	Wave Temperature	490° ($-$) vs. 510° ($+$)
Factor D	Conveyor Angle	5.0 ($-$) vs. 6.1 ($+$)
Factor E	Flux Type	A857 ($-$) vs. K192 ($+$)
Factor F	Direction of Boards	0 ($-$) vs. 90 ($+$)
Factor G	Wave Width	2.25 ($-$) vs. 3.00 ($+$)
Factor H	Conveyor Speed	3.5 ($-$) vs. 6.0 ($+$)

The generators F $\leftrightarrow$ $-$CD, G $\leftrightarrow$ $-$AD, and H $\leftrightarrow$ $-$ABCD were used to pick 32 different combinations of levels of these factors to run. For each combination, four special test printed circuit boards were soldered, and the lead shorts per board, y_1, and touch shorts per board, y_2, were counted, giving the accompanying data.

(The data here are exactly as given in the article, and this author has no explanation of the fact that some of the numbers do not seem to have come from division of a raw count by 4.)

Combination	y_1	y_2
(1)	6.00	13.00
agh	10.00	26.00
bh	10.00	12.00
abg	8.50	14.00
cfh	1.50	18.75
acfg	.25	16.25
bcf	1.75	25.75
abcfgh	4.25	18.50
dfgh	6.50	6.50
adf	.75	.00
bdfg	3.50	1.00
abdfh	3.25	6.50
cdg	6.00	7.25
acdh	9.50	11.25
bcdgh	6.25	10.00
abcd	6.75	12.50
e	20.00	29.25
aegh	16.50	31.25
beh	17.25	28.75
abeg	19.50	41.25
cefh	9.67	21.33
acefg	2.00	10.75
bcef	5.67	28.67
abcefgh	3.75	35.75
defgh	6.00	22.70
adef	7.30	25.70
bdefg	8.70	30.00
abdefh	9.00	29.70
cdeg	19.30	32.70
acdeh	26.70	25.70
bcdegh	17.70	45.30
abcde	10.30	37.00

(a) Verify that the 32 combinations of levels of the factors A through H listed here are in fact those that are prescribed by the choice of generators. (For each combination of levels of the factors A through E, determine what levels of F, G, and H are prescribed by the generators and check that such a combination is listed.)

(b) Use the generators given here and write out the whole defining relation for this study. (You will end with I aliased with seven other strings of letters.) What is the resolution of the design used in this study? According to Table 10-21, what was possible in terms of resolution for a 2^{8-3} study? Could the engineers in charge here have done better at containing the ambiguity that unavoidably follows from use of a fractional factorial study?

(c) The first accompanying printout (see page 688) is from the use of a "homegrown" microcomputer program for the analysis of 2^p (and therefore also 2^{p-q} fractional) factorial data sets, on the response y_1. (A Fortran listing of the program can be found in a paper by Crowder, Jensen, Stephenson, and Vardeman (*Journal of Quality Technology*, 1988).) Based on the normal plot, which estimated linear combinations of 2^8 factorial effects appear to be statistically detectable here? (Notice that the combinations are listed above in Yates standard order as regards factors A through E.) If possible, give a simple interpretation of this in light of the alias structure specified by the defining relation you found in part (b).

(d) The second accompanying printout (page 690) is from the use of the same program mentioned in (c) on the response y_2. Based on the normal plot, which estimated linear combinations of 2^8 factorial effects appear to be statistically detectable? If possible, give a simple interpretation of this in light of the alias structure specified by the defining relation you found in part (b).

(e) In light of your answers to (c) and (d), the signs of the fitted linear combinations of effects, and a desire to reduce both y_1 and y_2 to the minimum values possible, what combination of levels of the factors do you tentatively recommend here? Is the combination of levels that you see as promising one that is among the 32 tested? If it is not, how would you recommend proceeding in the real manufacturing scenario? (Would you, for example, order that any permanent process changes necessary to the use of your promising combination be adopted immediately?)

The original article reported a decrease in solder defects by nearly a factor of 10 in this process as a result of what was learned from this experiment.

10-19. In the situation of Exercise 10-18, the 32 different combinations of levels of factors A through H were run in the order listed. In fact, the first 16 runs were made by one shift of workers, and the the last 16 were made by a second shift.

(a) In light of the material of Chapter 2 on experiment planning and the formal notion of confounding, what risk of a serious logical flaw did the engineers run in the execution of their experiment? (How would possible shift-to-shift differences show up in the data from an experiment run like this? One of the main things learned from the experiment was that Factor E was very important. Did the engineers run the risk of clouding their view of this important fact?) Explain.

(b) Devise an alternative plan that could have been used to collect data in the situation of Exercise 10-18 without completely confounding the effects of Flux and Shift. Continue to use the 32 combinations of the original factors listed in Exercise 10-18, but give a better assignment of 16 of them to each shift. (*Hint:* Think of Shift as a ninth factor, pick a sensible generator, and use it to put half of the 32 combinations in each shift. There are a variety of possibilities here.)

(c) Discuss in qualitative terms how you would do data analysis if your suggestion in (b) were to be followed.

10-20. (§10-3) *Plackett-Burman designs* provide orthogonal array plans for two-level fractional factorial experiments with numbers of combinations other than the powers of 2 that are characteristic of the 2^{p-q} designs discussed in Sections 10-1 and 10-2. There are, for example, Plackett-Burman designs of 12, 20, and 24 runs. (The 12-run Plackett-Burman design can be used, for example, to create a $\frac{3}{4}$ fraction of the standard 2^4 factorial.) The article "High Cycle Fatigue of Weld Repaired Cast Ti-6Al-4V" by Hunter, Hodi, and Eagar (*Metallurgical Transactions*, 1982) employed a Plackett-Burman design to study the effects of seven two-level factors, using 12 of the 128 possible combinations of levels of the factors. The study involved the response

y = fatigue life of welded titanium castings (log cycles)

and the experimental factors and levels studied were as listed in the accompanying table.

Factor A	Initial Structure of Specimens	as-received ($-$) vs. β-treat ($+$)
Factor B	Weld Bead Size	small ($-$) vs. large ($+$)
Factor C	Pressure Treatment	none ($-$) vs. HIP ($+$)
Factor D	Heat Treatment	anneal ($-$) vs. solution treat and age ($+$)
Factor E	Cooling Rate	slow ($-$) vs. fast ($+$)
Factor F	Polish	chemical ($-$) vs. mechanical ($+$)
Factor G	Final Treatment	none ($-$) vs. peen ($+$)

The values of y reported by the authors are listed in the next table.

Combination	A	B	C	D	E	F	G	y
abdef	+	+	−	+	+	+	−	6.058
acde	+	−	+	+	+	−	−	4.733
bcd	−	+	+	+	−	−	−	4.625
abcg	+	+	+	−	−	−	+	5.899
abf	+	+	−	−	−	+	−	7.000*
aeg	+	−	−	−	+	−	+	5.752
dfg	−	−	−	+	−	+	+	5.682
cef	−	−	+	−	+	+	−	6.607
bdeg	−	+	−	+	+	−	+	5.818
acdfg	+	−	+	+	−	+	+	5.917
bcefg	−	+	+	−	+	+	+	5.863
(1)	−	−	−	−	−	−	−	4.809

(The starred value is actually only a lower bound for y under this set of conditions. The fatigue test was terminated without specimen failure. In the language of Appendix B-5, this value is a *censored* value. Data analysis methods that take account of this censoring do exist, but they are beyond present purposes. Proceed here as if this value were the actual fatigue life observed under conditions abf.)

(a) Verify that the experimental plan above satisfies Definition 10-5. How many times does a level of one factor appear in combination with a level of each other factor?

(b) Compute estimated main effects for each of the factors A through G under the assumption that there are no interactions.

(c) Compute s_{NI} and its associated degrees of freedom. Use half of the plus-or-minus part of formula (10-28) of Section 10-3 and 95% two-sided confidence to judge (still under an assumption of no interactions) which of the main effects estimated in part (b) are statistically detectable.

(d) Use the dummy variable idea discussed in Section 9-4 to fit a "main effects only" model to these data. Can answers to parts (b) and (c) of this exercise be read off your regression printout? Exactly how?

(e) Now use the dummy variable idea of Section 9-4 to fit an "F and D main effects plus FG interaction" model to these data. How does the fit to the data compare with that in part (d)? (E.g., how do the R^2 values compare?) How would the subject matter interpretations of the models compare? How would predictions of fatigue life for other combinations of levels of factors A through G (besides the ones above) compare for the two different fitted models?

PRINTOUT 1
for Exercise 10-18

```
                                    YATES ALGORITHM

CHOOSE OPTION FOR ENTERING DATA:

        (1)  ENTER DATA FROM KEYBOARD
        (2)  FETCH STORED DATA FILE

ENTER NUMBER OF SELECTION:   1
ENTER NUMBER OF FACTORS (UP TO 7):   5

ENTER CELL MEAN FOR SPECIFIED TREATMENT COMBINATION:

(1)       = 6.00
a         = 10.00
b         = 10.00
ab        = 8.50
c         = 1.50
ac        = .25
bc        = 1.75
abc       = 4.25
d         = 6.50
ad        = .75
bd        = 3.50
abd       = 3.25
cd        = 6.00
acd       = 9.50
bcd       = 6.25
abcd      = 6.75
e         = 20.00
ae        = 16.50
be        = 17.25
abe       = 19.50
ce        = 9.67
ace       = 2.00
bce       = 5.67
abce      = 3.75
de        = 6.00
ade       = 7.30
bde       = 8.70
abde      = 9.00
cde       = 19.30
acde      = 26.70
bcde      = 17.70
abcde     = 10.30

ESTIMATED EFFECTS CALCULATED BY YATES ALGORITHM:

I     =     8.8778
A     =     -.2341
B     =     -.3703
AB    =     -.1109
C     =     -.6691
AC    =     -.0372
BC    =     -.7859
ABC   =     -.4078
D     =      .3409
AD    =      .2091
BD    =     -.6672
ABD   =     -.7203
CD    =     4.2628
```

(continued)

PRINTOUT 1

for Exercise 10-18
(continued)

```
ACD     =      .5622
BCD     =     -.7391
ABCD    =     -.9859
E       =     3.5809
AE      =     -.3434
BE      =     -.6047
ABE     =     -.1578
CE      =      .0966
ACE     =     -.5841
BCE     =     -.7703
ABCE    =     -.4547
DE      =      .3253
ADE     =      .5684
BDE     =     -.0578
ABDE    =     -.9859
CDE     =     1.6847
ACDE    =     -.1409
BCDE    =     -.5047
ABCDE   =      .1234


CHOOSE OPTION FOR FURTHER ANALYSIS:

        1.   CONSTRUCT A NORMAL PLOT OF FITTED EFFECTS
        2.   CONSTRUCT A HALF-NORMAL PLOT OF ABSOLUTE FITTED
             EFFECTS
        3.   USE REVERSE YATES ALGORITHM
        4.   SAVE INPUT CELL MEANS IN A DISK FILE
        5.   EXIT FROM THE PROGRAM

ENTER NUMBER OF SELECTION:   1

CHOOSE OPTION REGARDING GRAND MEAN:

        1.   INCLUDE GRAND MEAN IN THE PLOT
        2.   DO NOT INCLUDE GRAND MEAN IN THE PLOT

ENTER NUMBER OF SELECTION:   2


             PLOT OF Y=NORMAL QUANTILE VS. X=FITTED EFFECT
```

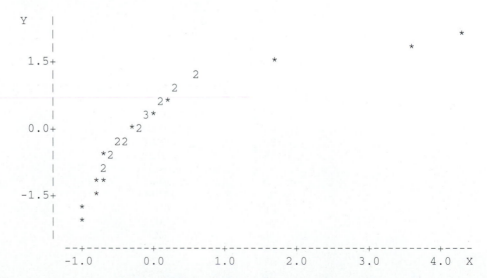

PRINTOUT 2
for Exercise 10-18

```
                              YATES ALGORITHM

CHOOSE OPTION FOR ENTERING DATA:

      (1) ENTER DATA FROM KEYBOARD
      (2) FETCH STORED DATA FILE

ENTER NUMBER OF SELECTION:  1
ENTER NUMBER OF FACTORS (UP TO 7):  5

ENTER CELL MEAN FOR SPECIFIED TREATMENT COMBINATION:

(1)      = 13.00
a        = 26.00
b        = 12.00
ab       = 14.00
c        = 18.75
ac       = 16.25
bc       = 25.75
abc      = 18.50
d        = 6.50
ad       = .00
bd       = 1.00
abd      = 6.50
cd       = 7.25
acd      = 11.25
bcd      = 10.00
abcd     = 12.50
e        = 29.25
ae       = 31.25
be       = 28.75
abe      = 41.25
ce       = 21.33
ace      = 10.75
bce      = 28.67
abce     = 35.75
de       = 22.70
ade      = 25.70
bde      = 30.00
abde     = 29.70
cde      = 32.70
acde     = 25.70
bcde     = 45.30
abcde    = 37.00

ESTIMATED EFFECTS CALCULATED BY YATES ALGORITHM:

I        =    21.0953
A        =      .2859
B        =     2.4466
AB       =      .5722
C        =     1.2453
AC       =    -1.6641
BC       =     1.8966
ABC      =      .0597
D        =    -2.1078
AD       =     -.7297
BD       =      .0659
ABD      =     -.2034
CD       =     2.4797
```

(continued)

PRINTOUT 2

for Exercise 10-18
(continued)

```
ACD     =     1.0078
BCD     =     -.9216
ABCD    =     -.7784
E       =     8.6422
AE      =     -.3859
BE      =     2.3684
ABE     =      .9003
CE      =    -1.3328
ACE     =     -.5859
BCE     =      .3184
ABCE    =      .5128
DE      =     3.4703
ADE     =     -.7453
BDE     =     -.4809
ABDE    =    -1.8441
CDE     =     1.6828
ACDE    =    -1.0078
BCDE    =      .2816
ABCDE   =      .4559
```

```
CHOOSE OPTION FOR FURTHER ANALYSIS:

        1.   CONSTRUCT A NORMAL PLOT OF FITTED EFFECTS
        2.   CONSTRUCT A HALF-NORMAL PLOT OF ABSOLUTE FITTED
             EFFECTS
        3.   USE REVERSE YATES ALGORITHM
        4.   SAVE INPUT CELL MEANS IN A DISK FILE
        5.   EXIT FROM THE PROGRAM

ENTER NUMBER OF SELECTION:   1

CHOOSE OPTION REGARDING GRAND MEAN:

        1.   INCLUDE GRAND MEAN IN THE PLOT
        2.   DO NOT INCLUDE GRAND MEAN IN THE PLOT

ENTER NUMBER OF SELECTION:   2

            PLOT OF Y=NORMAL QUANTILE VS. X=FITTED EFFECT
```

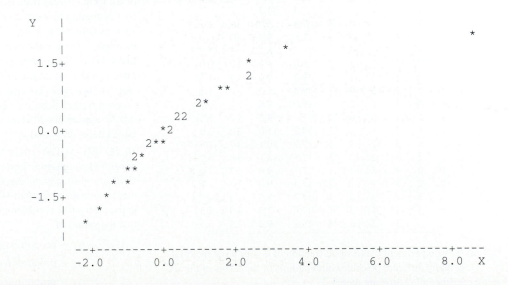

10-21. (§10-3) The article "Banbury Mixing Optimization Using Taguchi Design of Experiments Technique," by J. Rulon (*Rubber World*, 1987) discusses a mixed two- and three-level orthogonal array experiment done to study the effects of six process variables on several physical properties of a raw rubber compound mixed in a Banbury mixer. The factors and levels studied were as shown in the accompanying table.

Factor		Level 1	Level 2	Level 3
A	Mixing Speed	high	low	
B	Banbury Water Temperature	$-40°F$	current	$+40°F$
C	Batch Size	-20 lbs	current	$+20$ lbs
D	Ram Air Pressure	-20 psi	current	$+20$ psi
E	Drop Temperature	$-20°F$	current	$+20°F$
F	Polymer Breakdown Time	current	-1 min	$+1$ min

The response variables in the study were

$y_1 =$ Mooney viscosity, the amount of torque required to rotate the rubber in an enclosed die cavity (in. lb)

$y_2 =$ load deflection at .55, the amount of force required to compress the final product to 55% of its original height (lb)

$y_3 =$ a smoothness score for a specimen run through a laboratory extruder (small values are good)

$y_4 =$ a surface imperfection rating for a piece of final product made from rubber (small values are good)

The responses were measured for three specimens taken from a single batch of rubber processed under each of 18 different combinations of levels of the factors. The average values of the responses for the 18 batches are given in the following table.

A	B	C	D	E	F	y_1	y_2	y_3	y_4
1	1	1	1	1	1	16.3	10.0	2.0	1.0
1	1	2	2	2	2	17.3	11.7	2.7	.7
1	1	3	3	3	3	17.3	9.6	3.3	1.0
1	2	1	1	2	3	16.8	9.2	2.3	0.0
1	2	2	2	3	1	18.3	8.7	.3	5.0
1	2	3	3	1	2	20.0	8.9	3.7	0.0
1	3	1	2	1	2	19.2	9.0	4.7	1.3
1	3	2	3	2	3	18.3	9.0	3.0	1.7
1	3	3	1	3	1	19.0	8.8	3.0	.7
2	1	1	3	3	2	17.3	9.2	4.7	3.3
2	1	2	1	1	3	17.2	11.0	2.0	0.0
2	1	3	2	2	1	17.7	10.1	3.7	1.3
2	2	1	2	3	3	17.5	8.1	3.0	5.7
2	2	2	3	1	1	17.7	9.0	2.7	1.0
2	2	3	1	2	2	17.3	9.0	2.7	.3
2	3	1	3	2	1	17.7	8.5	2.3	.7
2	3	2	1	3	2	17.8	7.5	4.0	2.3
2	3	3	2	1	3	19.2	9.2	2.3	.3

(a) This experiment was a small fraction of a very large factorial. How many different possible combinations of the levels of the six experimental factors are there?

(b) Suppose that (in light of the fact that the objective of the original experiment was process optimization, and considering the discussion of Section 9-3) one had treated all of the experimental variables as quantitative and considered fitting full quadratic response surfaces to the responses. It turns out that for six factors, it suffices to use only a $\frac{1}{2}$ fraction of a 2^6 for the 2^p part of a central composite design (instead of the full 2^6 factorial). (Actually, it appears that even a $\frac{1}{4}$ fraction would suffice, although that possibility for some reason doesn't seem to be recommended in the statistical literature.) How many different combinations of the six factors would then suffice to fit full quadratic response surfaces? How does this number compare to the 18 combinations actually used here? In view of the number of terms in a full quadratic expression in six variables, is there any hope of fitting a full quadratic response surface to the experimenters' data?

(c) Why is there no hope of fitting even the "all main effects and 2-factor interactions" model to the data above? (If you were to try to use the dummy variable idea of Section 9-3, how many terms would there need to be in your regression equation? How does this number compare to the 18 combinations actually studied?)

(d) Parts (a) through (c) above are intended to convince you that the information carried by the experimenters' data on the nature of the joint impact of the six experimental variables on the responses is relatively limited. Explain which of the following are most seriously degraded by the fractional nature of the data here.

(i) one's ability to predict responses for those combinations of levels run in the experiment

(ii) one's ability to find a description of the physical system that can be reasonably assumed to hold somewhat globally

(iii) one's ability to predict responses for combinations of levels not run in the experiment

(e) Use either the hand-calculation methods of Section 10-3 or the dummy variable idea of Section 9-4 to fit "main effects only"/no-interaction models to all of the four responses. For each response, which factors have detectable differences between main effects under a no-interaction model? Use 90% two-sided confidence limits of the form (10-28) from Section 10-3 to make your judgments.

(f) Continuing to work with the no-interaction model for the responses here, write a report to your engineering manager recommending which combination(s) of levels of the factors you would like to try out in an attempt to find a combination that will minimize y_3 and y_4 in future operations of this process. In view of the fact that the major concern, as regards responses y_1 and y_2, is consistency (rather than, e.g., large mean), which process variables does it appear need the most careful watching because of their large effects on y_1 and y_2?

10-22. (§10-5) The article "Experiments with Mixtures of Components Having Lower Bounds," by I. Kurotori (*Industrial Quality Control*, 1966), contains an example of the application of the mixture techniques of Section 10-5 to the formulation of a rocket propellant. Physical considerations in the original situation required that the values of

$x_1 =$ proportion of binder in the mixture

$x_2 =$ proportion of oxidizer in the mixture

$x_3 =$ proportion of fuel in the mixture

satisfy the constraints

$$x_1 \geqslant .20, \qquad x_2 \geqslant .40, \qquad \text{and} \qquad x_3 \geqslant .20$$

The objective of the study conducted was to find a mixture of the components with

$$y = \text{modulus of elasticity}$$

of at least 3000, and possessing minimum x_1. With these constraints and goals in mind, a simplex centroid experiment augmented with three extra interior points was planned and executed. The resulting data are given here.

x_1	x_2	x_3	y
.40	.40	.20	2350
.20	.60	.20	2450
.20	.40	.40	2650
.30	.50	.20	2400
.30	.40	.30	2750
.20	.50	.30	2950
.266	.466	.266	3000
.333	.433	.233	2690
.233	.533	.233	2770
.233	.433	.333	2980

(a) Plot, in triangular coordinates, the locations of the 10 points (x_1, x_2, x_3) used in this experiment. On the plot, show clearly the boundary of the region defined by the constraints on x_1, x_2, and x_3. Do the design points adequately cover the region of interest?

(b) Fit the cubic equation

$$y \approx \beta_1 x_1 + \beta_2 x_2 + \beta_3 x_3 + \beta_4 x_1 x_2 + \beta_5 x_1 x_3 + \beta_6 x_2 x_3 + \beta_7 x_1 x_2 x_3$$

to the data. How does the estimated standard deviation of response compare to the original investigators' prior experience that measurement error standard deviation is somewhere between 10 and 15? (You should find $b_1 = 50381$, $b_2 = 17057$, $b_3 = 35850$, $b_4 = -154245$, $b_5 = -282954$, $b_6 = -114138$, $b_7 = 770378$, and $s_{SF} = 10.50$.)

(c) Using your equation from part (b), make (by hand if you don't have appropriate software) a contour plot for fitted y on the (x_1, x_2, x_3) region of interest. Make contours for at least $\hat{y} = 2950$, 3000, and 3050. (If you are computing by hand, eliminate x_3 from your fitted equation using the fact that $x_3 = 1 - x_1 - x_2$. Then for various values of x_1, you can solve quadratic equations in x_2 for values giving desired $\hat{y}$'s.)

(d) Use your plot from (c) and make a recommendation for the solution of the original rocket propellant problem.

10-23. The article "Computer Control of a Butane Hydrogenolysis Reactor" by Tremblay and Wright (*The Canadian Journal of Chemical Engineering*, 1974) contains an interesting data set concerned with the effects of $p = 3$ process variables on the performance of a chemical reactor. The factors and their levels were as follows.

Factor A	Total Feed Flow (cc/sec at STP)	50 (−) vs. 180 (+)
Factor B	Reactor Wall Temperature (°F)	470 (−) vs. 520 (+)
Factor C	Feed Ratio (Hydrogen/Butane)	4 (−) vs. 8 (+)

The data had to be collected over a four-day period, and two combinations of the levels of factors A, B, and C above were run each day along with a center point — a data point with Total Feed Flow 115, Reactor Wall Temperature 495, and Feed Ratio 6. The response variable was

$$y = \text{percent conversion of butane}$$

and the data in the accompanying table were collected.

Day	Feed Flow	Wall Temp.	Feed Ratio	Combination	y
1	115	495	6	—	78
1	50	470	4	(1)	99
1	180	520	8	abc	87
2	50	520	4	b	98
2	180	470	8	ac	18
2	115	495	6	—	87
3	50	520	8	bc	95
3	180	470	4	a	59
3	115	495	6	—	90
4	50	470	8	c	76
4	180	520	4	ab	92
4	115	495	6	—	89

(a) Suppose that to begin with, one ignores the fact that these data were collected over a period of four days and simply treats the data as a complete 2^3 factorial augmented with a repeated center point. Analyze these data using the methods of Chapter 8. (Compute s_P from four center points. Use the Yates algorithm and the eight corner points to compute fitted 2^3 factorial effects. Then judge the statistical significance of these using appropriate 95% two-sided confidence limits based on s_P.) Is any simple interpretation of the experimental results in terms of factorial effects obvious?

According to the authors, there was the possibility of "process drift" during the period of experimentation. The one-per-day center points were added to the 2^3 factorial at least in part to provide some check on that possibility, and the allocation of two ABC combinations to each day was very carefully done in order to try to minimize the possible confounding introduced by any Day/Block effects. The rest of this problem considers analyses that might be performed on the experimenters' data above, in recognition of the possibility of process drift.

(b) Plot the four center points against the number of the day on which they were collected. What possibility is at least suggested by your plot? Would the plot be particularly troubling if your experience with this reactor told you that a standard deviation of around 5(%) was to be expected for values of y from consecutive runs of the reactor under fixed operating conditions on a given day? Would the plot be troubling if your experience with this reactor told you that a standard deviation of around 1(%) was to be expected for values of y from consecutive runs of the reactor under fixed operating conditions on a given day?

(c) The four-level factor Day can be formally thought of in terms of two extra two-level factors — say, D and E. Consider the choice of generators D ↔ AB and E ↔ BC for a 2^{5-2} fractional factorial. Verify that the eight combinations of levels of A through E prescribed by these generators divide the eight possible combinations of levels of A through C up into the four groups of two corresponding to the four days of experimentation. (To begin with, note that both A low, B low, C low and A high, B high, C high correspond to D high and E high. That is, the first level of Day can be thought of as the D high and E high combination.)

(d) The choice of generators in (c) produces the defining relation I $\leftrightarrow$ ABD $\leftrightarrow$ BCE $\leftrightarrow$ ACDE. Write out, on the basis of this defining relation, the list of eight groups of aliased 2^5 factorial effects. Any effect involving factors A, B, or C with either of the letters D (δ) or E (ϵ) in its name represents some kind of interaction with Days. Explain what it means for there to be no interactions with Days. Make out a list of eight smaller groups of aliased effects that are appropriate supposing that there are no interactions with Days.

(e) Allowing for the possibility of Day (Block) effects, it does not make sense to use the center points to compute s_P. However, one might normal-plot (or half-normal-plot) the fitted effects from (a). Do so. Interpret your plot, supposing that one is willing to assume that there were no interactions with Days in the reactor study. How do your conclusions differ (if at all) from those in (a)?

(f) One possible way of dealing with the possibility of Day effects in this particular study is to use the center point on each day as a sort of baseline and express each other response as a deviation from that baseline. (If on day i there is a day effect γ_i, and on day i the mean response for any combination of levels of factors A through C is $\mu_{comb} + \gamma_i$, the mean of the difference $y_{comb} - y_{center}$ is $\mu_{comb} - \mu_{center}$; one can therefore hope to see 2^3 factorial effects uncontaminated by additive Day effects using such differences in place of the original responses.) For each of the four days, subtract the response at the center point from the other two responses and apply the Yates algorithm to the eight differences. Normal-plot the fitted effects on the (difference from the center point mean) response. Is there any substantial difference between the result of this analysis and that for the others suggested in this problem?

10-24. The text *Design of Experiments: A Realistic Approach*, by Anderson and McLean, contains the the results of a $k = 4$-component mixture experiment, where the object was to optimize

$$y = \text{illumination provided (1000 candles)}$$

by choice of the values of

$$x_1 = \text{proportion of magnesium}$$
$$x_2 = \text{proportion of sodium nitrate}$$
$$x_3 = \text{proportion of strontium nitrate}$$
$$x_4 = \text{proportion of binder}$$

(by weight) in a formula for type of flare. Physical considerations introduced the constraints

$$.40 \le x_1 \le .60, \qquad .10 \le x_2 \le .50,$$
$$.10 \le x_3 \le .50, \qquad \text{and} \qquad .03 \le x_4 \le .08$$

and data were collected at the eight extreme vertices plus various centroids of the region defined by these constraints. The data follow.

x_1	x_2	x_3	x_4	y
.40	.10	.47	.03	75
.40	.10	.42	.08	180
.60	.10	.27	.03	195
.60	.10	.22	.08	300
.40	.47	.10	.03	145
.40	.42	.10	.08	230
.60	.27	.10	.03	220
.60	.22	.10	.08	350
.50	.10	.345	.055	220
.50	.345	.10	.055	260
.40	.2725	.2725	.055	190
.60	.1725	.1725	.055	310
.50	.2350	.2350	.03	260
.50	.21	.21	.08	410
.50	.2225	.2225	.055	425

(a) Fit a full quadratic model to these data. What fraction of the raw variation in y is accounted for in the fitting process? What do you obtain for s_{SF}? What does this quantity intend to measure? What is a fairly obvious weakness in the present experimental design? (You should obtain $\hat{y} = 14298 - 22398x_1 - 33564x_2 - 33768x_3 + 6453x_1^2 + 16915x_2^2 + 17044x_3^2 + 31756x_1x_2 + 31661x_1x_3 + 37172x_2x_3$, and $s_{SF} = 59.9$.)

(b) Anderson and McLean report that use of a quadratic programming algorithm with the fitted response from (a) and the original physical constraints produced a point of maximum $\hat{y}$ when $x_4 = .08$. Set $x_4 = .08$ in your fitted response function from (a) and then make a contour plot (in triangular coordinates) of the fitted response for $x_4 = .08$ as a function of $x_1, x_2,$ and x_3 (summing to .92), similar to the plots in Figure 10-19. On this plot, mark the constraints for each of $x_1, x_2,$ and x_3. What combination of values of $x_1, x_2,$ and x_3 do you recommend for maximum $\hat{y}$, subject to the constraints and assuming that x_4 will be set to .08?

10-25. The text *Statistical Theory and Methodology in Science and Engineering*, by K.A. Brownlee, contains a large hierarchical data set that arose from the analysis of the amounts of solids suspended in batches of a slurry. $J = 4$ physical samples were taken from each batch, and by splitting each sample and analyzing the halves separately, $K = 2$ measurements were obtained on each sample. Part of Brownlee's data set is reproduced in the accompanying table. (Brownlee's book contains data on ten other batches of slurry, but the data from $I = 4$ samples will serve present purposes.)

(a) Brownlee noted that for the raw solids measurements (in the entire 14-batch data set), the batch average of the ranges of the two measurements for the samples from the batch tended to increase (roughly linearly) with the batch average, indicating a nonconstant σ, when measurement is done on the original scale. Why does this cause problems if one wishes to carry out the kind of analysis discussed in Section 10-4?

(b) In response to the difficulty noted in (a), Brownlee suggested that analysis of the solids content data be done on the log scale — hence the last column in the accompanying table. Using the logged values, make an ANOVA table for this part of Brownlee's data set.

Batch	Sample	Analysis	Solids, y	$\ln(y)$
1	1	1	76	4.33
1	1	2	85	4.44
1	2	1	69	4.23
1	2	2	82	4.41
1	3	1	72	4.28
1	3	2	78	4.36
1	4	1	75	4.32
1	4	2	84	4.43
2	1	1	110	4.70
2	1	2	109	4.69
2	2	1	119	4.78
2	2	2	106	4.66
2	3	1	120	4.79
2	3	2	121	4.80
2	4	1	111	4.71
2	4	2	119	4.78
3	1	1	130	4.87
3	1	2	140	4.94
3	2	1	143	4.96
3	2	2	121	4.80
3	3	1	141	4.95
3	3	2	147	4.99
3	4	1	129	4.86
3	4	2	140	4.94
4	1	1	62	4.13
4	1	2	67	4.20
4	2	1	50	3.91
4	2	2	61	4.11
4	3	1	71	4.26
4	3	2	74	4.30
4	4	1	66	4.19
4	4	2	67	4.20

(c) Find estimates of the three variance components σ^2, σ_β^2, and σ_α^2 in this balanced hierarchical study. State carefully what each of these variance components means in the context of the problem.

(d) Use your estimates from (c) and compute fractions of the variance in log solids contents that could be attributed to each of Batches, Samples (within a batch), and Analyses (within a sample). Does it appear that the solids measurement process is good enough to detect differences between batches?

(e) Use formulas (10-53) and (10-54) of Section 10-4 to give 90% two-sided confidence intervals for σ^2 and the ratio σ_β^2/σ^2.

10-26. Return to the fire retardant flame test study of Exercise 8-26. The original study, summarized in that exercise, was a full 2^4 factorial study.

(a) If you have not done so previously, use the (four-cycle) Yates algorithm and compute the fitted 2^4 factorial effects for the study. Normal-plot these. What subject matter interpretation of the data is suggested by the normal plot?

Now suppose that instead of a full factorial study, only the $\frac{1}{2}$ fraction with generator D $\leftrightarrow$ ABC had been conducted.

(b) Which eight of the 16 treatment combinations would have been run? List these combinations in Yates standard order as regards factors A, B, and C and use the (three-cycle) Yates algorithm to compute the eight estimated sums of effects that it is possible to derive from these eight treatment combinations. Verify that each of these eight estimates is the sum of two of your fitted effects from part (a). (For example, you should find that the first estimated sum here is $\bar{y}_{....} + abcd_{2222}$ from part (a).)

(c) Normal-plot the last seven of the estimated sums from (b). Interpret this plot. If you had only the data from this 2^{4-1} fractional factorial, would your subject matter conclusions be the same as those reached in part (a), based on the full 2^4 data set?

Putting It Together:
An Industrial Case Study

This book has described what may seem an almost dazzling variety of statistical concepts and methods. As these have been introduced, an effort has been made to elucidate where each one fits in the grand spectrum of tools available for use in engineering problem solving. But there remains a danger that the tools exposited here will seem like so many trees in the proverbial forest, each important in its own right, but isolated and of unknown relationship to the others or to the surrounding landscape. To reduce this danger, it seems prudent to discuss an important real industrial project, whose success story involves the integration of many of the tools presented in this text.

The project that will be discussed is one that was carried out in a division of Dow Chemical Company during the first half of a single year. This project has had a positive economic impact that is estimated to be several hundred thousand dollars annually. This author is grateful to Dow Chemical Company for permission to incorporate this case study as the final chapter of this book. He particularly thanks Mr. Robert Kasprzyk and Dr. Don Brattesani for supplying the information necessary to write the chapter and securing the company's permission to use it here.

In the interest of corporate security, the data that appear in this chapter have been "sanitized." In structure, they are like the data originally collected in the study but are not themselves the original data. Note also that the details of every application of statistical tools may not have been exactly as presented here, although this author has tried to be accurate at least in terms of generalities, if not specific details unknown to him. And finally, as is always true in summaries like this, it is possible that some important aspects of the engineering project will not be featured and perhaps not even mentioned. Nonetheless, the hope is to present a reasonably complete summary of what was done from the perspective of engineering statistics.

Background

The circumstances that led to the Dow engineering project were as follows. For many years, an important company product ("Product Z") had been produced by Dow by means of a batch chemical process. A large new demand for high-purity Product Z created a need for increased attention to the making of the product. A large amount of product of at least 95% purity was needed, and the existing Product Z process was averaging about 88% purity, at 43% yield in January, with large fluctuations apparent in both purity and yield variables.

The first wave of new orders for high-purity Product Z were met, but at a high cost. A high-level engineering project team, operating primarily on the basis of expert opinion, had taken on the task of improving process performance. Through some manipulation of process variables and a large rework rate, the team was able to direct use of the existing Product Z process to meet initial demand for the high-purity product. But it also became apparent that serious process improvement efforts were needed so as to create a permanent, sizable increase in both routine mean purity and mean yield for the process and also to increase the consistency of the purity and yield figures from batch to batch.

In April, a second, broader project team was assembled to attack the problem of improving the ongoing production of high-purity Product Z. Managers, engineers, chemists, plant operators, and a statistician participated in a joint process improvement effort. This case study is a description of some aspects of their work, which in less than three months increased product mean purity to 97% (from 88%), increased mean yield to 59% (from 43%), and radically reduced batch-to-batch variation in both purity and yield.

Measurement Consistency

A fundamental issue that had to be settled, before there could be any hope of successful data-directed process improvement efforts, was the matter of the ability to measure product purity. Product Z was being made in batches adequate to fill five drums. The standard method of measuring purity was to take a physical sample of product from the first, third, and fifth drums, mix these, and send the mixture to a laboratory for analysis. There, standard procedure was to make three separate determinations of purity from the (single composite) sample. At least two different aspects of this purity measurement process were considered.

For one thing, discussions with process operators revealed that no consistent protocol existed for the taking of physical samples from the drums. Each operator did things his or her own way. In order to eliminate this unnecessary source of variability in measured purity, a detailed sampling protocol was written and implemented to ensure that all operators would draw their samples in the same way. (See again Sections 1-3 and 2-1 of this book.)

To determine whether the analytical results were precise enough to allow investigators to see changes in purity from batch to batch, they conducted a hierarchical study quite similar to the one used as an example in Section 10-4. $I = 7$ different high-purity batches of Product Z were studied. $J = 2$ different physical samples were taken from each batch (in the manner samples are taken from a single drum), and each one of these individual samples was measured $K = 2$ times for purity, by the standard laboratory methods. The data in Table 11-1 are representative of those obtained in the study. Table 11-2 is a (nested/hierarchical) ANOVA table for the data of Table 11-1.

From this table and formulas (10-47), (10-48), and (10-49) of Section 10-4, estimates of batch, sample, and determination variance components σ_α^2, σ_β^2, and σ^2 can be

TABLE 11-1

Measured Purities from Seven Batches of Product Z

Batch	Sample	Measured Purity (%)
1	1	98.4, 99.3
	2	98.6, 98.2
2	1	97.6, 98.1
	2	97.4, 97.4
3	1	97.9, 98.9
	2	98.4, 97.9
4	1	98.7, 98.3
	2	99.6, 99.3
5	1	95.7, 96.0
	2	94.6, 94.9
6	1	95.0, 95.3
	2	95.1, 94.9
7	1	96.9, 96.3
	2	96.7, 96.7

TABLE 11-2

ANOVA Table for the Product Z Purities

ANOVA TABLE

Source	SS	df	MS	EMS
Batches	59.774	6	9.9623	$\sigma^2 + 2\sigma_\beta^2 + 4\sigma_\alpha^2$
Samples (Batches)	2.613	7	.3732	$\sigma^2 + 2\sigma_\beta^2$
Determinations	1.695	14	.1211	σ^2
Total	64.082	27		

seen to be (respectively) 2.397, .126, and .121. These represent 90.6%, 4.8%, and 4.6% of their total, respectively. (In the original data, these three estimated variance components accounted for 83%, 5%, and 12% of their total, respectively.) Two facts are relevant in considering the sizes of these estimated variance components. 1) The mixing of three different physical samples to produce a single composite sample amounts to a kind of averaging that (roughly speaking) ought to reduce the impact of the sample-to-sample variance by a factor of 3. 2) Arithmetic averaging of three different determinations on each composite sample ought to reduce the impact of determination-to-determination variance by a factor of 3 as well. In light of these facts, these figures indicate that the purity measurement system is adequate for purposes of detecting changes from batch to batch. This is true even for this homogeneous group of high-purity batches. In any subsequent experimentation, the problem would be to see the effects of process changes above a background noise level established primarily by batch-to-batch variation, not by measurement process variation.

Establishing a Baseline

In any process improvement study, it is important to know from where one is beginning. The data in Table 11-3 are representative of purity and yield figures for the process as of January. Figures 11-1 and 11-2 are plots (respectively) of purity, y_1, and yield, y_2, against time.

TABLE 11-3

Product Z Purities and Yields at the Beginning of the Study

Batch	y_1 Purity (%)	y_2 Yield (%)
1	92.9	55.4
2	90.8	49.9
3	90.6	44.8
4	87.2	43.7
5	83.5	34.9
6	83.6	40.3
7	76.9	38.0
8	90.7	51.7
9	85.5	41.1
10	93.2	42.1
11	96.7	48.9
12	80.6	32.9
13	74.8	33.4
14	98.9	55.5
15	82.6	38.7
16	76.4	28.7
17	99.9	46.6
18	96.1	48.8
19	94.2	48.3
20	86.3	44.0
21	88.0	33.8
22	93.1	44.9
23	92.2	44.0
24	99.6	55.9
25	90.1	46.9
26	75.4	35.9
27	80.4	40.0
28	86.4	39.5
29	93.0	45.4
30	80.2	33.8

FIGURE 11-1

Plot of Purity vs. Batch Number at the Beginning of the Study

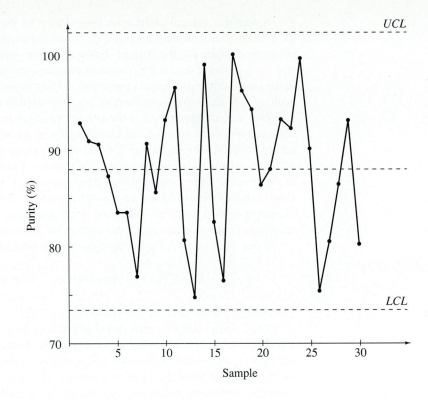

FIGURE 11-2

Plot of Yield vs. Batch Number at the Beginning of the Study

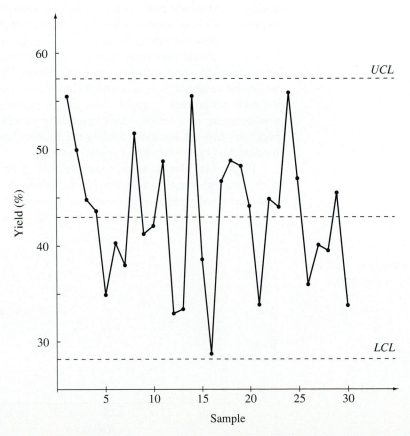

Notice that control limits have been drawn on the two plots in Figures 11-1 and 11-2. Even after reviewing Section 7-5, it may not be at all clear how such limits would be derived, since for the data of Table 11-2, the natural sample (or *subgroup*) size is $m = 1$. Using a reasonably common practice, the (retrospective) limits for the two figures were computed based on average *moving ranges*. That is, retrospective control limits (7-93) or (7-94) from Section 7-5 were used, with $m = 2$, d_2 based on sample size $m = 2$, and $\bar{R}$ computed as the average of absolute differences between successive observations. (This practice amounts roughly to averaging regular $\bar{R}$ values from two sets of artificial samples of size 2, made up by pairing adjacent observations beginning with the first and then with the second observation. The thinking is that artificial samples of consecutive values may exhibit slightly too much variation to use in estimating a short-term σ, because of day-to-day variation in μ. But at least as a conservative way of setting control limits on individual observations, using this measure of variation in artificial samples to estimate σ is better than having no way to proceed.) The reader may check that (for example) for the purity values, $\bar{R}$ based on the moving ranges is

$$\bar{R} = \frac{1}{29}(2.1 + .2 + 3.4 + \cdots + 6.6 + 12.8) = 7.64$$

The corresponding value for the yields is $\bar{R} = 7.82$.

The original control charts (represented here by Figures 11-1 and 11-2) portrayed a production process behaving in a fairly stable fashion but operating at mean levels for both y_1 and y_2 that were much below desired values, and with baseline variability far in excess of what the engineers running the process hoped for. (Notice that (for example) as regards yield, large variability in y_2 could make for chaotic production scheduling, since one can't predict with any certainty how many batches would be needed to produce a given amount of product.) The stability shown in plots like Figures 11-1 and 11-2 significantly dims the chances of finding a quick fix for unsatisfactory process behavior. It would have been simpler if a few instances of obvious lack of control had been identified and localized to particular times. Then a search for obvious physical, assignable causes at those times might have led to the discovery of circumstances out of the norm for process operation. Instead, the production process as initially configured was behaving in a stable, albeit practically unacceptable, fashion. Evidently, it was necessary to work on understanding the process better, which might point the way to fundamental improvements in operating policy.

Figure 11-3 shows an interesting feature of the data in Table 11-3, which is true to the original situation and provides some comfort that the twin goals of improving both purity and yield might possibly be accomplished simultaneously. It is clear from the simple scatterplot in Figure 11-3 that the purity and yield figures are positively

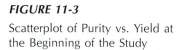

FIGURE 11-3

Scatterplot of Purity vs. Yield at the Beginning of the Study

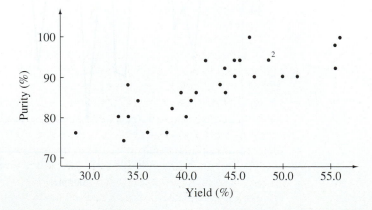

correlated. This fact provides hope that measures taken to increase yield might also increase purity.

Goal Setting, Brainstorming, and Attention Upstream

Beginning from the "stable but unsatisfactory" nature of initial process performance, the project team adopted a systematic, comprehensive approach to improving the process. The fact that some batches of Product Z were made with large purities and yields suggested that if one could identify and control the physical variables causing fluctuation in y_1 and y_2, it might well be possible to make both y_1 and y_2 consistently large. As a first step in process improvement efforts, the project team adopted the concrete goals of finding a way of running the process that would

> **1.** increase mean purity, y_1, of Product Z to at least 96.5% and reduce the batch-to-batch variability in purity,
> **2.** increase mean yield, y_2, of Product Z to at least 59%, and
> **3.** allow the process to meet certain minimum monthly production goals.

The project team brainstormed together regarding variables that might possibly be influencing process performance. (Many of the tools of Section 2-5 of this text were used in that effort.) They reached consensus on which relatively few of the many potential sources of variation in process purity and yield deserved highest priority and would be studied in detail. A number of work groups were then formed to investigate the impacts of various "upstream" factors on process performance. Measures were taken to improve the consistency of a number of process inputs, with the expectation that such improvements would be reflected in improved consistency of process output. In addition, an experimental study was commissioned to quantify (and hopefully exploit) the effects of four process variables that were generally agreed to affect performance. The remainder of this chapter is primarily a description of that experimental work.

Understanding the Effects of Four Process Variables: Two $\frac{1}{2}$ Fractions of a 2^4 Factorial

The experimentation team set to work planning and executing a series of studies aimed at quantifying the impacts of four factors (which will be called A, B, C, and D) — primarily on the yield variable, y_2, and secondarily on the purity variable, y_1. These factors were the concentrations or mole ratios of three reactants and the run time of the process. Initial experimental limits on these variables were chosen as in Table 11-4, with a view to adequately covering the ranges of feasible values without producing conditions that seemed unworkable, based on practical experience with the process.

TABLE 11-4

Initial Experimental Limits on Four Process Variables

Factor	Low Level	High Level
Reactant A Mole Ratio	1.0	2.5
Reactant B Mole Ratio	1.0	1.8
Reactant C Mole Ratio	1.0	2.5
Time (D)	2.0 (hr)	5.0 (hr)

An initial set of $n = 12$ experimental runs was planned. Eight of those constituted the standard $\frac{1}{2}$ fraction of a 2^4 factorial defined by the generator D $\leftrightarrow$ ABC. (See Section 10-1.) The remaining four observations were gathered at the center point of the set of ABCD conditions defined in Table 11-4. These four provided the experimenters with pooled estimates of response variation σ for both responses and at least some ability to check for curvature (as described in Section 9-3). The values in Table 11-5 are representative of the results obtained in the first phase of experimentation.

For the time being, attention will be focused on the yield variable, y_2. The value of s_P derived from the last four y_2 data points in Table 11-5 is $s_P = 1.13$. Table 11-6 shows the (three-cycle) Yates algorithm applied to the first eight y_2 data points in Table 11-5.

TABLE 11-5

Data Collected in a First Phase of Experimentation

x_1 A Level	x_2 B Level	x_3 C Level	x_4 D Level	2^4 Name	y_1	y_2
1.00	1.0	1.00	2.0	(1)	62.1	35.1
2.50	1.0	1.00	5.0	ad	92.2	45.9
1.00	1.8	1.00	5.0	bd	7.0	4.0
2.50	1.8	1.00	2.0	ab	84.0	46.0
1.00	1.0	2.50	5.0	cd	61.1	41.4
2.50	1.0	2.50	2.0	ac	91.6	51.2
1.00	1.8	2.50	2.0	bc	9.0	10.0
2.50	1.8	2.50	5.0	abcd	83.7	52.8
1.75	1.4	1.75	3.5	—	87.7	54.7
1.75	1.4	1.75	3.5	—	89.8	52.8
1.75	1.4	1.75	3.5	—	86.5	53.3
1.75	1.4	1.75	3.5	—	87.3	52.0

TABLE 11-6

The Yates Algorithm Applied to the 2^{4-1} Part of the Yield Data from the First Phase of Experimentation

Combination	y	Cycle 1	Cycle 2	Cycle 3	Cycle 3 ÷ 8	Sum Estimated
(1)	35.1	81.0	131.0	286.4	35.80	I + ABCD
ad	45.9	50.0	155.4	105.4	13.18	A + BCD
bd	4.0	92.6	52.8	−60.8	−7.60	B + ACD
ab	46.0	62.8	52.6	64.2	8.03	AB + CD
cd	41.4	10.8	−31.0	24.4	3.05	C + ABD
ac	51.2	42.0	−29.8	−.2	−.03	AC + BD
bc	10.0	9.8	31.2	1.2	.15	BC + AD
abcd	52.8	42.8	33.0	1.8	.23	ABC + D

Since s_P from the four center points has $\nu = 3$ associated degrees of freedom, the $m = 1$, $p = 3$ version of formula (8-46) of Section 8-3 shows that for 95% individual two-sided confidence intervals, precisions of roughly

$$\pm 3.182 \frac{1.13}{\sqrt{1 \cdot 2^3}}$$

that is,

$$\pm 1.27 \ (\%)$$

could be attached to each of the estimated sums of effects in the next-to-last column of Table 11-6. By this standard, each of the first five estimated sums is statistically detectable, while the last three are not. There is thus some hint that factor D, the run time, may be fairly inert, having minimal impact on yield, and that C may act on yield separately from A and B.

From the point of view of potential surface fitting, there is strong evidence in the data of Table 11-5 that a surface with a fair amount of curvature would be required to describe yield in terms of x_1, x_2, x_3, and x_4. For one thing (as discussed in Section 9-4), interaction effects represent a kind of curvature, and the fourth estimated sum in Table 11-6 represents some kind of interaction or sum of interactions. For another thing, the first eight y_2's in Table 11-5 have mean $\bar{y}_{\text{corners}} = 35.8$, while the four center points have mean $\bar{y}_{\text{center}} = 53.2$. In a manner similar to that leading to equations (9-84) and (9-85) of Section 9-3, the measure of curvature

$$\bar{y}_{\text{center}} - \bar{y}_{\text{corners}} = 53.2 - 35.8 = 17.4$$

is then clearly detectable above its measure of precision

$$\pm t(1.13)\sqrt{\frac{1}{4} + \frac{1}{2^3}}$$

And most conclusively, Printout 11-1 shows that fitting the linear expression

$$y_2 \approx \beta_0 + \beta_1 x_1 + \beta_2 x_2 + \beta_3 x_3 + \beta_4 x_4$$

to the $n = 12$ data points of Table 11-5 results in an R^2 value of only about .59 and $s_{SF} = 13.8$, which is way out of line with the value of $s_P = 1.13$.

After analyzing their first experimental data set and finding no definitive simple picture of yield as a function of the four process variables, the Dow experimentation team planned and executed a second experimental phase. This time, $n = 8$ experimental runs were made, constituting the other standard $\frac{1}{2}$ fraction of a 2^4 factorial, defined by the generator $D \leftrightarrow -ABC$. The values in Table 11-7 are representative of the results obtained in the second phase of this experimentation.

The eight process conditions represented in Table 11-7 are different from any included in the first experimental phase. Although such was not done in the Dow study, it might have been a good idea also to collect a few extra data points, again at the center

PRINTOUT 11-1

```
MTB > print c1-c5

ROW      x1      x2      x3      x4    yield

  1    1.00    1.0    1.00    2.0    35.1
  2    2.50    1.0    1.00    5.0    45.9
  3    1.00    1.8    1.00    5.0     4.0
  4    2.50    1.8    1.00    2.0    46.0
  5    1.00    1.0    2.50    5.0    41.4
  6    2.50    1.0    2.50    2.0    51.2
  7    1.00    1.8    2.50    2.0    10.0
  8    2.50    1.8    2.50    5.0    52.8
  9    1.75    1.4    1.75    3.5    54.7
 10    1.75    1.4    1.75    3.5    52.8
 11    1.75    1.4    1.75    3.5    53.3
 12    1.75    1.4    1.75    3.5    52.0

MTB > regress c5 4 c1-c4

The regression equation is
yield = 29.8 + 17.6 x1 - 19.0 x2 + 4.07 x3 + 0.15 x4

Predictor        Coef       Stdev      t-ratio         p
Constant        29.82       26.32         1.13     0.295
x1             17.567       6.490         2.71     0.030
x2             -19.00       12.17        -1.56     0.162
x3              4.067       6.490         0.63     0.551
x4              0.150       3.245         0.05     0.964

s = 13.77      R-sq = 59.2%      R-sq(adj) = 35.9%

Analysis of Variance

SOURCE         DF           SS           MS           F         p
Regression      4       1925.6        481.4        2.54     0.133
Error           7       1326.6        189.5
Total          11       3252.2
```

TABLE 11-7

Data Collected in a Second Phase of Experimentation

x_1 A Level	x_2 B Level	x_3 C Level	x_4 D Level	2^4 Name	y_1	y_2
1.00	1.0	1.00	5.0	d	64.0	35.3
2.50	1.0	1.00	2.0	a	91.9	47.2
1.00	1.8	1.00	2.0	b	6.5	3.9
2.50	1.8	1.00	5.0	abd	86.4	45.9
1.00	1.0	2.50	2.0	c	63.9	39.5
2.50	1.0	2.50	5.0	acd	93.1	51.6
1.00	1.8	2.50	5.0	bcd	6.8	9.2
2.50	1.8	2.50	2.0	abc	84.6	54.3

of the experimental region. A second sample of, say, four observations at the set of conditions $x_1 = 1.75$, $x_2 = 1.4$, $x_3 = 1.75$, and $x_4 = 3.5$ would have provided a rough check against the possibility of an undetected systematic Time or Block effect creeping in between the experimental periods. For example, formula (6-35) of Section 6-3 could have been used to compare means from two center point samples. (Had no detectable change been observed between the experimental phases, the $4 + 4 = 8$ center points could then have been used to produce a ($\nu = 7$ degrees of freedom) pooled estimate of σ for use in subsequent analysis.)

Upon combining the two $\frac{1}{2}$ fraction parts of the yield values in Tables 11-5 and 11-7 into a complete 2^4 factorial data set and applying the Yates algorithm, the fitted effects in Table 11-8 can be derived. Once again appealing to formula (8-46) of Section 8-3, 95% individual two-sided confidence intervals for the 2^4 factorial effects can be made using the fitted effects of Table 11-8 and plus-or-minus figures of

$$\pm 3.182 \frac{1.13}{\sqrt{1 \cdot 2^4}}$$

that is,

$$\pm .90 \, (\%)$$

TABLE 11-8

Fitted Effects from the 2^4 Part of the Yield Data

Fitted Effect	Value
$\bar{y}_{....}$	35.83
a_2	13.53
b_2	-7.57
ab_{22}	7.96
c_2	2.92
ac_{22}	.19
bc_{22}	.39
abc_{222}	.29
d_2	$-.07$
ad_{22}	$-.24$
bd_{22}	$-.22$
abd_{222}	.13
cd_{22}	.07
acd_{222}	$-.03$
bcd_{222}	$-.36$
$abcd_{2222}$	$-.03$

This value makes it clear that only the A, B, and C main effects and AB 2-factor interaction effects on yield are large enough to see above the background noise in these data. This confirms the appropriateness of the simplest possible interpretation of the first 2^{4-1} study.

In the Dow process improvement context, it was significant that factor D, the run time, had minimal impact on yield (either separately or in combination with other factors). This had the pleasing practical implication that production could apparently be speeded up (or at least didn't need to be slowed down to improve yield). The fact that factor C had minimal interactions with A and B also seemed to promise some simplification: in any optimization efforts, changes in reactant C's concentration would have the same impact for any combination of concentrations of A and B. But in terms of meeting the project team's 59% yield goal, the 2^4 factorial analysis was not enough to do the job. None of the combination means in Tables 11-5 and 11-7 appear to hold any promise of consistently giving 59% yields. (Indeed, no fitted y_2 value based on the grand mean, A, B, and C main effects, and AB interaction exceeds 53%.) But there is hope in the fact that the curvature indicated earlier in the first experimental phase is as present as ever in the combined data set. Perhaps somewhere in the interior of the experimental region (rather than at its corners), an area of improved mean yield might be found. (For example, repeat the $\bar{y}_{\text{center}} - \bar{y}_{\text{corners}}$ curvature check, based now on all of the data from both tables, and note that its result is qualitatively unchanged.) To find a region of improved mean yield (if it exists) by exploiting curvature, the nature of that curvature must be identified. As was seen in Section 9-3, that requires more data than simply a 2^4 configuration augmented by center points.

Understanding the Effects of Four Process Variables: Augmentation to a Central Composite Design

The data in Tables 11-5 and 11-7 are not sufficient to allow the fitting of a general quadratic response surface for yield in the four process variables x_1, x_2, x_3, and x_4. Recognizing this, the Dow experimentation team planned and executed a third phase of experimentation, aimed at providing the additional information necessary to allow such fitting. In the language and notation of Definition 9-4, eight star points derived using (at least approximately) $\alpha = 1.414$ were used. The data in Table 11-9 are representative of what the team found in their third experimental phase as they completed their central composite design.

TABLE 11-9

Data Collected in a Third Phase of Experimentation

x_1 A Level	x_2 B Level	x_3 C Level	x_4 D Level	y_1	y_2
.6895	1.4	1.75	3.5	20.8	13.0
2.8105	1.4	1.75	3.5	95.9	54.3
1.75	.8344	1.75	3.5	99.9	62.4
1.75	1.9656	1.75	3.5	65.9	41.2
1.75	1.4	.6895	3.5	64.4	32.7
1.75	1.4	2.8105	3.5	64.8	40.3
1.75	1.4	1.75	1.379	88.1	52.7
1.75	1.4	1.75	5.621	88.9	50.5

Printout 11-2 is from the fitting of a full quadratic response surface for yield in the four variables x_1, x_2, x_3, and x_4 to the central composite yield data set made up from Tables 11-5, 11-7, and 11-9. The R^2 value on the printout is huge, presumably larger than that originally obtained by the experimentation team. But the general features of the fitted equation, as well as the fact that $s_{\text{SF}} = 1.22$ is in essential agreement with $s_{\text{P}} = 1.13$, are true to the original situation.

Notice that interpreting a fitted equation like

$$\hat{y}_2 = 15.4 + 37.9x_1 - 66.2x_2 + 48.8x_3 + .97x_4 - 16.1x_1^2 - .03x_2^2$$
$$- 13.6x_3^2 - .046x_4^2 + 26.5x_1x_2 + .344x_1x_3 - .217x_1x_4 + 1.31x_2x_3 \qquad \textbf{(11-1)}$$
$$- .365x_2x_4 + .061x_3x_4$$

is not a particularly inviting task. But the earlier 2^4 factorial analysis and the nature of the table of t ratios for the fitted coefficients in the printout suggest the possibility of finding a much simpler-looking equation that might fit the yield data nearly as well as equation (11-1). Consider the conclusions of the earlier 2^4 factorial analysis: no effects involving D, and C acting separately from A and B. Consider also the many small t values corresponding to coefficients in the full equation for terms involving x_4 (the run time variable D), product terms involving x_3 (the reactant C variable), and the x_2^2 term.

PRINTOUT 11-2

```
MTB > print c1-c5

ROW      x1        x2        x3       x4    yield

  1   1.0000    1.0000    1.0000   2.000    35.1
  2   2.5000    1.0000    1.0000   5.000    45.9
  3   1.0000    1.8000    1.0000   5.000     4.0
  4   2.5000    1.8000    1.0000   2.000    46.0
  5   1.0000    1.0000    2.5000   5.000    41.4
  6   2.5000    1.0000    2.5000   2.000    51.2
  7   1.0000    1.8000    2.5000   2.000    10.0
  8   2.5000    1.8000    2.5000   5.000    52.8
  9   1.7500    1.4000    1.7500   3.500    54.7
 10   1.7500    1.4000    1.7500   3.500    52.8
 11   1.7500    1.4000    1.7500   3.500    53.3
 12   1.7500    1.4000    1.7500   3.500    52.0
 13   1.0000    1.0000    1.0000   5.000    35.3
 14   2.5000    1.0000    1.0000   2.000    47.2
 15   1.0000    1.8000    1.0000   2.000     3.9
 16   2.5000    1.8000    1.0000   5.000    45.9
 17   1.0000    1.0000    2.5000   2.000    39.5
 18   2.5000    1.0000    2.5000   5.000    51.6
 19   1.0000    1.8000    2.5000   5.000     9.2
 20   2.5000    1.8000    2.5000   2.000    54.3
 21   0.6895    1.4000    1.7500   3.500    13.0
 22   2.8105    1.4000    1.7500   3.500    54.3
 23   1.7500    0.8344    1.7500   3.500    62.4
 24   1.7500    1.9656    1.7500   3.500    41.2
 25   1.7500    1.4000    0.6895   3.500    32.7
 26   1.7500    1.4000    2.8105   3.500    40.3
 27   1.7500    1.4000    1.7500   1.379    52.7
 28   1.7500    1.4000    1.7500   5.621    50.5

MTB > regress c5 14 c1-c4 c6-c15
* NOTE *     x2 is highly correlated with other predictor variables
* NOTE *     x2sq is highly correlated with other predictor variables

The regression equation is
yield = 15.4 + 37.9 x1 - 66.2 x2 + 48.8 x3 + 0.97 x4 - 16.1 x1sq
             - 0.03 x2sq - 13.6 x3sq - 0.046 x4sq + 26.5 x1x2
             + 0.344 x1x3 - 0.217 x1x4 + 1.31 x2x3 - 0.365 x2x4
             + 0.061 x3x4
```

(continued)

PRINTOUT 11-2
(continued)

Predictor	Coef	Stdev	t-ratio	p
Constant	15.418	7.219	2.14	0.052
x1	37.864	3.215	11.78	0.000
x2	-66.245	7.771	-8.52	0.000
x3	48.816	3.215	15.19	0.000
x4	0.966	1.607	0.60	0.558
x1sq	-16.1456	0.7214	-22.38	0.000
x2sq	-0.026	2.536	-0.01	0.992
x3sq	-13.6115	0.7214	-18.87	0.000
x4sq	-0.0463	0.1804	-0.26	0.801
x1x2	26.521	1.017	26.09	0.000
x1x3	0.3444	0.5422	0.64	0.536
x1x4	-0.2167	0.2711	-0.80	0.439
x2x3	1.312	1.017	1.29	0.219
x2x4	-0.3646	0.5083	-0.72	0.486
x3x4	0.0611	0.2711	0.23	0.825

s = 1.220   R-sq = 99.7%   R-sq(adj) = 99.5%

Analysis of Variance

SOURCE	DF	SS	MS	F	p
Regression	14	7718.81	551.34	370.41	0.000
Error	13	19.35	1.49		
Total	27	7738.16			

These facts suggest the possibility of fitting an equation of the form

$$y_2 \approx \beta_0 + \beta_1 x_1 + \beta_2 x_2 + \beta_3 x_3 + \beta_4 x_1^2 + \beta_5 x_3^2 + \beta_6 x_1 x_2$$

to the central composite yield data. Printout 11-3 is from such a regression run.

Notice that the R^2 value for this second regression run is nearly as large as the first one, and the s_{SF} value is again well in line with s_P. As a matter of fact, fitted values for

PRINTOUT 11-3

```
MTB > regress c5 6 c1 c2 c3 c6 c8 c10

The regression equation is
yield = 13.8 + 37.8 x1 - 65.3 x2 + 51.6 x3 - 16.2 x1sq
        - 13.6 x3sq + 26.5 x1x2
```

Predictor	Coef	Stdev	t-ratio	p
Constant	13.782	3.337	4.13	0.000
x1	37.809	2.585	14.63	0.000
x2	-65.297	1.711	-38.16	0.000
x3	51.571	2.247	22.96	0.000
x1sq	-16.1745	0.6351	-25.47	0.000
x3sq	-13.6404	0.6351	-21.48	0.000
x1x2	26.5208	0.9131	29.05	0.000

s = 1.096   R-sq = 99.7%   R-sq(adj) = 99.6%

Analysis of Variance

SOURCE	DF	SS	MS	F	p
Regression	6	7712.9	1285.5	1070.78	0.000
Error	21	25.2	1.2		
Total	27	7738.2			

the equation

$$\hat{y}_2 = 13.8 + 37.8x_1 - 65.3x_2 + 51.6x_3 - 16.2x_1^2 - 13.6x_3^2 + 26.5x_1x_2 \qquad \textbf{(11-2)}$$

differ from those for the full quadratic fitted equation by no more than about 1.4 across the entire set of central composite design points (x_1, x_2, x_3, x_4), and most differences are less than 1.0.

Fitted equation (11-2) (which for most purposes is essentially equivalent to the full fitted quadratic (11-1) over the experimental region) has an interesting nature. Since x_3 enters the equation separately from x_1 and x_2, the basic shape of the fitted response as a function of x_1 and x_2 is the same for each x_3. Changing x_3 simply moves the whole (x_1, x_2, y) response surface up or down, without warping it. x_4 doesn't enter equation (11-2) at all. So as far as getting a picture of fitted yield goes, it is sufficient to pick a single value of x_3 and make a contour plot of $\hat{y}_2$ as a function of x_1 and x_2. Figure 11-4 shows such a plot for $x_3 = 1.75$, the center value of x_3 in experimentation and in fact the

FIGURE 11-4

Contour Plot of Fitted Yield for $x_3 = 1.75$

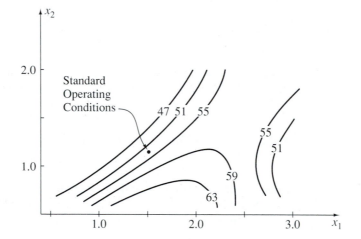

FIGURE 11-5

Three-Dimensional Perspective Plot of Fitted Yield for $x_3 = 1.75$

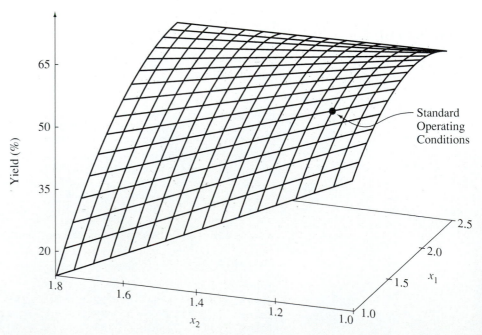

FIGURE 11-6

Plots of Fitted Yield vs. x_1 for Three Values of x_2 When $x_3 = 1.75$

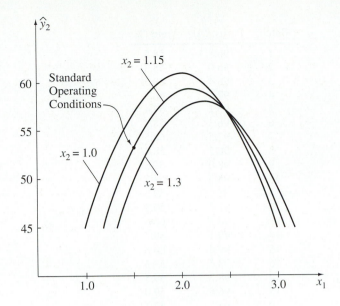

standard operating condition for x_3. Figure 11-5 is an equivalent figure, which attempts to portray the shape of the $x_3 = 1.75$ response surface as a three-dimensional object. Figure 11-6 is yet a third portrayal of the $x_3 = 1.75$ fitted yield, this time in terms of plots of $\hat{y}_2$ versus x_1 for several values of x_2. Standard operating conditions of $x_1 = 1.5$, $x_2 = 1.15$, and $x_3 = 1.75$ and the corresponding fitted yield $\hat{y}_2 = 53.2$ are marked on each of these.

Armed with an equation like (11-2) and plots like Figures 11-4, 11-5, and 11-6, the Dow engineering team was prepared to consider the question of how the reactant mole ratios x_1, x_2, and x_3 might be changed in order to improve process yield. To understand what was done, first consider the effect on yield of x_3, the C mole ratio (and the variable not directly represented in the figures). Notice that the part of the right side of equation (11-2) involving x_3, namely,

$$51.6x_3 - 13.6x_3^2$$

is a quadratic function of x_3, having a maximum at roughly $x_3 = 1.90$. This suggests that a change in x_3 from 1.75 would increase mean yield by at most

$$\left(51.6(1.90) - 13.6(1.90)^2\right) - \left(51.6(1.75) - 13.6(1.75)^2\right) = .29\,(\%)$$

regardless of the x_1 and x_2 values under consideration. Reactant C mole ratio changes simply didn't present any promise of increasing mean yield enough to meet team goals.

Consider then the joint effects of changing x_1 and/or x_2 from standard values like $x_1 = 1.5$ and $x_2 = 1.15$. It is clear from Figures 11-4, 11-5, and 11-6 that there are ways of changing x_1 and/or x_2 in order to increase $\hat{y}_2$ to the goal of 59% yield. An increase in x_1 and/or decrease in x_2 might be used to improve $\hat{y}_2$. Changing x_2 alone produces a linear change in $\hat{y}_2$ in equation (11-2), while $\hat{y}_2$ varies quadratically in x_1 for fixed x_2. What was ultimately done by the experimentation team was to recommend moving the standard value of x_1 alone, changing it to a larger value. (This choice of process change was made partially on the basis of consideration of the original analog of equation (11-2) and partially on the basis of other operating constraints for the Product Z production process.) For discussion, consider a new set of operating conditions $x_1 = 2.0$, $x_2 = 1.15$, and $x_3 = 1.75$. The reader is invited to verify that the corresponding fitted yield is then $\hat{y}_2 = 59.1$, which meets the team's mean yield goal.

TABLE 11-10

Product Z Purities and
Yields After the Process
Improvement Project

Batch	y_1 Purity (%)	y_2 Yield (%)
1	96.8	58.5
2	96.0	58.3
3	98.4	60.3
4	97.5	60.5
5	95.1	57.4
6	96.3	58.2
7	97.5	60.5
8	96.7	58.4
9	98.0	59.8
10	95.3	58.7
11	98.1	59.3
12	97.7	59.3
13	98.3	60.6
14	99.1	59.7
15	99.4	60.8
16	93.5	57.3
17	98.2	59.5
18	96.6	59.0
19	94.5	57.1
20	97.2	58.9
21	97.0	60.1
22	97.6	58.9
23	97.5	59.3
24	97.6	59.8
25	96.5	58.1
26	96.6	58.8
27	95.7	58.4
28	98.0	60.3
29	97.8	59.3
30	97.6	58.9

The data in Table 11-10 are representative of routine batch purity and yield values obtained in June, after making an adjustment to x_1 similar to that just suggested. The same values are control-charted in Figures 11-7 and 11-8.

The contrasts between the original control charts in Figures 11-1 and 11-2 and those in Figures 11-7 and 11-8 are startling. Not only are the means much higher for this second set, but the basic natural variation portrayed in this second set is much smaller than before the beginning of the process improvement project. Some of the improvement in consistency was no doubt due to better upstream control over (and therefore reduced variation in) process inputs, achieved by the general efforts of the project team. But an important part of the improvement in consistency was also due specifically to the change in x_1 that was instituted on the basis of the response surface experimentation aimed primarily at increasing mean yield. To understand how this came about, consider again Figures 11-4 through 11-6. The operating conditions $x_1 = 1.5$, $x_2 = 1.15$, $x_3 = 1.75$ put one on a "steep slope" of the yield response surface, where small (unintentional) variations in x_1 cause big variations in yield. The move to $x_1 = 2.0$, suggested by consideration of equation (11-2), puts one on a flatter part of the response surface, where changes in x_1 have a much smaller impact on $\hat{y}_2$.

In analytic terms, the reader can check that from equation (11-2), for $x_1 = 1.5$ and $x_2 = 1.15$,

$$\frac{\partial \hat{y}_2}{\partial x_1} = 37.8 - 32.4(1.5) + 26.5(1.15) = 19.7$$

FIGURE 11-7

Plot of Purity vs. Batch Number at the End of the Project

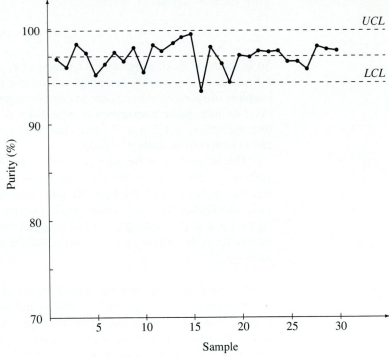

FIGURE 11-8

Plot of Yield vs. Batch Number at the End of the Project

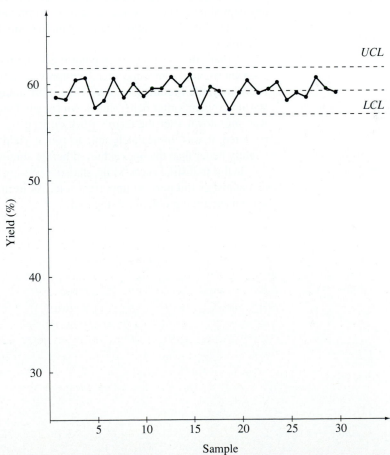

while for $x_1 = 2.0$ and $x_2 = 1.15$,

$$\frac{\partial \hat{y}_2}{\partial x_1} = 37.8 - 32.4(2.0) + 26.5(1.15) = 3.5$$

So in terms of yield, unintentional changes in x_1 around $x_1 = 1.5$ are magnified 5 or 6 times as much (roughly speaking) as similar changes in x_1 around $x_1 = 2.0$. (This coupling of variation in a response to derivatives and variation in inputs should bring to mind the material on propagation of error that was discussed in Section 5-5.) A change like one from $x_1 = 1.5$ to $x_1 = 2.0$ can thus produce benefits both in terms of mean yield and in terms of consistency of yield.

This discussion of the project team's experimentation has focused on the yield variable, y_2. The charge to the experimentation group was in fact to improve yield, and it is thus particularly pleasant that the same measures that improved yield average and yield consistency also had similar impacts on purity average and purity consistency. At least part of the explanation of this happy circumstance is that a quadratic response surface for y_1 in terms of x_1, x_2, x_3, and x_4 had the same qualitative character as the one fit to y_2.

Impact of the Project

It was remarked earlier that the annual impact of the Dow project was estimated to be several hundred thousand dollars. Before the project, only about 20% of batches of Product Z were (by chance) high-purity batches. After the project, essentially all were. Before the project, process yield averaged 43%; after the project, the average yield was over 59%. Thus, a large increase in effective production capacity was achieved without any capital investment. At the same time, improved consistency of production was also realized — which can allow better planning, smaller lead times, and other important benefits.

Perhaps the most revealing point regarding the effectiveness of the project was not even known until some time after the improved yield and purity benefits were recognized. Previously, the Product Z process (like most industrial processes) had had its share of startup problems after being shut down for any length of time. The first time the process was shut down after the team's work on it, it was down for $2\frac{1}{2}$ months. When it was restarted, it was immediately able to run at yield and purity levels equal to and even slightly better than the ones achieved before shutdown.

In this industrial success tale, statistics was not the whole story. But it should be clear that statistics did play an important role. As in many engineering situations, statistics was an essential problem solving tool.

A

The Matrix Formulation of Multiple Regression

T he treatment of curve- and surface-fitting in Chapters 4 and 9 contains very few explicit formulas, except for the very simple case of fitting a line to (x, y) data. The method of exposition used there was to rely instead on information provided by commonly available multiple linear regression programs. If one is willing to use matrix notation, it is possible to write fairly explicit formulas for the quantities that were so useful in Chapters 4 and 9. Partly for the sake of completeness with regard to regression material already introduced, and partly because there are some important types of inferences in multiple regression that couldn't be discussed in Chapters 4 and 9 due to the lack of explicit formulas, this section now provides the matrix formulas that underlie both descriptive and more formal analyses in curve- and surface-fitting.

The Matrix Representation of Multiple Regression Data and Least Squares Surface-Fitting

The gist of multiple regression is to take n data points $(x_{1i}, x_{2i}, \ldots, x_{ki}, y_i)$ and process them in such a way as to quantify an approximate relationship between y and the system variables $x_1, x_2, \ldots, x_k$. In most of the exposition of this text, the generic relationship fit to such data has been

$$y_i \approx \beta_0 + \beta_1 x_{1i} + \beta_2 x_{2i} + \cdots + \beta_k x_{ki} \tag{A-1}$$

However, in the mixture context of Section 10-5, it was also advantageous to consider no-intercept relationships like

$$y_i \approx \beta_1 x_{1i} + \beta_2 x_{2i} + \cdots + \beta_k x_{ki} \tag{A-2}$$

It turns out to be possible to think of the fitting of either of the two relationships (A-1) and (A-2) in the same framework, if one conceives of equation (A-1) as

$$y_i \approx \beta_0(1) + \beta_1 x_{1i} + \beta_2 x_{2i} + \cdots + \beta_k x_{ki} \tag{A-3}$$

That is, one thinks of there being an extra system variable x_0 involved when one fits an equation with intercept — a variable one that always has the value 1.

In this section, in order to write one set of formulas for the fitting of both intercept and no-intercept relationships like (respectively) equations (A-1) or (A-3) and (A-2) in k real system variables $x_1, x_2, \ldots, x_k$, the following will be done. The symbol $\boldsymbol{k}$ will be used to stand for the number of coefficients β appearing in the relationship to be fitted and the fitting of

$$y_i \approx \beta_1 x_{1i} + \beta_2 x_{2i} + \cdots + \beta_{\boldsymbol{k}} x_{\boldsymbol{k}i} \tag{A-4}$$

to a data set will be considered. In the familiar intercept context of relationship (A-1) or (A-3), this amounts to letting $\boldsymbol{k} = k + 1$, introducing the extra system variable $x_{\boldsymbol{k}} = x_{k+1} = 1$ and thinking of the intercept term as standing last, rather than first, in the list of terms to be fitted. In the less common no-intercept context of relationship (A-2), this notation will simply mean that $\boldsymbol{k} = k$, and no mental adjustments are necessary.

The basic fact that gets one started in writing matrix formulas for multiple regression is that relationship (A-4) is really standing for n equations, one for each data point. After appropriately defining matrices/vectors of responses, settings of system variables, and coefficients, the system of n equations can be written compactly in matrix form. That is, define an $n \times 1$ vector of responses

$$y = \begin{bmatrix} y_1 \\ y_2 \\ \vdots \\ y_n \end{bmatrix} \tag{A-5}$$

and an $n \times k$ matrix of corresponding settings of system variables

$$X = \begin{bmatrix} x_{11} & x_{21} & \cdots & x_{k1} \\ x_{12} & x_{22} & \cdots & x_{k2} \\ \vdots & \vdots & & \vdots \\ x_{1n} & x_{2n} & \cdots & x_{kn} \end{bmatrix} \tag{A-6}$$

and a $k \times 1$ matrix of coefficients

$$\beta = \begin{bmatrix} \beta_1 \\ \beta_2 \\ \vdots \\ \beta_k \end{bmatrix} \tag{A-7}$$

Then, in matrix notation using formulas (A-5) through (A-7), the relationship (A-4) can be written as

$$y \approx X\beta \tag{A-8}$$

EXAMPLE A-1

(Example 9-2 revisited) **The Nitrogen Plant Analysis in Matrix Notation.** Return again to the nitrogen plant example and the fitting of the relationship

$$y_i \approx \beta_0 + \beta_1 x_{1i} + \beta_2 x_{2i} + \beta_3 x_{1i}^2 \tag{A-9}$$

to the $n = 17$ data points of Table 4-9. Notice that for present purposes, relationship (A-9) can be rewritten in form (A-4) as

$$y_i \approx \beta_1 x_{1i} + \beta_2 x_{2i} + \beta_3 x_{3i} + \beta_4 x_{4i} \tag{A-10}$$

where $k = 4$, $x_3 = x_1^2$, and $x_4 = 1$. Then from equation (A-5), using the order of observations originally employed in Table 4-9, one creates the 17×1 vector

$$y = \begin{bmatrix} 37 \\ 18 \\ \vdots \\ 9 \\ 15 \end{bmatrix}$$

From formula (A-6), again using the numbering of observations laid out in Table 4-9, one has the 17×4 matrix

$$X = \begin{bmatrix} 80 & 27 & 6400 & 1 \\ 62 & 22 & 3844 & 1 \\ 62 & 23 & 3844 & 1 \\ \vdots & \vdots & \vdots & \vdots \\ 56 & 20 & 3136 & 1 \end{bmatrix}$$

Finally, one defines the 4×1 vector

$$\boldsymbol{\beta} = \begin{bmatrix} \beta_1 \\ \beta_2 \\ \beta_3 \\ \beta_4 \end{bmatrix}$$

and with this notation notes that the $n = 17$ different equations (A-10) may be written in the form (A-8),

$$y \approx X\boldsymbol{\beta}$$

EXAMPLE A-2

(Example 10-24 revisited) **The Pipe Compound Mixture Analysis in Matrix Notation.** To illustrate a no-intercept application of notations (A-5) through (A-7), consider again the ABS pipe compound mixture example from Section 10-5, particularly the influence of basic proportions x_1, x_2, and x_3 on the impact strength response variable, y_1. The data for this example were given in Table 10-45.

Consider first the fitting of a linear relationship

$$y_i \approx \beta_1 x_{1i} + \beta_2 x_{2i} + \beta_3 x_{3i} \tag{A-11}$$

to the strength data in Table 10-45. (This is not the relationship actually used in Section 10-5.) Here $k = k = 3$, and employing the order of data points given in Table 10-45, from formula (A-5) one first defines the 10×1 vector of strengths:

$$y = \begin{bmatrix} 6.1 \\ 2.1 \\ \vdots \\ 6.1 \\ 7.4 \end{bmatrix}$$

The corresponding 10×3 matrix of values of the system variables is

$$X = \begin{bmatrix} .45 & .55 & 0 \\ .45 & .425 & .125 \\ .45 & .30 & .25 \\ \vdots & \vdots & \vdots \\ .70 & .30 & 0 \end{bmatrix}$$

and the 3×1 matrix of coefficients is

$$\boldsymbol{\beta} = \begin{bmatrix} \beta_1 \\ \beta_2 \\ \beta_3 \end{bmatrix}$$

With this notation, the $n = 10$ different equations (A-11) can be written in form (A-8).

If the fitting of the (mixture study) full quadratic relationship

$$y_i \approx \beta_1 x_{1i} + \beta_2 x_{2i} + \beta_3 x_{3i} + \beta_4 x_{1i} x_{2i} + \beta_5 x_{1i} x_{3i} + \beta_6 x_{2i} x_{3i} \tag{A-12}$$

is of interest (as it was in Section 10-5), then $k = 6$ is called for. y is as before, but X becomes a 10×6 matrix

$$X = \begin{bmatrix} .45 & .55 & 0 & (.45)(.55) & 0 & 0 \\ .45 & .425 & .125 & (.45)(.425) & (.45)(.125) & (.425)(.125) \\ \vdots & \vdots & \vdots & \vdots & \vdots & \vdots \\ .70 & .30 & 0 & (.70)(.30) & 0 & 0 \end{bmatrix}$$

and $\boldsymbol{\beta}$ becomes a 6×1 vector

$$\boldsymbol{\beta} = \begin{bmatrix} \beta_1 \\ \beta_2 \\ \vdots \\ \beta_6 \end{bmatrix}$$

With these abbreviations, the mixture study quadratic relationship (A-12) also fits into form (A-8).

Once one has equations (A-5) through (A-8) and employs some matrix algebra and vector calculus, the process of least squares fitting of the coefficients $\beta_1, \beta_2, \ldots, \beta_k$ is straightforward. First, one defines the sum of squares

$$S(\boldsymbol{\beta}) = \sum_{i=1}^{n} \left(y_i - (\beta_1 x_{1i} + \beta_2 x_{2i} + \cdots + \beta_k x_{ki}) \right)^2$$

$$= (\boldsymbol{y} - \boldsymbol{X\beta})'(\boldsymbol{y} - \boldsymbol{X\beta})$$

which is the object of the minimization efforts. Taking k partial derivatives of $S(\boldsymbol{\beta})$ with respect to the entries of $\boldsymbol{\beta}$, setting that vector equal to $\boldsymbol{0}$ (the $k \times 1$ vector of 0's), and doing a small bit of matrix algebra yields the set of normal equations written in matrix notation,

$$\boldsymbol{X'X\beta} = \boldsymbol{X'y} \tag{A-13}$$

Then, provided that $\boldsymbol{X'X}$ is nonsingular (which is equivalent to $\boldsymbol{X}$ having full rank k, which in practical terms means that one has enough different combinations of $x_1, x_2, \ldots, x_k$ to allow the unambiguous fitting of the equation of interest), equation (A-13) leads to the matrix formula for the vector of fitted coefficients $b_1, b_2, \ldots, b_k$,

$$\boldsymbol{b} = \begin{bmatrix} b_1 \\ b_2 \\ \vdots \\ b_k \end{bmatrix} = (\boldsymbol{X'X})^{-1}\boldsymbol{X'y} \tag{A-14}$$

Fitted coefficients led, in the main exposition of regression analysis, directly to fitted responses, residuals, and sums of squares that are useful in the multiple regression context. Using formula (A-14), it is straightforward to produce matrix formulas for these. If one lets

$$\hat{\boldsymbol{y}} = \begin{bmatrix} \hat{y}_1 \\ \hat{y}_2 \\ \vdots \\ \hat{y}_n \end{bmatrix}$$

be an $n \times 1$ vector of (least squares) fitted values corresponding to equation (A-8), then from formulas (A-8) and (A-14),

$$\hat{\boldsymbol{y}} = \boldsymbol{Xb} = \boldsymbol{X}(\boldsymbol{X'X})^{-1}\boldsymbol{X'y} \tag{A-15}$$

An $n \times 1$ vector of residuals

$$\boldsymbol{e} = \begin{bmatrix} e_1 \\ e_2 \\ \vdots \\ e_n \end{bmatrix}$$

is (from formulas (A-5) and (A-15)),

$$e = y - \hat{y} = (I - X(X'X)^{-1}X')y \tag{A-16}$$

One may combine the fact that the error sum of squares in a regression context is the sum of squared residuals — that is,

$$SSE = e'e \tag{A-17}$$

with fact (A-16) to arrive at a value for SSE via matrix calculations. Then in those cases where it makes sense to do so (essentially where X either explicitly or implicitly (due, for example, to a mixture constraint) contains a column of 1's), one may subtract SSE so calculated from $SSTot$ to arrive at SSR, and then produce a value for R^2 according to Definitions 4-3 and 9-2.

EXAMPLE A-3

Regression Analysis of the First Phase of Dow Experimentation in Matrix Notation. The yield data in Table A-1 are repeated from Table 11-5 and will be used in this appendix to illustrate the use of matrix calculations in multiple regression.

TABLE A-1

Yield Data Collected in the First Phase of Dow Experimentation

x_1	x_2	x_3	x_4	y
1.00	1.0	1.00	2.0	35.1
2.50	1.0	1.00	5.0	45.9
1.00	1.8	1.00	5.0	4.0
2.50	1.8	1.00	2.0	46.0
1.00	1.0	2.50	5.0	41.4
2.50	1.0	2.50	2.0	51.2
1.00	1.8	2.50	2.0	10.0
2.50	1.8	2.50	5.0	52.8
1.75	1.4	1.75	3.5	54.7
1.75	1.4	1.75	3.5	52.8
1.75	1.4	1.75	3.5	53.3
1.75	1.4	1.75	3.5	52.0

Printout A-1 shows the use of Minitab in the least squares fitting of the equation

$$y_i \approx \beta_0 + \beta_1 x_{1i} + \beta_2 x_{2i} + \beta_3 x_{3i} + \beta_4 x_{4i}$$

to the data of Table A-1 in two different ways. In the first place, the usual "REGRESS" command was used to do this fitting. Then Minitab's facility to do matrix calculations was used in conjunction with formulas (A-14) through (A-17).

This is an example where $n = 12$ and $\mathbf{k} = k + 1 = 4 + 1 = 5$. The reader is encouraged to verify that the estimated coefficients, fitted responses, residuals, and sums of squares in the regression output agree with those produced through the less automatic matrix calculations.

The Matrix Formulation of the Multiple Regression Model and Inference for Linear Combinations of Coefficients

Section 9-2 discussed a wide range of formal inference methods that can be based on the multiple regression model summarized in the model equation

$$y_i = \beta_0 + \beta_1 x_{1i} + \beta_2 x_{2i} + \cdots + \beta_k x_{ki} + \epsilon_i \tag{A-18}$$

The study of mixture methods in Section 10-5 made use of the no-intercept version of the model, which can be summarized as

$$y_i = \beta_1 x_{1i} + \beta_2 x_{2i} + \cdots + \beta_k x_{ki} + \epsilon_i \tag{A-19}$$

PRINTOUT A-1

```
MTB > print c1-c6

ROW     C1      C2      C3      C4      C5     C6

  1    1.00    1.0    1.00    2.0    35.1    1
  2    2.50    1.0    1.00    5.0    45.9    1
  3    1.00    1.8    1.00    5.0     4.0    1
  4    2.50    1.8    1.00    2.0    46.0    1
  5    1.00    1.0    2.50    5.0    41.4    1
  6    2.50    1.0    2.50    2.0    51.2    1
  7    1.00    1.8    2.50    2.0    10.0    1
  8    2.50    1.8    2.50    5.0    52.8    1
  9    1.75    1.4    1.75    3.5    54.7    1
 10    1.75    1.4    1.75    3.5    52.8    1
 11    1.75    1.4    1.75    3.5    53.3    1
 12    1.75    1.4    1.75    3.5    52.0    1

MTB > regress c5 4 c1-c4

The regression equation is
C5 = 29.8 + 17.6 C1 - 19.0 C2 + 4.07 C3 + 0.15 C4

Predictor       Coef        Stdev     t-ratio          p
Constant       29.82        26.32        1.13      0.295
C1            17.567         6.490        2.71      0.030
C2           -19.00         12.17       -1.56      0.162
C3             4.067         6.490        0.63      0.551
C4             0.150         3.245        0.05      0.964

s = 13.77      R-sq = 59.2%      R-sq(adj) = 35.9%

Analysis of Variance

SOURCE        DF          SS          MS         F         p
Regression     4      1925.6       481.4      2.54     0.133
Error          7      1326.6       189.5
Total         11      3252.2

SOURCE        DF      SEQ SS
C1             1      1388.6
C2             1       462.1
C3             1        74.4
C4             1         0.4

Obs.       C1          C5       Fit   Stdev.Fit   Residual   St.Resid
  1      1.00       35.10     32.75       10.51       2.35       0.26
  2      2.50       45.90     59.55       10.51     -13.65      -1.54
  3      1.00        4.00     18.00       10.51     -14.00      -1.58
  4      2.50       46.00     43.90       10.51       2.10       0.24
  5      1.00       41.40     39.30       10.51       2.10       0.24
  6      2.50       51.20     65.20       10.51     -14.00      -1.58
  7      1.00       10.00     23.65       10.51     -13.65      -1.54
  8      2.50       52.80     50.45       10.51       2.35       0.26
  9      1.75       54.70     41.60        3.97      13.10       0.99
 10      1.75       52.80     41.60        3.97      11.20       0.85
 11      1.75       53.30     41.60        3.97      11.70       0.89
 12      1.75       52.00     41.60        3.97      10.40       0.79
```

(continued)

PRINTOUT A-1
(continued)

```
MTB > copy c1-c4 c6 m1
MTB > print m1
 MATRIX M1
```

$$\begin{bmatrix}
1.00 & 1.00 & 1.00 & 2.00 & 1.00 \\
2.50 & 1.00 & 1.00 & 5.00 & 1.00 \\
1.00 & 1.80 & 1.00 & 5.00 & 1.00 \\
2.50 & 1.80 & 1.00 & 2.00 & 1.00 \\
1.00 & 1.00 & 2.50 & 5.00 & 1.00 \\
2.50 & 1.00 & 2.50 & 2.00 & 1.00 \\
1.00 & 1.80 & 2.50 & 2.00 & 1.00 \\
2.50 & 1.80 & 2.50 & 5.00 & 1.00 \\
1.75 & 1.40 & 1.75 & 3.50 & 1.00 \\
1.75 & 1.40 & 1.75 & 3.50 & 1.00 \\
1.75 & 1.40 & 1.75 & 3.50 & 1.00 \\
1.75 & 1.40 & 1.75 & 3.50 & 1.00
\end{bmatrix} = X$$

```
MTB > transpose m1 m2
MTB > print m2
 MATRIX M2
```

$$\begin{bmatrix}
1.00 & 2.50 & 1.00 & 2.50 & 1.00 & 2.50 & 1.00 & 2.50 & 1.75 & 1.75 & 1.75 & 1.75 \\
1.00 & 1.00 & 1.80 & 1.80 & 1.00 & 1.00 & 1.80 & 1.80 & 1.40 & 1.40 & 1.40 & 1.40 \\
1.00 & 1.00 & 1.00 & 1.00 & 2.50 & 2.50 & 2.50 & 2.50 & 1.75 & 1.75 & 1.75 & 1.75 \\
2.00 & 5.00 & 5.00 & 2.00 & 5.00 & 2.00 & 2.00 & 5.00 & 3.50 & 3.50 & 3.50 & 3.50 \\
1.00 & 1.00 & 1.00 & 1.00 & 1.00 & 1.00 & 1.00 & 1.00 & 1.00 & 1.00 & 1.00 & 1.00
\end{bmatrix} = X'$$

```
MTB > mult m2 m1 m3
MTB > print m3
 MATRIX M3
```

$$\begin{bmatrix}
41.25 & 29.40 & 36.75 & 73.50 & 21.00 \\
29.40 & 24.80 & 29.40 & 58.80 & 16.80 \\
36.75 & 29.40 & 41.25 & 73.50 & 21.00 \\
73.50 & 58.80 & 73.50 & 165.00 & 42.00 \\
21.00 & 16.80 & 21.00 & 42.00 & 12.00
\end{bmatrix} = X'X$$

```
MTB > inverse m3 m4
MTB > print m4
 MATRIX M4
```

$$\begin{bmatrix}
0.22222 & -0.00000 & 0.00000 & 0.00000 & -0.38889 \\
-0.00000 & 0.78125 & -0.00000 & -0.00000 & -1.09375 \\
0.00000 & -0.00000 & 0.22222 & 0.00000 & -0.38889 \\
0.00000 & -0.00000 & 0.00000 & 0.05556 & -0.19444 \\
-0.38889 & -1.09375 & -0.38889 & -0.19444 & 3.65625
\end{bmatrix} = (X'X)^{-1}$$

(continued)

In keeping with the notation in this appendix and the desire to provide formulas for both intercept and no-intercept analyses in a single format, consider formal inference based on a model equation of the form

$$y_i = \beta_1 x_{1i} + \beta_2 x_{2i} + \cdots + \beta_k x_{ki} + \epsilon_i \tag{A-20}$$

Here x_k can, for example, be an extra system variable always having the value 1, so equation (A-20) includes both equations (A-18) and (A-19). In equation (A-20), the x's

PRINTOUT A-1
(continued)

```
MTB > copy c5 m5
MTB > print m5
 MATRIX M5
```

$$\begin{bmatrix} 35.1 \\ 45.9 \\ 4.0 \\ 46.0 \\ 41.4 \\ 51.2 \\ 10.0 \\ 52.8 \\ 54.7 \\ 52.8 \\ 53.3 \\ 52.0 \end{bmatrix} = y$$

```
MTB > mult m2 m5 m6
MTB > mult m4 m6 m6
MTB > print m6
 MATRIX M6
```

$$\begin{bmatrix} 17.5667 \\ -19.0000 \\ 4.0667 \\ 0.1500 \\ 29.8166 \end{bmatrix} = b$$

```
MTB > mult m1 m6 m7
MTB > print m7
 MATRIX M7
```

$$\begin{bmatrix} 32.7500 \\ 59.5500 \\ 18.0001 \\ 43.9001 \\ 39.3000 \\ 65.2000 \\ 23.6501 \\ 50.4501 \\ 41.6000 \\ 41.6000 \\ 41.6000 \\ 41.6000 \end{bmatrix} = \hat{y}$$

(continued)

are (as always) known settings of system variables, the β's are unknown parameters, the ϵ's are unobservable normal $(0, \sigma^2)$ random variables (where σ^2 is an unknown parameter), and the y's are observable responses.

Notice that much of the matrix formalism needed in formal inference under model (A-20) has already been prepared in discussing least squares fitting of equation (A-4). For example, using the fact that

$$s_{\text{SF}}^2 = \frac{1}{n-k} SSE \tag{A-21}$$

PRINTOUT A-1
(continued)

```
MTB > sub m7 m5 m8
MTB > print m8
 MATRIX M8
```

$$
\begin{bmatrix}
2.3500 \\
-13.6500 \\
-14.0001 \\
2.0999 \\
2.1000 \\
-14.0000 \\
-13.6501 \\
2.3499 \\
13.1000 \\
11.2000 \\
11.7000 \\
10.4000
\end{bmatrix} = e
$$

```
MTB > transpose m8 m9
MTB > mult m9 m8 m10
ANSWER =       1326.6097  = SSE = e'e
```

and the expressions leading to equation (A-17), one has the usual surface-fitting sample variance. F tests of hypotheses that sets of β's are all 0 depend on sums of squares that follow from $SSTot$ and SSE given in equation (A-17). The entries of b given in equation (A-14) serve as estimates of β's in equation (A-20), and $\hat{y}$ values that are needed to center confidence intervals for mean responses, prediction intervals, and tolerance intervals are provided by equation (A-15). The ingredients of the inference formulas of Section 9-2 that are not already covered by formulas (A-14) through (A-17) have to do with computing variances or estimated variances for the entries of b, $\hat{y}$, and e. Matrix formulas for such quantities will now be given, considering first a device that will cover the cases of b and $\hat{y}$ — and do more besides.

A useful way to think of both individual coefficients β and theoretical mean responses $\mu_{y|x_1, x_2, \ldots, x_k}$ is as linear combinations of the entries of $\boldsymbol{\beta}$ — that is, as of the form

$$
L = c_1 \beta_1 + c_2 \beta_2 + \cdots + c_k \beta_k \tag{A-22}
$$

for constants $c_1, c_2, \ldots, c_k$. For example, the lth coefficient, β_l, is thought of in terms (A-22) by taking $c_l = 1$ and all other c's equal to 0. The average response for values $x_1, x_2, \ldots, x_k$ of the system variables is thought of in terms (A-22) by taking $c_1 = x_1, c_2 = x_2, \ldots, c_k = x_k$.

If one lets

$$
c = \begin{bmatrix} c_1 \\ c_2 \\ \vdots \\ c_k \end{bmatrix}
$$

then equation (A-22) can be written in matrix notation as

$$
L = c' \boldsymbol{\beta}
$$

and it turns out to be straightforward to give inference methods for such linear combinations of β's based on model (A-20). That is, the linear combination of fitted coefficients from formula (A-14) corresponding to $L = c'\boldsymbol{\beta}$,

$$\hat{L} = c'\boldsymbol{b} = c_1 b_1 + c_2 b_2 + \cdots + c_k b_k \tag{A-23}$$

can be shown to be normally distributed with

$$E\hat{L} = L$$

and

$$Var\,\hat{L} = \sigma^2 \cdot c'(X'X)^{-1}c \tag{A-24}$$

and to be independent of s_{SF}^2. It is then plausible (and correct) that under model (A-20), the hypothesis $H_0: L = \#$ can be tested using a test statistic

$$T = \frac{\hat{L} - \#}{s_{SF}\sqrt{c'(X'X)^{-1}c}} \tag{A-25}$$

and a t_{n-k} reference distribution. Further, the values

$$\hat{L} \pm t s_{SF}\sqrt{c'(X'X)^{-1}c} \tag{A-26}$$

can be used as individual confidence limits for L, where t is chosen based on the desired confidence level and the t_{n-k} distribution.

It is worth pausing to carefully consider the relationship of formulas (A-25) and (A-26) to formulas seen in Section 9-2 and to consider other methods derivable from them. Take first the case of formula (A-22) where

$$L = \beta_l$$

(the lth regression coefficient) is a quantity of practical interest. In such a case, the product

$$c'(X'X)^{-1}c$$

appearing in formula (A-24) and making its way into formulas (A-25) and (A-26) turns out to be fairly simple. It is the lth diagonal entry of $(X'X)^{-1}$. That is, inference formulas (9-49) and (9-50) of Section 9-2 for β_l are special cases of (A-25) and (A-26) here, where $\hat{L} = b_l$ and d_l from Section 9-2 is

$$d_l = c'(X'X)^{-1}c = \text{the } l\text{th diagonal entry of } (X'X)^{-1}$$

So, where in Section 9-2 estimated standard deviations for individual fitted coefficients could only be read off a regression printout, it should now be clear how they are derived.

As a second instance of the relevance of this material to things discussed in Section 9-2, consider the case of formula (A-22) where

$$L = \mu_{y|x_1,x_2,\ldots,x_k} = \beta_1 x_1 + \beta_2 x_2 + \cdots + \beta_k x_k + \beta_{k+1} \cdot 1$$

in a regression with intercept application of this material. In such a situation,

$$c = \begin{bmatrix} x_1 \\ x_2 \\ \vdots \\ x_k \\ 1 \end{bmatrix} \tag{A-27}$$

and

$$\hat{L} = c'\boldsymbol{b} = b_1 x_1 + b_2 x_2 + \cdots + b_k x_k + b_{k+1} = \hat{y}$$

which is the fitted or predicted response at the set of conditions $x_1, x_2, \ldots, x_k$. Using the notation of formulas (9-52), (9-55), and (9-56) of Section 9-2, with c as in formula (A-27) and

$$A^2 = c'(X'X)^{-1}c$$

it is the case that

$$s_{\text{SF}}\sqrt{c'(X'X)^{-1}c} = s_{\text{SF}} \cdot A$$

is an estimated standard deviation for $\hat{y}$, for use in making confidence intervals for $\mu_{y|x_1,\ldots,x_k}$ as well as for use in prediction and tolerance interval making at the set of conditions $x_1, x_2, \ldots, x_k$.

These two specializations do not exhaust the usefulness of the idea of inference for the quantity L in display (A-22). Another important application that could not be made in Section 9-2 (or 9-3, 9-4, or 10-5 for that matter), because of the limitations in standard output of common regression programs, is inference for a *difference in mean responses* at two different combinations of system variables. That is, suppose a difference

$$
\begin{aligned}
L &= \mu_{y|x_1, x_2, \ldots, x_k} - \mu_{y|x_1', x_2', \ldots, x_k'} \\
&= (x_1 - x_1')\beta_1 + (x_2 - x_2')\beta_2 + \cdots + (x_k - x_k')\beta_k
\end{aligned}
\tag{A-28}
$$

is of practical interest. (This might represent a change in mean system response, to be experienced as a result of a move in standard operating conditions as described by the x variables.) It is easy to see that in a no-intercept regression context,

$$
c = \begin{bmatrix} x_1 - x_1' \\ x_2 - x_2' \\ \vdots \\ x_k - x_k' \end{bmatrix}
\tag{A-29}
$$

is used to produce formula (A-28) from formula (A-22), while in a context of regression with intercept,

$$
c = \begin{bmatrix} x_1 - x_1' \\ x_2 - x_2' \\ \vdots \\ x_k - x_k' \\ 0 \end{bmatrix}
\tag{A-30}
$$

is used. But in both situations,

$$\hat{L} = (x_1 - x_1')b_1 + (x_2 - x_2')b_2 + \cdots + (x_k - x_k')b_k$$

and appropriate use of formula (A-29) or (A-30) in relationship (A-25) or (A-26) provides an important method of inference for a change in mean response.

As yet one more instance of the potential usefulness of the ability to do inference for linear combinations L specified by formula (A-22), recall that in Section 9-4 it was useful to treat unbalanced factorial data sets through the use of regression analysis with dummy variables. In that context, it was possible to think of a difference in effects (e.g., a difference in levels 2 and 3 B main effects) in terms of a difference in particular regression coefficients. So there are applications of this idea where a version of formula (A-22) of the form

$$L = \beta_l - \beta_{l'}$$

would be of interest. In such cases, a choice of c with one entry of 1, one entry of -1, and $k - 2$ entries of 0 would produce

$$\hat{L} = b_l - b_{l'}$$

and a way to use formula (A-25) or (A-26) to do inference for a difference of factorial effects (based on even unbalanced data) under reduced models.

EXAMPLE A-3
(continued)

Printout A-2 is from a continuation of the Minitab session that produced Printout A-1 for the fitting of the model

$$y_i = \beta_1 x_{1i} + \beta_2 x_{2i} + \beta_3 x_{3i} + \beta_4 x_{4i} + \beta_5 \cdot 1 + \epsilon_i$$

to the data of Table A-1.

A number of applications of inference for linear combinations of the form given in formula (A-22) are enabled by Printout A-2. To begin with, since $(X'X)^{-1}$ has been printed out, it is possible to get estimated standard deviations for the fitted coefficients from the diagonal entries of this matrix. These are collected in Table A-2, which should be compared to the table of estimated regression coefficients on Printout A-1.

TABLE A-2

Estimated Regression Coefficients and Their Estimated Standard Deviations Derived from Printout A-2

Model Term	Estimated Coefficient	Estimated Standard Deviation of Estimate
x_1	$b_1 = 17.567$	$s_{SF}\sqrt{d_1} = 13.77\sqrt{.22222} = 6.49$
x_2	$b_2 = -18.999$	$s_{SF}\sqrt{d_2} = 13.77\sqrt{.78125} = 12.17$
x_3	$b_3 = 4.0667$	$s_{SF}\sqrt{d_3} = 13.77\sqrt{.22222} = 6.49$
x_4	$b_4 = .1500$	$s_{SF}\sqrt{d_4} = 13.77\sqrt{.05556} = 3.25$
$x_5 = 1$	$b_5 = 29.8165$	$s_{SF}\sqrt{d_5} = 13.77\sqrt{3.65625} = 26.33$

Next notice that Printout A-2 shows the result of the matrix calculation

$$[1, 1, 1, 2, 1](X'X)^{-1}\begin{bmatrix} 1 \\ 1 \\ 1 \\ 2 \\ 1 \end{bmatrix} = .5833$$

This in turn means that for the first set of conditions listed in Table A-1 (namely, the set specified by $x_1 = 1.0$, $x_2 = 1.0$, $x_3 = 1.0$, $x_4 = 2.0$, and $x_5 = 1.0$), the fitted response $\hat{y} = 32.75$ has estimated standard deviation

$$s_{SF} \cdot \sqrt{A^2} = 13.77\sqrt{.5833} = 10.51$$

which agrees with the value of this estimated standard deviation produced by the Minitab "REGRESS" command on Printout A-1.

Finally, note that with

$$c = \begin{bmatrix} 1.0 - 2.5 \\ 1.0 - 1.0 \\ 1.0 - 1.0 \\ 2.0 - 5.0 \\ 1.0 - 1.0 \end{bmatrix} = \begin{bmatrix} -1.5 \\ 0 \\ 0 \\ -3.0 \\ 0 \end{bmatrix}$$

the product

$$c'(X'X)^{-1}c = 1.0000$$

PRINTOUT A-2

```
MTB > print m4
 MATRIX M4
```

$$\begin{bmatrix} 0.22222 & -0.00000 & 0.00000 & 0.00000 & -0.38889 \\ -0.00000 & 0.78125 & -0.00000 & -0.00000 & -1.09375 \\ 0.00000 & -0.00000 & 0.22222 & 0.00000 & -0.38889 \\ 0.00000 & -0.00000 & 0.00000 & 0.05556 & -0.19444 \\ -0.38889 & -1.09375 & -0.38889 & -0.19444 & 3.65625 \end{bmatrix} = (X'X)^{-1}$$

```
MTB > read 1 5 m11
DATA> 1 1 1 2 1
      1 ROWS READ
MTB > print m11
 MATRIX M11


  1   1   1   2   1
```

```
MTB > transpose m11 m12
MTB > mult m11 m4 m13
MTB > mult m13 m12 m13          A² = c'(X'X)⁻¹c
ANSWER =          0.5833
MTB > read 1 5 m14
DATA> 2.5 1 1 5 1
      1 ROWS READ
MTB > print m14
 MATRIX M14


  2.5   1.0   1.0   5.0   1.0
```

$A^2 = c'(X'X)^{-1}c$

```
MTB > transpose m14 m15
MTB > sub m14 m11 m11
MTB > print m11
 MATRIX M11


 -1.5   0.0   0.0  -3.0   0.0
```

```
MTB > sub m15 m12 m12
MTB > print m12
 MATRIX M12


 -1.5
  0.0
  0.0
 -3.0
  0.0
```

```
MTB > mult m11 m4 m13
MTB > mult m13 m12 m13          c'(X'X)⁻¹c
ANSWER =          1.0000
```

$c'(X'X)^{-1}c$

is also given on the printout. This fact allows one (for example) to give a confidence interval comparing the mean response when $x_1 = 1.0, x_2 = 1.0, x_3 = 1.0, x_4 = 2.0$, and $x_5 = 1.0$ to that when $x_1 = 2.5, x_2 = 1.0, x_3 = 1.0, x_4 = 5.0$, and $x_5 = 1.0$. (These are in fact the first two sets of conditions listed in Table A-1.) Using the fact that for the first set of circumstances, $\hat{y} = 32.75$, and for the second, $\hat{y} = 59.55$, confidence limits for the difference in mean responses become

$$(32.75 - 59.55) \pm t(13.77)\sqrt{1.0000}$$

where (as always) $\nu = n - k - 1 = n - \mathbf{k} = 12 - 5 = 7$ degrees of freedom are associated with $s_{\text{SF}} = 13.77$ and should therefore be used in the choice of an appropriate value for t.

EXAMPLE A-4

(Example 9-9 revisited) **Matrices and the Analysis of Unbalanced 3 × 2 Wood Joint Strength Data.** Consider again the unbalanced factorial wood joint strength study of Kotlers, MacFarland, and Tomlinson that has been used as an illustration several times previously. Section 9-4 discussed how it was possible to fit a no-interaction model to the 3×2 unbalanced factorial obtained from the original data set by restricting attention to pine and oak woods, via the device of regression with dummy variables.

Consider here the problem of comparing Joint Type (factor A) main effects for butt and beveled joints (levels 1 and 2 of factor A). Using dummy variables x_1^A, x_2^A, and x_1^B as in Section 9-4, recall that the no-interaction model for the data of Table 9-22 was written in regression notation as

$$y = \beta_0 + \beta_1 x_1^A + \beta_2 x_2^A + \beta_3 x_1^B + \epsilon$$

where the difference $\beta_1 - \beta_2$ is the quantity of interest. This is a $\mathbf{k} = 4$ regression with intercept version of the problem discussed in this section, and the difference in main effects can be represented in form (A-22) if one uses

$$\mathbf{c} = \begin{bmatrix} 1 \\ -1 \\ 0 \\ 0 \end{bmatrix}$$

Printout A-3 shows the calculation of

$$\mathbf{c}'(\mathbf{X}'\mathbf{X})^{-1}\mathbf{c} = .5937$$

Combined with the results (from Printout 9-7) that estimated regression coefficients are $b_1 = -264.48$ and $b_2 = 292.86$, this leads via formula (A-26) to confidence limits for the difference in butt and beveled main effects of the form

$$(-264.48 - 292.86) \pm t(154.7)\sqrt{.5937}$$

where as in Section 9-4, $\nu = n - k - 1 = n - \mathbf{k} = 11 - 4 = 7$ degrees of freedom are appropriate for use in choosing t.

Even with the help of a computer program for matrix calculation, the computations surrounding the use of formula (A-25) or (A-26) can seem somewhat tedious, especially if they amount to an add-on, typically done only after one has already used a multiple regression program. It is thus encouraging that many commercially available regression programs have an option for outputting the important matrix $(\mathbf{X}'\mathbf{X})^{-1}$ (or the multiple

PRINTOUT A-3

```
MTB > print c1-c5

 ROW    C1    C2    C3     C4    C5

   1     1     0     1    829    1
   2     1     0     1    596    1
   3     1     0    -1   1169    1
   4     0     1     1   1348    1
   5     0     1     1   1207    1
   6     0     1    -1   1518    1
   7     0     1    -1   1927    1
   8    -1    -1     1   1000    1
   9    -1    -1     1    859    1
  10    -1    -1    -1   1295    1
  11    -1    -1    -1   1561    1

MTB > copy c1 c2 c3 c5 m1
MTB > transpose m1 m2
MTB > mult m2 m1 m3
MTB > inverse m3 m4
MTB > print m4
MATRIX M4
```

$$\begin{bmatrix} 0.208333 & -0.104167 & -0.020833 & 0.020833 \\ -0.104167 & 0.177083 & 0.010417 & -0.010417 \\ -0.020833 & 0.010417 & 0.093750 & -0.010417 \\ 0.020833 & -0.010417 & -0.010417 & 0.093750 \end{bmatrix} = (X'X)^{-1}$$

```
MTB > read 1 4 m5
DATA> 1 -1 0 0
       1 ROWS READ
MTB > print m5
 MATRIX M5

   1    -1     0     0

MTB > transpose m5 m6
MTB > mult m5 m4 m7
MTB > mult m7 m6 m8                    c'(X'X)^{-1}c
ANSWER =              0.5937
```

$c'(X'X)^{-1}c$

of it, $s_{\text{SF}}^2(X'X)^{-1}$). However, there is one important difference between the discussion given here and the way most packages operate in this matter. In cases of regression with intercept, the variable $x_k = x_{k+1} = 1$ is usually placed as the *first* rather than the *last* column of X. Therefore, in using an $(X'X)^{-1}$ matrix produced by most standard statistical packages, one must permute the entries of c (relative to the treatment here) and use $c^{*\prime}(X'X)^{-1}c^*$ where

$$c^* = \begin{bmatrix} c_k \\ c_1 \\ c_2 \\ \vdots \\ c_k \end{bmatrix}$$

in place of $c'(X'X)^{-1}c$ in formulas (A-25) and (A-26).

Standardized Residuals in Multiple Regression

The only inference-related matter discussed in Section 9-2 and not yet given a matrix formulation is the standardization of residuals. It has been noted several times that the residuals in regression contexts typically do not have equal variances. Related to formulas (A-15) and (A-16) is a matrix-based formula that produces those variances (and then allows computation of standardized residuals).

The matrix

$$H = X(X'X)^{-1}X'$$

appearing in formulas (A-15) and (A-16) is sometimes called the *hat matrix*, since multiplication of y by H produces the "y hats." If one lets h_i stand for the ith diagonal entry of H, it turns out that $0 < h_i < 1$, and under model (A-20),

$$Var\,\hat{y}_i = h_i\sigma^2 \tag{A-31}$$

and

$$Var\,e_i = (1 - h_i)\sigma^2 \tag{A-32}$$

(Differently put, the entries on the diagonal of the $n \times n$ matrix $\sigma^2 H$ are variances of the fitted values $\hat{y}_i$, while the entries on the diagonal of the $n \times n$ matrix $\sigma^2(I - H)$ are the variances of the residuals e_i.) In the notation of Section 9-2, one has from formula (A-31) that for values of $x_1, x_2, \ldots, x_k$ in the data set, A^2 can be found as a diagonal entry of H.

The relationship (A-32) correctly suggests that estimated variances for the residuals e_i can be calculated as

$$(1 - h_i)s_{\text{SF}}^2 \tag{A-33}$$

So, recalling Definition 7-2, one sees from expression (A-33) that under model (A-20), the n standardized residuals e_i^* are of the form

$$e_i^* = \frac{e_i}{s_{\text{SF}}\sqrt{1 - h_i}}$$

EXAMPLE A-3
(continued)

Printout A-4 is from a continuation of the Minitab session that produced Printouts A-1 and A-2 for the fitting of the model

$$y_i = \beta_1 x_{1i} + \beta_2 x_{2i} + \beta_3 x_{3i} + \beta_4 x_{4i} + \beta_5 \cdot 1 + \epsilon_i$$

to the data of Table A-1.

PRINTOUT A-4

```
MTB > print m1
 MATRIX M1
```

$$\begin{bmatrix} 1.00 & 1.00 & 1.00 & 2.00 & 1.00 \\ 2.50 & 1.00 & 1.00 & 5.00 & 1.00 \\ 1.00 & 1.80 & 1.00 & 5.00 & 1.00 \\ 2.50 & 1.80 & 1.00 & 2.00 & 1.00 \\ 1.00 & 1.00 & 2.50 & 5.00 & 1.00 \\ 2.50 & 1.00 & 2.50 & 2.00 & 1.00 \\ 1.00 & 1.80 & 2.50 & 2.00 & 1.00 \\ 2.50 & 1.80 & 2.50 & 5.00 & 1.00 \\ 1.75 & 1.40 & 1.75 & 3.50 & 1.00 \\ 1.75 & 1.40 & 1.75 & 3.50 & 1.00 \\ 1.75 & 1.40 & 1.75 & 3.50 & 1.00 \\ 1.75 & 1.40 & 1.75 & 3.50 & 1.00 \end{bmatrix} = X$$

(continued)

PRINTOUT A-4

(continued)

```
MTB > print m4
 MATRIX M4
```

$$
\begin{bmatrix}
0.22222 & -0.00000 & -0.00000 & -0.00000 & -0.38889 \\
-0.00000 & 0.78125 & -0.00000 & -0.00000 & -1.09375 \\
-0.00000 & -0.00000 & 0.22222 & -0.00000 & -0.38889 \\
-0.00000 & -0.00000 & -0.00000 & 0.05556 & -0.19444 \\
-0.38889 & -1.09375 & -0.38889 & -0.19444 & 3.65624
\end{bmatrix} = (X'X)^{-1}
$$

```
MTB > print m2
 MATRIX M2
```

$$
\begin{bmatrix}
1.00 & 2.50 & 1.00 & 2.50 & 1.00 & 2.50 & 1.00 & 2.50 & 1.75 & 1.75 & 1.75 & 1.75 \\
1.00 & 1.00 & 1.80 & 1.80 & 1.00 & 1.00 & 1.80 & 1.80 & 1.40 & 1.40 & 1.40 & 1.40 \\
1.00 & 1.00 & 1.00 & 1.00 & 2.50 & 2.50 & 2.50 & 2.50 & 1.75 & 1.75 & 1.75 & 1.75 \\
2.00 & 5.00 & 5.00 & 2.00 & 5.00 & 2.00 & 2.00 & 5.00 & 3.50 & 3.50 & 3.50 & 3.50 \\
1.00 & 1.00 & 1.00 & 1.00 & 1.00 & 1.00 & 1.00 & 1.00 & 1.00 & 1.00 & 1.00 & 1.00
\end{bmatrix} = X'
$$

```
MTB > mult m1 m4 m5
MTB > mult m5 m2 m6
MTB > print m6
 MATRIX M6
```

h_1

First 7 Columns of H:

$$
\begin{bmatrix}
\boxed{0.583333} & 0.083333 & 0.083334 & 0.083334 & 0.083333 & 0.083333 & 0.083334 \\
0.083333 & 0.583333 & 0.083333 & 0.083333 & 0.083333 & 0.083333 & -0.416667 \\
0.083334 & 0.083333 & 0.583333 & 0.083333 & 0.083333 & -0.416667 & 0.083333 \\
0.083334 & 0.083333 & 0.083333 & 0.583333 & -0.416667 & 0.083333 & 0.083333 \\
0.083333 & 0.083333 & 0.083333 & -0.416667 & 0.583333 & 0.083333 & 0.083333 \\
0.083333 & 0.083333 & -0.416667 & 0.083333 & 0.083333 & 0.583333 & 0.083333 \\
0.083334 & -0.416667 & 0.083333 & 0.083333 & 0.083333 & 0.083333 & 0.583333 \\
-0.416666 & 0.083334 & 0.083333 & 0.083333 & 0.083334 & 0.083334 & 0.083333 \\
0.083333 & 0.083333 & 0.083333 & 0.083333 & 0.083333 & 0.083333 & 0.083333 \\
0.083333 & 0.083333 & 0.083333 & 0.083333 & 0.083333 & 0.083333 & 0.083333 \\
0.083333 & 0.083333 & 0.083333 & 0.083333 & 0.083333 & 0.083333 & 0.083333 \\
0.083333 & 0.083333 & 0.083333 & 0.083333 & 0.083333 & 0.083333 & 0.083333
\end{bmatrix}
$$

Last 5 Columns of H:

$$
\begin{bmatrix}
-0.416666 & 0.083333 & 0.083333 & 0.083333 & 0.083333 \\
0.083334 & 0.083333 & 0.083333 & 0.083333 & 0.083333 \\
0.083333 & 0.083333 & 0.083333 & 0.083333 & 0.083333 \\
0.083333 & 0.083333 & 0.083333 & 0.083333 & 0.083333 \\
0.083334 & 0.083333 & 0.083333 & 0.083333 & 0.083333 \\
0.083333 & 0.083333 & 0.083333 & 0.083333 & 0.083333 \\
0.083333 & 0.083333 & 0.083333 & 0.083333 & 0.083333 \\
0.583333 & 0.083333 & 0.083333 & 0.083333 & 0.083333 \\
0.083333 & 0.083333 & 0.083333 & 0.083333 & 0.083333 \\
0.083333 & 0.083333 & 0.083333 & 0.083333 & 0.083333 \\
0.083333 & 0.083333 & 0.083333 & 0.083333 & 0.083333 \\
0.083333 & 0.083333 & 0.083333 & 0.083333 & 0.083333
\end{bmatrix}
$$

The reader is invited to verify that for the first data point (having $x_1 = 1.0$, $x_2 = 1.0$, $x_3 = 1.0$, $x_4 = 2.0$, and $x_5 = 1.0$),

$$h_1 = .583333$$

so the first standardized residual is

$$
\begin{aligned}
e_1^* &= \frac{e_1}{s_{SF}\sqrt{1 - h_1}} \\
&= \frac{35.1 - 32.75}{13.77\sqrt{1.0 - .58333}} \\
&= .26
\end{aligned}
$$

exactly as shown (for example) on Printout A-1.

B

More on Probability and Model Fitting

In keeping with this book's announced intention to emphasize data collection and analysis rather than mathematical theory, the introduction to probability theory in Chapter 5 was relatively brief. But there are, of course, important engineering applications of probability that require more background in the subject than was provided in Chapter 5. So this appendix gives a few more details and discusses some additional uses of the theory that are reasonably elementary, particularly in the contexts of reliability analysis and life data analysis.

The appendix begins by discussing the formal/axiomatic basis for the mathematics of probability and several of the most useful simple consequences of the basic axioms. It then applies those simple theorems of probability to the prediction of reliability for series, parallel, and combination series-parallel systems. A brief section treats principles of counting (permutations and combinations) that are sometimes useful in engineering applications of probability. There follows a section on special probability concepts used with life length (or time to failure) variables. The appendix concludes with a discussion of the use of maximum likelihood methods in drawing formal statistical inferences based on possibly nonstandard statistical models involving (for example) censored life-length data.

B-1 More Elementary Probability

Like any other mathematical theory or system, probability theory is built on a few basic definitions and some rules of the game called *axioms*. Logic is applied to determine what consequences (or theorems) follow from the definitions and axioms. These, in turn, can prove to be helpful guides as an engineer seeks to understand and predict the behavior of physical systems that involve chance.

For the sake of logical completeness, this section gives the formal axiomatic basis for probability theory and states and then illustrates the use of some simple theorems that follow from this base. Conditional probability and the independence of events are then defined, and a simple theorem related to these concepts is stated and its use illustrated.

Basic Definitions and Axioms

As was illustrated somewhat informally in Chapter 5, the practical usefulness of probability theory is in assigning sensible likelihoods of occurrence to possible happenings in chance situations. The basic, irreducible, potential results in such a chance situation are called *outcomes* belonging to a *sample space*.

DEFINITION B-1

A single potential result of a chance situation is called an **outcome**. All outcomes of a chance situation taken together make up a **sample space** for the situation. A script capital *S* is often used to stand for a sample space.

Mathematically, outcomes are points in a universal set that is the sample space, and notions of simple set theory become relevant. For one thing, subsets of *S* containing more than one outcome can often be of interest.

DEFINITION B-2

A collection of outcomes (a subset of *S*) is called an **event**. Capital letters near the beginning of the alphabet are sometimes used as symbols for events, as are English phrases describing the events.

Once one has defined events, the standard set-theoretic operations of complementation, union, and intersection can be applied to them. However, rather than using the typical ""^{c}," "$\cup$," and "$\cap$" mathematical notation for these operations, it is common in probability theory to substitute the use of the words "not," "or," and "and," respectively.

DEFINITION B-3

For event A and event B, subsets of some sample space S,

    1. *not A* is an event consisting of all outcomes not belonging to A;

    2. *A or B* is an event consisting of all outcomes belonging to one, the other, or both of the two events; and

    3. *A and B* is an event consisting of all outcomes belonging simultaneously to the two events.

EXAMPLE B-1

A Redundant Inspection System for Detecting Metal Fatigue Cracks. Consider a redundant inspection system for the detection of fatigue cracks in metal specimens. Suppose the system involves the making of a fluorescent penetrant inspection (FPI) and also a (magnetic) eddy current inspection (ECI). When a metal specimen is to be tested using this two-detector system, a potential sample space consists of four outcomes corresponding to the possible combinations of what can happen at each detector. That is, a possible sample space is specified in a kind of set notation as

$$S = \{(\text{FPI signal and ECI signal}), (\text{no FPI signal and ECI signal}),$$
$$(\text{FPI signal and no ECI signal}), (\text{no FPI signal and no ECI signal})\} \tag{B-1}$$

and in tabular and pictorial forms as in Table B-1 and Figure B-1.

TABLE B-1

A List of the Possible Outcomes for Two Inspections

Possible Outcome	FPI Detection Signal?	ECI Detection Signal?
1	yes	yes
2	no	yes
3	yes	no
4	no	no

FIGURE B-1

Graphical Representation of Four Outcomes of Two Inspections

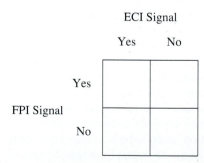

Notice that Figure B-1 can be treated as a kind of *Venn diagram* — the big square standing for S and the four smaller squares making up S standing for events that each consist of one of the four different possible outcomes.

Using this four-outcome sample space to describe experience with a metal specimen, one can define several events of potential interest and illustrate the use of the notation described in Definition B-3. That is, let

$$A = \{(\text{FPI signal and ECI signal}), (\text{FPI signal and no ECI signal})\} \tag{B-2}$$

$$B = \{(\text{FPI signal and ECI signal}), (\text{no FPI signal and ECI signal})\} \tag{B-3}$$

Then in words,

$$A = \text{the FPI detector signals}$$
$$B = \text{the ECI detector signals}$$

Part 1) of Definition B-3 means, for example, that using notations (B-1) and (B-2),

$$not\,A = \{(\text{no FPI signal and ECI signal}), (\text{no FPI signal and no ECI signal})\}$$
$$= \text{the FPI detector doesn't signal}$$

Part 2) of Definition B-3 means, for example, that using notations (B-2) and (B-3),

$$A\,or\,B = \{(\text{FPI signal and ECI signal}), (\text{FPI signal and no ECI signal}),$$
$$(\text{no FPI signal and ECI signal})\}$$
$$= \text{at least one of the two detectors signals}$$

And part 3) of Definition B-3 means that again using (B-2) and (B-3), one has

$$A\,and\,B = \{(\text{FPI signal and ECI signal})\}$$
$$= \text{both of the two detectors signal}$$

not A, *A or B*, and *A and B* are shown in Venn diagram fashion in Figure B-2.

FIGURE B-2

Graphical Representations of *A*, *B*, *not A*, *A or B*, and *A and B*

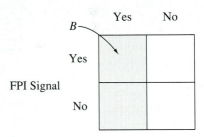
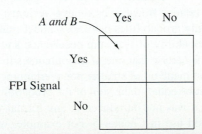
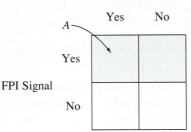
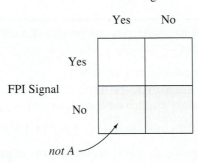
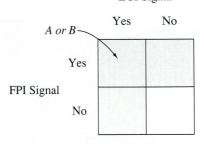

Elementary set theory allows the possibility that a set can be empty — that is, have no elements. Such a concept is also needed in probability theory.

DEFINITION B-4

The **empty event** is an event containing no outcomes. The symbol $\varnothing$ is typically used to stand for the empty event.

In the context of probability theory, $\varnothing$ has the interpretation that none of the possible outcomes of a chance situation occur. The way in which $\varnothing$ is most useful in probability is in describing the relationship between two events that have no outcomes in common, and thus cannot both occur. There is special terminology for this eventuality (that *A and B* = $\varnothing$).

DEFINITION B-5

If event *A* and event *B* have no outcomes in common (i.e., *A and B* = $\varnothing$), then the two events are called **disjoint** or **mutually exclusive**.

EXAMPLE B-1
(continued)

From Figure B-2 it is quite clear that, for example, the event *A* and the event *not A* are disjoint. And the event *A and B* and the event *not(A or B)*, for example, are also mutually exclusive events.

Manipulation of events using complementation, union, intersection, etc. is necessary background, but it is hardly the ultimate goal of probability theory. The goal is assignment of likelihoods to events. In order to guarantee that such assignments are internally coherent, probabilists have devised what seem to be intuitively sensible axioms (or rules of operation) for probability models. Assignment of likelihoods in conformance to those rules guarantees that (at a minimum) the assignment is logically consistent. (Whether it is realistic or useful is a separate question.) The axioms of probability are laid out next.

DEFINITION B-6

A **system of probabilities** is an assignment of numbers (probabilities) $P[A]$ to events *A* in such a way that

1. for each event *A*, $0 \leq P[A] \leq 1$,
2. $P[S] = 1$ and $P[\varnothing] = 0$, and
3. for mutually exclusive events $A_1, A_2, A_3, \ldots$,

$$P[A_1 \text{ or } A_2 \text{ or } A_3 \text{ or} \ldots] = P[A_1] + P[A_2] + P[A_3] + \cdots$$

The relationships 1), 2), and 3) are the **axioms of probability theory**.

Definition B-6 is meant to be perfectly plausible and in agreement with the ways that empirical relative frequencies behave. Axiom 1) says that, as in the case of relative frequencies, only probabilities between 0 and 1 make sense. Axiom 2) says that if one interprets a probability of 1 as sure occurrence and a probability of 0 as no chance of occurrence, it is certain that one of the outcomes in *S* will occur. Axiom 3) says that if an event can be made up of smaller nonoverlapping pieces, the probability assigned to that event must be equal to the sum of the probabilities assigned to the pieces.

Although it was not introduced in any formal way, the third axiom of probability was put to good use in Chapter 5. For example, when concluding that for a Poisson random variable X

$$P[2 \le X \le 5] = P[X = 2] + P[X = 3] + P[X = 4] + P[X = 5]$$
$$= f(2) + f(3) + f(4) + f(5)$$

one is really using the third axiom with

$$A_1 = \{X = 2\}$$
$$A_2 = \{X = 3\}$$
$$A_3 = \{X = 4\}$$
$$A_4 = \{X = 5\}$$

It is only in very simple situations that one would ever try to make use of Definition B-6 by checking that an entire candidate set of probabilities satisfies the axioms of probability. It is more common to assign probabilities (totaling to 1) to individual outcomes and then simply declare that the third axiom of Definition B-6 will be followed in making up any other probabilities. (This strategy guarantees that subsequent probability assignments will be logically consistent.)

EXAMPLE B-2

A System of Probabilities for Describing a Single Inspection of a Metal Part. As an extremely simple illustration, consider the result of a single inspection of a metal part for fatigue cracks using fluoride penetrant technology. With a sample space

$$S = \{\text{crack signaled, crack not signaled}\}$$

there are only four possible events:

$$S$$
$$\{\text{crack signaled}\}$$
$$\{\text{no crack signaled}\}$$
$$\varnothing$$

An assignment of probabilities that can be seen to conform to Definition B-6 is

$$P[S] = 1$$
$$P[\text{crack signaled}] = .3$$
$$P[\text{no crack signaled}] = .7$$
$$P[\varnothing] = 0$$

Since they conform to Definition B-6, these values make up a mathematically valid system of probabilities. Whether or not they constitute a *realistic* or *useful* model is a separate question that can really be answered only on the basis of empirical evidence.

EXAMPLE B-1
(continued)

Returning to the situation of redundant inspection of metal parts using both fluoride penetrant and eddy current technologies, suppose that via extensive testing it is possible to verify that for cracks of depth .005 in., the following four values are sensible:

$$P[\text{FPI signal and ECI signal}] = .48 \tag{B-4}$$

$$P[\text{FPI signal and no ECI signal}] = .02 \tag{B-5}$$

$$P[\text{no FPI signal and ECI signal}] = .32 \tag{B-6}$$

$$P[\text{no FPI signal and no ECI signal}] = .18 \tag{B-7}$$

ECI Signal

		Yes	No
	Yes	.48	.02
FPI Signal	No	.32	.18

This assignment of probabilities to the basic outcomes in S is illustrated in Figure B-3.

Since these four potential probabilities do total to 1, one can adopt them together with provision 3) of Definition B-6 and have a mathematically consistent assignment. Then simple addition gives appropriate probabilities for all other events. For example, with event A and event B as defined earlier,

$$P[A] = P[\text{the FPI detector signals}]$$
$$= P[\text{FPI signal and ECI signal}] + P[\text{FPI signal and no ECI signal}]$$
$$= .48 + .02$$
$$= .50$$

And further,

$$P[A \, or \, B] = P[\text{at least one of the two detectors signals}]$$
$$= P[\text{FPI signal and ECI signal}] + P[\text{FPI signal and no ECI signal}]$$
$$+ P[\text{no FPI signal and ECI signal}]$$
$$= .48 + .02 + .32$$
$$= .82$$

It is clear that to find the two values, one simply adds the numbers that appear in Figure B-3 in the regions that are shaded in Figure B-2 delimiting the events in question.

Simple Theorems of Probability Theory

The preceding discussion is typical of probability analyses, in that the probabilities for all possible events are not explicitly written down. Rather, probabilities for *some* events, together with logic and the basic rules of the game (the probability axioms), are used to deduce appropriate values for probabilities of *other* events that are of particular interest. This enterprise is often facilitated by the existence of a number of simple theorems. These are general statements that are logical consequences of the axioms in Definition B-6 and thus govern the assigning of probabilities for all probability models.

One such simple theorem concerns the relationship between $P[A]$ and $P[not \, A]$.

PROPOSITION B-1

For any event A, $P[not \, A] = 1 - P[A]$.

This fact is again one that was used freely in Chapter 5 without explicit reference. For example, in the context of independent, identical success-failure trials, the fact that the probability of at least one success (i.e., $P[X \geq 1]$ for a binomial random variable X) is 1 minus the probability of 0 successes (i.e., $1 - P[X = 0] = 1 - f(0)$) is really a consequence of Proposition B-1.

EXAMPLE B-1
(continued)

Once one has found, via the addition of probabilities for individual outcomes given in displays (B-4) through (B-7), that the assignment

$$P[A] = P[\text{the FPI detector signals}]$$
$$= .50$$

is appropriate, Proposition B-1 then immediately implies that

$$P[not\,A] = P[\text{the FPI detector doesn't signal}]$$
$$= 1 - P[A]$$
$$= 1 - .50$$
$$= .50$$

is also appropriate. (Of course, if the point here weren't to illustrate the use of Proposition B-1, this value could just as well have been gotten by adding .32 and .18.)

A second simple theorem of probability theory can be thought of as a variation on axiom 3 of Definition B-6 for two events that are not necessarily disjoint. It is sometimes called the *addition rule of probability*.

PROPOSITION B-2

(The Addition Rule of Probability) For any two events, event A and event B,

$$P[A\,or\,B] = P[A] + P[B] - P[A\,and\,B] \tag{B-8}$$

Note that when dealing with mutually exclusive events, the last term in equation (B-8) is $P[\varnothing] = 0$. Therefore, equation (B-8) simplifies to a two-event version of part 3) of Definition B-6. When the event A and the event B are not mutually exclusive, the simple addition $P[A] + P[B]$ (so to speak) counts $P[A\,and\,B]$ twice, and the subtraction in equation (B-8) corrects for this in the computing of $P[A\,or\,B]$.

The practical usefulness of an equation like (B-8) is that when furnished with any three of the four terms appearing in it, the fourth can be gotten by using simple arithmetic.

EXAMPLE B-3

Describing the Dual Inspection of a Single Cracked Part. Suppose that two different inspectors, both using a fluoride penetrant inspection technique, are to inspect a metal part actually possessing a crack .007 in. deep. Suppose further that some relevant probabilities in this context are

$$P[\text{inspector 1 detects the crack}] = .50$$
$$P[\text{inspector 2 detects the crack}] = .45$$
$$P[\text{at least one inspector detects the crack}] = .55$$

Then using equation (B-8), one could reason that

$$P[\text{at least one inspector detects the crack}] = P[\text{inspector 1 detects the crack}]$$
$$+ P[\text{inspector 2 detects the crack}] - P[\text{both inspectors detect the crack}]$$

Thus,

$$.55 = .50 + .45 - P[\text{both inspectors detect the crack}]$$

so

$$P[\text{both inspectors detect the crack}] = .40$$

Of course, the .40 value is only as good as the three others used to produce it. But it is at least logically consistent with the given probabilities, and if they have practical relevance, so does the .40 value.

A third simple theorem of probability concerns cases where the basic outcomes in a sample space are judged to be equally likely.

PROPOSITION B-3

If the outcomes in a finite sample space S all have the same probability, then for any event A,

$$P[A] = \frac{\text{the number of outcomes in } A}{\text{the number of outcomes in } S}$$

Proposition B-3 shows that if one is clever or fortunate enough to be able to conceive of a sample space where an *equally likely outcomes* assignment of probabilities is sensible, the assessment of probabilities can be reduced to a simple counting problem. This fact is particularly useful in enumerative contexts where one is drawing random samples from a finite population.

EXAMPLE B-4

Equally Likely Outcomes in a Random Sampling Scenario. Suppose that a storeroom holds, among other things, four integrated circuit chips of a particular type and that two of these are needed in the fabrication of a prototype of an advanced electronic instrument. Suppose further that one of these chips is defective. Consider assigning a probability that both of two chips selected on the first trip to the storeroom are good chips. One way to find such a value (there are others) is to use Proposition B-3. Naming the three good chips G1, G2, and G3 and the single defective chip D, one can invent a sample space made up of ordered pairs, the first entry naming the first chip selected and the second entry naming the second chip selected. This is given in set notation as follows.

$$S = \{(G1, G2), (G1, G3), (G1, D), (G2, G1), (G2, G3), (G2, D), (G3, G1),$$
$$(G3, G2), (G3, D), (D, G1), (D, G2), (D, G3)\}$$

A pictorial representation of S is given in Figure B-4.

FIGURE B-4

Graphical Representation of 12 Possible Outcomes When Selecting Two of Four IC Chips

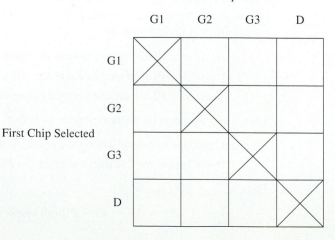

Then, noting that the 12 outcomes in this sample space are reasonably thought of as equally likely and that 6 of them do not have D listed either first or second, Proposition B-3 suggests the assessment

$$P[\text{two good chips}] = \frac{6}{12} = .50$$

Conditional Probability and the Independence of Events

Chapter 5 discussed the notion of independence for random variables; in that discussion, the idea of assigning probabilities for one variable conditional on the value of another was essential. The concept of conditional assignment of probabilities of events is spelled out next.

DEFINITION B-7

For event A and event B, provided event B has nonzero probability, the **conditional probability of A given B** is the ratio

$$\frac{P[A \, and \, B]}{P[B]} \tag{B-9}$$

The notation $P[A \mid B]$ will be used to stand for this quantity.

The ratio (B-9) ought to make reasonable intuitive sense. If, for example, $P[A \, and \, B] = .3$ and $P[B] = .5$, one might reason that "B occurs only 50% of the time, but of those times B occurs, A also occurs $\frac{3}{5} = 60\%$ of the time. So .6 is a sensible assessment of the likelihood of A knowing that indeed B occurs."

EXAMPLE B-4
(continued)

Return to the situation of selecting two integrated circuit chips at random from four residing in a storeroom, one of which is defective. Consider using expression (B-9) and evaluating

$$P[\text{the second chip selected is defective} \mid \text{the first chip selected is good}]$$

Simple counting in the 12-outcome sample space leads to the assignments

$$P[\text{the first chip selected is good}] = \frac{9}{12} = .75$$

$$P[\text{first chip selected is good and second is defective}] = \frac{3}{12} = .25$$

So using Definition B-7,

$$P[\text{the second chip selected is defective} \mid \text{the first selected is good}] = \frac{\dfrac{3}{12}}{\dfrac{9}{12}} = \frac{1}{3}$$

Of the nine equally likely outcomes in S for which the first chip selected is good, there are three for which the second chip selected is defective. If one thinks of the nine outcomes for which the first chip selected is good as a kind of reduced sample space (brought about by the partial restriction that the first chip selected is good), then the $\frac{3}{9}$ figure above is a perfectly plausible value for the likelihood that the second chip is defective.

There are sometimes circumstances that make it obvious how a conditional probability ought to be assigned. For example, in the context of Example B-4, one might argue that it is obvious that

$$P[\text{the second chip selected is defective} \mid \text{the first selected is good}] = \frac{1}{3}$$

because if the first is good, when the second is to be selected, the storeroom will contain three chips, one of which is defective.

When one does have a natural value for $P[A \mid B]$, the relationship between this and the probabilities $P[A \text{ and } B]$ and $P[B]$ can sometimes be exploited to evaluate one or the other of them. This notion is important enough that the relationship

$$P[A \mid B] = \frac{P[A \text{ and } B]}{P[B]}$$

is often rewritten by multiplying both sides by the quantity $P[B]$ and calling the result the *multiplication rule of probability*.

PROPOSITION B-4

(The Multiplication Rule of Probability) Provided $P[B] > 0$, so that $P[A \mid B]$ is defined,

$$P[A \text{ and } B] = P[A \mid B] \cdot P[B] \tag{B-10}$$

EXAMPLE B-5

The Multiplication Rule of Probability and a Probabilistic Risk Assessment. A probabilistic risk assessment of the solid rocket motor field joints used in space shuttles prior to the Challenger disaster was made in "Risk Analysis of the Space Shuttle: Pre-Challenger Prediction of Failure" (*Journal of the American Statistical Association*, 1989) by Dalal, Fowlkes, and Hoadley. They estimated that for each field joint (at 31° and 200 psi),

$$P[\text{primary O-ring erosion}] = .95$$

$$P[\text{primary O-ring blow-by} \mid \text{primary O-ring erosion}] = .29$$

Combining these two values according to rule (B-10), one then sees that the authors' assessment of the failure probability for each primary O-ring was

$$P[\text{primary O-ring erosion and blow-by}] = (.29)(.95) = .28$$

Typically, the numerical values of $P[A \mid B]$ and $P[A]$ are different. The difference can be thought of as reflecting the change in one's assessed likelihood of occurrence of A brought about by knowing that B's occurrence is certain. In cases where there is no difference, the terminology of *independence* is used.

DEFINITION B-8

If event A and event B are such that

$$P[A \mid B] = P[A]$$

they are said to be **independent**. Otherwise, they are called **dependent**.

EXAMPLE B-1
(continued)

Consider again the example of redundant fatigue crack inspection with probabilities given in Figure B-3. Since

$$P[\text{ECI signal}] = .80$$

$$P[\text{ECI signal} \mid \text{FPI signal}] = \frac{.48}{.50} = .96$$

the events {ECI signal} and {FPI signal} are dependent events.

Of course, different probabilities assigned to individual outcomes in this example can lead to the conclusion that the two events are independent. For example, the probabilities in Figure B-5 give

$$P[\text{ECI signal}] = .4 + .4 = .8$$

$$P[\text{ECI signal} \mid \text{FPI signal}] = \frac{.4}{.4 + .1} = .8$$

so with these probabilities, the two events would be independent.

FIGURE B-5

A Second Assignment of Probabilities to Four Possible Outcomes of Two Inspections

		ECI Signal	
		Yes	No
FPI Signal	Yes	.4	.1
	No	.4	.1

Independence is the mathematical formalization of the qualitative notion of *unrelatedness*. One way in which it is used in engineering applications is in conjunction with the multiplication rule. If one has values for $P[A]$ and $P[B]$ and judges the event A and the event B to be unrelated, then independence allows one to replace $P[A \mid B]$ with $P[A]$ in formula (B-10) and evaluate $P[A \text{ and } B]$ as $P[A] \cdot P[B]$. (This fact was behind the scenes in Section 5-1 when sequences of independent identical success-failure trials and the binomial and geometric distributions were discussed.)

EXAMPLE B-5 (continued)

In their probabilistic risk assessment of the pre-Challenger space shuttle solid rocket motor field joints, Dalal, Fowlkes, and Hoadley arrived at the figure

$$P[\text{failure}] = .023$$

for a single field joint in a shuttle launch at 31°F. A shuttle's two solid rocket motors have a total of six such field joints, and it is perhaps plausible to think of their failures as independent events.

If one does adopt a model of independence, it is possible to calculate as follows.

$$\begin{aligned} P[\text{joint 1 and joint 2 are both effective}] &= P[\text{joint 1 is effective}]P[\text{joint 2 is effective}] \\ &= (1 - .023)(1 - .023) \\ &= .96 \end{aligned}$$

And in fact, considering all six joints,

$$\begin{aligned} P[\text{at least one joint fails}] &= 1 - P[\text{all 6 joints are effective}] \\ &= 1 - (1 - .023)^6 \\ &= .13 \end{aligned}$$

B-2 *Applications of Simple Probability to System Reliability Prediction*

Sometimes engineering systems are made up of identifiable components or subsystems that operate reasonably autonomously, and for which fairly accurate reliability information is available. In such cases, it is typically of interest to try to predict overall system reliability from the available component reliabilities. This section considers how the simple probability material from Section B-1 can be used to help do this for series, parallel, and combination series-parallel systems.

Series Systems

DEFINITION B-9

A system consisting of components $C_1, C_2, C_3, \ldots, C_k$ is called a **series system** if its proper functioning requires the functioning of all k components.

Figure B-6 is a representation of a series system made up of $k = 3$ components. The interpretation to be attached to a diagram like Figure B-6 is that the system will function provided there is a path from point 1 to point 2 that crosses no failed component. (It is tempting, but *not* a good idea, to interpret a system diagram as a flow diagram or like an electrical circuit schematic. The flow diagram interpretation is often inappropriate because there need be no sequencing, time progression, communication, or other such relationship between components in a real series system. The circuit schematic notion often fails to be relevant, and even when it might seem to be, the independence assumptions typically used in arriving at a system reliability figure are of questionable practical appropriateness for electrical circuits.)

FIGURE B-6

Three-Component
Series System

If it is sensible to model the functioning of the individual system components as independent, then the overall system reliability is easily deduced from component reliabilities via simple multiplication. For example, for a two-component series system

$$P[\text{the system functions}] = P[C_1 \text{ functions and } C_2 \text{ functions}]$$
$$= P[C_2 \text{ functions} \mid C_1 \text{ functions}] \cdot P[C_1 \text{ functions}]$$
$$= P[C_2 \text{ functions}] \cdot P[C_1 \text{ functions}]$$

where the last step depends on the independence assumption. And in general, if the reliability of component C_i (i.e., $P[C_i \text{ functions}]$) is r_i, then assuming that the k components in a series system behave independently, the (series) system reliability (say, R_S), becomes

$$R_S = r_1 \cdot r_2 \cdot r_3 \cdot \cdots \cdot r_k \tag{B-11}$$

EXAMPLE B-6

(Example B-5 revisited) **Space Shuttle Solid Rocket Motor Field Joints as a Series System.** The probabilistic risk assessment of Dalal, Fowlkes, and Hoadley put the reliability (at 31°F) of pre-Challenger solid rocket motor field joints at .977 apiece. Since the proper functioning of six such joints is necessary for the safe operation of the solid rocket motors, the reliability of the system of joints is then

$$R_S = (.977)(.977)(.977)(.977)(.977)(.977) = .87$$

as in Example B-5. (One might well consider the .87 figure to be optimistic with regard to the entire solid rocket motor system, as it doesn't take into account any potential problems other than those involving field joints.)

Since typically each r_i is less than 1.0, formula (B-11) shows (as intuitively it should) that system reliability decreases as components are added to a series system. And system reliability is no better (larger) than the worst (smallest) component reliability.

Parallel Systems

In contrast to series system structure is *parallel* system structure.

DEFINITION B-10

A system consisting of components $C_1, C_2, C_3, \ldots, C_k$ is called a **parallel system** if its proper functioning requires only the functioning of at least one component.

Figure B-7 is a representation of a parallel system made up of $k = 3$ components. This diagram is interpreted in a manner similar to Figure B-6 (i.e., the system will function provided there is a path from point 1 to point 2 that crosses no failed component).

FIGURE B-7

Three-Component Parallel System

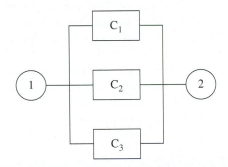

The fact that made it easy to develop formula (B-11) for the reliability of a series system is that for a series system to function, all components must function. The corresponding fact for a parallel system is that for a parallel system to *fail*, all components must *fail*. So if it is sensible to model the functioning of the individual components in a parallel system as independent, if r_i is the reliability of component i, and if R_P is the (parallel) system reliability, one has

$$1 - R_P = P[\text{the system fails}]$$
$$= P[\text{all components fail}]$$
$$= (1 - r_1)(1 - r_2)(1 - r_3) \cdots (1 - r_k) \qquad \textbf{(B-12)}$$

EXAMPLE B-7

Parallel Redundancy and Critical Safety Systems. The principle of parallel redundancy is often employed to improve the reliability of critical safety systems. For example, two physically separate automatic shutdown subsystems might be called for in the design of a nuclear power plant. The hope would be that in a rare overheating emergency, at least one of the two would successfully trigger reactor shutdown.

In such a case, if the shutdown subsystems are truly physically separate (so that independence could reasonably be used in a model of their emergency operation), relationship (B-12) might well describe the reliability of the overall safety system. And if, for example, each subsystem is 90% reliable, the overall reliability becomes

$$R_\text{P} = 1 - (.10)(.10) = 1 - .01 = .99$$

Expression (B-12) is perhaps a bit harder to absorb than expression (B-11). But it turns out that the formula functions the way one would intuitively expect. System reliability increases as components are added to a parallel system and is no worse (smaller) than the best (largest) component reliability.

One useful type of calculation that is sometimes done using expression (B-12) is to determine how many equally reliable components of a given reliability r are needed in order to obtain a desired system reliability, R_P. Substitution of r for each r_i in formula (B-12) gives

$$R_\text{P} = 1 - (1 - r)^k$$

and this can be solved for an approximate number of components required, giving

$$k \approx \frac{\ln(1 - R_\text{P})}{\ln(1 - r)} \qquad \textbf{(B-13)}$$

Using (for the sake of example) the values $r = .80$ and $R_\text{P} = .98$, expression (B-13) gives $k \approx 2.4$, so rounding *up* to an integer, one has that 3 components of individual 80% reliability will be required to give a parallel system reliability of at least 98%.

Combination Series-Parallel Systems

Real engineering systems rarely have purely series or purely parallel structure. However, it is sometimes possible to conceive of system structure as a combination of these two basic types. When this is the case, formulas (B-11) and (B-12) can be used to analyze successively larger subsystems until finally an overall reliability prediction is obtained.

EXAMPLE B-8

Predicting Reliability for a System with a Combination of Series and Parallel Structure. In order for an electronic mail message from individual A to reach individual B, the main computers at both A's site and B's site must be functioning, and at least one of three separate switching devices at a communications hub must be working. If the reliabilities for A's computer, B's computer, and each switching device are (respectively) .95, .99, and .90, a plausible figure for the reliability of the A-to-B electronic mail system can be determined as follows.

An appropriate system diagram is given in Figure B-8, with C_A, C_B, C_1, C_2, and C_3 standing (respectively) for the A site computer, the B site computer, and the three

FIGURE B-8

System Diagram for an Electronic Mail System

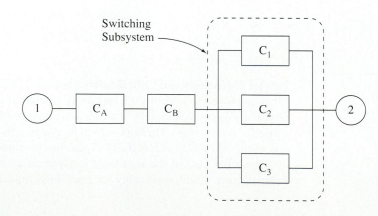

switching devices. Notice that although this system is neither purely series nor purely parallel, if one mentally replaces components C_1, C_2, and C_3 with a single "switching subsystem" block, there would be a purely series structure. So one is led to calculate

$$C_1, C_2, \text{ and } C_3 \text{ parallel subsystem reliability} = 1 - (1 - .90)^3 = .999$$

via formula (B-12). Then using formula (B-11),

$$\text{System reliability} = (.95)(.99)(.999) = .94$$

It is clear that the weak link in this communications system is at site A, rather than at B or at the communications hub.

B-3   Counting

Proposition B-3 and Example B-4 illustrate that when one can conceive of a model for a chance situation that consists of a finite sample space S with outcomes judged to be equally likely, the computation of probabilities for events of interest is conceptually a very simple matter. One simply counts up the number of outcomes in the event and divides by the total number of outcomes in the whole sample space. However, in most realistic applications of this simple idea to engineering problems, the process of actually writing down all outcomes in S and doing the counting involved would be most tedious indeed, and often far beyond the realm of practicality. Fortunately, there are some simple principles of counting that can often be applied to shortcut the process, allowing one to mentally count up outcomes. The purpose of this section is to present those counting techniques.

This section presents a multiplication principle of counting, the notion of permutations and how to count them, and the idea of combinations and how to count them, along with a few examples. It should be recognized that this material is on the very fringe of what is appropriate for inclusion in this book. It is not statistics, nor even really probability, but rather a piece of discrete mathematics that has some engineering implications. It is included here for two reasons. First is the matter of tradition. Counting has traditionally been part of most elementary expositions of probability, because games of chance (cards, coins, and dice) are often assumed to be *fair* and thus describable in terms of sample spaces with equally likely outcomes. And for better or worse, games of chance have been a principal source of examples in elementary probability. A second and perhaps more appealing reason for including the material is that it does have engineering applications (regardless of whether they are central to the particular mission of this text). Ultimately, the reader should take this short section for what it is: a digression from the book's main story line that can be quite helpful in engineering problems on occasion.

A Multiplication Principle, Permutations, and Combinations

The fundamental insight of this section is a multiplication principle that is simply stated but wide-ranging in its implications. As a means of emphasizing the principle, it will be stated in the form of a proposition.

PROPOSITION B-5

Suppose a complex action can be thought of as composed of r component actions, the first of which can be performed in n_1 different ways, the second of which can subsequently be performed in n_2 different ways, the third of which can subsequently be performed in n_3 different ways, etc. Then the total number of ways in which the complex action can be performed is

$$n = n_1 \cdot n_2 \cdot \ \cdots \ \cdot n_r$$

In graphical terms, this proposition is just a statement that a tree diagram that has n_1 first-level nodes, each of which leads to n_2 second-level nodes, and so on, must end up having a total of $n_1 \cdot n_2 \cdot \cdots \cdot n_r$ rth-level nodes.

EXAMPLE B-9

The Multiplication Principle and Counting the Number of Treatment Combinations in a Full Factorial. The familiar calculation of the number of different possible treatment combinations in a full factorial statistical study is an example of the use of Proposition B-5. Consider, for example, a $3 \times 4 \times 2$ study in the factors A, B, and C. One may think of the process of writing down a combination of levels for A, B, and C as consisting of $r = 3$ component actions. There are $n_1 = 3$ different ways of choosing a level for A, subsequently $n_2 = 4$ different ways of choosing a level for B, and then subsequently $n_3 = 2$ different ways of choosing a level for C. There are thus

$$n_1 \cdot n_2 \cdot n_3 = 3 \cdot 4 \cdot 2 = 24$$

different ABC combinations.

EXAMPLE B-10

The Multiplication Principle and Counting the Number of Ways of Assigning Four out of 100 Pistons to Four Cylinders. Suppose that four out of a production run of 100 pistons are to be installed in a particular engine block. Consider the number of possible placements of (distinguishable) pistons into the four (distinguishable) cylinders. One may think of the installation process as composed of $r = 4$ component actions. There are $n_1 = 100$ different ways of choosing a piston for installation into cylinder 1, subsequently $n_2 = 99$ different ways of choosing a piston for installation into cylinder 2, subsequently $n_3 = 98$ different ways of choosing a piston for installation into cylinder 3, and finally, subsequently $n_4 = 97$ different ways of choosing a piston for installation into cylinder 4. There are thus

$$100 \cdot 99 \cdot 98 \cdot 97 = 94{,}109{,}400$$

different ways of placing four of 100 (distinguishable) pistons into four (distinguishable) cylinders. (Notice that the job of actually making a list of the different possibilities is not one that is practically doable.)

Example B-10 is actually of a generic type of enough importance that there is some special terminology and notation associated with it. The general problem (of which the piston example is a special case) is that of choosing an ordering for r out of n distinguishable objects, or equivalently, placing r out of n distinguishable objects into r distinguishable positions. The application of Proposition B-5 shows that the number of different ways in which this placement can be accomplished is

$$n(n - 1)(n - 2) \cdots (n - r + 1) \tag{B-14}$$

since at each stage of sequentially placing objects into positions, there is one less object available for placement. The special terminology and notation for this are next.

DEFINITION B-11

If r out of n distinguishable objects are to be placed in an order 1 to r (or equivalently, placed into r distinguishable positions), each such potential arrangement is called a **permutation**. The number of such permutations possible will be symbolized as $P_{n,r}$, read "the number of permutations of n things r at a time."

In the notation of Definition B-11, one has (from expression (B-14)) that

$$P_{n,r} = n(n-1)(n-2)\cdots(n-r+1)$$

$$= \frac{n!}{(n-r)!} \qquad \textbf{(B-15)}$$

EXAMPLE B-10
(continued)

In the special permutation notation, the number of different ways of installing the four pistons is

$$P_{100,4} = \frac{100!}{96!}$$

EXAMPLE B-11

Permutations and Counting the Number of Possible Circular Arrangements of 12 Turbine Blades. The permutation idea of Definition B-11 can be used not only in straightforward ways, as in the previous example, but in slightly more subtle ways as well. To illustrate, consider a situation where 12 distinguishable turbine blades are to be placed into a central disk or hub at successive 30° angles, as sketched in Figure B-9. Notice that if one of the slots into which these blades fit is marked on the front face of the hub (and one therefore thinks of the blade positions as completely distinguishable), there are

$$P_{12,12} = 12 \cdot 11 \cdot 10 \cdot \cdots \cdot 2 \cdot 1$$

different possible arrangements of the blades.

FIGURE B-9

Hub with 12 Slots for Blade Installation

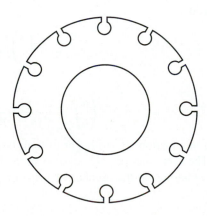

But now also consider the problem of counting all possible (circular) arrangements of the 12 blades if relative position is taken into account but absolute position is not. (Moving each blade 30° counterclockwise after first installing them would not create an arrangement different from the first, with this understanding.) The permutation idea can be used here as well, if one thinks as follows. Placing blade 1 anywhere in the hub essentially establishes a point of reference and makes the remaining 11 positions distinguishable (relative to the point of reference). One then has 11 distinguishable blades to place in 11 distinguishable positions. Thus, there must be

$$P_{11,11} = 11 \cdot 10 \cdot 9 \cdot \cdots \cdot 2 \cdot 1$$

such circular arrangements of the 12 blades, where only relative position is considered.

A second generic counting problem is related to the permutation idea and turns out to be particularly relevant in describing simple random sampling. That is the problem of choosing an unordered collection of r out of n distinguishable objects. The special terminology and notation associated with this generic problem are as follows.

DEFINITION B-12

If an unordered collection of r out of n distinguishable objects is to be made, each such potential collection is called a **combination**. The number of such combinations possible will be symbolized as $\binom{n}{r}$, read "the number of combinations of n things r at a time."

The alert reader will detect in Definition B-12 a slight conflict in terminology with other usage in this text. The "combination" in Definition B-12 is not the same as the "treatment combination" terminology used in connection with multifactor statistical studies to describe a set of conditions under which a sample is taken. (The "treatment combination" terminology has been used in this very section in Example B-9.) But this conflict rarely causes problems, since the intended meaning of the word *combination* is essentially always clear from context.

Appropriate use of Proposition B-5 and formula (B-15) makes it possible to develop a formula for $\binom{n}{r}$ as follows. One can think of creating a permutation of r out of n distinguishable objects as a two-step process. One might first select a combination of r out of the n objects and then place those selected objects in an order. This thinking suggests that $P_{n,r}$ can be written as

$$P_{n,r} = \binom{n}{r} \cdot P_{r,r}$$

But this means that

$$\frac{n!}{(n-r)!} = \binom{n}{r}\frac{r!}{0!}$$

that is, that

$$\binom{n}{r} = \frac{n!}{r!\,(n-r)!} \tag{B-16}$$

The ratio in equation (B-16) ought to look familiar to readers who have studied Section 5-1. The multiplier of $p^x(1-p)^{n-x}$ in the binomial probability function is of the form $\binom{n}{x}$, counting up the number of ways of placing x successes in a series of n trials.

EXAMPLE B-12

(Example 3-10 revisited) **Combinations and Counting the Numbers of Possible Samples of Cable Connectors with Prescribed Defect Class Compositions.** In the cable connector inspection scenario of Delva, Lynch, and Stephany, 3,000 inspections of connectors produced 2,985 connectors classified as having no defects, 1 connector classified as having only minor defects, and 14 others classified as having moderately serious, serious, or very serious defects. Suppose that in an effort to audit the work of the inspectors, a sample of 100 of the 3,000 previously inspected connectors is to be reinspected.

Then notice that directly from expression (B-16), there are in fact

$$\binom{3,000}{100} = \frac{3,000!}{100!\,2,900!}$$

different (unordered) possible samples for reinspection. Further, there are

$$\binom{2,985}{100} = \frac{2,985!}{100!\,2,885!}$$

different samples made up of only connectors judged to be defect-free. If (for some reason) the connectors to be reinspected were to be chosen as a simple random sample of the 3,000, the ratio

$$\frac{\binom{2,985}{100}}{\binom{3,000}{100}}$$

would then be a sensible figure to use for the probability that the sample is composed entirely of connectors initially judged to be defect-free.

It is instructive to take this example one step further and combine the use of Definition B-12 and Proposition B-5. So consider the problem of counting up the number of different samples containing 96 connectors initially judged defect-free, 1 judged to have only minor defects, and 3 judged to have moderately serious, serious, or very serious defects. To solve this problem, one may conceive of the creation of such a sample as a three-step process. In the first, 96 nominally defect-free connectors are chosen from 2,985; in the second, 1 connector nominally having minor defects only is chosen from 1; and finally, 3 connectors are chosen from the remaining 14. There are thus

$$\binom{2,985}{96} \cdot \binom{1}{1} \cdot \binom{14}{3}$$

different possible samples of this rather specialized type.

EXAMPLE B-13

An Application of Counting Principles to the Calculation of a Probability in a Scenario of Equally Likely Outcomes. As a final example in this section, most of the ideas discussed here can be applied to the computation of a probability in another situation of equally likely outcomes where one wouldn't really want to write out a list of the possible outcomes.

Consider a hypothetical situation where 15 manufactured devices of a particular kind are to be sent five apiece to three different testing labs. Suppose further that three of the seemingly identical devices are defective. Consider evaluating the probability that each lab receives one defective device, if the assignment of devices to labs is done at random.

The total number of possible assignments of devices to labs can be computed by thinking first of choosing 5 of 15 to send to Lab A, then 5 of the remaining 10 to send to Lab B, then sending the remaining 5 to Lab C. There are thus

$$\binom{15}{5} \cdot \binom{10}{5} \cdot \binom{5}{5}$$

such possible assignments of devices to labs.

On the other hand, if each lab is to receive one defective device, there are $P_{3,3}$ ways to assign defective devices to labs and then subsequently $\binom{12}{4} \cdot \binom{8}{4} \cdot \binom{4}{4}$ possible ways of completing the three shipments. So ultimately, an appropriate probability assignment for the event that each lab receives one defective device is

$$\frac{P_{3,3} \cdot \binom{12}{4} \cdot \binom{8}{4} \cdot \binom{4}{4}}{\binom{15}{5} \cdot \binom{10}{5} \cdot \binom{5}{5}} = \frac{3 \cdot 2 \cdot 1 \cdot 12! \cdot 8! \cdot 5! \cdot 10! \cdot 5! \cdot 5!}{15! \cdot 10! \cdot 4! \cdot 8! \cdot 4! \cdot 4!}$$

$$= \frac{3 \cdot 2 \cdot 1 \cdot 5 \cdot 5 \cdot 5}{15 \cdot 14 \cdot 13}$$

$$= .27$$

B-4 *Probabilistic Concepts Useful in Survival Analysis*

Section B-2 is meant to give readers the most elementary insights into how ideas of probability might prove useful in the context of reliability modeling and prediction. The ideas discussed in that section are of an essentially "static" nature, most appropriate when considering the likelihood of a system performing adequately at a single point in time — for example, at its installation, or at the end of its warranty period.

Reliability engineers also concern themselves with matters possessing a more dynamic flavor, having to do with the modeling and prediction of life-length variables associated with engineering systems and their components. It is outside the intended scope of this text to provide anything like a serious introduction to the large body of methods available in the engineering statistics literature for probability modeling and subsequent formal inference for such variables. But what will be done here is to provide some material (supplementary to that found in Section 5-2), that is part of the everyday jargon and intellectual framework of reliability engineering. This section will consider several descriptions and constructs related to continuous random variables that, like system or component life lengths, take only positive values.

Survivorship and Force-of-Mortality Functions

In this section, T will stand for a continuous random variable taking only nonnegative values. The reader may think of T as being the time till failure of an engineering component. In Section 5-2, continuous random variables X (or more properly, their distributions) were described through their probability densities $f(x)$ and cumulative probability functions $F(x)$. In the present context of lifetime random variables, there are several other more popular ways of conveying the information carried by $f(t)$ or $F(t)$. Two of these devices are introduced next.

DEFINITION B-13

The **survivorship function** for a nonnegative random variable T is the function

$$S(t) = P[T > t] = 1 - F(t)$$

The survivorship function is also sometimes known as the *reliability function*; it specifies the probability that the component being described survives beyond time t.

EXAMPLE B-14

The Survivorship Function and Diesel Engine Fan Blades. Data on 70 diesel engines of a single type (given in Table 1.1 of Nelson's *Applied Life Data Analysis*) indicate that lifetimes in hours of a certain fan on such engines could be modeled using an exponential distribution with mean $\alpha \approx 27{,}800$. So from Definition 5-17, to describe a fan lifetime T, one could use the density

$$f(t) = \begin{cases} 0 & \text{if } t < 0 \\ \dfrac{1}{27{,}800} e^{-t/27{,}800} & \text{if } t > 0 \end{cases}$$

or the cumulative probability function

$$F(t) = \begin{cases} 0 & \text{if } t \leq 0 \\ 1 - e^{-t/27{,}800} & \text{if } t > 0 \end{cases}$$

or from Definition B-13, the survivorship function

$$S(t) = \begin{cases} 1 & \text{if } t \leq 0 \\ e^{-t/27{,}800} & \text{if } t > 0 \end{cases}$$

The probability of a fan surviving at least 10,000 hours is then

$$S(10{,}000) = e^{-10{,}000/27{,}800} = .70$$

A second way of specifying the distribution of a life-length variable (unlike anything discussed in Section 5-2), is through a function giving a kind of "instantaneous rate of death of survivors."

DEFINITION B-14

The **force-of-mortality function** for a nonnegative continuous random variable T is, for $t > 0$, the function

$$h(t) = \frac{f(t)}{S(t)}$$

$h(t)$ is sometimes called the *hazard function* for T, but such usage tends to perpetuate unfortunate confusion with the entirely different concept of "hazard rate" for repairable systems. (The important difference between the two concepts is admirably exposited in the paper "On the Foundations of Reliability" by W.A. Thompson (*Technometrics*, 1981) and in the book *Repairable Systems Reliability* by Ascher and Feingold.) This author's strong preference is thus to stick to the term *force of mortality*.

The force-of-mortality function can be thought of heuristically as

$$h(t) = \frac{f(t)}{S(t)} = \lim_{\Delta \to 0} \frac{P[t < T < t + \Delta]/\Delta}{P[t < T]} = \lim_{\Delta \to 0} \frac{P[t < T < t + \Delta \mid t < T]}{\Delta}$$

which is indeed a sort of "death rate of survivors at time t."

EXAMPLE B-14
(continued)

The force-of-mortality function for the diesel engine fan example is, for $t > 0$,

$$h(t) = \frac{\dfrac{1}{27{,}800} e^{-t/27{,}800}}{e^{-t/27{,}800}} = \frac{1}{27{,}800}$$

The exponential mean $\alpha = 27{,}800$ model for fan life implies a constant $\left(\frac{1}{27{,}800}\right)$ force of mortality.

The property of the fan life model shown in the previous example is characteristic of exponential distributions. That is, a distribution has constant force of mortality exactly when that distribution is exponential. So having a constant force of mortality is equivalent to possessing the *memoryless* property of the exponential distributions discussed in Section 5-2. If the lifetime of an engineering component is described using a constant force of mortality, there is no (mathematical) reason to replace such a component before it fails, since the distribution of its remaining life from any point in time is the same as the distribution of the time till failure of a new component of the same type.

Potential probability models for lifetime random variables are often classified according to the nature of their force-of-mortality functions, and these classifications are taken into account when selecting models for reliability engineering applications. If $h(t)$ is increasing in t, the corresponding distribution is called an *increasing force-of-mortality* (IFM) distribution, and if $h(t)$ is decreasing in t, the corresponding distribution is called a *decreasing force-of-mortality* (DFM) distribution. The reliability engineering implications of an IFM distribution being appropriate for modeling the lifetimes of a particular type of component are often that (as a form of preventative maintenance) such components are retired from service once they reach a particular age, even if they have not failed.

EXAMPLE B-15

The Weibull Distributions and Their Force-of-Mortality Functions. The Weibull family of distributions discussed in Section 5-2 is commonly used in reliability engineering contexts. Using formulas (5-26) and (5-27) of Section 5-2, the Weibull force-of-mortality function for shape parameter β and scale parameter α is

$$h(t) = \frac{\beta t^{\beta-1}}{\alpha^\beta} \qquad \text{for } t > 0$$

which for $\beta = 1$ (the exponential distribution case) is constant, for $\beta < 1$ is decreasing in t, and for $\beta > 1$ is increasing in t. The Weibull distributions with $\beta < 1$ are DFM distributions, and the ones with $\beta > 1$ are IFM distributions.

EXAMPLE B-16

Force-of-Mortality Function for a Uniform Distribution. As an artificial but instructive example, consider the use of a uniform distribution on the interval $(0, 1)$ as a life-length model. With

$$f(t) = \begin{cases} 1 & \text{if } 0 < t < 1 \\ 0 & \text{otherwise} \end{cases}$$

the survivorship function is

$$S(t) = \begin{cases} 1 & \text{if } t < 0 \\ 1 - t & \text{if } 0 \le t < 1 \\ 0 & \text{if } 1 \le t \end{cases}$$

so

$$h(t) = \frac{1}{1 - t} \qquad \text{if } 0 < t < 1$$

Figure B-10 shows plots of both $f(t)$ and $h(t)$ for the uniform model. $h(t)$ is clearly increasing for $0 < t < 1$ (quite drastically so, in fact, as one approaches $t = 1$). And well it should be. Knowing that (according to the uniform model) life will certainly end by $t = 1$, one's nervousness about impending death should certainly skyrocket as one nears $t = 1$.

FIGURE B-10

Probability Density and Force-of-Mortality Function for a Uniform Distribution

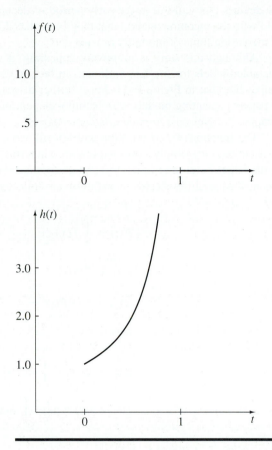

Conventional wisdom in reliability engineering is that many kinds of manufactured devices have life distributions that ought to be described by force-of-mortality functions qualitatively similar to the hypothetical one sketched in Figure B-11. The shape in Figure B-11 is often referred to as the *bathtub curve* shape. It includes an early region of decreasing force of mortality, a long central period of relatively constant force of

FIGURE B-11

A "Bathtub Curve" Force-of-Mortality Function

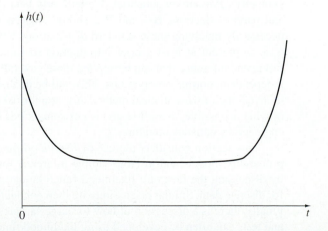

mortality, and a late period of rapidly increasing force of mortality. Devices with lifetimes describable as in Figure B-11 are sometimes subjected to a *burn-in* period, to eliminate the devices that will fail in the early period of decreasing force of mortality, and then sold with the recommendation that they be replaced before the onset of the late period of increasing force of mortality or *wear-out*.

Although this story is intuitively appealing, it should be admitted that the most tractable models for life length do not, in fact, have force-of-mortality functions with shapes like that in Figure B-11. For a further discussion of this matter and references to papers presenting models with bathtub-shaped force-of-mortality functions, refer to Chapter 2 of Nelson's *Applied Life Data Analysis*.

The functions $f(t)$, $F(t)$, $S(t)$, and $h(t)$ all carry the same information about a life distribution; they simply express it in different terms. Given one of them, the derivation of the others is (at least in theory) straightforward. Some of the relationships that exist among the four different characterizations are collected here for the reader's convenience. For $t > 0$,

$$F(t) = \int_0^t f(x)\,dx$$

$$f(t) = \frac{d}{dt}F(t)$$

$$S(t) = 1 - F(t)$$

$$h(t) = \frac{f(t)}{S(t)}$$

$$S(t) = \exp\left(-\int_0^t h(x)\,dx\right)$$

$$f(t) = h(t)\exp\left(-\int_0^t h(x)\,dx\right)$$

B-5 *Maximum Likelihood Fitting of Probability Models and Related Inference Methods*

The model fitting and inference methods discussed in this text are, for the most part, methods for independent, normally distributed observations. This is in spite of the fact that there are strong hints in Chapter 5 and this appendix that other kinds of probability models often prove useful in engineering problem solving. (For example, binomial, geometric, Poisson, exponential, Weibull, and beta distributions have been discussed, and parts of Sections B-1 and B-2 should suggest that probability models not even necessarily involving these standard distributions will often be helpful.) It thus seems wise to present at least a brief introduction to some principles of probability-model fitting and inference that can be applied more generally than to only scenarios involving independent, normal observations. This will be done, not so much to encourage readers to invent their own statistical methods for nonnormal data, but to give at least the flavor of what is possible, as well as an idea of some kinds of things likely to be found in the engineering statistics literature.

This section considers the use of *likelihood functions* in the fitting of parametric probability models and in large-sample inference for the model parameters. It begins by discussing the idea of a likelihood function and *maximum likelihood* model fitting for discrete data. Similar discussions are then conducted for continuous and mixed data. Finally, there is a discussion of how for large samples, approximate confidence regions and tests can often be developed using likelihood functions.

Likelihood Functions
for Discrete Data and
Maximum Likelihood
Model Fitting

To begin with, consider scenarios where the outcome of a chance situation can be described in terms of a data vector of jointly discrete random variables (or a single discrete random variable) Y, whose probability function f depends on some (unknown) vector of parameters (or single parameter) Θ. To make the dependence of f on Θ explicit, this section will use the notation

$$f_{\Theta}(y)$$

for the (joint) probability function of Y.

Chapter 5 made heavy use of particular parametric probability functions, primarily thinking of them as functions of y. In this section, it will be very helpful to shift perspective. With data $Y = y$ in hand, think of

$$f_{\Theta}(y) \tag{B-17}$$

or (often more conveniently) its natural logarithm

$$L(\Theta) = \ln(f_{\Theta}(y)) \tag{B-18}$$

as functions of Θ, specifying for various possible vectors of parameters "how likely" it would be to observe the particular data in hand. With this perspective, the function (B-17) is often called the *likelihood function* and function (B-18) the *log likelihood function* for the problem under consideration.

EXAMPLE B-17

(Example 5-4 revisited) **The Log Likelihood Function for the Number Conforming in a Sample of Hexamine Pellets.** In the pelletizing machine example used in Chapter 5 and earlier, it is possible to argue that under stable conditions,

X = the number of conforming pellets produced in a batch of 100 pellets

might well be modeled using a binomial distribution with $n = 100$ and p some unknown parameter. The corresponding probability function is thus

$$f(x) = \begin{cases} \dfrac{100!}{x!\,(100-x)!}\,p^x(1-p)^{100-x} & x = 0, 1, \ldots, 100 \\ 0 & \text{otherwise} \end{cases}$$

Should one observe $X = 66$ conforming pellets in a batch, the material just introduced says that the function of p

$$L(p) = \ln(f(66)) = \ln(100!) - \ln(66!) - \ln(34!) + 66\ln(p) + 34\ln(1-p) \tag{B-19}$$

is an appropriate log likelihood function. Figure B-12 is a sketch of $L(p)$ for this problem. Notice that (in an intuitively appealing fashion) the value of p maximizing $L(p)$ is

$$\hat{p} = \frac{66}{100} = .66$$

That is, $p = .66$ makes the chance of observing the particular data in hand ($X = 66$) as large as possible.

EXAMPLE B-18

The Log Likelihood Function for n Independent Poisson Observations. As a second, somewhat more abstract example of the idea of a likelihood function, suppose that $X_1, X_2, \ldots, X_n$ are independent Poisson random variables, X_i with mean $k_i\lambda$ for $k_1, k_2, \ldots, k_n$ known constants, and λ an unknown parameter. Such a model might,

FIGURE B-12

Plot of the Log Likelihood
Function Based on 66
Conforming Tablets out of 100

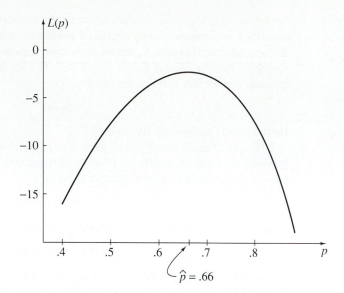

for example, be appropriate in a quality monitoring context, where at time i, k_i standard-sized units of product are inspected, X_i defects are observed, and λ is a constant mean defects per unit value.

The joint probability function for $X_1, X_2, \ldots, X_n$ is

$$f(x_1, x_2, \ldots, x_n) = \begin{cases} \displaystyle\prod_{i=1}^{n} \frac{e^{-k_i\lambda}(k_i\lambda)^{x_i}}{x_i!} & \text{for each } x_i \text{ a nonnegative integer} \\ 0 & \text{otherwise} \end{cases}$$

The log likelihood function in the present context is thus

$$L(\lambda) = -\lambda \sum_{i=1}^{n} k_i + \sum_{i=1}^{n} x_i \ln(k_i) + \sum_{i=1}^{n} x_i \ln(\lambda) - \sum_{i=1}^{n} \ln(x_i!) \qquad \textbf{(B-20)}$$

The likelihood functions in Examples B-17 and B-18 are for individual (univariate) parameters. The next example involves two parameters.

EXAMPLE B-19

A Log Likelihood Function Based on Pre-Challenger Space Shuttle O-Ring Failure Data.
Table B-2 contains pre-Challenger data on field joint primary O-ring failures on 23 (out of 24) space shuttle flights. (On one flight, the rocket motors were lost at sea, so no data are available.) The failure counts $x_1, x_2, \ldots, x_{23}$ are the numbers (out of 6 possible) of primary O-rings showing evidence of erosion or blow-by in post-flight inspections of the solid rocket motors, and $t_1, t_2, \ldots, t_{23}$ are the corresponding temperatures at the times of launch.

In "Risk Analysis of the Space Shuttle: Pre-Challenger Prediction of Failure" (*Journal of the American Statistical Association*, 1989), Dalal, Fowlkes, and Hoadley considered several analyses of the data in Table B-2 (and other pre-Challenger shuttle failure data). In one of their analyses of the data given here, Dalal, Fowlkes, and Hoadley used a likelihood approach based on the observations

$$y_i = \begin{cases} 1 & \text{if } x_i \geq 1 \\ 0 & \text{if } x_i = 0 \end{cases}$$

TABLE B-2

Pre-Challenger Field Joint
Primary O-Ring Failure Data

Flight Date	x Number of Field Joint Primary O-Ring Incidents	t Temperature at Launch (°F)
4/12/81	0	66
11/12/81	1	70
3/22/82	0	69
11/11/82	0	68
4/4/83	0	67
6/18/83	0	72
8/30/83	0	73
11/28/83	0	70
2/3/84	1	57
4/6/84	1	63
8/30/84	1	70
10/5/84	0	78
11/8/84	0	67
1/24/85	2	53
4/12/85	0	67
4/29/85	0	75
6/17/85	0	70
7/29/85	0	81
8/27/85	0	76
10/3/85	0	79
10/30/85	2	75
11/26/85	0	76
1/12/86	1	58

which indicate which flights experienced primary O-ring incidents. (They also considered a likelihood approach based on the counts x_i themselves. But here only the slightly simpler analysis based on the y_i's will be discussed.) The authors modeled $Y_1, Y_2, \ldots, Y_{23}$ as *a priori* independent variables and treated the probability of at least one O-ring incident on flight i,

$$p_i = P[Y_i = 1] = P[X_i \geq 1]$$

as a function of t_i. The particular form of dependence of p_i on t_i used by the authors was a "linear (in t) log odds" form

$$\ln\left(\frac{p}{1-p}\right) = \alpha + \beta t \qquad \text{(B-21)}$$

for α and β some unknown parameters. Equation (B-21) can be solved for p to produce the function of t

$$p(t) = \frac{1}{1 + e^{-(\alpha + \beta t)}} \qquad \text{(B-22)}$$

From either equation (B-21) or (B-22), it is possible to see that if $\beta > 0$, the probability of at least one O-ring incident is increasing in t (low-temperature launches are best). On the other hand, if $\beta < 0$, p is decreasing in t (high-temperature launches are best).

The joint probability function for $Y_1, Y_2, \ldots, Y_{23}$ employed by Dalal, Fowlkes, and Hoadley was then

$$f(y_1, y_2, \ldots, y_{23}) = \begin{cases} \displaystyle\prod_{i=1}^{23} p(t_i)^{y_i}\big(1 - p(t_i)\big)^{1-y_i} & \text{for each } y_i = 0 \text{ or } 1 \\ 0 & \text{otherwise} \end{cases}$$

The log likelihood function is then (using equations (B-21) and (B-22))

$$
\begin{aligned}
L(\alpha, \beta) &= \sum_{i=1}^{23} y_i \ln\left(\frac{p(t_i)}{1 - p(t_i)}\right) + \sum_{i=1}^{23} \ln(1 - p(t_i)) \\
&= \sum_{i=1}^{23} y_i(\alpha + \beta t_i) + \sum_{i=1}^{23} \ln\left(\frac{e^{-(\alpha + \beta t_i)}}{1 + e^{-(\alpha + \beta t_i)}}\right) \\
&= 7\alpha + \beta(70 + 57 + 63 + 70 + 53 + 75 + 58) \\
&\quad + \ln\left(\frac{e^{-(\alpha + 66\beta)}}{1 + e^{-(\alpha + 66\beta)}}\right) + \ln\left(\frac{e^{-(\alpha + 70\beta)}}{1 + e^{-(\alpha + 70\beta)}}\right) \\
&\quad + \cdots + \ln\left(\frac{e^{-(\alpha + 58\beta)}}{1 + e^{-(\alpha + 58\beta)}}\right)
\end{aligned}
\tag{B-23}
$$

where the sum abbreviated in equation (B-23) is over all 23 t_i's. Figure B-13 is a contour plot of $L(\alpha, \beta)$ given in equation (B-23).

FIGURE B-13

Contour Plot of the Dallal, Fowlkes, and Hoadley Log Likelihood Function

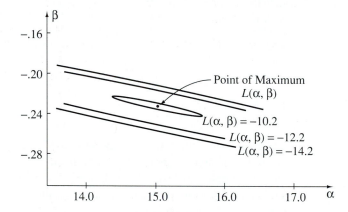

It is interesting (and sadly, of great engineering importance) that the region of (α, β) pairs making the data of Table B-2 most likely is in the $\beta < 0$ part of the (α, β)-plane — that is, where $p(t)$ is decreasing in t (i.e., increases as t falls). (Remember that the tragic Challenger launch was made at $t = 31°$.)

The binomial and Poisson examples of discrete-data likelihoods given thus far have arisen from situations that are most naturally thought of as intrinsically discrete. However, the details of how engineering data are collected sometimes lead to the production of essentially discrete data from intrinsically continuous variables. For example, consider a life test of some electrical components, where one begins a test by connecting 50 devices to a power source, goes away, and then returns every 10 hours to note which devices are still functioning. The details of data collection produce only discrete data (which 10-hour period produces failure) from the intrinsically continuous life lengths of the 50 devices. The next example shows how the likelihood idea might be used in another situation where the underlying phenomenon is continuous.

EXAMPLE B-20

A Log Likelihood Function for a Crudely Gaged Normally Distributed Dimension of Five Machined Metal Parts. In many contexts where industrial process monitoring involves relatively stable processes and relatively crude gaging, intrinsically continuous product characteristics are measured and recorded as essentially discrete data. For example, Table B-3 gives values (in units of .0001 in. over nominal) of a critical dimension measured on a sample of $n = 5$ consecutive metal parts produced by a CNC lathe.

TABLE B-3

Measurements of a Critical
Dimension on Five Metal Parts
Produced on a CNC Lathe

Part	Measured Dimension, y
1	4
2	3
3	3
4	2
5	3

It might make sense to model underlying values of this critical dimension as normal, with some (unknown) mean μ and some (unknown) standard deviation σ, but nonetheless want to explicitly recognize the discreteness of the recorded data. One way of doing so in this context is to think of the observed values as arising (after coding) from rounding normally distributed dimensions to the nearest integer. For a single metal part, this would mean that for any integer y,

$P[\text{the value recorded is } y] = P[\text{the actual dimension is between } y - .5 \text{ and } y + .5]$

$$= \Phi\left(\frac{y + .5 - \mu}{\sigma}\right) - \Phi\left(\frac{y - .5 - \mu}{\sigma}\right) \tag{B-24}$$

So treating $n = 5$ consecutive recorded dimensions as independent, equation (B-24) leads to the joint probability function

$$f(y_1, y_2, \ldots, y_5) = \prod_{i=1}^{5}\left\{\Phi\left(\frac{y_i + .5 - \mu}{\sigma}\right) - \Phi\left(\frac{y_i - .5 - \mu}{\sigma}\right)\right\}$$

and log likelihood function for the data in Table 5-3

$$\left.\begin{aligned}
L(\mu, \sigma) = {} & \ln\left(\Phi\left(\frac{2 + .5 - \mu}{\sigma}\right) - \Phi\left(\frac{2 - .5 - \mu}{\sigma}\right)\right) \\
& + 3\ln\left(\Phi\left(\frac{3 + .5 - \mu}{\sigma}\right) - \Phi\left(\frac{3 - .5 - \mu}{\sigma}\right)\right) \\
& + \ln\left(\Phi\left(\frac{4 + .5 - \mu}{\sigma}\right) - \Phi\left(\frac{4 - .5 - \mu}{\sigma}\right)\right)
\end{aligned}\right\} \tag{B-25}$$

Figure B-14 is a contour plot of $L(\mu, \sigma)$.

FIGURE B-14

Contour Plot of the "Rounded
Normal Data" Log Likelihood
for the Data of Table B-3

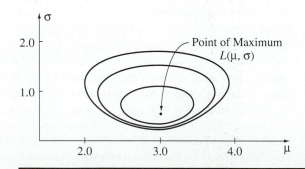

Contemplation of a likelihood function $f_\Theta(y)$ or its log version $L(\Theta)$ can be thought of as a way of assessing how consonant various probability models indexed by Θ are with the data in hand, $Y = y$. Different parameter vectors Θ having the same value of $L(\Theta)$ can be viewed as equally consonant with data in hand. A value of Θ maximizing $L(\Theta)$ might then be considered to be as consonant with the observed data as is possible; this value is often termed the *maximum likelihood estimate* of the parameter vector Θ. Finding maximum likelihood estimates of parameters is a very common method of fitting probability models to data. In simple situations, calculus can sometimes be employed to see how to maximize $L(\Theta)$, but in most nonstandard situations, numerical or graphical methods are required.

EXAMPLE B-17
(continued)

In the pelletizing example, simple investigation of Figure B-12 shows

$$\hat{p} = \frac{66}{100}$$

to maximize $L(p)$ given in display (B-19) and thus to be the maximum likelihood estimate of p. The reader is encouraged to verify that by differentiating $L(p)$ with respect to p, setting the result equal to 0, and solving for p, this maximizing value can also be found analytically.

EXAMPLE B-18
(continued)

Differentiating the log likelihood (B-20) with respect to λ, one obtains

$$\frac{d}{d\lambda}L(\lambda) = -\sum_{i=1}^{n} k_i + \frac{1}{\lambda}\sum_{i=1}^{n} x_i$$

Setting this derivative equal to 0 and solving for λ produces

$$\lambda = \frac{\sum_{i=1}^{n} x_i}{\sum_{i=1}^{n} k_i} = \hat{u}$$

which is the total number of defects observed divided by the total number of units inspected. Since the second derivative of $L(\lambda)$ is easily seen to be negative for all λ, $\hat{u}$ is the unique maximizer of $L(\lambda)$ — that is, the maximum likelihood estimate of λ.

EXAMPLE B-19
(continued)

Careful examination of contour plots like Figure B-13, or use of a numerical search method for the (α, β) pair maximizing $L(\alpha, \beta)$, produces maximum likelihood estimates of α and β of approximately

$$\hat{\alpha} = 15.043$$
$$\hat{\beta} = -.2322$$

based on the pre-Challenger data. Figure B-15 is a plot of $p(t)$ given in display (B-22) for these values of α and β. Notice the disconcerting fact that the corresponding estimate of $p(31)$ (the probability of at least one O-ring failure in a 31° launch) exceeds .99. ($t = 31$ is clearly a huge extrapolation away from any t values in Table B-2, but even so, this kind of analysis conducted before the Challenger launch could well have helped cast legitimate doubt on the advisability of a low-temperature launch.)

FIGURE B-15

Plot of Fitted Probability of
at Least One O-Ring Failure
as a Function of Shuttle
Launch Temperature

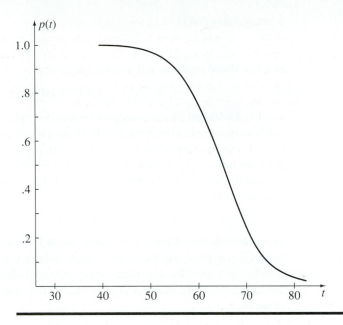

EXAMPLE B-20
(continued)

Examination of the contour plot in Figure B-14 shows maximum likelihood estimates of μ and σ based on the rounded normal data model and the data in Table B-3 to be approximately

$$\hat{\mu} = 3.0$$
$$\hat{\sigma} = .55$$

It is worth noting that for these data, $s = .71$, which is noticeably larger than $\hat{\sigma}$. This illustrates a fairly well established piece of statistical folklore. It is fairly well known that to ignore rounding of intrinsically continuous data will typically have the effect of inappropriately inflating one's perception of the spread of the underlying distribution.

**Likelihood Functions for
Continuous and Mixed Data
and Maximum Likelihood
Model Fitting**

The likelihood function ideas discussed thus far depend on treating the Θ *probability* of discrete data in hand, $Y = y$, as a function of Θ. When one comes to the analysis of data using continuous distributions, a slight logical snag is therefore encountered: If a continuous model is employed, the probability associated with observing any particular exact realization y is always 0, for every Θ.

To understand how one employs likelihood methods in continuous models, it is then useful to consider the probability of observing a value of Y "near" y as a function of Θ. That is, suppose that

$$f_{\Theta}(y)$$

is a joint probability density for Y depending on an unknown parameter vector Θ. Then in rough terms, if Δ is a small positive number and $y = (y_1, y_2, \ldots, y_n)$,

$$P[\text{each } Y_i \text{ is within } \tfrac{\Delta}{2} \text{ of } y_i] \approx f_{\Theta}(y)\Delta^n \qquad \textbf{(B-26)}$$

But in expression (B-26), Δ^n doesn't depend on Θ — that is, the approximate probability is proportional to the function of Θ, $f_{\Theta}(y)$. It is therefore plausible to use the joint density with data plugged in,

$$f_{\Theta}(y) \tag{B-27}$$

as a likelihood function and to use its logarithm,

$$L(\Theta) = \ln(f_{\Theta}(y)) \tag{B-28}$$

as a log likelihood for data modeled as jointly continuous. (Formulas (B-27) and (B-28) are formally identical to formulas (B-17) and (B-18), but they involve a different type of data.) Contemplation of formula (B-27) or (B-28) can be thought of as a way of assessing the consonance of different parameter vectors Θ with continuous data, y. And as for the discrete case, a vector Θ maximizing $L(\Theta)$ is often termed a *maximum likelihood estimate* of the parameter vector.

EXAMPLE B-21

Maximum Likelihood Estimation Based on iid Exponential Data. The exponential distribution is a popular model for life-length variables. The following are hypothetical life lengths (in hours) for $n = 4$ nominally identical electrical components, which will be assumed to have been *a priori* adequately described as iid exponential variables with mean α,

$$75.4, \quad 39.4, \quad 3.7, \quad 4.5 \tag{B-29}$$

If $Y_1, Y_2, Y_3,$ and Y_4 are iid exponential variables with means α, an appropriate joint probability density is

$$f(y) = \begin{cases} \prod_{i=1}^{4} \dfrac{1}{\alpha} e^{-y_i/\alpha} & \text{for each } y_i > 0 \\[2ex] 0 & \text{otherwise} \end{cases}$$

So with the data of display (B-29) in hand, the log likelihood function becomes

$$L(\alpha) = -4\ln(\alpha) - \frac{1}{\alpha}(75.4 + 39.4 + 3.7 + 4.5) \tag{B-30}$$

FIGURE B-16

Plot of a Log Likelihood Based on Four iid Exponential Observations

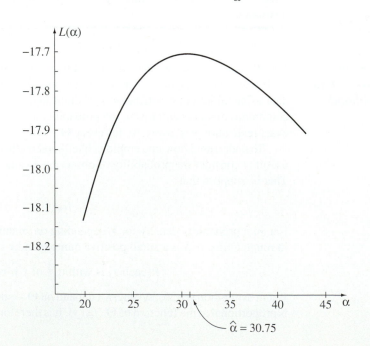

$\hat{\alpha} = 30.75$

It is easy to verify (using calculus and/or simply looking at the plot of $L(\alpha)$ in Figure B-16) that $L(\alpha)$ is maximized for

$$\hat{\alpha} = 30.75 = \frac{75.4 + 39.4 + 3.7 + 4.5}{4} = \bar{y}$$

This fact is a particular instance of the general result that the maximum likelihood estimate of an exponential mean is the sample average of the observations.

Example B-21 is fairly simple, in that only one parameter is involved and calculus can be used to find an explicit formula for the maximum likelihood estimator. The reader might be interested in working through the somewhat more complicated (two-parameter) situation involving n iid normal random variables with means μ and standard deviations σ. Two-variable calculus can be used to show that maximum likelihood estimates of the parameters based on observations $x_1, x_2, \ldots, x_n$ turn out to be, respectively,

$$\hat{\mu} = \bar{x}$$

$$\hat{\sigma} = \sqrt{\frac{n-1}{n}}\, s$$

The next example concerns an important continuous situation where no explicit formulas for maximum likelihood estimates seem to exist.

EXAMPLE B-22

Maximum Likelihood Estimation Based on iid Weibull Steel Specimen Failure Times. The data in Table B-4 are $n = 10$ ordered failure times for hardened steel specimens that were subjected to a particular rolling fatigue test. These data appeared originally in the paper of J.I. McCool, "Confidence Limits for Weibull Regression With Censored Data" (*IEEE Transactions on Reliability*, 1980).

TABLE B-4

Ten Ordered Failure Times of Steel Specimens

.073, .098, .117, .135, .175, .262, .270, .350, .386, .456

The Weibull probability plot of these data in Figure B-17 suggests the appropriateness of fitting a Weibull model to them (and indicates that β near 2 and α near .25 may be appropriate parameters for such a fitted model).

Notice that the joint density function of $n = 10$ iid Weibull random variables $Y_1, Y_2, \ldots, Y_{10}$ with parameters α and β is

$$f(\mathbf{y}) = \begin{cases} \displaystyle\prod_{i=1}^{10} \frac{\beta}{\alpha^{\beta}} y_i^{\beta-1} e^{-(y_i/\alpha)^{\beta}} & \text{for each } y_i > 0 \\[2mm] 0 & \text{otherwise} \end{cases}$$

So using the data of Table B-4, the log likelihood

$$L(\alpha, \beta) = 10\ln(\beta) - 10\beta\ln(\alpha) + (\beta - 1)(\ln(.073) + \ln(.098) + \cdots + \ln(.456))$$

$$- \frac{1}{\alpha^{\beta}}((.073)^{\beta} + (.098)^{\beta} + \cdots + (.456)^{\beta})$$

$$= 10\ln(\beta) - 10\beta\ln(\alpha) - 16.267(\beta - 1) - \frac{1}{\alpha^{\beta}}((.073)^{\beta} + (.098)^{\beta}$$

$$+ \cdots + (.456)^{\beta})$$

FIGURE B-17

Weibull Probability Plot of
McCool's Steel Specimen
Failure Times

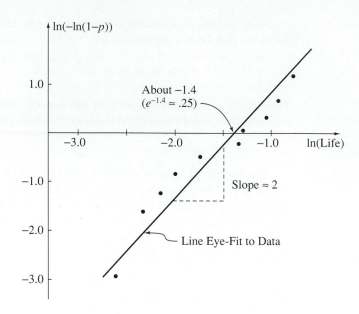

is indicated. Figure B-18 shows a contour plot of $L(\alpha, \beta)$ and indicates that maximum likelihood estimates of α and β are indeed in the vicinity of $\hat{\beta} = 2.0$ and $\hat{\alpha} = .26$.

FIGURE B-18

Contour Plot of a Weibull Log
Likelihood for McCool's Steel
Specimen Failure Times

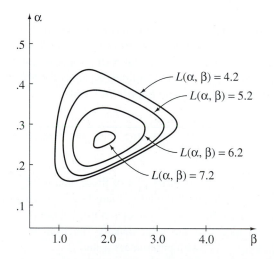

Analytical attempts to locate the maximum likelihood estimates for this kind of iid Weibull data situation are only partially fruitful. Setting partial derivatives of $L(\alpha, \beta)$ equal to 0, followed by some algebra, does lead to the two equations

$$\beta = \left(\frac{\sum y_i^\beta \ln(y_i)}{\sum y_i^\beta} - \frac{\sum \ln(y_i)}{n} \right)^{-1}$$

$$\alpha = \left(\frac{\sum y_i^\beta}{n} \right)^{1/\beta}$$

which maximum likelihood estimates must satisfy, but these must be solved numerically.

Discrete and continuous likelihood methods have thus far been discussed separately. However, particularly in life data analysis contexts, statistical engineering studies occasionally yield data that are *mixed* — in the sense that some parts are discrete, while other parts are continuous. If it is sensible to think of the two parts as independent, a combination of things already said here can lead to an appropriate likelihood function and then, for example, to maximum likelihood parameter estimates.

That is, suppose that one has available discrete data, $Y_1 = y_1$, and continuous data, $Y_2 = y_2$, which can be thought of as independently generated — Y_1 from a discrete joint distribution with joint probability function

$$f_{\Theta}^{(1)}(y_1)$$

and Y_2 from a continuous joint distribution with joint probability density

$$f_{\Theta}^{(2)}(y_2)$$

Then a sensible likelihood function becomes

$$f_{\Theta}^{(1)}(y_1) \cdot f_{\Theta}^{(2)}(y_2) \tag{B-31}$$

with corresponding log likelihood

$$L(\Theta) = \ln(f_{\Theta}^{(1)}(y_1)) + \ln(f_{\Theta}^{(2)}(y_2)) \tag{B-32}$$

Armed with equation (B-31) or (B-32), assessments of the consonance of different parameter vectors Θ with the data in hand and maximum likelihood model fitting can proceed just as for purely discrete or purely continuous cases.

EXAMPLE B-23

Maximum Likelihood Estimation of a Mean Insulating Fluid Breakdown Time Using Censored Data. Table 2.1 of Nelson's *Applied Life Data Analysis* gives some data on times to breakdown (in seconds) of an insulating fluid at several different voltages. The results of $n = 12$ tests made at 30 kV are repeated here in Table B-5.

TABLE B-5

12 Insulating Fluid
Breakdown Times

50, 134, 187, 882, 1450, 1470, 2290, 2930, 4180, 15800, > 29200, > 86100

The last two entries in Table B-5 mean that two tests were terminated at (respectively) 29,200 seconds and 86,100 seconds without failures having been observed. In common statistical jargon, these last two data values are *censored* (at the times 29,200 and 86,100, respectively).

Nelson remarks in his book that exponential distributions are often used to model life length for such fluids. Therefore, consider fitting an exponential distribution with mean α to the data of Table B-5. Notice that the first ten pieces of data in Table B-5 are continuous "exact" failure times, while the last two are essentially discrete pieces of information. Considering first the discrete part of the overall likelihood, the probability that two independent exponential variables exceed 29,200 and 86,100, respectively, is

$$f_{\alpha}^{(1)}(y_1) = e^{-29,200/\alpha} \cdot e^{-86,100/\alpha}$$

Then considering the continuous part of the likelihood, the joint density of ten independent exponential variables with mean α is

$$f_{\alpha}^{(2)}(y_2) = \begin{cases} \dfrac{1}{\alpha^{10}} e^{-\sum y_i/\alpha} & \text{for each } y_i > 0 \\ 0 & \text{otherwise} \end{cases}$$

Putting these two pieces together via equation (B-32), one has that the log likelihood function appropriate here is

$$L(\alpha) = -10\ln(\alpha) - \frac{1}{\alpha}(50 + 134 + 187 + \cdots + 15{,}800 + 29{,}200 + 86{,}100)$$

$$= -10\ln\alpha - \frac{1}{\alpha}(144{,}673) \tag{B-33}$$

This function of α is easily seen via elementary calculus to be maximized at

$$\hat{\alpha} = \frac{144{,}673}{10} = 14{,}467.3 \text{ sec}$$

which has the intuitively appealing interpretation of the ratio of the total time on test to the number of failures observed during testing.

Likelihood-Based Large-Sample Inference Methods

One of the appealing things about the likelihood function idea is that in many situations, it is possible to base large-sample significance testing and confidence region methods on the likelihood function. Intuitively, it would seem that those parameter vectors Θ "most consonant" with the data in hand ought to form a sensible confidence set for Θ. And in significance-testing terms, if a hypothesized value of Θ (say, Θ_0) has a corresponding value of the likelihood function far smaller than the maximum possible, that circumstance ought to produce a small p-value — that is, strong evidence against $H_0\colon \Theta = \Theta_0$.

To make this thinking precise, let

$$L^* = \max_{\Theta} L(\Theta)$$

that is, L^* is the largest possible value of the log likelihood. (If $\hat{\Theta}$ is a maximum likelihood estimate of Θ, then $L^* = L(\hat{\Theta})$.) An intuitively appealing way to make a confidence set for the parameter vector Θ is to use the set of all Θ's with log likelihood not too far below L^*,

$$\{\Theta \mid L(\Theta) > L^* - c\} \tag{B-34}$$

for an appropriate number c. And a plausible way of deriving a p-value for testing

$$H_0\colon \Theta = \Theta_0 \tag{B-35}$$

is by trying to identify a sensible probability distribution for

$$L^* - L(\Theta_0) \tag{B-36}$$

when H_0 holds, and using the upper-tail probability beyond an observed value of variable (B-36) as a p-value.

The practical gaps in this thinking are two: how to choose c in display (B-34) to get a desired confidence level and what kind of distribution to use to describe variable (B-36) under hypothesis (B-35). There are no general exact answers to these questions, but statistical theory does provide at least some indication of approximate answers that are often adequate for practical purposes when large samples are involved. That is, statistical theory (for iid observations from sufficiently "nice" distributions) suggests that in many large-sample situations, if Θ is of dimension k, choosing

$$c = \frac{1}{2}U \tag{B-37}$$

for U the γ quantile of the χ_k^2 distribution, produces a confidence set (B-34) of confidence level roughly γ. And similar reasoning suggests that in many large-sample situations, if Θ is of dimension k, the hypothesis (B-35) can be tested using the test statistic

$$2(L^* - L(\Theta_0)) \tag{B-38}$$

and a χ_k^2 approximate reference distribution, where large values of the test statistic (B-38) count as evidence against H_0.

EXAMPLE B-23
(continued)

Consider the problem of setting confidence limits on the mean time till breakdown of Nelson's insulating fluid tested at 30 kV. In this problem, Θ is $k = 1$-dimensional. So, for example, making use of the facts that the .9 quantile of the χ_1^2 distribution is 2.706 and that the maximum likelihood estimate of α turned out to be 14,467.3, displays (B-33), (B-34), and (B-37) suggest that those α having

$$L(\alpha) > -10\ln(14{,}467.3) - \frac{1}{14{,}467.3}(144{,}673) - \frac{1}{2}(2.706)$$

that is,

$$-10\ln(\alpha) - \frac{1}{\alpha}(144{,}673) > -107.15$$

form an approximate 90% confidence set for α. Figure B-19 shows a plot of the log likelihood (B-33) cut at the level -107.15 and the corresponding interval of α's. (Numerical solution of the equation

$$-10\ln(\alpha) - \frac{1}{\alpha}(144{,}673) = -107.15$$

shows the interval in question to extend from 8963 sec to 25,572 sec.)

FIGURE B-19

Plot of the Log Likelihood for Nelson's Insulating Fluid Breakdown Time Data and Approximate Confidence Limits for α

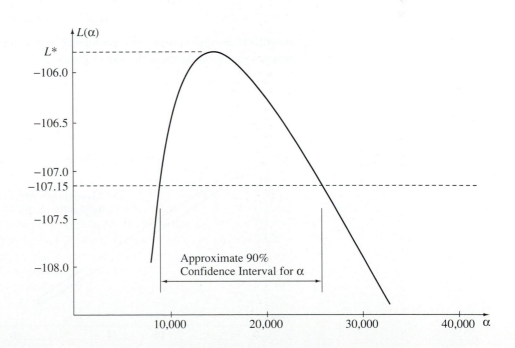

The $n = 12$ pieces of data in Table B-5 do not constitute an especially large sample, so the 90% approximate confidence level associated with the interval (8,963, 25,572) should be treated as very approximate. But even so, this interval does give one some feeling about the precision with which α is known based on the data of Table B-5. There is clearly substantial uncertainty associated with the estimate $\hat{\alpha} = 14{,}467.3$.

It is not a trivial matter to verify that the χ_k^2 approximations suggested here are adequate for a particular nonstandard probability model. In engineering situations where fairly exact confidence levels and/or p-values are critical, this author strongly suggests that readers seek genuinely expert statistical advice before placing too much faith in the χ_k^2 approximations. But for purposes of engineering problem solving requiring a rough, working quantification of uncertainty associated with parameter estimates, the use of the χ_k^2 approximation is certainly preferable to operating without any such quantification.

The insulating fluid example involved only a single parameter. As an example of a $k = 2$-parameter application, consider once again the space shuttle O-ring failure example.

EXAMPLE B-19
(continued)

Again use the log likelihood (B-23) and the fact that maximum likelihood estimates of α and β in equation (B-21) or (B-22) are $\hat{\alpha} = 15.043$ and $\hat{\beta} = -.2322$. These produce corresponding log likelihood -10.158. This, together with the fact that the .9 quantile of the χ_2^2 distribution is 4.605, gives one (from displays (B-34) and (B-37)), that the set of (α, β) pairs with

$$L(\alpha, \beta) > -10.158 - \frac{1}{2}(4.605)$$

that is,

$$L(\alpha, \beta) > -12.4605$$

constitutes an approximate 90% confidence region for (α, β). This set of possible parameter vectors is shown in the plot in Figure B-20.

FIGURE B-20

Likelihood-Based Approximate Confidence Region for the Parameters of the O-Ring Failure Model

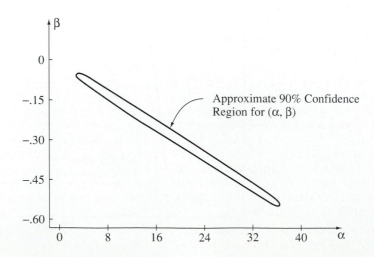

Notice that one message conveyed by the contour plot is that β is pretty clearly negative. Low-temperature launches are more prone to O-ring failure than moderate- to high-temperature launches.

The approximate inference methods represented in displays (B-34) through (B-38) concern the entire parameter vector Θ in cases where it is multidimensional. It is reasonably common, however, to desire inferences only for particular parameters individually. (For example, in the case of the O-rings, it is the parameter β that determines whether $p(t)$ is increasing, constant, or decreasing in t, and for many purposes β is of primary interest.) It is thus worth mentioning that the likelihood ideas discussed here can be adapted to provide inference methods for a *part* of a parameter vector Θ of individual interest. An exposition of these adaptations will not be attempted here, but be aware of their existence. For details, refer to more complete expositions of likelihood methods (such as that in Nelson's *Applied Life Data Analysis* text).

◼ EXERCISES

B-1. Return to the situation of Exercise 5-54, where measured diameters of a turned metal part were coded as Green, Yellow, or Red, depending upon how close they are to a mid-specification. Suppose that the probabilities that a given diameter falls into the various zones are .6247 for the Green Zone, .3023 for the Yellow Zone, and .0730 for the Red Zone. Suppose further (as in the problem in Chapter 5) that the lathe turning the parts is checked once per hour according to the following rules: One diameter is measured, and if it is in the Green Zone, no further action is needed that hour. If it is in the Red Zone, the process is immediately stopped. If it is in the Yellow Zone, a second diameter is measured. If the second diameter is in the Green Zone, no further action is necessary, but if it is not, the process is stopped immediately. Suppose further that the lathe is physically stable, so that it makes sense to think of successive color codes as independent.
(a) Show that the probability that the process is stopped in a given hour is .1865.
(b) Given that the process is stopped, what is the conditional probability that the first diameter was in the Yellow Zone?

B-2. A lot of 100 machine parts contains 10 with diameters that are out of specifications on the low side, 20 with diameters that are out of specifications on the high side, and 70 that are in specifications.
(a) How many different possible samples of $n = 10$ of these parts are there?
(b) How many different possible samples of size $n = 10$ are there that each contain exactly one part with diameter out of specifications on the low side, two parts with diameters out of specifications on the high side, and seven parts with diameters that are in specifications?
(c) Based on your answers to (a) and (b), what is the probability that a simple random sample of $n = 10$ of these contains exactly one part with diameter out of specifications on the low side, two parts with diameters out of specifications on the high side, and seven parts with diameters that are in specifications?

B-3. The lengths of bolts produced in a factory are checked with two "go–no go" gages and the bolts sorted into piles of short, OK, and long bolts. Suppose that of the bolts produced, about 20% are short, 30% are long, and 50% are OK.
(a) Find the probability that among the next ten bolts checked, the first three are too short, the next three are OK, and the last four are too long.
(b) Find the probability that among the next ten bolts checked, there are three that are too short, three that are OK, and four that are too long. (*Hint:* In how many ways can one choose three of the group to be short, three to be OK, and four to be long? Then use your answer to (a).)

B-4. A bin of nuts is mixed, containing 30% $\frac{1}{2}$ in. nuts and 70% $\frac{9}{16}$ in. nuts. A bin of bolts has 40% $\frac{1}{2}$ in. bolts and 60% $\frac{9}{16}$ in. bolts. Suppose that one bolt and one nut are selected (independently and at random) from the two bins.
(a) What is the probability that the nut and bolt match?
(b) What is the conditional probability that the nut is a $\frac{9}{16}$ in. nut, given that the nut and bolt match?

B-5. A physics student is presented with six unmarked specimens of radioactive material. She knows that two are of substance A and four are of substance B. Further, she knows that when tested with a Geiger counter, substance A will produce an average of 3 counts per second, while substance B will produce an average of 4 counts per second. (Use Poisson models for the counts per time period.)
(a) Suppose the student selects a sample at random and makes a one-second check of radioactivity. If 1 count is observed, how should the student assess the (conditional) probability that the specimen is of substance A?
(b) Suppose the student selects a sample at random and makes a ten-second check of radioactivity. If 10 counts are observed, how should the student assess the (conditional) probability that the specimen is of substance A?
(c) Are your answers to (a) and (b) the same? How should this be understood?

B-6. User names on a computer system consist of three letters A through Z, followed by two digits 0 through 9.

(a) How many user names of this type are there?

(b) Suppose that Joe has user name TPK66, but unfortunately he's forgotten it. Joe remembers only the format of the user names and that the letters K, P, and T appear in his name. If he picks a name at random from those consistent with his memory, what's the probability that he selects his own?

(c) If Joe in part (b) also remembers that his digits match, what's the probability that he selects his own?

B-7. A lot contains ten pH meters, three of which are miscalibrated. A technician selects these meters one at a time, at random without replacement, and checks their calibration.

(a) What is the probability that among the first four meters selected, exactly one is miscalibrated?

(b) What is the probability that the technician discovers his second miscalibrated meter when checking his fifth one?

B-8. At final inspection of certain integrated circuit chips, 20% of the chips are in fact defective. An automatic testing device does the final inspection. Its characteristics are such that 95% of good chips test as good. Also, 10% of the defective chips test as good.

(a) What is the probability that the next chip is good and tests as good?

(b) What is the probability that the next chip tests as good?

(c) What is the (conditional) probability that the next chip that tests as good is in fact good?

B-9. In the process of producing piston rings, the rings are subjected to a first grind. Those rings whose thicknesses remain above an upper specification are reground. The history of the grinding process has been that on the first grind,

> 60% of the rings meet specifications (and are done processing)
>
> 25% of the rings are above the upper specification (and are reground)
>
> 15% of the rings are below the lower specification (and are scrapped)

The history has been that after the second grind,

> 80% of the reground rings meet specifications
>
> 20% of the reground rings are below the lower specification (and are scrapped)

A ring enters the grinding process today.

(a) Evaluate P[the ring is ground only once].

(b) Evaluate P[the ring meets specifications].

(c) Evaluate P[the ring is ground only once | the ring meets specifications].

(d) Are the events {the ring is ground only once} and {the ring meets specifications} independent events? Explain.

(e) Describe any two mutually exclusive events in this situation.

B-10. A lot of machine parts is checked piece by piece for Brinell hardness and diameter, with the resulting counts as shown in the accompanying table.

A single part is selected at random from this lot.

(a) What is the probability that it is more than 1.005 in. in diameter?

(b) What is the probability that it is more than 1.005 in. in diameter and has Brinell hardness of more than 210?

		DIAMETER		
		< 1.000 in.	1.000 to 1.005 in.	> 1.005 in.
BRINELL HARDNESS	< 190	154	98	48
	190–210	94	307	99
	> 210	33	72	95

(c) What is the probability that it is more than 1.005 in. in diameter or has Brinell hardness of more than 210?

(d) What is the conditional probability that it has a diameter over 1.005 in., given that its Brinell hardness is over 210?

(e) Are the events {Brinell hardness over 210} and {diameter over 1.005 in.} independent? Explain.

(f) Name any two mutually exclusive events in this situation.

B-11. An engineer begins a series of presentations to his corporate management with a working bulb in his slide projector and (an inferior quality) Brand W replacement bulb in his briefcase. Suppose that the random variables

> X = the number of hours of service given by the bulb in the projector
>
> Y = the number of hours of service given by the spare bulb

may be modeled as independent exponential random variables with respective means 15 and 5. The number of hours that the engineer may operate without disaster is $X + Y$.

(a) Find the mean and standard deviation of $X + Y$ using Proposition 5-1.

(b) Find, for $t > 0$, $P[X + Y \leq t]$.

(c) Use your answer to (b) and find the probability density for $T = X + Y$.

(d) Find the survivorship and force-of-mortality functions for T. What is the nature of the force-of-mortality function? Is it constant like those of X and Y?

B-12. Widgets produced in a factory can be classified as defective, marginal, or good. At present, a machine is producing about 5% defective, 15% marginal, and 80% good widgets. An engineer plans the following method of checking on the machine's adjustment: Two widgets will be sampled initially, and if either is defective, the machine will be immediately adjusted. If both are good, testing will cease without adjustment. If neither of these first two possibilities occurs, an additional three widgets will be sampled. If all three of these are good, or two are good and one is marginal, testing will cease without machine adjustment. Otherwise, the machine will be adjusted.

(a) Evaluate P[only two widgets are sampled and no adjustment is made].

(b) Evaluate P[only two widgets are sampled].

(c) Evaluate P[no adjustment is made].

(d) Evaluate P[no adjustment is made | only two widgets are sampled].

(e) Are the events {only two widgets are sampled} and {no adjustment is made} independent events? Explain.

(f) Describe any two mutually exclusive events in this situation.

C

Answers to Selected Exercises

CHAPTER 1

1-6. The relationship between the variables x and y.

1-8. Full factorial data structure: (delta, construction, with clip), (t-wing, construction, with clip), (delta, typing, with clip), (t-wing, typing, with clip), (delta, construction, without clip), (t-wing, construction, without clip), (delta, typing, without clip), (t-wing, typing, without clip).

Fractional factorial data structure — one possibility is to choose the following "$\frac{1}{2}$ fraction": (delta, construction, without clip), (t-wing, construction, with clip), (delta, typing, with clip), (t-wing, typing, without clip).

1-9. You might be interested in monitoring the strength of the concrete. If you study 4 batches of concrete, taking from each batch 3 specimens and then making 2 strength measurements on each specimen, the data would have a nested/hierarchical structure.

1-15. (A, B, C) combinations: (1, 1, 1), (2, 1, 1), (3, 1, 1), (1, 2, 1), (2, 2, 1), (3, 2, 1), (1, 1, 2), (2, 1, 2), (3, 1, 2), (1, 2, 2), (2, 2, 2), (3, 2, 2).

1-16. If each alloy specimen is measured for hardness before and after heat treating, the data would be paired (according to specimen).

1-22. (a) Rockwell hardness: paired, quantitative data. Flatness: univariate, qualitative data.

(b) There are many possibilities. If you choose Vendor (1 vs. 2), Heating Time (short vs. long), and Cooling Method (1 vs. 2), the factor-level combinations are: (1, short, 1), (2, short, 1), (1, long, 1), (2, long, 1), (1, short, 2), (2, short, 2), (1, long, 2), (2, long, 2).

CHAPTER 2

2-4. Label the 38 runout values consecutively, 1 . . . 38, in the order given in Table 1-1 (smallest to largest). 1st sample labels: {12, 15, 05, 09, 11}; 1st sample runout values: {11, 11, 9, 10, 11}. 2nd sample labels: {34, 31, 36, 02, 14}; 2nd sample runout values: {17, 15, 18, 8, 11}. 3rd sample labels: {10, 35, 12, 27, 30}; 3rd sample runout values: {10, 17, 11, 14, 15}. 4th sample labels: {15, 05, 19, 11, 08}; 4th sample runout values: {11, 9, 12, 11, 10}. The samples are not identical. Note: the population mean is 12.63; the sample means are 10.4, 13.8, 13.4, and 10.6.

2-6. Advantage: may reduce baseline variation (background noise) in the response, making it easier to see the effects of factors. Disadvantage: the variable may fluctuate in the real world, so controlling it makes the experiment more artificial — it will be harder to generalize conclusions from the experiment to the real world.

2-8. List the tests for Juanita in the same order given for Exercise 1-8. Then list the tests for Tom after Juanita, again in the same order. Label the tests consecutively 1 . . . 16, in the order listed. Let the digits 01–05 refer to test 1, 06–10 to test 2, . . . , and 76–

80 to test 16. Move through Table D-1 choosing two digits at a time. Ignore previously chosen test labels or numbers between 81 and 00. Order the tests in the same order that their corresponding two-digit numbers are chosen from the table. Using this method (and starting from the upper left of the table), the test labeled 3 (Juanita, delta, typing, with clip) would be first, followed by the tests labeled 13, 9, 1, 2, 7, 10, 8, 14, 11, 6, 15, 4, 16, 12, and 5.

2-13. With (delta, construction) corresponding to the upper-left corner of the square, put Tom on the upper-left to lower-right diagonal: (Tom, delta, construction), (Tom, t-wing, typing), (Juanita, delta, typing), (Juanita, t-wing, construction).

2-17. Blocking is a way of supervising an extraneous variable: within each block, there may be less baseline variation (background noise) in the response than there would be if the variable were not supervised. This makes it easier to see the effects of the factors of interest within each block. Any effect of the extraneous variable can be isolated and distinguished from the effects of the factors of interest. Compared to holding the variable constant throughout the experiment, blocking also results in a more realistic experiment.

2-18. Replication is used to estimate the magnitude of baseline variation (background noise, experimental error) in the response. It thus helps sharpen and validate conclusions drawn from data.

2-21. (a) Label the gears 1 . . . 20. Let the digits 01–05 refer to gear 1, 06–10 to gear 2, . . . , and 96–00 to gear 20. Move through Table D-1 choosing two digits at a time, until 10 different gears are chosen. These gears will be laid; the remaining 10 gears will be hung. Using this method, the gears to be laid are the ones labeled 3, 20, 13, 9, 1, 2, 7, 10, 8, and 17.

(b) This will guard against bias. Unbeknownst to the experimenter, a biased method of assignment may surely assign most of the "good" gears to one group and most of the "bad" gears to the other. There is only a small chance that this type of assignment will result from randomization (but it is possible). Randomization attempts to average between samples for treatments of interest the effects of variables that are not managed.

2-29. (a) See Exercise 1-22(b). Two of the possible responses would be flatness and concentricity. Replication dictates that at least one of the 8 factor-level combinations given in 1-22(b) be run at least twice. One possibility is to run each factor-level combination twice, for a total of 16 runs.

CHAPTER 3

3-1. One choice of intervals for the frequency table and histogram is 65.5–66.4, 66.5–67.4, . . . , 73.5–74.4. For this choice, the frequencies are 3, 2, 9, 5, 8, 6, 2, 3, 2; the relative frequencies are .075, .05, .225, .125, .2, .15, .05, .075, .05; the cumulative relative frequencies are .075, .125, .35, .475, .675, .825, .875, .95, 1. The plots reveal a fairly symmetric, bell-shaped distribution.

3-2. The plot shows that the depths for 200 grain bullets are larger and have less variability than those for the 230 grain bullets.

3-3. (a) There are no obvious patterns.
(b) The differences are -15, 0, -20, 0, -5, 0, -5, 0, -5, 20, -25, -5, -10, -20, and 0. The dot diagram shows that most of the differences are negative and truncated at 0. The exception is the 10th piece of equipment, with a difference of 20. This point does not fit in with the distribution of the rest of the differences, so it is an outlier. Since most of the differences are negative, the bottom bolt generally required more torque than the top bolt.

3-4. (a) For the lengthwise sample: median $= .895$, $Q(.25) = .870$, $Q(.75) = .930$, $Q(.37) = .880$. For the crosswise sample: median $= .775$, $Q(.25) = .690$, $Q(.75) = .800$, $Q(.37) = .738$.
(b) On the whole, the impact strengths are larger and more consistent for lengthwise cuts. Each method also produced an unusual impact strength value (outlier).
(c) The nonlinearity of the Q-Q plot indicates that the overall shapes of these two data sets are not the same. The lengthwise cuts had an unusually large data point ("long right tail"), whereas the crosswise cuts had an unusually small data point ("long left tail"). Without these two outliers, the data sets would have similar shapes, since the rest of the Q-Q plot is fairly linear.

3-5. Use the $\frac{i-.5}{n}$ quantiles for the smaller data set. The plotted points have coordinates: $(.370, .907)$, $(.520, 1.22)$, $(.650, 1.47)$, $(.920, 1.70)$, $(2.89, 2.45)$, $(3.62, 5.89)$.

3-6. The first 3 plotted points are: $(65.6, -2.33)$, $(65.6, -1.75)$, $(66.2, -1.55)$. The normal plot is quite linear, indicating that the distribution is very bell-shaped.

3-7. For the lengthwise cuts: $\bar{x} = .919$, median $= .895$, $R = .310$, $IQR = .060$, $s = .088$. For the crosswise cuts: $\bar{x} = .743$, median $= .775$, $R = .430$, $IQR = .110$, $s = .120$. The sample means and medians show that the center of the distribution for lengthwise cuts is higher than the center for crosswise cuts. The sample ranges, interquartile ranges, and sample standard deviations show that there is less spread in the lengthwise data than in the crosswise data.

3-8. These values are statistics. They are summarizations of two samples of data; they do not represent exact summarizations of larger populations or theoretical (long-run) distributions.

3-9. In the first case, the sample mean and median increase by 1.3, but none of the measures of spread change; in the second case, all of the measures double.

3-15. (c) Even if you ignore the outlier, the distribution is right-skewed (not bell-shaped).
(d) median $= 119$, $Q(.25) = 87$, $Q(.75) = 182$, $Q(.58) = 122.5$.
(e) The median is closer to $Q(.25)$ than $Q(.75)$, and the upper whisker is longer than the lower whisker. Both of these features indicate that the distribution is right-skewed. The point for the 20th time period lies above $Q(.75) + 1.5(IQR)$ and should be plotted separately as an outlier.
(f) The coordinates of the first 3 plotted points are: $(30, -2.05)$, $(30, -1.55)$, $(60, -1.28)$. The plot is quite nonlinear, indicating that the data distribution is not bell-shaped. The data are more bunched up at low values and more spread out at high values than

they would be if the distribution were bell-shaped. This again shows that the distribution is right-skewed.
(h) For the original data: $\bar{x} = 142.65$, $R = 481$, $s = 98.20$. For the transformed data: $\bar{x} = 4.779$, $R = 2.835$, $s = .631$. No. Since the transformed data are more symmetric, their mean is not "pulled up" as much as the mean of the original data.

3-26. (a) median $= 5.53$, $Q(.25) = 5.48$, $Q(.75) = 5.59$, $\bar{x} = 5.514$, $s = .123$. Both plots reveal an outlier.
(b) The first 3 plotted points are: $(5.22, -1.75)$, $(5.37, -1.13)$, $(5.45, -.81)$. The plot is roughly linear, but it would be more linear if the lower left point were pushed to the right. This means that the lower tail of the distribution is longer than it would be for a bell-shaped distribution.

3-27. (a) median $= 21$, $Q(.25) = 17$, $Q(.75) = 25$, $Q(.64) = 22.9$.
(b) The points are: $(14, -1.65)$, $(16, -1.04)$, $(17, -.67)$, $(18, -.39)$, $(20, -.13)$, $(22, .13)$, $(23, .39)$, $(25, .67)$, $(27, 1.04)$, $(28, 1.65)$. The plot is fairly linear, but there is some indication that the data distribution has shorter tails than a bell-shaped distribution (the smallest and largest points would need to be "moved out" to make a perfectly straight line).
(c) $\bar{x} = 21$, $R = 14$, $s = 4.78$.
(e) For the Lab 2 data: median $= 29.5$, $Q(.25) = 29$, $Q(.75) = 31$.
(f) Both plots show that there is less spread in the Lab 2 data, so it produced the more precise results.
(g) No. You would need to know the "true" long-run average lifetime of this type of specimen in order to determine which lab is more accurate.

3-30. (a) For Method A: median $= 80.03$, $Q(.25) = 80.015$, $Q(.75) = 80.04$. For Method B: median $= 79.97$, $Q(.25) = 79.96$, $Q(.75) = 80.00$. There does not seem to be any important difference in the precisions of the two methods, but Method A generally produced larger values than Method B. Since there is some fixed, true, theoretical latent heat for the fusion of ice, at least one of the methods must be somewhat inaccurate.
(b) $\bar{x}_A = 80.021$, $s_A = .024$, $\bar{x}_B = 79.979$, $s_B = .031$. The sample standard deviations are similar, as reflected by the similar magnitudes of spread in the boxplots. $\bar{x}_A > \bar{x}_B$, as reflected by the locations of the boxes on the plot.

CHAPTER 4

4-1. (a) $\hat{y} = 9.4 - 1.0x$. (b) $r = -.945$.
(c) $r = .945$. This is the negative of the r in part (b), since the $\hat{y}$'s are perfectly negatively correlated with the x's.
(d) $R^2 = .893 = r^2$ (from both (b) and (c)).
(e) $-.4$, $.6$, $-.4$, $.6$, $-.4$. These are the vertical distances from each data point to the least squares line.

4-3. $b = \dfrac{\sum_{i=1}^{n} x_i y_i}{\sum_{i=1}^{n} x_i^2} = 1.13$; $\hat{y} = 1.13x$.

4-5. (a) $R^2 = .994$. (b) $\hat{y} = -3174.6 + 23.50x$. 23.5.
(c) Residuals: 105.36, -21.13, -60.11, -97.58, 16.95, 14.48, 42.00, $.02$.

(d) There is no replication (multiple experimental runs at a particular pot temperature).

(e) For $x = 188°C$, $\hat{y} = 1243.1$. For $x = 200°C$, $\hat{y} = 1525.1$. It would not be wise to make a similar prediction at $x = 70°C$, because there is no evidence that the fitted relationship is correct for pot temperatures as low as $x = 70°C$. Some data should be obtained around $x = 70°C$ before making such a prediction.

4-6. (a) The scatterplot is not linear, so the given straight-line relationship does not seem appropriate. $R^2 = .723$.

(b) This scatterplot is much more linear, and a straight-line relationship seems appropriate for the transformed variables. $R^2 = .965$.

(c) $\widehat{\ln y} = 34.344 - 5.1857 \ln x$. For $x = 550$, $\widehat{\ln y} = 1.6229$, so $\hat{y} = e^{1.6229} = 5.07$ min. The implied relationship between x and y is $y = e^{\beta_0} x^{\beta_1}$.

4-7. $\hat{y} = -1315 + 5.6x + .04212x^2$. $R^2 = .996$. For the quadratic model, at $x = 200°C$, $\hat{y} = 1487.2$, which is relatively close to 1525.1 from 4-5 (e).

4-8. (a) $\hat{y} = 6.0483 + .14167x_1 - .016944x_2$. $b_1 = .14167$ means that as x_1 increases by 1% (holding x_2 constant), y increases by roughly $.142$ cm^3/g. $b_2 = -.016944$ means that as x_2 increases by 1 min (holding x_1 constant), y decreases by roughly $.017$cm^3/g. $R^2 = .807$.

(b) The residuals are $-.015, .143, .492, -.595, -.457, -.188, .695, .143, -.218$.

(c) For $x_2 = 30$, the equation is $\hat{y} = 5.53998 + .14167x_1$. For $x_2 = 60$, the equation is $\hat{y} = 5.03166 + .14167x_1$. For $x_2 = 90$, the equation is $\hat{y} = 4.52334 + .14167x_1$. The fitted responses do not match up well, because the relationship between y and x_1 is not linear for any of the x_2 values.

(d) At $x_1 = 10\%$ and $x_2 = 70$ min, $\hat{y} = 6.279$ cm^3/g. It would not be wise to make a similar prediction at $x_1 = 10\%$ and $x_2 = 120$ min, because there is no evidence that the fitted relationship is correct under these conditions. Some data should be obtained around $x_1 = 10\%$ and $x_2 = 120$ min before making such a prediction.

(e) $\hat{y} = 4.98 + .260x_1 + .00081x_2 - .00197x_1x_2$, and $R^2 = .876$. The increase in R^2 from .807 to .876 is not very large; using the more complicated equation may not be desirable (this is subjective).

(f) For $x_2 = 30$, the equation is $\hat{y} = 5.0076 + .20084x_1$. For $x_2 = 60$, the equation is $\hat{y} = 5.0319 + .14168x_1$. For $x_2 = 90$, the equation is $\hat{y} = 5.0562 + .08252x_1$. The new model allows there to be a different slope for different values of x_2, so these lines fit the data better than the lines in part (c). But they still do not account for the nonlinearity between x_1 and y. An x_1^2 term should be added to the model.

(g) There is no replication (multiple experimental runs at a particular NaOH-Time combination).

4-9. (a) Labeling x_1 as A and x_2 as B, $a_1 = -.643$, $a_2 = -.413$, $a_3 = 1.057$, $b_1 = .537$, $b_2 = -.057$, $b_3 = -.480$, $ab_{11} = -.250$, $ab_{12} = -.007$, $ab_{13} = .257$, $ab_{21} = -.210$, $ab_{22} = .013$, $ab_{23} = .197$, $ab_{31} = .460$, $ab_{32} = -.007$, $ab_{33} = -.453$. The fitted interactions ab_{31} and ab_{33} are large (relative to fitted main effects), indicating that the effect on y of changing NaOH from 9% to 15%

depends on the Time (nonparallelism in the plot). It would not be wise to use the fitted main effects alone to summarize the data, since there may be an importantly large interaction.

(b) $\hat{y}_{11} = 6.20$, $\hat{y}_{12} = 5.61$, $\hat{y}_{13} = 5.18$, $\hat{y}_{21} = 6.43$, $\hat{y}_{22} = 5.84$, $\hat{y}_{23} = 5.41$, $\hat{y}_{31} = 7.90$, $\hat{y}_{32} = 7.31$, $\hat{y}_{33} = 6.88$. Like the plot in 4-8(c), and unlike the plot in 4-8(f), the fitted values for each level of B (x_2) must be parallel; no interactions are allowed. However, unlike 4-8(c) or 4-8(f), the current model allows these fitted values to be nonlinear in x_1.

(c) $R^2 = .914$.

4-10. (a) $\bar{y}_{...} = 20.792$, $a_2 = .113$, $b_2 = -13.807$, $ab_{22} = -.086$, $c_2 = 7.081$, $ac_{22} = -.090$, $bc_{22} = -6.101$, $abc_{222} = .118$. Other fitted effects can be obtained by appropriately changing the signs of the above. The simplest possible interpretation is that Diameter, Fluid, and their interaction are the only effects on time.

(b) $\bar{y}_{...} = 2.699$, $a_2 = .006$, $b_2 = -.766$, $ab_{22} = -.003$, $c_2 = .271$, $ac_{22} = -.003$, $bc_{22} = -.130$, $abc_{222} = .007$. Yes, but the Diameter $\times$ Fluid interaction still seems to be important.

(c) In standard order, the fitted values are 3.19, 3.19, 1.66, 1.66, 3.74, 3.74, 2.20, 2.20. $R^2 = .974$. For a model with all factorial effects ($\widehat{\ln y_{ijk}} = \overline{\ln y_{ijk}}$), $R^2 = .995$.

(d) $b_1 - b_2 = 1.532 \ln$ (sec) decrease; divide the .188 raw drain time by $e^{1.532}$ to get the .314 drain time. This suggests that (.188 drain time/.314 drain time) $= e^{1.532} = 4.63$; the theory predicts this ratio to be 7.78.

4-15. (a) $\hat{y} = 4345.9 - 3160.0x$.

(b) $r = -.984$. Since r is close to -1 and negative, there is a strong negative linear relationship between Water/Cement Ratio and 14-Day Compressive Strength.

(c) $R^2 = .968$.

(d) The residuals are 30.11, -10.89, $-.089$, -22.89, 13.11, -26.89, 44.11, $-.89$, -24.90. The normal plot of residuals is fairly linear; this implies that the distribution of residuals is roughly bell-shaped.

(e) $\hat{y} = 2829.09$ psi.

4-16. (a) There is no replication (multiple experimental runs at a particular jetting size).

(b) $\hat{y}_{line} = 13.731 + .01414x$. $\hat{y}_{quad} = 103.989 - 2.5343x + .017946x^2$.

(c) $R^2_{line} = .066$; $R^2_{quad} = .969$.

(d) $x_{opt} = 70.61$.

4-17. (a) A full factorial data structure. There is no replication (multiple experimental runs at a particular Temp-Time combination).

(b) $\hat{y} = -515.15 + .35667x_1 + .01069x_2$; $\hat{y} = -528.46 + .35667x_1 + 3.711 \ln x_2$; $\hat{y} = -42.4 + .0311x_1 - 93.72 \ln x_2 + .06526x_1 \ln x_2$. R^2 values: .886, .889, .962.

(d) For $x_1 = 1443$, the equation is $\hat{y} = 2.4773 + .45018 \ln x_2$. For $x_1 = 1493$, the equation is $\hat{y} = 4.0323 + 3.71318 \ln x_2$. For $x_1 = 1543$, the equation is $\hat{y} = 5.5873 + 6.97618 \ln x_2$. The 3rd equation allows there to be a different slope for different values of x_1; a similar plot for the 2nd equation would have parallel lines, because the 2nd equation does not allow for different slopes.

(e) 30.16 μm.

(f) The nonparallelism in the plot indicates that there may be an interaction between Temp and Time. The effect on y of changing Time depends on the Temp. It would not be wise to use the fitted main effects alone to summarize the data, since there may be an importantly large interaction. Labeling x_1 as A and x_2 as B, $a_1 = -15.89$, $a_2 = -3.89$, $a_3 = 19.78$, $b_1 = -6.56$, $b_2 = -2.22$, $b_3 = 8.78$, $ab_{11} = 4.89$, $ab_{12} = 1.56$, $ab_{13} = -6.44$, $ab_{21} = 1.89$, $ab_{22} = .56$, $ab_{23} = -2.44$, $ab_{31} = -6.78$, $ab_{32} = -2.11$, $ab_{33} = 8.89$.

4-18. (a) There is no replication (multiple experimental runs at a particular factor-level combination).
(b) $\bar{y}_{...} = 11.01$, $a_2 = 3.96$, $b_2 = 1.01$, $ab_{22} = .46$, $c_2 = .11$, $ac_{22} = .66$, $bc_{22} = -.59$, $abc_{222} = -.24$. The fitted main effects for A and B are much larger than the rest. The positive signs of a_2 and b_2 indicate that setting A and B at their ($+$) levels (1.2 and 40 mol %, respectively) results in the largest impact strength.
(c) The fitted values are 7.050 for combinations with A ($-$) and 14.975 for combinations with A ($+$). $R^2 = .882$.
(d) $\hat{y} = 3.088 + 9.906x$. The fitted values and R^2 are the same as in (c); since there are only 2 levels for A, this model fits the same as the one in (c).

4-23. (a) $\bar{y}_{....} = 17.625$, $a_2 = 11.50$, $b_2 = -6.0$, $ab_{22} = -4.625$, $c_2 = -.625$, $ac_{22} = -1.50$, $bc_{22} = -2.25$, $abc_{222} = -1.625$, $d_2 = -10.250$, $ad_{22} = -7.125$, $bd_{22} = 2.875$, $abd_{222} = 2.50$, $cd_{22} = 3.750$, $acd_{222} = 3.625$, $bcd_{222} = -.625$, $abcd_{2222} = -1.250$. The dominant effects seem to be the A, B, and D main effects and the A $\times$ D and A $\times$ B interactions.
(b) Using the fitted effects and interactions identified in (a), the fitted responses are (in standard order): 10.6, 57.1, 7.9, 35.9, 10.6, 57.1, 7.9, 35.9, 4.4, 22.4, 1.6, 1.1, 4.3, 22.4, 1.6, 1.1.
(c) To minimize the number of pull-outs for the above model (based on the smallest fitted responses), set A at glazed ($+$), B at 1.5 times normal ($+$), D at no clean ($+$), and C at its cheapest level.

4-26. (a) $\widehat{\ln y} = 18.750 - 5.1209 \ln x_1 - 3.7379 \ln x_2$. $R^2 = .960$. $\hat{\alpha}_1 = -b_1 = 5.1209$, $\hat{\alpha}_2 = -b_2 = 3.7379$, and $\hat{C} = e^{b_0} = 1.39 \times 10^8$.
(c) $\widehat{\ln y} = 1.7789$, so $\hat{y} = e^{1.7789} = 5.92$ min.

4-28. (a) $\widehat{\ln y} = 20.539 - 1.2060 \ln x_1 - 1.3988 \ln x_2$. $R^2 = .782$. $\hat{\alpha}_1 = -b_1 = 1.2060$, $\hat{\alpha}_2 = -b_2 = 1.3988$, and $\hat{C} = e^{b_0} = 8.317 \times 10^8$. For $x_2 = 3000$, the equation is $y = 11381 x_1^{-1.2060}$; for $x_2 = 6000$, the equation is $y = 4316 x_1^{-1.2060}$; for $x_2 = 10,000$, the equation is $y = 2112 x_1^{-1.2060}$.
(c) $\widehat{\ln y} = 4.043$, so $\hat{y} = 57.00$ hr. (d) $\widehat{\ln y} = 7.043$, so $\hat{y} = 1145$ hr.

4-34. (a) A full factorial data structure. There is no replication (multiple experimental runs at a particular (x_1, x_2) combination).
(b) $\hat{y}_1 = 96.621 - .01570x_1$; $\hat{y}_1 = 100.61 - .1917x_2$; $\hat{y}_1 = 115.788 - .01570x_1 - .1917x_2$. R^2 values: .537, .227, .764.
(d) For $x_1 = 1350$, the equation is $\hat{y}_1 = 94.593 - .19167x_2$. For $x_1 = 950$, the equation is $\hat{y}_1 = 100.873 - .19167x_2$. For $x_1 = 600$, the equation is $\hat{y}_1 = 106.368 - .19167x_2$.
(e) $\hat{y}_1 = 79.00\%$.
(f) $\hat{y}_1 = -17.92 + .01952x_1 + 2.235x_2 - .00001746x_1^2 - .012083x_2^2 - .0000104x_1 x_2$. $R^2 = .915$. For $x_1 = 1350$, the equa-

tion is $\hat{y}_1 = -23.37535 + 2.22096x_2 - .012083x_2^2$. For $x_1 = 950$, the equation is $\hat{y}_1 = -15.12415 + 2.22512x_2 - .012083x_2^2$. For $x_1 = 600$, the equation is $\hat{y}_1 = -12.4876 + 2.22876x_2 - .012083x_2^2$.
(g) The plot shows that the effect on y_1 of changing Speed depends on the Concentration (there seems to be an interaction). It would not be wise to use the fitted main effects alone to summarize the data, since there may be an importantly large interaction. Labeling x_1 as A and x_2 as B, $a_1 = -6.78$, $a_2 = 1.89$, $a_3 = 4.89$, $b_1 = 2.22$, $b_2 = 3.22$, $b_3 = -5.44$, $ab_{11} = .11$, $ab_{12} = 2.11$, $ab_{13} = -2.22$, $ab_{21} = -2.56$, $ab_{22} = .44$, $ab_{23} = 2.11$, $ab_{31} = 2.44$, $ab_{32} = -2.56$, $ab_{33} = .11$.
(h) No. The 3rd equation does not allow the effect on y_1 of changing x_2 to depend on x_1, as can be seen from the plot in (d). The equation in part (f) does allow this, since the plotted curves in (f) are (slightly) nonparallel.

CHAPTER 5

5-1. For $p = .1$, $f(0)–f(5)$ are .59, .33, .07, .01, .00, .00; $\mu = np = .5$; $\sigma = \sqrt{np(1-p)} = .67$. For $p = .3$, $f(0)–f(5)$ are .17, .36, .31, .13, .03, .00; $\mu = 1.5$; $\sigma = 1.02$. For $p = .5$, $f(0)–f(5)$ are .03, .16, .31, .31, .16, .03; $\mu = 2.5$; $\sigma = 1.12$. For $p = .7$, $f(0)–f(5)$ are .00, .03, .13, .31, .36, .17; $\mu = 3.5$; $\sigma = 1.02$. For $p = .9$, $f(0)–f(5)$ are .00, .00, .01, .07, .33, .59; $\mu = 4.5$; $\sigma = .67$.

5-2. (a) .531. (b) .984. (c) .016. (d) 5.4. (e) .54; .735.

5-4. For $\lambda = .5$, $f(0), f(1), \ldots$ are .61, .30, .08, .01, .00, .00, $\ldots$; $\mu = \lambda = .5$; $\sigma = \sqrt{\lambda} = .71$. For $\lambda = 1.0$, $f(0), f(1), \ldots$ are .37, .37, .18, .06, .02, .00, .00, $\ldots$; $\mu = 1.0$; $\sigma = 1.0$. For $\lambda = 2.0$, $f(0), f(1), \ldots$ are .14, .27, .27, .18, .09, .04, .01, .00, .00, $\ldots$; $\mu = 2.0$; $\sigma = 1.41$. For $\lambda = 4.0$, $f(0), f(1), \ldots$ are .02, .07, .15, .20, .20, .16, .10, .06, .03, .01, .01, .00, $\ldots$; $\mu = 4.0$; $\sigma = 2.0$.

5-5. (a) .323. (b) .368.

5-6. (a) .0067. (b) Y is binomial ($n = 4, p = .0067$); .00027.

5-8. (a) $\frac{2}{9}$. (c) .5.

(d) $F(x) = \begin{cases} 0 & \text{for } x \leq 0 \\ \dfrac{10x - x^2}{9} & \text{for } 0 < x < 1 \\ 1 & \text{for } x \geq 1 \end{cases}$

(e) $\frac{13}{27}$; .288.

5-9. (a) .2676. (b) .1446. (c) .3393. (d) .3616. (e) .3524. (f) .9974. (g) 1.28. (h) 1.645. (i) 2.17.

5-10. (a) .7291. (b) .3594. (c) .2794. (d) .4246. (e) .6384. (f) 48.922. (g) 44.872. (h) 7.056.

5-11. (a) .4938. (b) Set μ to the midpoint of the specifications: $\mu = 2.0000$; .7888. (c) .0002551.

5-12. (a) $\mu \approx 69.6$; $\sigma \approx 1/\text{slope} = 2.1$.

5-13. (a) The coordinates of the first 3 points on the normal plot of the raw data: $(17.88, -2.05)$, $(28.92, -1.48)$, $(33.00, -1.23)$. The normal plot is not linear, so a Gaussian (normal) distribution does not seem to fit these data. First 3 points on the normal plot of the natural log of the data: $(2.884, -2.05)$, $(3.365, -1.48)$, $(3.497, -1.23)$. This normal plot is fairly linear, indicating that a lognormal distribution fits the data well. $\mu \approx 4.15$, $\sigma \approx .54$. 3.26, 26.09.

(b) The first 3 points on the Weibull plot are $(2.88, -3.82)$, $(3.36, -2.70)$, $(3.50, -2.16)$. The Weibull plot is fairly linear, indicating that a Weibull distribution might be used to describe bearing load life. $\alpha \approx 81.12$, $\beta \approx 2.30$; 21.31.

5-15. mean $= .75$ in.; standard deviation $= .0037$ in.

5-16. (a) Propagation of error formula gives 1.415×10^{-6}.
(b) The lengths.

5-18. (a) $\frac{13}{27}$; .0576.
(b) $\bar{X}$ is normal with mean $\frac{13}{27}$ and std. deviation .0576.
(c) .3745. **(d)** .2736.
(e) $\frac{13}{27}$, .0288; $\bar{X}$ is normal with mean $\frac{13}{27}$ and std. deviation .0288; .2611; .5098.

5-19. .7888, .9876, 1.0000.

5-20. (b) 4.1; 1.136.

5-26. 3.148332.

5-28. (a) 8.792×10^{-9}. **(b)** Length.
(c) Changing lab conditions (especially those that affect temperature) may distort the relationship, creating sources of variability other than measurement error.

5-29. 37 mm; .0577 mm.

5-32. (a) .4972. **(b)** .9992. **(c)** .809.

5-44. (a) 2,311,904. **(b)** L_0 and L_1.
(c) It predicts that they will, because the effect on ΔL of changing Force will depend on the Diameter used.

5-47. $\sigma < .00102$ cm.

5-50. (b) The exponential plot is fairly linear, indicating that an exponential distribution fits the data well. Since a line on the plot indicates that $Q(0) \approx 0$, no need for a threshold parameter greater than 0 is indicated.
(c) .3935; .1353. **(d)** 51.29; 2302.56.

5-52. (a) The first 3 points are $(2.30, -3.51)$, $(2.48, -2.38)$, $(2.71, -1.84)$. The plot is pretty linear, so the Weibull model seems reasonable.
(b) $\alpha \approx 23.83$, $\beta \approx 3.70$. **(c)** 12.97.

5-55. (a) $f_X(x) = \begin{cases} 2x & \text{for } 0 \leq x \leq 1 \\ 0 & \text{otherwise} \end{cases}$
$f_Y(y) = \begin{cases} 2(1-y) & \text{for } 0 \leq y \leq 1 \\ 0 & \text{otherwise} \end{cases}$
(b) Yes, since $f(x, y) = f_X(x) f_Y(y)$. **(c)** .7083.

5-56. (a) For $y = 1, 2, 3, 4$, $f_{Y|X}(y \mid 0) = 0, 0, 0, 1$ and $f_{Y|X}(y \mid 1) = .25, .25, .25, .25$. $f(0, 1) = f(0, 2) = f(0, 3) = 0$, $f(0, 4) = p, f(1, 1) = f(1, 2) = f(1, 3) = f(1, 4) = .25(1 - p)$.
(b) $2.5 + 1.5p$. **(c)** $p > .143$.

5-58. (a) For $x = 0, 1, 2$, $f_X(x) = .5, .4, .1$. For $y = 0, 1, 2, 3, 4$, $f_Y(y) = .21, .19, .26, .21, .13$.
(b) No, since $f(x, y) \neq f_X(x) f_Y(y)$.
(c) .6; .44. **(d)** 1.86; 1.74.
(e) For $y = 0, 1, 2, 3, 4$, $f_{Y|X}(y \mid 0) = .3, .2, .2, .2, .1$; 1.6.
(f) 4.86. **(g)** .77. **(h)** $\frac{1}{.77} = 1.30$.

5-60. (a) $f(x, y) = f_X(x) f_Y(y) = \begin{cases} e^{-x} e^{-y} & \text{if } x \geq 0 \text{ and } y \geq 0 \\ 0 & \text{otherwise} \end{cases}$
(b) $e^{-2t}, t > 0$.
(c) $f_T(t) = \begin{cases} 2e^{-2t} & \text{for } t \geq 0 \\ 0 & \text{otherwise} \end{cases}$

This is an exponential distribution with mean .5.
(d) $(1 - e^{-t})^2, t > 0$.
(e) $f_T(t) = \begin{cases} 2e^{-t}(1 - e^{-t}) & \text{for } t > 0 \\ 0 & \text{otherwise} \end{cases}$
$E(T) = 1.5$.

5-66. (b) .274; .1968. **(c)** .107. **(d)** .235. **(e)** 9.3.

5-69. Binomial distribution: $n = 8, p = .20$.
(a) .147. **(b)** .797. **(c)** $np = 1.6$. **(d)** $np(1 - p) = 1.28$.
(e) 1.13.

5-70. Geometric distribution: $p = .20$. **(a)** .08. **(b)** .59.
(c) $\frac{1}{p} = 5$. **(d)** $\frac{1-p}{p^2} = 20$. **(e)** 4.47.

5-75. (a) .5899. **(b)** Binomial; .2483. **(c)** Geometric; .3818.
(d) .9050. **(e)** $n \geq 98$.

▮ CHAPTER 6

6-1. (a) $[8480, 9684]$; 8594. **(b)** $[7085, 11079]$; 7464.
(c) $[5347, 12818]$; 5731. **(d)** $[579, 1537]$; 614.
(e) $H_0: \mu = 9500$; $H_a: \mu < 9500$; $t = -1.57$ on 9 df; $.05 < p\text{-value} < .10$.
(f) $H_0: \sigma^2 = 160,000$; $H_a: \sigma^2 > 160,000$; $x^2 = 39.87$ on 9 df; $p\text{-value} < .005$.

6-2. (a) $H_0: \mu_H - \mu_L = 0$; $H_a: \mu_H - \mu_L > 0$; $z = 4.18$; $p\text{-value} < .0002$.
(b) $[3.22, 7.41]$; 3.69. **(c)** $[11.60, 13.66]$. **(d)** There is an outlier. 97.4%; 85.8%.

6-3. (a) $H_0: \mu_{4000} - \mu_{2000} = 0$; $H_a: \mu_{4000} - \mu_{2000} > 0$; $t = 11.99$ on 4 df; $p\text{-value} < .0005$. **(b)** .062.
(c) $H_0: \frac{\sigma_{2000}^2}{\sigma_{4000}^2} = 1$; $H_a: \frac{\sigma_{2000}^2}{\sigma_{4000}^2} \neq 1$; $f = .40$ on 2, 2 df; $p\text{-value} > .50$. **(d)** $[.146, 2.770]$.
(e) Independence within and between samples; normal distribution for individual data points; for (a) and (b), needed to assume that $\sigma_{2000} = \sigma_{4000}$.

6-4. (a) Independence among differences; normal distribution for differences.
(b) $H_0: \mu_d = 0$; $H_a: \mu_d < 0$ (where differences are Top − Bottom); $t = -2.10$ on 14 df; $.025 < p\text{-value} < .05$.
(c) $[-13.49, 1.49]$.

6-5. (a) Conservative method: $[.562, .758]$; .578. Other method: $[.567, .753]$; .582.
(b) $H_0: p = .55$; $H_a: p > .55$; $z = 2.21$; $p\text{-value} = .0136$.
(c) Conservative method: $[-.009, .269]$. Other method: $[-.005, .265]$.
(d) $H_0: p_S - p_L = 0$; $H_a: p_S - p_L \neq 0$; $z = 1.87$; $p\text{-value} = .0614$.

6-6. (a) $[111.0, 174.4]$. **(b)** $[105.0, 180.4]$. **(c)** 167.4.
(d) 174.4.

6-8. 92.6%; 74.9%.

6-9. (a) $[3.42, 6.38]$; $[30.6, 589.1]$.
(b) $[3.87, 5.93]$; $[48.1, 375.0]$.

6-10. Conservative method: $[.50, .89]$. Other method: $[.52, .87]$.

6-14. (b) $H_0: \mu_T - \mu_S = 0$; $H_a: \mu_T - \mu_S \neq 0$; $t = 2.49$ on 10 df; $.02 < p\text{-value} < .05$.

(c) [2.65, 47.35].

6-16. (a) [9.60, 37.73].   **(b)** 57.58.

(c) H_0: $\frac{\sigma_T^2}{\sigma_S^2} = 1$; H_a: $\frac{\sigma_T^2}{\sigma_S^2} \neq 1$; $f = .64$ on $5, 5$ df; p-value $> .50$.

(d) [.36, 1.80].

6-17. (a) 1.476.   **(b)** -1.476.   **(c)** 14.067.   **(d)** 2.167.
(e) 6.04.   **(f)** $\frac{1}{3.84} = .26$.

6-19. Conservative method: [.22, .35]. Other method: [.23, .34].

6-20. H_0: $p_1 - p_2 = 0$; H_a: $p_1 - p_2 \neq 0$; $z = -.97$; p-value $= .3320$.

6-26. 9604.

6-33. (a) 45.   **(b)** [102.3, 103.3].

6-34. (a) 98.5%.   **(b)** 96.4%.

6-36. (a) Independence among observations; normal distribution for individual measurements.
(c) (i) [.0391, .0513];   **(ii)** [.0249, .0655];   **(iii)** [.0150, .0754];
(iv) [.00771, .01739].
(e) The sample standard deviations would not be based on two independent samples.
(f) Independence among observations and normal distribution for individual measurements within each sample; common standard deviation; independence between samples.
(h) (i) H_0: $\mu_{1op} - \mu_{1or} = 0$; H_a: $\mu_{1op} - \mu_{1or} < 0$; $t = -2.86$ on 18 df; $.005 < p$-value $< .01$.   **(ii)** .0068.

6-50. Conservative method: [.24, .52]. Other method: [.25, .52].

▮ CHAPTER 7

7-1. (b) .02057. This measures the magnitude of baseline variation in any of the 5 treatments, assuming it is the same for all 5 treatments. [.01521, .03277].
(c) .02646.   **(d)** .03742.   **(e)** [.214, .262].
(f) [2.516, 2.622]; [2.716, 2.822].   **(g)** 2.4835; 2.6828.
(h) 75%.   **(i)** .0368.   **(j)** .05522.   **(k)** .04854.
(l) H_0: $\mu_1 = \mu_2 = \mu_3 = \mu_4 = \mu_5$; H_a: at least two of the means are not equal; $h = 17.70$; p-value $< .001$.
(m) $SSTr = .285135$, $MSTr = .071284$, df $= 4$; $SSE = .00423$, $MSE = .000423$, df $= 10$; $SSTot = .289365$, df $= 14$; $f = 168.52$ on 4, 10 df; p-value $< .001$. $R^2 = .985$.

7-2. (b) $SSTr = 9310.5$, $MSTr = 1862.1$, df $= 5$; $SSE = 194.0$, $MSE = 16.2$, df $= 12$; $SSTot = 9505.5$, df $= 17$; $f = 115.8$ on 5, 12 df. $\hat{\sigma} = 4.025$ measures variation in y from repeated measurements of the same rail; $\hat{\sigma}_\tau = 615.3$ measures the variation in y from differences among rails.
(c) [11.99, 178.98].

7-3. (a) Center line$_{\bar{x}} = 21.0$, $UCL_{\bar{x}} = 22.73$, $LCL_{\bar{x}} = 19.27$. Center line$_R = 1.693$, $UCL_R = 4.358$, no LCL_R.
(b) Center line$_s = .8862$, $UCL_s = 2.276$, no LCL_s.
(c) 1.3585; 1.3645; $s_p = 1.315$.
(d) Center line$_{\bar{x}} = 21.26$, $UCL_{\bar{x}} = 23.61$, $LCL_{\bar{x}} = 18.91$. Center line$_R = 2.3$, $UCL_R = 5.9202$, no LCL_R.

(e) Center line$_{\bar{x}} = 21.26$, $UCL_{\bar{x}} = 23.62$, $LCL_{\bar{x}} = 18.90$. Center line$_s = 1.21$, $UCL_s = 3.105$, no LCL_s.

7-4. (a) Center line$_{\hat{p}} = .02$, $UCL_{\hat{p}} = .0438$, no $LCL_{\hat{p}}$.
(b) Center line$_{\hat{p}} = .0234$, $UCL_{\hat{p}} = .0491$, no $LCL_{\hat{p}}$.

7-5. Center line$_{\hat{u}_i} = .138$ for all i, $UCL_{\hat{u}_i} = .138 + 3\sqrt{\frac{.138}{k_i}}$, no $LCL_{\hat{u}_i}$ for all i.

7-6. (a) Independence and normal distribution for individual measurements within each sample; independence among samples; equal variances for all lead types.
(b) 31.58; 5.62.
(c) [52.64, 63.60], [93.66, 104.62], [59.74, 70.70].
(d) $[-48.76, -33.28]$, $[-14.84, .64]$, [26.18, 41.66].
(e) [6.70, 20.12].
(f) [51.17, 65.07], [92.19, 106.09], [58.27, 72.17].
(g) $[-50.49, -31.54]$, $[-16.57, 2.37]$, [24.45, 43.39].
(h) H_0: $\mu_{4H} = \mu_H = \mu_B$; H_a: at least two of the means are not equal; $h = 12.17$; p-value $< .001$.
(i) $f = 76.10$ on 2,12 df; p-value $< .001$.
(j) $SSTr = 4806.0$, $MSTr = 2403.0$, df $= 2$; $SSE = 378.9$, $MSE = 31.6$, df $= 12$; $SSTot = 5185.0$, df $= 14$.

7-10. (b) If there are many technicians, and 5 of these were randomly chosen to be in the study, then interest is in the variation among all technicians, not just the 5 chosen for the study.
(c) $\hat{\sigma} = .00155$ in.; $\hat{\sigma}_\tau = .00071$ in.

7-12. (a) [0, 2193.1].
(b) H_0: $\mu_{pine} - \mu_{oak} = 0$; H_a: $\mu_{pine} - \mu_{oak} \neq 0$; $t = -3.31$ on 2 df; $.05 < p$-value $< .10$.
(c) [.042, 6.73].   **(d)** 155.44.
(e) Butt/Pine: 289.3; Butt/Oak: 570.6; Lap/Pine: 506.3; Lap/Oak: 1004.8.
(f) 264.6.   **(g)** $[-629.0, 153.0]$.

7-13. (a) Center line$_{\bar{x}} = .35080$, $UCL_{\bar{x}} = .351157$, $LCL_{\bar{x}} = .350443$. Center line$_R = .00019$, $UCL_R = .000621$, no LCL_R.
(b) Specifications apply to individual measurements; control limits apply to $\bar{x}$'s. Specifications are *external* standards used to judge quality; control limits are based on process history and are used to monitor process stability.

7-20. (a) Center line$_{\hat{p}} = .0001$, $UCL_{\hat{p}} = .0031$, no $LCL_{\hat{p}}$.
(b) .02.
(c) Even with large sample sizes, attributes data will often not provide enough information to detect even large changes in p.

7-26. (a) Center line$_{\hat{u}_i} = .714$ for all i, $UCL_{\hat{u}_i} = .714 + 3\sqrt{\frac{.714}{k_i}}$, no $LCL_{\hat{u}_i}$ for all i. The process seems to be stable.
(b) (i) if $k_i = 1$, .0078; if $k_i = 2$, .0033.   **(ii)** if $k_i = 1$, .0959; if $k_i = 2$, .1133.

7-30. (a) $SSTr = 1052.39$, $MSTr = 45.756$, df $= 23$; $SSE = 353.2$, $MSE = 3.679$, df $= 96$; $f = 12.44$.
(b) The focus is on variability among the many orders that are run over time, not just the 24 orders used in the study.
(c) $\hat{\sigma} = 1.918$ measures within-order variability in skewness of boxes; $\hat{\sigma}_\tau = 2.901$ measures between-order variability.
(d) [1.147, 2.086].
(e) Most of the variability seems to be coming from differences among orders (setups). The manufacturer should try to reduce

variability in setups before trying to improve within-setup precision of the equipment.

7-33. Center line$_{\hat{p}}$ = .0939, $UCL_{\hat{p}}$ = .1492, $LCL_{\hat{p}}$ = .0385.

7-35. (a) Center line$_R$ = .0072, UCL_R = .0164, no LCL_R.
(b) Center line$_s$ = .00339, UCL_s = .00768, no LCL_s.
(c) Based on $\bar{R}$: Center line$_{\bar{x}}$ = −.00159, $UCL_{\bar{x}}$ = .00158, $LCL_{\bar{x}}$ = −.00476. Based on $\bar{s}$: Center line$_{\bar{x}}$ = −.00159, $UCL_{\bar{x}}$ = .00161, $LCL_{\bar{x}}$ = −.00479.
(d) This caused the within-bundle variability to be much smaller than the between-bundle variability. The estimates of σ used above were "too small," because they only measured within-bundle variability.
(e) Based on $\bar{R}$: .00211; based on $\bar{s}$: .00213. These measure within-bundle variability, assuming it is the same for the last 10 bundles.
(f) Essentially 1.
(g) Due to bundling, the within-sample variability (as measured in (e)) is much smaller than the overall variability within and between bundles (as measured by s).

CHAPTER 8

8-1. (a) Error bars: $\bar{y}_{ij} \pm 23.54$.
(b) $a_1 = 21.78$, $a_2 = -21.78$, $b_1 = -41.61$, $b_2 = 16.06$, $b_3 = 25.56$, $ab_{11} = -1.94$, $ab_{12} = 1.39$, $ab_{13} = .56$, $ab_{21} = 1.94$, $ab_{22} = -1.39$, $ab_{23} = -.56$. Interactions: $ab_{ij} \pm 9.52$. A main effects: $a_i \pm 6.73$. B main effects: $b_j \pm 9.52$. Interactions are not detectable, but main effects for both A and B are.
(c) $\bar{y}_{j} - \bar{y}_{j'} \pm 20.18$.

8-2. (a) $s_{NI} = 12.2458$; close to $s_P = 13.113$.
(b) Using no-interaction assumption: [121.23, 145.99]. Using general method: [115.17, 148.16]. The additional assumption provides additional "information," leading to a shorter interval.
(c) Using no-interaction assumption: [104.57, 162.65]. Using general method: [98.67, 164.66].
(d) Using no-interaction assumption: $\bar{y}_{j} - \bar{y}_{j'} \pm 18.50$.

8-3. (a) and **(b)** $SSTr = 24427$; $SSA = 8537$, $MSA = 8537$, df = 1, $f = 49.63$ on 1, 12 df, p-value < .001; $SSB = 15854$, $MSB = 7927$, df = 2, $f = 46.09$ on 2, 12 df, p-value < .001; $SSAB = 36$, $MSAB = 18$, df = 2, $f = .10$ on 2, 12 df, p-value > .25; $SSE = 2063$, $MSE = 172$, df = 12.

8-4. (a) $\hat{E} \pm .014$. B and C main effects, BC interaction.
(b) $s_{FE} = .0314$ with 20 df; close to $s_P = .0329$.
(c) Using few-effects model: [3.037, 3.091]. Using general method: [3.005, 3.085].
(d) Using few-effects model: [2.993, 3.135]; [19.945, 22.989]. Using general method: [2.964, 3.126]; [19.375, 22.783].

8-5. (a) Only the main effect for A plots "off the line."
(b) Since the D main effect is almost as big (in absolute value) as the main effect for A, you might choose to include it. For the "A and D main effects only" model, the fitted values are (in standard order): 16.375, 39.375, 16.375, 39.375, 16.375, 39.375, 16.375, 39.375, −4.125, 18.875, −4.125, 18.875, −4.125, 18.875, −4.125, 18.875.
(c) Set A low (unglazed) and D high (no clean). [0, 9.09].

(d) [0, 29.13]. **(e)** 30.89.

8-6. (a) $s_P = 33.25$ measures baseline variation in y for each factor-level combination, assuming it is the same for all factor-level combinations.
(b) Error bars: $\hat{y}_{ij} \pm 27.36$.
(d) $a_1 = -2.77$, $a_2 = -17.4$, $a_3 = 20.17$, $b_1 = -13.33$, $b_2 = -1.20$, $b_3 = 14.53$, $ab_{11} = .033$, $ab_{12} = -5.40$, $ab_{13} = 5.37$, $ab_{21} = -2.13$, $ab_{22} = -.567$, $ab_{23} = 2.70$, $ab_{31} = 2.100$, $ab_{32} = 5.97$, $ab_{33} = -8.07$.
(e) 18.24. No.

8-7. Coordinates of points for normal plot: $(-.59, -1.48)$, $(-.24, -.81)$, $(.11, -.36)$, $(.46, 0)$, $(.66, .36)$, $(1.01, .81)$, $(3.96, 1.48)$. Main effect for A is detectable.

8-11. (b) $E(MSE) = \sigma^2$, $E(MSAB) = 3\sigma_{\alpha\beta}^2 + \sigma^2$, $E(MSA) = 9\sigma_{\alpha}^2 + 3\sigma_{\alpha\beta}^2 + \sigma^2$, $E(MSB) = 30\sigma_{\beta}^2 + 3\sigma_{\alpha\beta}^2 + \sigma^2$. $\hat{\sigma}^2 = .122$, $\hat{\sigma}_{\alpha\beta}^2 = 0$, $\hat{\sigma}_{\alpha}^2 = 1.405$, $\hat{\sigma}_{\beta}^2 = 0$.
(c) $\hat{\sigma} = .349 \times 10^{-3}$ in. = .000349. The micrometer is precise enough. **(d)** 0 in.

8-12. (a) $s_P = .12235$ with 108 df. **(c)** .065. **(d)** .0825.
(e) Weight main effects — yes; Brand main effects — no.

8-14. (a) $s_P = 2.606$ with 32 df. **(b)** $\bar{y}_{ij} \pm 3.38$.
(c) $a_1 = 2.89$, $a_2 = -2.89$, $b_1 = 3.77$, $b_2 = -1.475$, $b_3 = -1.66$, $b_4 = -.635$, $ab_{11} = -.05$, $ab_{12} = .125$, $ab_{13} = -.06$, $ab_{14} = -.015$, $ab_{21} = .05$, $ab_{22} = -.125$, $ab_{23} = .06$, $ab_{24} = .015$.
(d) 1.454. Interactions are not detectable. **(e)** [−7.46, −4.10].

8-15. Yes, since the Ratio × Temp interaction changes as you go from Type I to Type III concrete.

8-16. (a) $s_{NI} = 2.493$; $s_p = 2.606$.
(b) Based on no-interaction model: [40.30, 43.88]. Based on general method: [39.765, 44.515].
(c) Based on no-interaction model: [36.72, 47.46]. Based on general method: [36.32, 47.96].
(d) Based on no-interaction model: 37.124. Based on general method: 36.52.

8-18. (a) Plotted points: (.625, .05), (.625, .13), (1.25, .20), (1.50, .31), (1.625, .39), (2.25, .47), (2.50, .58), (2.875, .67), (3.625, .77), (3.750, .92), (4.625, 1.04), (6.000, 1.18), (7.125, 1.41), (10.250, 1.65), (11.500, 2.05). The absolute values of the main effects for A and D plot slightly off the line; in Exercise 8-5, it appeared that only A was off the line.
(b) Plotted points: (.1125, .10), (.2375, .28), (.4625, .47), (.5875, .67), (.6625, .92), (1.0125, 1.23), (3.9625, 1.75). Yes; only the main effect for A plots off the line.

8-19. (a) $\bar{y}_{...} = .68125$, $a_2 = .11875$, $b_2 = -.31875$, $ab_{22} = .01875$, $c_2 = -.01875$, $ac_{22} = .01875$, $bc_{22} = .03125$, $abc_{222} = -.03125$.
(b) $s_p = .12247$. **(c)** .09184. **(d)** Main effects for A and B.

8-26. $\bar{y}_{...} = 3.594$, $a_2 = -.8064$, $b_2 = .156$, $ab_{22} = -.219$, $c_2 = -.056$, $ac_{22} = -.031$, $bc_{22} = .081$, $abc_{222} = .031$, $d_2 = -.056$, $ad_{22} = -.156$, $bd_{22} = .006$, $abd_{222} = -.119$, $cd_{22} = -.031$, $acd_{222} = -.056$, $bcd_{222} = -.044$, $abcd_{2222} = .006$.
(c) To minimize y, use A(+) (monk's cloth) and B(+) (treatment Y).

CHAPTER 9

9-1. (a) $s_{LF} = 67.01$ measures the baseline variation in Average Molecular Weight for any particular Pot Temperature.
(b) Standardized residuals: 2.013, $-.3719$, $-.9998$, -1.562, $.2715$, $.2394$, $.7450$, $.0004$.
(c) $[22.08, 24.91]$.   **(d)** $[1761, 1853]$, $[2630, 2770]$.
(e) $[1745, 1869]$, $[2605, 2795]$.   **(f)** 1705; 2590.
(g) 1625; 2497.
(h) $SSR = 4{,}676{,}798, MSR = 4{,}676{,}798, df = 1; SSE = 26941$, $MSE = 4490, df = 6; SSTot = 4{,}703{,}739, df = 7; f = 1041.58$ on $1, 6$ df; p-value $< .001$.

9-2. (a) $s_{SF} = .04677$ measures baseline in Elapsed Time for any particular Jetting Size.
(b) Standardized residuals: $-.181$, $.649$, $-.794$, $-.747$, 1.55, -1.26.
(c) $[81.32, 126.66]$; $[-3.17, -1.89]$; $[.01344, .02245]$.
(d) $[14.462, 14.596]$; $[14.945, 15.145]$.
(e) $[14.415, 14.644]$; $[14.875, 15.215]$.
(f) 14.440; 14.942.   **(g)** 14.32; 14.81.
(h) $SSR = .20639, MSR = .10319, df = 2; SSE = .00656$, $MSE = .00219, df = 3; SSTot = .21295, df = 5; f = 42.17$ on $2, 3$ df; p-value $= .005$. H_0 means that Elapsed Time is not related to Jetting Size.
(i) $t = 9.38$; p-value $= .003$. $H_0 : y \approx \beta_0 + \beta_1 x + 0$; i.e., Elapsed Time is related to Jetting Size only linearly (no curvature).

9-3. (a) $s_{SF} = .4851$ measures baseline variation in y for any (x_1, x_2) combination.
(b) Standardized residuals: $-.041$, $.348$, 1.36, -1.44, -1.00, $-.457$, 1.92, $.348$, $-.604$.
(c) $[5.036, 7.060]$; $[.0775, .2058]$; $[-.0298, -.0041]$.
(d) $[5.992, 6.622]$; $[5.933, 6.625]$.
(e) $[5.798, 6.816]$; $[5.720, 6.838]$.   **(f)** 5.571; 5.535.
(g) 4.999; 4.951.
(h) $SSR = 5.8854, MSR = 2.9427, df = 2; SSE = 1.4118$, $MSE = .2353, df = 6; SSTot = 7.2972, df = 8; f = 12.51$ on $2, 6$ df; p-value $= .007$.

9-4. (a) $b_0 = 4345.9$, $b_1 = -3160.0$, $s_{LF} = 26.76$ (close to $s_P = 26.89$).
(b) Standardized residuals: 1.32, $-.48$, $-.04$, $-.91$, $.52$, -1.07, 1.94, $-.04$, -1.09.
(c) $[-357.4, -274.64]$.
(d) $t = -14.47$ on 7 df, p-value $< .001$; or $f = 209.24$ on $1, 7$ df, p-value $< .001$.
(e) $[2744.8, 2787.0]$.   **(f)** $[2699.2, 2832.6]$.   **(g)** 2697.6.

9-5. (a) See Exercise 4-17(b). $s_{SF} = 4.401$.
(b) See Exercise 4-17(d). Since this complex model was fitted based on a small amount of data (and no replication), the fitted model may give very inaccurate predictions. (There is no way to check the validity of the fitted model.)
(c) $[18.01, 25.60]$.   **(d)** $[9.87, 33.74]$.   **(e)** $[25.46, 34.86]$.
(f) H_0: Grain Size is not at all related to Time or Temp. $f = 42.37$ on $3, 5$ df. p-value $= .001$.
(g) H_0: No interaction between Time and Temp. $t = 3.12$ on 5 df. p-value $= .026$.

9-6. (a) $\hat{y} = 31.40 + 7.430 \ln x_1 - .08101 x_2 - .2760 (\ln x_1)^2 + .00004792 x_2^2 - .006596 x_2 \ln x_1$. $R^2 = .724$. $s_{SF} = 1.947$. $s_P = 2.136$, which is greater than s_{SF}, so there is no indication that the model is inappropriate.
(b) Factor-level combinations have fitted values which differ by as much as $.77$.
(d) (i) $[.128, 2.781]$.   **(ii)** $[-2.693, 5.601]$.   **(iii)** -2.332.

9-9. (a) There is no replication, so one cannot compute s_P.
(b) $f = 1.00$ on $1, 6$ df; p-value $> .25$.   **(c)** $f = .29$ on $10, 32$ df; p-value $> .25$.

9-10. (a) Estimate of $\mu_{...} = .67407$; estimate of $\alpha_2 = .12407$; estimate of $\beta_2 = -.30926$.
(b) Residuals: $.141$, $-.107$, $.093$, $-.041$, $.011$, $.041$, $-.159$, $-.007$, $-.041$, $.059$, $.011$.
(c) $s_{FE} = .09623$. $s_P = .12247$. No; $s_{FE} < s_P$.

9-11. (b) For y_1: max $\hat{y}_1$ − min $\hat{y}_1 = 20.41$, which is greater than $4\sqrt{\dfrac{6(3.321)^2}{9}}$, so this response surface seems to be adequately determined.
(c) For y_1: stationary point at $x_1 = 531.802$ and $x_2 = 92.256$; eigenvalues are $-.0120830$, $-.0000174$. The surface is bowl-shaped down with a maximum at the stationary point.

CHAPTER 10

10-1. (a) I $\leftrightarrow$ ABCDF $\leftrightarrow$ ABCEG $\leftrightarrow$ DEFG.
(b) ABDF, ABEG, CDEFG.   **(c)** $+, +; -, -$.
(d) That only A, F, and their interaction are important in describing y.

10-2. Since A $\leftrightarrow$ BCDE, if both are large but opposite in sign, their estimated sum will be small.

10-3. 3.264.

10-4. (a) I $\leftrightarrow$ ABCE $\leftrightarrow$ BCDF $\leftrightarrow$ ADEF.   **(b)** $-, -; +, -$.
(c) $.489$.

10-5. (a) 8.23, $.379$, $.256$, $-.056$, $.344$, $-.069$, $-.081$, $-.093$, $-.406$, $.181$, $.269$, $-.344$, $-.094$, $-.156$, $-.069$, $.019$.
(b) $.312$. The sums $\alpha_2 + \beta\gamma\delta\epsilon_{2222}$, $\gamma_2 + \alpha\beta\delta\epsilon_{2222}$, $\delta_2 + \alpha\beta\gamma\epsilon_{2222}$, and $\alpha\beta\delta_{222} + \gamma\epsilon_{22}$ are detectable. Simplest explanation: A, C, D main effects and CE interaction are responsible for these large sums.
(c) A $(+)$, C $(+)$, D $(-)$, and E $(-)$. The abc combination, which did have the largest observed bond strength.

10-10. (a) d, a, b, abd, cd, ac, bc, abcd.
(b) I $\leftrightarrow$ ABD, A $\leftrightarrow$ BD, B $\leftrightarrow$ AD, AB $\leftrightarrow$ D, C $\leftrightarrow$ ABCD, AC $\leftrightarrow$ BCD, BC $\leftrightarrow$ ACD, ABC $\leftrightarrow$ CD.
(c) A, B, and D main effects are dominant; A and B main effects and AB interaction are dominant; A and D main effects and AD interaction are dominant; B and D main effects and BD interaction are dominant.
(d) For D $\leftrightarrow$ ABC, main effects are aliased with 3-factor interactions, which are hopefully small, allowing the main effects to be detected less ambiguously. For D $\leftrightarrow$ AB, the main effects for A, B, and D are aliased with 2-factor interactions, which are often not small.

10-12. (a) $\bar{y}_{...} = 66.50$, $a_2 = -3.25$, $b_2 = .75$, $ab_{22} = 0$, $c_2 = 1.00$, $ac_{22} = 2.25$, $bc_{22} = .75$, $abc_{222} = 1.00$.
(c) .519. A, B, C, AC, BC, and ABC are detectable.
(d) Main effect for A and AC (or BD) interaction are dominant.
(e) A and E main effects are dominant.

10-16. (a) Best possible resolution = 4. The students' plan has resolution 3, with some main effects aliased with 2-factor interactions. A resolution 4 design would have all main effects aliased with 3- or higher-factor interactions.
(b) $E \leftrightarrow BCD$ and $F \leftrightarrow ABC$. (There are others.)

10-17. (a) $\hat{\sigma}^2 = 4.7$ measures variation from different pieces of sliver cut from a particular roll; $\hat{\sigma}_\beta^2 = 9.8$ measures variation from different rolls produced by a particular machine; $\hat{\sigma}_\alpha^2 = 37.2$ measures variation from different machines.
(b) .72, .19, .09. Concentrate on reducing machine-to-machine variation.
(c) [3.566, 6.530]; [.990, 5.239].

10-20. (a) 3 times.
(b) $a_2^* = .1629$, $b_2^* = .1469$, $c_2^* = -.1229$, $d_2^* = -.2581$, $e_2^* = .0749$, $f_2^* = .4576$, $g_2^* = .0916$. Get estimated main effects of factors at their low levels by switching signs.
(c) $s_{NI} = .5929$ with 4 df. $\frac{1}{2}\Delta = .475$; none are detectable.
(d) Yes. $a_2^*, \ldots, g_2^*$ are the estimated regression coefficients, and s_{NI} is s_{SF} on the printout.
(e) $R^2_{\text{all main effects}} = .751$; $R^2_{F,G,FG} = .910$, a much better fit using only 2 of the factors. The two models could give very different predictions for combinations not used in the experiment.

10-21. (a) 486. (b) 45, which is much larger than 18. No.
(c) There would be 22 terms, which is greater than the number of data points, so one cannot estimate all of the parameters in the model.

(d) (i) Not seriously degraded, since there is information on these combinations in the data. If a few replications could be made to obtain an s_P, you'd be in pretty good shape. (ii) Seriously degraded. (iii) Very seriously degraded.

10-26. (b) (1), ad, bd, ab, cd, ac, bc, abcd. Estimated sums of effects: 3.600, $-.850$, .100, $-.250$, $-.175$, $-.025$, $-.075$, $-.025$.
(c) The estimate of $\alpha_2 + \beta\gamma\delta_{222}$ plots off the line. Still might conclude that this is due to the main effect for A, but the conclusion here would be a little more tentative.

APPENDIX B

B-2. (a) 1.7310×10^{13}. (b) 2.2777×10^{12}. (c) .1316.
B-3. (a) .0000081. (b) .03402.
B-4. (a) .54. (b) .78.
B-5. (a) .505. (b) .998.
B-6. (a) 1,757,600. (b) .00167. (c) .0167.
B-7. (a) .5. (b) .167.
B-8. (a) .76. (b) .78. (c) .974.
B-11. (a) 20; 15.81. (b) $1 - \frac{3}{2}\exp\left(-\frac{t}{15}\right) + \frac{1}{2}\exp\left(-\frac{t}{5}\right)$.
(c) $f_T(t) = \frac{1}{10}\left(\exp\left(-\frac{t}{15}\right) - \exp\left(-\frac{t}{5}\right)\right)$.
(d) $S_T(t) = \frac{3}{2}\exp\left(-\frac{t}{15}\right) - \frac{1}{2}\exp\left(-\frac{t}{5}\right)$.

$$h_T(t) = \frac{1}{5}\left(\frac{\exp\left(-\frac{t}{15}\right) - \exp\left(-\frac{t}{5}\right)}{3\exp\left(-\frac{t}{15}\right) - \exp\left(-\frac{t}{5}\right)}\right)$$

$h_T(t)$ is not constant. It starts at 0 and increases to an asymptote of $\frac{1}{15}$. For $t > 30$ hr, the function is nearly constant.

D

Tables

TABLE D-1   Random Digits

12159	66144	05091	13446	45653	13684	66024	91410	51351	22772
30156	90519	95785	47544	66735	35754	11088	67310	19720	08379
59069	01722	53338	41942	65118	71236	01932	70343	25812	62275
54107	58081	82470	59407	13475	95872	16268	78436	39251	64247
99681	81295	06315	28212	45029	57701	96327	85436	33614	29070
27252	37875	53679	01889	35714	63534	63791	76342	47717	73684
93259	74585	11863	78985	03881	46567	93696	93521	54970	37601
84068	43759	75814	32261	12728	09636	22336	75629	01017	45503
68582	97054	28251	63787	57285	18854	35006	16343	51867	67979
60646	11298	19680	10087	66391	70853	24423	73007	74958	29020
97437	52922	80739	59178	50628	61017	51652	40915	94696	67843
58009	20681	98823	50979	01237	70152	13711	73916	87902	84759
77211	70110	93803	60135	22881	13423	30999	07104	27400	25414
54256	84591	65302	99257	92970	28924	36632	54044	91798	78018
36493	69330	94069	39544	14050	03476	25804	49350	92525	87941
87569	22661	55970	52623	35419	76660	42394	63210	62626	00581
22896	62237	39635	63725	10463	87944	92075	90914	30599	35671
02697	33230	64527	97210	41359	79399	13941	88378	68503	33609
20080	15652	37216	00679	02088	34138	13953	68939	05630	27653
20550	95151	60557	57449	77115	87372	02574	07851	22128	39189
72771	11672	67492	42904	64647	94354	45994	42538	54885	15983
38472	43379	76295	69406	96510	16529	83500	28590	49787	29822
24511	56510	72654	13277	45031	42235	96502	25567	23653	36707
01054	06674	58283	82831	97048	42983	06471	12350	49990	04809
94437	94907	95274	26487	60496	78222	43032	04276	70800	17378
97842	69095	25982	03484	25173	05982	14624	31653	17170	92785
53047	13486	69712	33567	82313	87631	03197	02438	12374	40329
40770	47013	63306	48154	80970	87976	04939	21233	20572	31013
52733	66251	69661	58387	72096	21355	51659	19003	75556	33095
41749	46502	18378	83141	63920	85516	75743	66317	45428	45940
10271	85184	46468	38860	24039	80949	51211	35411	40470	16070
98791	48848	68129	51024	53044	55039	71290	26484	70682	56255
30196	09295	47685	56768	29285	06272	98789	47188	35063	24158
99373	64343	92433	06388	65713	35386	43370	19254	55014	98621
27768	27552	42156	23239	46823	91077	06306	17756	84459	92513
67791	35910	56921	51976	78475	15336	92544	82601	17996	72268
64018	44004	08136	56129	77024	82650	18163	29158	33935	94262
79715	33859	10835	94936	02857	87486	70613	41909	80667	52176
20190	40737	82688	07099	65255	52767	65930	45861	32575	93731
82421	01208	49762	66360	00231	87540	88302	62686	38456	25872

Reprinted from *A Million Random Digits with 100,000 Normal Deviates*, RAND (New York: The Free Press, 1955). Copyright © 1955 and 1983 by RAND. Used by permission.

TABLE D-2   Control Chart Constants

m	d_2	d_3	c_4	A_2	A_3	B_3	B_4	B_5	B_6	D_1	D_2	D_3	D_4
2	1.128	0.853	0.7979	1.880	2.659		3.267		2.606		3.686		3.267
3	1.693	0.888	0.8862	1.023	1.954		2.568		2.276		4.358		2.574
4	2.059	0.880	0.9213	0.729	1.628		2.266		2.088		4.698		2.282
5	2.326	0.864	0.9400	0.577	1.427		2.089		1.964		4.918		2.115
6	2.534	0.848	0.9515	0.483	1.287	0.030	1.970	0.029	1.874		5.078		2.004
7	2.704	0.833	0.9594	0.419	1.182	0.118	1.882	0.113	1.806	0.205	5.204	0.076	1.924
8	2.847	0.820	0.9650	0.373	1.099	0.185	1.815	0.179	1.751	0.387	5.307	0.136	1.864
9	2.970	0.808	0.9693	0.337	1.032	0.239	1.761	0.232	1.707	0.546	5.394	0.184	1.816
10	3.078	0.797	0.9727	0.308	0.975	0.284	1.716	0.276	1.669	0.687	5.469	0.223	1.777
11	3.173	0.787	0.9754	0.285	0.927	0.321	1.679	0.313	1.637	0.812	5.534	0.256	1.744
12	3.258	0.778	0.9776	0.266	0.886	0.354	1.646	0.346	1.610	0.924	5.592	0.284	1.716
13	3.336	0.770	0.9794	0.249	0.850	0.382	1.618	0.374	1.585	1.026	5.646	0.308	1.692
14	3.407	0.762	0.9810	0.235	0.817	0.406	1.594	0.399	1.563	1.121	5.693	0.329	1.671
15	3.472	0.755	0.9823	0.223	0.789	0.428	1.572	0.421	1.544	1.207	5.737	0.348	1.652
20	3.735	0.729	0.9869	0.180	0.680	0.510	1.490	0.504	1.470	1.548	5.922	0.414	1.586
25	3.931	0.709	0.9896	0.153	0.606	0.565	1.435	0.559	1.420	1.804	6.058	0.459	1.541

This table is from *Quality Control and Industrial Statistics* by A.J. Duncan and was originally adapted from the *A.S.T.M. Manual on Quality Control of Materials*, Table B2, and the ASQC Standard A1, Table 1. Reprinted by permission of Richard D. Irwin, Inc.

TABLE D-3 Standard Normal Cumulative Probabilities

$$\Phi(z) = \int_{-\infty}^{z} \frac{1}{\sqrt{2\pi}} \exp\left(-\frac{t^2}{2}\right) dt$$

z	.00	.01	.02	.03	.04	.05	.06	.07	.08	.09
0.0	.5000	.5040	.5080	.5120	.5160	.5199	.5239	.5279	.5319	.5359
0.1	.5398	.5438	.5478	.5517	.5557	.5596	.5636	.5675	.5714	.5753
0.2	.5793	.5832	.5871	.5910	.5948	.5987	.6026	.6064	.6103	.6141
0.3	.6179	.6217	.6255	.6293	.6331	.6368	.6406	.6443	.6480	.6517
0.4	.6554	.6591	.6628	.6664	.6700	.6736	.6772	.6808	.6844	.6879
0.5	.6915	.6950	.6985	.7019	.7054	.7088	.7123	.7157	.7190	.7224
0.6	.7257	.7291	.7324	.7357	.7389	.7422	.7454	.7486	.7517	.7549
0.7	.7580	.7611	.7642	.7673	.7704	.7734	.7764	.7794	.7823	.7852
0.8	.7881	.7910	.7939	.7967	.7995	.8023	.8051	.8078	.8106	.8133
0.9	.8159	.8186	.8212	.8238	.8264	.8289	.8315	.8340	.8365	.8389
1.0	.8413	.8438	.8461	.8485	.8508	.8531	.8554	.8577	.8599	.8621
1.1	.8643	.8665	.8686	.8708	.8729	.8749	.8770	.8790	.8810	.8830
1.2	.8849	.8869	.8888	.8907	.8925	.8944	.8962	.8980	.8997	.9015
1.3	.9032	.9049	.9066	.9082	.9099	.9115	.9131	.9147	.9162	.9177
1.4	.9192	.9207	.9222	.9236	.9251	.9265	.9279	.9292	.9306	.9319
1.5	.9332	.9345	.9357	.9370	.9382	.9394	.9406	.9418	.9429	.9441
1.6	.9452	.9463	.9474	.9484	.9495	.9505	.9515	.9525	.9535	.9545
1.7	.9554	.9564	.9573	.9582	.9591	.9599	.9608	.9616	.9625	.9633
1.8	.9641	.9649	.9656	.9664	.9671	.9678	.9686	.9693	.9699	.9706
1.9	.9713	.9719	.9726	.9732	.9738	.9744	.9750	.9756	.9761	.9767
2.0	.9773	.9778	.9783	.9788	.9793	.9798	.9803	.9808	.9812	.9817
2.1	.9821	.9826	.9830	.9834	.9838	.9842	.9846	.9850	.9854	.9857
2.2	.9861	.9864	.9868	.9871	.9875	.9878	.9881	.9884	.9887	.9890
2.3	.9893	.9896	.9898	.9901	.9904	.9906	.9909	.9911	.9913	.9916
2.4	.9918	.9920	.9922	.9925	.9927	.9929	.9931	.9932	.9934	.9936
2.5	.9938	.9940	.9941	.9943	.9945	.9946	.9948	.9949	.9951	.9952
2.6	.9953	.9955	.9956	.9957	.9959	.9960	.9961	.9962	.9963	.9964
2.7	.9965	.9966	.9967	.9968	.9969	.9970	.9971	.9972	.9973	.9974
2.8	.9974	.9975	.9976	.9977	.9977	.9978	.9979	.9979	.9980	.9981
2.9	.9981	.9982	.9983	.9983	.9984	.9984	.9985	.9985	.9986	.9986
3.0	.9987	.9987	.9987	.9988	.9988	.9989	.9989	.9989	.9990	.9990
3.1	.9990	.9991	.9991	.9991	.9992	.9992	.9992	.9992	.9993	.9993
3.2	.9993	.9993	.9994	.9994	.9994	.9994	.9994	.9995	.9995	.9995
3.3	.9995	.9995	.9996	.9996	.9996	.9996	.9996	.9996	.9996	.9997
3.4	.9997	.9997	.9997	.9997	.9997	.9997	.9997	.9997	.9997	.9998

z	.00	.01	.02	.03	.04	.05	.06	.07	.08	.09
−3.4	.0003	.0003	.0003	.0003	.0003	.0003	.0003	.0003	.0003	.0002
−3.3	.0005	.0005	.0005	.0004	.0004	.0004	.0004	.0004	.0004	.0003
−3.2	.0007	.0007	.0006	.0006	.0006	.0006	.0006	.0005	.0005	.0005
−3.1	.0010	.0009	.0009	.0009	.0008	.0008	.0008	.0008	.0007	.0007
−3.0	.0013	.0013	.0013	.0012	.0012	.0011	.0011	.0011	.0010	.0010
−2.9	.0019	.0018	.0018	.0017	.0016	.0016	.0015	.0015	.0014	.0014
−2.8	.0026	.0025	.0024	.0023	.0023	.0022	.0021	.0021	.0020	.0019
−2.7	.0035	.0034	.0033	.0032	.0031	.0030	.0029	.0028	.0027	.0026
−2.6	.0047	.0045	.0044	.0043	.0041	.0040	.0039	.0038	.0037	.0036
−2.5	.0062	.0060	.0059	.0057	.0055	.0054	.0052	.0051	.0049	.0048
−2.4	.0082	.0080	.0078	.0075	.0073	.0071	.0069	.0068	.0066	.0064
−2.3	.0107	.0104	.0102	.0099	.0096	.0094	.0091	.0089	.0087	.0084
−2.2	.0139	.0136	.0132	.0129	.0125	.0122	.0119	.0116	.0113	.0110
−2.1	.0179	.0174	.0170	.0166	.0162	.0158	.0154	.0150	.0146	.0143
−2.0	.0228	.0222	.0217	.0212	.0207	.0202	.0197	.0192	.0188	.0183
−1.9	.0287	.0281	.0274	.0268	.0262	.0256	.0250	.0244	.0239	.0233
−1.8	.0359	.0351	.0344	.0336	.0329	.0322	.0314	.0307	.0301	.0294
−1.7	.0446	.0436	.0427	.0418	.0409	.0401	.0392	.0384	.0375	.0367
−1.6	.0548	.0537	.0526	.0516	.0505	.0495	.0485	.0475	.0465	.0455
−1.5	.0668	.0655	.0643	.0630	.0618	.0606	.0594	.0582	.0571	.0559
−1.4	.0808	.0793	.0778	.0764	.0749	.0735	.0721	.0708	.0694	.0681
−1.3	.0968	.0951	.0934	.0918	.0901	.0885	.0869	.0853	.0838	.0823
−1.2	.1151	.1131	.1112	.1093	.1075	.1056	.1038	.1020	.1003	.0985
−1.1	.1357	.1335	.1314	.1292	.1271	.1251	.1230	.1210	.1190	.1170
−1.0	.1587	.1562	.1539	.1515	.1492	.1469	.1446	.1423	.1401	.1379
−0.9	.1841	.1814	.1788	.1762	.1736	.1711	.1685	.1660	.1635	.1611
−0.8	.2119	.2090	.2061	.2033	.2005	.1977	.1949	.1922	.1894	.1867
−0.7	.2420	.2389	.2358	.2327	.2297	.2266	.2236	.2206	.2177	.2148
−0.6	.2743	.2709	.2676	.2643	.2611	.2578	.2546	.2514	.2483	.2451
−0.5	.3085	.3050	.3015	.2981	.2946	.2912	.2877	.2843	.2810	.2776
−0.4	.3446	.3409	.3372	.3336	.3300	.3264	.3228	.3192	.3156	.3121
−0.3	.3821	.3783	.3745	.3707	.3669	.3632	.3594	.3557	.3520	.3483
−0.2	.4207	.4168	.4129	.4090	.4052	.4013	.3974	.3936	.3897	.3859
−0.1	.4602	.4562	.4522	.4483	.4443	.4404	.4364	.4325	.4286	.4247
−0.0	.5000	.4960	.4920	.4880	.4840	.4801	.4761	.4721	.4681	.4641

This table was generated using the "CDF" command in Minitab.

TABLE D-4   t Distribution Quantiles

ν	$Q(.9)$	$Q(.95)$	$Q(.975)$	$Q(.99)$	$Q(.995)$	$Q(.999)$	$Q(.9995)$
1	3.078	6.314	12.706	31.821	63.657	318.317	636.607
2	1.886	2.920	4.303	6.965	9.925	22.327	31.598
3	1.638	2.353	3.182	4.541	5.841	10.215	12.924
4	1.533	2.132	2.776	3.747	4.604	7.173	8.610
5	1.476	2.015	2.571	3.365	4.032	5.893	6.869
6	1.440	1.943	2.447	3.143	3.707	5.208	5.959
7	1.415	1.895	2.365	2.998	3.499	4.785	5.408
8	1.397	1.860	2.306	2.896	3.355	4.501	5.041
9	1.383	1.833	2.262	2.821	3.250	4.297	4.781
10	1.372	1.812	2.228	2.764	3.169	4.144	4.587
11	1.363	1.796	2.201	2.718	3.106	4.025	4.437
12	1.356	1.782	2.179	2.681	3.055	3.930	4.318
13	1.350	1.771	2.160	2.650	3.012	3.852	4.221
14	1.345	1.761	2.145	2.624	2.977	3.787	4.140
15	1.341	1.753	2.131	2.602	2.947	3.733	4.073
16	1.337	1.746	2.120	2.583	2.921	3.686	4.015
17	1.333	1.740	2.110	2.567	2.898	3.646	3.965
18	1.330	1.734	2.101	2.552	2.878	3.610	3.922
19	1.328	1.729	2.093	2.539	2.861	3.579	3.883
20	1.325	1.725	2.086	2.528	2.845	3.552	3.849
21	1.323	1.721	2.080	2.518	2.831	3.527	3.819
22	1.321	1.717	2.074	2.508	2.819	3.505	3.792
23	1.319	1.714	2.069	2.500	2.807	3.485	3.768
24	1.318	1.711	2.064	2.492	2.797	3.467	3.745
25	1.316	1.708	2.060	2.485	2.787	3.450	3.725
26	1.315	1.706	2.056	2.479	2.779	3.435	3.707
27	1.314	1.703	2.052	2.473	2.771	3.421	3.690
28	1.313	1.701	2.048	2.467	2.763	3.408	3.674
29	1.311	1.699	2.045	2.462	2.756	3.396	3.659
30	1.310	1.697	2.042	2.457	2.750	3.385	3.646
40	1.303	1.684	2.021	2.423	2.704	3.307	3.551
60	1.296	1.671	2.000	2.390	2.660	3.232	3.460
120	1.289	1.658	1.980	2.358	2.617	3.160	3.373
∞	1.282	1.645	1.960	2.326	2.576	3.090	3.291

This table was generated using the "INVCDF" command in Minitab.

TABLE D-5   Chi-Square Distribution Quantiles

ν	Q(.005)	Q(.01)	Q(.025)	Q(.05)	Q(.1)	Q(.9)	Q(.95)	Q(.975)	Q(.99)	Q(.995)
1	0.000	0.000	0.001	0.004	0.016	2.706	3.841	5.024	6.635	7.879
2	0.010	0.020	0.051	0.103	0.211	4.605	5.991	7.378	9.210	10.597
3	0.072	0.115	0.216	0.352	0.584	6.251	7.815	9.348	11.345	12.838
4	0.207	0.297	0.484	0.711	1.064	7.779	9.488	11.143	13.277	14.860
5	0.412	0.554	0.831	1.145	1.610	9.236	11.070	12.833	15.086	16.750
6	0.676	0.872	1.237	1.635	2.204	10.645	12.592	14.449	16.812	18.548
7	0.989	1.239	1.690	2.167	2.833	12.017	14.067	16.013	18.475	20.278
8	1.344	1.646	2.180	2.733	3.490	13.362	15.507	17.535	20.090	21.955
9	1.735	2.088	2.700	3.325	4.168	14.684	16.919	19.023	21.666	23.589
10	2.156	2.558	3.247	3.940	4.865	15.987	18.307	20.483	23.209	25.188
11	2.603	3.053	3.816	4.575	5.578	17.275	19.675	21.920	24.725	26.757
12	3.074	3.571	4.404	5.226	6.304	18.549	21.026	23.337	26.217	28.300
13	3.565	4.107	5.009	5.892	7.042	19.812	22.362	24.736	27.688	29.819
14	4.075	4.660	5.629	6.571	7.790	21.064	23.685	26.119	29.141	31.319
15	4.601	5.229	6.262	7.261	8.547	22.307	24.996	27.488	30.578	32.801
16	5.142	5.812	6.908	7.962	9.312	23.542	26.296	28.845	32.000	34.267
17	5.697	6.408	7.564	8.672	10.085	24.769	27.587	30.191	33.409	35.718
18	6.265	7.015	8.231	9.390	10.865	25.989	28.869	31.526	34.805	37.156
19	6.844	7.633	8.907	10.117	11.651	27.204	30.143	32.852	36.191	38.582
20	7.434	8.260	9.591	10.851	12.443	28.412	31.410	34.170	37.566	39.997
21	8.034	8.897	10.283	11.591	13.240	29.615	32.671	35.479	38.932	41.401
22	8.643	9.542	10.982	12.338	14.041	30.813	33.924	36.781	40.290	42.796
23	9.260	10.196	11.689	13.091	14.848	32.007	35.172	38.076	41.638	44.181
24	9.886	10.856	12.401	13.848	15.659	33.196	36.415	39.364	42.980	45.559
25	10.520	11.524	13.120	14.611	16.473	34.382	37.653	40.647	44.314	46.928
26	11.160	12.198	13.844	15.379	17.292	35.563	38.885	41.923	45.642	48.290
27	11.808	12.879	14.573	16.151	18.114	36.741	40.113	43.195	46.963	49.645
28	12.461	13.565	15.308	16.928	18.939	37.916	41.337	44.461	48.278	50.994
29	13.121	14.256	16.047	17.708	19.768	39.087	42.557	45.722	49.588	52.336
30	13.787	14.953	16.791	18.493	20.599	40.256	43.773	46.979	50.892	53.672
31	14.458	15.655	17.539	19.281	21.434	41.422	44.985	48.232	52.192	55.003
32	15.134	16.362	18.291	20.072	22.271	42.585	46.194	49.480	53.486	56.328
33	15.815	17.074	19.047	20.867	23.110	43.745	47.400	50.725	54.775	57.648
34	16.501	17.789	19.806	21.664	23.952	44.903	48.602	51.966	56.061	58.964
35	17.192	18.509	20.569	22.465	24.797	46.059	49.802	53.204	57.342	60.275
36	17.887	19.233	21.336	23.269	25.643	47.212	50.998	54.437	58.619	61.581
37	18.586	19.960	22.106	24.075	26.492	48.364	52.192	55.668	59.893	62.885
38	19.289	20.691	22.878	24.884	27.343	49.513	53.384	56.896	61.163	64.183
39	19.996	21.426	23.654	25.695	28.196	50.660	54.572	58.120	62.429	65.477
40	20.707	22.164	24.433	26.509	29.051	51.805	55.759	59.342	63.691	66.767

This table was generated using the "INVCDF" command in Minitab.

For $\nu > 40$, the approximation $Q(p) \approx \nu \left(1 - \dfrac{2}{9\nu} + Q_z(p)\sqrt{\dfrac{2}{9\nu}} \right)^3$ can be used.

TABLE D-6-A   F Distribution .75 Quantiles

ν_1 (NUMERATOR DEGREES OF FREEDOM)

ν_2	1	2	3	4	5	6	7	8	9	10	12	15	20	24	30	40	60	120	∞
1	5.83	7.50	8.20	8.58	8.82	8.98	9.10	9.19	9.26	9.32	9.41	9.49	9.58	9.63	9.67	9.71	9.76	9.80	9.85
2	2.57	3.00	3.15	3.23	3.28	3.31	3.34	3.35	3.37	3.38	3.39	3.41	3.43	3.43	3.44	3.45	3.46	3.47	3.48
3	2.02	2.28	2.36	2.39	2.41	2.42	2.43	2.44	2.44	2.44	2.45	2.46	2.46	2.46	2.47	2.47	2.47	2.47	2.47
4	1.81	2.00	2.05	2.06	2.07	2.08	2.08	2.08	2.08	2.08	2.08	2.08	2.08	2.08	2.08	2.08	2.08	2.08	2.08
5	1.69	1.85	1.88	1.89	1.89	1.89	1.89	1.89	1.89	1.89	1.89	1.89	1.88	1.88	1.88	1.88	1.87	1.87	1.87
6	1.62	1.76	1.78	1.79	1.79	1.78	1.78	1.78	1.77	1.77	1.77	1.76	1.76	1.75	1.75	1.75	1.74	1.74	1.74
7	1.57	1.70	1.72	1.72	1.71	1.71	1.70	1.70	1.69	1.69	1.68	1.68	1.67	1.67	1.66	1.66	1.65	1.65	1.65
8	1.54	1.66	1.67	1.66	1.66	1.65	1.64	1.64	1.64	1.63	1.62	1.62	1.61	1.60	1.60	1.59	1.59	1.58	1.58
9	1.51	1.62	1.63	1.63	1.62	1.61	1.60	1.60	1.59	1.59	1.58	1.57	1.56	1.56	1.55	1.54	1.54	1.53	1.53
10	1.49	1.60	1.60	1.59	1.59	1.58	1.57	1.56	1.56	1.55	1.54	1.53	1.52	1.52	1.51	1.51	1.50	1.49	1.48
11	1.47	1.58	1.58	1.57	1.56	1.55	1.54	1.53	1.53	1.52	1.51	1.50	1.49	1.49	1.48	1.47	1.47	1.46	1.45
12	1.46	1.56	1.56	1.55	1.54	1.53	1.52	1.51	1.51	1.50	1.49	1.48	1.47	1.46	1.45	1.45	1.44	1.43	1.42
13	1.45	1.55	1.55	1.53	1.52	1.51	1.50	1.49	1.49	1.48	1.47	1.46	1.45	1.44	1.43	1.42	1.42	1.41	1.40
14	1.44	1.53	1.53	1.52	1.51	1.50	1.49	1.48	1.47	1.46	1.45	1.44	1.43	1.42	1.41	1.41	1.40	1.39	1.38
15	1.43	1.52	1.52	1.51	1.49	1.48	1.47	1.46	1.46	1.45	1.44	1.43	1.41	1.41	1.40	1.39	1.38	1.37	1.36
16	1.42	1.51	1.51	1.50	1.48	1.47	1.46	1.45	1.44	1.44	1.43	1.41	1.40	1.39	1.38	1.37	1.36	1.35	1.34
17	1.42	1.51	1.50	1.49	1.47	1.46	1.45	1.44	1.43	1.43	1.41	1.40	1.39	1.38	1.37	1.36	1.35	1.34	1.33
18	1.41	1.50	1.49	1.48	1.46	1.45	1.44	1.43	1.42	1.42	1.40	1.39	1.38	1.37	1.36	1.35	1.34	1.33	1.32
19	1.41	1.49	1.49	1.47	1.46	1.44	1.43	1.42	1.41	1.41	1.40	1.38	1.37	1.36	1.35	1.34	1.33	1.32	1.30
20	1.40	1.49	1.48	1.47	1.45	1.44	1.43	1.42	1.41	1.40	1.39	1.37	1.36	1.35	1.34	1.33	1.32	1.31	1.29
21	1.40	1.48	1.48	1.46	1.44	1.43	1.42	1.41	1.40	1.39	1.38	1.37	1.35	1.34	1.33	1.32	1.31	1.30	1.28
22	1.40	1.48	1.47	1.45	1.44	1.42	1.41	1.40	1.39	1.39	1.37	1.36	1.34	1.33	1.32	1.31	1.30	1.29	1.28
23	1.39	1.47	1.47	1.45	1.43	1.42	1.41	1.40	1.39	1.38	1.37	1.35	1.34	1.33	1.32	1.31	1.30	1.28	1.27
24	1.39	1.47	1.46	1.44	1.43	1.41	1.40	1.39	1.38	1.38	1.36	1.35	1.33	1.32	1.31	1.30	1.29	1.28	1.26
25	1.39	1.47	1.46	1.44	1.42	1.41	1.40	1.39	1.38	1.37	1.36	1.34	1.33	1.32	1.31	1.29	1.28	1.27	1.25
26	1.38	1.46	1.45	1.44	1.42	1.41	1.39	1.38	1.37	1.37	1.35	1.34	1.32	1.31	1.30	1.29	1.28	1.26	1.25
27	1.38	1.46	1.45	1.43	1.42	1.40	1.39	1.38	1.37	1.36	1.35	1.33	1.32	1.31	1.30	1.28	1.27	1.26	1.24
28	1.38	1.46	1.45	1.43	1.41	1.40	1.39	1.38	1.37	1.36	1.34	1.33	1.31	1.30	1.29	1.28	1.27	1.25	1.24
29	1.38	1.45	1.45	1.43	1.41	1.40	1.38	1.37	1.36	1.35	1.34	1.32	1.31	1.30	1.29	1.27	1.26	1.25	1.23
30	1.38	1.45	1.44	1.42	1.41	1.39	1.38	1.37	1.36	1.35	1.34	1.32	1.30	1.29	1.28	1.27	1.26	1.24	1.23
40	1.36	1.44	1.42	1.40	1.39	1.37	1.36	1.35	1.34	1.33	1.31	1.30	1.28	1.26	1.25	1.24	1.22	1.21	1.19
60	1.35	1.42	1.41	1.38	1.37	1.35	1.33	1.32	1.31	1.30	1.29	1.27	1.25	1.24	1.22	1.21	1.19	1.17	1.15
120	1.34	1.40	1.39	1.37	1.35	1.33	1.31	1.30	1.29	1.28	1.26	1.24	1.22	1.21	1.19	1.18	1.16	1.13	1.10
∞	1.32	1.39	1.37	1.35	1.33	1.31	1.29	1.28	1.27	1.25	1.24	1.22	1.19	1.18	1.16	1.14	1.12	1.08	1.00

ν_2 (DENOMINATOR DEGREES OF FREEDOM)

This table was generated using the "INVCDF" command in Minitab.

TABLE D-6-B   *F Distribution .90 Quantiles*

v_1 *(NUMERATOR DEGREES OF FREEDOM)*

	1	2	3	4	5	6	7	8	9	10	12	15	20	24	30	40	60	120	∞
1	39.86	49.50	53.59	55.84	57.24	58.20	58.90	59.44	59.85	60.20	60.70	61.22	61.74	62.00	62.27	62.53	62.79	63.05	63.33
2	8.53	9.00	9.16	9.24	9.29	9.33	9.35	9.37	9.38	9.39	9.41	9.42	9.44	9.45	9.46	9.47	9.47	9.48	9.49
3	5.54	5.46	5.39	5.34	5.31	5.28	5.27	5.25	5.24	5.23	5.22	5.20	5.18	5.18	5.17	5.16	5.15	5.14	5.13
4	4.54	4.32	4.19	4.11	4.05	4.01	3.98	3.95	3.94	3.92	3.90	3.87	3.84	3.83	3.82	3.80	3.79	3.78	3.76
5	4.06	3.78	3.62	3.52	3.45	3.40	3.37	3.34	3.32	3.30	3.27	3.24	3.21	3.19	3.17	3.16	3.14	3.12	3.10
6	3.78	3.46	3.29	3.18	3.11	3.05	3.01	2.98	2.96	2.94	2.90	2.87	2.84	2.82	2.80	2.78	2.76	2.74	2.72
7	3.59	3.26	3.07	2.96	2.88	2.83	2.78	2.75	2.72	2.70	2.67	2.63	2.59	2.58	2.56	2.54	2.51	2.49	2.47
8	3.46	3.11	2.92	2.81	2.73	2.67	2.62	2.59	2.56	2.54	2.50	2.46	2.42	2.40	2.38	2.36	2.34	2.32	2.29
9	3.36	3.01	2.81	2.69	2.61	2.55	2.51	2.47	2.44	2.42	2.38	2.34	2.30	2.28	2.25	2.23	2.21	2.18	2.16
10	3.28	2.92	2.73	2.61	2.52	2.46	2.41	2.38	2.35	2.32	2.28	2.24	2.20	2.18	2.16	2.13	2.11	2.08	2.06
11	3.23	2.86	2.66	2.54	2.45	2.39	2.34	2.30	2.27	2.25	2.21	2.17	2.12	2.10	2.08	2.05	2.03	2.00	1.97
12	3.18	2.81	2.61	2.48	2.39	2.33	2.28	2.24	2.21	2.19	2.15	2.10	2.06	2.04	2.01	1.99	1.96	1.93	1.90
13	3.14	2.76	2.56	2.43	2.35	2.28	2.23	2.20	2.16	2.14	2.10	2.05	2.01	1.98	1.96	1.93	1.90	1.88	1.85
14	3.10	2.73	2.52	2.39	2.31	2.24	2.19	2.15	2.12	2.10	2.05	2.01	1.96	1.94	1.91	1.89	1.86	1.83	1.80
15	3.07	2.70	2.49	2.36	2.27	2.21	2.16	2.12	2.09	2.06	2.02	1.97	1.92	1.90	1.87	1.85	1.82	1.79	1.76
16	3.05	2.67	2.46	2.33	2.24	2.18	2.13	2.09	2.06	2.03	1.99	1.94	1.89	1.87	1.84	1.81	1.78	1.75	1.72
17	3.03	2.64	2.44	2.31	2.22	2.15	2.10	2.06	2.03	2.00	1.96	1.91	1.86	1.84	1.81	1.78	1.75	1.72	1.69
18	3.01	2.62	2.42	2.29	2.20	2.13	2.08	2.04	2.00	1.98	1.93	1.89	1.84	1.81	1.78	1.75	1.72	1.69	1.66
19	2.99	2.61	2.40	2.27	2.18	2.11	2.06	2.02	1.98	1.96	1.91	1.86	1.81	1.79	1.76	1.73	1.70	1.67	1.63
20	2.97	2.59	2.38	2.25	2.16	2.09	2.04	2.00	1.96	1.94	1.89	1.84	1.79	1.77	1.74	1.71	1.68	1.64	1.61
21	2.96	2.57	2.36	2.23	2.14	2.08	2.02	1.98	1.95	1.92	1.87	1.83	1.78	1.75	1.72	1.69	1.66	1.62	1.59
22	2.95	2.56	2.35	2.22	2.13	2.06	2.01	1.97	1.93	1.90	1.86	1.81	1.76	1.73	1.70	1.67	1.64	1.60	1.57
23	2.94	2.55	2.34	2.21	2.11	2.05	1.99	1.95	1.92	1.89	1.84	1.80	1.74	1.72	1.69	1.66	1.62	1.59	1.55
24	2.93	2.54	2.33	2.19	2.10	2.04	1.98	1.94	1.91	1.88	1.83	1.78	1.73	1.70	1.67	1.64	1.61	1.57	1.53
25	2.92	2.53	2.32	2.18	2.09	2.02	1.97	1.93	1.89	1.87	1.82	1.77	1.72	1.69	1.66	1.63	1.59	1.56	1.52
26	2.91	2.52	2.31	2.17	2.08	2.01	1.96	1.92	1.88	1.86	1.81	1.76	1.71	1.68	1.65	1.61	1.58	1.54	1.50
27	2.90	2.51	2.30	2.17	2.07	2.00	1.95	1.91	1.87	1.85	1.80	1.75	1.70	1.67	1.64	1.60	1.57	1.53	1.49
28	2.89	2.50	2.29	2.16	2.06	2.00	1.94	1.90	1.87	1.84	1.79	1.74	1.69	1.66	1.63	1.59	1.56	1.52	1.48
29	2.89	2.50	2.28	2.15	2.06	1.99	1.93	1.89	1.86	1.83	1.78	1.73	1.68	1.65	1.62	1.58	1.55	1.51	1.47
30	2.88	2.49	2.28	2.14	2.05	1.98	1.93	1.88	1.85	1.82	1.77	1.72	1.67	1.64	1.61	1.57	1.54	1.50	1.46
40	2.84	2.44	2.23	2.09	2.00	1.93	1.87	1.83	1.79	1.76	1.71	1.66	1.61	1.57	1.54	1.51	1.47	1.42	1.38
60	2.79	2.39	2.18	2.04	1.95	1.87	1.82	1.77	1.74	1.71	1.66	1.60	1.54	1.51	1.48	1.44	1.40	1.35	1.29
120	2.75	2.35	2.13	1.99	1.90	1.82	1.77	1.72	1.68	1.65	1.60	1.55	1.48	1.45	1.41	1.37	1.32	1.26	1.19
∞	2.71	2.30	2.08	1.94	1.85	1.77	1.72	1.67	1.63	1.60	1.55	1.49	1.42	1.38	1.34	1.30	1.24	1.17	1.00

v_2 *(DENOMINATOR DEGREES OF FREEDOM)*

This table was generated using the "INVCDF" command in Minitab.

TABLE D-6-C   F Distribution .95 Quantiles

ν_1 (NUMERATOR DEGREES OF FREEDOM)

ν_2	1	2	3	4	5	6	7	8	9	10	12	15	20	24	30	40	60	120	∞
1	161.44	199.50	215.69	224.57	230.16	233.98	236.78	238.89	240.55	241.89	243.91	245.97	248.02	249.04	250.07	251.13	252.18	253.27	254.31
2	18.51	19.00	19.16	19.25	19.30	19.33	19.35	19.37	19.39	19.40	19.41	19.43	19.45	19.45	19.46	19.47	19.48	19.49	19.50
3	10.13	9.55	9.28	9.12	9.01	8.94	8.89	8.85	8.81	8.79	8.74	8.70	8.66	8.64	8.62	8.59	8.57	8.55	8.53
4	7.71	6.94	6.59	6.39	6.26	6.16	6.09	6.04	6.00	5.96	5.91	5.86	5.80	5.77	5.75	5.72	5.69	5.66	5.63
5	6.61	5.79	5.41	5.19	5.05	4.95	4.88	4.82	4.77	4.74	4.68	4.62	4.56	4.53	4.50	4.46	4.43	4.40	4.36
6	5.99	5.14	4.76	4.53	4.39	4.28	4.21	4.15	4.10	4.06	4.00	3.94	3.87	3.84	3.81	3.77	3.74	3.70	3.67
7	5.59	4.74	4.35	4.12	3.97	3.87	3.79	3.73	3.68	3.64	3.57	3.51	3.44	3.41	3.38	3.34	3.30	3.27	3.23
8	5.32	4.46	4.07	3.84	3.69	3.58	3.50	3.44	3.39	3.35	3.28	3.22	3.15	3.12	3.08	3.04	3.01	2.97	2.93
9	5.12	4.26	3.86	3.63	3.48	3.37	3.29	3.23	3.18	3.14	3.07	3.01	2.94	2.90	2.86	2.83	2.79	2.75	2.71
10	4.96	4.10	3.71	3.48	3.33	3.22	3.14	3.07	3.02	2.98	2.91	2.85	2.77	2.74	2.70	2.66	2.62	2.58	2.54
11	4.84	3.98	3.59	3.36	3.20	3.09	3.01	2.95	2.90	2.85	2.79	2.72	2.65	2.61	2.57	2.53	2.49	2.45	2.40
12	4.75	3.89	3.49	3.26	3.11	3.00	2.91	2.85	2.80	2.75	2.69	2.62	2.54	2.51	2.47	2.43	2.38	2.34	2.30
13	4.67	3.81	3.41	3.18	3.03	2.92	2.83	2.77	2.71	2.67	2.60	2.53	2.46	2.42	2.38	2.34	2.30	2.25	2.21
14	4.60	3.74	3.34	3.11	2.96	2.85	2.76	2.70	2.65	2.60	2.53	2.46	2.39	2.35	2.31	2.27	2.22	2.18	2.13
15	4.54	3.68	3.29	3.06	2.90	2.79	2.71	2.64	2.59	2.54	2.48	2.40	2.33	2.29	2.25	2.20	2.16	2.11	2.07
16	4.49	3.63	3.24	3.01	2.85	2.74	2.66	2.59	2.54	2.49	2.42	2.35	2.28	2.24	2.19	2.15	2.11	2.06	2.01
17	4.45	3.59	3.20	2.96	2.81	2.70	2.61	2.55	2.49	2.45	2.38	2.31	2.23	2.19	2.15	2.10	2.06	2.01	1.96
18	4.41	3.55	3.16	2.93	2.77	2.66	2.58	2.51	2.46	2.41	2.34	2.27	2.19	2.15	2.11	2.06	2.02	1.97	1.92
19	4.38	3.52	3.13	2.90	2.74	2.63	2.54	2.48	2.42	2.38	2.31	2.23	2.16	2.11	2.07	2.03	1.98	1.93	1.88
20	4.35	3.49	3.10	2.87	2.71	2.60	2.51	2.45	2.39	2.35	2.28	2.20	2.12	2.08	2.04	1.99	1.95	1.90	1.84
21	4.32	3.47	3.07	2.84	2.68	2.57	2.49	2.42	2.37	2.32	2.25	2.18	2.10	2.05	2.01	1.96	1.92	1.87	1.81
22	4.30	3.44	3.05	2.82	2.66	2.55	2.46	2.40	2.34	2.30	2.23	2.15	2.07	2.03	1.98	1.94	1.89	1.84	1.78
23	4.28	3.42	3.03	2.80	2.64	2.53	2.44	2.37	2.32	2.27	2.20	2.13	2.05	2.01	1.96	1.91	1.86	1.81	1.76
24	4.26	3.40	3.01	2.78	2.62	2.51	2.42	2.36	2.30	2.25	2.18	2.11	2.03	1.98	1.94	1.89	1.84	1.79	1.73
25	4.24	3.39	2.99	2.76	2.60	2.49	2.40	2.34	2.28	2.24	2.16	2.09	2.01	1.96	1.92	1.87	1.82	1.77	1.71
26	4.23	3.37	2.98	2.74	2.59	2.47	2.39	2.32	2.27	2.22	2.15	2.07	1.99	1.95	1.90	1.85	1.80	1.75	1.69
27	4.21	3.35	2.96	2.73	2.57	2.46	2.37	2.31	2.25	2.20	2.13	2.06	1.97	1.93	1.88	1.84	1.79	1.73	1.67
28	4.20	3.34	2.95	2.71	2.56	2.45	2.36	2.29	2.24	2.19	2.12	2.04	1.96	1.91	1.87	1.82	1.77	1.71	1.65
29	4.18	3.33	2.93	2.70	2.55	2.43	2.35	2.28	2.22	2.18	2.10	2.03	1.94	1.90	1.85	1.81	1.75	1.70	1.64
30	4.17	3.32	2.92	2.69	2.53	2.42	2.33	2.27	2.21	2.16	2.09	2.01	1.93	1.89	1.84	1.79	1.74	1.68	1.62
40	4.08	3.23	2.84	2.61	2.45	2.34	2.25	2.18	2.12	2.08	2.00	1.92	1.84	1.79	1.74	1.69	1.64	1.58	1.51
60	4.00	3.15	2.76	2.53	2.37	2.25	2.17	2.10	2.04	1.99	1.92	1.84	1.75	1.70	1.65	1.59	1.53	1.47	1.39
120	3.92	3.07	2.68	2.45	2.29	2.18	2.09	2.02	1.96	1.91	1.83	1.75	1.66	1.61	1.55	1.50	1.43	1.35	1.25
∞	3.84	3.00	2.60	2.37	2.21	2.10	2.01	1.94	1.88	1.83	1.75	1.67	1.57	1.52	1.46	1.39	1.32	1.22	1.00

ν_2 (DENOMINATOR DEGREES OF FREEDOM)

This table was generated using the "INVCDF" command in Minitab.

TABLE D-6-D   F Distribution .99 Quantiles

ν_1 *(NUMERATOR DEGREES OF FREEDOM)*

ν_2	1	2	3	4	5	6	7	8	9	10	12	15	20	24	30	40	60	120	∞
1	4052	4999	5403	5625	5764	5859	5929	5981	6023	6055	6107	6157	6209	6235	6260	6287	6312	6339	6366
2	98.51	99.00	99.17	99.25	99.30	99.33	99.35	99.38	99.39	99.40	99.41	99.43	99.44	99.45	99.47	99.47	99.48	99.49	99.50
3	34.12	30.82	29.46	28.71	28.24	27.91	27.67	27.49	27.35	27.23	27.05	26.87	26.69	26.60	26.51	26.41	26.32	26.22	26.13
4	21.20	18.00	16.69	15.98	15.52	15.21	14.98	14.80	14.66	14.55	14.37	14.20	14.02	13.93	13.84	13.75	13.65	13.56	13.46
5	16.26	13.27	12.06	11.39	10.97	10.67	10.46	10.29	10.16	10.05	9.89	9.72	9.55	9.47	9.38	9.29	9.20	9.11	9.02
6	13.75	10.92	9.78	9.15	8.75	8.47	8.26	8.10	7.98	7.87	7.72	7.56	7.40	7.31	7.23	7.14	7.06	6.97	6.88
7	12.25	9.55	8.45	7.85	7.46	7.19	6.99	6.84	6.72	6.62	6.47	6.31	6.16	6.07	5.99	5.91	5.82	5.74	5.65
8	11.26	8.65	7.59	7.01	6.63	6.37	6.18	6.03	5.91	5.81	5.67	5.52	5.36	5.28	5.20	5.12	5.03	4.95	4.86
9	10.56	8.02	6.99	6.42	6.06	5.80	5.61	5.47	5.35	5.26	5.11	4.96	4.81	4.73	4.65	4.57	4.48	4.40	4.31
10	10.04	7.56	6.55	5.99	5.64	5.39	5.20	5.06	4.94	4.85	4.71	4.56	4.41	4.33	4.25	4.17	4.08	4.00	3.91
11	9.65	7.21	6.22	5.67	5.32	5.07	4.89	4.74	4.63	4.54	4.40	4.25	4.10	4.02	3.94	3.86	3.78	3.69	3.60
12	9.33	6.93	5.95	5.41	5.06	4.82	4.64	4.50	4.39	4.30	4.16	4.01	3.86	3.78	3.70	3.62	3.54	3.45	3.36
13	9.07	6.70	5.74	5.21	4.86	4.62	4.44	4.30	4.19	4.10	3.96	3.82	3.66	3.59	3.51	3.43	3.34	3.25	3.17
14	8.86	6.51	5.56	5.04	4.69	4.46	4.28	4.14	4.03	3.94	3.80	3.66	3.51	3.43	3.35	3.27	3.18	3.09	3.00
15	8.68	6.36	5.42	4.89	4.56	4.32	4.14	4.00	3.89	3.80	3.67	3.52	3.37	3.29	3.21	3.13	3.05	2.96	2.87
16	8.53	6.23	5.29	4.77	4.44	4.20	4.03	3.89	3.78	3.69	3.55	3.41	3.26	3.18	3.10	3.02	2.93	2.84	2.75
17	8.40	6.11	5.18	4.67	4.34	4.10	3.93	3.79	3.68	3.59	3.46	3.31	3.16	3.08	3.00	2.92	2.83	2.75	2.65
18	8.29	6.01	5.09	4.58	4.25	4.01	3.84	3.71	3.60	3.51	3.37	3.23	3.08	3.00	2.92	2.84	2.75	2.66	2.57
19	8.19	5.93	5.01	4.50	4.17	3.94	3.77	3.63	3.52	3.43	3.30	3.15	3.00	2.92	2.84	2.76	2.67	2.58	2.49
20	8.10	5.85	4.94	4.43	4.10	3.87	3.70	3.56	3.46	3.37	3.23	3.09	2.94	2.86	2.78	2.69	2.61	2.52	2.42
21	8.02	5.78	4.87	4.37	4.04	3.81	3.64	3.51	3.40	3.31	3.17	3.03	2.88	2.80	2.72	2.64	2.55	2.46	2.36
22	7.95	5.72	4.82	4.31	3.99	3.76	3.59	3.45	3.35	3.26	3.12	2.98	2.83	2.75	2.67	2.58	2.50	2.40	2.31
23	7.88	5.66	4.76	4.26	3.94	3.71	3.54	3.41	3.30	3.21	3.07	2.93	2.78	2.70	2.62	2.54	2.45	2.35	2.26
24	7.82	5.61	4.72	4.22	3.90	3.67	3.50	3.36	3.26	3.17	3.03	2.89	2.74	2.66	2.58	2.49	2.40	2.31	2.21
25	7.77	5.57	4.68	4.18	3.85	3.63	3.46	3.32	3.22	3.13	2.99	2.85	2.70	2.62	2.54	2.45	2.36	2.27	2.17
26	7.72	5.53	4.64	4.14	3.82	3.59	3.42	3.29	3.18	3.09	2.96	2.81	2.66	2.58	2.50	2.42	2.33	2.23	2.13
27	7.68	5.49	4.60	4.11	3.78	3.56	3.39	3.26	3.15	3.06	2.93	2.78	2.63	2.55	2.47	2.38	2.29	2.20	2.10
28	7.64	5.45	4.57	4.07	3.75	3.53	3.36	3.23	3.12	3.03	2.90	2.75	2.60	2.52	2.44	2.35	2.26	2.17	2.06
29	7.60	5.42	4.54	4.04	3.73	3.50	3.33	3.20	3.09	3.00	2.87	2.73	2.57	2.49	2.41	2.33	2.23	2.14	2.03
30	7.56	5.39	4.51	4.02	3.70	3.47	3.30	3.17	3.07	2.98	2.84	2.70	2.55	2.47	2.39	2.30	2.21	2.11	2.01
40	7.31	5.18	4.31	3.83	3.51	3.29	3.12	2.99	2.89	2.80	2.66	2.52	2.37	2.29	2.20	2.11	2.02	1.92	1.80
60	7.08	4.98	4.13	3.65	3.34	3.12	2.95	2.82	2.72	2.63	2.50	2.35	2.20	2.12	2.03	1.94	1.84	1.73	1.60
120	6.85	4.79	3.95	3.48	3.17	2.96	2.79	2.66	2.56	2.47	2.34	2.19	2.03	1.95	1.86	1.76	1.66	1.53	1.38
∞	6.63	4.61	3.78	3.32	3.02	2.80	2.64	2.51	2.41	2.32	2.18	2.04	1.88	1.79	1.70	1.59	1.47	1.32	1.00

ν_2 (DENOMINATOR DEGREES OF FREEDOM)

This table was generated using the "INVCDF" command in Minitab.

TABLE D-6-E   F Distribution .999 Quantiles

ν_1 *(NUMERATOR DEGREES OF FREEDOM)*

	1	2	3	4	5	6	7	8	9	10	12	15	20	24	30	40	60	120	∞
1	405261	499996	540349	562463	576409	585904	592890	598185	602359	605671	610644	615766	620884	623544	626117	628724	631381	634002	636619
2	998.55	999.01	999.23	999.26	999.29	999.38	999.40	999.35	999.45	999.41	999.46	999.40	999.44	999.45	999.47	999.49	999.50	999.52	999.50
3	167.03	148.50	141.11	137.10	134.58	132.85	131.58	130.62	129.86	129.25	128.32	127.37	126.42	125.94	125.45	124.96	124.47	123.97	123.47
4	74.14	61.25	56.18	53.44	51.71	50.53	49.66	49.00	48.48	48.05	47.41	46.76	46.10	45.77	45.43	45.09	44.75	44.40	44.05
5	47.18	37.12	33.20	31.08	29.75	28.83	28.16	27.65	27.24	26.92	26.42	25.91	25.40	25.13	24.87	24.60	24.33	24.06	23.79
6	35.51	27.00	23.70	21.92	20.80	20.03	19.46	19.03	18.69	18.41	17.99	17.56	17.12	16.90	16.67	16.44	16.21	15.98	15.75
7	29.24	21.69	18.77	17.20	16.21	15.52	15.02	14.63	14.33	14.08	13.71	13.32	12.93	12.73	12.53	12.33	12.12	11.91	11.70
8	25.41	18.49	15.83	14.39	13.48	12.86	12.40	12.05	11.77	11.54	11.19	10.84	10.48	10.30	10.11	9.92	9.73	9.53	9.33
9	22.86	16.39	13.90	12.56	11.71	11.13	10.70	10.37	10.11	9.89	9.57	9.24	8.90	8.72	8.55	8.37	8.19	8.00	7.81
10	21.04	14.91	12.55	11.28	10.48	9.93	9.52	9.20	8.96	8.75	8.45	8.13	7.80	7.64	7.47	7.30	7.12	6.94	6.76
11	19.69	13.81	11.56	10.35	9.58	9.05	8.66	8.35	8.12	7.92	7.63	7.32	7.01	6.85	6.68	6.52	6.35	6.18	6.00
12	18.64	12.97	10.80	9.63	8.89	8.38	8.00	7.71	7.48	7.29	7.00	6.71	6.40	6.25	6.09	5.93	5.76	5.59	5.42
13	17.82	12.31	10.21	9.07	8.35	7.86	7.49	7.21	6.98	6.80	6.52	6.23	5.93	5.78	5.63	5.47	5.30	5.14	4.97
14	17.14	11.78	9.73	8.62	7.92	7.44	7.08	6.80	6.58	6.40	6.13	5.85	5.56	5.41	5.25	5.10	4.94	4.77	4.60
15	16.59	11.34	9.34	8.25	7.57	7.09	6.74	6.47	6.26	6.08	5.81	5.54	5.25	5.10	4.95	4.80	4.64	4.47	4.31
16	16.12	10.97	9.01	7.94	7.27	6.80	6.46	6.19	5.98	5.81	5.55	5.27	4.99	4.85	4.70	4.54	4.39	4.23	4.06
17	15.72	10.66	8.73	7.68	7.02	6.56	6.22	5.96	5.75	5.58	5.32	5.05	4.78	4.63	4.48	4.33	4.18	4.02	3.85
18	15.38	10.39	8.49	7.46	6.81	6.35	6.02	5.76	5.56	5.39	5.13	4.87	4.59	4.45	4.30	4.15	4.00	3.84	3.67
19	15.08	10.16	8.28	7.27	6.62	6.18	5.85	5.59	5.39	5.22	4.97	4.70	4.43	4.29	4.14	3.99	3.84	3.68	3.51
20	14.82	9.95	8.10	7.10	6.46	6.02	5.69	5.44	5.24	5.08	4.82	4.56	4.29	4.15	4.01	3.86	3.70	3.54	3.38
21	14.59	9.77	7.94	6.95	6.32	5.88	5.56	5.31	5.11	4.95	4.70	4.44	4.17	4.03	3.88	3.74	3.58	3.42	3.26
22	14.38	9.61	7.80	6.81	6.19	5.76	5.44	5.19	4.99	4.83	4.58	4.33	4.06	3.92	3.78	3.63	3.48	3.32	3.15
23	14.20	9.47	7.67	6.70	6.08	5.65	5.33	5.09	4.89	4.73	4.48	4.23	3.96	3.82	3.68	3.53	3.38	3.22	3.05
24	14.03	9.34	7.55	6.59	5.98	5.55	5.23	4.99	4.80	4.64	4.39	4.14	3.87	3.74	3.59	3.45	3.29	3.14	2.97
25	13.88	9.22	7.45	6.49	5.89	5.46	5.15	4.91	4.71	4.56	4.31	4.06	3.79	3.66	3.52	3.37	3.22	3.06	2.89
26	13.74	9.12	7.36	6.41	5.80	5.38	5.07	4.83	4.64	4.48	4.24	3.99	3.72	3.59	3.44	3.30	3.15	2.99	2.82
27	13.61	9.02	7.27	6.33	5.73	5.31	5.00	4.76	4.57	4.41	4.17	3.92	3.66	3.52	3.38	3.23	3.08	2.92	2.75
28	13.50	8.93	7.19	6.25	5.66	5.24	4.93	4.69	4.50	4.35	4.11	3.86	3.60	3.46	3.32	3.18	3.02	2.86	2.69
29	13.39	8.85	7.12	6.19	5.59	5.18	4.87	4.64	4.45	4.29	4.05	3.80	3.54	3.41	3.27	3.12	2.97	2.81	2.64
30	13.29	8.77	7.05	6.12	5.53	5.12	4.82	4.58	4.39	4.24	4.00	3.75	3.49	3.36	3.22	3.07	2.92	2.76	2.59
40	12.61	8.25	6.59	5.70	5.13	4.73	4.44	4.21	4.02	3.87	3.64	3.40	3.14	3.01	2.87	2.73	2.57	2.41	2.23
60	11.97	7.77	6.17	5.31	4.76	4.37	4.09	3.86	3.69	3.54	3.32	3.08	2.83	2.69	2.55	2.41	2.25	2.08	1.89
120	11.38	7.32	5.78	4.95	4.42	4.04	3.77	3.55	3.38	3.24	3.02	2.78	2.53	2.40	2.26	2.11	1.95	1.77	1.54
∞	10.83	6.91	5.42	4.62	4.10	3.74	3.47	3.27	3.10	2.96	2.74	2.51	2.27	2.13	1.99	1.84	1.66	1.45	1.00

ν_2 *(DENOMINATOR DEGREES OF FREEDOM)*

This table was generated using the "INVCDF" command in Minitab.

TABLE D-7-A **Factors for Two-Sided Tolerance Intervals for Normal Distributions**

	95% CONFIDENCE			99% CONFIDENCE		
n	p = .90	p = .95	p = .99	p = .90	p = .95	p = .99
2	31.092	36.519	46.944	155.569	182.720	234.877
3	8.306	9.789	12.647	18.782	22.131	28.586
4	5.368	6.341	8.221	9.416	11.118	14.405
5	4.291	5.077	6.598	6.655	7.870	10.220
6	3.733	4.422	5.758	5.383	6.373	8.292
7	3.390	4.020	5.241	4.658	5.520	7.191
8	3.156	3.746	4.889	4.189	4.968	6.479
9	2.986	3.546	4.633	3.860	4.581	5.980
10	2.856	3.393	4.437	3.617	4.294	5.610
11	2.754	3.273	4.282	3.429	4.073	5.324
12	2.670	3.175	4.156	3.279	3.896	5.096
13	2.601	3.093	4.051	3.156	3.751	4.909
14	2.542	3.024	3.962	3.054	3.631	4.753
15	2.492	2.965	3.885	2.967	3.529	4.621
16	2.449	2.913	3.819	2.893	3.441	4.507
17	2.410	2.868	3.761	2.828	3.364	4.408
18	2.376	2.828	3.709	2.771	3.297	4.321
19	2.346	2.793	3.663	2.720	3.237	4.244
20	2.319	2.760	3.621	2.675	3.184	4.175
25	2.215	2.638	3.462	2.506	2.984	3.915
30	2.145	2.555	3.355	2.394	2.851	3.742
35	2.094	2.495	3.276	2.314	2.756	3.618
40	2.055	2.448	3.216	2.253	2.684	3.524
50	1.999	2.382	3.129	2.166	2.580	3.390
60	1.960	2.335	3.068	2.106	2.509	3.297
80	1.908	2.274	2.988	2.028	2.416	3.175
100	1.875	2.234	2.936	1.978	2.357	3.098
150	1.826	2.176	2.859	1.906	2.271	2.985
200	1.798	2.143	2.816	1.866	2.223	2.921
500	1.737	2.070	2.721	1.777	2.117	2.783
1000	1.709	2.036	2.676	1.736	2.068	2.718
∞	1.645	1.960	2.576	1.645	1.960	2.576

This table was adapted from *Tables for Normal Tolerance Limits, Sampling Plans and Screening* by R. E. Odeh and D. B. Owen. Reprinted by permission of Marcel Dekker Inc.

TABLE D-7-B **Factors for One-Sided Tolerance Intervals for Normal Distributions**

	95% CONFIDENCE			99% CONFIDENCE		
n	p = .90	p = .95	p = .99	p = .90	p = .95	p = .99
2	20.581	26.260	37.094	103.029	131.426	185.617
3	6.155	7.656	10.553	13.995	17.370	23.896
4	4.162	5.144	7.042	7.380	9.083	12.387
5	3.407	4.203	5.741	5.362	6.578	8.939
6	3.006	3.708	5.062	4.411	5.406	7.335
7	2.755	3.399	4.642	3.859	4.728	6.412
8	2.582	3.187	4.354	3.497	4.285	5.812
9	2.454	3.031	4.143	3.240	3.972	5.389
10	2.355	2.911	3.981	3.048	3.738	5.074
11	2.275	2.815	3.852	2.898	3.556	4.829
12	2.210	2.736	3.747	2.777	3.410	4.633
13	2.155	2.671	3.659	2.677	3.290	4.472
14	2.109	2.614	3.585	2.593	3.189	4.337
15	2.068	2.566	3.520	2.521	3.102	4.222
16	2.033	2.524	3.464	2.459	3.028	4.123
17	2.002	2.486	3.414	2.405	2.963	4.037
18	1.974	2.453	3.370	2.357	2.905	3.960
19	1.949	2.423	3.331	2.314	2.854	3.892
20	1.926	2.396	3.295	2.276	2.808	3.832
25	1.838	2.292	3.158	2.129	2.633	3.601
30	1.777	2.220	3.064	2.030	2.515	3.447
35	1.732	2.167	2.995	1.957	2.430	3.334
40	1.697	2.125	2.941	1.902	2.364	3.249
50	1.646	2.065	2.862	1.821	2.269	3.125
60	1.609	2.022	2.807	1.764	2.202	3.038
80	1.559	1.964	2.733	1.688	2.114	2.924
100	1.527	1.927	2.684	1.639	2.056	2.850
150	1.478	1.870	2.611	1.566	1.971	2.740
200	1.450	1.837	2.570	1.524	1.923	2.679
500	1.385	1.763	2.475	1.430	1.814	2.540
1000	1.354	1.727	2.430	1.385	1.762	2.475
∞	1.282	1.645	2.326	1.282	1.645	2.326

This table was adapted from *Tables for Normal Tolerance Limits, Sampling Plans and Screening* by R. E. Odeh and D. B. Owen. Reprinted by permission of Marcel Dekker Inc.

TABLE D-8   Values of λ for Use in Computing 95% One-Sided Normal Tolerance Limits

ξ	ν									
	4	5	6	7	8	9	16	36	144	∞
1.0	1.636	1.648	1.655	1.660	1.662	1.6638	1.6665	1.6634	1.6559	1.6449
0.9	1.643	1.655	1.662	1.666	1.668	1.6695	1.6711	1.6667	1.6576	1.6449
0.8	1.650	1.662	1.668	1.672	1.674	1.6747	1.6751	1.6691	1.6586	1.6449
0.7	1.657	1.668	1.674	1.677	1.679	1.6796	1.6782	1.6707	1.6589	1.6449
0.6	1.664	1.675	1.680	1.682	1.684	1.6838	1.6804	1.6714	1.6587	1.6449
0.5	1.671	1.681	1.686	1.687	1.687	1.6871	1.6817	1.6709	1.6580	1.6449
0.4	1.679	1.687	1.690	1.691	1.691	1.6896	1.6816	1.6698	1.6568	1.6449
0.3	1.687	1.693	1.694	1.693	1.692	1.6902	1.6804	1.6677	1.6550	1.6449
0.2	1.693	1.697	1.696	1.694	1.692	1.6898	1.6779	1.6646	1.6529	1.6449
0.1	1.698	1.699	1.697	1.693	1.690	1.6874	1.6738	1.6606	1.6504	1.6449
0.0	1.703	1.699	1.695	1.690	1.686	1.6827	1.6682	1.6558	1.6477	1.6449
−0.1	1.702	1.695	1.689	1.684	1.679	1.6756	1.6611	1.6503	1.6447	1.6449
−0.2	1.698	1.688	1.680	1.674	1.669	1.6657	1.6525	1.6442	1.6417	1.6449
−0.3	1.687	1.676	1.667	1.661	1.657	1.6535	1.6427	1.6378	1.6388	1.6449
−0.4	1.670	1.658	1.650	1.645	1.642	1.6391	1.6322	1.6314	1.6359	1.6449
−0.5	1.646	1.636	1.630	1.627	1.624	1.6231	1.6213	1.6252	1.6334	1.6449
−0.6	1.615	1.610	1.607	1.606	1.606	1.6066	1.6108	1.6195	1.6313	1.6449
−0.7	1.582	1.583	1.585	1.587	1.589	1.5911	1.6019	1.6150	1.6299	1.6449
−0.8	1.551	1.559	1.565	1.571	1.575	1.5792	1.5954	1.6122	1.6291	1.6449
−0.9	1.531	1.544	1.553	1.561	1.567	1.5722	1.5925	1.6116	1.6292	1.6449
−1.0	1.528	1.543	1.554	1.563	1.569	1.5744	1.5952	1.6141	1.6307	1.6449

This table is from "Applications of the Noncentral t Distribution" by N. L. Johnson and B. L. Welch, *Biometrika*, *31*, pp. 362–389, 1940. Reprinted by permission of the Biometrika Trustees.

Johnson and Welch recommend that for $\nu > 9$, interpolation in this table be done linearly in $12/\sqrt{\nu}$. (Notice that $\nu = 9, 16, 36, 144$, and ∞ have values of $12/\sqrt{\nu}$ equal to 4, 3, 2, 1, and 0, respectively.)

TABLE D-9-A *Factors for Simultaneous 95% Two-Sided Confidence Limits for r Means*

				r				
ν	1	2	3	4	5	6	7	8
5	2.57	3.09	3.40	3.62	3.78	3.92	4.04	4.14
10	2.23	2.61	2.83	2.98	3.10	3.19	3.28	3.35
15	2.13	2.47	2.67	2.81	2.91	2.99	3.06	3.12
20	2.09	2.41	2.59	2.72	2.82	2.90	2.97	3.02
24	2.06	2.38	2.56	2.68	2.78	2.84	2.91	2.96
30	2.04	2.35	2.52	2.64	2.73	2.80	2.86	2.91
40	2.02	2.32	2.49	2.60	2.69	2.76	2.82	2.86
60	2.00	2.29	2.46	2.56	2.65	2.72	2.77	2.82
120	1.98	2.26	2.43	2.53	2.61	2.68	2.73	2.77
∞	1.96	2.23	2.39	2.49	2.57	2.64	2.69	2.73

This table is taken from "On the Distribution of the Ratio of the ith Observation in an Ordered Sample from a Normal Population to an Independent Estimate of the Standard Deviation" by K. C. S. Pillai and K. V. Ramachandran, *Annals of Mathematical Statistics*, *25*, pp. 565–572, 1954. Reprinted by permission of the Institute of Mathematical Statistics.

TABLE D-9-B *Factors for Simultaneous 95% One-Sided Confidence Limits for r Means*

				r				
ν	1	2	3	4	5	6	7	8
3	2.35	3.10	3.53	3.85	4.12	4.34	4.53	4.71
4	2.13	2.72	3.06	3.31	3.51	3.67	3.82	3.98
5	2.01	2.53	2.83	3.03	3.21	3.34	3.45	3.58
6	1.94	2.42	2.68	2.86	3.02	3.14	3.25	3.36
7	1.89	2.34	2.60	2.75	2.90	3.01	3.11	3.21
8	1.86	2.29	2.52	2.67	2.82	2.92	3.02	3.11
9	1.83	2.24	2.47	2.62	2.75	2.85	2.94	3.04
10	1.81	2.22	2.44	2.59	2.70	2.79	2.88	2.97
12	1.78	2.18	2.38	2.53	2.63	2.72	2.81	2.90
14	1.76	2.14	2.34	2.49	2.58	2.67	2.75	2.83
15	1.75	2.13	2.32	2.47	2.56	2.65	2.73	2.81
16	1.75	2.12	2.31	2.45	2.54	2.63	2.71	2.78
18	1.73	2.10	2.29	2.43	2.52	2.61	2.68	2.75
20	1.72	2.09	2.27	2.41	2.50	2.58	2.65	2.72
24	1.71	2.06	2.24	2.38	2.47	2.55	2.62	2.68
30	1.70	2.04	2.22	2.35	2.44	2.52	2.59	2.65
40	1.68	2.02	2.20	2.32	2.41	2.49	2.55	2.61
60	1.67	2.00	2.17	2.29	2.38	2.45	2.51	2.57
120	1.66	1.98	2.14	2.26	2.35	2.42	2.47	2.53
∞	1.64	1.96	2.12	2.23	2.32	2.39	2.44	2.49

This table is taken from "On the Distribution of the Ratio of the ith Observation in an Ordered Sample from a Normal Population to an Independent Estimate of the Standard Deviation" by K. C. S. Pillai and K. V. Ramachandran, *Annals of Mathematical Statistics*, *25*, pp. 565–572, 1954. Reprinted by permission of the Institute of Mathematical Statistics.

TABLE D-10-A   .95 Quantiles of the Studentized Range Distribution

NUMBER OF MEANS TO BE COMPARED

ν	2	3	4	5	6	7	8	9	10	11	12	13	15	20
5	3.64	4.60	5.22	5.67	6.03	6.33	6.58	6.80	6.99	7.17	7.32	7.47	7.72	8.21
6	3.46	4.34	4.90	5.30	5.63	5.90	6.12	6.32	6.49	6.65	6.79	6.92	7.14	7.59
7	3.34	4.16	4.68	5.06	5.36	5.61	5.82	6.00	6.16	6.30	6.43	6.55	6.76	7.17
8	3.26	4.04	4.53	4.89	5.17	5.40	5.60	5.77	5.92	6.05	6.18	6.29	6.48	6.87
9	3.20	3.95	4.41	4.76	5.02	5.24	5.43	5.59	5.74	5.87	5.98	6.09	6.28	6.64
10	3.15	3.88	4.33	4.65	4.91	5.12	5.30	5.46	5.60	5.72	5.83	5.93	6.11	6.47
11	3.11	3.82	4.26	4.57	4.82	5.03	5.20	5.35	5.49	5.61	5.71	5.81	5.98	6.33
12	3.08	3.77	4.20	4.51	4.75	4.95	5.12	5.27	5.39	5.51	5.61	5.71	5.88	6.21
13	3.06	3.73	4.15	4.45	4.69	4.88	5.05	5.19	5.32	5.43	5.53	5.63	5.79	6.11
14	3.03	3.70	4.11	4.41	4.64	4.83	4.99	5.13	5.25	5.36	5.46	5.55	5.71	6.03
15	3.01	3.67	4.08	4.37	4.59	4.78	4.94	5.08	5.20	5.31	5.40	5.49	5.65	5.96
16	3.00	3.65	4.05	4.33	4.56	4.74	4.90	5.03	5.15	5.26	5.35	5.44	5.59	5.90
17	2.98	3.63	4.02	4.30	4.52	4.70	4.86	4.99	5.11	5.21	5.31	5.39	5.54	5.84
18	2.97	3.61	4.00	4.28	4.49	4.67	4.82	4.96	5.07	5.17	5.27	5.35	5.50	5.79
19	2.96	3.59	3.98	4.25	4.47	4.65	4.79	4.92	5.04	5.14	5.23	5.31	5.46	5.75
20	2.95	3.58	3.96	4.23	4.45	4.62	4.77	4.90	5.01	5.11	5.20	5.28	5.43	5.71
24	2.92	3.53	3.90	4.17	4.37	4.54	4.68	4.81	4.92	5.01	5.10	5.18	5.32	5.59
30	2.89	3.49	3.85	4.10	4.30	4.46	4.60	4.72	4.82	4.92	5.00	5.08	5.21	5.47
40	2.86	3.44	3.79	4.04	4.23	4.39	4.52	4.63	4.73	4.82	4.90	4.98	5.11	5.36
60	2.83	3.40	3.74	3.98	4.16	4.31	4.44	4.55	4.65	4.73	4.81	4.88	5.00	5.24
120	2.80	3.36	3.68	3.92	4.10	4.24	4.36	4.47	4.56	4.64	4.71	4.78	4.90	5.13
∞	2.77	3.31	3.63	3.86	4.03	4.17	4.29	4.39	4.47	4.55	4.62	4.68	4.80	5.01

This table is taken from *Probability and Statistics for Engineering and the Sciences* by J.L. Devore and originally abridged from Table 29 of *Biometrika Tables for Statisticians*, E.S. Pearson and H.O. Hartley, editors (Third Edition, 1966). Reprinted by permission of the Biometrika Trustees.

TABLE D-10-B   *.99 Quantiles of the Studentized Range Distribution*

NUMBER OF MEANS TO BE COMPARED

v	2	3	4	5	6	7	8	9	10	11	12	13	15	20
5	5.70	6.98	7.80	8.42	8.91	9.32	9.67	9.97	10.24	10.48	10.70	10.89	11.24	11.93
6	5.24	6.33	7.03	7.56	7.97	8.32	8.61	8.87	9.10	9.30	9.48	9.65	9.95	10.54
7	4.95	5.92	6.54	7.01	7.37	7.68	7.94	8.17	8.37	8.55	8.71	8.86	9.12	9.65
8	4.75	5.64	6.20	6.62	6.96	7.24	7.47	7.68	7.86	8.03	8.18	8.31	8.55	9.03
9	4.60	5.43	5.96	6.35	6.66	6.91	7.13	7.33	7.49	7.65	7.78	7.91	8.13	8.57
10	4.48	5.27	5.77	6.14	6.43	6.67	6.87	7.05	7.21	7.36	7.49	7.60	7.81	8.23
11	4.39	5.15	5.62	5.97	6.25	6.48	6.67	6.84	6.99	7.13	7.25	7.36	7.56	7.95
12	4.32	5.05	5.50	5.84	6.10	6.32	6.51	6.67	6.81	6.94	7.06	7.17	7.36	7.73
13	4.26	4.96	5.40	5.73	5.98	6.19	6.37	6.53	6.67	6.79	6.90	7.01	7.19	7.55
14	4.21	4.89	5.32	5.63	5.88	6.08	6.26	6.41	6.54	6.66	6.77	6.87	7.05	7.39
15	4.17	4.84	5.25	5.56	5.80	5.99	6.16	6.31	6.44	6.55	6.66	6.76	6.93	7.26
16	4.13	4.79	5.19	5.49	5.72	5.92	6.08	6.22	6.35	6.46	6.56	6.66	6.82	7.15
17	4.10	4.74	5.14	5.43	5.66	5.85	6.01	6.15	6.27	6.38	6.48	6.57	6.73	7.05
18	4.07	4.70	5.09	5.38	5.60	5.79	5.94	6.08	6.20	6.31	6.41	6.50	6.65	6.97
19	4.05	4.67	5.05	5.33	5.55	5.73	5.89	6.02	6.14	6.25	6.34	6.43	6.58	6.89
20	4.02	4.64	5.02	5.29	5.51	5.69	5.84	5.97	6.09	6.19	6.28	6.37	6.52	6.82
24	3.96	4.55	4.91	5.17	5.37	5.54	5.69	5.81	5.92	6.02	6.11	6.19	6.33	6.61
30	3.89	4.45	4.80	5.05	5.24	5.40	5.54	5.65	5.76	5.85	5.93	6.01	6.14	6.41
40	3.82	4.37	4.70	4.93	5.11	5.26	5.39	5.50	5.60	5.69	5.76	5.83	5.96	6.21
60	3.76	4.28	4.59	4.82	4.99	5.13	5.25	5.36	5.45	5.53	5.60	5.67	5.78	6.01
120	3.70	4.20	4.50	4.71	4.87	5.01	5.12	5.21	5.30	5.37	5.44	5.50	5.61	5.83
∞	3.64	4.12	4.40	4.60	4.76	4.88	4.99	5.08	5.16	5.23	5.29	5.35	5.45	5.65

This table is taken from *Probability and Statistics for Engineering and the Sciences* by J.L. Devore and originally abridged from Table 29 of *Biometrika Tables for Statisticians*, E.S. Pearson and H.O. Hartley, editors (Third Edition, 1966). Reprinted by permission of the Biometrika Trustees.

TABLE D-11-A   *Factors for Two-Sided Simultaneous 95% Confidence Intervals for All Differences between a Control and r − 1 Treatment Means*

ν	1	2	3	4	5	6	7	8	9	10	11	12	15	20
							$r-1$							
5	2.57	$3.03^{2.3}$	$3.29^{3.6}$	$3.48^{4.6}$	$3.62^{5.4}$	$3.73^{5.9}$	$3.82^{6.4}$	$3.90^{6.8}$	$3.97^{7.2}$	$4.03^{7.5}$	$4.09^{7.8}$	$4.14^{8.0}$	$4.26^{8.7}$	$4.42^{9.4}$
6	2.45	$2.86^{2.1}$	$3.10^{3.4}$	$3.26^{4.3}$	$3.39^{5.0}$	$3.49^{5.6}$	$3.57^{6.0}$	$3.64^{6.4}$	$3.71^{6.8}$	$3.76^{7.1}$	$3.81^{7.4}$	$3.86^{7.6}$	$3.97^{8.2}$	$4.11^{9.0}$
7	2.36	$2.75^{2.0}$	$2.97^{3.2}$	$3.12^{4.1}$	$3.24^{4.8}$	$3.33^{5.3}$	$3.41^{5.7}$	$3.47^{6.1}$	$3.53^{6.5}$	$3.58^{6.7}$	$3.63^{7.0}$	$3.67^{7.2}$	$3.78^{7.8}$	$3.91^{8.6}$
8	2.31	$2.67^{2.0}$	$2.88^{3.1}$	$3.02^{3.9}$	$3.13^{4.5}$	$3.22^{5.1}$	$3.29^{5.5}$	$3.35^{5.9}$	$3.41^{6.2}$	$3.46^{6.5}$	$3.50^{6.7}$	$3.54^{6.9}$	$3.64^{7.5}$	$3.76^{8.2}$
9	2.26	$2.61^{1.9}$	$2.81^{3.0}$	$2.95^{3.8}$	$3.05^{4.4}$	$3.14^{4.9}$	$3.20^{5.3}$	$3.26^{5.6}$	$3.32^{5.9}$	$3.36^{6.2}$	$3.40^{6.5}$	$3.44^{6.7}$	$3.53^{7.2}$	$3.65^{7.9}$
10	2.23	$2.57^{1.8}$	$2.76^{2.9}$	$2.89^{3.6}$	$2.99^{4.2}$	$3.07^{4.7}$	$3.14^{5.1}$	$3.19^{5.4}$	$3.24^{5.7}$	$3.29^{6.0}$	$3.33^{6.2}$	$3.36^{6.5}$	$3.45^{7.0}$	$3.57^{7.7}$
11	2.20	$2.53^{1.8}$	$2.72^{2.8}$	$2.84^{3.5}$	$2.94^{4.1}$	$3.02^{4.6}$	$3.08^{4.9}$	$3.14^{5.3}$	$3.19^{5.6}$	$3.23^{5.8}$	$3.27^{6.1}$	$3.30^{6.3}$	$3.39^{6.8}$	$3.50^{7.5}$
12	2.18	$2.50^{1.7}$	$2.68^{2.7}$	$2.81^{3.4}$	$2.90^{4.0}$	$2.98^{4.4}$	$3.04^{4.8}$	$3.09^{5.1}$	$3.14^{5.4}$	$3.18^{5.7}$	$3.22^{5.9}$	$3.25^{6.1}$	$3.34^{6.6}$	$3.45^{7.3}$
13	2.16	$2.48^{1.7}$	$2.65^{2.7}$	$2.78^{3.4}$	$2.87^{3.9}$	$2.94^{4.3}$	$3.00^{4.7}$	$3.06^{5.0}$	$3.10^{5.3}$	$3.14^{5.5}$	$3.18^{5.8}$	$3.21^{6.0}$	$3.29^{6.5}$	$3.40^{7.1}$
14	2.14	$2.46^{1.7}$	$2.63^{2.6}$	$2.75^{3.3}$	$2.84^{3.8}$	$2.91^{4.2}$	$2.97^{4.6}$	$3.02^{4.9}$	$3.07^{5.2}$	$3.11^{5.4}$	$3.14^{5.6}$	$3.18^{5.8}$	$3.26^{6.3}$	$3.36^{7.0}$
15	2.13	$2.44^{1.7}$	$2.61^{2.6}$	$2.73^{3.2}$	$2.82^{3.8}$	$2.89^{4.2}$	$2.95^{4.5}$	$3.00^{4.8}$	$3.04^{5.1}$	$3.08^{5.3}$	$3.12^{5.5}$	$3.15^{5.7}$	$3.23^{6.2}$	$3.33^{6.8}$
20	2.09	$2.38^{1.6}$	$2.54^{2.4}$	$2.65^{3.0}$	$2.73^{3.5}$	$2.80^{3.9}$	$2.86^{4.2}$	$2.90^{4.5}$	$2.95^{4.7}$	$2.98^{4.9}$	$3.02^{5.1}$	$3.05^{5.3}$	$3.12^{5.7}$	$3.22^{6.3}$
30	2.04	$2.32^{1.5}$	$2.47^{2.3}$	$2.58^{2.8}$	$2.66^{3.2}$	$2.72^{3.6}$	$2.77^{3.9}$	$2.82^{4.1}$	$2.86^{4.3}$	$2.89^{4.5}$	$2.92^{4.7}$	$2.95^{4.8}$	$3.02^{5.2}$	$3.11^{5.8}$
40	2.02	$2.29^{1.4}$	$2.44^{2.2}$	$2.54^{2.7}$	$2.62^{3.1}$	$2.68^{3.4}$	$2.73^{3.7}$	$2.77^{3.9}$	$2.81^{4.1}$	$2.85^{4.3}$	$2.87^{4.5}$	$2.90^{4.6}$	$2.97^{5.0}$	$3.06^{5.5}$
60	2.00	$2.27^{1.4}$	$2.41^{2.1}$	$2.51^{2.6}$	$2.58^{3.0}$	$2.64^{3.3}$	$2.69^{3.5}$	$2.73^{3.7}$	$2.77^{3.9}$	$2.80^{4.1}$	$2.83^{4.2}$	$2.86^{4.4}$	$2.92^{4.7}$	$3.00^{5.1}$
120	1.98	$2.24^{1.3}$	$2.38^{2.0}$	$2.47^{2.5}$	$2.55^{2.8}$	$2.60^{3.1}$	$2.65^{3.3}$	$2.69^{3.5}$	$2.73^{3.7}$	$2.76^{3.8}$	$2.79^{4.0}$	$2.81^{4.1}$	$2.87^{4.4}$	$2.95^{4.8}$
∞	1.96	$2.21^{1.3}$	$2.35^{1.9}$	$2.44^{2.3}$	$2.51^{2.7}$	$2.57^{2.9}$	$2.61^{3.1}$	$2.65^{3.3}$	$2.69^{3.5}$	$2.72^{3.6}$	$2.74^{3.7}$	$2.77^{3.8}$	$2.83^{4.1}$	$2.91^{4.5}$

Reproduced from C.W. Dunnett, "New Tables for Multiple Comparisons with a Control." *Biometrics*, 20, pp. 482–491, 1964. With permission from The Biometric Society.

TABLE D-11-B Factors for Two-Sided Simultaneous 99% Confidence Intervals for All Differences between a Control and r − 1 Treatment Means

ν							$r-1$							
	1	2	3	4	5	6	7	8	9	10	11	12	15	20
5	4.03	$4.63^{1.8}$	$4.98^{3.0}$	$5.22^{3.9}$	$5.41^{4.6}$	$5.56^{5.2}$	$5.69^{5.7}$	$5.80^{6.1}$	$5.89^{6.5}$	$5.98^{6.9}$	$6.05^{7.2}$	$6.12^{7.4}$	$6.30^{8.1}$	$6.52^{9.0}$
6	3.71	$4.21^{1.6}$	$4.51^{2.7}$	$4.71^{3.5}$	$4.87^{4.1}$	$5.00^{4.6}$	$5.10^{5.1}$	$5.20^{5.5}$	$5.28^{5.8}$	$5.35^{6.1}$	$5.41^{6.4}$	$5.47^{6.7}$	$5.62^{7.3}$	$5.81^{8.1}$
7	3.50	$3.95^{1.5}$	$4.21^{2.4}$	$4.39^{3.1}$	$4.53^{3.7}$	$4.64^{4.2}$	$4.74^{4.6}$	$4.82^{5.0}$	$4.89^{5.3}$	$4.95^{5.6}$	$5.01^{5.8}$	$5.06^{6.1}$	$5.19^{6.7}$	$5.36^{7.4}$
8	3.36	$3.77^{1.3}$	$4.00^{2.2}$	$4.17^{2.9}$	$4.29^{3.4}$	$4.40^{3.9}$	$4.48^{4.2}$	$4.56^{4.6}$	$4.62^{4.9}$	$4.68^{5.1}$	$4.73^{5.4}$	$4.78^{5.6}$	$4.90^{6.1}$	$5.05^{6.9}$
9	3.25	$3.63^{1.2}$	$3.85^{2.1}$	$4.01^{2.7}$	$4.12^{3.2}$	$4.22^{3.6}$	$4.30^{3.9}$	$4.37^{4.2}$	$4.43^{4.5}$	$4.48^{4.8}$	$4.53^{5.0}$	$4.57^{5.2}$	$4.68^{5.7}$	$4.82^{6.4}$
10	3.17	$3.53^{1.2}$	$3.74^{1.9}$	$3.88^{2.5}$	$3.99^{3.0}$	$4.08^{3.4}$	$4.16^{3.7}$	$4.22^{4.0}$	$4.28^{4.2}$	$4.33^{4.5}$	$4.37^{4.7}$	$4.42^{4.9}$	$4.52^{5.4}$	$4.65^{6.0}$
11	3.11	$3.45^{1.1}$	$3.65^{1.8}$	$3.79^{2.4}$	$3.89^{2.8}$	$3.98^{3.2}$	$4.05^{3.5}$	$4.11^{3.8}$	$4.16^{4.0}$	$4.21^{4.2}$	$4.25^{4.4}$	$4.29^{4.6}$	$4.39^{5.1}$	$4.52^{5.7}$
12	3.05	$3.39^{1.1}$	$3.58^{1.7}$	$3.71^{2.3}$	$3.81^{2.7}$	$3.89^{3.0}$	$3.96^{3.3}$	$4.02^{3.6}$	$4.07^{3.8}$	$4.12^{4.0}$	$4.16^{4.2}$	$4.19^{4.4}$	$4.29^{4.8}$	$4.41^{5.4}$
13	3.01	$3.33^{1.0}$	$3.52^{1.7}$	$3.65^{2.2}$	$3.74^{2.6}$	$3.82^{2.9}$	$3.89^{3.2}$	$3.94^{3.4}$	$3.99^{3.6}$	$4.04^{3.8}$	$4.08^{4.0}$	$4.11^{4.2}$	$4.20^{4.6}$	$4.32^{5.2}$
14	2.98	$3.29^{1.0}$	$3.47^{1.6}$	$3.59^{2.1}$	$3.69^{2.5}$	$3.76^{2.8}$	$3.83^{3.0}$	$3.88^{3.3}$	$3.93^{3.5}$	$3.97^{3.7}$	$4.01^{3.9}$	$4.05^{4.0}$	$4.13^{4.4}$	$4.24^{5.0}$
15	2.95	$3.25^{0.9}$	$3.43^{1.5}$	$3.55^{2.0}$	$3.64^{2.4}$	$3.71^{2.7}$	$3.78^{2.9}$	$3.83^{3.2}$	$3.88^{3.4}$	$3.92^{3.6}$	$3.95^{3.7}$	$3.99^{3.9}$	$4.07^{4.3}$	$4.18^{4.8}$
20	2.85	$3.13^{0.8}$	$3.29^{1.4}$	$3.40^{1.7}$	$3.48^{2.1}$	$3.55^{2.3}$	$3.60^{2.5}$	$3.65^{2.7}$	$3.69^{2.9}$	$3.73^{3.1}$	$3.77^{3.2}$	$3.80^{3.4}$	$3.87^{3.7}$	$3.97^{4.2}$
30	2.75	$3.01^{0.7}$	$3.15^{1.2}$	$3.25^{1.5}$	$3.33^{1.8}$	$3.39^{2.0}$	$3.44^{2.2}$	$3.49^{2.3}$	$3.52^{2.5}$	$3.56^{2.6}$	$3.59^{2.7}$	$3.62^{2.8}$	$3.69^{3.1}$	$3.78^{3.5}$
40	2.70	$2.95^{0.7}$	$3.09^{1.1}$	$3.19^{1.4}$	$3.26^{1.6}$	$3.32^{1.8}$	$3.37^{2.0}$	$3.41^{2.1}$	$3.44^{2.3}$	$3.48^{2.4}$	$3.51^{2.5}$	$3.53^{2.6}$	$3.60^{2.8}$	$3.68^{3.2}$
60	2.66	$2.90^{0.6}$	$3.03^{1.0}$	$3.12^{1.3}$	$3.19^{1.5}$	$3.25^{1.6}$	$3.29^{1.8}$	$3.33^{1.9}$	$3.37^{2.0}$	$3.40^{2.1}$	$3.42^{2.2}$	$3.45^{2.3}$	$3.51^{2.5}$	$3.59^{2.8}$
120	2.62	$2.85^{0.6}$	$2.97^{0.9}$	$3.06^{1.1}$	$3.12^{1.3}$	$3.18^{1.5}$	$3.22^{1.6}$	$3.26^{1.7}$	$3.29^{1.8}$	$3.32^{1.9}$	$3.35^{2.0}$	$3.37^{2.1}$	$3.43^{2.2}$	$3.51^{2.5}$
∞	2.58	$2.79^{0.5}$	$2.92^{0.8}$	$3.00^{1.0}$	$3.06^{1.2}$	$3.11^{1.3}$	$3.15^{1.4}$	$3.19^{1.5}$	$3.22^{1.6}$	$3.25^{1.7}$	$3.27^{1.7}$	$3.29^{1.8}$	$3.35^{1.9}$	$3.42^{2.2}$

Reproduced from C.W. Dunnett, "New Tables for Multiple Comparisons with a Control." *Biometrics, 20*, pp. 482–491, 1964. With permission from The Biometric Society.

TABLE D-12-A   .90 Quantiles of the Distribution of the ANOM Statistic When All Means Are Equal

								NUMBER OF MEANS, r										
ν	3	4	5	6	7	8	9	10	11	12	13	14	15	16	17	18	19	20
3	3.16																	
4	2.81	3.10																
5	2.63	2.88	3.05															
6	2.52	2.74	2.91	3.03														
7	2.44	2.65	2.81	2.92	3.02													
8	2.39	2.59	2.73	2.85	2.94	3.02												
9	2.34	2.54	2.68	2.79	2.88	2.95	3.01											
10	2.31	2.50	2.64	2.74	2.83	2.90	2.96	3.02										
11	2.29	2.47	2.60	2.70	2.79	2.86	2.92	2.97	3.02									
12	2.27	2.45	2.57	2.67	2.75	2.82	2.88	2.93	2.98	3.02								
13	2.25	2.43	2.55	2.65	2.73	2.79	2.85	2.90	2.95	2.99	3.03							
14	2.23	2.41	2.53	2.63	2.70	2.77	2.83	2.88	2.92	2.96	3.00	3.03						
15	2.22	2.39	2.51	2.61	2.68	2.75	2.80	2.85	2.90	2.94	2.97	3.01	3.04					
16	2.21	2.38	2.50	2.59	2.67	2.73	2.79	2.83	2.88	2.92	2.95	2.99	3.02	3.05				
17	2.20	2.37	2.49	2.58	2.65	2.72	2.77	2.82	2.86	2.90	2.93	2.97	3.00	3.03	3.05			
18	2.19	2.36	2.47	2.56	2.64	2.70	2.75	2.80	2.84	2.88	2.92	2.95	2.98	3.01	3.03	3.06		
19	2.18	2.35	2.46	2.55	2.63	2.69	2.74	2.79	2.83	2.87	2.90	2.94	2.96	2.99	3.02	3.04	3.06	
20	2.18	2.34	2.45	2.54	2.62	2.68	2.73	2.78	2.82	2.86	2.89	2.92	2.95	2.98	3.00	3.03	3.05	3.07
24	2.15	2.32	2.43	2.51	2.58	2.64	2.69	2.74	2.78	2.82	2.85	2.88	2.91	2.93	2.96	2.98	3.00	3.02
30	2.13	2.29	2.40	2.48	2.55	2.61	2.66	2.70	2.74	2.77	2.81	2.84	2.86	2.89	2.91	2.93	2.96	2.98
40	2.11	2.27	2.37	2.45	2.52	2.57	2.62	2.66	2.70	2.73	2.77	2.79	2.82	2.85	2.87	2.89	2.91	2.93
60	2.09	2.24	2.34	2.42	2.49	2.54	2.59	2.63	2.66	2.70	2.73	2.75	2.78	2.80	2.82	2.84	2.86	2.88
120	2.07	2.22	2.32	2.39	2.45	2.51	2.55	2.59	2.62	2.66	2.69	2.71	2.74	2.76	2.78	2.80	2.82	2.84
∞	2.05	2.19	2.29	2.36	2.42	2.47	2.52	2.55	2.59	2.62	2.65	2.67	2.69	2.72	2.74	2.76	2.77	2.79

This table is from "Exact Critical Values for Use with the Analysis of Means" by L. S. Nelson, *Journal of Quality Technology*, *15*, pp. 40–44, 1983. Reprinted by permission of the American Society for Quality Control.

TABLE D-12-B   *.95 Quantiles of the Distribution of the ANOM Statistic When All Means Are Equal*

ν	NUMBER OF MEANS, r																	
	3	4	5	6	7	8	9	10	11	12	13	14	15	16	17	18	19	20
3	4.18																	
4	3.56	3.89																
5	3.25	3.53	3.72															
6	3.07	3.31	3.49	3.62														
7	2.94	3.17	3.33	3.45	3.56													
8	2.86	3.07	3.21	3.33	3.43	3.51												
9	2.79	2.99	3.13	3.24	3.33	3.41	3.48											
10	2.74	2.93	3.07	3.17	3.26	3.33	3.40	3.45										
11	2.70	2.88	3.01	3.12	3.20	3.27	3.33	3.39	3.44									
12	2.67	2.85	2.97	3.07	3.15	3.22	3.28	3.33	3.38	3.42								
13	2.64	2.81	2.94	3.03	3.11	3.18	3.24	3.29	3.34	3.38	3.42							
14	2.62	2.79	2.91	3.00	3.08	3.14	3.20	3.25	3.30	3.34	3.37	3.41						
15	2.60	2.76	2.88	2.97	3.05	3.11	3.17	3.22	3.26	3.30	3.34	3.37	3.40					
16	2.58	2.74	2.86	2.95	3.02	3.09	3.14	3.19	3.23	3.27	3.31	3.34	3.37	3.40				
17	2.57	2.73	2.84	2.93	3.00	3.06	3.12	3.16	3.21	3.25	3.28	3.31	3.34	3.37	3.40			
18	2.55	2.71	2.82	2.91	2.98	3.04	3.10	3.14	3.18	3.22	3.26	3.29	3.32	3.35	3.37	3.40		
19	2.54	2.70	2.81	2.89	2.96	3.02	3.08	3.12	3.16	3.20	3.24	3.27	3.30	3.32	3.35	3.37	3.40	
20	2.53	2.68	2.79	2.88	2.95	3.01	3.06	3.11	3.15	3.18	3.22	3.25	3.28	3.30	3.33	3.35	3.37	3.40
24	2.50	2.65	2.75	2.83	2.90	2.96	3.01	3.05	3.09	3.13	3.16	3.19	3.22	3.24	3.27	3.29	3.31	3.33
30	2.47	2.61	2.71	2.79	2.85	2.91	2.96	3.00	3.04	3.07	3.10	3.13	3.16	3.18	3.20	3.22	3.25	3.27
40	2.43	2.57	2.67	2.75	2.81	2.86	2.91	2.95	2.98	3.01	3.04	3.07	3.10	3.12	3.14	3.16	3.18	3.20
60	2.40	2.54	2.63	2.70	2.76	2.81	2.86	2.90	2.93	2.96	2.99	3.02	3.04	3.06	3.08	3.10	3.12	3.14
120	2.37	2.50	2.59	2.66	2.72	2.77	2.81	2.84	2.88	2.91	2.93	2.96	2.98	3.00	3.02	3.04	3.06	3.08
∞	2.34	2.47	2.56	2.62	2.68	2.72	2.76	2.80	2.83	2.86	2.88	2.90	2.93	2.95	2.97	2.98	3.00	3.02

This table is from "Exact Critical Values for Use with the Analysis of Means" by L. S. Nelson, *Journal of Quality Technology*, *15*, pp. 40–44, 1983. Reprinted by permission of the American Society for Quality Control.

TABLE D-12-C   .99 Quantiles of the Distribution of the ANOM Statistic When All Means Are Equal

ν	\multicolumn{18}{c}{NUMBER OF MEANS, r}
	3	4	5	6	7	8	9	10	11	12	13	14	15	16	17	18	19	20
3	7.51																	
4	5.74	6.21																
5	4.93	5.29	5.55															
6	4.48	4.77	4.98	5.16														
7	4.18	4.44	4.63	4.78	4.90													
8	3.98	4.21	4.38	4.52	4.63	4.72												
9	3.84	4.05	4.20	4.33	4.43	4.51	4.59											
10	3.73	3.92	4.07	4.18	4.28	4.36	4.43	4.49										
11	3.64	3.82	3.96	4.07	4.16	4.23	4.30	4.36	4.41									
12	3.57	3.74	3.87	3.98	4.06	4.13	4.20	4.25	4.31	4.35								
13	3.51	3.68	3.80	3.90	3.98	4.05	4.11	4.17	4.22	4.26	4.30							
14	3.46	3.63	3.74	3.84	3.92	3.98	4.04	4.09	4.14	4.18	4.22	4.26						
15	3.42	3.58	3.69	3.79	3.86	3.92	3.98	4.03	4.08	4.12	4.16	4.19	4.22					
16	3.38	3.54	3.65	3.74	3.81	3.87	3.93	3.98	4.02	4.06	4.10	4.14	4.17	4.20				
17	3.35	3.50	3.61	3.70	3.77	3.83	3.89	3.93	3.98	4.02	4.05	4.09	4.12	4.14	4.17			
18	3.33	3.47	3.58	3.66	3.73	3.79	3.85	3.89	3.94	3.97	4.01	4.04	4.07	4.10	4.12	4.15		
19	3.30	3.45	3.55	3.63	3.70	3.76	3.81	3.86	3.90	3.94	3.97	4.00	4.03	4.06	4.08	4.11	4.13	
20	3.28	3.42	3.53	3.61	3.67	3.73	3.78	3.83	3.87	3.90	3.94	3.97	4.00	4.02	4.05	4.07	4.09	4.12
24	3.21	3.35	3.45	3.52	3.58	3.64	3.69	3.73	3.77	3.80	3.83	3.86	3.89	3.91	3.94	3.96	3.98	4.00
30	3.15	3.28	3.37	3.44	3.50	3.55	3.59	3.63	3.67	3.70	3.73	3.76	3.78	3.81	3.83	3.85	3.87	3.89
40	3.09	3.21	3.29	3.36	3.42	3.46	3.50	3.54	3.58	3.60	3.63	3.66	3.68	3.70	3.72	3.74	3.76	3.78
60	3.03	3.14	3.22	3.29	3.34	3.38	3.42	3.46	3.49	3.51	3.54	3.56	3.59	3.61	3.63	3.64	3.66	3.68
120	2.97	3.07	3.15	3.21	3.26	3.30	3.34	3.37	3.40	3.42	3.45	3.47	3.49	3.51	3.53	3.55	3.56	3.58
∞	2.91	3.01	3.08	3.14	3.18	3.22	3.26	3.29	3.32	3.34	3.36	3.38	3.40	3.42	3.44	3.45	3.47	3.48

This table is from "Exact Critical Values for Use with the Analysis of Means" by L. S. Nelson, *Journal of Quality Technology*, *15*, pp. 40–44, 1983. Reprinted by permission of the American Society for Quality Control.

TABLE D-12-D   *.999 Quantiles of the Distribution of the ANOM Statistic When All Means Are Equal*

ν	NUMBER OF MEANS, r																	
	3	4	5	6	7	8	9	10	11	12	13	14	15	16	17	18	19	20
3	16.4																	
4	10.6	11.4																
5	8.25	8.79	9.19															
6	7.04	7.45	7.76	8.00														
7	6.31	6.65	6.89	7.09	7.25													
8	5.83	6.12	6.32	6.49	6.63	6.75												
9	5.49	5.74	5.92	6.07	6.20	6.30	6.40											
10	5.24	5.46	5.63	5.76	5.87	5.97	6.05	6.13										
11	5.05	5.25	5.40	5.52	5.63	5.71	5.79	5.86	5.92									
12	4.89	5.08	5.22	5.33	5.43	5.51	5.58	5.65	5.71	5.76								
13	4.77	4.95	5.08	5.18	5.27	5.35	5.42	5.48	5.53	5.58	5.63							
14	4.66	4.83	4.96	5.06	5.14	5.21	5.28	5.33	5.38	5.43	5.48	5.51						
15	4.57	4.74	4.86	4.95	5.03	5.10	5.16	5.21	5.26	5.31	5.35	5.39	5.42					
16	4.50	4.66	4.77	4.86	4.94	5.00	5.06	5.11	5.16	5.20	5.24	5.28	5.31	5.34				
17	4.44	4.59	4.70	4.78	4.86	4.92	4.98	5.03	5.07	5.11	5.15	5.18	5.22	5.25	5.28			
18	4.38	4.53	4.63	4.72	4.79	4.85	4.90	4.95	4.99	5.03	5.07	5.10	5.14	5.16	5.19	5.22		
19	4.33	4.47	4.58	4.66	4.73	4.79	4.84	4.88	4.93	4.96	5.00	5.03	5.06	5.09	5.12	5.14	5.17	
20	4.29	4.42	4.53	4.61	4.67	4.73	4.78	4.83	4.87	4.90	4.94	4.97	5.00	5.03	5.05	5.08	5.10	5.12
24	4.16	4.28	4.37	4.45	4.51	4.56	4.61	4.65	4.69	4.72	4.75	4.78	4.81	4.83	4.86	4.88	4.90	4.92
30	4.03	4.14	4.23	4.30	4.35	4.40	4.44	4.48	4.51	4.54	4.57	4.60	4.62	4.64	4.67	4.69	4.71	4.72
40	3.91	4.01	4.09	4.15	4.20	4.25	4.29	4.32	4.35	4.38	4.40	4.43	4.45	4.47	4.49	4.50	4.52	4.54
60	3.80	3.89	3.96	4.02	4.06	4.10	4.14	4.17	4.19	4.22	4.24	4.27	4.29	4.30	4.32	4.33	4.35	4.37
120	3.69	3.77	3.84	3.89	3.93	3.96	4.00	4.03	4.05	4.07	4.09	4.11	4.13	4.15	4.16	4.17	4.19	4.21
∞	3.58	3.66	3.72	3.76	3.80	3.84	3.87	3.89	3.91	3.93	3.95	3.97	3.99	4.00	4.02	4.03	4.04	4.06

This table is from "Exact Critical Values for Use with the Analysis of Means" by L. S. Nelson, *Journal of Quality Technology*, *15*, pp. 40–44, 1983. Reprinted by permission of the American Society for Quality Control.

Index

Note: **Boldface** page numbers indicate definitions.